Power System Analysis

POWER ENGINEERING

Series Editor
H. Lee Willis
Quanta Technology
Raleigh, North Carolina

Advisory Editor
Muhammad H. Rashid
University of West Florida
Pensacola, Florida

1. Power Distribution Planning Reference Book, *H. Lee Willis*
2. Transmission Network Protection: Theory and Practice *Y. G. Paithankar*
3. Electrical Insulation in Power Systems, *N. H. Malik, A. A. Al-Arainy, and M. I. Qureshi*
4. Electrical Power Equipment Maintenance and Testing, *Paul Gill*
5. Protective Relaying: Principles and Applications, Second Edition, *J. Lewis Blackburn*
6. Understanding Electric Utilities and De-Regulation *Lorrin Philipson and H. Lee Willis*
7. Electrical Power Cable Engineering, *William A. Thue*
8. Electric Systems, Dynamics, and Stability with Artificial Intelligence Applications, *James A. Momoh and Mohamed E. El-Hawary*
9. Insulation Coordination for Power Systems *Andrew R. Hileman*
10. Distributed Power Generation: Planning and Evaluation *H. Lee Willis and Walter G. Scott*
11. Electric Power System Applications of Optimization *James A. Momoh*
12. Aging Power Delivery Infrastructures, *H. Lee Willis, Gregory V. Welch, and Randall R. Schrieber*
13. Restructured Electrical Power Systems: Operation, Trading, and Volatility, *Mohammad Shahidehpour and Muwaffaq Alomoush*
14. Electric Power Distribution Reliability, *Richard E. Brown*
15. Computer-Aided Power System Analysis *Ramasamy Natarajan*
16. Power Transformers: Principles and Applications, *John J. Winders, Jr.*
17. Spatial Electric Load Forecasting: Second Edition, Revised and Expanded, *H. Lee Willis*
18. Dielectrics in Electric Fields, *Gorur G. Raju*
19. Protection Devices and Systems for High-Voltage Applications, *Vladimir Gurevich*

20. Electrical Power Cable Engineering, Second Edition
William Thue
21. Vehicular Electric Power Systems: Land, Sea, Air, and Space Vehicles, *Ali Emadi, Mehrdad Ehsani, and John Miller*
22. Power Distribution Planning Reference Book, Second Edition, *H. Lee Willis*
23. Power System State Estimation: Theory and Implementation, *Ali Abur*
24. Transformer Engineering: Design and Practice, *S.V. Kulkarni and S. A. Khaparde*
25. Power System Capacitors, *Ramasamy Natarajan*
26. Understanding Electric Utilities and De-regulation: Second Edition, *Lorrin Philipson and H. Lee Willis*
27. Control and Automation of Electric Power Distribution Systems, *James Northcote-Green and Robert G. Wilson*
28. Protective Relaying for Power Generation Systems
Donald Reimert
29. Protective Relaying: Principles and Applications, Third Edition
J. Lewis Blackburn and Thomas J. Domin
30. Electric Power Distribution Reliability, Second Edition
Richard E. Brown
31. Electrical Power Equipment Maintenance and Testing, Second Edition, *Paul Gill*
32. Electricity Pricing: Engineering Principles and Methodologies
Lawrence J. Vogt
33. Power System Analysis: Short-Circuit Load Flow and Harmonics, Second Edition, *J. C. Das*

Power System Analysis

Short-Circuit Load Flow and Harmonics

Second Edition

J. C. Das

Amec, Incorporated
Tucker, Georgia
USA

CRC Press
Taylor & Francis Group
Boca Raton London New York

CRC Press is an imprint of the
Taylor & Francis Group, an **informa** business

CRC Press
Taylor & Francis Group
6000 Broken Sound Parkway NW, Suite 300
Boca Raton, FL 33487-2742

First issued in paperback 2017

Version Date: 20110517

ISBN 13: 978-1-4398-2078-0 (hbk)
ISBN 13: 978-1-138-07504-7 (pbk)

Visit the Taylor & Francis Web site at
http://www.taylorandfrancis.com

and the CRC Press Web site at
http://www.crcpress.com

Contents

Series Introduction

Power engineering is the oldest and most traditional of the various areas within electrical engineering, yet no other facet of modern technology is currently undergoing a more dramatic revolution in both technology and industry structure. But none of these changes alter the basic complexity of electric power system behavior, or reduce the challenge that power system engineers have always faced in designing an economical system that operates as intended and shuts down in a safe and noncatastrophic mode when something fails unexpectedly. In fact, many of the ongoing changes in the power industry—deregulation, reduced budgets and staffing levels, and increasing public and regulatory demand for reliability among them—make these challenges all the more difficult to overcome.

Therefore, I am particularly delighted to see this latest addition to the Power Engineering series. J.C. Das's *Power System Analysis: Short-Circuit Load Flow and Harmonics* provides comprehensive coverage of both theory and practice in the fundamental areas of power system analysis, including power flow, short-circuit computations, harmonics, machine modeling, equipment ratings, reactive power control, and optimization. It also includes an excellent review of the standard matrix mathematics and computation methods of power system analysis in a readily usable format.

Of particular note, this book discusses both ANSI/IEEE and IEC methods, guidelines, and procedures for applications and ratings. Over the past few years, my work as vice president of technology and strategy for ABB's global consulting organization has given me an appreciation that the IEC and ANSI standards are not so much in conflict as they are slightly different but equally valid approaches to power engineering. There is much to be learned from each, and from the study of the differences between them.

As the editor of the Power Engineering series, I am proud to include *Power System Analysis* among this important group of books. Like all the volumes in the Power Engineering series, this book provides modern power technology in a context of proven, practical application. It is useful as a reference book as well as for self-study and advanced classroom use. The series includes books covering the entire field of power engineering, in all its specialties and subgenres, all aimed at providing practicing power engineers with the knowledge and techniques they need to meet the electric industry's challenges in the twenty-first century.

H. Lee Willis

Preface to the Second Edition

In recent times, two new aspects of power system analysis have emerged: (1) the arc flash hazard analysis and reduction of hazard risk category (HRC) in electrical systems and (2) the wind power generation and its integration in utility systems. Maintaining the structure and order of the first edition of the book, two new chapters, Chapters 21 and 22, have been added to address these new technologies. The ANSI/IEEE rating structures of the high-voltage circuit breakers have undergone many changes in an attempt to harmonize with IEC standards. Chapters 7 through 9 have been revised to reflect these changes and current ANSI/IEEE and IEC standards. New material has been added to practically each chapter, for example, Chapters 12 through 15 on load flow and Chapters 17 through 20 on harmonic analysis and harmonic filter designs. Errata of the first edition have been taken care of. New figures and supporting mathematical equations have been added where required.

This new edition should prove all the more popular with the academia and practicing power system engineers as it enhances the technical content and the presentation of the subjects covered in this book.

I would like to thank Nora Konopka of CRC Press for all her help and cooperation in publishing this second edition.

J.C. Das

Preface to the First Edition

Power system analysis is fundamental in the planning, design, and operating stages, and its importance cannot be overstated. This book covers the commonly required short-circuit, load flow, and harmonic analyses. Practical and theoretical aspects have been harmoniously combined. Although there is the inevitable computer simulation, a feel for the procedures and methodology is also provided, through examples and problems. *Power System Analysis: Short-Circuit Load Flow and Harmonics* should be a valuable addition to the power system literature for practicing engineers, those in continuing education, and college students.

Short-circuit analyses are included in chapters on rating structures of breakers, current interruption in ac circuits, calculations according to the IEC and ANSI/IEEE methods, and calculations of short-circuit currents in dc systems.

The load flow analyses cover reactive power flow and control, optimization techniques, and introduction to FACT controllers, three-phase load flow, and optimal power flow.

The effect of harmonics on power systems is a dynamic and evolving field (harmonic effects can be experienced at a distance from their source). The book derives and compiles ample data of practical interest, with the emphasis on harmonic power flow and harmonic filter design. Generation, effects, limits, and mitigation of harmonics are discussed, including active and passive filters and new harmonic mitigating topologies.

The models of major electrical equipment—i.e., transformers, generators, motors, transmission lines, and power cables—are described in detail. Matrix techniques and symmetrical component transformation form the basis of the analyses. There are many examples and problems. The references and bibliographies point to further reading and analyses. Most of the analyses are in the steady state, but references to transient behavior are included where appropriate.

A basic knowledge of per unit system, electrical circuits and machinery, and matrices is required, although an overview of matrix techniques is provided in Appendix A. The style of writing is appropriate for the upper-undergraduate level, and some sections are at graduate-course level.

Power Systems Analysis is a result of my long experience as a practicing power system engineer in a variety of industries, power plants, and nuclear facilities. Its unique feature is applications of power system analyses to real-world problems.

I thank ANSI/IEEE for permission to quote from the relevant ANSI/IEEE standards. The IEEE disclaims any responsibility or liability resulting from the placement and use in the described manner. I am also grateful to the International Electrotechnical Commission (IEC) for permission to use material from the international standards IEC 60660-1 (1997) and IEC 60909 (1988). All extracts are copyright IEC Geneva, Switzerland. All rights reserved. Further information on the IEC, its international standards, and its role is available at www.iec.ch. IEC takes no responsibility for and will not assume liability from the reader's misinterpretation of the referenced material due to its placement and context in this publication. The material is reproduced or rewritten with their permission.

Finally, I thank the staff of Marcel Dekker, Inc., and special thanks to Ann Pulido for her help in the production of this book.

J.C. Das

Author

J.C. Das is currently staff consultant, Electrical Power Systems, AMEC Inc., Tucker, Georgia. He has varied experience in the utility industry, industrial establishments, hydroelectric generation, and atomic energy. He is responsible for power system studies, including short circuit, load flow, harmonics, stability, arc-flash hazard, grounding, switching transients, and protective relaying. He conducts courses for continuing education in power systems and has authored or coauthored about 60 technical publications. He is the author of the book *Transients in Electrical Systems—Analysis Recognition and Mitigation*, McGraw-Hill, New York, 2010. His interests include power system transients, EMTP simulations, harmonics, power quality, protection, and relaying. He has also published 190 electrical power system study reports for his clients.

Das is a life fellow of the Institute of Electrical and Electronics Engineers, IEEE (United States), a member of the IEEE Industry Applications and IEEE Power Engineering societies, a fellow of the Institution of Engineering Technology (United Kingdom), a life fellow of the Institution of Engineers (India), a member of the Federation of European Engineers (France), and a member of CIGRE (France). He is a registered professional engineer in the states of Georgia and Oklahoma, a chartered engineer (CEng) in the United Kingdom, and a European engineer (Eur. Ing.).

He received his MSEE from Tulsa University, Tulsa, Oklahoma, and his BA (advanced mathematics) and BEE from Punjab University, India.

1

Short-Circuit Currents and Symmetrical Components

Short circuits occur in well-designed power systems and cause large *decaying* transient currents, generally much above the system load currents. These result in disruptive electrodynamic and thermal stresses that are potentially damaging. Fire risks and explosions are inherent. One tries to limit short circuits to the faulty section of the electrical system by appropriate switching devices capable of operating under short-circuit conditions without damage and isolating only the faulty section, so that a fault is not escalated. The faster the operation of sensing and switching devices, the lower is the fault damage, and the better is the chance of systems holding together without loss of synchronism.

Short circuits can be studied from the following angles:

1. Calculation of short-circuit currents
2. Interruption of short-circuit currents and rating structure of switching devices
3. Effects of short-circuit currents
4. Limitation of short-circuit currents, that is, with current-limiting fuses and fault current limiters
5. Short-circuit withstand ratings of electrical equipment like transformers, reactors, cables, and conductors
6. Transient stability of interconnected systems to remain in synchronism until the faulty section of the power system is isolated

We will confine our discussions to the calculations of short-circuit currents, and the basis of short-circuit ratings of switching devices, that is, power circuit breakers and fuses. As the main purpose of short-circuit calculations is to select and apply these devices properly, it is meaningful for the calculations to be related to current interruption phenomena and the rating structures of interrupting devices. The objectives of short-circuit calculations, therefore, can be summarized as follows:

- Determination of short-circuit duties on switching devices, that is, high-, medium-, and low-voltage circuit breakers and fuses
- Calculation of short-circuit currents required for protective relaying and coordination of protective devices
- Evaluations of adequacy of short-circuit withstand ratings of static equipment like cables, conductors, bus bars, reactors, and transformers
- Calculations of fault voltage dips and their time-dependent recovery profiles

The type of short-circuit currents required for each of these objectives may not be immediately clear but will unfold in the chapters to follow.

In a three-phase system, a fault may equally involve all three phases. A bolted fault means that the three phases are connected together with links of zero impedance prior to the fault, that is, the fault impedance itself is zero and the fault is limited by the system and machine impedances only. Such a fault is called a symmetrical three-phase bolted fault or a solid fault. Bolted three-phase faults are rather uncommon. Generally, such faults give the maximum short-circuit currents and form the basis of calculations of short-circuit duties on switching devices.

Faults involving one, or more than one, phase and ground are called unsymmetrical faults. Under certain conditions, the line-to-ground fault or double line-to-ground fault currents may exceed three-phase symmetrical fault currents, discussed in the chapters to follow. Unsymmetrical faults are more common as compared to three-phase faults, that is, a support insulator on one of the phases on a transmission line may start flashing to ground, ultimately resulting in a single line-to-ground fault.

Short-circuit calculations are, thus, the primary study whenever a new power system is designed or an expansion and upgrade of an existing system are planned.

1.1 Nature of Short-Circuit Currents

The transient analysis of the short circuit of a passive impedance connected to an alternating current (ac) source gives an initial insight into the nature of the short-circuit currents. Consider a sinusoidal time-invariant single-phase 60 Hz source of power, $E_m \sin \omega t$, connected to a single-phase short distribution line, $Z = (R + j\omega L)$, where Z is the complex impedance, R and L are the resistance and inductance, E_m is the peak source voltage, and ω is the angular frequency $= 2\pi f$, f being the frequency of the ac source. For a balanced three-phase system, a single-phase model is adequate, as we will discuss further. Let a short circuit occur at the far end of the line terminals. As an ideal voltage source is considered, that is, zero Thévenin impedance, the short-circuit current is limited only by Z, and its steady-state value is vectorially given by E_m/Z. This assumes that the impedance Z does not change with flow of the large short-circuit current. For simplification of empirical short-circuit calculations, the impedances of static components like transmission lines, cables, reactors, and transformers are assumed to be time invariant. Practically, this is not true, that is, the flux densities and saturation characteristics of core materials in a transformer may entirely change its leakage reactance. Driven to saturation under high current flow, distorted waveforms and harmonics may be produced.

Ignoring these effects and assuming that Z is time invariant during a short circuit, the transient and steady-state currents are given by the differential equation of the R–L circuit with an applied sinusoidal voltage:

$$L\frac{di}{dt} + Ri = E_m \sin(\omega t + \theta) \tag{1.1}$$

where θ is the angle on the voltage wave, at which the fault occurs. The solution of this differential equation is given by

$$i = I_m \sin(\omega t + \theta - \phi) - I_m \sin(\theta - \phi)e^{-Rt/L} \tag{1.2}$$

where I_m is the maximum steady-state current, given by E_m/Z, and the angle

$$\phi = \frac{\tan^{-1}(\omega L)}{R}.$$

In power systems $\omega L \gg R$. A 100 MVA, 0.85 power factor synchronous generator may have an X/R of 110, and a transformer of the same rating, an X/R of 45. The X/R ratios in low-voltage systems are on the order of 2–8. For present discussions, assume a high X/R ratio, that is, $\phi \approx 90°$.

If a short circuit occurs at an instant $t=0$, $\theta=0$ (i.e., when the voltage wave is crossing through zero amplitude on the X-axis), the instantaneous value of the short-circuit current, from Equation 1.2, is $2I_m$. This is sometimes called *the doubling effect.*

If a short circuit occurs at an instant when the voltage wave peaks, $t=0, =\pi/2$, the second term in Equation 1.2 is zero and there is no transient component.

These two situations are shown in Figure 1.1a and b. The voltage at the point of bolted fault will be zero. The voltage E shown in Figure 1.1a and b signifies that prior to fault and after the fault is cleared, the voltage remains constant.

A simple explanation of the origin of the transient component is that in power systems the inductive component of the impedance is high. The current in such a circuit is at zero value when the voltage is at peak, and for a fault at this instant no direct current (dc) component is required to satisfy the physical law that the current in an inductive circuit cannot change

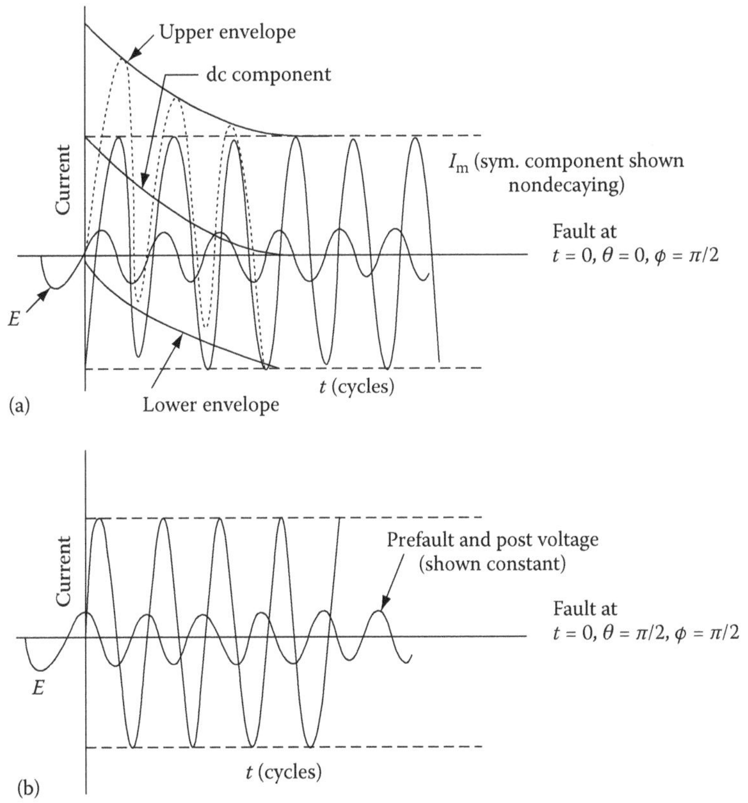

FIGURE 1.1
(a) Terminal short circuit of time-invariant impedance, current waveforms with maximum asymmetry; (b) current waveform with no dc component.

suddenly. When the fault occurs at an instant when $\theta = 0$, there has to be a transient current whose initial value is equal and opposite to the instantaneous value of the ac short-circuit current. This transient current, the second term of Equation 1.2 can be called a dc component and it decays at an exponential rate. Equation 1.2 can be simply written as

$$i = I_{\mathrm{m}} \sin \omega t + I_{\mathrm{dc}} e^{-Rt/L} \tag{1.3}$$

$$\text{where the initial value of } I_{\mathrm{dc}} = I_{\mathrm{m}} \tag{1.4}$$

The following inferences can be drawn from the above discussions:

1. There are two distinct components of a short-circuit current: (1) a non-decaying ac component or the steady-state component, and (2) a decaying dc component at an exponential rate, the initial magnitude of which is a maximum of the ac component and it depends on the time on the voltage wave at which the fault occurs.
2. The decrement factor of a decaying exponential current can be defined as its value any time after a short circuit, expressed as a function of its initial magnitude per unit. Factor L/R can be termed the time constant. The exponential then becomes $I_{\mathrm{dc}} e^{t/t'}$, where $t' = L/R$. In this equation, making $t = t' =$ time constant will result in a decay of approximately 62.3% from its initial magnitude, that is, the transitory current is reduced to a value of 0.368 per unit after an elapsed time equal to the time constant, as shown in Figure 1.2.
3. The presence of a dc component makes the fault current wave-shape envelope asymmetrical about the zero line and axis of the wave. Figure 1.1a clearly shows the profile of an asymmetrical waveform. The dc component always decays to zero in a short time. Consider a modest X/R ratio of 15, say, for a medium-voltage 13.8 kV system. The dc component decays to 88% of its initial value in five cycles. The higher the X/R ratio, the slower is the decay and the longer is the time for which the asymmetry in the total current will be sustained. The stored energy can be thought to be expanded in I^2R losses. After the decay of the dc component, only the symmetrical component of the short-circuit current remains.
4. Impedance is considered as time invariant in the above scenario. Synchronous generators and dynamic loads, that is, synchronous and induction motors are the major sources of short-circuit currents. The trapped flux in these rotating machines at the instant of short circuit cannot change suddenly and decays, depending on machine time constants. Thus, the assumption of constant L is not valid for rotating machines and decay in the ac component of the short-circuit current must also be considered.

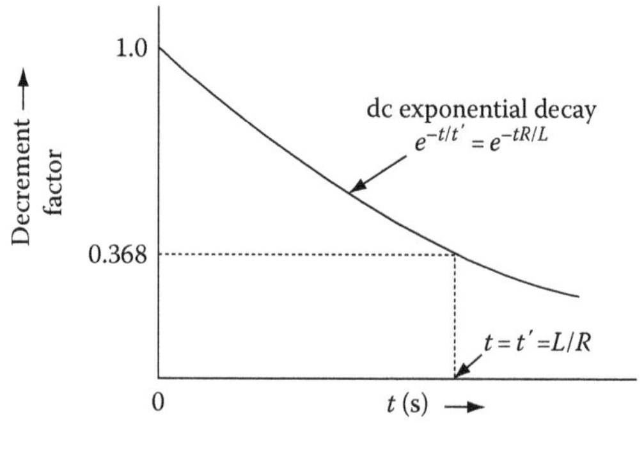

FIGURE 1.2
Time constant of dc-component decay.

5. In a three-phase system, the phases are time displaced from each other by 120 electrical degrees. If a fault occurs when the unidirectional component in phase *a* is zero, the phase *b* component is positive and the phase *c* component is equal in magnitude and negative. Figure 1.3 shows a three-phase fault current waveform. As the fault is symmetrical, $I_a + I_b + I_c$ is zero at any instant, where I_a, I_b, and I_c are the short-circuit currents in phases *a*, *b*, and *c*, respectively. For a fault close to a synchronous generator, there is a 120 Hz current also, which rapidly decays to zero. This gives rise to the characteristic non-sinusoidal shape of three-phase short-circuit currents observed in test oscillograms. The effect is insignificant, and ignored in the short-circuit calculations. This is further discussed in Chapter 6.
6. The load current has been ignored. Generally, this is true for empirical short-circuit calculations, as the short-circuit current is much higher than the load current. Sometimes the load current is a considerable percentage of the short-circuit current. The load currents determine the effective voltages of the short-circuit sources, prior to fault.

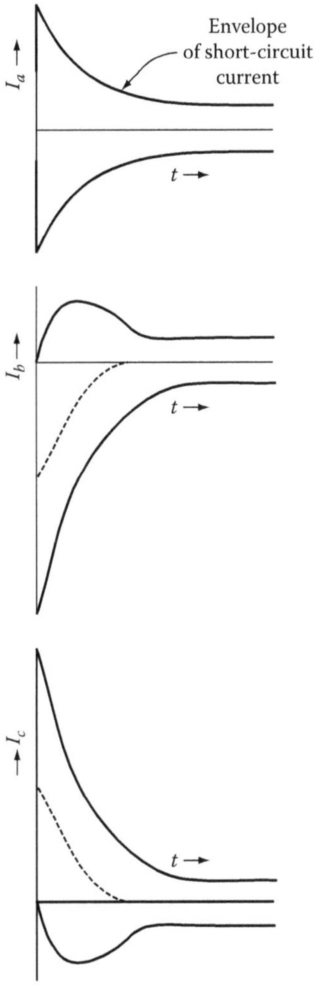

FIGURE 1.3
Asymmetries in phase currents in a three-phase short circuit.

The ac short-circuit current sources are synchronous machines, that is, turbogenerators and salient pole generators, asynchronous generators, and synchronous and asynchronous motors. Converter motor drives may contribute to short-circuit currents when operating in the inverter or regenerative mode. For extended duration of short-circuit currents, the control and excitation systems, generator voltage regulators, and turbine governor characteristics affect the transient short-circuit process.

The duration of a short-circuit current depends mainly on the speed of operation of protective devices and on the interrupting time of the switching devices.

1.2 Symmetrical Components

The method of symmetrical components has been widely used in the analysis of unbalanced three-phase systems, unsymmetrical short-circuit currents, and rotating electrodynamic machinery. The method was originally presented by Fortescue in 1918 and has been popular ever since.

Unbalance occurs in three-phase power systems due to faults, single-phase loads, untransposed transmission lines, or nonequilateral conductor spacings. In a three-phase balanced system, it is sufficient to determine the currents and voltages in one phase, and the currents and voltages in the other two phases are simply phase displaced. In an unbalanced system, the simplicity of modeling a three-phase system as a single-phase system is not valid. A convenient way of analyzing unbalanced operation is through symmetrical components. The three-phase voltages and currents, which may be unbalanced, are transformed into three sets of balanced voltages and currents, called symmetrical components. The impedances presented by various power system components, that is, transformers, generators, and transmission lines, to symmetrical components are *decoupled* from each other, resulting in independent networks for each component. These form a balanced set. This simplifies the calculations.

Familiarity with electrical circuits and machine theory, per unit system, and matrix techniques is required before proceeding with this book. A review of the matrix techniques in power systems is included in Appendix A. The notations described in this appendix for vectors and matrices are followed throughout the book.

The basic theory of symmetrical components can be stated as a mathematical concept. A system of three coplanar vectors is completely defined by six parameters, and the system can be said to possess six degrees of freedom. A point in a straight line being constrained to lie on the line possesses but one degree of freedom, and by the same analogy, a point in space has three degrees of freedom. A coplanar vector is defined by its terminal and length and therefore possesses two degrees of freedom. A system of coplanar vectors having six degrees of freedom, that is, a three-phase unbalanced current or voltage vectors, can be represented by three symmetrical systems of vectors each having two degrees of freedom. In general, a system of n numbers can be resolved into n sets of component numbers each having n components, that is, a total of n^2 components. Fortescue demonstrated that an unbalanced set on n phasors can be resolved into $n-1$ balanced phase systems of different phase sequence and one zero sequence system, in which all phasors are of equal magnitude and cophasial:

$$V_a = V_{a1} + V_{a2} + V_{a3} + \cdots + V_{an}$$
$$V_b = V_{b1} + V_{b2} + V_{b3} + \cdots + V_{bn} \tag{1.5}$$
$$V_n = V_{n1} + V_{n2} + V_{n3} + \cdots + V_{nn}$$

where

$V_a, V_b, \ldots, V_n$ are original n unbalanced voltage phasors
$V_{a1}, V_{b1}, \ldots, V_{n1}$ are the first set of n *balanced* phasors, at an angle of $2\pi/n$ between them
$V_{a2}, V_{b2}, \ldots, V_{n2}$ are the second set of n balanced phasors at an angle $4\pi/n$
$V_{an}, V_{bn}, \ldots, V_{nn}$ is the zero sequence set, all phasors at $n(2\pi/n) = 2\pi$, that is, cophasial

In a symmetrical three-phase balanced system, the generators produce balanced voltages, which are displaced from each other by $2\pi/3 = 120°$. These voltages can be called positive sequence voltages. If a vector operator a is defined, which rotates a unit vector through 120° in a counterclockwise direction, then $a = -0.5 + j0.866$, $a^2 = -0.5 - j0.866$, $a^3 = 1$, $1 + a^2 + a = 0$. Considering a three-phase system, Equation 1.5 reduces to

$$V_a = V_{a0} + V_{a1} + V_{a2}$$
$$V_b = V_{b0} + V_{b1} + V_{b2} \tag{1.6}$$
$$V_c = V_{c0} + V_{c1} + V_{c2}$$

We can define the set consisting of V_{a0}, V_{b0}, and V_{c0} as the zero sequence set, the set V_{a1}, V_{b1}, and V_{c1} as the positive sequence set, and the set V_{a2}, V_{b2}, and V_{c2} as the negative sequence set of voltages. The three original unbalanced voltage vectors give rise to nine voltage vectors, which must have constraints of freedom and are not totally independent. By definition of positive sequence, V_{a1}, V_{b1}, and V_{c1} should be related as follows, as in a normal balanced system:

$$V_{b1} = a^2 V_{a1}, \quad V_{c1} = aV_{a1}$$

Note that V_{a1} phasor is taken as the reference vector.

The negative sequence set can be similarly defined, but of opposite phase sequence:

$$V_{b2} = aV_{a2}, \quad V_{c2} = a^2 V_{a2}$$

Also, $V_{a0} = V_{b0} = V_{c0}$. With these relations defined, Equation 1.6 can be written as

$$\begin{vmatrix} V_a \\ V_b \\ V_c \end{vmatrix} = \begin{vmatrix} 1 & 1 & 1 \\ 1 & a^2 & a \\ 1 & a & a^2 \end{vmatrix} \begin{vmatrix} V_{a0} \\ V_{a1} \\ V_{a2} \end{vmatrix} \tag{1.7}$$

or in the abbreviated form

$$\overline{V}_{abc} = \overline{T}_s \overline{V}_{012} \tag{1.8}$$

where $\overline{T}_s$ is the transformation matrix. Its inverse will give the reverse transformation.

While this simple explanation may be adequate, a better insight into the symmetrical component theory can be gained through matrix concepts of similarity transformation, diagonalization, eigenvalues, and eigenvectors.

The discussions to follow show that

- Eigenvectors giving rise to symmetrical component transformation are the same though the eigenvalues differ. Thus, these vectors are not unique.
- The Clarke component transformation is based on the same eigenvectors but different eigenvalues.
- The symmetrical component transformation does not uncouple an initially unbalanced three-phase system. *Prima facie* this is a contradiction of what we said earlier, that the main advantage of symmetrical components lies in decoupling unbalanced systems, which could then be represented much akin to three-phase balanced systems. We will explain what is meant by this statement as we proceed.

1.3 Eigenvalues and Eigenvectors

The concept of eigenvalues and eigenvectors is related to the derivation of symmetrical component transformation. It can be briefly stated as follows.

Consider an arbitrary square matrix $\overline{A}$. If a relation exists so that

$$\overline{A}\overline{x} = \lambda\overline{x} \tag{1.9}$$

where

λ is a scalar called an eigenvalue, characteristic value, or root of the matrix $\overline{A}$

$\overline{x}$ is a vector called the eigenvector or characteristic vector of $\overline{A}$

Then, there are n eigenvalues and corresponding n sets of eigenvectors associated with an arbitrary matrix $\overline{A}$ of dimensions $n \times n$. The eigenvalues are not necessarily distinct, and multiple roots occur.

Equation 1.9 can be written as

$$[\overline{A} - \lambda I]\,[\overline{x}] = 0 \tag{1.10}$$

where I is the identity matrix. Expanding:

$$\begin{vmatrix} a_{11}-\lambda & a_{12} & a_{13} & \cdots & a_1n \\ a_{12} & a_{22}-\lambda & a_{23} & \cdots & a_2n \\ \cdots & \cdots & \cdots & \cdots & \cdots \\ a_{n1} & a_{n2} & a_{n3} & \cdots & a_{nn}-\lambda \end{vmatrix} \begin{vmatrix} x_1 \\ x_2 \\ \cdots \\ x_n \end{vmatrix} = \begin{vmatrix} 0 \\ 0 \\ \cdots \\ 0 \end{vmatrix} \tag{1.11}$$

This represents a set of homogeneous linear equations. Determinant $|A - \lambda I|$ must be zero as $\overline{x} \neq 0$.

$$|\overline{A} - \lambda I| = 0 \tag{1.12}$$

This can be expanded to yield an nth order algebraic equation:

$$a_n\lambda^n + a_n - I\lambda^n - 1 + \cdots + a_1\lambda + a_0 = 0, \text{ that is,}$$
$$(\lambda_1 - a_1)(\lambda_2 - a_2)\ldots(\lambda_n - a_n) = 0 \tag{1.13}$$

Equations 1.12 and 1.13 are called the characteristic equations of the matrix $\overline{A}$. The roots $\lambda_1, \lambda_2, \lambda_3, \ldots, \lambda_n$ are the eigenvalues of matrix $\overline{A}$. The eigenvector $\overline{x}_j$ corresponding to $\overline{\lambda}_j$ is found from Equation 1.10. See Appendix A for details and an example.

1.4 Symmetrical Component Transformation

Application of eigenvalues and eigenvectors to the decoupling of three-phase systems is useful when we define similarity transformation. This forms a diagonalization technique and decoupling through symmetrical components.

1.4.1 Similarity Transformation

Consider a system of linear equations:

$$\overline{A}\overline{x} = \overline{y} \tag{1.14}$$

A transformation matrix $\overline{C}$ can be introduced to relate the original vectors $\overline{x}$ and $\overline{y}$ to new sets of vectors $\overline{x}_n$ and $\overline{y}_n$ so that

$$\overline{x} = \overline{C}\overline{x}_n$$
$$\overline{y} = \overline{C}\overline{y}_n$$
$$\overline{A}\,\overline{C}\overline{x}_n = \overline{C}\overline{y}_n$$
$$\overline{C}^{-1}\overline{A}\,\overline{C}\overline{x}_n = \overline{C}^{-1}\overline{C}\overline{y}_n$$
$$\overline{C}^{-1}\overline{A}\,\overline{C}\overline{x}_n = \overline{y}_n$$

This can be written as

$$\begin{aligned}\overline{A}_n\overline{x}_n &= \overline{y}_n\\ \overline{A}_n &= \overline{C}^{-1}\overline{A}\,\overline{C}\end{aligned} \tag{1.15}$$

$\overline{A}_n\overline{x}_n = \overline{y}_n$ is distinct from $\overline{A}\overline{x} = \overline{y}$. The only restriction on choosing $\overline{C}$ is that it should be nonsingular. Equation 1.15 is a set of linear equations, derived from the original equations (Equation 1.14) and yet distinct from them.

If $\overline{C}$ is a nodal matrix $\overline{M}$, corresponding to the coefficients of $\overline{A}$, then

$$\overline{C} = \overline{M} = [x_1, x_2 \cdots x_n] \tag{1.16}$$

where $\overline{x}_i$ are the eigenvectors of the matrix $\overline{A}$, then

$$\begin{aligned}\overline{C}^{-1}\overline{A}\,\overline{C} &= \overline{C}^{-1}\overline{A}[x_1, x_2, \ldots, x_n] \\ &= \overline{C}^{-1}[\overline{A}x_1, \overline{A}x_2, \ldots, \overline{A}x_n] \\ &= \overline{C}^{-1}[\lambda_1 x_1, \lambda_2 x_2, \ldots, \lambda_n x_n] \\ &= C^{-1}[x_1, x_2, \ldots x_n]\begin{vmatrix}\lambda_1 & & & \\ & \lambda_2 & & \\ & & \cdot & \\ & & & \lambda_n\end{vmatrix} \\ &= \overline{C}^{-1}\overline{C}\begin{vmatrix}\lambda_1 & & & \\ & \lambda_2 & & \\ & & \cdot & \\ & & & \lambda_n\end{vmatrix} \\ &= \overline{\lambda}\end{aligned} \tag{1.17}$$

Thus, $\overline{C}^{-1}\,\overline{A}\,\overline{C}$ is reduced to a diagonal matrix $\overline{\lambda}$, called a spectral matrix. Its diagonal elements are the eigenvalues of the original matrix $\overline{A}$. The new system of equations is an uncoupled system. Equations 1.14 and 1.15 constitute a similarity transformation of matrix $\overline{A}$. The matrices $\overline{A}$ and $\overline{A}_n$ have the same eigenvalues and are called similar matrices. The transformation matrix $\overline{C}$ is nonsingular.

1.4.2 Decoupling a Three-Phase Symmetrical System

Let us decouple a three-phase transmission line section, where each phase has a mutual coupling with respect to ground. This is shown in Figure 1.4a. An impedance matrix of the three-phase transmission line can be written as

$$\begin{vmatrix}Z_{aa} & Z_{ab} & Z_{ac} \\ Z_{ba} & Z_{bb} & Z_{bc} \\ Z_{ca} & Z_{cb} & Z_{cc}\end{vmatrix} \tag{1.18}$$

where

Z_{aa}, Z_{bb}, and Z_{cc} are the self-impedances of the phases a, b, and c

Z_{ab} is the mutual impedance between phases a and b

Z_{ba} is the mutual impedance between phases b and a

Assume that the line is *perfectly symmetrical*. This means all the mutual impedances, that is, $Z_{ab} = Z_{ba} = M$ and all the self-impedances, that is, $Z_{aa} = Z_{bb} = Z_{cc} = Z$ are equal. This reduces the impedance matrix to

$$\begin{vmatrix}Z & M & M \\ M & Z & M \\ M & M & Z\end{vmatrix} \tag{1.19}$$

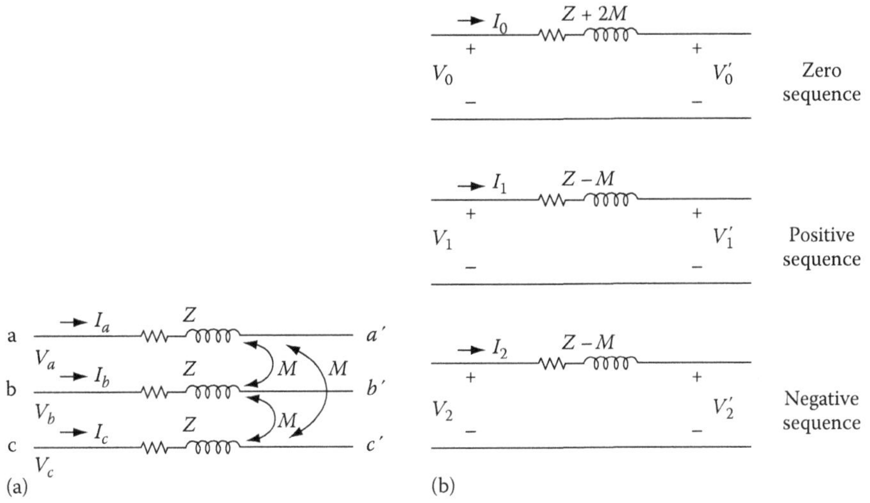

FIGURE 1.4
(a) Impedances in a three-phase transmission line with mutual coupling between phases; (b) resolution into symmetrical component impedances.

It is required to decouple this system using symmetrical components. First find the eigenvalues:

$$\begin{vmatrix} Z-\lambda & M & M \\ M & Z-\lambda & M \\ M & M & Z-\lambda \end{vmatrix} = 0 \tag{1.20}$$

The eigenvalues λ, can be

$$\begin{gathered} Z+2M \\ Z-M \\ Z-M \end{gathered}$$

The eigenvectors can be found by making $\lambda = Z + 2M$ and then $Z - M$. Substituting $\lambda = Z + 2M$:

$$\begin{vmatrix} Z-(Z+2M) & M & M \\ M & Z-(Z+2M) & M \\ M & M & Z-(Z+2M) \end{vmatrix} \begin{vmatrix} X_1 \\ X_2 \\ X_3 \end{vmatrix} = 0 \tag{1.21}$$

This can be reduced to

$$\begin{vmatrix} -2 & 1 & 1 \\ 0 & -1 & 1 \\ 0 & 0 & 0 \end{vmatrix} \begin{vmatrix} X_1 \\ X_2 \\ X_3 \end{vmatrix} = 0 \tag{1.22}$$

This gives $X_1 = X_2 = X_3 =$ any arbitrary constant k. Thus, one of the eigenvectors of the impedance matrix is

$$\begin{vmatrix} k \\ k \\ k \end{vmatrix} \quad (1.23)$$

It can be called the zero sequence eigenvector of the symmetrical component transformation matrix and can be written as

$$\begin{vmatrix} 1 \\ 1 \\ 1 \end{vmatrix} \quad (1.24)$$

Similarly for $\lambda = Z - M$:

$$\begin{vmatrix} Z-(Z-M) & M & M \\ M & Z-(Z-M) & M \\ M & M & Z-(Z-M) \end{vmatrix} \begin{vmatrix} X_1 \\ X_2 \\ X_3 \end{vmatrix} = 0 \quad (1.25)$$

which gives

$$\begin{vmatrix} 1 & 1 & 1 \\ 0 & 0 & 0 \\ 0 & 0 & 0 \end{vmatrix} \begin{vmatrix} X_1 \\ X_2 \\ X_3 \end{vmatrix} = 0 \quad (1.26)$$

This gives the general relation $X_1 + X_2 + X_3 = 0$. Any choice of X_1, X_2, X_3 that satisfies this relation is a solution vector. Some choices are shown below.

$$\begin{vmatrix} X_1 \\ X_2 \\ X_3 \end{vmatrix} = \begin{vmatrix} 1 \\ a^2 \\ a \end{vmatrix}, \begin{vmatrix} 1 \\ a \\ a^2 \end{vmatrix}, \begin{vmatrix} 0 \\ \sqrt{3}/2 \\ -\sqrt{3}/2 \end{vmatrix}, \begin{vmatrix} 1 \\ -1/2 \\ -1/2 \end{vmatrix} \quad (1.27)$$

where a is a unit vector operator, which rotates by 120° in the counterclockwise direction, as defined before.

Equation 1.27 is an important result and shows that, for perfectly symmetrical systems, the common eigenvectors are the same, although the eigenvalues are different in each system. The Clarke component transformation (described in Section 1.5) is based on this observation.

The symmetrical component transformation is given by solution vectors:

$$\begin{vmatrix} 1 \\ 1 \\ 1 \end{vmatrix} \begin{vmatrix} 1 \\ a \\ a^2 \end{vmatrix} \begin{vmatrix} 1 \\ a^2 \\ a \end{vmatrix} \quad (1.28)$$

A symmetrical component transformation matrix can, therefore, be written as

$$\overline{T}_s = \begin{vmatrix} 1 & 1 & 1 \\ 1 & a^2 & a \\ 1 & a & a^2 \end{vmatrix} \quad (1.29)$$

This is the same matrix as was arrived at in Equation 1.8. Its inverse is

$$\overline{T}_s^{-1} = \frac{1}{3}\begin{vmatrix} 1 & 1 & 1 \\ 1 & a & a^2 \\ 1 & a^2 & a \end{vmatrix} \tag{1.30}$$

For the transformation of currents, we can write

$$\overline{I}_{abc} = \overline{T}_s \overline{I}_{012} \tag{1.31}$$

where $\overline{I}_{abc}$, the original currents in phases *a*, *b*, and *c*, are transformed into zero sequence, positive sequence, and negative sequence currents, $\overline{I}_{012}$. The original phasors are subscripted *abc* and the sequence components are subscripted 012. Similarly, for transformation of voltages:

$$\overline{V}_{abc} = \overline{T}_s \overline{V}_{012} \tag{1.32}$$

Conversely,

$$\overline{I}_{012} = \overline{T}_s^{-1} \overline{I}_{abc}, \quad \overline{V}_{012} = \overline{T}_s^{-1} \overline{V}_{abc} \tag{1.33}$$

The transformation of impedance is not straightforward and is derived as follows:

$$\begin{aligned} \overline{V}_{abc} &= \overline{Z}_{abc} \overline{I}_{abc} \\ \overline{T}_s \overline{V}_{012} &= \overline{Z}_{abc} \overline{T}_s \overline{I}_{012} \\ \overline{V}_{012} &= \overline{T}_s^{-1} \overline{Z}_{abc} \overline{T}_s \overline{I}_{012} = \overline{Z}_{012} \overline{I}_{012} \end{aligned} \tag{1.34}$$

Therefore,

$$\overline{Z}_{012} = \overline{T}_s^{-1} \overline{Z}_{abc} \overline{T}_s \tag{1.35}$$

$$\overline{Z}_{abc} = \overline{T}_s \overline{Z}_{012} \overline{T}_s^{-1} \tag{1.36}$$

Applying the impedance transformation to the original impedance matrix of the three-phase symmetrical transmission line in Equation 1.19, the transformed matrix is

$$\begin{aligned} \overline{Z}_{012} &= \frac{1}{3}\begin{vmatrix} 1 & 1 & 1 \\ 1 & a & a^2 \\ 1 & a^2 & a \end{vmatrix} \begin{vmatrix} Z & M & M \\ M & Z & M \\ M & M & Z \end{vmatrix} \begin{vmatrix} 1 & 1 & 1 \\ 1 & a^2 & a \\ 1 & a & a^2 \end{vmatrix} \\ &= \frac{1}{3}\begin{vmatrix} Z+2M & 0 & 0 \\ 0 & Z-M & 0 \\ 0 & 0 & Z-M \end{vmatrix} \end{aligned} \tag{1.37}$$

The original three-phase coupled system has been decoupled through symmetrical component transformation. It is diagonal, and all off-diagonal terms are zero, meaning that there is no coupling between the sequence components. Decoupled positive, negative, and zero sequence networks are shown in Figure 1.4b.

1.4.3 Decoupling a Three-Phase Unsymmetrical System

Now consider that the original three-phase system is not completely balanced. Ignoring the mutual impedances in Equation 1.18, let us assume unequal phase impedances, Z_1, Z_2, and Z_3, that is, the impedance matrix is

$$\overline{Z}_{abc} = \begin{vmatrix} Z_1 & 0 & 0 \\ 0 & Z_2 & 0 \\ 0 & 0 & Z_3 \end{vmatrix} \tag{1.38}$$

The symmetrical component transformation is

$$\overline{Z}_{012} = \frac{1}{3}\begin{vmatrix} 1 & 1 & 1 \\ 1 & a & a^2 \\ 1 & a^2 & a \end{vmatrix}\begin{vmatrix} Z_1 & 0 & 0 \\ 0 & Z_2 & 0 \\ 0 & 0 & Z_3 \end{vmatrix}\begin{vmatrix} 1 & 1 & 1 \\ 1 & a^2 & a \\ 1 & a & a^2 \end{vmatrix}$$

$$= \frac{1}{3}\begin{vmatrix} Z_1+Z_2+Z_3 & Z_1+a^2Z_2+aZ_3 & Z_1+aZ_2+a^2Z_3 \\ Z_1+aZ_2+a^2Z_3 & Z_1+Z_2+Z_3 & Z_1+a^2Z_2+aZ_3 \\ Z_1+a^2Z_2+aZ_3 & Z_1+aZ_2+a^2Z_3 & Z_1+Z_2+Z_3 \end{vmatrix} \tag{1.39}$$

The resulting matrix shows that the *original unbalanced system is not decoupled*. If we start with equal self-impedances and unequal mutual impedances or vice versa, the resulting matrix is nonsymmetrical. It is a minor problem today, as nonreciprocal networks can be easily handled on digital computers. Nevertheless, the main application of symmetrical components is for the study of unsymmetrical faults. Negative sequence relaying, stability calculations, and machine modeling are some other examples. It is assumed that the system is perfectly symmetrical before an unbalance condition occurs. *The asymmetry occurs only at the fault point*. The symmetrical portion of the network is considered to be isolated, to which an unbalanced condition is applied at the fault point. In other words, the unbalance part of the network can be thought to be connected to the balanced system at the point of fault. Practically, the power systems are not perfectly balanced and some asymmetry always exists. However, the error introduced by ignoring this asymmetry is small. (This may not be true for highly unbalanced systems and single-phase loads.)

1.4.4 Power Invariance in Symmetrical Component Transformation

Symmetrical component transformation is power invariant. The complex power in a three-phase circuit is given by

$$S = V_aI_a^* + V_bI_b^* + V_cI_c^* = \overline{V}'_{abc}\overline{I}^*_{abc} \tag{1.40}$$

where I_a^* is the complex conjugate of I_a. This can be written as

$$S = [\overline{T}_s\overline{V}_{012}]\overline{T}_s^*\overline{I}_{012}^* = \overline{V}'_{012}\overline{T}'_s\overline{T}_s^*\overline{I}_{012}^* \tag{1.41}$$

The product $\overline{T}_s^t\overline{T}_s^*$ is given by (see Appendix A)

$$\overline{T}'_s\overline{T}_s^* = 3\begin{vmatrix} 1 & 0 & 0 \\ 0 & 1 & 0 \\ 0 & 0 & 1 \end{vmatrix} \tag{1.42}$$

Thus,

$$S = 3V_1I_1^* + 3V_2I_2^* + 3V_0I_0^* \tag{1.43}$$

This shows that complex power can be calculated from symmetrical components, and is the sum of the symmetrical component powers.

1.5 Clarke Component Transformation

It has been already shown that, for perfectly symmetrical systems, the component eigenvectors are the same, but eigenvalues can be different. The Clarke component transformation is defined as

$$\begin{vmatrix} V_a \\ V_b \\ V_c \end{vmatrix} = \begin{vmatrix} 1 & 1 & 0 \\ 1 & -\frac{1}{2} & \frac{\sqrt{3}}{2} \\ 1 & -\frac{1}{2} & -\frac{\sqrt{3}}{2} \end{vmatrix}\begin{vmatrix} V_0 \\ V_\alpha \\ V_\beta \end{vmatrix} \tag{1.44}$$

Note that the eigenvalues satisfy the relations derived in Equation 1.27, and

$$\begin{vmatrix} V_0 \\ V_\alpha \\ V_\beta \end{vmatrix} = \begin{vmatrix} \frac{1}{3} & \frac{1}{3} & \frac{1}{3} \\ \frac{2}{3} & -\frac{1}{3} & -\frac{1}{3} \\ 0 & \frac{1}{\sqrt{3}} & -\frac{1}{\sqrt{3}} \end{vmatrix}\begin{vmatrix} V_a \\ V_b \\ V_c \end{vmatrix} \tag{1.45}$$

Similar equations may be written for the current. Note that

$$\begin{vmatrix} 1 \\ 1 \\ 1 \end{vmatrix}, \quad \begin{vmatrix} 1 \\ -\frac{1}{2} \\ -\frac{1}{2} \end{vmatrix}, \quad \begin{vmatrix} 0 \\ \frac{\sqrt{3}}{2} \\ -\frac{\sqrt{3}}{2} \end{vmatrix}$$

are the eigenvectors of a perfectly symmetrical impedance.

The transformation matrices are

$$\overline{T}_c = \begin{vmatrix} 1 & 1 & 0 \\ 1 & -1/2 & \sqrt{3}/2 \\ 1 & 1/2 & -\sqrt{3}/2 \end{vmatrix} \tag{1.46}$$

$$\overline{T}_c^{-1} = \begin{vmatrix} 1/3 & 1/3 & 1/3 \\ 2/3 & -1/3 & -1/3 \\ 0 & 1/\sqrt{3} & -1/\sqrt{3} \end{vmatrix} \tag{1.47}$$

and as before:

$$\overline{Z}_{0\alpha\beta} = \overline{T}_c^{-1}\overline{Z}_{abc}\overline{T}_c \tag{1.48}$$

$$\overline{Z}_{abc} = \overline{T}_c\overline{Z}_{0\alpha\beta}\overline{T}_c^{-1} \tag{1.49}$$

The Clarke component expression for a perfectly symmetrical system is

$$\begin{vmatrix} V_0 \\ V_\alpha \\ V_\beta \end{vmatrix} = \begin{vmatrix} Z_{00} & 0 & 0 \\ 0 & Z_{\alpha\alpha} & 0 \\ 0 & 0 & Z_{\beta\beta} \end{vmatrix} \begin{vmatrix} I_0 \\ I_\alpha \\ I_\beta \end{vmatrix} \tag{1.50}$$

The same philosophy of transformation can also be applied to systems with two or more three-phase circuits in parallel. The instantaneous power theory, Chapter 20, and EMTP (electromagnetic transient program) modeling of transmission lines are based upon Clarke's transformation.

1.6 Characteristics of Symmetrical Components

Matrix equations (1.32) and (1.33) are written in the expanded form

$$\begin{aligned} V_a &= V_0 + V_1 + V_2 \\ V_b &= V_0 + a^2V_1 + aV_2 \\ V_c &= V_0 + aV_1 + a^2V_2 \end{aligned} \tag{1.51}$$

and

$$\begin{aligned} V_0 &= \frac{1}{3}(V_a + V_b + V_c) \\ V_1 &= \frac{1}{3}(V_a + aV_b + a^2V_c) \\ V_2 &= \frac{1}{3}(V_a + a^2V_b + aV_c) \end{aligned} \tag{1.52}$$

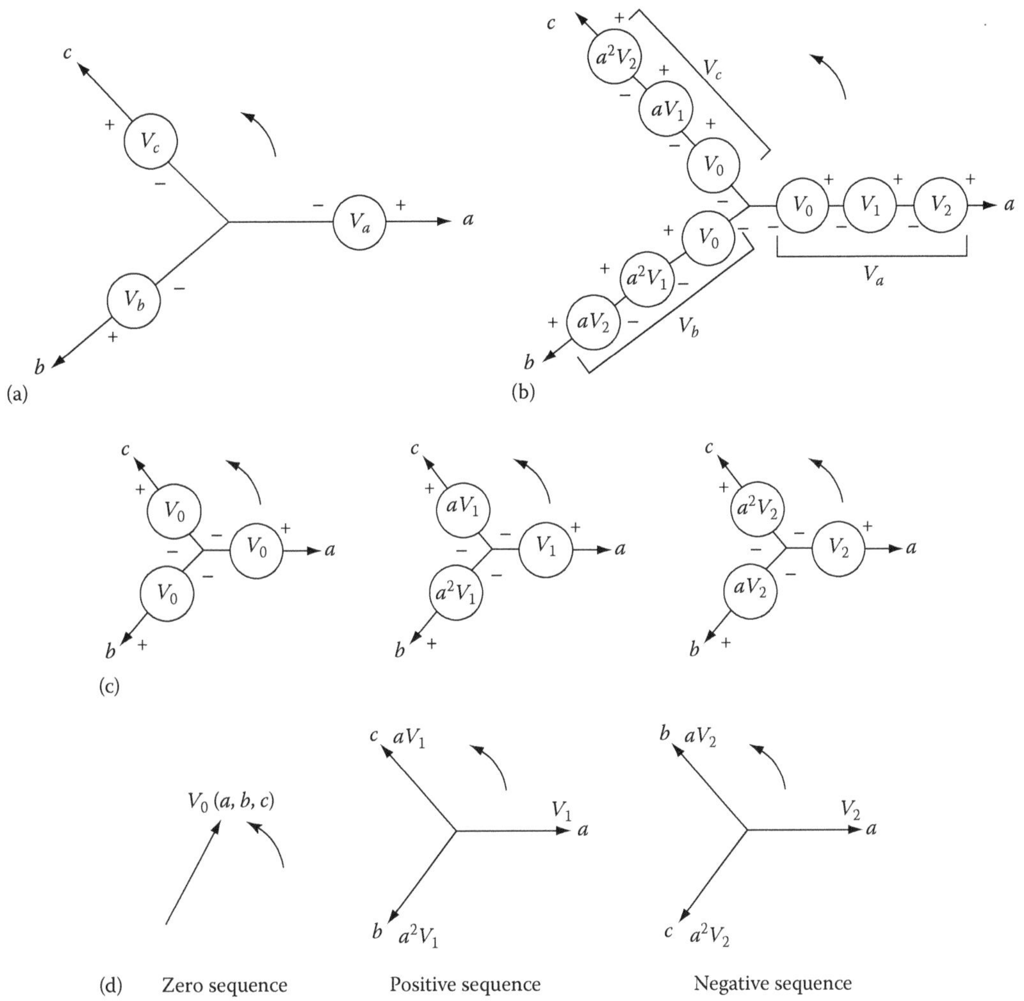

FIGURE 1.5
(a–d) Progressive resolution of voltage vectors into sequence components.

These relations are graphically represented in Figure 1.5, which clearly shows that phase voltages V_a, V_b, and V_c can be resolved into three voltages: V_0, V_1, and V_2, defined as follows:

- V_0 is the zero sequence voltage. It is of equal magnitude in all the three phases and is cophasial.
- V_1 is the system of balanced positive sequence voltages, *of the same phase sequence* as the original unbalanced system of voltages. It is of equal magnitude in each phase, but displaced by 120°, the component of phase *b* lagging the component of phase *a* by 120°, and the component of phase *c* leading the component of phase *a* by 120°.

- V_2 is the system of balanced negative sequence voltages. It is of equal magnitude in each phase, and there is a 120° phase displacement between the voltages, the component of phase *c* lagging the component of phase *a*, and the component of phase *b* leading the component of phase *a*.

Therefore, the positive and negative sequence voltages (or currents) can be defined as "the order in which the three phases attain a maximum value." For the positive sequence, the order is *abca* while for the negative sequence it is *acba*. We can also define positive and negative sequence by the order in which the phasors pass a *fixed point* on the vector plot. *Note that the rotation is counterclockwise for all three sets of sequence components, as was assumed for the original unbalanced vectors*, Figure 1.5d. Sometimes, this is confused and negative sequence rotation is said to be the *reverse of* positive sequence. *The negative sequence vectors do not rotate in a direction opposite to the positive sequence vectors*, though the negative phase sequence is opposite to the positive phase sequence.

Example 1.1

An unbalanced three-phase system has the following voltages:

$V_a = 0.9 < 0°$ per unit
$V_b = 1.25 < 280°$ per unit
$V_c = 0.6 < 110°$ per unit

The phase rotation is *abc*, counterclockwise. The unbalanced system is shown in Figure 1.6a. Resolve into symmetrical components and sketch the sequence voltages.

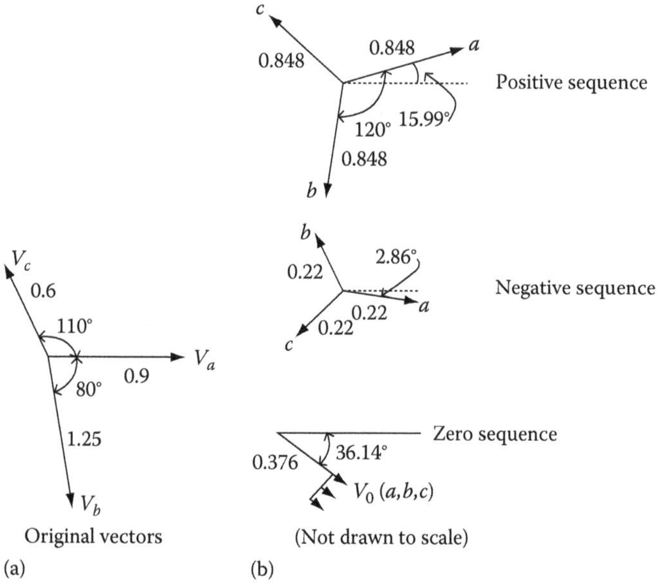

FIGURE 1.6
(a) Unbalanced voltage vectors; (b) resolution into symmetrical components.

Using the symmetrical component transformation, the resolution is shown in Figure 1.6b. The reader can verify this as an exercise and then convert back from the calculated sequence vectors into original *abc* voltages, graphically and analytically.

In a symmetrical system of three phases, the resolution of voltages or currents into a system of zero, positive, and negative components is equivalent to three separate systems. Sequence voltages act in isolation and produce zero, positive, and negative sequence currents, and the theorem of superposition applies. The following generalizations of symmetrical components can be made:

1. In a three-phase unfaulted system in which all loads are balanced and in which generators produce positive sequence voltages, only positive sequence currents flow, resulting in balanced voltage drops of the same sequence. There are no negative sequence or zero sequence voltage drops.
2. In symmetrical systems, the currents and voltages of different sequences do not affect each other, that is, positive sequence currents produce only positive sequence voltage drops. By the same analogy, the negative sequence currents produce only negative sequence drops, and zero sequence currents produce only zero sequence drops.
3. Negative and zero sequence currents are set up in circuits of unbalanced impedances only, that is, a set of unbalanced impedances in a symmetrical system may be regarded as a source of negative and zero sequence current. Positive sequence currents flowing in an unbalanced system produce positive, negative, and possibly zero sequence voltage drops. The negative sequence currents flowing in an unbalanced system produce voltage drops of all three sequences. The same is true about zero sequence currents.
4. In a three-phase three-wire system, no zero sequence currents appear in the line conductors. This is so because $I_0 = (1/3)(I_a + I_b + I_c)$ and, therefore, there is no path for the zero sequence current to flow. In a three-phase four-wire system with neutral return, the neutral must carry out-of-balance current, that is, $I_n = (I_a + I_b + I_c)$. Therefore, it follows that $I_n = 3I_0$. At the grounded neutral of a three-phase wye system, positive and negative sequence voltages are zero. The neutral voltage is equal to the zero sequence voltage or product of zero sequence current and three times the neutral impedance, Z_n.
5. From what has been said in point 4 above, phase conductors emanating from ungrounded wye- or delta-connected transformer windings cannot have zero sequence current. In a delta winding, zero sequence currents, if present, set up circulating currents in the delta winding itself. This is because the delta winding forms a closed path of low impedance for the zero sequence currents; each phase zero sequence voltage is absorbed by its own phase voltage drop and there are no zero sequence components at the terminals.

1.7 Sequence Impedance of Network Components

The impedance encountered by the symmetrical components depends on the type of power system equipment, that is, a generator, a transformer, or a transmission line. The sequence impedances are required for component modeling and analysis. We derived the sequence impedances of a symmetrical coupled transmission line in Equation 1.37. Zero sequence impedance of overhead lines depends on the presence of ground wires, tower footing resistance, and grounding. It may vary between two and six times the positive sequence impedance. The line capacitance of overhead lines is ignored in short-circuit calculations. Appendix B details three-phase matrix models of transmission lines, bundle conductors, and cables, and their transformation into symmetrical components. While estimating sequence impedances of power system components is one problem; constructing the zero, positive, and negative sequence impedance networks is the first step for unsymmetrical fault current calculations.

1.7.1 Construction of Sequence Networks

A sequence network shows how the sequence currents, if these are present, will flow in a system. Connections between sequence component networks are necessary to achieve this objective. The sequence networks are constructed as viewed from the *fault point*, which can be defined as the point at which the unbalance occurs in a system, that is, a fault or load unbalance.

The voltages for the sequence networks are taken as line-to-neutral voltages. The only active network containing the voltage source is the positive sequence network. Phase *a* voltage is taken as the reference voltage, and the voltages of the other two phases are expressed with reference to phase *a* voltage, as shown in Figure 1.5d.

The sequence networks for positive, negative, and zero sequence will have per phase impedance values, which may differ. Normally, the sequence impedance networks are constructed on the basis of per unit values on a common MVA base, and a base MVA of 100 is in common use. For nonrotating equipment like transformers, the impedance to negative sequence currents will be the same as for positive sequence currents. The impedance to negative sequence currents of rotating equipment will be different from the positive sequence impedance and, in general, for all apparatuses the impedance to zero sequence currents will be different from the positive or negative sequence impedances. For a study involving sequence components, the sequence impedance data can be (1) calculated by using subroutine computer programs, (2) obtained from manufacturers' data, (3) calculated by long-hand calculations, or (4) estimated from tables in published references.

The positive direction of current flow in each sequence network is *outward* at the faulted or unbalance point. This means that the sequence currents flow in the same direction in all three sequence networks.

Sequence networks are shown schematically in boxes in which the fault points from which the sequence currents flow outward are marked as F_1, F_2, and F_0, and the neutral buses are designated as N_1, N_2, and N_0, respectively, for the positive, negative, and zero sequence impedance networks. Each network forms a two-port network with Thévenin sequence voltages across sequence impedances. Figure 1.7 illustrates this basic formation. Note the direction of currents. The voltage across the sequence impedance rises from N to F. As stated before, only the positive sequence network has a voltage source, which is the Thevenin equivalent. With this convention, appropriate signs must be allocated to the sequence voltages:

$$
\begin{aligned}
V_1 &= V_a - I_1 Z_1 \\
V_2 &= -I_2 Z_2 \\
V_0 &= -I_0 Z_0
\end{aligned}
\tag{1.53}
$$

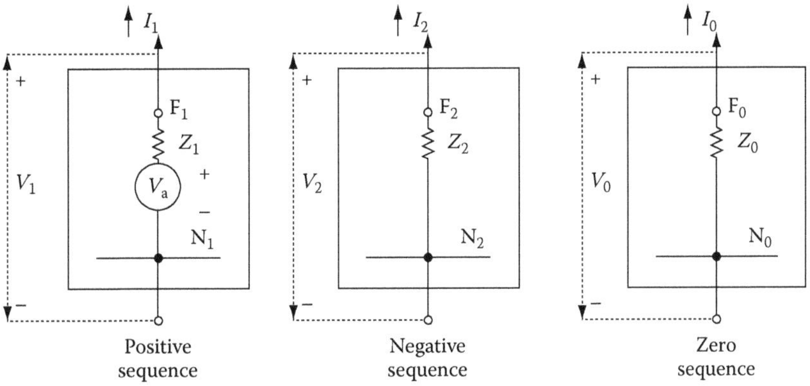

FIGURE 1.7
Positive, negative, and zero sequence network representation.

or in matrix form:

$$\begin{vmatrix} V_0 \\ V_1 \\ V_2 \end{vmatrix} = \begin{vmatrix} 0 \\ V_a \\ 0 \end{vmatrix} - \begin{vmatrix} Z_0 & 0 & 0 \\ 0 & Z_1 & 0 \\ 0 & 0 & Z_2 \end{vmatrix} \begin{vmatrix} I_0 \\ I_1 \\ I_2 \end{vmatrix} \tag{1.54}$$

Based on the discussions so far, we can graphically represent the sequence impedances of various system components.

1.7.2 Transformers

The positive and negative sequence impedances of a transformer can be taken to be equal to its leakage impedance. As the transformer is a static device, the positive or negative sequence impedances do not change with phase sequence of the applied balanced voltages. The zero sequence impedance can, however, vary from an open circuit to a low value, depending on the transformer winding connection, method of neutral grounding, and transformer construction, that is, core or shell type.

We will briefly discuss the shell and core form of construction, as it has a major impact on the zero sequence flux and impedance. Referring to Figure 1.8a, in a three-phase core-type transformer, the sum of the fluxes in each phase in a given direction along the cores is zero; however, the flux going up one leg of the core must return through the other two, that is, the magnetic circuit of a phase is completed through the other two phases in parallel. The magnetizing current per phase is that required for the core and part of the yoke. This means that in a three-phase core-type transformer the magnetizing current will be different in each phase. Generally, the cores are long compared to yokes and the yokes are of greater cross section. The yoke reluctance is only a small fraction of the core and the variation of magnetizing current per phase is not appreciable. However, consider now the zero sequence flux, which will be directed in one direction, in each of the core-legs. The return path lies, not through the core legs, but through insulating medium and tank.

In three separate single-phase transformers connected in three-phase configuration or in shell-type three-phase transformers, the magnetic circuits of each phase are complete in themselves and do not interact, Figure 1.8b. Due to the advantages in short-circuit and

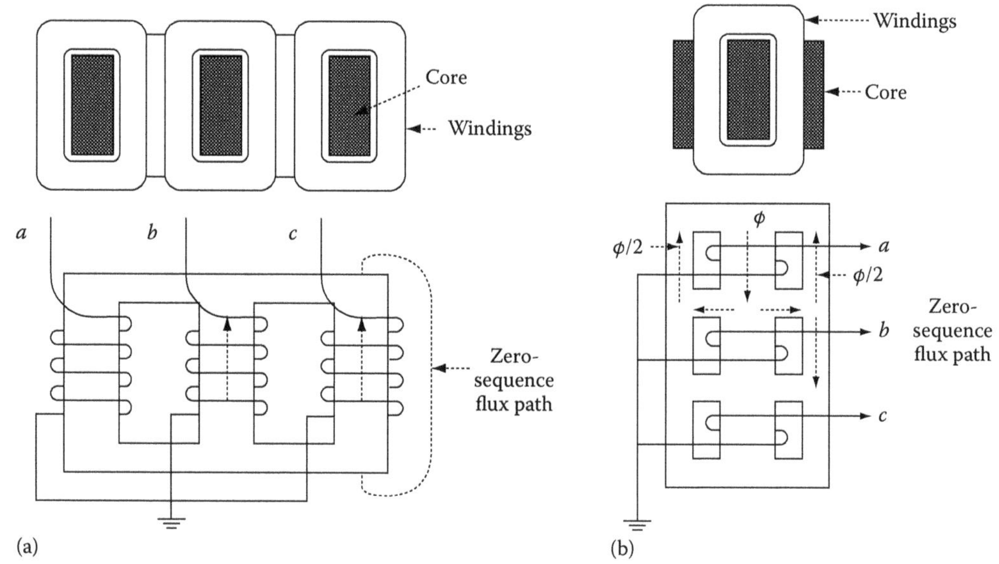

FIGURE 1.8
(a) Core form of three-phase transformer, flux paths for phase and zero sequence currents; (b) shell form of three-phase transformer.

transient voltage performance, the shell form is used for larger transformers. The variations in shell form have five- or seven-legged cores. Briefly, we can say that, in a core type, the windings surround the core, and in the shell type, the core surrounds the windings.

1.7.2.1 Delta–Wye or Wye–Delta Transformer

In a delta–wye transformer with the wye winding grounded, zero sequence impedance will be approximately equal to positive or negative sequence impedance, viewed from the wye connection side. Impedance to the flow of zero sequence currents in the core-type transformers is lower as compared to the positive sequence impedance. This is so, because there is no return path for zero sequence exciting flux in core type units except through insulating medium and tank, a path of high reluctance. In groups of three single-phase transformers or in three-phase shell-type transformers, the zero sequence impedance is higher.

The zero sequence network for a wye–delta transformer is constructed as shown in Figure 1.9a. The grounding of the wye neutral allows the zero sequence currents to return through the neutral and circulate in the windings to the source of unbalance. Thus, the circuit on the wye side is shown connected to the L side line. On the delta side, the circuit is open, as no zero sequence currents appear in the lines, though these currents circulate in the delta windings to balance the ampere turns in the wye windings. The circuit is open on the H side line, and the zero sequence impedance of the transformer seen from the high side is an open circuit. If the wye winding neutral is left isolated, Figure 1.9b, the circuit will be open on both sides, presenting an infinite impedance.

Three-phase current-flow diagrams can be constructed based on the convention that current always flows to the unbalance and that the ampere turns in primary windings must be balanced by the ampere turns in the secondary windings.

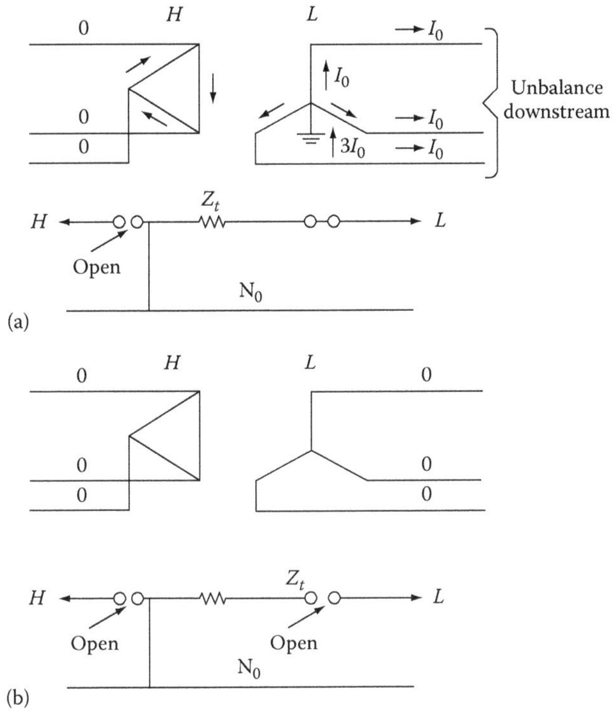

FIGURE 1.9
(a) Derivations of equivalent zero sequence circuit for a delta–wye transformer, wye neutral solidly grounded; (b) zero sequence circuit of a delta–wye transformer, wye neutral isolated.

1.7.2.2 Wye–Wye Transformer

In a wye–wye-connected transformer, with both neutrals isolated, no zero sequence currents can flow. The zero sequence equivalent circuit is open on *both* sides and presents an infinite impedance to the flow of zero sequence currents. When one of the neutrals is grounded, still no zero sequence currents can be transferred from the grounded side to the ungrounded side. With one neutral grounded, there are no balancing ampere turns in the ungrounded wye windings to enable current to flow in the grounded neutral windings. Thus, neither of the windings can carry a zero sequence current. Both neutrals must be grounded for the transfer of zero sequence currents.

A wye–wye-connected transformer with isolated neutrals is not used, due to the phenomenon of the oscillating neutral. This is discussed in Chapter 17. Due to saturation in transformers, and the flat-topped flux wave, a peak emf is generated, which does not balance the applied sinusoidal voltage and generates a resultant third (and other) harmonics. These distort the transformer voltages as the neutral oscillates at thrice the supply frequency, a phenomenon called the "oscillating neutral." A tertiary delta is added to circulate the third harmonic currents and stabilize the neutral. It may also be designed as a load winding, which may have a rated voltage distinct from high- and low-voltage windings. This is further discussed in Section 1.7.2.5. When provided for zero sequence current circulation and harmonic suppression, the terminals of the tertiary connected delta winding may not be brought out of the transformer tank. Sometimes core-type transformers are provided with five-legged cores to circulate the harmonic currents.

1.7.2.3 Delta–Delta Transformer

In a delta–delta connection, no zero currents will pass from one winding to another. The windings are shown connected to the reference bus, allowing the circulation of currents within the windings.

1.7.2.4 Zigzag Transformer

A zigzag transformer is often used to derive a neutral for grounding of a delta–delta connected system. This is shown in Figure 1.10. Windings a_1 and a_2 are on the same core

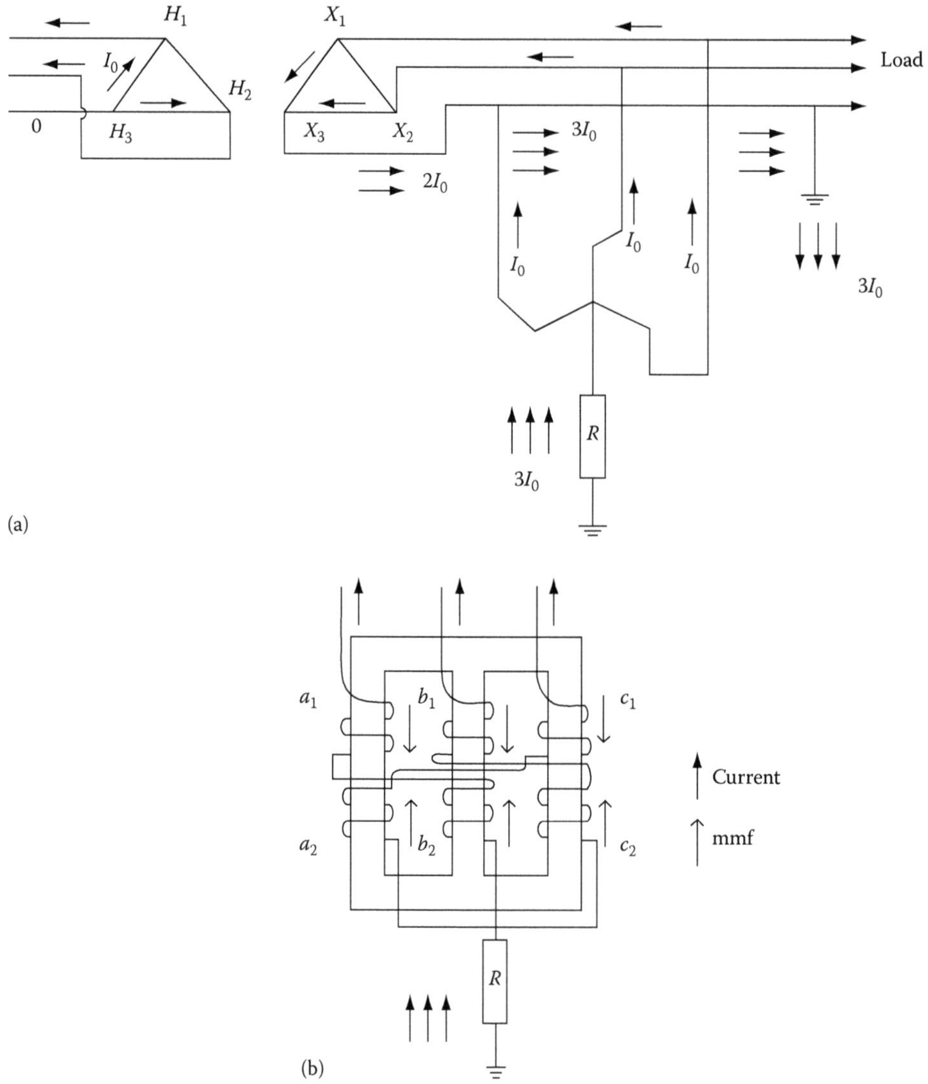

FIGURE 1.10
(a) Current distribution in a delta–delta system with zigzag grounding transformer for a single line-to-ground fault; (b) zigzag transformer winding connections.

leg and have the same number of turns but are wound in the opposite direction. The zero sequence currents in the two windings on the same core leg have canceling ampere turns. Referring to Figure 1.10b the currents in the winding sections a_1 and c_2 must be equal as these are in series. By the same analogy, all currents must be equal, balancing the mmfs in each leg:

$$i_{a1} = i_{a2} = i_{b1} = i_{b2} = i_{c1} = i_{c2}$$

The impedance to the zero sequence currents is that due to leakage flux of the windings. For positive or negative sequence currents, neglecting magnetizing current, the connection has infinite impedance. Figure 1.10a shows the distribution of zero sequence current and its return path for a single line-to-ground fault on one of the phases. The ground current divides equally through the zigzag transformer; one-third of the current returns directly to the fault point and the remaining two-thirds must pass through two phases of the delta connected windings to return to the fault point. Two phases and windings on the primary delta must carry current to balance the ampere turns of the secondary winding currents, Figure 1.10b. An impedance can be added between the artificially derived neutral and ground to limit the ground fault current.

Table 1.1 shows the sequence equivalent circuits of three-phase two-winding transformers. When the transformer neutral is grounded through an impedance Z_n, a term $3Z_n$ appears in the equivalent circuit. We have already proved that $I_n = 3I_0$. The zero sequence impedance of the high- and low-voltage windings is shown as Z_H and Z_L, respectively. The transformer impedance $Z_T = Z_H + Z_L$ on a per unit basis. This impedance is specified by the manufacturer as a percentage impedance on transformer MVA base, based on ONAN (self-cooled rating, windings and core immersed in insulating liquid with fire point less than or equal to 300°C) rating of the transformer. For example, a 138–13.8 kV transformer may be rated as follows:

40 MVA, ONAN ratings at 55°C rise

44.8 MVA, ONAN rating at 65°C rise

60 MVA, ONAF (forced air, i.e., fan cooled) rating at first stage of fan cooling, 65°C rise

75 MVA, ONAF second-stage fan cooling, 65°C rise

These ratings are normally applicable for an ambient temperature of 40°C, with an average of 30°C over a period of 24 h. The percentage impedance will be normally specified on a 40 MVA or possibly a 44.8 MVA base.

The difference between the zero sequence impedance circuits of wye–wye-connected shell- and core-form transformers in Table 1.1 is noteworthy. Connections 8 and 9 are for a core-type transformer and connections 7 and 10 are for a shell-type transformer. The impedance Z_M accounts for magnetic coupling between the phases of a core-type transformer.

1.7.2.5 Three-Winding Transformers

The theory of linear networks can be extended to apply to multiwinding transformers. A linear network having n terminals requires $(1/2)n(n+1)$ quantities to specify it completely for a given frequency and emf. Figure 1.11 shows the wye-equivalent circuit of a three-winding transformer. One method to obtain the necessary data is to designate the pairs of terminals as 1, 2, . . . , n. All the terminals are then short-circuited except terminal

TABLE 1.1

Equivalent Positive, Negative, and Zero Sequence Circuits for Two-Winding Transformers

No.	Winding Connections	Zero Sequence Circuit	Positive or Negative Sequence Circuit
1	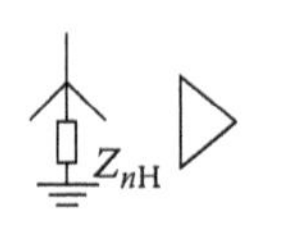	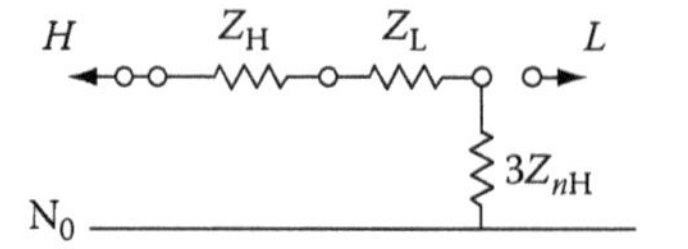	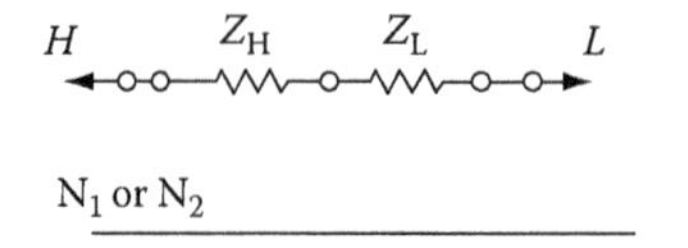
2	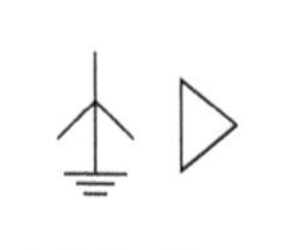	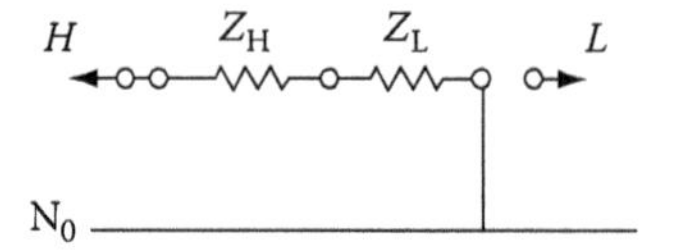	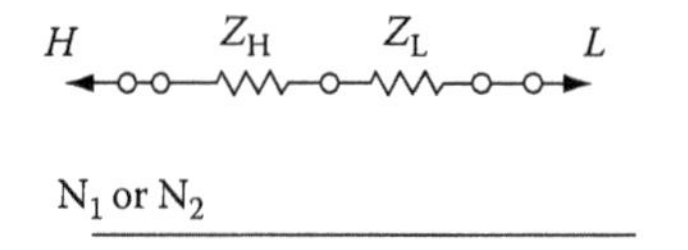
3	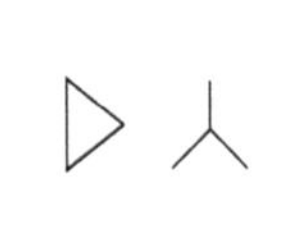	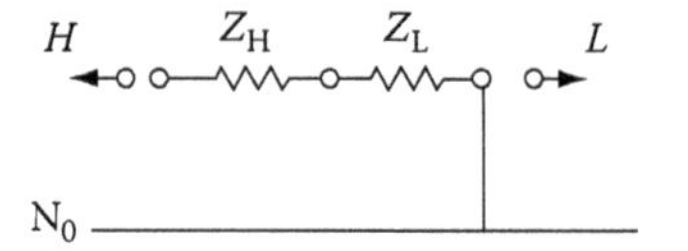	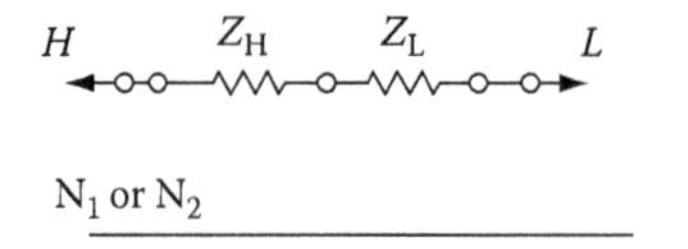
4	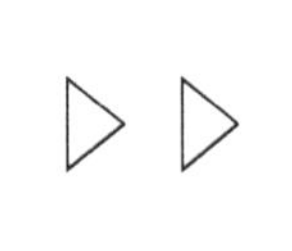	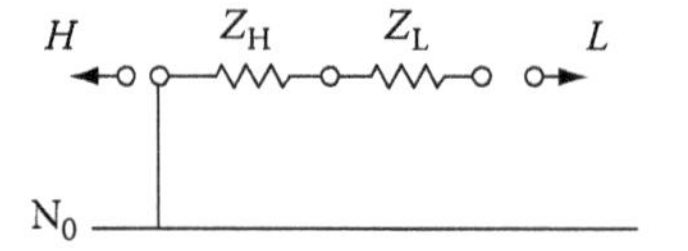	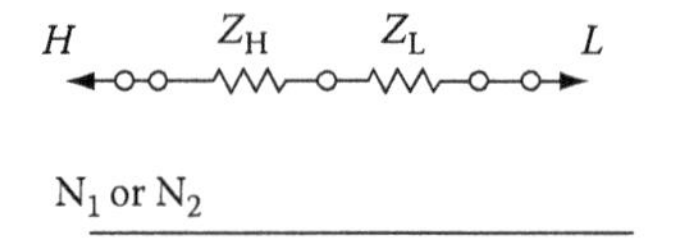
5		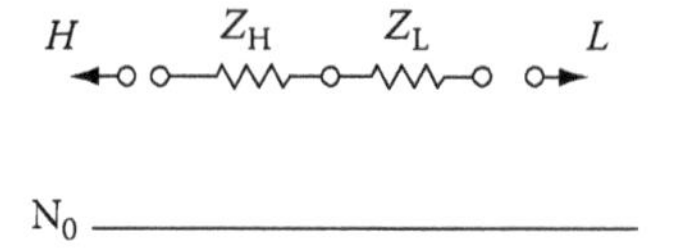	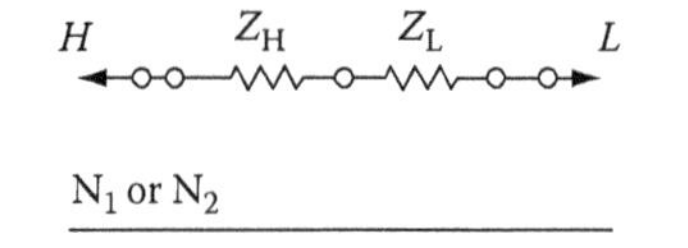
6	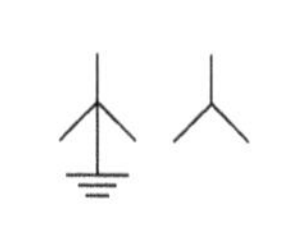	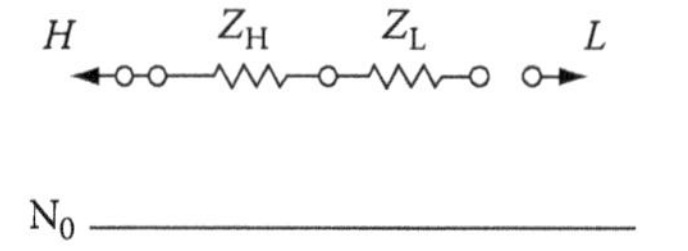	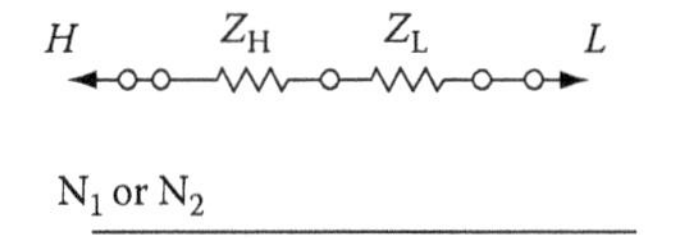
7	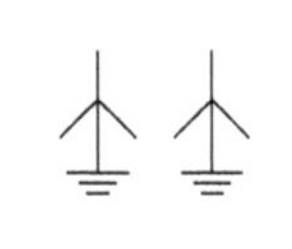	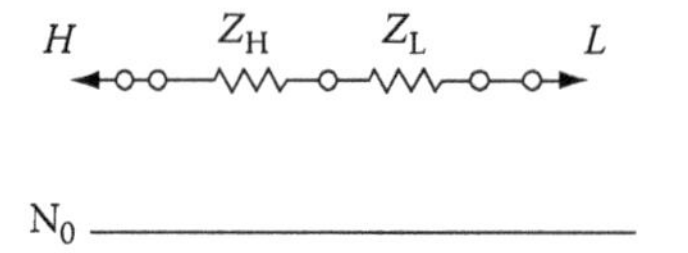	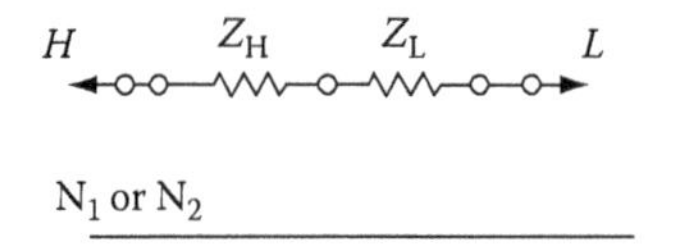
8	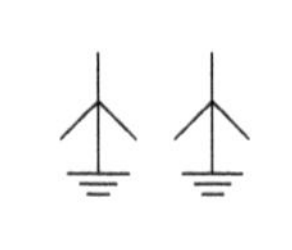	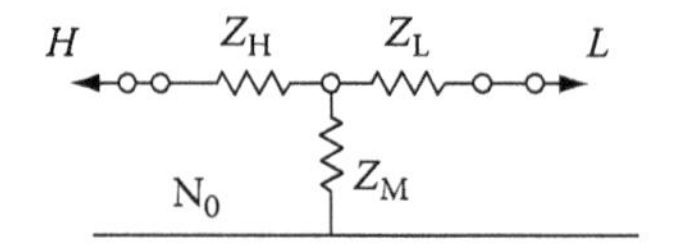	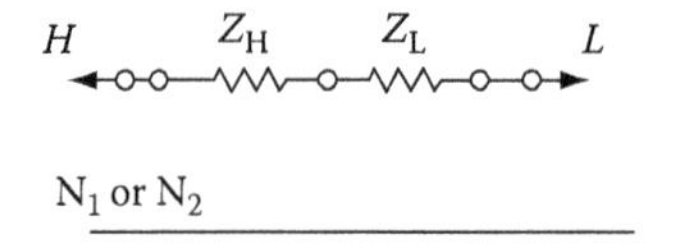
9	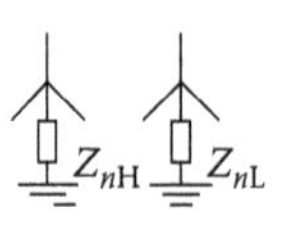	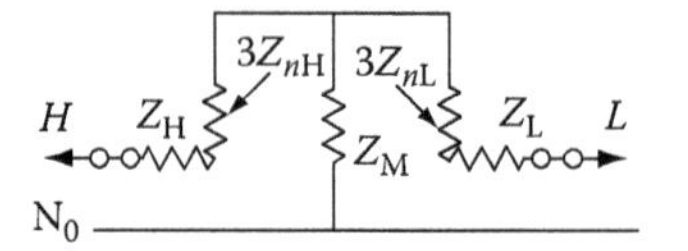	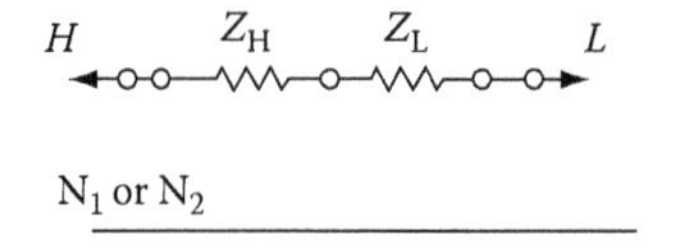
10	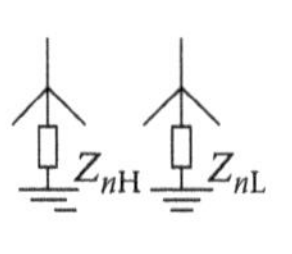	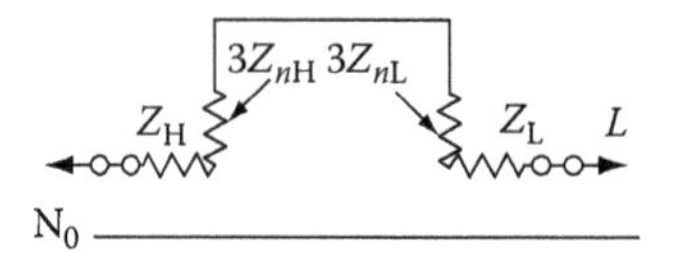	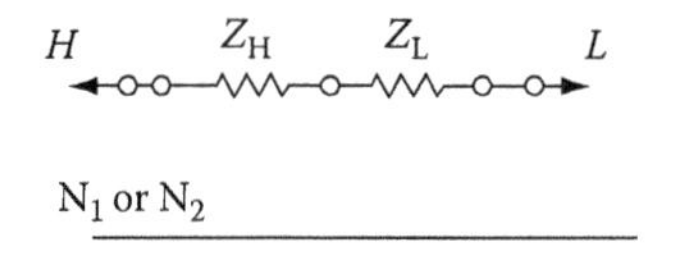

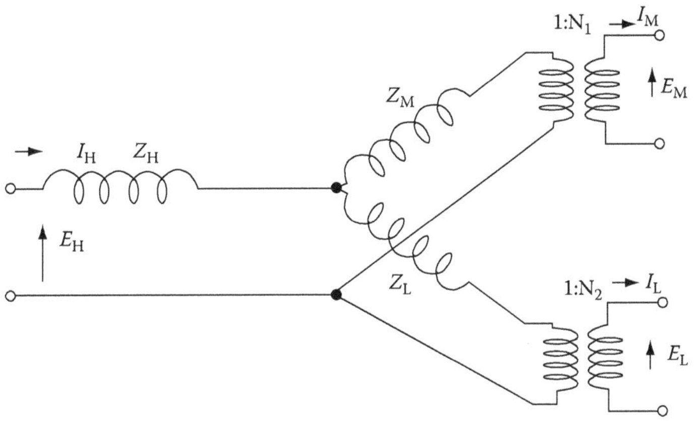

FIGURE 1.11
Wye-equivalent circuit of a three-winding transformer.

one and a suitable emf is applied across it. The current flowing in each pair of terminals is measured. This is repeated for all the terminals. For a three-winding transformer,

$$Z_H = \frac{1}{2}(Z_{HM} + Z_{HL} - Z_{ML})$$
$$Z_H = \frac{1}{2}(Z_{ML} + Z_{HM} - Z_{HL}) \quad (1.55)$$
$$Z_L = \frac{1}{2}(Z_{HL} + Z_{ML} - Z_{HM})$$

where

Z_{HM} is the leakage impedance between the H and M windings, as measured on the H winding with M winding short-circuited and L winding open-circuited

Z_{HL} is the leakage impedance between the H and L windings, as measured on the H winding with L winding short-circuited and M winding open-circuited

Z_{ML} is the leakage impedance between the M and L windings, as measured on the M winding with L winding short-circuited and H winding open-circuited

Equation 1.55 can be written as

$$\begin{vmatrix} Z_H \\ Z_M \\ Z_L \end{vmatrix} = \frac{1}{2}\begin{vmatrix} 1 & 1 & -1 \\ 1 & -1 & 1 \\ -1 & 1 & 1 \end{vmatrix}\begin{vmatrix} Z_{HM} \\ Z_{HL} \\ Z_{ML} \end{vmatrix}$$

We also see that

$$Z_{HL} = Z_H + Z_L$$
$$Z_{HM} = Z_H + Z_M \quad (1.56)$$
$$Z_{ML} = Z_M + Z_L$$

Table 1.2 shows the equivalent sequence circuits of a three-winding transformer.

TABLE 1.2

Equivalent Positive, Negative, and Zero Sequence Circuits for Three-Winding Transformers

No.	Winding Connections	Zero Sequence Circuit	Positive or Negative Sequence Circuit
1	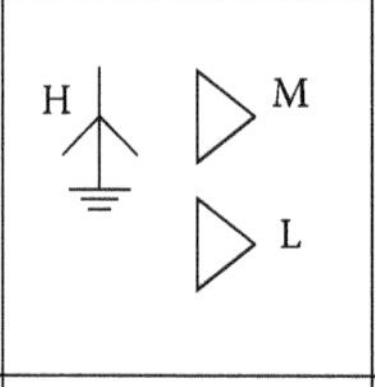	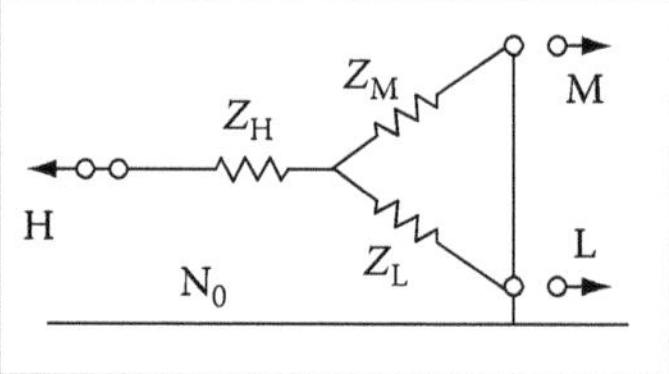	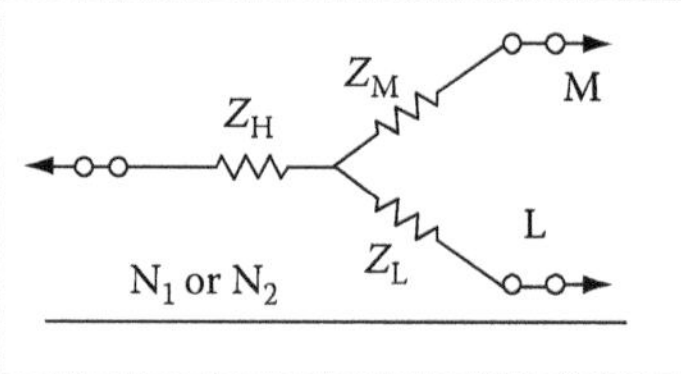
2	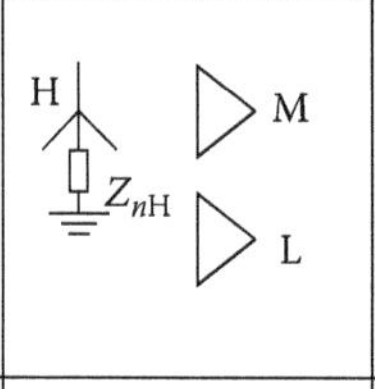	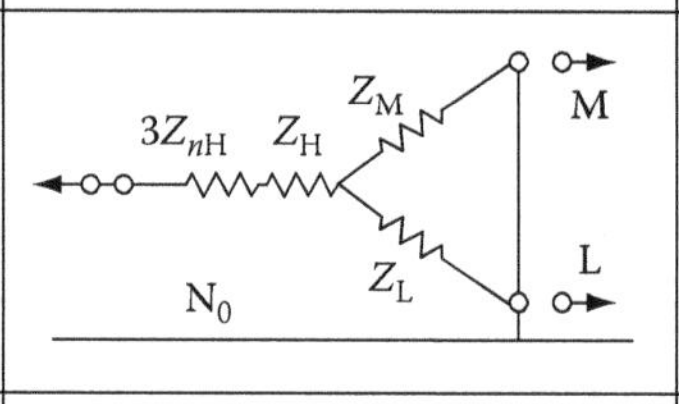	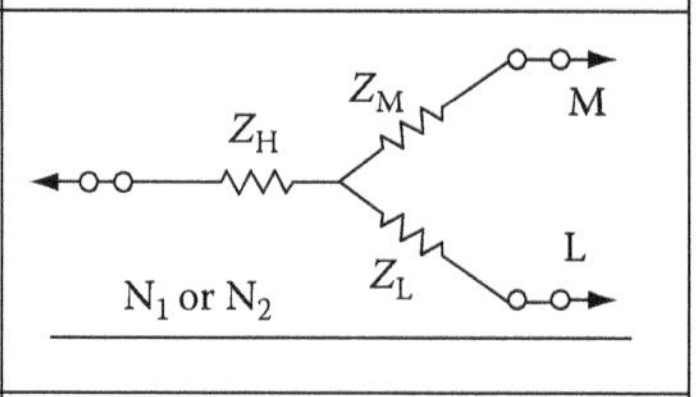
3	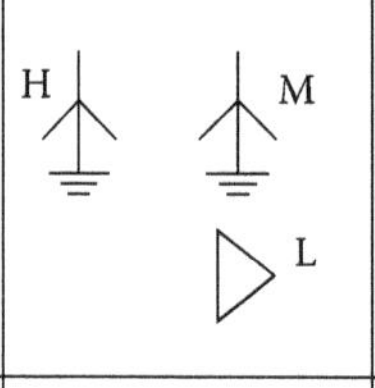	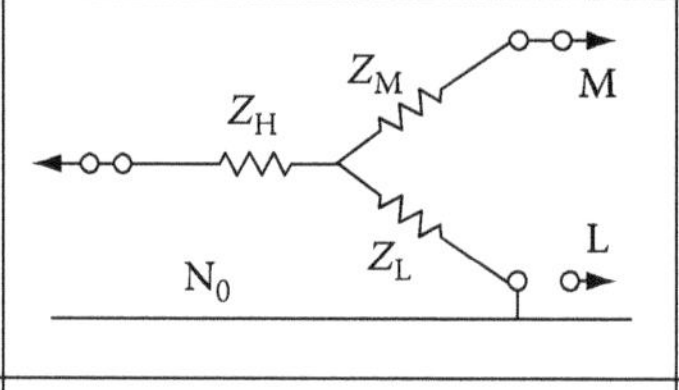	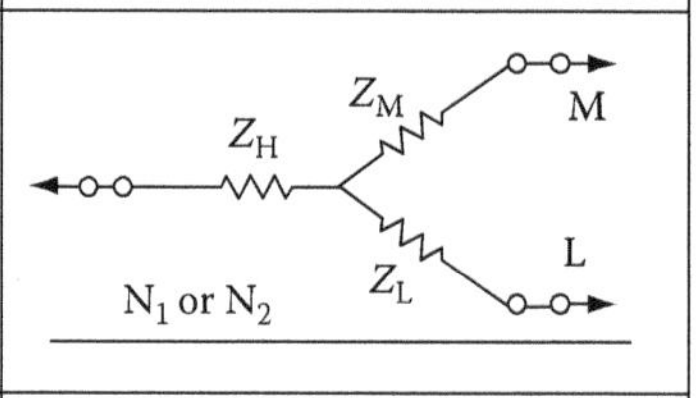
4	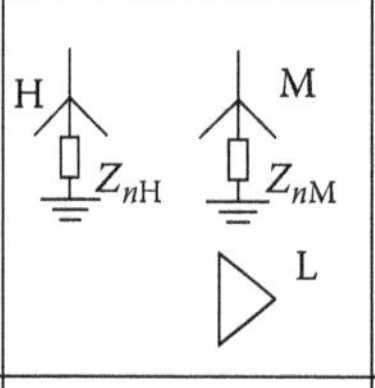	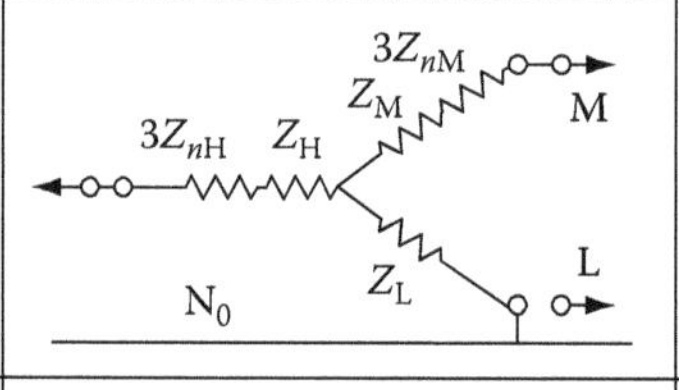	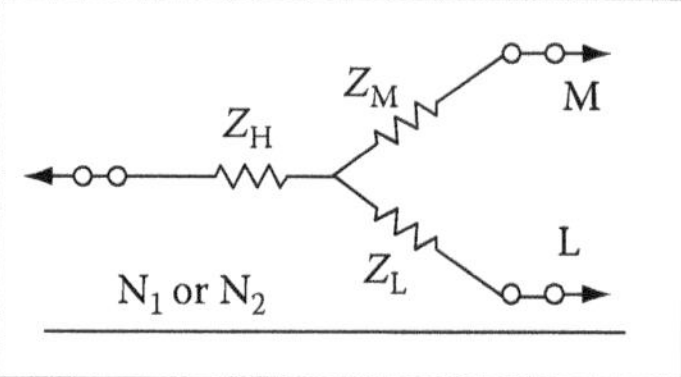
5	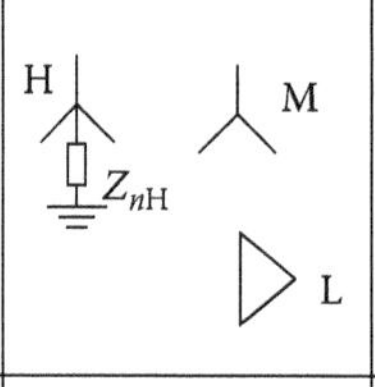	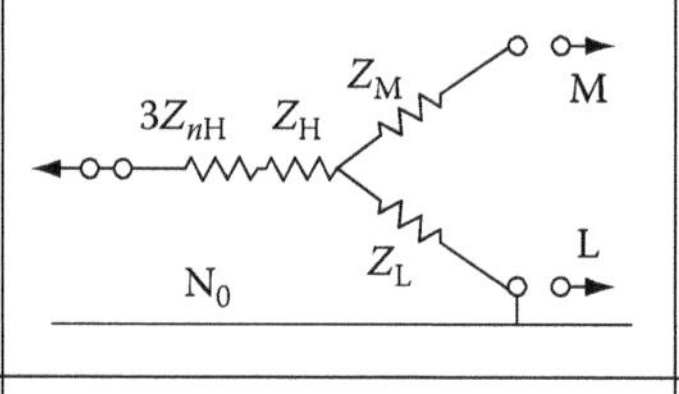	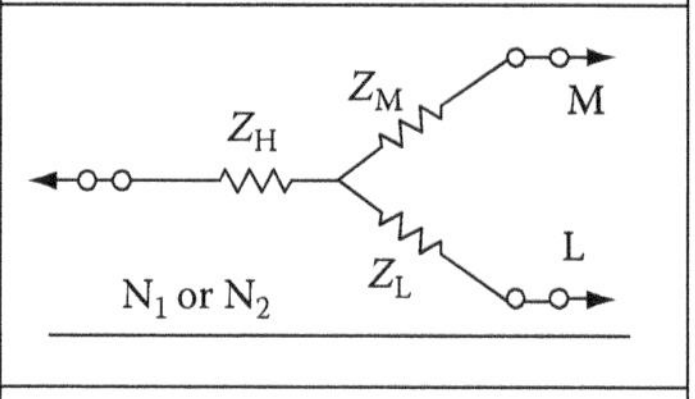
6	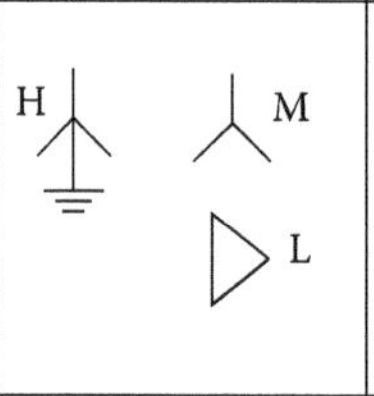	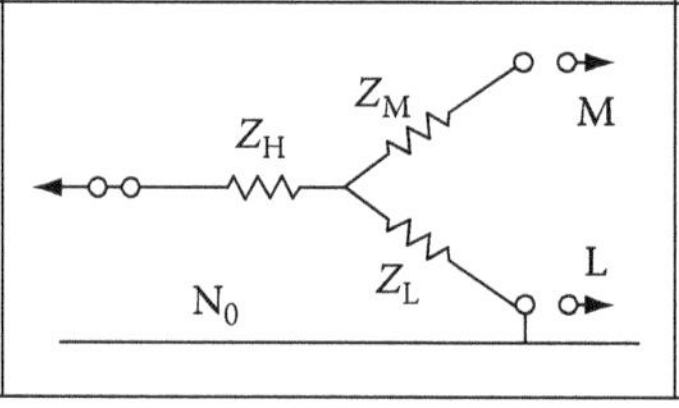	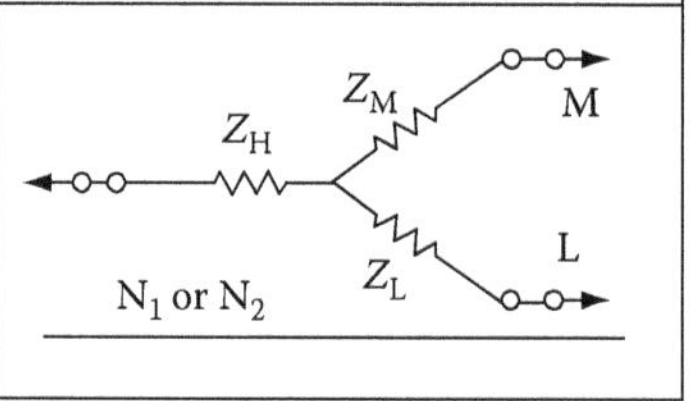

Example 1.2

A three-phase three-winding transformer nameplate reads as

High-voltage winding: 138 kV, wye connected rated at 60 MVA, medium-voltage winding 69 kV, wye connected rated at 50 MVA, 13.8 kV winding (tertiary winding), delta connected and rated 10 MVA. $Z_{HM} = 9\%$, $Z_{HL} = 6\%$, $Z_{ML} = 3\%$. Calculate Z_H, Z_M, and Z_L for representation in the circuit of Figure 1.11.

Before Equation 1.56 is applied, the percentage impedances given on respective MVA base of the windings are converted to a common base, say on the primary winding rating of 60 MVA. This gives $Z_{HM} = 9\%$. $Z_{HL} = 7.2\%$, $Z_{ML} = 18\%$. Then, $Z_H = -0.9\%$, $Z_M = 9.9\%$, $Z_L = 8.1\%$. Note that Z_H is negative. In a three-winding transformer, a negative impedance means that the voltage drop on load flow can be reduced or even made to rise.

1.7.3 Static Load

Consider a static three-phase load connected in a wye configuration with the neutral grounded through an impedance Z_n. Each phase impedance is Z. The sequence transformation is

$$\begin{vmatrix} V_a \\ V_b \\ V_c \end{vmatrix} = \begin{vmatrix} Z & 0 & 0 \\ 0 & Z & 0 \\ 0 & 0 & Z \end{vmatrix} \begin{vmatrix} I_a \\ I_b \\ I_c \end{vmatrix} + \begin{vmatrix} I_n Z_n \\ I_n Z_n \\ I_n Z_n \end{vmatrix}$$

$$T_s \begin{vmatrix} V_0 \\ V_1 \\ V_2 \end{vmatrix} = \begin{vmatrix} Z & 0 & 0 \\ 0 & Z & 0 \\ 0 & 0 & Z \end{vmatrix} T_s \begin{vmatrix} I_0 \\ I_1 \\ I_2 \end{vmatrix} + \begin{vmatrix} 3I_0 Z_n \\ 3I_0 Z_n \\ 3I_0 Z_n \end{vmatrix} \quad (1.57)$$

$$\begin{vmatrix} V_0 \\ V_1 \\ V_2 \end{vmatrix} = T_s^{-1} \begin{vmatrix} Z & 0 & 0 \\ 0 & Z & 0 \\ 0 & 0 & Z \end{vmatrix} T_s \begin{vmatrix} I_0 \\ I_1 \\ I_2 \end{vmatrix} + T_s^{-1} \begin{vmatrix} 3I_0 Z_n \\ 3I_0 Z_n \\ 3I_0 Z_n \end{vmatrix} \quad (1.58)$$

$$= \begin{vmatrix} Z & 0 & 0 \\ 0 & Z & 0 \\ 0 & 0 & Z \end{vmatrix} \begin{vmatrix} I_0 \\ I_1 \\ I_2 \end{vmatrix} + \begin{vmatrix} 3I_0 Z_n \\ 0 \\ 0 \end{vmatrix} = \begin{vmatrix} Z + 3Z_n & 0 & 0 \\ 0 & Z & 0 \\ 0 & 0 & Z \end{vmatrix} \begin{vmatrix} I_0 \\ I_1 \\ I_2 \end{vmatrix} \quad (1.59)$$

This shows that the load can be resolved into sequence impedance circuits. This result can also be arrived at by merely observing the symmetrical nature of the circuit.

1.7.4 Synchronous Machines

Negative and zero sequence impedances are specified for synchronous machines by the manufacturers on the basis of the test results. The negative sequence impedance is measured with the machine driven at rated speed and the field windings short-circuited. A balanced negative sequence voltage is applied and the measurements taken. The zero sequence impedance is measured by driving the machine at rated speed, field windings short-circuited, all three-phases in series, and a single-phase voltage applied to circulate a single-phase current. The zero sequence impedance of generators is low, while the negative sequence impedance is approximately given by

$$\frac{X_d'' + X_q''}{2} \tag{1.60}$$

where X_d'' and X_q'' are the direct axis and quadrature axis subtransient reactances. An explanation of this averaging is that the negative sequence in the stator results in a double-frequency negative component in the field. (Chapter 18 provides further explanation.) The negative sequence flux component in the air gap may be considered to alternate between poles and interpolar gap, respectively.

The following expressions can be written for the terminal voltages of a wye-connected synchronous generator, neutral grounded through an impedance Z_n:

$$\begin{aligned} V_a &= \frac{d}{dt}[L_{af}\cos\theta I_f - L_{aa}I_a - L_{ab}I_b - L_{ac}I_c] - I_aR_a + V_n \\ V_b &= \frac{d}{dt}[L_{bf}\cos(\theta - 120°)I_f - L_{ba}I_a - L_{bb}I_b - L_{bc}I_c] - I_aR_b + V_n \\ V_c &= \frac{d}{dt}[L_{cf}\cos(\theta - 240°)I_f - L_{ca}I_a - L_{cb}I_b - L_{cc}I_c] - I_aR_c + V_n \end{aligned} \tag{1.61}$$

The first term is the generator internal voltage, due to field linkages, and L_{af} denotes the field inductance with respect to phase A of stator windings and I_f is the field current. These internal voltages are displaced by 120° and may be termed E_a, E_b, and E_c. The voltages due to armature reaction, given by the self-inductance of a phase, that is, L_{aa}, and its mutual inductance with respect to other phases, that is, L_{ab} and L_{ac}, and the IR_a drop is subtracted from the generator internal voltage and the neutral voltage is added to obtain the line terminal voltage V_a.

For a symmetrical machine,

$$\begin{aligned} L_{af} &= L_{bf} = L_{cf} = L_f \\ R_a &= R_b = R_c = R \\ L_{aa} &= L_{bb} = L_{cc} = L \\ L_{ab} &= L_{bc} = L_{ca} = L' \end{aligned} \tag{1.62}$$

Thus,

$$\begin{vmatrix} V_a \\ V_b \\ V_c \end{vmatrix} = \begin{vmatrix} E_a \\ E_b \\ E_c \end{vmatrix} - j\omega \begin{vmatrix} L & L' & L' \\ L' & L & L' \\ L' & L' & L \end{vmatrix} \begin{vmatrix} I_a \\ I_b \\ I_c \end{vmatrix} - \begin{vmatrix} R & 0 & 0 \\ 0 & R & 0 \\ 0 & 0 & R \end{vmatrix} \begin{vmatrix} I_a \\ I_b \\ I_c \end{vmatrix} - Z_n \begin{vmatrix} I_n \\ I_n \\ I_n \end{vmatrix} \tag{1.63}$$

Transform using symmetrical components:

$$\begin{aligned} T_s \begin{vmatrix} V_0 \\ V_1 \\ V_2 \end{vmatrix} &= T_s \begin{vmatrix} E_0 \\ E_1 \\ E_2 \end{vmatrix} - j\omega \begin{vmatrix} L & L' & L' \\ L' & L & L' \\ L' & L' & L \end{vmatrix} T_s \begin{vmatrix} I_0 \\ I_1 \\ I_2 \end{vmatrix} - \begin{vmatrix} R & 0 & 0 \\ 0 & R & 0 \\ 0 & 0 & R \end{vmatrix} T_s \begin{vmatrix} I_0 \\ I_1 \\ I_2 \end{vmatrix} - 3Z_n \begin{vmatrix} I_0 \\ I_0 \\ I_0 \end{vmatrix} \\ \begin{vmatrix} V_0 \\ V_1 \\ V_2 \end{vmatrix} &= \begin{vmatrix} E_0 \\ E_1 \\ E_2 \end{vmatrix} - j\omega \begin{vmatrix} L_0 & 0 & 0 \\ 0 & L_1 & 0 \\ 0 & 0 & L_2 \end{vmatrix} \begin{vmatrix} I_0 \\ I_1 \\ I_2 \end{vmatrix} - \begin{vmatrix} R & 0 & 0 \\ 0 & R & 0 \\ 0 & 0 & R \end{vmatrix} \begin{vmatrix} I_0 \\ I_1 \\ I_2 \end{vmatrix} - \begin{vmatrix} 3I_0Z_n \\ 0 \\ 0 \end{vmatrix} \end{aligned} \tag{1.64}$$

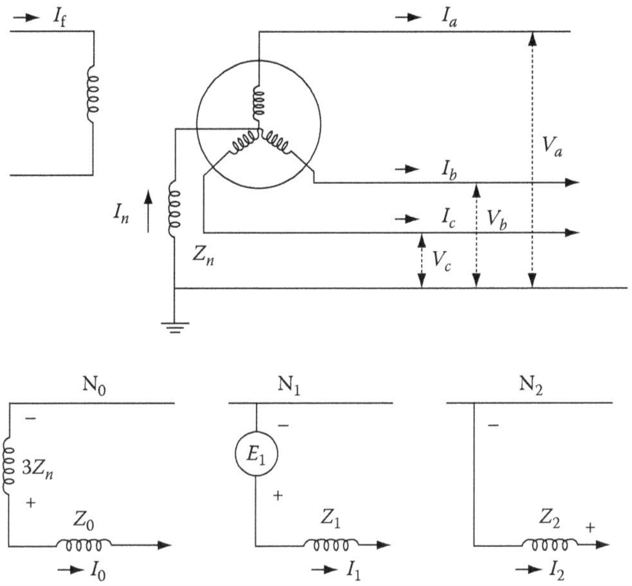

FIGURE 1.12
Sequence components of a synchronous generator impedances.

where

$$\begin{aligned} L_0 &= L + 2L' \\ L_1 &= L_2 + L - L' \end{aligned} \tag{1.65}$$

The equation may, thus, be written as

$$\begin{vmatrix} V_0 \\ V_1 \\ V_2 \end{vmatrix} = \begin{vmatrix} 0 \\ E_1 \\ 0 \end{vmatrix} - \begin{vmatrix} Z_0 + 3Z_n & 0 & 0 \\ 0 & Z_1 & 0 \\ 0 & 0 & Z_2 \end{vmatrix} \begin{vmatrix} I_0 \\ I_1 \\ I_2 \end{vmatrix} \tag{1.66}$$

The equivalent circuit is shown in Figure 1.12. This can be compared with the static three-phase load equivalents. Even for a cylindrical rotor machine, the assumption $Z_1 = Z_2$ is not strictly valid. The resulting generator impedance matrix is nonsymmetrical.

Example 1.3

Figure 1.13a shows a single line diagram, with three generators, three transmission lines, six transformers, and three buses. It is required to construct positive, negative, and zero sequence networks looking from the fault point marked F. Ignore the load currents.

The positive sequence network is shown in Figure 1.13b. There are three generators in the system, and their positive sequence impedances are clearly marked in Figure 1.13b. The generator impedances are returned to a common bus. The Thévenin voltage at the fault point is shown to be equal to the generator voltages, which are all equal. This has to be so as all load currents are neglected, that is, all the shunt elements representing loads are open-circuited. Therefore, the voltage magnitudes and phase angles of all three generators must be equal. When load flow is considered, generation voltages will differ in magnitude and phase, and the voltage vector at the chosen fault point, prior to the fault, can be calculated based on load flow. We have discussed that the load currents are normally ignored in short-circuit calculations. Fault duties of switching devices are calculated based on rated system voltage rather than the actual voltage, which varies with load flow. This is generally true, unless the prefault voltage at the fault point remains continuously above or below the rated voltage.

Figure 1.13c shows the negative sequence network. Note the similarity with the positive sequence network with respect to interconnection of various system components.

Figure 1.13d shows a zero sequence impedance network. This is based on the transformer zero sequence networks shown in Table 1.1. The neutral impedance is multiplied by a factor of 3.

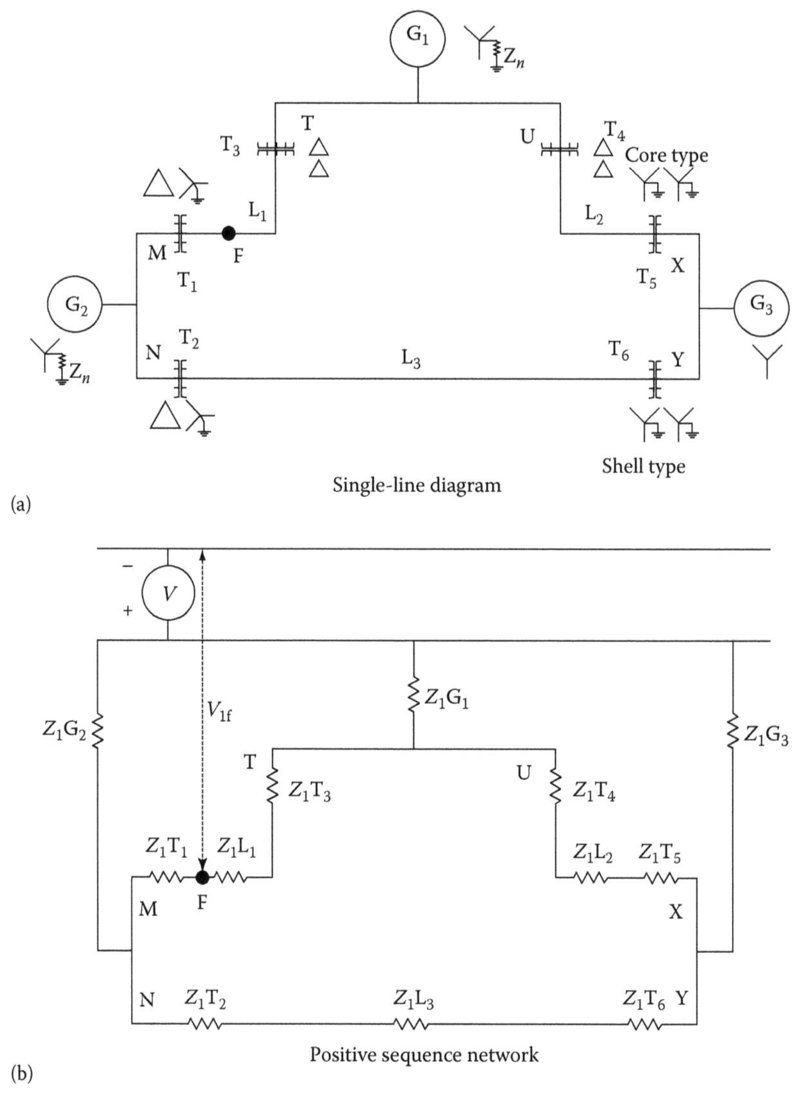

FIGURE 1.13
(a) A single line diagram of a distribution system; (b–d) positive, negative, and zero sequence networks of the distribution system in (a).

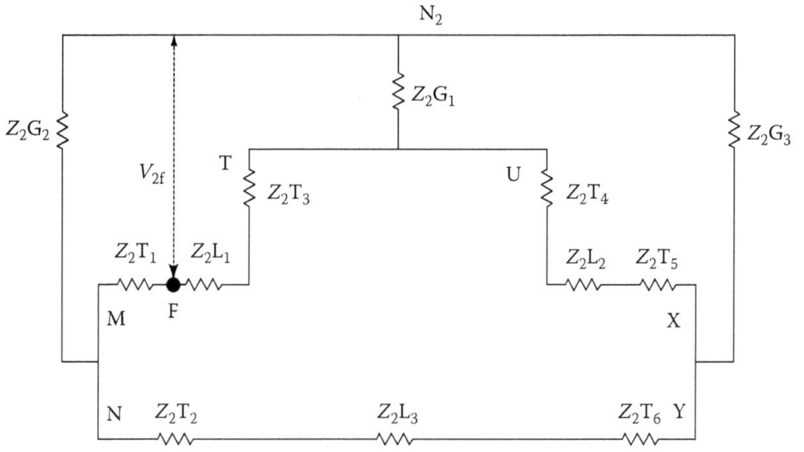

Negative sequence network

(c)

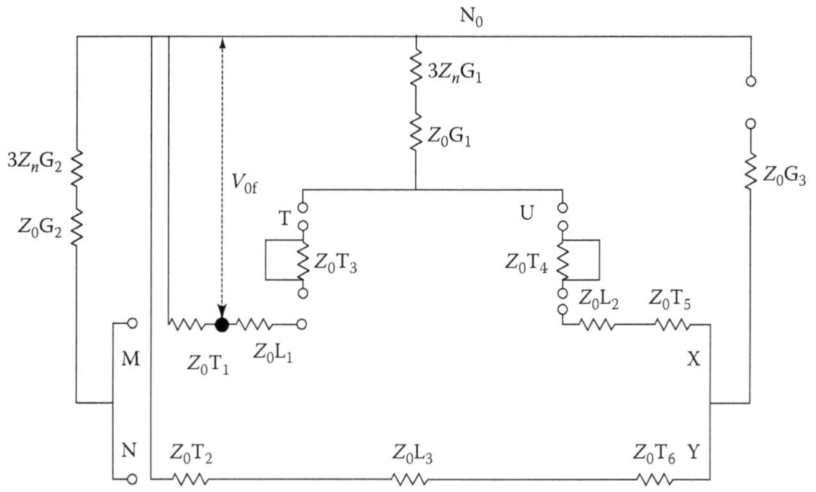

Zero sequence network

(d)

FIGURE 1.13 (continued)

Each of these networks can be reduced to a single impedance using elementary network transformations. Referring to Figure 1.14, wye-to-delta and delta-to-wye impedance transformations are given by

$$
\begin{aligned}
Z_1 &= \frac{Z_{12}Z_{31}}{Z_{12} + Z_{23} + Z_{31}} \\
Z_2 &= \frac{Z_{12}Z_{23}}{Z_{12} + Z_{23} + Z_{31}} \\
Z_3 &= \frac{Z_{23}Z_{31}}{Z_{12} + Z_{23} + Z_{31}}
\end{aligned} \tag{1.67}
$$

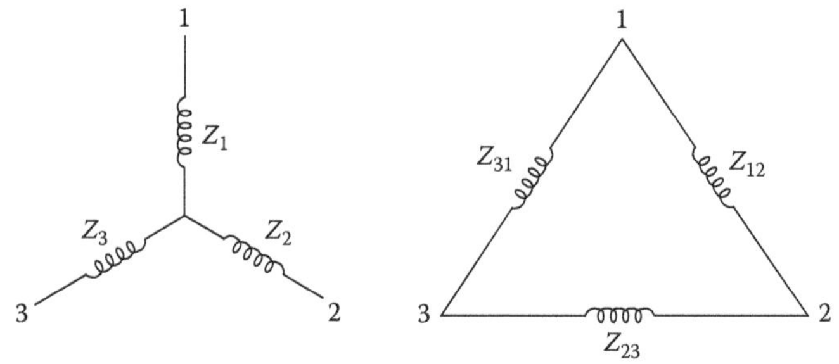

FIGURE 1.14
Wye–delta and delta–wye transformation of impedances.

and from wye to delta,

$$
\begin{aligned}
Z_{12} &= \frac{Z_1Z_2 + Z_2Z_3 + Z_3Z_1}{Z_3} \\
Z_{23} &= \frac{Z_1Z_2 + Z_2Z_3 + Z_3Z_1}{Z_1} \\
Z_{31} &= \frac{Z_1Z_2 + Z_2Z_3 + Z_3Z_1}{Z_2}
\end{aligned} \tag{1.68}
$$

1.8 Computer Models of Sequence Networks

Referring to the zero sequence impedance network of Figure 1.13d, a number of discontinuities occur in the network, depending on transformer winding connections and system grounding. These disconnections are at nodes marked T, U, M, and N. These make a node disappear in the zero sequence network, while it exists in the models of positive and negative sequence networks. The integrity of the nodes should be maintained in all the sequence networks for computer modeling. Figure 1.15 shows how this discontinuity can be resolved.

Figure 1.15a shows a delta–wye transformer, wye side neutral grounded through an impedance Z_n, connected between buses numbered 1 and 2. Its zero sequence network, when viewed from the bus 1 side, is an open circuit.

Two possible solutions in computer modeling are shown in Figure 1.15b and c. In Figure 1.15b a fictitious bus R is created. The positive sequence impedance circuit is modified by dividing the transformer positive sequence impedance into two parts:

Z_{TL} stands for the low-voltage winding and Z_{TH} for the high-voltage winding. An infinite impedance between the junction point of these impedances to the fictitious bus R is connected. In computer calculations, this infinite impedance will be simulated by a large value, that is, $999 + j9999$, on a per unit basis.

The zero sequence network is treated in a similar manner, that is, the zero sequence impedance is split between the windings and the equivalent grounding resistor $3R_N$ is connected between the junction point and the fictitious bus R.

Figure 1.15c shows another approach to the creation of a fictitious bus R to preserve the integrity of nodes in the sequence networks. For the positive sequence network, a large impedance is connected between bus 2 and bus R, while for the zero sequence network an impedance equal to $Z_{0TH} + 3R_N$ is connected between bus 2 and bus R.

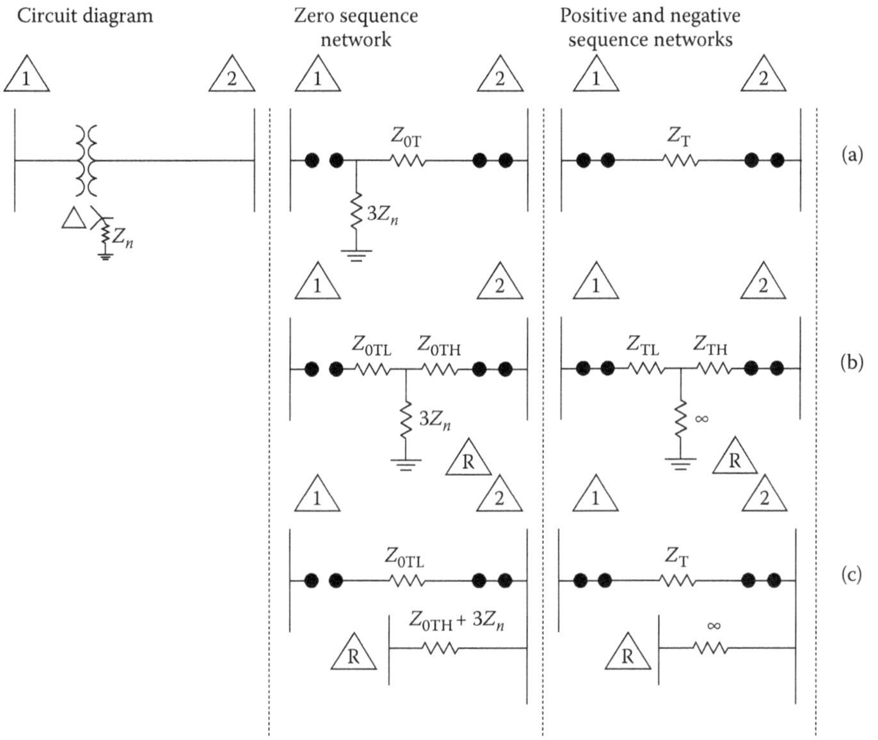

FIGURE 1.15
(a) Representation of a delta–wye transformer; (b) and (c) zero and positive and negative sequence network representation maintaining integrity of nodes.

1.9 Structure and Nature of Electrical Power Systems

Electrical power systems can be broadly classified into generation, transmission, subtransmission, and distribution systems. Individual power systems are organized in the form of electrically connected areas of *regional* grids, which are interconnected to form *national* grids and also international grids. Each area is interconnected to another area in terms of some contracted parameters like generation and scheduling, tie line power flow, and contingency operations. The business environment of power industry has entirely changed due to deregulation and decentralization, and more emphasis on economics and cost centers.

The irreplaceable sources of power generation are petroleum, natural gas, oil, and nuclear fuels. The fission of heavy atomic weight elements like uranium and thorium and fusion of lightweight elements, that is, deuterium offer almost limitless reserves. Replaceable sources are elevated water, pumped storage systems, solar, geothermal, wind, and fuel cells, which in recent times have received much attention driven by strategic planning for independence from foreign oil imports. Figure 1.16 shows the world's energy generation till 2020, and the type of energy source. The forecast projects 54% growth in electricity generation and consumption in 2015 based upon 1995 levels; though many efforts are directed toward conservation of energy. The social and environmental issues

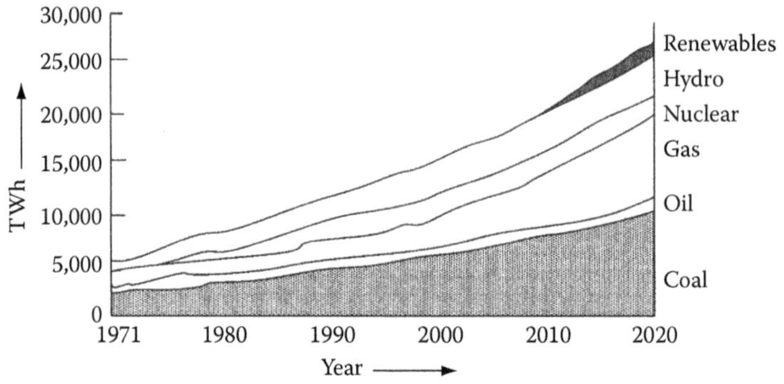

FIGURE 1.16
World electrical power generation growth.

are of importance; the increase in renewable energy sources compared to that shown in Figure 1.16 can be foreseen. Nuclear generation in the United States has much public concern due to a possible meltdown and disposal of nuclear waste, though France relies heavily on this source. U238 has a half-life of 4.5×10^{10} years. The disposal of nuclear waste products is one public deterrent against nuclear generation. The coal-based plants produce SO_2, nitrogen oxides, CO, CO_2, hydrocarbons, and particulates. Single-shaft steam units of 1500 MW are in operation, and superconducting single units of 5000 MW or more are a possibility. On the other hand, dispersed generating units integrated with grid may produce only a few kilowatt of power.

The generation voltage is low, 13.8–25 kV, because of problems of inter-turn and winding insulation at higher voltages in the generator stator slots, though a 110 kV Russian generator is in operation. The transmission voltages have risen to 765 kV or higher and many HVDC links around the world are in operation. Maintaining acceptable voltage profile and load frequency control are major issues. Synchronous condensers, shunt capacitors, static var compensators, and FACTS devices (Chapter 13) are employed to improve power system stability and enhance power handling capability of transmission lines. The sub-transmission voltage levels are 23 kV to approximately 69 kV, though for large industrial consumers, voltages of 230 and 138 kV are in use. Sub-transmission systems connect high-voltage substations to local distribution substations. The voltage is further reduced to 12.47 kV and several distribution lines and cables emanate from distribution stations. At the consumer level, the variety of load types, their modeling and different characteristics present a myriad of complexities. A pulp mill may use single synchronous motors of 30,000 hp or more and ship propulsion can use even higher ratings.

The electrical power systems are highly nonlinear and dynamic—perhaps the most nonlinear man-made systems of the highest order on earth. The systems are constantly subjected to internal switching transients (faults, closing and opening of breakers, charging of long transmission lines, varying generation in response to load demand, etc.) and of external origin, that is, lightning surges. Electromagnetic and electromechanical energy is constantly being redistributed amongst power system components. This cannot happen suddenly, and takes some time, which brings about the transient state. Electromagnetic transients are due to interaction between electrical fields of capacitance and magnetic fields of inductances, while electromechanical transients originate due to interaction between electrical energy stored in the system and the mechanical energy of the rotating machines.

This book does not provide an insight into the transient behavior of the power systems and confines the analyses to steady state. Short circuits are decaying transients and subject the system to severe stresses and stability problems. Some sporadic reference to transients may be seen in some chapters and appendices of the book.

1.9.1 Power System Component Models

The time duration (frequency) of the transient phenomena in the power system varies. CIGRE (International Council for Large Electrical Power Systems) classifies the transients with respect to frequency in four groups. These groups are low-frequency oscillations (0.1–3 kHz), slow front surges (50/60–20 kHz), fast front surges (10–3 MHz), and very fast front surges (100–50 MHz). It is very difficult to develop a power system component (say transmission lines) model that is accurate from low to very high frequencies. Thus, models are good for a certain frequency range. A model must reproduce the frequency variations, saturation, nonlinearity, surge arrester characteristics, power fuse and circuit breaker operations accurately for the frequency range considered. The models in this book are mostly steady-state models. For harmonic analysis, the impacts of higher frequencies involved on some component models are discussed in Chapter 19.

1.9.2 Smart Grids

In the years to come, power industry will undergo profound changes, need based—environmental compatibility, reliability, improved operational efficiencies, integration of renewable energy sources, and dispersed generation. The dynamic state of the system will be known at all times and under any disturbance. The technologies driving smart grids are RAS (remedial action schemes), SIPS (system integrated protection systems), WAMS (wide area measurement systems), FACTS (flexible ac transmission systems—chapter 13), EMS (energy management systems), PMUs (phasor measurement units), overlay of 750 kV lines, and HVDC links.

After the November 1965 great blackout, National Reliability Council was created in 1968 and later, its name changed to North American Reliability Council (NERC). It lays down guidelines and rules for utility companies to follow and adhere to. Yet it seems that adequate dynamic system studies are not carried out during the planning stage. This is attributed to lack of resources, data, or expertise (lack of models for dispersed generation, verified data and load models, models for wind generation). In a modern complex power system, the potential problem area may lay hidden and may not be possible to identify intuitively or with some power system studies. It is said that cascading-type blackouts can be minimized and time frame prolonged but these cannot be entirely eliminated.

1.9.3 Linear and Nonlinear Systems

It is pertinent to define mathematically linearity and nonlinearity, static, and dynamic systems.

Linearity implies two conditions:

1. Homogeneity
2. Superimposition

Consider the state of a system defined by

$$\dot{\mathbf{x}} = \mathbf{f}[\mathbf{x}(t), \mathbf{r}(t), t] \tag{1.69}$$

If $\mathbf{x}(t)$ is the solution to this differential equation with initial conditions $x(t_0)$ at $t = t_0$ and input $\mathbf{r}(t)$, $t > t_0$:

$$\mathbf{x}(t) = \varphi[\mathbf{x}(t_0), \mathbf{r}(t)] \tag{1.70}$$

Then homogeneity implies that

$$\varphi[\mathbf{x}(t_0), \alpha\mathbf{r}(t)] = \alpha\varphi[\mathbf{x}(t_0), \mathbf{r}(t)] \tag{1.71}$$

where α is a scalar constant. This means that $\mathbf{x}(t)$ with input $\alpha\mathbf{r}(t)$ is equal to α times $\mathbf{x}(t)$ with input $\mathbf{r}(t)$ for any scalar α.

Superposition implies that

$$\varphi[\mathbf{x}(t_0), \mathbf{r}_1(t) + \mathbf{r}_2(t)] = \varphi[\mathbf{x}(t_0), \mathbf{r}_1(t)] + \varphi[\mathbf{x}(t_0), \mathbf{r}_2(t)] \tag{1.72}$$

That is, $\mathbf{x}(t)$ with inputs $\mathbf{r}_1(t) + \mathbf{r}_2(t)$ is = sum of $\mathbf{x}(t)$ with input $\mathbf{r}_1(t)$ and $\mathbf{x}(t)$ with input $\mathbf{r}_2(t)$. Thus linearity is superimposition plus homogeneity.

1.9.3.1 Property of Decomposition

A system is said to be linear if it satisfies the decomposition property and the decomposed components are linear.

If $x'(t)$ is solution of Equation 1.69 when system is in zero state for all inputs $\mathbf{r}(t)$, that is,

$$\mathbf{x}'(t) = \varphi(\mathbf{0}, \mathbf{r}(t)) \tag{1.73}$$

and $\mathbf{x}''(t)$ is the solution when all states $x(t_0)$, the input $\mathbf{r}(t)$ is zero, that is,

$$\mathbf{x}''(t) = \varphi(\mathbf{x}(t_0), \mathbf{0}) \tag{1.74}$$

then, the system is said to have decomposition property, if

$$\mathbf{x}(t) = \mathbf{x}'(t) + \mathbf{x}''(t) \tag{1.75}$$

The zero input response and zero state response satisfy the property of homogeneity and superimposition with respect to initial states and initial inputs, respectively. If this is not true then the system is nonlinear.

For nonlinear systems general methods of solutions are not available and each system must be studied specifically. Yet, we apply linear techniques of solution to nonlinear systems over a certain time interval. Perhaps the system is not changing so fast, and for a certain range of applications linearity can be applied. Thus, the linear system analysis forms the very fundamental aspect of the study.

1.9.4 Static and Dynamic Systems

Consider a time-invariant, linear resistor element, across a voltage source. The output, that is, the voltage across the resistor is solely dependent upon the input voltage at that instant. We may say that the resistor does not have a *memory, and is a static system*. On the other hand, the voltage across a capacitor depends not only upon the input, but also upon its initial charge, that is, the past history of current flow. We say that the capacitor has a *memory* and is a dynamic system.

The state of the system with memory is determined by state variables that vary with time. The state transition from $x(t_1)$ at time t_1 to $x(t_2)$ at time t_2 is a dynamic process that can be described by differential equations. For a capacitor connected to a voltage source, the dynamics of the state variable $x(t) = e(t)$ can be described by

$$\dot{x} = \frac{1}{C}r(t), \quad r(t) = i(t) \tag{1.76}$$

1.10 Power System Studies

Power system analysis is a vast subject. Consider the following broad categories:

- Short-circuit studies
- Load flow studies
- Stability studies; large rotor angle, small disturbances, and voltage instability
- Motor starting studies
- Optimal load flow, contingency, and security analysis studies
- Harmonic analysis studies, harmonic mitigation, flicker mitigation, and harmonic filter designs studies
- Application of FACTS and power electronics studies
- Transmission and distribution lines planning and design studies, underground transmission and distribution studies
- Insulation coordination and application of surge protection studies
- Switching transient and transient analysis studies
- Power system reliability studies
- Cable ampacity calculation studies
- Torsional dynamics studies
- HVDC transmission studies, short circuits and load flow in dc systems studies
- Wind power and renewable energy sources, feasibility studies and their integration in the grid studies
- Power quality for sensitive and electronic equipment studies
- Ground mat (grid) design for safety studies, system grounding studies, grounding for electronic equipment studies
- Protective relaying and relay coordination studies
- Arc flash hazard analysis studies

In addition, specific studies may be required for a specific task, for example, transmission line designs or generating stations, transmission substation, and consumer load substations. This book is confined to short-circuit, load flow, and harmonic analyses. This second edition includes chapters on arc flash hazard analysis and wind power integration into utility systems. The arc flash hazard analyses have gained much importance in the recent times.

This chapter provides the basic concepts and an overall condensed picture of power systems. The discussions of symmetrical components, construction of sequence networks, and fault current calculations are carried over to Chapter 2.

Problems

1.1 A short transmission line of inductance 0.05 H and resistance 1 Ω is suddenly short-circuited at the receiving end, while the source voltage is $480(\sqrt{2})\sin(2\pi ft + 30°)$. At what instant of the short circuit will the dc offset be zero? At what instant will the dc offset be a maximum?

1.2 Figure 1.1 shows a non-decaying ac component of the fault current. Explain why this is not correct for a fault close to a generator.

1.3 Explain similarity transformation. How is it related to the diagonalization of a matrix?

1.4 Find the eigenvalues of the matrix

$$\begin{bmatrix} 6 & -2 & 2 \\ -2 & 3 & -1 \\ 2 & -2 & 3 \end{bmatrix}$$

1.5 A power system is shown in Figure P1.1. Assume that loads do not contribute to the short-circuit currents. Convert to a common 100 MVA base, and form sequence impedance networks. Redraw zero sequence network to eliminate discontinuities.

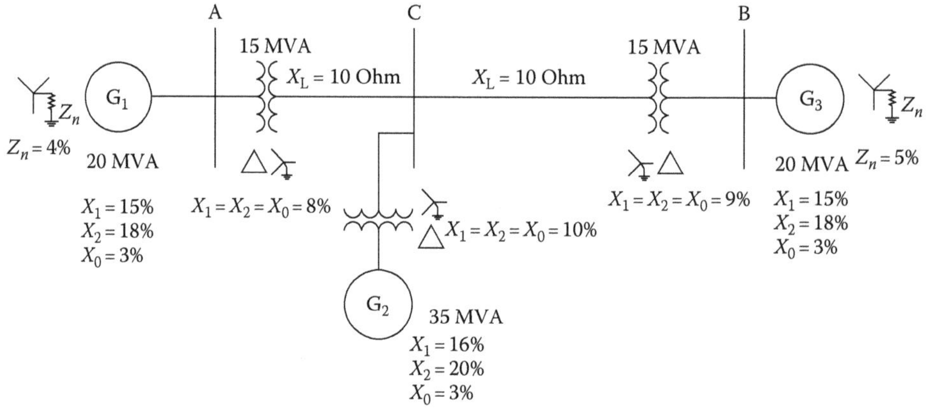

FIGURE P1.1
Power system with impedance data for Problem 1.5.

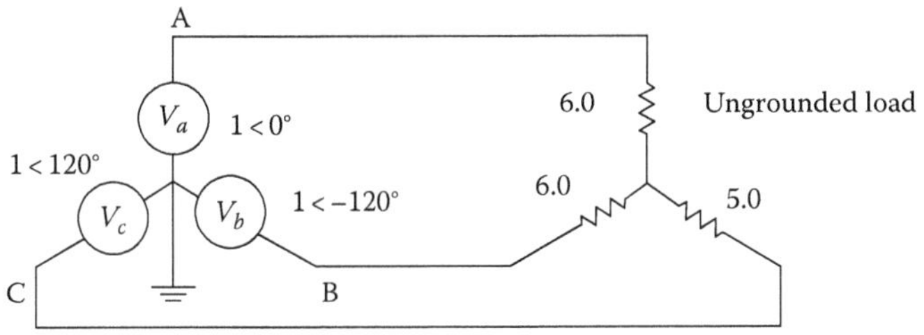

FIGURE P1.2
Network for Problem 1.8.

1.6 Three unequal load resistances of 10, 20, and 20 Ω are connected in delta, 10 Ω between lines *a* and *b*, 20 Ω between lines *b* and *c*, and 200 Ω between lines *c* and *a*. The power supply is a balanced three-phase system of 480 V rms between the lines. Find symmetrical components of line currents and delta currents.

1.7 In Figure 1.10, the zigzag transformer is replaced with a wye–delta-connected transformer. Show the distribution of the fault current for a phase-to-ground fault on one of the phases.

1.8 Resistances of 6, 6, and 5 Ω are connected in a wye configuration across a balanced three-phase supply system of line-to-line voltage of 480 V rms (Figure P1.2). The wye point of the load (neutral) is not grounded. Calculate the neutral voltage with respect to ground using symmetrical components and Clarke's components' transformation.

1.9 Based on the derivation of symmetrical component theory presented in this chapter, can another transformation system be conceived?

1.10 Write equations for a symmetrical three-phase fault in a three-phase wye-connected system, with balanced impedances in each line.

1.11 The load currents are generally neglected in short-circuit calculations. Do these have any effect on the dc component asymmetry: (1) increase it; (2) decrease it; (3) have no effect? Explain.

1.12 Write a 500 word synopsis on symmetrical components, without using equations or figures.

1.13 Figure 1.9a shows the zero sequence current flow for a delta–wye transformer, with the wye neutral grounded. Construct a similar diagram for a three-winding transformer, wye–wye-connected, with tertiary delta and both wye neutrals solidly grounded.

1.14 Convert the sequence impedance networks of Example 1.2 to single impedances as seen from the fault point. Use the following numerical values on a per unit basis (all on a common MVA base). Neglect resistances.

Generators G_1, G_2, and G_3: $Z_1 = 0.15$, $Z_2 = 0.18$, $Z_0 = 0.08$, Z_n (neutral grounding impedance) $= 0.20$;

Transmission lines L_1, L_2, and L_3: $Z_1 = 0.2$, $Z_2 = 0.2$;

Transformers T_1, T_2, T_3, T_4, T_5, and T_6: $Z_1 = Z_2 = 0.10$, transformer T_1: $Z_0 = 0.10$.

1.15 Repeat Problem 1.14 for a fault at the terminals of generator G_2.

Bibliography

ATP Rule Book. Portland, OR: ATP User Group, 1992.

Bird, J.O. *Electrical Circuit Theory and Technology*. Oxford, U.K.: Butterworth Heinemann, 1997.

Chen, M.S. and W.E. Dillon. Power system modeling. *Proc. IEEE*, 62: 901–915, July 1974.

Chowdhuri, P. *Electromagnetic Transients in Power Systems*. Somerset, U.K.: Research Studies Press Ltd., 1996.

Clarke, E. *Circuit Analysis of Alternating Current Power Systems*, Vol. 1. New York: Wiley, 1943.

Das, J.C. *Transients in Electrical Systems*. New York: McGraw-Hill, 2010.

Dommel, H.W. and W.S. Meyer. Computation of electromagnetic transients. *IEE Proc*. 62: 983–993, 1974.

EMTP web site: www.emtp.org

Fitzgerald, A.E., C. Kingsley, and A. Kusko. *Electric Machinery*, 3rd edn. New York: McGraw-Hill, 1971.

Fortescu, C.L. Method of symmetrical coordinates applied to the solution of polyphase networks. *AIEE* 37: 1027–1140, 1918.

Greenwood, A. *Electrical Transients in Power Systems*, 2nd edn. New York: Wiley, 1991.

IEEE Std. 399, IEEE Recommended Practice for Power Systems Analysis, 1997.

Lewis, W.E. and D.G. Pryce. *The Application of Matrix Theory to Electrical Engineering*. London, U.K.: E&FN Spon, 1965, Chapter 6.

Myatt, L.J. *Symmetrical Components*. Oxford, U.K.: Pergamon Press, 1968.

NERC web site: www.nerc.com

Stagg, G.W. and A. Abiad. *Computer Methods in Power Systems Analysis*. New York: McGraw-Hill, 1968.

Wagner, C.F. and R.D. Evans. *Symmetrical Components*. New York: McGraw-Hill, 1933.

Westinghouse Electric Transmission and Distribution Handbook, 4th edn. East Pittsburgh, PA: Westinghouse Electric Corp., 1964.

Worth, C.A. (ed.). *J. & P. Transformer Book*, 11th edn. London, U.K.: Butterworth, 1983.

2

Unsymmetrical Fault Calculations

Chapter 1 discussed the nature of sequence networks and how three distinct sequence networks can be constructed as seen from the fault point. Each of these networks can be reduced to a single Thévenin positive, negative, or zero sequence impedance. Only the positive sequence network is active and has a voltage source, which is the prefault voltage. For unsymmetrical fault current calculations, the three separate networks can be connected in a certain manner, depending on the type of fault.

Unsymmetrical fault types involving one or two phases and ground are as follows:

- A single line-to-ground fault
- A double line-to-ground fault
- A line-to-line fault

These are called shunt faults. A three-phase fault may also involve ground. The unsymmetrical series type faults are as follows:

- One conductor opens
- Two conductors open

The broken conductors may be grounded on one side or on both sides of the break. An open conductor fault can occur due to operation of a fuse in one of the phases.

Unsymmetrical faults are more common. The most common type is a line-to-ground fault. Approximately 70% of the faults in power systems are single line-to-ground faults.

While applying symmetrical component method to fault analysis, we will ignore the load currents. This makes the positive sequence voltages of all the generators in the system identical and equal to the prefault voltage.

In the analysis to follow, Z_1, Z_2, and Z_0 are the positive, negative, and zero sequence impedances as seen from the fault point; V_a, V_b, and V_c are the phase-to-ground voltages at the fault point, prior to fault, that is, *if the fault does not exist*; and V_1, V_2, and V_0 are corresponding sequence component voltages. Similarly, I_a, I_b, and I_c are the line currents and I_1, I_2, and I_0 their sequence components. A fault impedance of Z_f is assumed in every case. For a bolted fault $Z_f = 0$.

2.1 Line-to-Ground Fault

Figure 2.1a shows that phase *a* of a three-phase system goes to ground through an impedance Z_f. The flow of ground fault current depends on the method of system grounding. A solidly grounded system with zero ground resistance is assumed. There will be some impedance to flow of fault current in the form of impedance of the return ground conductor

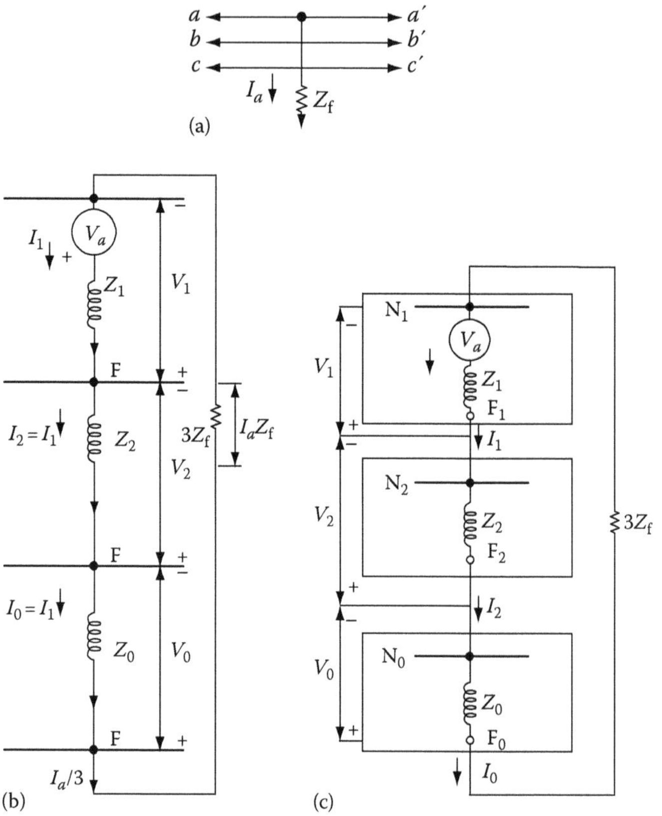

FIGURE 2.1
(a) Line-to-ground fault in a three-phase system; (b) line-to-ground fault equivalent circuit; (c) sequence network interconnections.

or the grounding grid resistance. A ground resistance can be added in series with the fault impedance Z_f. The ground fault current must have a return path through the grounded neutrals of generators or transformers. If there is no return path for the ground current, $Z_0=\infty$ and the ground fault current is zero. This is an obvious conclusion.

Phase a is faulted in Figure 2.1a. As the load current is neglected, currents in phases b and c are zero, and the voltage at the fault point $V_a=I_aZ_f$. The sequence components of the currents are given by

$$\begin{vmatrix} I_0 \\ I_1 \\ I_2 \end{vmatrix} = \frac{1}{3}\begin{vmatrix} 1 & 1 & 1 \\ 1 & a & a^2 \\ 1 & a^2 & a \end{vmatrix}\begin{vmatrix} I_a \\ 0 \\ 0 \end{vmatrix} = \frac{1}{3}\begin{vmatrix} I_a \\ I_a \\ I_a \end{vmatrix} \tag{2.1}$$

Also,

$$I_0 = I_1 = I_2 = \frac{1}{3}I_a \tag{2.2}$$

$$3I_0Z_f = V_0 + V_1 + V_2 = -I_0Z_0 + (V_a - I_1Z_1) - I_2Z_2 \tag{2.3}$$

which gives

$$I_0 = \frac{V_a}{Z_0 + Z_1 + Z_2 + 3Z_f} \tag{2.4}$$

The fault current I_a is

$$I_a = 3I_0 = \frac{3V_a}{(Z_1 + Z_2 + Z_0) + 3Z_f} \tag{2.5}$$

This shows that the equivalent fault circuit using sequence impedances can be constructed as shown in Figure 2.1b. In terms of sequence impedances' network blocks, the connections are shown in Figure 2.1c.

This result could also have been arrived at from Figure 2.1b:

$$(V_a - I_1Z_1) + (-I_2Z_2) + (-I_0Z_0) - 3Z_fI_0 = 0$$

which gives the same Equations 2.4 and 2.5. The voltage of phase b to ground under fault conditions is

$$\begin{aligned} V_b &= a^2V_1 + aV_2 + V_0 \\ &= V_a\frac{3a^2Z_f + Z_2(a^2 - a) + Z_0(a^2 - 1)}{(Z_1 + Z_2 + Z_0) + 3Z_f} \end{aligned} \tag{2.6}$$

Similarly, the voltage of phase c can be calculated.

An expression for the ground fault current for use in grounding grid designs and system grounding is as follows:

$$I_a = \frac{3V_a}{(R_0 + R_1 + R_2 + 3R_f + 3R_G) + j(X_0 + X_1 + X_2)} \tag{2.7}$$

where

R_f is the fault resistance
R_G is the resistance of the grounding grid
R_0, R_1, and R_2 are the sequence resistances
X_0, X_1, and X_2 are sequence reactances

2.2 Line-to-Line Fault

Figure 2.2a shows a line-to-line fault. A short-circuit occurs between phases b and c, through a fault impedance Z_f. The fault current circulates between phases b and c, flowing back to source through phase b and returning through phase c; $I_a = 0$, $I_b = -I_c$. The sequence components of the currents are

$$\begin{vmatrix} I_0 \\ I_1 \\ I_2 \end{vmatrix} = \frac{1}{3}\begin{vmatrix} 1 & 1 & 1 \\ 1 & a & a^2 \\ 1 & a^2 & a \end{vmatrix}\begin{vmatrix} 0 \\ -I_c \\ I_c \end{vmatrix} = \frac{I_c}{3}\begin{vmatrix} 0 \\ -a + a^2 \\ -a^2 + a \end{vmatrix} \tag{2.8}$$

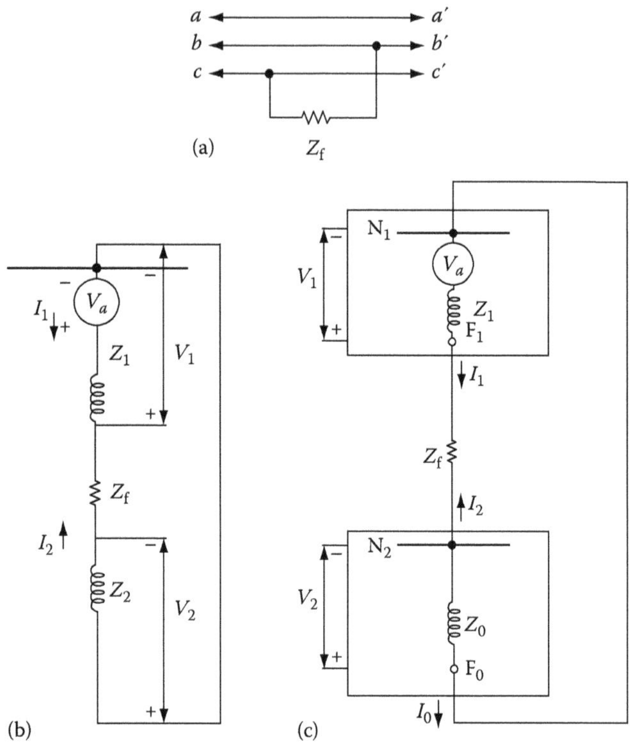

FIGURE 2.2
(a) Line-to-line fault in a three-phase system; (b) line-to-line fault equivalent circuit; (c) sequence network interconnections.

From Equation 2.8, $I_0 = 0$ and $I_1 = -I_2$.

$$V_b - V_c = |0 \quad 1 \quad -1| \begin{vmatrix} V_a \\ V_b \\ V_c \end{vmatrix} = |0 \quad 1 \quad -1| \begin{vmatrix} 1 & 1 & 1 \\ 1 & a^2 & a \\ 1 & a & a^2 \end{vmatrix} \begin{vmatrix} V_0 \\ V_1 \\ V_2 \end{vmatrix}$$

$$= |0 \quad a^2 - a \quad a - a^2| \begin{vmatrix} V_0 \\ V_1 \\ V_2 \end{vmatrix} \tag{2.9}$$

Therefore,

$$\begin{aligned} V_b - V_c &= (a^2 - a)(V_1 - V_2) \\ &= (a^2 I_1 + a I_2) Z_f \\ &= (a^2 - a) I_1 Z_f \end{aligned} \tag{2.10}$$

This gives

$$(V_1 - V_2) = I_1 Z_f \tag{2.11}$$

The equivalent circuit is shown in Figure 2.2b and c.

Also

$$I_b = (a^2 - a) I_1 = -j\sqrt{3} I_1 \tag{2.12}$$

and

$$I_1 = \frac{V_a}{Z_1 + Z_2 + Z_f} \tag{2.13}$$

The fault current is

$$I_b = -I_c = \frac{-j\sqrt{3}V_a}{Z_1 + Z_2 + Z_f} \tag{2.14}$$

2.3 Double Line-to-Ground Fault

A double line-to-ground fault is shown in Figure 2.3a. Phases b and c go to ground through a fault impedance Z_f. The current in the ungrounded phase is zero, that is, $I_a = 0$. Therefore, $I_1 + I_2 + I_0 = 0$.

$$V_b = V_c = (I_b + I_c)Z_f \tag{2.15}$$

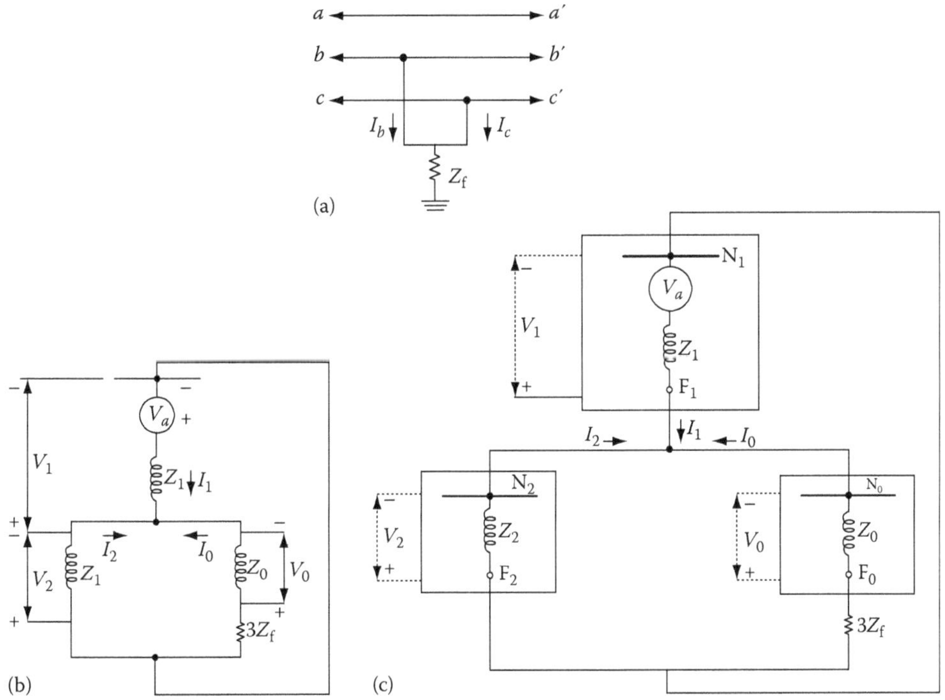

FIGURE 2.3
(a) Double line-to-ground fault in a three-phase system; (b) double line-to-ground fault equivalent circuit; (c) sequence network interconnections.

Thus,

$$\begin{vmatrix} V_0 \\ V_1 \\ V_2 \end{vmatrix} = \frac{1}{3}\begin{vmatrix} 1 & 1 & 1 \\ 1 & a & a^2 \\ 1 & a^2 & a \end{vmatrix}\begin{vmatrix} V_a \\ V_b \\ V_b \end{vmatrix} = \frac{1}{3}\begin{vmatrix} V_a + 2V_b \\ V_a + (a + a^2)V_b \\ V_a + (a + a^2)V_b \end{vmatrix} \tag{2.16}$$

which gives $V_1 = V_2$ and

$$\begin{aligned} V_0 &= \frac{1}{3}(V_a + 2V_b) \\ &= \frac{1}{3}[(V_0 + V_1 + V_2) + 2(I_b + I_c)Z_f] \\ &= \frac{1}{3}[(V_0 + 2V_1) + 2(3I_0)Z_f] \\ &= V_1 + 3Z_f I_0 \end{aligned} \tag{2.17}$$

This gives the equivalent circuit of Figure 2.3b and c.

The fault current is

$$\begin{aligned} I_1 &= \frac{V_a}{Z_1 + [Z_2 \| (Z_0 + 3Z_f)]} \\ &= \frac{V_a}{Z_1 + \dfrac{Z_2(Z_0 + 3Z_f)}{Z_2 + Z_0 + 3Z_f}} \end{aligned} \tag{2.18}$$

2.4 Three-Phase Fault

The three phases are short-circuited through equal fault impedances Z_f, in Figure 2.4a. The vectorial sum of fault currents is zero, as a symmetrical fault is considered and there is no path to ground.

$$I_0 = 0 \qquad I_a + I_b + I_c = 0 \tag{2.19}$$

As the fault is symmetrical:

$$\begin{vmatrix} V_a \\ V_b \\ V_c \end{vmatrix} = \begin{vmatrix} Z_f & 0 & 0 \\ 0 & Z_f & 0 \\ 0 & 0 & Z_f \end{vmatrix}\begin{vmatrix} I_a \\ I_b \\ I_c \end{vmatrix} \tag{2.20}$$

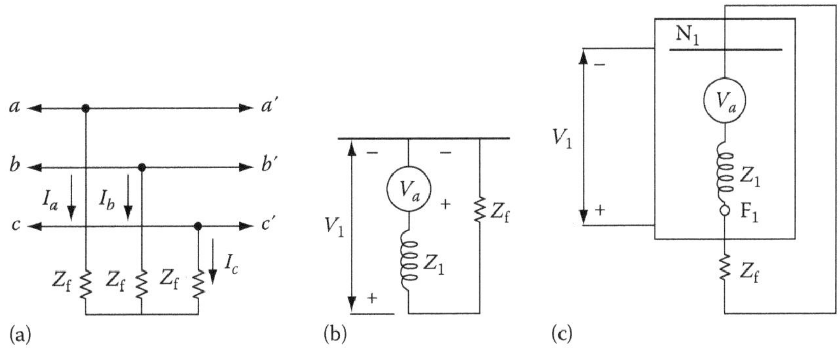

FIGURE 2.4
(a) Three-phase symmetrical fault; (b) equivalent circuit; (c) sequence network.

The sequence voltages are given by

$$\begin{vmatrix} V_0 \\ V_1 \\ V_2 \end{vmatrix} = [T_s]^{-1} \begin{vmatrix} Z_f & 0 & 0 \\ 0 & Z_f & 0 \\ 0 & 0 & Z_f \end{vmatrix} [T_s] \begin{vmatrix} I_0 \\ I_1 \\ I_2 \end{vmatrix} = \begin{vmatrix} Z_f & 0 & 0 \\ 0 & Z_f & 0 \\ 0 & 0 & Z_f \end{vmatrix} \begin{vmatrix} I_0 \\ I_1 \\ I_2 \end{vmatrix} \tag{2.21}$$

This gives the equivalent circuit of Figure 2.4b and c.

$$\begin{aligned} I_a &= I_1 = \frac{V_a}{Z_1 + Z_f} \\ I_b &= a^2 I_1 \\ I_c &= a I_1 \end{aligned} \tag{2.22}$$

2.5 Phase Shift in Three-Phase Transformers

2.5.1 Transformer Connections

Transformer windings can be connected in wye, delta, zigzag, or open delta. The transformers may be three-phase units, or three-phase banks can be formed from single-phase units. Autotransformer connections should also be considered. The variety of winding connections is, therefore, large [1]. It is not the intention to describe these connections completely. The characteristics of a connection can be estimated from the vector diagrams of the primary and secondary EMFs (Electromotive Force). There is a phase shift in the secondary voltages with respect to the primary voltages, depending on the connection. This is of importance when paralleling transformers. A vector diagram of the transformer connections can be constructed based on the following:

1. The voltages of primary and secondary windings on the same leg of the transformer are in opposition, while the induced EMFs are in the same direction. (Refer to Appendix C for further explanation.)
2. The induced EMFs in three phases are equal, balanced, and displaced mutually by a one-third period in time. These have a definite phase sequence.

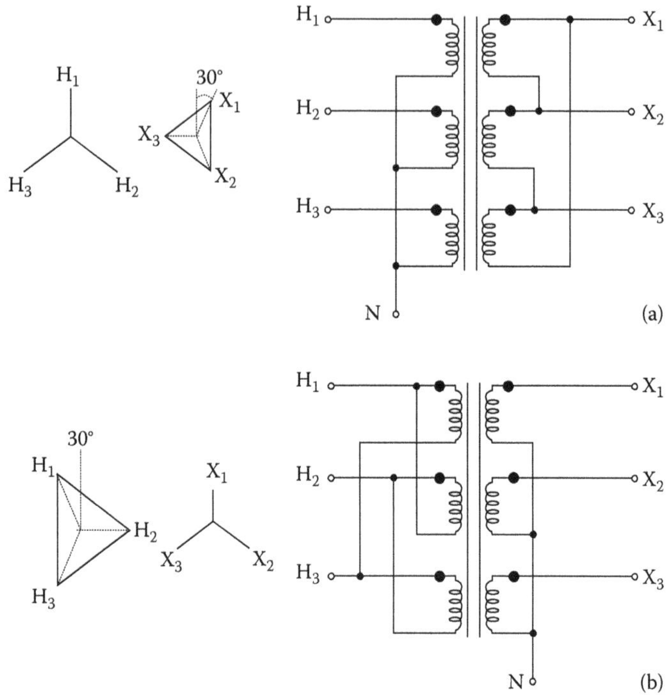

FIGURE 2.5
Winding connections and phase displacement for voltage vectors for delta–wye-connected transformers. (a) High-voltage winding: wye, Low-voltage winding: delta; (b) High-voltage winding: delta, Low-voltage winding: wye.

Delta–wye connections are discussed, as these are most commonly used. Figure 2.5 shows polarity markings and connections of delta–wye transformers. For all liquid-immersed transformers the polarity is subtractive according to American National Standard Institute (ANSI) standard [2] (refer to Appendix C for an explanation). Two-winding transformers have their windings designated as high voltage (H) and low voltage (X). Transformers with more than two windings have their windings designated as H, X, Y, and Z. External terminals are distinguished from each other by marking with a capital letter, followed by a subscript number, that is, H_1, H_2, and H_3.

2.5.2 Phase Shifts in Winding Connections

The angular displacement of a polyphase transformer is the time angle expressed in degrees between the line-to-neutral voltage of the reference identified terminal and the line-to-neutral voltage of the corresponding identified low-voltage terminal. In Figure 2.5a, wye-connected side voltage vectors lead the delta-connected side voltage vectors by 30°, for counterclockwise rotation of phasors. In Figure 2.5b the delta-connected side leads the wye-connected side by 30°. For transformers manufactured according to the ANSI/IEEE (Institute of Electrical and Electronics Engineers, Inc., USA) standard [3], the *low-voltage side, whether in wye or delta* connection, has a phase shift of 30° lagging with respect to the high-voltage side phase-to-neutral voltage vectors. Figure 2.6 shows ANSI/IEEE [3] transformer connections and a phasor diagram of the delta side and wye side voltages. These relations and phase displacements are applicable to positive sequence voltages.

The International Electrotechnical Commission (IEC) Standard for Power Transformers (IEC 60076-1) allocates vector groups, giving the type of phase connection and the *angle of*

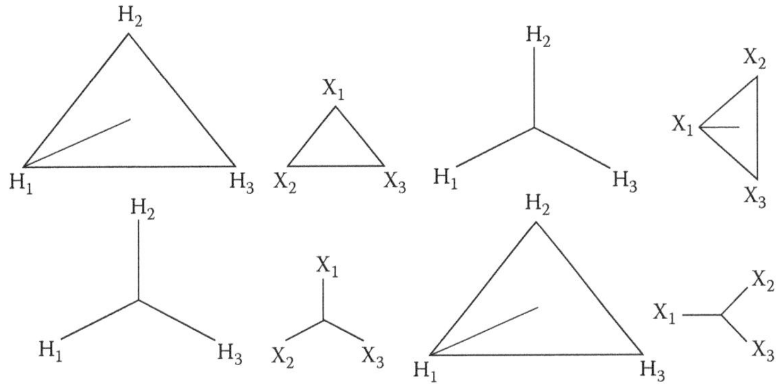

FIGURE 2.6
Phase designations of terminal markings in three-phase transformers according to ANSI/IEEE standard.

advance turned though in passing from the vector representing the high-voltage side EMF to that representing the low-voltage side EMF at the corresponding terminals. The angle is indicated much like the hands of a clock, the high-voltage vector being at 12 o'clock (zero) and the corresponding low-voltage vector being represented by the hour hand. The total rotation corresponding to hour hand of the clock is 360°. Thus, Dy11 and Yd11 symbols specify 30° lead (11 being the hour hand of the clock) and Dy1 and Yd1 signify 30° lag. Table 2.1 shows some IEC vector groups of transformers and their winding connections.

2.5.3 Phase Shift for Negative Sequence Components

The phase shifts described earlier are applicable to positive sequence voltages or currents. If a voltage of negative phase sequence is applied to a delta–wye-connected transformer, the phase angle displacement will be equal to the positive sequence phasors, but in the opposite direction. Therefore, when the positive sequence currents and voltages on one side lead the positive sequence current and voltages on the other side by 30°, the corresponding negative sequence currents and voltages will lag by 30°. In general, if the positive sequence voltages and currents on one side lag the positive sequence voltages and currents on the other side by 30°, the negative sequence voltages and currents will lead by 30°.

Example 2.1

Consider a balanced three-phase delta load connected across an unbalanced three-phase supply system, as shown in Figure 2.7. The currents in lines *a* and *b* are given.

The currents in the delta-connected load and also the symmetrical components of line and delta currents are required to be calculated. From these calculations, the phase shifts of positive and negative sequence components in delta windings and line currents can be established.

The line current in *c* is given by

$$
\begin{aligned}
I_c &= -(I_a + I_b) \\
&= -30 + j6.0 \text{ A}
\end{aligned}
$$

TABLE 2.1

Transformer Vector Groups, Winding Connections, and Vector Diagrams

Vector Group & Phase Shift	Winding Connections	Vector Diagram
Yy0 0°	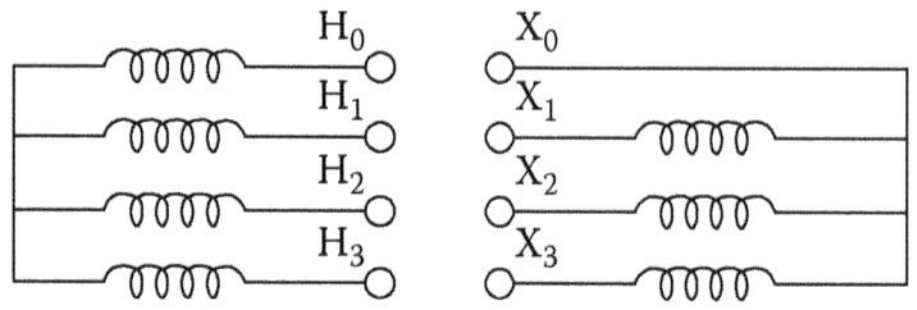	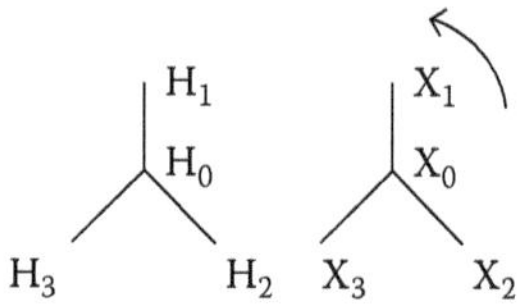
Yy6 180°	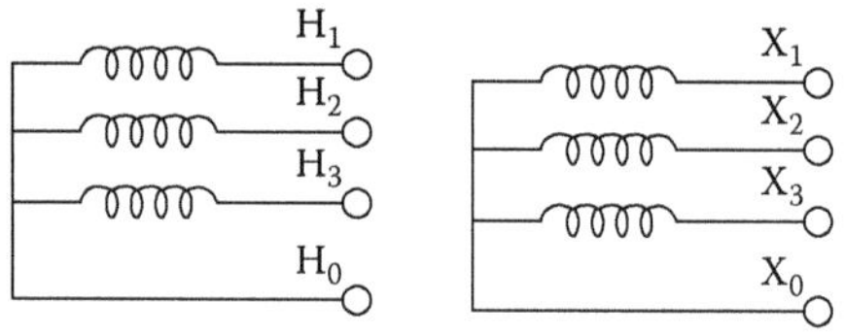	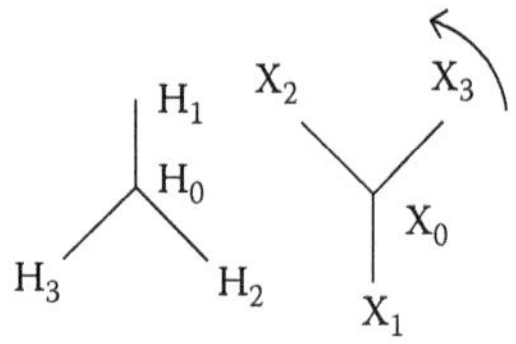
Dd0 0°	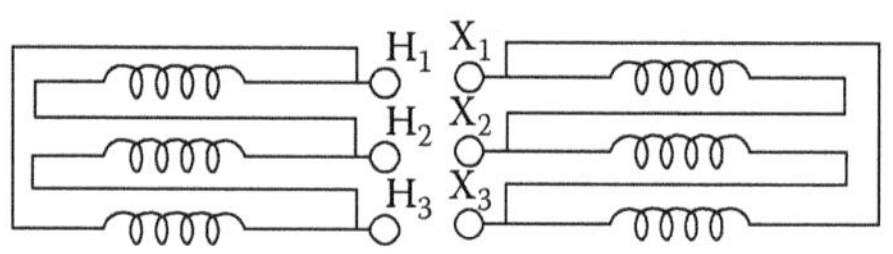	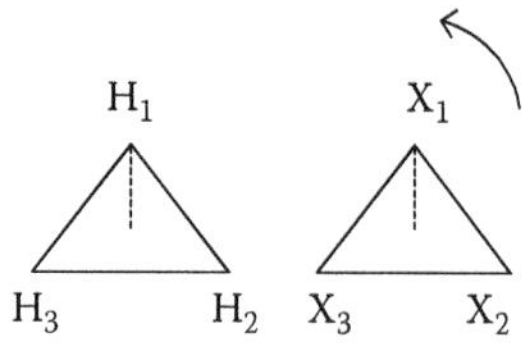
Dz0 0°	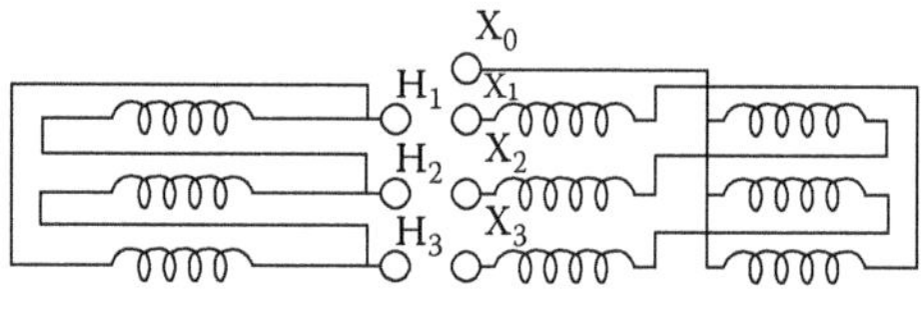	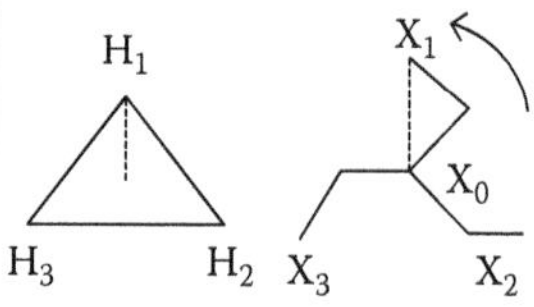
Dz6 180°	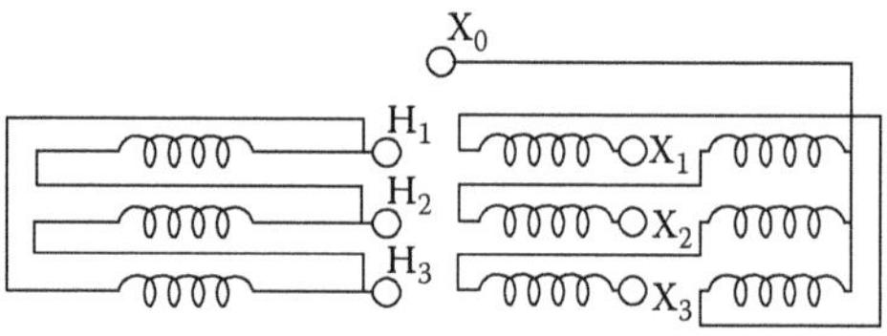	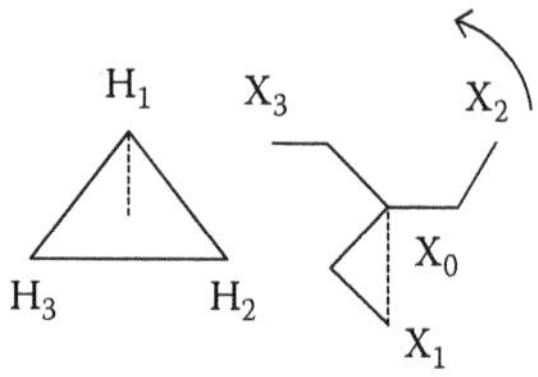
Dd6 180°	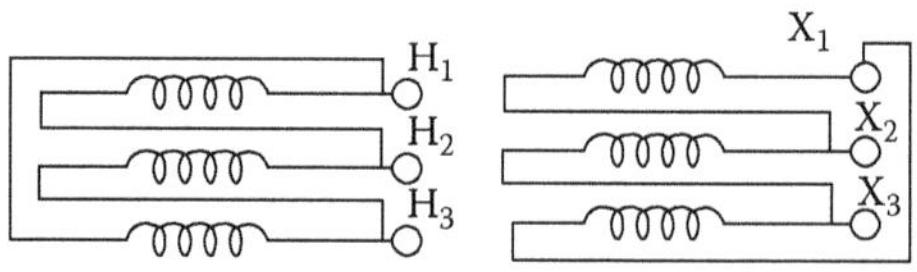	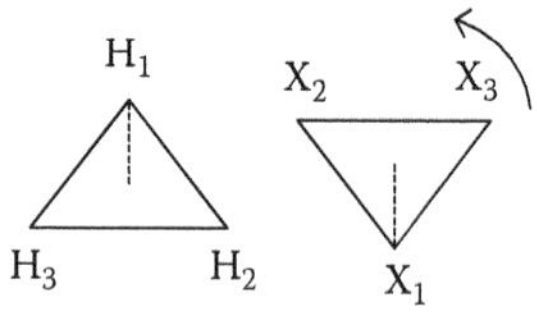

TABLE 2.1 (continued)

Transformer Vector Groups, Winding Connections, and Vector Diagrams

Vector Group & Phase Shift	Winding Connections	Vector Diagram
Dy1 −30°	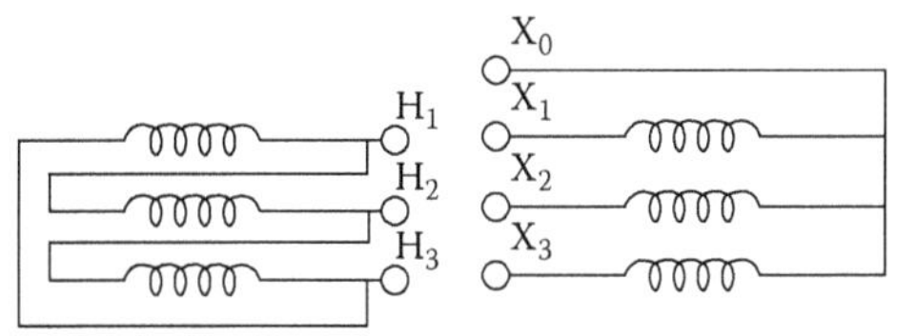	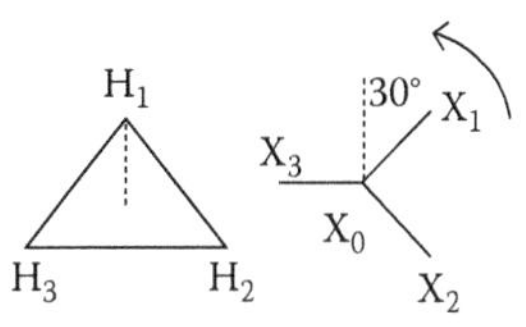
Yd1 −30°	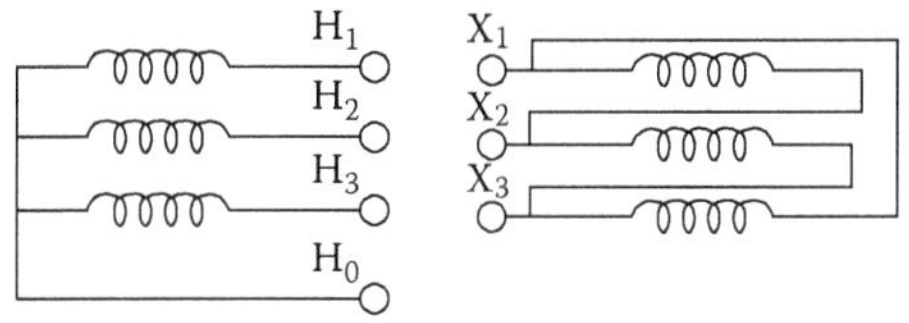	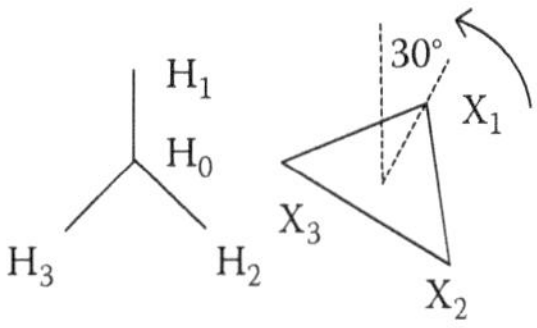
Dy11 +30°	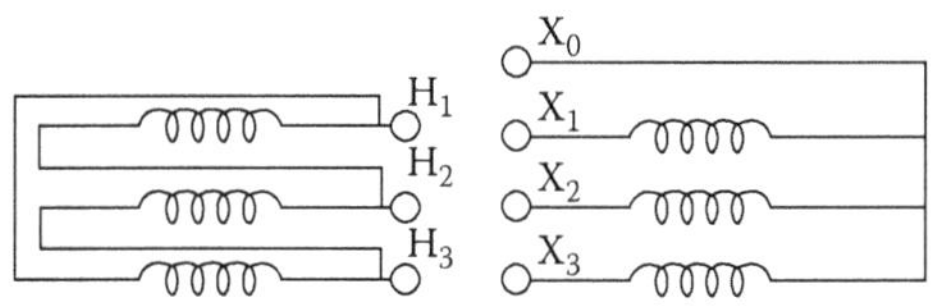	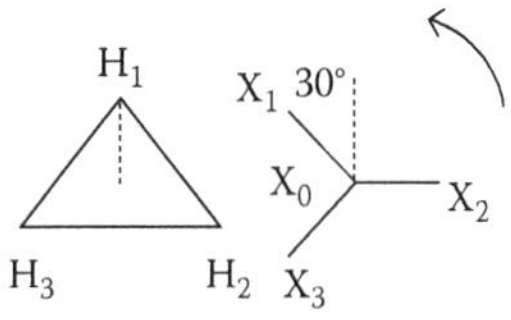
Yd11 +30°	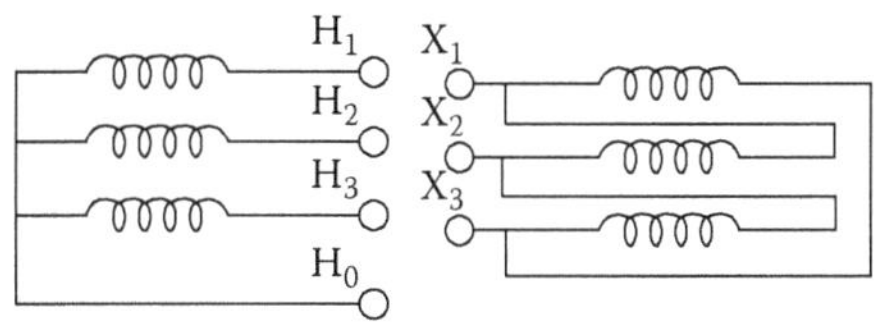	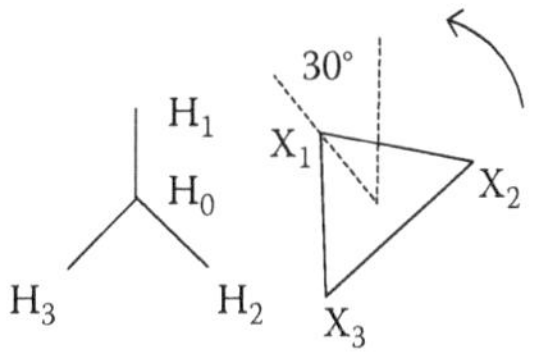
Yz1 −30°	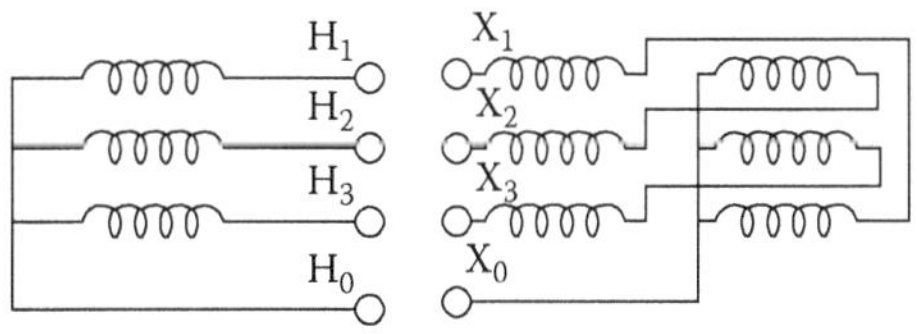	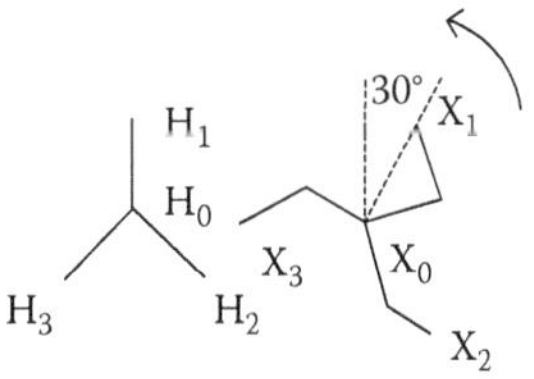
Yz11 +30°	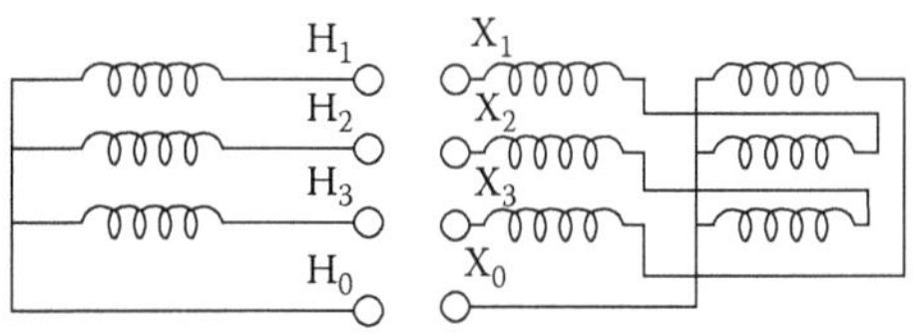	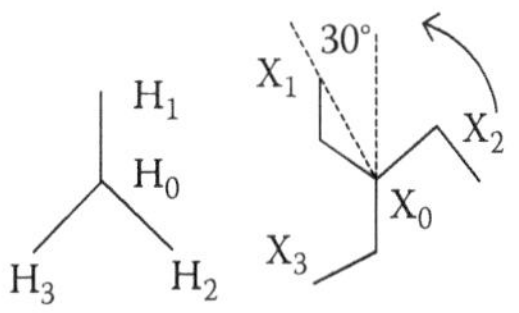

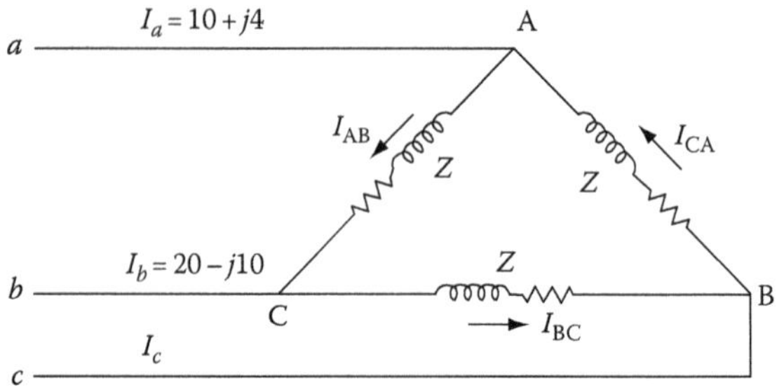

FIGURE 2.7
Balanced delta-connected load on an unbalanced three-phase power supply.

The currents in delta windings are

$$I_{AB} = \frac{1}{3}(I_a - I_b) = -3.33 + j4.67 = 5.735 < 144.51^\circ \text{ A}$$

$$I_{BC} = \frac{1}{3}(I_b - I_c) = 16.67 - j5.33 - 17.50 < -17.7^\circ \text{ A}$$

$$I_{CA} = \frac{1}{3}(I_c - I_a) = -13.33 + j0.67 = 13.34 < 177.12^\circ \text{ A}$$

Calculate the sequence component of the currents I_{AB}. This calculation gives

$$I_{AB1} = 9.43 < 89.57^\circ \text{ A}$$
$$I_{AB2} = 7.181 < 241.76^\circ \text{ A}$$
$$I_{AB0} = 0 \text{ A}$$

Calculate the sequence component of current I_a. This calculation gives

$$I_{a1} = 16.33 < 59.57^\circ \text{ A}$$
$$I_{a2} = 12.437 < 271.76^\circ \text{ A}$$
$$I_{a0} = 0 \text{ A}$$

This shows that the positive sequence current in the delta winding is $1/\sqrt{3}$ times the line positive sequence current, and the phase displacement is $+30^\circ$, that is,

$$I_{AB1} = 9.43 < 89.57^\circ = \frac{I_{a1}}{\sqrt{3}} < 30^\circ = \frac{16.33}{\sqrt{3}} < (59.57^\circ + 30^\circ) \text{ A}$$

The negative sequence current in the delta winding is $1/\sqrt{3}$ times the line negative sequence current, and the phase displacement is -30°, that is,

$$I_{AB2} = 7.181 < 241.76^\circ = \frac{I_{a2}}{\sqrt{3}} < -30^\circ = \frac{12.437}{\sqrt{3}} < (271.76^\circ - 30^\circ) \text{ A}$$

This example illustrates that the negative sequence currents and voltages undergo a phase shift that is the reverse of the positive sequence currents and voltages.

The relative magnitudes of fault currents in two-winding transformers for secondary faults are shown in Figure 2.8, on a per unit basis. The reader can verify the fault current flows shown in this figure.

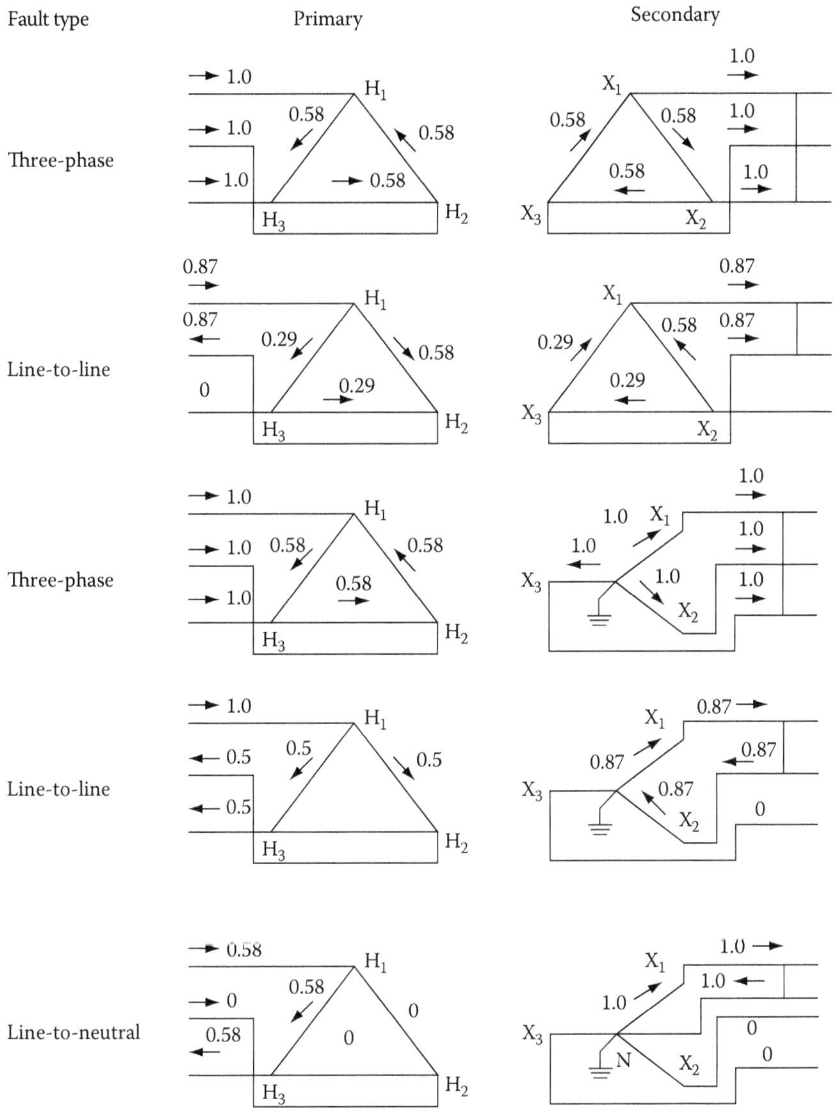

FIGURE 2.8
Three-phase transformer connections and fault current distribution for secondary faults.

(continued)

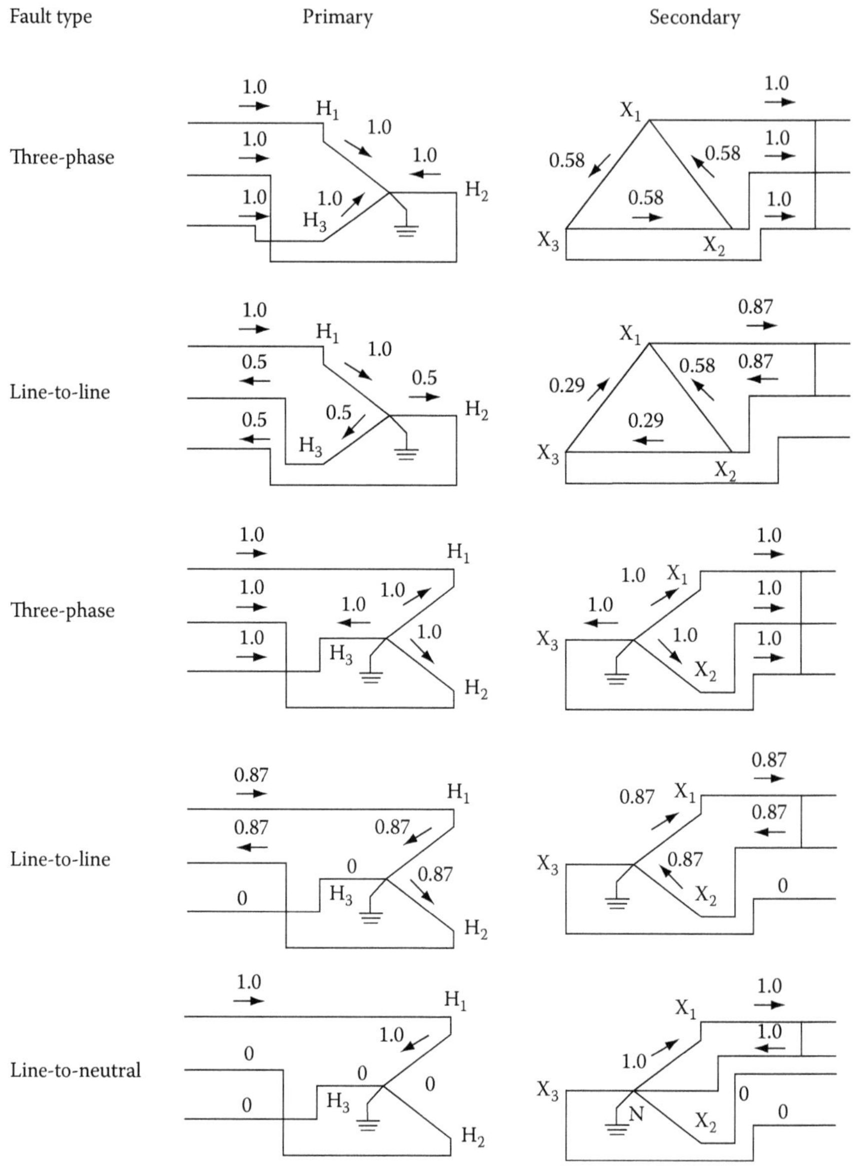

FIGURE 2.8 (continued)

2.6 Unsymmetrical Fault Calculations

Example 2.2

The calculations using symmetrical components can best be illustrated with an example. Consider a subtransmission system as shown in Figure 2.9. A 13.8 kV generator G_1 voltage is stepped up to 138 kV. At the consumer end the voltage is stepped down to 13.8 kV, and generator G_2 operates in

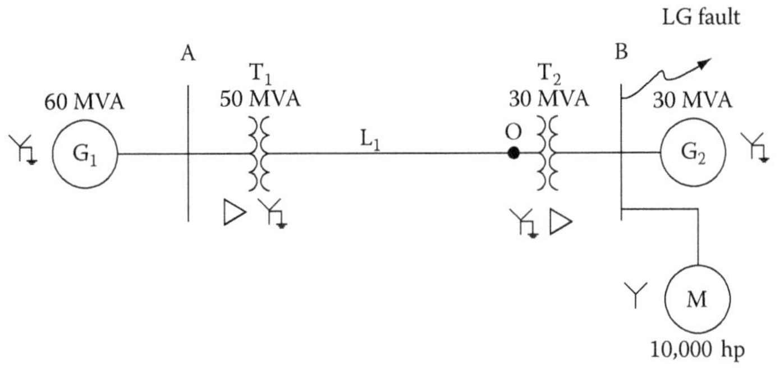

FIGURE 2.9
A single line diagram of power system for Example 2.2.

synchronism with the supply system. Bus B has a 10,000 hp motor load. A line-to-ground fault occurs at bus B. It is required to calculate the fault current distribution throughout the system and also the fault voltages. The resistance of the system components is ignored in the calculations.

Impedance Data

The impedance data for the system components are shown in Table 2.2. Generators G_1 and G_2 are shown solidly grounded, which will not be the case in a practical installation.

TABLE 2.2
Impedance Data for Example 2.2

Equipment	Description	Impedance Data	Per Unit Impedance 100 MVA Base
G_1	13.8 kV, 60 MVA, 0.85 power factor generator	Subtransient reactance = 15% Transient reactance = 20% Zero sequence reactance = 8% Negative sequence reactance = 16.8%	$X_1 = 0.25$ $X_2 = 0.28$ $X_0 = 0.133$
T_1	13.8–138 kV step-up transformer, 50/84 MVA, delta–wye-connected, wye neutral solidly grounded	Z = 9% on 50 MVA base	$X_1 = X_2 = X_0 = 0.18$
L_1	Transmission line, 5 mi. long, 266.8 KCMIL, ACSR	Conductors at 15 ft (4.57 m) equivalent spacing	$X_1 = X_2 = 0.04$ $X_0 = 0.15$
T_2	138–13.2 kV, 30 MVA step-down transformer, wye–delta-connected, high-voltage wye neutral solidly grounded	Z = 8%	$X_1 = X_2 = X_0 = 0.24$
G_2	13.8 kV, 30 MVA, 0.85 power factor generator	Subtransient reactance = 11% Transient reactance = 15% Zero sequence reactance = 6% Negative sequence reactance = 16.5%	$X_1 = 0.37$ $X_2 = 0.55$ $X_0 = 0.20$
M	10,000 hp induction motor load	Locked rotor reactance = 16.7% on motor base kVA (consider 1 hp ≈ 1 kVA)	$X_1 = 1.67$ $X_2 = 1.80$ $X_0 = \infty$

Resistances are neglected in the calculations. KCMIL: Kilo-circular mils, same as MCM. ACSR: Aluminum conductor steel reinforced.

A high-impedance grounding system is used by utilities for grounding generators in step-up transformer configurations. Generators in industrial facilities, directly connected to the load buses, are low-resistance grounded, and the ground fault currents are limited to 200–400 A. The simplifying assumptions in the example are not applicable to a practical installation, but clearly illustrate the procedure of calculations.

The first step is to examine the given impedance data. Generator-saturated subtransient reactance is used in the short-circuit calculations and this is termed positive sequence reactance; 138 kV transmission line reactance is calculated from the given data for conductor size and equivalent conductor spacing. The zero sequence impedance of the transmission line cannot be completely calculated from the given data and is estimated on the basis of certain assumptions, that is, a soil resistivity of 100 Ω-m.

Compiling the impedance data for the system under study from the given parameters, from manufacturers' data, or by calculation and estimation can be time consuming. Most computer-based analysis programs have extensive data libraries and companion programs for calculation of system impedance data and line constants, which has partially removed the onus of generating the data from step-by-step analytical calculations. Appendix B provides models of line constants for coupled transmission lines, bundle conductors, and line capacitances. Refs. [3,4] provide analytical and tabulated data.

Next, the impedance data are converted to a common MVA base. A familiarity with the per unit system is assumed. The voltage transformation ratio of transformer T_2 is 138–13.2 kV, while a bus voltage of 13.8 kV is specified, which should be considered in transforming impedance data on a common MVA base. Table 2.2 shows raw impedance data and their conversion into sequence impedances.

For a single line-to-ground fault at bus B, the sequence impedance network connections are shown in Figure 2.10, with the impedance data for components clearly marked. This figure is based on the fault equivalent circuit shown in Figure 2.1b, with fault impedance $Z_f = 0$. The calculation is carried out in per unit, and the units are not stated in every step of the calculation.

The positive sequence impedance to the fault point is

$$Z_1 = \frac{j(0.25 + 0.18 + 0.04 + 0.24) \times \dfrac{j0.37 \times j1.67}{j(0.37 + 1.67)}}{j(0.25 + 0.18 + 0.04 + 0.24) + \dfrac{j0.37 \times j1.67}{j(0.37 + 1.67)}}$$

This gives $Z_1 = j0.212$.

$$Z_2 = \frac{j(0.28 + 0.18 + 0.04 + 0.24) \times \dfrac{j0.55 \times j1.8}{j(0.55 + 1.8)}}{j(0.28 + 0.18 + 0.04 + 0.24) + \dfrac{j0.55 \times j1.8}{j(0.55 + 1.8)}}$$

This gives $Z_2 = j0.266$.

$Z_0 = j0.2$. Therefore,

$$I_1 = \frac{E}{Z_1 + Z_2 + Z_0} = \frac{1}{j0.212 + j0.266 + j0.2} = -j1.45 \text{ pu}$$

$$I_2 = I_0 = -j1.475$$

$$I_a = I_0 + I_1 + I_2 = 3(-j1.475) = -j4.425 \text{ pu}$$

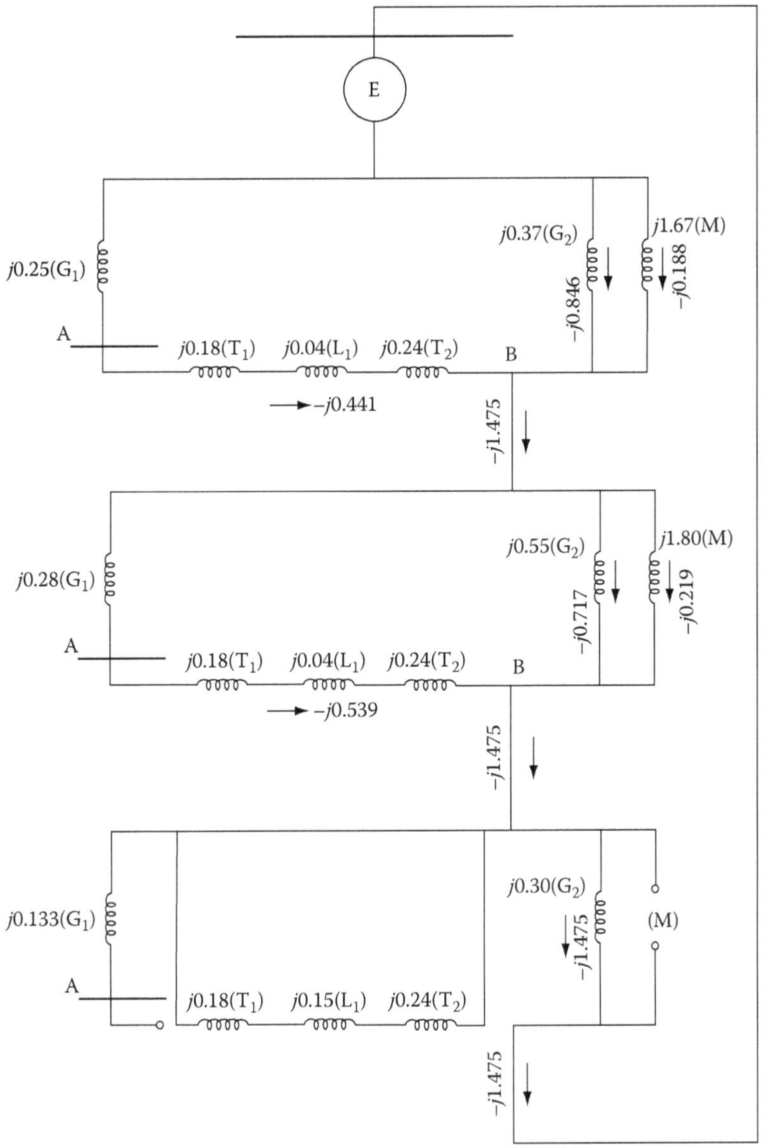

FIGURE 2.10
Sequence network connections for single line-to-ground fault (Example 2.2).

In terms of actual values this is equivalent to 1.851 kA. The fault currents in phases *b* and *c* are zero:

$$I_b = I_c = 0$$

The sequence voltages at a fault point can now be calculated:

$$V_0 = -I_0 Z_0 = j1.475 \times j0.2 = -0.295$$

$$V_2 = -I_2 Z_2 = j1.475 \times j0.266 = -0.392$$

$$V_1 = E - I_1 Z_1 = I_1(Z_0 + Z_2) = 1 - (-j1.475 \times j0.212) = 0.687$$

A check of the calculation can be made at this stage; the voltage of the faulted phase at fault point $B = 0$:

$$V_a = V_0 + V_1 + V_2 = -0.295 - 0.392 + 0.687 = 0$$

The voltages of phases b and c at the fault point are

$$\begin{aligned}
V_b &= V_0 + aV_1 + a^2V_2 \\
&= V_0 - 0.5(V_1 + V_2) - j0.866(V_1 - V_2) \\
&= -0.295 - 0.5(0.687 - 0.392) - j0.866(0.687 + 0.392) \\
&= -0.4425 - j0.9344 \\
|V_b| &= 1.034 \text{ pu}
\end{aligned}$$

Similarly,

$$\begin{aligned}
V_c &= V_0 - 0.5(V_1 + V_2) + j0.866(V_1 - V_2) \\
&= -0.4425 + j0.9344 \\
|V_c| &= 1.034 \text{ pu}
\end{aligned}$$

The distribution of the sequence currents in the network is calculated from the known sequence impedances. The positive sequence current contributed from the right side of the fault, that is, by G_2 and motor M is

$$-j1.475 \frac{j(0.25 + 0.18 + 0.04 + 0.24)}{j(0.25 + 0.18 + 0.04 + 0.24) + \dfrac{j0.37 \times j1.67}{j(0.37 + 1.67)}}$$

This gives $-j1.0338$. This current is composed of two components, one from the generator G_2 and the other from the motor M. The generator component is

$$(-j1.0338)\frac{j1.67}{j(0.37 + 1.67)} = -j0.8463$$

The motor component is similarly calculated and is equal to $-j0.1875$.

The positive sequence current from the left side of bus B is

$$-j1.475 \frac{\dfrac{j0.37 \times j1.67}{j(0.37 + 1.67)}}{j(0.25 + 0.18 + 0.04 + 0.24) + \dfrac{j0.37 \times j1.67}{j(0.37 + 1.67)}}$$

This gives $-j0.441$. The currents from the right side and the left side should sum to $-j1.475$. This checks the calculation accuracy.

The negative sequence currents are calculated likewise and are as follows:

In generator $G_2 = -j0.7172$
In motor $M = -j0.2191$
From left side, bus $B = -j0.5387$
From right side $= -j0.9363$

The results are shown in Figure 2.10. Again, verify that the vectorial summation at the junctions confirms the accuracy of calculations.

Currents in Generator G_2

$$\begin{aligned} I_a(G_2) &= I_1(G_2) + I_2(G_2) + I_0(G_2) \\ &= -j0.8463 - j0.7172 - j1.475 \\ &= -j3.0385 \\ |I_a(G_2)| &= 3.0385 \text{ pu} \end{aligned}$$

$$\begin{aligned} I_b(G_2) &= I_0 - 0.5(I_1 + I_2) - j0.866(I_1 - I_2) \\ &= -j1.475 - 0.5(-j0.8463 - j0.7172) - j0.866(-j0.8463 + j0.7172) \\ &= -0.1118 - j0.6933 \\ |I_b(G_2)| &= 0.7023 \text{ pu} \end{aligned}$$

$$\begin{aligned} I_c(G_2) &= I_0 - 0.5(I_1 + I_2) + j0.866(I_1 - I_2) \\ &= 0.1118 - j0.6933 \\ |I_c(G_2)| &= 0.7023 \text{ pu} \end{aligned}$$

This large unbalance is noteworthy. It gives rise to increased thermal effects due to negative sequence currents and results in overheating of the generator rotor. A generator will trip quickly on negative sequence currents.

Currents in Motor M

The zero sequence current in the motor is zero, as the motor wye-connected windings are not grounded as per industrial practice in the United States. Thus,

$$\begin{aligned} I_a(M) &= I_1(M) + I_2(M) \\ &= -j0.1875 - j0.2191 \\ &= -j0.4066 \\ |I_a(M)| &= 0.4066 \text{ pu} \end{aligned}$$

$$\begin{aligned} I_b(M) &= -0.5(-j0.4066) - j0.866(0.0316) = 0.0274 + j0.2033 \\ I_c(M) &= -0.0274 + j0.2033 \\ |I_b(M)| &= |I_c(M)| = 0.2051 \text{ pu} \end{aligned}$$

The summation of the line currents in the motor M and generator G_2 are

$$\begin{aligned} I_a(G_2) + I_a(M) &= -j3.0385 - j0.4066 = -j3.4451 \\ I_b(G_2) + I_b(M) &= -0.118 - j0.6993 + 0.0274 + j0.2033 = -0.084 - j0.490 \\ I_c(G_2) + I_c(M) &= 0.1118 - j0.6933 - 0.0274 + j0.2033 = 0.084 - j0.490 \end{aligned}$$

Currents from the left side of the bus B are

$$I_a = -j0.441 - j0.5387$$
$$= -j0.98$$
$$I_b = -0.5(-0.441 - j0.5387) - j0.866(-0.441 + j0.5387)$$
$$= 0.084 + j0.490$$
$$I_c = -0.084 + j0.490$$

These results are consistent as the sum of currents in phases *b* and *c* at the fault point from the right and left side is zero and the summation of phase *a* currents gives the total ground fault current at $b = -j4.425$. The distribution of currents is shown in a three-line diagram (Figure 2.11).

Continuing with the example, the currents and voltages in the transformer T_2 windings are calculated. We should correctly apply the phase shifts for positive and negative sequence components when passing from delta secondary to wye primary of the transformer. The positive and negative sequence current on the wye side of transformer T_2 are

$$I_{1(p)} = I_1 < 30^\circ = -j0.441 < 30^\circ = 0.2205 - j0.382$$
$$I_{2(p)} = I_2 < -30^\circ = -j0.5387 < -30^\circ = -0.2695 - j0.4668$$

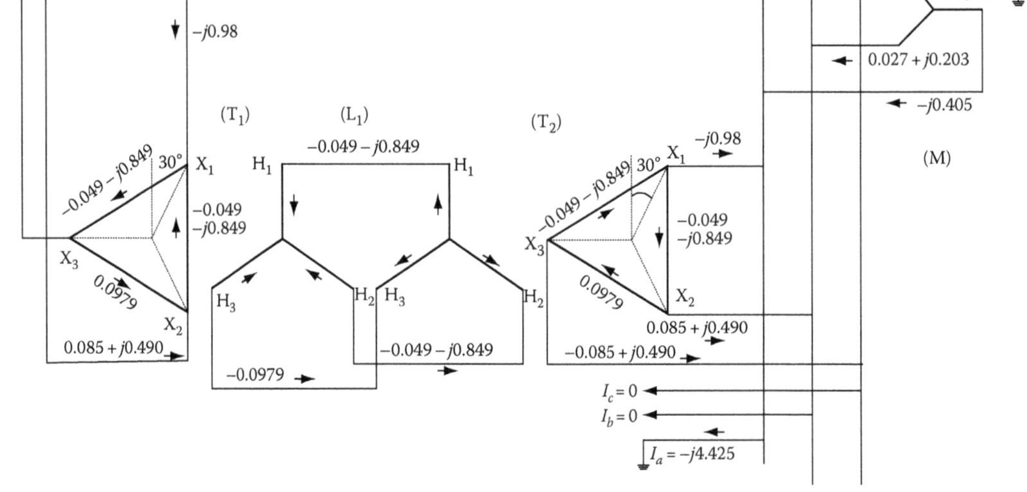

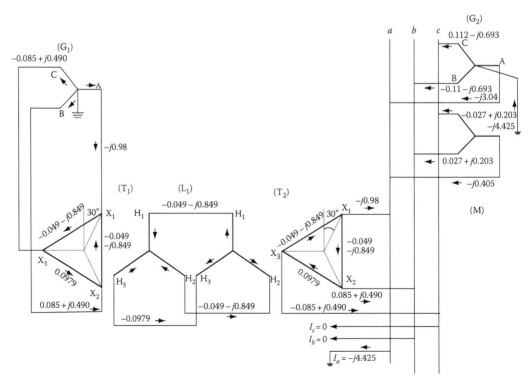

FIGURE 2.11
Three-line diagram of fault current distribution (Example 2.2).

Also, the zero sequence current is zero. The primary currents are

$$I_{a(p)} = I_{1(p)} + I_{2(p)}$$
$$= 0.441 < 30^\circ + 0.5387 < -30^\circ = -0.049 - j0.8487$$
$$I_{b(p)} = a^2 I_{1(p)} + a I_{2(p)} = -0.0979$$
$$I_{c(p)} = a I_{1(p)} + a^2 I_{2(p)} = -0.049 - j0.8487$$

Currents in the lines on the delta side of the transformer T_1 are similarly calculated. The positive sequence component, which underwent a 30° positive shift from delta to wye in transformer T_2, undergoes a −30° phase shift; as for an ANSI connected transformer it is the low-voltage vectors, which lag the high-voltage side vectors. Similarly, the negative sequence component undergoes a positive phase shift. The currents on the delta side of transformers T_1 and T_2 are identical in amplitude and phase. Note that 138 kV line is considered lossless. Figure 2.11 shows the distribution of currents throughout the distribution system.

The voltage on the primary side of transformer T_2 can be calculated. The voltages undergo the same phase shifts as the currents. Positive sequence voltage is the base fault positive sequence voltage, phase shifted by 30° (positive) minus the voltage drop in transformer reactance due to the positive sequence current:

$$V_1(p) = 1.0 < 30^\circ - jI_{1(p)}X_2$$
$$= 1.0 < 30^\circ - (j0.441 < 30^\circ)(j0.24)$$
$$= 0.9577 + j0.553$$
$$V_2(p) = 0 - I_{2(p)}X_2$$
$$= -(0.539 < -30^\circ)(j0.24)$$
$$= 0.112 - j0.0647$$

Thus,

$$V_{a(p)} = 0.9577 + j0.553 + 0.112 - j0.0647 = 1.0697 + j0.4883 = 1.17 < 24.5^\circ$$
$$V_{b(p)} = -0.5(V_{1(p)} + V_{2(p)}) - j0.866(V_{1(p)} - V_{2(p)})$$
$$= -j0.9763$$
$$V_{c(p)} = 0.5(V_{1(p)} + V_{2(p)}) - j0.866(V_{2(p)} - V_{1(p)})$$
$$= 1.0697 + j0.4883 = 1.17 < 155.5^\circ$$

Note the voltage unbalance caused by the fault.

2.7 System Grounding

System grounding refers to the electrical connection between the phase conductors and ground and dictates the manner in which the neutral points of wye-connected transformers and generators or artificially derived neutral systems through delta-wye or zigzag transformers are grounded. The equipment grounding refers to the grounding of the exposed metallic parts of the electrical equipment, which can become energized and create a potential to ground, say due to breakdown of insulation or fault, and can be a potential safety hazard. The safety of the personnel and human life is of importance. The safety grounding is to establish an equipotential surface in the work area to mitigate shock hazard. The utility systems at high-voltage transmission level, subtransmission level, and distribution level are solidly grounded. The utility generators connected through step-up transformers are invariably high-resistance grounded through a distribution transformer with secondary loading resistor. The industrial systems at medium-voltage level are low-resistance grounded. The implications of system grounding are as follows:

- Enough ground fault current should be available to selectively trip the faulty section with minimum disturbance to the system. Ground faults are cleared even faster than the phase faults to limit equipment damage.
- Line-to-ground fault is the most important cause of system temporary overvoltages, which dictates the selection of surge arresters.
- Grounding should prevent high overvoltages of ferroresonance, overvoltages of arcing type ground fault, and capacitive-inductive resonant couplings. If a generator neutral is left ungrounded, there is possibility of generating high voltages through inductive-capacitive couplings. Ferroresonance can also occur due to the presence of generator PTs (Potential Transformers). In ungrounded systems, a possibility of resonance with high voltage generation, approaching five times or more of the system voltage exists for values of X_0/X_1 between 0 and −40. For the first phase-to-ground fault, the continuity of operations can be sustained, though unfaulted phases have $\sqrt{3}$ times the normal line-to-ground voltage. All unremoved faults, thus, put greater than normal voltage on system insulation and increased level of conductor and motor insulation may be required. The grounding practices in the industry have withdrawn from this method of grounding.
- The relative magnitude of the fault currents depends upon sequence impedances.
- In industrial systems the continuity of processes is important. The current industry practice is high-resistance grounded systems for low-voltage distributions.

Figure 2.12 shows the system grounding methods. A brief discussion follows.

2.7.1 Solidly Grounded Systems

In a solidly grounded system, no intentional impedance is introduced between the system neutral and ground. These systems meet the requirement of "effectively grounded" systems in which the ratio X_0/X_1 is positive and less than 3.0 and the ratio R_0/X_0 is positive and less than 1, where X_1, X_0, and R_0 are the positive sequence reactance, zero sequence reactance, and zero sequence resistance, respectively. The coefficient of grounding (COG) is defined as a ratio of E_{LG}/E_{LL} in percentage, where E_{LG} is the highest rms voltage on a sound phase, at a selected location, during a fault affecting one or more phases to ground, and E_{LL} is the rms phase-to-phase power frequency voltage that is obtained at the same location with the fault removed. Calculations in Example 2.2 show that the fault voltage rises on unfaulted phases. Solidly grounded systems are characterized by a COG of 80%. By contrast, *for ungrounded systems*, definite values cannot be assigned to ratios X_0/X_1 and R_0/X_0. The ratio X_0/X_1 is negative and may vary from low to high values. The COG

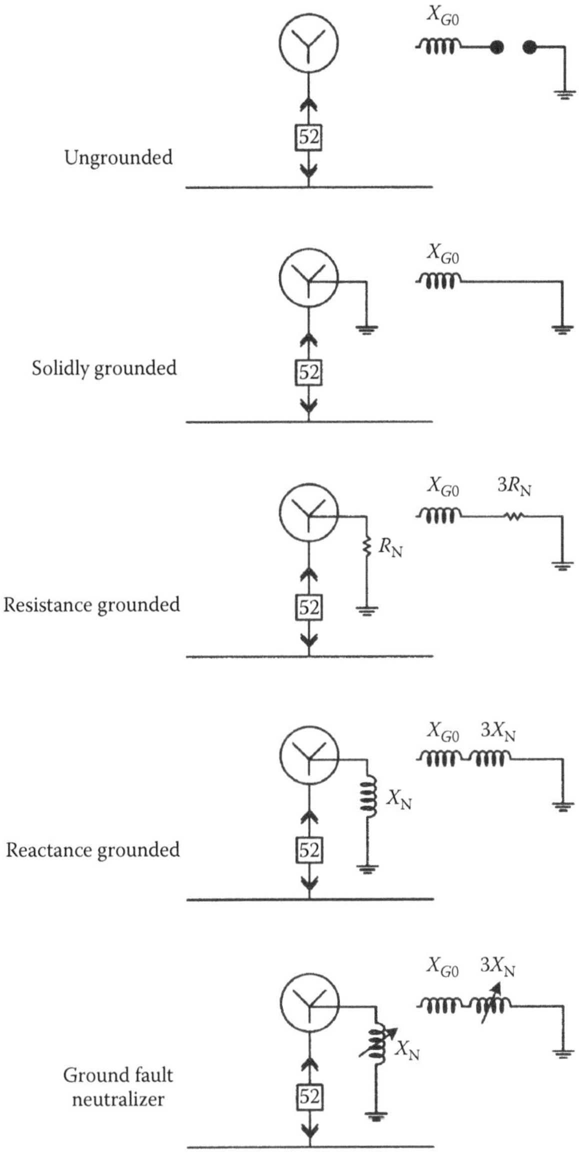

FIGURE 2.12
Methods of system grounding.

approaches 120%. For values of X_0/X_1 between 0 and −40, a possibility of resonance with consequent generation of high voltages exists. The overvoltages based on relative values of sequence impedances are plotted in Ref. [4].

The COG affects the selection of rated voltage of the surge arresters and stresses on the insulation systems. Solidly grounded systems are, generally, characterized by COG of 80%. Approximately, a surge arrester with its rated voltage calculated on the basis of the system voltage multiplied by 0.8 can be applied.

The low-voltage systems in industrial power distribution systems used to be solidly grounded. However, this trend is changing and high-resistance grounding is being adopted.

The solidly grounded systems have an advantage of providing effective control of overvoltages, which become impressed on or are self-generated in the power system by insulation breakdowns and restriking faults. Yet, these give the highest arc fault current and consequent damage and require immediate isolation of the faulty section. Single-line-to-ground fault currents can be higher than the three-phase fault currents. These high magnitudes of fault currents have twofold effect:

- Higher burning or equipment damage
- Interruption of the processes, as the faulty section must be selectively isolated without escalation of the fault to unfaulted sections

The arcing faults are caused by insulation failure, loose connections or accidents. The arc behavior is difficult to predict and model because of spasmodic nature of the arc fault. This is due to elongation and blowout effects and arc re-ignition. Arc travels from point to point and physical flexing of cables and structures can occur. Arcing faults can exhibit low levels because of the impedance of the arc fault circuit itself. Arcing faults can be discontinuous requiring a certain minimum voltage for re-ignition. The limits of the acceptable damage to material for arc fault currents of 3,000–26,000 A in 480 V systems have been established by testing [5,6] and are given by the following equation:

$$\text{Fault damage } \propto (I)^{1.5}t \tag{2.23}$$

where
I is the arc fault current
t is the duration in seconds

$$V_{\text{D}} = K_{\text{s}}(I)^{1.5}\, t\ (\text{in.})^3 \tag{2.24}$$

where
K_{s} is the burning rate of material in in.^{3}/As$^{1.5}$
V_{D} is acceptable damage to material in in.^{3}
I is the arc fault current
t is the duration of flow of fault current
K_{s} depends upon type of material and is given by

$$\begin{aligned} K_{\text{s}} &= 0.72 \times 10^{-6} \text{ for copper,} \\ &= 1.52 \times 10^{-6} \text{ for aluminum,} \\ &= 0.66 \times 10^{-6} \text{ for steel.} \end{aligned} \tag{2.25}$$

NEMA (National Electrical Manufacturers Association) [7] assumes a practical limit for the ground fault protective devices, so that

$$(I)^{1.5}\, t \lhd 250 I_{\text{r}} \tag{2.26}$$

I_{r} is the rated current of the conductor, bus, and disconnect or circuit breaker to be protected.

Combining these equations, we can write

$$V_D = 250K_sI_r \tag{2.27}$$

As an example, consider a circuit of 4000 A. Then the NEMA practical limit is 1.0×10^6 $(A)^{1.5}$ s and the permissible damage to copper, from (Equation 2.24) is 0.72 in.^{3}. To limit the arc fault damage to this value, the maximum fault clearing time can be calculated. Consider that the arc fault current is 20 kA. Then, the maximum fault clearing time including the relay operating time and breaker interrupting time is 0.35 s. It is obvious that vaporizing 0.72 in.^{3} of copper on a ground fault, which is cleared according to established standards, is still determinant to the operation of the equipment. A shutdown and repairs will be needed after the fault incidence.

The arc fault current is not of the same magnitude as the three-phase fault current, due to a voltage drop in the arc. In the low-voltage 480 V systems it may be 50%–60% of the bolted three-phase current, while for medium voltage systems it will approach three-phase bolted fault current, but somewhat lower.

Due to high arc fault damage and interruption of processes, the solidly grounded systems are not in much use in the industrial distribution systems. However, ac circuits of less than 50 V and circuits of 50–1000 V for supplying premises wiring systems and single-phase 120/240 V control circuits must be solidly grounded according to NEC (National Electrical Code) [8].

2.7.2 Resistance Grounding

An impedance grounded system has a resistance or reactance connected in the neutral circuit to ground, as shown in Figure 2.12b. In a low-resistance grounded system the resistance in the neutral circuit is so chosen that the ground fault is limited to approximately full load current or even lower, typically 200–400 A. The arc fault damage is reduced, and these systems provide effective control of the overvoltages generated in the system by resonant capacitive-inductive couplings and restriking ground faults. Though the ground fault current is much reduced, it cannot be allowed to be sustained and selective tripping must be provided to isolate the faulty section. For a ground fault current limited to 400 A, the pick up sensitivity of modern ground fault devices can be even lower than 5 A. Considering an available fault current of 400 A and a relay pickup of 5 A, approximately 98.75% of the transformer or generator windings from the line terminal to neutral are protected. This assumes a linear distribution of voltage across the winding. (Practically the pickup will be higher than the low set point of 5 A.) The incidence of ground fault occurrence toward the neutral decreases as square of the winding turns. Medium voltage distribution systems in industrial distributions are commonly low resistance grounded.

The low resistance grounded systems are adopted at medium voltages, 13.8, 4.16, and 2.4 kV for industrial distribution systems. Also industrial bus-connected generators were commonly low resistance grounded. A recent trend in industrial bus-connected medium voltage generator grounding is hybrid grounding systems [9].

2.7.2.1 High-Resistance Grounded Systems

High-resistance grounded systems limit the ground fault current to a low value, so that an immediate disconnection on occurrence of a ground fault is not required. It is well

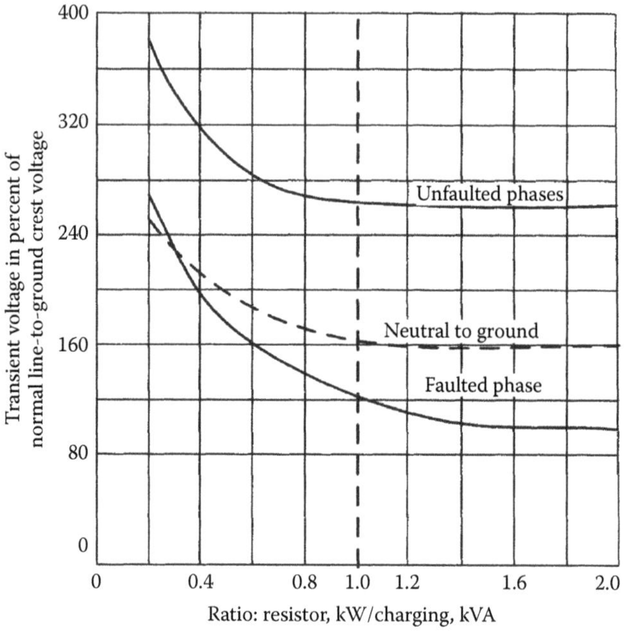

FIGURE 2.13
Overvoltages versus ratio of resistor kW/charging kVA.

documented that to control overvoltages in the high-resistance grounded systems, the grounding resistor should be so chosen that

$$R_n = \frac{V_{ln}}{3I_c} \tag{2.28}$$

where
V_{ln} is the line to neutral voltage
I_c is the stray capacitance current of each phase conductor

Figure 2.13 shows transient voltage in percent of normal line-to-ground crest voltage verses the resistor kW/charging capacitive kVA. The transients are a minimum when this ratio is unity [4]. This leads to the requirement of accurately calculating the stray capacitance currents in the system [10]. Cables, motors, transformers, surge arresters, and generators all contribute to the stray capacitance current. Surge capacitors–connected line to ground must be considered in the calculations. Once the system stray capacitance is determined, then, the charging current per phase, I_c is given by

$$I_c = \frac{V_{ln}}{X_{c0}} \tag{2.29}$$

where X_{c0} is the capacitive reactance of each phase, stray capacitance considered lumped together.

This can be illustrated with an example. A high-resistance grounding system for a wye-connected neutral of a 13.8–0.48 kV transformer is shown in Figure 2.14a. This shows that the stray capacitance current per phase of all the distribution systems connected to the

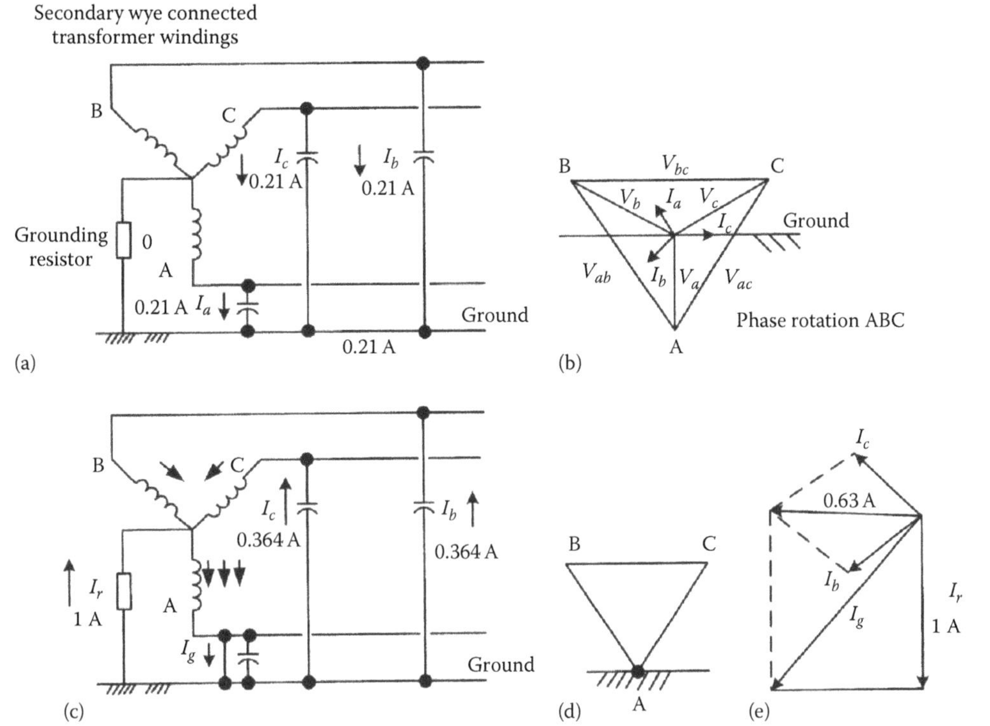

FIGURE 2.14
(a) and (b) stray currents under no fault conditions; (c) Flow of phase capacitance and ground currents for a phase-to-ground fault phase a; (d) Voltages to ground; (e) Summation of capacitance and resistor currents.

secondary of the transformer is 0.21 A per phase, assumed to be balanced in each phase. Generally for low-voltage distribution systems, a stray capacitance current of 0.1 A per MVA of transformer load can be taken, though this rule of thumb is no substitute for accurate calculations of stray capacitance currents. Figure 2.14a shows that under no fault condition, the vector sum of three capacitance currents is zero, as these are 90° displaced with respect to each voltage vector and therefore 120° displaced with respect to each other. Thus, the grounded neutral does not carry any current and the neutral of the system is held at the ground potential as in Figure 2.14b. As

$$I_{c1} + I_{c2} + I_{c3} = 0 \tag{2.30}$$

no capacitance current flows into the ground. On occurrence of a ground fault, say in phase *a*, the situation is depicted in Figure 2.14c and d. The capacitance current of faulted *a* phase is short-circuited to ground. The faulted phase, assuming zero fault resistance is at the ground potential as in Figure 2.14d and the other two phases have line-to-line voltages with respect to ground. Therefore, the capacitance current of the unfaulted phases *b* and *c* increases proportional to the voltage increase, that is, $\sqrt{3} \times 0.21 = 0.364$ A. Moreover, this current in phases *b* and *c* reverses, flows through the transformer windings, and sums up in the transformer winding of phase *a*. Figure 2.14e shows that this vector sum = 0.63 A.

Now consider that the ground current through the grounding resistor is limited to 1 A only. This is acceptable according to Equation 2.28 as the stray capacitance current is 0.63 A. This resistor ground current also flows through the transformer phase winding *a* to the fault and the total ground fault current is $I_g = \sqrt{1^2 + 0.63^2} = 1.182A$ as in Figure 2.14e.

The preceding analysis assumes a full neutral shift, ignores the fault impedance itself and assumes that the ground grid resistance and the system zero sequence impedances are zero. Practically, the neutral shift will vary.

Some obvious advantages are as follows:

- The resistance limits the ground fault current and, therefore, reduces burning and arcing effects in switchgear, transformers, cables, and rotating equipment.
- It reduces mechanical stresses in circuits and apparatus carrying fault current.
- It reduces arc blast and arc flash hazard to personnel who happen to be in close proximity of the ground fault.
- It reduces line-to-line voltage dips due to ground fault and three-phase loads can be served.
- Control of transient overvoltages is secured by proper selection of the resistor.

2.7.2.1.1 Limitations of High Resistance Grounded (HRG) Systems

The limitation of the system is that the capacitance current should not exceed approximately 10 A to prevent immediate shutdown. As the system voltage increases, so does the capacitance current. This limits the applications to systems of rated voltages of 4.16 kV and below.

Though immediate shutdown is prevented, the fault situation should not be prolonged; the fault should be localized and removed. There are three reasons for this:

1. Figure 2.14d shows that the unfaulted phases have voltage rise by a factor of $\sqrt{3}$ to ground. This increases the normal insulation stresses between phase-to-ground. This may be of special concern for low-voltage cables. If the time required to de-energize the system is indefinite, 173% insulation level for the cables must be selected [11]. However, NEC does not specify 173% insulation level and for 600 V cables insulation levels correspond to 100% and 133%. Also Ref. [6] specifies that the actual operating voltage on cables should not exceed 5% during continuous operation and 10% during emergencies. This is of importance when 600 V nominal three-phase systems are used for power distributions. The dc loads served through six-pulse converter systems will have a dc voltage of 648 and 810 V, respectively, for 480 and 600 V rms ac systems.
2. Low levels of fault currents if sustained for long time may cause irreparable damage. Though the burning rate is slow, but the heat energy released over the course of time can damage cores and windings of rotating machines even for ground currents as low as 3–4 A. This has been demonstrated in test conditions [12].
3. A first ground fault left in the system increases the probability of a second ground fault in another phase. If this happens, then it amounts to a two-phase-to ground-fault with some interconnecting impedance depending upon the fault location. The potentiality of equipment damage and burnout increases.

Example 2.3

Figure 2.15a shows a 5 MVA 34.5–2.4 kV delta–delta transformer serving industrial motor loads. It is required to derive an artificial neutral through a wye–delta-connected transformer and high-resistance ground the 2.4 kV secondary system through a grounding resistor. The capacitance charging current of the system is 8 A. Calculate the value of the resistor to limit the ground fault current through the resistor to 10 A. Neglect transformer resistance and source resistance.

The connection of sequence networks is shown in Figure 2.15b and the given impedance data reduced to a common 100 MVA base are shown in Table 2.3. Motor wye-connected neutrals are left ungrounded in the industrial systems, and therefore motor zero sequence impedance is infinite. (This contrasts with grounding practices in some European countries, where the motor neutrals are grounded.) The source zero sequence impedance can be calculated based on the assumption of equal positive and negative sequence reactances. The motor voltage is 2.3 kV, and, therefore, its per unit reactance on 100 MVA base is given by

$$\left(\frac{16.7}{1.64}\right)\left(\frac{2.3}{2.4}\right)^2 = 9.35$$

Similarly, the grounding transformer per unit calculations should be adjusted for correct voltages:

$$X_0 = \left(\frac{1.5}{0.06}\right)\left(\frac{2.4}{2.4/\sqrt{3}}\right)^2 = 75$$

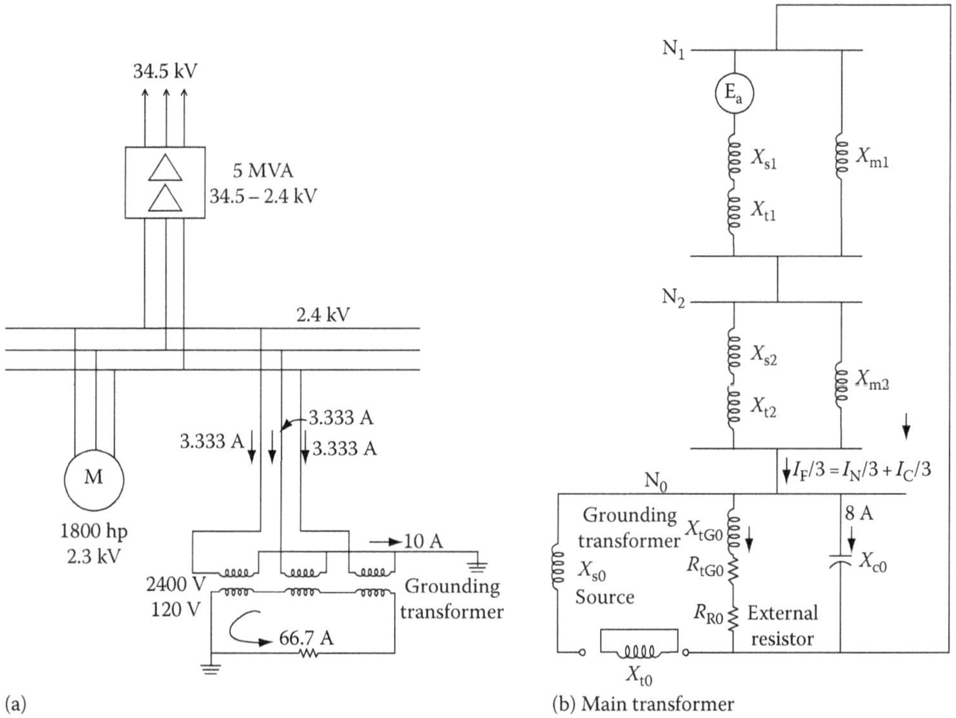

FIGURE 2.15
(a) Artificially derived neutral grounding in a delta–delta system through a wye–delta grounding transformer (Example 2.3); (b) connection of sequence impedances for high-resistance fault calculations (Example 2.3).

TABLE 2.3

Impedance Data for Example 2.3—High-Resistance Grounding

Equipment	Given Data	Per Unit Impedance on 100 MVA Base
34.5 kV source	Three-phase fault = 1500 MVA	$X_{s1} = X_{s2} = 0.067$
	Line-to-ground fault = 20 kA sym	$X_{s0} = 0.116$
34.5–2.4 kV, 5 MVA transformer, delta–delta-connected	$X_1 = X_2 = X_0 = 8\%$	$X_{t1} = X_{t2} = X_{t0} = 1.60$
2.3 kV 1800 hp (1640 kVA) induction motor load	Locked rotor reactance = 16.7% (on motor base kVA)	$X_{m1} = X_{m2} = 9.35$
Grounding transformer, 60 kVA, wye–delta-connected 2400:120 V	$X_0 = 1.5\%$	$X_0 = 75$
	$R_0 = 1.0\%$	$R_0 = 50$

The equivalent positive and negative sequence reactances are as follows: 1.41 per unit each. The zero sequence impedance of the grounding transformer is $50 + j75$ per unit. The total fault current should be limited to $10 - j8 = 12.80$ A. Thus, the required impedance is

$$Z_t = \left(\frac{2400/\sqrt{3}}{12.8/3}\right) = 324.8 \text{ ohm}$$

The base ohms (100 MVA base) = 0.0576. The required $Z_t = 324.8/\text{base ohms} = 5638.9$ per unit. This shows that the system positive and negative sequence impedances are low compared to the desired total impedance in the neutral circuit. The system positive and negative sequence impedances can, therefore, be neglected.

$I_{R0} = 10/3 = 3.33$ A. Therefore, $Z_{R0} = (2400/\sqrt{3})/3.33 = 416.09$ ohm = 416.09/base ohms = 7223.9 per unit. The additional resistor to be inserted is given by the expression:

$$\begin{aligned} R_{R0} &= \sqrt{Z_{R0} - X_{tG0}} - R_{tG0} \\ &= \sqrt{7223.9^2 - 75^2} - 50 \\ &= 7173.5 \text{ pu} \end{aligned}$$

where

R_{R0} is the added resistor

Z_{R0} is the total impedance in the ground circuit

X_{tG0} and R_{tG0} are reactance and resistance of the grounding transformer as in Figure 2.15b

Multiplying by base ohms, the required resistance = 413.2 ohm

These values are in symmetrical component equivalents. In actual values, referred to 120 V secondary, the resistance value is

$$R_R = \left(\frac{120}{2400}\right)^2 413.2 \times 3 = 3.1 \text{ ohm}$$

If we had ignored all the sequence impedances, including that of the grounding transformer, the calculated value is 3.12 ohm. This is often done in the calculations for grounding resistance for high-resistance grounded systems, and all impedances including that of the grounding transformer can be ignored without appreciable error in the final results. The grounding transformer should be rated to permit continuous operation, with a ground fault on the system. The per phase grounding transformer kVA requirement is 2.4 (kV) × 3.33 A = 8 kVA, *that is, a total of* 8 × 3 = 24 kVA. The grounding transformer of the example is, therefore, adequately rated.

2.7.2.2 Coefficient of Grounding

Simplified equations can be applied for calculation of COG. Sometimes we define EFF (IEC standards, earth fault factor). It is simply

$$\text{EFF} = \sqrt{3}\text{COG} \tag{2.31}$$

COG can be calculated by the following equations and more rigorously by the sequence component matrix methods as illustrated earlier.

Single line-to-ground fault

$$\begin{aligned} \text{COG}(\textit{phase } b) &= -\frac{1}{2}\left(\frac{\sqrt{3}k}{2+k} + j1\right) \\ \text{COG}(\textit{phase } c) &= -\frac{1}{2}\left(\frac{\sqrt{3}k}{2+k} - j1\right) \end{aligned} \tag{2.32}$$

Double line-to-ground fault

$$\text{COG}(\textit{phase } a) = \frac{\sqrt{3}k}{1+2k} \tag{2.33}$$

where k is given by

$$k = \frac{Z_0}{Z_1} \tag{2.34}$$

To take fault resistance into account, k is modified as follows:

Single line-to-ground fault

$$k = \frac{(R_0 + R_f + jX_0)}{(R_1 + R_f + jX_1)} \tag{2.35}$$

For double line-to-ground fault

$$k = \frac{(R_0 + 2R_f + jX_0)}{(R_1 + 2R_f + jX_1)} \tag{2.36}$$

If R_0 and R_1 are zero, then the earlier equations reduce to

For single line-to-ground fault:

$$\text{COG} = \frac{\sqrt{k^2 + k + 1}}{k+2} \tag{2.37}$$

For double line-to-ground fault:

$$COG = \frac{k}{2k+1} \tag{2.38}$$

$$\text{where } k \text{ is now } = \frac{X_0}{X_1} \tag{2.39}$$

In general, fault resistance will reduce COG, except in low resistance systems. Ref. [4] provides curves for direct reading of COG for *any type of fault* in percent of unfaulted line-to-line voltage for the area bonded by the curve and the axes. All impedances must be on the same MVA base.

Example 2.4

Calculate the COG at the faulted bus B in Example 2.2. If generator G_2 is grounded through a 400 A resistor, what is the COG?

In Example 2.2, all resistances are ignored. A voltage of 1.034 pu was calculated on the unfaulted phases, which gives a COG of 0.597.

If the generator is grounded through 400 A resistor, then $R_0 = 19.19$ ohm, the positive sequence reactance is 0.4 ohm, and the zero sequence reactance is 0.38 ohm, which is much smaller than R_0. In fact, in a resistance grounded or high-resistance grounded system, the sequence components are relatively small and the ground fault current can be calculated based upon the grounding resistor alone. The total ground fault current at bus 2 will reduce to approximately 400 A. This gives a COG of approximately 100%. This means that phase-to-ground voltage on unfaulted phases will be equal to line-to-line voltage.

2.8 Open Conductor Faults

Symmetrical components can also be applied to the study of open conductor faults. These faults are in series with the line and are called series faults. One or two conductors may be opened, due to mechanical damage or by operation of fuses on unsymmetrical faults.

2.8.1 Two-Conductor Open Fault

Consider that conductors of phases *b* and *c* are open-circuited. The currents in these conductors then go to zero.

$$I_b = I_c = 0 \tag{2.40}$$

The voltage across the unbroken phase conductor is zero, at the point of break as in Figure 2.16a.

$$\begin{aligned} V_{a0} &= V_{a01} + V_{a02} + V_{a0} = 0 \\ I_{a1} &= I_{a2} = I_{a0} = \frac{1}{3} I_a \end{aligned} \tag{2.41}$$

This suggests that sequence networks can be connected in series as shown in Figure 2.16b.

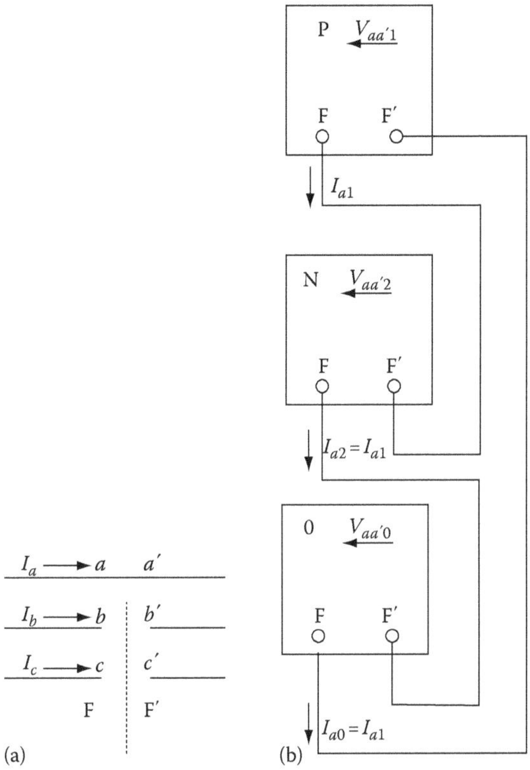

FIGURE 2.16
(a) Two-conductor open series fault; (b) connection of sequence networks.

2.8.2 One-Conductor Open Fault

Now consider that phase *a* conductor is broken as in Figure 2.17a

$$I_a = 0 \qquad V_{b0} - V_{c0} = 0 \tag{2.42}$$

Thus,

$$\begin{aligned} V_{ao1} = V_{ao2} = V_{ao0} = \frac{1}{3} V_{ao} \\ I_{a1} + I_{a2} + I_{a0} = 0 \end{aligned} \tag{2.43}$$

This suggests that sequence networks are connected in parallel as in Figure 2.17b.

Example 2.5

Consider that one conductor is broken on the high-voltage side at the point marked O in Figure 2.9. The equivalent circuit is shown in Figure 2.18.

An induction motor load of 10,000 hp was considered in the calculations for a single-line-to-ground fault in Example 2.2. All other *static loads*, that is, lighting and resistance heating loads, were ignored, as these do not contribute to short-circuit currents. Also, all drive system loads,

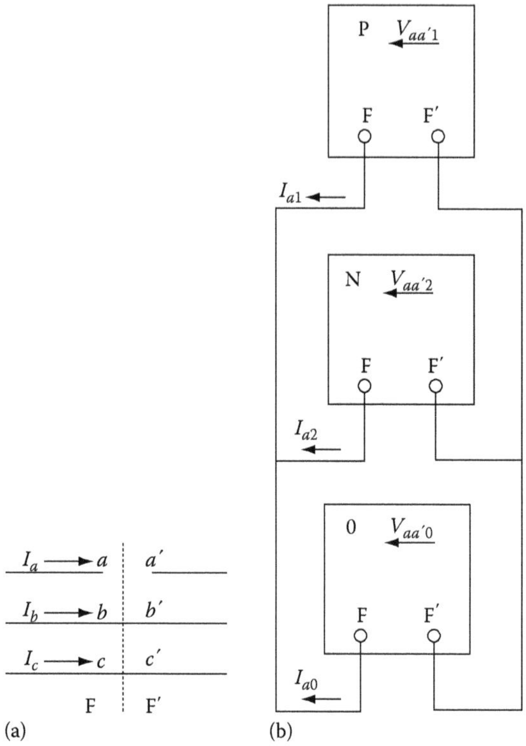

FIGURE 2.17
(a) One-conductor open series fault; (b) connection of sequence networks.

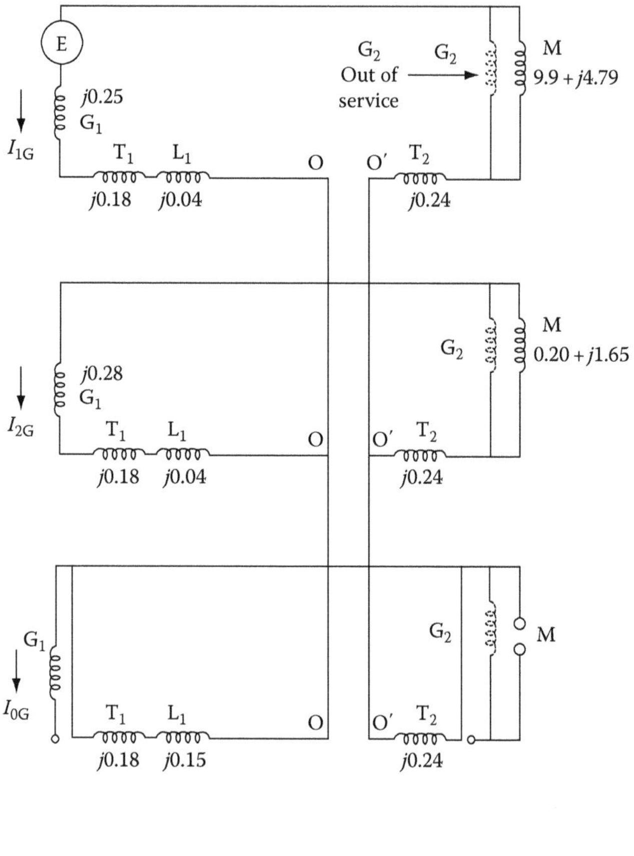

FIGURE 2.18
Equivalent circuit of an open conductor fault (Example 2.5).

connected through converters are ignored, unless the drives are in a regenerative mode. If there are no loads and a broken conductor fault occurs, no load currents flow.

Therefore, for broken conductor faults all loads, irrespective of their types, should be modeled. For simplicity of calculations, again consider that a 10,000 hp induction motor is the only load. Its positive and negative sequence impedances for load modeling will be entirely different from the impedances used in short-circuit calculations.

Chapter 12 discusses induction motor equivalent circuits for positive and negative sequence currents. The range of induction motor impedances per unit (based upon motor kVA base) are

$$\begin{aligned} X_{1r} &= 0.14 - 0.2 < 83^\circ - 75^\circ \\ X^+_{load} &= 0.9 - 0.95 < 20^\circ - 26^\circ \\ X^-_{load} &\approx X_{lr} \end{aligned} \tag{2.44}$$

where

X_{1r} is the induction motor locked rotor reactance at its rated voltage
X^+_{load} is the positive sequence load reactance
X^-_{load} is the negative sequence load reactance

The load impedances for the motor are as shown in Figure 2.18. For an open conductor fault as shown in this figure, the load is not interrupted. Under normal operating conditions, the motor load is served by generator G_2, and in the system of Figure 2.18 no current flows in the transmission line L. If an open conductor fault occurs, generator G_2, operating in synchronism, will trip on operation of negative sequence current relays. To proceed with the calculation assume that G_2 is out of service, when the open conductor fault occurs.

The equivalent impedance across an open conductor is

$$\begin{aligned} &[j0.25 + j0.18 + j0.04 + j0.24 + 9.9 + j4.79]_{pos} \\ &\quad + \{ [j0.28 + j0.18 + j0.04 + j0.24 + 0.20 + j1.65]_{neg} \\ &\qquad \text{in parallel with } [j0.18 + j0.15 + j0.24]_{zero}\} \\ &= 9.918 + j6.771 = 12.0 < 34.32^\circ \end{aligned}$$

The motor load current is

$$0.089 < -25.84^\circ \text{ pu (at 0.9 power factor [PF])}$$

The load voltage is assumed as the reference voltage, thus the generated voltage is

$$\begin{aligned} V_g &= 1 < 0^\circ + (0.089 < -25.84^\circ)(j0.25 + j0.18 + j0.04 + j0.24) \\ &= 1.0275 + j0.0569 = 1.0291 < 3.17^\circ \end{aligned}$$

The positive sequence current is

$$\begin{aligned} I_{1g} = V_G/Z_t &= \left[\frac{1.0291 < 3.17^\circ}{12.00 < 34.32^\circ}\right] \\ &= 0.0857 < 31.15^\circ = 0.0733 + j0.0443 \end{aligned}$$

The negative sequence and zero sequence currents are

$$I_{2g} = -I_{1g}\frac{Z_0}{Z_2 + Z_0}$$

$$= (0.0857 < 31.15°)\left[\frac{0.53 < 90°}{2.897 < 86.04°}\right]$$

$$0.0157 < 215.44°$$

$$I_{0g} = -I_{1g}\frac{Z_2}{Z_2 + Z_0}$$

$$= 0.071 < 210.33°$$

Current $3I_{0g}$ flows through the ground.

Calculate line currents:

$$I_{ag} = I_{1g} + I_{2g} + I_{0g} = 0$$

$$I_{bg} = a^2 I_{1g} + aI_{2g} + I_{0g} = 0.1357 < 250.61°$$

$$I_{cg} = 0.1391 < -8.78°$$

The line currents in the two phases are increased by 52%, indicating serious overheating. A fully loaded motor will stall. The effect of negative sequence currents in the rotor is simulated by the equation:

$$I^2 = I_1^2 + kI_2^2 \quad (2.45)$$

where k can be as high as 6. The motors are disconnected from service by anti-single-phasing devices and protective relays.

The "long way" of calculation using symmetrical components, illustrated by the examples, shows that, even for simple systems, the calculations are tedious and lengthy. For large networks consisting of thousands of branches and nodes these are impractical. There is an advantage in the hand calculations, in the sense that verification is possible at each step and the results can be correlated with the expected final results. For large systems, matrix methods and digital simulation of the systems are invariable. This gives rise to an entirely new challenge for analyzing the cryptical volumes of analytical data, which can easily mask errors in system modeling and output results.

Problems

2.1 A double line-to-ground fault occurs on the high-voltage terminals of transformer T_2 in Figure 2.9. Calculate the fault current distribution and the fault voltages throughout the system, similar to Example 2.2.

2.2 Repeat Problem 2.1 for a line-to-line fault and then for a line-to-ground fault.

2.3 Calculate the percentage reactance of a 60/100 MVA, 13.8–138 kV transformer in Figure P2.1 to limit the three-phase fault current at bus A to 28 kA symmetrical, for a

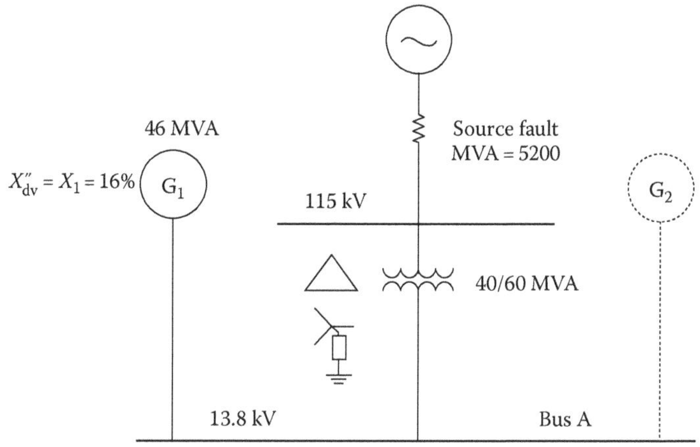

FIGURE P2.1
Distribution system for Problem 2.3.

three-phase symmetrical fault at the bus. Assume only non-decaying ac component of the short-circuit current and neglect resistances.

2.4 In Problem 2.3, another similar generator is to be added to bus A. What is the new short-circuit current? What can be done to limit the three-phase short-circuit level at this bus to 36 kA?

2.5 Calculate the three-phase and single line-to-ground fault current for a fault at bus C, for the system shown in Figure P1.1. As all the generators are connected through delta–wye transformers, delta windings block the zero-sequence currents. Does the presence of generators (1) increase, (2) decrease, or (3) have no effect on the single line-to-ground fault at bus C?

2.6 In Problem 2.5 list the fault types in the order of severity, that is, the magnitude of the fault current.

2.7 Calculate the three-phase symmetrical fault current at bus 3 in the system configuration of Figure P2.2. Neglect resistances.

2.8 Figure P2.3 shows an industrial system motor load being served from a 115 kV utility's system through a step-down transformer. A single line-to-ground fault occurs at the secondary of the transformer terminals. Specify the sequence network considering the motor load. Consider a load operating power factor of 0.85 and an overall efficiency of 94%. Calculate the fault current.

2.9 A wye–wye-connected transformer, with neutral isolated and a tertiary close-circuited delta winding serves a single-phase load between two phases of the secondary windings. A 15 ohm resistance is connected between two lines. For a three-phase balanced supply voltage of 480 V between the primary windings, calculate the distribution of currents in all the windings. Assume a unity transformation ratio.

2.10 Calculate the COG factor in Example 2.2 for all points where the fault voltages have been calculated.

2.11 Why is it permissible to ignore all the sequence impedances of the system components and base the fault current calculations only on the system voltage and resistance to be inserted in the neutral circuit when designing a high-resistance grounded system?

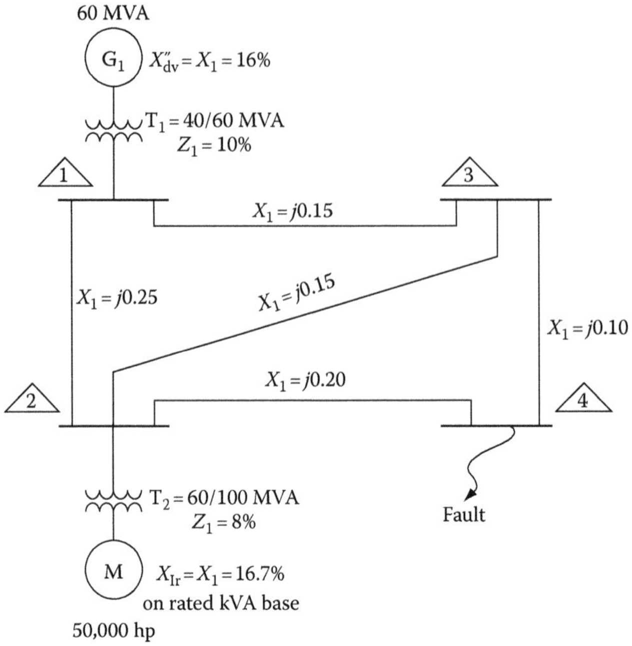

FIGURE P2.2
System configuration for Problem 2.7.

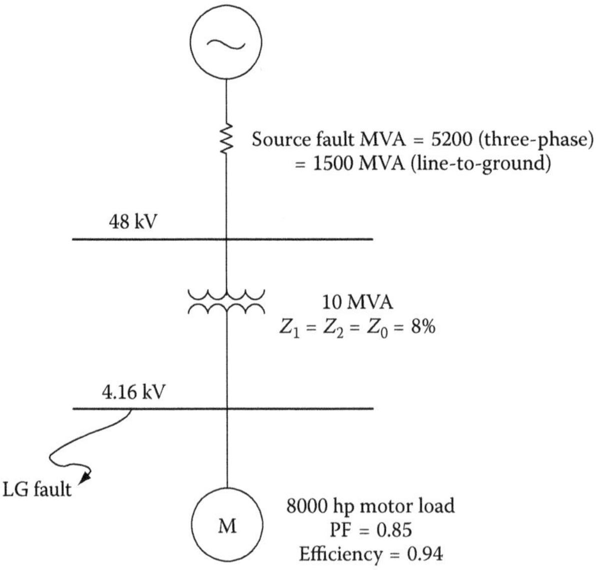

FIGURE P2.3
System configuration for Problem 2.8.

References

1. *Transformer Connections (Including Auto-Transformer Connections)*. Publication no. GET-2H. Pittsfield, MA: General Electric, 1967.
2. ANSI/IEEE. General Requirements of Liquid Immersed Distribution, Power and Regulating Transformers. Standard C57.12.00-2006.
3. ANSI. Terminal Markings and Connections for Distribution and Power Transformers. Standard C57.12.70-2000.
4. *Electrical Transmission and Distribution Reference Book*, 4th edn. East Pittsburgh, PA: Westinghouse Electric Corp., 1964.
5. H.L. Stanback. Predicting damage from 277 volt single-phase-to-ground arcing faults. *IEEE Trans. IA* 13(4): 307–314, July/August 1977.
6. R.H. Kaufman and J.C. Page. Arcing fault protection for low voltage power distribution systems—Nature of the problem. *IEEE Trans. IA* 79: 160–167, June 1960.
7. NEMA PB1-2. Application Guide for Ground Fault Protective Devices for Equipment, 1977.
8. ANSI/NFPA 70, National Electric Code, 2008.
9. J.C. Das. Ground fault protection of bus-connected generators in an interconnected 13.8 kV system. *IEEE Trans. IA* 43(2): pp. 453–461, March/April 2007.
10. D.S. Baker. Charging current data for guess work-free design of high resistance grounded systems. *IEEE Trans. IA* 15(2): 136–140, March/April 1979.
11. ICEA Pub. S-61-40, NEMA WCS, Thermoplastic Insulated Wire for Transmission and Distribution of Electrical Energy, 1979.
12. J.R. Dunki-Jacobs. The reality of high resistance grounding. *IEEE Trans. IA* 13: 469–475, September/October 1977.

Bibliography

Blackburn, J.L. *Symmetrical Components for Power Systems Engineering*. New York: Marcel Dekker, 1993.

Calabrase, G.O. *Symmetrical Components Applied to Electric Power Networks*. New York: Ronald Press Group, 1959.

Das, J.C. Grounding of AC and DC low-voltage and medium-voltage drive systems. *IEEE Trans. IA* 34(1): 295–216, January/February 1998.

Gross, C.A. *Power System Analysis*. New York: Wiley, 1979.

IEEE Std. 142, IEEE Recommended Practice for Grounding of Industrial and Commercial Power Systems, 1991.

Myatt, L.J. *Symmetrical Components*. Oxford, U.K.: Paragon Press, 1968.

Robertson, W.F. and J.C. Das. Grounding medium voltage mobile or portable equipment. *Ind. Appl. Mag.* 6(3): 33–42, May/June 2000.

Smith, D.R. Digital simulation of simultaneous unbalances involving open and faulted conductors. *IEEE Trans. PAS* 89(8): 1826–1835, 1970.

Stevenson, W.D. *Elements of Power System Analysis*. 4th ed. New York: McGraw-Hill, 1982.

3

Matrix Methods for Network Solutions

Calculations for the simplest of power systems, i.e., Example 2.2, are involved and become impractical. For speed and accuracy, modeling on digital computers is a must. The size of the network is important. Even the most powerful computer may not be able to model all the generation, transmission, and consumer connections of a national grid, and the network of interest is "islanded" with boundary conditions represented by current injection or equivalent circuits. Thus, for performing power system studies on a digital computer, the first step is to construct a suitable mathematical model of the power system network and define the boundary conditions. As an example, for short-circuit calculations in industrial systems, the utility's connection can be modeled by sequence impedances, which remain invariant. This generalization may not, however, be valid in every case. For a large industrial plant, with cogeneration facilities, and the utility's generators located close to the industrial plant, it will be necessary to extend the modeling into the utility's system. The type of study also has an effect on the modeling of the boundary conditions. For the steady-state analysis, this model describes the characteristics of the individual elements of the power system and also the interconnections.

A transmission or distribution system network is an assemblage of a linear, passive, bilateral network of impedances connected in a certain manner. The points of connections of these elements are described as buses or nodes. The term bus is more prevalent and a bus may be defined as a point where shunt elements are connected between line potential and ground, though it is not a necessary requirement. A bus may be defined in a series circuit as the point at which a system parameter, i.e., current or voltage, needs to be calculated. The generators and loads are also connected to buses or nodes.

Balanced three-phase networks can be described by equivalent positive sequence elements with respect to a neutral or ground point. An infinite conducting plane of zero impedance represents this ground plane, and all voltages and currents are measured with reference to this plane. If the ground is not taken as the reference plane, a bus known as a slack or swing bus is taken as the reference bus and all the variables are measured with reference to this bus.

3.1 Network Models

Mathematically, the network equations can be formed in the bus (or nodal) frame of reference, in the loop (or mesh) frame of reference, or in the branch frame of reference. The bus frame of reference is important. The equations may be represented using either impedance or admittance parameters.

In the bus frame of reference, the performance is described by $n-1$ linear independent equations for n number of nodes. As stated earlier, the reference node, which is at ground potential, is always neglected. In the admittance form, the performance equation can be written as

$$\overline{I}_{\mathrm{B}} = \overline{Y}_{\mathrm{B}}\overline{V}_{\mathrm{B}} \tag{3.1}$$

where $\overline{I}_{\mathrm{B}}$ is the vector of injection bus currents. The usual convention for the flow of current is that it is positive when flowing toward the bus, and negative when flowing away from the bus. $\overline{V}_{\mathrm{B}}$ is the vector of bus or nodal voltages measured from the reference node, and $\overline{Y}_{\mathrm{B}}$ is the bus admittance matrix. Expanding Equation 3.1,

$$\begin{vmatrix} I_1 \\ I_2 \\ \cdot \\ I_{(n-1)} \end{vmatrix} = \begin{vmatrix} Y_{11} & Y_{12} & \cdot & Y_{1,n-1} \\ Y_{21} & Y_{22} & \cdot & Y_{2,n-1} \\ \cdot & \cdot & \cdot & \cdot \\ Y_{(n-1),1} & Y_{(n-1),2} & \cdot & Y_{(n-1),(n-1)} \end{vmatrix} \begin{vmatrix} V_1 \\ V_2 \\ \cdot \\ V_{n-1} \end{vmatrix} \tag{3.2}$$

$\overline{Y}_{\mathrm{B}}$ is a nonsingular square matrix of order $(n-1)(n-1)$. It has an inverse:

$$\overline{Y}_{\mathrm{B}}^{-1} = \overline{Z}_{\mathrm{B}} \tag{3.3}$$

where $\overline{Z}_{\mathrm{B}}$ is the bus impedance matrix. Equation 3.3 shows that this matrix can be formed by inversion of the bus admittance matrix; $\overline{Z}_{\mathrm{B}}$ is also of the order $(n-1)(n-1)$. It also follows that

$$\overline{V}_{\mathrm{B}} = \overline{Z}_{\mathrm{B}}\overline{I}_{\mathrm{B}} \tag{3.4}$$

3.2 Bus Admittance Matrix

We note the similarity of the bus impedance and admittance matrices; however, there are differences in their formation and application as we will examine. In the impedance matrix, the voltage equations are written in terms of known constant voltage sources, known impedances, and unknown loop currents. In the admittance matrix, current equations are written in terms of known admittances and unknown node voltages. The voltage source of Thévenin branch equivalent acting through a series impedance Z is replaced with a current source equal to EY, in parallel with an admittance $Y=1/Z$, according to Norton's current equivalent. The two circuits are essentially equivalent, and the terminal conditions remain unaltered. These networks deliver at their terminals a specified current or voltage irrespective of the state of the rest of the system. This may not be true in every case. A generator is neither a true current nor a true voltage source.

The formation of a bus admittance matrix for a given network configuration is straightforward. Figure 3.1a shows a five-node impedance network with three voltage sources, E_x, E_y, and E_z; Figure 3.1b shows the admittance equivalent network derived from the impedance network. The following current equations can be written for each of the nodes 1–5.

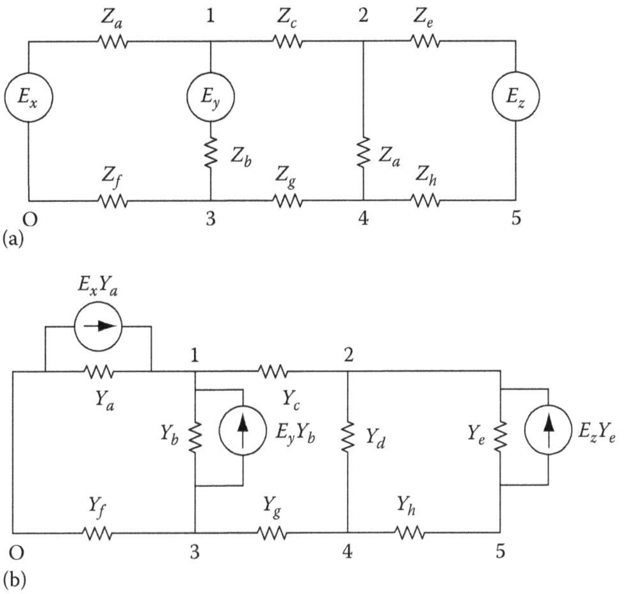

FIGURE 3.1
(a) Network with voltage sources; (b) identical network with Norton equivalent current sources.

Note that node 0 is the reference node. Five independent node-pair voltages are possible, measured from node 0 to the other nodes. As node 0 is taken as the reference node, there is one node–voltage pair less than the number of the nodes. The current equation at node 1 is

$$\begin{aligned} E_xY_a + E_yY_b &= V_{01}Y_a + (V_{01} - V_{03})Y_b + (V_{01} - V_{02})Y_c \\ &= V_{01}(Y_a + Y_b + Y_c) - V_{02}Y_c - V_{03}Y_b \end{aligned} \tag{3.5}$$

Similarly, for node 2:

$$E_zY_e = V_{02}(Y_c + Y_d + Y_e) - V_{01}Y_c - V_{04}\ Y_d - V_{05}Y_e \tag{3.6}$$

and equations for nodes 3–5 are

$$\text{Node 3:}\ -E_yY_b = V_{03}(Y_b + Y_f + Y_g) - V_{01}\ Y_b - V_{04}\ Y_g \tag{3.7}$$

$$\text{Node 4:}\ 0 = V_{04}(Y_d + Y_g + Y_h) - V_{02}\ Y_d - V_{03}\ V_g - V_{05}\ Y_h \tag{3.8}$$

$$\text{Node 5:}\ -E_z\ Y_e = V_{05}(Y_h + Y_e) - V_{02}\ Y_e - V_{04}\ Y_h \tag{3.9}$$

In writing these equations, the direction of current flow must be properly accounted for by change of sign. If a source current arrow is directed away from the node, a minus sign is associated with the term. The above equations can be written in the matrix form:

$$\begin{vmatrix} E_xY_a + E_yY_b \\ E_ZY_e \\ -E_yY_b \\ 0 \\ -E_zY_e \end{vmatrix} = \begin{vmatrix} (Y_a+Y_b+Y_c) & -Y_c & -Y_b & 0 & 0 \\ -Y_c & (Y_e+Y_d+Y_c) & 0 & -Y_d & -Y_e \\ -Y_b & 0 & (Y_b+Y_f+Y_g) & -Y_g & 0 \\ 0 & -Y_d & -Y_g & (Y_d+Y_g+Y_h) & -Y_h \\ 0 & -Y_e & 0 & -Y_h & (Y_h+Y_e) \end{vmatrix} \begin{vmatrix} V_{01} \\ V_{02} \\ V_{03} \\ V_{04} \\ V_{05} \end{vmatrix} \tag{3.10}$$

For a general network with $n+1$ nodes,

$$\overline{Y} = \begin{vmatrix} Y_{11} & Y_{12} & \cdot & Y_{1n} \\ Y_{21} & Y_{22} & \cdot & Y_{2n} \\ \cdot & \cdot & \cdot & \cdot \\ Y_{n1} & Y_{n2} & \cdot & Y_{nn} \end{vmatrix} \tag{3.11}$$

where each admittance Y_{ii} ($i=1, 2, 3, 4, \ldots$) is the self-admittance or driving point admittance of node i, given by the diagonal elements, and it is equal to an algebraic sum of all admittances terminating in that node. Y_{ik} ($i,k=1, 2, 3, 4, \ldots$) is the mutual admittance between nodes i and k or transfer admittance between nodes i and k and is equal to the negative of the sum of all admittances directly connected between those nodes. The current entering a node is given by

$$I_k = \sum_{n=1}^{n} Y_{kn} V_n \tag{3.12}$$

To find an element, say, Y_{22}, the following equation can be written:

$$I_2 = Y_{21}V_1 + Y_{22}V_2 + Y_{23}V_3 + \cdots + Y_{2n}V_n \tag{3.13}$$

The self-admittance of a node is measured by shorting all other nodes and finding the ratio of the current injected at that node to the resulting voltage (Figure 3.2):

$$Y_{22} = \frac{I_2}{V_2}(V_1 = V_3 = \cdots V_n = 0) \tag{3.14}$$

Similarly, the transfer admittance is

$$Y_{21} = \frac{I_2}{V_1}(V_2 = V_3 = \cdots V_n = 0) \tag{3.15}$$

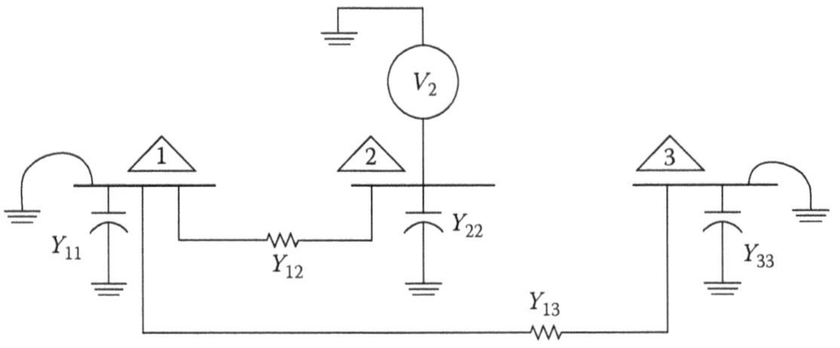

FIGURE 3.2
Calculations of self-admittance in a network, with unit voltage applied at a bus and other buses short-circuited to ground or reference node.

Example 3.1

Figure 3.3a shows a simple network of three buses with series and shunt elements. Numerical values of circuit elements are impedances. The shunt resistors may represent unity power factor loads. Write the bus admittance matrix by examination and by use of Equations 3.14 and 3.15.

The bus admittance matrix is formed by inspection. At node 1, the self-admittance Y_{11} is $1+1/j0.2=1-j5$, and the transfer admittance between node 1 and 2, $Y_{12}=-(1/j0.2)=j5$. Similarly, the other admittance elements are easily calculated:

$$\overline{Y}_B = \begin{vmatrix} Y_{11} & Y_{12} & Y_{13} \\ Y_{21} & Y_{22} & Y_{23} \\ Y_{31} & Y_{32} & Y_{33} \end{vmatrix}$$

$$= \begin{vmatrix} 1-j5 & j5 & 0 \\ j5 & 0.5-j8.33 & j3.33 \\ 0 & j3.33 & 0.33-j3.33 \end{vmatrix} \tag{3.16}$$

Alternatively, the bus admittance matrix can be constructed by use of Equations 3.14 and 3.15. Apply unit voltages, one at a time, to each bus while short-circuiting the other bus voltage sources to ground. Figure 3.3b shows unit voltage applied to bus 1, while buses 2 and 3 are short-circuited to ground. The input current I to bus 1 gives the driving point admittance. This current is given by

$$\frac{V}{1.0}+\frac{V}{j0.2}=1-j5=Y_{11}$$

Bus 2 is short-circuited to ground; the current flowing to bus 2 is

$$\frac{-V}{j0.2}=Y_{12}=Y_{21}$$

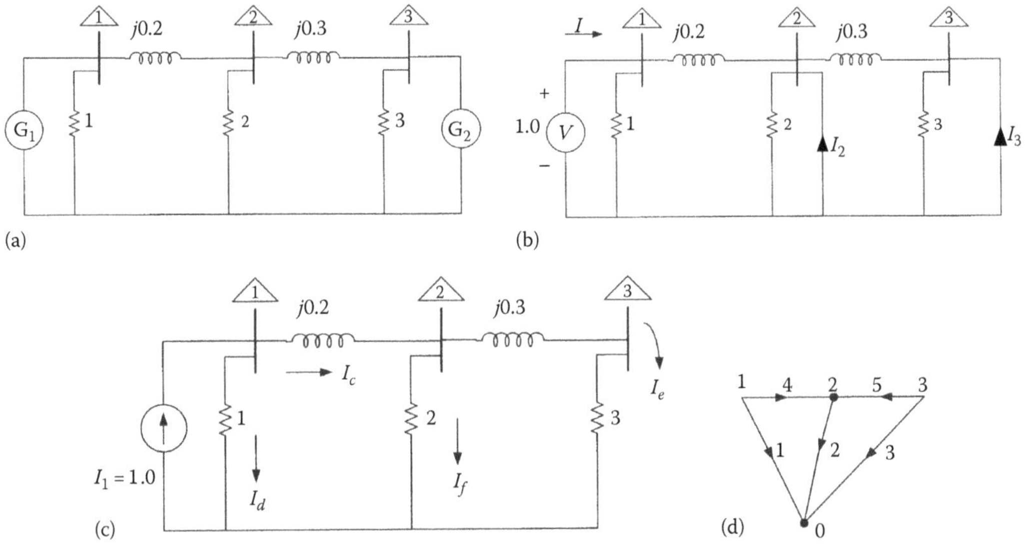

FIGURE 3.3
(a) Network for Examples 3.1 and 3.2; (b) calculation of admittance elements by unit voltage injection; (c) calculation of impedance elements by unit current injection; (d) diagram of network.

Only buses directly connected to bus 1 have current contributions. The current flowing to bus 3 is zero:

$$Y_{13} = Y_{31} = 0$$

The other elements of the matrix can be similarly found and the result obtained is the same as in Equation 3.16.

3.3 Bus Impedance Matrix

The bus impedance matrix for $(n+1)$ nodes can be written as

$$\begin{vmatrix} V_1 \\ V_2 \\ \cdot \\ V_m \end{vmatrix} = \begin{vmatrix} Z_{11} & Z_{12} & \cdot & Z_{1m} \\ Z_{21} & Z_{22} & \cdot & Z_{2m} \\ \cdot & \cdot & \cdot & \cdot \\ Z_{m1} & Z_{m2} & \cdot & Z_{mm} \end{vmatrix} \begin{vmatrix} I_1 \\ I_2 \\ \cdot \\ I_m \end{vmatrix} \quad (3.17)$$

Unlike the bus admittance matrix, the bus impedance matrix cannot be formed by simple examination of the network circuit. The bus impedance matrix can be formed by the following methods:

- Inversion of the admittance matrix
- By open-circuit testing
- By step-by-step formation
- From graph theory

Direct inversion of the Y matrix is rarely implemented in computer applications. Certain assumptions in forming the bus impedance matrix are

1. The passive network can be shown within a closed perimeter (Figure 3.4). It includes the impedances of all the circuit components, transmission lines, loads,

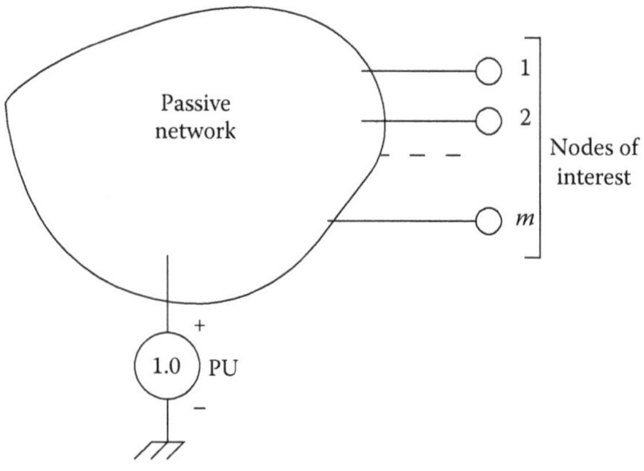

FIGURE 3.4
Representation of a network as passive elements with loads and faults excluded. The nodes of interest are pulled out of the network and unit voltage is applied at the common node.

transformers, cables, and generators. The nodes of interest are brought out of the bounded network, and it is excited by a unit generated voltage.

2. The network is passive in the sense that no circulating currents flow in the network. Also, the load currents are negligible with respect to the fault currents. For any currents to flow, an external path (a fault or load) must exist.
3. All terminals marked 0 are at the same potential. All generators have the same voltage magnitude and phase angle and are replaced by one equivalent generator connected between 0 and a node. For fault current calculations, a unit voltage is assumed.

3.3.1 Bus Impedance Matrix from Open-Circuit Testing

Consider a passive network with m nodes as shown in Figure 3.5 and let the voltage at node 1 be measured when unit current is injected at bus 1. Similarly, let the voltage at bus 1 be measured when unit current is injected at bus 2. All other currents are zero and the injected current is 1 per unit. The bus impedance matrix equation (3.17) then becomes

$$\begin{vmatrix} V_1 \\ V_2 \\ \cdot \\ V_m \end{vmatrix} = \begin{vmatrix} Z_{11} \\ Z_{21} \\ \cdot \\ Z_{m1} \end{vmatrix} \tag{3.18}$$

where Z_{11} can be defined as the voltage at bus 1 when one per unit current is injected at bus 1. This is the *open-circuit driving point impedance.* Z_{12} is defined as voltage at bus 1 when one per unit current is injected at bus 2. This is the *open-circuit transfer impedance* between

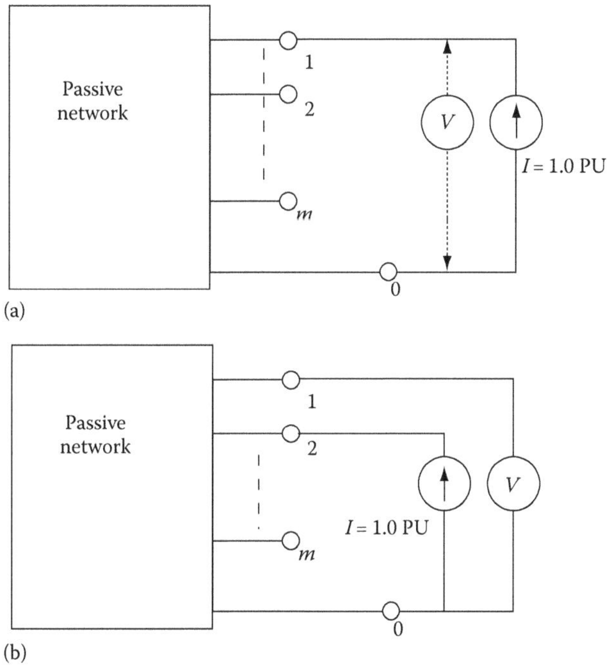

FIGURE 3.5
(a, b) Equivalent networks for calculations of Z_{11} and Z_{12} for formation of bus impedance matrix.

buses 1 and 2. Z_{21} is defined as voltage at bus 2 when one per unit current is injected at bus 1. This is the open-circuit transfer impedance between buses 2 and 1. Generally,

$$Z_{12} = Z_{21} \tag{3.19}$$

To summarize, the open-circuit driving point impedance of bus i is determined by injecting a unit current between bus i and the reference node and keeping all other buses open circuited. This gives the diagonal elements of the bus impedance matrix. The open-circuit transfer impedance between buses i and j is found by applying a unit current between bus i and the reference node and measuring the voltage at bus j, while keeping all other buses open circuited. This gives the off-diagonal elements.

Example 3.2

Find the bus impedance matrix of Example 3.1 by inversion and also by open-circuit testing.

The inversion of the admittance matrix in Equation 3.16, calculated in Example 3.1, gives

$$\bar{Z}_B = \begin{vmatrix} 0.533 + j0.05 & 0.543 - j0.039 & 0.533 - j0.092 \\ 0.543 - j0.039 & 0.55 + j0.069 & 0.552 + j0.014 \\ 0.533 - j0.092 & 0.522 + j0.014 & 0.577 + j0.258 \end{vmatrix} \tag{3.20}$$

From Equation 3.20 we note that the zero elements of the bus admittance matrix get populated in the bus impedance matrix. As we will discover, the admittance matrix for power system networks is sparse and this sparsity is lost in the impedance matrix.

The same bus impedance matrix can be constructed from the open-circuit test results. Unit currents are injected, one at a time, at each bus and the other current sources are open circuited. Figure 3.3c shows unit current injected at bus 1 with bus 3 current source open circuited. Z_{11} is given by voltage at node 1 divided by current I_1 (=1.0 per unit):

$$1\|[(2 + j0.02)\|(3 + j0.03)] = Z_{11} = 0.533 + j0.05$$

The injected current divides as shown in Figure 3.3c. Transfer impedance $Z_{12}=Z_{21}$ at bus 2 is the potential at bus 2. Similarly, the potential at bus 3 gives $Z_{13}=Z_{31}$. The example shows that this method of formation of bus impedance matrix is tedious.

3.4 Loop Admittance and Impedance Matrices

In the loop frame of reference,

$$\bar{V}_L = \bar{Z}_L \bar{I}_L \tag{3.21}$$

where

$\bar{V}_L$ is the vector of loop voltages

$\bar{I}_L$ is the vector of unknown loop currents

$\bar{Z}_L$ is the loop impedance matrix of order $l \times l$

$\overline{Z}_L$ is a nonsingular square matrix and it has an inverse

$$\overline{Z}_L^{-1} = \overline{Y}_L \tag{3.22}$$

where $\overline{Y}_B$ is the loop admittance matrix. We can write

$$\overline{I}_L = \overline{Y}_L \overline{V}_L \tag{3.23}$$

It is important to postulate the following:

- The loop impedance matrix can be constructed by examination of the network. The diagonal elements are the self-loop impedances and are equal to the sum of the impedances in the loop. The off-diagonal elements are the mutual impedances and are equal to the impedance of the elements common to a loop.
- The loop admittance matrix can only be constructed by inversion of the loop impedance matrix. It has no direct relation with the actual network components.

Compare the formation of bus and loop impedance and admittance matrices
The loop impedance matrix is derived from basic loop-impedance equations. It is based on Kirchoff's voltage law which states that voltage around a closed loop sums to zero. A potential rise is considered positive and a potential drop, negative. Consider the simple network of Figure 3.6. Three independent loops can be formed as shown in this figure, and the following equations can be written:

$$\begin{aligned} E_1 &= I_1(Z_1 + Z_2) - I_2Z_2 \\ 0 &= -I_1Z_2 + I_2(Z_2 + Z_3) - I_3Z_4 \\ -E_2 &= 0 - I_2Z_4 + I_3(Z_4 + Z_5) \end{aligned} \tag{3.24}$$

In the matrix form, these equations can be written as

$$\begin{vmatrix} E_1 \\ 0 \\ -E_2 \end{vmatrix} = \begin{vmatrix} Z_1 + Z_2 & -Z_2 & 0 \\ -Z_2 & Z_2 + Z_3 & -Z_4 \\ 0 & -Z_4 & Z_4 + Z_5 \end{vmatrix} \begin{vmatrix} I_1 \\ I_2 \\ I_3 \end{vmatrix} \tag{3.25}$$

The impedance matrix in the above example can be written without writing the loop equations and by examining the network. As stated before, the diagonal elements of the matrix are the self-impedances around each loop, while off-diagonal elements are the

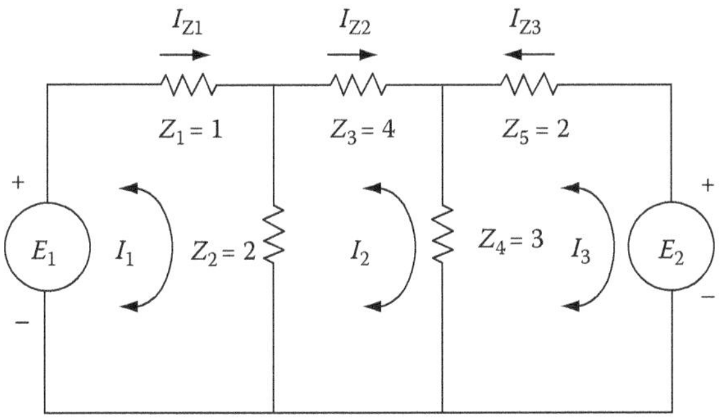

FIGURE 3.6
Network with correct choice of loop currents.

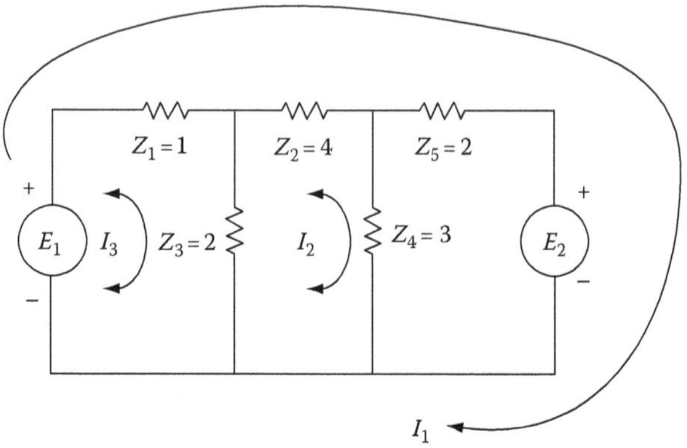

FIGURE 3.7
Incorrect choice of loop currents in the network when only two loops I_1 and I_2 are selected. I_3 must be selected.

impedances common to coupled loops. An off-diagonal element is negative when it carries a current in a loop in the opposite direction to the current in the coupled loop.

3.4.1 Selection of Loop Equations

The selection of loop equations is arbitrary, yet there are certain limitations in forming these equations. The selection should result in a sufficient number of independent voltage equations. As an example, in Figure 3.7 the selection of loop currents, I_1 and I_2 is not adequate. An additional loop current marked I_3 must be selected.

3.5 Graph Theory

Linear network graphs help in the assembly of a network. The problem for large networks can be stated that a minimum number of linearly independent equations of zero redundancy must be selected to provide sufficient information for the solution of the network.

A topographical graph or map of the network is provided by shorting branch EMFs, opening branch current sources, and considering all branch impedances as zero. The network is, thus, replaced by simple lines joining the nodes. A linear graph depicts the geometric interconnections of the elements of a network. A graph is said to be *connected* only if there is a path between every pair of nodes. If each element is assigned a direction, it is called an *oriented* graph. The direction assignment follows the direction assumed for the current in the element.

Consider the network of Figure 3.8a. It is required to construct a graph. First, the network can be redrawn as shown in Figure 3.8b. Each source and the shunt admittance across it are represented by a single element. Its graph is shown in Figure 3.8c. It has a total of nine branches, which are marked from 1 through 9.

The *node* or *vertex* of a graph is the end point of a branch. In Figure 3.8c, the nodes are marked as 0–4. Node 0 is at ground potential. A route traced out through a linear graph which goes through a node no more than one time is called a *path*.

The *tree-link* concept is useful when large systems are involved. A tree of the network is formed, which includes all the nodes of a graph, but no closed paths. Thus, a tree in a subgraph of a given connected graph, which has the following characteristics:

- It is connected.
- It contains all the nodes of the original graph.
- It does not contain any closed paths.

Figure 3.8d shows a tree of the original graph. The elements of a tree are called *tree branches*. The number of branches B is equal to the number of nodes minus 1, which is the reference node:

$$B = n - 1 \tag{3.26}$$

The number of tree branches in Figure 3.8d is four, one less than the number of nodes. The elements of the graph that are not included in the tree are called links or link branches and they form a *subgraph*, which may or may not be connected. This graph is called a *cotree*. Figure 3.8e shows the cotree of the network. It has five *links*. Each *link* in the tree will close a *new* loop and a corresponding loop equation is written. Whenever a link closes a loop,

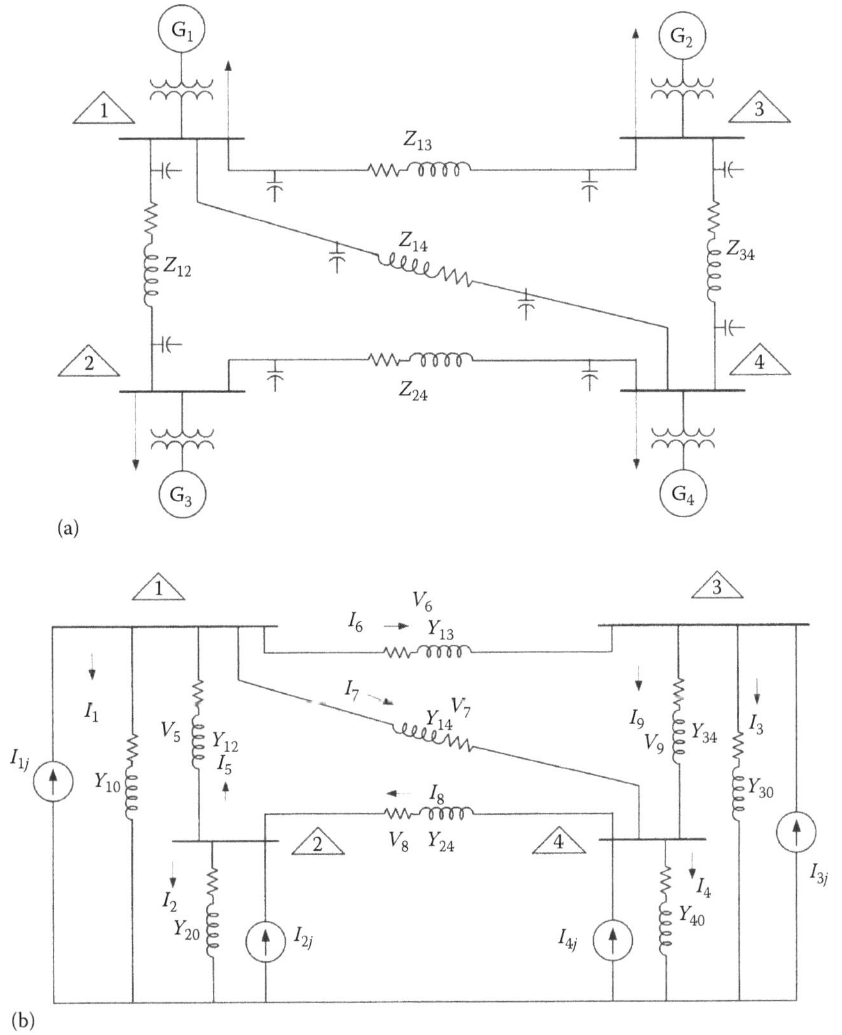

FIGURE 3.8
(a) Equivalent circuit of a network; (b) network redrawn with lumped elements and current injections at the nodes; *(continued)*

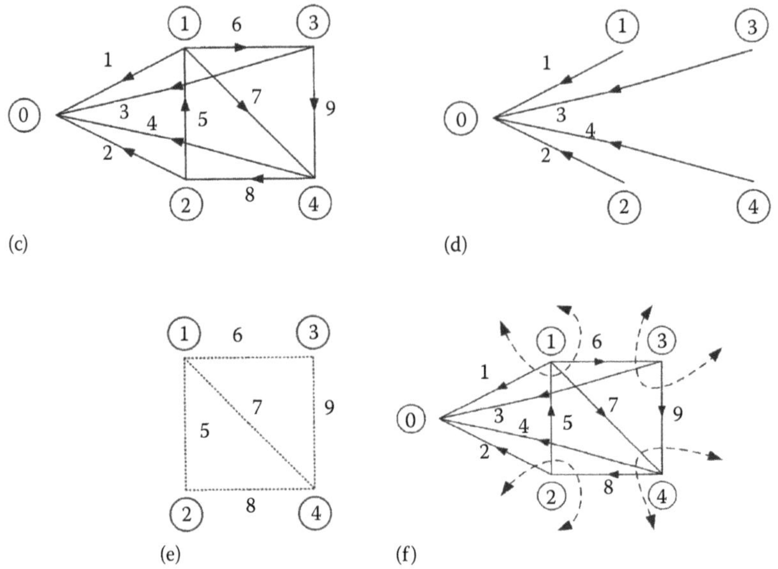

FIGURE 3.8 (continued)
(c) oriented connected graph of the network; (d) tree of the oriented network; (e) cotree of the oriented network; (f) cut-sets of the network.

a *tie-set* is formed, which contains one link and one or more branches of the established tree. A tree and cotree of a graph are not unique. A set of branches of a connected graph is called a *cut-set*, if the following properties hold:

- If the branches of the cut-set are removed, the graph is split into two parts, which are not connected.
- No subset of the above set has this property. This ensures that no more than the minimum number of branches are included in the cut-set.

Basic cut-sets are those which contain only one branch and, therefore, the number of basic cut-sets is equal to the number of branches. An admittance matrix can be formed from the cut-sets. The network is divided into pieces of cut-sets and Kirchoff's current law must be satisfied for each cut-set line. No reference node need be given. The various cut-sets for the tree in Figure 3.8d are as shown in Figure 3.8f.

A loop which is formed by closing only one link is called a basic loop. The number of basic loops is equal to the number of links. In a metallically coupled network, loops L are given by

$$L = e - B \tag{3.27}$$

where

B is the number of tree branches
e is the number of nodes

The graph of Figure 3.8c has nine elements and there are four tree branches. Therefore, the number of loops is equal to five.

If any two loops, say loops 1 and 2, have a mutual coupling, the diagonal term of loop 1 will have all the self-impedances, while off-diagonal terms will have mutual impedance between 1 and 2.

Elimination of loop currents can be done by matrix partitioning. In a large network, the currents in certain loops may not be required and these can be eliminated. Consider a partitioned matrix:

$$\begin{vmatrix} \overline{E}_x \\ \overline{E}_y \end{vmatrix} = \begin{vmatrix} \overline{Z}_1 & \overline{Z}_2 \\ \overline{Z}_3 & \overline{Z}_4 \end{vmatrix} \begin{vmatrix} \overline{I}_x \\ \overline{I}_y \end{vmatrix} \tag{3.28}$$

The current given by array $\overline{I}_x$ is only of interest. This is given by

$$\overline{I}_x = \left[\overline{Z}_1 - \overline{Z}_2\overline{Z}_4^{-1}\overline{Z}_3\right]^{-1}\overline{E}_x \tag{3.29}$$

3.6 Bus Admittance and Impedance Matrices by Graph Approach

The bus admittance matrix can be found by graph approach:

$$\overline{Y}_{\mathrm{B}} = \overline{A}\,\overline{Y}_{\mathrm{p}}\overline{A}' \tag{3.30}$$

where

$\overline{A}$ is the bus incidence matrix
$\overline{Y}_{\mathrm{p}}$ is the primitive admittance matrix
$\overline{A}'$ is the transpose of the bus incident matrix

Similarly, the loop impedance matrix can be formed by

$$\overline{Z}_{\mathrm{L}} = \overline{B}\,\overline{Z}_{\mathrm{p}}\overline{B}' \tag{3.31}$$

where

$\overline{B}$ is the basic loop incidence matrix
$\overline{B}'$ is its transpose
$\overline{Z}_{\mathrm{p}}$ is the primitive bus impedance matrix

3.6.1 Primitive Network

A network element may contain active and passive components. Figure 3.9 shows impedance and admittance forms of a network element, and equivalence can be established between these two. Consider the impedance form shown in Figure 3.9a. The nodes P and Q have voltages V_{p} and V_{q} and let $V_{\mathrm{p}} > V_{\mathrm{q}}$. Self-impedance, Z_{pq}, has a voltage source, e_{pq}, in series. Then

$$V_{\mathrm{p}} + e_{\mathrm{pq}} - Z_{\mathrm{pq}}i_{\mathrm{pq}} = V_{\mathrm{q}}$$

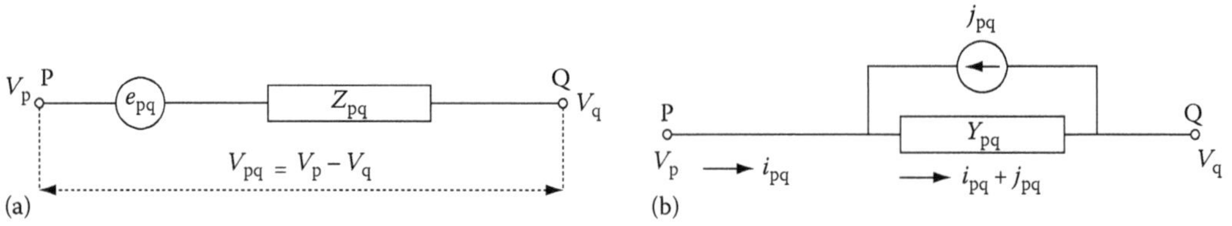

FIGURE 3.9
(a, b) Primitive network, impedance, and admittance forms.

or

$$V_{pq} + e_{pq} = Z_{pq} V_{pq} \tag{3.32}$$

where
$V_{pq} = V_p - V_q$ = voltage across P–Q
i_{pq} is current through P–Q

In the admittance form and referring to Figure 3.9b

$$i_{pq} + j_{pq} = Y_{pq} V_{pq}$$

Also

$$j_{pq} = -Y_{pq} e_{pq} \tag{3.33}$$

where j_{pq} is current source in parallel with P–Q.

The performance equations of the primitive network can be derived from Equations 3.32 and 3.33. For the entire network, the variables become column vector and parameters become matrices. The performance equations in impedance and admittance forms are

$$\overline{V} + \bar{e} = \overline{Z}_p \bar{i}$$

$$\bar{i} + \bar{j} = \overline{Y}_p \overline{V}$$

Also

$$\overline{Y}_p = \overline{Z}_p^{-1} \tag{3.34}$$

The diagonal elements are the impedances or admittances of the element $p-q$, and the off-diagonal elements are mutual impedances/admittances. When there is no coupling between the elements, the matrices are diagonal.

Example 3.3

Write the primitive admittance matrix of the network of Figure 3.8b.

The matrix is written by simply examining the figure. There are no mutual couplings between the elements. The required matrix is

	1	2	3	4	5	6	7	8	9
	0–1	0–2	0–3	0–4	1–2	1–3	1–4	2–4	3–4
0–1	y_{10}								
0–2		y_{20}							
0–3			y_{30}						
0–4				y_{40}					
1–2					y_{12}				
1–3						y_{13}			
1–4							y_{14}		
2–4								y_{24}	
3–4									y_{34}

The top and left-side identifications of the elements between nodes are helpful in the formation of the matrix. It is a diagonal matrix. If there are mutual couplings between elements of the network, the appropriate off-diagonal elements in the matrix are populated.

3.6.2 Incidence Matrix from Graph Concepts

Consider a graph with n nodes and e elements. The matrix $\overline{A}$ of Equation 3.30 has n rows which correspond to the n nodes and e columns which correspond to the e elements. This matrix is known as an *incidence matrix*. The matrix elements can be formed as follows:

$a_{ij} = 1$, if the jth element is incident to and directed away from the node i.

$a_{ij} = -1$, if the jth element is incident to but directed *towards* the node i.

$a_{if} = 0$, if the jth element is not incident to the node.

When a node is taken as the reference node, then the matrix $\overline{A}$ is called a *reduced incidence matrix* or *bus incidence matrix*.

Example 3.4

Form an incidence and reduced incidence matrix from the graph of the network in Figure 3.8a.

The bus incidence matrix is formed from the graph of the network in Figure 3.8c:

e/n	1	2	3	4	5	6	7	8	9
1	1				−1	1	1		
2		1			1			−1	
3			1			−1			1
4				1			−1	1	−1
0	−1	−1	−1	−1					

This matrix is singular. The matrix without the last row pertaining to the reference node, which is shown shaded, is a reduced incidence matrix. It can be partitioned as follows:

e/(n−1)	Branches	Links
Buses	A_b	A_L

We can therefore write:

$$A_{(n-1),e} = (A_b)_{(n-1),\,(n-1)} \times (A_L)_{(n-1),l} \tag{3.35}$$

Figure 3.8b ignores the shunt capacitance of the lines. Referring to the graph in Figure 3.8c, we can write

$$\begin{vmatrix} 1 & 0 & 0 & 0 & -1 & 1 & 1 & 0 & 0 \\ 0 & 1 & 0 & 0 & 1 & 0 & 0 & -1 & 0 \\ 0 & 0 & 1 & 0 & 0 & -1 & 0 & 0 & 1 \\ 0 & 0 & 0 & 1 & 0 & 0 & -1 & 1 & -1 \end{vmatrix} \begin{vmatrix} I_1 \\ I_2 \\ I_3 \\ I_4 \\ I_5 \\ I_6 \\ I_7 \\ I_8 \\ I_9 \end{vmatrix} = \begin{vmatrix} I_{1j} \\ I_{2j} \\ I_{3j} \\ I_{4j} \end{vmatrix} \tag{3.36}$$

where I_{1j}, I_{2j} are the injected currents at nodes I, 2,…
In abbreviated form

$$\bar{A}\bar{I}_{pr} = \bar{I} \tag{3.37}$$

where
$\bar{I}_{pr}$ is the column vector of the branch currents
$\bar{I}$ is column vector of the injected currents

With respect to voltages,

$$\begin{vmatrix} V_1 \\ V_2 \\ V_3 \\ V_4 \\ V_5 \\ V_6 \\ V_7 \\ V_8 \\ V_9 \end{vmatrix} = \begin{vmatrix} 1 & 0 & 0 & 0 \\ 0 & 1 & 0 & 0 \\ 0 & 0 & 1 & 0 \\ 0 & 0 & 0 & 1 \\ -1 & 1 & 0 & 0 \\ 1 & 0 & -1 & 0 \\ 1 & 0 & 0 & -1 \\ 0 & -1 & 0 & 1 \\ 0 & 0 & 1 & 1 \end{vmatrix} \begin{vmatrix} V_1 \\ V_2 \\ V_3 \\ V_4 \end{vmatrix} \tag{3.38}$$

Referring to Figure 3.8b,

$$
\begin{aligned}
V_5 &= V_2 - V_1 \\
V_6 &= V_1 - V_3 \\
V_7 &= V_1 - V_4 \\
V_8 &= V_4 - V_2 \\
V_9 &= V_3 - V_4
\end{aligned} \tag{3.39}
$$

Equation 3.38 can be written as

$$\overline{V}_{pr} = \overline{A}^t V \tag{3.40}$$

where
$\overline{V}_{pr}$ is the column vector of the primitive branch voltage drops
$\overline{V}$ is the vector of the bus voltages measured to the common node

From these relations a reader can prove Equation 3.30. Using this equation, the bus admittance matrix is

$$\overline{A}\overline{Y}_p\overline{A}^t = \overline{Y}_B = \begin{vmatrix} Y_{10}+Y_{12}+Y_{13}+Y_{14} & -Y_{12} & -Y_{13} & -Y_{14} \\ -Y_{12} & Y_{20}+Y_{12}+Y_{24} & 0 & -Y_{24} \\ -Y_{13} & 0 & Y_{30}+Y_{13}+Y_{43} & -Y_{43} \\ -Y_{14} & -Y_{24} & -Y_{43} & Y_{40}+Y_{14}+Y_{24}+Y_{43} \end{vmatrix} \tag{3.41}$$

This matrix could be formed by simple examination of the network.

Example 3.5

Form the primitive impedance and admittance matrix of the network in Figure 3.3a and then the bus incidence matrix. From these calculate the bus admittance matrix.

The graph of the network is shown in Figure 3.3d. The primitive impedance matrix is

$$\vec{Z}_P = \begin{vmatrix} 1 & 0 & 0 & 0 & 0 \\ 0 & 2 & 0 & 0 & 0 \\ 0 & 0 & 3 & 0 & 0 \\ 0 & 0 & 0 & j0.2 & 0 \\ 0 & 0 & 0 & 0 & j0.3 \end{vmatrix}$$

Then the primitive admittance matrix is

$$\overline{Y}_\mathrm{p} = \begin{vmatrix} 1 & 0 & 0 & 0 & 0 \\ 0 & 0.5 & 0 & 0 & 0 \\ 0 & 0 & 0.333 & 0 & 0 \\ 0 & 0 & 0 & -j5 & 0 \\ 0 & 0 & 0 & 0 & -j3.33 \end{vmatrix}$$

None of the diagonal elements are populated.

From the graph of Figure 3.3d, the reduced bus incidence matrix (ignoring node 0) is

$$\overline{A} = \begin{vmatrix} 1 & 0 & 0 & 1 & 0 \\ 0 & 1 & 0 & -1 & -1 \\ 0 & 0 & 1 & 0 & 1 \end{vmatrix}$$

The bus admittance matrix is

$$\overline{Y}_\mathrm{B} = \overline{A}\,\overline{Y}_\mathrm{p}\overline{A}' = \begin{vmatrix} 1-j5 & j5 & 0 \\ j5 & 0.5-j8.33 & j3.33 \\ 0 & j3.33 & 0.33-j3.33 \end{vmatrix}$$

This is the same matrix as calculated before.

The loop impedance matrix can be similarly formed from the graph concepts. There are five basic loops in the network of Figure 3.8b. The basic loop matrix $\overline{B}_\mathrm{L}$ of Equation 3.31 is constructed with its elements as defined below:

$b_{ij} = 1$, if *j*th element is incident to *j*th basic loop and is oriented in the same direction.

$b_{ij} = -1$, if *j*th element is incident to the *j*th basic loop and is oriented in the opposite direction.

$b_{ij} = 0$, if *j*th basic loop does not include the *j*th element.

Branch Frame of Reference

The equations can be expressed as follows:

$$\begin{aligned} \overline{V}_\mathrm{BR} &= \overline{Z}_\mathrm{BR}\overline{I}_\mathrm{BR} \\ I_\mathrm{BR} &= Y_\mathrm{BR}\overline{V}_\mathrm{BR} \end{aligned} \tag{3.42}$$

The branch impedance matrix is the matrix of the branches of the tree of the connected power system network and has the dimensions, $b \times b$; $\overline{V}_\mathrm{BR}$ is the vector of branch voltages and $\overline{I}_\mathrm{BR}$ is the vector of currents through the branches.

Example 3.6

In Example 3.1, unit currents $(1 < \overset{\circ}{0})$ are injected at nodes 1 and 3. Find the bus voltages.

From Example 3.1,

$$\begin{vmatrix} 1-j5 & j5 & 0 \\ j5 & 0.5-j8.33 & j3.33 \\ 0 & j3.33 & 0.33-j3.33 \end{vmatrix} \begin{vmatrix} 1 \\ 0 \\ 1 \end{vmatrix} = \begin{vmatrix} V_1 \\ V_2 \\ V_3 \end{vmatrix}$$

Taking inverse of the Y matrix, the equation is

$$\begin{vmatrix} V_1 \\ V_2 \\ V_3 \end{vmatrix} = \begin{vmatrix} 0.552+j0.05 & 0.542-j0.039 & 0.532-j0.092 \\ 0.542-j0.039 & 0.55+j0.069 & 0.551+j0.014 \\ 0.532-j0.092 & 0.551+j0.014 & 0.577+j0.257 \end{vmatrix} \begin{vmatrix} 1 \\ 0 \\ 1 \end{vmatrix} = \begin{vmatrix} 1.084-j0.042 \\ 1.093-j0.025 \\ 1.109+j0.164 \end{vmatrix}$$

Example 3.7

Consider that there is a mutual coupling of $Z_M = j0.4$ between the branch elements 4 and 5 as shown in Figure 3.4. Recalculate the bus admittance matrix.

The primitive impedance matrix is

$$\vec{Z}_P = \begin{vmatrix} 1 & 0 & 0 & 0 & 0 \\ 0 & 2 & 0 & 0 & 0 \\ 0 & 0 & 3 & 0 & 0 \\ 0 & 0 & 0 & j0.2 & j0.4 \\ 0 & 0 & 0 & j0.4 & j0.3 \end{vmatrix}$$

For the coupled branch,

$$\overline{Y}_{jj} = \frac{1}{\overline{Z}_{jj}}$$

Therefore,

$$\begin{vmatrix} j0.2 & j0.4 \\ j0.4 & j0.3 \end{vmatrix}^{-1} = \begin{vmatrix} -j3 & j4 \\ j4 & -j2 \end{vmatrix}$$

Then the primitive Y matrix is

$$\overline{Y}_P = \begin{vmatrix} 1 & 0 & 0 & 0 & 0 \\ 0 & 0.5 & 0 & 0 & 0 \\ 0 & 0 & 0.333 & 0 & 0 \\ 0 & 0 & 0 & j3 & -j4 \\ 0 & 0 & 0 & -j4 & j2 \end{vmatrix}$$

Then the Y bus matrix is

$$\overline{Y}_B = \begin{vmatrix} 1+j3 & j & -4j \\ j & 0.5-j3 & j2 \\ -j4 & 2j & 0.333+j2 \end{vmatrix}$$

If similar currents as in Example 3.1 are injected at buses 1 and 3, then

$$\begin{vmatrix} V_1 \\ V_2 \\ V_3 \end{vmatrix} = \begin{vmatrix} 0.549 - j0.015 & 0.54 - j0.033 & 0.543 - j0.093 \\ 0.54 - j0.033 & 0.563 + j0.147 & 0.537 - j0.123 \\ 0.543 - j0.093 & 0.537 - j0.123 & 0.565 - j0.096 \end{vmatrix} \begin{vmatrix} 1 \\ 0 \\ 1 \end{vmatrix} = \begin{vmatrix} 1.092 - j0.079 \\ 1.077 - j0.155 \\ 1.108 + j0.003 \end{vmatrix}$$

3.6.3 Node Elimination in Y-Matrix

It is possible to eliminate a group of nodes form the nodal matrix equations. Consider a five-node system, and assume that nodes 4 and 5 are to be eliminated. Then the Y-matrix can be partitioned as follows:

$$\begin{vmatrix} I_1 \\ I_2 \\ I_3 \\ - \\ I_4 \\ I_5 \end{vmatrix} = \begin{vmatrix} Y_{11} & Y_{12} & Y_{13} & | & Y_{14} & Y_{15} \\ Y_{21} & Y_{22} & Y_{23} & | & Y_{24} & Y_{25} \\ Y_{31} & Y_{32} & Y_{33} & | & Y_{34} & Y_{35} \\ -- & -- & -- & -- & -- & -- \\ Y_{41} & Y_{42} & Y_{43} & | & Y_{44} & Y_{45} \\ Y_{51} & Y_{52} & Y_{53} & | & Y_{54} & Y_{55} \end{vmatrix} \begin{vmatrix} V_1 \\ V_2 \\ V_3 \\ - \\ V_4 \\ V_5 \end{vmatrix} \tag{3.43}$$

This can be written as

$$\begin{vmatrix} \overline{I}_x \\ \overline{I}_y \end{vmatrix} = \begin{vmatrix} \overline{Y}_1 & \overline{Y}_2 \\ \overline{Y}_3 & \overline{Y}_4 \end{vmatrix} \begin{vmatrix} \overline{V}_x \\ \overline{V}_y \end{vmatrix} \tag{3.44}$$

The current $\overline{I}_y = 0$ for the eliminated nodes. Therefore,

$$\overline{I}_y = 0 = \overline{Y}_3\overline{V}_x + \overline{Y}_4\overline{V}_y \tag{3.45}$$

or

$$\overline{V}_y = -\overline{Y}_4^{-1}\overline{Y}_3\overline{V}_x \tag{3.46}$$

Also

$$\overline{I}_x = \overline{Y}_1\overline{V}_x + \overline{Y}_2\overline{V}_y \tag{3.47}$$

Eliminate $\overline{V}_y$ from Equation 3.47. This gives

$$\overline{I}_x = [\overline{Y}_1 - \overline{Y}_2\overline{Y}_4^{-1}\overline{Y}_3]\overline{V}_x \tag{3.48}$$

And the reduced matrix is

$$\overline{Y}_{\text{reduced}} = \overline{Y}_1 - \overline{Y}_2\overline{Y}_4^{-1}\overline{Y}_3 \tag{3.49}$$

3.7 Algorithms for Construction of Bus Impedance Matrix

The $\overline{Z}$ matrix can be formed step by step from basic building concepts. This method is suitable for large power systems and computer analysis.

Consider a passive network with m independent nodes. The bus impedance matrix of this system is given by

$$\begin{vmatrix} Z_{11} & Z_{12} & \cdot & Z_{1m} \\ Z_{21} & Z_{22} & \cdot & Z_{2m} \\ \cdot & \cdot & \cdot & \cdot \\ Z_{m1} & Z_{m2} & \cdot & Z_{mm} \end{vmatrix} \tag{3.50}$$

The buildup of the impedance matrix can start with an arbitrary element between a bus and a common node and then adding branches and links, one by one. The following procedure describes the building blocks of the matrix.

3.7.1 Adding a Tree Branch to an Existing Node

Figure 3.10a shows that a branch pk is added at node p. This increases the dimensions of the primitive bus impedance matrix by 1:

$$\overline{Z} = \begin{vmatrix} Z_{11} & Z_{12} & \cdot & Z_{1m} & Z_{1k} \\ \cdot & \cdot & \cdot & \cdot & \cdot \\ Z_{m1} & Z_{m2} & \cdot & Z_{mm} & Z_{mk} \\ Z_{k1} & Z_{k2} & \cdot & Z_{km} & Z_{kk} \end{vmatrix} \tag{3.51}$$

This can be partitioned as shown:

$$\begin{vmatrix} \overline{Z}_{xy,xy} & \overline{Z}_{xy,pk} \\ \overline{Z}_{pk,xy} & \overline{Z}_{pk,pk} \end{vmatrix} \begin{vmatrix} \overline{I}_{xy} \\ \overline{I}_{pk} \end{vmatrix} = \begin{vmatrix} \overline{V}_{xy} \\ \overline{V}_{pk} \end{vmatrix} \tag{3.52}$$

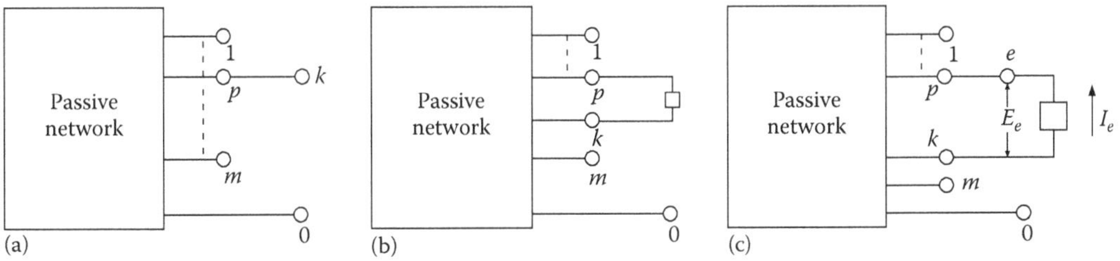

FIGURE 3.10
(a) Adding a tree-branch; (b) and (c) adding a link, in the step-by-step formation of bus impedance matrix.

where $\overline{Z}_{xy,xy}$ = primitive bus impedance matrix. This remains unchanged as addition of a new branch does not change the voltages at the nodes in the primitive matrix. $\overline{Z}_{xy,pk}$ = mutual impedance matrix between the original primitive matrix and element pq. $\overline{Z}_{pk,xy} = \overline{Z}_{xy,pk}$ and $\overline{Z}_{pk,pk}$ = impedance of new element.

From Equation 3.52,

$$\begin{vmatrix} \overline{I}_{xy} \\ \overline{I}_{pk} \end{vmatrix} = \begin{vmatrix} \overline{Y}_{xy,xy} & \overline{Y}_{xy,pk} \\ \overline{Y}_{pk,xy} & \overline{Y}_{pk,pk} \end{vmatrix} \begin{vmatrix} \overline{V}_{xy} \\ \overline{V}_{pk} \end{vmatrix} \tag{3.53}$$

where

$$\begin{vmatrix} \overline{Y}_{xy,xy} & \overline{Y}_{xy,pk} \\ \overline{Y}_{pk,xy} & \overline{Y}_{pk,pk} \end{vmatrix} = \begin{vmatrix} \overline{Z}_{xy,xy} & \overline{Z}_{xy,pk} \\ \overline{Z}_{pk,xy} & \overline{Z}_{pk,pk} \end{vmatrix}^{-1} \tag{3.54}$$

The matrix $\overline{Z}$ has therefore to be inverted.

From Equation 3.53,

$$\overline{I}_{pk} = \overline{Y}_{pk,xy}\overline{V}_{xy} + \overline{Y}_{pk,pk}\overline{V}_{pk} \tag{3.55}$$

If 1 per unit current is injected at any bus other than k and all other currents are zero then $I_{pk} = 0$. This gives

$$\overline{V}_{pk} = -\frac{\overline{Y}_{pk,xy}\overline{V}_{xy}}{\overline{Y}_{pk,pk}}$$

$$\overline{V}_p - \overline{V}_k = -\frac{\overline{Y}_{pk,xy}(\overline{V}_x - \overline{V}_y)}{\overline{Y}_{pk,pk}} \tag{3.56}$$

$$\overline{Z}_{kj} - \overline{Z}_{pj} = \frac{\overline{Y}_{pk,xy}(\overline{Z}_{xj} - \overline{Z}_{yj})}{\overline{Y}_{pk,pk}}$$

Thus, Z_{kj} ($j = 1, 2, \ldots, m, j \neq k$) can be found from the following equation:

$$\overline{Z}_{kj} = \overline{Z}_{pj} + \frac{\overline{Y}_{pk,xy}(\overline{Z}_{xj} - \overline{Z}_{yj})}{\overline{Y}_{pk,pk}} \tag{3.57}$$

If a per unit current is injected at bus k and all other currents are zero, then Equation 3.55 becomes

$$-1 = \overline{Y}_{pk,xy}\overline{V}_{xy} + \overline{Y}_{pk,pk}\overline{V}_{pk} \tag{3.58}$$

This gives Z_{kk}:

$$\overline{Z}_{kk} = \overline{Z}_{pk} + \frac{1 + \overline{Y}_{pk,xy}(\overline{Z}_{xk} - \overline{Z}_{yk})}{\overline{Y}_{pk,pk}} \tag{3.59}$$

2. If there is no coupling between pk and any existing branch xy, then

$$\overline{Z}_{kj} = \overline{Z}_{pj} \quad j = 1, 2, \ldots, m, \quad j \neq k \tag{3.60}$$

$$\overline{Z}_{xj} = \overline{Z}_{pk} + \overline{Z}_{pk,pk} \tag{3.61}$$

3. If the new branch is added between p and the reference node 0, then

$$\overline{Z}_{pj} = 0 \tag{3.62}$$

$$\overline{Z}_{kj} = 0 \tag{3.63}$$

$$\overline{Z}_{kk} = \overline{Z}_{pk,pk} \tag{3.64}$$

3.7.2 Adding a Link

A link can be added as shown in Figure 3.10b. As k is not a new node of the system, the dimensions of the bus impedance matrix do not change; however, the elements of the bus impedance matrix change. To retain the elements of the primitive impedance matrix let a new node e be created by breaking the link pk, as shown in Figure 3.10c. If E_e is the voltage of node e with respect to node k, the following equation can be written:

$$\begin{vmatrix} V_1 \\ V_2 \\ \cdot \\ V_p \\ \cdot \\ V_m \\ E_e \end{vmatrix} = \begin{vmatrix} Z_{11} & Z_{12} & \cdot & Z_{1m} & Z_{1e} \\ Z_{21} & Z_{22} & \cdot & Z_{2m} & Z_{2e} \\ \cdot & \cdot & \cdot & \cdot & \cdot \\ Z_{p1} & Z_{p2} & \cdot & Z_{pm} & Z_{pe} \\ \cdot & \cdot & \cdot & \cdot & \cdot \\ Z_{m1} & Z_{m2} & \cdot & Z_{mm} & Z_{me} \\ Z_{e1} & Z_{e2} & \cdot & Z_{em} & Z_{ee} \end{vmatrix} \begin{vmatrix} I_1 \\ I_2 \\ \cdot \\ I_p \\ \cdot \\ I_m \\ I_e \end{vmatrix} \tag{3.65}$$

In this case the primitive impedance matrix does not change, as the new branch pe can be treated like the addition of a branch from an existing node to a new node, as discussed above. The impedances bearing a subscript e have the following definitions:

Z_{1e} = voltage at bus 1 with respect to the reference node when unit current is injected at k.

Z_{ei} = voltage at bus e *with reference to* k when unit current is injected at bus 1 from reference bus.

Z_{ee} = voltage at bus e *with respect to* k when unit current is injected to bus e from k.

Equation 3.65 is partitioned as shown:

$$\begin{vmatrix} \bar{I}_{xy} \\ \bar{I}_{pe} \end{vmatrix} = \begin{vmatrix} \overline{Y}_{xy,xy} & \overline{Y}_{xy,pe} \\ \overline{Y}_{pe,xy} & \overline{Y}_{pe,pe} \end{vmatrix} \begin{vmatrix} \overline{V}_{xy} \\ \overline{V}_{pe} \end{vmatrix} \tag{3.66}$$

Thus,

$$\overline{I}_{pe} = \overline{Y}_{pe,xy}\overline{V}_{xy} + \overline{Y}_{pe,pe}\overline{V}_{pe} \tag{3.67}$$

If unit current is injected at any node, except node e, and all other currents are zero:

$$0 = \overline{Y}_{pe,xy}\overline{V}_{xy} + \overline{Y}_{pe,pe}\overline{V}_{pe} \tag{3.68}$$

This gives

$$\overline{Z}_{ej} = \overline{Z}_{pj} - \overline{Z}_{kj} + \frac{\overline{Y}_{pe,xy}(\overline{Z}_{xj} - \overline{Z}_{yj})}{\overline{Y}_{pe,pe}} \quad j = 1, 2, \ldots, m \quad j \neq e \tag{3.69}$$

If I_e is 1 per unit and all other currents are zero:

$$-1 = \overline{Y}_{pe,xy}\,\overline{V}_{xy} + \overline{Y}_{pe,pe}\,\overline{V}_{pe} \tag{3.70}$$

This gives

$$\overline{Z}_{ee} = \overline{Z}_{pe} - \overline{Z}_{ke} + \frac{1 + \overline{Y}_{pe,xy}(\overline{Z}_{xe} - \overline{Z}_{ye})}{\overline{Y}_{pe,pe}} \tag{3.71}$$

Thus, this treatment is similar to that of adding a link. If there is no mutual coupling between pk and other branches and p is the reference node:

$$\overline{Z}_{pj} = 0 \tag{3.72}$$

$$\overline{Z}_{ej} = \overline{Z}_{pj} - \overline{Z}_{kj} \tag{3.73}$$

$$\overline{Z}_{ee} = \overline{Z}_{pk,pk} - \overline{Z}_{ke} \tag{3.74}$$

The artificial node can be eliminated by letting voltage at node $e = 0$:

$$\begin{vmatrix} \overline{V}_{\text{bus}} \\ 0 \end{vmatrix} = \begin{vmatrix} \overline{Z}_{\text{bus}} & \overline{Z}_{ej} \\ \overline{Z}_{ej} & \overline{Z}_{ee} \end{vmatrix} \begin{vmatrix} \overline{I}_{\text{bus}} \\ I_e \end{vmatrix} \tag{3.75}$$

$$\overline{Z}_{\text{bus,modified}} = \overline{Z}_{\text{bus,primitive}} - \frac{\overline{Z}_{je}\overline{Z}'_{je}}{\overline{Z}_{ee}} \tag{3.76}$$

3.7.3 Removal of an Uncoupled Branch

An uncoupled branch can be removed by adding a branch in parallel with the branch to be removed, with an impedance equal to the negative of the impedance of the uncoupled branch.

3.7.4 Changing Impedance of an Uncoupled Branch

The bus impedance matrix can be modified by adding a branch in parallel with the branch to be changed with its impedance given by

$$Z_n = Z \| \left(\frac{Z.Z_n}{Z + Z_n} \right) \tag{3.77}$$

where
Z_n is the required new impedance
Z is the original impedance of the branch

3.7.5 Removal of a Coupled Branch

A branch with mutual coupling M can be modeled as shown in Figure 3.11. We calculated the elements of bus impedance matrix by injecting a current at a bus and measuring the voltage at the other buses. The voltage at buses can be maintained if four currents as shown in Figure 3.11 are injected at either side of the coupled branch:

$$\begin{vmatrix} I_{gh} \\ I_{pk} \end{vmatrix} = \begin{vmatrix} Y_{gh,gh} & Y_{gh,pk} \\ Y_{pk,gh} & Y_{pk,pk} \end{vmatrix} \begin{vmatrix} V_g - V_h \\ V_p - V_k \end{vmatrix} \tag{3.78}$$

This gives

$$\begin{vmatrix} I \\ -I \\ I' \\ -I' \end{vmatrix} = \begin{vmatrix} Y_{gh,gh} & -Y_{gh,gh} & Y_{gh,pk} & -Y_{gh,pk} \\ -Y_{gh,gh} & Y_{gh,gh} & -Y_{gh,pk} & Y_{gh,pk} \\ Y_{pk,gh} & -Y_{pk,gh} & Y_{pk,pk} - 1/Z_{pk,pk} & Y_{pk,pk} + 1/Z_{pk,pk} \\ -Y_{pk,gh} & Y_{pk,gh} & -Y_{pk,pk} + 1/Z_{pk,pk} & Y_{pk,pk} - 1/Z_{pk,pk} \end{vmatrix} \begin{vmatrix} V_g \\ V_h \\ V_p \\ V_k \end{vmatrix} \tag{3.79}$$

which is written in abbreviated form as

$$\bar{I}_w = \overline{K}\,\overline{V}_w \tag{3.80}$$

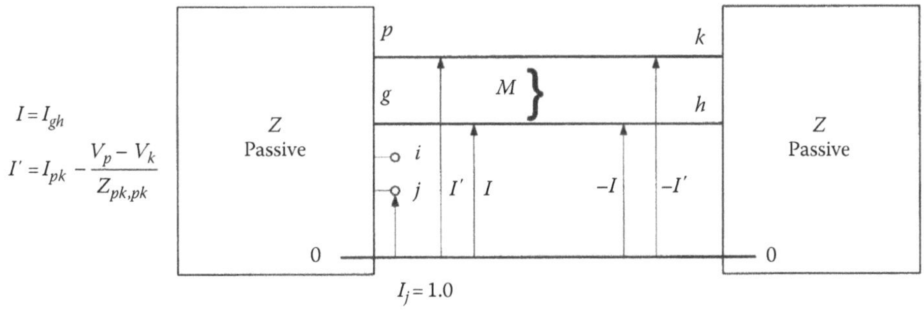

FIGURE 3.11
Adding a coupled branch in the step-by-step formation of impedance matrix.

Partition the original matrix, separating out the coupled portion as follows:

$$\begin{vmatrix} V_1 \\ \cdot \\ V_j \\ \cdot \\ V_{g-1} \\ V_{h+1} \\ \cdot \\ V_{p-1} \\ V_{k+1} \\ \cdot \\ V_m \\ -- \\ V_g \\ V_h \\ V_p \\ V_k \end{vmatrix} = \begin{vmatrix} Z_{11} & | & \cdot & \cdot & \cdot & \cdot \\ \cdot & | & \cdot & \cdot & \cdot & \cdot \\ Z_{j1} & | & \cdot & \cdot & \cdot & \cdot \\ \cdot & | & \cdot & \cdot & \cdot & \cdot \\ \cdot & | & \cdot & \cdot & \cdot & \cdot \\ \cdot & | & \cdot & \cdot & \cdot & \cdot \\ \cdot & | & \cdot & \cdot & \cdot & \cdot \\ \cdot & | & \cdot & \cdot & \cdot & \cdot \\ \cdot & | & \cdot & \cdot & \cdot & \cdot \\ \cdot & | & \cdot & \cdot & \cdot & \cdot \\ \cdot & | & \cdot & \cdot & \cdot & \cdot \\ - & - & - & - & - & - \\ Z_{g1} & | & Z_{gg} & Z_{gh} & Z_{gp} & Z_{kk} \\ Z_{h1} & | & \cdot & \cdot & \cdot & \cdot \\ Z_{p1} & | & \cdot & \cdot & \cdot & \cdot \\ Z_{k1} & | & Z_{kg} & Z_{kh} & Z_{kp} & Z_{kk} \end{vmatrix} \begin{vmatrix} 0 \\ \cdot \\ 1 \\ \cdot \\ \cdot \\ \cdot \\ \cdot \\ \cdot \\ \cdot \\ \cdot \\ \cdot \\ -- \\ I \\ -I \\ I' \\ -I' \end{vmatrix} \tag{3.81}$$

This is written as

$$\begin{vmatrix} \overline{V}_u \\ \overline{V}_w \end{vmatrix} \begin{vmatrix} \overline{A} & \overline{B} \\ \overline{C} & \overline{D} \end{vmatrix} \begin{vmatrix} \overline{I}_u \\ \overline{I}_w \end{vmatrix} \tag{3.82}$$

From Equations 3.80 and 3.82:

$$\text{If } \overline{I}_u = 0$$

$$\overline{V}_w = (I - \overline{D}\overline{K})^{-1}\overline{D}\,\overline{I}_j \tag{3.83}$$

$$\overline{V}_u = \overline{B}\,\overline{K}\,\overline{V}_w + \overline{B}\,\overline{I}_j \tag{3.84}$$

$$\text{If } \overline{I}_u \neq 0$$

$$\overline{V}_w = (I - \overline{D}\overline{K})^{-1}\overline{C}\,\overline{I}_u \tag{3.85}$$

$$\overline{V}_u = \overline{A}\,\overline{I}_u + \overline{B}\,\overline{K}\,\overline{V}_w \tag{3.86}$$

If the transformed matrix is defined as

$$\begin{vmatrix} \overline{A}' & \overline{B}' \\ \overline{C}' & \overline{D}' \end{vmatrix} \tag{3.87}$$

then Equations 3.83 through 3.86 give $\overline{C}'$, $\overline{D}'$, $\overline{A}'$, and $\overline{B}'$, respectively.

Example 3.8

Consider a distribution system with four buses, whose positive and negative sequence networks are shown in Figure 3.12a and zero sequence network in Figure 3.12b. The positive and negative sequence networks are identical and rather than $r + jx$ values, numerical values are shown for ease of hand calculations. There is a mutual coupling between parallel lines in the zero sequence network. It is required to construct bus impedance matrices for positive and zero sequence networks.

The primitive impedance or admittance matrices can be written by examination of the network. First consider the positive or negative sequence network of Figure 3.12a. The following steps illustrate the procedure:

1. The buildup is started with branches 01, 02, and 03 that are connected to the reference node, Figure 3.13a. The primitive impedance matrix can be simply written as

$$\begin{vmatrix} 0.05 & 0 & 0 \\ 0 & 0.2 & 0 \\ 0 & 0 & 0.05 \end{vmatrix}$$

2. Next, add link 1–2, Figure 3.13b. As this link has no coupling with other branches of the system,

$$Z_{ej} = Z_{pj} - Z_{kj}$$
$$Z_{ee} = Z_{pk,pk} + Z_{pe} - Z_{ke}$$
$$p = 1, \quad k = 2.$$

$$Z_{e1} = Z_{11} - Z_{21} = 0.05$$
$$Z_{e2} = Z_{12} - Z_{22} = 0 - 0.2 = -0.2$$
$$Z_{e3} = Z_{13} - Z_{23} = 0$$
$$Z_{ee} = Z_{12,12} + Z_{1e} - Z_{2e} = 0.04 + 0.05 + 0.2 = 0.29$$

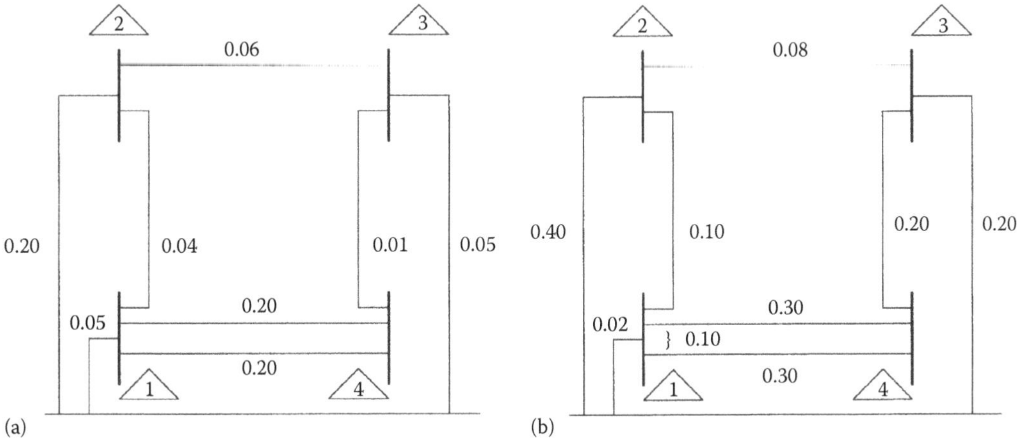

FIGURE 3.12
(a) Positive and negative sequence network for Example 3.6; (b) zero sequence network for Example 3.6.

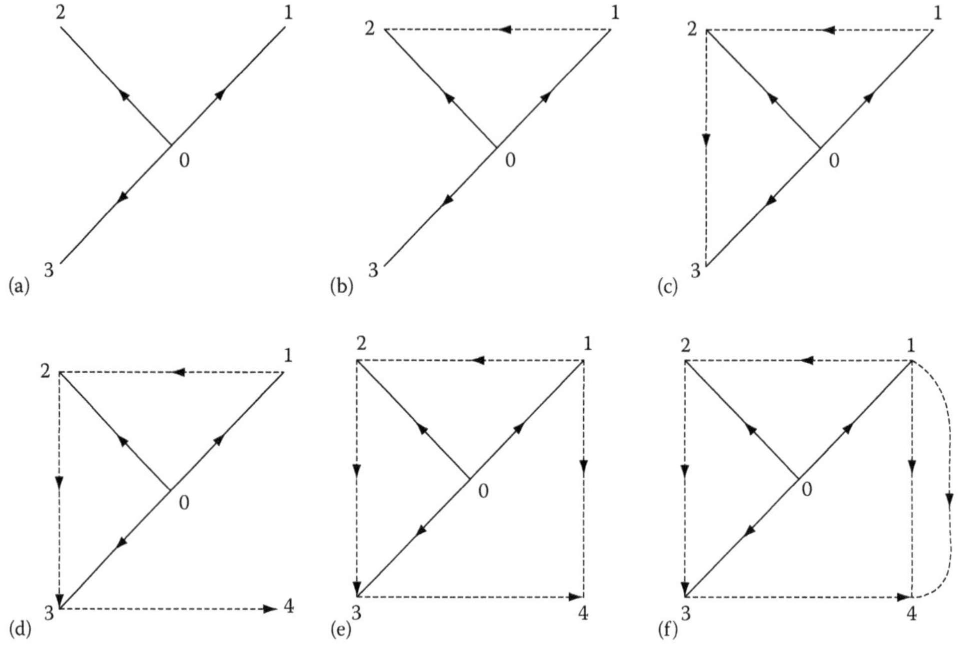

FIGURE 3.13
(a–f) Step-by-step formation of impedance matrix.

The augmented matrix is

$$\begin{vmatrix} 0.05 & 0 & 0 & 0.05 \\ 0 & 0.2 & 0 & -0.2 \\ 0 & 0 & 0.05 & 0 \\ 0.05 & -0.2 & 0 & 0.29 \end{vmatrix}$$

3. Eliminate the last row and last column by using Equation 3.76. This gives

$$\begin{vmatrix} 0.0414 & 0.0345 & 0 \\ 0.0345 & 0.0621 & 0 \\ 0 & 0 & 0.05 \end{vmatrix}$$

4. Add link 2–3, Figure 3.13c:
   $p = 2$, $k = 3$ and there is no mutual coupling with other branches.
   This gives

$$Z_{e1} = Z_{21} - Z_{31} = 0.345 - 0 = 0.345$$

$$Z_{e2} = Z_{22} - Z_{32} = 0.0621 - 0 = 0.0621$$

$$Z_{e3} = Z_{23} - Z_{33} = 0 - 0.05 = -0.05$$

$$Z_{ee} = Z_{23,23} + Z_{2e} - Z_{3e} = 0.06 + 0.0621 - (-0.05) = 0.1721$$

The augmented matrix is

$$\begin{vmatrix} 0.0414 & 0.0345 & 0 & 0.0345 \\ 0.0345 & 0.0621 & 0 & 0.0621 \\ 0 & 0 & 0.05 & -0.05 \\ 0.0345 & 0.0621 & -0.05 & 0.1721 \end{vmatrix}$$

5. Eliminate last row and column. The modified matrix is

$$\begin{vmatrix} 0.0345 & 0.0221 & 0.010 \\ 0.0221 & 0.0397 & 0.018 \\ 0.010 & 0.0180 & 0.355 \end{vmatrix}$$

6. Add branch 3 to 4, $p=3$, $k=4$:

$$Z_{41} = Z_{31} = 0.01$$
$$Z_{42} = Z_{32} = 0.018$$
$$Z_{43} = Z_{33} = 0.0355$$
$$Z_{44} = Z_{34} + Z_{34,34} = 0.0355 + 0.1 = 0.1355$$

The augmented matrix is

$$\begin{vmatrix} 0.0345 & 0.0221 & 0.01 & 0.01 \\ 0.0221 & 0.0397 & 0.018 & 0.018 \\ 0.01 & 0.018 & 0.0355 & 0.0355 \\ 0.01 & 0.018 & 0.0355 & 0.1355 \end{vmatrix}$$

7. Add first parallel link 1–4, Figure 3.13d:

$$p = 1, \quad k = 4$$

$$Z_{e1} = Z_{1e} = 0.0345 - 0.01 = 0.0245$$
$$Z_{e2} = Z_{2e} = 0.0221 - 0.018 = 0.0041$$
$$Z_{e3} = Z_{3e} = 0.01 - 0.0355 = -0.0255$$
$$Z_{e4} = Z_{4e} = 0.01 - 0.1355 = 0.1255$$
$$Z_{ee} = 0.2 + 0.245 - (-0.1255) = 0.350$$

This gives the matrix

$$\begin{vmatrix} 0.0345 & 0.0221 & 0.01 & 0.01 & 0.0245 \\ 0.0221 & 0.0397 & 0.018 & 0.018 & 0.0041 \\ 0.01 & 0.018 & 0.0355 & 0.0355 & -0.0255 \\ 0.01 & 0.018 & 0.0355 & 0.1355 & -0.1255 \\ 0.0245 & 0.0041 & -0.0255 & -0.1255 & 0.349 \end{vmatrix}$$

8. Eliminate last row and last column using Equation 3.76:

$$\begin{vmatrix} 0.0328 & 0.0218 & 0.0118 & 0.0188 \\ 0.0218 & 0.0397 & 0.0183 & 0.0195 \\ 0.0118 & 0.0183 & 0.0336 & 0.0264 \\ 0.0188 & 0.0195 & 0.0264 & 0.0905 \end{vmatrix}$$

9. Finally, add second parallel link 1–4, Figure 3.13e:

$$Z_{e1} = Z_{1e} = 0.0328 - 0.0188 = 0.014$$
$$Z_{e2} = Z_{2e} = 0.0218 - 0.0195 = 0.0023$$
$$Z_{e3} = Z_{3e} = 0.0118 - 0.0264 = -0.0146$$
$$Z_{e4} = Z_{4e} = 0.0188 - 0.0905 = -0.0717$$
$$Z_{ee} = 0.2 + 0.014 - (-0.0717) = 0.2857$$

This gives

$$\begin{vmatrix} 0.0328 & 0.0218 & 0.0118 & 0.0188 & 0.014 \\ 0.0218 & 0.0397 & 0.0183 & 0.0195 & 0.0023 \\ 0.0118 & 0.0183 & 0.0336 & 0.0264 & -0.0146 \\ 0.0188 & 0.0195 & 0.0264 & 0.0905 & -0.0717 \\ 0.014 & 0.0023 & -0.0146 & -0.0717 & 0.2857 \end{vmatrix}$$

10. Eliminate last row and column:

$$\begin{vmatrix} 0.0328 & 0.0218 & 0.0118 & 0.0188 \\ 0.0218 & 0.0397 & 0.0183 & 0.0195 \\ 0.0118 & 0.0183 & 0.0336 & 0.0264 \\ 0.0188 & 0.0195 & 0.0264 & 0.0905 \end{vmatrix}$$

$$- \frac{\begin{vmatrix} 0.014 \\ 0.0023 \\ -0.0416 \\ -0.0717 \end{vmatrix} \begin{vmatrix} 0.0014 & 0.0023 & -0.0146 & -0.0717 \end{vmatrix}}{0.2857}$$

This gives the final positive or negative sequence matrix as

$$\overline{Z}^{+}, \overline{Z}^{-} = \begin{vmatrix} 0.0321 & 0.0217 & 0.0125 & 0.0223 \\ 0.217 & 0.0397 & 0.0184 & 0.0201 \\ 0.0125 & 0.0184 & 0.0329 & 0.0227 \\ 0.0223 & 0.0201 & 0.0227 & 0.0725 \end{vmatrix}$$

Zero Sequence Impedance Matrix

The zero sequence impedance matrix is similarly formed, until the last parallel coupled line between buses 1 and 4 is added. The zero sequence impedance matrix, until the coupled branch is required to be added, is formed by a step-by-step procedure as outlined for the positive sequence matrix:

$$\begin{vmatrix} 0.0184 & 0.0123 & 0.0098 & 0.0132 \\ 0.0123 & 0.0670 & 0.0442 & 0.314 \\ 0.0098 & 0.0442 & 0.0806 & 0.0523 \\ 0.0132 & 0.0314 & 0.0523 & 0.1567 \end{vmatrix}$$

Add parallel coupled lines between buses 1 and 4, $p=1$, $k=4$.

The coupled primitive impedance matrix is

$$Z_{pr} = \begin{vmatrix} 0.3 & 0.1 \\ 0.1 & 0.3 \end{vmatrix}$$

Its inverse is given by

$$Z_{pr}^{-1} = \begin{vmatrix} 0.3 & 0.1 \\ 0.1 & 0.03 \end{vmatrix}^{-1} = \begin{vmatrix} 3.750 & -1.25 \\ -1.25 & 3.750 \end{vmatrix}$$

$$Y_{pe,pe} = 3.75$$

$$Y_{pe,xy} = -1.25$$

$p=1$, $k=4$ coupled with 2–3. Thus,

$$Z_{x1} = 0.0184, \quad Z_{x2} = 0.0123, \quad Z_{x3} = 0.0098, \quad Z_{x4} = 0.0132$$
$$Z_{y1} = 0.0132, \quad Z_{y2} = 0.0314, \quad Z_{y3} = 0.0523, \quad Z_{y4} = 0.1567$$

This gives

$$Z_{e1} = Z_{11} - Z_{41} + \frac{Y_{pe,pe}(Z_{x1} - Z_{y1})}{Y_{pe,ep}}$$
$$= 0.0184 - 0.0132 + \frac{(-1.25)(0.0184) - (0.00132)}{3.75} = -0.0035$$

Similarly,

$$Z_{e2} = 0.0123 - 0.314 + \left(\frac{-1.25}{3.75}\right)(0.0123 - 0.314) = -0.0127$$

$$Z_{e3} = 0.0098 - 0.523 + \left(\frac{-1.25}{3.75}\right)(0.0098 - 0.0523) = -0.0283$$

$$Z_{e4} = 0.0132 - 0.1567 + \left(\frac{-1.25}{3.75}\right)(0.0132 - 0.1567) = -0.0957$$

Z_{ee} is given by

$$Z_{ee} = Z_1e - Z_4e + \frac{1 + (-1.25)(Z_{e1} - Z_{e4})}{3.75}$$
$$= 0.0035 - (-0.0957) + \frac{1 + (-25)(0.0035 - (-0.0957))}{3.75} = 0.3328$$

The modified impedance matrix is now

$$\begin{vmatrix} 0.0184 & 0.0123 & 0.0098 & 0.0132 & 0.0035 \\ 0.0123 & 0.0670 & 0.0442 & 0.0314 & -0.0127 \\ 0.0098 & 0.0442 & 0.0806 & 0.0523 & -0.0283 \\ 0.0132 & 0.0324 & 0.0523 & 0.1567 & -0.0957 \\ 0.0035 & -0.0127 & -0.0283 & -0.0957 & 0.3328 \end{vmatrix}$$

Finally, eliminate the last row and column:

$$\begin{vmatrix} 0.0184 & 0.0123 & 0.0098 & 0.132 \\ 0.0123 & 0.0670 & 0.0442 & 0.0314 \\ 0.0098 & 0.0442 & 0.0806 & 0.0523 \\ 0.0132 & 0.0314 & 0.0523 & 0.1567 \end{vmatrix} - \frac{\begin{vmatrix} 0.0035 \\ -0.0127 \\ -0.0283 \\ -0.0957 \end{vmatrix} \begin{vmatrix} 0.0035 & -0.0127 & -0.0283 & -0.0957 \end{vmatrix}}{0.3328}$$

This gives the final zero sequence bus impedance matrix:

$$\begin{vmatrix} 0.0184 & 0.0124 & 0.0101 & 0.0142 \\ 0.0124 & 0.0665 & 0.0431 & 0.0277 \\ 0.0101 & 0.0431 & 0.0782 & 0.0442 \\ 0.0142 & 0.0277 & 0.0442 & 0.1292 \end{vmatrix}$$

Note that at each last row and column elimination, $Z_{21}=Z_{12}$, $Z_{23}=Z_{32}$, etc. Also, in the final matrix there are no negative elements. If a duplex reactor is modeled (Appendix C), some elements may be negative; the same is true, in some cases, for modeling of three-winding transformers. Other than that, there is no check on correct formation.

3.8 Short-Circuit Calculations with Bus Impedance Matrix

Short-circuit calculations using bus impedance matrices follow the same logic as developed in Chapter 2. Consider that the positive, negative, and zero sequence bus impedance matrices Z^1_{ss}, Z^2_{ss} and Z^0_{ss} are known and a single line-to-ground fault occurs at the rth bus. The positive sequence is then injected only at the rth bus and all other currents in the positive sequence current vector are zero. The positive sequence voltage at bus r is given by

$$V^1_r = -Z^1_{rr}I^1_r \tag{3.88}$$

Similarly, the negative and zero sequence voltages are

$$
\begin{aligned}
V_r^2 &= -Z_{rr}^2 I_r^2 \\
V_r^0 &= -Z_{rr}^0 I_r^0
\end{aligned}
\tag{3.89}
$$

From the sequence network connections for a line-to-ground fault,

$$
I_r^1 = I_r^2 = I_r^0 = \frac{1.0}{Z_{rr}^1 + Z_{rr}^2 + Z_{rr}^0 + 3Z_f} \tag{3.90}
$$

This shows that the following equations can be written for a shorted bus *s*.

3.8.1 Line-to-Ground Fault

$$
I_s^0 = I_s^1 = I_s^2 = \frac{1}{Z_{ss}^1 + Z_{ss}^2 + Z_{ss}^0 + 3Z_f} \tag{3.91}
$$

3.8.2 Line-to-Line Fault

$$
I_s^1 = -I_s^2 = \frac{1}{Z_{ss}^1 + Z_{ss}^2 + 3Z_f} \tag{3.92}
$$

3.8.3 Double Line-to-Ground Fault

$$
I_s^1 = \frac{1}{Z_{ss}^1 + \frac{Z_{ss}^2\left(Z_{ss}^0 + 3Z_f\right)}{Z_{ss}^2 + \left(Z_{ss}^0 + 3Z_f\right)}} \tag{3.93}
$$

$$
I_s^0 = -\frac{Z_{ss}^2}{Z_{ss}^2 + \left(Z_{ss}^0 + 3Z_f\right)} I_s^1 \tag{3.94}
$$

$$
I_s^2 = \frac{\left(-Z_{ss}^0 + 3Z_f\right)}{Z_{ss}^2 + \left(Z_{ss}^0 + 3Z_f\right)} I_s^1 \tag{3.95}
$$

The phase currents are calculated by

$$
I_s^{abc} = T_s I_s^{012} \tag{3.96}
$$

The voltage at bus j of the system is

$$\begin{vmatrix} V_j^0 \\ V_j^1 \\ V_j^2 \end{vmatrix} = \begin{vmatrix} 0 \\ 1 \\ 0 \end{vmatrix} - \begin{vmatrix} Z_{js}^0 & 0 & 0 \\ 0 & Z_{js}^1 & 0 \\ 0 & 0 & Z_{js}^2 \end{vmatrix} \begin{vmatrix} I_s^0 \\ I_s^1 \\ I_s^2 \end{vmatrix} \quad (3.97)$$

where $j = 1, 2, \ldots, s, \ldots, m$.

The fault current from bus x to bus y is given by

$$\begin{vmatrix} I_{xy}^0 \\ I_{xy}^1 \\ I_{xy}^2 \end{vmatrix} = \begin{vmatrix} Y_{xy}^0 & 0 & 0 \\ 0 & Y_{xy}^1 & 0 \\ 0 & 0 & Y_{xy}^2 \end{vmatrix} \begin{vmatrix} V_x^0 - V_y^0 \\ V_x^1 - V_y^1 \\ V_x^2 - V_y^2 \end{vmatrix} \quad (3.98)$$

where

$$I_{xy}^0 = \begin{vmatrix} I_{12}^0 \\ I_{13}^0 \\ \cdot \\ I_{mm}^0 \end{vmatrix} \quad (3.99)$$

and

$$\overline{Y}_{xy}^0 = \begin{vmatrix} Y_{12,12}^0 & Y_{12,13}^0 & \cdot & Y_{12,mn}^0 \\ Y_{13,12}^0 & Y_{13,13}^0 & \cdot & Y_{13,mn}^0 \\ \cdot & \cdot & \cdot & \cdot \\ Y_{mn,12}^0 & Y_{mn,13}^0 & \cdot & Y_{mn,mn}^0 \end{vmatrix} \quad (3.100)$$

where Y_{xy}^0 is the inverse of the primitive matrix of the system. Similar expressions apply to positive sequence and negative sequence quantities.

Example 3.9

The positive, negative, and zero sequence matrices of the system in Figure 3.12 are calculated in Example 3.8. A double line-to-ground fault occurs at bus 4. Using the matrices already calculated, it is required to calculate

- Fault current at bus 4
- Voltage at bus 4
- Voltage at buses 1, 2, and 3
- Fault current flows from buses 3 to 4, 1 to 4, 2 to 3
- Current flow in node 0 to bus 3

The fault current at bus 4 is first calculated as follows:

$$I_4^1 = \frac{1}{Z_{4s}^1 + \dfrac{Z_{4s}^2 \times Z_{4s}^0}{Z_{4s}^2 + Z_{4s}^0}}$$

$$= \frac{1}{0.0725 + \dfrac{0.0725 \times 0.1292}{0.0725 + 0.1292}}$$

$$= 8.408$$

$$I_4^0 = \frac{-Z_{4s}^2}{Z_{4s}^2 + Z_{4s}^0} I_4^1$$

$$= \frac{-0.0725 \times 8.408}{0.0725 + 0.1292}$$

$$= -3.022$$

$$I_4^2 = \frac{-Z_{4s}^0}{Z_{4s}^2 + Z_{4s}^0} I_4^1$$

$$= \frac{-0.1292 \times 8.408}{0.0727 + 0.1292}$$

$$= -5.386$$

The line currents are given by

$$I_4^{abc} = T_s I_4^{012}$$

i.e.,

$$\begin{vmatrix} I_4^a \\ I_4^b \\ I_4^c \end{vmatrix} = \begin{vmatrix} 1 & 1 & 1 \\ 1 & a^2 & a \\ 1 & a & a^2 \end{vmatrix} \begin{vmatrix} I_4^0 \\ I_4^1 \\ I_4^2 \end{vmatrix}$$

$$= \begin{vmatrix} 1 & 1 & 1 \\ 1 & a^2 & a \\ 1 & a & a^2 \end{vmatrix} \begin{vmatrix} -3.022 \\ 8.408 \\ -5.386 \end{vmatrix}$$

$$= \begin{vmatrix} 0 \\ -4.533 - j11.946 \\ -4.533 + j11.946 \end{vmatrix} = \begin{vmatrix} 0 \\ 12.777 < 249.2^\circ \\ 12.777 < 110.78^\circ \end{vmatrix}$$

Sequence voltages at bus 4 are given by

$$\begin{vmatrix} V_4^0 \\ V_4^1 \\ V_4^2 \end{vmatrix} = \begin{vmatrix} 0 \\ 1 \\ 0 \end{vmatrix} - \begin{vmatrix} Z_{4s}^0 & & \\ & Z_{4s}^1 & \\ & & Z_{4s}^2 \end{vmatrix} \begin{vmatrix} I_4^0 \\ I_4^1 \\ I_4^2 \end{vmatrix}$$

$$= \begin{vmatrix} 0 \\ 1 \\ 0 \end{vmatrix} - \begin{vmatrix} 0.1292 & 0 & 0 \\ 0 & 0.0725 & 0 \\ 0 & 0 & 0.0725 \end{vmatrix} \begin{vmatrix} -3.022 \\ 8.408 \\ -5.386 \end{vmatrix}$$

$$= \begin{vmatrix} 0.3904 \\ 0.3904 \\ 0.3904 \end{vmatrix}$$

Line voltages are, therefore,

$$\begin{vmatrix} V_4^a \\ V_4^b \\ V_4^c \end{vmatrix} = \begin{vmatrix} 1 & 1 & 1 \\ 1 & a^2 & a \\ 1 & a & a^2 \end{vmatrix} \begin{vmatrix} V_4^0 \\ V_4^1 \\ V_4^2 \end{vmatrix}$$

$$= \begin{vmatrix} 1 & 1 & 1 \\ 1 & a^2 & a \\ 1 & a & a^2 \end{vmatrix} \begin{vmatrix} 0.3904 \\ 0.3904 \\ 0.3904 \end{vmatrix} = \begin{vmatrix} 1.182 \\ 0 \\ 0 \end{vmatrix}$$

Similarly,

$$\overline{V}_1^{0,1,2} = \begin{vmatrix} 0 \\ 1 \\ 0 \end{vmatrix} - \begin{vmatrix} 0.0142 & 0 & 0 \\ 0 & 0.0223 & 0 \\ 0 & 0 & 0.0223 \end{vmatrix} \begin{vmatrix} -3.022 \\ 8.408 \\ -5.386 \end{vmatrix}$$

$$= \begin{vmatrix} 0.0429 \\ 0.8125 \\ 0.1201 \end{vmatrix}$$

Sequence voltages at buses 2 and 3, similarly calculated, are

$$\overline{V}_2^{0,1,2} = \begin{vmatrix} 0.0837 \\ 0.8310 \\ 0.1083 \end{vmatrix}$$

$$\overline{V}_3^{0,1,2} = \begin{vmatrix} 0.1336 \\ 0.8091 \\ 0.1223 \end{vmatrix}$$

The sequence voltages are converted into line voltages:

$$\overline{V}_1^{abc} = \begin{vmatrix} 0.976 < 0° \\ 0.734 < 125.2° \\ 0.734 < 234.8° \end{vmatrix} \quad \overline{V}_2^{abc} = \begin{vmatrix} 1.023 < 0° \\ 0.735 < 121.5° \\ 0.735 < 238.3° \end{vmatrix} \quad \overline{V}_3^{abc} = \begin{vmatrix} 1.065 < 0° \\ 0.681 < 119.2° \\ 0.681 < 240.8° \end{vmatrix}$$

The sequence currents flowing between buses 3 and 4 are given by

$$\begin{vmatrix} I_{34}^0 \\ I_{34}^1 \\ I_{34}^2 \end{vmatrix} = \begin{vmatrix} 1/0.2 & 0 & 0 \\ 0 & 1/0.1 & 0 \\ 0 & 0 & 1/0.1 \end{vmatrix} \begin{vmatrix} 0.1336 - 0.3904 \\ 0.8091 - 0.3904 \\ 0.1223 - 0.3904 \end{vmatrix}$$

$$= \begin{vmatrix} -1.284 \\ 4.1870 \\ -2.681 \end{vmatrix}$$

Similarly, the sequence currents between bus 3 and 2 are

$$\begin{vmatrix} I_{32}^0 \\ I_{32}^1 \\ I_{32}^2 \end{vmatrix} = \begin{vmatrix} 1/0.08 & 0 & 0 \\ 0 & 1/0.06 & 0 \\ 0 & 0 & 1/0.06 \end{vmatrix} \begin{vmatrix} 0.1336 - 0.0837 \\ 0.8091 - 0.8310 \\ 0.1223 - 0.1083 \end{vmatrix} = \begin{vmatrix} 0.624 \\ -0.365 \\ 0.233 \end{vmatrix}$$

This can be transformed into line currents.

$$\bar{I}_{32}^{abc} = \begin{vmatrix} 0.492 \\ 0.69 + j0.518 \\ 0.69 + j0.518 \end{vmatrix} \quad \bar{I}_{34}^{abc} = \begin{vmatrix} 0.222 \\ -2.037 + j5.948 \\ -2.037 + j5.948 \end{vmatrix}$$

The lines between buses 1 and 4 are coupled in the zero sequence network. The $\overline{Y}$ matrix between zero sequence coupled lines is

$$\overline{Y}_{14}^0 = \begin{vmatrix} 3.75 & -1.250 \\ -1.25 & 3.75 \end{vmatrix}$$

Therefore, the sequence currents are given by

$$\begin{vmatrix} I_{14a}^0 \\ I_{14b}^0 \\ I_{14a}^1 \\ I_{14b}^1 \\ I_{14a}^2 \\ I_{14b}^2 \end{vmatrix} = \begin{vmatrix} 3.75 & -1.25 & 0 & 0 & 0 & 0 \\ -1.25 & 3.75 & 0 & 0 & 0 & 0 \\ 0 & 0 & 5 & 0 & 0 & 0 \\ 0 & 0 & 0 & 5 & 0 & 0 \\ 0 & 0 & 0 & 0 & 5 & 0 \\ 0 & 0 & 0 & 0 & 0 & 5 \end{vmatrix} \begin{vmatrix} 0.0429 - 0.3904 \\ 0.0429 - 0.3904 \\ 0.8125 - 0.3904 \\ 0.8125 - 0.3904 \\ 0.1201 - 0.3904 \\ 0.1201 - 0.3904 \end{vmatrix} = \begin{vmatrix} -0.8688 \\ -0.8688 \\ 2.1105 \\ 2.1105 \\ -1.3515 \\ -1.3515 \end{vmatrix}$$

Each of the lines carries sequence currents:

$$\bar{I}_{14a}^{012} = \bar{I}_{14b}^{012} = \begin{vmatrix} -0.8688 \\ 2.1105 \\ -1.3515 \end{vmatrix}$$

Converting into line currents,

$$\bar{I}_{14a}^{012} = \bar{I}_{14b}^{012} = \begin{vmatrix} -0.11 \\ -1.248 - j2.998 \\ -1.248 + j2.998 \end{vmatrix}$$

Also the line currents between buses 3 and 4 are

$$\bar{I}_{34}^{abc} = \begin{vmatrix} -0.222 \\ -2.037 - j5.948 \\ -2.037 + j5.948 \end{vmatrix}$$

Within the accuracy of calculation, the summation of currents (sequence components as well as line currents) at bus 4 is zero. This is a verification of the calculation. Similarly, the vectorial sum of currents at bus 3 should be zero. As the currents between 3 and 4 and 3 and 2 are already known, the currents from node 0 to bus 3 can be calculated.

Example 3.10

Reform the positive and zero sequence impedance matrices of Example 3.8 after removing one of the parallel lines between buses 1 and 4.

Positive Sequence Matrix
The positive sequence matrix is

$$\begin{vmatrix} 0.0321 & 0.0217 & 0.0125 & 0.0223 \\ 0.0217 & 0.0397 & 0.0184 & 0.0201 \\ 0.0125 & 0.0184 & 0.0329 & 0.0277 \\ 0.0223 & 0.0201 & 0.0227 & 0.0725 \end{vmatrix}$$

Removing one of the parallel lines is similar to adding an impedance of −0.2 between 1 and 4:

$$Z_{1e} = Z_{e1} = Z_{11} - Z_{41} = 0.0321 - 0.0223 = 0.0098$$

$$Z_{2e} = Z_{e2} = Z_{12} - Z_{42} = 0.0217 - 0.0201 = 0.0016$$

$$Z_{3e} = Z_{e3} = Z_{13} - Z_{43} = 0.0125 - 0.0227 = -0.0102$$

$$Z_{4e} = Z_{e4} = Z_{14} - Z_{44} = 0.0223 - 0.0725 = 0.0502$$

$$Z_{ee} = -0.2 + 0.0098 - (-0.0502) = -0.1400$$

The augmented impedance matrix is, therefore,

$$\begin{vmatrix} 0.0321 & 0.0217 & 0.0125 & 0.0223 & 0.0098 \\ 0.0217 & 0.0397 & 0.0184 & 0.0201 & 0.0016 \\ 0.0125 & 0.0184 & 0.0329 & 0.0227 & -0.0102 \\ 0.0223 & 0.0201 & 0.0227 & 0.0725 & -0.0502 \\ 0.0098 & 0.0016 & -00102 & -0.0502 & -0.1400 \end{vmatrix}$$

Eliminate the last row and column:

$$\begin{vmatrix} 0.0321 & 0.0217 & 0.0125 & 0.0223 \\ 0.0217 & 0.0397 & 0.0184 & 0.0201 \\ 0.0125 & 0.0184 & 0.0329 & 0.0227 \\ 0.0223 & 0.0201 & 0.0227 & 0.0725 \end{vmatrix} - \frac{\begin{vmatrix} 0.0098 \\ 0.0016 \\ -0.0102 \\ -0.0502 \end{vmatrix} \begin{vmatrix} 0.0998 & 0.0016 & -0.0102 & -0.0502 \end{vmatrix}}{-0.14}$$

This is equal to

$$\begin{vmatrix} 0.0328 & 0.0218 & 0.0118 & 0.0188 \\ 0.218 & 0.0397 & 0.0183 & 0.0195 \\ 0.0118 & 0.0183 & 0.0336 & 0.0264 \\ 0.0188 & 0.0195 & 0.0264 & 0.0905 \end{vmatrix}$$

The results can be checked with those of Example 3.8. This is the same matrix that was obtained before the last link between buses 1 and 4 was added.

Zero Sequence Matrix

Here, removal of a coupled link is involved. The zero sequence matrix from Example 3.8 is

	1	2	3	4
1	0.0184	0.0124	0.0101	0.0142
2	0.0124	0.0665	0.0431	0.0277
3	0.0101	0.0431	0.0782	0.0442
4	0.0142	0.0277	0.0442	0.1292

Rewrite this matrix as

	2	3	1	4
2	0.0665	0.0431	0.0124	0.0277
3	0.0431	0.0782	0.0101	0.0442
1	0.0124	0.0101	0.0184	0.0142
4	0.0277	0.0442	0.0142	0.1292

Therefore,

$$\begin{vmatrix} V_2 \\ V_3 \\ V_1 \\ V_4 \end{vmatrix} = \begin{vmatrix} 0.0665 & 0.0431 & 0.0124 & 0.0277 \\ 0.0431 & 0.0782 & 0.0101 & 0.0442 \\ 0.0124 & 0.0101 & 0.0184 & 0.0142 \\ 0.0277 & 0.0442 & 0.0142 & 0.1292 \end{vmatrix} \begin{vmatrix} 0 \\ 0 \\ I+I' \\ -I-I' \end{vmatrix}$$

or

$$\begin{vmatrix} \overline{V}_u \\ \overline{V}_w \end{vmatrix} = \begin{vmatrix} \overline{A} & \overline{B} \\ \overline{C} & \overline{D} \end{vmatrix} \begin{vmatrix} \overline{I}_u \\ \overline{I}_w \end{vmatrix}$$

From adding the coupled link in Example 3.8, we have

$$\begin{vmatrix} I_{14} \\ I'_{14} \end{vmatrix} = \begin{vmatrix} 3.75 & -1.25 \\ -1.25 & 3.75 \end{vmatrix} \begin{vmatrix} V_1 - V_4 \\ V_1 - V_4 \end{vmatrix}$$

The matrix $\overline{K}$ is

$$\overline{K} = \begin{vmatrix} Y_{14,14} & -Y_{14,14} & -Y_{14,14a} & -Y_{14,14a} \\ -Y_{14,14} & Y_{14,14} & -Y_{14,14a} & Y_{14,14a} \\ Y_{14a,14} & -Y_{14a,14} & Y_{14a,14a} - 1/Z_{14a,14a} & Y_{14a,14a} + 1/Z_{14a,14a} \\ -Y_{14a,14} & -Y_{14a,14a} & -Y_{14a,14a} + 1/Z_{14a,14a} & Y_{14a,14a} - 1/Z_{14a,14a} \end{vmatrix}$$

Substituting the values,

$$\overline{K} = \begin{vmatrix} 3.75 & 3.75 & -1.25 & 1.25 \\ -3.75 & 3.75 & 1.25 & -1.25 \\ -1.25 & 1.25 & 0.4167 & -0.4167 \\ 1.25 & -1.25 & -0.4167 & 0.4167 \end{vmatrix}$$

From Equation 3.80,

$$\overline{I}_w = \begin{vmatrix} I \\ -I \\ I' \\ -I' \end{vmatrix} = \overline{K} \begin{vmatrix} V_1 \\ V_4 \\ V_1 \\ V_4 \end{vmatrix}$$

This gives

$$\begin{vmatrix} I+I' \\ -I-I' \end{vmatrix} = \begin{vmatrix} 1.6667 & -1.6667 \\ -1.6667 & 1.6667 \end{vmatrix} \begin{vmatrix} V_1 \\ V_4 \end{vmatrix}$$

Therefore, from equivalence of currents, we apply Equations 3.83 through 3.86.

$$\text{New } \overline{C}', \overline{I}_u \neq 0$$

$$\overline{V}_w = (1 - \overline{D}\,\overline{K})^{-1}\overline{C}\,\overline{I}_u$$

$$\begin{vmatrix} V_1 \\ V_4 \end{vmatrix}_2 = \left[\begin{vmatrix} 1 & 0 \\ 0 & 1 \end{vmatrix} - \begin{vmatrix} 0.0184 & 0.0142 \\ 0.0142 & 0.1292 \end{vmatrix}\begin{vmatrix} 1.6667 & -1.6667 \\ -1.6667 & 1.6667 \end{vmatrix}\right]^{-1} \times \begin{vmatrix} 0.0124 & 0.0101 \\ 0.0277 & 0.0442 \end{vmatrix}\begin{vmatrix} 1 \\ 0 \end{vmatrix}$$

This is equal to

$$\begin{vmatrix} 1.0087 & -0.0087 \\ -0.2392 & 1.2392 \end{vmatrix}\begin{vmatrix} 0.0124 \\ 0.0277 \end{vmatrix} = \begin{vmatrix} 0.0123 \\ 0.0314 \end{vmatrix}$$

Similarly,

$$\begin{vmatrix} V_1 \\ V_4 \end{vmatrix}_3 = \left[\begin{vmatrix} 1 & 0 \\ 0 & 1 \end{vmatrix} - \begin{vmatrix} 0.0184 & 0.0142 \\ 0.0142 & 0.1292 \end{vmatrix}\begin{vmatrix} 1.6667 & -1.6667 \\ -1.6667 & 1.6667 \end{vmatrix}\right]^{-1}$$

$$\times \begin{vmatrix} 0.0124 & 0.0101 & 0.0277 & 0.0442 \end{vmatrix}\begin{vmatrix} 0 \\ 1 \end{vmatrix}$$

$$= \begin{vmatrix} 1.0087 & -0.0087 \\ -0.2392 & 1.2392 \end{vmatrix}\begin{vmatrix} 0.0101 \\ 0.0442 \end{vmatrix} = \begin{vmatrix} 0.0098 \\ 0.0523 \end{vmatrix}$$

Therefore, the new $\overline{C}^t$ is

$$\overline{C}' = \begin{vmatrix} 0.0123 & 0.0098 \\ 0.0314 & 0.0523 \end{vmatrix}$$

$$\text{New } \overline{A}',\ \overline{I}_u \neq 0$$

$$\overline{V}_u = \overline{A}\,\overline{I}_u + \overline{B}\,\overline{K}\,\overline{V}_w$$

Thus,

$$\begin{vmatrix} V_2 \\ V_3 \end{vmatrix}_2 = \begin{vmatrix} 0.0665 & 0.0431 \\ 0.0431 & 0.0782 \end{vmatrix}\begin{vmatrix} 1 \\ 0 \end{vmatrix} + \begin{vmatrix} 0.0124 & 0.0277 \\ 0.0101 & 0.0442 \end{vmatrix}\begin{vmatrix} 1.6667 & -1.6667 \\ -1.6667 & 1.6667 \end{vmatrix}\begin{vmatrix} 0.0123 \\ 0.0314 \end{vmatrix} = \begin{vmatrix} 0.067 \\ 0.0442 \end{vmatrix}$$

Similarly,

$$\begin{vmatrix} V_2 \\ V_3 \end{vmatrix}_3 = \begin{vmatrix} 0.0442 \\ 0.0806 \end{vmatrix}$$

Thus, the new $\overline{A}'$ is

$$\overline{A}' = \begin{vmatrix} 0.0670 & 0.0442 \\ 0.0442 & 0.0806 \end{vmatrix}$$

$$\text{New } \overline{D}',\ \overline{I}_u = 0$$

$$\overline{V}_u = (1 - \overline{D}\overline{K})^{-1}\overline{D}\overline{I}_j$$

$$\begin{vmatrix} V_1 \\ V_4 \end{vmatrix}_1 = \left[\begin{vmatrix} 1 & 0 \\ 0 & 1 \end{vmatrix} - \begin{vmatrix} 0.0184 & 0.0142 \\ 0.0142 & 0.1292 \end{vmatrix}\begin{vmatrix} 1.6667 & -1.6667 \\ -1.6667 & 1.6667 \end{vmatrix}\right]^{-1}\begin{vmatrix} 0.0184 & 0.0142 \\ 0.0142 & 0.1292 \end{vmatrix}\begin{vmatrix} 1 \\ 0 \end{vmatrix}$$

$$= \begin{vmatrix} 0.0184 \\ 0.0132 \end{vmatrix}$$

Similarly,

$$\begin{vmatrix} V_1 \\ V_4 \end{vmatrix}_4 = \begin{vmatrix} 0.0132 \\ 0.1567 \end{vmatrix}$$

Therefore, $\overline{D}'$ is

$$\overline{D}' = \begin{vmatrix} 0.0184 & 0.0132 \\ 0.0132 & 0.1567 \end{vmatrix}$$

$$\text{New } \overline{B}',\ \overline{I}_u = 0$$

$$\overline{V}_u = \overrightarrow{B}\overline{K}\,\overline{V}_w + \overline{B}\,\overline{I}_j$$

$$\begin{vmatrix} V_2 \\ V_3 \end{vmatrix}_1 = \begin{vmatrix} 0.01214 & 0.0277 \\ 0.0101 & 0.0442 \end{vmatrix}\begin{vmatrix} 1.6667 & -1.6667 \\ -1.6667 & 1.6667 \end{vmatrix}\begin{vmatrix} 0.0184 \\ 0.0132 \end{vmatrix} + \begin{vmatrix} 0.0124 \\ 0.0101 \end{vmatrix}$$

$$= \begin{vmatrix} 0.0123 \\ 0.0098 \end{vmatrix}$$

Similarly,

$$\begin{vmatrix} V_2 \\ V_3 \end{vmatrix}_4 = \begin{vmatrix} 0.0314 \\ 0.0523 \end{vmatrix}$$

The new $\overline{B}'$ is

$$\overline{B}' = \begin{vmatrix} 0.0123 & 0.0314 \\ 0.0098 & 0.0523 \end{vmatrix}$$

Substituting these values, the impedance matrix after removal of the coupled line is

	2	3	1	4
2	0.0670	0.0442	0.0123	0.0314
3	0.0442	0.0806	0.0098	0.0523
1	0.0123	0.0098	0.0184	0.0132
4	0.0314	0.0523	0.0132	0.1567

Rearranging in the original form:

	1	2	3	4
1	0.0184	0.0123	0.0098	0.0132
2	0.0123	0.0670	0.0442	0.0314
3	0.0098	0.0442	0.0806	0.0523
4	0.0132	0.0314	0.0523	0.1567

Referring to Example 3.8, this is the same matrix before the coupled link between 1 and 4 was added. This verifies the calculation.

3.9 Solution of Large Network Equations

The bus impedance method demonstrates the ease of calculations throughout the distribution system, with simple manipulations. Yet it is a full matrix and requires storage of each element. The *Y*-matrix of a large network is very sparse and has a large number of zero elements. In a large system the sparsity may reach 90%, because each bus is connected to only a few other buses. The sparsity techniques are important in matrix manipulation and are covered in Appendixes A and D. Some of these matrix techniques are

- Triangulation and factorization: Crout's method, bifactorization, and product form.
- Solution by forward–backward substitution.
- Sparsity and optimal ordering.

A matrix can be factored into lower, diagonal, and upper form called LDU form. This is of special interest. This formation always requires less computer storage. The sparse techniques exhibit a distinct advantage in computer time required for the solution of a network and can be adapted to system changes, without rebuilding these at every step.

Problems

3.1 For the network in Figure P3.1, draw its graph and specify the total number of nodes, branches, buses, basic loops, and cut-sets. Form a tree and a cotree. Write the bus admittance matrix by direct inspection. Also, form a reduced bus incidence admittance matrix, and form a bus admittance matrix using Equation 3.30. Write the basic loop incidence impedance matrix and form loop impedance matrix, using Equation 3.31.

3.2 For the network shown in Figure P3.2, draw its graph and calculate all the parameters, as specified in Problem 3.1. The self-impedances are as shown in Figure P3.2. The mutual impedances are as follows:

Buses 1–3: 0.2 ohm; buses 2–5 = 0.3 ohm

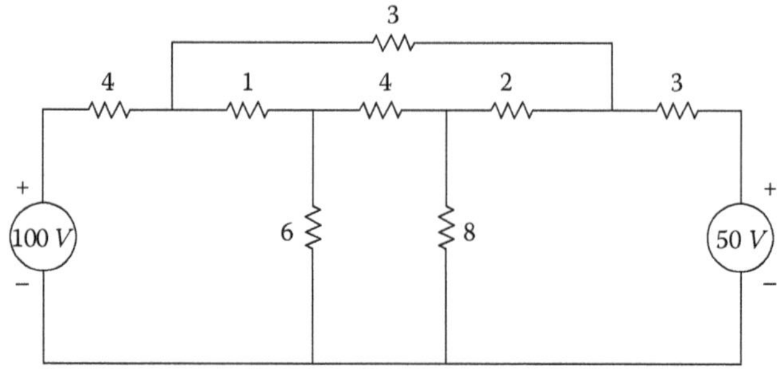

FIGURE P3.1
Network for Problem 3.1.

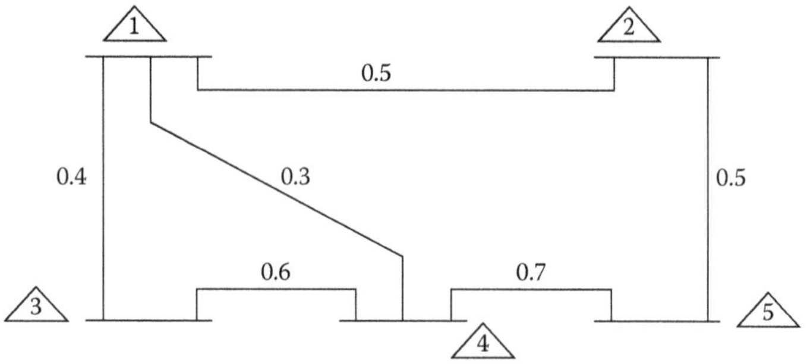

FIGURE P3.2
Network for Problem 3.2.

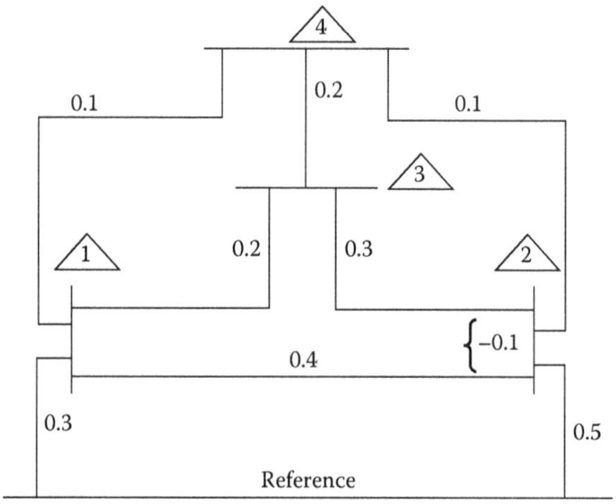

FIGURE P3.3
Network for Problems 3.3 through 3.8.

3.3 Figure P3.3 shows the positive and negative sequence network of a power system. Form the bus impedance matrix by a step-by-step buildup process, as illustrated in Example 3.8.

3.4 A double line-to-ground fault occurs at bus 2 in the network of Figure P3.3. Find the fault current. Assume for the simplicity of hand calculations that the zero sequence impedance network is the same as the positive and negative sequence impedance.

3.5 In Problem 3.4, calculate all the bus voltages.

3.6 In Problem 3.4, calculate the fault currents flowing between nodes 1–2, 4–2, 3–1, and a 3–4, also the current flowing between node 0 and bus 1.

3.7 Remove the coupled element of 0.4 between buses 3 and 4 and reform the bus impedance matrix.

3.8 A single line-to-ground fault occurs at bus 4 in Problem 3.7, after the impedance matrix is reformed. Calculate the fault current and fault voltages at buses 1, 2, 3, and 4, and the currents flowing between all buses and the current in the ground circuit from buses 1 and 2.

Bibliography

Anderson, P.M. *Analysis of Faulted Power Systems*. Ames, IA: Iowa State Press, 1973.

Arrillaga, J., C.P. Arnold, and B.J. Harker. *Computer Modeling of Electrical Power Systems*. New York: Wiley, 1983.

Bergen, R. and V. Vittal. *Power System Analysis*, 2nd edn. Upper Saddle River, NJ: Prentice Hall, 1999.

Brown, H.E. *Solution of Large Networks by Matrix Methods*. New York: Wiley, 1975.

Brown, H.E. and C.E. Parson. Short-circuit studies of large systems by the impedance matrix method. *Proc. PICA* 1967, pp. 335–346.

Dong, Z., P. Zhang, J. Ma, and J. Zhao. *Engineering Techniques in Power Systems Analysis*. Springer, 2010.

Heydt, G.T. *Computer Analysis Methods for Power Systems*. New York: Macmillan, 1986.

Maron, W.J. *Numerical Analysis*. New York: Macmillan, 1987.

Stagg, G.W. and A.H. El-Abiad. *Computer Methods in Power Systems Analysis*. New York: McGraw-Hill, 1968.

4

Current Interruption in AC Networks

Current interruption in high-voltage ac networks has been intensively researched since the introduction of high-voltage transmission lines. These advances can be viewed as an increase in the breaker-interrupting capacity per chamber or a decrease in the weight with respect to interrupting capacity. Fundamental electrical phenomena occurring in the electrical network and the physical aspects of arc interruption processes need to be considered simultaneously. The phenomena occurring in an electrical system and the resulting demands on the switchgear can be well appreciated and explained theoretically, yet no well-founded and generally applicable theory of the processes in a circuit breaker itself exists. Certain characteristics have a different effect under different conditions, and care must be applied in generalizations.

The interruption of short circuits is not the most frequent duty a circuit breaker has to perform, but this duty subjects it to the greatest stresses for which it is designed and rated. As a short circuit represents a serious disturbance in the electrical system, a fault must be eliminated rapidly by isolating the faulty part of the system. Stresses imposed on the circuit breaker also depend on the system configuration and it is imperative that the fault isolation takes place successfully and that the circuit breaker itself is not damaged during fault interruption and remains serviceable.

This chapter explores the various fault types, their relative severity, effects on the electrical system, and the circuit breaker itself. The basic phenomenon of interruption of short-circuit currents is derived from the electrical system configurations and the modifying effect of the circuit breaker itself. This chapter shows that short-circuit calculations according to empirical methods in IEC or ANSI/IEEE standards do not and cannot address all the possible applications of circuit breakers. It provides a background in terms of circuit interruption concepts and paves a way for better understanding of the calculation methods to follow in the following chapters.

The two basic principles of interruption are (1) high-resistance interruption or an ideal rheostatic circuit breaker and (2) low-resistance or zero-point arc extinction. Direct current circuit breakers also employ high-resistance arc interruption; however, emphasis in this chapter is on ac current interruption.

4.1 Rheostatic Breaker

An ideal rheostatic circuit breaker inserts a constantly increasing resistance in the circuit until the current to be interrupted drops to zero. The arc is extinguished when the system voltage can no longer maintain the arc, because of high-voltage drops. The arc length is increased and the arc resistance acquires a high value. The energy stored in the system is gradually dissipated in the arc. The volt–ampere characteristic of a *steady* arc is given by

$$V_{arc} = \text{Anode voltage} + \text{cathode voltage} + \text{voltage across length of arc}$$

$$A + \frac{C}{I_{arc}} + \left(B + \frac{D}{I_{arc}}\right)d \tag{4.1}$$

where
I_{arc} is the arc current
V_{arc} is the voltage across the arc
d is the length of the arc
A, B, C, and D are constants

For small arc lengths, the voltage across the arc length can be neglected:

$$V_{arc} = A + \frac{C}{I_{arc}} \tag{4.2}$$

The voltage across the arc reduces as the current increases. The energy dissipated in the arc is

$$E_{arc} = \int_0^t iv\, dt \tag{4.3}$$

where
$i = i_m \sin \omega t$ is current
$v = ir$ voltage in the arc

Equation 4.3 can be written as

$$E_{arc} = \int_0^t i_m^2 r \sin^2 \omega t\, dt \tag{4.4}$$

The approximate variation of arc resistance, r, with time, t, is obtained for different parameters of the arc by experimentation and theoretical analysis.

In a rheostatic breaker, if the arc current is assumed to be constant, the arc resistance can be increased by increasing the arc voltage. Therefore, the arc voltage and the arc resistance can be increased by increasing the arc length. The arc voltage increases until it is greater than the voltage across the contacts. At this point, the arc is extinguished. If the arc voltage remains lower, the arc will continue to burn until the contacts are destroyed.

Figure 4.1 shows a practical design of the arc-lengthening principle. The arc originates at the bottom of the arc chutes and is blown upward by the magnetic force. It is split by arc splitters, which may consist of resin-bonded plates of high-temperature fiber glass, placed perpendicular to the arc path. Blowout coils, in some breaker designs, subject the arc to a strong magnetic field, forcing it upward in the arc chutes.

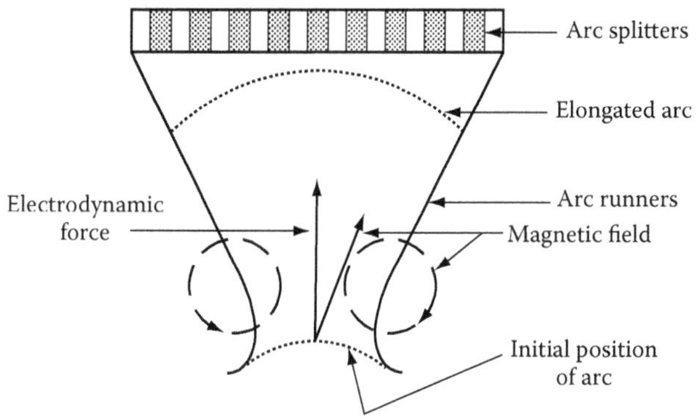

FIGURE 4.1
Principle of a rheostatic breaker and arc elongation.

This principle has been successfully employed in some commercial designs of medium-voltage breakers, and is stated here for reference only. All high-voltage breakers are current-zero breakers and further discussions are confined to this method of arc interruption.

4.2 AC Arc Interruption

The ionization of gaseous mediums and the contact space of circuit breakers depends upon the following:

- Thermal ionization.
- Ionization by collision.
- Thermal emission from contact surfaces.
- Secondary emission from contact surface.
- Photoemission and field emission.
- The arc models in circuit breakers are further influenced by the contact geometry, turbulence, gas pressure, and arc-extinguishing devices [1].

Deionization can take place by recombination, diffusion, and conduction heat. The arc behaves like a nonlinear resistor with arc voltage in phase with the arc current. At low temperature the arc has falling volt–ampere characteristics. At higher currents voltage gradient levels out, practically becoming independent of current. At high currents, arc temperature is of the order of 20,000 K and tends to be limited by radiant heat transfer. The heat transfer and electrical conductivity have nearly constant values within the arc column.

It is rather perplexing that when the arc is cooled the temperature increases. This happens due to the reduction in the diameter of the core, which results in higher current density (of the order of several thousand amperes/cm^2).

High- and low-pressure arcs can be distinguished. High-pressure arcs exist at pressures above atmosphere and appear as a bright column, characterized by a small burning core at high temperatures, of the order of 20,000 K. In low pressure (vacuum arcs), the arc voltages are low, of the order of 40 V, and the positive column of the arc is solely influenced by the electrode material, while in high-pressure arcs it is made up of ionized gasses from the arc's surrounding medium.

4.2.1 Arc Interruption Theories

There are a number of arc interruption theories, and the Cassie Mayer theory briefly discussed below seem to be practical and employed in circuit-breaker arc interruption models.

4.2.1.1 Cassie's Theory

A differential equation describing the behavior of arc was presented by Cassie in 1939 [2]. It assumes a constant temperature across the arc diameter. As the current varies, so does the arc cross section, but not the temperature inside the arc column. Under given assumptions the conductance G of the model is proportional to current, so that the steady state voltage gradient E_0 is fixed. Define the time constant:

$$\theta = \frac{Q}{N} \tag{4.5}$$

where
Q is the energy storage capability
N is the finite rate of energy loss

The following is simplified Cassie's equation:

$$\frac{\mathrm{d}}{\mathrm{d}t}(G^2) = \frac{2}{\theta}\left(\frac{I}{E_0}\right)^2 \tag{4.6}$$

For high-current region, there is good agreement with the model, but, for current-zero region, agreement is good only for high rates of current decay.

4.2.1.2 Mayr's Theory

O. Mayr considered an arc column where arc diameter is constant and where arc temperature varies with time and radial dimensions. The decay of the temperature of arc is assumed due to thermal conduction and electrical conductivity of arc is dependent upon temperature:

$$\frac{1}{G}\frac{\mathrm{d}G}{\mathrm{d}t} = \frac{1}{\theta}\left(\frac{EI}{N_0} - 1\right) \tag{4.7}$$

where

$$\theta = \frac{Q_0}{N_0} \tag{4.8}$$

The validity of the theory during current-zero periods is acknowledged.

4.2.1.3 Cassie–Mayr Theory

Mayr assumed arc temperature of 6000 K, but it is recognized to be in excess of 20,000 K. At these high temperatures there is a linear increase in gas conductivity. Assuming that before current zero the current is defined by driving circuit and that after current zero, the voltage across the gap is determined by arc circuit, we write the following two equations:

1. Cassie's period prior to current zero:

$$\frac{\mathrm{d}}{\mathrm{d}t}\left(\frac{1}{R^2}\right)+\frac{2}{\theta}\left(\frac{1}{R^2}\right)=\frac{2}{\theta}\left(\frac{1}{E_0}\right)^2 \tag{4.9}$$

2. Mayr's period around current zero:

$$\frac{\mathrm{d}R}{\mathrm{d}t}-\frac{R}{\theta}=-\frac{e^2}{\theta N_0} \tag{4.10}$$

4.3 Current-Zero Breaker

The various arc-quenching mediums in circuit breakers for arc interruption have different densities, thermal conductivities, dielectric strengths, arc time constants, etc. In a current-zero circuit breaker, the interruption takes place during the passage of current through zero. At the same time the electrical strength of the break-gap increases so that it withstands the recovery voltage stress. All high-voltage breakers, and high-interrupting capacity breakers, whatever may be the arc-quenching medium (oil, air, or gas), use current-zero interruption. In an ideal circuit breaker, with no voltage drop before interruption, the arc energy is zero. Modern circuit breakers approach this ideal on account of short arc duration and low arc voltage.

The circuit breakers are called upon to interrupt currents of the order of tens of kilo amperes or even more. Generator circuit breakers having an interrupting capability of 250 kA rms symmetrical are available. At the first current zero, the electrical conductivity of the arc is fairly high and since the current-zero period is very short, the reignition cannot be prevented. The time lag between current and temperature is commonly called arc hysteresis.

Figure 4.2b shows a typical short-circuit current waveform in an inductive circuit of Figure 4.2a; the short circuit is applied at $t=0$. The short-circuit current is limited by the impedance $R+j\omega L$ and is interrupted by breaker B. The waveform shows asymmetry as discussed in Chapter 1. At $t=t_2$, the contacts start parting. The time t_2-t_0 is termed the contact-parting time, since it takes some finite time for the breaker-operating mechanism to set in motion and the protective relaying to signal a fault condition. The contact-parting time is the sum of tripping delay, which is taken in the standards as half a cycle (t_1-t_0) and contact-parting time (t_2-t_1).

As the contacts start parting, an arc is drawn, which is intensely cooled by the quenching medium (air, SF_6, or oil). The arc current varies sinusoidally for a short duration. As the contacts start parting, the voltage across these increases. This voltage is the voltage drop across the arc during the arcing period and is shown exaggerated in Figure 4.2c for clarity. The arc is mostly resistive and the voltage in the arc is in phase with the current.

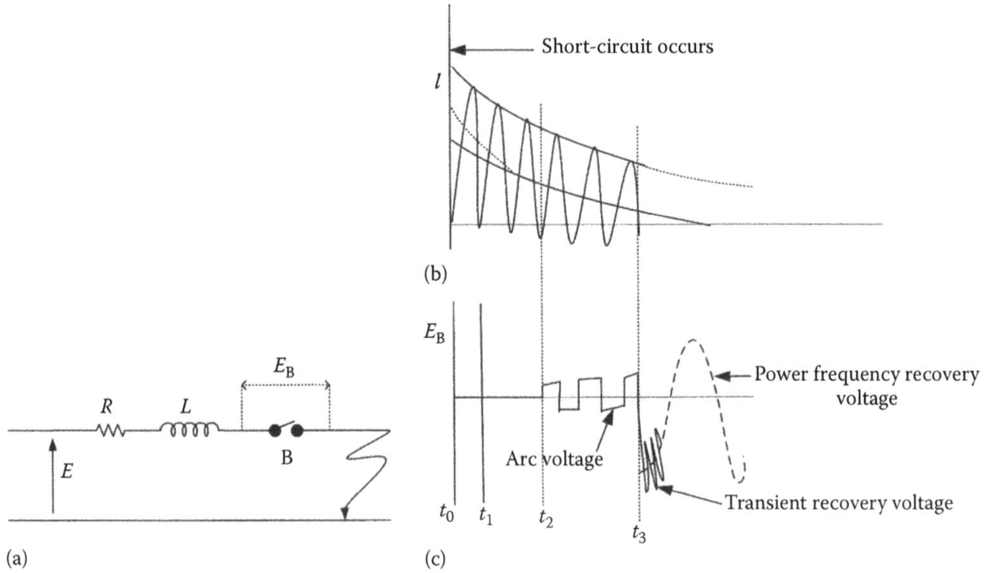

FIGURE 4.2
Current interruption in a current-zero breaker: (a) circuit diagram showing a mainly inductive circuit and short circuit at terminals; (b) current waveform; (c) voltage during interruption and after fault clearance.

The peculiar shape of the arc voltage shown in Figure 4.2c is the result of the volt–ampere characteristic of the arc, and at a particular current zero, the dielectric strength of the arc space is sufficiently high for the arcing to cease. The contacts part at high speed, to prevent continuation of the arc. When the dielectric strength builds and a current zero occurs, the current is interrupted. In Figure 4.2c it occurs at $t = t_3$, and the interval $t_3 - t_2$ is the arcing time. The interval $t_3 - t_0$ is the total interrupting time. With modern high-voltage breakers, it is as low as two cycles or even lower, based on the system power frequency. The ANSI standard [3] defines the rated interrupting time of a circuit breaker as the time between trip circuit energization and power arc interruption on an opening operation, and it is used to classify breakers of different speeds. The rated interrupting time may be exceeded at low values of current and for close–open operations; also, the time for interruption of the resistor current for interrupters equipped with resistors may exceed the rated interrupting time. The increase in interrupting time on close–open operation may be important from the standpoint of line damage or possible instability.

The significance of interrupting time on breaker-interrupting duty can be appreciated from Figure 4.2b. As the short-circuit current consists of decaying ac and dc components, the *faster is the breaker, the greater the asymmetry, and the interrupted current.*

The interrupting current at final arc extinction is asymmetrical in nature, consisting of an ac component and a dc component. The rms value of this asymmetrical current is given by

$$I_i = \sqrt{(\text{rms of ac component})^2 + (\text{dc component})^2} \tag{4.11}$$

IEC specifications term this as the breaking current.

4.4 Transient Recovery Voltage

The essential problem of current interruption consists in rapidly establishing an adequate electrical strength across the break after current zero, so that restrikes are eliminated. Whatever may be the breaker design, it can be said that it is achieved in most interrupting mediums, i.e., oil, air blast, or SF_6 by an intense blast of gas. The flow velocities are always governed by aerodynamic laws. However, there are other factors that determine the rate of recovery of the dielectric medium: nature of the quenching gases, mode of interaction of pressure and velocity of the arc, arc control devices, contact shape, and number of breaks per phase. Interruption in vacuum circuit breakers is entirely different and discussed in Section 4.16.

At the final arc interruption, a high-frequency oscillation superimposed on the power frequency appears across the breaker contacts. A short-circuit current loop is mainly inductive, and the power frequency voltage has its peak at current zero; however, a sudden voltage rise across the contacts is prevented by the inherent capacitance of the system, and in the simplest cases a transient of the order of some hundreds to 10,000 c/s occurs. It is termed the natural frequency of the circuit. Figure 4.3 shows the recovery voltage profile after final current extinction. The two components of the recovery voltage, (1) a high-frequency damped oscillation, and (2) the power frequency recovery voltages, are shown. The high-frequency component is called the transient recovery voltage (TRV) and sometimes the restriking voltage. Its frequency is given by

$$f_n = \frac{1}{2\pi\sqrt{LC}} \tag{4.12}$$

where

f_n is the natural frequency

L and C are equivalent inductance and capacitance of the circuit

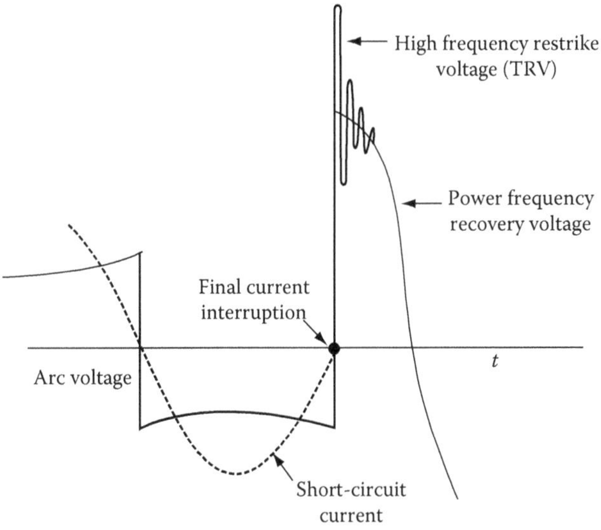

FIGURE 4.3
Final current interruption and the resulting recovery voltages.

If the contact space breaks down within a period of a quarter of a cycle of initial arc extinction, the phenomena are called reignitions and if the breakdown occurs after a quarter cycle, the phenomena are called restrikes.

The transient oscillatory component subsides in a few microseconds and the power frequency component continues.

4.4.1 First Pole to Clear Factor

TRV refers to the voltage across the first pole to clear, because it is generally higher than the voltage across the other two poles of the breaker, which clear later. Consider a three-phase ungrounded fault; the voltage across the breaker phase, first to clear, is 1.5 times the phase voltage (Figure 4.4). The arc interruption in three phases is not simultaneous, as the three phases are mutually 120° apart. Thus, theoretically, the power frequency voltage of the first pole to clear is 1.5 times the phase voltage. It may vary from 1.5 to 2, rarely exceeding 3, and can be calculated using symmetrical components. The first pole to clear factor is defined as the ratio of rms voltage between the faulted phase and unfaulted phase and phase-to-neutral voltage with the fault removed. Figure 4.4 shows first pole to clear factors for three-phase terminal faults. The first pole to clear factor for a three-phase fault with ground contact is calculated as

$$1.5\frac{2X_0/X_1}{1+2X_0/X_1} \tag{4.13}$$

where X_1 and X_2 are the positive and zero sequence reactances of the source side. Also, in Figure 4.4, Y_1 and Y_2 are the sequence reactances of the load side. Figure 4.5 illustrates the slopes of tangents to three TRV waveforms of different frequencies. As the

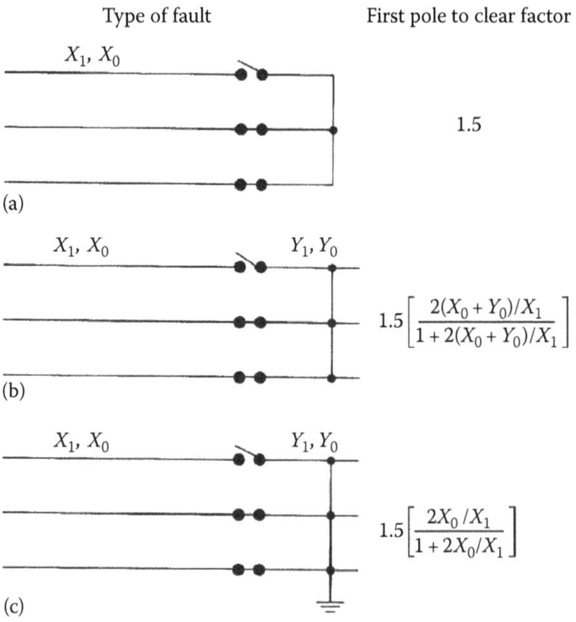

FIGURE 4.4
First pole to clear factor for three-phase faults: (a) a three-phase terminal fault with no connection to ground; (b) three-phase fault with no contact to ground and an extension of the load-side circuit; (c) three-phase fault with ground contact.

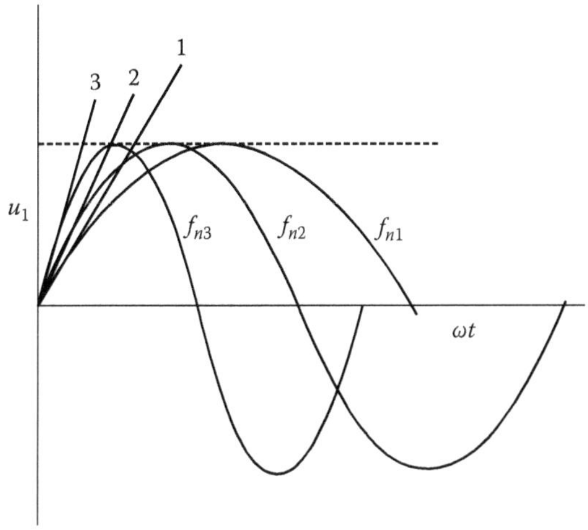

FIGURE 4.5
Effect of frequency of transient recovery voltage (TRV) on the rate of rise of recovery voltage (RRRV).

natural frequency rises, the rate of rise of recovery voltage (RRRV) increases. Therefore, it can be concluded as follows:

1. Voltage across breaker contacts rises slowly, as RRRV decreases.
2. There is a swing beyond the recovery voltage value, the amplitude of which is determined by the circuit breaker and breaker damping.
3. The higher is the natural frequency of the circuit, the lower is the breaker-interrupting rating.
4. Interrupting capacity/frequency characteristics of the breaker should not fall below that of the system.

The TRV is affected by many factors, among which the power factor of the current being interrupted is important. At zero power factor, maximum voltage is impressed across the gap at the instant of current zero, which tends to reignite the arc in the hot arc medium.

TRV can be defined by specifying the crest and the time to reach the crest, and alternatively, by defining the segments of lines which envelop the TRV waveform.

The steepest rates of rise in a system are due to short circuits beyond transformers and reactors which are connected to a system of high short-circuit power. In these cases, the capacitance which retards the voltage rise is small; however, the breaking capacity of the breaker for such faults need only be small compared to the short-circuit power of the system, as the current is greatly reduced by the reactance of the transformers and reactors. It means that, in most systems, high short-circuit levels of current and high natural frequencies may not occur simultaneously.

The interrupting capacity of every circuit breaker decreases with an increase in natural frequency. It can, however, be safely said that the interrupting (or breaking) capacity for the circuit breakers decreases less rapidly with increasing natural frequency than the short-circuit power of the system. The simplest way to make the breaking capacity independent of the natural frequency is to influence the RRRV across breaker contacts by resistors, which is discussed further. Yet, there may be special situations where the interrupting rating of a breaker may have to be reduced or a breaker of higher interrupting capacity may be required.

A circuit of a single-frequency transient occurs for a terminal fault in a power system composed of distributed capacitance and inductances. A terminal fault is defined as a fault close to the circuit breaker, and the reactance between the fault and the circuit breaker is negligible. TRV can vary from low to high values, in the range of 20–10,000 Hz.

A circuit with inductance and capacitance on both sides of the circuit breaker gives rise to a double-frequency transient. After fault interruption, both circuits oscillate at their own frequencies and a composite double-frequency transient appears across the circuit-breaker contacts. This can occur for a short-line fault. Recovery may feature traveling waves, depending on the faulted components in the network.

In the analyses to follow, we will describe IEC [4] methods of estimating the TRV wave shape. The ANSI/IEEE methods are described in the rating structure of ANSI-rated breakers in Chapter 5. The IEC methods are simpler for understanding the basic principles and these are well defined in the IEC standard.

Figure 4.6 shows the basic parameters of the TRV for a terminal fault in a simplified network. Figure 4.6a shows power system constants, i.e., resistance, inductance, and capacitances. The circuit shown may well represent the π model of a transmission line (see Chapter 10). Figure 4.6b shows the behavior of the recovery voltage, transient component, and power frequency component. The amplitude of the power frequency component is given by

$$\alpha\sqrt{2}u_0 \tag{4.14}$$

where

α depends on the type of fault and the network

u_0 is the rated system rms voltage

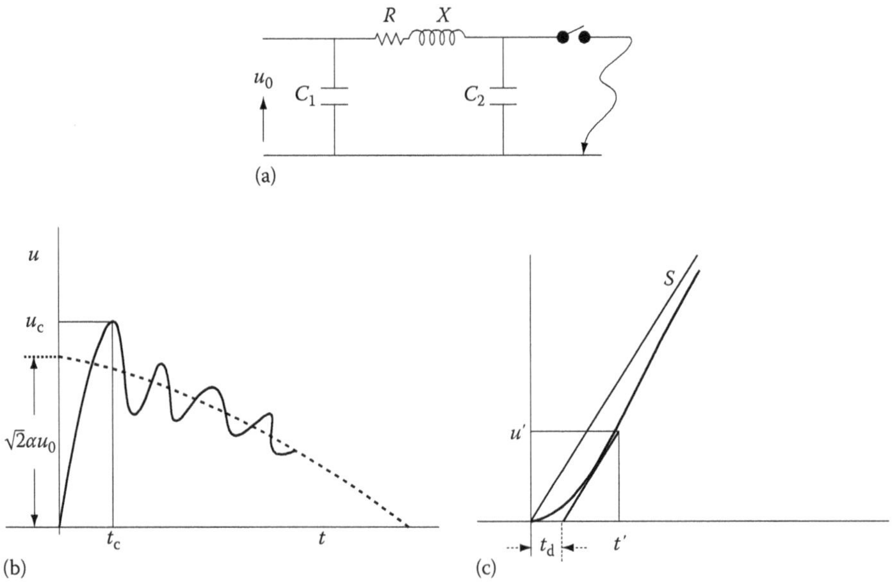

FIGURE 4.6
Basic parameters of the recovery voltage profile in a simplified terminal fault: (a) system configuration—R, X, C_1, and C_2 are system resistance, reactance, and shunt capacitances, respectively; (b) recovery voltage profile; (c) initial TRV curve, delay line, and RRRV shown as S.

The rate of rise of recovery voltage (RRRV = S) is the tangent to the TRV starting from the zero point of the unaffected or inherent transient recovery voltage (ITRV). This requires some explanation. The TRV can change by the circuit breaker design and operation. The TRV measured across terminals of two circuit breakers can be different. The power system characteristics are calculated, ignoring the effect of the breakers. This means that an ideal circuit breaker has zero terminal impedance when carrying its rated current, and when interrupting short-circuit current its terminal impedance changes from zero to infinity instantaneously at current interruption. The TRV is then called inherent transient recovery voltage (ITRV).

Figure 4.6c shows an enlarged view of the slope. Under a few microseconds, the voltage behavior may be described by the time delay t_d, which is dependent on the ground capacitance of the circuit. The time delay t_d in Figure 4.6c is approximated by

$$t_d = CZ_0 \tag{4.15}$$

where
C is the ground capacitance
Z_0 is the surge impedance

Measurements show that a superimposed high-frequency oscillation often appears. IEC specifications recommend a linear voltage rise with a surge impedance of 450 ohm and no time delay, when the faulted phase is the last to be cleared in a single line-to-ground fault. This gives the maximum TRV. It is practical to distinguish between terminal faults and short-line faults for these phenomena.

4.5 Terminal Fault

This is a fault in the immediate vicinity of a circuit breaker, which may or may not involve ground. This type of fault gives the maximum short-circuit current. There are differences in the system configuration, and the TRV profile differs widely. Two- and four-parameter methods are used in the IEC standard.

4.5.1 Four-Parameter Method

Figure 4.7 shows a representation of the TRV wave by the four-parameter method. In systems above 72.5 kV clearing terminal faults higher than 30% of the rating will result in TRV characteristics that have a four-parameter envelope. The wave has an initial period of high rise, followed by a low rate of rise. Such waveforms can be represented by the four-parameter method:

u_1 = first reference voltage (kV)

t_1 = time to reach u_1, in μs

u_c = second reference voltage, peak value of TRV

t_2 = time to reach u_c, in μs

IEC specifies values of u_1, u_c, t_1, and t_2 for the circuit breakers. The interrupting current tests are carried out on circuit breakers with specified TRVs. The segments can be plotted,

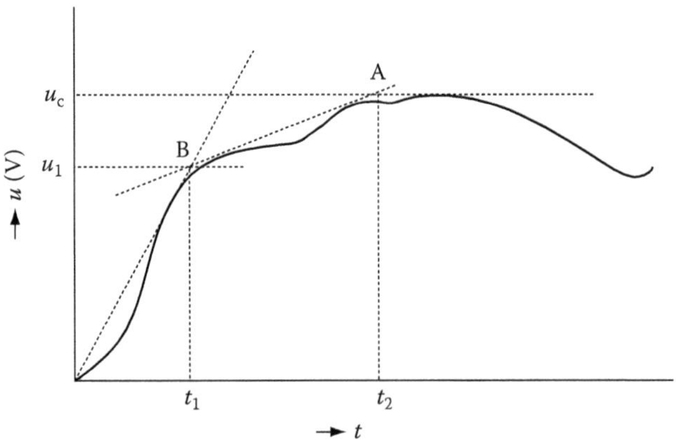

FIGURE 4.7
IEC four-parameter representation of TRV.

as shown in Figure 4.7, based on the breaker data. A table in IEEE Std. C37.011 [5] provides factors for calculating rated TRV profiles for terminal faults, based upon IEC standards.

$$u_c = k_{af} \times k_{pp} \times \sqrt{\frac{2}{3}} U_r \tag{4.16}$$

$$u_c(T10) = u_c(T100)ku_c \tag{4.17}$$

where U_r is the breaker rated voltage. The amplitude factor k is defined as the ratio of the peak recovery voltage and the power frequency voltage. $T10$ and $T100$ are percentages of interrupting capability. See Ref. [5] for further details.

$$k = \frac{u_c}{u_1} \tag{4.18}$$

The natural frequency is given by

$$f_n = \frac{10^3}{2t_2} \text{kHz} \tag{4.19}$$

4.5.2 Two-Parameter Representation

Figure 4.8 shows representation of TRV wave by two-parameter method. For terminal faults between 10% and 30% for systems above 72.5 kV and for terminal faults for systems of 72.5 kV and below, TRV can be approximately represented by a single-frequency transient [6].

u_c = peak of TRV wave (kV)
t_3 = time to reach peak (μs)

The initial rate of rise of TRV is contained within segments drawn according to the two- or four-parameter method by specifying the delay line, as shown in Figure 4.6. The rate of rise of TRV can be estimated from

$$S = 2\pi f \sqrt{2} I_k Z_r \tag{4.20}$$

where
I_k is short-circuit current
Z_r is the resultant surge impedance

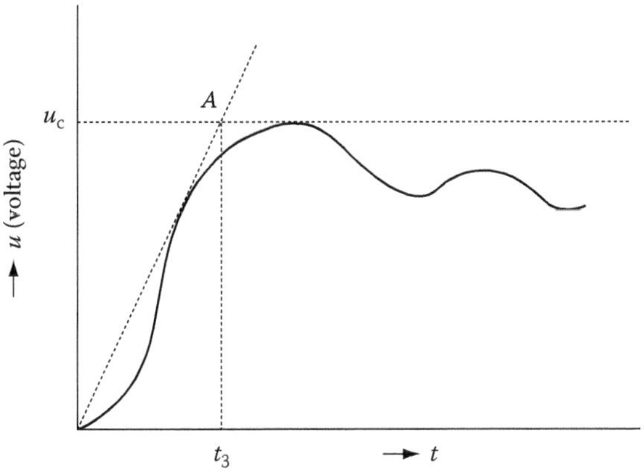

FIGURE 4.8
IEC two-parameter representation of TRV.

Z_r can be found from sequence impedances. For a three-phase ungrounded fault, and with n equal outgoing lines,

$$Z_r = 1.5\left(\frac{Z_1}{n}\right)\frac{2Z_0/Z_1}{1 + 2Z_0/Z_1} \tag{4.21}$$

where
Z_1 and Z_0 are the surge impedances in positive and zero sequence of individual lines
n is the number of lines emanating from the substation

Factor 1.5 in Equation 4.21 can be 1.3. For single-frequency transients, with concentrated components, IEC tangent (rate of rise) can be estimated from

$$S = \frac{2\sqrt{2}kf_n u_0}{0.85} \tag{4.22}$$

where
f_n is the natural frequency
k is the amplitude factor

The peak value u_c in the four-parameter method cannot be found easily, due to many variations in the system configurations. The traveling waves are partially or totally reflected at the points of discontinuity of the surge impedance. A superimposition of all forward and reflected traveling waves gives the overall waveform. IEEE unapproved draft proposal [6] adopts IEC representation of the TRV.

4.6 Short-Line Fault

Faults occurring between a few and some hundreds of kilometers from the breaker are termed short-line faults. A small length of the line lies between the breaker and the fault location, Figure 4.9a. After the short-circuit current is interrupted, the breaker terminal at the line end assumes a sawtooth oscillation shape, as shown in Figure 4.9c. The rate of rise of voltage is directly proportional to the effective surge impedance (which can vary between 35

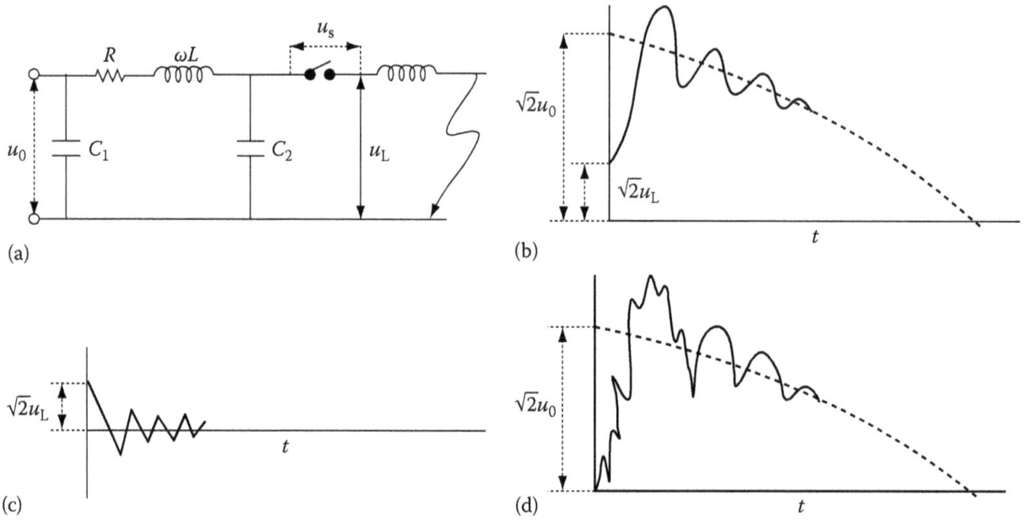

FIGURE 4.9
Behavior of TRV in a short-line fault: (a) system equivalent circuit; (b) recovery voltage on source side; (c) recovery voltage on load side; (d) voltage across the breaker contacts.

and 450 ohm, the lower value being applicable to cables) and to the rate of rise of current at current zero. The component on the supply side exhibits the same waveform as for a terminal fault (Figure 4.9b). The circuit breaker is stressed by the difference between these two voltages (Figure 4.9d). Because of the high-frequency oscillation of the line side terminal, the TRV has a very steep initial rate of rise. In many breaker designs, the short-line fault may become a limiting factor of the current-interrupting capability of the breaker.

It is possible to reduce the voltage stresses of TRV by incorporating resistances or capacitances in parallel with the contact break. SF_6 circuit breakers of single-break design up to 345 kV and 50 kA interrupting rating have been developed.

4.7 Interruption of Low Inductive Currents

A circuit breaker is required to interrupt low inductive currents of transformers at no load, high-voltage reactors or locked rotor currents of motors. On account of arc instability in a low-current region, *current chopping* can occur, irrespective of the breaker-interrupting medium. In a low-current region, the characteristics of the arc decrease, corresponding to a negative resistance which lowers the damping of the circuit. This sets up a high-frequency oscillation, depending on the LC of the circuit.

Figure 4.10a shows the circuit diagram for interruption of low inductive currents. The inductance L_2 and capacitance C_2 on the load side can represent transformers and motors. As the arc becomes unstable at low currents, the capacitances C_1 and C_2 partake in an oscillatory process of frequency:

$$f_3 = \frac{1}{2\pi\sqrt{L\dfrac{C_1C_2}{C_1+C_2}}} \tag{4.23}$$

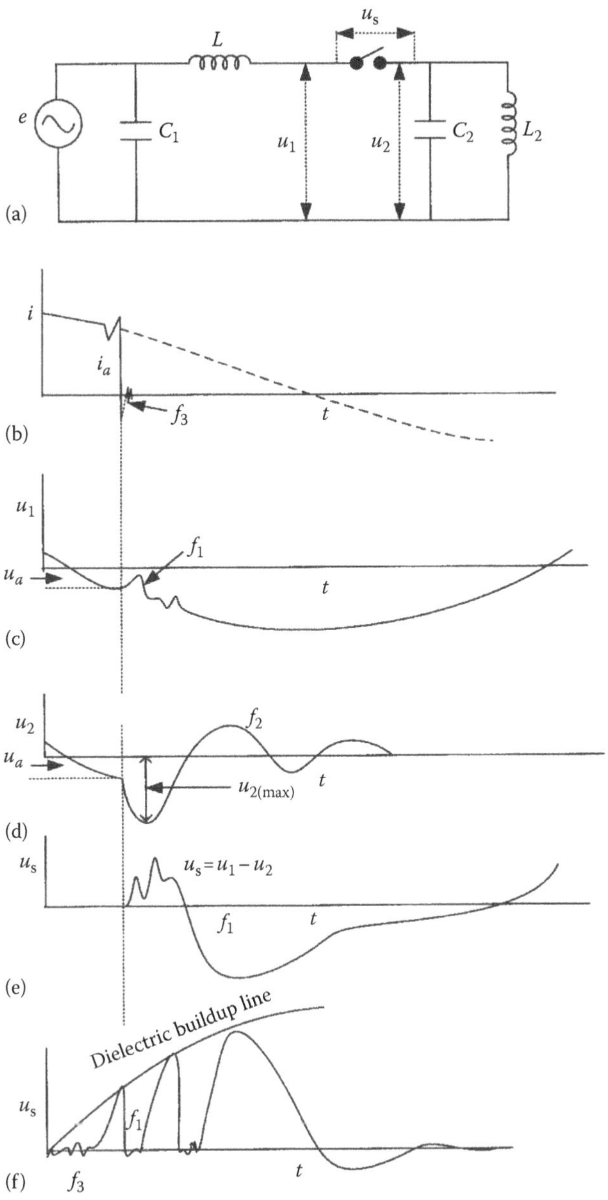

FIGURE 4.10
Interruption of low inductive currents: (a) the equivalent circuit diagram; (b) chopped current i_a at u_a; (c) source-side voltage; (d) load-side voltage; (e) voltage across breaker contacts; (f) phenomena of repeated restrikes.

Practically no current flows through the load inductance L_2. This forces a current zero, before the natural current zero, and the current is interrupted (Figure 4.10b). This interrupted current i_a is the chopped current at voltage u_a (Figure 4.10c). Thus, the chopped current is affected not only by the circuit breaker but also by the properties of the circuit. The energy stored in the load at this moment is

$$i_a^2 \frac{L_2}{2} + u_a^2 \frac{C_2}{2} \tag{4.24}$$

which oscillates at the natural frequency of the disconnected circuit:

$$f_2 = \frac{1}{2\pi\sqrt{L_2C_2}} \tag{4.25}$$

This frequency may vary between 200 and 400 Hz for a transformer. The maximum voltage on the load side occurs when all the inductive energy is converted into capacitive energy:

$$u_{2\,\max}^2 \frac{C_2}{2} = u_2^2 \frac{C_2}{2} + i_a^2 \frac{L_2}{2} \tag{4.26}$$

The source-side voltage builds up with the frequency:

$$f_1 = \frac{1}{2\pi\sqrt{L_1C_1}} \tag{4.27}$$

The frequency f lies between 1 and 5 kHz. This is shown in Figure 4.10c.

The linear expression of magnetic energy at the time of current chopping is not strictly valid for transformers and should be replaced with

$$\text{Volume of transformer core} \times \int_0^{B_m} H\,dB \tag{4.28}$$

where

B is the magnetic flux density

H the magnetic field intensity (B–H hysteresis curve)

The load-side voltage decays to zero, on account of system losses. The maximum load-side overvoltage is of concern; from Figure 4.10d and from the simplified relationship (Equation 4.26), it is given by

$$u_{2\,\max} = \sqrt{u_a^2 + i_a^2 \frac{L_2}{C_2}} \tag{4.29}$$

A similar expression applies for the supply-side voltage.

Thus, the overvoltage is dependent on the chopped current. If the current is chopped at its peak value, the voltage is zero. The chopped currents in circuit breakers have been reduced with better designs and arc control. The voltage across the supply side of the break, neglecting the arc voltage drop, is u_s and it oscillates at the frequency given by L and C_1. The voltage across the breaker contacts is $u_s = u_2 - u_1$. The supply-side frequency is generally between 1 and 5 kHz.

If the circuit-breaker voltage intersects the dielectric recovery characteristics of the breaker, reignition occurs and the process is repeated a new (Figure 4.10f). With every reignition, the energy stored is reduced, until the dielectric strength is large enough and further reignitions are prevented. Overvoltages of the order of two to four times may be produced on disconnection of inductive loads.

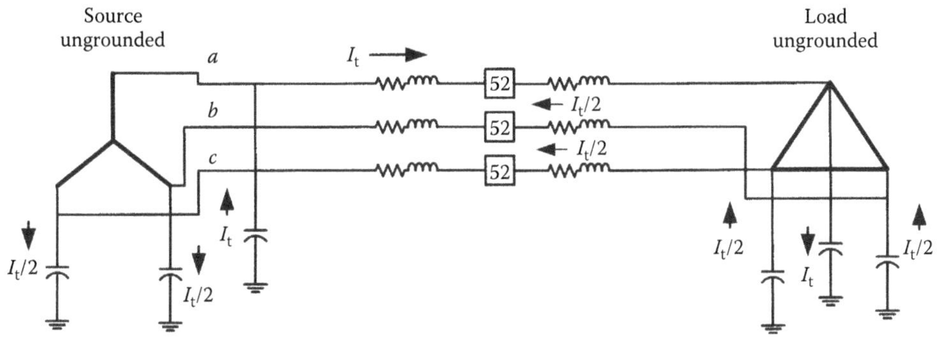

FIGURE 4.11
Circuit illustrating virtual current chopping due to reignition, restrike, or prestrike.

4.7.1 Virtual Current Chopping

This term is frequently used and needs some explanation. Reignition, restrike, or prestrike in one phase can cause high-frequency currents to flow which can get coupled to the other two phases and this phenomenon is called virtual current chopping [7,8]. This is clearly illustrated in Figure 4.11. At the instant of reignition in phase *a*, the high-frequency current flows to ground through the load capacitance of phase *a* and divides into phases *b* and *c* equally, assuming a balanced three-phase system. The power frequency current in phases *b* and *c* is approximately 0.87 times the peak value of the 60 Hz current. If the magnitude of the high-frequency currents in phases *b* and *c* ($I_t/2$) is greater than the power frequency current, the high-frequency current forces the 60 Hz current to zero. This forced current phenomenon is virtual current chopping and gives rise to overvoltages.

4.8 Interruption of Capacitive Currents

A breaker may be used for line dropping and interrupt charging currents of cables open at the far end or shunt capacitor currents. These duties impose voltage stresses on the breaker. Consider the single-phase circuit of Figure 4.12a. The distributed line capacitance is represented by a lumped capacitance C_2, or C_2 may be a power capacitor. The current and voltage waveforms of capacitance current interruption in a single pole of a circuit breaker under the following three conditions are shown:

1. Without restrike
2. With restrike
3. With restrike and current chopping

After interruption of the capacitive current, the voltage across the capacitance C_2 remains at the peak value of the power frequency voltage:

$$u_2 = \frac{\sqrt{2}u_n}{\sqrt{3}} \tag{4.30}$$

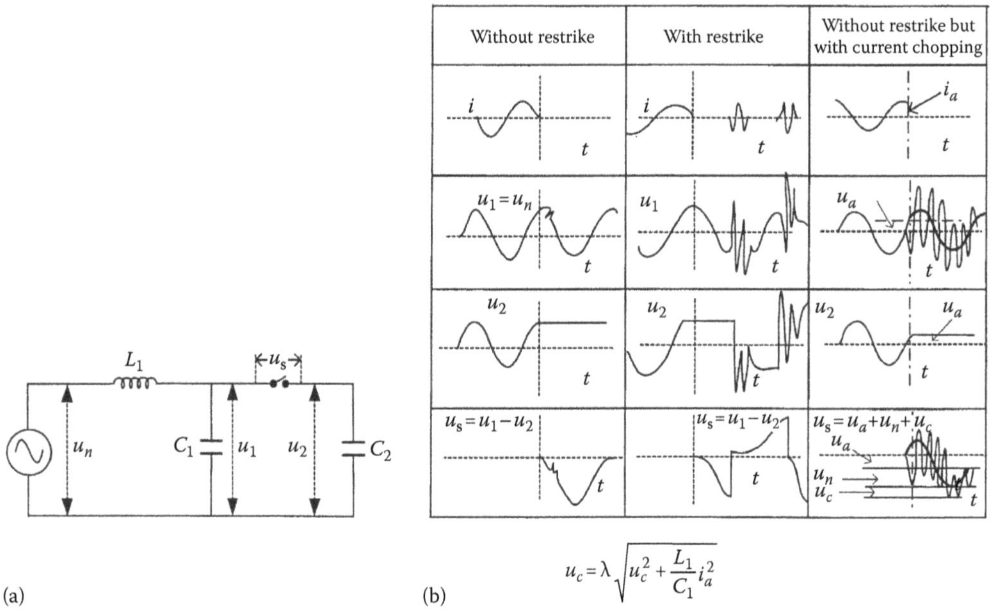

FIGURE 4.12
Interruption of capacitance current: (a) the equivalent circuit diagram; (b) current and voltage waveforms without restrike, with restrike, and with current chopping. i_a is chopping current and u_a is voltage on the disconnected side on current chopping. λ = damping factor < 1.

The voltage at the supply side oscillates at a frequency given by supply side C_1 and L_1, about the driving voltage u_n. The difference between these two voltages appears at the breaker pole. This can be more than double the rated voltage, with no prior charge on the capacitors.

If the gap across poles of a circuit breaker has not recovered enough dielectric strength, restrike may occur. As the arc bridges the parting contacts, the capacitor being disconnected is again reconnected to the supply system. This results in a frequency higher than that of the natural frequency of the source-side system being superimposed on the 60 Hz system voltage. The current may be interrupted at a zero crossing in the reignition process. Thus, the high-frequency voltage at its crest is trapped on the capacitors. Therefore, after half a cycle following the restrike, the voltage across the breaker poles is the difference between the supply side and the disconnected side, which is at the peak voltage of the equalizing process, and a second restrike may occur. Multiple restrikes can occur, pumping the capacitor voltage to 3, 5, 7, ... times the system voltage at each restrike. The multiple restrikes can terminate in two ways: (1) these may cease as the breaker parting contacts increase the dielectric strength, and (2) these may continue for a number of cycles, until these are damped out.

A distinction should be made between reignitions in less than 5 ms of current zero and reignitions at 60 Hz power frequency. Reignitions in less than 5 ms have a low voltage across the circuit breaker gap and do not lead to overvoltages.

Disconnecting a three-phase capacitor circuit is more complex. The instant of current interruption and trapped charge level depends on the circuit configuration. In an

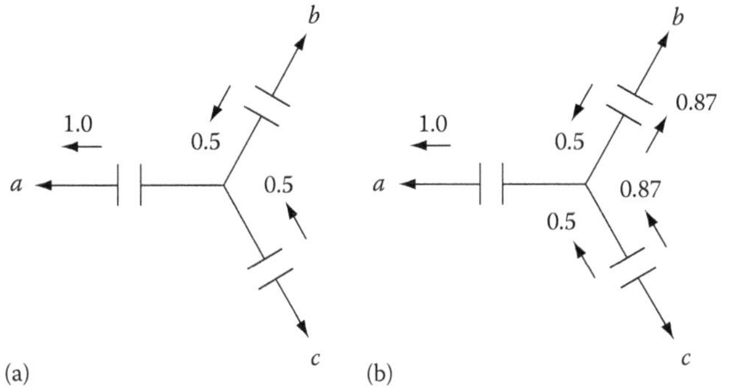

FIGURE 4.13
Sequence of creating trapped charges in a three-phase ungrounded capacitor bank: (a) first phase *a* clears; (b) phases *b* and *c* clear in series.

ungrounded three-phase wye-connected bank, commonly applied at medium- and high-voltage levels, let phase *a* current be first interrupted. This will occur when the voltage of phase *a* is at its peak. Figure 4.13a shows that phase *a* is interrupted first. The charge trapped in phase *a* is 1 per unit and that trapped in phases *b* and *c* is 0.5 per unit.

The interruption of phase *a* changes the circuit configuration and connects the capacitors in phases *b* and *c* in series. These capacitors are charged with equal and opposite polarities. The current in phases *b* and *c* will interrupt simultaneously as soon as the phase-to-phase current becomes zero. This will occur at 90° after the current interruption in phase *a*, at the crest of the phase-to-phase voltage, so that an additional charge of $\sqrt{3/2}$ is stored in the capacitors, as shown in Figure 4.13b. These charges will add to those already trapped on the capacitors in Figure 4.13a and thus voltages across capacitor terminals are

$$
\begin{aligned}
E_{ab} &= 0.634 \\
E_{bc} &= 1.73 \text{ per unit} \\
E_{ac} &= 2.37 \text{ per unit}
\end{aligned} \tag{4.31}
$$

Further escalation of voltages occurs if the phases *b* and *c* are not interrupted after 90° of current interruption in phase *a*. It is hardly possible to take into account all forms of three-phase interruptions with restrikes.

4.9 TRV in Capacitive and Inductive Circuits

The TRV on interruption of capacitive and inductive circuits is illustrated in Figure 4.14. In a capacitive circuit when the current passes through zero (Figure 4.14a), the system voltage is trapped on the capacitors. The recovery voltage, the difference between the source side and the load side of the breaker reaches a maximum of 2.0 per unit after half a cycle of current interruption (Figure 4.14b and c). The TRV oscillations are practically absent, as large capacitance suppresses the oscillatory frequency (Equation 4.12) and the rate of rise of the TRV is low. This may prompt circuit breaker contacts to interrupt when there is not enough separation between them and precipitate restrikes.

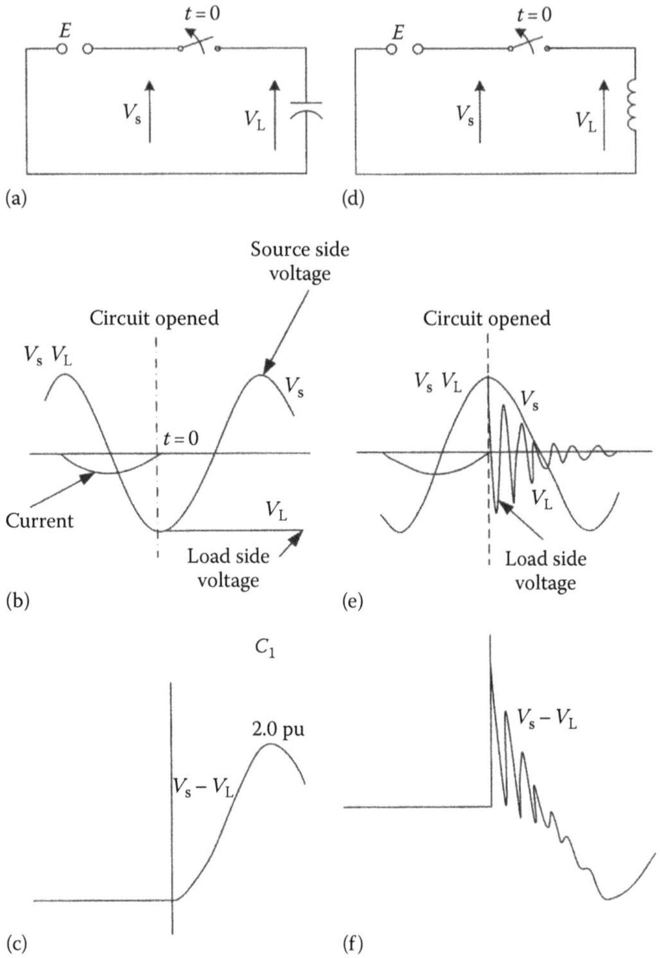

FIGURE 4.14
TRV, interruption of capacitive and inductive currents: (a), (b), and (c) interruption of capacitive current; (d), (e), and (f) interruption of inductive current.

This is not the case when disconnecting an inductive load (Figure 4.14d through f). The capacitance on the disconnected side is low and the frequency of the isolated circuit is high. The TRV is therefore oscillatory. The rate of rise of TRV after disconnection is fairly high.

Reverting to TRV on disconnection of a capacitance, IEEE standard [8] specifies that voltage across capacitor neutral, not ground, during an opening operation should not exceed 3.0 per unit for a general purpose breaker. For the definite purpose breaker rated 72.5 kV and lower, this limit is 2.5 per unit. For breakers rated 121 kV and above capacitor banks are normally grounded, and the voltage is limited to 2.0 per unit. It was observed that when interrupting capacitance current 2.0 pu voltage occurs at small contact separation and this can lead to restrikes. With no damping, the voltage will reach 3.0 pu. This suggests that one restrike is acceptable for general purpose breaker, but not for definite purpose breakers.

Some voltage controlled methods are resistance switching and surge arresters (see Section 4.13).

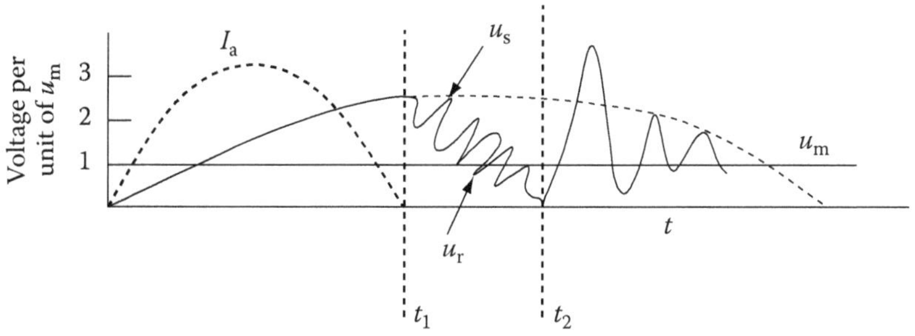

FIGURE 4.15
Voltages due to prestrike at $t = t_1$ and $t = t_2$. The inrush current is interrupted at $t = t_1$.

4.10 Prestrikes in Breakers

A prestrike may occur on closing a circuit breaker, establishing the current flow, before the contacts physically close. A prestrike occurs in a current flow at a frequency given by the inductances and capacitances of the supply circuit and the circuit being closed. In Figure 4.15, this high-frequency current is interrupted at $t = t_1$. Assuming no trapped charge on the capacitors, the voltage rises to approximately 2 per unit. A high-frequency voltage given by source reactance and stray capacitance is superimposed on the recovering bus voltage. If a second prestrike occurs at $t = t_2$, a further escalation of the bus voltage occurs. Thus, the transient conditions are similar as for restrikes; however, the voltage tends to decrease as the contacts come closer in a closing operation. In Figure 4.15, u_m is the maximum system voltage, u_r is the recovery voltage, and u_s is the voltage across the breaker contacts.

4.11 Overvoltages on Energizing High-Voltage Lines

The highest overvoltages occur when unloaded high-voltage transmission lines are energized and reenergized and this imposes voltage stresses on circuit breakers.

Figure 4.16a shows the closing of a line of surge impedance Z_0 and of length l, open at the far end. Before the breaker is closed, the voltage on the supply side of the breaker terminal is equal to the power system voltage, while the line voltage is zero. At the moment of closing the voltage at the sending end must rise from zero to the power frequency voltage. This takes place in the form of a traveling wave on the line with its peak at u_m interacting with the supply system parameters. As an unloaded line has capacitive impedance, the steady-state voltage at the supply end is higher than the system voltage and, due to the Ferranti effect, the receiving end voltage is higher than the sending end (see Chapter 10 for further discussions). Overvoltage factor can be defined as follows:

$$\text{Total overvoltage factor} = \text{OV}_{\text{tot}} = \frac{u_m}{u_n} \tag{4.32}$$

where

u_m is the highest voltage peak at a given point
u_n is the power frequency voltage at the supply side of breaker before switching

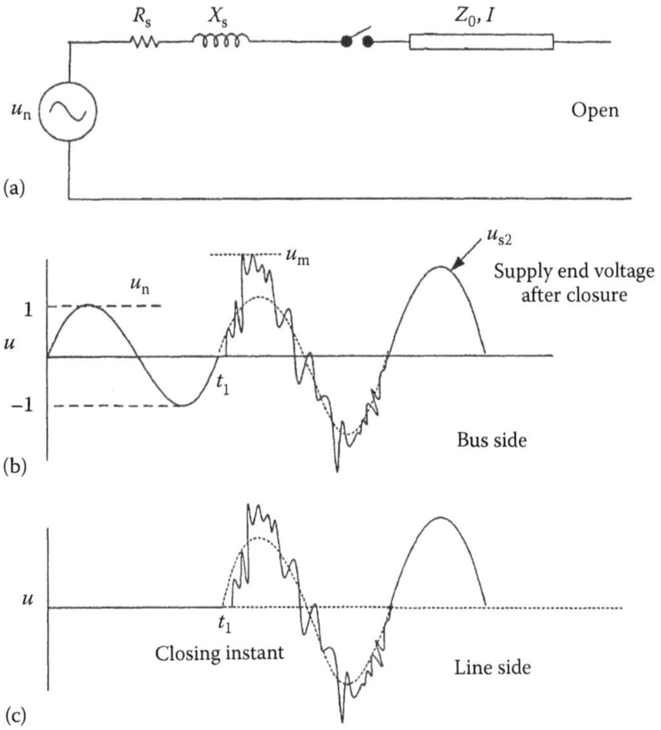

FIGURE 4.16
Overvoltages to ground on closing a transmission line: (a) basic circuit; (b) and (c) voltages on the source and line side and superimposition of traveling waves occurs at $t = t_1$.

$$\text{Power frequency overvoltage factor} = \text{OV}_{\text{pf}} = \frac{u_{\text{pf}}}{u_{\text{n}}} \tag{4.33}$$

This is the ratio of the power frequency voltage u_{pf} after closure at a point and power frequency voltage u_{n} on the supply side before closing.

$$\text{Transient overvoltage factor} = \text{OV}_{\text{tr}} = \frac{u_{\text{m}}}{u_{\text{pf}}} \tag{4.34}$$

The power frequency overvoltage factor can be calculated by known line parameters. This is given by

$$\frac{1}{\cos\alpha 1 - X_s/Z_0 \sin\alpha l} \tag{4.35}$$

where the surge impedance and α are given by

$$Z_0 = \sqrt{\frac{L_1}{C_1}} \tag{4.36}$$

$$\alpha = 2\pi f \sqrt{L_1 C_1} \tag{4.37}$$

The relationship between sending and receiving end voltages is $1/\cos\alpha l$.

This shows that the increase in power frequency voltage depends considerably on the line length. The transient voltage is not so simple to determine and depends on the phase angle at the closing instant (Figure 4.16). At the instant $t = t_1$, the maximum superposition of the transient and power frequency voltages occurs.

Trapped charges occur on the transmission lines in three-pole autoclosure operations. Contact making of three poles of a circuit breaker is nonsimultaneous. Consider breakers at the sending and receiving ends of a line and a transient ground fault, which needs to be cleared by an auto-reclosure operation. The opening of the two breakers is nonsimultaneous and the one which opens later must clear two line phases at no load. These two phases can, therefore, remain charged at the peak of the power frequency voltage, which is still present when the closure takes place. After the dead time, one breaker has to close with two phases still charged. If the closing instant happens to be such that the trapped charge and the power frequency voltage are of opposite polarity, maximum transient overvoltage will occur.

Switching overvoltages do not cause flashovers to the same extent as caused by lightning. The frequency, point of occurrence, amplitude, and characteristics govern the selection of equipment, insulation level, and economical design. The switching surges gain more importance as the system voltage rises and in EHV networks, it is the switching surges which are of primary importance in insulation coordination. The switching surges have, generally, a power frequency and a transient component, which bear certain relation to each other and may be of equal amplitude in the absence of damping. The nonsimultaneous closing of breaker poles can increase the transient component. The EHV installations are primarily concerned with the stresses imposed on the insulation by switching surges and the coordination of the insulation is based upon these values. To lower the costs, it is desirable to reduce the insulation levels at high voltages. On the transmission line towers, the flashover voltage cannot be allowed to increase along with service voltage, as beyond a certain point the electrical strength of air gaps can no longer be economically increased by increasing the clearances [9]. Thus, while the external lightning voltages are limited, the switching over voltages become of primary concern.

All equipment for operating voltages above 300 kV is tested for switching impulses. No switching surge values are assigned to equipment below 60 kV.

A reader may refer to Chapter 5, and Table 5.7 which shows rated line closing switching surge factors for circuit breakers specifically designed to control line switching closing surge. Note that the parameters are applicable for testing with standard reference transmission lines, as shown in this table.

The lightning surges are responsible for approximately 10% of all short-circuits in substations and almost 50% of short-circuits on lines and systems above 300 kV. Less than 1% short-circuits are caused by switching surges [10].

4.11.1 Overvoltage Control

The power frequency component of the overvoltage is controlled by connecting high-voltage reactors from line to ground at the sending and receiving ends of the transmission lines. The effect of the trapped charge on the line can be eliminated if the closing takes place during that half cycle of the power frequency voltage, which has the same polarity as the trapped charge. The high-voltage circuit breakers may be fitted with devices for polarity-dependent closing. Controlling overvoltages with switching resistors is yet another method.

Lines with trapped charge and no compensation and no switching resistors in breakers may have overvoltages greater than three times the rated voltage. Without trapped charge this overvoltage will be reduced to 2.0–2.8 times the rated voltage.

With single-stage closing resistors and compensated line, overvoltages are reduced to less than twice the rated voltage. With two-stage closing resistors or compensated lines with optimized closing resistors, the overvoltage factor is 1.5.

4.11.2 Synchronous Operation

A breaker can be designed to open or close with reference to the system voltage sensing and zero crossing. An electronic control monitors the zero crossing of the voltage wave and controls the shunt release of the breaker. The contacts can be made to touch at voltage zero or voltage crest. In the opening operation, the current zero occurs at a definite contact gap. As the zeroes in three-phase voltages will be displaced, the three poles of the breaker must have independent operating mechanisms. Independent pole operation is a standard feature for breakers of 550 kV. Though it can be fitted to breakers of even lower voltages, say 138 kV. A three-pole device is commercially available for staggering the pole-operating sequence.

Switching of unloaded transformers, transmission lines, shunt reactors, and capacitor banks can benefit from the synchronous operation. The operating characteristics of the breaker must be matched with that of the electrical system. Variations can occur due to aging ambient temperature, level of energy stored in the operating mechanism, and control voltage levels. Deviations in the closing time of 1–2 cycles and in the opening time of 2.9 ms can occur. Resistance switching can achieve the same objectives and is less prone to the mechanical variations—a point often put in the forefront by the advocates of resistance switching. Currently, IEC circuit-breaker standard IEC 62271-100, Ref. [11] excludes circuit breakers with intentional nonsimultaneous pole operation, but this operation may be soon added to the specifications. The resistance switching and controlled switching are not used in combination.

RRDS (rate of rise of dielectric strength) is defined as the circuit-breaker characteristic, which describes the rate of voltage-withstand at opening of a circuit breaker. The value defines the maximum arcing time needed for reignition-free interruption of inductive loads. The RRDS determination is done at no-load opening operations by determining the flashover limits at different contact distances, which translate into different times after contact separation by applying a rapidly increasing voltage. The manufacturers specify their specific breaker designs with staggered pole closing controllers for controlled switching. Upon closing operation, the RRDS will rapidly decrease and is zero when the contacts touch.

4.11.3 Synchronous Capacitor Switching

Though the restrikes in vacuum circuit breakers are greatly reduced (but not eliminated), synchronous switching can eliminate higher frequency component of the voltage swings. The contacts are closed at a nominal voltage zero.

Energizing a capacitor bank means that there should be zero voltage across the breaker contacts when the contacts touch. A number of studies show that the overvoltages can be controlled if the contacts touch 1 ms before or after the voltage zero. As shown in Figure 4.17, the closing can be set slightly after voltage zero in order to minimize the influence of statistical variations.

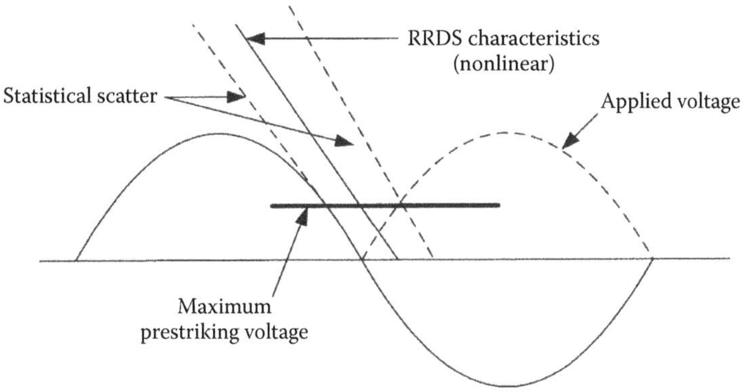

FIGURE 4.17
Synchronous closing operation with respect to voltage zero and RDDS.

The closing operation of grounded and ungrounded capacitor banks must be distinguished. When the capacitors are grounded, then each pole must close independently at the voltage zero of that phase. When the capacitor neutrals are ungrounded, two poles can close at phase-to-phase voltage of zero, and the last pole will close at 4.2 ms. This is shown in Figure 4.18a and b.

4.11.4 Shunt Reactors

Shunt reactors normally have iron cores with integrated air gaps. Due to air gaps, the iron core will not significantly saturate and during energizing the nonlinearity is not of concern. A neutral reactor is sometimes connected between neutral and ground, to increase the zero

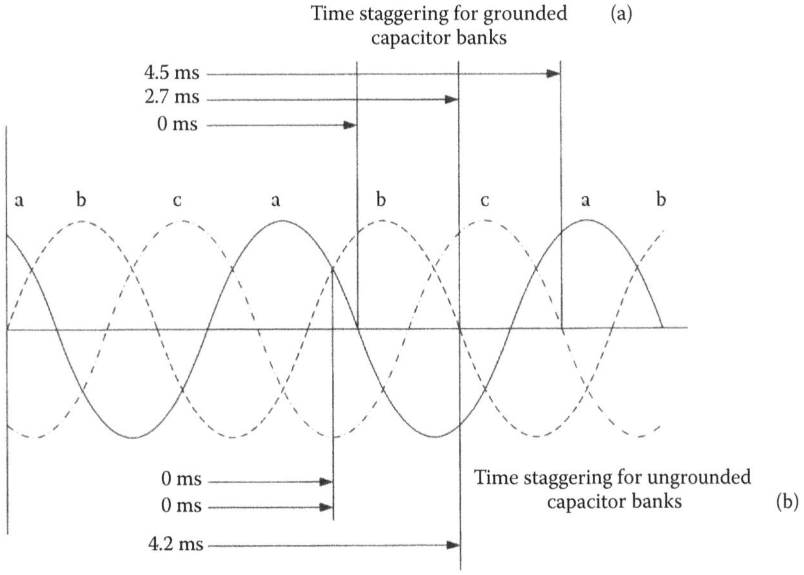

FIGURE 4.18
Synchronous closing of shunt capacitor banks: (a) grounded banks and (b) ungrounded banks.

sequence impedance of the transmission line. When opening the reactor, the voltage across the reactor will be oscillatory, assuming no reignition. For modern SF_6 circuit breakers, the chopping overvoltages are of the order of 1.2–2.0 pu. Due to oscillating reactor voltage, if the circuit breaker contacts reignite, very steep voltage transients caused by reignition will be unevenly distributed across the reactor windings. This may puncture the reactor winding insulations. By controlling the contact separation in such a manner that short arcing times do not occur, reignition will be eliminated.

4.12 Out-of-Phase Closing

Figure 4.19 shows two interconnected systems which are totally out of phase. In Figure 4.19a, a voltage equal to three times the system peak voltage appears across the breaker pole, while in Figure 4.19b, a ground fault exists on two different phases at the sending and receiving ends (rather an unusual condition). The maximum voltage across a breaker pole is $2 \times \sqrt{3}$ times the normal system peak voltage. The present-day high-speed relaying has reduced the tripping time and, thus, the divergence of generator rotors on fast closing is reduced. Simultaneously, the high-speed auto-reclosing to restore service and remove faults increases the possibility of out-of-phase closing, especially under fault conditions. The effect of the increased recovery voltage when the two systems have drifted apart can be stated in terms of the short-circuit power that needs to be interrupted. If the interrupting capacity of a circuit breaker remains unimpaired up to double the rated voltage, it will perform all events satisfactorily as a tie-line breaker when the two sections of the system are completely out of synchronism. The short-circuit power to be interrupted under out-of-step conditions is approximately equal to the total short-circuit power of the entire system, but reaches this level only if the two systems which are out-of-phase have the same capacity. In Figure 4.20, P_1 is the interrupting capacity under completely out-of-phase conditions of two interconnected systems, and P_2 is the total short-circuit capacity; X_a and X_b are the short-circuit reactances of the two systems.

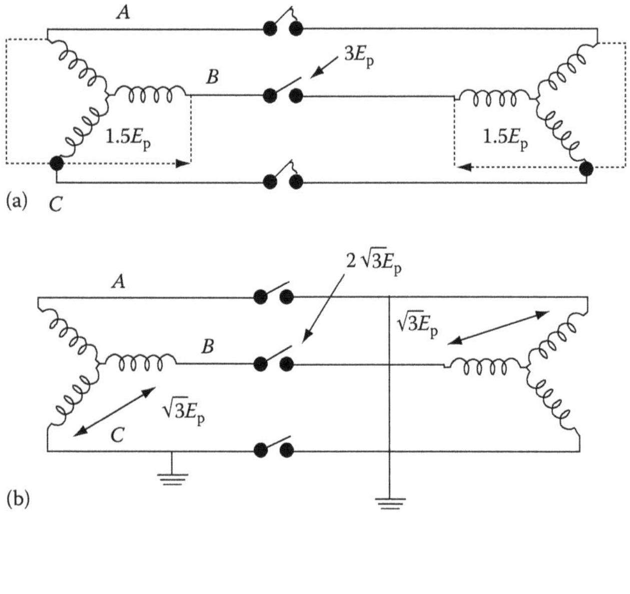

FIGURE 4.19
Overvoltage due to two interconnected systems totally out of phase: (a) unfaulted condition, the maximum voltage equal to three times the peak system voltage; (b) ground fault on different phases on the load and source sides, the maximum voltage equal to $2\sqrt{3}$ times the peak system voltage.

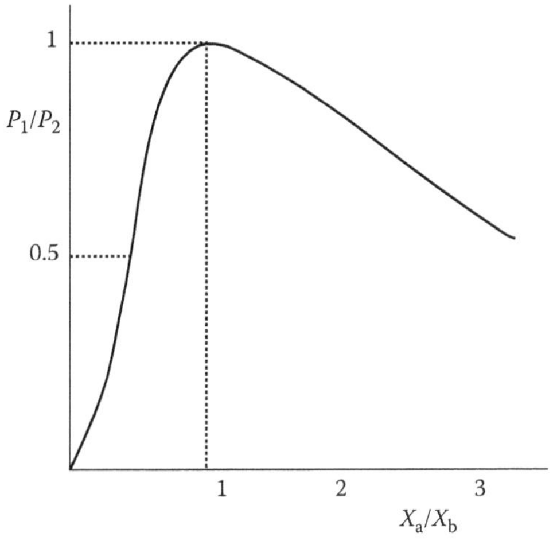

FIGURE 4.20
Interrupting capacity P_1 in the case of out-of-step operation as a function of short-circuit reactances X_a/X_b and P_2 the total short-circuit capacity.

4.13 Resistance Switching

Circuit-breaker resistors can be designed and arranged for the following functions:

- To reduce switching surges and overvoltages
- For potential control across multi-breaks per phase
- To reduce natural frequency effects

Figure 4.21 shows a basic circuit of resistance switching. A resistor r is provided in parallel with the breaker pole, and R, L, and C are the system parameters on the source side of the break. Consider the current loops in this figure. The following equations can be written:

$$u_n = iR + L\frac{di}{dt} + \frac{1}{C}\int i_c\,dt \tag{4.38}$$

$$\frac{1}{C}\int i_c\,dt = i_r r \tag{4.39}$$

$$i = i_r + i_c \tag{4.40}$$

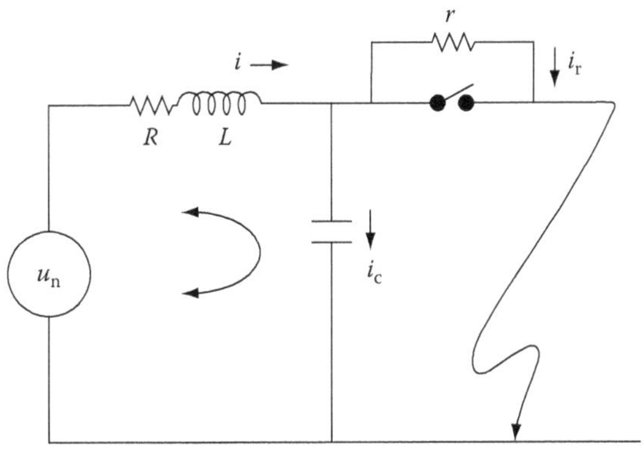

FIGURE 4.21
Resistance r connected in parallel with the circuit-breaker contacts (resistance switching) on a short-circuit interruption.

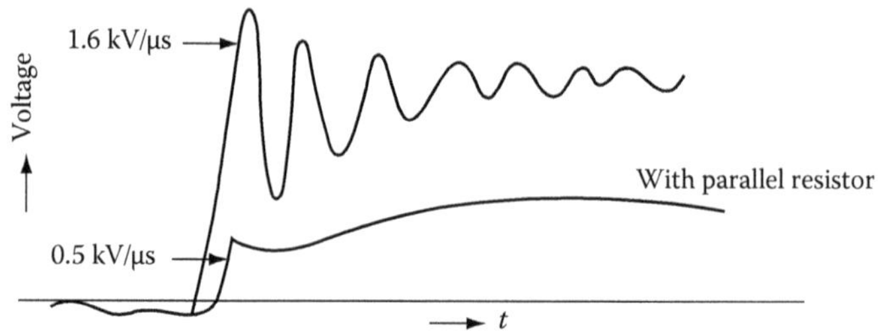

FIGURE 4.22
Reduction of RRRV with parallel resistor switching.

This gives

$$\frac{d^2 i_r}{dt^2} + \left(\frac{R}{L} + \frac{1}{rC}\right)\frac{di_r}{dt} + \left(\frac{1}{LC} + \frac{R}{rLC}\right) i_r = 0 \tag{4.41}$$

The frequency of the transient is given by

$$f_n = \frac{1}{2\pi}\sqrt{\frac{1}{LC} - \frac{1}{4}\left(\frac{R}{L} - \frac{1}{rC}\right)^2} \tag{4.42}$$

In power systems, R is $\ll L$. If a parallel resistor across the contacts of value $r < (1/2)\sqrt{L/C}$ is provided, the frequency reduces to zero. The value of r at which frequency reduces to zero is called the critical damping resistor. The critical resistance can be evaluated in terms of the system short-circuit current, I_{sc}:

$$r = \frac{1}{2}\sqrt{\frac{u}{I_{sc}\omega C}} \tag{4.43}$$

Figure 4.22 shows the effect of the resistors on the recovery voltage. Opening resistors are also called switching resistors and are in parallel with the main break and in series with an auxiliary resistance break switch. On an opening operation, the resistor switch remains closed and opens with a certain delay after the main contacts have opened. The resistance switch may be formed by the moving parts in the interrupter or striking of an arc, dependent on the circuit breaker design.

Figure 4.23 shows the sequence of opening and closing in a circuit breaker provided with both opening and closing resistors. The closing resistors control the switching overvoltage on energization of, say, long lines. An interrupting and closing operation is shown. The main break is shown as SB, the breaking resistor as RB. On an opening operation, as the main contacts start arcing, the current is diverted through the resistor RB, which is interrupted by contacts SC. In Figure 4.23d, the breaker is in open position. Figure 4.23e and f show the closing operation. Finally, the closed breaker is shown in Figure 4.23a.

4.14 Failure Modes of Circuit Breakers

In ac circuit breakers, the phenomena of arc interruption are complex. Arc plasma temperatures of the order of 25,000 to 5,000 K are involved, with conductivity changing a billion times as fast as temperature in the critical range associated with thermal ionization.

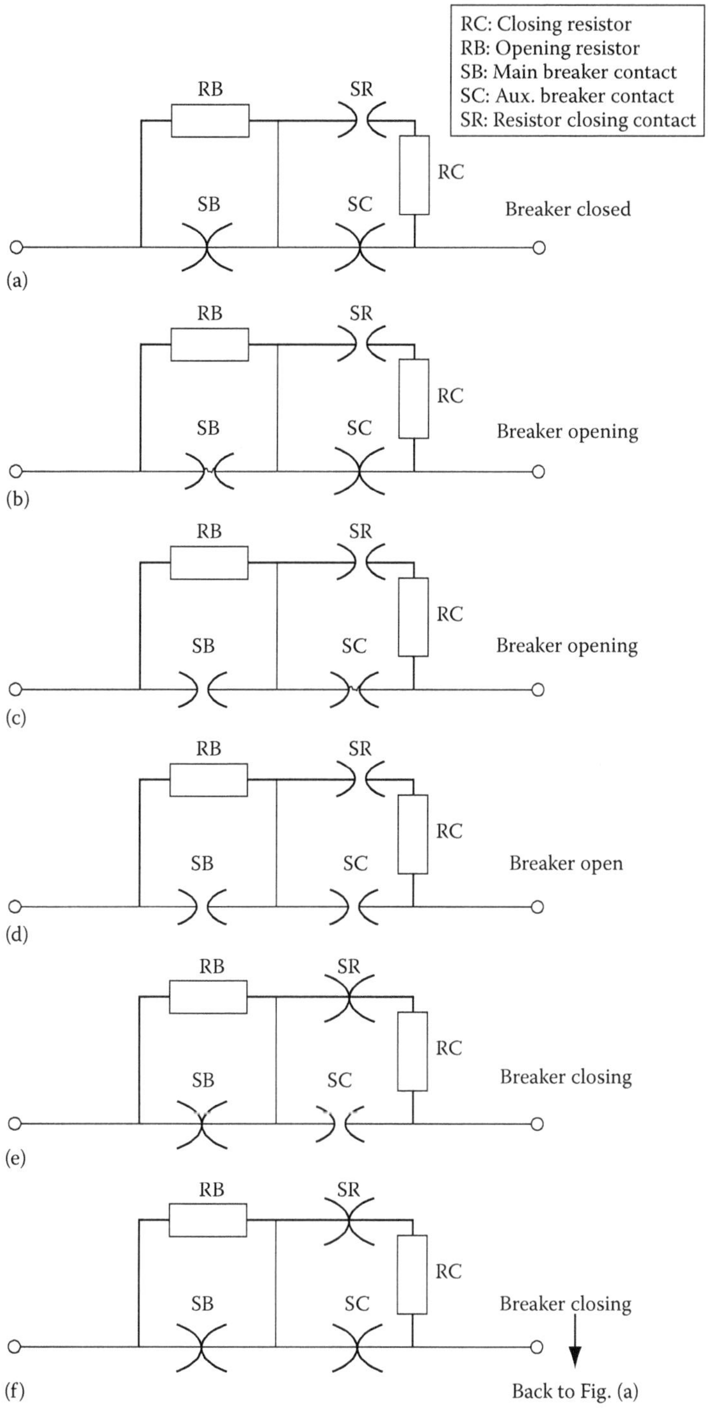

FIGURE 4.23
Sequence of closing and opening operation of a high-voltage circuit breaker provided with opening and closing resistors. Figures (a) through (f)—Breaker opening & then closing.

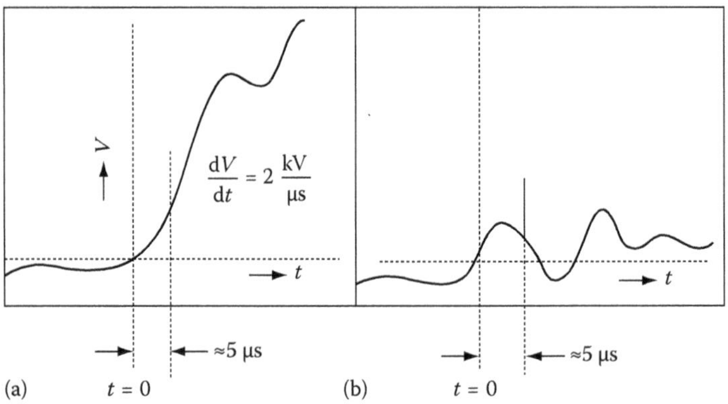

FIGURE 4.24
Thermal failure mode of a circuit breaker, while opening: (a) successful interruption; (b) failure in the thermal mode.

Supersonic turbulent flow occurs in changing flow and contact geometry at speeds from 100 to 1000 m/s in the arc. The contact system should accelerate from a stationary condition to high speeds in a very short time.

With parabolic pressure distribution in the contact zone of a double-nozzle configuration, a cylindrical arc with temperatures nearing 25,000 K exists. Due to the low density of gas at high temperatures, the plasma is strongly accelerated by an axial pressure gradient. A so-called thermal arc boundary at a temperature of 300–2000 K exists. The arc downstream expands into nozzles and in this region the boundary layer between arc and gas is turbulent with formation of vortices.

Two types of failures can occur: (1) dielectric failure which is usually coupled with a terminal fault and (2) thermal failure which is associated with a short-line fault. If after a current zero, the RRRV is greater than a critical value, the decaying arc channel is reestablished by ohmic heating. This period, which is controlled by the energy balance in the arc, is called the thermal interruption mode. Figure 4.24 shows successful thermal interruption and a thermal failure. Within 2 μs after interruption, the voltage deviates from TRV. It decreases and approaches the arc voltage.

Following the thermal mode, a hot channel exists at temperatures from 300 to 5000 K and a gas zone adjacent to the arc diminishes at a slow rate. The recovering system voltage distorts and sets the dielectric limits. After successful thermal interruption, if the TRV can reach such a high peak value that the circuit breaker gap fails, it is called a dielectric failure mode. This is shown in Figure 4.25. Figure 4.25a shows successful interruption, and Figure 4.25b shows dielectric failure at the peak of the recovery voltage, and rapid voltage decay.

The limit curves for circuit breakers can be plotted on a log u and log I basis, as shown in Figure 4.26. In this figure, u is the system voltage and I the short-circuit current. The portion in thick lines shows dielectric limits, while the vertical portion in thin lines shows thermal limits. In thermal mode, (1) metal vapor production from contact surfaces; (2) di/dt, i.e., the rate of decrease of the current at current zero; (3) arc constrictions in the nozzle due to finite velocity; (4) nozzle configurations; (5) presence of parallel capacitors and resistors; and (6) type of quenching medium and pressures are of importance. In the dielectric mode, the generation of electrons in an electric field is governed by Townsend's equation:

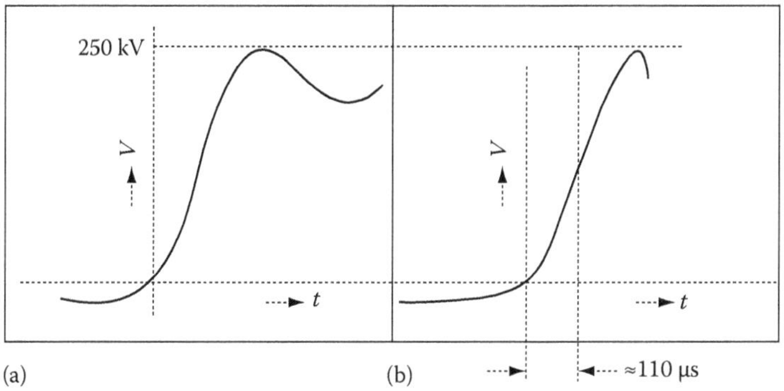

FIGURE 4.25
Dielectric failure mode of a high-voltage circuit breaker: (a) successful interruption; (b) failure at the peak of TRV.

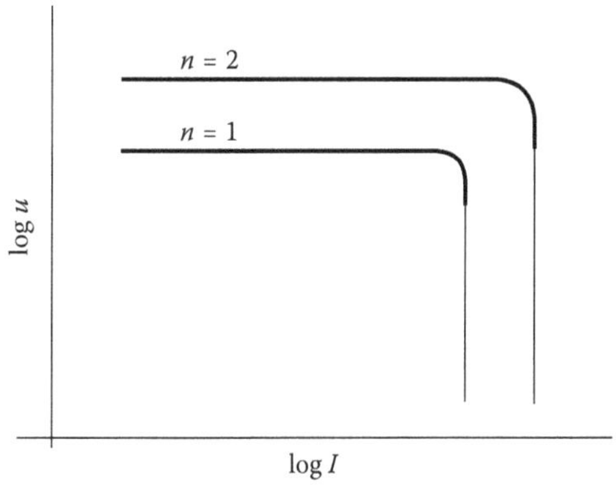

FIGURE 4.26
General form of limiting curve in a circuit breaker plotted as log u versus log I, where V is the rated voltage of the breaker and I is the short-circuit current; n indicates the number of interrupting chambers in series. Thick lines show dielectric mode and thin lines thermal mode of possible failure.

$$\frac{\partial n_e}{\partial t} = -\frac{\partial n_e V_e}{\partial d} + (\alpha - \eta) n_s V_e \tag{4.44}$$

where

- n_e is the number of electrons
- α is the Townsend coefficient (number of charged particles formed by negatively charged ions)
- d is the spacing of electrodes
- η is the attachment coefficient
- V_e is the electron drift velocity

4.15 Operating Mechanisms-SF_6 Breakers

The operating energy mechanisms of circuit breakers have evolved into energy-efficient and lighter designs, made possible by computer simulations and laboratory models. With respect to SF_6 breakers, the earlier designs of double pressure breakers are obsolete, and "puffer type" designs were introduced in 1967. The gas is sealed in the interrupting chamber at a low pressure and the contact movement itself generates a high pressure—part of the moving contact forms a cylinder and it moves against a stationary piston, thus the terminology puffer type, meaning blowing out the arc with a puff. The last 20 years have seen the development of "self-blast technology" applicable to 800 kV. A valve was introduced between the expansion and compression volume. When the interrupting current is low, the valve opens under effect of overpressure and the interruption phenomena is similar to puffer design. At high current interruption, the arc energy produces a high overpressure in the expansion volume and the valve remains closed. The overpressure for breaking is obtained by optimal use of thermal effect and nozzle-clogging effect to heat the gas and raise its pressure. The cross section of the arc significantly reduces the exhaust of gas in the nozzle. The self-blast technology is further optimized by double motion principle; this consists of displacing the two arcing contacts in opposite directions. This leads to further reduction in the operating energy. In "thermal blast chamber with arc-assisted opening" arc energy is used to generate the blast by thermal expansion and also to accelerate the moving part of the circuit breaker when interrupting high currents [12,13].

4.16 Vacuum Interruption

Vacuum circuit breakers are extensively applied in the voltage range of 5–38 kV, though vacuum breakers have been built to higher voltages. Vacuum contactors are used for motor starting, tap changing—they must be suitable for repeatedly switching the motor load currents, though the interrupting currents are low in 5–10 kA. The vacuum breakers are currently available to 63 kA interrupting at 13.8 kV and 40 kA at 38 kV.

Pressures below 10^{-3} mm of mercury are considered high vacuum. The charged particles moving from one electrode to another electrode at such a low pressure are unlikely to cause collision with a residual gas molecule, though electrons can be emitted from cold metal surfaces due to high electrical field intensity, called field emission. This electron current may vary according to

$$I = Ce^{-B/E} \tag{4.45}$$

where B and C are constants and E is electrical field intensity. As the contacts separate, at a lower gap the electrical field intensity is high. As the voltage is increased, the current suddenly increases and gap breaks down, the phenomena called vacuum spark. This depends upon the surface conditions and material of the electrodes. Secondary emission takes place by bombardment of high energy on electrode surfaces. The current leaves the electrodes from a few bright spots, and the current densities are high. The core of the arc has high temperatures of the order of 6,000–15,000 K. At such temperatures thermal emission takes place. The dielectric recovery after sparkover takes place at a rate of 5–10 kV/μs.

An arc cannot persist in ideal vacuum, but separation of contacts causes vapor to be released from the contact surface, giving rise to plasma. At low currents there are several parallel paths, each originating and sinking in a hot spot, a diffused arc. At high currents above 15 kA, the arc becomes concentrated on a small region and becomes self-sustaining. The transition from diffused arc to concentrated arc depends upon contact material and geometry. As the current wave approaches zero, rate of release of vapor is reduced, the arc becomes diffused, and the medium tends to regain the dielectric strength, provided vapor density around the contacts is substantially reduced. Thus, extinction process is related to the material and shape of the contacts and techniques used to condense the metal vapor. Contact geometry is designed so that the root of the arc keeps moving. The axial magnetic field decreases the arc voltage.

The vacuum interrupters interrupt the small current before natural current zero, causing current chopping. Again this is a function of contact material and the chopping current of modern circuit breakers has been reduced to 3 A, 50% cumulative probability. The interrupting capability depends upon the contact material and the size and type of the magnetic field produced.

A number of restrikes and insulation failures were reported during the initial development of vacuum interrupter technology. With innovations in contact materials and configurations these are rare in the modern designs. However, we must distinguish between the vacuum circuit breakers and contactors; the latter are more prone to restrikes. An area of possible multiple ignitions occurs with respect to locked rotor current of the motor (for application of motors up to 6.6 kV). The frequency shown in this figure is given by [14]

$$f_n = \frac{1}{2\pi\sqrt{LC}} \tag{4.46}$$

where L is the ungrounded motor inductance at locked rotor condition, l is the length of the cable connection from the contactor to the motor, C_v is the capacitance of the cable per unit length. ($C = lC_v$). Essentially Equation 4.46 is a resonance formula of cable capacitance with motor-locked rotor reactance. Therefore, a certain length of cable if exceeded can lead to reignition. This shows that (1) motor insulation is more vulnerable to stresses on account of lower BIL, and (2) possibility of restrikes in the switching devices. As an example, for a 500 hp motor overvoltage protection device is needed, if #4 shielded cable length, used to connect the motor to the switching vacuum contactor, exceeds approximately 470 ft. This is based upon the cable data.

Further descriptions of constructional features, interruptions in other mediums are not discussed. It is sufficient to highlight that the interruption of currents in ac circuits in not an independent phenomena depending upon the circuit conditions alone, it is seriously modified by the physical and electrical properties of the interrupting mediums in circuit breakers and the system to which these are connected.

The failure rate of circuit breakers all over the world is decreasing on account of better designs and applications.

4.17 Stresses in Circuit Breakers

The stresses in a circuit breaker under various operating conditions are summarized in Figure 4.27. These stresses are shown in terms of three parameters, current, voltage, and du/dt, in a three-dimensional plane. Let the current stress be represented along the x axis, the du/dt, stress along the y axis, and the voltage stress along the z axis. We see that a short-line fault (A_1, A_2, A_3) gives the maximum RRRV stress, though the voltage stress is low. A terminal fault (B_1, B_2, B_3) results in the maximum interrupting current, while capacitor switching (C) and out-of-phase switching (D) give the maximum voltage stresses. All the stresses do not occur simultaneously in an interrupting process.

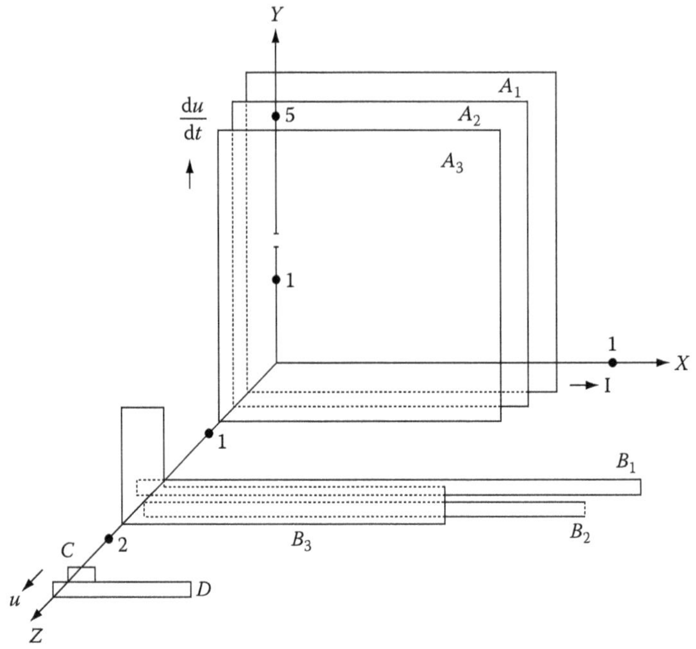

FIGURE 4.27
Stresses in a high-voltage circuit breaker in terms of short-circuit current, RRRV, and maximum overvoltage. A_1, A_2, and A_3 short-line faults; B_1, B_2, and B_3 terminal faults; C capacitance currents; D out-of-phase or asynchronous conditions.

Problems

4.1 Distinguish between reignitions, restrikes, and current chopping in high-voltage circuit breakers.

4.2 What is a delay line in TRV representation by two- and four-parameter representation? Describe the parameters on which it depends and its calculation.

4.3 Describe two accepted failure modes of circuit breakers. Categorize the fault types which can lead to each of these two modes.

4.4 Find the recovery voltage across the breaker contacts while interrupting the 4 A (peak) magnetizing current of a 138 kV, 20 MVA transformer. Assume a capacitance of 4 nF to ground and an inductance of 4 H.

4.5 What is the value of a switching resistor to eliminate the restriking transient in Problem 4.4?

4.6 On the source side of a generator breaker, $L = 1.5$ mH and $C = 0.005$ μF. The breaker interrupts a current of 20 kA. Find (a) RRRV, (b) time to reach peak recovery voltage, and (c) frequency of oscillation.

4.7 Explain the influence of power factor and first pole to clear on TRV. What is the effect of frequency of TRV and load current on the interrupting duty of the circuit breaker?

4.8 A synchronous breaker is required to control a large shunt capacitor bank. Overvoltages can be reduced by closing the breaker at (1) peak of the voltage, and (2) zero crossing of the voltage. Which of the two statements is correct? Assume that the capacitors do not have a residual charge. Create a timing diagram for synchronous switching of grounded and ungrounded shunt capacitors.

4.9 Comment on the correctness of these statements: (1) interrupting an asymmetrical current gives rise to higher TRV than interrupting a symmetrical current; (2) as the current to be interrupted reduces, so does the initial rate of rise of the recovery voltage; (3) the thermal mode of a failure of breaker is excited when interrupting a capacitor current, due to higher TRV; (4) an oscillatory TRV occurs for a fault on a transformer connected to a transmission line; (5) selecting a breaker of higher interrupting rating is an assurance that, in general, its TRV capability is better.

4.10 Describe a simple circuit element to control the RRRV, when interrupting a highly magnetizing current.

4.11 Why is there a high probability of flashover across breaker contacts when interrupting capacitive currents? How can this be avoided?

4.12 What are the common measures to control the effect of restrikes in vacuum contactors applied for switching long cables to motors?

References

1. K. Ragaller. *Current Interruption in High Voltage Networks*. New York: Plenum Press, 1978.
2. A.M. Cassie. Arc rupture and circuit theory. CIGRE Report No. 102, Paris, France, 1939.
3. ANSI/IEEE Std. C37.010. *IEEE Application Guide for High-Voltage Circuit Breakers Rated on Symmetrical Current Basis*, 1999.
4. IEC. *High Voltage Alternating Current Circuit Breakers*. Standard 62271-100, 2010.
5. IEEE Std. C37.011, *IEEE Application Guide for Transient Recovery Voltage for AC High Voltage Circuit Breakers*, 2005.
6. J. Panek and K.G. Fehrie. Overvoltage phenomena associated with virtual current chopping in three-phase circuits. *IEEE Trans*. PAS-94: 1317–1325, 1975.
7. J.F. Perkins. Evaluation of switching surge overvoltages on medium voltage power systems. *IEEE Trans*. PAS-101 (6): 1727–1734, 1982.
8. IEEE Std. C37.012. *Application Guide for Capacitor Current Switching for AC High Voltage Circuit Breakers*, 2005.
9. H. Glavitsch. Problems associated with switching surges in EHV networks. *BBC Rev*. 53: 267–277, April/May 1966.
10. CIGRE Working Group 13-02, Switching surge phenomena in EHV systems. Switching overvoltages in transmission lines with special reference to closing and reclosing transmission lines. *Electra* 30: 70–122, 1973.
11. D. Dufournet, F. Sciullo, J. Ozil, and A. Ludwig. New interrupting and drive techniques to increase high voltage circuit breakers performance and reliability. CIGRE Session, paper 13-104, Paris, France, 1998.
12. H.E. Spindle, T.F. Garrity, and C.L. Wagner, Development of 1200 kV circuit breaker for GIS system—Requirements and circuit breaker parameters. CIGRE, Rep. 13-07, Paris, France, 1980.
13. S.F. Frag and R.G. Bartheld. Guidelines for application of vacuum contactors. *IEEE Trans*. IA, 22 (1): 102–108, January/February 1986.
14. T. Itoh, T. Muri, T. Ohkura, and T. Yakami. Voltage escalation in switching of motor control circuit by vacuum contactors. *IEEE Trans*. PAS-91: 1897–1903, 1972.

Bibliography

Braun, A., A. Eidinger, and E. Rouss. Interruption of short-circuit currents in high-voltage AC networks. *Brown Boveri Rev.* 66: 240–245, April 1979.

Browne, T.E. Jr. (ed.). *Circuit Interruption—Theory and Techniques*. New York: Marcel Dekker, 1984.

CIGRE Working Group 13-02. Switching overvoltages in EHV and UHV systems with special reference to closing and reclosing transmission lines. *Electra* 30: 70–122, 1973.

CIGRE WG A3.12. Failure survey on circuit breaker controls systems. *Electra* 251: 17–31, April 2007.

Das, J.C. SF6 in high-voltage outdoor switchgear. *Proc. IEEE* 61 (pt. EL2): 1–7, October 1980.

Garzon, R.D. *HV Circuit Breakers-Design and Applications*, 2nd edn. New York: Marcel Dekker, 2002.

Greenwood, A. *Electrical Transients in Power Systems*. New York: Wiley Interscience, 1991.

Hermann, W. and K. Ragaller. Theoretical description of current interruption in gas blast circuit breakers. *IEEE Trans.* PAS-96: 1546–1555, 1977.

Koschik, V., S.R. Lambert, R.G. Rocamora, C.E. Wood, and G. Worner. Long line single phase switching transients and their effect on station equipment. *IEEE Trans.* PAS-97: 857–964, 1978.

Loeb, L.B. *Fundamental Processes of Electrical Breakdown in Gases*. New York: John Wiley, 1975.

Slamecka, E. Interruption of small interrupting currents, CIGRE WG 13.02. *Electra* 72: 73–103, October 1980, and *Electra* 5–30, March 1981.

Sluis, L.V.D. and A.L.J. Jansen. Clearing faults near shunt capacitor banks. *IEEE Trans.* PD 5 (3): 1346–1354, July 1990.

Toda, H., Y. Ozaki, and I. Miwa. Development of 800-kV gas insulated switchgear. *IEEE Trans.* PD **7**: 316–322, 1992.

Ushio, T., I. Shimura, and S. Tominaga. Practical problems of SF6 gas circuit breakers. *IEEE Trans.* PAS-90 (5): 2166–2174, 1971.

5

Application and Ratings of Circuit Breakers and Fuses according to ANSI Standards

In Chapter 4, we discussed current interruption in ac circuits and the stresses that can be imposed on the circuit breakers, depending on the nature of fault or the switching operation. We observed that the system modulates the arc interruption process and the performance of a circuit breaker. In this chapter, we will review the ratings of circuit breakers and fuses according to ANSI, mainly from the short circuit and switching point of view, and examine their applications. While a general-purpose circuit breaker may be adequate in most applications, higher duties may be imposed in certain systems, which should be carefully considered. This chapter also forms a background to the short-circuit calculation procedures in Chapter 7.

There has been attempt to harmonize ANSI/IEEE standards with IEC. IEEE unapproved draft standard IEEE PC37.06/D11 for AC high-voltage circuit breakers rated on symmetrical current basis—preferred ratings and required related capabilities for voltage above 1000 V [1] may be referred to. A joint task force group was established to harmonize requirements of TRV with that in IEC62271-100 [2] including amendments 1 and 2. This draft standard also publishes new rating tables and completely revises the ratings and the nomenclature of the circuit breakers, which has been in use in the United States for many years:

- Classes S1 and S2 are used to denote traditional terms of "indoor" and "outdoor, respectively." The class S1 circuit breaker is for cable systems, and indoor circuit breakers are predominantly used with cable distribution systems. Class S2 is for overhead line systems.
- The term "peak" is used—the term "crest" has been dropped form the usage. All tables show "prospective" or "inherent" characteristics of the current and voltages. The word "prospective" is used in conformance to International standard.
- Two and four parameter of representations of TRV is adopted in line with IEC standards
- For class C0, general purpose circuit breakers, no ratings are assigned for back-to-back capacitor switching. For class C0, exposed to transient currents for nearby capacitor banks during fault conditions, the capacitance transient current on closing shall not exceed lesser of either 1.41 times rated short-circuit current or 50 kA peak. The product of transient inrush current peak and frequency shall not exceed 20 kA kHz. Definite purpose circuit breakers are now identified as class C1 and class C2. Here, the manufacturer shall specify the inrush current and frequency at which class C1 or C2 performance is met.

5.1 Total and Symmetrical Current Rating Basis

Prior to 1964, high-voltage circuit breakers were rated on a total current basis [3,4]. At present, these are rated on a symmetrical current basis [5–8]. Systems of nominal voltage less than 1000 V are termed low voltage, from 1,000 to 100,000 V as medium voltage, and from 100 to 230 kV as high voltage. Nominal system voltages from 230 to 765 kV are extrahigh voltage (EHV) and higher than that ultrahigh voltage (UHV). ANSI covers circuit breakers rated above 1–800 kV. The difference in total and symmetrical ratings depends on how the asymmetry in short-circuit current is addressed. The symmetrical rating takes asymmetry into account in the rating structure of the breaker itself. The asymmetrical profile of the short-circuit current shown in Figure 5.1 is the same as that of Figure 1.1, except that a decaying ac component is shown. The rms value of a symmetrical sinusoidal wave, at any instant, is equal to the peak-to-peak value divided by 2.828. The rms value of an asymmetrical wave shape at any instant is given by Equation 4.11. For circuit breakers rated on a total current rating basis, short-circuit interrupting capabilities are expressed in total rms current. For circuit breakers rated on a symmetrical current basis, the interrupting capabilities are expressed as rms symmetrical current at contact parting time. The symmetrical capacity for polyphase is the highest value of the symmetrical component of the short-circuit current in rms ampères at the instant of primary arcing contact separation, which the circuit breaker will be required to interrupt at a specified operating voltage on the standard operating duty and *with a direct current component of less than 20% of the current value of the symmetrical component* [6, 1999 revision]. Note that 1978 edition of this standard stated "irrespective of the total dc component of the total short-circuit current."

Figure 5.2 shows an asymmetrical current waveform, where t is the instant of contact parting. The peak-to-peak value of the asymmetrical ac current is given by ordinate A. This is the sum of two ordinates A' and B', called the major ordinate and minor ordinates, respectively, as measured from the zero line. The dc component D is

$$D = \frac{A' - B'}{2} \tag{5.1}$$

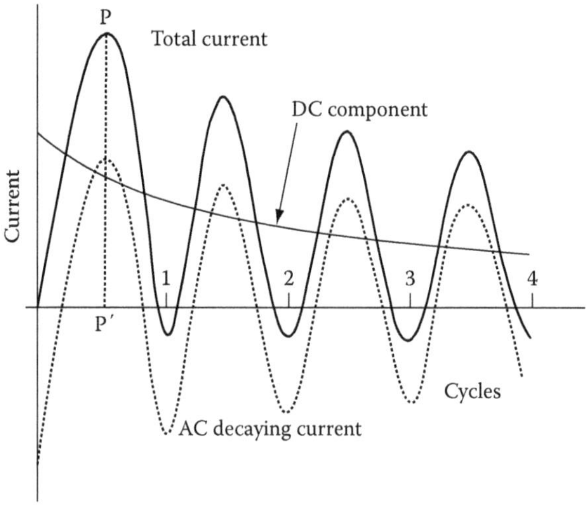

FIGURE 5.1
Asymmetrical current wave with decaying ac component.

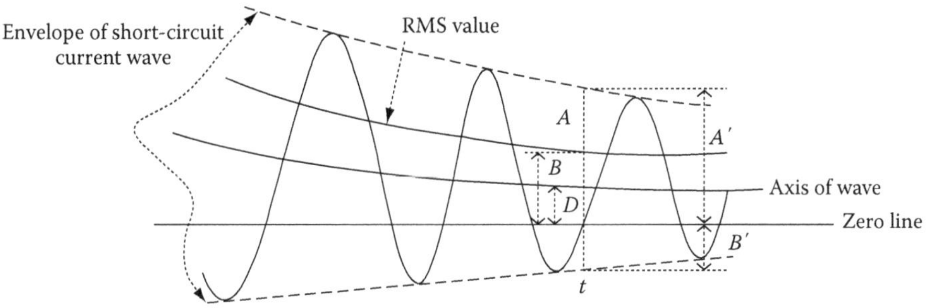

FIGURE 5.2
Evaluation of rms value of an offset wave.

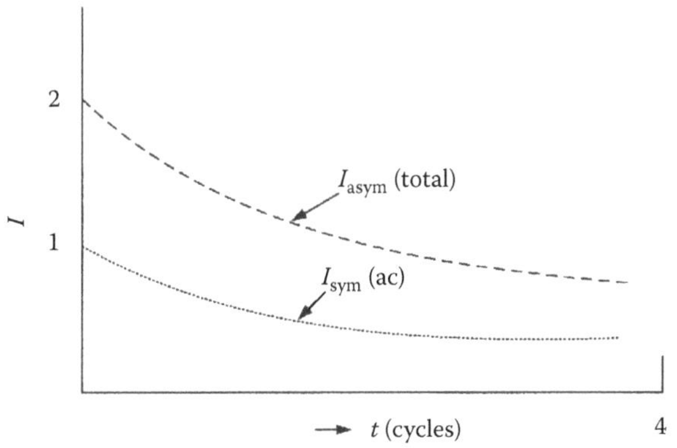

FIGURE 5.3
Symmetrical ac and total current profiles, rms values.

The rms value of ac component B is

$$B = \frac{A' + B'}{2.828} \tag{5.2}$$

Thus, total interrupting current in rms is

$$\sqrt{B^2 + D^2} \tag{5.3}$$

This is an asymmetrical current. Equation 5.3 forms the total current rating basis, and Equation 5.2 forms the symmetrical current basis. Figure 5.3 shows that the total rms asymmetrical current is higher than the rms of the ac component alone. This does not mean that the effect of the dc component at the contact parting time is ignored in the symmetrical rating of the breakers. It is considered in the testing and rating of the breaker. Breakers rated on a symmetrical current basis are discussed in the rest of this chapter.

5.2 Asymmetrical Ratings

5.2.1 Contact Parting Time

The asymmetry in polyphase and phase-to-phase interrupting is accounted for by testing the breaker at higher total current (asymmetrical current) based on a minimum contact parting time. This includes the *tripping delay*. The contact parting is the sum of the tripping delay plus breaker opening time. The ANSI standard [6, 1999 revision] specifies that the primary contact parting time will be considered equal to the sum of one-half cycle (tripping delay) and the *lesser of* (1) the actual opening time of the particular breaker or (2) 1.0, 1.5, 2.5, or 3.5 cycles for breakers having a rated interrupting time of 2, 3, 5, or 8 cycles, respectively. This means that 2-, 3-, 5-, and 8-cycle breakers have contact parting times of 1.5, 2, 3, and 4 cycles, respectively, unless the test results show a *lower* opening time plus a half-cycle tripping delay.

The asymmetrical rating in 1979 standard was expressed as a ratio S, which is the required asymmetrical interrupting capability per unit of the symmetrical interrupting capability. Ratio S is found by multiplying the symmetrical interrupting capability of the breaker determined for the operating voltage by an appropriate factor. The value of S was specified as 1.4, 1.3, 1.2, 1.1, and 1.0 for breakers having a contact parting time of 1, 1.5, 2, 3, or 4 or more cycles, respectively.

The factor S curve is now replaced by the curve shown in Figure 5.4. This curve is from [5, revision 2005]. The percentage dc (%dc) is given by the expression:

$$\%\text{dc} = 100e^{-t/45} \tag{5.4}$$

where t is the contact parting time in ms.

The required asymmetrical capability for the three-phase faults is the value of total rms current at the instant of arcing contact separation that the circuit breaker will be required to interrupt at a specified operating voltage, on the standard operating duty cycle.

$$I_\text{t} = I_\text{sym}\sqrt{1 + 2\left(\frac{\%\text{dc}}{100}\right)^2} \tag{5.5}$$

Consider a breaker, interrupting time 5 cycles and therefore contact parting time = 3 cycles. From Figure 5.4, the %dc = 32.91%. Then

$$\frac{I_\text{t}}{I_\text{sym}} = \sqrt{1 + 2(0.3291)^2} \approx 1.1$$

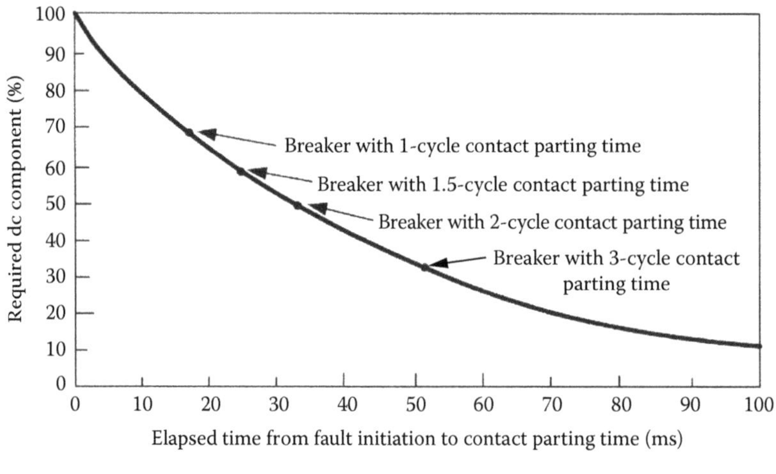

FIGURE 5.4
DC component at contact parting time.

As per earlier (1979) edition of [5], *S* was specified equal to 1.1 for 5-cycle breakers. Thus, the new revised standards have converted the *S* curve to %dc curve. The following observations are pertinent:

- If the actual relay time is less than 0.5 cycles, the fact should be considered in calculations of asymmetrical ratings. The curve in Figure 5.5 is based upon half-cycle tripping delay.
- With X/R greater than 17 at 60 Hz, the decrement rate of dc will be slower than that incorporated in rating structures in standards, Refs. [6,7]. E/X method of ac and dc adjustments can be used, Chapter 7.
- The longer dc time constants can cause problems with some types of SF_6 puffer breakers (see Chapter 7). The E/X adjustment can be used, provided the X/R does not exceed 45 at 60 Hz (dc time constant of 120 ms).
- For time constants beyond 120 ms, consultation with manufacturer is recommended.
- In some application of large generators, current zeros may not be obtained due to high asymmetry (see Chapter 7).

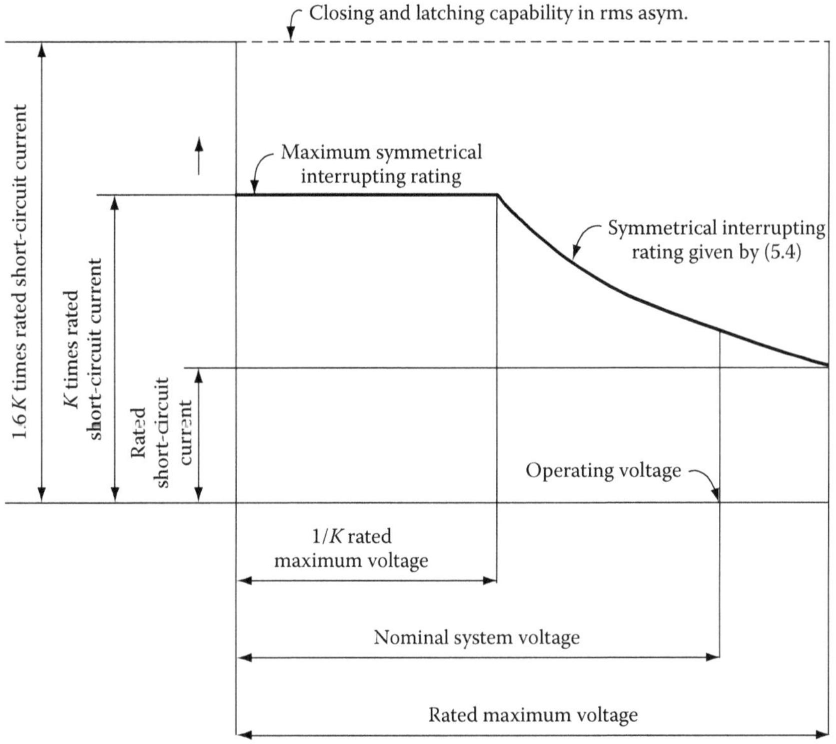

FIGURE 5.5
Relationship of interrupting capacity, closing and latching capability, and carrying capability to rated short-circuit current. (From ANS1/IEEE Standard C37.04, Rating Structure for AC High Voltage Circuit Breakers, 1999 (revision of 1979).)

5.3 Voltage Range Factor K

The 2000 revision of ANSI/IEEE standard [8] has changed the K factor of indoor oil-less circuit breakers to 1 (Table 5.1). Prior to that, K factor was higher than 1, Table 5.2. As the K factor rated breakers have been extensively used in the industry over the past years and may continue to be in service, manufacturer's can still supply K factor rated breakers, though retrofitting with newly rated $K=1$ breakers is a possibility; if replacements and upgrades are required.

Note that both Tables 5.1 and 5.2 show a rated interrupting time of 5 cycles; however, 3-cycle breakers are available and can be used to advantage faster fault clearing time, for example, improving the transient stability limits and reducing arc flash hazard, Chapter 21. IEEE draft standard [1] shows both 5- and 3-cycle breakers, interrupting times of 83 and 50 ms, respectively.

The maximum symmetrical interrupting capability of a circuit breaker is K times the rated short-circuit current. Between the rated maximum voltage and $1/K$ times the rated maximum voltage, the symmetrical interrupting capacity is defined as

$$\text{Rated short circuit current} \times \frac{\text{Rated maximum voltage}}{\text{Operating voltage}} \tag{5.6}$$

This is illustrated in Figure 5.5. The interrupting rating at lower voltage of application cannot exceed the rated short-circuit current multiplied by K. With new $K=1$ rated breakers, a 15 kV breaker rated at 40 kA interrupting duty can be applied at 2.4 kV, and the interrupting rating will still be 40 kA.

Example 5.1

A 15 kV circuit breaker (maximum rated voltage) has a K factor of 1.30 and rated short-circuit current of 37 kA rms. What is its interrupting rating at 13.8 kV? What is the breaker interrupting rating if applied at 2.4 kV?

The rated interrupting current at 13.8 kV is given by Equation 5.6 and is $37 \times (15/13.8) = 40.2$ kA. The maximum symmetrical interrupting capability of this breaker is 37 K = 48 kA rms. Thus, the interrupting rating if applied at 2.4 kV will be 48 kA.

Compare this to a 15 kV 40 kA rated breaker $K=1$. The interrupting rating if applied at 13.8 kV or 2.4 kV is 40 kA.

5.4 Circuit Breaker Timing Diagram

Figure 5.6 shows the operating time diagram of a circuit breaker. We have already defined half-cycle relay time, contact opening time, contact parting time, arcing time, and interrupting time. The interrupting time is not a constant parameter, and variations occur depending upon the fault. For a line-to-ground fault, it may increase by 0.1 cycle, and, for asymmetrical faults, it may exceed rated interrupting time by 0.2 cycles.

High speed reclosing is applied to radial and tie lines to (1) clear a ground fault of the transient nature, (2) improve system stability through fast reclosing; the phase angles and

TABLE 5.1

Preferred Ratings of Indoor Oil-Less Circuit Breakers, $K = 1$

Rated Maximum Voltage (kV, rms)	Rated Voltage Range Factor (K)	Rated Continuous Current at 60 Hz (A, rms)	Rated Short-Circuit Current at Rated Maximum kV (kA, rms)	Rated TRV: Rated Peak Voltage E_2 (kV, Peak)	Rated TRV: Rated Time to Peak T_2 (μs)	Rated Interrupting Time (ms)	Closing and Latching Current (kA, Peak)
4.76	1.0	1200, 2000	31.5	8.9	50	83	82
4.76	1.0	1200, 2000	40	8.9	50	83	104
4.76	1.0	1200, 2000, 3000	50	8.9	50	83	130
8.25	1.0	1200, 2000, 3000	40	15.5	60	83	104
15	1.0	1200, 2000	20	28	75	83	52
15	1.0	1200, 2000	25	28	75	83	65
15	1.0	1200, 2000	31.5	28	75	83	82
15	1.0	1200, 2000, 3000	40	28	75	83	104
15	1.0	1200, 2000, 3000	50	28	75	83	130
15	1.0	1200, 2000, 3000	63	28	75	83	164
27	1.0	1200	16	51	105	83	42
27	1.0	1200, 2000	25	51	105	83	65
38	1.0	1200	16	71	125	83	42
38	1.0	1200, 2000	25	71	125	83	65
38	1.0	1200, 2000, 3000	31.5	71	125	83	82
38	1.0	1200, 2000, 3000	40	71	125	83	104

Source: ANSI/IEEE Standard C37.010, Application Guide for AC High-Voltage Circuit Breakers Rated on a Symmetrical Current Basis, 1999 (R-2005).

TABLE 5.2

Preferred Ratings of Indoor Oil-Less Circuit Breakers, $K > 1$

Rated Maximum Voltage (kV, rms)	Rated Voltage Range Factor (K)	Rated Continuous Current at 60 Hz (A, rms)	Rated Short-Circuit Current at Rated Maximum kV (kA, rms)	Rated Interrupting Time Cycles	Rated Maximum Voltage Divided by K (kV, rms)	Maximum Symmetrical Interrupting Capability and Rated Short-Time Current (kA, rms)	Closing and Latching Capability 2.7 K Times Rated Short-Circuit Current (kA, peak)
4.76	1.36	1200	8.8	5	3.5	12	32
4.76	1.24	1200, 2000	29	5	3.85	36	97
4.76	1.19	1200, 2000, 3000	41	5	4.0	49	132
8.25	1.25	1200, 2000	33	5	6.6	41	111
15.0	1.3	1200, 2000	18	5	11.5	23	62
15.0	1.3	1200, 2000	28	5	11.5	36	97
15.0	1.3	1200, 2000, 3000	37	5	11.5	48	130
38.0	1.65	1200, 2000, 3000	21	5	23.0	35	95
38.0	1.0	1200, 3000	41	5	38.0	40	108

Source: ANSI Standard C37.06, AC High-Voltage Circuit Breakers Rated on a Symmetrical Current Basis—Preferred Ratings and Related Capabilities, 2000 (revision of 1987).

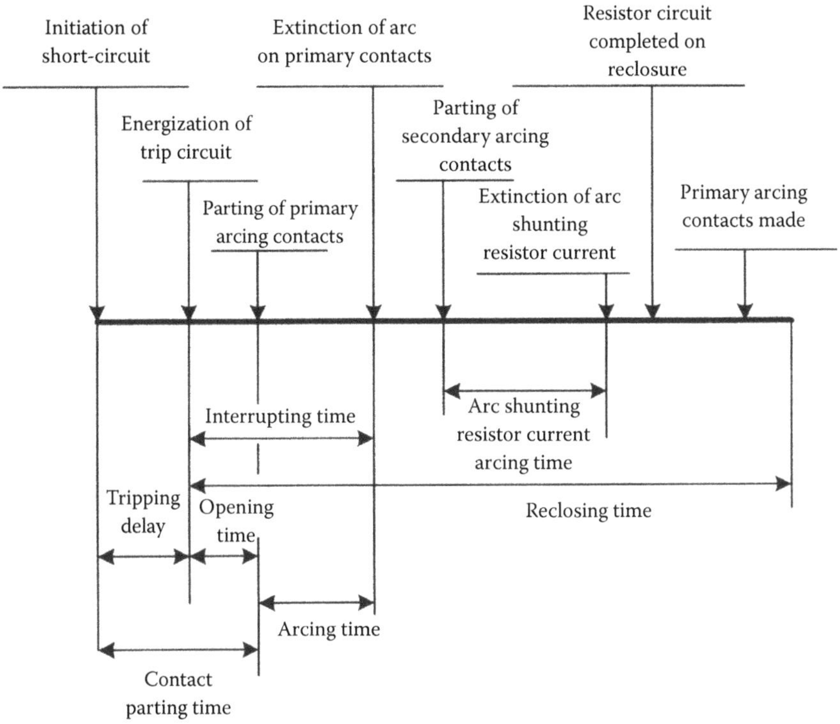

FIGURE 5.6
Operating time diagram of circuit breakers. (From ANSI/IEEE Standard C37.010, Application Guide for AC High-Voltage Circuit Breakers Rated on a Symmetrical Current Basis, 1999 (R-2005).)

voltages across breaker contacts may not reach beyond the capability to maintain synchronism, and (3) during single-pole reclosing, some synchronizing power is transferred and stability improved. Before a circuit is reenergized, there has to be some dead time in the circuit breaker for arc path to become deionized. A dead time of 135 ms is normally required at 115–138 kV for breakers without resistors across the interrupters. A manufacturer will supply this data for the specific breaker type. The rated interrupting capability has to be derated for reclosing according to [5].

For utility services to industrial consumers having induction and synchronous motor loads, the reclosing is not used. A sudden uncontrolled return of power can damage the motors. Synchronous motors are not suitable for reclosing and for induction motor transfer schemes special electronic, and control devices are available to affect (1) fast transfer, (2) in-phase transfer, and (3) residual voltage transfer, not discussed.

5.5 Maximum Peak Current

The asymmetrical peak *does not* occur at half cycle of the 60 Hz wave, considering that the fault occurs at zero point on the voltage wave (see Chapter 7). The peak will occur at half cycle for purely inductive circuits only. IEEE standard, Ref. [9] tabulates per unit peak currents using exact calculations, IEC equations, half cycle values, and also the equations recommended in this standard. The peak given at 0.5 cycles by the following equation is *nonconservative*:

$$Halfcycle_{\text{peak}} = I_{\text{ac peak}}\left(1 + e^{-\pi/\left(\frac{X}{R}\right)}\right) \tag{5.7}$$

See Chapter 7 for further discussions.

Example 5.2

Consider a circuit breaker of 15 kV, $K=1.3$, Table 5.2, rated short-circuit current of 37 kA applied at 13.8 kV. Its interrupting duty at voltage of application is 40.2 kA, and peak close and latch capability is 130 kA. The breakers thus rated had a peak close and latch = 2.7 K times the short-circuit rating = $2.7 \times 1.3 \times 37 = 128.87$ kA ≈ 130 kA, as read from Table 5.2.

Now consider a 15 kV, 50 kA interrupting rating breaker, $K=1$, Table 5.1. Its close and latch are also 130 kA. This is so because in the new standards, the close and latch have been reduced to 2.6 times the rated short-circuit current = $2.6 \times 50 = 130$ kA.

This factor should be considered when applying circuit breakers according to the revised standards. A circuit breaker of higher interrupting rating has a corresponding lower close and latch capability. Close to the generating stations, due to high-X/R ratios, the close and latch capability may be the limiting factor in selecting the ratings of the circuit breakers. In fact, manufacturers marketed breakers of close and latch capability higher than the interrupting rating. For example, referring to Table 5.2, a 15.5 kV breaker of rated short-circuit rating of 28 kA, but close and latch capability of 130 kA peak, was available. Also see Chapter 7.

5.6 Permissible Tripping Delay

The values of maximum tripping delay Y are specified in an ANSI standard [8], i.e., for circuit breakers of 72.5 kV and below, the rated permissible delay is 2 s and for circuit breakers of 121 kV and above it is 1 s. The tripping delay at lower values of current can be increased and, within a period of 30 min, should not increase the value given by

$$\int_0^t i^2 dt = Y[\text{rated short-circuit current}]^2 \tag{5.8}$$

It will then be capable of interrupting any short-circuit current, which at the instant of primary arcing contact separation has a symmetrical value not exceeding the required asymmetrical capability.

5.7 Service Capability Duty Requirements and Reclosing Capability

ANSI rated circuit breakers are specified to have the following interrupting performance:

1. Between 85% and 100% of asymmetrical interrupting capability at operating voltage three standard duty cycles.

O – 15 s – CO – 3 min – CO or O – 0.3 s–
CO – 3 min – CO for circuit breakers for rapid reclosing
For generator breakers: CO–30 min–CO

2. Between rated continuous current and 85% of required asymmetrical capability, a number of operations in which the sum of interrupted currents is a minimum of 800% of the required asymmetrical interrupting capability of the breaker at rated voltage.
   CO: close–open

Whenever a circuit breaker is applied having more operations or a shorter time interval between operations, other than the standard duty cycle, the rated short-circuit current and related required capabilities are reduced by a reclosing capability factor R, determined as follows:

$$R = (100 - D)\% \tag{5.9}$$

$$D = d_1(n - 2) + d_1\frac{(15 - t_1)}{15} + d_1\frac{(15 - t_2)}{15} + \cdots \tag{5.10}$$

where
D is the total reduction factor in percent
d_1 is specified in an ANSI standard [5]
n is the total number of openings
t_1 is the first time interval (<15 s)
t_2 is the second time interval (<15 s)

Interrupting duties thus calculated are subject to further qualifications. These should be adjusted for X/R ratios. All breakers are not rated for reclosing duties. Breakers rated more than 1200 A and below 100 kV are not intended for reclosing operations. Breakers rated 100 kV and above have reclosing capabilities irrespective of the current ratings. Table 5.3 shows preferred ratings for outdoor circuit breakers 121 kV and above.

5.7.1 Transient Stability on Fast Reclosing

High-speed reclosing is used to improve transient stability and voltage conditions in a grid system. On a step variation of the shaft power (increase), the torque angle of the synchronous machine (Chapter 6) will overshoot, which may pass the peak of the stability curve. It will settle down to *new torque angle demanded by the changed conditions* after a series of oscillations. If these oscillations damp out, we say that the stability will be achieved; if these oscillations diverge, then the stability will be lost.

The basic concept of equal area criteria of stability is illustrated with reference to Figure 5.7. Note that the acceleration area due to variation of the kinetic energy of the rotating masses is

$$A_{\text{accelerating}} = ABC = \int_{\sigma_1}^{\sigma_2} (T_{\text{shaft}} - T_e)\, d\sigma \tag{5.11}$$

TABLE 5.3

Preferred Ratings for Outdoor Circuit Breakers 121 kV and Above Including Circuit Breakers Applied in Gas-Insulated Substations

Rated Maximum Voltage (kV)	Rated Voltage Range Factor (K)	Rated Continuous Current at 60 Hz (A, rms)	Rated Short-Circuit Current at Rated Maximum Voltage (kA, rms)	Rated Time to Point P (T_2, μs)	Rated Rate R (kV/μs)	Rated Delay Time (T_1, μs)	Rated Interrupting Time Cycles	Maximum Permissible Tripping Delay	Closing and Latching Capability, 2.6 K Times Rated Short-Circuit Current (kA, Peak)
123	1.0	1200, 2000	31.5	260	2.0	2	3 (50 ms)	1	82
123	1.0	1600, 2000, 3000	40	260	2.0	2	3	1	104
123	1.0	2000, 3000	63	260	2.0	2	3	1	164
145	1.0	1200, 2000	31.5	330	2.0	2	3	1	82
145	1.0	2000, 3000	63	310	2.0	2	3	1	164
145	1.0	2000, 3000	80	310	2.0	2	3	1	208
170	1.0	1600, 2000, 3000	31.5	360	2.0	2	3	1	82
170	1.0	2000	40	360	2.0	2	3	1	104
170	1.0	2000	63	360	2.0	2	3	1	164
245	1.0	1600, 2000, 3000	31.5	520	2.0	2	3	1	82
245	1.0	2000, 3000	63	520	2.0	2	3	1	164
362	1.0	2000, 3000	40	775	2.0	2	2 (33 ms)	1	104
362	1.0	2000, 3000	63	775	2.0	2	2	1	164
550	1.0	2000, 3000	50	1325	2.0	2	2	1	130
550	1.0	3000, 4000	63	1325	2.0	2	2	1	164
800	1.0	2000, 3000	40	1530	2.0	2	2	1	104
800	1.0	3000, 4000	63	1530	2.0	2	2	1	164

Source: ANSI Standard C37.06, AC High-Voltage Circuit Breakers Rated on a Symmetrical Current Basis—Preferred Ratings and Related Capabilities, 2000 (revision of 1987).

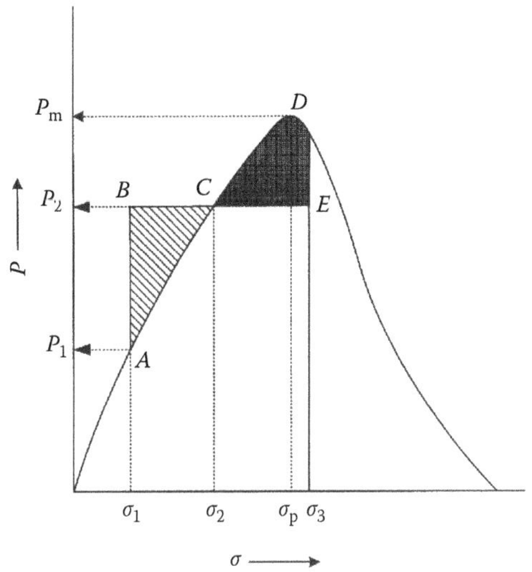

FIGURE 5.7
To illustrate the concept of equal area criteria of stability.

The area *CDE* is the de-acceleration area:

$$B_{\text{deaccelerating}} = CDE = \int_{\sigma_2}^{\sigma_3} (T_{\text{shaft}} - T_{\text{e}})\,\mathrm{d}\sigma \tag{5.12}$$

In accelerating and decelerating the machine passes, the equilibrium point given by δ_2 giving rise to oscillations, which will either increase or decrease. If the initial impact is large enough, the machine will be instable in the very first swing.

If,

$$A_{\text{accelerating}} = B_{\text{deaccelerating}} \tag{5.13}$$

Then, there are chances of the generator remaining in synchronism [10]. The asynchronous torque produced by the dampers has been neglected in this analysis; also, the synchronizing power is assumed to remain constant during the disturbance. It is clear that at point *C*, the accelerating power is zero, and assuming a generator connected to an infinite bus, the speed continues to increase. It is more than the speed of the infinite bus and at point *E*, the relative speed is zero, and the torque angle ceases to increase, but the output is more than the input and the torque angle starts decreasing; rotor decelerates. But, for damping, these oscillations can continue. This is the concept of "equal area criterion of stability."

Figure 5.8 illustrates the effect of high-speed single-phase reclosing on transient stability using equal area criteria of stability. A transient single line-to-ground fault occurs on the tie line; the tie line breaker opens and then closes within a short-time delay, called the dead time of the breaker. Some synchronizing power flows through two unfaulted phases, during a single line-to-ground fault, and no power flows during the dead time. The dead time of the circuit breaker is σ_1–σ_2, and synchronous motors and power capacitors tend to prolong the arcing time. Applying equal area criteria of stability, if shaded area $1 + C$ is equal to shaded area 2; stability is possible.

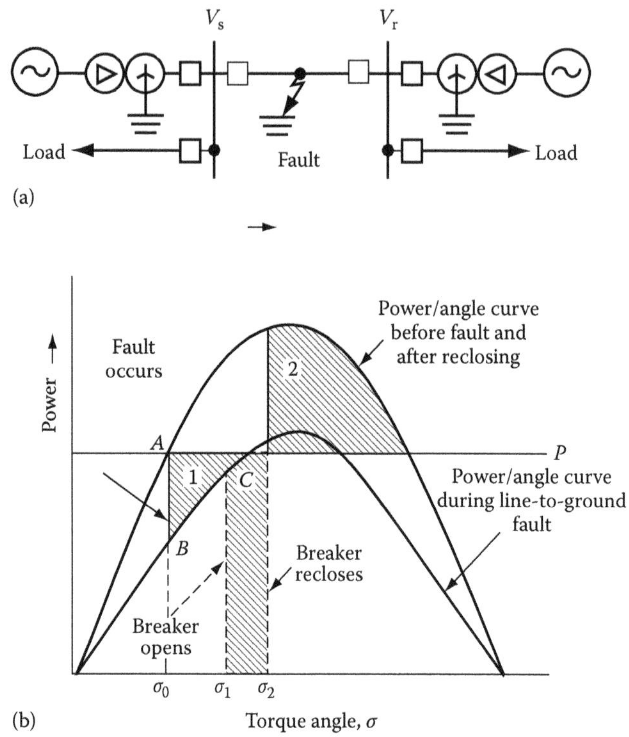

FIGURE 5.8
Transient stability of a tie line circuit with fast reclosing on a single line-to-ground fault: (a) equivalent system representation; (b) equal area criteria of stability. Fault occurs at torque angle σ_0; breaker opens at σ_1 and recloses at σ_2 to remove the transient fault.

5.8 Capacitance Current Switching

The capacitance switching is encountered when switching overhead lines and cables at no load and when switching shunt capacitor banks and filters. With respect to the application of circuit breakers for capacitance switching consider (1) type of application that is overhead line, cable or capacitor bank, or shunt capacitor filter switching, (2) power frequency and system grounding, and (3) presence of one-or-two-phase-to-ground faults.

ANSI ratings [8] distinguish between general-purpose and *definite-purpose* circuit breakers, and there is a vast difference between their capabilities for capacitance current switching. Definite-purpose breakers may have different constructional features, i.e., a heavy-duty closing and tripping mechanism and testing requirements. Table 5.4 shows the capacitance current switching ratings. General-purpose breakers do not have any back-to-back capacitance current switching capabilities. A 123 kV general-purpose circuit breaker of rated current 2 kA and 63 kA symmetrical short circuit has overhead line charging current or isolated capacitor switching current capability of 50 A rms and no back-to-back switching capability.

Rated transient inrush current is the highest magnitude, which the circuit breaker will be required to close at any voltage up to the rated maximum voltage and will be measured by the system and unmodified by the breaker. Rated transient inrush frequency is the highest natural frequency that the circuit breaker is required to close at 100% of its rated back-to-back shunt capacitor or cable switching current.

For systems below 72.5 kV, shunt capacitors may be grounded or ungrounded and for systems above 121 kV, both the shunt capacitors and the systems will be solidly grounded.

TABLE 5.4

Preferred Capacitance Current Switching Rating for Outdoor Circuit Breakers 121 kV and Above Including Circuit Breakers Applied in Gas-Insulated Substations

					Definite-Purpose Breakers Rated Capacitance Switching Current Shunt Capacitor Bank or Cable				
							Back-to-Back Switching		
								Inrush Current	
Rated Maximum Voltage (kV)	Rated Short-Circuit Current at Rated Maximum Voltage (kA, rms)	Rated Continuous Current at 60 Hz (A, rms)	General-Purpose Circuit Breakers Rated Overhead Line Current (A, rms)	General-Purpose Circuit Breakers, Rated Isolated Current (A, rms)	Overhead Line Current (A, rms)	Rated Isolated Current (A, rms)	Current (A, rms)	Peak Current (kA)	Frequency (Hz)
123	31.5	1200, 2000	50	50	160	315	315	16	4250
123	40	1600, 2000, 3000	50	50	160	315	315	16	4250
123	63	2000, 3000	50	50	160	315	315	16	4250
145	31.5	1200, 2000	80	80	160	315	315	16	4250
145	40	1600, 2000, 3000	80	80	160	315	315	16	4250
145	63	2000, 3000	80	80	160	315	315	16	4250
145	80	2000, 3000	80	80	160	315	315	16	4250
170	31.5	1600, 2000	100	100	160	400	400	20	4250
170	40	2000, 3000	100	100	160	400	400	20	4250
170	50	2000, 3000	100	100	160	400	400	20	4250
170	63	2000, 3000	100	100	160	400	400	20	4250
245	31.5	1600, 2000, 3000	160	160	200	400	400	20	4250
245	40	2000, 3000	160	160	200	400	400	20	4250
245	50	2000, 3000	160	160	200	400	400	20	4250
245	63	2000, 3000	160	160	200	400	400	20	4250
362	40	2000, 3000	250	250	315	500	500	25	4250
362	63	2000, 3000	250	250	315	500	500	25	4250
550	40	2000, 3000	400	400	500	500	500	25	4250
550	63	3000, 4000	400	400	500	500	500	25	4250
800	40	2000, 3000	500	500	500	500	—	—	—
800	63	3000, 4000	500	500	500	500	—	—	—

Source: ANSI Standard C37.06, AC High-Voltage Circuit Breakers Rated on a Symmetrical Current Basis—Preferred Ratings and Related Capabilities, 2000 (revision of 1987).

If the neutral of the system, the capacitor bank, or both are ungrounded, the manufacturer should be consulted for circuit breaker application. The first phase to interrupt affects the recovery voltage.

The following definitions are applicable:

1. Rated open wire line charging current is the highest line charging current that the circuit breaker is required to switch at any voltage up to the rated voltage.
2. Rated isolated cable charging and isolated shunt capacitor bank switching current are the highest isolated cable or shunt capacitor current that the breaker is required to switch at any voltage up to the rated voltage.
3. The cable circuits and switched capacitor bank are considered isolated if the rate of change of transient inrush current, $\mathrm{d}i/\mathrm{d}t$ does not exceed the maximum rate of change of symmetrical interrupting capability of the circuit breaker at the applied voltage [11]:

$$\left(\frac{\mathrm{d}i}{\mathrm{d}t}\right)_{\max} = \sqrt{2}\omega I \quad (5.14)$$

   where I is the rated short-circuit current in ampères.
4. Cable circuits and shunt capacitor banks are considered switched back to back if the highest rate of change of inrush current on closing exceeds that for which the cable or shunt capacitor can be considered isolated.

The oscillatory current on back-to-back switching is limited only by the impedance of the capacitor bank and the circuit between the energized bank and the switched bank.

The inrush current and frequency on capacitor current switching can be calculated by the solution of the following differential equation:

$$iR + L\frac{\mathrm{d}i}{\mathrm{d}t} + \int \frac{i\,\mathrm{d}t}{C} = E_{\mathrm{m}} \sin \omega t \quad (5.15)$$

The solution to this differential equation is discussed in many texts and is of the form

$$i = A \sin(\omega t + \alpha) + Be^{-Rt/2L} \sin(\omega_0 t - \beta) \quad (5.16)$$

where

$$\omega_0 = \sqrt{\frac{1}{LC} - \frac{R^2}{4L^2}} \quad (5.17)$$

The first term is a forced oscillation, which in fact is the steady-state current, and the second term represents a free oscillation and has a damping component $e^{-rt/2L}$. Its frequency is given by $\omega_0/2\pi$. Resistance can be neglected, and this simplifies the solution. The maximum inrush current is given at an instant of switching when $t = \sqrt{(LC)}$. For the purpose of evaluation of switching duties of circuit breakers, the maximum inrush current on switching an isolated bank is

$$I_{\text{peak}} = \frac{\sqrt{2}}{\sqrt{3}} E_{\text{rms}} \sqrt{\frac{C}{L}} \tag{5.18}$$

where
E is the line-to-line voltage
C and L are in H and F, respectively

The inrush frequency is

$$f_{\text{inrush}} = \frac{1}{2\pi\sqrt{LC}} \tag{5.19}$$

For back-to-back switching, i.e., energizing a bank on the same bus when another energized bank is present, the inrush current is entirely composed of interchange of currents between the two banks. The component supplied by the source is of low frequency and can be neglected. This will not be true if the source impedance is comparable to the impedance between the banks being switched back to back. The back-to-back switching current is given by

$$I_{\text{inrush}} = \frac{\sqrt{2}}{\sqrt{3}} E_{\text{rms}} \sqrt{\frac{C_1 C_2}{(C_1 + C_2)(L_{\text{eq}})}} \tag{5.20}$$

where L_{eq} is the equivalent reactance between the banks being switched. The inrush frequency is

$$f_{\text{inrush}} = \frac{1}{2\pi\sqrt{L_{\text{eq}} C_1 C_2/(C_1 + C_2)}} \tag{5.21}$$

The 2005 revision to IEEE standard [11,12] and also IEEE standard [1] classify the breakers for capacitance switching as class C0, C1, and C2 as well as mechanical endurance class M1 or M2 are assigned. These classes have specific type-testing duties, coordinated as per standards [13,14]. Table 5.4 from IEEE standard [8] shows the capacitance current switching ratings of breakers from 121 to 800 kV, while Table 5.5 shows similar ratings for classes C0, C1, and C2 from standard [1]. These two tables can be compared.

The presence of single- and two-phase faults is an important factor that determines the recovery voltage across the breaker contacts, see Chapter 4. Another phenomena that has been observed primarily in vacuum circuit breakers is called NSDD (nonsustained disruptive discharge), which is defined as the disruptive discharge associated with current interruption that does not result in current at natural frequency of the circuit. For other types of breakers, this can be ignored.

Class C1 circuit breaker is acceptable for medium voltage circuit breakers and for circuit breakers applied for infrequent switching of transmission lines and cables. Class C2 is recommended for frequent switching of transmission lines and cables [11]. An important consideration is the transient overvoltages that may be generated by restrikes during opening operation (see Figure 4.12). The effect of these transients will be local as well as remote.

TABLE 5.5

Preferred Capacitance Current Switching Ratings for Circuit Breakers Rated 100 kV and Above Including Circuit Breakers in Gas Insulated Substations

		Class C0 Circuit Breakers		Class C1 or C2 Circuit Breakers											
						Back to Back Capacitor Bank Switching									
							Rated Inrush Current								
							Preferred Ratings		Alternate 1 Rating		Alternate 2 Rating		Alternate 3 Rating		
Rated Maximum Voltage	Rated Continuous Current (Applicable to All Continuous Current Ratings) (A rms)	Rated Overhead Line Current (A rms)	Rated Isolated Capacitor Bank or Cable Current (A rms)	Rated Isolated Capacitor Bank or Cable Current (A rms)	Rated Overhead Line Current (A rms)	Rated Current (A rms)	Peak Value (kA)	Frequency (kHz)	Peak Value (kA)	Frequency (kHz)	Peak Value (kA)	Frequency (kHz)	Peak Value (kA)	Frequency (kHz)
123	9	50	50	1200	160	700	16	4.3	6	2	25	13	60	8.5
145	9	80	80	1200	160	700	16	4.3	6	2	25	13	60	8.5
170	9	100	100	1200	160	700	20	4.3	6	2	25	13	60	8.5
245	9	160	160	1200	200	700	20	4.3	6	2	25	13	60	8.5
362	9	250	250	1200	315	800	25	4.3	6	2	20	21	65	8.5
550	9	400	400	1000	500	800	25	4.3	6	2	20	21	65	8.5
800	9	900	500	1000	900	800	25	4.3	6	2	20	21	65	8.5

Source: IEEE PC37.06/D11, Draft Standard AC High-Voltage Circuit Breakers Rated on Symmetrical Current Basis-Preferred Ratings and Related Required Capabilities for Voltages Above 1000 Volts, 2008.

5.8.1 Switching of Cables

The cable charging currents depend upon the length of cable, cable construction, system voltage, and insulation dielectric constants. Much akin to switching devices for capacitor banks, a cable is considered isolated if the maximum rate of change with respect to time of transient inrush current on energizing an uncharged cable does not exceed the rate of change associated with maximum symmetrical interrupting current of the switching device.

The cables may be switched back to back, much akin to capacitor banks. Transient currents of high magnitude and initial high rate of change flow between cables when the switching circuit breaker is closed or on restrikes on opening.

For isolated cable switching, we can use the following expression:

$$i = \frac{u_m - u_t}{Z}\left[1 - \exp\left(-\frac{Z}{L}t\right)\right] \tag{5.22}$$

where

u_m is the supply system voltage
u_t is the trapped voltage on the cable being switched
Z is the cable surge impedance
L is the source inductance

From Equation 5.22, the maximum inrush current is

$$i_p = \frac{u_m - u_t}{Z} \tag{5.23}$$

Equation 5.23 can be modified for back-to-back switching

$$i = \frac{u_m - u_t}{Z_1 + Z_2}\left[1 - \exp\left(-\frac{Z_1 + Z_2}{L}t\right)\right] \tag{5.24}$$

where Z_1 and Z_2 are the surge impedances of cable 1 and cable 2, respectively.

Again, akin to back-to-back switching of capacitors, the source reactance can be ignored. The peak current is

$$i_p = \frac{u_m - u_t}{Z_1 + Z_2} \tag{5.25}$$

And its frequency is given by

$$f_{eq} = f\left[\frac{u_m - u_t}{\omega(L_1 + L_2)I_{ir}}\right] \tag{5.26}$$

where

f_{eq} is the inrush frequency
I_{ir} is the charging current of one cable
L_1 and L_2 are the inductances of cables 1 and 2, respectively

The switching of single cable forms a series circuit, and the transient will be oscillatory, damped, or critically damped. Following equations can be written for the inrush current:

$$
\begin{aligned}
&Z = \sqrt{4L/C} \qquad i_{\mathrm{p}} = 0.368\frac{u_{\mathrm{m}} - u_{\mathrm{t}}}{Z}\\
&Z < \sqrt{4L/C} \qquad i_{\mathrm{p}} = \frac{u_{\mathrm{m}} - u_{\mathrm{t}}}{\sqrt{L/C - Z^2/4}}\left[\exp\left(-\frac{Z\pi}{4L}\frac{1}{\sqrt{L/C - Z^2/4}}\right)\right]\\
&Z > \sqrt{4L/C} \qquad i_{\mathrm{p}} = \frac{u_{\mathrm{m}} - u_{\mathrm{t}}}{\sqrt{L/C - Z^2/4}}[\exp(-\alpha t_{\mathrm{m}}) - \exp(-\beta t_{\mathrm{m}})]
\end{aligned}
\tag{5.27}
$$

where

$$
\begin{aligned}
t_{\mathrm{m}} &= \frac{\ln \alpha/\beta}{\alpha - \beta}\\
\alpha &= \frac{Z}{L} - \sqrt{\frac{Z^2}{L^2} - \frac{4}{LC}}\\
\beta &= \frac{Z}{L} + \sqrt{\frac{Z^2}{L^2} - \frac{4}{LC}}
\end{aligned}
\tag{5.28}
$$

The magnetic fields due to high inrush currents during back-to-back switching can induce voltage in control cables by capacitive and magnetic couplings. This is minimized by shielding the control cables.

Example 5.3

Consider the system of Figure 5.9a. Two capacitor banks C_1 and C_2 are connected on the same 13.8 kV bus. The inductances in the switching circuit are calculated in Table 5.6. Let C_1 be first switched. The inrush current is mostly limited by the source inductance, which predominates. The inrush current magnitude and frequency, using the expressions in Equations 5.18 and 5.19, are 5560 A peak and 552.8 Hz, respectively. The maximum rate of change of current is

$$2\pi(552.81)(5560) \times 10^{-6} = 19.31\ \mathrm{A}/\mu\mathrm{s}$$

Consider that a definite-purpose indoor 15 kV breaker of 2 kA continuous rating and 40.2 kA interrupting at 13.8 kA. From Equation 5.14, the breaker di/dt is:

$$2\pi(60)\sqrt{2}(40{,}200)10^{-6} = 21.43\ \mathrm{A}/\mu\mathrm{s}$$

This is more than 19.283 A/μs as calculated above. Thus, the capacitor bank can be considered isolated.

Now calculate the inrush current and frequency on back-to-back switching, i.e., capacitor C_2 is switched when C_1 is already connected to the bus. The equivalent inductance on back-to-back switching consists of a small length of bus between the banks and their cable connections. The source inductance is ignored, as practically no current is contributed from the source. The inductance in the back-to-back switching circuit is 14.46 μH, Figure 5.9b. From Equations 5.20 and 5.21, the inrush current is 22.7 kA, and the inrush frequency is 5653 Hz. As per data from [8], the intended 15.5 kV definite-purpose breaker to be used has a maximum peak inrush current of 18 kA and an inrush frequency of 2.4 kHz. In this example, even a definite-purpose circuit breaker

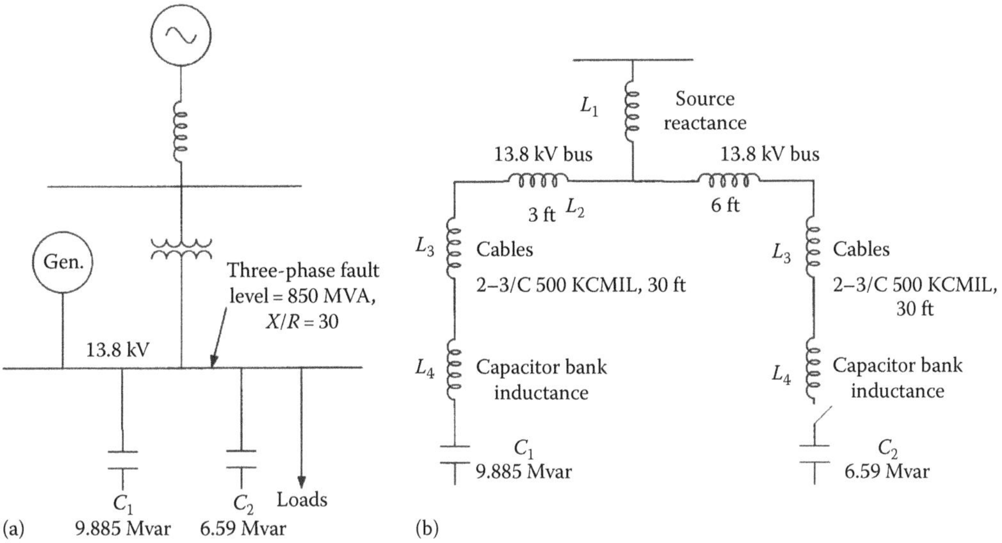

FIGURE 5.9
(a) Connection diagram for capacitor bank switching; (b) equivalent reactance diagram (Example 5.3).

TABLE 5.6

Capacitor Switching (Example 5.3): Calculation of Inductances and Capacitances

No.	System Data	Calculated Inductance or Capacitance
1	Three-phase short-circuit level at 13.8 kV bus, 850 MVA, $X/R = 30$	L_1 source = 593.97 μH
2	3′ of 13.8 kV bus	L_2 bus = 0.63 μH
3	30′ of 2–3/C 500 KCMIL cables	L_3 cable = 1.26 μH
4	Inductance of the bank itself	L_4 bank = 5 μH
5	Total inductance, when capacitor C_1 is switched = $L_1 + L_2 + L_3 + L_4$	600.86 μH
6	Capacitance of bank C_1, consisting of 9 units in parallel, one series group, wye connected, rated voltage 8.32 kV, 400 kvar each, total three-phase kvar at 13.8 kV = 9.885 Mvar	$C_1 = 0.138 \times 10^{-3}$ F
7	Total inductance when C_2 is switched and C_1 is already energized = inductance of 9 feet of 13.8 kV bus, 60 feet of cables and inductances of banks themselves	14.42 μH
8	Capacitance of bank C_2, consisting of 6 units in parallel, one series group, wye connected, rated voltage 8.32 kV, 400 kvar each, total three-phase kvar at 13.8 kV = 6.59 Mvar	$C_2 = 0.092 \times 10^{-3}$ F

will be applied beyond its rating. In order to reduce the inrush current and frequency, an additional reactance should be introduced into the circuit. An inductance of 70 μH will reduce the inrush current to 9.4 kA and the frequency to 2332 Hz. Inrush current-limiting reactors are generally required when the capacitor banks are switched back to back on the same bus. In the case where power capacitors are applied as shunt-tuned filters, the filter reactors will reduce the inrush current and its frequency, so that the breaker duties are at acceptable levels.

An electromagnetic transient program (EMTP) simulation of the back-to-back switching of inrush current in Example 5.3 is shown in Figure 5.10.

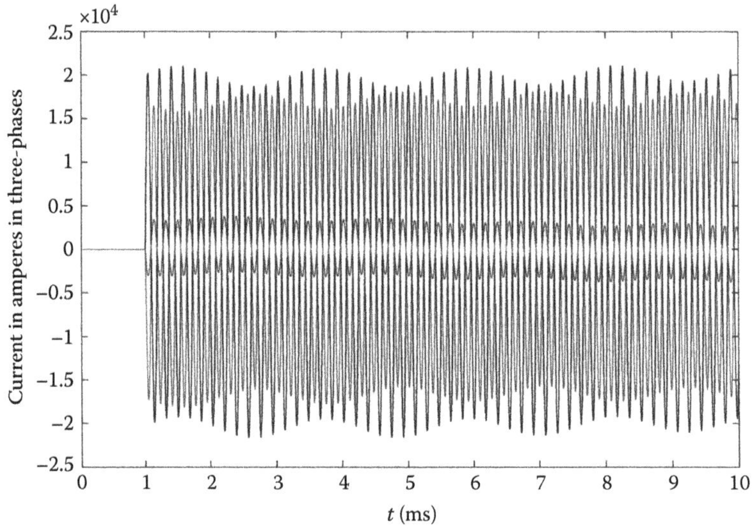

FIGURE 5.10
Results of EMTP simulation for back-to-back switching of Example 5.3.

5.8.2 Effects of Capacitor Switching

The switching of capacitor currents is associated with restrikes, Chapter 4. The overvoltage control is also discussed in Chapter 4. Special applications may exist where the circuit-breaker duties need to be carefully evaluated [8,11]. These applications may be:

1. Switching through a transformer of turns ratio greater than one will have the effect of increasing the switching current. Dropping EHV and UHV lines through low-voltage circuit breakers can increase the effective line charging current in the 750–1000 A range.
2. The effect of the capacitive discharge currents on voltage induced in the secondary of the bushing type current transformer should be considered. In certain system configurations, i.e., when a number of capacitors are connected to a bus, for a fault on a feeder circuit, all bus-connected capacitors will discharge into the fault. The BCT (bushing current transformer) secondary voltage may reach high values. This secondary voltage can be estimated from

$$\begin{gathered}\left(\frac{1}{\text{BCT ratio}}\right)(\text{Crest transient current})\\ \left((\text{Relay reactance})\left(\frac{\text{Transient frequency}}{\text{System frequency}}\right)\right)\end{gathered} \tag{5.29}$$

   Higher than the normal inrush currents are possible on fast reclosing of power capacitors. Reclosing is, generally, not attempted on power capacitor banks. Capacitors over 600 V are provided with internal discharge devices (resistors) to reduce the residual charge to 50 V or less within 5 min.
3. When parallel banks of capacitors are located on a bus section, caution must be applied in fault switching sequence, so that the last circuit breaker to clear the fault is not subjected to a capacitive switching duty beyond its capability.

4. Switching capacitor banks under faulted conditions give rise to high recovery voltages, depending on the grounding and fault type. A phase-to-ground fault produces the most severe conditions when the source is ungrounded and the bank neutral is grounded. If an unfaulted phase is first to clear, the current may reach 1.73 times the rated current and the recovery voltage $3.46E_{max}$, phase-to-ground. When the faulted phase is first to interrupt, the current is 3.06 times the rated current and the recovery voltage $3.0E_{max}$.
5. Switching of transformers and capacitors can bring harmonic resonance and increase the inrush current and its time duration.

5.9 Line-Closing Switching Surge Factor

The rated line-closing switching surge factors are specified in ANSI for breakers of 362 kV and above specifically designed to control the switching overvoltages and are shown in Table 5.7 [8]. The rating designates that the breaker is capable of controlling the switching surge voltages so that the probability of not exceeding the rated overvoltage factor is 98% or higher when switching the standard *reference transmission line* from a *standard reference source* [13].

Switching surge overvoltages are discussed in Section 4.11. ANSI takes a statistical approach. Random closing of circuit breaker will produce line-closing switching surge maximum voltages, which vary in magnitude according to the instantaneous value of the source voltage, the parameters of the connected system, and the time difference between completions of a circuit path by switching traveling waves in each phase. These variations will be governed by laws of probability, and the highest and lowest overvoltages will occur infrequently.

The assumptions are that the circuit breaker connects the overhead line directly to a power source, open at the receiving end and not connected to terminal apparatus such as a power transformer, though it may be connected to an open switch or circuit breaker. The system does not include surge arresters, shunt reactors, potential transformers, or series or shunt capacitors.

TABLE 5.7

Rated Line Closing Switching Surge Factors for Circuit Breakers Specifically Designed to Control Line Closing Switching Surge Maximum Voltage and Parameters of Standard Reference Transmission Lines

Rated Maximum Voltage (kV, rms)	Rated Line Closing Switching Surge Factor	Line Length (mi)	Percentage Shunt Capacitance Divided Equally at Line Ends	L_1	L_0/L_1	R_1	R_0	C_1	C_1/C_0
362	2.4	150	0	1.6	3	0.05	0.5	0.02	1.5
500	2.2	200	0	1.6	3	0.03	0.5	0.02	1.5
800	2.0	200	60	1.4	3	0.02	0.5	0.02	1.5

Source: ANSI/IEEE Standard C37.010, Application Guide for AC High-Voltage Circuit Breakers Rated on a Symmetrical Current Basis, 1999 (R-2005).

L_1 = positive and negative sequence inductance in mH per mi, L_0 = zero sequence inductance in mH per mi, R_1 = positive and negative sequence resistance in ohms per mi, R_0 = zero sequence resistance in ohms per mi, C_1 = positive and negative sequence capacitance in microfarads per mi, C_0 = zero sequence capacitance in microfarads per mi.

The reference power source is a three-phase wye-connected voltage source with the neutral grounded and with each of the three-phase voltages in series with an inductive reactance, which represents the short-circuit capability of the source. The maximum source voltage, line to line, is the rated voltage of the circuit breaker. The series reactance is that which produces the rated short-circuit current of the circuit breaker, both three phase and single phase at rated maximum voltage with the short-circuit applied at the circuit breaker terminals.

The standard transmission line is a perfectly transposed three-phase transmission line with parameters as listed in the ANSI/IEEE standard [13]. Any power system that deviates too greatly from the standard reference power system may require that a simulated study be made.

5.9.1 Switching of Transformers

Certain switching operations can excite transformer windings high ringing frequencies. This can cause excessive stresses on the transformer interturn windings. Traditional surge arresters applied at the circuit breakers or transformer terminals are ineffective to counteract this phenomena. Resistance and capacitance "snubber" [5] networks connected across the transformer windings, phase to ground are shown to be effective.

The term "sympathetic inrush" is applied when a transformer is switched on the same bus to which another transformer is connected. The transformer that is on line will experience inrush currents when the parallel transformer on the same bus is switched.

5.10 Out-of-Phase Switching Current Rating

The assigned out-of-phase switching rating is the maximum out-of-phase current that can be switched at an out-of-phase recovery voltage specified in ANSI and under prescribed conditions. If a circuit breaker has an out-of-phase switching current rating, it will be 25% of the maximum short-circuit current in kiloampères, unless otherwise specified. The duty cycles are specified in ANSI/IEEE standard [13]. The conditions for out-of-phase switching currents are

1. Opening and closing operations in conformity with manufacturers' instructions, closing angle limited to a maximum out-of-phase angle of 90° whenever possible
2. Grounding conditions of the neutral corresponding to that for which the circuit breaker is tested
3. Frequency within ±20% of the rated frequency of the breaker
4. Absence of fault on either side of the circuit breaker

Where frequent out-of-phase operations are anticipated, the actual system recovery voltages should be evaluated (see Section 4.12). A special circuit breaker, or one rated at a higher voltage, may sometimes be required. As an alternative solution, the severity of out-of-phase switching can be reduced in several systems by using relays with coordinated impedance sensitive elements to control the tripping instant, so that interruption will occur substantially after or substantially before the instant the phase angle is 180°. Polarity sensing and synchronous breakers are discussed in Chapter 4.

5.11 Transient Recovery Voltage

As discussed in Chapter 4, the interrupting capability of the circuit breaker is related to transient recovery voltage (TRV). If the specified TRV withstand boundary is exceeded in any application, a different circuit breaker should be used, or the system should be modified. The addition of capacitors to a bus or line is one method of improving the recovery voltage characteristics. For proper application:

$$\mathrm{TRV}_{\mathrm{breaker}} > \mathrm{TRV}_{\mathrm{system}}$$

To calculate system TRV, dynamic simulation is required though simplified equations can be used for hand calculations. IEEE standard C.37.011 [15] is the application guide for TRV for ac high-voltage circuit breakers.

Breakers Rated Below 100 kV

For circuit breakers rated below 100 kV, the rated transient voltage is defined as the envelope formed by a 1 – cosine curve using the values of E_2 and T_2 defined in the ANSI standard [8]; E_2 is the peak of TRV, and its value is 1.88 times the maximumrated voltage. This value of 1.88 considers a first pole to clear factor of 1.5, as the systems below 100 kV may be ungrounded. The time T_2, specified in microseconds, to reach the peak is variable, depending on short-circuit type, circuit breaker, and voltage rating. For indoor oil-less circuit breakers up to 38 kV, T_2 varies from 50 to 125 μs, see Ref. [8], 2000 and Table 5.8.

The plot of this response curve for first half-cycle of the oscillatory component of TRV is shown in Figure 5.11. The supply voltage is considered at its peak during this interval and is represented by a straight line in Figure 5.11, i.e., the power frequency component of TRV is constant. This definition of TRV by two parameters, E_2 and T_2 is akin to that of Figures 5.4 through 5.8, i.e., IEC representation by two parameters. The curve of Figure 5.11 is called *one-minus-cosine* curve.

Breakers Rated 100 kV and Above

For breakers rated 100 kV and above, the rated TRV is defined by higher of an exponential waveform and 1 – cosine wavefrom, Figure 5.12 For systems above 100 kV, most, if not all, of the systems will be grounded, and a first pole to clear factor of 1.3 is considered. $E_1 = 1.06 \times V$ and $E_2 = 1.49$ V. Envelope formed by the exponential cosine curve obtained by using the rated values of E_1 R, T_1, E_2, and T_2 from the standards, and applying these values at the rated short-circuit current of the breaker, Table 5.8, R is defined as the rated TRV rate, ignoring the effect of the bus side-lumped capacitance, at which the recovery voltage rises across the terminals of a first-pole-to-interrupt for a three-phase, ungrounded load-side terminal fault (TF) under the specified rated conditions. The rate is a close approximation of the maximum de/dt in the rated envelope, but is slightly higher, because the bus side capacitance is ignored.

TABLE 5.8

Preferred Ratings of Prospective TRV for Class S1 Circuit Breakers Rated Below 100 kV, for Cable Systems Noneffectively Grounded (T100, T60, T10 Test Duties. TRV Representation by Two-Parameter Method)

Rated Maximum Voltage (kV, rms)	Test Duty	First Pole to Clear Factor k_{pp} (pu)	Amplitude Factor k_{af} (pu)	TRV Peak Value u_c	Time t_3 (μs)	Time Delay t_d (μs)	Reference Voltage u' (kV)	Time t' (μs)	RRRV u_c/t_3 (kV/μs)
4.76	TF	1.5	1.4	8.2	44	7	2.7	21	0.19
	OS	2.5	1.25	12.1	88	13	4.0	43	0.14
8.25	TF	1.5	1.4	14.1	52	8	4.7	25	0.27
	OS	2.5	1.25	21.1	104	16	7.0	50	0.20
15	TF	1.5	1.4	25.7	66	10	8.6	32	0.39
	OS	2.5	1.25	38.3	132	20	12.8	64	0.29
27	TF	1.5	1.4	46.3	92	14	15.4	45	0.50
	OS	2.5	1.25	68.9	184	28	23.0	90	0.37
38	TF	1.5	1.4	65.2	109	16	21.7	53	0.60
	OS	2.5	1.25	97.0	218	33	32.3	105	0.45
72.5	TF	1.5	1.4	124	165	25	41.4	80	0.75
	OS	2.5	1.25	185	330	50	61.7	160	0.56

Source: IEEE PC37.06/D11, Draft Standard AC High-Voltage Circuit Breakers Rated on Symmetrical Current Basis-Preferred Ratings and Related Required Capabilities for Voltages above 1000 Volts, 2008.

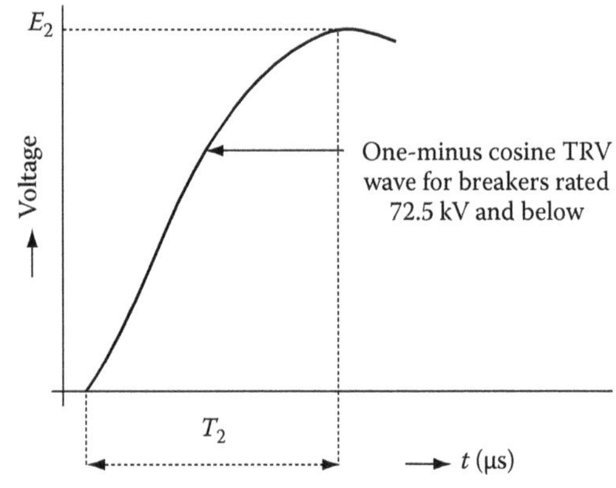

FIGURE 5.11
One-minus-cosine TRV wave for breakers rated 72.5 kV and below.

The exponential cosine envelope is defined by whichever of e_1 and e_2 is larger:

$$e_1 = E_1(1 - e^{-t/\tau}) \quad \text{with a time delay } T_1 \ \mu\text{s} \tag{5.30}$$

$$\tau = \frac{E_1}{R} \tag{5.31}$$

$$e_2 = \frac{E_2}{2}\left(1 - \cos\left(\frac{\pi t}{T_2}\right)\right) \tag{5.32}$$

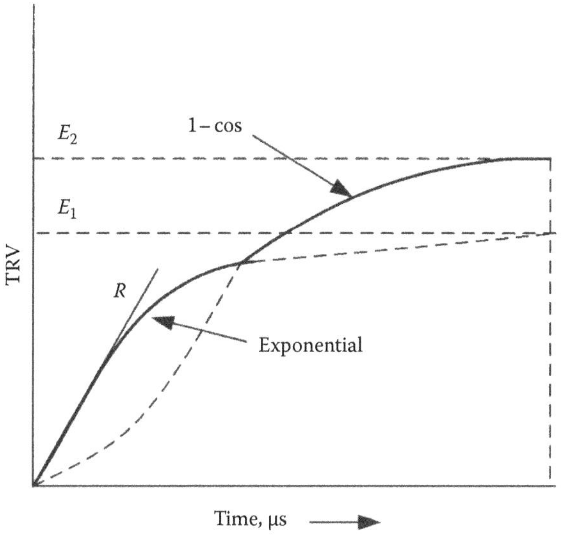

FIGURE 5.12
TRV profile for system voltages of 100 kV and above.

Example 5.4

Consider a 550 kV breaker. The ratings are

K factor = 1
Current rating = 2 kA
Rated short-circuit current = 40 kA
Rated time to point P, T_2 μs = 1325
R, rate of rise of recovery voltage = 2 kV/μs
Rated time delay $T_1 = 2$ μs $E_2 = 1.49 \times$ rated maximum voltage
$E_1 = 1.06 \times$ rated maximum voltage

Then

$$E_1 = 583\text{ kV}$$
$$E_2 = 819.5\text{ kV}$$
$$\tau = E_1$$
$$R = 583$$
$$2 = 291.5\ \mu\text{s}$$

Substituting in Equations 5.19 and 5.21:

$$e_1 = 673.6(1 - e^{-t/291.5})$$
$$e_2 = 409.80(1 - \cos 0.1358t^0)$$

The calculated TRV, for a rated fault current, is shown in Figure 5.13.

Fault Currents Other Than Rated

Circuit breakers are required to interrupt short-circuit currents that are less than the rated short-circuit currents. This increases E_2, and T_2 time is shorter. The adjustment factors are given in curves in Figure 5.14 from 1979 edition of Ref. [6]. The revision to this standard does not reproduce these curves but refers to 1979 edition.

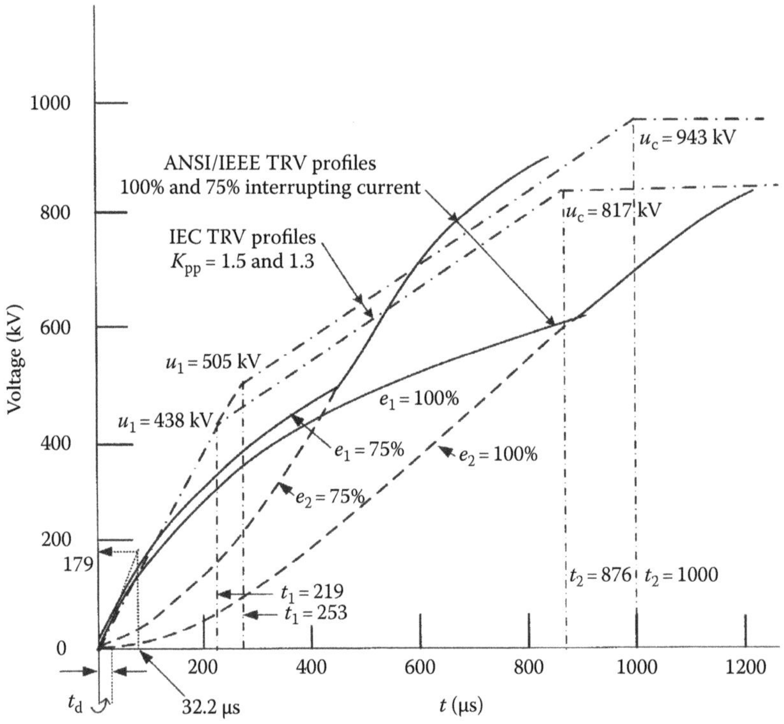

FIGURE 5.13
Calculated TRV waveshapes (Examples 5.4 and 5.5).

Example 5.5

Now consider that the TRV is required to be calculated for 75% of the rated fault current. This requires calculation of adjustment factors from Figure 5.9.

$K_r = 1.625$ (rate of rise multiplying factor)
$K_1 = 1.044$ (E_2 multiplying factor)
$K_t = 1.625$ (T_2 dividing factor)

The adjusted parameters are

$E_1 = 583$ kV
$E_2 = (819.5)(K_1) = 855.5$ kV
$R = (2)(K_r) = 3.25$ kV/μs
$T_2 = (1325)/(K_t) = 815.4$ μs
$T_1 = 2$ μs
$\tau = E_1/R = (583)/3.25 = 179.4$ μs

The TRV for 75% interrupting fault duty is superimposed in Figure 5.13 for comparison, and it is higher than the TRV for 100% interrupting current. TRV at lower short-circuit current is calculated to 10% of the rated short-circuit current.

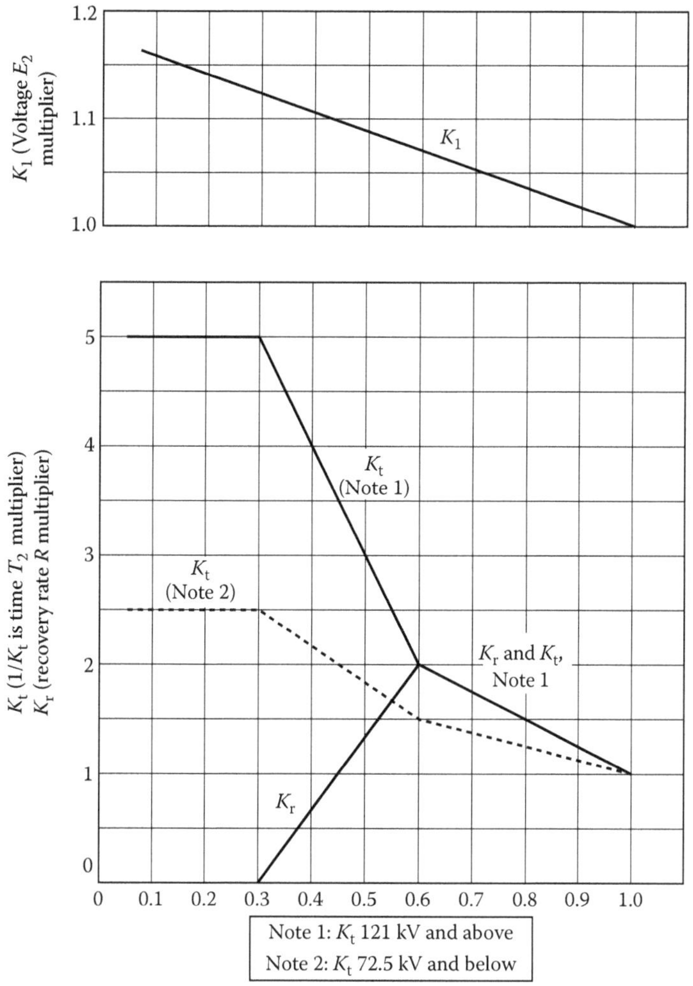

FIGURE 5.14
TRV rate and voltage multipliers for fractions of rated interrupting current. (From ANS1/IEEE Standard C37.04, Rating Structure for AC High Voltage Circuit Breakers, 1999 (revision of 1979).)

5.11.1 Short-Line Faults

We discussed sawtooth TRV waveform for a short-line fault in Section 4.6. Initial TRV can be defined as an initial ramp and plateau of voltage added to the initial front of an exponential cosine wave shape. This TRV is due to relatively close inductance and capacitance associated with substation work. For breakers installed in gas-insulated substations, the initial TRV can be neglected because of low-bus surge impedance and small distance to the first major discontinuity. However, for other systems at low levels of fault current, the initial rate of TRV may exceed the envelope defined by the standards. In such cases, the short-line initial TRV capability can be superimposed on the calculated TRV curve and the results examined. The saw tooth voltage at the input line terminals can be calculated by ladder diagrams.

The circuit breaker will be capable of interrupting single-phase line faults at any distance from the circuit breaker on a system in which

- The TRV on a TF is within the rated or related transient voltage envelope
- The voltage on the first ramp of the sawtooth wave is equal to or less than that in an ideal system in which surge impedance and amplitude constant are as follows:

$$242\,\text{kV and below, single conductor line: } Z = 450,\ d = 1.8 \tag{5.33}$$

$$362\,\text{kV and above, bundled conductors: } Z = 360,\ d = 1.6 \tag{5.34}$$

The amplitude constant d is the peak of the ratio of the sawtooth component, which will appear across the circuit breaker terminal at the instant of interruption. The short-line fault (SLF) TRV capability up to the first peak of TRV is defined as

$$e = e_L + e_S$$

where
e_L is the line side contribution to TRV
e_S is the source side contribution to TRV

$$e_L = d(1 - M)\sqrt{\frac{2}{3}}E_{max}$$

$$e_S = 2M(t_L - t_d) \tag{5.35}$$

$$R_L = \sqrt{2}\omega MIZ \times 10^{-6}\,\text{kV}/\mu\text{s} \tag{5.36}$$

$$t_L = \frac{e}{R_L}\,\mu\text{s} \tag{5.37}$$

where
R_L is the rate of rise
t_L is the time to peak
M is the ratio of fault current to rated short-circuit current
I is the rated short-circuit current in kA
V is the rated voltage
Z is the surge impedance
e is the peak voltage in kV

There is a delay of 0.5 μs for circuit breakers rated 245 kV and above and 0.2 μs for circuit breakers rated below 245 kV. It is not necessary to calculate SLF TRV, as long as TF TRV are within rating and transmission line parameter are within those specified in Table 5.7.

Example 5.6

For a 550 kV breaker, whose TRV wave shapes are plotted in Figure 5.13, plot the short-line capability for a 75% short circuit current and a surge impedance of 360 ohm.

$M = 0.75$, $I = 40$ kA, $V = 550$, $d = 1.6$, and $Z = 360$ ohm. This gives

$$e = 1.6(1 - 0.75)(550)\frac{\sqrt{2}}{\sqrt{3}} = 179.6\,\text{kV}$$

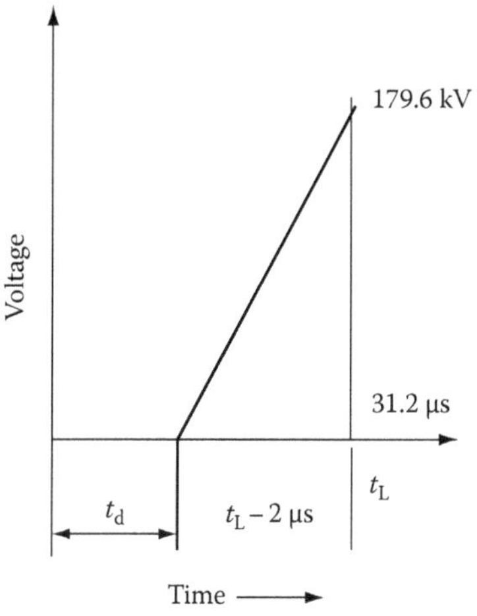

FIGURE 5.15
TRV profile, first peak only, Example 5.6.

Also,

$$R_L = \sqrt{2} \times 377 \times 0.75 \times 40 \times 360 \times 10^{-6} = 5.75\ \text{kV}/\mu\text{s}$$

and $t_L = 31.2\ \mu$s.

This is shown in Figure 5.15.

5.11.2 Oscillatory TRV

Figure 5.16 shows an example of an underdamped TRV, where the system TRV exceeds the breaker TRV capability curve. Such a waveform can occur when a circuit breaker clears a low-level three-phase-ungrounded fault, limited by a transformer on the source side or a reactor, Figure 5.16a and b. Figure 5.16c shows that the circuit breaker TRV capability is exceeded. Where this happens, the following choices exist:

1. Use a breaker with higher interrupting rating.
2. Add capacitance to the circuit breaker terminals to reduce the rate of rise of TRV.
3. Consult the manufacturer concerning the application.

A computer simulation using EMTP of the TRV will be required.

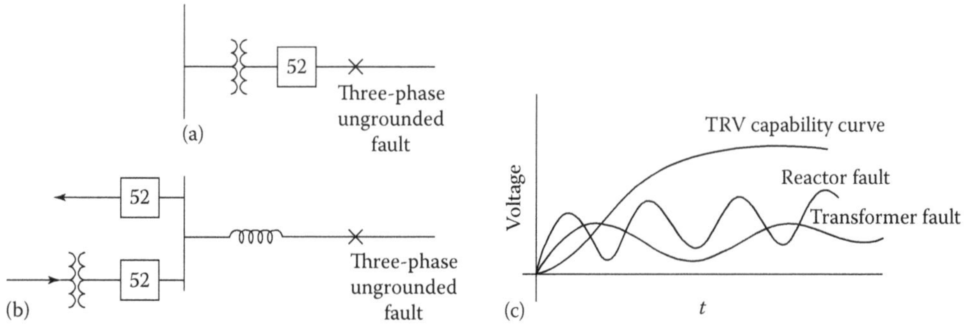

FIGURE 5.16
(a and b) Power system configurations where TRV may exceed the breaker capabilities for a fault limited by the transformer or reactor; (c) oscillatory TRV.

5.11.3 Initial TRV

Circuit beakers rated 100 kV and short-circuit rating of 31.5 kV and above will have an initial TRV capability for phase-to-ground fault. This rises linearly from origin to first peak voltage E_i and time T_i [6]. The E_i is given by the following expression:

$$E_i = \omega \times \sqrt{2} \times I \times Z_b \times T_i \times 10^{-6}\,\text{kV} \tag{5.38}$$

where

Z_b is the bus surge impedance = 450 ohm for outdoor substations, phase-to-ground faults

I is in the fault current in kA

The term initial TRV refers to the conditions during the first microsecond or so, following a current interruption, when the recovery voltage on the source side of the circuit breaker is influenced by proximity of buses, capacitors, isolators, etc. Akin to short-line fault voltage oscillation is produced, but this oscillation has a lower voltage peak magnitude. The traveling wave will move down the bus where the first discontinuity occurs. The initial slope depends upon the surge impedance and di/dt, and the peak of ITRV appears at a time equal to twice the traveling wave time. There can be as many variations of ITRV, as the station layouts. T_i in μs, time to peak is given in IEEE standard [6] with respect to maximum system voltage: For 121, 145, 169, 362, 350, and 500, it is 0.3, 0.4, 0.5, 0.6, 0.8, 1.0, and 1.1, respectively.

5.11.4 Adopting IEC TRV Profiles in IEEE Standards

In the draft revision of IEEE standard [1], the TRV capability is defined by two-parameter and four-parameter envelope, much akin to IEC standards (Chapter 4). These revisions are in the draft form and not yet approved, at the time of writing this edition of the book. The TRV profiles from this draft standard are shown in Tables 5.8 and 5.9. Note that TRV profiles are distinct for the type of operation. TF signifies terminal fault, OS, out of step switching, and SLF, short-line fault.

Interestingly, the TRV capability envelope of a 550 kV breaker at 100% of its short-circuit current rating in Figure 8 of standard [11] is superimposed upon the calculated curves in Figure 5.13. This standard contains example of TRV calculations in a practical electrical system, which an interested reader may refer to. The calculations involve the following:

1. Start with a circuit configuration
2. Put a fault on the system, to be studied, for example, a three-phase-to-ground fault.
3. Draw the circuit on each side (line and source side) of the breaker poles.
4. Reduce this circuit to an equivalent circuit. The transmission lines and cables can be modeled with their surge impedances. The capacitances on either side of the breaker are important.
5. An equivalent circuit with positive zero and negative sequence impedances can be constructed.
6. Calculate the fault current at the circuit breaker.
7. Consider first pole to clear factor.

TABLE 5.9

Preferred Ratings of Prospective TRV for Class S1 Circuit Breakers Rated 100 kV and Above Including Circuit Breakers Applied in Gas Insulated Substations, for Effectively Grounded Systems and Grounded Faults with a First Pole to Clear Factor of 1.3, T100

Rated Maximum Voltage (kV, rms)	Test Duty	First Pole to Clear Factor k_{pp} (pu)	Amplitude Factor k_{af} (pu)	First Ref. Voltage u_1 (kV)	Time t_1 (μs)	TRV Peak Value (kV)	Time T_2 (μs)	Time Delay T_d (μs)	Voltage U' (kV)	Time t' (μs)	RRRV u_c/t_3 (kV/μs)
123	TF	1.3	1.4	98	49	183	196	2	49	27	2
	SLF	1.0	1.4	75	38	141	152	2	38	21	2
	OS	2.0	1.25	151	98	251	392	2	75	51	1.54
145	TF	1.3	1.4	115	58	215	232	2	58	31	2
	SLF	1.0	1.4	89	44	166	176	2	44	24	2
	OS	2.0	1.25	178	116	296	464	2	89	60	1.54
170	TF	1.3	1.4	135	68	253	272	2	68	36	2
	SLF	1.0	1.4	104	52	194	208	2	52	28	2
	OS	2.0	1.25	208	136	347	544	2	104	70	1.54
245	TF	1.3	1.4	195	98	364	392	2	98	51	2
	SLF	1.0	1.4	150	75	280	300	2	75	40	2
	OS	2.0	1.25	300	196	500	784	2	150	99	1.54
362	TF	1.3	1.4	288	144	538	576	2	144	74	2
	SLF	1.0	1.4	222	111	414	444	2	111	57	2
	OS	2.0	1.25	443	288	739	1152	2	222	146	1.54
550	TF	1.3	1.4	438	219	817	876	2	219	112	2
	SLF	1.0	1.4	337	168	629	672	2	168	86	2
	OS	2.0	1.25	674	438	1120	1752	2	337	221	1.54
800	TF	1.3	1.4	637	318	1190	1272	2	319	161	2
	SLF	1.0	1.4	490	245	914	980	2	245	124	2
	OS	2.0	1.25	980	636	1630	2544	2	490	326	1.54

Source: IEEE PC37.06/D11, Draft Standard AC High-Voltage Circuit Breakers Rated on Symmetrical Current Basis-Preferred Ratings and Related Required Capabilities for Voltages above 1000 Volts, 2008.

The TRV profiles can be calculated based upon the simplified equations in [15]. The standard also provides typical values of surge impedances and first pole to clear factor for various fault types. The TRV at reduced short-circuit currents can be calculated using adjustment factors provided in this standard. A rigorous calculation is through EMTP and similar transient simulation programs [16].

5.11.5 Definite Purpose TRV Breakers

ANSI standard C37.06.1 [17] is for "definite purpose circuit breakers for fast transient recovery voltages." This is somewhat akin to specifications of definite purpose breakers for capacitor switching. The standard qualifies the following:

1. No fast T_2 values or tests are proposed for fault currents >30% of the rated short-circuit current.
2. The proposed T_2 values are chosen to meet 90% of the known TRV circuits, but even these fast values do not meet the requirements of all fast TRV applications.
3. A circuit breaker that meets the requirements of definite purpose for fast TRV may or may not meet the requirements of definite purpose circuit breakers for capacitor switching.

5.12 Generator Circuit Breakers

There is no other national or international standard on generator circuit breakers, other than IEEE standard C37.013 [18,19], revision 1997 of this standard are adapted to international practice. The specified ratings of the generator breakers are as follows:

Required symmetrical interrupting capability for three-phase faults: For a three-phase fault, the generator breaker shall be capable of interrupting the rated three-phase symmetrical short-circuit current for the rated duty cycle, irrespective of the direct current component of the total short-circuit current, at the instant of primary arcing contact separation for operating voltages equal to rated maximum voltage.

Required asymmetrical interrupting capability for three-phase faults: The required asymmetrical *system source* interrupting rating, at maximum operating voltage and rated duty cycle is composed of the rms symmetrical current and the percent dc component. This value of the dc component in percent of peak value of the symmetrical short-circuit current is given by Figure 5.17 for primary arcing contact parting time in ms. This figure is applicable for system short-circuit currents. The curve is based upon time constant of decay of dc component of 133 ms. The primary contact parting time is considered equal to 0.5 cycle plus minimum opening time of the particular breaker. In Section 5.2.1, for a 5-cycle breaker with contact parting time of 3 cycle, dc component of 32.91% was calculated. From Figure 5.17, for a generator breaker, the dc component is 54.7%. For time constants different from 133 ms, the following expressions are used:

$$\alpha = \frac{I_{dc}}{I_{ac\ peak}} \tag{5.39}$$

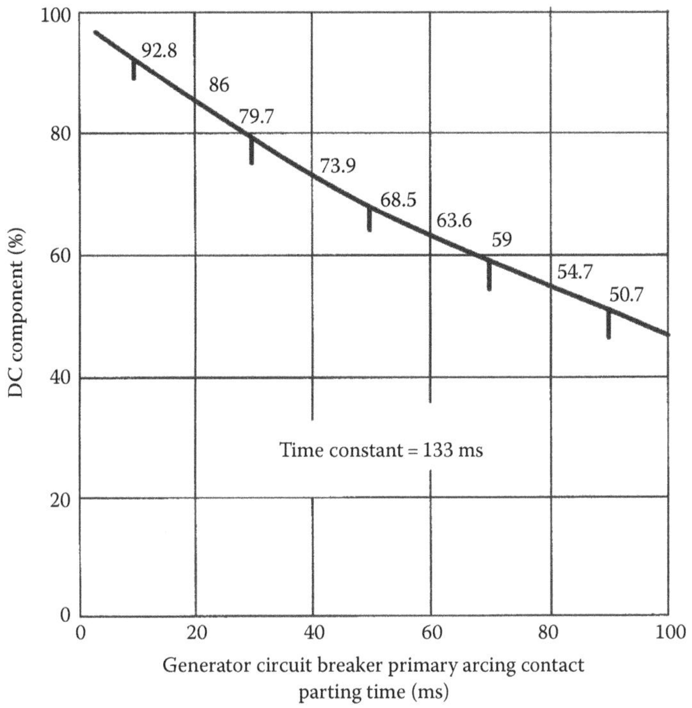

FIGURE 5.17
Asymmetrical interrupting capability of generator circuit breakers: dc component in percentage of peak value of the symmetrical three-phase system-source short-circuit currents.

where α is the degree of asymmetry calculated at the contact part time t_{cp} (in ms) of the breaker, and the dc component is

$$I_{dc} = I_{ac\ peak} e^{-t_{cp}/133} \tag{5.40}$$

Generator source symmetrical interrupting capability for three-phase faults: No specific value is assigned to generator source symmetrical interrupting short-circuit current for three-phase faults, because its maximum value is usually less than the short-circuit current from the power system. If a rating is assigned, the generator breaker shall be tested for the three-phase short-circuit interrupting current solely contributed by the generator source.

Generator source asymmetrical interrupting capability for three-phase faults: The required generator source asymmetrical interrupting capability for three-phase faults at rated maximum voltage and duty cycle is composed of rms generator source symmetrical current and a dc component. A dc component value of 110% of the peak value of the symmetrical generator source short-circuit current is specified.

Generator source asymmetrical interrupting capability for maximum required degree of asymmetry: The maximum required degree of asymmetry of the current required for the maximum required degree of asymmetry is 130% of the peak value of symmetrical current for this condition. The symmetrical component of the current for this condition of maximum asymmetry is only 74% of the value of required generator source symmetrical interrupting capability. This is further discussed in Chapter 7, with an example of calculation.

Required interrupting capability for single-phase-to-ground faults: The generator circuit breakers are designed for use on high-resistance-grounded systems where single phase-to-ground short-circuit current will not exceed 50 A.

The first pole-to-clear factor shall be 1.5, and the amplitude factor shall be 1.5

Closing and latching: (a) The generator circuit breaker is capable of closing and latching any power frequency current whose crest does not exceed 2.74 times rated symmetrical short-circuit current or the maximum peak (peak making current) and the generator source short-circuit current, whichever is higher. No numerical value is given for generator-source peak current as it depends upon the generator characteristic data. (b) The circuit breaker shall carry the short-circuit current for 0.25 s.

Short-time current carrying capability: The short-time carrying is for a period of 1 s, any short-circuit current determined from the envelope of current wave at the time of maximum peak, whose value does not exceed 2.74 times the rated short-circuit current:

$$I = \sqrt{\int_0^1 i^2 \, dt} \tag{5.41}$$

TRV: The standard specifies TRV values for system source faults, generator source faults, load current switching, and out-of phase current switching. Both power frequency recovery voltage and inherent TRV (unmodified by the presence of generator breaker) should be considered. The power frequency recovery voltage across generator breaker contacts consists of sum of voltage variations at each side of the generator breaker. Maximum voltage is

$$1.5\frac{V}{\sqrt{3}} \times \left(X''_{\mathrm{d}} + X_{\mathrm{t}}\right) \tag{5.42}$$

where

V is rated maximum voltage in pu
X''_{d} and X_{t} are generator and transformer per unit reactance, respectively

Their sum does not exceed 0.5 pu even for large machines, and therefore recovery voltage that appears across generator breaker contacts after a short-circuit current interruption is standardized at 0.43 V, from Equation 5.42.

The TRV for load current switching is normally a dual frequency curve, field tests for accurate estimate rather than theoretical calculations.

Short-circuit currents with delayed current zero: A generator breaker will be required to interrupt generator source currents with delayed current zeros. The magnitude of these currents is considerably lower than the rated short-circuit currents. The standard recommends that capability to interrupt delayed current zeros can be ascertained by computations that consider effect of arc voltage on prospective short-circuit current.

Out of phase current switching capability: When out-of-phase switching capability is assigned, it is based upon an out-of-phase angle of 90° at rated maximum voltage. The maximum out-of-phase current will be 50% of the symmetrical system short-circuit current. The out-of-phase switching current can be calculated from the expression

$$I_{\text{oph}} = \frac{\delta I_n}{X_d'' + X_t + X_s} \tag{5.43}$$

where X_s is the system reactance. All reactances are based on generator rated MVA in pu. I_n is the rated generator current and $\delta = 1.4$ for a 90° out of phase angle and 2 for 180° out of phase angle.

A generator breaker may be specified for out-of phase current rating and also capacitance switching capability, which should be demonstrated by tests.

In the utility systems, the synchronous generators are connected directly through a step-up transformer to the transmission systems, Figure 5.18. In this figure, the ratings of generators and transformers are not shown for generality. Note that generators 1 and 4 do not have a generator breaker, and the generator and transformer are protected as a unit with overlapping zones of differential protection (not discussed here). Generators 2 and 3

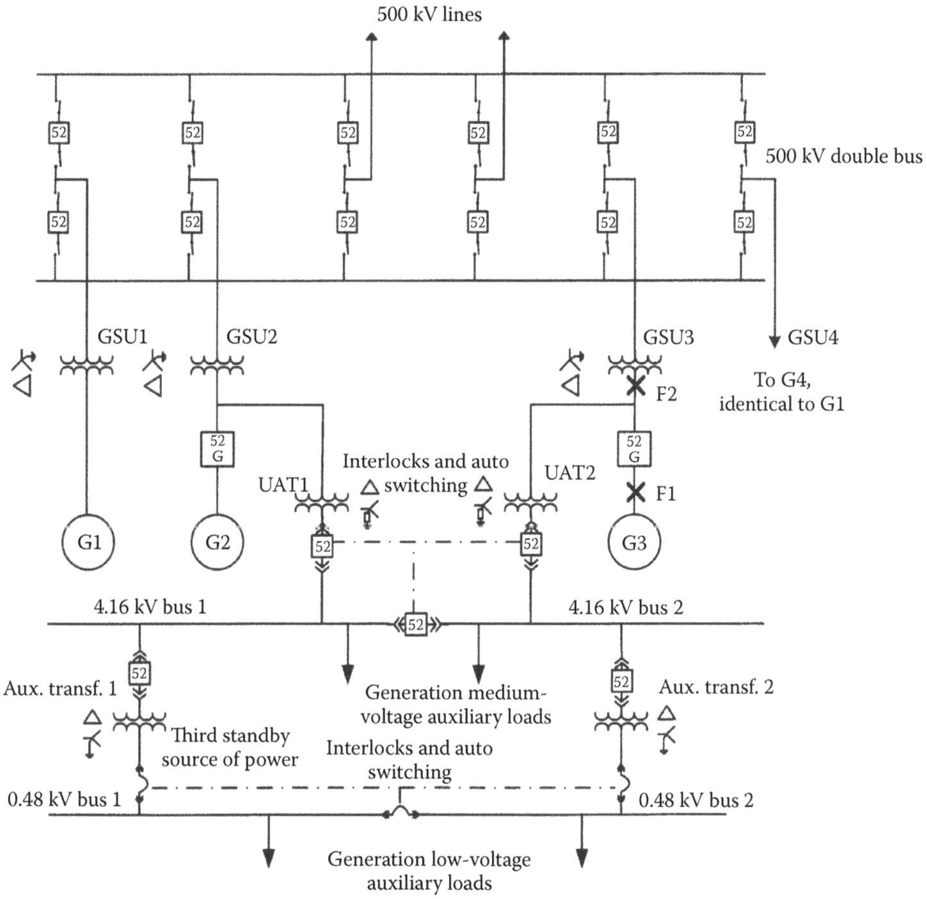

FIGURE 5.18
Interconnection of large generators in utility systems, with and without a generator circuit breaker.

have generator breakers, which allow generators step-up transformers to be used as step-up down transformers during cold start-up. The power to the generation auxiliary loads is duplicated with autoswitching of bus section breaker, and also there is a third standby power source. Utilizing the generator step-up (GSU) transformer as generator, step-down transformer may result in cost savings; the addition of generator breaker may be more economical than a dedicated high-voltage step-down transformer. In-line generator breakers of 100 kA, an advantage is that generator and transformer can be protected by separate differential zones of protection compared to protecting the two as a unit. A disadvantage that is noted is as follows:

1. During cold start and when the transformer is used as a step-down transformer, with generator breaker open, the adjustment of load-voltage profiles requires negative tap adjustment on the transformer primary windings to counteract the voltage drops on load flow in the auxiliary distribution.
2. After the start-up with generator synchronized, positive transformer taps are required to utilize the generator reactive power capability.

These are two conflicting requirement. Sometimes, the generator reactive power may remain trapped due to this limitation. GSU transformers are sometimes provided additional positive tap adjustment range, see Chapter 13, reactive power flow and control.

Figure 5.19 shows a generator of 81.82 MVA, 12.47 kV, 0.85 power factor directly connected to a 12.47 kV bus, which is also powered by a 30/40/50 MVA, 115–12.47 kV

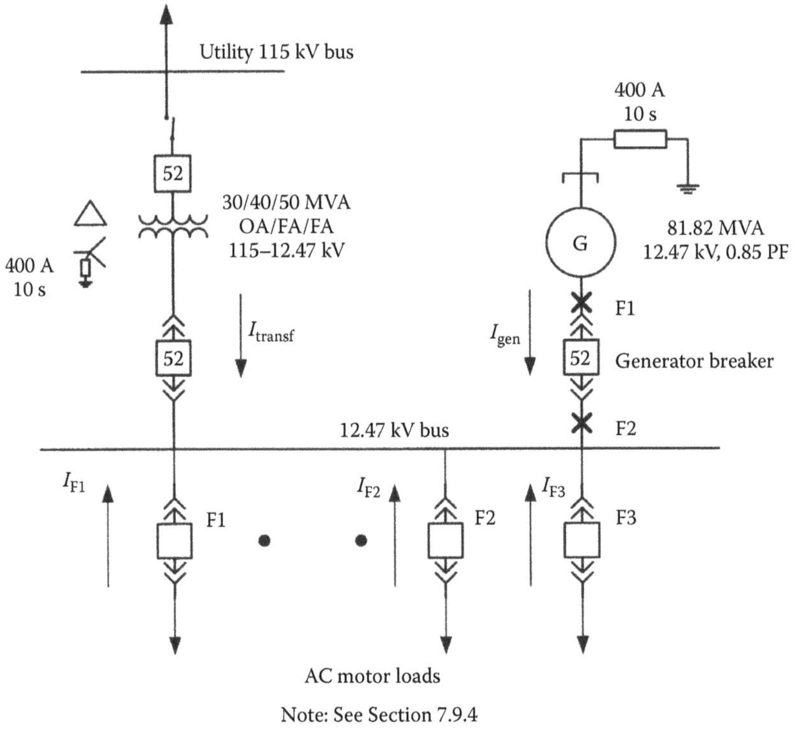

FIGURE 5.19
Interconnection of a large bus connected generator in an industrial system.

utility transformer. The two sources are run in synchronism, and the plant running load is 45 MVA; the excess generated power is supplied into the utility system. The size of a generator that can be bus connected in an industrial distribution, primary distribution voltage of 13.8 kV, is approximately limited to 100 MVA, as an acceptable level of short-circuit should be maintained at the medium voltage switchgear and the downstream distributions.

5.13 Specifications of High-Voltage Circuit Breakers

The discussions in the above sections are limited to current interrupting performance of the high-voltage circuit breakers. There are a number of other relevant specifications, which depend upon a specific installation. As an example, ambient temperature, short-time overload capability, location in environmentally adverse conditions, outdoor installations in high seismic zones, dielectric strength of external and internal insulation systems should form a part of the comprehensive specifications. Standards specify power frequency tests, dry and wet; impulse tests, full wave and chopped wave, switching impulse tests (voltages 362 kV and higher), minimum creepage distances to ground, etc. Current transformers of relaying and metering accuracy can be located in the circuit breakers (for outdoor installations in the outdoor bushings, for indoor installation bus mounted or window type on draw-out contacts spouts of metal-clad switchgear). A breaker may be fitted with more than one trip coil. Specifications of power operating mechanism and host of safety interlocks are required. For SF_6 breakers, a gas density monitor is normally provided. Breakers requiring single pole operation at higher voltages must be specified for this application, and manufacturer should supply the expected pole scatter.

A circuit breaker manufactured according to ANSI/IEEE must comply with the rating structure and other specifications laid down in the standard, yet specific applications will require additional data to be made available to the manufacturers. The short-circuit ratings are an important parameter and must be calculated accurately (Chapters 7 and 8). Electrical power systems invariably expand, and a conservative margin in short-circuit ratings can save much higher costs of replacements when the circuit breakers short-circuit ratings become inadequate due to system growth. The capacitance current switching performance, TRV performance will require definite purpose breakers, specified based upon the study results. Out-of-phase switching duty, single-pole closing capability, limiting switching overvoltages, synchronous, and resistance switching will require specific calculations. These switching transients can be calculated based upon EMTP-type programs [16] and are not discussed in this book.

5.14 Low-Voltage Circuit Breakers

The three classifications of low-voltage circuit breakers are (1) molded case circuit breakers, (2) insulated case circuit breakers, and (3) low-voltage power circuit breakers (LVPCBs) [20–25].

5.14.1 Molded Case Circuit Breakers

In molded case circuit breakers (MCCBs), the current carrying parts, mechanism, and trip devices are completely contained in a molded case insulating material, and these breakers are not maintainable. Available frame sizes range from 15 to 6000 A, interrupting ratings from 10 to 100 kA symmetrical without current-limiting fuses and to 200 kA symmetrical with current-limiting fuses. These can be provided with electronic trip units and have limited short-time delay and ground fault sensing capability. When provided with thermal magnetic trips, the trips may be adjustable or nonadjustable and are instantaneous in nature. Motor circuit protectors (MCPs) may be classed as a special category of MCCBs and are provided with instantaneous trips only. MCPs do not have an interrupting rating by themselves and are tested in conjunction with motor starters. All MCCBs are fast enough to limit the amount of prospective current let-through, and some are fast enough to be designated as current-limiting circuit breakers. For breakers claimed to be current limiting, peak current and I^2t are tabulated for the threshold of current-limiting action.

5.14.2 Insulated Case Circuit Breakers

Insulated case circuit breakers (ICCBs) utilize characteristics of design from both the power and MCCBs, are not fast enough to qualify as current-limiting type, and are partially field maintainable. These can be provided with electronic trip units and have short-time ratings and ground fault sensing capabilities. These utilize stored energy mechanisms similar to LVPCBs.

MCCBs and ICCBs are rated and tested according to UL 489 standard [24]. Both MCCBs and ICCBs are tested in the open air without enclosure and are designed to carry 100% of their current rating in open air. When housed in an enclosure, there is 20% derating, though some models and frame sizes may be listed for application at 100% of their continuous current rating in an enclosure. MCCBs are fixed mounted in switchboards and bolted to bus bars. ICCBs can be fixed mounted or provided in draw-out design.

5.14.3 Low-Voltage Power Circuit Breakers

Low-voltage power circuit breakers (LVPCBs) are rated and tested according to ANSI C37.13 [21,22] and are used primarily in draw-out switchgear. These are the largest in physical size and are field maintainable. Electronic trip units are almost standard with these circuit breakers, and these are available in frame sizes from 800 to 6000 A, interrupting ratings, 40–100 kA sym. without current-limiting fuses.

All the three types of circuit breakers have different ratings, short-circuit test requirements, and applications. The short-circuit ratings and fault current calculation considerations are of interest here.

The symmetrical interrupting rating of the circuit breaker takes into account the initial current offset due to circuit X/R ratio. The value of the standard X/R ratio is used in the test circuit. For LVPCBs, this standard is $X/R = 6.6$, corresponding to a 15% power factor. Table 5.10 shows the multiplying factor (MF) for other X/R ratios. The recommended MFs for unfused circuit breakers are based on highest peak current and can be calculated from

$$\mathrm{MF} = \frac{\sqrt{2}[1 + e^{-\pi/(X/R)}]}{2.29} \tag{5.44}$$

TABLE 5.10

Multiplying Factors for Low-Voltage LVPCBs

System Short-Circuit Power Factor (%)	System X/R Ratio	Multiplying Factors for the Calculated Current	
		Unfused Circuit Breakers	Fused Circuit Breakers
20	4.9	1.00	1.00
15	6.6	1.00	1.07
12	8.27	1.04	1.12
10	9.95	1.07	1.15
8.5	11.72	1.09	1.18
7	14.25	1.11	1.21
5	20.0	1.14	1.26

Source: IEEE Standard Ref. [21].

The MF for the fused breaker is based on the total rms current (asymmetrical) and is calculated from:

$$\text{MF} = \frac{\sqrt{1 + 2e^{-2\pi/(X/R)}}}{1.25} \tag{5.45}$$

In general, when X/R differs from the test power factor, the MF can be approximated by

$$\text{MF} = \frac{1 + e^{-\pi(X/R)}}{1 + e^{-\pi/\tan\phi}} \tag{5.46}$$

where ϕ is the test power factor.

MCCBs and ICCBs are tested in the prospective fault test circuit according to UL 489 [24]. Power factor values for the test circuit are different from LVPCBs and are given in Table 5.11. If a circuit has an X/R ratio, which is equal to or lower than the test circuit, no corrections to interrupting rating are required. If the X/R ratio is higher than the test circuit X/R ratio, the interrupting duty requirement for that application is increased by a MF from Table 5.12. The MF can be interpreted as a ratio of the offset peak of the calculated system peak (based on X/R ratio) to the test circuit offset peak.

While testing the breakers, the actual trip unit type installed during testing should be the one represented by referenced specifications and time-current curves. The short-circuit ratings may vary with different trip units, i.e., a short-time trip only (no instantaneous) may result in reduced short-circuit interrupting rating compared to testing with instantaneous

TABLE 5.11

Test Power Factors of MCCBs

Interrupting Rating (kA, rms sym.)	Test Power Factor Range	X/R
10 or less	0.45–0.50	1.98–1.73
10–20	0.25–0.30	3.87–3.18
Over 20	0.15–0.20	6.6–4.9

TABLE 5.12

Short-Circuit Multiplying Factors for MCCBs and ICCBs

		Interrupting Rating Multiplying Factor		
Power Factor (%)	X/R Ratio	10 kA or Less	10–20 kA	>20 kA
5	19.97	1.59	1.35	1.22
6	16.64	1.57	1.33	1.20
7	14.25	1.55	1.31	1.18
8	12.46	1.53	1.29	1.16
9	11.07	1.51	1.28	1.15
10	9.95	1.49	1.26	1.13
13	7.63	1.43	1.21	1.09
15	6.59	1.39	1.18	1.06
17	5.80	1.36	1.15	1.04
20	4.90	1.31	1.11	1.00
25	3.87	1.24	1.05	1.00
30	3.18	1.18	1.00	1.00
35	2.68	1.13	1.00	1.00
40	2.29	1.08	1.00	1.00
50	1.98	1.04	1.00	1.00

trips. The trip units may be rms sensing or peak sensing, electronic or electromagnetic, and may include ground fault trips.

IEC standards do not directly correspond to the practices and standards in use in North America for single-pole duty, thermal response, and grounding. A direct comparison is not possible.

5.14.3.1 Single-Pole Interrupting Capability

A single-pole interruption connects two breaker poles in series, and the maximum fault current interrupted is 87% of the full three-phase fault current. The interrupting duty is less severe as compared to a three-phase interruption test, where the first-pole-to-clear factor can be 1.5. Therefore, the three-phase tests indirectly prove the single-pole interrupting capability of three-pole circuit breakers. For the rated X/R, every three-pole circuit breaker intended for operation on a three-phase circuit can interrupt a bolted single-phase fault. LVPCBs are single-pole tested with maximum line-to-line voltage impressed across the single pole and at the theoretical maximum single-phase fault current level of 87% of maximum three-phase bolted fault current. Generally, single-pole interrupting is not a consideration. Nevertheless, all MCCBs and ICCBs do not receive the same 87% test at full line-to-line voltage. In a corner-grounded delta system (not much used in the industry), a single line-to-ground fault on the load side of the circuit breaker will result in single-phase fault current flowing through only one pole of the circuit breaker, but full line-to-line voltage impressed across that pole. A rare fault situation in ungrounded or high-resistance grounded systems can occur with two simultaneous bolted faults on the line side and load side of a circuit breaker and may require additional considerations. Some manufacturers market circuit breakers rated for a corner-grounded systems.

Thus, normally, the three-phase faults, calculated at the point of application, give the maximum short-circuit currents on which the circuit breaker rating can be based, adjusted for fault point X/R. But, in certain cases, a line-to-ground fault in solidly grounded system can exceed the three-phase symmetrical fault. Care needs to be exercised in such applications.

5.14.3.2 Short-Time Ratings

MCCBs, generally, do not have short-time ratings. These are designed to trip and interrupt high-level faults without intentional delays. When provided with electronic trip units, capabilities of these breakers are utilized for short-delay tripping. ICCBs do have some short-time capability, typically 15 cycles. Yet these are provided with high-set instantaneous trips. LVPCBs are designed to have short-time capabilities, typically 30 cycles, and can withstand short-time duty cycle tests.

Short-time rating becomes of concern when two devices are to be coordinated in series, and these see the same magnitude of fault current. If an upstream device has a short-time withstand capability, a slight delay in the settings can ensure coordination. This is an important concept from time-current coordination point of view (see Chapter 21).

For an unfused LVPCB, the rated short-time current is the designated limit of prospective current at which it will be required to perform its short-time duty cycle of two periods of 0.5 s current flow separated by 15 s intervals of zero current at rated maximum voltage under prescribed test conditions. This current is expressed in rms symmetrical ampères. The unfused breakers will be capable of performing the short-time current duty cycle with all degrees of asymmetry produced by three-phase or single-phase circuits having a short-circuit power factor of 15% or greater. Fused circuit breakers do not have a short-time current rating, though the unfused circuit breaker element has a short-time rating as described above.

5.14.3.3 Series Connected Ratings

Series connection of MCCBs or MCCBs and fuses permits a downstream circuit breaker to have an interrupting rating less than the calculated fault duty, and the current-limiting characteristics of the upstream device "protects" the downstream lower-rated devices. Series combination is recognized for application by testing only. The upstream device is fully rated for the available short-circuit current and protects a downstream device, which is not fully rated for the available short-circuit current by virtue of its current-limiting characteristics. The series rating of the two circuit breakers makes it possible to apply the combination as a single device, the interrupting rating of the combination being that of the higher-rated device. As an example, a single upstream incoming breaker of 65 kA interrupting may protect a number of downstream feeder breakers of 25 kA interrupting, and the complete assembly will be rated for 65 kA interrupting. The series rating should not be confused with cascading arrangement. IEC also uses this term for their series rated breakers [25]. A method of cascading, which is erroneous and has been in use in the past, is shown in Figure 5.20.

Consider a series combination of an upstream current-limiting fuse of 1200 A and a downstream MCCB. The available short-circuit current is 50 kA sym., while the MCCB is rated for 25 kA. Figure 5.20 shows the let-through characteristics of the fuse. The required interrupting capability of the system, i.e., 50 kA is entered at the point A, and moving upward the vertical line is terminated at the 1200 A fuse let-through characteristics. Moving horizontally, the point C is intercepted and then moving vertically down the point D is located. The symmetrical current given by D is read off, which in Figure 5.20

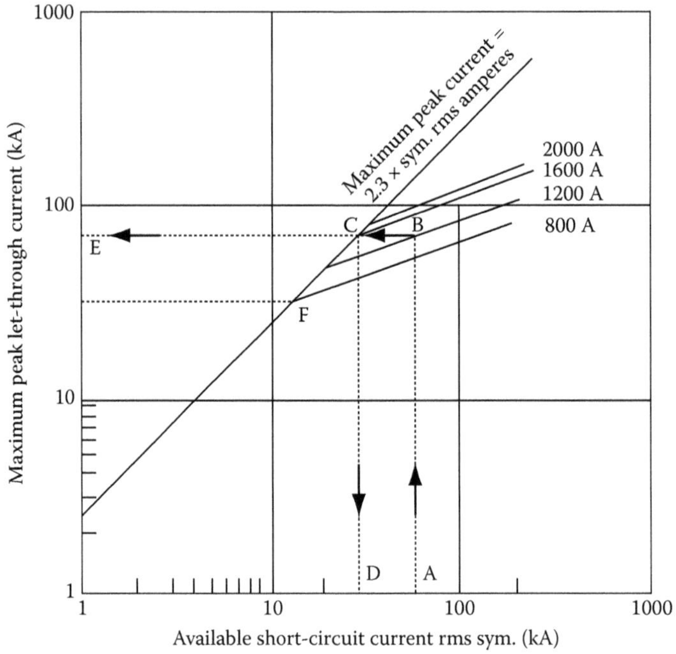

FIGURE 5.20
Let-through curves of current-limiting fuses.

is 19 kA. As this current is less than the interrupting rating of the downstream device to be protected, the combination is considered safe. This method can lead to erroneous results, as the combination may not be able to withstand the peak let-through current given by point E in Figure 5.20 on the y axis. Calculations of series ratings are not permissible, and these can only be established by testing.

A disadvantage of series combination is lack of selective co-ordination. On a high fault current magnitude, both the line side and load side circuit breakers will trip. A series combination should not be applied if motors or other loads that contribute to short-circuit current are connected between the line-side and load-side MCCBs. NEC (240.86 (c)) [26] specifies that series rating will not be used where

- Motors are connected on the load side of the higher-rated overcurrent device and on the line side of the lower-rated overcurrent device.
- The sum of motor full load currents exceeds 1% of the interrupting rating of the lower-rated circuit breaker.

5.15 Fuses

Fuses are fault sensing and interrupting devices, while circuit breakers must have protective relays as sensing devices before these can operate to clear short-circuit faults. Fuses are direct acting, single-phase devices, which respond to magnitude and duration of current.

Electronically actuated fuses are a recent addition, and these incorporate a control module that provides current sensing, electronically derived time-current characteristics, energy to initiate tripping, and an interrupting module, which interrupts the current.

5.15.1 Current-Limiting Fuses

A current-limiting fuse is designed to reduce equipment damage by interrupting the rising fault current before it reaches its peak value. Within its current-limiting range, the fuse operates within 1/4 to 1/2 cycle. The total clearing time consists of melting time, sometimes called the prearcing time and the arcing time. This is shown in Figure 5.21. The let-through current can be much lower than the prospective fault current peak, and the rms symmetrical available current can be lower than the let-through current peak. The prospective fault current can be defined as the current that will be obtained if the fuse was replaced with a bolted link of zero impedance. By limiting the rising fault current, the I^2t let-through to the fault is reduced because of two counts: (1) high speed of fault clearance in 1/4 cycle typically in the current-limiting range and (2) fault current limitation. This reduces the fault damage.

Current-limiting fuses have a fusible element of nonhomogeneous cross-section. It may be perforated or notched and while operating it first melts at the notches because of reduced cross-sectional area. Each melted notch forms an arc that lengthens and disperses the element material into the surrounding medium. When it is melted by current in the specified current-limiting range, it abruptly introduces a high resistance to reduce the current magnitude and duration. It generates an internal arc voltage, much greater than the system voltage, to force the current to zero, before the natural current zero crossing. Figure 5.22 shows the current interruption in a current-limiting fuse. Controlling the arcs in series controls the rate of rise of arc voltage and its magnitude. The arc voltages must be controlled to levels specified in the standards [27], i.e., for 15.5 kV fuses of 0.5–12 A, the maximum arc voltage is 70 kV peak, and for fuses >12 A, the arc voltage is 49 kV peak.

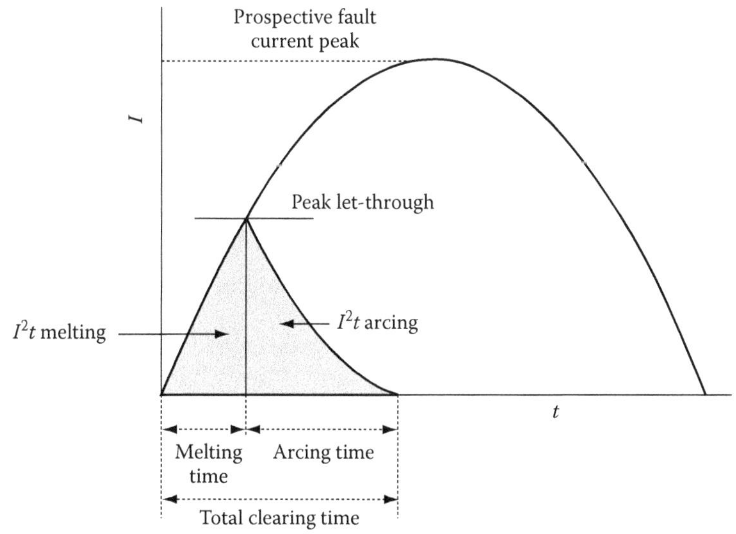

FIGURE 5.21
Current interruption by a current-limiting fuse.

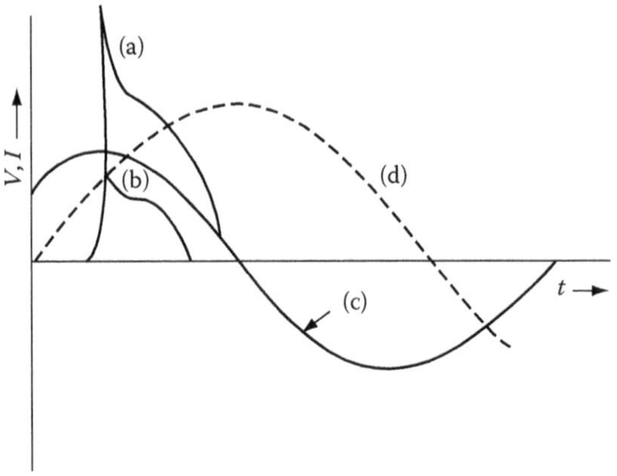

FIGURE 5.22
Arc voltage generated by a current-limiting fuse during interruption: (a) arc voltage; (b) interrupted current; (c) system voltage; (d) perspective fault current.

The current-limiting action of a fuse becomes effective only at a certain magnitude of the fault current called the critical current or threshold current. It can be defined as the first peak of a fully asymmetrical current wave at which the current-limiting fuse will melt. This can be determined by the fuse let-through characteristics and is given by the inflection point on the curve where the peak let-through current begins to increase less steeply with increasing short-circuit current, i.e., point F in Figure 5.20 for a 800 A fuse. The higher is the rated current of the fuse, the greater is the value of the threshold current at which the current-limiting action starts.

5.15.2 Low-Voltage Fuses

Low-voltage fuses can be divided into two distinct classes, current-limiting type and non-current-limiting type. The current-limiting class fuses are types CC, T, K, G, J, L, and R. Noncurrent-limiting fuses, i.e., class H fuses, have a low interrupting rating of 10 kA, are not in much use in industrial power systems and are being replaced with current-limiting fuses. Current-limiting fuses have interrupting capabilities up to 200 kA rms symmetrical. The various classes of current-limiting fuses are designed for specific applications, have different sizes and mounting dimensions, and are not interchangeable. As an example, classes J, RK1, and RK5 may be used for motor controllers, control transformers, and backup protection. Class L (available in current ratings up to 6 kA) is commonly used as a current-limiting device in series rated circuits. Class T is a fast-acting fuse that may be applied to load-center, panel-board, and circuit-breaker backup protection.

5.15.3 High-Voltage Fuses

High-voltage fuses [30,31] can be divided into two distinct categories: distribution fuse cutouts and power fuses. Distribution cutouts are meant for outdoor pole or cross arm mounting (except distribution oil cutouts), have basic insulation levels (BILs) at distribution levels, and are primarily meant for distribution feeders and circuits. These are available in voltage ratings up to 34.5 kV. The interrupting ratings are relatively low, 5.00 kA rms sym. at 34.5 kV. The power fuses are adapted to station and substation mounting, have BILs at power levels, and are meant primarily for applications in stations and substations. These are of two types: expulsion-type fuses and current-limiting fuses. Expulsion-type fuses can again be of two types: (1) fiber-lined fuses having voltage ratings up to 169 kV

TABLE 5.13
Short-Circuit Interrupting Ratings of High-Voltage Fuses

Fuse Type	Current Ratings	Nominal Voltage Rating in kV-Maximum Short-Circuit Interrupting Rating (kA rms Symmetrical)
Distributions fuse cutouts	Up to 200 A	4.8–12.5, 7.2–15, 14.4–13.2, 25–8, 34.5–5
Solid-material boric acid fuses	Up to 300 A	17.0–14.0, 38–33.5, 48.3–31.5, 72.5–25, 121–10.5, 145–8.75
Current-limiting fuses	Up to 1350 A for 5.5 kV, up to 300 A for 15.5 kV, and 100 A for 25.8 and 38 kV	5.5–50, 15.5–50 (85 sometimes), 25.8–35, 38.0–35

Note: 4.8–12.5 signifies 4.8 kV, and maximum short-circuit interrupting rating of 12.5 kA rms symmetrical.

and (2) solid boric acid fuses, which have voltage ratings up to 145 kV. The solid boric acid fuse can operate without objectionable noise or emission of flame and gases. High-voltage current-limiting fuses are available up to 38 kV, and these have comparatively much higher interrupting ratings. Table 5.13 shows comparative interrupting ratings of distribution cutouts, solid boric acid, and current-limiting fuses. While the operating time of the current-limiting fuses is typically one-quarter of a cycle in the current-limiting range, the expulsion-type fuses will allow the maximum peak current to pass through and interrupt in more than 1 cycle. This can be a major consideration in some applications where a choice exists between the current-limiting and expulsion-type fuses.

Class E fuses are suitable for the protection of voltage transformers, power transformers, and capacitor banks, while class R fuses are applied for medium-voltage motor starters. All class E rated fuses are not current limiting; E rating merely signifies that class E rated power fuses in ratings of 100E or less will open in 300 s at currents between 200% and 240% of their E rating. Fuses rated above 100E open in 600 s at currents between 220% and 264% of their E ratings.

5.15.4 Interrupting Ratings

The interrupting ratings relate to the maximum rms asymmetrical current available in the first half-cycle after fault, which the fuse must interrupt under the specified conditions. The interrupting rating itself has no direct bearing on the current-limiting effect of the fuse. Currently, the rating is expressed in maximum rms symmetrical current, and thus the fault current calculation based on an E/Z basis can be directly used to compare the calculated fault duties with the short-circuit ratings. Many power fuses and distribution cutouts were earlier rated on the basis of maximum rms asymmetrical currents; rms asymmetrical rating represents the maximum current that the fuse has to interrupt because of its fast-acting characteristics. For power fuses the rated asymmetrical capability is 1.6 times the symmetrical current rating. The asymmetrical rms factor can exceed 1.6 for high-X/R ratios or low-power factors short-circuit currents. Figure 5.23 from [28] relates rms multiplying factors and peak multiplying factors.

Note that the test X/R ratio is 25 only for expulsion type and current-limiting type fuses [29]. For distribution class fuse cutouts interrupting tests (except current-limiting and open-link cutouts), the minimum X/R ratio varies, 1.5–15 [28]. It is important to calculate the interrupting duty based upon the actual system X/R and apply proper adjustment factors.

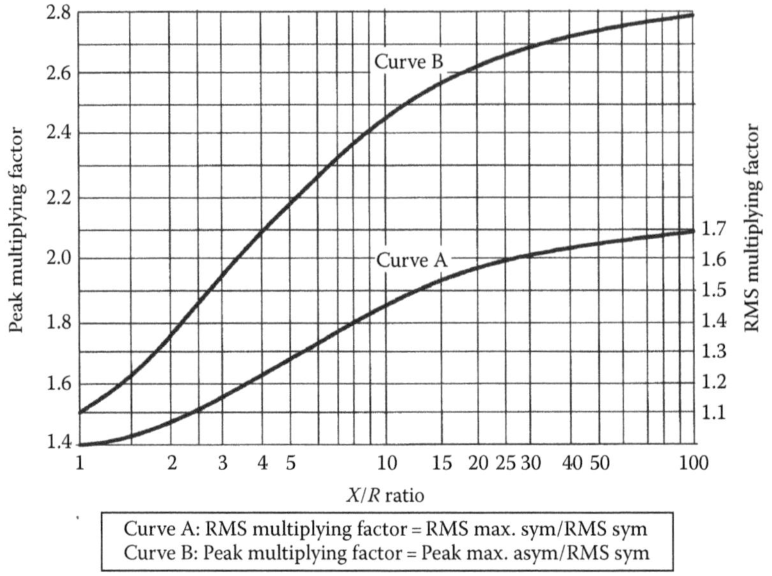

FIGURE 5.23
Relation of X/R to rms and peak multiplying factors. (From IEEE Standard C37.41, IEEE Standard Design Tests for High-Voltage (>1000 V) Fuses, Fuse and Disconnecting Cutouts, Distribution Enclosed Single-Pole Air Switches, Fuse Disconnecting Switches, and Fuse Links and Accessories Used with These Devices, 2008.)

A basic understanding of the ratings and problems of application of circuit breakers and fuses for short-circuit and switching duties can be gained from this chapter. The treatment is not exhaustive, and an interested reader will like to explore further.

Problems

5.1 A 4.76 kV rated breaker has a rated short-circuit current of 41 kA sym. and a K factor of 1.19. Without referring to tables calculate its (1) maximum symmetrical interrupting capability, (2) short-time current rating for 3 s, and (3) close and latch capability in asymmetrical rms and peak. If the breaker is applied at 4.16 kV, what is its interrupting capability and close and latch capability? How will these values change if the breaker is applied at 2.4 kV? Calculate similar ratings for 40 kA, $K=1$ rated circuit breaker according to ANSI year 2000 revision.

5.2 The breaker of Problem 5.1 has a rated interrupting time of 5 cycles. What is its symmetrical and asymmetrical rating for phase faults, when applied at 4.16 and 2.4 kV, respectively?

5.3 A 15 kV circuit breaker applied at 13.8 kV has a rated short-circuit current of 28 kA rms, K factor $=1.3$, and permissible tripping delay $Y=2$ s. What is its permissible delay for a short-circuit current of 22 kA?

5.4 In Example 5.3, reduce all reactances by 10% and increase all capacitances by 10%. Calculate the inrush current and frequency on (1) isolated capacitor bank switching, and (2) back-to-back switching. Find the value of reactor to be added to limit the inrush current magnitude and frequency to acceptable levels for a definite-purpose breaker.

5.5 Provide two examples of power system configurations, where TRV is likely to increase the standard values. Why can it be reduced by adding capacitors?

5.6 Plot the TRV characteristics of a 121 kV breaker, from the data in Table 5.9, at (1) a rated interrupting current of 40 kA and (2) a 50% short-circuit current. Also, plot the initial profile of TRV for a short-line fault.

5.7 Why is the initial TRV not of concern for gas-insulated substations?

5.8 A LVPCB, ICCB, and MCCB are similarly rated at 65 kA sym. interrupting. What other short-circuit rating is important for their application and protection coordination?

5.9 Each type of breaker in Problem 5.8 is subjected to a fault current of 50 kA, $X/R=7.0$. Calculate the interrupting duty multiplying factors from the tables in this chapter.

5.10 What are the advantages and disadvantages of current-limiting fuses as compared to relayed circuit breakers for short-circuit interruption? How do these compare with expulsion-type fuses?

5.11 Explain the series interrupting ratings of two devices. What are the relative advantages and disadvantages of this configuration? Why should the series rating of two devices not be calculated?

5.12 Construct a table and show the major ratings of a 13.8 kV general purpose breaker, definite purpose breaker for capacitor switching, definite purpose breaker for TRV, and a generator circuit breaker.

References

1. IEEE PC37.06/D11. Draft Standard AC High-Voltage Circuit Breakers Rated on Symmetrical Current Basis-Preferred Ratings and Related Required Capabilities for Voltages above 1000 Volts, 2008.
2. IEC Std. 62271-100. High Voltage alternating Current Circuit Breakers, 2001.
3. ANSI Standard C37.010. Application Guide for High-Voltage Circuit Breakers, 1961.
4. ANSI/IEEE Standard C37.5. Guide for Calculation of Fault Currents for Application of AC High-Voltage Circuit Breakers Rated on a Total Current Basis, 1979.
5. ANSI/IEEE Standard C37.010. Application Guide for AC High-Voltage Circuit Breakers Rated on a Symmetrical Current Basis, 1999 (R-2005).
6. ANS1/IEEE Standard C37.04. Rating Structure for AC High Voltage Circuit Breakers, 1999 (revision of 1979).
7. IEEE Standard PC37.04-1999/Cor 1/Draft B3. Draft IEEE Standard Rating Structure for AC High Voltage Circuit Breakers-Corrigendum 1, May 2009.
8. ANSI Standard C37.06. AC High-Voltage Circuit Breakers Rated on a Symmetrical Current Basis—Preferred Ratings and Related Capabilities, 2000 (revision of 1987).
9. IEEE Standard 551 (Violet Book). IEEE Recommended Practice for Calculating Short-Circuit Currents in Industrial and Commercial Power Systems, 2006.
10. Kundur, P. *Power System Stability and Control*. New York: McGraw Hill, 1993.
11. ANSI/IEEE Standard C37.012. Application Guide for Capacitance Current Switching for AC High-Voltage Circuit Breakers Rated on a Symmetrical Current Basis, 2005 (revision of 1979).

12. IEEE standard C37.04a. IEEE Standard Rating Structure for AC High-Voltage Circuit Breakers Rated on Symmetrical Current Basis, Amendment 1: Capacitance Current Switching, 2003.
13. IEEE Standard C37.09. Test Procedure for AC High-Voltage Circuit Breakers Rated on a Symmetrical Current Basis, 1999 (R2007).
14. IEEE Standard C37.09. Test Procedure for AC High-Voltage Circuit Breakers Rated on a Symmetrical Current Basis. Corrigendum 1, 2007.
15. IEEE Standard C37.011. Application Guide for Transient Recovery Voltage for AC High-Voltage Circuit Breakers, 1995.
16. J.C. Das. *Transients in Electrical Systems—Analysis, Recognition and Mitigation*. New York: McGraw-Hill, 2010.
17. IEEE Standard C37.06.1. Guide for High Voltage Circuit Breakers Rated on Symmetrical Current Basis Designated, "Definite Purpose for Fast Transient Voltage Recovery Times," 2000.
18. IEEE Standard C37.013. IEEE Standard for AC High-Voltage Generator Circuit Breakers Rated on a Symmetrical Current Basis, 1997.
19. IEEE Standard C37.013. IEEE Standard for AC High-Voltage Generator Circuit Breakers Rated on a Symmetrical Current Basis Amendment 1: Supplement for Use with Generators Rated 10-100 MVA, 2007.
20. IEEE Standard 1015. Applying Low-Voltage Circuit Breakers Used in Industrial and Commercial Power Systems, 1997.
21. ANSI/IEEE. Standard C37.13. Standard for Low-Voltage AC Power Circuit Breakers Used in Enclosures, 2008.
22. IEEE Standard C37.13.1. IEEE Standard for Definite–Purpose Switching Devices for use in Metal-Enclosed Low-Voltage Power Circuit Breaker Switchgear, 2006.
23. NEMA. Molded Case Circuit Breakers and Molded Case Switches, 1993, Standard AB-1.
24. UL Standard 489. Molded Case Circuit Breakers and Circuit-Breaker Enclosures, 1991.
25. IEC Standard 60947-2. Low-voltage Switchgear and Control Gear-Part 2: Circuit Breakers, 2009.
26. NEC. National Electric Code, NFPA 70, 2008.
27. ANSI Standard C37.46. American National Standard for High-Voltage Expulsion and Current Limiting Type Power Class Fuses and Fuse Disconnecting Switches, 2000.
28. IEEE Standard C37.41. IEEE Standard Design Tests for High-Voltage (>1000 V) Fuses, Fuse and Disconnecting Cutouts, Distribution Enclosed Single-Pole Air Switches, Fuse Disconnecting Switches, and Fuse Links and Accessories Used with These Devices, 2008.
29. ANSI Standard C37.47. American National Standard for High-Voltage Current-Limiting Type Distribution Class Fuses and Fuse Disconnecting Switches, 2000.
30. NEMA Standard SG2. High-Voltage Fuses, 1981.
31. IEEE Standard C37.42. IEEE Standard Specifications for High-Voltage (>1000 V) Expulsion-Type Distribution-Class Fuses, Fuse and Disconnecting Cutouts, Fuse Disconnecting Switches, and Fuse Links, and Accessories Used with These Devices, 2008.

6

Short Circuit of Synchronous and Induction Machines

A three-phase short circuit on the terminals of a generator has twofold effects. One, large disruptive forces are brought into play in the machine itself and the machine should be designed to withstand these forces. Two, short circuits should be removed quickly to limit fault damage and improve stability of the interconnected systems. The circuit breakers for generator application sense a fault current of high asymmetry and must be rated to interrupt successfully the short circuit currents. This is discussed in Chapters 7 and 8.

According to NEMA [1] specifications, a synchronous machine shall be capable of withstanding, without injury, a 30 s, three-phase short circuit at its terminals when operating at rated kVA and power factor, at 5% overvoltage, with fixed excitation. With a voltage regulator in service, the allowable duration t, in seconds, is determined from the following equation, where the regulator is designed to provide a ceiling voltage continuously during a short circuit:

$$t = \left(\frac{\text{Norminal field voltage}}{\text{Exciter ceiling voltage}}\right)^2 \times 30\,\text{s} \tag{6.1}$$

The generator should also be capable of withstanding without injury *any other* short circuit at its terminals for 30 s, provided that

$$I_2^2 t \preceq 40 \quad \text{for salient-pole machines} \tag{6.2}$$

$$I_2^2 t \preceq 30 \quad \text{for air-cooled cylindrical rotor machines} \tag{6.3}$$

and the maximum current is limited by external means so as not to exceed the three-phase fault; I_2 is the negative sequence current due to unsymmetrical faults.

Synchronous generators are major sources of short-circuit currents in power systems. The fault current depends on

1. The instant at which the short circuit occurs
2. The load and excitation of the machine immediately before the short circuit
3. The type of short circuits, i.e., whether three phases or one or more than one phase and ground are involved
4. Constructional features of the machine, especially leakage and damping.
5. The interconnecting impedances between generators

An insight into the physical behavior of the machine during a short circuit can be made by considering the theorem of constant flux linkages. For a closed circuit with resistance r and inductance L, $ri + L\ di/dt$ must be zero. If resistance is neglected, $L\ di/dt = 0$, i.e., the flux linkage *Li* must remain constant. In a generator, the resistance is small in comparison with the inductance, the field winding is closed on the exciter, and the stator winding is closed due to the short circuit. During the initial couple of cycles following a short circuit, the flux linkages with these two windings must remain constant. On a terminal fault, the generated EMF acts on a closed circuit of stator windings and is analogous to an EMF being suddenly applied to an inductive circuit. Dynamically, the situation is more complex, i.e., the lagging stator current has a demagnetizing effect on the field flux, and there are time constants associated with the penetration of the stator flux and decay of short-circuit current.

6.1 Reactances of a Synchronous Machine

The following definitions are applicable.

6.1.1 Leakage Reactance X_l

The leakage reactance can be defined but cannot be tested. It is the reactance due to flux setup by armature windings, but not crossing the air gap. It can be divided into end-winding leakage and slot leakage. A convenient way of picturing the reactances is to view these in terms of permeances of various magnetic paths in the machine, which are functions of dimensions of iron and copper circuits and independent of the flux density or the current loading. The permeances thus calculated can be multiplied by a factor to consider the flux density and current. For example, the leakage reactance is mainly given by the slot permeance and the end-coil permeance.

6.1.2 Subtransient Reactance X''_d

Subtransient reactance equals the leakage reactance plus the reactance due to the flux setup by stator currents crossing the air gap and penetrating the rotor as far as the damper windings in a laminated pole machine or as far as the surface damping currents in a solid pole machine. The subtransient conditions last for 1–5 cycles on a 60 Hz basis.

6.1.3 Transient Reactance X'_d

Transient reactance is the reactance after all damping currents in the rotor surface or amortisseur windings have decayed, but before the damping currents in the field winding have decayed. The transient reactance equals the leakage reactance plus the reactance due to flux setup by the armature that penetrates the rotor to the field windings. Transient conditions last for 5–200 cycles on a 60 Hz basis.

6.1.4 Synchronous Reactance X_d

Synchronous reactance is the steady-state reactance after all damping currents in the field windings have decayed. It is the sum of leakage reactance and a fictitious armature reaction reactance, which is much larger than the leakage reactance. Ignoring resistance, the per unit synchronous reactance is the ratio of per unit voltage on an open circuit divided by per unit armature current on a short circuit for a given field excitation. This gives *saturated* synchronous reactance. The unsaturated value of the synchronous reactance is given by the per unit voltage on air-gap open circuit line divided by per unit armature current on short circuit. If 0.5 per unit field excitation produces full-load armature current on short circuit, the saturated synchronous reactance is 2.0 per unit. The saturated value may be only 60%–80% of the unsaturated value.

6.1.5 Quadrature Axis Reactances X''_q, X'_q, and X_q

Quadrature axis reactances are similar to direct axis reactances, except that they involve the rotor permeance encountered when the stator flux enters one pole tip, crosses the pole, and leaves the other pole tip. The direct axis permeance is encountered by the flux crossing the air gap to the center of one pole, then crossing from one pole to the other pole, and entering the stator from that pole. Figure 6.1 shows the armature reaction components. The total armature reaction F can be divided into two components, F_{ad} and F_{aq}; F_{ad} is directed across the direct axis and F_{aq} across the quadrature axis. As these MMFs (magneto-motive forces) act on circuits of different permeances, the flux produced varies. If damper windings across pole faces are connected X''_q is nearly equal to X''_d.

6.1.6 Negative Sequence Reactance X_2

The negative sequence reactance is the reactance encountered by a voltage of reverse-phase sequence applied to the stator, with the machine running. Negative sequence flux revolves opposite to the rotor and is at twice the system frequency. Negative sequence reactance is

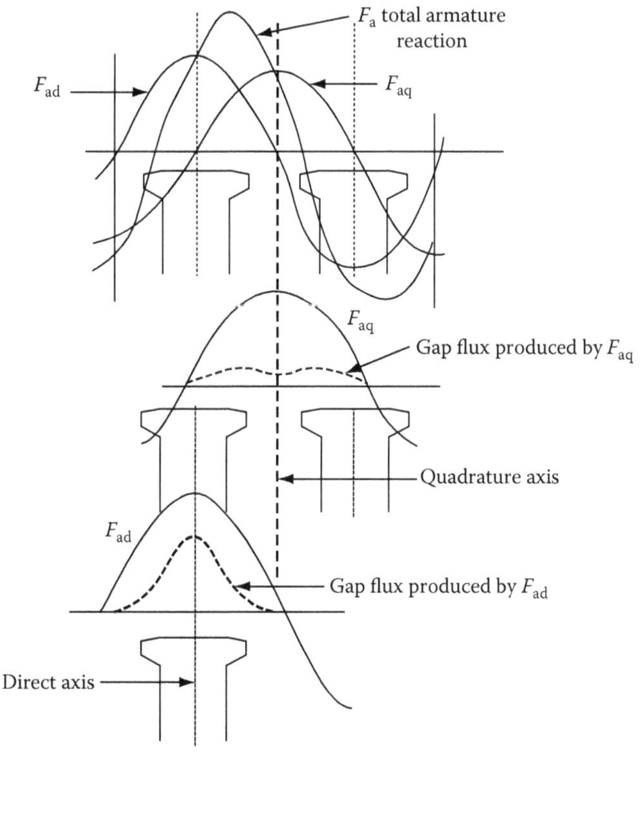

FIGURE 6.1
Armature reaction components in a synchronous machine in the direct and quadrature axes.

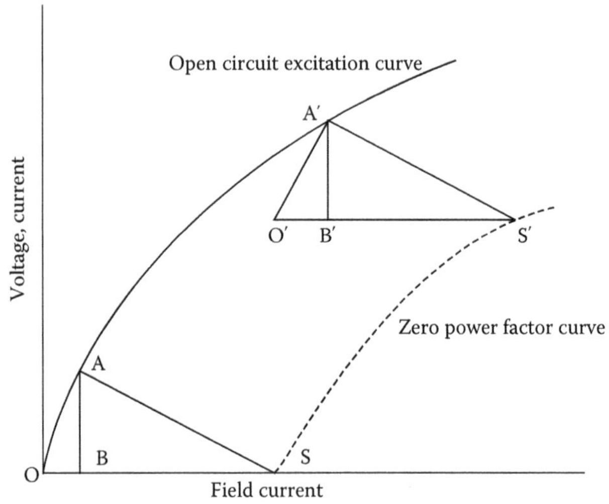

FIGURE 6.2
Open circuit, zero-power factor curves, and Potier triangle and reactance.

practically equal to the subtransient reactance as the damping currents in the damper windings or solid pole rotor surface prevent the flux from penetrating farther. The negative sequence reactance is generally taken as the average of subtransient direct axis and quadrature axis reactances, Equation 1.60.

6.1.7 Zero-Sequence Reactance X_0

The zero-sequence reactance is the reactance effective when rated frequency currents enter all three terminals of the machine simultaneously and leave at the neutral of the machine. It is approximately equal to the leakage reactance of the machine with full-pitch coils. With two-thirds-pitch stator coils, the zero-sequence reactance will be a fraction of the leakage reactance.

6.1.8 Potier Reactance X_p

Potier reactance is a reactance with numerical value between transient and subtransient reactances. It is used for the calculation of field current when open circuit and zero power factor curves are available. Triangle ABS in Figure 6.2 is a Potier triangle. As a result of the different slopes of open circuit and zero power factor curves, A′ B′ in Figure 6.2 is slightly larger than AB, and the value of reactance obtained from it is known as the Potier reactance.

6.2 Saturation of Reactances

Saturation varies with voltage, current, and power factor. *For short-circuit calculations according to ANSI/IEEE methods described in Chapter 7, saturated subtransient reactance must be considered.* The saturation factor is usually applied to transient and synchronous reactances, though all other reactances change, though slightly, with saturation. The saturated and unsaturated synchronous reactances are already defined above. In a typical machine, transient reactances may reduce from 5% to 25% on saturation. Saturated reactance is sometimes called the rated voltage reactance and is denoted by subscript "v" added to the "d" and "q" subscript axes, i.e., X''_{dv} and X_{qv} denote saturated subtransient reactances in direct and quadrature axes, respectively.

6.3 Time Constants of Synchronous Machines

6.3.1 Open Circuit Time Constant T'_{do}

The open circuit time constant expresses the rate of decay or buildup of field current when the stator is open circuited and there is zero resistance in the field circuit.

6.3.2 Subtransient Short-Circuit Time Constant T''_{d}

The subtransient short-circuit time constant expresses the rate of decay of the subtransient component of current under a bolted (zero resistance), three-phase short circuit at the machine terminals.

6.3.3 Transient Short-Circuit Time Constant T'_{d}

The transient short-circuit time constant expresses the rate of decay of the transient component of the current under a bolted (zero resistance), three-phase short circuit at the machine terminals.

6.3.4 Armature Time Constant T_a

The armature time constant expresses the rate of decay of the dc component of the short-circuit current under the same conditions.

Table 6.1 shows electrical data, reactances, and time constants of a 13.8 kV, 112.1 MVA 0.85 power factor generator.

6.4 Synchronous Machine Behavior on Terminal Short Circuit

The time immediately after a short circuit can be divided into three periods:

- The subtransient period lasting from 1 to 5 cycles.
- The transient period that may last up to 100 cycles or more.
- The final or steady-state period. Normally, the generator will be removed from service by protective relaying, much before the steady-state period is reached.

In the subtransient period, the conditions can be represented by the flux linking the stator and rotor windings. Any sudden change in the load or power factor of a generator produces changes in the MMFs, both in direct and quadrature axes. A terminal three-phase

TABLE 6.1

Generator Data

Description	Symbol	Data
Generator		
112.1 MVA, 2-pole, 13.8 kV, 0.85 PF, 95.285 MW, 4690 A, 0.56 SCR, 235 field V, wye connected		
Per unit reactance data, direct axis		
Saturated synchronous	X_{dv}	1.949
Unsaturated synchronous	X_d	1.949
Saturated transient	X'_{dv}	0.207
Unsaturated transient	X'_d	0.278
Saturated subtransient	X''_{dv}	0.164
Unsaturated subtransient	X''_d	0.193
Saturated negative sequence	X_{2v}	0.137
Unsaturated negative sequence	X_{2I}	0.185
Saturated zero sequence	X_{0v}	0.092
Leakage reactance, overexcited	X_{0I}	0.111
Leakage reactance, underexcited	$X_{LM,OXE}$	0.164
	$X_{LM,UXE}$	0.164
Per unit reactance data, quadrature axis		
Saturated synchronous	X_{qv}	1.858
Unsaturated synchronous	X_q	1.858
Unsaturated transient	X'_q	0.434
Saturated subtransient	X''_{qv}	0.140
Unsaturated subtransient	X''_q	0.192
Field time constant data, direct axis		
Open circuit	T'_{d0}	5.615
Three-phase short-circuit transient	T'_{d3}	0.597
Line-to-line short-circuit transient	T'_{d2}	0.927
Line-to-neutral short-circuit transient	T'_{d1}	1.124
Short-circuit subtransient	T''_d	0.015
Open circuit subtransient	T''_{d0}	0.022
Field time constant data quadrature axis		
Open circuit	T'_{q0}	0.451
Three-phase short-circuit transient	T'_q	0.451
Short-circuit subtransient	T''_q	0.015
Open circuit subtransient	T''_{q0}	0.046
Armature dc component time constant data		
Three-phase short circuit	T_{a3}	0.330
Line-to-line short circuit	T_{a2}	0.330
Line-to-neutral short circuit	T_{a1}	0.294

short circuit is a large disturbance. At the moment of short circuit, the flux linking the stator from the rotor is trapped to the stator, giving a stationary replica of the main-pole flux. The rotor poles may be in a position of maximum or minimum flux linkage, and as these rotate, the flux linkages tend to change. This is counteracted by a current in the stator windings. The short-circuit current is, therefore, dependent on rotor angle. As the energy stored can be considered as a function of armature and field linkages, the torque fluctuates and reverses cyclically. The dc component giving rise to asymmetry is caused by the flux trapped in the stator windings at the instant of short circuit, which sets up a dc transient in the armature circuit. This dc component establishes a component field in the air gap, which is stationary in space, and which, therefore, induces a fundamental frequency voltage and current in the synchronously revolving rotor circuits. Thus, an increase in the stator current is followed by an increase in the field current. The field flux has superimposed on it a new flux pulsating with respect to field windings at normal machine frequency. The single-phase-induced current in the field can be resolved into two components, one stationary with respect to the stator, which counteracts the dc component of the stator current, and the other component travels at twice the synchronous speed with respect to the stator and induces a second harmonic in it.

The armature and field are linked through the magnetic circuit, and the ac component of lagging current creates a demagnetizing effect. However, some time must elapse before it starts becoming effective in reducing the field current and the steady-state current is reached. The protective relays will normally operate to open the generator breaker and simultaneously the field circuit for suppression of generated voltage.

The above is rather an oversimplification of the transient phenomena in the machine on short circuit. In practice, a generator will be connected in an interconnected system. The machine terminal voltage, rotor angle, and frequency all change depending on the location of the fault in the network, the network impedance, and the machine parameters. The machine output power will be affected by the change in the rotor winding EMF and the rotor position in addition to any changes in the impedance seen at the machine terminals. For a terminal fault, the voltage at the machine terminals is zero and, therefore, power supplied by the machine to load reduces to zero, while the prime mover output cannot change suddenly. Thus, the generator accelerates. In a multimachine system with interconnecting impedances, the speeds of all machines change, so that these generate their share of synchronizing power in the overall impact, as these strive to reach a mean retardation through oscillations due to stored energy in the rotating masses.

In a dynamic simulation of a short circuit, the following may be considered:

- Network before, during, and after the short circuit
- Induction motors' dynamic modeling, with zero excitation
- Synchronous machine dynamic modeling, considering saturation
- Modeling of excitation systems
- Turbine and governor models

Figure 6.3 shows the transients in an interconnected system on a three-phase short circuit lasting for 5 cycles. Figure 6.3a shows the torque angle swings of two generators, which are stable after the fault, Figure 6.3b shows speed transients, and Figure 6.3c shows the field voltage response of a high-response excitation system. A system having an excitation system voltage response of 0.1 s or less is defined as the high-response excitation system [2]. The excitation systems may not affect the first cycle momentary currents, but are of consideration for interrupting duty and 30 cycle currents. The reactive and active power transients are shown in Figure 6.3d and e, respectively. The voltage dip and recovery characteristics are shown in Figure 6.3f. A fault voltage dip of more than 60% occurs. Though the generators are stable after the fault removal, the large voltage dip can precipitate shutdown of consumer loads, i.e., the magnetic contactors in motor controllers can drop out during the first-cycle voltage dip. This is of major consideration in continuous process plants. Figure 6.3 is based on system transient stability study. Transient analysis programs such as EMTP [3] can be used for dynamic simulation of the short-circuit currents.

For practical calculations, the dynamic simulations of short-circuit currents are rarely carried out. The generator is replaced with an equivalent circuit of voltage and certain impedances intended to represent the worst conditions, after the fault (Chapter 7).

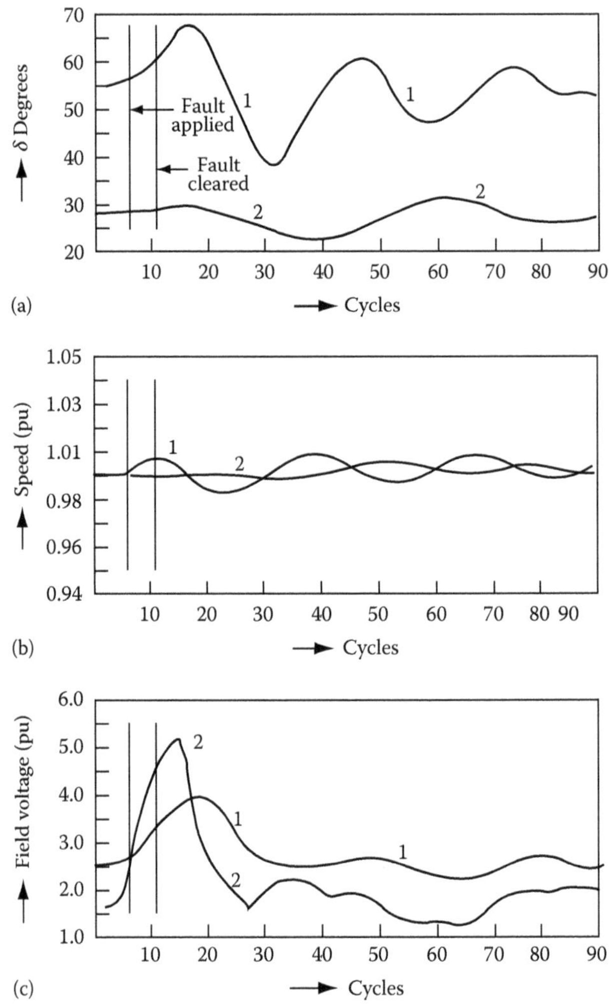

FIGURE 6.3
Transient behavior of two generators in an interconnected system for a three-phase fault cleared in 5 cycles: (a) torque angle swings; (b) speed transients; (c) field voltage;

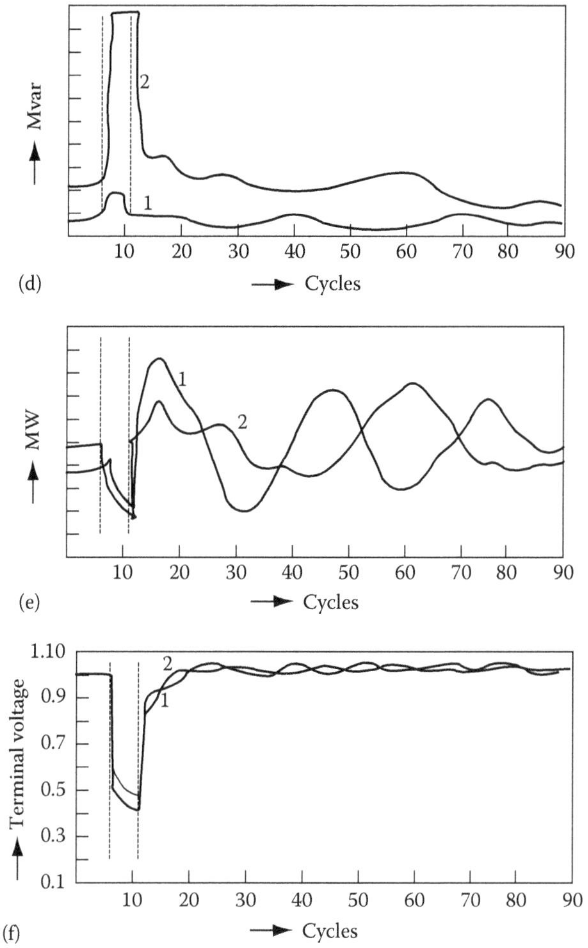

FIGURE 6.3 (continued)
(d) and (e) reactive and active power swings; (f) voltage dip and recovery profile.

The speed change is ignored. The excitation is assumed to be constant, and the generator load is ignored. That this procedure is safe and conservative has been established in the industry by years of applications, testing, and experience.

6.4.1 Equivalent Circuits during Fault

Figure 6.4 shows an envelope of decaying ac component of the short-circuit current wave, neglecting the dc component. The extrapolation of the current envelope to zero time gives the peak current. Note that, immediately after the fault, the current decays rapidly and then more slowly.

Transformer equivalent circuits of a salient pole synchronous machine in the direct and quadrature axis at the instant of short circuit and during subsequent time delays help to derive the short-circuit current equations and explain the decaying ac component of Figure 6.4. Based on the above discussions, these circuits are shown in Figure 6.5 in the subtransient, transient, and steady-state periods. As the flux penetrates into the rotor, and

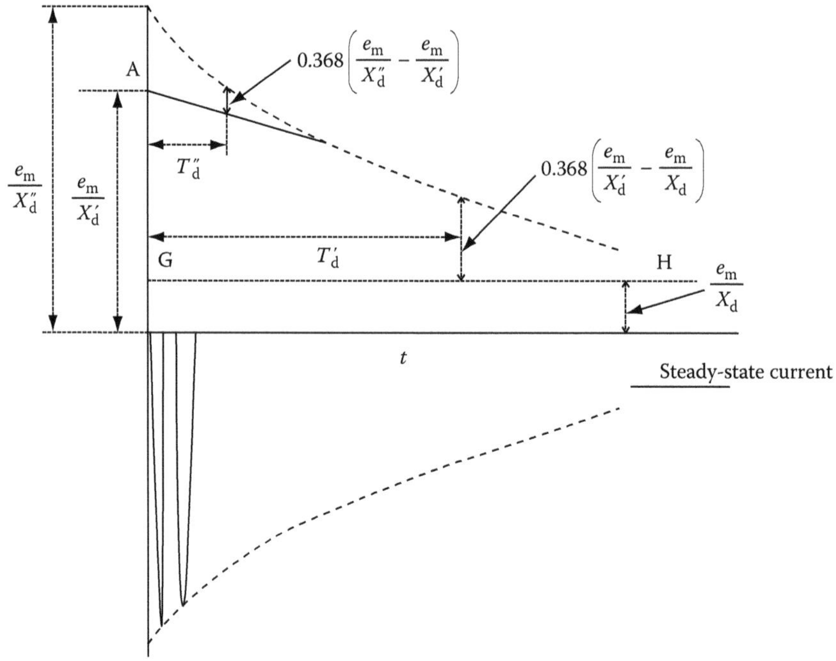

FIGURE 6.4
Decaying ac component of the short-circuit current and subtransient, transient, and steady-state currents.

the currents die down, it is equivalent to opening a circuit element, i.e., from subtransient to transient state, the damper circuits are opened.

The direct axis subtransient reactance is given by

$$X''_d = X_l + \frac{1}{1/X_{ad} + 1/X_f + 1/X_{kD}} \tag{6.4}$$

where

X_{ad} is the reactance corresponding to the fundamental space wave of the armature in the direct axis
X_f is the reactance of the field windings
X_{kD} is the damper windings in the direct axis
X_l is the leakage reactance

X_f and X_{kD} are also akin to leakage reactances. Similarly, the quadrature axis subtransient reactance is given by

$$X''_q = X_l + \frac{1}{1/X_{aq} + 1/X_{kQ}} \tag{6.5}$$

where

X_{aq} is the reactance corresponding to the fundamental space wave in the quadrature axis
X_{kQ} is the reactance of the damper winding in the quadrature axis
X_l is identical in the direct and quadrature axes

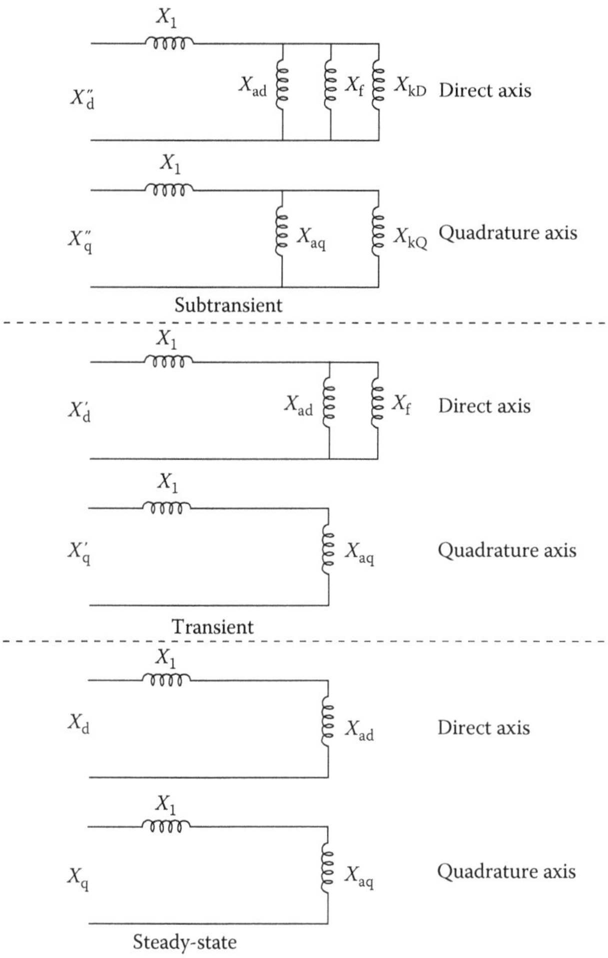

FIGURE 6.5
Equivalent transformer circuits of a synchronous generator during subtransient, transient, and steady-state periods, after a terminal fault.

The quadrature axis rotor circuit does not carry a field winding, and this circuit is composed of damper bars or rotor iron in the interpolar axis of a cylindrical rotor machine.

The direct axis and quadrature axis short-circuit time constants associated with decay of the subtransient component of the current are

$$T''_{\mathrm{d}} = \frac{1}{\omega r_{\mathrm{D}}}\left[\frac{X_{\mathrm{ad}}X_{\mathrm{f}}X_{\mathrm{l}}}{X_{\mathrm{ad}}X_{\mathrm{f}} + X_{\mathrm{f}}X_{\mathrm{l}} + X_{\mathrm{ad}}X_{\mathrm{l}}} + X_{\mathrm{kD}}\right] \tag{6.6}$$

$$T''_{\mathrm{q}} = \frac{1}{\omega r_{\mathrm{Q}}}\left[\frac{X_{\mathrm{aq}}X_{\mathrm{l}}}{X_{\mathrm{aq}} + X_{\mathrm{l}}} + X_{\mathrm{kQ}}\right] \tag{6.7}$$

where r_{D} and r_{Q} are resistances of the damper windings in the direct and quadrature axis, respectively. When the flux has penetrated the air gap, the effect of the eddy currents in the pole face cease after a few cycles is given by short-circuit subtransient time constants. The resistance of the damper circuit is much higher than that of the field windings.

This amounts to opening of the damper winding circuits and the direct axis and quadrature axis transient reactances are given by

$$X'_{\mathrm{d}} = X_{\mathrm{l}} + \frac{1}{1/X_{\mathrm{ad}} + 1/X_{\mathrm{f}}} = \left[\frac{X_{\mathrm{ad}}X_{\mathrm{f}}}{X_{\mathrm{ad}} + X_{\mathrm{f}}} + X_{\mathrm{l}}\right] \tag{6.8}$$

$$X'_{\mathrm{q}} = X_{\mathrm{l}} + X_{\mathrm{aq}} \tag{6.9}$$

The direct axis transient time constant associated with this decay is

$$T'_{\mathrm{d}} = \frac{1}{\omega r_{\mathrm{F}}}\left[\frac{X_{\mathrm{ad}}X_{\mathrm{l}}}{X_{\mathrm{ad}} + X_{\mathrm{l}}} + X_{\mathrm{f}}\right] \tag{6.10}$$

where r_{F} is the resistance of the field windings.

Finally, when the currents in the field winding have also died down, given by the transient short-circuit time constant, the steady-state short-circuit current is given by the synchronous reactance:

$$X_{\mathrm{d}} = X_{\mathrm{l}} + X_{\mathrm{ad}} \tag{6.11}$$

$$X_{\mathrm{q}} = X_{\mathrm{l}} + X_{\mathrm{aq}} \tag{6.12}$$

Equations 6.9 and 6.12 show that X'_{q} is equal to X_{q}. The relative values of X''_{q}, X'_{q}, and X_{q} depend on machine construction. For cylindrical rotor machines, $X_{\mathrm{q}} \gg X'_{\mathrm{q}}$. Sometimes, one or two damper windings are modeled in the q axis.

Reverting to Figure 6.4, the direct axis transient reactance determines the initial value of the symmetrical transient envelope, and the direct axis short-circuit time constant T'_{d} determines the decay of this envelope. The direct axis time constant T'_{d} is the time required by the transient envelope to decay to a point where the difference between it and the steady-state envelope GH is $1/e(=0.368)$ of the initial difference GA. A similar explanation applies to the decay of the subtransient component in Figure 6.4.

6.4.2 Fault Decrement Curve

Based on Figure 6.4, the expression for a decaying ac component of the short-circuit current of a generator can be written as

$$\begin{aligned} i_{\mathrm{ac}} &= \text{Decaying subtransient component} + \text{decaying transient component} \\ &\quad + \text{steady-state component} \\ &= \left(i''_{\mathrm{d}} - i'_{\mathrm{d}}\right)e^{-t/T''_{\mathrm{d}}} + \left(i'_{\mathrm{d}} - i_{\mathrm{d}}\right)e^{-t/T'_{\mathrm{d}}} + i_{\mathrm{d}} \end{aligned} \tag{6.13}$$

The subtransient current is given by

$$i''_{\mathrm{d}} = \frac{E''}{X''_{\mathrm{d}}} \tag{6.14}$$

where E'' is the generator internal voltage behind subtransient reactance:

$$E'' = V_a + X''_d \sin\phi \tag{6.15}$$

where
V_a is the generator terminal voltage
ϕ is the load power factor angle, prior to fault

Similarly, the transient component of the current is given by

$$i'_d = \frac{E'}{X'_d} \tag{6.16}$$

where E' is the generator internal voltage behind transient reactance:

$$E' = V_a + X'_d \sin\phi \tag{6.17}$$

The steady-state component is given by

$$i_d = \frac{V_a}{X_d}\left(\frac{i_F}{i_{Fg}}\right) \tag{6.18}$$

where
i_F is the field current at given load conditions (when regulator action is taken into account)
i_{Fg} is the field current at no-load rated voltage

The dc component is given by

$$i_{dc} = \sqrt{2} i''_d e^{-t/T_a} \tag{6.19}$$

where T_a is the armature short-circuit time constant, given by

$$T_a = \frac{1}{\omega r}\left[\frac{2X''_d X''_q}{X''_d + X''_q}\right] \tag{6.20}$$

where r is the stator resistance.

The open circuit time constant describes the decay of the field transient; the field circuit is closed and the armature circuit is open:

$$T''_{do} = \frac{1}{\omega r_D}\left[\frac{X_{ad} X_f}{X_{ad} + X_f} + X_{kD}\right] \tag{6.21}$$

and the quadrature axis subtransient open circuit time constant is

$$T''_{qo} = \frac{1}{\omega r_Q}(X_{aq} + X_{kQ}) \tag{6.22}$$

The open circuit direct axis transient time constant is

$$T'_{do} = \frac{1}{\omega r_F}(X_{ad} + X_f) \tag{6.23}$$

The short-circuit direct axis transient time constant can be expressed as

$$T'_d = T'_{do}\left[\frac{X'_d}{(X_{ad} + X_l)}\right] = T'_{do}\frac{X'_d}{X_d} \tag{6.24}$$

It may be observed that the resistances have been neglected in the above expressions. In fact, these can be included, i.e., the subtransient current is

$$i''_d = \frac{E''}{r_D + X''_d} \tag{6.25}$$

where r_D is defined as the resistance of the armortisseur windings on salient pole machines and analogous body of cylindrical rotor machines. Similarly, the transient current is

$$i'_d = \frac{E'}{r_f + X'_d} \tag{6.26}$$

Example 6.1

Consider a 13.8 kV, 100 MVA 0.85 power factor generator. Its rated full-load current is 4184 A. Other data are

Saturated subtransient reactance $X''_{dv} = 0.15$ per unit
Saturated transient reactance $X'_{dv} = 0.2$ per unit
Synchronous reactance $X_d = 2.0$ per unit
Field current at rated load $i_f = 3$ per unit
Field current at no-load rated voltage $i_{fg} = 1$ per unit
Subtransient short circuit time constant $T''_d = 0.012$ s
Transient short circuit time constant $T'_d = 0.35$ s
Armature short circuit time constant $T_a = 0.15$ s
Effective resistance* = 0.0012 per unit
Quadrature axis synchronous reactance* = 1.8 per unit

A three-phase short circuit occurs at the terminals of the generator, when it is operating at its rated load and power factor. It is required to construct a fault decrement curve of the generator for (1) the ac component, (2) dc component, and (3) total current. Data marked with an asterisk are intended, e.g., 6.5.

From Equation 6.15, the voltage behind subtransient reactance at the generator rated voltage, load, and power factor is

$$E'' = V + X''_d \sin\phi = 1 + (0.15)(0.527) = 1.079\,\text{pu}$$

From Equation 6.14, the subtransient component of the current is

$$i''_{d} = \frac{E''}{X''_{dv}} = \frac{1.079}{0.15} \text{ per unit} = 30.10\,\text{kA}$$

Similarly, from Equation 6.17, E', the voltage behind transient reactance is 1.1054 per unit and, from Equation 6.16, the transient component of the current is 23.12 kA.

From Equation 6.18, current i_d at constant excitation is 2.09 kA rms. For a ratio of $i_f/i_{Fg} = 3$, current $i_d = 6.28$ kA rms. Therefore, the following equations can be written for the ac component of the current:

With constant excitation:

$$i_{ac} = 6.98^{-t/0.012} + 20.03e^{-t/0.35} + 2.09\,\text{kA}$$

With full load excitation:

$$i_{ac} = 6.98e^{-t/0.012} + 16.84e^{-t/0.35} + 6.28\,\text{kA}$$

The ac decaying component of the current can be plotted from these two equations, with the lowest value of $t = 0.01$–1000 s. This is shown in Figure 6.5. The dc component is given by Equation 6.19:

$$i_{dc} = \sqrt{2}i''_{d}e^{-t/T_a} = 42.57e^{-t/0.15}\,\text{kA}$$

This is also shown in Figure 6.6. At any instant, the total current is

$$\sqrt{i_{ac}^2 + i_{dc}^2}\ \text{kA rms}$$

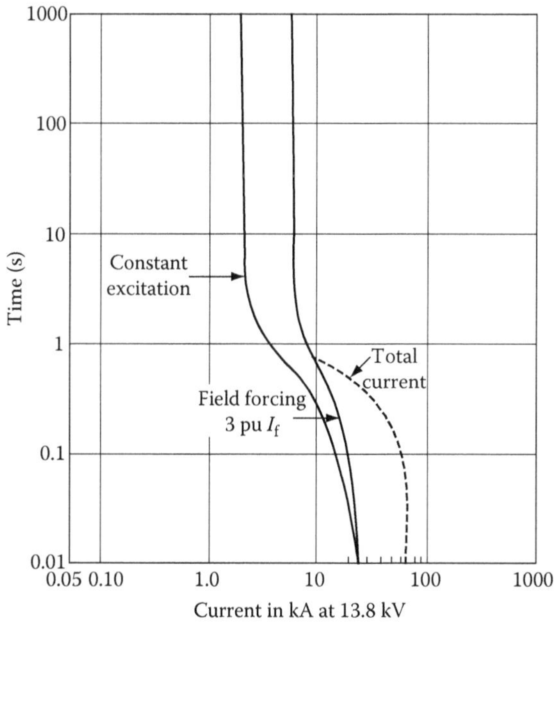

FIGURE 6.6
Calculated fault decrement curves of generator (Example 6.1).

The fault decrement curves are shown in Figure 6.6. Short-circuit current with constant excitation is 50% of the generator full-load current. This can occur for a stuck voltage regulator condition. Though this current is lower than the generator full-load current, it cannot be allowed to be sustained. Voltage restraint or voltage-controlled overcurrent generator backup relays (ANSI/IEEE device number 51V) or distance relays (device 21) are set to pick up on this current. The generator fault decrement curve is often required for appropriate setting and coordination of these relays with the system relays.

6.5 Circuit Equations of Unit Machines

The behavior of machines can be analyzed in terms of circuit theory, which makes it useful not only for steady-state performance but also for transients like short circuits. The circuit of machines can be simplified in terms of coils on the stationary (stator) and rotating (rotor) parts and these coils interact with each other according to fundamental electromagnetic laws. The circuit of a unit machine can be derived from consideration of generation of EMF in coupled coils due to (1) transformer EMF, also called pulsation EMF, and (2) the EMF of rotation.

Consider two magnetically coupled, stationary, coaxial coils as shown in Figure 6.7. Let the applied voltages be v_1 and v_2 and the currents i_1 and i_2, respectively. This is, in fact, the circuit of a two-winding transformer, the primary and secondary being represented by single-turn coils. The current in the primary coil (any coil can be called a primary coil) sets up a total flux linkage Φ_{11}. Change of current in this coil induces an EMF is given by

$$e_{11} = -\frac{d\Phi_{11}}{dt} = -\frac{d\Phi_{11}}{di_1}\cdot\frac{di_1}{dt} = -L_{11}\frac{di_1}{dt} = -L_{11}pi_1 \tag{6.27}$$

where $L_{11} = -d\Phi_{11}/di_1$ is the total primary self-inductance and the operator $p = d/dt$. If Φ_1 is the leakage flux and Φ_{12} is the flux linking with the secondary coil, then the variation of current in the primary coil induces in the secondary coil an EMF:

$$e_{12} = -\frac{d\Phi_{12}}{dt} = -\frac{d\Phi_{12}}{di_1}\cdot\frac{di_1}{dt} = -L_{12}\frac{di_1}{dt} = -L_{12}pi_1 \tag{6.28}$$

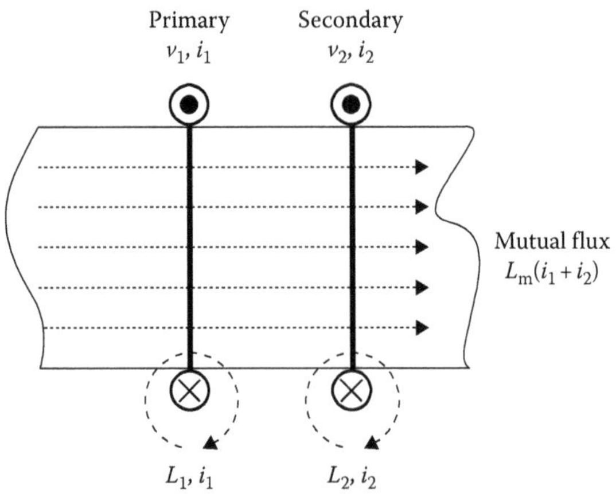

FIGURE 6.7
Representation of magnetically coupled coils of a two-winding transformer.

where $L_{12} = -d\Phi_{12}/di_1$ is the mutual inductance of the primary coil winding with the secondary coil winding. Similar equations apply for the secondary coil winding. All the flux produced by the primary coil winding does not link with the secondary. The leakage inductance associated with the windings can be accounted for as

$$L_{11} = L_{12} + L_1 \tag{6.29}$$

$$L_{22} = L_{21} + L \tag{6.30}$$

The mutual inductance between coils can be written as

$$L_{12} = L_{21} = L_m = \sqrt{[(L_{11} - L_1)(L_{22} - L_2)]} = k\sqrt{L_{11}L_{22}} \tag{6.31}$$

Thus, the equations of a unit transformer are

$$v_1 = r_1 i_1 + (L_m + L_1)pi_1 + L_m p i_2$$
$$v_2 = r_2 i_2 + (L_m + L_2)pi_2 + L_m p i_1$$

Or in the matrix form

$$\begin{vmatrix} v_1 \\ v_2 \end{vmatrix} = \begin{vmatrix} r_1 + (L_1 + L_m)p & L_m p \\ L_m p & r_2 + (L_2 + L_m)p \end{vmatrix} \begin{vmatrix} i_1 \\ i_2 \end{vmatrix} \tag{6.32}$$

If the magnetic axis of the coupled coils is at right angles, no mutually induced pulsation or transformer EMF can be produced by variation of currents in either of the windings. However, if the coils are free to move, the coils with magnetic axes at right angles have an EMF of rotation, e_r, induced when the winding it represents rotates:

$$e_r = \omega_r \Phi \tag{6.33}$$

where
 ω_r is the angular speed of rotation
 Φ is the flux

This EMF is a maximum when the two coils are at right angles to each other and zero when these are cophasial.

To summarize, a pulsation EMF is developed in two coaxial coils and there is no rotational EMF. Conversely, a rotational EMF is developed in two coils at right angles, but no pulsation EMF. If the relative motion is at an angle θ, the EMF of rotation is multiplied by $\sin\theta$:

$$e_r = \omega_r \Phi \sin\theta \tag{6.34}$$

The equations of a unit machine may be constructed based on the above simple derivation of EMF production in coils. Consider a machine with direct and quadrature axis coils as shown in Figure 6.8. Note that the armature is shown rotating and has two coils D and Q at right angles in the *d*–*q* axes. The field winding F and the damper winding KD are shown stationary in the direct axis. All coils are single turn coils. In the direct axis, there are three

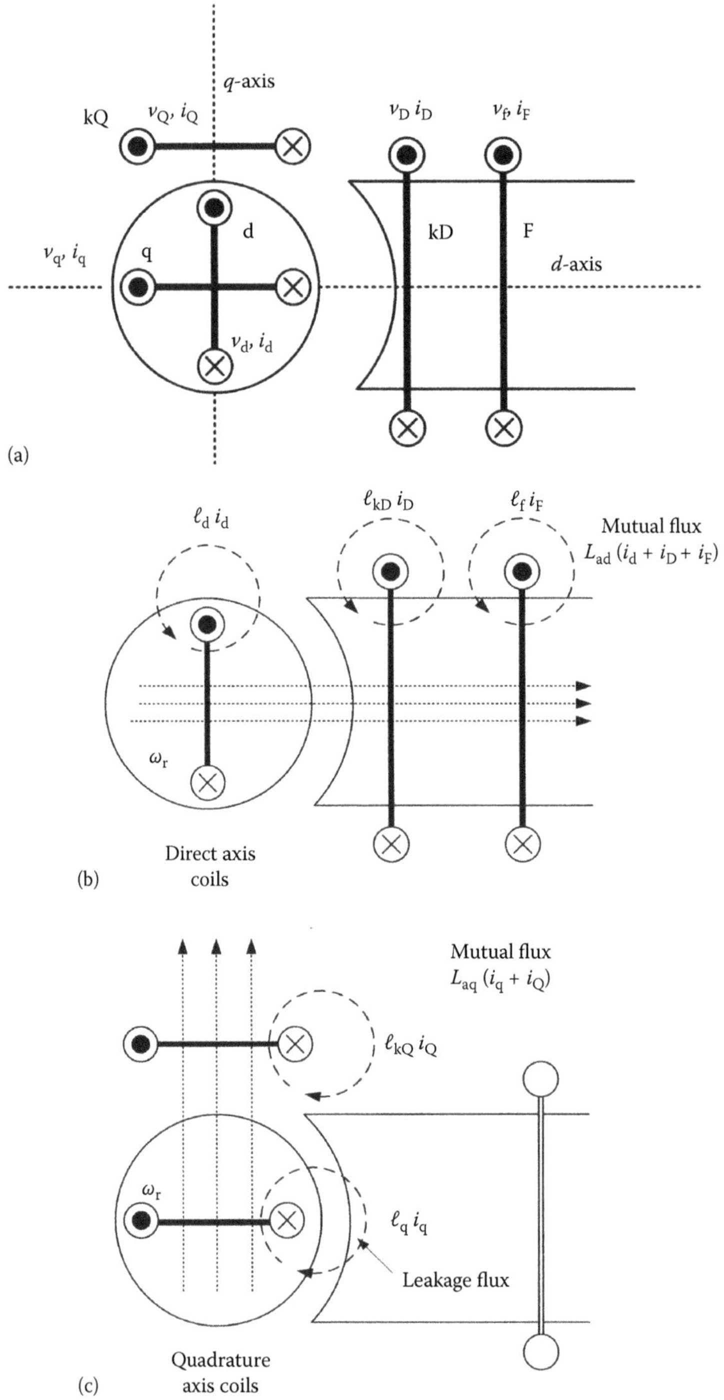

FIGURE 6.8
(a) Development of the circuit of a unit machine; (b) and (c) flux linkages in the direct and quadrature axes' coils.

mutual inductances, i.e., of D with KD, KD with F, and F with D. A simplification is to consider these equal to inductance L_{ad}. Each coil has a leakage inductance of its own. Consequently, the total inductances of the coils are

$$\begin{aligned} &\text{Coil D:} && (l_d + L_{ad}) \\ &\text{Coil KD:} && (l_{kD} + L_{ad}) \\ &\text{Coil F:} && (l_f + L_{ad}) \end{aligned} \tag{6.35}$$

The mutual linkage with armature coil D when all three coils carry currents is

$$\Phi_d = L_{ad}(i_F + i_D + i_d) \tag{6.36}$$

where i_F, i_D, and i_d are the currents in the field, damper, and direct axis coils. Similarly, in the quadrature axis,

$$\Phi_q = L_{aq}(i_q + i_Q) \tag{6.37}$$

The EMF equations in each of the coils can be written on the basis of these observations.

Field coil: no q-axis circuit will affect its flux, nor do any rotational voltages appear. The applied voltage v_f is

$$v_f = r_F i_F + (L_{ad} + l_f)pi_F + L_{ad}pi_D + L_{ad}pi_d \tag{6.38}$$

Stator coil KD is located similarly to coil F:

$$v_D = r_D i_D + (L_{ad} + l_{kD})pi_D + L_{ad}pi_F + L_{ad}pi_d \tag{6.39}$$

Coil KQ has no rotational EMF, but will be affected magnetically by any current i_Q in coil Q:

$$v_Q = r_Q i_Q + (L_{aq} + l_{kQ})pi_Q + L_{aq}pi_q \tag{6.40}$$

Armature coils D and Q have the additional property of inducing rotational EMF:

$$v_d = r_d i_d + (L_{ad} + l_d)pi_d + L_{ad}pi_F + L_{ad}pi_D + L_{aq}\omega_r i_Q + (L_{aq} + l_q)\omega_r i_q \tag{6.41}$$

$$v_q = r_q i_q + (L_{aq} + l_q)pi_q + L_{aq}pi_Q - L_{ad}\omega_r i_F - L_{ad}\omega_r i_D - (L_{ad} + l_d)\omega_r i_d \tag{6.42}$$

These equations can be written in a matrix form

$$\begin{vmatrix} v_f \\ v_D \\ v_Q \\ v_d \\ v_q \end{vmatrix} = \begin{vmatrix} r_F + (L_{ad} + l_f)p & L_{ad}p & \cdot & L_{ad}p & \cdot \\ L_{ad}p & r_D + (L_{ad} + l_{kD})p & \cdot & L_{ad}p & \cdot \\ \cdot & \cdot & r_Q + (L_{aq} + l_{kQ})p & \cdot & L_{aq}p \\ L_{ad}p & L_{ad}p & L_{aq}\omega_r & r_d + (L_{ad} + l_d)p & (L_{aq} + l_q)\omega_r \\ -L_{ad}\omega_r & -L_{ad}\omega_r & L_{aq}p & -(L_{ad} + l_d)\omega_r & r_q + (L_{aq} + l_q)p \end{vmatrix} \begin{vmatrix} i_F \\ i_D \\ i_Q \\ i_d \\ i_q \end{vmatrix} \tag{6.43}$$

6.6 Park's Transformation

Park's transformation [4,5] greatly simplifies the mathematical model of synchronous machines. It describes a new set of variables, such as currents, voltages, and flux linkages, obtained by transformation of the actual (stator) variables in three axes: 0, *d*, and *q*. The *d* and *q* axes are already defined, the 0 axis is a stationary axis.

6.6.1 Reactance Matrix of a Synchronous Machine

Consider the normal construction of a three-phase synchronous machine, with three-phase stationary ac windings on the stator, and the field and damper windings on the rotor (Figure 6.9). The stator inductances vary, depending on the relative position of the stator and rotor. Consider that the field winding is cophasial with the direct axis and also that the direct axis carries a damper winding. The *q* axis also has a damper winding. The phase windings are distributed, but are represented by single turn coils aa, bb, and cc in Figure 6.9. The field flux is directed along the *d* axis, and, therefore, the machine-generated voltage is at right angles to it, along the *q* axis. For generator action, the generated voltage vector *E* leads the terminal voltage vector *V* by an angle δ, and, from basic machine theory, we know that δ is the torque angle. At $t = 0$, the voltage vector *V* is located along the axis of phase a, which is the reference axis in Figure 6.9. The *q* axis is at an angle δ and the *d* axis is at an angle $\delta + \pi/2$. For $t > 0$, the reference axis is at an angle $\omega_r t$ with respect to the axis of phase a. The *d* axis of the rotor is, therefore, at

$$\theta = \omega_r t + \delta + \frac{\pi}{2} \tag{6.44}$$

For synchronous operation, $\omega_r = \omega_0 =$ constant.

Consider phase a inductance, it is a combination of its own self-inductance L_{aa} and its mutual inductances L_{ab} and L_{bc} with phases b and c. All three inductances vary with the relative position of the rotor with respect to the stator because of saliency of the air gap. When the axis of phase a coincides with the direct axis (Figure 6.9), i.e., $\theta = 0$ or π, the resulting flux of coil aa is maximum in the horizontal direction and its self-inductance

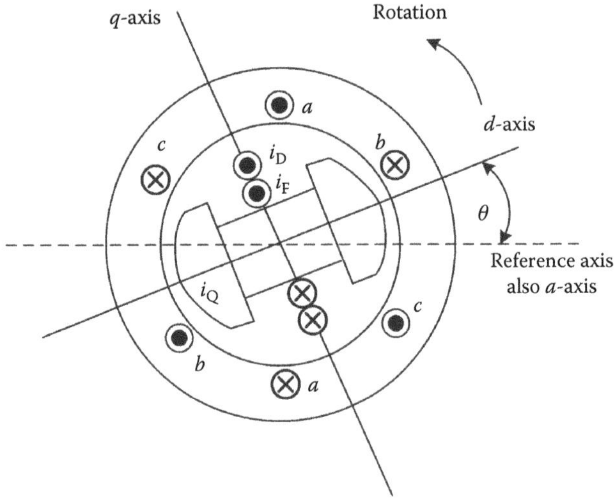

FIGURE 6.9
Representation of *d–q* axes, reference axes, and windings in a synchronous machine.

is a maximum. When at right angles to the d axis, $\theta = \pi/2$ or $3\pi/2$ its inductance is a minimum. Thus, L_{aa} fluctuates twice per revolution and can be expressed as

$$L_{aa} = L_s + L_m \cos 2\theta \tag{6.45}$$

Similarly, self-inductance of phase b is maximum at $\theta = 2\pi/3$ and of phase c at $\theta = -2\pi/3$:

$$L_{bb} = L_s + L_m \cos 2\left(\theta - 2\frac{\pi}{3}\right) \tag{6.46}$$

$$L_{cc} = L_s + L_m \cos 2\left(\theta + 2\frac{\pi}{3}\right) \tag{6.47}$$

Phase-to-phase mutual inductances are also a function of θ; L_{ab} is negative and is maximum at $\theta = -\pi/6$. This can be explained as follows: for the direction of currents shown in coils aa and bb, L_{ab} is negative. When the angle is $-\pi/3$, the current in the phase b coil generates the maximum flux, but the linkage with the phase a coil is better when the angle is 0°. However, at this angle, the flux is reduced. The maximum flux linkage can be considered to take place at an angle, which is an average of these two angles, i.e., $\pi/6$:

$$L_{ab} = -\left[M_s + L_m \cos 2\left(\theta + \frac{\pi}{6}\right)\right] \tag{6.48}$$

$$L_{bc} = -\left[M_s + L_m \cos 2\left(\theta - \frac{\pi}{2}\right)\right] \tag{6.49}$$

$$L_{ca} = -\left[M_s + L_m \cos 2\left(\theta + 5\frac{\pi}{6}\right)\right] \tag{6.50}$$

Stator-to-rotor mutual inductances are the inductances between stator windings and field windings, between stator windings and direct axis damper windings, and between stator windings and quadrature axis damper windings. These reactances are

From stator phase windings to field windings

$$L_{aF} = M_F \cos\theta \tag{6.51}$$

$$L_{bF} = M_F \cos\left(\theta - 2\frac{\pi}{3}\right) \tag{6.52}$$

$$L_{cF} = M_F \cos\left(\theta + 2\frac{\pi}{3}\right) \tag{6.53}$$

From stator phase windings to direct axis damper windings

$$L_{aD} = M_D \cos\theta \tag{6.54}$$

$$L_{bD} = M_D \cos\left(\theta - 2\frac{\pi}{3}\right) \tag{6.55}$$

$$L_{cD} = M_D \cos\left(\theta + 2\frac{\pi}{3}\right) \tag{6.56}$$

From stator phase windings to damper windings in the quadrature axis

$$L_{aQ} = M_Q \sin\theta \tag{6.57}$$

$$L_{bQ} = M_Q \sin\left(\theta - 2\frac{\pi}{3}\right) \tag{6.58}$$

$$L_{cQ} = M_Q \sin\left(\theta + 2\frac{\pi}{3}\right) \tag{6.59}$$

The rotor self-inductances are L_F, L_D, and L_Q. The mutual inductances are

$$L_{DF} = M_R \qquad L_{FQ} = 0 \qquad L_{DQ} = 0 \tag{6.60}$$

The mutual inductance between field windings and direct axis damper windings is constant and does not vary. Also, the d and q axes are displaced by 90°, and the mutual inductances between the field and direct axis damper windings and quadrature axis damper windings are zero.

The inductance matrix can therefore be written as

$$\overline{L} = \begin{vmatrix} \overline{L}_{aa} & \overline{L}_{aR} \\ \overline{L}_{Ra} & \overline{L}_{RR} \end{vmatrix} \tag{6.61}$$

where $\overline{L}_{aa}$ is a stator-to-stator inductance matrix:

$$\overline{L}_{aa} = \begin{vmatrix} L_s + L_m \cos 2\theta & -M_s - L_m \cos 2(\theta + \pi/6) & -M_s - L_m \cos 2(\theta + 5\pi/6) \\ -M_s - L_m \cos 2(\theta + \pi/6) & L_s + L_m \cos 2(\theta - 2\pi/3) & -M_s - L_m \cos 2(\theta - \pi/2) \\ -M_s - L_m \cos 2(\theta + 5\pi/6) & -M_s - L_m \cos 2(\theta - \pi/2) & L_s + L_m \cos 2(\theta + 2\pi/3) \end{vmatrix} \tag{6.62}$$

$\overline{L}_{aR} = \overline{L}_{Ra}$ is the stator-to-rotor inductance matrix:

$$\overline{L}_{aR} = \overline{L}_{Ra} = \begin{vmatrix} M_F \cos\theta & M_D \cos\theta & M_Q \sin\theta \\ M_F \cos(\theta - 2\pi/3) & M_D \cos(\theta - 2\pi/3) & M_Q \sin(\theta - 2\pi/3) \\ M_F \cos(\theta + 2\pi/3) & M_D \cos(\theta + 2\pi/3) & M_Q \sin(\theta + 2\pi/3) \end{vmatrix} \tag{6.63}$$

$\overline{L}_{RR}$ is the rotor-to-rotor inductance matrix

$$\overline{L}_{RR} = \begin{vmatrix} L_F & M_R & 0 \\ M_R & L_D & 0 \\ 0 & 0 & L_Q \end{vmatrix} \tag{6.64}$$

The inductance matrix of Equation 6.61 shows that the inductances vary with the angle θ. By referring the stator quantities to rotating rotor dq axes through Park's transformation, this dependence on θ is removed and a constant reactance matrix emerges.

6.6.2 Transformation of Reactance Matrix

Park's transformation describes a new set of variables, such as currents, voltages, and flux linkages in $0dq$ axes. The stator parameters are transferred to the rotor parameters. For the currents, this transformation is

$$\begin{vmatrix} i_0 \\ i_d \\ i_q \end{vmatrix} = \sqrt{\frac{2}{3}} \begin{vmatrix} \frac{1}{\sqrt{2}} & \frac{1}{\sqrt{2}} & \frac{1}{\sqrt{2}} \\ \cos\theta & \cos\left(\theta - 2\frac{\pi}{3}\right) & \cos\left(\theta + 2\frac{\pi}{3}\right) \\ \sin\theta & \sin\left(\theta - 2\frac{\pi}{3}\right) & \sin\left(\theta + 2\frac{\pi}{3}\right) \end{vmatrix} \begin{vmatrix} i_a \\ i_b \\ i_c \end{vmatrix} \tag{6.65}$$

Using matrix notation,

$$\bar{i}_{0dq} = \bar{P}\bar{i}_{abc} \tag{6.66}$$

Similarly,

$$\bar{v}_{0dq} = \bar{P}\bar{v}_{abc} \tag{6.67}$$

$$\bar{\lambda}_{0dq} = \bar{P}\bar{\lambda}_{abc} \tag{6.68}$$

where $\bar{\lambda}$ is the flux linkage vector. The a–b–c currents in the stator windings produce a synchronously rotating field, stationary with respect to the rotor. This rotating field can be produced by constant currents in the fictitious rotating coils in the *dq* axes; *P* is nonsingular and $\bar{P}^{-1} = \bar{P}'$:

$$\bar{P}^{-1} = \bar{P}' = \sqrt{\frac{2}{3}}\begin{vmatrix} \frac{1}{\sqrt{2}} & \cos\theta & \sin\theta \\ \frac{1}{\sqrt{2}} & \cos\left(\theta - \frac{2\pi}{3}\right) & \sin\left(\theta - \frac{2\pi}{3}\right) \\ \frac{1}{\sqrt{2}} & \cos\left(\theta + \frac{2\pi}{3}\right) & \sin\left(\theta + \frac{2\pi}{3}\right) \end{vmatrix} \tag{6.69}$$

To transform the stator-based variables into rotor-based variables, define a matrix as follows:

$$\begin{vmatrix} i_0 \\ i_d \\ i_q \\ i_F \\ i_D \\ i_Q \end{vmatrix} = \begin{vmatrix} \bar{P} & \bar{0} \\ \bar{0} & \bar{1} \end{vmatrix} \begin{vmatrix} i_a \\ i_b \\ i_c \\ i_F \\ i_D \\ i_Q \end{vmatrix} = \bar{B}\bar{i} \tag{6.70}$$

where $\bar{1}$ is a 3×3 unity matrix and $\bar{0}$ is a 3×3 zero matrix. The original rotor quantities are left unchanged. The time-varying inductances can be simplified by referring all quantities to the rotor frame of reference:

$$\begin{vmatrix} \bar{\lambda}_{odq} \\ \bar{\lambda}_{FDQ} \end{vmatrix} = \begin{vmatrix} \bar{P} & \bar{0} \\ \bar{0} & \bar{1} \end{vmatrix} \begin{vmatrix} \bar{\lambda}_{abc} \\ \bar{\lambda}_{FDQ} \end{vmatrix} = \begin{vmatrix} \bar{P} & \bar{0} \\ \bar{0} & \bar{1} \end{vmatrix} \begin{vmatrix} \bar{L}_{aa} & \bar{L}_{aR} \\ \bar{L}_{Ra} & \bar{L}_{RR} \end{vmatrix} \begin{vmatrix} \bar{P}^{-1} & \bar{0} \\ \bar{0} & \bar{1} \end{vmatrix} \begin{vmatrix} \bar{P} & \bar{0} \\ \bar{0} & \bar{1} \end{vmatrix} \begin{vmatrix} \bar{i}_{abc} \\ \bar{i}_{FDQ} \end{vmatrix} \tag{6.71}$$

This transformation gives

$$\begin{vmatrix} \lambda_0 \\ \lambda_d \\ \lambda_q \\ \lambda_F \\ \lambda_D \\ \lambda_Q \end{vmatrix} = \begin{vmatrix} L_0 & 0 & 0 & 0 & 0 & 0 \\ 0 & L_d & 0 & kM_F & kM_D & 0 \\ 0 & 0 & L_q & 0 & 0 & kM_q \\ 0 & kM_F & 0 & L_F & M_R & 0 \\ 0 & kM_D & 0 & M_R & L_D & 0 \\ 0 & 0 & kM_Q & 0 & 0 & L_Q \end{vmatrix} \begin{vmatrix} i_0 \\ i_d \\ i_q \\ i_F \\ i_D \\ i_Q \end{vmatrix} \tag{6.72}$$

Define

$$L_d = L_s + M_s + \frac{3}{2}L_m \tag{6.73}$$

$$L_q = L_s + M_s - \frac{3}{2}L_m \tag{6.74}$$

$$L_0 = L_s - 2M_s \tag{6.75}$$

$$k = \sqrt{\frac{3}{2}} \tag{6.76}$$

The inductance matrix is sparse, symmetric, and constant. It decouples the $0dq$ axes as will be illustrated further.

6.7 Park's Voltage Equation

The voltage equation [4,5] in terms of current and flux linkages is

$$\bar{v} = -\bar{R}\bar{i} - \frac{d\bar{\lambda}}{dt} \tag{6.77}$$

or

$$\begin{vmatrix} v_a \\ v_b \\ v_c \\ v_F \\ v_D \\ v_Q \end{vmatrix} = \begin{vmatrix} r & 0 & 0 & 0 & 0 & 0 \\ 0 & r & 0 & 0 & 0 & 0 \\ 0 & 0 & r & 0 & 0 & 0 \\ 0 & 0 & 0 & r_F & 0 & 0 \\ 0 & 0 & 0 & 0 & r_D & 0 \\ 0 & 0 & 0 & 0 & 0 & r_Q \end{vmatrix} \begin{vmatrix} i_a \\ i_b \\ i_c \\ i_F \\ i_D \\ i_Q \end{vmatrix} - \frac{di}{dt} \begin{vmatrix} \lambda_a \\ \lambda_b \\ \lambda_c \\ \lambda_F \\ \lambda_D \\ \lambda_Q \end{vmatrix} \tag{6.78}$$

This can be partitioned as

$$\begin{vmatrix} \bar{v}_{abc} \\ \bar{v}_{FDQ} \end{vmatrix} = -\begin{vmatrix} \bar{r}_{s} & \\ & \bar{r}_{FDQ} \end{vmatrix} \begin{vmatrix} \bar{i}_{abc} \\ \bar{i}_{FDQ} \end{vmatrix} - \frac{di}{dt}\begin{vmatrix} \bar{\lambda}_{abc} \\ \bar{\lambda}_{FDQ} \end{vmatrix} \tag{6.79}$$

The transformation is given by

$$\bar{B}^{-1}\bar{v}_{B} = -\bar{R}\bar{B}^{-1}\bar{i}_{B} - \frac{d}{dt}\left(\bar{B}^{-1}\bar{\lambda}_{B}\right) \tag{6.80}$$

where

$$\begin{vmatrix} \bar{P} & \bar{0} \\ \bar{0} & \bar{1} \end{vmatrix} = \bar{B}, \quad \bar{B}\begin{vmatrix} \bar{i}_{abc} \\ \bar{i}_{FDQ} \end{vmatrix} = \begin{vmatrix} \bar{i}_{odq} \\ \bar{i}_{FDQ} \end{vmatrix} = \bar{i}_{B}, \quad \bar{B}\begin{vmatrix} \bar{\lambda}_{abc} \\ \bar{\lambda}_{FDQ} \end{vmatrix} = \begin{vmatrix} \bar{\lambda}_{odq} \\ \bar{\lambda}_{FDQ} \end{vmatrix} = \bar{\lambda}_{B}, \quad \bar{B}\begin{vmatrix} \bar{v}_{abc} \\ \bar{v}_{FDQ} \end{vmatrix} = \begin{vmatrix} \bar{v}_{0dq} \\ \bar{v}_{FDQ} \end{vmatrix} = \bar{v}_{B} \tag{6.81}$$

Equation 6.80 can be written as

$$v_{B} = -\bar{B}R\bar{B}^{-1} - \bar{B}\frac{d}{dt}(\bar{B}^{-1}\bar{\lambda}_{B}) \tag{6.82}$$

First, evaluate

$$\bar{B}\frac{d\bar{B}^{-1}}{d\theta} = \begin{vmatrix} \bar{P} & \bar{0} \\ \bar{0} & \bar{1} \end{vmatrix} \begin{vmatrix} \dfrac{d\bar{P}^{-1}}{d\theta} & \bar{0} \\ \bar{0} & \bar{0} \end{vmatrix} = \begin{vmatrix} \bar{P}\dfrac{d\bar{P}^{-1}}{d\theta} & \bar{0} \\ \bar{0} & \bar{0} \end{vmatrix} \tag{6.83}$$

where it can be shown that

$$\bar{P}\frac{d\bar{P}^{-1}}{d\theta} = \begin{vmatrix} 0 & 0 & 0 \\ 0 & 0 & 1 \\ 0 & -1 & 0 \end{vmatrix} \tag{6.84}$$

As we can write

$$\frac{d\bar{B}^{-1}}{dt} = \frac{d\bar{B}^{-1}}{d\theta}\frac{d\theta}{dt} \tag{6.85}$$

$$\bar{B}\frac{d\bar{B}^{-1}}{d\theta} = \begin{vmatrix} 0 & 0 & 0 & 0 & 0 & 0 \\ 0 & 0 & 1 & 0 & 0 & 0 \\ 0 & -1 & 0 & 0 & 0 & 0 \\ 0 & 0 & 0 & 0 & 0 & 0 \\ 0 & 0 & 0 & 0 & 0 & 0 \\ 0 & 0 & 0 & 0 & 0 & 0 \end{vmatrix} \tag{6.86}$$

The voltage equation becomes

$$\bar{v}_B = \bar{R}\bar{i}_B - \frac{d\theta}{dt}\begin{vmatrix} 0 \\ \lambda_q \\ -\lambda_d \\ 0 \\ 0 \\ 0 \end{vmatrix} - \frac{d\bar{\lambda}_B}{dt} \tag{6.87}$$

When the shaft rotation is uniform $d\theta/dt$ is a constant, and Equation 6.87 is linear and time invariant.

6.8 Circuit Model of Synchronous Machines

From the above treatment, the following decoupled voltage equations can be written as

Zero sequence

$$v_0 = ri_0 - \frac{d\lambda_0}{dt} \tag{6.88}$$

Direct axis

$$v_d = -ri_d - \frac{d\theta}{dt}\lambda_q - \frac{d\lambda_d}{dt} \tag{6.89}$$

$$v_F = r_F i_F + \frac{d\lambda_F}{dt} \tag{6.90}$$

$$v_D = r_D i_D + \frac{d\lambda_D}{dt} = 0 \tag{6.91}$$

Quadrature axis

$$v_q = -ri_q + \frac{d\theta}{dt}\lambda_d - \frac{d\lambda_q}{dt} \tag{6.92}$$

$$v_Q = r_Q i_Q + \frac{d\lambda_Q}{dt} = 0 \tag{6.93}$$

The decoupled equations relating to flux linkages and currents are

Zero sequence

$$\lambda_0 = L_0 i_0 \tag{6.94}$$

Direct axis

$$\begin{vmatrix} \lambda_d \\ \lambda_F \\ \lambda_D \end{vmatrix} = \begin{vmatrix} L_d & kM_F & kM_D \\ kM_F & L_F & M_R \\ kM_D & M_R & L_D \end{vmatrix} \begin{vmatrix} i_d \\ i_F \\ i_D \end{vmatrix} \tag{6.95}$$

Quadrature axis

$$\begin{vmatrix} \lambda_q \\ \lambda_Q \end{vmatrix} = \begin{vmatrix} L_q & kM_Q \\ kM_Q & L_Q \end{vmatrix} \begin{vmatrix} i_q \\ i_Q \end{vmatrix} \tag{6.96}$$

This decoupling is shown in equivalent circuits in Figure 6.10. Note that ld, lf, and lD are the self inductances in armature, field, and damper circuits and Lad is defined as

$$L_{ad} = L_D - l_D = L_d - l_d = L_F - l_f = kM_F = kM_D = M_R$$

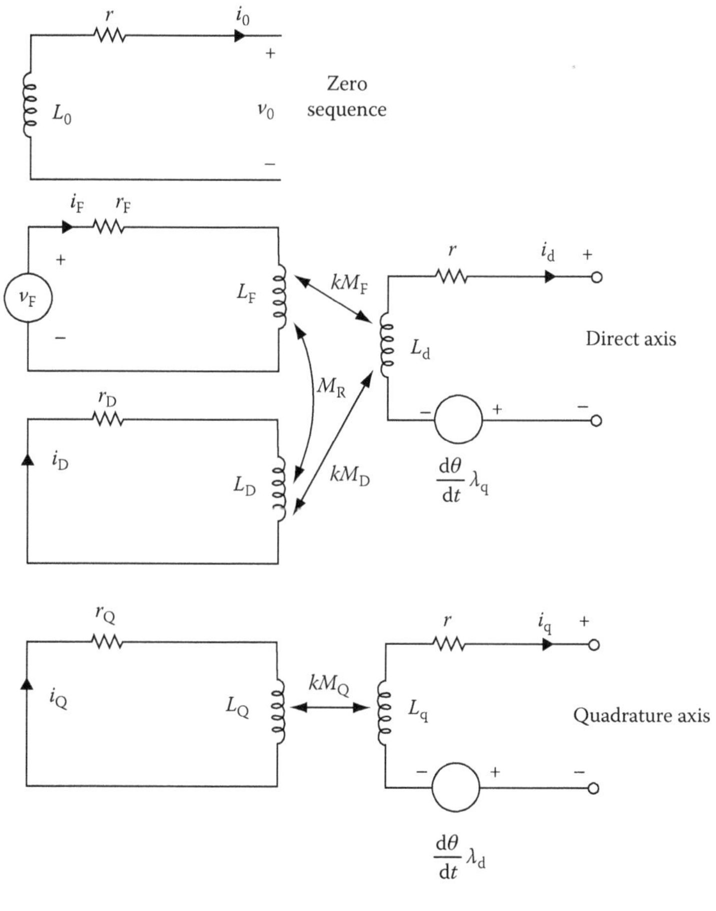

FIGURE 6.10
Synchronous generator decoupled circuits in *d–q* axes.

6.9 Calculation Procedure and Examples

There are three steps involved:

1. The problem is normally defined in stator parameters, which are of interest. These are transformed into $0dq$ axes variables.
2. The problem is solved in $0dq$ axes, parameters, using Laplace transform or other means.
3. The results are transformed back to a–b–c variables of interest.

These three steps are inherent in any calculation using transformations. For simpler problems, it may be advantageous to solve directly in stator parameters.

Example 6.2

Calculate the time variation of the direct axis, quadrature axis voltages, and field current, when a step function of field voltage is suddenly applied to a generator at no load. Neglect damper circuits.

As the generator is operating without load, $i_{abc}=i_{0dq}=0$. Therefore, from Equations 6.94 through 6.96

$$\lambda_0 = 0 \qquad \lambda_d = kM_F i_F \qquad \lambda_F = L_F i_F \qquad \lambda_q = 0$$

From Equations 6.88 through 6.92,

$$v_0 = 0$$

$$v_d = -\frac{d\lambda_d}{dt} = -kM_F\frac{di_E}{dt}$$

$$v_F = r_F i_F + L_F\frac{di_F}{dt}$$

Therefore, as expected, the time variation of field current is

$$i_F = \frac{1}{r_F}\left(1 - e^{(-r_F/L_F)t}\right)$$

The direct axis and quadrature axis voltages are given by

$$v_d = -\frac{kM_F}{L_F}e^{-(r_F/L_F)t}$$

$$v_q = -\frac{\omega kM_F}{r_F}v_F\left(1 - e^{-(r_F/L_F)t}\right)$$

The phase voltages can be calculated using Park's transformation.

Example 6.3

A generator is operating with balanced positive sequence voltage of

$$v_a = \sqrt{2}|V|\cos(\omega_0 t + \angle V)$$

across its terminals. The generator rotor is described by

$$\theta = \omega_1 t + \frac{\pi}{2} + \delta$$

Find v_0, v_d, and v_q.

This is a simple case of transformation using (6.65):

$$\begin{vmatrix} v_0 \\ v_d \\ v_q \end{vmatrix} = \frac{2|V|}{\sqrt{3}} \begin{vmatrix} \frac{1}{\sqrt{2}} & \frac{1}{\sqrt{2}} & \frac{1}{\sqrt{2}} \\ \cos\theta & \cos\left(\theta - \frac{2\pi}{3}\right) & \cos\left(\theta + \frac{2\pi}{3}\right) \\ \sin\theta & \sin\left(\theta - \frac{2\pi}{3}\right) & \sin\left(\theta + \frac{2\pi}{3}\right) \end{vmatrix} \begin{vmatrix} \cos(\omega_0 t + \angle V) \\ \cos\left(\omega_0 t + \angle V - \frac{2\pi}{3}\right) \\ \cos\left(\omega_0 t + \angle V - \frac{4\pi}{3}\right) \end{vmatrix}$$

A solution of this equation gives

$$\begin{aligned} v_d &= \sqrt{3}|V|\sin[(\omega_0 - \omega_1)t + < V - \delta] \\ v_q &= \sqrt{3}|V|\cos[(\omega_0 - \omega_1)t + < V - \delta] \end{aligned} \tag{6.97}$$

These relations apply equally well to the derivation of i_d, i_q, λ_d, and λ_q.

For synchronous operation $\omega_1 = \omega_0$ and the equations reduce to

$$\begin{aligned} v_q &= \sqrt{3}|V|\cos(\angle V - \delta) \\ v_d &= \sqrt{3}|V|\sin(\angle V - \delta) \end{aligned} \tag{6.98}$$

Note that v_q and v_d are now constant and do not have the slip frequency term $\omega_1 - \omega_0$. We can write

$$v_q + jv_d = \sqrt{3}|V|e^{j(\angle V - \delta)} = \sqrt{3}V_a e^{-j\delta} \tag{6.99}$$

Therefore, V_a can be written as

$$V_a = \left(\frac{v_q}{\sqrt{3}} + j\frac{v_d}{\sqrt{3}}\right)e^{j\delta} = (V_q + jV_d)e^{j\delta} \tag{6.100}$$

where

$$V_q = \frac{v_q}{\sqrt{3}} \quad \text{and} \quad V_d = \frac{v_d}{\sqrt{3}} \tag{6.101}$$

This is shown in the phasor diagram of Figure 6.11.

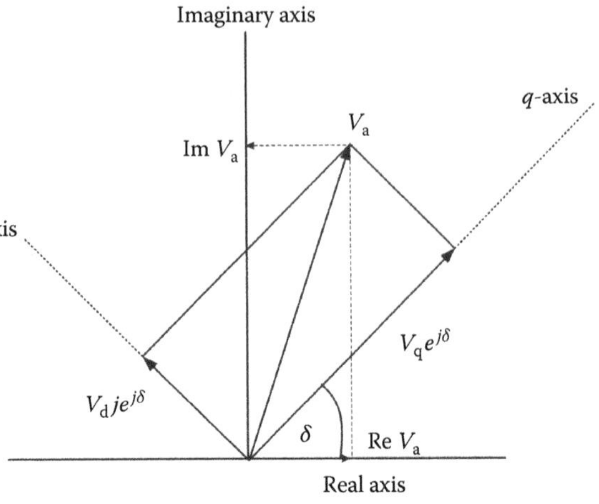

FIGURE 6.11
Vector diagram illustrating relationship of direct axes and quadrature axes' voltages to the terminal voltage.

We can write these equations in the following form:

$$\begin{vmatrix} \text{Re} & V_a \\ \text{Im} & V_a \end{vmatrix} = \begin{vmatrix} \cos\delta & -\sin\delta \\ \sin\delta & \cos\delta \end{vmatrix} \begin{vmatrix} V_q \\ V_d \end{vmatrix}$$
$$\begin{vmatrix} V_q \\ V_d \end{vmatrix} = \begin{vmatrix} \cos\delta & \sin\delta \\ -\sin\delta & \cos\delta \end{vmatrix} \begin{vmatrix} \text{Re} & V_a \\ \text{Im} & V_a \end{vmatrix} \quad (6.102)$$

Example 6.4

Steady-State Model of Synchronous Generator

Derive a steady-state model of a synchronous generator and its phasor diagram.

In the steady state, all the currents and flux linkages are constant. Also, $i_0 = 0$ and rotor damper currents are zero. Equations 6.89 through 6.94 reduce to

$$\begin{aligned} v_d &= -ri_d - \omega_0\lambda_q \\ v_q &= -ri_q + \omega_0\lambda_d \\ v_F &= r_F i_F \end{aligned} \quad (6.103)$$

where

$$\begin{aligned} \lambda_d &= L_d i_d + kM_F i_F \\ \lambda_F &= kM_F i_d + L_F i_F \\ \lambda_q &+ L_q i_q \end{aligned} \quad (6.104)$$

Substitute values of λ_d and λ_q from Equation 6.104 to 6.103; then, from Example 6.3, and then from Equations 6.100 to 6.101, we can write the following equation:

$$V_a = -r(I_q + jI_d)e^{j\delta} + \omega_0 L_d I_d e^{j\delta} - j\omega_0 L_q I_q e^{j\delta} + \tfrac{1}{\sqrt{2}}\omega_0 M_F i_F e^{j\delta} \quad \text{where } i_d = \sqrt{3}I_d \text{ and } i_q = \sqrt{3}I_q.$$

Define

$$\sqrt{2}E = \omega_0 M_F i_F e^{j\delta} \quad (6.105)$$

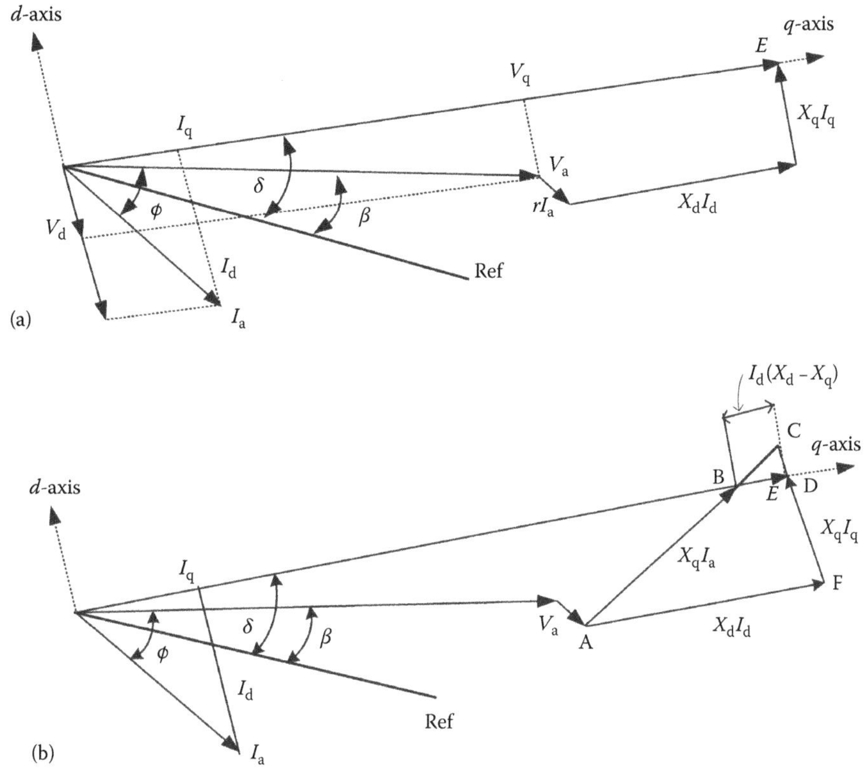

FIGURE 6.12
(a) Phasor diagram of a synchronous generator operating at lagging power factor; (b) to illustrate the construction of phasor diagram from known stator parameters.

This is the no-load voltage or the open-circuit voltage with generator current = 0. We can then write

$$E = V_a + rI_a + jX_d I_d e^{j\delta} + jX_q I_q e^{j\delta} \tag{6.106}$$

The phasor diagram is shown in Figure 6.12a. The open circuit voltage on no load is a q-axis quantity and is equal to the terminal voltage.

As the components I_d and I_q are not initially known, the phasor diagram is constructed by first laying out the terminal voltage V_a and line current I_a at the correct phase angle ϕ, then adding the resistance drop and reactance drop IX_q. At the end of this vector, the quadrature axis is located. Now the current is resolved into direct axis and quadrature axis components. This allows the construction of vectors I_qX_q and I_qX_q. This is shown in Figure 6.12b.

Example 6.5

Consider the generator data of Example 6.1. Calculate the direct and quadrature axis components of the currents and voltages and machine voltage E when the generator is delivering its full-load rated current at its rated voltage. Also calculate all the angles shown in the phasor diagram, Figure 6.12a. If this generator is connected to an infinite bus through an impedance of $0.01 + j0.1$ per unit (100 MVA base), what is the voltage of the infinite bus?

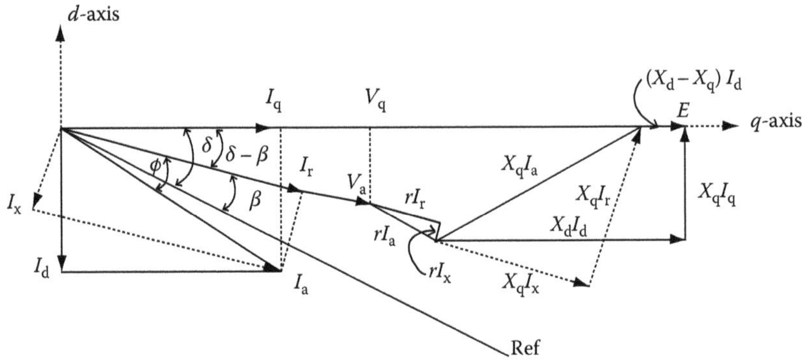

FIGURE 6.13
Phasor diagram of the synchronous generator for Example 6.5.

The generator operates at a power factor of 0.85 at its rated voltage of 1.0 per unit. Therefore, $\phi = 31.8°$. The generator full-load current is 4183.8 A = 1.0 per unit. The terminal voltage vector can be drawn to scale, the Ir drop (= 0.0012 per unit) is added, and the vector $IX_q = 1.8$ per unit is drawn to locate the q axis. Current I can be resolved into direct axis and quadrature axis components, and the phasor diagram is completed as shown in Figure 6.12b and the values of V_d, I_d, V_q, I_q, and E are read from it. The analytical solution is as follows:

The load current is resolved into active and reactive components $I_r = 0.85$ per unit and $I_x = 0.527$ per unit, respectively. Then, from the geometric construction shown in Figure 6.13:

$$(\delta - \beta) = \tan^{-1}\left(\frac{X_q I_r + r I_x}{V_a + r I_r - X_q I_x}\right)$$

$$= \tan^{-1}\left(\frac{(1.8)(0.85) + (0.0012)(0.527)}{1 + (0.0012)(0.85) + (1.8)(0.527)}\right) = 38.14° \quad (6.107)$$

From the above calculation, resistance can even be ignored without an appreciable error. Thus, $(\delta - \beta + \phi) = 69.93°$; this is the angle of the current vector with the q axis. Therefore,

$$I_q = I_a \cos(\delta - \beta - \phi) = 0.343\,\text{pu}, \quad i_q = 0.594\,\text{pu}$$
$$I_d = -I_a \sin(\delta - \beta - \phi) = -0.939\,\text{pu}, \quad i_d = -1.626\,\text{pu}$$
$$V_q = V_a \cos(\delta - \beta) = 0.786\,\text{pu}, \quad v_q = 1.361\,\text{pu}$$
$$V_d = -V_a \sin(\delta - \beta) = -0.618\,\text{pu}, \quad v_d = 1.070\,\text{pu}$$

The machine generated voltage is

$$E = V_q + rI_q - X_d I_d = 2.66\,\text{pu}$$

The infinite bus voltage is simply the machine terminal voltage less the IZ drop subtracted vectorially:

$$V_\infty = V_a\angle 0° - I_a\angle 31.8° Z\angle 84.3° = 0.94\angle -4.8°$$

The infinite bus voltage lags the machine voltage by 4.8°. Practically, the infinite bus voltage will be held constant and the generator voltage changes, depending on the system impedance and generator output.

Example 6.6

Symmetrical Short Circuit of a Generator at No Load

Derive the short-circuit equations of a synchronous generator for a balanced three-phase short circuit at its terminals. Ignore damper circuit and resistances and neglect the change in speed during short circuit. Prior to short circuit, the generator is operating at no load. Ignoring the damper circuit means that the subtransient effects are ignored. As the generator is operating at no load, $i_{abc} = i_{odq} = 0$, prior to the fault.

From Equations 6.89 through 6.94,

Zero sequence

$$v_0 = -L_0 \frac{di_0}{dt} = 0$$

Direct axis

$$v_d = -\omega_0 \lambda_q - \frac{d\lambda_d}{dt} = 0$$
$$v_f = \frac{d\lambda_F}{dt}$$

Quadrature axis

$$v_q = \omega_0 \lambda_d - \frac{d\lambda_q}{dt} = 0$$

The flux linkages can be expressed in terms of currents by using Equations 6.95 and 6.96:

$$\omega_0 L_q i_q + L_d \frac{di_d}{dt} + kM_F \frac{di_F}{dt} = 0$$
$$v_F = kM_F \frac{di_d}{dt} + L_F \frac{di_F}{dt}$$
$$-\omega_0 L_d i_d - \omega_0 kM_F i_F + L_q \frac{di_q}{dt} = 0$$

These equations can be solved using Laplace transformation. The initial conditions must be considered. In Example 6.4, we demonstrated that at no load, prior to fault, the terminal voltage is equal to the generated voltage and this voltage is

$$\sqrt{2}E = \omega_0 M_F i_F e^{j\delta}$$

and this is a quadrature axis quantity. Also, $v_d = 0$. The effect of short circuit is, therefore, to reduce v_q to zero. This is equivalent to applying a step function of $-v_q 1$ to the q axis. The transient currents can then be superimposed on the currents prior to fault. Except for current in the field coil, all these currents are zero. The solution for i_F will be superimposed on the existing current i_{F0}.

If we write

$$kM_F = L_{ad} = \frac{X_{ad}}{\omega}$$
$$L_F = \frac{(X_f + X_{ad})}{\omega}$$

The expressions converted into reactances and using Laplace transform, $dx/dt = sX(s) - x(0^-)$ [where $X(s)$ is Laplace transform of $x(t)$] reduce to

$$0 = \left(\frac{1}{\omega}\right)(X_{ad} + X_f)si_F + \left(\frac{1}{\omega}\right)X_{ad}si_d \quad (6.108)$$

$$0 = \left(\frac{1}{\omega}\right)(X_d)si_d + \left(\frac{1}{\omega}\right)X_{ad}si_f + (X_q)i_q \quad (6.109)$$

$$-v_q = \left(\frac{1}{\omega}\right)X_q si_q - X_{ad}i_F - X_d i_d \quad (6.110)$$

The field current from Equation 6.108 is

$$i_F = \frac{-i_d X_{ad}}{(X_{ad} + X_f)} \quad (6.111)$$

The field current is eliminated from Equations 6.109 to 6.110 by substitution. The quadrature axis current is

$$\begin{aligned} i_q &= \frac{1}{X_{aq} + X_l}\left[\frac{X_{aq}X_f}{X_{aq} + X_f} + X_l\right]\frac{s}{\omega}i_d \\ &= -\left(\frac{X'_d}{X_q}\right)\frac{s}{\omega}i_d \end{aligned}$$

and

$$i_d = \frac{\omega^2}{X'_d}\left[\frac{1}{s^2 + \omega^2}\right]v_q$$

Solving these equations gives

$$i_d = \frac{(\sqrt{3}|E|)}{X'_d}(1 - \cos \omega t)$$

$$i_q = -\frac{(\sqrt{3}|E|)}{X_q}\sin \omega t$$

Note that $k = \sqrt{3/2}$. Apply

$$\bar{i}_{abc} = \bar{P}\bar{i}_{0dq}$$

with $\theta = \omega t + \pi/2 + \delta$, the short-circuit current in phase a is

$$i_a = \sqrt{2}|E|\left[\left(\frac{1}{X'_d}\right)\sin(\omega t + \delta) + \frac{X_q + X'_d}{2X'_d X_q}\sin\delta - \frac{X_q - X'_d}{2X'_d X_q}\sin(2\omega t + \delta)\right] \quad (6.112)$$

The first term is normal-frequency short-circuit current, the second is constant asymmetric current, and the third is double-frequency short-circuit current. The 120 Hz component imparts a nonsinusoidal characteristic to the short-circuit current waveform. It rapidly decays to zero and is ignored in the calculation of short-circuit currents.

When the damper winding circuit is considered, the short-circuit current can be expressed as

$$i_a = \sqrt{2}E\left[\left(\frac{1}{X_d}\right)\sin(\omega t+\delta)+\left(\frac{1}{X'_d}-\frac{1}{X_d}\right)e^{-t/T'_d}\sin(\omega t+\delta)+\left(\frac{1}{X''_d}-\frac{1}{X'_d}\right)e^{-t/T''_d}\sin(\omega t+\delta)\right.$$
$$\left.-\frac{\left(X''_d+X''_q\right)}{2X''_dX''_q}e^{-t/T_a}\sin\delta-\frac{\left(X''_d-X''_q\right)}{2X''_dX''_q}e^{-t/T_a}\sin(2\omega t+\delta)\right] \quad (6.113)$$

- The first term is final steady-state short-circuit current.
- The second term is normal-frequency decaying transient current.
- The third term is normal-frequency decaying subtransient current.
- The fourth term is asymmetric decaying dc current.
- The fifth term is double-frequency decaying current.

Example 6.7

Calculate the component short-circuit currents at the instant of three-phase terminal short circuit of the generator (particulars as shown in Table 6.1). Assume that phase a is aligned with the field at the instant of short circuit, maximum asymmetry, i.e., $\delta = 0$. The generator is operating at no load prior to short circuit.

The calculations are performed by substituting the required numerical data from Table 6.1 into (6.113):

Steady-state current = 2.41 kA rms
Decaying transient current = 20.24 kA rms
Decaying subtransient current = 5.95 kA rms
Decaying DC component = 43.95 kA
Decaying second-harmonic component = 2.35 kA rms

Note that the second-harmonic component is zero if the direct axis and quadrature axis subtransient reactances are equal. Also, the dc component in this case is 40.44 kA.

6.9.1 Manufacturer's Data

The relationship between the various inductances and the data commonly supplied by a manufacturer for a synchronous machine is not obvious. The following relations hold:

$$L_{ad} = L_d - l_l = kM_F = kM_D = M_R \quad (6.114)$$

$$L_{aq} = L_q - l_l = kM_Q \quad (6.115)$$

Here, l_l is the leakage reactance corresponding to X_l and L_{ad} is the mutual inductance between the armature and rotor = mutual inductance between field and rotor = mutual inductance between damper and rotor. Similar relations are applicable in the quadrature axis. Some texts have used different symbols. Field leakage reactance l_f is

$$l_f = \frac{L_{ad}(L'_d - l_l)}{(L_d - L'_d)} \tag{6.116}$$

and

$$L_F = l_f + L_{ad} \tag{6.117}$$

The damper leakage reactance in the direct axis is

$$l_{kD} = \frac{L_{ad} l_f (L''_d - l_l)}{L_{ad} l_f - L_F (L''_d - l_l)} \tag{6.118}$$

and

$$L_D = l_{kD} + L_{ad} \tag{6.119}$$

$$l_{kQ} = \frac{L_{aq}\left(L''_q - l_l\right)}{\left(L_q - L''_q\right)} \tag{6.120}$$

In the quadrature axis, the damper leakage reactance is

$$L_Q = l_{kQ} + L_{aq} \tag{6.121}$$

The field resistance is

$$r_F = \frac{L_F}{T'_{do}} \tag{6.122}$$

The damper resistances in direct axis can be obtained from

$$T''_d = \frac{(L_D L_F - L^2_{ad})}{r_D L_F}\left(\frac{L''_d}{L'_d}\right) \tag{6.123}$$

and in the quadrature axis

$$T''_q = \frac{L''_q L_Q}{L_a r_Q} \tag{6.124}$$

Example 6.8

Using the manufacturer's data in Table 6.1, calculate the machine parameters in the d–q axes. Applying the equations in Section 6.9.1:

$$L_{ad} = L_d - l_l = 1.949 - 0.164 = 1.785\,\text{pu} = kM_F = kM_D = M_R$$

$$L_{aq} = L_q - l_l = 1.858 - 0.164 = 1.694\,\text{pu} = kM_Q$$

$$l_f = (1.758)(0.278 - 0.164)/(1.964 - 0.278) = 0.121\,\text{pu}$$

$$L_F = 0.121 + 1.785 = 1.906\,\text{pu}$$

$$l_{kd} = (1.785)(0.121)(0.193 - 0.164)/\{(1.785)(0.164) - (1.096)(0.193 - 0.164)\} = 0.026\,\text{pu}$$

$$L_D = 0.026 + 1.785 = 1.811\,\text{pu}$$

$$l_{kq} = (1.694)(0.192 - 0.164)/(1.858 - 0.192) = 0.028\,\text{pu}$$

$$L_Q = 0.028 + 1.694 = 1.722\,\text{pu}$$

$$T'_{do} = 5.615s = 2116.85\,\text{rad}$$

$$r_F = 1.906/2116.85 = 1.005 \times 10^{-5}\,\text{pu}$$

$$r_D = \frac{(1.811)(1.906) - 1.785^2}{(0.015)(377)(1.906)}\left(\frac{0.193}{0.278}\right) = 0.0131\,\text{pu}$$

$$r_Q = \left(\frac{0.192}{1.858}\right)\left(\frac{1.722}{0.015 \times 377}\right) = 0.031\,\text{pu}$$

Note that in Table 6.1, as the data is in per unit, we consider $X_d = L_d$, $X_l = l_l$, etc. The per unit system for synchronous machines is not straightforward and variations in the literature exist. Reference [6] provides further reading.

6.10 Short Circuit of Synchronous Motors and Condensers

The equations derived above for synchronous generators can be applied to synchronous motors and condensers. Synchronous condensers are used for reactive power support in transmission systems and sometimes in industrial systems also and are not connected to any load. Large synchronous motors are used to drive certain loads in the industry, e.g., single units of 30,000–40,000 hp have been used in pulp industry. Figure 6.14 shows V curves of the synchronous machines, generators, and motors, which are obtained with machines operating at different excitations. A phasor diagram of synchronous motor under steady-state operation can be drawn similar to that of a synchronous generator.

With respect to short-circuit calculations according to empirical methods of ANSI/IEEE, the excitation systems and prior loads are ignored—Chapter 7. The subtransient and transient reactances of synchronous motors are much higher compared to that of synchronous generators of the same ratings. Manufacturers may not supply all the reactances and time constants of the synchronous motors, unless specifically requested. Thus, for the same ratings, the synchronous motors contribute smaller magnitude of short-circuit currents. Also, these currents decay faster.

6.11 Induction Motors

A rotating machine can be studied in the steady state depending upon the specific type of the machine, e.g., synchronous, induction, or dc machine. Generalized machine theory [7,8] attempts to unify the piecemeal treatment of rotating machines—after all each machine type consists of some coils on the stationary and rotating part, which interact with each other.

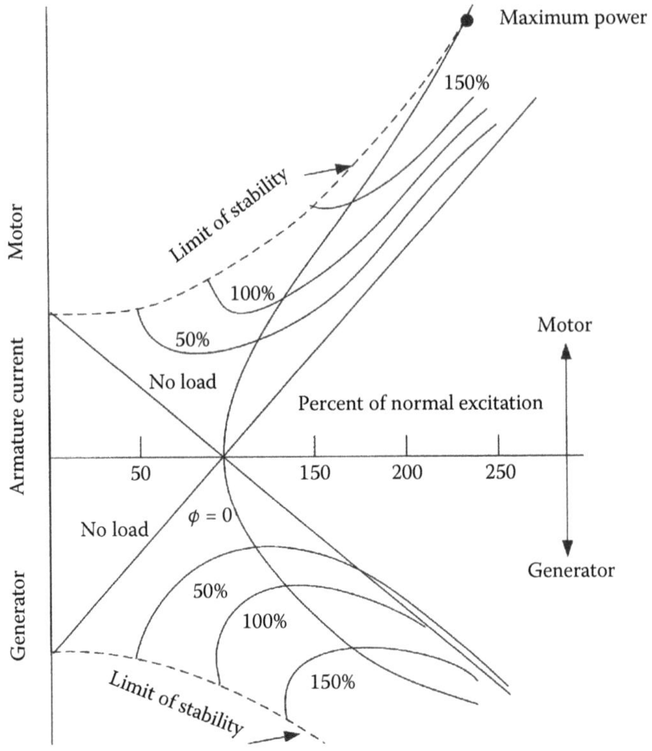

FIGURE 6.14
V-curves of synchronous machines.

The treatment of synchronous machine above amply demonstrates this concept. The ideas were further developed by Kron and others and treatment of electrical machines using tensors or matrices and linear transformation is the basis of the generalized machine theory. Though there are some limitations that saturation, commutation effects, surge phenomena, and eddy current losses cannot be accounted for and also the theory is not applicable if there is saliency in both rotor and stator pole construction, it is a powerful analytical tool for machine modeling.

The following transformations are used to derive the model of an induction motor:

$$3-\varphi|_{\mathrm{abc}} \Leftrightarrow 2-\varphi|_{\alpha\beta0} \Leftrightarrow 2-\varphi|_{\mathrm{dq0}} \tag{6.125}$$

In terms of d–q axes, both the stator and rotor are cylindrical and symmetrical. The d axis is chosen arbitrarily as the axis of the stator phase a. The three-phase winding is converted into a two-phase winding, so that the axis of the second phase becomes the q axis. The two stator phases are then the fixed axis coils, 1D and 1Q, respectively, Figure 6.15.

The rotor circuit, whether of wound type or cage type, is also represented by d- and q-axis coils, though a squirrel cage rotor is more complex and space harmonics are neglected. These coils are 2D and 2Q. The impedance matrix can be set up as for a synchronous machine, as follows:

$$\begin{vmatrix} v_{1\mathrm{d}} \\ v_{1\mathrm{q}} \\ v_{2\mathrm{d}} \\ v_{2\mathrm{q}} \end{vmatrix} = \begin{vmatrix} r_1+(L_{\mathrm{m}}+L_1)p & & -L_{\mathrm{m}}p & \\ & r_1+(L_{\mathrm{m}}+L_1)p & & L_{\mathrm{m}}p \\ L_{\mathrm{m}}p & -L_{\mathrm{m}}\omega_{\mathrm{r}} & -r_2-(L_{\mathrm{m}}+L_2)p & (L_{\mathrm{m}}+L_2)\omega_{\mathrm{r}} \\ -L_{\mathrm{m}}\omega_{\mathrm{r}} & L_{\mathrm{m}}p & -(L_{\mathrm{m}}+L_2)\omega_{\mathrm{r}} & -r_2-(L_{\mathrm{m}}+L_2)p \end{vmatrix} \begin{vmatrix} i_{1\mathrm{d}} \\ i_{1\mathrm{q}} \\ i_{2\mathrm{d}} \\ i_{2\mathrm{q}} \end{vmatrix} \tag{6.126}$$

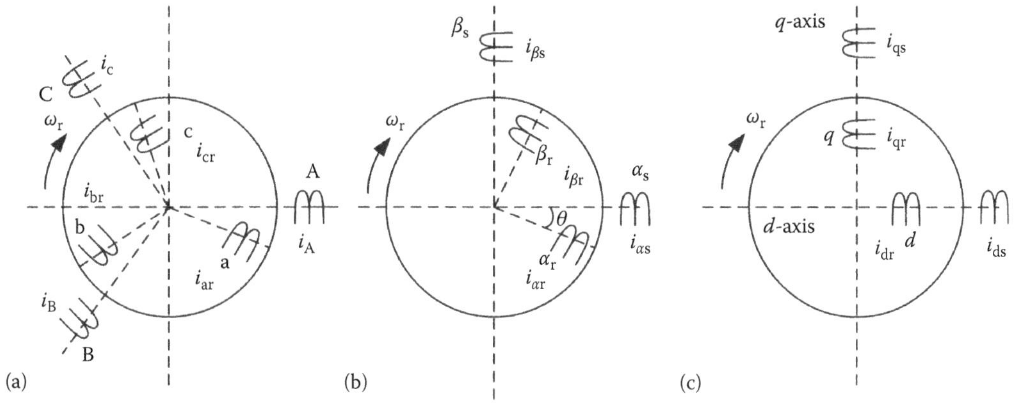

FIGURE 6.15
Transformation of an induction machine.

where
r_1 and r_2 are stator and rotor resistances, respectively
L_1 and L_2 are stator and rotor reactances, respectively

1D and 1Q have mutual inductance L_m, which is also the mutual inductance between 1Q and 2Q. Here, $\omega_r = (1 - s_1)\,\omega$, where s_1 is the motor slip. The rotor is short circuited on itself; therefore, $v_{2d} = v_{2q} = 0$. Equation 6.126 can be written as

$$\begin{vmatrix} v_{1d} \\ v_{1q} \\ 0 \\ 0 \end{vmatrix} = \begin{vmatrix} r_1 + X_s & & -jX_m & \\ & r_s + jX_s & & -jX_m \\ jX_m & -(1-s)X_m & -(r_2 + jX_r) & (1-s)X_m \\ (1-s)X_m & jX_m & -(1-s)X_r & -(r_2 + jX_r) \end{vmatrix} \begin{vmatrix} i_{1d} \\ i_{1q} \\ i_{2d} \\ i_{2q} \end{vmatrix}$$

There is no difference in the d and q axes except for time, and it is possible to write

$$\begin{aligned} v_{1d} &= v_1 \quad & v_{1q} &= -jv_1 \\ i_{1d} &= i_1 \quad & ti_{1q} &= -ji_1 \\ i_{2d} &= i_2 \quad & i_{2q} &= -ji_2 \end{aligned} \tag{6.127}$$

For a terminal fault, whether an induction machine is operating as an induction generator or motor, the machine feeds into the fault due to trapped flux linkage with the rotor. This fault current will be a decaying transient. A dc decaying component also occurs to maintain the flux linkage constant.

By analogy with a synchronous machine, reactances X_a, X_f, and X_{ad} are equivalent to X_1, X_2, and X_m in the induction machine. The transient reactance of the induction machine is

$$X' = X_1 + \frac{X_m X_2}{X_m + X_2} \tag{6.128}$$

This is also the motor-locked rotor reactance. The equivalent circuit of the induction motor is shown in Chapter 12. The open-circuit transient time constant is

$$T_0' = \frac{X_2 + X_m}{\omega r_2} \tag{6.129}$$

The short-circuit transient time constant is

$$T' = T'_0 \frac{X'}{X_1 + X_m} \tag{6.130}$$

This is approximately equal to

$$T' = \frac{X'}{\omega r_2} \tag{6.131}$$

and the time constant for the decay of dc component is

$$T_{dc} = \frac{X'}{\omega r_1} \tag{6.132}$$

AC symmetrical short-circuit current is

$$i_{ac} = \frac{E}{X'} e^{-t/T'} \tag{6.133}$$

and dc current is

$$i_{dc} = \sqrt{2} \frac{E}{X'} e^{-t/T_{dc}} \tag{6.134}$$

where E is the prefault voltage behind the transient reactance X'. At no load, E is equal to the terminal voltage.

Example 6.9

Consider a 4 kV, 5000 hp four-pole motor, with full-load kVA rating = 4200. The following parameters in per unit on motor-base kVA are specified:

$$r_1 = 0.0075$$
$$r_2 = 0.0075$$
$$X_1 = 0.0656$$
$$X_2 = 0.0984$$
$$R_m = 100$$
$$X_m = 3.00$$

Calculate the motor short-circuit current equations for a sudden terminal fault.

The following parameters are calculated using Equations 6.128 through 6.132:

$$X' = 0.1608\,\text{pu}$$
$$T' = 0.057\,\text{s}$$
$$T_{dc} = 0.057\,\text{s}$$
$$T_0 = 1.09\,\text{s}$$

The ac component of the short-circuit current is

$$i_{ac} = 6.21e^{-t/0.057}$$

At $t = 0$, the ac symmetrical short-circuit current is 6.21 times the full-load current.

The dc component of the short-circuit current is

$$i_{dc} = 8.79e^{-t/0.057}$$

The nature of short-circuit currents is identical to that of synchronous machines; however, the currents decay more rapidly. The decay is a function of the motor rating, inertia, and load–torque characteristics. Typically, the effect of short-circuit currents from induction machines can be ignored after 6 cycles.

6.12 Practical Short-Circuit Calculations

For practical short-circuit calculations, dynamic simulation or analytical calculations are rarely carried out. Chapter 7 describes the ANSI empirical calculation procedures and shows that the machine models are simple and represented by a voltage behind impedance, which changes with the type of calculations. The detailed machine models and calculation of time variation of short-circuit currents are sometimes required to validate the empirical results [9]. These form a background to the empirical methods to be discussed in Chapters 7 and 8.

Problems

6.1 Calculate the fault decrement curves of the generator, data as given in Table 6.1. Calculate (1) the ac decaying component, (2) dc component, and (3) total current. Plot the results in a similar manner to those of Figure 6.6.

6.2 Consider the system and data shown in Figure P6.1. Calculate (1) prefault voltages behind reactances X_d, X'_d, and X''_d for faults at G and F, and (2) the largest possible dc component for faults at G and F.

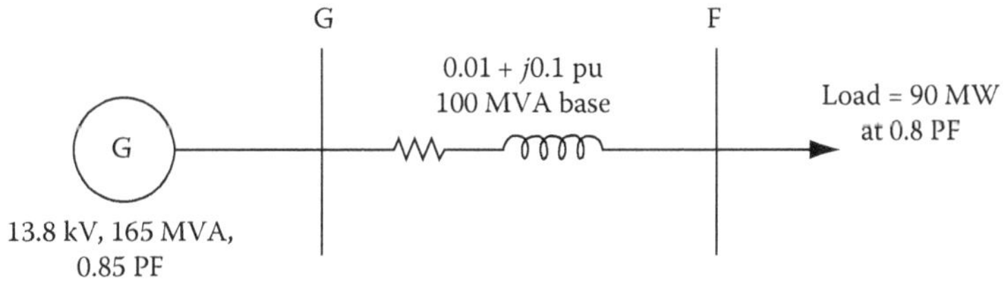

$X''_{dv} = 16\%$	$X''_d = 18\%$
$X'_d = 20\%$	$X_d = 200\%$
$X_l = 15\%$	$X_q = 190\%$
$X'_q = 26\%$	$X''_q = 17.8\%$
$r_D = 1.5\%$	$T'_{do} = 4.8$ s
$r_Q = 2.5\%$	$X/R = 110$

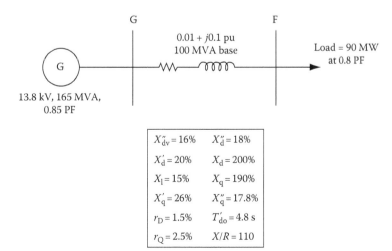

FIGURE P6.1
Circuit diagram and system data for Problem 6.2.

6.3 Calculate the field current in Problem 6.2 on application of short circuit.

6.4 Calculate three-phase short-circuit subtransient and transient time constants in Problem 6.2 for a fault at F.

6.5 Write a numerical expression for the decaying ac component of the current for faults at G and F in Problem 6.2. What is the ac component of the fault current at 0.05 and 0.10 s?

6.6 Transform the calculated direct axis and quadrature axis voltages derived in Problem 6.2 into stator voltages using Park's transformation.

6.7 Draw a general steady-state phasor diagram of a synchronous motor operating at (1) leading power factor and (2) lagging power factor.

6.8 Construct a simplified dynamic phasor diagram (ignoring damper circuit and resistances) of a synchronous generator using Park's transformations. How does it differ from the steady-state phasor diagram?

6.9 Show that first column of P' is an eigenvector of L_{11} corresponding to eigenvalue $L_d = L_s - 2M_S$.

6.10 Form an equivalent circuit similar to Figure 6.10 with numerical values using the generator data from Table 6.1.

6.11 A 13.8 kV, 10,000 hp four-pole induction motor has a full load efficiency of 96% and a power factor of 0.93. The locked rotor current is six times the full-load current at a power factor of 0.25. Calculate the time variation of ac and dc components of the current. Assume equal stator and rotor resistances and reactances. The magnetizing resistance and reactance are equal to 130 and 3.0 per unit, respectively, on machine MVA base.

References

1. NEMA. Large Machines—Synchronous Generators. MG-1, Part 22.
2. IEEE, Standard 421A. IEEE Guide for Identification and Evaluation of Dynamic Response of Excitation Control Systems, 1978.
3. *ATP Rule Book*. Portland, OR: Canadian/American EMTP Group, 1987–1992.
4. R.H. Park. Two reaction theory of synchronous machines, Part I. *AIEE Trans.* 48: 716–730, 1929.
5. R.H. Park. Two reaction theory of synchronous machines, Part II. *AIEE Trans.* 52: 352–355, 1933.
6. P.M. Anderson and A. Fouad. *Power System Control and Stability*. New York: IEEE Press, 1991.
7. C.V. Jones. *The Unified Theory of Electrical Machines*. Oxford, U.K.: Pergamon Press, 1964.
8. A.T. Morgan. *General Theory of Electrical Machines*. London, U.K.: Heyden & Sons Ltd., 1979.
9. J.R. Dunki-Jacobs, P. Lam, and P. Stratford. A comparison of ANSI-based and dynamically rigorous short-circuit current calculation procedures. *Trans. IEEE Ind. Appl.* 24: 1180–1194, 1988.

Bibliography

Adkins, B. *The General Theory of Electrical Machines*. London, U.K.: Chapman & Hall, 1964.

Anderson, P.M. *Analysis of Faulted Power Systems*. Ames, IA: Iowa State University Press, 1973.

Boldea, I. *Synchronous Generators*. Boca Raton, FL: CRC Press, 2005.

Concordia, C. *Synchronous Machines*. New York: Wiley, 1951.

Fitzgerald Jr., A.E. Umans, S.D., and Kingsley, C. *Electrical Machinery*. New York: McGraw Hill Higher Education, 2002.

Hancock, N.N. *Matrix Analysis of Electrical Machinery*. Oxford, U.K.: Pergamon Press, 1964.

IEEE Committee Report. Recommended phasor diagram for synchronous machines. *IEEE Trans. PAS* 88: 1593–1610, 1963.

Park, R.H. Two reaction theory of synchronous machines, Part-1. *AIEEE Trans.* 48: 716–730, 1929.

7

Short-Circuit Calculations according to ANSI Standards

ANSI methods of short-circuit calculations are used all over North America and are accepted in many other countries. These standards have been around for a much longer time than any other standard in the world. The IEC [1] standard for short-circuit calculation was published in 1988, which is now revised Ref. [2] and the calculation procedures according to IEC are discussed in Chapter 8. A VDE [3] (Deutsche Electrotechnische Kommission) standard has been around since 1971. There has been a thrust for analog methods too. Nevertheless, for all equipment manufactured and applied in industry in the United States, ANSI/IEEE standards prevail. Most foreign equipment for use in the U.S. market has been assigned ANSI ratings.

We will confine our discussions to ANSI/IEEE symmetrical rating basis of the circuit breakers. The United States is the only country that rates the circuit breakers on symmetrical current basis. The interpretations, theory, and concepts governing short-circuit calculations according to the latest ANSI/IEEE standards are discussed with illustrative examples.

7.1 Types of Calculations

In a multivoltage system, four types of short-circuit calculations may be required. These are

1. First-cycle (momentary) duties for fuses and low-voltage circuit breakers
2. First-cycle (momentary) duties for medium- or high-voltage circuit breakers
3. Contact parting (interrupting) duties for high-voltage circuit breakers (circuit breakers rated above 1 kV)
4. Short-circuit currents for time-delayed relaying devices

The first-cycle, momentary duty, or (close and latch duty) are all synonymous. However, the close and latch term is applicable to high-voltage circuit breakers only, while fist-cycle term can be applied to low-voltage breakers and fuses too. Irrespective of the type of fault current calculation, the power system is reduced to single Thevénin equivalent impedance behind the source voltage.

7.1.1 Assumptions—Short-Circuit Calculations

The source voltage or prefault voltage is the system rated voltage, though a higher or lower voltage can be used in the calculations. The worst short-circuit conditions occur at maximum loads, because the rotating loads contribute to the short-circuit currents. It is unlikely that the operating voltage will be above the rated voltage at maximum loading. Under light load conditions, the operating voltage may be higher, but the load contributions to the short-circuit currents will also be reduced. The effect of higher voltage at a reduced load is offset by the reduced contributions from the loads. Therefore, the short-circuit calculations are normally carried out at the rated voltage. Practically, the driving voltage will not remain constant; it will be reduced and vary with the machine loading and time subsequent to short circuit. The fault current source is assumed sinusoidal, and all harmonics and saturation are neglected. All circuits are linear; the nonlinearity associated with rotating machines, transformer, and reactor modeling is neglected. As the elements are linear, the theorem of superimposition is applicable.

Loads prior to the short circuit are neglected; the short circuit occurs at zero crossing of the voltage wave. At the instant of fault, the dc current value is equal in magnitude to the ac fault current value but opposite in sign.

7.1.2 Maximum Peak Current

The maximum peak current *does not occur* at half cycle in the phase that has maximum initial dc component, unless short circuit occurs on purely inductive circuits with resistance equal to zero. The maximum peak occurs before half cycle and before the symmetrical current peak. The IEC equation for peak is in Chapter 8, Equation 8.3. IEEE standard, Ref. [4], provides the following equations that give better approximation to peak, as compared to IEC equations:

$$I_{\text{peak}} = I_{\text{ac peak}} + I_{\text{dc}} = \sqrt{2}I_{\text{ac,rms}}\left(1 + e^{-2\pi\tau/(X/R)}\right) = I_{\text{ac peak}}\left(1 + e^{-2\pi\tau/(X/R)}\right) \tag{7.1}$$

where

$$\tau = 0.49 - 0.1e^{-(X/R)/3} \tag{7.2}$$

For the rms first-cycle rms current, the equation

$$I_{\text{rms}} = \sqrt{I_{\text{ac,rms}}^2 + I_{\text{dc}}^2} \tag{7.3}$$

is only correct if dc is constant, but it is not and decays exponentially. Many times, dc component is calculated at half cycle, though this does not correspond to the peak value [5]. Following equations are provided in Ref. [4]:

$$\text{IEC}_{\text{rms}} = I_{\text{ac,rms}}\sqrt{1 + 2(1.02 + 0.98e^{-3/(X/R)})^2} \tag{7.4}$$

$$\text{Half cycle}_{\text{rms}} = I_{\text{ac,rms}}\sqrt{1 + 2(e^{-\pi/(X/R)})^2} \tag{7.5}$$

The more exact equation from Ref. [4] is

$$I_{\text{rms,total}} = I_{\text{ac rms}}\sqrt{(1 + 2e^{-4\pi\tau/(X/R)})} \tag{7.6}$$

Reference [4] tabulated the peak and rms value calculations using the above equations. Consider an $X/R = 10$, then the peak values in pu are

Exact calculations: Time to peak = 0.4735 cycles, dc = 0.7368, maximum peak = 1.7368
IEC equations: Peak = 1.7460, % error = 0.53% [see chapter 8]
Half cycle equations: Peak = 1.7304, error = −0.37% [Equation 5.7]
Reference [4] equations: Peak = 1.7367, error = −0.01% [Equation 7.1].

A peak multiplier of 2.6 is often used for simplicity, when calculating the duties of medium and high-voltage circuit breakers above 1 kV. This 2.6 factor corresponds to $X/R = 17$ for 60 Hz systems or equivalently a dc component decay governed by L/R time constant of 45 ms for 60 Hz system ($X/R = 14$ for 50 Hz system). When larger X/R ratios are encountered, higher multipliers result. *When selecting a breaker, it is important that the peak is adjusted for the actual fault point X/R ratio.*

For older high-voltage circuit breakers, calculate asymmetrical rms current based upon half cycle peak, given

$$I_{\mathrm{asym}} = I_{\mathrm{sym}}\sqrt{1 + 2e^{-2\pi/(X/R)}} \tag{7.7}$$

This essentially calculated total asymmetrical current at half cycle. An asymmetrical multiplier of 1.6 has been used, which corresponds to $X/R = 25$. Higher factors will be obtained with higher X/R. *When selecting a breaker, it is important that the peak is adjusted for the actual fault point X/R ratio.*

7.2 Accounting for Short-Circuit Current Decay

We have amply discussed that short-circuit current are decaying transients. In the short-circuit calculations according to standards, a step-by-step account of the decay cannot be considered. Depending on the type of calculation, the ANSI/IEEE standards consider that the dynamic (rotating equipment) reactances are multiplied by factors given in Table 7.1 [6,7]. The static equipment impedances are assumed to be time invariant, i.e., harmonics and saturation are neglected. Maintaining a constant emf and artificially increasing the equivalent impedance to model a machine during short circuit has the same effect as the decay of the flux trapped in the rotor circuit. In Table 7.1, manufacturer's data for the transient and subtransient and locked rotor reactances should be used in the calculations. Some industries may use large motors of the order of thousands of hp rating, i.e., the pulp mill industry may use a single synchronous motor of the order of 40,000 hp for a refiner application. Reactances and time constants need to be accurately modeled for such large machines (Example 7.6). The decay of short-circuit currents of the motors depends upon the size of the motors. For induction machine, a simulation will show high initial current decay followed by fairly rapid decay to zero. For synchronous machines, there is a high initial decay followed by a slower rate of decay to a steady-state value, Chapter 6.

TABLE 7.1

Impedance Multiplier Factors for Rotating Equipment for Short-Circuit Calculations

Type of Rotating Machine	Positive Sequence Reactance for Calculating: Interrupting Duty (Per Unit)	Positive Sequence Reactance for Calculating: Closing and Latching Duty (Per Unit)
All turbogenerators, all hydrogenerators with amortisseur windings, and all condensers	$1.0X''_d$	$1.0X''_d$
Hydrogenerators without amortisseur windings	$0.75X'_d$	$0.75X'_d$
All synchronous motors	$1.5X''_d$	$1.0X''_d$
Induction motors		
Above 1000 hp at 1800 r/min or less	$1.5X''_d$	$1.0X''_d$
Above 250 hp at 3600 r/min	$1.5X''_d$	$1.0X''_d$
From 50 to 1000 hp at 1800 r/min or less	$3.0X''_d$	$1.2X''_d$
From 50 to 250 hp at 3600 r/min	$3.0X''_d$	$1.2X''_d$
Neglect all three-phase induction motors below 50 hp and all single-phase motors		
Multivoltage level calculations including low-voltage systems		
Induction motors of above 50 hp	$3.0X''_d$	$1.2X''_d$
Induction motors below 50 hp	∞	$1.67X''_d$

Source: Ref. [6,7].

X''_d of synchronous rotating machines is the rated voltage (saturated) direct-axis subtransient reactance. X'_d of synchronous rotating machines is rated-voltage (saturated) direct-axis transient reactance. X''_d of induction motors equals 1.00 divided by per unit locked rotor current at rated voltage.

7.2.1 Low-Voltage Motors

For the calculation of short-circuit duties on low-voltage systems, a modified subtransient reactance for group of low-voltage induction and synchronous motors fed from a low-voltage substation can be used. If the total motor horse-power rating is approximately equal to or less than the self-cooled rating in kVA of the substation transformer, a reactance equal to 0.25 per unit, based on the transformer self-cooled rating, may be used as a single impedance to represent the motors. This means that the combined motor loads contribute a short-circuit current equal to four times the rated current. This estimate is based on the low-voltage circuit breaker application guide, IEEE Std. C37.13 [8]. It assumes a typical motor group having 75% induction motors, which contribute short-circuit currents equal to 3.6 times their rated current, and 25% synchronous motors, which contribute 4.8 times the rated current. At present, low-voltage synchronous motors are not in much use in industrial distribution systems; however, higher rated induction motors in a group may contribute a higher amount of short-circuit current, compensating for the absence of synchronous motors. Overall, four times the full load contribution can be retained.

For calculations of short-circuit duties for comparison with medium- or high-voltage circuit breakers closing and latching capabilities (or momentary ratings according to pre-1964 basis, now no longer in use), motors smaller than 50 hp can be ignored. However, these are required to be considered for low-voltage circuit breaker applications. To simplify multivoltage distribution system short-circuit calculations and obviate the necessity of running two first-cycle calculations, one for the low-voltage circuit breakers and the other for medium- and high-voltage circuit breakers, a single first-cycle network can replace the two networks [7]. This network is constructed by

1. Including all motors <50 hp using a multiplying factor of 1.67 for subtransient reactance or an estimate of first-cycle impedance of 0.28 per unit based on the motor rating
2. Including all motors ≥50 hp and using a multiplying factor of 1.2 for the subtransient reactance or an estimate of first-cycle impedance of 0.20 per unit based on the motor rating

This single-combination first-cycle network adds conservatism to both low- and high-voltage calculations. A typical short-circuit contribution for a terminal fault of an induction motor is six times the full load current. Thus, an estimate of 0.20 per unit for larger low-voltage motors or a multiplying factor of 1.2 is equivalent to a fault current contribution of 4.8 times the rated full-load current. Similarly, for motors ≥50 hp, a multiplying factor of 1.67 or an estimate of 0.28 per unit impedance means a short-circuit contribution of approximately 3.6 times the rated current. These factors are shown in Table 7.1.

Though this simplification can be adopted, where the medium- and high-voltage breakers are applied close to their first-cycle ratings, it is permissible to ignore all low-voltage motors rated <50 hp. Depending on the extent of low-voltage loads, this may permit retaining the existing medium- or high-voltage breakers in service, if these are overdutied from the close and latch capability considerations. Close to a generating station, the close and latch capabilities may be the limiting factor in the application of circuit breakers.

7.3 Rotating Machines Model

The rotating machine model for the short-circuit calculations is shown in Figure 7.1. The machine reactances are modeled with suitable multiplying factors from Table 7.1. The multiplying factors are applicable to resistances as well as reactances, so that the X/R ratio remains the same. The voltage behind the equivalent transient reactance at no load will be equal to the terminal voltage, i.e., $V_s = V_t$, as no prefault currents need be considered in the calculations. A justification of neglecting the prefault currents is indirectly discussed in

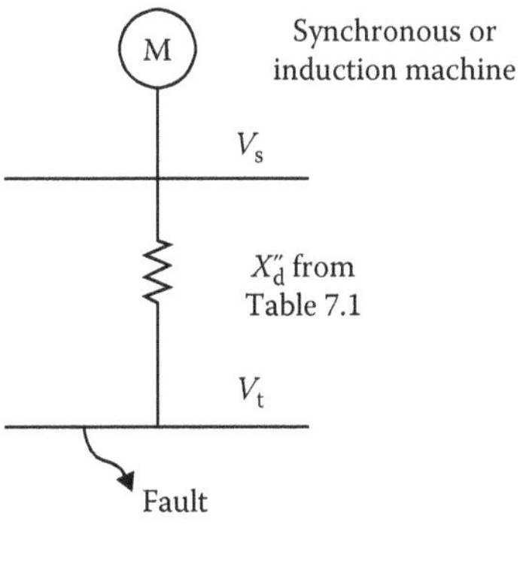

FIGURE 7.1
Equivalent machine model for short-circuit calculations.

Chapter 6, i.e., the concept of constant flux linkages. The total current before and after the transition (pre- and post-fault) should not change. The dc component is equal in magnitude to the ac component, but of the opposite polarity. Thus, the ac and dc components of the current summate to zero to maintain a constant flux linkage, i.e., no load conditions. If preloading is assumed, this balance is no longer valid.

7.4 Types and Severity of System Short Circuits

The short-circuit currents in the decreasing order of severity are

- Three-phase ungrounded
- Three-phase grounded
- Phase-to-phase ungrounded
- Phase-to-phase grounded
- Single-phase to ground

This order is also applicable to transient stability studies, a three-phase fault giving the worst situation. If the system is shown to hold together for a three-phase fault, it can be concluded that it will be stable for *other fault types cleared in the same time duration* as the three-phase fault.

A three-phase power system may be subjected to symmetrical and unsymmetrical faults. Generally, three-phase ungrounded faults impose the most severe duty on a circuit breaker, since the first phase to interrupt has a normal frequency recovery voltage of approximately 87% of the system phase-to-phase voltage. A single phase-to-ground fault current can be higher than the three-phase current. This condition exists when the zero sequence impedance at the fault point is less than the positive sequence impedance. Depending on the relative magnitude of sequence impedances, it may be necessary to investigate other types of faults. For calculations of short-circuit duties, the fault resistance is ignored. This gives conservatism to the calculations.

7.5 Calculation Methods

Two calculation procedures are

1. E/X or E/Z simplified method
2. E/X or E/Z method with adjustments for ac and dc decrements

7.5.1 Simplified Method $X/R \leq 17$

The results of the E/X calculation can be directly compared with the circuit breaker symmetrical interrupting capability, provided that the circuit X/R ratio for three-phase faults and $(2X_1+X_0)/(2R_1+R_0)$ for single line-to-line ground fault is 17 or less. (For single-phase to-ground faults X_1 is assumed equal to X_2.) This is based on the rating structure of the breakers. When the circuit X/R ratio is 17 or less, the asymmetrical short-circuit duty never exceeds the symmetrical short-circuit duty by a proportion greater than that by which the circuit breaker asymmetrical rating exceeds the symmetrical capability. It may only be slightly higher at four-cycle contact parting time.

7.5.2 Simplified Method $X/R > 17$

A further simplification of the calculations is possible when the X/R ratio exceeds 17. For X/R ratios higher than 17, the dc component of the short-circuit current may increase the short-circuit duty beyond the compensation provided in the rating structure of the breakers. A circuit breaker can be immediately applied without calculation of system resistance, X/R ratio, or remote/local considerations, if the E/X calculation does not exceed 80% of the breaker symmetrical interrupting capability. This means that resistance component of the system need not be determined.

7.5.3 E/X Method for AC and DC Decrement Adjustments

Where a closer calculation is required, and the current exceeds 80% of the breaker symmetrical rating, ac and dc decrement adjustments should be considered. This method is also recommended when a single line-to-ground fault supplied predominantly by generators, at generator voltage, exceeds 70% of the circuit breaker interrupting capability for single line-to-ground faults. For calculations using this method, the fault point X/R ratio is necessary. Two separate networks are constructed; (1) a resistance network, with complete disregard of the reactance, and (2) a reactance network with complete disregard of the resistance. The fault point X/R ratio is calculated by reducing these networks to an equivalent resistance and an equivalent reactance at the fault point. This gives more accurate results than any other reasonably simple procedure, including the phasor representation at the system frequency.

The resistance values for various system components are required, and, for accuracy of calculations, these should be obtained from the manufacturer's data. In the absence of these data, Table 7.2 and Figures 7.2 through 7.4 provide typical resistance data. The variations in X/R ratio between an upper and lower bound are shown in the ANSI/IEEE standard [6].

TABLE 7.2

Resistance of System Components for Short-Circuit Calculations

System Component	Approximate Resistance
Turbine generators and condensers	Effective resistance
Salient pole generators and motors	Effective resistance
Induction motors	1.2 times the dc armature resistance
Power transformers	AC load loss resistance (not including no-load losses or auxiliary losses)
Reactors	AC resistance
Lines and cables	AC resistance

Source: Ref. [6].

The effective resistance $= X_{2v}/(2\pi f T_{a3})$, where X_{2v} is the rated-voltage negative-sequence reactance and T_{a3} is the rated voltage generator armature time constant in seconds.

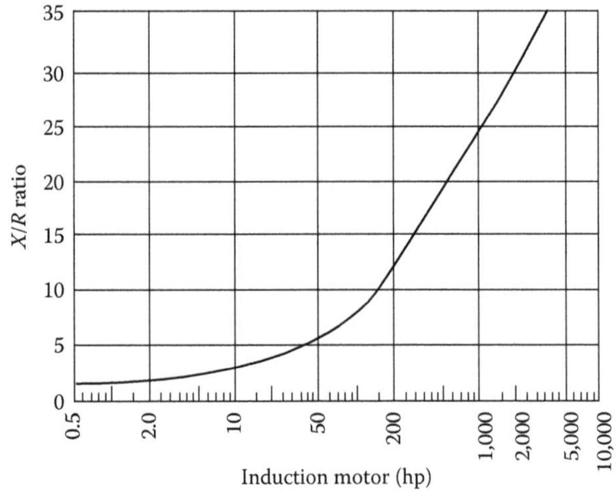

FIGURE 7.2
Typical X/R ratios for induction motors based on induction motor hp rating.

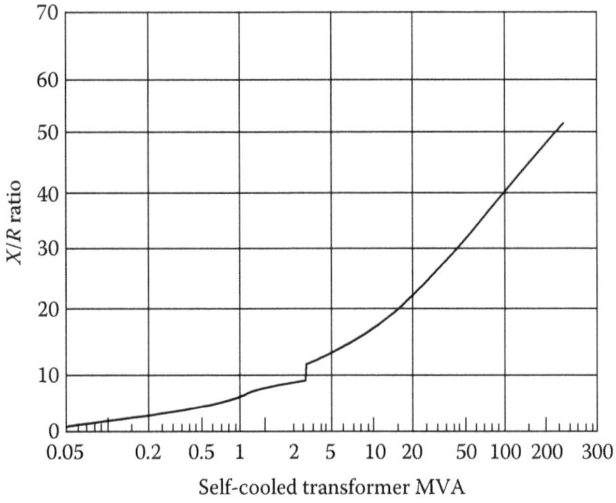

FIGURE 7.3
Typical X/R ratios for transformers based on transformer self-cooled MVA rating.

Once the E/X calculation is made and the X/R ratio is known, the interrupting duty on the high-voltage circuit breakers can be calculated by multiplying the calculated short-circuit currents with an appropriate multiplying factor. This multiplying factor is based on

1. Contact parting time of the circuit breaker
2. Calculated X/R ratio
3. Effects of dc decay (remote sources) and effects of ac and dc decay (local sources)

7.5.4 Fault Fed from Remote Sources

If the short-circuit current is fed from generators through (1) two and more transformations or (2) a per unit reactance external to the generator that is equal to or exceed 1.5 times the generator per unit subtransient reactance on a common MVA base, i.e., it supplies less than 40% of its terminal short-circuit current, it is considered a remote source. In this case, the effect of ac decay need not be considered, and the curves of multiplying factors include only dc decay. These curves are shown in Figure 7.5. The decrement factor for the standard

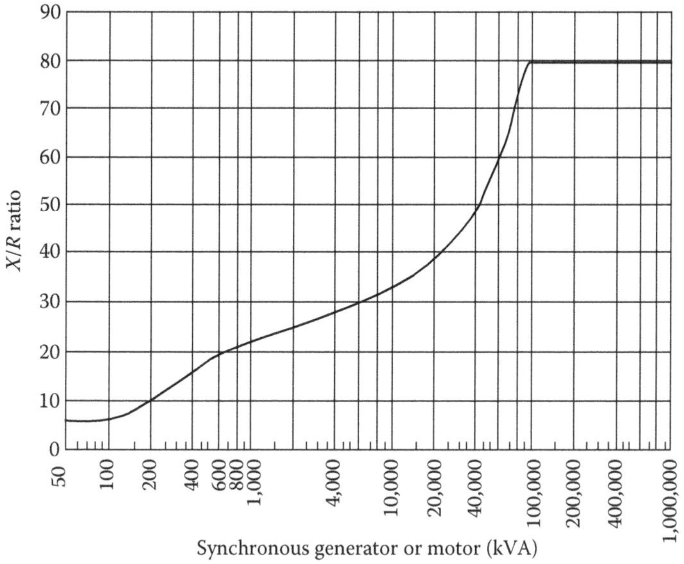

FIGURE 7.4
Typical X/R ratios for synchronous generators and synchronous motors based on their kVA rating.

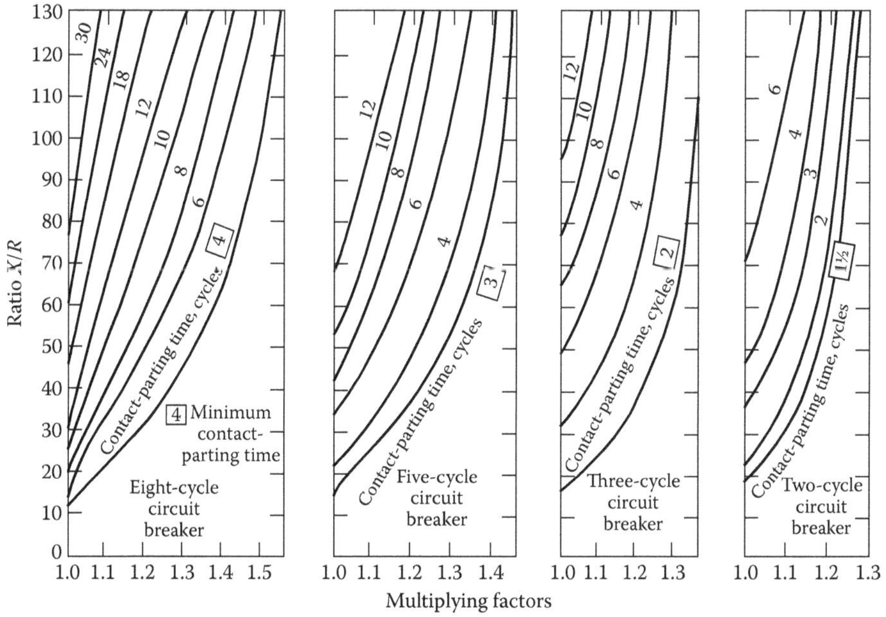

FIGURE 7.5
Three-phase and line-to-ground faults, E/X multiplying factors, dc decrement only (remote sources). (*Source:* Ref. [6].)

contact parting time of the breakers is shown within a rectangle, which includes a half-cycle tripping delay. Factors for higher contact parting time, applicable when the tripping delay is increased above a half-cycle, are also shown. Interpolation between the curves is possible. The multiplying factor for the remote curves is calculable and is given by

$$MF_{remote} = \frac{I_{sym}}{I_{asym}} \times \sqrt{1 + 2e^{-4\pi C/(X/R)}} \tag{7.8}$$

where C is the contact parting time in cycles at 60 Hz. As an example, the remote multiplying factor for five-cycle breaker, which has a contact parting time of three cycles, and for a fault point X/R of 40, from Equation 7.8, is 1.21, which can also be read from the curves in Figure 7.5.

7.5.5 Fault Fed from Local Sources

When the short-circuit current is predominantly fed through no more than one transformation or a per unit reactance external to the generator, which is less than 1.5 times the generator per unit reactance on the same MVA base, i.e., it supplies more than 40% of its maximum terminal fault current, it is termed a local source. The effect of ac and dc decrements should be considered. The multiplying factors are applied from separate curves, reproduced in Figures 7.6 and 7.7.

The asymmetrical multiplying factors for the *local* curves are not a known equation. A number of sources may contribute to a fault through varying impedances. Each of these contributions has a different ac and dc decay rate. The impedance through which a fault is fed determines whether it is considered a local or remote source. The ac decay in

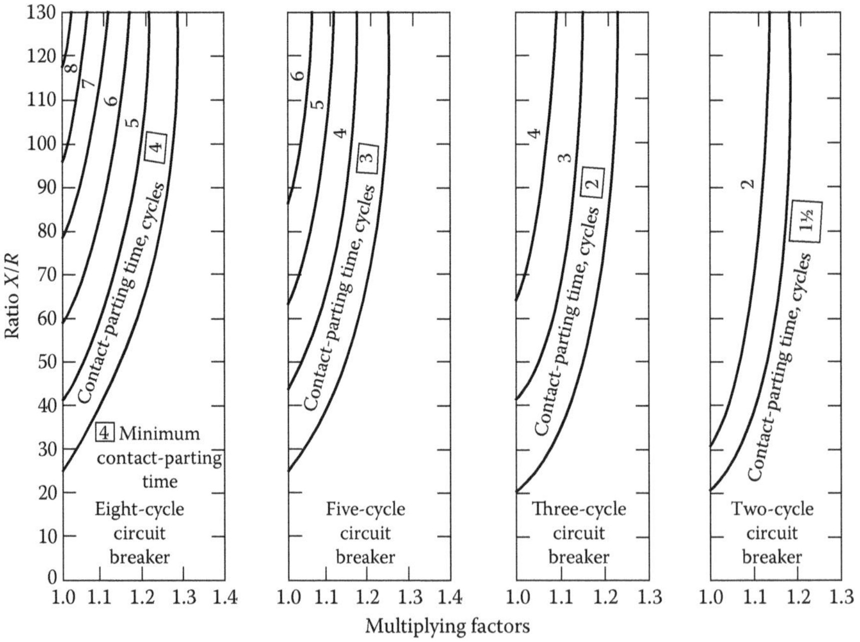

FIGURE 7.6
Three-phase faults, E/X multiplying factors, ac and dc decrement (local sources). (*Source*: Ref. [6].)

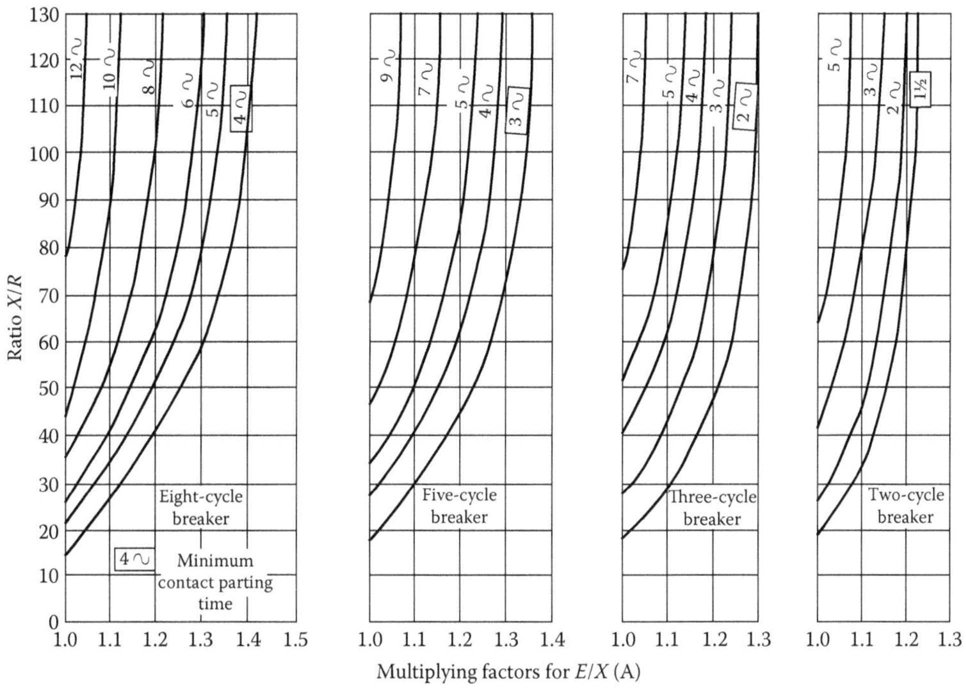

FIGURE 7.7
Line-to-ground faults, E/X multiplying factors, ac and dc decrement (local sources). (*Source*: Ref. [6].)

electrically remote sources is slower, as compared to the near sources. The time constant associated with the ac decay is a function of rotor resistance and added external reactance prolongs it (Chapter 6). An explanation and derivation of the multiplying factors for ac and dc decrements, provided in Ref. [6], is as follows.

Figure 7.8 shows the relationship of fault current $(I_{asym}/I_{sym})_{nacd}$ (the subscript "nacd" means that there is no ac decrement), as a function of X/R ratio for various contact parting times. The curves of this figure are modified, so that the decrement of the symmetrical component of the fault current is taken into consideration. Figure 7.9a shows the general relationship of X/R to the ac decrement as the fault location moves away from the generating station. This empirical relationship is shown as a band, based on small to large machines of various manufacturers. Figure 7.9b shows the decay of the symmetrical component (ac component) of the fault current at various times after fault initiation, as a function of the contact parting time and the type of fault. Figure 7.9c establishes reduction factors that can be applied to $(I_{asym}/I_{sym})_{nacd}$ to obtain this effect. The reduction factor is obtained from the following relationship:

$$\text{Reduction factor} = \frac{\sqrt{I_{ac}^2 + I_{dc}^2}/(E/X)}{(I_{asym}/I_{sym})_{nacd}} \tag{7.9}$$

As an example, consider an X/R ratio of 80 and contact parting time of three cycles; the factor $(I_{sym}/I_{sym})_{nacd}$, as read from Figure 7.8, is 1.5. Enter curve in Figure 7.9a at X/R of 80, follow down to contact parting time curve in Figure 7.9b, and go across to Figure 7.9c, curve labeled

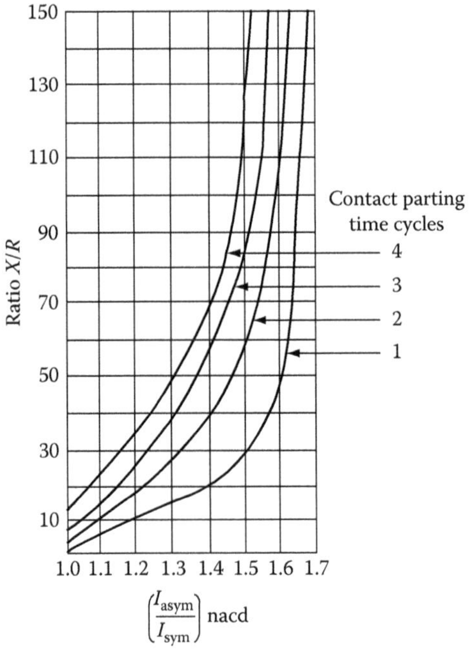

FIGURE 7.8
Ratio $(I_{asym}/I_{sym})_{nacd}$ versus X/R ratio for breaker contact parting times. (From VDE, Standard 0102, Parts 1/11.71 and 2/11.75, Calculations of Short-Circuit Currents in Three-Phase Systems, 1971 and 1975.)

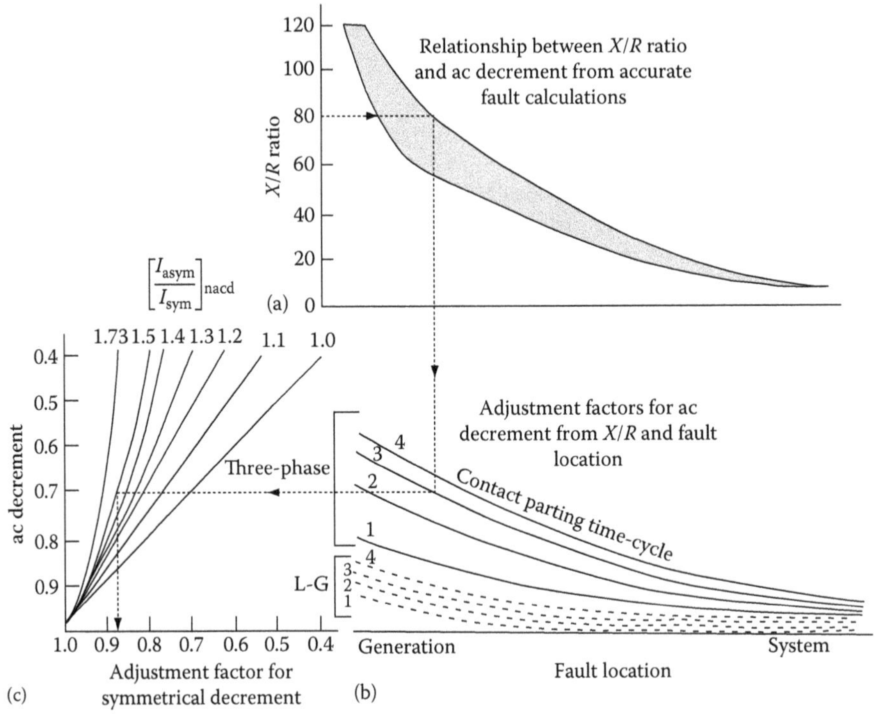

FIGURE 7.9
(a) ac decrement for faults away from the sources; (b) adjustment factors for ac decrement from X/R and fault location for breaker contact parting times (solid lines: three-phase faults; dotted lines: line-to-ground faults); (c) adjustment factors for ac decrement. (*Source*: Ref. [6].)

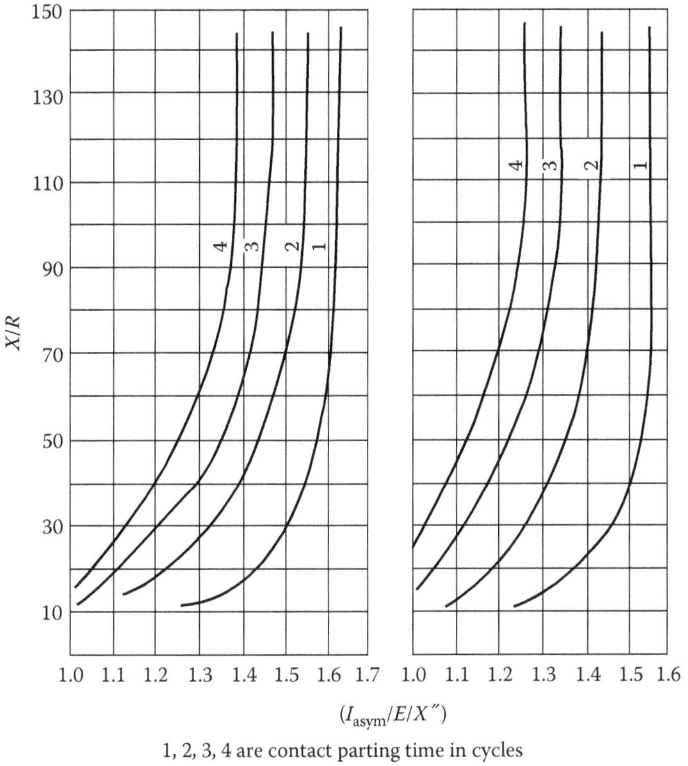

FIGURE 7.10
Relationship of I_{asym}/I_{sym} to X/R for several breaker contact parting times, ac decrement included. (*Source*: Ref. [6].)

$(I_{asym}/I_{asym})_{nacd} = 1.5$. A reduction factor of 0.885 is obtained. The modifier $(I_{sym}/I_{sym})_{nacd}$ ratio for an X/R of 80 is calculated as $0.885 \times 1.5 = 1.33$, and this establishes one point on a three-phase modified decrement curve, shown in Figure 7.10. This curve is constructed by following the procedure outlined above. Finally, E/X multipliers for the breaker application are obtained through the use of a modified X/R decrement curve and the breaker capability. Continuing with the above calculation, the breaker asymmetric capability factor for a three-cycle parting time is 1.1. The E/X multiplier required to ensure sufficient breaker capability is, therefore, $1.33/1.1 = 1.21$. This establishes one point in Figure 7.6.

In a digital computer-based calculation, the matrix equations can be used to calculate voltages at buses other than the faulted bus, and current contributions from individual sources can be calculated (Chapter 3). These currents can then be labeled as remote or local.

In certain breaker applications, a breaker contact parting time in excess of the contact parting time, with a half-cycle tripping delay assumed for the rating structure, may be used. If a breaker with a minimum contact parting time of two cycles is relayed, such that it actually parts contacts after four cycles after fault initiation, the E/X multiplier for breaker selection can be reduced to account for the fault current decay during the two-cycle period. This will reduce the interrupting duty because of decaying nature of short-circuit currents. Sometimes, it can prevent costly replacement of circuit breakers, which are applied close to their ratings and short-circuit duties increase because of system growth [9]. However, the related aspects of delayed opening of the breaker by introducing addition relay time delay has other adverse effects on the system: increase in fault damage and jeopardizing the transient stability in some cases. This does not impact the closing and latch duty.

7.5.6 Weighted Multiplying Factors

For a system with several short-circuit sources, which may include generators that may be classified local or remote, depending on the interconnecting impedances, neither the remote nor the local multiplying factors can be exclusively applied. It is logical to make use of both local and remote multiplying factors in a weighting process. This weighting consists of applying remote multiplying factor to that part of the E/X symmetrical short-circuit current that is contributed by remote sources. Similarly, the local multiplying factor is applied to the local component of the fault current contribution. The fraction of interrupting current that is contributed by remote sources is identified as the NACD ratio:

$$\text{NACD ratio} = \frac{\sum \text{NACD source currents}}{E/X \text{ for the interrupting network}} \tag{7.10}$$

This computation requires additional calculations of remote and total current contributed at the fault point from various sources and is facilitated by digital computers. Figure 7.11 shows interpolated multiplying factors for various NACD ratios [10].

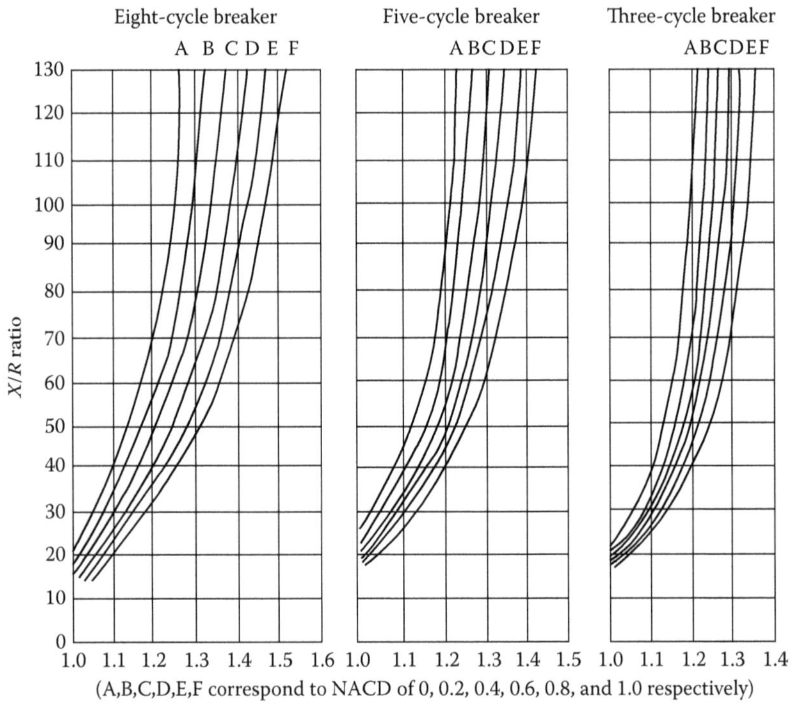

FIGURE 7.11
Multiplying factors for E/X amperes, three-phase faults, for various NACD ratios and breaker interrupting times. (*Source*: Ref. [10].)

For the short-circuit current contribution from motors, *irrespective of their type and rating and location in the system*, the ac decay is built into the premultiplying impedance factors in Table 7.1. Thus, it is assumed that the motors, howsoever remote in the system, will continue contributing to the fault. In an actual system, the postfault recovering voltage may return the motor to normal motoring function. The magnetic contactors controlling the motors may drop in the first cycle of the voltage dip on 30%–70% of their rated voltage, disconnecting the motors from service. This is conservative from the short-circuit calculation point of view but may give increased arc flash hazard calculations (see Chapter 21).

7.6 Network Reduction

Two methods of network reduction are as follows:

1. Short-circuit current can be determined by complex impedance network reduction, this gives the E/X complex method. It is permissible to replace E/X with E/Z. In fact, in all commercially available software, the default calculations are based upon E/Z method, though E/X method can be selected.
2. Short-circuit current can be determined from R and X calculations from separate networks and treating them as a complex impedance at the fault point.

In either case, the X/R ratio is calculated from separate resistance and reactance networks. The X/R ratio thus calculated is used to ascertain multiplying factors and also for calculation of asymmetry factors for the first-cycle calculation. The X/R ratio for single line-to-ground faults is $(X_1+X_2+X_0)/(R_1+R_2+R_0)$.

The E/Z calculation from separate networks is conservative, and results of sufficient accuracy are obtained. Branch current flows and angles may have greater variation as compared to the complex network solution. Contributions from the adjacent buses may not correlate well. However, this procedure results in much simpler computing algorithms.

There could be a difference of 5%–6% in the calculated results of short-circuit currents between the complex impedance reduction method and the calculations from the separate R and X networks. The separate R and jX calculations give higher values, as compared to the $R+jX$ complex calculation.

7.6.1 E/X or E/Z Calculation

The E/X calculation will give conservative results, and, as the X/R at the fault point is high, there may not be much difference between E/X and E/Z calculations. This may not be always true. For low-voltage systems, it is appropriate to perform E/Z calculations, as the X/R ratios are low, and the difference in the results between E/Z and E/X calculations can be significant. Generally, E/Z calculations using the complex method are the standard in industry.

7.7 Breaker Duty Calculations

Once the results of E/X or E/Z calculation for the interrupting network are available and the weighted multiplying factor is ascertained, the adequacy of the circuit breaker for interrupting duty application is given by

$$\begin{aligned}&MF \times E/X(E/Z)(\text{interrupting network})\\&< \text{Breaker interrupting rating kA sym.}\end{aligned} \tag{7.11}$$

For calculations of close and latch capability or the first-cycle calculations, *no considerations of local and remote are required*. Old high-voltage breakers are rated at 1.6 K times the rated short-circuit current in rms asym. And new breakers are rated at 2.6 times the rated short-circuit current in peak. The peak current based upon X/R is given by Equation 7.1. The results are compared with the breaker close and latch capability. The adequacy of the breaker for first-cycle or close and latch capability is given by

$$\text{Calculated } I_{\text{peak}} \text{ at the fault point } X/R < \text{Breaker close and latch capability} \tag{7.12}$$

Reference [6] cautions that E/X method of calculations with ac and dc adjustments described in Section 7.5.3 can be applied provided X/R ratio does not exceed 45 at 60 Hz (dc time constant not greater than 120 ms). For higher X/R ratios, it recommends consulting the manufacturer. The interruption process can be affected, and the interruption window, which is the time difference between the minimum and maximum arcing times of SF_6 puffer breakers, may exceed due to delayed current zero. We will examine in the calculations to follow that the current zeros may be altogether missing for a number of cycles. This qualification has been ignored in the commercially available short-circuit calculations programs, claiming to follow IEEE standards.

7.8 Generator Source Short-Circuit Current Asymmetry

For a generator circuit breaker, the highest value of asymmetry occurs, when prior to fault the generator is operating underexcited with a leading power factor and the generator breaker sees a fault current contribution of the generator, that is, a fault on the utility side of the generator breaker. Also see Chapter 8. The dc component may be higher than the symmetrical component of the short-circuit current and may lead to delayed current zeros. An analysis of a large number of generators resulted in a maximum asymmetry of 130% of the actual generator current [11,12]. The symmetrical component of the short-circuit current is 74% of generator current. Consequently, the ratio of the asymmetrical to symmetrical short-circuit current rating is 1.55.

α is a factor of asymmetry given by

$$\alpha = \frac{I_{dc}}{\sqrt{2}I_{sym}} \tag{7.13}$$

From Equation 7.13

$$\frac{I_{rms\ asym}}{I_{rms\ sym}} = \sqrt{2\alpha^2 + 1}$$

Thus, for $\alpha = 1.3$, $I_{asym}/I_{sym} = 2.09$. For $I_{sym} = 0.74I_{gen}$, ratio I_{asym}/I_{gen} can be written as

$$\frac{I_{asym}}{I_{gen}} = \left(\frac{I_{asym}}{I_{sym}}\right)\left(\frac{I_{sym}}{I_{gen}}\right) = 1.55 \tag{7.14}$$

The asymmetry can be calculated by considering the dc component at the contact parting time. Depending on generator subtransient and transient short-circuit time constants in the direct and quadrature axes and armature time constant T_a, the ac component may decay faster than the dc component, leading to delayed current zeros.

Additional resistance in series with the armature resistance forces the dc component to decay faster. The time constant with added resistance is

$$T_a = \frac{X''_d}{2\pi f(r + R_e)} \tag{7.15}$$

where R_e is the external resistance (see Equation 6.20). If there is an arc at the fault point, the arc resistance further reduces the time constant of the dc component. Figure 7.12 shows that at the contact parting time, the dc component changes suddenly due to the influence of the arc voltage of the generator circuit breaker, and a current zero is obtained within one cycle.

As we have seen generator characteristics, X/R ratio, time constants, and the subtransient component of the current influence the degree of asymmetry (Figure 6.4). Also see Refs. [13,14].

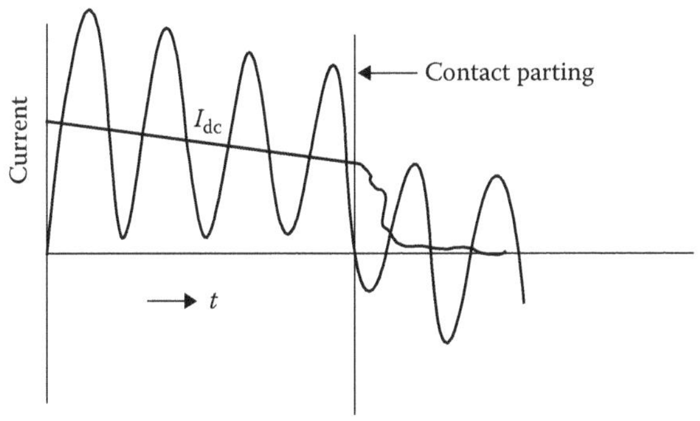

FIGURE 7.12
Fault of high asymmetry: effect of arc resistance to force a current zero at contact separation.

The interruption of ac current without a current zero is equivalent to the interruption of dc current, and HV breakers have very limited capabilities to interrupt dc currents. Delaying the opening of contacts may ultimately bring a current zero, but note that *both* ac and dc components are decaying and a number of cycles may elapse before current zero is obtained. Practically, delaying the opening of contacts is not implemented.

The generator breakers are designed and tested to interrupt currents of high asymmetry. The arc interruption medium and arc control devices, i.e., arc rotation have an effect on the interruption process and introducing an arc resistance to force current zero (Figure 7.12). Large generators and transformers are often protected as a unit and generator circuit breaker eliminated. When required, a generator circuit breaker should be carefully selected and applied. The actual asymmetry can be analytically calculated, Example 7.2.

Generators circuit breakers (in-line circuit breakers) mounted on phase-isolated buses connecting large generators to step-up transformers are available with short-circuit capabilities of 250 kA rms symmetrical and continuous current rating of 100 kA. The manufacturer's confirm compliance to IEEE Std. C37-013 [11] and can supply test certificates for asymmetrical current interruption with 130% asymmetry factor.

7.9 Calculation Procedure

The calculation procedure is described for hand calculations, which is instructive. Reproduction of computer printouts is avoided. Not much data preparation is required for present-day computer-based calculations. Most programs will accept *raw* impedance data, through a graphic user interface; the calculation algorithms tag it with respect to the short-circuit type and apply appropriate impedance multiplying factors, depending on the calculation type. Depending upon the in-built libraries, positive, negative, and zero sequence impedances for each component are generated and converted to a common MVA base. Transformer ratios and tap settings can be accounted for and the adjusted impedances calculated. Matrix equations are solved and results presented in a user-friendly format. These include fault point complex impedance, X/R ratio, and magnitude and phase angles of all the contributions to a bus from the adjacent buses. Remote and local components of the currents, fault voltages, NACD ratio, and interrupting duty and asymmetrical multiplying factors are tabulated for each faulted bus. Based on the input short-circuit ratings of the switching devices, all devices, which are overdutied from short-circuit considerations, can be flagged and the percentage over duty factors plotted. A variety of user-friendly output format are available including tabulation of all input data, conversions to common MVA base, all generator and motor data, their subtransient reactances, outputs of sequence impedances, NACD ratios, and calculated weighted multiplying factors for breakers of different contact parting times.

7.9.1 Necessity of Gathering Accurate Data

Though the digital computer-based calculations have become user friendly and are necessary for large systems consisting of thousands of buses and machines, the accuracy of the input data must be ensured. As an example, omitting even short lengths of cables in low-voltage systems can seriously impact the results. Gathering accurate impedance data may be time consuming. For example ascertaining correct X/R ratio for the current limiting reactors is important. These rectors have high X/R ratios, and there can be large differences with respect to the actual manufacturer's data and the built in computer database. This can impact the short-circuit asymmetry and the calculated results.

7.9.2 Calculations—Step by Step

This can be summarized in the following steps:

1. A single-line diagram of the system to be studied is required as a first step. It identifies impedances of all system components as pertinent to the short-circuit calculations. For hand calculation, a separate impedance diagram may be constructed, which follows the pattern of a single-line diagram with impedances and their X/R ratios calculated on a common MVA base. The transformer voltage ratios may be different from the base voltages considered for data reduction. The transformer impedance can be adjusted for transformer voltage adjustment taps and voltage ratios, see Chapter 11, modeling of transformers for tap ratios.
2. Appropriate impedance multiplying factors are applied from Table 7.1, depending on the type of calculation. For high-voltage breakers, at least two networks are required to be constructed, one for the first-cycle calculations and the other for the interrupting duty calculations.
3. A fault point impedance positive sequence network (for three-phase faults) is then constructed, depending on the location of the fault in the system. Both resistances and reactances can be shown in this network, or two separate networks, one for resistance and the other for reactance, can be constructed.
4. For unsymmetrical faults, similarly, zero sequence and negative sequence impedance networks can be constructed and reduced to single impedance at the fault point, Chapters 2 and 3. For rotating machines, positive and negative sequence impedances are not equal, Chapters 2 and 12. Computer programs will calculate these networks based upon raw equipment data inputs, based upon the built in databases, though a user can overwrite the numbers with better/manufacturer's data.
5. For E/Z complex calculation, the fault point positive sequence network is reduced to single impedance using complex phasors. Alternatively, the resistance and reactance values obtained by reducing separate resistance and reactance networks to a single-point network to calculate the fault point X/R ratio can also be used for E/Z calculation. This considerably simplifies hand calculations, compared to complex impedance reduction.
6. If there are many sources in the network, NACD is required to be calculated and that sets a limit to the complexity of networks, which can be solved by hand calculations. The currents from NACD sources have to be traced throughout the system to the faulty node to apply proper weighting factors, and this may not be easy in interconnected networks. The calculation of the first-cycle duty does not require considerations of remote or local.
7. The adjusted currents, thus calculated, can be used to compare with the short-circuit ratings of the existing switching equipment or selection of new equipment.

7.9.3 Analytical Calculation Procedure

7.9.3.1 Hand Calculations

In simple systems, with limited buses not more than 6–8, the currents from the various sources can be vectorially summed at the fault point. No preimpedance multiplying factors and postmultiplying duty factors are required. The dc component of the short-circuit current is calculated. The time constants associated with ac and dc decay are required to calculate the currents at the contact parting time of the breaker. This may not be always easy. Once each of the components is calculated, the theorem of superimposition applies, and the total currents can be calculated for circuit-breaker duties.

7.9.3.2 Dynamic Simulation

Alternative methods of calculations of short-circuit currents that give accurate results are recognized. The short-circuit calculations can be conducted on EMTP or other digital computer programs, which could emulate the transient behavior of machines during short-circuit conditions. In this sense, these simulations become more like transient stability analyses carried out in the time domain. The behavior of machines as influenced by their varying electrical and mechanical characteristics can be modeled, and the calculation accuracy is a function of the machine models, static and dynamic elements as well as the assumptions. Synchronous machines can be modeled in much greater details, IEEE Standard [15] details out the synchronous machine models for transient analysis studies. This may result in a number of differential equations, so that harmonics and dc offset effects can be accounted for. The inclusion of voltage regulators and excitation systems is possible. Dynamic simulation puts heavy demand on the computing resources as well as on accurate data entries, yet this forms an important verification tool for the empirical calculations.

Variations in the results of the short-circuit calculations have been studied by various authors [16–18]. An example is provided in Chapter 8. Table 7.3 shows these variations for

TABLE 7.3

Comparison of Short-Circuit Calculations, ANSI/IEC/EMTP, and Dynamic Simulation Methods

Bus kV	ANSI	IEC	EMTP/Dynamic Simulation	Remarks	Reference
13.8	23.39		17.81[a]	Breaking/interrupting current at 50 ms ac + dc	[16]
	24.87		18.52[a]		
	25.49		20.16[a]		
4.16	46.58		45.70[a]		
	35.37		36.44[a]		
13.8	15.45	16.31		E/Z symmetrical current	[17]
	15.36	15.21			
2.4	21.47	19.09			
220	20.99	22.57	20.30[b]	E/Z symmetrical current	[18]
21	74.35	74.43	60.23[b]		
10	25.98	23.68	17.75[b]		

All currents are in kA rms symmetrical.

[a] The dynamic simulation results.

[b] EMTP simulation results.

the same system configurations using different methods of calculations. The table shows that generally ANSI/IEEE calculations are more conservative when compared with dynamic simulation or EMTP. This table shows that in some cases ANSI calculations are 30% higher than the corresponding results obtained with dynamic simulation. Due to large variations documented in this table, the following question arises: Can the calculated overduty by a certain percentage be ignored? This will not be a prudent approach, unless rigorous alternate techniques are explored and are acceptable to the owners of an establishment.

7.9.4 Devices with Sources on Either Side

Circuit breakers are often applied when there are short-circuit sources on either side of the circuit breaker. In such cases, short-circuit duty need be calculated for the maximum through fault current. This is illustrated with reference to Figures 5.18 and 5.19.

Figure 5.18 shows two fault locations F1 and F2. Fault at location F1 is fed from (1) the generator, (2) from transformer GSU_3, and also (3) from transformer UAT2. But the generator breaker sees only the components (2) and (3) and not its own contribution to fault at F1. For a fault at location F2, again, there are three components of the fault current as before, but the generator breaker does not see components (2) and (3) and sees only its own component. Thus, its duty should be based upon the highest of these components in either direction. A similar explanation applies to faults at F1 and F2 in Figure 5.19. Faults on either side of a tie breaker should be considered and the first cycle and interrupting duty based upon the largest of the fault currents. Apart from the absolute magnitude of the currents the X/R ratio is also important. It is possible that smaller of the two currents, which has a high-X/R ratio can give a higher duty compared to the higher magnitude short-circuit current with a lower X/R ratio.

Figure 5.19 shows a number of feeder circuit breakers connected to 12.47 kV bus, which serve motor loads. Then, for a fault on the bus, and applying superimposition theorem, the total short-circuits current is:

$$\vec{I}_{t,bus} = \vec{I}_{gen} + \bar{I}_{trans} + \vec{I}_{F1} + \vec{I}_{F2} + \cdots \tag{7.16}$$

A feeder circuit breaker selected on this basis is conservatively applied, and this method is appropriate for the new installations. When evaluating an existing system, where the circuit breakers are applied close to their short-circuit ratings, the duties should be calculated based upon maximum through fault current. Say for calculation of the interrupting duty of a feeder breaker, the entire component short-circuit currents should be considered, less the short-circuit current contributed by the loads connected to that breaker itself:

$$\bar{I}_{F1\,max} = \bar{I}_{t,bus} - \bar{I}_{F1} \tag{7.17}$$

A highly loaded feeder circuit breaker will have a lower short-circuit duty as compared to a lightly loaded feeder circuit breaker. Practically, for a switchgear lineup, a uniform ratings of switching devices will be selected, based upon the maximum short-circuit duty that any one breaker may experience. Generally, 10%–15% additional margin is allowed in the selection of the ratings for future growth. Utility source impedances invariably decrease over the course of time.

However, for the first cycle duty, entire short-circuit components on the bus should be considered, as this bus runs common through the entire lineup of the circuit breakers. Some available commercial programs will calculate the short-circuit duties on individual breakers on a common bus depending upon the maximum duty imposed by the through fault currents in either direction. Hand calculations can also be made, but these will be laborious.

7.9.5 Switching Devices without Short-Circuit Interruption Ratings

A switching device may not be designed for short-circuit current interruption. For example, high-voltage disconnect switches that are designed to interrupt the transformer magnetizing currents but not the short-circuit currents. These are interlocked with the appropriate circuit breakers. Another example in industrial distribution systems is NEMA type E2 motor starters for medium voltage motors. These are fused with current limiting type R fuses, which have a symmetrical interrupting rating of 50 kA, while the contactor itself has a short-circuit interrupting rating of only a few kA. The fuse characteristics are coordinated to protect the contactor. The transformer primary fused load-break switches can interrupt the rated load current, but not the short-circuit current. These are protected by the current limiting fuses in the same enclosure. Now consider a transformer load break switch without fuses. It should still have the first cycle or momentary rating, though does not have an interrupting rating. It may be closed on to a fault and should have fault closing rating for certain duration. Calculations of first cycle duties are still required for such devices.

Similarly, bus ducts, cables, and transformers can be cited as examples of the equipment that should withstand through fault currents. Withstand capability curves of transformers category I through IV are in ANSI/IEEE standard [19]. The short-circuit protection of transformers is achieved through proper relaying practices, not discussed in this book.

7.9.6 Capacitor and Static Converter Contributions to Short-Circuit Currents

A number of dynamic simulations of capacitors are presented in [4]. During fault, the capacitor discharge takes place in a very short time of 1/30–1/8 cycles. The electromagnetically induced forces of discharge current are instantaneously proportional to square of the current. The capacitor discharge currents will have no effect on breaker parting or clearing operations. Some stresses may be imposed for breaker closing and latching duty. The standard also considers rectifier, inverter, cycloconverter, and chopper circuits and concludes that for faults in the ac systems, dc systems provide fault currents:

1. When dc system has a fault current source such as motors, batteries, or photovoltaic cells.
2. When the converter operates as an inverter.
3. For the first cycle, until the grid protection system operates and magnitude is approximately three times the ac three-phase apparent power input of the converter transformer at rated machine dc load and an X/R ratio of 10. Also, see Chapter 8 for IEC specifications.

7.10 Examples of Calculations

Example 7.1

Example 7.1 is for calculations in a multivoltage level distribution system (Figure 7.13). Short-circuit duties are required to verify the adequacy of ratings of the switching devices shown in this single-line diagram. These devices are

1. 13.8 kV circuit breakers at buses 1, 2, and 3
2. 138 kV circuit breakers
3. 4.16 kV circuit breakers at bus 6
4. Primary switches of transformers T_3 and T_4
5. Primary fused switches of transformers T_5 and T_6
6. Type R fuses for medium-voltage motor starters, buses 7 and 8
7. Low-voltage power circuit breakers, bus 10
8. ICCB at bus 10 (10F5)
9. Molded case circuit breakers at low-voltage motor control center, bus 14

Also, calculate bus bracings, withstand capability of 13.8 kV #4 ACSR overhead line conductors connected to feeder breaker 2F3 with transformer T_7 and #4/0 cable C_1 connected to feeder breaker 2F4.

In this example, emphasis is upon evaluation of calculated short-circuit duties with respect to the equipment ratings. Three-phase fault calculations are required to be performed.

Calculation of Short-Circuit Duties

The impedance data reduced to a common 100 MVA base are shown in Table 7.4. Note that no impedance multiplying factors are applied—this table is merely a conversion of the raw equipment impedance data to a common 100 MVA base. Depending upon the type of calculation, the impedance multiplying factors from Table 7.1 are applied to the calculated impedances in Table 7.4. The fault point networks for various faulted buses can be constructed, one at a time, and reduced to a single network. As an example, the fault network for the 13.8 kV bus 2 is shown in Figure 7.14. Reducing it to single impedance requires wye-delta impedance transformation. The simplicity, accuracy, and speed of computer methods of solution can be realized from this exercise. The reduced complex impedance for interrupting duty calculations for bus 2 fault is $Z = 0.003553 + j0.155536$, and X/R from separate networks is 47.9; $E/Z = 26.892 < -88.7°$. All the generator contribution of 14.55 kA is a local source as the generator is directly connected to the bus. The remote (utility) source contributes to the fault at bus 2 through transformers T_1 and T_2 and synchronizing bus reactors. The utility's contributions through transformers T_1 and T_2 and synchronizing bus reactors are summed up. This gives 9.01 kA. The remote/total ratio, i.e., NACD ratio is 0.335. The multiplying factor is 1.163, and the interrupting duty is 31.28 for a five-cycle symmetrical breaker. If the calculation is based on a separate R–X method, the fault point impedance is $0.003245 + j0.15521$. This gives $E/Z = 26.895$ kA. There is not much difference in the calculations by using the two methods, though a difference up to 5% can occur. The results of calculations are shown in Tables 7.5 through 7.11.

Table 7.5 shows the interrupting and close and latch rating of the breakers according to revised IEEE standards, $K = 1$. The peak first cycle current is calculated using Equation 7.1

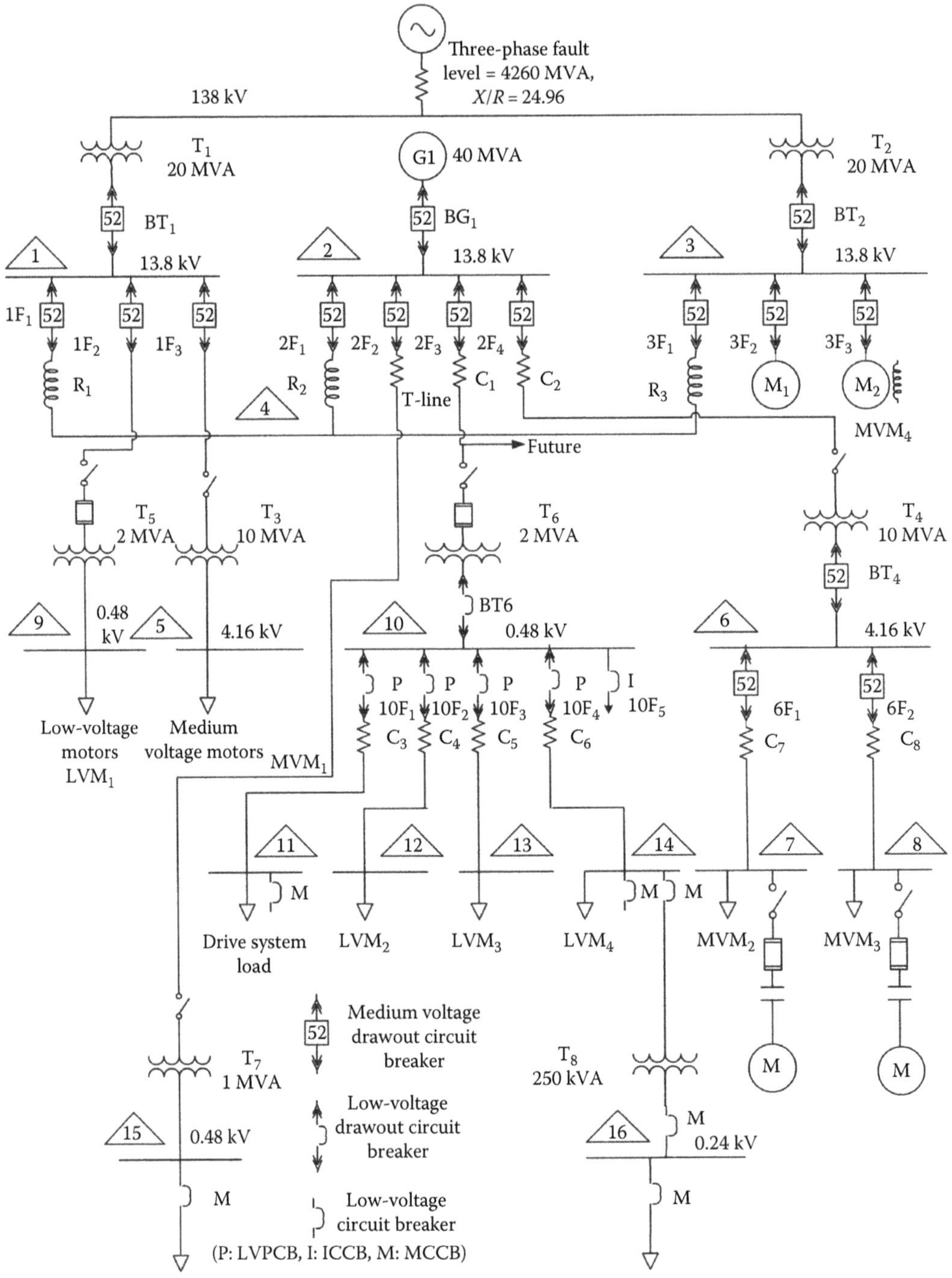

FIGURE 7.13
Single-line diagram of a multivoltage level distribution system for short-circuit calculations (Example 7.1).

TABLE 7.4

Impedance Data (100 MVA Base) Distribution System (Example 7.1)

Symbol	Equipment Description	Per Unit Resistance	Per Unit Reactance
U_1	Utility source, 138 kV, 4260 MVA, $X/R=25$	0.00094	0.02347
G_1	Synchronous generator, 13.8 kV, 40 MVA, 0.85 power factor, saturated subtransient reactance = 11.5%, saturated transient = 15%, $X/R=56.7$	0.00507	0.28750
R_1, R_2, R_3	Reactors, 13.8 kV, 2 kA, 0.25 ohm, 866 KVA, $X/R=88.7$	0.00148	0.13127
T_1, T_2	20/33.4 MVA, ONAN/ONAF, 138–13.8 kV, delta-wye transformers, $Z=8.0\%$ on 20 MVA ONAN rating, $X/R=21.9$, wye winding low resistance grounded through 400 A, 10 s resistor	0.01827	0.39958
T_3, T_4	10/14 MVA, ONAN/ONAF, 13.8–4.16 kV, delta-wye transformer, $Z=5.5\%$, $X/R=15.9$, wye-winding low-resistance grounded through 200 A 10 s resistor	0.03452	0.54892
T_5, T_6	2/2.58 MVA, ONAN/ONAF, 13.8–0.48 kV, delta-wye transformer, $Z=5.75\%$, $X/R=6.3$, wye-winding high-resistance grounded	0.44754	2.83995
T_7	1/1.29 MVA, AA/FA, 13.8–0.48 kV delta-wye transformer, $Z=5.75\%$, $X/R=5.3$, wye-winding high-resistance grounded	1.06494	5.65052
T_8	250 KVA, AA, 0.48–0.24 kV delta-wye transformer, $Z=4\%$, $X/R=2.7$, solidly grounded	5.49916	15.02529
C_1	1–3/C #4/0 15 kV grade shielded, MV-90, IAC (interlocked armor), XLPE cable laid in aluminum tray, 200 ft	0.00645	0.00377
C_2	1–3/C 500 KCMIL, 15 kV grade shielded, MV-90, IAC, XLPE cable laid in aluminum tray, 400 ft	0.00586	0.00666
C_7, C_8	2–3/C, 350 KCMIL, 5 kV grade, shielded, MV-90, XLPE, IAC cable, laid in aluminum tray, 100 ft	0.01116	0.00790
C_3, C_4, C_5, C_6	3–3/C, 500 KCMIL, 0.6 kV grade, THNN, 90°C cables, laid in tray, 60 ft	0.23888	0.23246
M_1	1 × 12,000 hp, squirrel cage induction motor, 2 pole (10,800 kVA), locked rotor reactance = 16.7%, $X/R=46$	0.0336	1.54630
M_2	1 × 10,000 hp synchronous motor, 8-pole, 0.8 power factor (10,000 KVA), $X=20\%$, $X/R=34.4$	0.05822	2.0000
MVM_1	2 × 1500 hp squirrel cage induction, 2-pole,	0.21667	6.1851 (2 × 1500 hp)
	5 × 300 hp induction motors, 4-pole, and	0.77958	11.719 (5 × 300 hp)
	1 × 3500 hp, 12 pole, 0.8 power factor synchronous motor	0.20324	5.7142 (3500 hp)
MVM_2, MVM_3	1 × 1800 hp, 12 pole, 0.8 power factor synchronous and 3 × 1100 hp, 6-pole induction motors	0.45904	11.111 (1800 hp)
		0.21675	5.6229 (3 × 1100 hp)
LVM_1	Grouped 1350 hp induction motors ≥50 hp and 640 hp induction motors <50 hp	1.4829	12.370 (>50 hp)
		5.4362	26.093 (<50 hp)
LVM_2,	Grouped 300 hp induction motors ≥50 hp and	6.0664	50.606 (>50 hp)
LVM_3, LVM_4	172 hp motors, <50 hp	20.227	97.093 (<50 hp)
L1	5200 ft long, GMD = 4 ft, ACSR conductor A-AA class (SWAN), #4 AWG [24]	1.41718	0.41718

2–3/C: Abbreviated for two three-conductor cables per phase. MV-90, XLPE, THNN: Cable insulation types. See Ref. [20]. The impedance of a cable is also a function of its construction and method of installation.

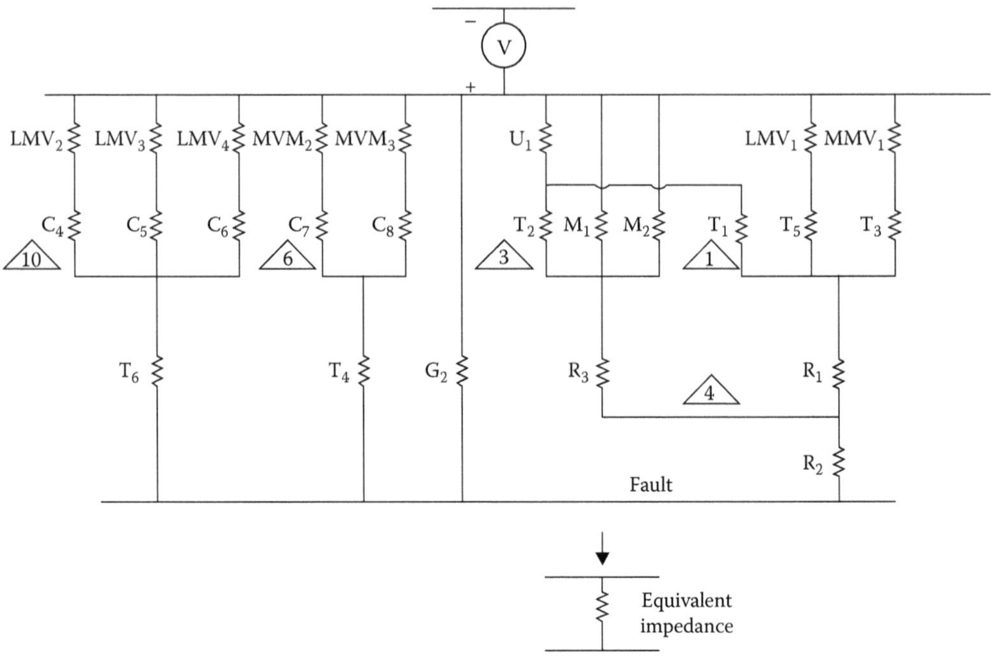

FIGURE 7.14
Positive sequence impedance diagram for a fault at bus 2 (Example 7.1).

for the calculated X/R ratio. *The interrupting duty is based upon bus fault current*. Table 7.6 shows contributions to this bus from all other buses. The vector sum of these currents gives faulted bus 2 total current.

15 kV circuit breakers. If old ratings of the breakers with $K = 1.3$ is considered Ref. [21], then for a standard five-cycle symmetrical breaker rating of 15 kV, rated short-circuit current is 28 kA, and close and latch is 97 kA peak or 58 kA rms asymmetrical. The interrupting duty current at the voltage of application, 13.8 kV will be 30.43 kA rms symmetrical. The calculated interrupting duty currents shown in Table 7.5, the multiplying factors, the X/R ratios, NACD ratio remains the same as newly rated breakers. The first cycle duty at the calculated X/R gives a multiplying factor of 1.659, and the first cycle duty of 46.48 kA rms asymmetrical. This shows that for these K-rated breakers, interrupting duties on bus 2 exceed the breaker ratings by 2.79%, though the close and latch duty of 46.48 kA rms asym is much below the breaker ratings. It would be hasty to suggest that the entire bus 2 switchgear be replaced.

As the duties are calculated for a bus fault, it is necessary to calculate the duties on individual breakers on this bus, by dropping the short-circuit contributions from the loads connected to each of these breakers. Neglecting the load contribution means that the fault point X/R ratio and duty multiplying factors will change. The load current component of the feeder breaker can be vectorially subtracted from the bus fault current. Table 7.7 is compiled on this basis and shows the interrupting duties on the feeder breakers connected to bus 2. It is observed that the feeder breaker 2F2 to the transmission line and the feeder breaker 2F3 to the 2 MVA transformer are overdutied. As the feeder breaker 2F2 has no rotating loads, its interrupting duty is the same as that for a bus fault. It is generally possible to retrofit these two breakers with breakers of higher interrupting rating, depending on the manufacturer and age of the equipment. The newly rated $K = 1$ breakers can usually be fitted in the same breaker cubicles with minor modifications. Increasing bus tie

TABLE 7.5

Calculated Duties on 13.8 kV Breakers (Example 7.1)

				Calculate First-Cycle Duties (Close and Latch Capabilities)				Calculate Interrupting Duties			
Bus No.	Breaker Type	Breaker Close and Latch Capability (kA, peak)	Breaker Interrupting Rating (kA sym rms)	Fault Point Z Based on Complex Impedance Reduction	Fault Point X/R Ratio Based on Separate Networks	Multiplying Factor Based on X/R	Calculated Duty (kA peak)	Fault Point Z Based on Complex Impedance Reduction	Fault Point X/R Ratio	NACD/ Weighted Multiplying Factor	Calculated Duty (kA rms sym)
1	15 kV, indoor	82	31.5	0.005767 + j0.175930	41.61	2.727	64.81	0.005831 + j0.18426	41.73	0.6/1.169	26.53
2	Oil-less, five-cycle	Ref. [21], $k=1$		0.003578 + j0.149318	47.84	2.740	76.75	0.003553 + j0.155536	47.92	0.335/1.163	31.28
3	Sym $K=1$			0.004903 + j0.159347	41.05	2.726	71.54	0.005291 + j0.171520	41.16	0.571/1.162	28.33

TABLE 7.6

Flow of Component Interrupting Duty Short-Circuit Currents from Connected Buses to 13.8 kV Bus 2 (Example 7.1)

From Bus	Current Amplitude and Angle	X/R
Gen. 2	$24.550 < -88.988°$	56.61
Bus 4	$11.021 < -88.491°$	37.96
Bus 6	$1.244 < -87.338°$	21.51
Bus 10	$0.078 < -82.965°$	8.10
Bus 15	$0 < 0°$	—
Vector sum of currents	$26.9 < -88.805$	47.92

TABLE 7.7

Calculated Duties on Feeder Circuit Breakers rated $K > 1$ on 13.8 kV Bus 2 (Example 7.1)

Breaker ID	Breaker Service	Breaker Rating Interrupting (kA, sym rms)	Breaker Calculated Interrupting Duty (kA sym rms)
BG2	Generator breaker	30.43	16.94
2F1	Synchronous bus		18.68
2F2	13.8 kV distribution line		31.28 (overexposure = 2.79%)
2F3	2 MVA transformer		31.18 (overexposure = 2.46%)
2F4	10 MVA transformer		30.03

reactor impedance by providing another reactor in series with the existing reactor could be another solution, especially if a system expansion is planned. Some remedial measures to the short-circuit problems are

- Retrofitting overdutied breakers with new breakers
- Replacing with new $K = 1$ breakers or entirely new switchgear of higher ratings
- Adding current-limiting reactors
- Redistribution of loads and reorganization of distribution system
- Short-circuit current limiters [22]
- Duplex reactors (see Appendix C)
- Series connected devices for low-voltage systems (see Section 5.14.3.3)

A detailed discussion of these topics is not covered.

4.16 kV circuit breakers and motor starters. Table 7.8 shows 4.16 kV metal-clad circuit breaker ratings and calculated duties for K-factor-rated breakers, and Table 7.9 shows a similar comparison of R-type fuses in medium-voltage motor starters, NEMA type E2. These devices are applied much below their short-circuit ratings. Note that, for the fuses in the medium voltage starters, it is not necessary to construct the interrupting duty network.

Transformer primary switches and fused switches. Short-circuit ratings of transformer primary switches (without fuses) are specified in terms of asymmetrical kA rms and 10-cycle fault closing. The former rating indicates the maximum asymmetrical withstand current capability, and the latter signifies that the switch can be closed onto a fault for 10 cycles, with maximum fault limited to specified asymmetrical fault closing rating. The upstream

TABLE 7.8

Calculated Duties on 4.16 kV Breakers (Example 7.1)

				Calculate First-Cycle Duties (Close and Latch Capabilities)				Calculate Interrupting Duties			
Bus No.	Breaker Type	Breaker Close and Latch Capability (kA, asym rms)	Breaker Interrupting Rating at Voltage of Application (kA sym rms)	Fault Point Z Based on Complex Impedance Reduction	Fault Point X/R Ratio	Multiplying Factor Based on X/R	Calculated Duty (kA rms asym)	Fault point Z Based on Complex Impedance Reduction	Fault Point X/R Ratio	NACD/ Weighted Multiplying Factor	Calculated Duty (kA rms sym)
6	4.16 kV, indoor, oil-less, five cycle sym	58 (=97 kA crest) $K=1.24$, Ref. [21]	30.43	$0.029041 + j0.517160$	18.41	1.649	41.70	$0.032647 + j0.572088$	18.03	0.475/1.00	24.22

TABLE 7.9

Short-Circuit Duties on 4.16 kV MCC (Example 7.1)

			Calculated First-Cycle Duties (Sym and Asym)			
Bus No.	Motor Starter Fuse Type	Fuse Interrupting Rating (kA, rms sym/asym)	Fault Point Z Based on Complex Impedance Reduction	Fault Point X/R Ratio	Multiplying Factor Based X/R	Calculated Duty (kA rms sym/asym)
7, 8	Current-limiting type R	50/80	$0.037043 + j0.522930$	15.09	1.523	26.47/40.31

protective devices must isolate the fault within 10 cycles. The short-circuit ratings on power fuses are discussed in Chapter 5. First-cycle calculations are required. Table 7.10 shows the comparison and that the equipment is applied within its short-circuit ratings.

Low-voltage circuit breakers. The switching devices in low-voltage distribution should be categorized into low-voltage power circuit breakers (LVPCBs), insulated case circuit breakers (ICCBs), and molded case circuit breakers (MCCBs), as discussed in Chapter 5. These have different test power factors, depending on their type and ratings, and interrupting duty multiplying factors are different. First-cycle calculation is required for ascertaining the duties. Table 7.11 shows these calculations. It is observed that the short-circuit duties on MCCBs at buses 11 and 16 exceed the ratings. A reactor in the incoming service to these buses can be provided. Alternatively, the underrated MCCBs can be replaced, or series rated devices (Chapter 5) can be considered, as there are no downstream short-circuit contributions to these buses.

Bus bracings. Bus bracings are, generally, specified in peak and symmetrical rms ampères and are indicative of mechanical strength under short-circuit conditions. The mechanical stresses are proportional to I^2/d, where d is the phase-to-phase spacing. First-cycle symmetrical current is, therefore, used to compare with the specified bus bracings. In terms of asymmetrical current, the bus bracings are 1.6 times the symmetrical current. Both the symmetrical and asymmetrical calculated values should be lower than the ratings. It is sometimes possible to raise the short-circuit capability of the buses in metal-clad switchgear by adding additional bus supports.

TABLE 7.10

Example 7.1: Calculated Duties 13.8 kV Transformer Primary Switches and Fuses

Transformer	Transformer Primary Switch/Fused Switch	Short-Circuit Ratings	Fault Point Impedance (100 MVA Base)	Fault Point X/R	Symmetrical Current (kA rms)	Asymmetrical Current (kA rms)
10 MVA	Switch only	Switch 61 kA rms asym Fault closing 10 cycles = 61 kA rms asym	$0.008709 + j0.153013$	18.48	26.926	41.92
2 MVA	Fused switch, with current-limiting type class E fuse	Fuse: interrupting = 50 kA rms, sym = 80 kA rms, asym	$0.009926 + j0.153013$	16.04	27.285	41.84

TABLE 7.11

Short-Circuit Duties on Low-Voltage Circuit Breakers (Example 7.1)

Breaker Identification	Breaker Interrupting Rating (kA sym)	Fault point Z per Unit (100 MVA Base)	Fault Point X/R	Multiplying Factor	E/Z	Calculated Duty (kA Sym)
Bus 10, LVPCB	50	0.37625 + j2.491190	6.65	1.002	47.73 < −81.36°	47.82
Bus 10, ICCB	65	0.37625 + j2.491190	6.65	1.063	47.73 < −81.36°	50.73
Bus 14, $MCCB_1$	65	0.588317 + j2.697419	4.68	1	43.57 < −77.70°	43.57
Bus 11, MCCB	35	0.617505 + j2.723650	4.44	1	43.07 < −77.23°	43.07
Bus 15, MCCB	35	2.480748 + j6.217019	2.51	1	17.97 < −68.25°	17.97
Bus 16, 240 V MCCB	10	6.087477 + j17.72271	2.92	1	12.84 < 71.04°	12.84

The short-circuit ratings (rated short-circuit withstand current and rated short-time current) of nonphase segregated, phase-segregated, and isophase metal-enclosed buses are specified in IEEE standard C37.23 [23]. The rated short-circuit withstand current is specified in kA asym for 167 ms duration. Rated short-time current for isophase buses is specified in kA sym for 1 s duration.

Power cables. Power cables should be designed to withstand short-circuit currents, so that these are not damaged within the total fault clearing time of the protective devices. During short circuit, approximately, all heat generated is absorbed by the conductor metal, and the heat transfer to insulation and surrounding medium can be ignored. An expression relating the size of copper conductor, magnitude of fault current, and duration of current flow is

$$\left(\frac{I}{CM}\right)^2 tF_{\text{ac}} = 0.0297\,\log_{10}\frac{T_{\text{f}} + 234}{T_0 + 234} \tag{7.18}$$

where

I is the magnitude of fault current in A
CM is the conductor size in circular mils
F_{ac} is the skin effect ratio or ac resistance/dc resistance ratio of the conductor
T_{f} is the final permissible short-circuit conductor temperature, depending on the type of insulation
T_0 is the initial temperature prior to current change

For aluminum conductors, this expression is

$$\left(\frac{I}{CM}\right)^2 tF_{\text{ac}} = 0.00125\,\log_{10}\frac{T_{\text{f}} + 228}{T_0 + 228} \tag{7.19}$$

where F_{ac} is given in Table 7.12 [25]. The short-circuit withstand capability of 4/0 (211600 CM) copper conductor cable of 13.8 kV breaker 2F4 has a short-circuit withstand capability

TABLE 7.12

AC/DC Resistance Ratios: Copper and Aluminum Conductors at 60 Hz and 65°C

Conductor Size (KCMIL or AWG)	5–15 kV Nonleaded Shielded Power Cable, 3 Single Concentric Conductors in Same Metallic Conduit	
	Copper	Aluminum
1000	1.36	1.17
900	1.30	1.14
800	1.24	1.11
750	1.22	1.10
700	1.19	1.09
600	1.14	1.07
500	1.10	1.05
400	1.07	1.03
350	1.05	1.03
300	1.04	1.02
250	1.03	1.01
4/0	1.02	1.01
3/0	1.01	<1%
2/0	1.01	<1%

of 0.238 s. This is based on an initial conductor temperature of 90°C, a final short-circuit temperature for XLPE (cross-linked polyethylene) insulation of 250°C, and a fault current of 31.18 kA sym (F_{ac} from Table 7.12 = 1.02). The breaker interrupting time is five cycles, which means that the protective relays must operate in less than nine cycles to clear the fault. Major cable circuits in industrial distribution systems are sized, so that these are not damaged even if the first zone of protective relays (instantaneous) fails to operate and the fault has to be cleared in the time-delay zone of the backup device. From these criteria, the cable may be undersized.

Overhead line conductors. Calculations of short-circuit withstand ratings for overhead line conductors must also receive similar considerations as cables, i.e., these should be sized not only for load current and voltage drop consideration, but also from short-circuit considerations. For ACSR (aluminum conductor steel reinforced) conductors, a temperature of 100°C (60°C rise over 40°C ambient) is frequently used for normal loading conditions, as the strands retain approximately 90% of rated strength after 10,000 h of operation. Under short circuit, 340°C may be selected as the maximum temperature for all aluminum conductors and 645°C for ACSR, with a sizable steel content. An expression for safe time duration based on this criterion and no heat loss during short circuit for ACSR is given in [24]:

$$t = \left(0.0862\frac{CM}{I}\right)^2 \tag{7.20}$$

where

t is the duration in s

CM is the area of conductor in circular mils

I is the current in A, rms

From Equation 7.20 # 4 (41740 CM) ACSR of the transmission line connected to breaker 2F2 has a short-circuit withstand capability of 0.013 s for a symmetrical short-circuit current of 31.28 kA close to bus 2. The conductors, though adequately sized for the load current of a 1 MVA transformer, are grossly undersized from short-circuit considerations.

Example 7.2

This example explores the problems of application of a generator circuit breaker for high-X/R ratio fault, when the natural current zero is not obtained because of high asymmetry. The calculation is based upon IEEE standard [11]. Figure 7.15 shows a generating station with auxiliary distribution system. A 112.1 MVA generator is connected through a step-up transformer to supply power to a 138 kV system. The generator data are the same as presented in Table 6.1. Auxiliary transformers of 7.5 MVA, 13.8–4.16 kV, and 1.5 MVA, 4.16–0.48 kV supply medium- and low-voltage motor loads. The generator neutral is high-resistance grounded, through a distribution transformer, the 4.16 kV system is low-resistance grounded, and the 480 V system is high-resistance grounded. Thus, breaker duties are based on three-phase fault currents. The generator auxiliary distribution systems are designed with double-ended substations, for reliability and redundancy. Here, we ignore the alternate source of power.

Three-Phase Fault at F1 (138 kV), E/Z Method

The fault at F1 is fed by three sources: utility source, generator, and the motor loads from the auxiliary distribution. The utility contribution predominates and for the selection of 138 kV breaker, the generator contribution and motor contribution can be ignored. However, this breaker will also be required to interrupt the generator and motor contribution currents through the step-up transformer and TRV can be of concern. Table 7.13 gives the impedance data for all the system components broken down in per unit R and X on a common 100 MVA base. The calculation is carried out on per unit basis, and the units are not stated at each step. The effective generator resistance is calculated from the following expression in Table 7.2:

$$\text{Effective resistance} = \frac{X_{2\text{v}}}{2\pi f T_{\text{a3}}} \tag{7.21}$$

Using the generator data from Table 6.1, $X/R = 130$. Correct input of X/R ratios for generators and large reactors is important, as it may have a pronounced effect on ac and dc decay.

Appropriate impedance multiplying factors from Table 7.1 are used before constructing the positive sequence fault point network. The impedance multiplying factor for calculation of first cycle or interrupting duties is 1 for all turbogenerators. The motor impedances after appropriate multiplying factors for first cycle and interrupting duty calculation are shown in Table 7.14. A positive sequence impedance network to the fault point under consideration may now be constructed, using modified impedances. Low-voltage motors of <50 hp are ignored for interrupting duty calculations. Figure 7.16 shows the interrupting duty network for a fault at F1.

The result of reduction of impedance of the network is shown in Figure 7.16 with complex phasors gives an interrupting duty impedance of $Z = 0.001426 + j0.020318$ per unit.

The X/R ratio is calculated by separate resistance and reactance networks, i.e., the resistance network is constructed by dropping out the reactances in Figure 7.16, and the

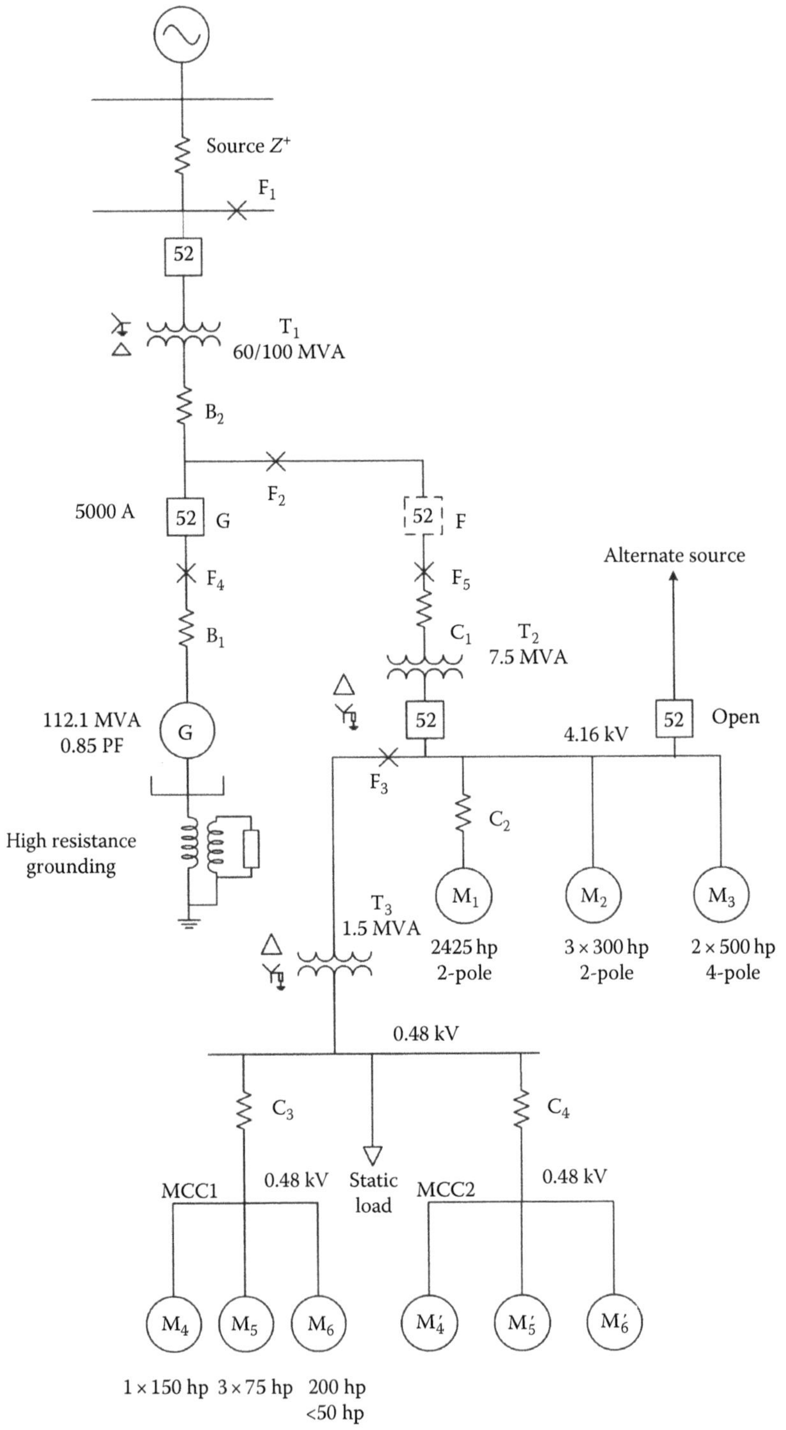

FIGURE 7.15
A generating station single-line diagram for short-circuit calculations (Example 7.2).

TABLE 7.13

Impedance Data (Example 7.2)

Description of Equipment	Per Unit Resistance on a 100 MVA Base	Per Unit Reactance on a 100 MVA Base
Utility's 138 kV source, three-phase fault level = 4556 MVA, X/R ratio = 13.4	0.00163	0.02195
112.1 MVA generator, saturated sub transient = 16.4% data in Table 6.1	0.001133	0.14630
Transformer T_1, 60/100 MVA, Z = 7.74%, X/R = 32	0.00404	0.12894
Transformer T_2, 7.5 MVA, Z = 6.75%, X/R = 14.1	0.06349	0.89776
Transformer T_3, 1.5 MVA, Z = 5.75, X/R = 5.9%	0.63909	3.77968
13.8 kV cable C_1, 2-1/C per phase, 1000 KCMIL, in steel conduit, 80 ft	0.00038	0.00101
4.16 kV cable C_2, 1-1/C per phase, 500 KCMIL, in steel conduit, 400 ft	0.06800	0.10393
0.48 kV cables C_3 and C_4, 3-1/C per phase, 750 KCMIL, in steel conduit, 150 ft	0.46132	0.85993
M_1, 2425 hp, 2-pole induction motor	0.23485	7.6517
M_2, 300 hp, 2-pole induction motors, 3 each	1.2997	19.532
M_3, 500 hp, 2-pole induction motors, 2 each	0.90995	17.578
M_4 and M_4', 150 hp, 4-pole induction motor	11.714	117.19
M_5 and M_5', 75 hp, 4-pole, induction motor, 3 each	10.362	74.222
M_6 and M_6', 200 hp induction motors, lumped, < 50 hp	20.355	83.458
B_1, 5 kA bus duct, phase-segregated, 40 ft	0.00005	0.00004
B_2, 5 kA bus duct, phase-segregated, 80 ft	0.00011	0.00008

TABLE 7.14

Example 7.2: Motor Impedances After Multiplying Factors in PU at 100 MVA Base

Motor ID	Quantity	Interrupting Duty MF	First-Cycle MF	Interrupting Duty Z per Unit, 100 MVA Base	First-Cycle Z per Unit, 100 MVA Base
M_1	1	1.5	1.0	$0.352 + j11.477$	$0.2348 + j7.6517$
M_2	3	1.5	1.0	$1.949 + j29.298$	$1.2997 + j19.532$
M_3	2	3.0	1.2	$2.7298 + j52.734$	$1.0919 + j21.094$
M_4, M_4'	1	3.0	1.2	$35.142 + j140.628$	$14.057 + j140.628$
M_5, M_5'	3	3.0	1.2	$31.086 + j89.066$	$12.2134 + j89.066$
M_6, M_6'	Group < 50 hp	∞	1.67	∞	$33.99 + j139.380$

reactance network is constructed by dropping out the resistances in Figure 7.16. This gives an X/R ratio of 16.28. The interrupting duty fault current is, therefore, $E/Z = 20.54$ kA symmetrical at 138 kV. To calculate the interrupting duty, NACD is required. This being a radial system, it is easy to calculate the local (generator) contribution through the transformer impedance, which is equal to 1.51 kA. The utility source is considered remote, and it contributes 19.01 kA; NACD = 0.925. A 138 kV breaker will be a three-cycle breaker, with a contact parting time of two cycles. The multiplying factor from Figure 7.11 is, therefore, equal to 1.

The first-cycle network will be similar to that shown in Figure 7.14, except that the motor impedances will change. This gives a complex impedance of $Z = 0.001426 + j0.020309$. The X/R ratio is 16.27, and the first-cycle symmetrical current is 20.55 kA, very close to the

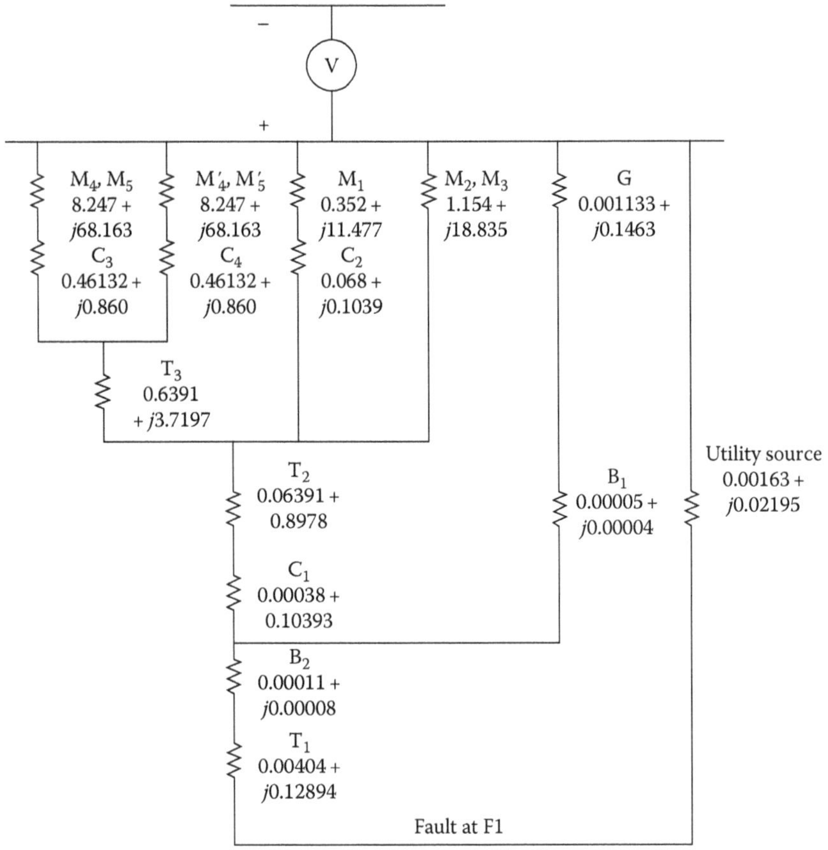

FIGURE 7.16
Positive sequence impedance diagram for a fault at F1 (Example 7.2).

interrupting duty current. The peak asymmetrical current is calculated from Equation 7.1 and is equal to 53.1 kA.

The effect of motor loads in this case is small. If the motor loads are dropped and calculations repeated, the interrupting duty current = first-cycle current = 20.525 kA symmetrical, i.e., a difference of only 0.13%. This is because the low-voltage motors contribute through impedances of transformers T_1, T_2, and T_3 in series, and medium-voltage motors contribute through two-stage transformations. The current contributed by small induction motors, and small synchronous motors in utility systems can, usually, be ignored except station service supply systems or at substations supplying industrial distribution systems or locations close to large motors, or both. Motor contributions increase half-cycle current more than the symmetrical interrupting current at the contact parting time.

Generator Source Symmetrical Short-Circuit Current

The generator has a high X/R ratio, and according to qualifications of E/Z method, that calculations with adjustments of ac and dc adjustment factors can be applied provided X/R does not exceed 45, an analytical calculation based upon Example in IEEE standard [11] is carried out. No premultiplying impedance factors and postmultiplying duty factors are applicable. The calculation should account for the short-circuit current decay at the contact parting time. The symmetrical current at the contact parting time (three cycles contact parting time = 50.0 ms) can be calculated using Equation 6.13. When an external reactance is added to the generator circuit, the subtransient component of the current is given by

$$i''_d = \frac{e''}{X''_d + X_e} \tag{7.22}$$

where X_e is the external reactance. Similar expressions apply for i'_d and i'_d. The time constants are also changed, i.e., the short-circuit transient time constant in Equation 6.24 becomes

$$T'_d = T'_{do} \frac{X'_d + X_e}{X_d + X_e} \tag{7.23}$$

This means that adding an external reactance is equivalent to increasing the armature leakage reactance. The calculations use the generator data in Table 6.1. The short-circuit subtransient time constant is 0.015 s. From Table 7.3, the reactance of the bus duct B_1 is 0.00004 per unit. Its effect on the subtransient time constant from is

$$T''_d = T''_{do} \frac{X''_d + X_e}{X'_d + X_e} = (0.22)\left(\frac{0.193 + 0.00004}{0.278 + 0.00004}\right) = 0.015 \tag{7.24}$$

As the bus B_1 reactance is small, there is not much change in the subtransient time constant. However, this illustrates the procedure. The transient time constant is also practically unchanged at 0.597 s.

The generator voltage behind subtransient or transient reactances is equal to its rated terminal voltage, as the generator is considered at no load and constant excitation. The procedure of calculation is similar to that illustrated in Example 6.1 for calculation of a fault decrement curve. The subtransient current is

$$i''_d = \frac{E''}{X''_d + X_e} = \frac{1}{0.14630 + 0.00004}\,\text{pu} = 28.59\,\text{kA} \tag{7.25}$$

We need not drop the resistance. The generator X/R is high, and the bus duct B_1 has a resistance of 0.00005 ohm. The calculated fault point impedance including resistance is:

$$0.001183 + j0.14634 \text{ per unit}$$

Therefore, to be more precise, this gives a current of 28.588 KA at $<-89.543°(X/R = 123.7)$.

Similarly, the transient and steady-state components of the currents are 20.17 and 2.15 kA, respectively. The following equation can, therefore, be written for symmetrical current in kA:

$$i_{ac} = 8.42e^{-t/0.015} + 18.02e^{-t/0.597} + 2.15 \tag{7.26}$$

At contact parting time, this gives the symmetrical current component as 18.95 kA.

Generator Source Asymmetrical Current

The asymmetrical current at contact parting can also be calculated from Equation 6.113. As the generator subtransient reactances in the direct and quadrature axes are approximately equal, the second frequency term in this equation can be neglected, and the dc component at contact parting time needs to be calculated. The decaying dc component is given by

$$i_{\mathrm{dc}} = \sqrt{2} i''_{\mathrm{d}} e^{-t/T_{\mathrm{a}}} = 40.43 e^{-t/T_{\mathrm{a}}} \tag{7.27}$$

The effect of external resistance T_a should be considered according to Equation 7.23. Table 6.1 shows an armature time constant of 0.33 s; considering the bus duct B_1 resistance of 0.00005 per unit, the time constant is reduced to approximately 0.32 s. At contact parting time, the dc component has decayed to 34.58 kA. The asymmetry factor α at contact parting time is calculated as follows:

$$\begin{aligned} &\sqrt{2} \text{ times ac current at contact parting time} = \sqrt{2}(18.95) = 26.80\,\text{kA} \\ &\text{DC current at contact parting} = 34.58\,\text{kA} \end{aligned} \tag{7.28}$$

Factor $\alpha = 34.58/26.80 = 1.29$, i.e., the asymmetry at contact parting time is approximately 129%, and the current zero is not obtained. The total rms asymmetrical breaking current at contact parting time is therefore

$$\sqrt{18.95^2 + (35.58)^2} = 39.43\,\text{kA}$$

The example in Appendix A of IEEE standard C37.013 [11] does not calculate the generator source asymmetrical current analytically, though it provides the mathematical equation on which above calculation is based. The computer simulation shows reduced asymmetry factor by approximately 3%. Further simulation with arc voltage shows a reduction of asymmetry factor to 68%. A free burning arc in air has a voltage of 10 V/cm. Further, demonstrating the capability of a generator breaker to interrupt short-circuit currents with delayed current zeros may be difficult and limited in high-power testing stations. In contrast, note that IEEE standard C 37.010 [6] considers bolted fault currents for ascertaining the duties of circuit breakers and any arc fault resistance is neglected. A fault resistance may introduce enough arc resistance to force the current to zero after the contact parting time (Figure 7.12). Simplified empirical/analytical calculations in such cases have limitations, and a dynamic simulation is recommended.

As stated in Chapter 5, symmetrical short-circuit rating of a generator breaker is not specified, but the generator source asymmetrical capability states that the generator breaker should be capable of interrupting up to 110% dc component, based upon the peak value of the symmetrical current. Based upon the above calculation results, the dc component is 34.58 kA. Therefore, the three-phase symmetrical interrupting must be at least 22.2 kA rms symmetrical.

System Source Symmetrical Short-Circuit Current, Fault at F4

We can now conduct analytical calculation for the fault at F4. Utility source, transformer T_1, and bus duct B_2 impedance in series gives $Z = 0.00577 + j0.15097$; $X/R = 26.16$ from the separate X and R networks. This gives a short-circuit current of 27.692 kA at an angle of $< -87.81°$.

The equivalent impedance of low-voltage motors through cables and a 1.5 MVA transformer plus medium-voltage motor loads through a 7.5 MVA transformer, as seen from the fault point F4, is $0.307 + j4.356$. No impedance multiplying factors to adjust motor impedance from Table 7.1 are used in this calculation. This gives a short-circuit current of 0.958 kA at $<-85.968°$; $X/R = 14.2$. The time constant for the auxiliary distribution is therefore

$$\frac{1}{2\pi f}\left(\frac{X}{R}\right) = 37.66\,\text{ms} \tag{7.29}$$

The ac symmetrical component of the auxiliary system decays and at contact parting time it can be *assumed* to be 0.7–0.8 times the initial short-circuit current. Therefore, the contribution from the auxiliary system is = 0.766 kA. No decay is applicable to the utility source connected through the transformer. The total system source symmetrical current is

$$I_{\text{sym}} = 27.692 + 0.766 = 28.458\,\text{kA rms}$$

The total fault point X/R calculated from separate R and X networks is 26.

System Source Asymmetrical Short-Circuit Current, Fault at F4

The asymmetrical current is now calculated. The time constant of the utility's contribution for an X/R of 26.16 is 0.0694 s. The time constant of the auxiliary distribution was calculated as 37.66 ms in Equation 7.29. Total dc current at the contact parting time is the sum of components from the utility's source and auxiliary distribution. Thus, the total dc current is given by

$$I_{\text{dc}} = \sqrt{2}[27.692e^{-50.0/69.4} + 0.958e^{-50.0/37.66}]$$

This gives $I_{\text{dc}} = 19.14$ kA. Asymmetrical current is given by

$$\begin{aligned} I_{\text{asym}} &= \sqrt{I_{\text{sym}} + I_{\text{dc}}} \\ &= 34.44\,\text{kA rms} \end{aligned}$$

The asymmetry factor $\alpha = 19.14/(\sqrt{2}\,28.458) = 0.476$. There is no problem of not obtaining a current zero. This means that the dc component at the primary arcing contact parting time is 47.6% of the peak value of the symmetrical system-source short-circuits current. Referring to Figure 5.17, a five-cycle symmetrical rated generator breaker with contact parting time of three cycles has a dc interrupting capability = 68.5%.

Required Closing Latching Capabilities

The ratio of the maximum asymmetrical short-circuit peak current at half cycle to the rated short-circuit current of the generator breaker is determined from the following equation from Ref. [11]:

$$\frac{I_{\text{peak}}}{I_{\text{sym}}} = \sqrt{2} \times \left(1 + e^{t/133}\right) = 2.74 \tag{7.30}$$

where time t can be approximately taken as half cycle.

We can calculate the close and latch current using Equation 7.1, which requires X/R ratio, based upon separate R and X networks. For the generator source current $X/R = 127.7$, symmetrical interrupting current is 22.2 kA and for the system source current $X/R=26$ and symmetrical interrupting current 28.485 kA. These give close and latch current of

62.04 kA peak, generator source current

76.05 kA peak system source current

The higher of the two values, that is, 76.05 kA, should be selected.

Selection of the Generator Breaker

Based upon above calculations, a generator breaker of 30 kA symmetrical interrupting capability can be selected. This will have $30 \times 2.74 = 82.2$ kA close and latch capability, five-cycle symmetrical rating, three-cycle contact parting time, dc component interrupting capability = 68.5%, generator source three-phase fault interrupting capability = 25 kA, generator source dc interrupting capability = 110% of the peak of the generator source three-phase fault capability = 38.9 kA. Rated continuous current = 5000 A, short-time current for 1 s = 30 kA. The TRV profile for a fault at F1 should be calculated and simulated. Other specifications like out of phase switching capabilities are required depending upon the system conditions.

While these specifications may seem adequate, specifying a certain asymmetry is no guarantee that the breaker will be able to interrupt the nonzero currents at the calculated asymmetry. Though the standards state that practically, the asymmetry factor will be reduced on account of arc voltage, it is prudent to add to the specifications the calculated asymmetry. Generator breakers meeting 130% asymmetry requirements are commercially available.

Example 7.3

What are the limitations of using a circuit breaker shown dotted for the primary protection of 7.5 MVA unit auxiliary transformer in Figure 7.15?

For application of a breaker for primary protection of the UAT, through fault currents for faults at F2 and F5 can be considered. Fault at F5 gives higher duties. Only the short-circuit contributions from distribution system rotating loads can be ignored, and the short-circuit contributions from utility and generator source should be summed up. Based on E/Z calculation of, though $X/R=75.81$, gives calculated interrupting duty = 71.69 kA, which is beyond the ratings of any commercially available oil-less metal clad or cubical type circuit breaker at 13.8 kV.

Generally, in the generating stations, no breaker is used to selectively isolate the UAT faults. The differential protection is arranged to take care of this fault condition, and both the generator breaker and utility 13.8 kV tie breakers are tripped.

Example 7.4

Construct bus admittance and impedance matrices of the network for Example 7.2 for interrupting duty calculations. Compare the calculated values of self-impedances with the values arrived at in Example 7.2.

The system in terms of admittances can be modeled as shown in Figure 7.17. The bus admittance matrix can be written by examination as follows:

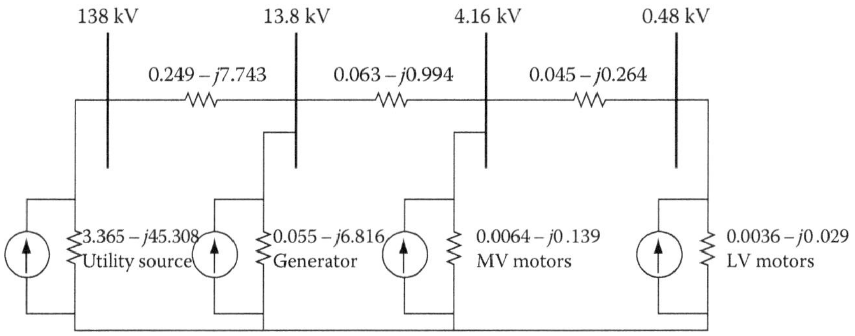

FIGURE 7.17
Equivalent admittance diagram of system of Figure 7.15 (Example 7.2).

$$Y_{bus} = \begin{vmatrix} 3.614 - j53.051 & -0.249 + j7.743 & 0 & 0 \\ -0.249 + j7.743 & 0.367 - j15.553 & -0.063 + j0.994 & 0 \\ 0 & -0.063 + j0.994 & 0.1144 - j1.397 & -0.045 + j0.264 \\ 0 & 0 & -0.045 + j0.264 & 0.0486 - j0.293 \end{vmatrix}$$

The bus impedance matrix is

$$Z_{bus} = \begin{vmatrix} 0.0014 + j0.0203 & 0.0006 + j0.0107 & 0.0005 + j0.0092 & 0.00045 + j0.0083 \\ 0.0006 + j0.0107 & 0.0017 + j0.0736 & 0.0014 + j0.0632 & 0.0010 + j0.0570 \\ 0.0005 + j0.0092 & 0.0014 + j0.06321 & 0.0551 + j0.9135 & 0.0460 + j0.8239 \\ 0.00045 + j0.0083 & 0.0010 + j0.0570 & 0.0460 + j0.8239 & 0.5892 + j4.0647 \end{vmatrix}$$

The self-impedances Z_{11}, Z_{22}, etc., compare well with the values calculated from positive sequence impedances of the fault point networks, reduced to a single impedance. The X/R ratio should not be calculated from complex impedances and separate R and X matrices are required. Zero values of elements are not acceptable.

7.10.1 Deriving an Equivalent Impedance

A section of a network can be reduced to single equivalent impedance. This tool can prove useful when planning a power system. Consider that in Figure 7.13, the 13.8 kV system is of interest for optimizing the bus tie reactors or developing a system configuration. A number of computer runs may be required for this purpose. The distribution connected to each of the 13.8 kV buses can be represented by equivalent impedances, one for interrupting duty calculation and the other for first-cycle calculation. Figure 7.18 shows these equivalent impedances. These can be derived by a computer calculation or from the vectorial summation of the short-circuit currents contributed to the buses.

This concept can be used in subdividing a large network into sections with interfaced impedances at their boundaries, representing the contributions of the connected systems. Attention can then be devoted to the section of interest, with less computer running time and saved efforts in modeling and analyzing the output results. Once the system of interest is finalized, its impact on the interfaced systems can be evaluated by detailed modeling.

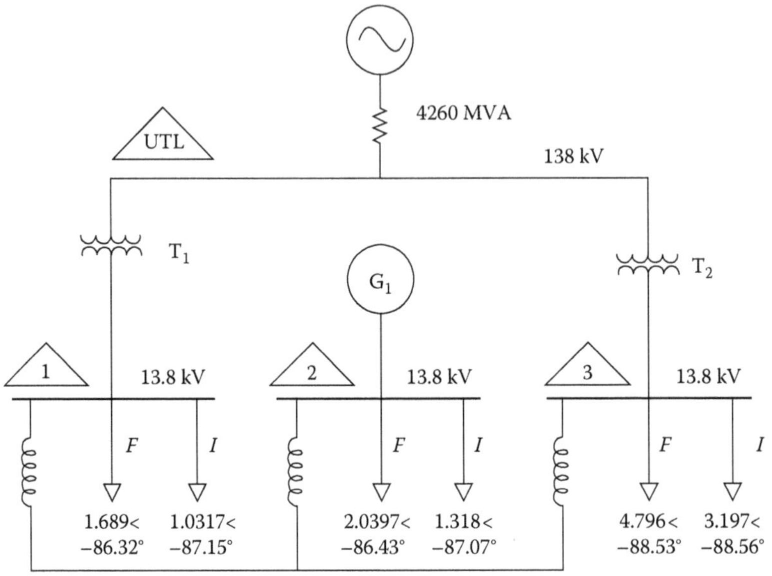

FIGURE 7.18
Equivalent impedances of the distribution system connected to 13.8 kV buses in the distribution system of Figure 7.16. *F*, first-cycle impedance; *I*, interrupting-duty impedance.

Example 7.5

In the majority of cases, three-phase short-circuit currents give the maximum short-circuit duties; however, in some cases, a single line-to-ground fault may give a higher short-circuit duty. Consider the large generating station shown in Figure 7.19. Fault duties are required at point *F*. The source short-circuit MVA at 230 kV for a three-phase and single line-to-ground fault is the same, 4000 MVA, $X/R = 15$ (10.04 kA $< -86.19°$). Each generator is connected through a delta-wye step-up transformer. The high-voltage wye neutral is solidly grounded. This is the usual connection of a step-up transformer in a generating station, as the generator is high-impedance grounded. With the impedance data for the generators and transformers shown in this figure, the positive sequence fault point impedance $= 0.00067 + j0.0146$ per unit, 100 MVA base. Thus, the three-phase short-circuit current at $F = 17.146$ kA sym. Each generator contributes 2.31 kA of short-circuit current. The zero-sequence impedance at the fault point $= 0.00038 + j0.00978$ and the single line-to-ground fault current at $F = 19.277$ kA, approximately 12.42% higher than the three-phase short-circuit current.

Example 7.6

Figure 7.20a shows three motors of 10,000 hp, each connected to a 13.8 kV bus, fed from a 138 kV source through a single step-down transformer. For a fault on load side of breaker 52 calculate the motor contributions at the 13.8 kV bus, and the first-cycle and interrupting duty currents by simplified ANSI methods, and by analytical calculations for a three-cycle breaker contact parting time. Repeat the calculations with series impedances added in the motor circuit as shown in Figure 7.20b.

ANSI/IEEE Calculations

The source impedance in series with the transformer impedance is same for the first-cycle and interrupting duty calculations. This gives $Z = 0.0139 + 0.2954$ per unit on a 100 MVA base. The source contribution through the transformer is, therefore, 14.147 kA at $< -87.306°$.

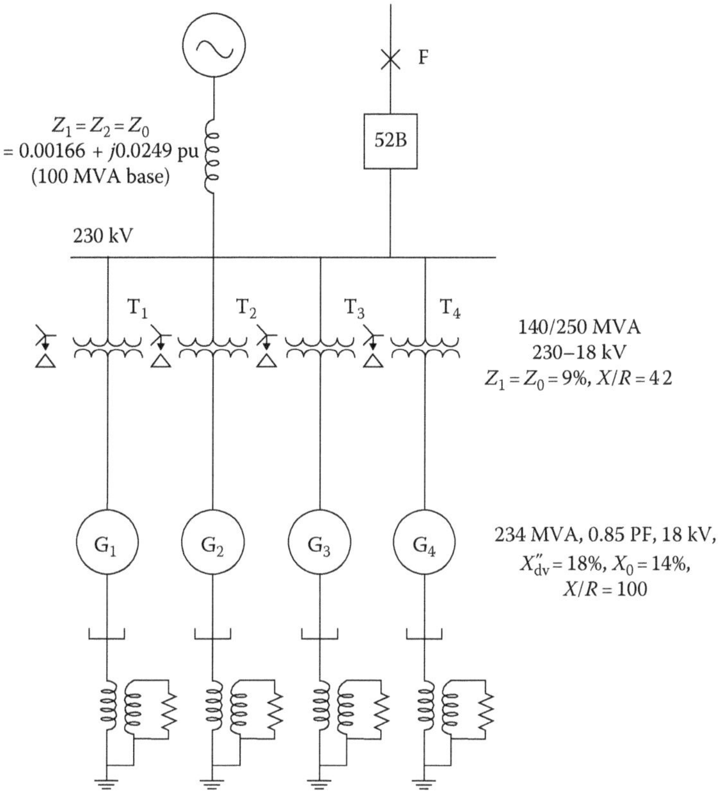

FIGURE 7.19
Single-line diagram of a large generating station for calculation of fault currents (Example 7.4).

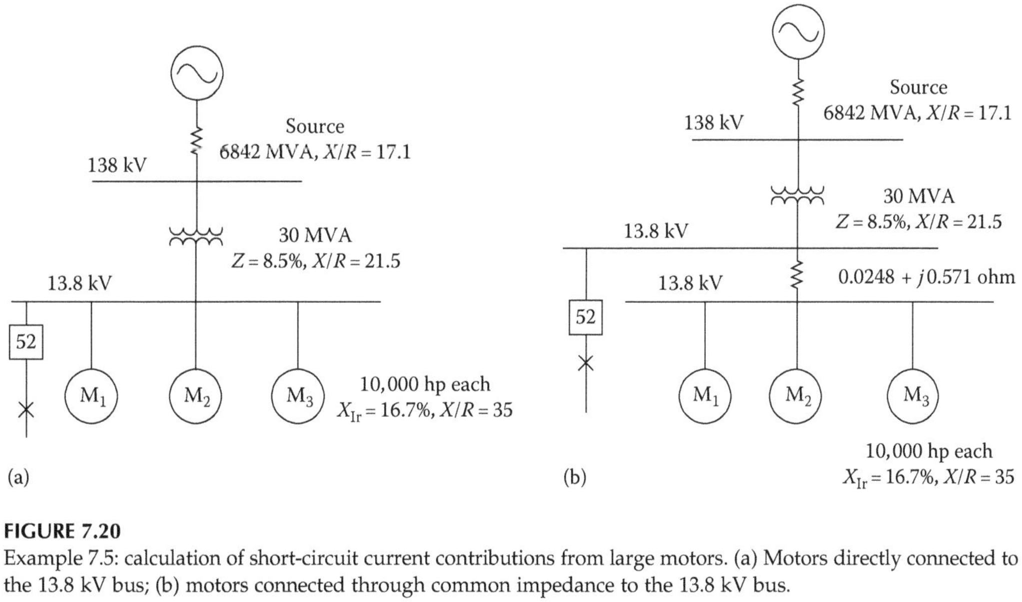

FIGURE 7.20
Example 7.5: calculation of short-circuit current contributions from large motors. (a) Motors directly connected to the 13.8 kV bus; (b) motors connected through common impedance to the 13.8 kV bus.

The first-cycle impedance multiplying factor for a 10,000 hp motor is unity. The motor kVA based on a power factor of 0.93 and efficiency of 0.96 = 8375; $X/R = 35$. Therefore, the per unit locked rotor impedance for three motors in parallel is $0.0189 + j0.0667$. This gives a first-cycle current of 6.270 kA at <−88.375°. The total first-cycle current is 20.416 kA sym at <−87.63°; $X/R = 25.56$. This gives an asymmetrical current of 32.69 kA.

For interrupting duty, the impedance multiplying factor is 1.5. The equivalent motor impedance of three motors in parallel is $0.02835 + j1.0005$ per unit. This gives a motor current contribution of 4.180 kA at <−88.375°. The total current is 18.327 kA <−87.55°; $X/R = 24.45$; NACD = 0.772. The E/X multiplying factor for dc decay at a three-cycle contact parting time is 1.027. The calculated interrupting duty current is 18.821 kA.

When a series impedance of 0.015 + 0.30 per unit (100 MVA base) is introduced into the circuit, the following are the results of the calculation:

First-cycle motor contribution = 4.324 kA <−87.992°

Total first-cycle current = 18.47 kA sym < −87.4°; $X/R = 22.95$

Asymmetry multiplying factor = 1.5888

Asymmetrical current = 29.33 kA rms

Interrupting duty motor contribution = 3.215 < −88.09°

Total current = 17.326 < −87.45°; $X/R = 22.87$

NACD = 0.816

E/X multiplying factor = 1.035

Calculated interrupting duty current = 17.93 kA

Analytical Calculation

We will recall Equations 6.133 and 6.134 and Example 6.9. The equations for ac and dc components of the short-circuit currents are reproduced as follows:

$$i_{ac} = \frac{E}{X'} e^{-t/T'}$$

$$i_{dc} = \sqrt{2}\frac{E}{X'} e^{-t/T_{dc}}$$

We showed that X' is the locked rotor reactance of the motor. This can be replaced with Z'. $T_{dc} = X'/\omega r_1$ (Equation 6.132), where r_1 is the stator resistance. Therefore, X'/r_1 is in fact the ANSI X/R ratio. When external impedance is added:

$$T_{dc} = \frac{X' + X_e}{\omega(r_1 + R_e)} \tag{7.31}$$

where R_e and X_e are the external resistances and reactances. Similarly, the time constant T' becomes:

$$T' = \frac{X' + X_e}{\omega r_2} \tag{7.32}$$

We will first calculate the motor first-cycle and interrupting current contributions, without series impedance in the motor circuit; $X' = 0.0189 + j0.667$ as before. The time constants of large motors are generally specified by the manufacturers. For example, $T_{dc} = 0.093$ and $T' = 0.102$. The ac and dc current equations for the motor are

$$i_{acm} = 6.27\varepsilon^{-t/0.102} \qquad i_{dcm} = 8.87\varepsilon^{-t/0.093}$$

At half-cycle, the ac and dc currents are 5.28 and 8.10 kA, respectively. The source impedance in series with the transformer is 0.0139 + j0.2954 per unit. Therefore, $X/R = 21.25$ and the symmetrical current is 14.147 kA. The equation for the dc current from the source is

$$i_{\mathrm{dcs}} = 20e^{-t/0.0564}$$

At half-cycle, the dc component from the source is 17.26 kA. Therefore, the total symmetrical current is 5.28 + 14.147 = 19.427 kA and the dc current is 25.36 kA. This gives an asymmetrical current of 31.94 kA. We calculated 32.69 kA by ANSI/IEEE method.

The interrupting currents are calculated at three cycles. The source dc component is 8.24 kA, the motor dc component is 5.17 kA, the source ac component is 14.147 kA, and the motor ac component is 3.84 kA. This gives a symmetrical current of 17.987 kA and an asymmetrical current of 22.44 kA. The interrupting duty to compare with a five-cycle breaker is 22.444/1.1 = 20.39 kA. We calculated 18.821 kA by the ANSI/IEEE method.

Now consider the effect of series resistance in the motor circuit. The motor impedance plus series impedance is $(0.0189 + j0.0667) + (0.015 + j0.30) = 0.0239 + j0.967$ per unit. This gives a current of 4.32 kA.

The time constants are modified to consider external impedance according to Equations 7.31 and 7.32:

$$T_{\mathrm{dc}} = \frac{0.667 + 0.30}{\omega(0.0189 + 0.015)} = 0.0757$$

Also, T' is

$$T' = \frac{0.667 + 0.30}{\omega(0.0173)} = 0.0148$$

The equations for motor ac and dc components of currents are then:

$$i_{\mathrm{acm}} = 4.32e^{-t/0.148} \qquad i_{\mathrm{dcm}} = 6.13e^{-t/0.0757}$$

This gives a total first-cycle current of 29.14 kA rms (versus 29.33 kA by the ANSI/IEEE method) and a symmetrical interrupting current of 17.23 kA; the dc component at contact parting time is 11.4 kA. The total asymmetrical current is 20.65 kA. The current for comparison with a five-cycle circuit breaker is 20.65/1.1 = 18.77 kA. We calculated 17.93 kA with the simplified calculations. ANSI/IEEE calculations do not consider decay of short-circuit currents of motors and there are variations.

7.11 Thirty-Cycle Short-Circuit Currents

Thirty-cycle short-circuit currents are required for overcurrent devices co-coordinated on a time-current basis. The 30-cycle short-circuit current is calculated on the following assumptions:

- The contributions from the utility sources remain unchanged.
- The dc component of the short-circuit current decays to zero.
- The contribution from the synchronous and induction motors decays to zero.
- The generator subtransient reactance is replaced with transient reactance or a value higher than the subtransient reactance.

The 30-cycle current of the synchronous generators and motors will vary depending upon the excitation systems. If the power for excitation system is taken from the same bus to

TABLE 7.15

30-Cycle Currents, Distribution System (Figure 7.13)

Bus Identification	30-Cycle Currents in kA Sym
13.8 kV buses	
1	19.91 < −88.12°
2	21.11 < −88.68°
3	19.91 < −88.12°
4	22.73 < −88.30°
4.16 kV buses	
5	18.26 < −86.88°
6	18.38 < −86.59°
7	18.17 < −85.79°
8	18.17 < −85.79°
Low-voltage buses	
9	39.01 < −81.53°
10	39.10 < −81.43°
11	35.93 < −77.98°
12	35.93 < −77.98°
15	17.85 < −68.39°
16	12.45 < −71.29°

which the motor is connected, then for a fault close the motor, the voltage reduces practically to zero, and the field current will quickly decay depending upon machine time constants. If the excitation power is taken from an independent source, not affected by the short circuit, the excitation will be maintained. This can be an important factor for machine stability. For brushless excitation systems of synchronous motors, the excitation power required is small, which can be provided by a battery source or UPS (uninterruptible power system) enhancing the motor stability.

Example 7.7

Calculate 30-cycle currents in the distribution system of Figure 7.13 (Example 7.1).

The generator reactance is changed to transient reactance, and all motor contributions are dropped. Table 7.15 shows the results, which can be compared with interrupting duty currents. The decay varies from 8% to 28%. The buses that serve the motor loads show the highest decay.

Problems

(The problems in this section are constituted, so that these can be solved by hand calculations.)

7.1 A circuit breaker is required for 13.8 kV application. The E/X calculation gives 25 kA sym, and the X/R ratio is below 17. Select a suitable breaker from the tables in Chapter 5.

7.2 The X/R ratio in Problem 7.1 is 25. Without making a calculation, what is the minimum interrupting rating of a circuit breaker for safe application?

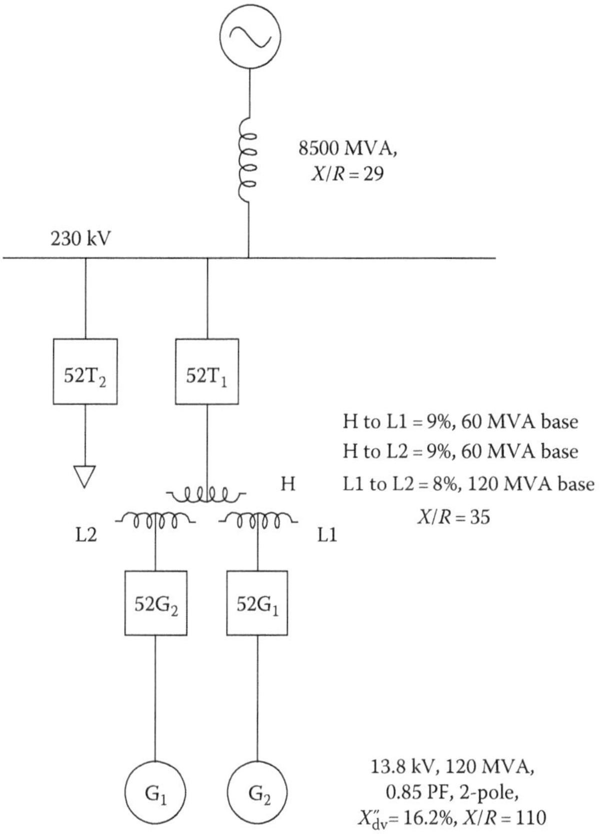

FIGURE P7.1
Power system for Problem 7.3.

7.3 A generating station is shown in Figure P7.1. The auxiliary distribution loads are omitted. Calculate the short-circuit duties on the generator breaker and the 230 kV breaker.

7.4 Indicate whether the following sources will be considered remote or local in the ANSI calculation method. What is the NACD ratio? Which E/X multiplying curve shown in this chapter will be used for interrupting duty currents?

a. A 50 MVA 0.85 power factor generator; saturated subtransient reactance = 16%, $X/R = 60$.

b. The above generator connected through a 0.4 ohm reactor in series with the generator.

c. The generator is connected through a step-up transformer of 40/60 MVA; transformer impedance on a 40 MVA base = 10%, $X/R = 20$.

d. A 10,000 hp synchronous motor, operating at 0.8 power factor leading.

e. A 10,000 hp induction motor, operating at 0.93 power factor lagging.

7.5 Figure P7.2 shows the single-line diagram of a multilevel distribution system. Calculate first-cycle and interrupting currents using the ANSI calculation and analytical methods for faults at F1, F2, and F3. Use the impedance data and X/R ratios specified in Table P7.1.

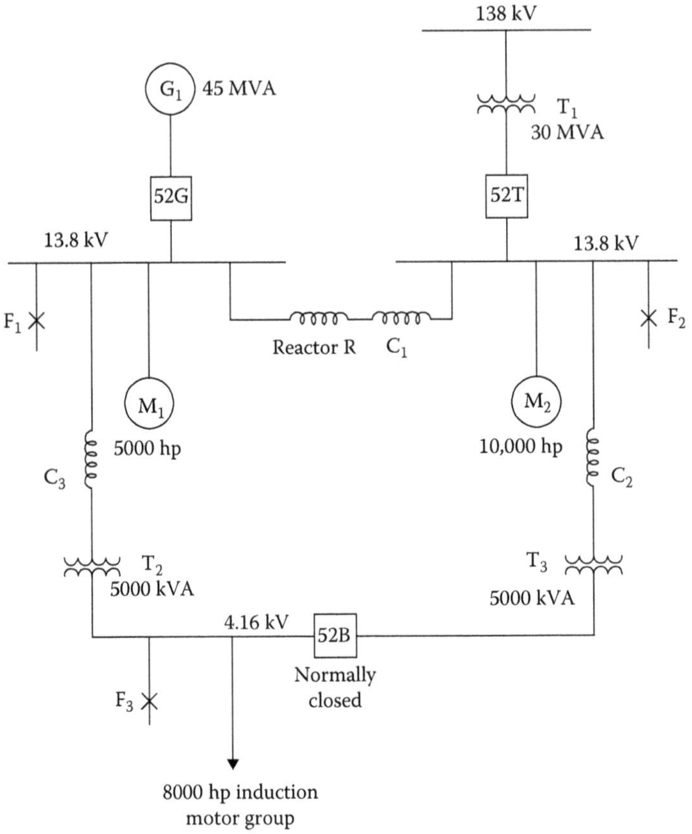

FIGURE P7.2
Power system for Problems 7.5 and 7.6.

TABLE P7.1

Impedance Data (Problems 7.5 and 7.6)

Equipment Identification	Impedance Data
138 kV utility's source	3600 MVA, $X/R = 20$
Generator	13.8 kV, 45 MVA, 0.85 pF, 2-pole subtransient-saturated reactance = 11.7%, $X/R = 100$, transient saturated = 16.8%, synchronous = 205% on generator MVA base, constant excitation, three-phase subtransient and transient time constants equal to 0.016 and 0.46 s, respectively, armature time constant = 0.18 s
Induction motor M_1	13.2 kV, 5000 hp, 2-pole-locked rotor reactance = 16.7%, $X/R = 55$
Synchronous motor M_2	13.8 kV, 10,000 hp, 4-pole, $X_d'' = 20\%$, $X/R = 38$
Motor group	4 kV, 200–1000 hp, 4-pole induction motors, locked rotor reactance = 17%, consider an average $X/R = 15$, total installed kVA = 8000
Transformer T_1	30/50 MVA, 138–13.8 kV, $Z = 10\%$, $X/R = 25$, data–wye, secondary neutral resistance grounded.
Transformers T_2 and T_3	5000 kVA, 13.8–4.16 kV, $Z = 5.5\%$, $X/R = 11$, transformer taps are set to provide 5% secondary voltage boost.
Reactor	0.3 ohm, $X/R = 100$
Cable C_1	$0.002 + j0.018$ ohm
Cables C_2 and C_3	$0.055 + j0.053$ ohm

7.6 Calculate the duties on circuit breakers marked 52G and 52T in Figure P7.2. Use separate $R \sim X$ method for E/Z calculation. Each of these breakers is five-cycle symmetrical rated.

7.7 Calculate the short-circuit withstand rating of a 250 KCMIL cable; conductor temperature 90°C, maximum short-circuit temperature 250°C, ambient temperature 40°C, and short-circuit duration 0.5 s.

7.8 The multiplying factors for short-circuit duty calculations for LVPCB and MCCB are 1.095 and 1.19, respectively, though the available short circuit current is 52.3 kA in each case. What is the fault point X/R?

7.9 In Example 7.1, construct a fault impedance network for the fault at Bus 3. Calculate fault point impedance and the X/R ratio based on separate R-X networks. How does this impedance differ from the values arrived at in Table 7.5 from complex impedance reduction calculation?

7.10 Write five lines on each to describe how the following devices are rated from a short-circuit consideration? LVPCB, ICCB, and MCCB, power cables, overhead ACSR conductors, transformer primary fused and unfused switches, bus-bars in switchgear and switchboard enclosures, phase segregated, and nonsegregated metal-enclosed buses and transformers and reactors.

7.11 Explain the problems of high-X/R faults. How does this affect the duties on the high-voltage generator circuit breakers?

7.12 Which of the following locations of breakers in the same system and on the same bus will have the maximum fault duties: bus tie breaker, incoming breaker, feeder breaker fully loaded, and feeder breaker lightly loaded?

7.13 Represent all the downstream distributions in Figure P7.2 at the 13.8 kV bus with a single interrupting duty and first-cycle impedance.

7.14 Calculate 30-cycle currents in Figure P7.2 at all buses.

7.15 Form a bus admittance and bus impedance matrix of the system in Figure P7.2.

References

1. IEC. Standard 909. Short Circuit Calculations in Three-Phase AC Systems, 1988.
2. IEC. Standard 60909-0. Short Circuit Calculations in Three-Phase AC Systems. 0—Calculation of Currents, 2001–07.
3. VDE. Standard 0102, Parts 1/11.71 and 2/11.75. Calculations of Short-Circuit Currents in Three-Phase Systems, 1971 and 1975.
4. IEEE. Standard 551, Violet Book. IEEE Recommended Practice for Calculating Short-Circuit Currents in Industrial and Commercial Power Systems, 2006.
5. C.N. Hartman. Understanding asymmetry. *IEEE Trans. Ind. Appl.* 21: 4–85, 1985.
6. ANSI/IEEE. Standard C37.010. Application Guide for AC High-Voltage Circuit Breakers Rated on a Symmetrical Current Basis, 1999 (revision of 1979 standard).
7. IEEE. Standard 141. IEEE Recommended Practice for Electrical Power Distribution for Industrial Plants, 1993.
8. ANSI/IEEE. Standard C37.13. Standard for Low-Voltage AC Power Circuit Breakers in Enclosures, 1997.

9. J.C. Das. Reducing interrupting duties of high voltage circuit breakers by increasing contact parting time. *IEEE Trans. Ind. Appl.* 44: 1027–1034, 2008.
10. W.C. Huening Jr. Interpretation of new American standards for power circuit breaker applications. *IEEE Trans. Ind. Gen. Appl.* 5: 121–143, 1969.
11. IEEE. Standard C37-013. IEEE Standard for AC High-Voltage Generator Breakers Rated on Symmetrical Current Basis, 1999.
12. IEEE. Standard C37-013a. IEEE Standard for AC High-Voltage Generator Breakers Rated on Symmetrical Current Basis, Amendment 1: Supplement for Use with Generators Rated 10–100 MVA, 2007.
13. I.M. Canay and L. Warren. Interrupting sudden asymmetric short-circuit currents without zero transition. *Brown Boveri Rev.* 56: 484–493, 1969.
14. I.M. Canay. Comparison of generator circuit breaker stresses in test laboratory and real service condition. *IEEE Trans. Power Deliv.* 16: 415–421, 2001.
15. IEEE. Standard 1110. Guide for Synchronous Generator Modeling Practice and Applications in Power System Stability Analysis, 2002.
16. J.R. Dunki-Jacobs, B.P. Lam, and R.P. Stafford. A comparison of ANSI-based and dynamically rigorous short-circuit current calculation procedures. *IEEE Trans. Ind. Appl.* 24: 1180–1194, 1988.
17. A. Berizzi, S. Massucco, A. Silvestri, and D. Zaninelli. Short-circuit current calculations—A comparison between methods of IEC and ANSI standards using dynamic simulation as reference. *IEEE Trans. Ind. Appl.* 30: 1099–1106, 1994.
18. A.J. Rodolakis. A comparison of North American (ANSI) and European (IEC) fault calculation guidelines. *IEEE Trans. Ind. Appl.* 29: 515–521, 1993.
19. IEEE. Standard C37.91. IEEE Guide for Protecting Power Transformers, 2008.
20. NFPA 70 (National Fire Protection Association) National Electric Code.
21. ANSI Standard C37.06. AC High-Voltage Circuit Breakers Rated on a Symmetrical Current Basis—Preferred Ratings and Related Capabilities, 2000 (revision of 1987).
22. J.C. Das. Limitations of fault current limiters for expansion of electrical distribution systems. *IEEE Trans. Ind. Appl.* 33: 1073–1082, 1997.
23. ANSI/IEEE. Standard C37.23. Guide for Metal-Enclosed Bus and Calculating Losses in Isolated-Phase Bus, 1987.
24. The Aluminum Association. *Aluminum Electrical Conductor Handbook*, 2nd edn. Washington, DC: The Aluminum Association, 1982.
25. D.G. Fink and J.M. Carroll (eds). *Standard Handbook for Electrical Engineers*, 10th edn. New York: McGraw-Hill, 1968 (see Section 4, Table 4-8).

8

Short-Circuit Calculations according to IEC Standards

Since the publication of IEC 909 for the calculation of short-circuit currents in 1988 and its subsequent revisions [1], it has attracted much attention and the different methodology compared to ANSI methods of calculation has prompted a number of discussions and technical papers. The examples of short-circuit calculations, which were included in 1988 standard, have been removed in the revised standard. It is recommended that this standard [1] is read in conjunction with Refs. [2–7]. The IEC standard showing examples of short-circuit calculations part-4 is yet to be published. The IEC standard [1] qualifies that systems at highest voltages of 550 kV and above with long transmission lines need special considerations. Short-circuit currents and impedances can also be determined by system tests, by measurement on a network analyzer or with a digital computer. Calculations of short-circuit currents in installations on board ships and airplanes are excluded.

This chapter analyzes and compares the calculation procedures in IEC and ANSI standards. Using exactly the same system, configurations and impedance data comparative results are arrived at, which show considerable differences in the calculated results by the two methods. Some explanation of these variances is provided based on different procedural approaches. Neither standard precludes alternative methods of calculation, which give equally accurate results.

8.1 Conceptual and Analytical Differences

The short-circuit calculations in IEC and ANSI standards are conceptually and analytically different and the rating structure of the circuit breakers is not identical. There are major differences in the duty cycles, testing, temperature rises, and recovery voltages, though there have been attempts to harmonize ANSI/IEEE standards with IEC, Chapters 5 and 7. For the purpose of IEC short-circuit calculations, we will confine our attention to the specifications of interest. Entirely different terminology is used in IEC [1,6] to describe the same phenomena in circuit breakers. The following overview provides a broad picture and correlation with ANSI.

8.1.1 Breaking Capability

The rated breaking capability of a circuit breaker corresponds to the rated voltage and to a reference restriking voltage, equal to the rated value, expressed as (1) rated symmetrical breaking current that each pole of the circuit breaker can break, and (2) rated asymmetrical breaking capability that any pole of the circuit breaker can break. The breaking capacity is expressed in MVA for convenience, which is equal to the product of the rated breaking current in kA and rated voltage multiplied by an appropriate factor, depending on the type of circuit: one for a single-phase circuit, two for a two-phase circuit, and $\sqrt{3}$ for a three-phase circuit.

This is equivalent to the interrupting capability in ANSI standards. In IEC calculations, the asymmetry at the contact parting time must be calculated to ascertain the asymmetrical rating of the breaker. As discussed in Chapter 5, ANSI breakers are rated on a symmetrical current basis, and the asymmetry is allowed in the rating structure and postfault correction factors. Unlike ANSI, IEC does not recommend any postmultiplying factors to account for asymmetry in short-circuit currents.

8.1.2 Rated Restriking Voltage

The rated restriking voltage is the reference restriking voltage to which the breaking capacity of the circuit breaker is related. It is recommended that the nameplate of the circuit breaker be marked with the amplitude factor and either the rate of rise of the restriking voltage in volts per microsecond or natural frequency in kilohertz per second be stated [6].

8.1.3 Rated Making Capacity

The rated making capacity corresponds to rated voltages and is given by $1.8 \times \sqrt{2}(=2.55)$ times the rated symmetrical breaking capacity. The making capacity in ampères is inversely proportional to the voltage, when the circuit breaker is dual-voltage rated. For voltages below the lower rated voltage, the making capacity has a constant value corresponding to the lower rated voltage, and, for voltages higher than the rated voltage, no making capacity is guaranteed. This is equivalent to the close and latch capability of ANSI standards.

8.1.4 Rated Opening Time and Break Time

The rated opening time up to separation of contacts is the opening time, which corresponds to rated breaking capacity. The rated total breaking time is the total break time, which corresponds to the rated breaking capacity. It may be different, depending on whether it refers to symmetrical or asymmetrical breaking capacity.

The minimum time delay t_m is the sum of the shortest possible operating time of the instantaneous relay (ANSI tripping delay, equal to one half-cycle) and the shortest opening time of the circuit breaker. Thus, the IEC breaking time is equivalent to the ANSI interrupting time, and the IEC minimum time delay t_m is equivalent to the ANSI contact parting time.

8.1.5 Initial Symmetrical Short-Circuit Current

IEC defines I_k'', the initial symmetrical short-circuit current as the ac symmetrical component of a prospective (available) short-circuit current applicable at the instant of short circuit if the impedance remains at zero-time value. This is approximately equal to ANSI first-cycle current in rms symmetrical, obtained in the first cycle at the maximum asymmetry in one of the phases. Note the difference in the specifications. The prospective (available) short-circuit current is defined as the current that will flow if the short circuit was replaced with an ideal connection of negligible impedance. This is the "bolted" fault current. IEEE adopts the definition of prospective current.

8.1.6 Peak Making Current

The peak making current, i_p, is the first major loop of the current in a pole of a circuit breaker during the transient period following the initiation of current during a making operation. This includes the dc component. This is the highest value reached in a phase in a polyphase circuit. It is the maximum value of the prospective (available) short-circuit current. The rated peak withstand current is equal to the rated short-circuit making current. This can be reasonably compared with ANSI close and latch capability, though there are differences in the rating structure. Revision of factor 2.7 to 2.6 for 60 Hz circuit breakers and 2.5 for 50 Hz circuit breakers in ANSI standards (see Section 7.1.2) brings these two standards closer, though there are differences. Also, IEC does not have any requirement, similar to that of ANSI, for latching and carrying a current before interrupting.

8.1.7 Breaking Current

The rated short-circuit breaking current, I_{basym}, is the highest short-circuit current that the circuit breaker shall be capable of breaking (this term is equivalent to ANSI, "interrupting") under the conditions of use and behavior prescribed in IEC, in a circuit having a power frequency recovery voltage corresponding to the rated voltage of the circuit breaker and having a transient recovery voltage equal to the rated value specified in the standards. The breaking current is characterized by (1) the ac component and (2) the dc component. The rms value of the ac component is termed the rated short-circuit current. The symmetrical short-circuit breaking current is defined as the rms value of an integral cycle of symmetrical ac component of the prospective short-circuit current at the instant of contact separation of the fist pole to open of a switching device. The standard values in IEC are 6.3, 8, 10, 12.5, 16, 25, 31.5, 40, 50, 63, 80, and 100 kA. The dc component is calculated at minimum time delay t_m. This is entirely different from ANSI symmetrical ratings and calculations (Chapter 7).

8.1.8 Steady-State Current

The calculations of steady-state fault currents from generators and synchronous motors according to IEC take into consideration the generator excitation, the type of synchronous machine, salient or cylindrical generators, and the excitation settings. The fault current contributed by the generator becomes a function of its rated current using multiplying factors from curves parameterized against saturated synchronous reactance of the generator, excitation settings, and the machine type.

This calculation is more elaborate and departs considerably from ANSI-based procedures for the calculation of 30-cycle currents. For the purpose of short-circuit calculations, Table 8.1 shows the equivalence between IEC and ANSI duties, though qualifications apply.

The specimen IEC ratings of a typical dual-voltage rated circuit breaker are as follows:

Voltage: 10 kV/11.5 kV

Frequency: 50 Hz

Symmetrical breaking capacity: 20.4 kA/17.8 kA

Asymmetrical breaking capacity: 25.5 kA/22.2 kA

Rated making capacity: 52 kA/45 kA

TABLE 8.1

Equivalence between ANSI and IEC Short-Circuit Calculation Types

ANSI Calculation Type	IEC Calculation Type
Closing–latching duty current, crest	Peak current (making current), i_p
Interrupting duty current	Breaking current I_{bsym} and I_{basym} (symmetrical and asymmetrical)
Time-delayed 30-cycle current	Steady-state current, I_k

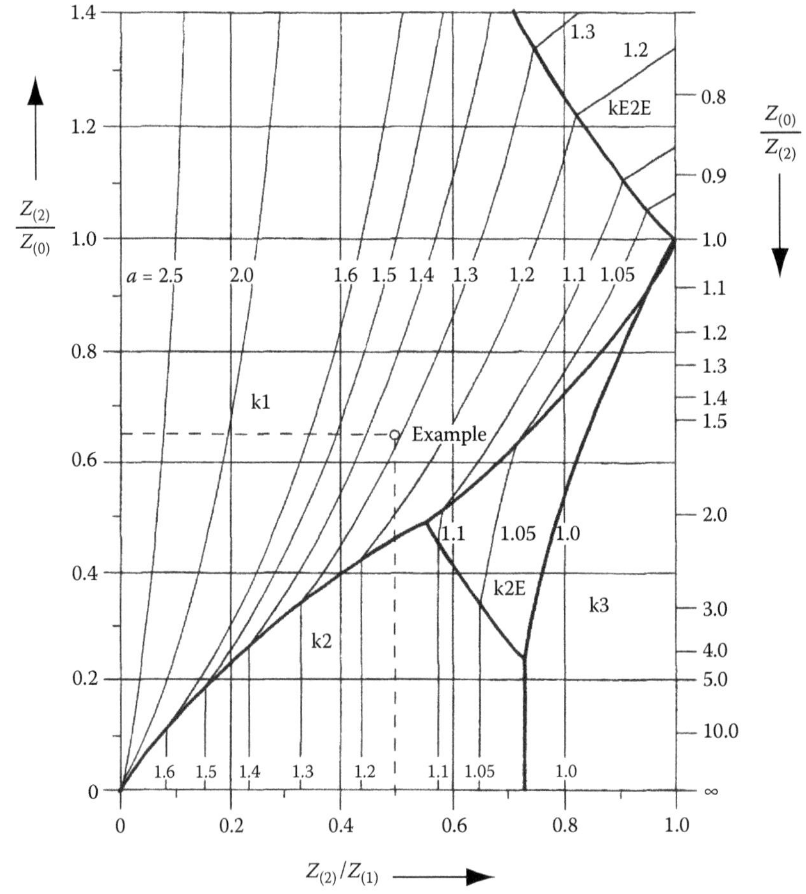

FIGURE 8.1
Diagram to determine the short-circuit type for the highest short-circuit current referred to symmetrical three-phase short-circuit current at the short-circuit location when the impedance angles of positive, negative, and zero-sequence impedances are identical.

Rated restriking voltage:
Amplitude factor: 1.25
Rate of rise: 500 V/μs
Rated short-time current (1 s): 20.4 kA
Rated operating duty: O-3m-CO-3m-CO

8.1.9 Highest Short-Circuit Currents

The three-phase, single line-to-ground, double line-to-ground, and phase-to-phase fault currents are to be considered. Based on the sequence impedances, Figure 8.1 shows which type of short circuit leads to the highest short-circuit currents. All the sequence impedances have the same impedance angle. This figure is useful for information but should not be used instead of calculations. As an example if $Z_2/Z_1 = 0.5$ and with ac decay $Z_2/Z_0 = 0.65$, the single-line-to-ground fault is the highest.

8.2 Prefault Voltage

IEC defines an equivalent voltage source given in Table 8.2 and states that the operational data on the static loads of consumers, position of tap changers on transformers, excitation of generators, etc., are dispensable; additional calculations about all the different possible load flows at the moment of short circuit are superfluous. The equivalent voltage source is the only active voltage in the system, and all network feeders and synchronous and asynchronous machines are replaced by their internal impedances. This equivalent voltage source is derived by multiplying the nominal system voltage by a factor c given in Table 8.2.

ANSI uses a prefault voltage equal to the system rated voltage, though a higher or lower voltage is permissible, depending on the operating conditions. IEC requires that in *every case*, the system voltage be multiplied by factor c from Table 8.2. We will again revert to this c factor.

TABLE 8.2
IEC Voltage Factor c

	Voltage Factor c for Calculation of	
Nominal Voltage (U_n)	**Maximum Short-Circuit Current (c_{max})**	**Minimum Short-Circuit Current (c_{min})**
Low voltage (100–1000 V)		
(a) 230 V/400 V	1.00	0.95
(b) Other voltages	1.05	1.00
Medium voltage (>1–35 kV)	1.10	1.00
High voltage (>35–230 kV)	1.10	1.00

Source: IEC, Short-Circuit Calculations in Three-Phase AC Systems, 1st edn., 1988, Now revised IEC 60909-0, Short-Circuit Currents in Three-Phase AC Systems, 0—Calculation of Currents, 2001–2007.

8.3 Far-from-Generator Faults

A "far-from-generator" short circuit is defined as a short circuit during which the magnitude of the symmetrical ac component of the prospective (available) current remains essentially constant. These systems have no ac component decay. For the duration of a short circuit, there is no change in the voltage or voltages that caused the short circuit to develop, nor any significant change in the impedance of the circuit, i.e., impedances, is considered constant and linear. Far from generator is equivalent to ANSI *remote* sources, i.e., no ac decay.

The following equation is supported:

$$I_b = I_k'', \quad I_{b2} = I_{k2}'', \quad I_{b2E} = I_{k2E}'', \quad I_{b1} = I_{k1}'' \tag{8.1}$$

where

I_b is the symmetrical breaking current
I_k'' is the initial symmetrical short-circuit current

The subscripts k1, k2, and k2E are line-to-earth short circuit, line-to-line short circuit, and line-to-line short circuit with earth connection. For a single-fed short-circuit current, as shown in Figure 8.2, I_k'' is given by

$$I_k'' = \frac{cU_n}{\sqrt{3}\sqrt{R_k^2 + X_k^2}} = \frac{cU_n}{\sqrt{3}Z_k} \tag{8.2}$$

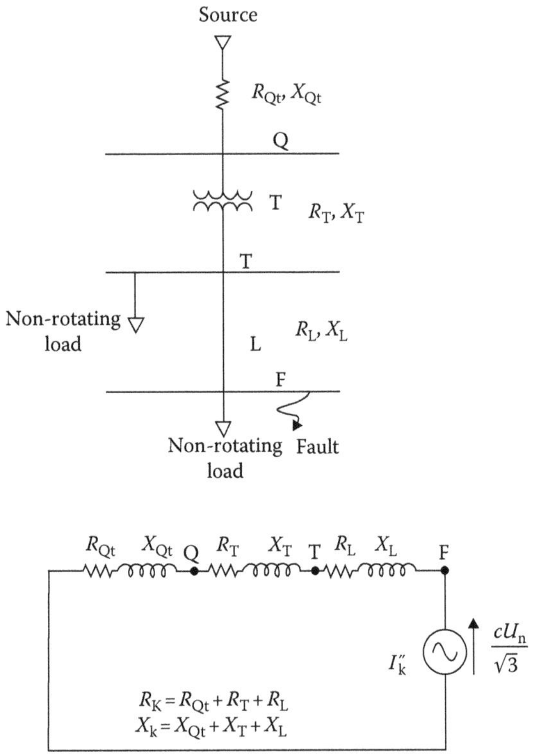

FIGURE 8.2
Calculation of initial short-circuit current, with equivalent voltage source.

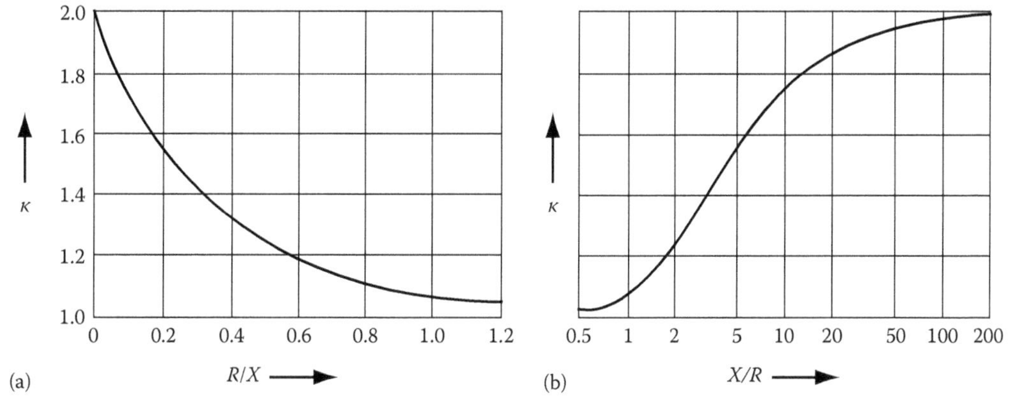

FIGURE 8.3
Factor χ for the calculation of peak current. (From IEC, Short-Circuit Calculations in Three-Phase AC Systems, 1st edn., 1988, Now revised IEC 60909-0, Short-Circuit Currents in Three-Phase AC Systems, 0—Calculation of Currents, 2001–2007.)

where

U_n is the normal system phase-to-phase voltage in V

I_k'' is in A

R_k and X_k are in ohms and are the sum of the source, transformer, and line impedances, as shown in Figure 8.2

The peak short-circuit current is given by

$$i_p = \chi\sqrt{2}I_k'' \quad (8.3)$$

where χ can be ascertained from the X/R ratio from the curves in Figure 8.3 or calculated from the expression

$$\chi = 1.02 + 0.98e^{-3R/X} \quad (8.4)$$

The peak short-circuit current, fed from sources that are not meshed with one another, is the sum of the partial short-circuit currents:

$$i_p = \sum i_{pi} \quad (8.5)$$

8.3.1 Nonmeshed Sources

IEC distinguishes between the types of networks. For nonmeshed sources (Figure 8.4), the initial short-circuit current, the symmetrical breaking current, and the steady-state short-circuit current at fault location F are composed of various separate branch short-circuit currents, which are *independent of each other*. The branch currents are calculated and summed to obtain the total fault current:

$$I_k'' = I_{kT1}'' + I_{kT2}'' \quad (8.6)$$

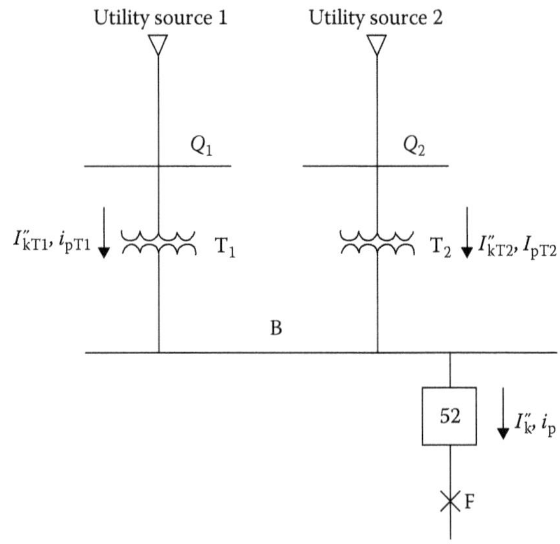

FIGURE 8.4
Short circuit fed from various sources that are independent of each other.

That is, the theorem of superimposition applies and the initial short-circuit current is the phasor sum of the individual short-circuit currents.

For calculating the short-circuit currents in Figure 8.5 in case of a power station unit (PSU) with *on-load tap changer, the equations for partial initial currents are*:

$$I''_{kG} = \frac{cU_{rG}}{\sqrt{3}K_{G,S}Z_G} \tag{8.7}$$

with

$$K_{G,S} = \frac{c_{max}}{1 + x''_d \sin\phi_{rG}} \tag{8.8}$$

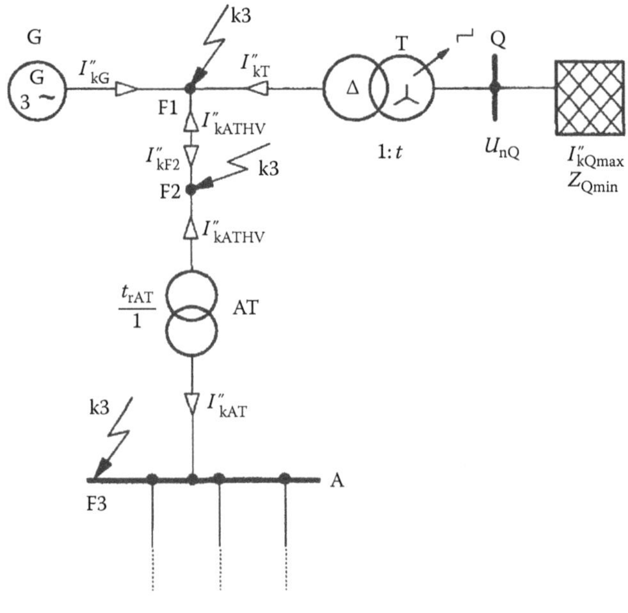

FIGURE 8.5
Short-circuit currents and parallel short-circuit currents for three-phase short circuits between generator and unit transformer with or without load tap changer or at connection of the auxiliary transformer of the PSU and at auxiliary busbar A.

and

$$I''_{\mathrm{KT}} = \frac{cU_{\mathrm{rG}}}{\sqrt{3}\left|Z_{\mathrm{TLV}} + \dfrac{1}{t_{\mathrm{T}}^{2}} Z_{\mathrm{Qmin}}\right|} \tag{8.9}$$

For the short-circuit current I''_{kF2} feeding into short-circuit location F2 at the high voltage side of the auxiliary transformer AT, Figure 8.5:

$$I''_{\mathrm{kF2}} = \frac{cU_{\mathrm{rG}}}{\sqrt{3}}\left[\frac{1}{K_{\mathrm{G,S}}Z_{\mathrm{G}}} + \frac{1}{K_{\mathrm{T,S}}Z_{\mathrm{TLV}} + \dfrac{1}{t_{\mathrm{T}}^{2}} Z_{\mathrm{Qmin}}}\right] = \frac{cU_{\mathrm{rG}}}{\sqrt{3}Z_{\mathrm{rsl}}} \quad \text{with} \quad K_{\mathrm{T,S}} = \frac{c_{\max}}{1 - x_{\mathrm{T}}\sin\phi_{\mathrm{rG}}} \tag{8.10}$$

where
Z_g = subtransient impedance of the generator $= R_{\mathrm{G}} + jX''_{\mathrm{d}}$
x''_{d} = subtransient reactance referred to the rated impedance
Z_{TLV} = short-circuit impedance of the transformer referred to low-voltage side
t_{T} = rated transformation ratio
Z_{Qmin} = minimum value of the impedance of the network feeder

The equations are in actual units, ohms. The calculations can be carried out in per unit system.

For power stations without on-load tap changers, the equations are similar except that the modified factors are

$$K_{\mathrm{G,SO}} = \frac{1}{1 + p_{\mathrm{G}}} \frac{c_{\max}}{1 + x''_{\mathrm{d}} \sin\phi_{\mathrm{rG}}}$$

and

$$K_{\mathrm{T,SO}} = \frac{1}{1 + p_{\mathrm{G}}} \frac{c_{\max}}{1 - x_{\mathrm{T}} \sin\phi_{\mathrm{rG}}} \tag{8.11}$$

where p_{G} is the range of generator voltage regulation.

8.3.2 Meshed Networks

For calculation of i_{p} in meshed networks, Figure 8.6, three methods (A, B, and C) are described.

8.3.2.1 Method A: Uniform Ratio R/X or X/R Ratio Method

The factor χ in Equation 8.4 is determined from the smallest ratio of R/X of all branches of the network. Only the branches that carry the partial short-circuit currents at the nominal

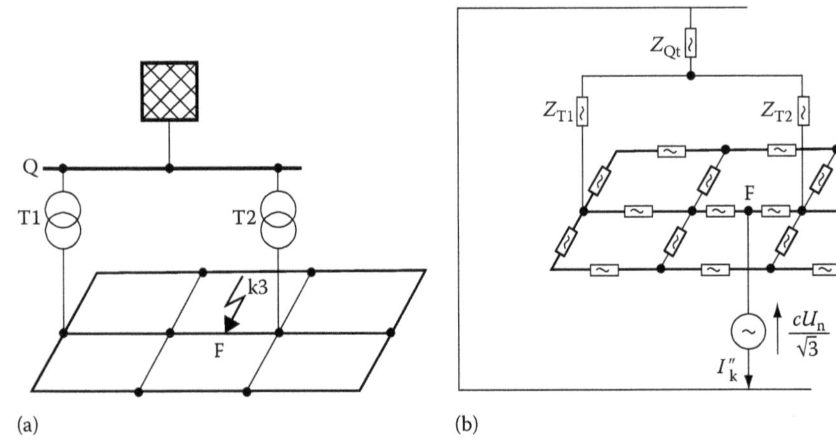

FIGURE 8.6
Calculation of initial short-circuit current in a meshed network. (a) The system diagram; (b) equivalent circuit diagram with equivalent voltage source. (From IEC, Short-Circuit Calculations in Three-Phase AC Systems, 1st edn., 1988, Now revised IEC 60909-0, Short-Circuit Currents in Three-Phase AC Systems, 0—Calculation of Currents, 2001–2007.)

voltage corresponding to the short-circuit location and branches with transformers adjacent to the short-circuit locations. Any branch may be a series combination of several elements.

8.3.2.2 Ratio R/X or X/R at the Short-Circuit Location

The factor $\chi = 1.15\chi_b$, where factor 1.15 is a safety factor to cover inaccuracies caused by using X/R from a meshed network reduction with complex impedances, and χ_b is calculated from curves in Figure 8.3 or mathematically from Equation 8.4. In the low-voltage networks, the product of $1.15\chi_b$ is limited to 1.8 and in the high-voltage networks to 2.0. As long as R/X remains smaller than 0.3 in all branches, it is not necessary to use factor 1.15.

8.3.2.3 Method C: Equivalent Frequency Method

This method provides the equivalent frequency approach. A source of 20 Hz for 50 Hz systems and 24 Hz for 60 Hz systems is considered to excite the network at the fault point. The X/R at the fault point is then

$$\frac{X}{R} = \left(\frac{X_c}{R_c}\right)\left(\frac{f}{f_c}\right) \tag{8.12}$$

where
- f is the system frequency
- f_c is the excitation frequency
- $Z_c = R_c + jX_c$ at the excitation frequency

The factor $\chi = \chi_c$ is used in the calculations for the peak current.

8.4 Near-to-Generator Faults

A "near-to-generator" fault is a short circuit to which at least one synchronous machine contributes a prospective initial symmetrical short-circuit current that is more than twice the generator's rated current or a short circuit to which synchronous and asynchronous motors contribute more than 5% of the initial symmetrical short-circuit current I_k'', calculated without motors. These fault types have ac decay. This is equivalent to ANSI *local* faults.

The factor c is applicable to prefault voltages as in the case of far-from-generator faults. The impedances of the generators and power station transformers are modified by additional factors, depending on their connection in the system.

8.4.1 Generators Directly Connected to Systems

When generators are directly connected to the systems, their positive sequence impedance is modified by a factor K_G:

$$Z_{GK} = K_G(R_G + jX_d'') \tag{8.13}$$

K_G is given by

$$K_G = \frac{U_n}{U_{rG}}\left(\frac{c_{max}}{1 + X_d'' \sin\phi_{rG}}\right) \tag{8.14}$$

where

U_{rG} is the rated voltage of the generator
U_n is the nominal system voltage
ϕ_{rG} is the phase angle between the generator current I_{rG} and generator voltage U_{rG}
X_d'' is the subtransient reactance of the generator, at a generator-rated voltage on a generator MVA base

Figure 8.7 shows the applicable phasor diagram.

If the generator voltage is different from U_{rG}, use

$$U_G = U_{rG}(1 + p_G) \tag{8.15}$$

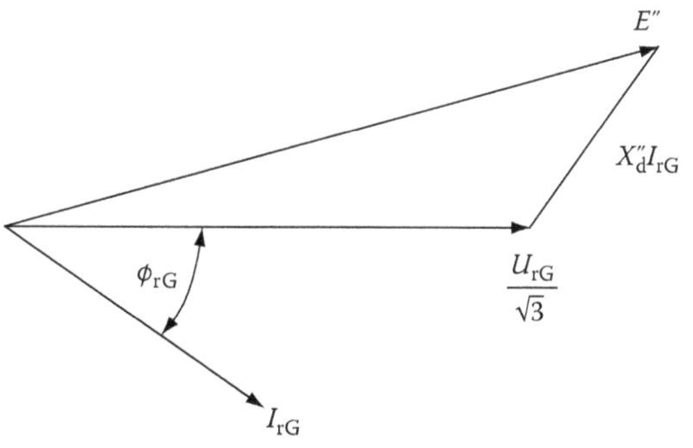

FIGURE 8.7
Phasor diagram of a synchronous generator at rated load and power factor.

The generator resistance R_G with sufficient accuracy is given by the following expressions:

$$\begin{aligned} R_{Gf} &= 0.05X''_d \quad \text{for generators with } U_{rG} > 1\text{ kV and } S_{rG} \geq 100\text{ MVA} \\ R_{Gf} &= 0.07X''_d \quad \text{for generators with } U_{rG} > 1\text{ kV and } S_{rG} < 100\text{ MVA} \\ R_{Gf} &= 0.15X''_d \quad \text{for generators with } U_{rG} \leq 1\text{ kV} \end{aligned} \tag{8.16}$$

These values of R_{Gf} are not to be used for the calculations of the peak short-circuit current. These values cannot be used when calculating the aperiodic component I_{dc} of the short-circuit current. The effective stator resistance of synchronous machines is much below the value of R_{Gf}, and manufacturer's value of R_G should be used.

This was a limitation in IEC standard of year 1988 that the above values of R_{Gf} gave a depressed value of I_{dc} as compared to other methods of calculations. This anomaly has now been rectified.

8.4.2 Generators and Unit Transformers of Power Station Units

For generators and unit transformers of power stations, the generator and the transformer are considered as a single unit. The following equation is used for the impedance of the whole power station unit (PSU) for the short circuit on the high side of the unit transformer, *with on-load tap changer* Figure 8.5:

$$Z_s = K_s\left(t_r^2 Z_G + Z_{THV}\right) \tag{8.17}$$

with

$$K_s = \frac{U_{nQ}^2}{U_{rG}^2}\frac{U_{rTLV}^2}{U_{rTHV}^2}\frac{c_{max}}{1 + |x''_d - x_T|\sin\phi_{rG}} \tag{8.18}$$

Here, U_{nq} is the nominal system voltage at the feeder connection point Q of the power station unit.

For calculations *without on-load tap changers*, the following equation can be used for the short circuit on the high side of the transformer unit.

$$Z_s = K_{SO}\left(t_t^2 Z_G + Z_{THV}\right) \quad \text{with } K_{SO} = \frac{U_{nQ}}{U_{rG}(1 + p_G)}\frac{U_{rTLV}}{U_{rTHV}}(1 \pm p_T)\frac{c_{max}}{1 + x''_d \sin\phi_{rG}} \tag{8.19}$$

$(1 \pm p_T)$ is introduced if the unit transformer has off-load taps and if one of the taps is permanently used. The highest short-circuit current will be given by $1 - p_T$.

8.4.3 Motors

For calculations of I''_k, synchronous motors and synchronous compensators are treated as synchronous generators. The impedance Z_M of asynchronous motors is determined from their locked rotor currents.

The following ratios of resistance to reactance of the motors apply with sufficient accuracy:

$\frac{R_M}{X_M} = 0.10$ with $X_M = 0.995\ Z_M$ for high-voltage motors with powers P_{rM} per pair of poles ≥ 1 MW

$\frac{R_M}{X_M} = 0.15$ with $X_M = 0.989\ Z_M$ for high-voltage motors with powers P_{rM} per pair of poles < 1 MW

$\frac{R_M}{X_M} = 0.42$ with $X_M = 0.922\ Z_M$ for low-voltage motor groups with connection cables (8.20)

8.4.4 Short-Circuit Currents Fed from One Generator

The initial short-circuit current is given by Equation 8.2. The peak short-circuit current is calculated as for far-from-generator faults, considering the type of network. For generator, corrected resistance $K_G R_G$ and corrected reactance $K_G X''_d$ are used.

8.4.4.1 Breaking Current

The symmetrical short-circuit breaking current for single fed or nonmeshed systems is given by

$$I_b = \mu I''_k \tag{8.21}$$

where factor μ accounts for ac decay. The following values of μ are applicable for medium-voltage turbine generators, salient pole generators, and synchronous compensators excited by rotating exciters or by static exciters, provided that for the static exciters the minimum time delay is less than 0.25 s, and the maximum excitation voltage is less than 1.6 times the rated excitation voltage. For all other cases μ is taken to be 1, if the exact value is not known.

When there is a unit transformer between the generator and short-circuit location, the partial short-circuit current at the high side of the transformer is calculated.

$$\begin{aligned} \mu &= 0.84 + 0.26e^{-0.26 I''_{kG}/I_{rG}} \quad \text{for } t_{min} = 0.02 \text{ s} \\ \mu &= 0.71 + 0.51e^{-0.30 I''_{kG}/I_{rG}} \quad \text{for } t_{min} = 0.05 \text{ s} \\ \mu &= 0.62 + 0.72e^{-0.32 I''_{kG}/I_{rG}} \quad \text{for } t_{min} = 0.10 \text{ s} \\ \mu &= 0.56 + 0.94e^{-0.38 I''_{kG}/I_{rG}} \quad \text{for } t_{min} \geq 0.25 \text{ s} \end{aligned} \tag{8.22}$$

If the ratio of the initial short-circuit current and the machine-rated current is equal to or less than 2, then the following relation holds:

$$\frac{I''_{kG}}{I_{rG}} \leq 2 \quad \mu = 1 \quad \text{for all values of } t_{min} \tag{8.23}$$

In the case of asynchronous motors, replace

$$\frac{I''_{kG}}{I_{rG}} \text{ by } \frac{I''_{kM}}{I_{rM}} \tag{8.24}$$

The equations can also be used for compound excited low-voltage generators with a minimum time delay not >0.1 s. The calculations of low-voltage breaking currents for a time duration >0.1 s are not included in the IEC standard.

8.4.4.2 Steady-State Current

The maximum and minimum short-circuit currents are calculated from

$$I_{k\,max} = \lambda_{max} I_{rG} \tag{8.25}$$

$$I_{k\,min} = \lambda_{min} I_{rG} \tag{8.26}$$

where λ_{max} and λ_{min} for turbine generators are calculated from the graphs in [1]. Figure 8.8 shows these values for cylindrical rotor generators. In this figure, X_{dsat} is the reciprocal of the short-circuit ratio. We have not yet defined the short-circuit ratio of a generator. It is given by

$$R_{sc} = \frac{\text{per unit excitation at normal voltage on open circuit}}{\text{per unit excitation for rated armature current on short circuit}} \tag{8.27}$$

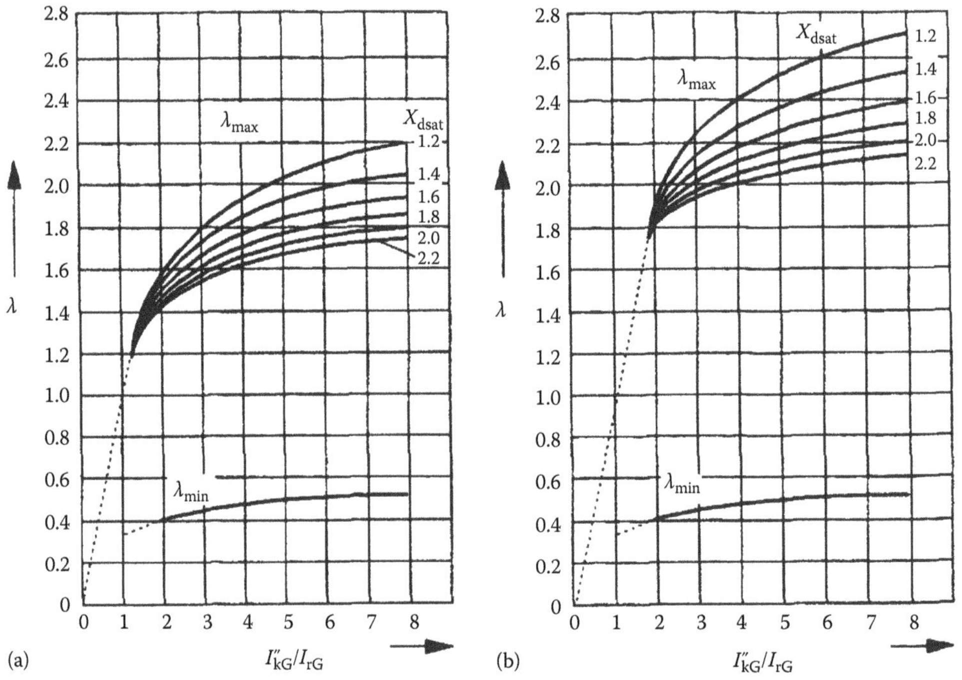

FIGURE 8.8
(a, b) Factors λ_{max} and λ_{min} for cylindrical rotor generators; Series One and Series Two are defined in the text. (From IEC, Short-Circuit Calculations in Three-Phase AC Systems, 1st edn., 1988, Now revised IEC 60909-0, Short-Circuit Currents in Three-Phase AC Systems, 0—Calculation of Currents, 2001–2007.)

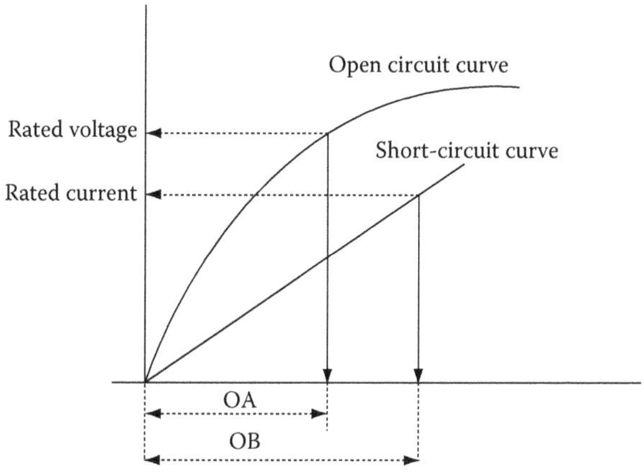

FIGURE 8.9
Open-circuit magnetization and short-circuit curves of a synchronous generator to illustrate short-circuit ratio.

Referring to the open circuit and short-circuit curves of the generator, shown in Figure 8.9, the short-circuit ratio is OA/OB. Chapter 6 defined saturated synchronous reactance as the ratio of per unit voltage on open circuit to per unit armature current on short circuit. In Figure 8.9, OA is related to the normal rated voltage, and OB is proportional to the rated current of the machine. The short-circuit ratio is, therefore, the reciprocal of synchronous reactance. It is a measure of the stiffness of the machine, and modern generators tend to have lower short-circuit ratios compared to those of their predecessors, of the order of 0.5 or even lower.

For the static excitation systems fed from generator terminals and a short circuit at the terminals, the field voltage collapses with the terminal voltage and therefore take $\lambda_{max} = \lambda_{min} = 0$. Maximum λ curves for Series One are based on the highest possible excitation voltage according to either 1.3 times the rated excitation at rated load and power factor for turbine generators or 1.6 times the rated excitation for a salient-pole machine. The maximum λ curves for Series Two are based on the highest excitation voltage according to either 1.6 times the rated excitation at rated load and power factor for turbine generators or 2.0 times the rated excitation for salient-pole machines. The graphs for the salient pole machines are similar and not shown.

8.4.5 Short-Circuit Currents in Nonmeshed Networks

The procedure is the same as that described for far-from-generator faults. The modified impedances are used. The branch currents are superimposed, as shown in Figure 8.10.

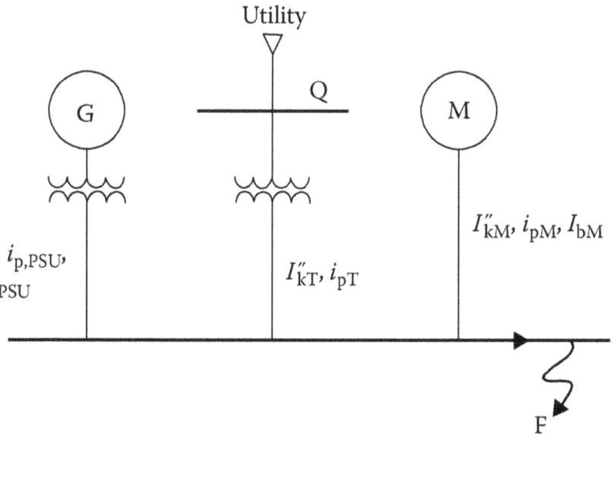

FIGURE 8.10
Calculation of I''_k, i_p, I_b, and I_k for a three-phase short-circuit fed from nonmeshed sources.

$$\begin{aligned} I_k'' &= I_{k,PSU}'' + I_{kT}'' + I_{kM}'' + \cdots \\ i_p &= i_{p,PSU} + i_{pT} + i_{pM} + \cdots \\ I_b &= I_{b,PSU} + I_{kT}'' + I_{bM} + \cdots \\ I_K &= I_{b,PSU} + I_{kT}'' + \cdots \end{aligned} \tag{8.28}$$

8.4.6 Short-Circuit Currents in Meshed Networks

Figure 8.11 shows that the initial short-circuit currents in meshed networks can be calculated by using modified impedances and the prefault voltage at the fault point. The peak current i_p is calculated as for far-from-generator faults. Methods A, B, and C for meshed

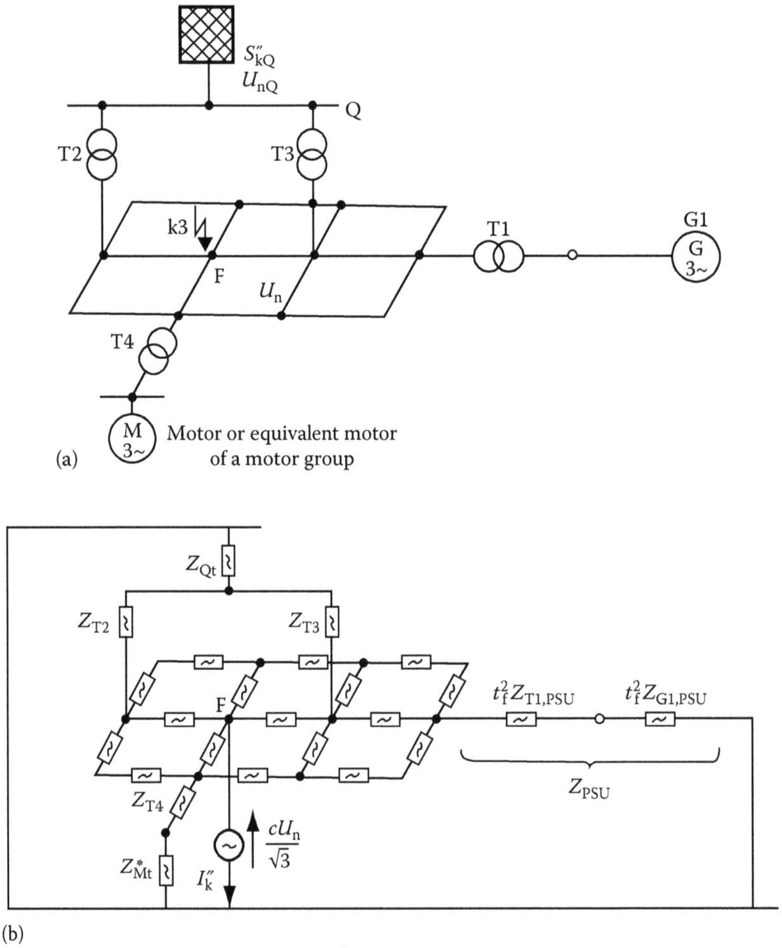

FIGURE 8.11
Calculation of initial short-circuit current in a meshed network fed from several sources. (a) The system diagram; (b) equivalent circuit diagram with equivalent voltage source. (From IEC, Short-Circuit Calculations in Three-Phase AC Systems, 1st edn., 1988, Now revised IEC 60909-0, Short-Circuit Currents in Three-Phase AC Systems, 0—Calculation of Currents, 2001–2007.)

networks are applicable. The symmetrical short-circuit breaking current for meshed networks is conservatively given by

$$I_b = I''_k \tag{8.29}$$

A more accurate expression is provided as follows:

$$I_b = I''_k - \sum_i \left(\frac{\Delta U''_{Gi}}{cU_n/\sqrt{3}}\right)(1-\mu_i)I''_{kGi} - \sum_j \left(\frac{\Delta U''_{Mj}}{cU_n/\sqrt{3}}\right)(1-\mu_j q_j)I''_{kMj} \tag{8.30}$$

$$\begin{aligned} \Delta U''_{Gi} &= jX''_{di}I''_{kGi} \\ \Delta U''_{Mj} &= jX''_{Mj}I''_{kMj} \end{aligned} \tag{8.31}$$

where

$\Delta U''_{Gi}$ and $\Delta U''_{Mj}$ are the initial voltage differences at the connection points of the synchronous machine i, and the asynchronous motor j

I''_{kGi} and I''_{kMj} are the parts of the initial symmetrical short-circuit currents of the synchronous machine i and the asynchronous motor j

μ is defined in Equation 8.22 and q is defined in Equation 8.36

For the steady-state current, the effect of motors is neglected. It is given by

$$I_k = I''_{k,M} \tag{8.32}$$

8.5 Influence of Motors

Synchronous motors and synchronous compensators contribute to I''_k, i_p, I_b, and I_k. Asynchronous motors contribute to I''_k, i_p, and I_b and for unbalanced faults to I_k. Low-voltage motors in public power supply systems can be neglected. High- and low-voltage motors that are connected through a two-winding transformer can be neglected if

$$\frac{\sum P_{rM}}{\sum S_{rT}} \leq \frac{0.8}{\left|\frac{c100\sum S_{rT}}{S''_{kQ}} - 0.3\right|} \tag{8.33}$$

where

$\sum P_{rM}$ is the sum of rated active power of the motors

$\sum S_{rT}$ is the sum of rated apparent power of the transformers

S''_{kQ} is the symmetrical short-circuit power at the connection point without the effect of motors

This expression is not valid for three-winding transformers.

8.5.1 Low-Voltage Motor Groups

For simplifications of the calculations, groups of low-voltage motors including their connecting cables can be combined into an equivalent motor:

$$
\begin{aligned}
&I_{\mathrm{rM}} = \text{sum of rated currents of all motors in the group}\\
&\text{Ratio of locked rotor current of full load current} = \frac{I_{\mathrm{LR}}}{I_{\mathrm{rM}}} = 5\\
&\frac{R_{\mathrm{M}}}{X_{\mathrm{M}}} = 0.42,\ \chi_{\mathrm{a}} = 1.3, \text{ and } m = 0.05 \text{ MW if nothing definite is known}\\
&\text{where } m = P_{\mathrm{rm}}/\mathrm{p},\quad \mathrm{p} = \text{number of pair of poles}
\end{aligned}
\tag{8.34}
$$

The partial short-circuit current of low-voltage motors is neglected if the rated current of the equivalent motor (sum of the ratings of group of motors) is <0.01% of the initial symmetrical short-circuit current at the low-voltage bus to which these motors are directly connected, without the contributions from the motors:

$$I_{\mathrm{rM}} \le 0.01 I''_{\mathrm{k,M}} \tag{8.35}$$

8.5.2 Calculations of Breaking Currents of Asynchronous Motors

For the calculation of breaking short-circuit current from asynchronous motors, another factor q (in addition to μ) is introduced: $q = 1$ for synchronous machines. The factor q is given by

$$
\begin{aligned}
q &= 1.03 + 0.12 \ln m \quad \text{for } t_{\min} = 0.02 \text{ s}\\
&= 0.79 + 0.12 \ln m \quad \text{for } t_{\min} = 0.05 \text{ s}\\
&= 0.57 + 0.12 \ln m \quad \text{for } t_{\min} = 0.10 \text{ s}\\
&= 0.26 + 0.10 \ln m \quad \text{for } t_{\min} \ge 0.25 \text{ s}
\end{aligned}
\tag{8.36}
$$

where m is the rated active power of motors per pair of poles.

Therefore, the breaking current of asynchronous machines is given by

$$I_{\mathrm{b,sym}} = \mu q I''_{\mathrm{k}} \tag{8.37}$$

8.5.3 Static Converter Fed Drives

Static converters for drives as in rolling mills (reversible converter fed drives) contribute to I''_{k} and i_{p} only if the rotational masses of the motors and the static equipment provide reverse transfer of energy for deceleration (a transient inverter operation) at the time of short circuit. These do not contribute to I_{b}. Nonrotating loads and capacitors (parallel or series) do not contribute to the short-circuit currents. Static power converter devices are treated in a similar manner as asynchronous motors with following parameters:

U_{rM} = rated voltage of the static converter transformer on the network side or rated voltage of the static converter if no transformer is present

I_{rM} = rated current of the static converter transformer on the network side or rated voltage of the static converter if no transformer is present

$$\frac{I_{LR}}{I_{rM}} = 3$$
$$\frac{R_M}{X_M} = 0.10 \quad \text{with } X_M = 0.995\ Z_M \tag{8.38}$$

8.6 Comparison with ANSI Calculation Procedures

The following comparison of calculation procedures underlines the basic philosophies in ANSI and IEC calculations [8].

1. IEC requires calculation of initial symmetrical short-circuit current, I_k'', in each contributing source. These component I_k'' currents form the basis of further calculations. Thus, tracking each contributing source current throughout the network is necessary. Each of these component currents is a function of machine characteristics, R/X ratio, type of network (meshed or radial), type of excitation system for synchronous generators, contact parting time (minimum time delay), and the determination whether contribution is near to or far from the short-circuit location. Multiplying factors on generators, unit transformers, and PSUs are applicable before the calculation proceeds. The PSU consisting of a generator and a transformer is considered as a single entity, and separate procedures are applicable for calculation of whether the fault is on the high- or low-voltage side of the transformer. IEC treats each of above factors differently for each contributing source.

   This approach is conceptually different from that of ANSI, which makes no distinction between the type of network, and the network is reduced to a single Thévenin impedance at the fault point, using complex reduction, or from separate R and X networks, though prior impedance multiplying factors are applicable to account for ac decay. ANSI states that there is no completely accurate way of combining two parallel circuits with different values of X/R ratios into a single circuit with one X/R ratio. The current from the several circuits will be the sum of the decaying terms, usually with different exponents, while from a single circuit, it contains just one such term. The standard then advocates separate X and R networks and states that the error for practical purposes is on the conservative side.

   In all IEC calculations, therefore, the initial short-circuit current is first required to be calculated as all other currents are based on this current. These initial short-circuit currents from sources must be tracked throughout the distribution system. The peak current and breaking current are then calculated based on factors to be applied to initial short-circuit current. In ANSI calculations, interrupting duty and first-cycle networks can be independently formed with prior impedance multiplying factors. In an application of circuit breakers, one of the two duties, i.e., the first cycle or interrupting may only be the limiting factor.

2. Both standards recognize ac decay, though the treatment is different. ANSI standards model motors with prior multipliers. Where the contribution of the large motors is an appreciable portion of the short-circuit current, substitution of tabulated multipliers with more accurate data based on manufacturer's time constants is recommended, but ANSI treats these multipliers on a *global basis*, and these do

not change with the location of fault points or the contact parting time of the breaker. IEC treats each motor individually and decay must be calculated on the basis of contact parting time, machine type, and its size, speed, and proximity to the fault. Contributions from motors can be ignored in certain cases in IEC, while ANSI considers motor contributions throughout, except as discussed in Section 7.2.1, though effect of motor loads can be insignificant in some cases, Example 7.2.

3. ANSI makes no distinction between remoteness of induction and synchronous motors for the short-circuit calculations. Impedance multiplying factors (Table 7.1) of the motors are considered to account for ac decay, irrespective of their location. IEC considers generators and *motors* as near to or far from the fault location for breaking and steady-state current calculations. Asynchronous machines are considered near if the sum of all motors I_k'' is >5% of the total I_k'' without motors; otherwise, these are considered remote. Synchronous machines are considered near if their I_k'' is more than twice the rated current.
4. Both standards recognize the rapid decay of the dc component of the fault current and add a half-cycle of tripping time to arrive at the contact parting time (IEC minimum time delay).
5. IEC calculations require that the dc component be calculated at the contact parting time to calculate the asymmetrical breaking current, i.e.,

$$i_{dc} = \sqrt{2} I_k'' \varepsilon^{-\omega t/X/R} \tag{8.39}$$

where X/R is computed differently for radial or meshed networks. In radial networks, i_{dc} is the sum of the dc currents calculated with X/R ratios in each of the contributing elements. See the qualification with respect to X/R ratio of generators in Section 8.4.1. For meshed networks, the ratio R/X should be determined by method (c). The equivalent f_c should be used as shown in Table 8.3. Methods B and C are applicable for meshed networks for calculation of i_p.

Calculations of the dc component at the contact parting time are not required in ANSI. The dc decay is built into the postfault calculations with E/X or E/Z multipliers from the curves. Also, the rating structure of ANSI takes into account certain asymmetry, depending on the contact parting time, Figure 5.4.

6. ANSI uses a prefault voltage equal to the rated system voltage, unless operating conditions show otherwise. IEC considers a prior voltage multiplying factor c from Table 8.2 *irrespective of the system conditions*. This factor seems to be imported from VDE (Deutsche Electrotechnische Kommission) and is approximately given by:

$$cU_n = U_{rG}\left(1 + \frac{I_{rG}\sqrt{3}X_d'' \sin\phi_{rG}}{U_{rG}}\right) \approx 1.1 \tag{8.40}$$

TABLE 8.3

Equivalent Frequency f_c for Meshed Networks, Method (c)

$f \cdot t$	<1	<2.5	<5	<12.5
f_c/f	0.27	0.15	0.092	0.055

$f \cdot t$ is product of frequency and time.

7. The calculations of steady-state current are materially different and more involved in IEC calculations.
8. IEC specifically details the calculation of i_p and I_k'' from static converter-fed drive systems and so does ANSI/IEEE specification, see Section 7.9.6.
9. IEC calculations are more demanding on the computing resources and require a larger database.

8.7 Examples of Calculations and Comparison with ANSI Methods

Example 8.1

Calculate the fault current contributions of the following synchronous machines, directly connected to a bus, using ANSI and IEC methods. Calculate the first-cycle (IEC peak) and interrupting (IEC breaking, symmetrical, and asymmetrical) currents for contact parting times of two cycles and three cycles (IEC minimum time delay = 0.03 and 0.05 s, approximately for a 60 Hz system). Compare the results.

- 110 MVA, 13.8 kV, 0.85 power factor generator, $X''_{dv} = 16\%$ on generator MVA base
- 50 MVA, 13.8 kV, 0.85 power factor generator, $X''_{dv} = 11\%$ on generator MVA base
- 2000 hp, 10-pole, 2.3 kV, 0.8 power factor synchronous motor, $X''_{lr} = 20\%$
- 10,000 hp, 4-pole, 4 kV synchronous motor, 0.8 power factor, $X''_{lr} = 15\%$

Table 8.4 shows the results of ANSI calculations. The first-cycle peak is calculated using Equation 7.1 for new ratings of the breakers. These calculations have already been discussed in Chapter 7, and the description is not repeated here. Table 8.5 shows all the steps in IEC calculations. We will go through these steps for a sample calculation for a 110 MVA generator.

The percentage subtransient reactances for all the machines are the same in both calculations. The X/R ratio for ANSI calculations is estimated from Figure 7.4, while for IEC calculations, it is based on Equation 8.16. For a 110 MVA generator, $R_{Gf} = 0.05\ X''_d$, i.e., $X/R = 20$.

Next, factor K_G is calculated from Equation 8.14. This is based on a rated power factor of 0.85 of the generator:

$$K_G = \frac{U_n}{U_{rG}}\left(\frac{c_{max}}{1 + X''_d \sin\phi_{rG}}\right) = \frac{1.10}{1 + 0.16 \times 0.526} = 1.05$$

In the above calculation, X''_d is in per unit on a machine MVA base at machine rated voltage and $U_n = U_{rG}$. From Equation 8.13, the modified generator impedance is

$$Z_{GK} = K_G\left(R_G + jX''_d\right) = 1.05(0.0073 + j0.1455) \text{ per unit 100 MVA base}$$

The initial short-circuit current from Equation 8.2 is

$$\left|I''_k\right| = \frac{c_{max}}{|Z_{GK}|} = \frac{1.1}{0.153} = 7.191 \text{ per unit} = 30.09 \text{ kA}$$

TABLE 8.4

ANSI Fault Current Calculations from Synchronous Generators and Motors Directly Connected to a Bus

			Impedance Multiplying Factors		First-Cycle Calculations		Interrupting Duty Calculations	
Description	**Percentage X''_d on Equipment MVA Base**	X/R	**First Cycle**	**Interrupting**	**First-Cycle Current (kA sym.)**	**First-Cycle Current (kA peak)**	**3-Cycle Contact Parting Time (kA rms)**	**2-Cycle Contact Parting Time (kA rms)**
110 MVA, 0.85 pF, 13.8 kV generator	16	80.0	1	1	28.76	79.81	34.62 MF = 1.204	34.20 MF = 1.189
50 MVA, 0.85 pF, 13.8 kV generator	11	65.0	1	1	19.02	52.55	22.30 MF = 1.173	22.35 MF = 1.175
2000 hp, 10-pole, 2.3 kV synchronous motor, 0.8 pF (2000 kVA)	20	25.0	1	1.5	2.51	6.67	1.67	1.67
10,000 hp, 4-pole, 4 kV synchronous motor, 0.8 pF (10,000 kVA)	15	34.4	1	1.5	9.62	25.99	6.41	6.41

TABLE 8.5

IEC Fault Current Calculations from Synchronous Generators and Motors Directly Connected to a Bus, Three-Phase Short-Circuit

Equipment	110 MVA, 0.85 pF, 13.8 kV Generator	50 MVA, 0.85 pF, 13.8 kV Generator	2000 hp, 10-pole, 0.8 pF, 2.4 kV Synchronous Motor	10,000 hp, 4-pole, 0.8 pF, 4 kV Synchronous Motor
Percentage X''_d on equipment kVA base	16	11	20	15
R_G or R_M	$0.05\ X''_d$	$0.07\ X''_d$	$0.07\ X''_d$	$0.07\ X''_d$
X/R for dc component	80	65	—	—
c_{max}	1.1	1.1	1.1	1.1
K_G or K_M	1.015	1.040	0.982	1.010
I''_{kG} or I''_{kM} kA rms	30.09	20.06	2.68	10.45
κ	1.863	1.814	1.814	1.814
i_p kA peak	79.27	51.46	6.88	26.80
μ (0.05 s)	0.78	0.74	0.81	0.73
M (0.03 s)	0.85	0.82	0.85	0.81
i_{bsym} (0.05 s)	23.47	14.84	2.17	7.63
i_{bsym} (0.03 s)	25.57	16.45	2.28	8.46
I_{DC} (0.05 s) kA	34.52	21.22	1.01	3.95
I_{DC} (0.03 s) kA	37.93	23.80	1.72	6.70
i_{basym} (0.05 s) kA	41.74	25.89	2.39	8.50
i_{basym} (0.03 s) kA	45.74	28.90	2.86	10.79

Here, we are interested in magnitude only, as the calculations are from a single source and summations are not involved. For the calculation of peak current, the factor χ is calculated from Equation 8.4:

$$\chi = 1.02 + 0.98e^{-3R/X} = 1.02 + 0.98 \times 0.861 = 1.863$$

The peak current from Equation 8.3 is therefore

$$i_p = \chi\sqrt{2}I''_k = 1.863 \times \sqrt{2} \times 30.09 = 79.27\ \text{kA}$$

The breaking current factor μ is calculated from Equation 8.22. For 0.05 minimum time delay:

$$\mu = 0.71 + 0.51e^{-0.30I''_{kG}/I_{rG}} = 0.71 + 0.51e^{-0.30\times 30.09/4.60} = 0.78$$

The calculation for 0.03 minimum time delay is not given directly by Equations 8.22, and interpolation is required. Alternatively, the factor can be estimated from the graphs in the IEC standard.

The symmetrical interrupting current at 0.05 s minimum time delay is

$$i_{bsym} = \mu I''_k = 0.78 \times 30.09 = 23.47\ \text{kA}$$

The dc component is calculated from Equation 8.39. However, to calculate X/R ratio R_G as calculated above is not used, as per qualification stated in section. Using an X/R of 80, the same as for the ANSI/IEEE calculation, gives

$$i_{dc} = \sqrt{2}I''_k e^{-\omega t/X/R} = \sqrt{2} \times 30.9 \times e^{-377\times 0.05/80} = 34.52\ \text{kA}$$

The asymmetrical breaking current at 0.05 s parting time is

$$i_{\text{basym}} = \sqrt{i_{\text{bsym}}^2 + i_{\text{dc}}^2} = 41.74 \text{ kA rms}$$

Table 8.5 is compiled similarly for other machines. Synchronous motors are treated as synchronous generators for the calculations. A comparison of the results by two methods of calculation shows some differences. ANSI first-cycle current and IEC peak currents are comparable, with a difference within 3%. ANSI interrupting duty symmetrical currents for generators are higher than IEC breaking currents, i.e., for a 110 MVA generator, the ANSI current at 2-cycle contact parting time is 28.76 kA (without multiplying factor), while IEC current is 25.57 kA.

For comparison with IEC asymmetrical breaking current for 5-cycle breaker, from Table 8.5, $i_{\text{basym}} = 41.74$ kA, and, from Table 8.4, ANSI calculation, the interrupting sym rating is calculated as 34.62 kA. As per the rating structure of the breakers, the asymmetrical interrupting current is $1.1 \times 34.62 = 38.08$ kA. Thus, IEC calculation gives higher asymmetrical breaking current. Similarly, for 3-cycle breaker, $i_{\text{basym}} = 45.74$ kA, and as per ANSI calculation the corresponding number to compare is $1.2 \times 34.20 = 41.04$. In ANSI calculations, the short-circuit currents from the motors for various contact parting times do not change, and no remote or local multiplying factors are applicable to these currents.

Example 8.2

Calculate the fault current contributions of the following asynchronous machines, directly connected to a bus, using ANSI and IEC methods. Calculate first-cycle (IEC peak) currents and the interrupting (breaking) currents at contact parting times of two and three cycles, 60 Hz basis, and IEC minimum time delays of 0.03 and 0.05 s approximately. Compare the results.

- 320 hp, 2-pole, 2.3 kV induction motor, $X_{\text{lr}} = 16.7\%$
- 320 hp, 4-pole, 2.3 kV induction motor, $X_{\text{lr}} = 16.7\%$
- 1560 hp, 4-pole, 2.3 kV induction motor, $X_{\text{lr}} = 16.7\%$

These results of ANSI calculations are shown in Table 8.6, while those of IEC calculations are shown in Table 8.7. Most of the calculation steps for asynchronous motors are in common with those for synchronous motors, as illustrated in Example 8.1. The motor-locked rotor reactance of $X_{\text{lr}} = 16.7\%$ on a motor kVA base is used in both calculation methods; however, the resistances are based on recommendations in each standard. Factor q must also be calculated for asynchronous motors and it is given by Equation 8.36. This requires m equal to the motor-rated power in megawatts per *pair* of poles to be calculated on the basis of motor power factor and efficiency. The symmetrical breaking current is then given by (8.37).

A comparison of results again shows divergence in the calculated currents. In ANSI calculations, the interrupting duty current for a 320 hp two-pole motor is twice that of the four-pole motor, 0.28 kA versus 0.14 kA. This is so because a prior impedance multiplying factor of 1.5 is applicable to a two-pole 320 hp motor, and this factor is 3 for a four-pole 320 hp motor. IEC calculation results for a two-pole motor is only slightly higher, 0.235 kA. The IEC calculations consider decay of currents from asynchronous motors, while ANSI/IEEE does not. There is no direct comparison of the asymmetrical motor currents in Table 8.7.

Example 8.3

Example 7.2 is repeated with the IEC method of calculation. All the impedance data remain unchanged, except that the resistance components are estimated from IEC equations.

TABLE 8.6

ANSI Fault Current Calculations from Asynchronous Motors Directly Connected to a Bus

			Impedance Multiplying Factors		First-Cycle Calculations		Interrupting Duty Calculations	
Description	X_{lr} on Equipment MVA Base	X/R	First Cycle	Interrupting	First-Cycle Current (kA sym.)	First-Cycle Current (kA peak)	3-Cycle Contact Parting Time (kA rms)	2-Cycle Contact Parting Time (kA rms)
320 hp, 2-pole induction motor, 2.3 kV (kVA = 285)	16.7	15	1	1.5	0.427	1.095	0.28	0.28
320 hp, 4-pole induction motor, 2.3 kV (kVA = 285)	16.7	15	1.2	3	0.356	0.909	0.14	0.14
1560 hp, 4-pole, 2.3 kV induction motor (kVA = 1350)	16.7	28.5	1	1.5	2.028	5.441	1.35	1.35

TABLE 8.7

IEC Fault Current Calculations from Asynchronous Motors Directly Connected to a Bus, Three-Phase Short-Circuit

Description	300 hp, 2-Pole, 2.3 kV Induction Motor (kVA = 285), Power Factor = 0.9, Efficiency = 0.93	300 hp, 4-Pole, 2.3 kV Induction Motor (kVA = 285), Power Factor = 0.9, Efficiency = 0.93	1500 hp, 4-Pole, 2.3 kV Induction Motor (kVA = 1350), Power Factor = 0.92, Efficiency = 0.94
I_{LR}/I_{rM}	6	6	6
$I''_K I_{rM}$	6.6	6.6	6.6
m (electric power per pair of poles) (MW)	0.256	0.128	0.621
R_M/X_M	0.15	0.15	0.15
k_M	1.65	1.65	1.65
μ (0.05 s)	0.79	0.79	0.79
μ (0.03 s)	0.83	0.83	0.83
q (0.05 s)	0.63	0.54	0.73
q (0.03 s)	0.79	0.66	0.86
I''_{KM} (kA rms)	0.473	0.473	2.229
i_{pM} (kA crest)	1.104	1.104	5.201
i_{bsym} (0.05 s) kA	0.235	0.202	1.285
i_{bsym} (0.03 s) kA	0.310	0.259	1.591
I_{DC} (0.05 s) kA	0.029	0.029	0.136
I_{DC} (0.03 s) kA	0.101	0.101	0.479
i_{basym} (0.05 s)	0.237	0.204	1.292
i_{basym} (0.03 s)	0.326	0.278	1.661

Three-Phase Short-Circuit at F1

For a fault at Fl, the generator and transformer are considered as a PSU. The ANSI/IEEE calculations in Example 7.2 are performed at rated system voltages. This means that the adjustments of taps due to tap changing of transformers, variations in the source voltage and generation voltage are neglected. Equations 8.18 and 8.19 consider variation of the voltages at the fault point due to tap changing, generator voltage regulation, or off-load tap settings. As the purpose of the example is to compare results on the same basis, rated voltages are considered. This means that the 60/100 MVA transformer is at rated voltage tap ratio, which is also the nominal system voltage on the high-voltage side and the generator-rated voltage on the low-voltage side. For a fault on the high-voltage side, the correction factor, given by Equation 8.18, becomes

$$K_S = \frac{c_{max}}{1 + \left|x''_d - x_T\right| \sin\phi_{rG}}$$

The ANSI calculations in Example 7.2 are performed with generator at no load. Here, consider a power factor of 0.85 lagging. Then:

$$K_S = \frac{1.1}{1 + (0.164 - 0.0774)(0.527)} = 1.052$$

The generator and transformer per unit impedances are on their respective MVA base, as shown in Table 7.13. The modified power station impedance is then

$$Z_s = K_S(Z_G + Z_T)$$

The generator resistance calculated is 0.05 times the subtransient reactance. Therefore:

$$Z_s = 1.052(0.007315 + j0.14630 + 0.00404 + j0.12894)$$
$$= 0.01195 + j0.28955$$

in per unit on a 100 MVA base. It is not necessary to consider the asynchronous motor contributions. The initial short-circuit current is then the sum of the source and PSU components, given by

$$I''_k = I''_{k,PSU} + I''_{kQ}$$

The utility's contribution, based on the source impedance in per unit from Table 7.13:

$$I''_{kQ} = \frac{1.1}{Z_{kQ}} = \frac{1.1}{(0.00163 + j0.02195)} = 3.7187 - j49.78 \text{ pu}$$
$$= |20.90| \text{ kA at } 13.8 \text{ kV}$$

PSU contribution

$$I''_{k,PSU} = \frac{1.1}{Z_{PSU}} = \frac{1.1}{0.01195 + j0.28955} = 0.1565 - j3.7925 \text{ pu}$$
$$= |1.59| \text{ kA at } 138 \text{ kV}$$

Therefore:

$$I''_k = 3.7187 - j49.78 + 0.1565 - j3.793 = j3.8875 - j53.573 \text{ pu} = |22.47| \text{ kA}$$

The peak current i_p is

$$i_p = i_{p,PSU} + i_{pQ}$$

Power station $R/X = 0.01195/0.28955 = 0.04128$. Therefore, χ_{PSU} from Equation 8.4 = 1.886. This gives

$$i_{p,PSU} = \chi_{PSU}\sqrt{2}I''_{k,PSU} = (1.886)(\sqrt{2})(1.59) = 4.24 \text{ kA}$$

Utility system $R/X = 0.0746$, $\chi_{PQ} = 1.803$:

$$i_p = 53.29 + 4.24 = 57.53 \text{ kA peak.}$$

The symmetrical breaking current is the summation of two currents—one from the PSU and the other from the source:

$$I_b = I_{b,PSU} + I_{bQ} = I_{b,PSU} + I''_{kQ}$$

Considering a contact parting time of 0.03 s, for the PSU contribution, μ is calculated from Equation 8.22. This gives $\mu = 0.94$.

$$I_{b,PSU} = 0.94 \times 1.59 = 1.495 \text{ kA}$$

Total breaking current = 20.90 + 1.495 = 22.395 kA symmetrical.

The dc components at contact parting time are calculated using Equation 8.39, which is repeated as follows:

$$i_{dc} = \sqrt{2}I''_{k}e^{-2\pi ftR/X}$$

The dc component of the utility's source, based on $X/R = 14.3$, from Table 7.13, is

$$\sqrt{2}(20.90)e^{-377(0.03)(0.0746)} = 12.71 \text{ kA}$$

Similarly, the dc component from the PSU, based on $X/R = 24.23$ from $Z_{PSU} = 1.41$ kA; the total dc current is 14.12 kA.

Finally, the asymmetrical breaking current is

$$\begin{aligned} I_{basym} &= \sqrt{I_b^2 + (i_{dc})^2} \\ &= \sqrt{(22.395)^2 + (14.12)^2} \\ &= 26.47 \text{ kA} \end{aligned}$$

Three-Phase Short Circuit at F2

The adjustment factor $Z_{G,PSU}$ for the low-voltage side faults for PSUs is given by Equation 8.13:

$$Z_G = R_G + jX''_d$$

$$R_G = 0.05\, X''_d \text{ for generator } U_{rG} > 1 \text{ kV} \quad \text{and} \quad S_{rG} \geq 100 \text{ MVA}$$

$$R_G = 0.05 \times 0.14630 = 0.007315$$

$$Z_G = 0.007315 + j0.14630$$

$$\begin{aligned} K_{G,PSU} &= \frac{c_{max}}{1 + X''_d \sin\phi_{rG}} \\ &= \frac{1.1}{(1 + 0.164 \times 0.52)} = 1.0136 \end{aligned}$$

Therefore:

$$\begin{aligned} Z_{G,PSU} = K_{G,PSU}Z_G &= 1.0136(0.007315 + j0.14630) \\ &= 0.0074 + j0.1483 \end{aligned}$$

We have ignored small impedances of bus ducts B1 and B2 in the above calculations. We will also ignore contributions from *all* motors. The initial short-circuit currents are the sum of the partial short-circuit currents from the generator and source through transformer T1. The partial short-circuit current from the generator is

$$\begin{aligned} I''_{kG} &= \frac{1.1}{Z_{G,PSU}} = \frac{1.1}{(0.0074 + j0.1483)} \\ &= 0.369 - j7.399 \text{ pu} \\ &= |30.99| \text{ kA} \end{aligned}$$

To calculate the partial short-circuit current of a utility's system through the transformer T1, the transformer impedance is modified. From the Equation 8.11:

$$K_{T,S} = \frac{c_{max}}{1 - x_T \sin \phi_{rG}} = \frac{1.1}{(1 - 0.0774 \times 0.5267)} = 1.147$$

The partial short-circuit current is

$$\begin{aligned} I''_{kT} &= \frac{1.1}{1.147(0.00404 + j0.12894) + (0.00163 + j0.02195)} \\ &= \frac{1}{0.00626 + j0.16984} = 0.2167 - j5.8799 \text{ pu} \\ &= |24.62| \text{ kA} \end{aligned}$$

By summation, the initial short-circuit current is then

$$\begin{aligned} I''_{kF2} &= I''_{kG} + I''_{kT} \\ &= 0.369 - j7.399 + 0.2167 - j5.8799 \\ &= 0.5857 - j13.279 \text{ pu} \\ &= |55.56| \text{ kA} \end{aligned}$$

As resistance is low, this current could have been calculated using reactances only. The peak current i_p is the sum of the component currents:

$$i_p = i_{pG} + i_{pT}$$

Factor χ for the generator, based on $R_G/X''_d = 0.05$, from Equation 8.4 = 1.86. Thus, the peak current contributed by the generator is

$$i_{pG} = \chi_G \sqrt{2} I''_{kG} = 1.86 \times \sqrt{2} \times 30.99 = 81.52 \text{ kA}$$

Similarly, calculate i_{pT}:

$$\begin{aligned} \frac{R}{X} &= 0.00607/0.16378 = 0.0371 \\ \chi_T &= 1.90 \\ i_{pT1} &= 1.90(\sqrt{2})(24.60) = 66.09 \text{ kA} \end{aligned}$$

Therefore, the total peak current is 147.61 kA. The symmetrical breaking current I_b is calculated at a minimum time delay of 0.05 s. It is the summation of the currents from the source through T1, which is equal to the initial short-circuit current and the symmetrical breaking current from the generator: $I_b = I_{bG} + I_{bT} = I_{bG} + I''_{kT}$.

$$\begin{aligned} I_{bG} &= \mu I''_{kG} \\ \frac{I''_{kG}}{I_{RG}} &= \frac{30.99}{4.690} = 6.60 \\ \mu &= 0.78, \quad q = 1 \\ I_{bG} &= (0.78)(30.99) = 24.17 \text{ kA} \end{aligned}$$

Total breaking current symmetrical = 48.79 kA.

TABLE 8.8

Examples 8.3 and 8.4: Comparative Results of Three-Phase Short-Circuit Calculations

Fault Location	Calculation Method	First-Cycle Current kA asym. Peak (ANSI) or Peak Current i_p (IEC)	Interrupting Duty Current (ANSI) or I_{basym}(IEC)
Fl (138 kV)	ANSI calculation	53.1	20.54 × 1.2 = 24.65 (Example 7.2)
	IEC calculation	57.53	26.47
F2(13.8 kV)	ANSI calculation	155.69	71.69 × 1.1 = 78.86 (Example 7.3)
	IEC calculation	147.61	74.21
F3 (4.16 kV)	ANSI calculation	46.23	16.35 × 1.1 = 17.99 (Note)
	IEC calculation	53.11	18.98

Note: Not calculated in chapter 7, a reader can verify the calculations shown.

For asymmetrical breaking, current dc components at a minimum time delay of 0.05 s are calculated. Consider a generator X/R of 125.

Generator component = 37.96 kA

Transformer component = 17.96 kA

Total dc component = 55.92 kA

This gives a total asymmetrical breaking current of 74.21 kA.

These results are compared with the calculations in Example 7.2 and are shown in Table 8.8. IEC currents are higher for a fault at F1 and lower for a fault at F2. Generally, for calculations involving currents contributed mainly by generators, ANSI interrupting currents are higher than IEC currents. IEC uses an artificially high generator resistance, which is further multiplied by factor $K_{G,PSU}$. As a result, the fault currents have comparatively much reduced in magnitude and asymmetry. The calculation for faults at F1 is higher in IEC, because of factor c, which increases the source contribution by 10%, while the generator contribution through transformer T1 is comparatively small.

Generator Source Short-Circuit Current

In Example 7.2, we calculated the generator fault current for a fault at F2 and noted high asymmetry. The current zeros are not obtained at the contact parting time, and an asymmetry factor of 129% is calculated. The calculation is repeated with IEC methods, the generator X/R ratio for the dc component is the same as for ANSI calculations = 125. The calculations give

$$I''_{kG} = 30.99 \text{ kA}$$
$$i_{pG} = 81.52 \text{ kA}$$
$$i_{bGsym} = 24.20 \text{ kA}$$
$$i_{GDC} = 37.96 \text{ kA}$$
$$i_{bGasym} = 45.02 \text{ kA}$$

The asymmetry factor is $\alpha = (37.96)/(\sqrt{2}\cdot 24.20) = 111.0\%$.

The results of the calculations are shown in Table 8.9. IEC standard [1] does not discuss the asymmetry at the contact parting time of the breaker. Short-circuit current profiles for "far from" and "near to" the generator are shown in Figures 1 and 2 *of this standard*, respectively. There is no discussion of not obtaining a current zero at the contact parting time of the breaker. The IEC standard showing the examples of short-circuit calculations, part-4 is yet to be published. IEC may adopt IEEE standard [2] for the generator breakers.

TABLE 8.9

Generator Fault Currents: Example 8.4

Calculation Type	Symmetrical Current (kA rms)	dc Component (kA)	Asymmetrical Current	Asymmetry Factor α
ANSI	18.95	34.58	39.43	1.29
IEC, generator X/R same as in ANSI calculation	24.20	37.96	45.02	1.11

Example 8.4

The effect of motors is neglected in Example 8.3. Calculate the partial currents from the motors at 13.8 kV. Do these motor contributions need to be considered in IEC calculations for a fault at F2? Also, calculate the peak current and the asymmetrical breaking current for a fault at F3 on the 4.16 kV bus.

Effect of Motor Contribution at 13.8 kV Bus, Fault F2

For a fault at the 13.8 kV bus F2, an equivalent impedance of the motors through transformers and cables is calculated. The partial currents from medium- and low-voltage motors are calculated in Tables 8.10 and 8.11, respectively. The equivalent impedance of low-voltage motors of two identical groups, from Table 8.11, is $6.45 + j15.235$ per unit.

TABLE 8.10

Partial Short-Circuit Currents from Asynchronous Medium-Voltage Motors: Example 8.4

Parameter	2425 hp	300 hp	500 hp	Sum ($\sum$)
Power output, P_{rm} (MW)	1.81	0.224	0.373	
Quantity	1	3	2	
Power factor (cos ϕ)	0.93	0.92	0.92	
Efficiency (η_r)	0.96	0.93	0.94	
Ratio, locked rotor current to full load current (I_{LR}/I_{rM})	6	6	6	
Pair of poles (p)	1	1	2	
Sum of MVA (S_{rM})	2.03	0.78	0.86	
Sum, rated current (I_{rM})	0.28	0.11	0.12	
I''_K/I_{rM}	6.6	6.6	6.6	
Power per pole pair (m)	1.81	0.223	0.186	
R_M/X_M	0.10	0.15	0.15	
κ_m	1.75	1.65	1.65	
μ	0.78	0.78	0.78	
q	0.86	0.61	0.59	
I''_{KM}	1.68	0.66	0.72	$\sum = 3.06$
i_{pM}	4.18	1.54	1.67	$\sum = 7.39$
i_{bM}	1.13	0.32	0.33	$\sum = 1.78$
Z_M (pu)	8.23	21.41	19.42	
X_M (pu)	$0.995\ Z_M = 8.189$	$0.989\ Z_M = 21.17$	$0.989\ Z_M = 19.21$	
R_M (pu)	$0.1\ X_M = 0.82$	$0.15\ X_M = 3.18$	$0.15\ X_M = 2.88$	
Cable C2	$0.068 + j0.104$			
$\sum$ MV motors and cable				$0.58 + j4.55$

TABLE 8.11

Low-Voltage Motors, Partial Short-Circuit Current Contributions: Example 8.4

Parameter	Motors M4, M5, and M6 (or Identical Group of Motors M4′, M5′, and M6)	Remarks
P_{rm} (MW)	0.43	Calculated active power rating of the motor group
Sum of MVA (S_{rM})	0.52	Active power rating divided by power factor
R_M/X_M	0.42	From Equation 8.20 for group of motors connected through cables
χ_m	1.3	From Equation 8.4
Ratio, locked rotor current to full load current (I_{LR}/I_{rM})	6	
Z_M in per unit 100 MVA base	32.12	
X_M in per unit 100 MVA base	$0.922\ Z_M = 29.61$	Equation 8.20
R_M in per unit 100 MVA base	$0.42\ X_M = 12.44$	Equation 8.20
Cables C3 or C4 in per unit 100 MVA base	$0.46 + j0.860$	Table 7.13

The per unit impedance of transformer T3 from Table 7.13 is $0.639 + j3.780$. Therefore, the low-voltage motor impedance through transformer T3, seen from the 4.16 kV bus, is $7.089 + j19.015$ per unit.

From Table 8.10, the equivalent impedance of medium-voltage motors is $0.58 + j4.55$ pu. The equivalent impedances of low- and medium-voltage motors in parallel are

$$(7.089 + j19.015) \parallel (0.58 + j4.55) = 0.637 + j3.707$$

To this, add the impedance of cable C1 and transformer T2 from Table 7.13, which gives $0.701 + j4.606$ pu. This is the equivalent impedance as seen from the 13.8 kV bus. Thus, the initial short-circuit current from the motor contribution is $1.1/(0.701 + j4.606)$ per unit $= |0.99|$ kA.

The effect of motors in this example can be ignored, and the above calculation of currents from motor contributions is not necessary. From Equation 8.33, ΣP_{rm} = sum of the active powers of all medium- and low-voltage motors = 0.86 MW. Also, ΣS_{rT} = rated apparent power of the transformer = 7.5 MVA. The left-hand side of Equation 8.33 = 0.1147. Symmetrical short-circuit power at the point of connection, without effect of motors, is

$$S''_{kQ} = \sqrt{3} I''_k U_n = \sqrt{3}(56.5)(13.8) = 1350.7 \text{ MVA}$$

The right-hand side of Equation 8.33 gives 2.571, and the identity in Equation 8.33 is satisfied. The effect of motors can be ignored for a fault at 13.8 kV.

If the calculation reveals that motor contributions should be considered, we have to modify i_p at the fault point. This requires calculation of χ, which is not straightforward. High-voltage motors have $\chi = 1.75$ or 1.65, and low-voltage motors have $\chi = 1.65$. For a combination load, $\chi = 1.7$ can be used to calculate i_p approximately.

Three-phase short-circuit at F3. For a fault at F3, we will first calculate the motor contributions. The low-voltage motor impedance plus transformer T3 impedance is $7.089 + j19.015$, as calculated above. The initial short-circuit contribution from the low-voltage motor contribution is $1.1/(7.089 + j19.015) = 0.019 - j0.051$ pu or $|I''_k| = 0.76$ kA. Medium-voltage impedance, from Table 8.10, is $0.58 + j4.55$ per unit. The medium-voltage motor contribution is $0.028 - j0.239$ pu or $|I''_k| = 3.43$ kA.

To calculate the generator and utility source contributions, the impedances $Z_{G,PSU}$ are in $\|$ with $(Z_{T,PSU} + Z_Q)$, i.e., $0.0074 + j0.1483$ in $\|$ with $0.006074 + j0.16378$. This gives $0.0034 + j0.0778$ pu. Add transformer T2 impedance $(0.06349 + j0.89776)$ and cable C1 impedance $(0.00038 + j0.00101)$ from Table 7.13. This gives an equivalent impedance of $0.0673 + j0.976$ per unit. Thus, the initial short-circuit current is $0.077 - j1.122$ per unit or $|I''_K| = 15.61$ kA. The total initial symmetrical current, considering low- and medium-voltage motor contributions, is 19.80 kA. To calculate i_p, χ must be calculated for the component currents.

For contribution through transformer T2, using Equation 8.4:

$$\begin{aligned}\chi_{AT} &= 1.02 + 0.98\varepsilon^{-3(0.06895)} \\ &= 1.82\end{aligned}$$

As this is calculated from a meshed network, a safety factor of 1.15 is applicable, i.e., $\chi = 1.15$ multiplied by $1.82 = 2.093$. However, for high-voltage systems, χ is not > 2.0. This gives $i_{pAT} = (2)(\sqrt{2})(15.65) = 44.27$. For medium-voltage motors, χ can be calculated from Table 8.10:

$$\chi_{MV} = \frac{i_p}{\sqrt{2}I''_K} = \frac{7.39}{\sqrt{2}(3.06)} = 1.71$$

For low-voltage motors through transformer T3:

$$\chi_{LVT} = 1.02 + 0.98e^{-3(6.97/19.02)} = 1.346$$

$$i_{pLV} = (1.346)(\sqrt{2})(0.76) = 1.45 \text{ kA}$$

Total peak current by summation $= (44.27 + 7.39 + 1.45) = 53.11$ kA.

The breaking current is the summation of individual breaking currents:

Breaking current through transformers at 4.16 kV $= |I''_K| = 15.65$ kA
Breaking current medium-voltage motors from Table 8.10 $= 1.78$ kA

For low-voltage motors with $I''_{KM}/I_{RM} = 6.6$, $\mu = 0.78$, q can be conservatively calculated for $m \leq 0.3$ and $p = 2$. This gives $q = 0.64$. The component breaking current from low-voltage motors is therefore, $0.78 \times 0.64 \times 0.76 = 0.38$ kA. Total symmetrical breaking current $= 17.81$ kA.

To calculate the asymmetrical breaking current, the dc components of the currents should be calculated:

The dc component of the low-voltage motor contribution is practically zero.

The dc component of the medium-voltage motors at contact parting time of 0.05 s $= 0.5$ kA.

The dc component of current through transformer T2 $= 6.07$ kA.

Total dc current at contact parting time $= 6.57$ kA; this gives asymmetrical breaking current of 18.98 kA.

The results are shown in Table 8.8.

Example 8.5 Steady-State Currents

Calculate the steady-state currents on the 13.8 kV bus, fault-point F2, in the system of Example 7.2, according to IEC and ANSI methods.

IEC Method

The steady-state current is the summation of the source current through the transformer and the generator steady-state current:

$$I_k = I_{k,PSU} + I_{kG} = I''_{K,PSU} + \lambda_{max} I_{rG}$$

Substitution of λ_{min} gives the minimum steady-state current; λ_{max} and λ_{min} are calculated from Figure 8.8; and $I_{rG} = 4.69$ kA, $I''_{kG} = 30.99$ kA, and ratio $I''_{kG}/I_{rG} = 6.61$. Also from Table 6.1, the generator $X_{d\text{-sat}} = 1.949$. From Series One curves in Figure 8.8, $\lambda_{min} = 0.5$ and $\lambda_{max} = 1.78$. Therefore, the maximum generator steady-state current is 8.35 kA, and the minimum generator steady-state current is 2.345 kA. The source steady-state current is equal to the initial short-circuit current (25.53 kA); therefore, the total steady-state short-circuit current is 33.88 kA maximum and 27.88 kA minimum.

ANSI Method

The source current from Example 7.2 is 27.69 kA. From Table 6.1, the generator transient reactance is 0.278 on machine MVA base. The generator contribution is, therefore, 16.87 kA, and the total steady-state current is 44.56 kA.

Example 8.6

This example demonstrates the generator source short-circuit calculations using ANSI/IEEE, IEC analytical methods, and rigorous EMTP simulation. The generator is 234 MVA, 2-pole, 18 kV, 0.85 PF, 198.9 MW, 7505 amperes stator current 60 Hz, 0.56 SCR, 350 V field volts, and wye connected. The manufacturer's data of the generator are shown in Table 8.12. All the data

TABLE 8.12

Manufacturer's Generator Data, 234 MVA, 0.85 PF, 198.9 MW, 7505 A, 60 Hz 0.56 SCR

Description	Symbol	Data	Description	Symbol	Data
Saturated synchronous	X_{dv}	2.120	Saturated negative sequence	X_{2v}	0.150
Unsaturated synchronous	X_d	2.120	Unsaturated negative sequence	X_{2I}	0.195
Saturated transient	X_{dv}	0.230	Saturated zero sequence	X_{0v}	0.125
Unsaturated transient	X'_d	0.260	Unsaturated zero sequence	X_{0I}	0.125
Saturated subtransient	X''_{dv}	0.150	Leakage reactance, overexcited	$X_{LM,OXE}$	0.135
Unsaturated subtransient	X'_d	0.195	Leakage reactance, under excited	$X_{LM,UEX}$	0.150
Saturated synchronous	X_{qv}	1.858	Saturated subtransient	X''_{qv}	0.140
Unsaturated synchronous	X_q	1.858	Unsaturated subtransient	X'_q	0.192
Unsaturated transient	X'_q	0.434	Field time constants open circuit	T'_{do}	5.615
Three-phase short-circuit transient	T'_{d3}	0.597	Line-to-neutral short-circuit transient	T'_{d1}	1.124
Line-to-line short-circuit transient	T'_{d2}	0.927	Short-circuit subtransient	T''_d	0.015
Open circuit	T'_{qo}	0.451	Open circuit subtransient	T''_{do}	0.022
Three-phase short-circuit transient	T'_q	0.451	Short-circuit subtransient	T''_q	0.015
Three-phase short-circuit	T_{a3}	0.330	Open circuit subtransient	T''_{qo}	0.046
Line-to-line short-circuit	T_{a2}	0.330	Effective X/R	X/R	125

d, direct axis; q, quadrature axis. All reactance's in per unit on machine MVA base. All time constants in seconds.

TABLE 8.13

Generator Parameters in *d-q* Axis, Park's Transformation

Parameter	Symbol	Value in Ohms	Parameter	Symbol	Value in Ohms
Self-inductance *d*-axis	L_f	3.463E+02	Mutual inductance *q*-axis, circuit 2 to armature	L_{akq}	2.822E+01
Mutual inductance field and armature	L_{af}	3.000E+01	Self-inductance of circuit 2 of *q*-axis.	L_{kq}	3.106E+02
Field damper mutual inductance *d*-axis	L_{fkd}	3.275E+02	Zero-sequence inductance	L_0	1.731E−01
Self-inductance armature *d*-axis	L_d	2.935E+00	Zero sequence resistance	R_0	3.946E+03
Mutual inductance armature *d*-axis damper	L_{akd}	3.000E+01	Resistance *d*-axis field winding	R_f	1.518E−01
Self-inductance of *d*-axis damper winding	L_{kd}	3.303E+02	Resistance of armature	R_a	1.523E−03
Self-inductance circuit 1 of the *q* axis	L_g	3.741E+02	Resistance of *d*-axis damper winding	R_{kd}	1.272E+00
Mutual inductance, *q*-axis circuit 1 to armature	L_{ag}	2.822E+01	Resistance circuit 1 of the *q*-axis	R_g	1.679E+00
Mutual inductance circuit 1 to circuit 2	L_{gkq}	3.080E+02	Resistance circuit 2 of *q*-axis	R_{kq}	1.914E+00
Self-inductance *q*-axis armature winding	L_q	2.772E+00			

shown in this table are not required for calculations according to Equation 6.113. The generator breaker is a 5-cycle breaker with contact parting time = 50 ms. EMTP routines can accept the manufacturer's data, which are converted to dq0 axis using Park's transformation. The results of the calculations are shown in Table 8.13. Figure 8.12 shows the three-phase terminal short-circuit current, and Figure 8.13 shows the current in phase c only for clarity. Zero crossing is not obtained

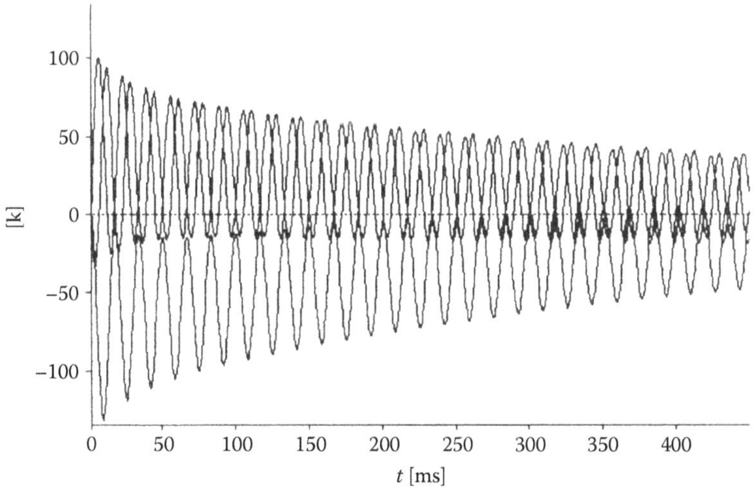

FIGURE 8.12
EMTP simulation results of the terminal three-phase short circuit of 234 MVA generator (Example 8.6).

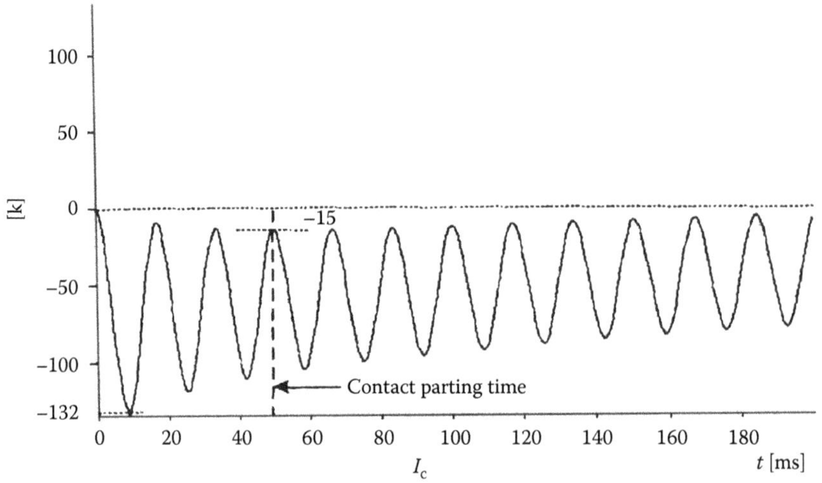

FIGURE 8.13
Short-circuit current profile in phase c with maximum asymmetry, showing delayed current zeros.

TABLE 8.14
Comparison of Calculations Using IEEE/IEC Standards and EMTP Simulations

Calculated Parameter	IEEE	IEC	EMTP
Close and latch, kA peak (IEC peak short-circuit current)	112.2	131.60	132.05
Generator source interrupting kA sym. RMS (IEC symmetrical breaking current i_{bsym}	30.90	38.50	33.59
Dc component, kA	59.22	60.73	62.50
Total asymmetrical, kA RMS (IEC i_{basym})	66.80	71.90	70.90
Asymmetry factor	135%	112%	131%

for a number of cycles after the contact parting time of the breaker [9–11]. The results of the comparative calculations are in Table 8.14.

Effect of Power Factor

The load power factor (lagging) in IEC equations does not change the asymmetry at the contact parting time. However, the asymmetry *does* change with the power factor and prior load. This is clearly shown in Figure 8.14, with generator absorbing reactive power, i.e., operating at leading power factor of 0.29. Generator load prior to short circuit = 28 MW, 92.4 Mvar. It is seen that the current zeros are further delayed, as compared to generator operating at no-load; compare with Figure 8.13. The prior leading power factor loading further delays the occurrence of current zeros. The example of calculation demonstrates that asymmetry at contact parting time can be even 130% or more. The delayed current zeros can also occur on short circuits in large industrial systems, with co-generation facilities.

The calculations in this chapter demonstrate that there are differences in results obtained by ANSI and IEC methods, and one or the other calculation method can give higher or lower results. The predominant differences are noted in the contributions from motors and generators, which are the major sources of short-circuit currents in power systems. These differences vary with the

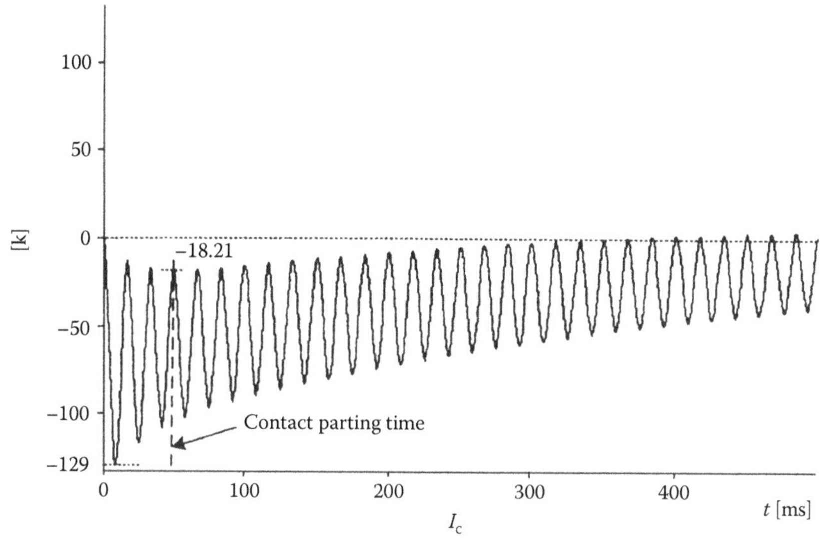

FIGURE 8.14
Effect of generator operating power factor (generator loaded to 28 MVA at 0.29 leading power factor) prior to short circuit on asymmetry. Current zeros are further delayed.

contact parting times. The factor c in IEC calculations makes the source contributions higher, and this generally results in higher currents. It seems appropriate to follow the calculations and rating structures of breakers in these standards in their entity, i.e., use IEC calculations for IEC-rated breakers and ANSI calculations for ANSI-rated breakers. Example 8.6 demonstrates the comparative results of calculations with EMTP simulation. Other investigators have reached the same conclusions, Table 7.3.

Problems

(The problems in this section are constituted, so that they can be solved by hand calculations.)

8.1 A 13.8 kV, 60 MVA, two-pole 0.8 power factor synchronous generator has a subtransient reactance of 11%. Calculate its corrected impedance for a bus fault. If this generator is connected through a step-up transformer of 60 MVA, 13.8–138 kV, transformer $Z = 10\%$, $X/R = 35$, what are the modified impedances for a fault on the 138 kV side and 13.8 kV side?

8.2 In Problem 8.1, the 138 kV system has a three-phase fault level of 6500 MVA, $X/R = 19$. Calculate initial symmetrical short-circuit current, peak current, symmetrical breaking current, and asymmetrical breaking currents for a three-phase fault on the 138 kV side and 13.8 kV side. The minimum time delay for fault on the 138 kV side is 0.03 s for a fault on the 13.8 kV side is 0.05 s.

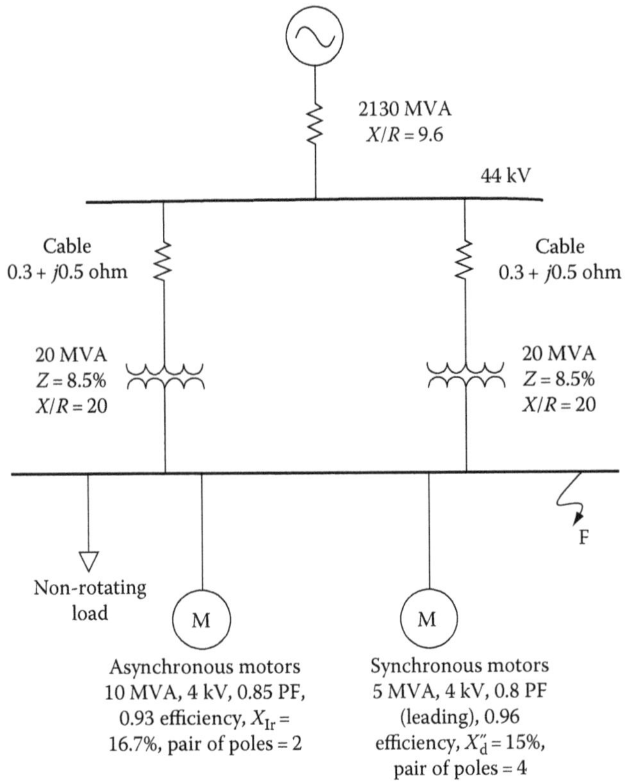

FIGURE P8.1
System configuration for Problem 8.3.

8.3 Figure P8.1 shows a double-ended substation with parallel running transformers. A three-phase fault occurs at *F*. Calculate I_k'', i_p, $I_{b\,sym}$, $I_{b\,asym}$, I_{dc}, and *k*.

8.4 Figure P8.2 shows a generating station with auxiliary loads. Calculate three-phase fault currents for faults at Fl, F2, and F3. Calculate all component currents I_k'', i_p, $i_{b\,sym}$, $I_{b\,asym}$, I_{dc}, and I_k in each case. The system data are shown in Table P8.1.

8.5 Repeat Problem 8.4 with typical X/R ratios from Figures 7.2 and 7.4.

8.6 In Example 8, using the manufacturer's data of the generator in Table 8.11, mathematically derive the dq0 axis parameters shown in Table 8.12 (see Chapter 6).

8.7 In Example 8, verify the results of ANSI/IEEE and IEC short-circuit calculations.

8.8 A 12.47 kV, 81.82 MVA, 0.85 power factor generator is directly connected to a 12.47 kV bus. The following generator parameters are applicable:

$$X_d'' = 0.162, X_d' = 0.223, X_d = 2.01, X_q'' = 0.159, T_d'' = 0.015, T_d' = 0.638, T_a = 0.476$$

Considering a 5-cycle breaker analytically calculate the asymmetry factor using ANSI/IEEE calculations.

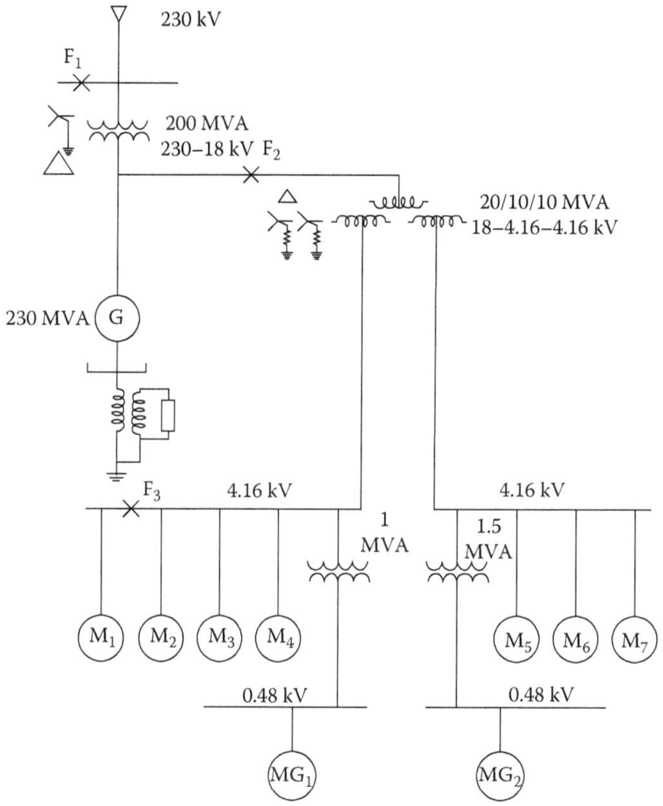

FIGURE P8.2
System configuration for Problems 8.4 and 8.5.

TABLE P8.1

Impedance Data for Problems 8.4 and 8.5

System Component	Description and Impedance Data
230 kV utility source	Three-phase short-circuit = 8690 MV A, $X/R = 16.9$
Step-up transformer	230–18 kV, 200 MVA, $Z = 10\%$, $X/R = 35$, HV tap at 23,5750 V
Generator	18 kV, 230 MVA, 0.85 power factor, subtransient reactance = 16%
Three-winding auxiliary transformer	18–4.16–4.16 kV, 20/10/10 MVA. impedance 18 kV to each 4.16 kV winding = 8% 20 MVA base, impedance 4.16 kV winding to 4.16 kV winding = 15%, 20 MVA base
1 MVA transformer	4.16–0.48 kV, $Z = 5.75\%$, $X/R = 5.75$
1.5 MVA transformer	4.16–0.48 kV, $Z = 5.75\%$, $X/R = 7.5$
Medium-voltage motors	$M_1 = 5000$ hp, 6-pole, synchronous, 0.8 power factor leading
	$M_2 = 2500$ hp, 8-pole, induction
	$M_3 = 2500$ hp, 2-pole, induction
	$M_4 = 2500$ hp, 12-pole, synchronous, 0.8 power factor leading
	$M_5 = 2000$ hp, 4-pole, synchronous, unity power factor
	$M_6 = 1000$ hp, 2-pole induction
	$M_7 = 1000$ hp, 6-pole induction
Low-voltage motor groups	$MG_1 = 2 \times 100$ hp and 6×40 hp induction 4-pole induction
	$MG_2 = 4$ groups 150 hp and 8×75 hp induction 6-pole induction

References

1. IEC. Short-Circuit Calculations in Three-Phase AC Systems, 1st edn., 1988. Now revised IEC 60909-0, Short-Circuit Currents in Three-Phase AC Systems, 0—Calculation of Currents, 2001–2007.
2. IEC. Factors for Calculation of Short-Circuit Currents in Three-Phase AC Systems According to IEC 60909-0, 1991.
3. IEC. Electrical Equipment-Data for Short-Circuit Current Calculations in Accordance with IEC 60909, IEC-60909-2, 1992.
4. IEC. Short-Circuit Current Calculations in AC Systems-Part 3: Currents During Two Separate Simultaneous Single-Phase Line-to-Earth Short-Circuits and Partial Short-Circuit Currents Following Through Earth, 60909-3, 1995.
5. IEC. Short-Circuit Calculations in Three-Phase Systems Part-4: Examples of Calculations of Short-Circuit Currents, 60909-4 (under publication).
6. IEC. High Voltage Alternating Current Circuit Breakers, 60056, 1987.
7. IEC 60060-1. High Voltage Test Techniques-Part 1: General definitions and Test Requirements, 1989.
8. J.C. Das. Short-circuit calculations-ANSI/IEEE & IEC methods, similarities and differences. In: *Proceedings of 8th International Symposium on Short-Circuit Currents in Power Systems*, Brussels, Belgium, 1998.
9. J.C. Das. Study of generator source short-circuit currents with respect to interrupting duty of the generator circuit breakers, EMTP simulation, ANSI/IEEE and IEC methods. *Int. J. Emerg. Electrical Power Syst*. 9(3): 1–23, 2008.
10. I.M. Canay and L. Warren. Interrupting sudden asymmetrical short-circuit currents without zero transition. *BBC Rev*. 56: 484–493, 1969.
11. D. Dufournet, J.M. Willieme, and G.F. Montillet, Design and Implementation of a SF6 interrupting chamber applied to low range generator breakers suitable for interrupting currents having a non-zero passage. *IEEE Trans. Power Delivery* 17: 963–969, Oct. 2002.

9

Calculations of Short-Circuit Currents in DC Systems

The calculations of short-circuit currents in dc systems are essential for the design and application of distribution and protective apparatuses used in these systems. The dc systems include dc motor drives and controllers, battery power applications, emergency power supply systems for generating stations, data-processing facilities, and computer-based dc power systems and transit systems.

Maximum short-circuit currents should be considered for selecting the rating of electrical equipment like cables, buses, and their supports. The high-speed dc protective devices may interrupt the current, before the maximum value is reached. It becomes necessary to consider the rate of rise of the current, along with interruption time, in order to determine the maximum current that will be actually obtained. Lower speed dc protective devices may permit the maximum value to be reached, before current interruption.

Though the simplified procedures for dc short-circuit current calculation are documented in some publications, these are not well established. There is no ANSI/IEEE standard for calculation of short-circuit currents in dc systems. A General Electric Company publication [1] and ANSI/IEEE standard C37.14 [2] provide some guidelines. IEC standard 61660-1 [3], published in 1997, is the only comprehensive document available on the subject. This standard addresses calculations of short-circuit currents in dc auxiliary installations in power plants and substations and does not include calculations in other large dc power systems, such as electrical railway traction and transit systems.

The IEC standard describes quasi-steady-state methods for dc systems. The time variation of the characteristics of major sources of dc short-circuit current from initiation to steady state is discussed and appropriate estimation curves and procedures are outlined.

A dynamic simulation is an option, however, akin to short-circuit current calculations in ac systems; the simplified methods are easy to use and apply, though these should be verified rigorously by an actual simulation.

9.1 DC Short-Circuit Current Sources

Four types of dc sources can be considered:

- Lead acid storage batteries
- DC motors
- Converters in three-phase bridge configuration
- Smoothing capacitors

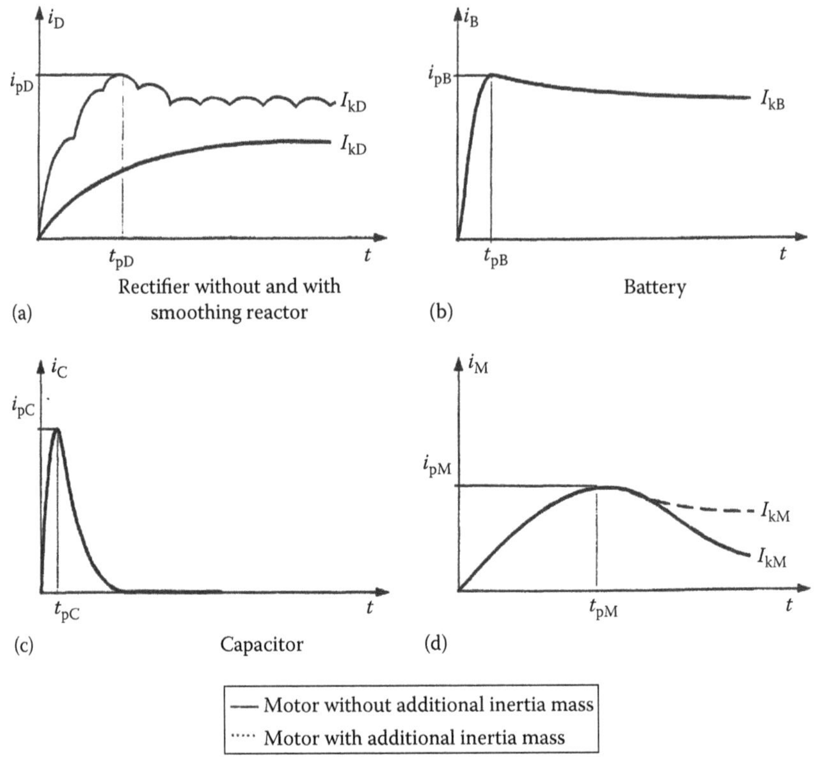

FIGURE 9.1
Short-circuit current–time profile of various dc sources: (a) rectifier without and with smoothing reactor; (b) battery; (c) capacitor; (d) dc motor with and without additional inertia mass. (From IEC, Short-Circuit Currents in DC Auxiliary Installations in Power Plants and Substations, Standard 61660-1, 1997. Copyright 1997 IEC. Reproduced with permission.)

Figure 9.1 shows the typical short-circuit current–time profiles of these sources, and Figure 9.2 shows the standard approximate function assumed in the IEC standard [3]. The following definitions apply:

I_k = Quasi-steady-state short-circuit current
i_p = Peak short-circuit current
T_k = Short-circuit duration
t_p = Time to peak
τ_1 = Rise time constant
τ_2 = Decay time constant

The function is described by

$$i_1(t) = i_p \frac{1 - e^{-t/\tau_1}}{1 - e^{-t_p/\tau_1}} \tag{9.1}$$

$$i_2(t) = i_p \left[(1-\alpha)e^{-(t-t_p)/\tau_2} + \alpha\right] \quad t \geq t_p \tag{9.2}$$

$$\alpha = \frac{I_k}{i_p} \tag{9.3}$$

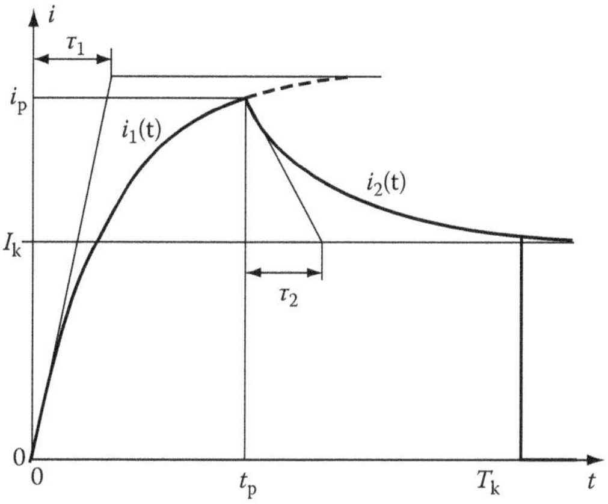

FIGURE 9.2
Standard approximation of short-circuit function. (From IEC, Short-Circuit Currents in DC Auxiliary Installations in Power Plants and Substations, Standard 61660-1, 1997. Copyright 1997 IEC. Reproduced with permission.)

The quasi-steady-state current I_k is conventionally assumed as the value at 1 s after the beginning of short circuit. If no definite maximum is present, as shown in Figure 9.1a for the converter current, then the function is given by Equation 9.1 only.

9.2 Calculation Procedures

9.2.1 IEC Calculation Procedure

Figure 9.3 shows a hypothetical dc distribution system, which has all the four sources of short-circuit current, i.e., a storage battery, a charger, a smoothing capacitor, and a dc motor. Two locations of short-circuit are shown: (1) at F_1, without a common branch, and (2) at F_2, through resistance and inductance, R_y and L_y of the common branch. The short-circuit current at F_1 is the summation of short-circuit currents of the four sources, as if these were acting alone through the series resistances and inductances. Compare this to the IEC method of ac short-circuit calculations in nonmeshed systems, discussed in Chapter 8.

For calculation of the short-circuit current at F_2, the short-circuit currents are calculated as for F_1 but adding R_y and L_y to the series circuit in each of the sources. Correction factors are introduced and the different time functions are added to the time function of the total current.

Whether it is the maximum or minimum short-circuit current calculation, the loads are ignored (i.e., no shunt branches) and the fault impedance is considered to be zero. For the maximum short-circuit current, the following conditions are applicable:

- The resistance of joints (in bus bars and terminations) is ignored.
- The conductor resistance is referred to 20°C.
- The controls for limiting the rectifier current are not effective.
- The diodes for the decoupling part are neglected.
- The battery is fully charged.
- The current-limiting effects of fuses or other protective devices are taken into account.

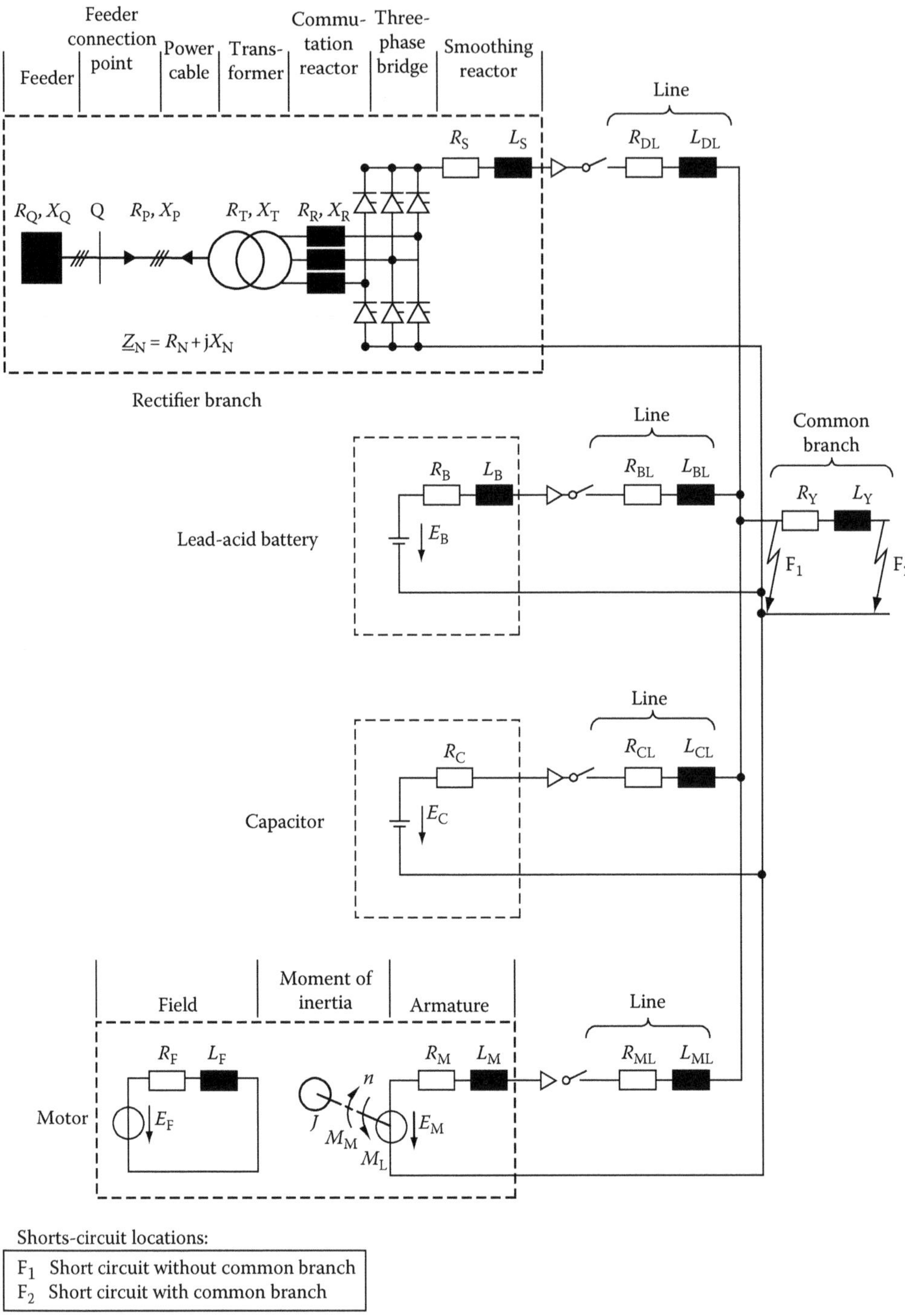

FIGURE 9.3
A dc distribution system for calculation of short-circuit currents. (From IEC, Short-Circuit Currents in DC Auxiliary Installations in Power Plants and Substations, Standard 61660-1, 1997. Copyright 1997 IEC. Reproduced with permission.)

For calculation of the minimum short-circuit current, the following conditions are applicable:

- The conductor resistance is referred to maximum temperature.
- The joint resistance is taken into account.
- The contribution of the rectifier is its rated short-circuit current.
- The battery is at the final voltage as specified by the manufacturer.
- Any diodes in the decoupling parts are taken into account.
- The current-limiting effects of fuses or other protective devices are taken into account.

9.2.2 Matrix Methods

Matrix methods contrast with superimposition techniques. In an example of calculation in Ref. [1], three sources of current, i.e., a generator, a rectifier, and a battery, are considered in parallel. The inductances and resistances of the system components are calculated and separate resistance and inductance networks are constructed, much akin to the ANSI/IEEE method for short-circuit current calculations in ac systems. These networks are reduced to a single resistance and inductance and then the maximum short-circuit current is simply given by the voltage divided by the equivalent resistance and its rate of rise by the equivalent time constant, which is equal to the ratio of equivalent inductance over resistance. This procedure assumes that all sources have the same voltage. When the source voltages differ, then the partial current of each source can be calculated and summed. This is rather a simplification. For calculation of currents from rectifier sources, an iterative procedure is required, as the resistance to be used in a Thévenin equivalent circuit at a certain level of terminal voltage during a fault needs to be calculated. This will be illustrated with an example.

9.3 Short Circuit of a Lead Acid Battery

The battery short-circuit model is shown in Figure 9.4; R_B is the internal resistance of the battery, E_B is the internal voltage, R_C is the resistance of cell connectors, L_{CC} is the inductance of the cell circuit in H, and L_{BC} is the inductance of the battery cells considered as bus bars. The *internal* inductance of the cell itself is zero. The line resistance and

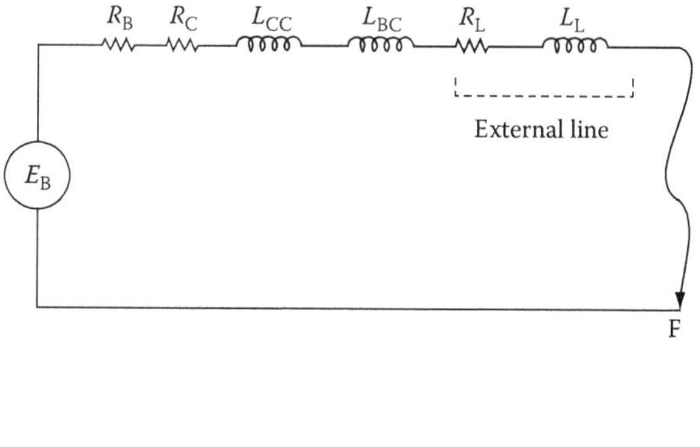

FIGURE 9.4
Equivalent circuit of the short-circuit of a battery through external resistance and inductance.

inductance are R_L and L_L, respectively. The equivalent circuit is that of a short circuit of a dc source through equivalent resistance and inductance, i.e.,

$$iR + L\frac{di}{dt} = E_B \tag{9.4}$$

The solution of this equation is

$$i = \frac{E_B}{R}\left(1 - e^{-(R/L)t}\right) \tag{9.5}$$

The maximum short-circuit current is then

$$I_{Bsc} = \frac{E_B}{R} \tag{9.6}$$

and the initial maximum rate of rise of the current is given by di/dt at $t = 0$, i.e.,

$$\frac{di_B}{dt} = \frac{E_B}{L} \tag{9.7}$$

Referring to Figure 9.4 all the resistance and inductance elements in the battery circuit are required to be calculated. The battery internal resistance R_B by the IEEE method [4] is given by

$$R_B = R_{cell}N = \frac{R_p}{N_p} \tag{9.8}$$

where
- R_{cell} is the resistance/per cell
- N is the number of cells
- R_p is the resistance per positive plate
- N_p is the number of positive plates in a cell

R_p is given by

$$R_p = \frac{V_1 - V_2}{I_2 - I_1}\,\Omega/\text{positive plate} \tag{9.9}$$

where
- V_1 is the cell voltage
- I_1 is the corresponding rated discharge current per plate

Similarly, V_2 is the cell voltage and I_2 is the corresponding rated discharge current per plate at V_2.

The following equation for the internal resistance is from Ref. [1]:

$$R_B = \frac{E_B}{100 \times I_{8h}}\,\Omega \tag{9.10}$$

where $I_{8\,h}$ is the 8 h ampere rating of the battery to 1.75 V per cell at 25°C. R_B is normally available from manufacturers' data; it is not a constant quantity and depends on the charge state of the battery. A discharged battery will have much higher cell resistance.

Example 9.1

A 60-cell 120 V sealed, valve regulated, lead acid battery has the following electrical and installation details:

Battery rating = 200 Ah (ampere hour), 8 h rate of discharge to 1.75 V per cell. Each cell has the following dimensions: height = 7.9 in. (= 200 mm), length = 10.7 in. (= 272 mm), and width = 6.8 in. (= 173 mm). The battery is rack mounted, 30 cells per row, and the configuration is shown in Figure 9.5. Cell inter-connectors are 250 KCMIL, diameter = 0.575 in.

Calculate the short-circuit current at the battery terminals. If the battery is connected through a cable of approximately 100 ft length to a circuit breaker, cable resistance 5 mΩ, and inductance 14 μH, calculate the short-circuit current at breaker terminals.

The battery resistance according to Equation 9.10 and considering a cell voltage of 1.75 V per cell, is

$$R_B = \frac{E_B}{100 \times I_{8-h}} = \frac{120}{100 \times 200} = 6\ \text{m}\Omega$$

The manufacturer supplies the following equation for calculating the battery resistance:

$$R_B = \frac{31 \times E_B}{I_{8\,h}}\ \text{m}\Omega \tag{9.11}$$

Substituting the values, this gives a battery resistance of 18.6 mΩ. There is three times the difference in these values, and the manufacturer's data should be used.

From Figure 9.5, battery connectors have a total length of 28 ft, size 250 KCMIL. Their resistance from the conductor resistance data is 1.498 mΩ at 25°C. The total resistance in the battery circuit is $R_B + R_C = 20.098$ mΩ (this excludes the external cable connections). Therefore, the maximum short-circuit current is $120/(20.098 \times 10^{-3}) = 5970$ A.

The inductance L_c of the battery circuit is sum of the inductances of the cell circuit L_{CC} plus the inductance of the battery cells, L_{CB}. The inductance of two round conductors of radius r, spaced at a distance d, is given by the expression

$$L = \frac{\mu_0}{\pi}\left(0.25 + \ln\frac{d}{r}\right) \tag{9.12}$$

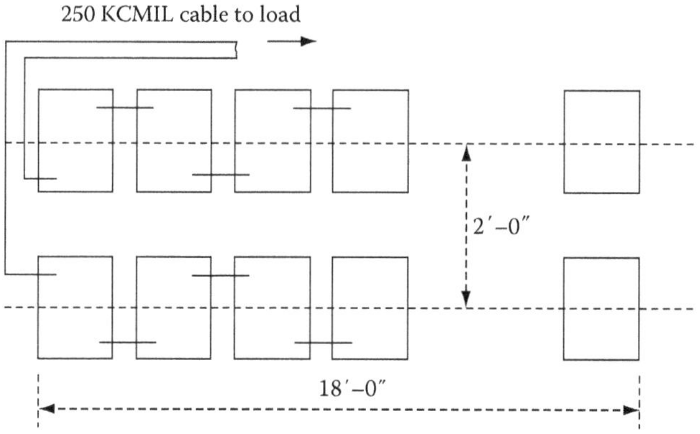

FIGURE 9.5
Battery system layout for calculation of short-circuit current (Examples 9.1 and 9.2).

where μ_0 is the permeability in vacuum $= 4\pi 10^{-7}$ H/m. From Figure 9.5, the distance $d = 24$ in. and r, the radius of 250 KCMIL conductor, is 0.2875 in. Substituting the values in Equation 9.12 the inductance is 1.87 μH/m for the *loop* length. Therefore, for an 18 ft loop length in Figure 9.5, the inductance $L_{CC} = 10.25$ μH.

The inductance of battery cells can be determined by treating each row of cells like a bus bar. Thus, the two rows of cells are equivalent to parallel bus bars at a spacing $d = 24$ in., the height of the bus bar $h =$ height of the cell $= 7.95$ in. and the width of the bus bars $w =$ width of the cell $= 6.8$ in. The expression for inductance of the bus bars in this configuration is

$$L = \frac{\mu_0}{\pi}\left(\frac{3}{2} + \ln\frac{d}{h + w}\right) \tag{9.13}$$

This gives inductance in H per meter loop length. Substituting the values, for an 18 ft loop length, inductance $L_{BC} = 4.36$ μH. The total inductance is, therefore, 14.61 μH. The initial rate of rise of the short-circuit current is given by

$$\frac{E_B}{L_C} = \frac{120}{14.61 \times 10^{-6}} = 8.21 \times 10^6 \text{A/s}$$

The time constant is

$$\frac{L_C}{R_B + R_C} = \frac{14.61 \times 10^{-6}}{20.01 \times 10^{-3}} = 0.73\text{ms}$$

The current reaches $0.63 \times 5970 = 3761$ A in 0.73 ms and in 1.46 ms it will be $0.87 \times 5970 = 5194$ A.

The cable resistance and inductance can be added to the values calculated above, i.e., total resistance $= 25.01$ mΩ and total inductance is 28.61 μH. The maximum short-circuit current is, therefore, 4798 A, and the time constant changes to 1.14 ms. The current profiles can be plotted.

9.3.1 IEC Method of Short-Circuit of a Lead Acid Battery

To calculate the maximum short-circuit current or the peak current according to IEC, the battery cell resistance R_B is multiplied by a factor 0.9. All other resistances in Figure 9.4 remain unchanged. Also, if the open-circuit voltage of the battery is unknown then use $E_B = 1.05U_{nB}$, where $U_{nB} = 2.0$ V/cell for lead acid batteries. The peak current is given by

$$i_{pB} = \frac{E_B}{R_{BBr}} \tag{9.14}$$

where

i_{pB} is the peak short-circuit current from the battery
R_{BBr} is the total equivalent resistance in the battery circuit, with R_B multiplied by a factor of 0.9

The time to peak and the rise time are read from curves in Figure 9.6, based on $1/\delta$, which is defined as follows:

$$\frac{1}{\delta} = \frac{2}{\frac{R_{BBr}}{L_{BBr}} + \frac{1}{T_B}} \tag{9.15}$$

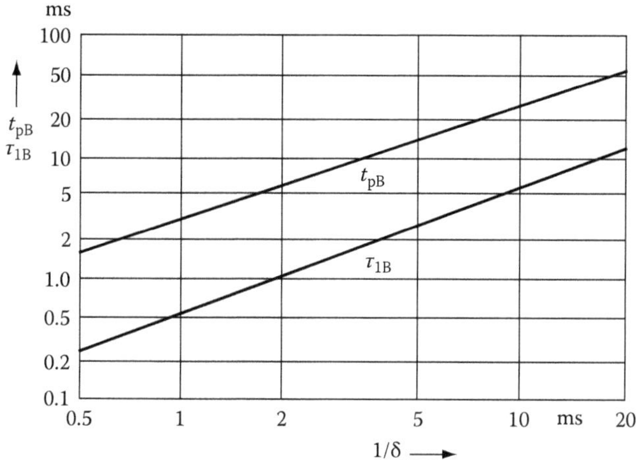

FIGURE 9.6
Time to peak t_{pB} and rise time constant τ_{1B} for short circuit of a battery. (From IEC, Short-Circuit Currents in DC Auxiliary Installations in Power Plants and Substations, Standard 61660-1, 1997. Copyright 1997 IEC. Reproduced with permission.)

The time constant T_B is specified as equal to 30 ms and L_{BBr} is the total equivalent inductance to the fault point in the battery circuit. The decay time constant τ_{2B} is considered to be 100 ms. The quasi-steady-state short-circuit current is given by

$$I_{kB} = \frac{0.95E_B}{R_{BBr} + 0.1R_B} \tag{9.16}$$

This expression considers that the battery voltage falls and the internal cell resistance increases after the short circuit. Note that all equations from IEC are in MKS units.

Example 9.2

Calculate the short-circuit current of the battery in Example 9.1, by the IEC method.

The total resistance in the battery circuit, without external cable, is $0.9 \times 18.6 + 1.498 = 18.238$ mΩ. The battery voltage of 120 V is multiplied by factor 1.05. Therefore, the peak short-circuit current is

$$i_{pB} = \frac{E_B}{R_{BBr}} = \frac{1.05 \times 120}{18.238 \times 10^{-3}} = 6908.6 \text{ A}$$

This is 15.7% higher compared to the calculation in Example 9.1: $1/\delta$ is calculated from Equation 9.15.

$$\frac{1}{\delta} = \frac{2}{\dfrac{18.238 \times 10^{-3}}{14.61 \times 10^{-6}} + \dfrac{1}{30 \times 10^{-3}}} = 1.56 \text{ ms}$$

From Figure 9.6, the time to peak = 4.3 ms and the rise time constant is 0.75 ms. The quasi-steady-state short-circuit current is

$$I_{kB} = \frac{0.95 \times 120 \times 10^3}{18.238 + 0.1(18.6)} = 5956 \text{ A}$$

The calculations with external cable added are similarly carried out. The cable resistance is 5 mΩ and inductance is 14 μH. Therefore, $R_{BBr} = (0.9)(18.6) + 1.498 + 5 = 23.24$ mΩ. This gives a peak current of 5422 A; $1/\delta = 2.40$ ms and time to peak is 5.4 ms. The rise time constant is 1.3 ms, and the quasi-steady-state short-circuit current is 4796 A. The short-circuit current profile is plotted in Figure 9.7a.

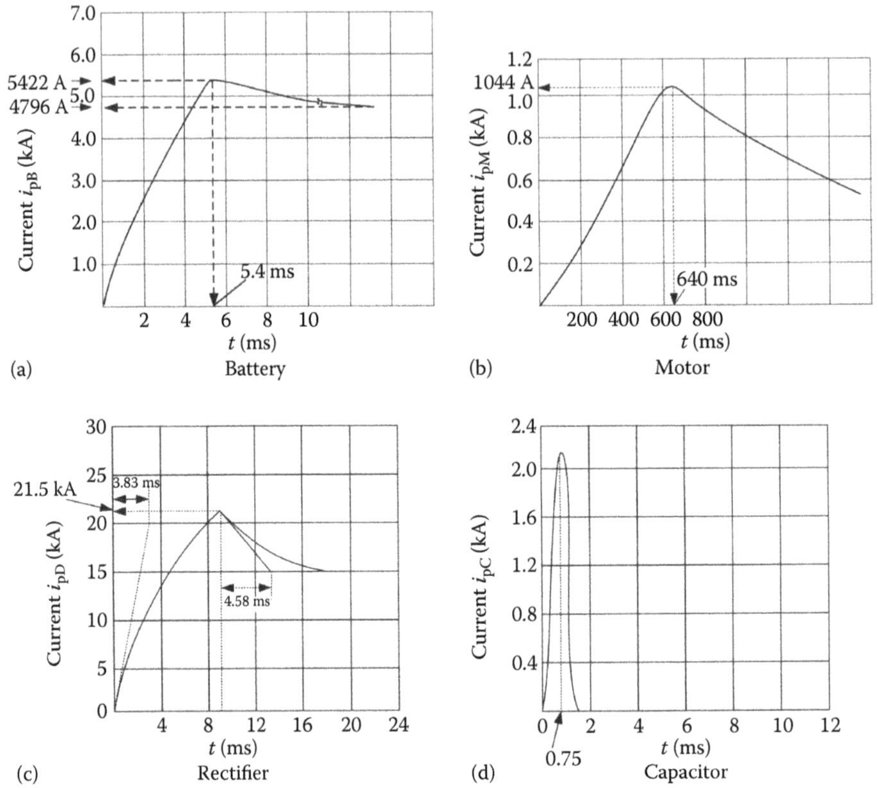

FIGURE 9.7
Calculated short-circuit current–time profiles: (a) battery (Example 9.2); (b) dc motor (Example 9.4); (c) rectifier (Example 9.6); (d) capacitor (Example 9.7).

9.4 Short-Circuit Current of DC Motors and Generators

An expression for the short-circuit from dc generators and motors [1] is

$$i_a = \frac{e_0}{r'_d}\left(1 - e^{-\sigma_a t}\right) - \left(\frac{e_0}{r'_d} - \frac{e_0}{r_d}\right)\left(1 - e^{\sigma_f t}\right) \tag{9.17}$$

where

i_a = per unit current
e_0 = internal emf prior to short-circuit in per unit
r_d = steady-state effective resistance of the machine in per unit
r'_d = transient effective resistance of the machine in per unit
σ_a = armature circuit decrement factor
σ_f = field circuit decrement factor

The first part of the equation has an armature time constant, which is relatively short and controls the buildup and peak of the short-circuit current; the second part is determined by the shunt field excitation and it controls the decay of the peak value. The problem of calculation is that the time constants in this equation are not time invariant. Saturation

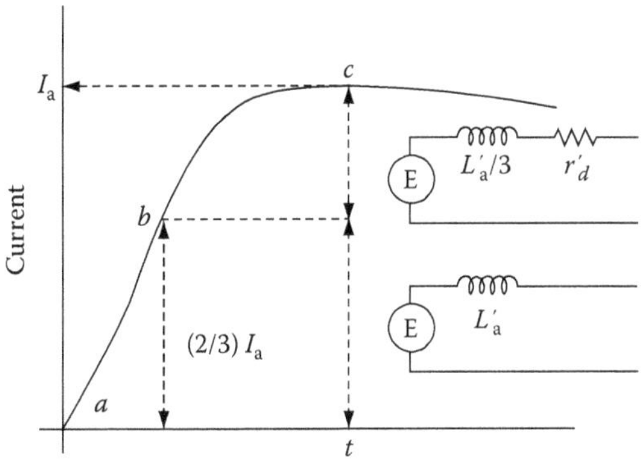

FIGURE 9.8
Short-circuit current–time curve for a dc motor or generator showing two distinct time constants.

causes the armature circuit decrement factor to increase as the motor saturates. Approximate values suggested for saturated conditions are 1.5–3.0 times the unsaturated value and conservatively a value of 3.0 can be used. The unsaturated value is applicable at the start of the short-circuit current and the saturated value at the maximum current. Between these two extreme values the decrement is changing from one value to another. Figure 9.8 shows the approximate curve of the short-circuit current and its equivalent circuit. For the first two-thirds of the curve, the circuit is represented by machine unsaturated inductance L'_a, and for the last one-third L'_a is reduced to one-third with series transient resistance. The peak short-circuit current in per unit is given by

$$i'_a = \frac{e_0}{r'_d} \tag{9.18}$$

The transient resistance r'_d in per unit requires some explanation. It is the effective internal resistance:

$$r'_d = r_w + r'_b + r'_x \tag{9.19}$$

where
- r_w is the total resistance of the windings in the armature circuit
- r'_x is the equivalent to flux reduction in per unit
- r'_b is the transient resistance equal to reactance voltage and brush contact resistance in per unit

The flux reduction and distortion are treated as ohmic resistance. The values of transient resistance r'_d in per unit are given graphically in the AIEE Committee Report of 1950 [5], depending on the machine rating, voltage, and speed. The transient resistance is not constant and there is a variation band. The machine load may also affect the transient resistance [6]. There does not seem to be any later publication on the subject. Similarly, the steady-state resistance is defined as

$$r_d = r_w + r_b + r_x \tag{9.20}$$

where
- r_b is the steady-state resistance equivalent to reactance voltage and brush contact in per unit
- r_x is the steady-state resistance equivalent to flux reduction in per unit

The maximum rate of rise of the current is dependent on armature unsaturated inductance. The unit inductance is defined as

$$L_{a1} = \frac{V_1}{I_a} \frac{2 \times 60}{2\pi P N_1} \tag{9.21}$$

The per unit inductance is the machine inductance L'_a divided by the unit inductance:

$$C_x = \frac{L'_a}{L_{a1}} = \frac{P N_1 L'_a}{19.1} \frac{I_a}{V_1} \tag{9.22}$$

This can be written as

$$L'_a = \frac{19.1 C_x V_1}{P N_1 I_a} \tag{9.23}$$

where
P is the number of poles
N_1 is the base speed
V_1 is the rated voltage
I_a is the rated machine current
C_x varies with the type of machine

Charts of initial inductance plotted against unit inductance show a linear relationship for a certain group of machines. For this purpose the machines are divided into four broad categories as follows:

Motors: $C_x = 0.4$ for motors without pole face windings.
Motors: $C_x = 0.1$ for motors with pole face windings.
Generators: $C_x = 0.6$ for generators without pole face windings.
Generators: $C_x = 0.2$ for generators with pole face windings.

The armature circuit decrement factor is

$$\sigma_a = \frac{r'_d 2\pi f}{C_x} \tag{9.24}$$

The maximum rate of rise of current in amperes per second is given by

$$\frac{di_a}{dt} = \frac{V_1 e_0}{L'_a} \tag{9.25}$$

The rate of current rise can also be expressed in terms of per unit rated current:

$$\frac{di_a}{dt} = \frac{P N_1 e_0}{19.1 C_x} \tag{9.26}$$

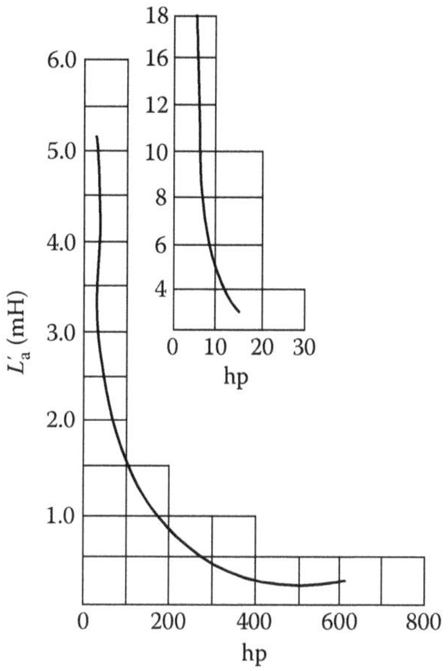

FIGURE 9.9
Inductance of dc motors in mH versus motor hp.

In Equations 9.25 and 9.26, e_0 can be taken as equal to unity without appreciable error. More accurately, e_0 can be taken as 0.97 per unit for motors and 1.03 for generators. The inductance L'_a is given in tabular and graphical forms in an AIEE publication [7] of 1952. Figure 9.9 shows L'_a values in mH for certain motor sizes, without pole face windings. Again, there does not seem to be a later publication on the subject.

Example 9.3

Calculate the short-circuit current of a 230 V, 150 hp, 1150-rpm motor, for a short circuit at motor terminals. The motor armature current is 541 A.

From graphical data in Ref. [1], the transient resistance is 0.068 per unit. Also, the inductance L'_a from the graphical data in Figure 9.9 is 1.0 mH. Then, the peak short-circuit current is

$$I'_a = \frac{I_a}{r'_d} = \frac{541}{0.068} = 7956 \text{ A}$$

and the initial rate of rise of the current is

$$\frac{di_a}{dt} = \frac{V_1}{L'_a} = \frac{230}{1 \times 10^{-3}} = 230 \text{ kA/s}$$

As shown in Figure 9.8, the time constant changes at point b.

$$\frac{L'_a}{3r'_d}$$

The base ohms are $V_a/I_a = 230/541 = 0.425$. Therefore, r'_d in ohms $= (0.425) \times (0.068) = 0.0289\ \Omega$. This gives a time constant of 11.52 ms. The short-circuit profile can now be plotted.

9.4.1 IEC Method of Short-Circuit of DC Machines

The resistance and inductance network of the short circuit of a dc machine with a separately excited field is shown in Figure 9.3. The equivalent resistance and inductance are

$$\begin{aligned} R_{\mathrm{MB}} &= R_{\mathrm{M}} + R_{\mathrm{ML}} + R_y \\ L_{\mathrm{MB}} &= L_{\mathrm{M}} + L_{\mathrm{ML}} + L_y \end{aligned} \tag{9.27}$$

where

R_{M} and L_{M} are the resistance and inductance of the armature circuit including the brushes

R_{ML} and L_{ML} are the resistance and inductance of the conductor in the motor circuit

R_y and L_y are the resistance and inductance of the common branch, if present

The time constant of the armature circuit up to the short-circuit location, τ_{M}, is given by

$$\tau_{\mathrm{M}} = \frac{L_{\mathrm{MB}}}{R_{\mathrm{MB}}} \tag{9.28}$$

The quasi-steady-state short-circuit current is given by

$$I_{\mathrm{KM}} = \frac{L_{\mathrm{F}}}{L_{0\mathrm{F}}} \frac{U_{\mathrm{rM}} - I_{\mathrm{rM}} R_{\mathrm{M}}}{R_{\mathrm{MB}}} \quad \text{when } n = n_{\mathrm{n}} = \text{constant} \tag{9.29}$$

where

L_{F} = equivalent saturated inductance of the field circuit on short-circuit
$L_{0\mathrm{F}}$ = equivalent unsaturated inductance of the field circuit at no load
U_{rM} = rated voltage of the motor
I_{rM} = rated current of the motor
n = motor speed
n_{n} = rated motor speed

The peak short-circuit current of the motor is given by

$$i_{\mathrm{pM}} = \kappa_{\mathrm{M}} \frac{U_{\mathrm{rM}} - I_{\mathrm{rM}} R_{\mathrm{M}}}{R_{\mathrm{MB}}} \tag{9.30}$$

At normal speed or decreasing speed with $\tau_{\mathrm{mec}} \geq 10\tau_{\mathrm{F}}$, the factor $\kappa_{\mathrm{M}} = 1$, where τ_{mec} is the mechanical time constant, given by

$$\tau_{\mathrm{mec}} = \frac{2\pi J n_0 R_{\mathrm{MB}} I_{\mathrm{rM}}}{M_{\mathrm{r}} U_{\mathrm{rM}}} \tag{9.31}$$

where

J is the moment of inertia

M_{r} is the rated torque of the motor

The field circuit time constant τ_{F} is given by

$$\tau_{\mathrm{F}} = \frac{L_{\mathrm{F}}}{R_{\mathrm{F}}} \tag{9.32}$$

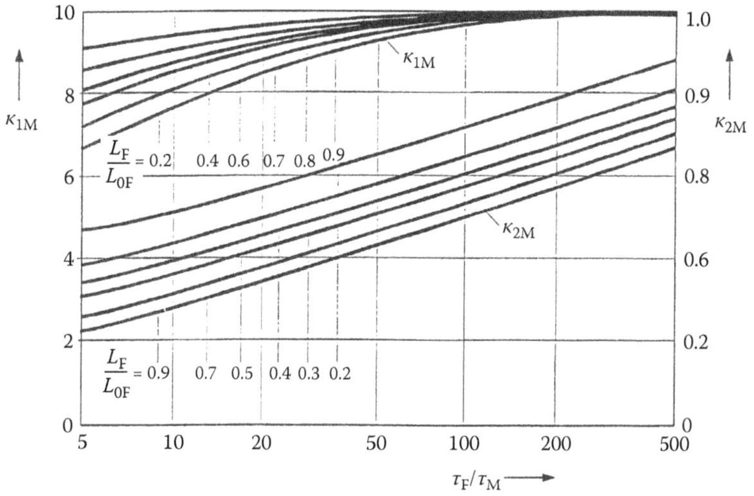

FIGURE 9.10
Factors κ_{1M} and κ_{2M} for determining the time to peak t_{pM} and the rise time constant τ_{1M} for normal and decreasing speed with $\tau_{mec} \geq 10\tau_F$; short circuit of a dc motor. (From IEC, Short-Circuit Currents in DC Auxiliary Installations in Power Plants and Substations, Standard 61660-1, 1997. Copyright 1997 IEC. Reproduced with permission.)

For $\tau_{mec} \geq 10\tau_F$, the time to peak and time constant are given by

$$
\begin{aligned}
t_{pM} &= \kappa_{1M}\tau_M \\
\tau_{1M} &= \kappa_{2M}\tau_M
\end{aligned}
\tag{9.33}
$$

The factor κ_{1M} and κ_{2M} are taken from Figure 9.10 and are dependent on τ_F/τ_M and L_F/L_{OF}. For decreasing speed with $\tau_{mec} < 10\tau_F$, the factor κ_M is dependent on $1/\delta = 2\tau_M$ and ω_0:

$$
\omega_0 = \sqrt{\frac{1}{\tau_{mec}\tau_M}\left(1 - \frac{I_{rM}R_M}{U_{rM}}\right)} \tag{9.34}
$$

where

ω_0 is the undamped natural angular frequency
δ is the decay coefficient
κ_M is derived from the curves in the IEC standard [3]

For decreasing speed with $\tau_{mec} < 10\tau_F$, the time to peak τ_M is read from a curve in the IEC standard, and the rise time constant is given by

$$
\tau_{1M} = \kappa_{3M}\tau_M \tag{9.35}
$$

where the factor κ_{3M} is again read from a curve in the IEC standard [3], not reproduced here.

Decay time constant τ_{2M}
For nominal speed or decreasing speed with $\tau_{mec} \geq 10\tau_F$:

$$
\begin{aligned}
\tau_{2M} &= \tau_F \quad \text{when } n = n_n = \text{const.} \\
\tau_{2M} &= \frac{L_{0f}}{L_F}\kappa_{4M}\tau_{mec} \quad \text{when } n \to 0 \text{ with } \tau_{mec} \geq 10\tau_F
\end{aligned}
\tag{9.36}
$$

For decreasing speed with $\tau_{mec} < 10\tau_F$:

$$\tau_{2M} = \kappa_{4M}\tau_{mec} \tag{9.37}$$

where κ_{4M} is again read from a curve in the IEC standard not reproduced here. Thus, the IEC calculation method requires extensive motor data and use of a number of graphical relations in the standard. The rise and decay time constants are related to $\tau_{mec} < 10\tau_F$ or $\tau_{mec} \geq 10\tau_F$.

Example 9.4

Calculate the short-circuit current for a terminal fault on a 115 V, 1150-rpm, six-pole, 15 hp motor. The armature current = 106 A, the armature and brush circuit resistance = 0.1 Ω, and the inductance in the armature circuit = 8 mH; $\tau_F = 0.8$ s, $\tau_{mec} > 10\tau_F$, $L_{OF}/L_F = 0.5$, and $\tau_{mec} = 20$ s.

There is no external resistance or inductance in the motor circuit. Therefore, $R_{MBr} = R_M = 0.10\ \Omega$. IEC is not specific about the motor circuit resistance, or how it should be calculated or ascertained.

The time constant is

$$\tau_M = \frac{L_M}{R_M} = \frac{8 \times 10^{-3}}{0.10} = 80 \text{ ms}$$

The quasi-steady-state current from Equation 9.29 is

$$0.5\left(\frac{115 - (0.10)(106)}{0.10}\right) = 522 \text{ A}$$

From Equation 9.30, the peak current is 1044 A, because for $\tau_{mec} > 10\tau_F$, factor κ_M in Equation 9.30 = 1. The time to peak and time constant are given by Equation 9.33. From Figure 9.10, and for $\tau_F/\tau_M = 10$ and $L_F/L_{OF} = 0.5$, factor $\kappa_{1M} = 8.3$ and $\kappa_{2M} = 3.7$. Therefore, the time to peak is 640 ms and the time constant $\kappa_{1M} = 296$ ms.

The short-circuit profile is plotted in Figure 9.7b.

9.5 Short-Circuit Current of a Rectifier

The typical current–time curve for a rectifier short circuit is shown in Figure 9.11. The maximum current is reached at one half-cycle after the fault occurs. The peak at half-cycle is caused by the same phenomenon that creates a dc offset in ac short-circuit calculations. The magnitude of this peak is dependent on X/R ratio, the ac system source reactance, rectifier transformer impedance, and the resistance and reactance through which the current flows in the dc system. The addition of resistance or inductance to the dc system reduces this peak and, depending on the magnitude of these components, the peak may be entirely eliminated, with a smoothing dc reactor as shown in Figure 9.1a. The region A in Figure 9.11 covers the initial rise of current, the peak current occurs in region B, and region C covers the time after one cycle until the current is interrupted.

The initial rate of rise of the dc short-circuit current for a bolted fault varies with the magnitude of the sustained short-circuit current. The addition of inductance to the dc circuit tends to decrease the rate of rise.

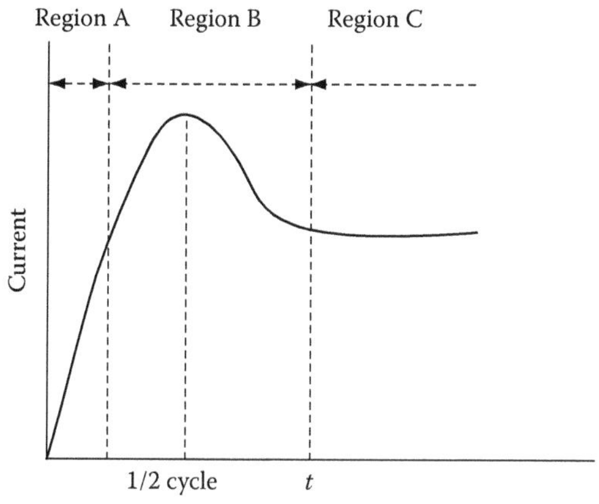

FIGURE 9.11
Short-circuit current profile of a rectifier.

An equivalent circuit of the rectifier short-circuit current is developed with a voltage source and equivalent resistance and inductance. The equivalent resistance varies with rectifier terminal voltage, which is a function of the short-circuit current. The equivalent resistance is determined from the rectifier regulation curve by an iterative process, which can admirably lend itself to iterative computer solution. The equivalent inductance is determined from a sustained short-circuit current for a bolted fault and rated system voltage. The magnitude of the peak current is determined from ac and dc system impedance characteristics [1].

The following step-by-step procedure can be used:

Calculate total ac system impedance $Z_C = R_C + X_C$ in ohms. Convert to per unit impedance z_C, which may be called the commutating impedance in per unit on a rectifier transformer kVA base. This is dependent on the type of rectifier circuit. For a double-wye, six-phase circuit, the conversion is given by [1]

$$Z_C = z_C \times 0.6 \times \frac{E_D}{I_D} \Omega \tag{9.38}$$

where E_D and I_D are rectifier rated dc voltage and rated dc current.

Assume a value of rectifier terminal voltage e_{da} under faulted condition and obtain factor K_2 from Figure 9.12. The *preliminary* calculated value of the sustained short-circuit current is then given by

$$I_{da} = \frac{K_2}{z_C} I_D \tag{9.39}$$

The equivalent rectifier resistance is then given by

$$R_R = \frac{(E_D - E_{da})}{I_{da}} \Omega \tag{9.40}$$

where E_{da} is the assumed rectifier terminal voltage in pu under fault conditions.

The sustained value of the fault current is

$$I_{dc} = \frac{E_D}{R_R + R_D} \text{A} \tag{9.41}$$

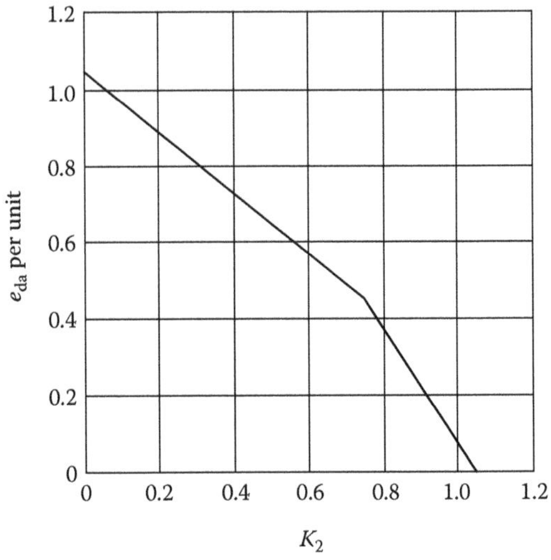

FIGURE 9.12
Sustained fault current factor versus rectifier terminal voltage.

where R_D is the resistance external to the rectifier. The rectifier terminal voltage in volts is

$$E_{dc} = E_D - I_{dc}R_R \tag{9.42}$$

The value of $E_{da} = e_{da} \times E_D$ should be within 10% of the calculated value E_{dc}, the rectifier terminal voltage under sustained short-circuit current. The iterative process is repeated until the desired tolerance is achieved.

Example 9.5

Consider a 100 kW source at 125 V dc. The dc resistance of the feeder cable, R_D, is 0.004 Ω. Let the ac source and rectifier transformer impedance $z_C = 0.05$ pu and $I_D = 800$ A. Calculate the rectifier resistance for a fault at the end of the cable.

Assume $e_{da} = 0.5$ per unit, i.e., $E_{da} = 62.5$ V; K_2 from Figure 9.12 = 0.70. Therefore,

$$I_{da} = \frac{K_2}{z_C} I_D = \frac{0.70}{0.05} \times 800 = 11{,}200 \text{ A}$$

Then R_R is

$$R_R = \frac{E_D - E_{da}}{I_{da}} = \frac{62.5}{11{,}200} = 0.00558\ \Omega$$

$$I_{dc} = \frac{E_D}{R_R + R_D} = \frac{125}{0.00558 + 0.004} = 13{,}048 \text{ A}$$

$$E_{dc} = 125 - (13{,}048)(0.00558) = 72.18 \text{ V}$$

We can iterate once more for a closer estimate of R_R:

$$E_{dc} = 72.18 \text{ V}$$

$$e_{da} = 0.392 \text{ pu}$$

$$K = 0.77$$

$$I_{da} = 12{,}320 \text{ A}$$

$$R_R = 0.006169\ \Omega$$
$$I_{dc} = 12{,}292.5\ \text{A}$$
$$E_{dc} = 49.17\ \text{V}$$

This is close enough an estimate as $E_{dc} = 0.392 \times 125\ \text{V} = 49\ \text{V}$. To calculate rate of rise of current, the rectifier inductance is required. This is given by

$$L_R = \frac{E_D}{360 \times I_{ds}}\ \text{H} \quad \text{(for 60 Hz)}$$

where I_{ds} is the terminal short-circuit current of the rectifier. On a terminal short-circuit voltage is zero and from Figure 9.12, $K_2 = 1.02$. This gives a short-circuit current of 15,320 A. Thus, inductance is 0.0213 mH. Add to it the inductance of the cable = 0.00796 mH. Then the rate of rise of current is 125/(0.00796 + 0.213) = 4.27 kA/s. The peak current can be 1.4–1.6 times the sustained current depending upon the system parameters [1].

9.5.1 IEC Method of Short-Circuit of a Rectifier

The equivalent short-circuit diagram is shown in Figure 9.13. The maximum dc short-circuit current is given by the minimum impedance $Z_{Q\min}$, which is obtained from the maximum short-circuit current $I''_{KQ\max}$ of the ac system:

$$Z_{Q\min} = \frac{cU_n}{\sqrt{3}I''_{KQ\max}} \tag{9.43}$$

The minimum dc current is given by

$$Z_{Q\max} = \frac{cU_n}{\sqrt{3}I''_{KQ\min}} \tag{9.44}$$

In Figure 9.13 the resistance and inductances on the ac side are

$$\begin{aligned} R_N &= R_Q + R_P + R_T + R_R \\ X_N &= X_Q + X_P + X_T + X_R \end{aligned} \tag{9.45}$$

where

R_Q and X_Q are the short-circuit resistance and reactance of the ac source referred to the secondary of the rectifier transformer

R_P and X_P are the short-circuit resistance and reactance of the power supply cable referred to the secondary side of the transformer

R_T and X_T are the short-circuit resistance and reactance of the rectifier transformer referred to the secondary side of the transformer

R_R and X_R are the short-circuit resistance and reactance of the commutating reactor, if present

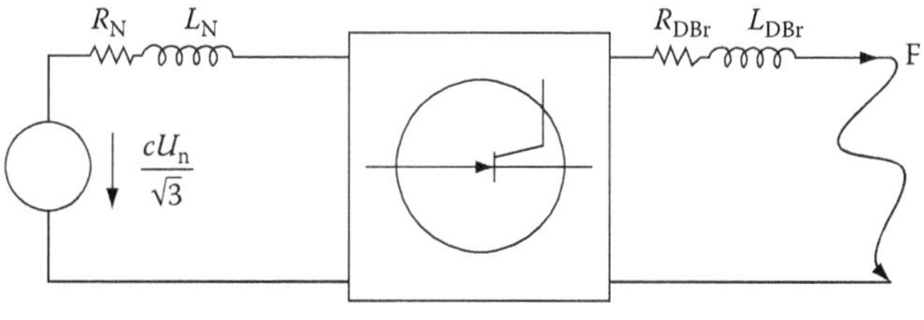

FIGURE 9.13
Equivalent circuit for short-circuit current calculations of a rectifier.

Similarly, on the dc side,

$$R_{DBr} = R_S + R_{DL} + R_y$$
$$L_{DBr} = L_S + L_{DL} + L_y \tag{9.46}$$

where

R_S, R_{DL}, and R_y are the resistances of the dc saturated smoothing reactor, the conductor in the rectifier circuit, and the common branch, respectively

L_S, L_{DL}, and L_y are the corresponding inductances

The quasi-steady-state short-circuit current is

$$i_{kD} = \lambda_D \frac{3\sqrt{2}}{\pi} \frac{cU_n}{\sqrt{3}Z_N} \frac{U_{rTLV}}{U_{rTHV}} \tag{9.47}$$

where Z_N is the impedance on the ac side of three-phase network. The factor λ_D as a function of R_N/X_N and R_{DBr}/R_N is estimated from the curves in the IEC standard [3]. Alternatively, it is given by the following equation:

$$\lambda_D = \sqrt{\frac{1 + (R_N/X_N)^2}{1 + (R_N/X_N)^2(1 + 0.667(R_{DBr}/R_N))^2}} \tag{9.48}$$

The peak short-circuit current is given by

$$i_{pD} = \kappa_D I_{kD} \tag{9.49}$$

where the factor κ_D is dependent on

$$\frac{R_N}{X_N}\left[1 + \frac{2R_{DBr}}{3R_N}\right] \quad \text{and} \quad \frac{L_{DBr}}{L_N} \tag{9.50}$$

It is estimated from the curves in the IEC standard [3] (not reproduced) or from the following equation:

$$\kappa_D = \frac{i_{pD}}{I_{kD}} = 1 + \frac{2}{\pi} e^{-\left(\frac{\pi}{3}+\phi_D\right)\cot\phi_D} \sin\phi_D \left(\frac{\pi}{2} - \arctan\frac{L_{DBr}}{L_N}\right) \tag{9.51}$$

where

$$\phi_D = \arctan\frac{1}{\frac{R_N}{X_N}\left(1 + \frac{2}{3}\frac{R_{DBr}}{R_N}\right)} \tag{9.52}$$

Time to peak t_{pD}, when $\kappa_D \geq 1.05$, is given by

$$i_{pD} = (3\kappa_D + 6)\ \text{ms} \quad \text{when} \quad \frac{L_{DBr}}{L_N} \leq 1$$
$$i_{pD} = \left[(3\kappa_D + 6) + 4\left(\frac{L_{DBr}}{L_N} - 1\right)\right]\ \text{ms} \quad \text{when} \quad \frac{L_{DBr}}{L_N} > 1 \tag{9.53}$$

If $\kappa_D < 1.05$, the maximum current, compared with the quasi-steady-state short-circuit current, is neglected, and $t_{pD} = T_k$ is used.

The rise time constant for 50 Hz is given by

$$\tau_{1D} = \left[2 + (\kappa_D - 0.9)\left(2.5 + 9\frac{L_{DBr}}{L_N}\right)\right] \text{ms} \quad \text{when } \kappa_D \geq 1.05$$
$$\tau_{1D} = \left[0.7 + \left[7 - \frac{R_N}{X_N}\left(1 + \frac{2}{3}\frac{L_{DBr}}{L_N}\right)\right]\left(0.1 + 0.2\frac{L_{DBr}}{L_N}\right)\right] \text{ms} \quad \text{when } \kappa_D < 1.05 \tag{9.54}$$

For simplification:

$$\tau_{1D} = \frac{1}{3}t_{pD} \tag{9.55}$$

The decay time constant τ_{2D} for 50 Hz is given by

$$\tau_{2D} = \frac{2}{\frac{R_N}{X_N}\left(0.6 + 0.9\frac{R_{DBr}}{R_N}\right)} \text{ms} \tag{9.56}$$

The time constants for a 60 Hz power system are not given in the IEC standard.

Example 9.6

A three-phase rectifier is connected on the ac side to a three-phase, 480–120 V, 100 kVA transformer of percentage $Z_T = 3\%$, $X/R = 4$. The 480 V source short-circuit MVA is 30, and the X/R ratio $= 6$. The dc side smoothing inductance is 5 μH and the resistance of the cable connections is 0.002 Ω. Calculate and plot the short-circuit current profile at the end of the cable on the dc side.

Based on the ac side data, the source impedance in series with the transformer impedance referred to the secondary side of the rectifier transformer is

$$R_Q + jX_Q = 0.00008 + j0.00048\,\Omega$$
$$R_T + jX_T = 0.001 + j0.00419\,\Omega$$

Therefore,

$$R_N + jX_N = 0.0011 + j0.004671\,\Omega$$

On the dc side:

$$R_{DBr} = 0.002\,\Omega \quad \text{and} \quad L_{DBr} = 5\,\mu\text{H}$$

This gives

$$\frac{R_N}{X_N} = 0.24 \quad \text{and} \quad \frac{R_{DBr}}{R_N} = 2.0$$

Calculate λ_D from Equation 9.48:

$$\lambda_D = \sqrt{\frac{1 + (0.24)^2}{1 + (0.24)^2(1 + 0.667)(2.0)^2}} = 0.897$$

The quasi-steady-state current is, therefore, from Equation 9.47,

$$I_{kD} = (0.897)\left(\frac{3\sqrt{2}}{\pi}\right)\left(\frac{1.05 \times 480}{\sqrt{3} \times 0.0048}\right)\left(\frac{120}{480}\right) = 18.36 \text{ kA}$$

To calculate the peak current, calculate the ratios,

$$\frac{R_N}{X_N}\left(1 + \frac{2}{3}\frac{R_{DBr}}{R_N}\right) = (0.24)(1 + 0.667 \times 2) = 0.56$$

$$\frac{L_{DBr}}{L_N} = \frac{5 \times 10^{-6}}{0.0128 \times 10^{-3}} = 0.392$$

Calculate κ_D from Equations 9.51 and 9.52. From Equation 9.52,

$$\phi_D = \tan^{-1}\left(\frac{1}{1 + 0.667(2.0)}\right) = 60.75°$$

and from Equation 9.51, $\kappa_D = 1.204$. Thus, the peak short-circuit current is

$$i_{pD} = \kappa_D I_{kD} = 1.204 \times 18.36 = 22.10 \text{ kA}$$

The time to peak is given by Equation 9.53 and is equal to

$$t_{pD} = (3\kappa_D + 6)\text{ms} = (3 \times 1.204 + 6) = 9.62 \text{ ms}$$

The rise time constant is given by Equation 9.54 and is equal to 3.83 ms, and the decay time constant is given by Equation 9.56 and is equal to 4.58 ms.

The current profile is plotted in Figure 9.7c, which shows the calculated values. The intermediate shape of the curve can be correctly plotted using Equations 9.1 and 9.2. Note that, in this example, the IEC equations are for a 50 Hz system. For a 60 Hz system, the peak will occur around 8.3 ms.

9.6 Short Circuit of a Charged Capacitor

9.6.1 IEC Method

The resistance and inductance in the capacitor circuit from Figure 9.3 are

$$\begin{aligned} R_{CBr} &= R_C + R_{CL} + R_y \\ L_{CBr} &= L_{CL} + L_y \end{aligned} \tag{9.57}$$

where

R_C is the equivalent dc resistance of the capacitor

R_{CL} and L_{CL} are the resistance and inductance of a conductor in the capacitor circuit

The steady-state short-circuit current of the capacitor is zero and the peak current is given by

$$i_{pC} = \kappa_C \frac{E_C}{R_{CBr}} \tag{9.58}$$

where
E_C is the capacitor voltage before the short circuit
κ_C is read from curves in the IEC standard [3], based on

$$\frac{1}{\delta} = \frac{2L_{CBr}}{R_{CBr}}$$
$$\omega_0 = \frac{1}{\sqrt{L_{CBr}C}} \qquad (9.59)$$

If $L_{CBr} = 0$, then $\kappa_C = 1$.

The time to peak t_{pC} is read from curves in the IEC standard [3]. If $L_{CBr} = 0$, then $t_{pC} = 0$. The rise time constant is

$$\tau_{1C} = \kappa_{1C} t_{pC} \qquad (9.60)$$

where κ_{1C} is read from curves in IEC. The decay time constant is

$$\tau_{2C} = \kappa_{2C} R_{CBr} C \qquad (9.61)$$

where κ_{2C} is read from curves in IEC standard [3]. The curves for these factors are not reproduced.

Example 9.7

A 120 V, 100 μF capacitor has $R_{CBr} = 0.05\,\Omega$ and $L_{CBr} = 10\,\text{mH}$. Calculate the terminal short-circuit profile.

From Equation 9.59,

$$\frac{1}{\delta} = \frac{2 \times 10 \times 10^{-3}}{0.05} = 0.4$$

Also,

$$\omega_0 = \frac{1}{\sqrt{10 \times 10^{-3} \times 100 \times 10^{-6}}} = 1000$$

From curves in the IEC standard, $\kappa_c = 0.92$. The peak current from Equation 9.58 is then $(0.92) \times (120/0.05) = 2208$ A. The time to peak from curves in IEC = 0.75 ms, and $\kappa_{IC} = 0.58$. From Equation 9.60 the rise time constant is $(0.58) \times (0.75) = 0.435$ ms. Also, $\kappa_{2C} = 1$, and, from Equation 9.61, the decay time constant is 5 μs. The short-circuit current profile is plotted in Figure 9.7d.

9.7 Total Short-Circuit Current

The total short-circuit current at fault F_1 (Figure 9.3) is the sum of the partial short-circuit currents calculated from the various sources. For calculation of the total short-circuit current at F_2 (Figure 9.3), the partial currents from each source should be calculated by adding the resistance and inductance of the common branch to the equivalent circuit. A correction factor is then applied. The correction factors for every source are obtained from

$$i_{\text{pcor}j} = \sigma_j i_{\text{p}j}$$
$$I_{\text{kcor}j} = \sigma_j I_{\text{k}j} \tag{9.62}$$

where the correction factor σ_j is described in the IEC standard [3].

Example 9.8

The sources (rectifier, battery, motor, and capacitor) in Examples 9.2, 9.4, 9.6, and 9.7 are connected together in a system configuration as shown in Figure 9.3. Plot the total short-circuit current.

The profiles of partial currents shown in Figure 9.7 are summed. As the time to peak, magnitudes, and decay time constants are different in each case, a graphical approach is taken and the total current profile is shown in Figure 9.14. The peak current is approximately 27.3 kA and the peak occurs at approximately 9 ms after the fault.

The short-circuit current from the rectifier predominates. The short-circuit current from the capacitor is a high rise pulse, which rapidly decays to zero. The dc motor short-circuit current rises slowly. Smaller dc motors have higher armature inductance (Figure 9.9), resulting in a slower rate of current rise. The rectifier current peaks approximately in one half-cycle of the power system frequency. The relative magnitudes of the partial currents can vary substantially, depending on the system configuration. This can give varying profiles of total current and time to peak.

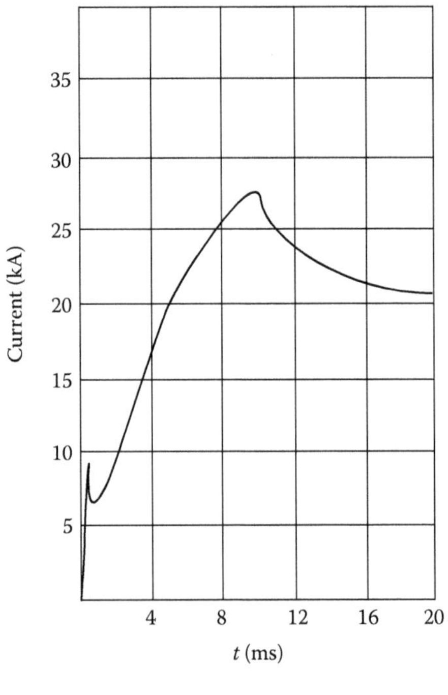

FIGURE 9.14
Total fault current profile of the partial short-circuit currents calculated in Examples 9.1, 9.4, 9.6, and 9.7.

9.8 DC Circuit Breakers

The general-purpose low-voltage dc power circuit breakers do not limit the peak available current and these breakers are assigned peak, short-time, and short-circuit ratings. A high-speed breaker during interruption limits the peak to a value less than the available (perspective) current and these breakers have a short-circuit rating and short-time rating. A semi-high-speed circuit breaker does not limit the short-circuit current on circuits with minimal inductance, but becomes current limiting for highly inductive circuits. This design also requires a peak rating. Rectifier circuit breakers are a class in themselves and these carry the normal current output of one rectifier, and during fault conditions function to withstand or interrupt abnormal currents as required. This breaker requires a short-circuit current rating for $n-1$ rectifiers and a short-time current rating for its own rectifier.

The dc breakers may have thermal magnetic or electronic trip devices, i.e., general-purpose circuit breakers of 2 kA or lower are provided with instantaneous tripping elements set to operate at 15 times the rated continuous current, and breakers rated >2 kA have instantaneous trips set to operate at 12 times the rated current. The rectifier circuit breakers have a reverse-current tripping element set to operate at no more than 50% of the continuous currents rating.

Two or three poles of a breaker may be connected in series for enhanced interrupting rating. The interrupting capacity of a breaker decreases with increasing dc voltage. The maximum inductance for full interrupting rating in microhenries is specified and the reduced interrupting rating for higher values of inductance can be calculated. When the breakers are rated for ac as well as dc systems, the interrupting rating on dc systems is much lower.

With general-purpose dc circuit breaker for the fault location F_1, the fault current profile shown in Figure 9.14 should have a peak rating of 40 kA and a short-circuit rating of higher than 25 kA.

9.8.1 High-Voltage DC Circuit Breakers

In Chapter 4, we discussed the arc-lengthening principle for arc extinguishing. Figure 9.15 shows the dc arc characteristics for different arc lengths. The resistance characteristic is shown by a straight line. For arc length given by curve a, the arc voltage is lower than the

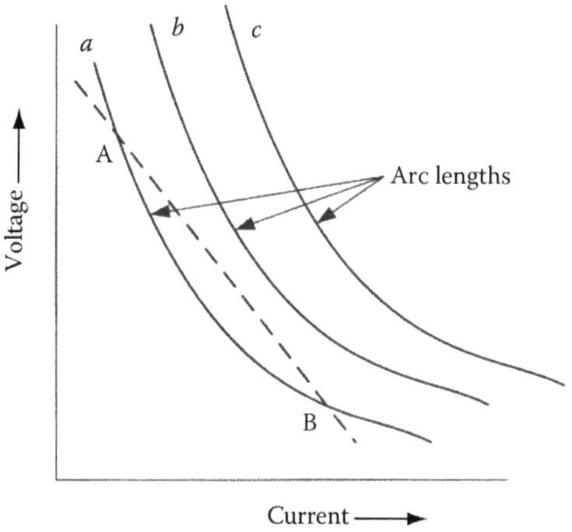

FIGURE 9.15
DC arc characteristics for different lengths.

supply system voltage by an amount IR. Only at the points of intersection it is possible to have a stable arc. For the arc of length in curve a, it intersects at point A. The arc voltage is less than the supply voltage and the arc will continue to burn. As the arc length is increased, the arc voltage increases above the supply voltage and the arc is extinguished. During the arcing time, the supply source continues to give up energy (energy stored in the inductance of the system); the longer the arc burns the greater is the energy. Thus, the entire interruption process is a question of energy balance.

This principle has been successfully employed in some commercial designs of medium-voltage ac breakers also. In these breaker designs, the current can be easily extinguished at current zero, there is practically no current chopping, and the breaker mechanism is very light, except for a large arc chute with splitters. In case of dc some chopping can be expected as there is no current zero.

Consider a dc current flowing in a parallel LC circuit. If this current is suddenly interrupted (instantaneously), the recovery voltage situation can be simulated by forcing an equal and opposite current through the interrupting breaker. The recovery voltage in this case will be given by the following differential equation written in Laplace transform:

$$V_r(s) = \frac{i}{C}\left(\frac{1}{s^2 + \frac{1}{LC}}\right) \tag{9.63}$$

And its solution is

$$V_r(t) = i\sqrt{\frac{L}{C}}\sin\frac{t}{\sqrt{LC}} \tag{9.64}$$

In case a ramp is used, i.e., the current is gradually brought to zero, the rate of rise of the recovery voltage can be slowed, i.e., the slower the chopping rate, the slower will be the recovery voltage. A ramp signal in Laplace transform can be written as

$$i = -\left[\frac{a}{s^2} - \frac{a}{s^2}e^{-t_c s}\right] \tag{9.65}$$

Where $a = i_o/t_c$ = slope of the ramp.
and the solution with this signal is

$$V_r(t) = \frac{-iL}{t_c}\left[\left(t - \cos\frac{t}{\sqrt{LC}}\right) - \left(1 - \cos\frac{t - t_c}{\sqrt{LC}}\right)u(t - t_c)\right] \tag{9.66}$$

For HVDC transmission, high-voltage DC circuit breakers are not necessary and are not used. Thyristor control takes care of the overcurrent, the short circuits and abnormal conditions in HVDC converters. However, MRTB (metallic return transfer breaker is used in bipolar HVDC transmission and the bipolar mode is changed to monopolar mode during a fault on one of the poles.

In high-voltage dc circuit breakers, the energy stored in the system can be high, 0.5 LI_d^2 = 10–25 MJ. Artificial current zeros are produced by a reverse flow of current. The main breaker contacts are in parallel with a charged capacitor, reactor, and vacuum gap.

This circuit is called a commutating circuit. When the main breaker contacts open, the vacuum gap is triggered. The precharged capacitor discharges violently through main circuit, causing discharge current to flow in opposition to the main current. The main current drops down and oscillates with current zeros. The current is interrupted by the main circuit breaker at one of the current zeros. Vacuum gap seals in and the commutating circuit is opened.

The energy in the main dc pole is partly absorbed by the commutating circuit and partly by the zinc oxide surge arrester which can absorb up to 20 kJ/kV. The arc in the main circuit is quenched by interrupting medium, air, or SF_6. The stresses are much higher because of higher energy associated with dc current interruption, and higher TRV appears across the breaker pole.

The HVDC breaker types identified by CIGRE working group [8] are classified on the following basis:

- Switching time
- Current to be interrupted
- TRV
- Switching energy

The classification according to CIGRE is as follows:

- Type A: It is a fast breaker, less than 15 ms and does not wait for converter action to reduce dc voltage. The TRV is the highest. Current capability is up to 1.25 I_d.
- Type B: It is slow and waits for converter action to reduce dc line current to low value. It is classified into two types B1 and B2. Type B1 has higher TRV compared to type B2, the operating time of type B2 is longer, 90–120 ms, while that of type B1 is 60–90 ms. Type A operating time is less than 15 ms.

Problems

9.1 Calculate and plot the short-circuit current profile for a battery system with details as follows: lead acid battery, 240 V, 120 cells, 400 Ah rating at an 8 h rate of 1.75 V per cell at 25°C. Each cell has a length of 15.5 in., width 6.8 in., and height 10 in. The cells are arranged in two-tier configuration in four rows, 30 cells per row. Intercell connectors are of l in. × 1/2 in. cross section, resistance 0.0321 mΩ/ft. Calculate the battery internal resistance by using Equation 9.11. The battery is connected through a cable of resistance 0.002 Ω and inductance 15 μH. The fault occurs at the end of the battery cable. Repeat with the IEC calculation method, as in Example 9.2.

9.2 Calculate and plot the terminal short-circuit current of a dc motor of 50 hp, 230 V, 690 rpm, armature current 178 A, and transient resistance 0.07 Ω, using the method of Example 9.3. Repeat the calculations using the IEC method. Additional motor data: $\tau_F = 1.0$ s, $J = 2$ kg/m^2, $L_{OF}/L_F = 0.3$.

9.3 Calculate and plot the short-circuit current profile for a fault on the dc side of a rectifier system in the following configuration: 480 V, three-phase ac source fault levels 20 kA,

$X/R = 5$, 480–230 V, three-phase 300 kVA rectifier transformer, $Z = 3.5\%$, $X/R = 5$, the dc side equivalent resistance and inductance equal to 0.001 Ω and 3 μH, respectively.

9.4 Sum the partial fault currents calculated in Problems 9.1 through 9.3 and calculate the maximum short-circuit current and time to peak. What should be the peak short-circuit rating and interrupting rating of a general-purpose dc circuit breaker?

References

1. General Electric Company. *GE Industrial Power System Data Book*. Schenectady, NY, 1978
2. ANSI/IEEE. Standard for Low-Voltage DC Power Circuit Breakers Used in Enclosures, 1979. Standard C37.14.
3. IEC. Short-Circuit Currents in DC Auxiliary Installations in Power Plants and Substations, 1997. Standard 61660-1.
4. IEEE. DC Auxiliary Power Systems for Generating Stations, 1992. Standard 946.
5. AIEE Committee Report. Maximum short-circuit current of DC motors and generators. Transient characteristics of DC motors and generators. *AIEE Trans*. 69:146–149, 1950.
6. A.T. McClinton, E.L. Brancato, and R. Panoff. Maximum short-circuit current of DC motors and generators. Transient characteristics of DC motors and generators. *AIEE Trans*. 68:1100–1106, 1949.
7. A.G. Darling and T.M. Linville. Rate of rise of short-circuit current of DC motors and generators. *AIEE Trans*. 71:314–325, 1952.
8. CIGRE Joint Working Group 13/14-08. Circuit breakers for meshed multi-terminal HVDC systems, Part 1: DC side substation switching under normal and fault conditions. *Electra* 163, December 1995. 98–122.

10

Load Flow over Power Transmission Lines

Load flow is a solution of the steady-state operating conditions of a power system. It presents a "frozen" picture of a scenario with a given set of conditions and constraints. This can be a limitation, as the power system's operations are dynamic. In an industrial distribution system, the load demand for a specific process can be predicted fairly accurately and a few load flow calculations will adequately describe the system. For bulk power supply, the load demand from hour to hour is uncertain, and winter and summer load flow situations, though typical, are not adequate. A moving picture scenario could be created from static snapshots, but it is rarely adequate in large systems having thousands of controls and constraints. Thus, the spectrum of load flow (power flow) embraces a large area of calculations, from calculating the voltage profiles and power flows in small systems to problems of online energy management and optimization strategies in interconnected large power systems.

Load flow studies are performed using digital computer simulations. These address operation, planning, running, and development of control strategies. Applied to large systems for optimization, security, and stability, the algorithms become complex and involved. While the treatment of load flow, and finally optimal power flow, will unfold in the following chapters, it can be stated that there are many load flow techniques and there is a historical background to the development of these methods.

In this chapter, we will study the power flow over power transmission lines, which is somewhat distinct and a problem by itself. The characteristics and performance of transmission lines can vary over wide limits, mainly dependent on their length. Maintaining an acceptable voltage profile at various nodes with varying power flow is a major problem. We will consider two-port networks, i.e., a single transmission line, to appreciate the principles and complexities involved.

10.1 Power in AC Circuits

The concepts of instantaneous power, average power, apparent power, and reactive power are fundamental and are briefly discussed here. Consider a lumped impedance $Ze^{j\theta}$, excited by a sinusoidal voltage $E < 0°$ at constant frequency. The linear load draws a sinusoidal current. The time-varying power can be written as

$$p(t) = \mathrm{Re}(\sqrt{2}E\varepsilon^{j\omega t})\,\mathrm{Re}(\sqrt{2}I\varepsilon^{j(\omega t-\theta)})$$

$$= 2EI\cos\omega t \cdot \cos(\omega t - \theta) \tag{10.1}$$

$$= EI\cos\theta + EI\cos(2\omega t - \theta) \tag{10.2}$$

where E and I are the rms voltage and current, respectively. The first term is the average time-dependent power, when the voltage and current waveforms consist only of fundamental components. The second term is the magnitude of power swing. Equation 10.2 can be written as

$$p(t) = EI\cos\theta(1+\cos 2\omega t) + EI\sin\theta\cdot\sin 2\omega t \tag{10.3}$$

The first term is the power actually exhausted in the circuit and the second term is power exchanged between the source and circuit, but not exhausted in the circuit. The active power is measured in watts and is defined as

$$P = EI\cos\theta(1+\cos 2\omega t) \approx EI\cos\theta \tag{10.4}$$

The reactive power is measured in var and is defined as

$$Q = EI\sin\theta\ \sin 2\omega t \approx EI\sin\theta \tag{10.5}$$

These relationships are shown in Figure 10.1; $\cos\theta$ is called the *power factor* (PF) of the circuit, and θ is the PF angle.

The apparent power in VA (volt-amperes) is given by

$$S = \sqrt{P^2+Q^2} = \sqrt{P^2+(P\tan\theta)^2} = P\sqrt{(1+\tan^2\theta)} = P\sec\theta = P\cos\theta \tag{10.6}$$

The PF angle is generally defined as

$$\theta = \tan^{-1}\left(\frac{Q}{P}\right) \tag{10.7}$$

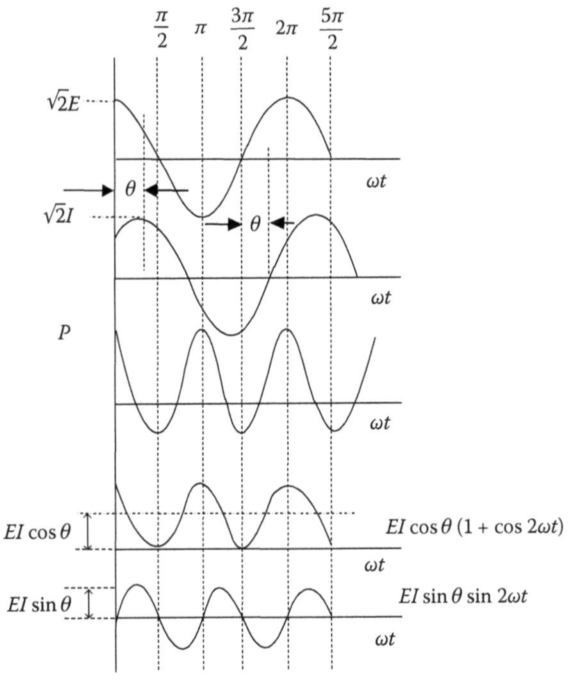

FIGURE 10.1
Active and reactive power in ac circuits.

If $\cos\theta = 1$, $Q = 0$. Such a load is a unity PF load. Except for a small percentage of loads, i.e., resistance heating and incandescent lighting, the industrial, commercial, or residential loads operate at lagging PF. As the electrical equipment is rated on a kVA basis, a lower PF derates the equipment and limits its capacity to supply active power loads. The reactive power flow and control is one important aspect of power flow and Chapter 13 is devoted to it. The importance of PF (reactive power) control can be broadly stated as follows:

- Improvement in the active power handling capability of transmission lines.
- Improvement in voltage stability limits.
- Increasing capability of existing systems: the improvement in PF for release of a certain per unit kVA capacity can be calculated from Equation 10.6:

$$\mathrm{PF_{imp}} = \frac{\mathrm{PF_{ext}}}{1 - \mathrm{kVA_{ava}}} \tag{10.8}$$

where
$\mathrm{PF_{imp}}$ is improved PF
$\mathrm{PF_{ext}}$ is existing PF
$\mathrm{kVA_{ava}}$ is kVA made available as per unit of existing kVA

- Reduction in losses: the active power losses are reduced as these are proportional to the square of the current. With PF improvement, the current per unit for the same active power delivery is reduced. The loss reduction is given by the expression

$$\mathrm{Loss_{red}} = 1 - \left(\frac{\mathrm{PF_{ext}}}{\mathrm{PF_{imp}}}\right)^2 \tag{10.9}$$

where $\mathrm{Loss_{red}}$ is reduction in losses in per unit with improvement in PF from $\mathrm{PF_{ext}}$ to $\mathrm{PF_{imp}}$. An improvement of PF from 0.7 to 0.9 reduces the losses by 39.5%.
- Improvement of transmission line regulation: the PF improvement improves the line regulation by reducing the voltage drops on load flow.

All these concepts may not be immediately clear and are further developed.

The active power can also be written as

$$P = \frac{1}{T}\int_0^T p(t)\mathrm{d}t = EI\cos\theta = S\cos\theta \tag{10.10}$$

That is the active power, i.e., the average value of instantaneous power over one period T.

The source current can be expressed as

$$\begin{aligned} I &= \frac{P}{E}\sin\omega t + \frac{Q}{E}\cos\omega t \\ &= \sqrt{\left(\frac{P}{E}\right)^2 + \left(\frac{Q}{E}\right)^2} \end{aligned} \tag{10.11}$$

In a three-phase circuit the currents and voltages are represented as column vectors. The three-phase apparent power can be expressed as

$$
\begin{aligned}
S &= E_a I_a + E_b I_b + E_c I_c \\
S &= \sqrt{P^2_{\text{three-phase}} + Q^2_{\text{three-phase}}} \\
S &= \sqrt{E_a^2 + E_b^2 + E_c^2} \cdot \sqrt{I_a^2 + I_b^2 + I_c^2}
\end{aligned}
\tag{10.12}
$$

where

I_a, I_b, and I_c are the currents in the three phases
E_a, E_b, and E_c are the voltages

So long as the voltages are sinusoidal and loads are *linear* and balanced, all the three equations of S will give the same results.

10.1.1 Complex Power

If the voltage vector is expressed as $A+jB$ and the current vector as $C+jD$, then by convention the volt-amperes in ac circuits are vectorially expressed as

$$
\begin{aligned}
EI^* &= (A + jB)(C - jD) \\
&= AC + BD + j(BC - AD) \\
&= P + jQ
\end{aligned}
\tag{10.13}
$$

where

$P = AC + BD$ is the active power
$Q = BC - AD$ is the reactive power
I^* is the conjugate of I

This convention makes the imaginary part representing reactive power negative for the leading current and positive for the lagging current. *This is the convention used by power system engineers.* If a conjugate of voltage, instead of current, is used, the reactive power of the leading current becomes positive. The PF is given by

$$\cos\theta = \frac{AC + BD}{\sqrt{A^2 + B^2}\sqrt{C^2 + D^2}} \tag{10.14}$$

10.1.2 Conservation of Energy

The conservation of energy concept (Tellegen's theorem) is based on Kirchoff laws and states that the power generated by the network is equal to the power consumed by the network (inclusive of load demand and losses). If $i_1, i_2, i_3, \ldots, i_n$ are the currents and $v_1, v_2, v_3, \ldots, v_n$ the voltages of n single-port elements connected in any manner,

$$\sum_{k=1}^{k=n} V_k I_k = 0 \tag{10.15}$$

This is an obvious conclusion.

Also, in a linear system of passive elements, the complex power, active power, and reactive power should summate to zero:

$$\sum_{k=1}^{k=n} S_n = 0 \tag{10.16}$$

$$\sum_{k=1}^{k=n} P_n = 0 \tag{10.17}$$

$$\sum_{k=1}^{k=n} Q_n = 0 \tag{10.18}$$

10.2 Power Flow in a Nodal Branch

The modeling of transmission lines is unique in the sense that capacitance plays a significant role and cannot be ignored, except for short lines of length less than approximately 50 mi (80 km). Let us consider power flow over a short transmission line. As there are no shunt elements, the line can be modeled by its series resistance and reactance, load, and terminal conditions. Such a system may be called a nodal branch in load flow or a two-port network. The sum of the sending end and receiving end active and reactive powers in a nodal branch is not zero, due to losses in the series admittance Y_{sr} (Figure 10.2). Let us define Y_{sr}, the admittance of the series elements $= g_{sr} - jb_{sr}$ or $Z = zl = l(r_{sr} + jx_{sr}) = R_{sr} + X_{sr} = 1/Y_{sr}$, where l is the length of the line. The sending end power is

$$S_{sr} = V_s I_s^* \tag{10.19}$$

where I_s^* is conjugate of I_s. This gives

$$\begin{aligned} S_{sr} &= V_s[Y_{sr}(V_s - V_r)]^* \\ &= [V_s^2 - V_s V_r \varepsilon^{j(\theta_s - \theta_r)}](g_{sr} - jb_{sr}) \end{aligned} \tag{10.20}$$

where sending-end voltage is $V_s < \theta_s$ and receiving-end voltage is $V_r < \theta_r$. The complex power in Equation 10.20 can be divided into active and reactive power components; at the sending end,

$$P_{sr} = [V_s^2 - V_s \cos(\theta_s - \theta_r)]g_{sr} - [V_s V_r \sin(\theta_s - \theta_r)]b_{sr} \tag{10.21}$$

$$Q_{sr} = [-V_s V_r \sin(\theta_s - \theta_r)]g_{sr} - [V_s^2 - V_s V_r \cos(\theta_s - \theta_r)]b_{sr} \tag{10.22}$$

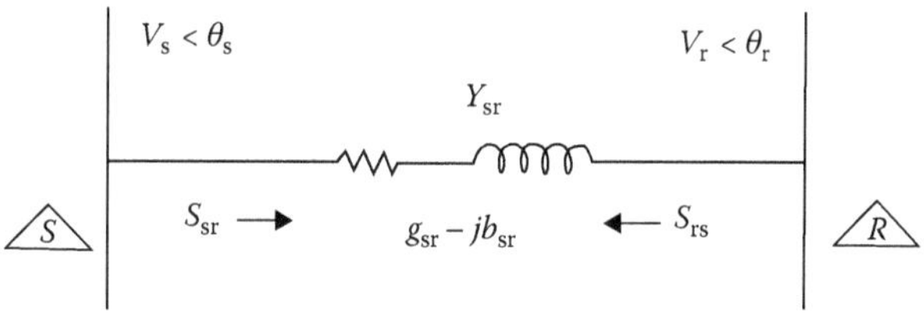

FIGURE 10.2
Power flow over a two-port line.

and, at the receiving end,

$$P_{rs} = [V_r^2 - V_r V_s \cos(\theta_r - \theta_s)]g_{sr} - [V_r V_s \sin(\theta_r - \theta_s)]b_{sr} \tag{10.23}$$

$$Q_{rs} = [-V_r V_s \sin(\theta_r - \theta_s)]g_{sr} - [V_r^2 - V_r V_s \cos(\theta_r - \theta_s)]b_{sr} \tag{10.24}$$

If g_{sr} is neglected,

$$P_{rs} = \frac{|V_s||V_r|\sin\delta}{X_{sr}} \tag{10.25}$$

$$Q_{rs} = \frac{|V_s||V_r|\cos\delta - |V_r|^2}{X_{sr}} \tag{10.26}$$

where δ is the difference between the sending-end and receiving-end voltage vector angles $= (\theta_s - \theta_r)$. For small values of delta, the reactive power equation can be written as

$$Q_{rs} = \frac{|V_r|}{X_{sr}}(|V_s| - |V_r|) = \frac{|V_r|}{X_{sr}}|\Delta V| \tag{10.27}$$

where $|\Delta V|$ is the voltage drop. For a short line it is

$$|\Delta V| = I_r Z = (R_{sr} + jX_{sr})\frac{(P_{rs} - jQ_{rs})}{V_r} \approx \frac{R_{sr}P_{rs} + X_{sr}Q_{rs}}{|V_r|} \tag{10.28}$$

Therefore, the transfer of real power depends on the angle δ, called the *transmission* angle, and the relative magnitudes of the sending and receiving end voltages. As these voltages will be maintained close to the rated voltages, it is mainly a function of δ. The maximum power transfer occurs at $\delta = 90°$ (steady-state stability limit).

The reactive power flows is in the direction of lower voltage and it is independent of δ. The following conclusions can be drawn:

1. For small resistance of the line, the real power flow is proportional to $\sin\delta$. It is a maximum at $\delta = 90°$. For stability considerations, the value is restricted to below 90°. The real power transfer rises with the rise in the transmission voltage.
2. The reactive power flow is proportional to the voltage drop in the line, and is independent of δ. The receiving-end voltage falls with increase in reactive power demand.

10.2.1 Simplifications of Line Power Flow

Generally, the series conductance is less than the series susceptance, the phase angle difference is small, and the sending-end and receiving-end voltages are close to the rated voltage:

$$\begin{aligned} &g_{sr} \prec b_{sr} \\ &\sin(\theta_s - \theta_r) \approx \theta_s - \theta_r \\ &\cos(\theta_s - \theta_r) \approx 1 \\ &V_s \approx V_r \approx 1 \text{ per unit} \end{aligned} \tag{10.29}$$

If these relations are used,

$$\begin{aligned} P_{sr} &\approx (\theta_s - \theta_r)b_{sr} \\ Q_{sr} &\approx (V_s - V_r)b_{sr} \\ P_{rs} &\approx -(\theta_r - \theta_s)b_{sr} \\ Q_{rs} &\approx -(V_r - V_s)b_{sr} \end{aligned} \quad (10.30)$$

10.2.2 Voltage Regulation

The voltage regulation is defined as the rise in voltage at the receiving end, expressed as a percentage of full-load voltage when the full load at a specified PF is removed. The sending-end voltage is kept constant. The voltage regulation is expressed as a percentage or as per unit of the receiving-end full-load voltage:

$$\text{VR} = \frac{V_{rnl} - V_{rfl}}{V_{rfl}} \times 100 \quad (10.31)$$

where
V_{rnl} is the no-load receiving-end voltage
V_{rfl} is the full-load voltage at a given PF

10.3 *ABCD* Constants

A transmission line of any length can be represented by a four-terminal network, Figure 10.3a. In terms of A, B, C, and D constants, the relation between sending- and receiving-end voltages and currents can be expressed as

$$\begin{vmatrix} V_s \\ I_s \end{vmatrix} = \begin{vmatrix} A & B \\ C & D \end{vmatrix} \begin{vmatrix} V_r \\ I_r \end{vmatrix} \quad (10.32)$$

In the case where sending-end voltages and currents are known, the receiving-end voltage and current can be found by

$$\begin{vmatrix} V_r \\ I_r \end{vmatrix} = \begin{vmatrix} D & -B \\ -C & A \end{vmatrix} \begin{vmatrix} V_s \\ I_s \end{vmatrix} \quad (10.33)$$

Also,

$$AD - BC = 1 \quad (10.34)$$

The significance of these constants can be stated as follows:

$A = V_s/V_r$, when $I_r = 0$, i.e., the receiving end is open-circuited. It is the ratio of two voltages and is dimensionless

$B = V_r/I_r$, when $V_r = 0$, i.e., the receiving end is short-circuited. It has the dimensions of an impedance and specified in ohms

$C = I_r/V_r$, when the receiving end is open-circuited and I_r is zero. It has the dimensions of an admittance

$D = I_s/I_r$, when $V_r = 0$, that is the receiving and is short-circuited. It is the ratio of two currents and dimensionless

(10.35)

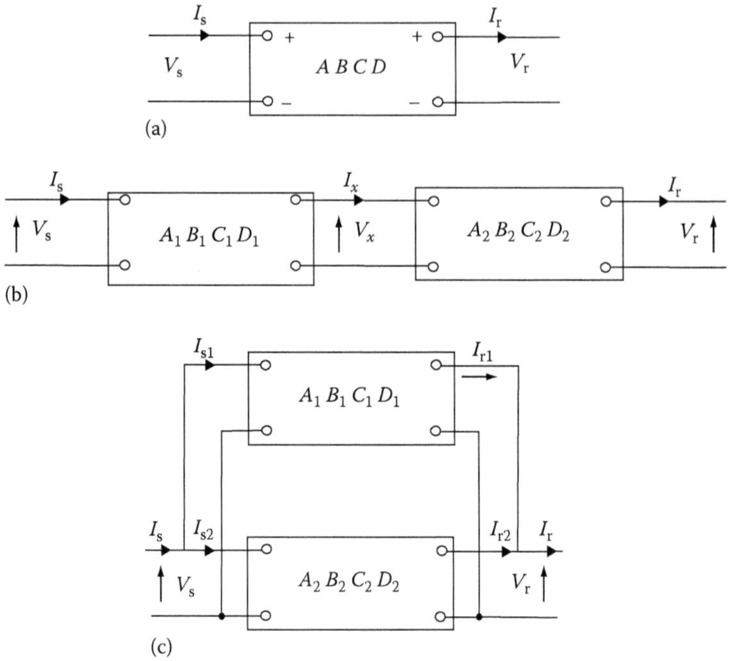

FIGURE 10.3
(a) Schematic representation of a two-terminal network using *ABCD* constants; (b) two networks in series; (c) two networks in parallel.

Two *ABCD* networks in series, in Figure 10.3b, can be reduced to a single equivalent network:

$$\begin{vmatrix} V_s \\ I_s \end{vmatrix} = \begin{vmatrix} A_1 & B_1 \\ C_1 & D_1 \end{vmatrix} \begin{vmatrix} A_2 & B_2 \\ C_2 & D_2 \end{vmatrix} \begin{vmatrix} V_r \\ I_r \end{vmatrix} = \begin{vmatrix} A_1A_2 + B_1C_2 & A_1B_2 + B_1D_2 \\ C_1A_2 + D_1C_2 & C_1B_2 + D_1D_2 \end{vmatrix} \begin{vmatrix} V_r \\ I_r \end{vmatrix} \tag{10.36}$$

For parallel *ABCD* networks, as in Figure 10.3c, the combined *ABCD* constants are

$$\begin{aligned} A &= \frac{(A_1B_2 + A_2B_1)}{(B_1 + B_2)} \\ B &= \frac{B_1B_2}{(B_1 + B_2)} \\ C &= \frac{(C_1 + C_2) + (A_1 - A_2)(D_2 - D_1)}{(B_1 + B_2)} \\ D &= \frac{(B_2D_1 + B_1D_2)}{(B_1 + B_2)} \end{aligned} \tag{10.37}$$

ABCD parameters with proper definitions and analytical expressions have also been applied for load flow through three-phase transformers.

Example 10.1

Calculate the *ABCD* constants of a short transmission line, voltage regulation, and load PF for zero-voltage regulation.

In a short transmission line, the sending end current is equal to the receiving end current. The sending end voltage can be vectorially calculated by adding the *IZ* drop to the receiving-end voltage. Considering a receiving-end voltage $V_r < 0°$ and current $I_r < \phi$,

$$V_s = V_r < 0° + ZI_r < \phi$$

Therefore, from Equation 10.32, $A = 1$, $B = Z$, $C = 0$, and $D = 1$:

$$\begin{vmatrix} V_s \\ V_r \end{vmatrix} = \begin{vmatrix} 1 & Z \\ 0 & 1 \end{vmatrix} \begin{vmatrix} V_r \\ I_r \end{vmatrix}$$

The equation can be closely approximated as

$$V_s = V_r + IR_{sr}\cos\phi + IX_{sr}\sin\phi \tag{10.38}$$

For a short line, the no-load receiving-end voltage equals the sending-end voltage:

$$V_s = AV_r + BI_r$$

At no load, $I_r = 0$ and $A = 1$; therefore, the no-load receiving-end voltage = sending-end voltage = V_s/A. Therefore, the regulation is

$$\frac{V_s/A - V_r}{V_r} = \frac{IR_{sr}\cos\phi + IX_{sr}\sin\phi}{V_r} \tag{10.39}$$

The voltage regulation is negative for a leading PF.

Example 10.2

A short three-phase, 13.8 kV line supplies a load of 10 MW at 0.8 lagging PF. The line resistance is 0.25 ohm and its reactance is 2.5 ohm. Calculate the sending-end voltage, regulation, and value of a capacitor to be connected at the receiving end to reduce the calculated line regulation by 50%.

The line current at 10 MW and 0.8 PF = 523 < −36.87°. The sending-end voltage is

$$V_s = V_r < 0^{\circ} + \sqrt{3} \times 523(0.8 - j0.6)(0.25 + j2.5)$$

or this is closely approximated by Equation 10.38, which gives a sending-end voltage of 15.34 kV. From Equation 10.39, the line regulation is 11.16%. This is required to be reduced by 50%, i.e., to 5.58%. The sending-end voltage is given by

$$\frac{|V_s| - 13.8}{13.8} = 0.0538 \quad \text{or} \quad |V_s| = 14.54 \text{ kV}$$

The line voltage drop must be reduced to 742 V. This gives two equations: $428 = I_n(0.25\cos ø_n + 2.5\sin ø_n)$ and $I_n = 418/\cos ø_n$ (10 MW of power at three-phase, 13.8 kV, and unity PF = 418 A), where I_n is the new value of the line current and $ø_n$ the improved PF angle. From these two equations, $ø_n = 17.2°$, i.e., the PF should be improved to approximately 0.955. The new current $I_n = 437.5 < -17.2°$. Therefore, the current supplied by the intended capacitor bank is

$$I - I_n = (417.9 - j129.37) - (418.0 - j313.8) = -j184 \text{ A (leading)}$$

Within the accuracy of calculation, the active part of the current should cancel out, as the capacitor bank supplies only a reactive component. The capacitor reactance to be added per phase = 13,800/($\sqrt{3}$.184) = 43.30 ohm, which is equal to 61.25 μF.

10.4 Transmission Line Models

10.4.1 Medium Long Transmission Lines

For transmission lines, in the range 50–150 mi (80–240 km), the shunt admittance cannot be neglected. There are two models in use, the nominal Π- and nominal T-circuit models. In the T-circuit model the shunt admittance is connected at the midpoint of the line, while in the Π model, it is equally divided at the sending end and the receiving end. The Π equivalent circuit and phasor diagram are shown in Figure 10.4a and b. The nominal T-circuit model and phasor diagram are shown in Figure 10.4c and d. The *ABCD* constants are shown in Table 10.1.

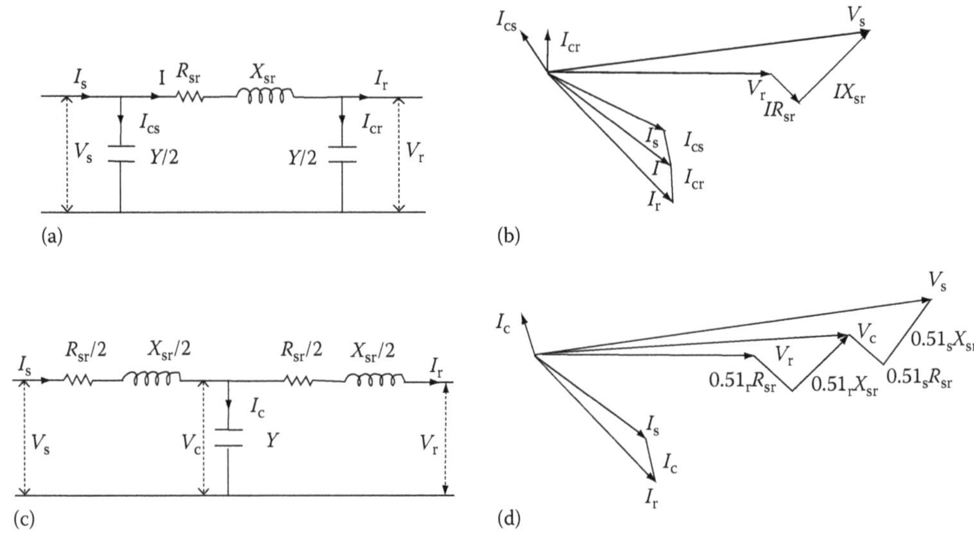

FIGURE 10.4
(a) Π representation of a transmission line; (b) phasor diagram of Π representation; (c) T representation of a transmission line; (d) phasor diagram of T-representation.

TABLE 10.1

ABCD Constants of Transmission Lines

Line Length	Equivalent Circuit	*A*	*B*	*C*	*D*
Short	Series impedance only	1	Z	0	1
Medium	Nominal Π, Figure 10.4a	$1+\frac{1}{2}YZ$	Z	$Y[1+(1/4)(YZ)]$	$1+\frac{1}{2}YZ$
Medium	Nominal T, Figure 10.4b	$1+\frac{1}{2}YZ$	$Z[1+(1/4)(YZ)]$	Y	$1+\frac{1}{2}YZ$
Long	Distributed parameters	$\cosh \gamma l$	$Z_0 \sinh \gamma l$	$(\sin \gamma l)/Z_0$	$\cos \gamma l$

Example 10.3

Calculate the *ABCD* constants of the Π model of the transmission line shown in Table 10.1.

The sending-end current is equal to the receiving-end current, and the current through the shunt elements $Y/2$ at the receiving and sending ends is

$$I_s = I_r + \frac{1}{2}V_rY + \frac{1}{2}V_sY$$

The sending-end voltage is the vector sum of the receiving-end voltage and the drop through the series impedance Z is

$$V_s = V_r + \left(I_r + \frac{1}{2}V_rY\right)Z = V_r\left(1 + \frac{1}{2}YZ\right) + I_rZ$$

The sending-end current can, therefore, be written as

$$I_s = I_r + \frac{1}{2}V_rY + \frac{1}{2}Y\left[V_r\left(1 + \frac{1}{2}YZ\right) + I_rZ\right]$$

$$V_rY\left(1 + \frac{1}{4}YZ\right) + I_r\left(1 + \frac{1}{2}YZ\right)$$

or in matrix form

$$\begin{vmatrix} V_s \\ I_s \end{vmatrix} = \begin{vmatrix} \left(1 + \frac{1}{2}YZ\right) & Z \\ Y\left(1 + \frac{1}{4}YZ\right) & \left(1 + \frac{1}{2}YZ\right) \end{vmatrix} \begin{vmatrix} V_r \\ I_r \end{vmatrix}$$

10.4.2 Long Transmission Line Model

Lumping the shunt admittance of the lines is an approximation and for line lengths over 150 mi (240 km), the distributed parameter representation of a line is used. Each elemental section of line has a series impedance and shunt admittance associated with it. The operation of a long line can be examined by considering an elemental section of impedance z per unit length, and admittance y per unit length. The impedance

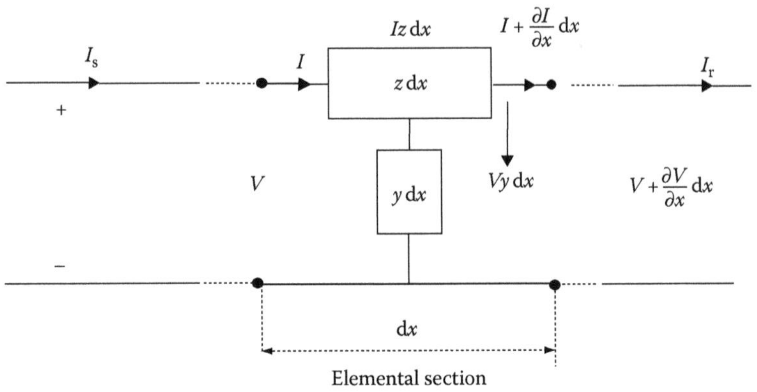

FIGURE 10.5
Model of an elemental section of a long transmission line.

for the elemental section of length dx is z dx and the admittance is y dx. Referring to Figure 10.5, by Kirchoff's voltage law,

$$V = Iz\,\mathrm{d}x + V + \frac{\partial V}{\partial x}\mathrm{d}x$$
$$\frac{\partial V}{\partial x} = -IZ \tag{10.40}$$

Similarly, from the current law,

$$I = Vy\,\mathrm{d}x + I + \frac{\partial I}{\partial x}\mathrm{d}x$$
$$\frac{\partial I}{\partial x} = -Vy \tag{10.41}$$

Differentiating Equations 10.40 and 10.41,

$$\frac{\partial^2 V}{\partial x^2} = yzV \tag{10.42}$$

$$\frac{\partial^2 I}{\partial x^2} = yzI \tag{10.43}$$

These differential equations have solutions of the form

$$V = V_1 e^{\gamma x} + V_2 e^{-\gamma x} \tag{10.44}$$

where γ is the *propagation constant*. It is defined as

$$\gamma = \sqrt{zy} \tag{10.45}$$

Define the following shunt elements, to distinguish from line series elements: g_{sh} is the shunt conductance, b_{sh} is the shunt susceptance, x_{sh} is the shunt capacitive reactance, and

$Y = yl$, the shunt admittance, where l is the length of the line. The shunt conductance is small and ignoring it gives $y = j/x_{sh}$.

$$\begin{aligned} y &= jb_{sh} \\ yz &= -b_{sh}x_{sc} + jr_{sc}b_{sh} \\ |\gamma| &= \sqrt{b_{sh}}(r_{sc}^2 + x_{sc}^2)^{1/4} \end{aligned} \tag{10.46}$$

The complex propagation constant can be written as

$$\gamma = \alpha + j\beta \tag{10.47}$$

where α is defined as the *attenuation constant*. Common units are nepers per mile or per km.

$$\alpha = |\gamma| \cos\left[\frac{1}{2}\tan^{-1}\left(\frac{-r_{sc}}{x_{sc}}\right)\right] \tag{10.48}$$

$$\beta = |\gamma| \sin\left[\frac{1}{2}\tan^{-1}\left(-\frac{r_{sc}}{x_{sc}}\right)\right] \tag{10.49}$$

where β is the *phase constant*. Common units are radians per mile. The *characteristic impedance* is

$$Z_0 = \sqrt{\frac{z}{y}} \tag{10.50}$$

Again neglecting shunt conductance,

$$\begin{aligned} \frac{z}{y} &= \frac{r_{sc} + jx_{sc}}{jb_{sh}} = x_{sc}x_{sh} - jx_{sh}r_{sc} \\ Z_0 &= \sqrt{x_{sh}}(r_{sc}^2 + x_{sc}^2)^{1/4} \end{aligned} \tag{10.51}$$

$$<Z_0 = \frac{1}{2}\tan^{-1}\left(-\frac{r_{sc}}{x_{sc}}\right) \tag{10.52}$$

The voltage at any distance χ can be written as

$$V_x = \left|\frac{V_r + Z_0 I_r}{2}\right| e^{\alpha x + j\beta x} + \left|\frac{V_r - Z_0 I_r}{2}\right| e^{-\alpha x - j\beta x} \tag{10.53}$$

These equations represent traveling waves. The solution consists of two terms, each of which is a function of two variables, time and distance. At any instant of time the first term, the incident wave is distributed sinusoidally along the line, with amplitude increasing exponentially from the receiving end. After a time interval Δt, the distribution advances in phase by $\omega\Delta t/\beta$, and the wave is traveling toward the receiving end. The second term is the reflected wave, and after time interval Δt, the distribution retards in phase by $\omega\Delta t/\beta$, the wave traveling from the receiving end to the sending end.

A similar explanation holds for the current:

$$I_x = \left|\frac{V_r/Z_0 + I_r}{2}\right| e^{\alpha x + j\beta x} - \left|\frac{V_r/Z_0 - I_r}{2}\right| e^{-\alpha x - j\beta x} \tag{10.54}$$

These equations can be written as

$$\begin{aligned} V_x &= V_r\left(\frac{e^{\gamma x} + e^{-\gamma x}}{2}\right) + I_r Z_0\left(\frac{e^{\gamma x} - e^{-\gamma x}}{2}\right) \\ I_x &= \frac{V_r}{Z_0}\left(\frac{e^{\gamma x} - e^{-\gamma x}}{2}\right) + I_r\left(\frac{e^{\gamma x} + e^{-\gamma x}}{2}\right) \end{aligned} \tag{10.55}$$

or in matrix form

$$\begin{vmatrix} V_s \\ I_s \end{vmatrix} = \begin{vmatrix} \cosh \gamma l & Z_0 \sinh \gamma l \\ \frac{1}{Z_0} \sinh \gamma l & \cosh \gamma l \end{vmatrix} \begin{vmatrix} V_r \\ I_r \end{vmatrix} \tag{10.56}$$

These *ABCD* constants are shown in Table 10.1.

10.4.3 Reflection Coefficient

The relative values of sending-end and receiving-end voltages, V_1 and V_2 depend on the conditions at the terminals of the line. The reflection coefficient at the load end is defined as the ratio of the amplitudes of the backward and forward traveling waves. For a line terminated in a load impedance Z_L,

$$V_2 = \left(\frac{Z_L - Z_0}{Z_L + Z_0}\right) V_1 \tag{10.57}$$

Therefore, the voltage reflection coefficient at the load end is

$$\rho_L = \frac{Z_L - Z_0}{Z_L + Z_0} \tag{10.58}$$

The current reflection coefficient is negative of the voltage reflection coefficient. For a short-circuited line, the current doubles and for an open-circuit line the voltage doubles. Figure 10.6 shows the traveling wave phenomenon. The reflected wave at an impedance discontinuity is a mirror image of the incident wave moving in the opposite direction. Every point in the reflected wave is the corresponding point on the incident wave multiplied by the reflection coefficient, but a mirror image. At any time, the total voltage is the sum of the incident and reflected waves. Figure 10.6b shows the reinforcement of the incident and reflected waves. The reflected wave moves toward the source and is again reflected. The source reflection coefficient, akin to the load reflection coefficient, can be defined as

$$\rho_s = \frac{Z_s - Z_0}{Z_s + Z_0} \tag{10.59}$$

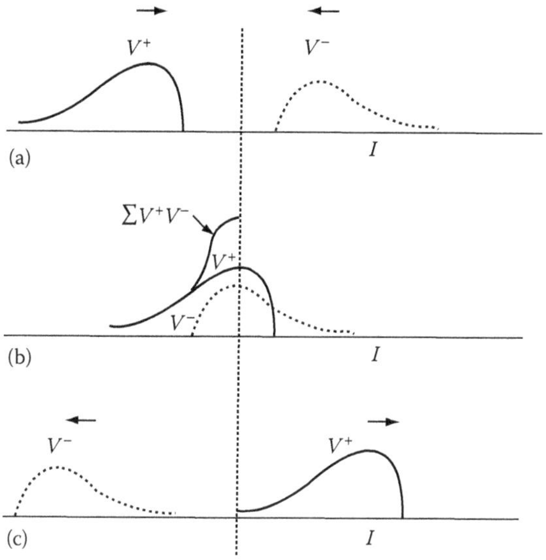

FIGURE 10.6
(a) Incident and reflected waves at an impedance change in a long transmission line; (b) reinforcement of incident and reflected waves; (c) incident and reflected waves crossing each other.

A forward traveling wave originates at the source, as the backward traveling wave originates at the load. At any time, the voltage or current at any point on the line is the sum of all voltage or current waves existing at the line at that point and time.

Consider that the line is terminated with a reactor, and a step voltage is applied. Calculate the expressions for transmitted and reflected voltages as follows:

We can write $Z_2 = Ls$ in Laplace transform and let $Z_1 = Z$

Then the reflection coefficient is

$$\rho_L = \frac{(s - Z/L)}{(s + Z/L)} \tag{10.60}$$

The reflected wave is

$$V'(s) = \frac{(s - Z/L)}{(s + Z/L)}\frac{V}{s} = \left[-\frac{1}{s} + \frac{2}{s + Z/L}\right] \tag{10.61}$$

Taking inverse transform,

$$V' = [-1 + 2e^{-(Z/L)t}] \tag{10.62}$$

Voltage of the transmitted wave is

$$V''(s) = (1 + \rho_L)V(s) = \frac{2V}{(s + Z/L)} \tag{10.63}$$

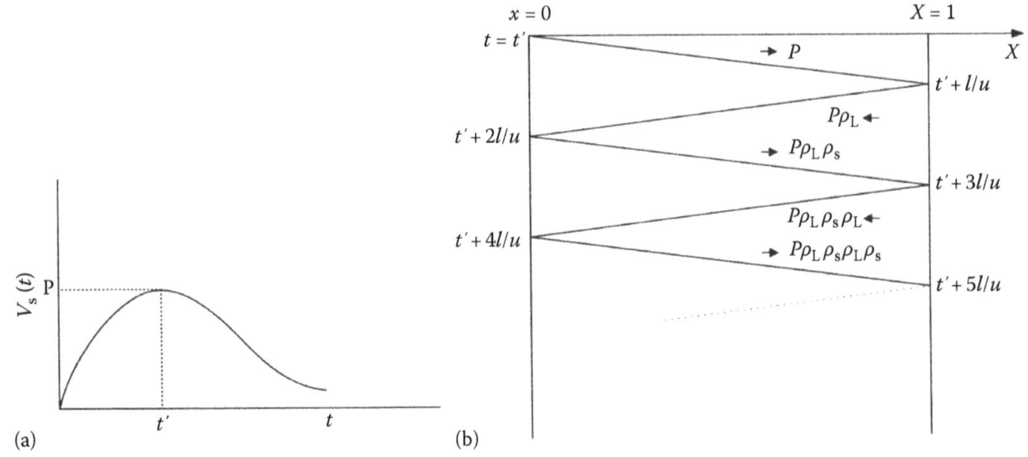

FIGURE 10.7
(a) A point P at time t' on a pulse signal applied to the sending end of a transmission line; (b) lattice diagram.

Thus,

$$V'' = 2Ve^{-(Z/L)t} \tag{10.64}$$

The voltage across the inductor rises initially to double the incident voltage and decays exponentially.

10.4.4 Lattice Diagrams

The forward and backward traveling waves can be shown on a lattice diagram (Figure 10.7). The horizontal axis is the distance of the line from the source and the vertical axis is labeled in time increments, each increment being the time required for the wave to travel the line in one direction, i.e., source to the load. Consider a point P in the pulse shape of Figure 10.7b at time t'. The time to travel in one direction is l/u, where u is close to the velocity of light, as we will discuss further. The point P then reaches the load end at $t' + l/us$ and is reflected back. The corresponding point on the reflected wave is $P\rho_L$. At the sending end it is re-reflected as $\rho_L\rho_s$, Figure 10.7b.

Example 10.4

A lossless transmission line has a surge impedance of 300 ohm. It is terminated in a resistance of 600 ohm. A 120 V dc source is applied to the line at $t=0$ at the sending end. Considering that it takes t seconds to travel the voltage wave in one direction, draw the lattice diagram from $t=0$ to $7t$. Plot the voltage profile at the load terminals.

The sending-end reflection coefficient from Equation 10.59 is -1 and the load reflection coefficient from Equation 10.58 is 0.33. The lattice diagram is shown in Figure 10.8a. At first reflection at the load, 120 V is reflected as $0.33 \times 120 = 39.6$ V, which is re-reflected from the sending end as $-1 \times 39.6 = -39.6$ V. The voltage at the receiving end can be plotted from the lattice diagram in Figure 10.8a and is shown in Figure 10.8b.

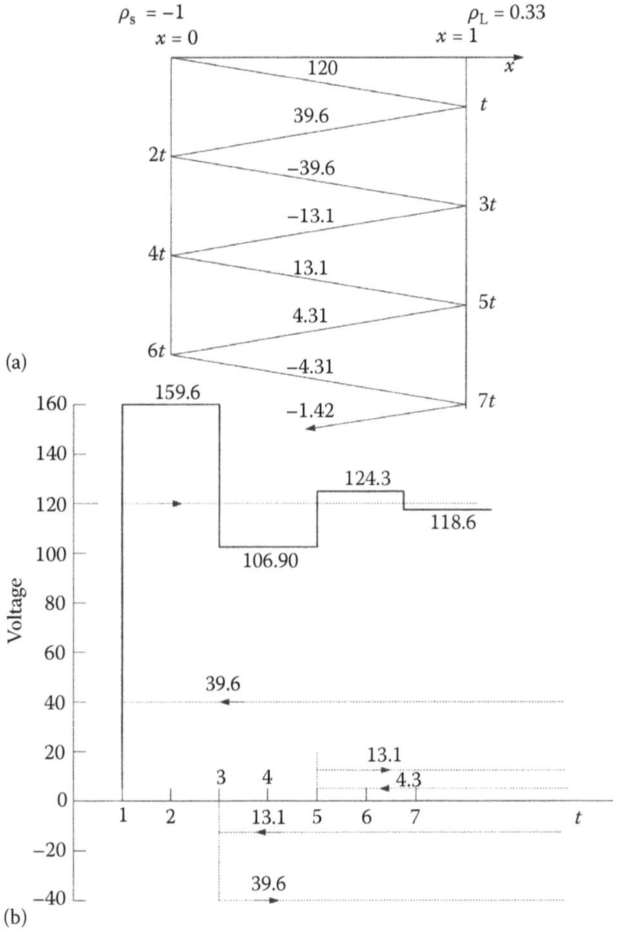

FIGURE 10.8
(a) Lattice diagram of Example 10.4; (b) the receiving-end voltage profile.

10.4.5 Infinite Line

When the line is terminated in its characteristic load impedance, i.e., $Z_L = Z_0$, the reflected wave is zero. Such a line is called an *infinite line* and the incident wave cannot distinguish between a termination and the continuation of the line.

The characteristic impedance is also called the surge impedance. It is approximately 400 ohm for overhead transmission lines and its phase angle may vary from 0° to 15°. For underground cables, the surge impedance is much lower, approximately 1/10 that of overhead lines.

10.4.6 Surge Impedance Loading

The surge impedance loading (SIL) of the line is defined as the power delivered to a purely resistive load equal in value to the surge impedance of the line:

$$\text{SIL} = \frac{V_r^2}{Z_0} \tag{10.65}$$

For 400 ohm surge impedance, SIL in kW is 2.5 multiplied by the square of the receiving-end voltage in kV. The surge impedance is a real number and, therefore, the PF along the line is unity, i.e., no reactive power compensation (see Chapter 13) is required. The SIL loading is also called the natural loading of the transmission line.

10.4.7 Wavelength

A complete voltage or current cycle along the line, corresponding to a change of 2π rad in angular argument of β is defined as the wavelength λ. If β is expressed in rad/mi,

$$\lambda = \frac{2\pi}{\beta} \tag{10.66}$$

For a lossless line,

$$\beta = \omega\sqrt{LC} \tag{10.67}$$

Therefore,

$$\lambda = \frac{1}{f\sqrt{LC}} \tag{10.68}$$

and the velocity of propagation of the wave

$$v = f\lambda = \frac{1}{\sqrt{LC}} \approx \frac{1}{\sqrt{\mu_0 k_0}} \tag{10.69}$$

where
$\mu_0 = 4\pi \times 10^{-7}$ is the permeability of the free space
$k_0 = 8.854 \times 10^{-12}$ is the permittivity of the free space

Therefore, $1/[\sqrt{(\mu_0 k_0)}] = 3 \times 10^{10}$ cm/s or 186,000 mi/s = velocity of light. We have considered a lossless line in developing the above expressions. The actual velocity of the propagation of the wave along the line is somewhat less than the speed of light.

10.5 Tuned Power Line

In the long transmission line model, if the shunt conductance and series resistance are neglected, then

$$\gamma = \sqrt{YZ} = j\omega\sqrt{LC} \tag{10.70}$$

$$\cosh \gamma l = \cosh j\omega l\sqrt{LC} = \cos \omega l\sqrt{LC} \tag{10.71}$$

$$\sinh \gamma l = \sinh j\omega l\sqrt{LC} = j \sin \omega l\sqrt{LC} \tag{10.72}$$

where l is the length of the line. This simplifies *ABCD* constants and the following relationship results:

$$\begin{vmatrix} V_s \\ I_s \end{vmatrix} = \begin{vmatrix} \cos \omega l\sqrt{LC} & jZ_0 \sin \omega l\sqrt{LC} \\ \frac{j}{Z_0} \sin \omega l\sqrt{LC} & \cos \omega l\sqrt{LC} \end{vmatrix} \begin{vmatrix} V_r \\ I_r \end{vmatrix} \tag{10.73}$$

If

$$\omega l\sqrt{LC} = n\pi \quad (n = 1, 2, 3 \ldots) \tag{10.74}$$

then

$$|V_s| = |V_r| \qquad |I_s| = |I_r| \tag{10.75}$$

This will be an ideal situation to operate a transmission line. The receiving-end voltage and currents are equal to the sending-end voltage and current. The line has a flat voltage profile.

$$\beta = \frac{2\pi f}{v} = \frac{2\pi}{\lambda} \tag{10.76}$$

As $1/\sqrt{LC} \approx$ velocity of light, the line length λ is 3100, 6200, ... mi or $\beta = 0.116°$ per mi. The quantity β_1 is called the *electrical length* of the line. The length calculated above is too long to avail this ideal property and suggests that power lines can be tuned with series capacitors to cancel the effect of inductance and with shunt inductors to neutralize the effect of line capacitance. This compensation may be done by sectionalizing the line. For power lines, series and shunt capacitors for heavy load conditions and shunt reactors under light load conditions are used to improve power transfer and line regulation. We will revert to this in Chapter 13.

10.6 Ferranti Effect

As the transmission line length increases, the receiving-end voltage rises above the sending-end voltage, due to line capacitance. This is called the Ferranti effect. In a long line model, at no load ($I_R = 0$), the sending-end voltage is

$$V_s = \frac{V_r}{2} e^{\alpha l} e^{j\beta l} + \frac{V_r}{2} e^{-\alpha l} e^{-j\beta l} \tag{10.77}$$

At $l = 0$, both incident and reflected waves are equal to $V_r/2$. As l increases, the incident wave increases exponentially, while the reflected wave decreases. Thus, the receiving-end voltage rises. Another explanation of the voltage rise can be provided by considering that the line capacitance is lumped at the receiving end. Let this capacitance be Cl; then, on open circuit, the sending-end current is

$$I_s = \frac{V_s}{\left(j\omega Ll - \dfrac{1}{j\omega Cl}\right)} \tag{10.78}$$

C is small in comparison with L. Thus, ωLl can be neglected. The receiving-end voltage can then be written as

$$\begin{aligned} V_r &= V_s - I_s(j\omega Ll) \\ &= V_s + V_s\omega^2 CLl^2 \\ &= V_s(1 + \omega^2 CLl^2) \end{aligned} \tag{10.79}$$

This gives a voltage rise at the receiving end of

$$|V_s|\omega^2 CLl^2 = |V_s|\omega^2 l^2/v^2 \tag{10.80}$$

where v is the velocity of propagation. Considering that v is constant, the voltage rises with the increase in line length.

Also, from Equation 10.73 the voltage at any distance x terms of the sending-end voltage, with the line open-circuited and resistance neglected, is

$$V_x = V_s \frac{\cos\beta(l-x)}{\cos\beta l} \tag{10.81}$$

and the current is

$$I_x = j\frac{V_s}{Z_0}\frac{\sin\beta(l-x)}{\cos\beta l} \tag{10.82}$$

Example 10.5

A 230 kV three-phase transmission line has 795 kcmil, ACSR conductors, one per phase. Neglecting resistance, $z = j0.8\ \Omega$/mi and $y = j5.4 \times 10^{-6}$ S (siemens, mho) per mi. Calculate the voltage rise at the receiving end for a 400 mi long line.

Using the expressions developed above,

$$Z_0 = \sqrt{\frac{z}{y}} = \sqrt{\frac{j0.8}{j5.4 \times 10^{-6}}} = 385\ \Omega$$

$\beta = \sqrt{zy} = 2.078 \times 10^{-3}$ rad/mi $= 0.119°$/mi; $\beta_l = 0.119 \times 400 = 47.6°$. The receiving-end voltage rise from Equation 10.81:

$$V_r = V_s \frac{\cos(l-l)}{\cos 47.6°} = \frac{V_s}{0.674} = 1.483V_s$$

The voltage rise is 48.3% and at 756 mi, one-quarter wavelength, it will be infinite.

Even a voltage rise of 10% at the receiving end is not acceptable as it may give rise to insulation stresses and affect the terminal regulating equipment. Practically, the voltage rise will be more than that calculated above. As the load is thrown off, the sending-end voltage will rise before the generator voltage regulators and excitation systems act to reduce the voltage, further increasing the voltages on the line. This points to the necessity of compensating the transmission lines.

The sending-end charging current from Equation 10.82 is 1.18 per unit and falls to zero at the receiving end. This means that the charging current flowing in the line is 118% of the line natural load.

10.6.1 Approximate Long Line Parameters

Regardless of voltage, conductor size, or spacing of a line, the series reactance is approximately 0.8 Ω/mi and the shunt-capacitive reactance is 0.2 MΩ/mi. This gives a β of 1.998×10^{-3}/mi or 0.1145°/mi.

10.7 Symmetrical Line at No Load

If we consider a symmetrical line at no load, with the sending-end and receiving-end voltages maintained the same, these voltages have to be in phase as no power is transferred. Half the charging current is supplied from each end and the line is equivalent to two equal halves connected back-to-back. The voltage rises at the midpoint, where the charging current falls to zero and reverses direction. The synchronous machines at the sending end absorb leading reactive power, while the synchronous machines at the receiving end generate lagging reactive power Figure 10.9a through c. The midpoint voltage is, therefore, equal to the voltage as if the line was of half the length.

On loading, the vector diagram shown in Figure 10.9d is applicable. By symmetry, the midpoint voltage vector exactly bisects the sending- and receiving-end voltage vectors, and the sending-end and receiving-end voltages are equal. The PF angle at both ends are equal

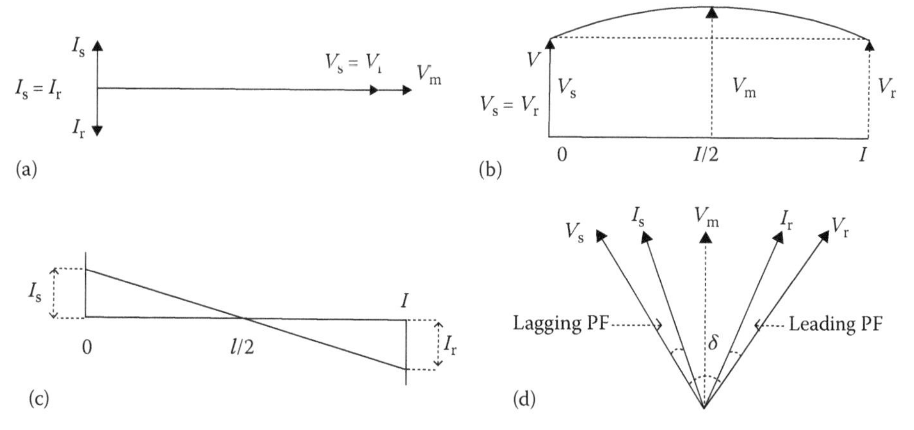

FIGURE 10.9
(a) Phasor diagram of a symmetrical line at no load; (b) the voltage profile of a symmetrical line at no load; (c) the charging current profile of a symmetrical line at no load; (d) the phasor diagram of a symmetrical line at load.

but of opposite sign. Therefore, the receiving-end voltage on a symmetric line of length $2l$ is the same as that of line of length l at unity PF load. From Equation 10.73, the equations for the sending-end voltage and current for a symmetrical line can be written with βl replaced by $\beta l/2 = \theta/2$.

$$V_s = V_m \cos\left(\frac{\theta}{2}\right) + jZ_0 I_m \sin\left(\frac{\theta}{2}\right) \tag{10.83}$$

$$I_s = j\frac{V_m}{Z_0}\sin\left(\frac{\theta}{2}\right) + I_m \cos\left(\frac{\theta}{2}\right) \tag{10.84}$$

At the midpoint,

$$P_m + jQ_m = V_m I_m^* = P$$

$$Q_s = \text{Img}\, V_s I_s^* \tag{10.85}$$

$$= j\frac{\sin\theta}{2}\left[Z_0 I_m^2 - \frac{V_m^2}{Z_0}\right] \tag{10.86}$$

where P is the transmitted power. No reactive power flows past the midpoint and it is supplied by the sending end.

10.8 Illustrative Examples

Example 10.6

Consider a line of medium length represented by a nominal Π circuit. The impedances and shunt susceptances are shown in Figure 10.10 in per unit on a 100 MVA base. Based on the given data calculate *ABCD* constants. Consider that one per unit current at 0.9 PF lagging is required to be supplied to the receiving end. Calculate the sending-end voltage, currents, power, and losses. Compare the results with approximate power flow relations derived in Equation 10.30:

$$D = A = 1 + \frac{1}{2}YZ$$
$$= 1 + 0.5[j0.0538][0.0746 + j0.394]$$
$$= 0.989 + j0.002$$

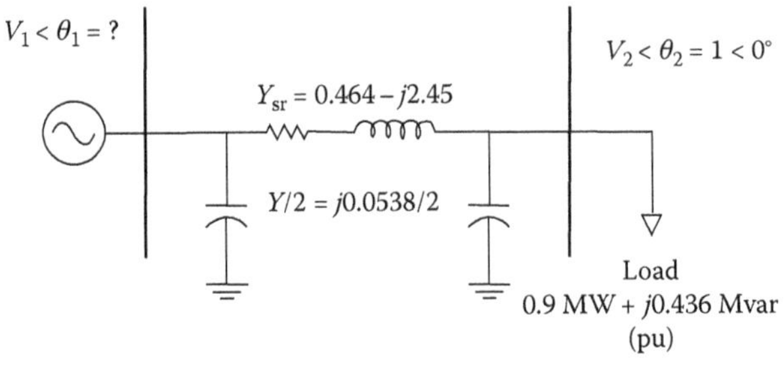

FIGURE 10.10
Transmission line and load parameters for Examples 10.6 and 10.7

and

$$C = Y\left(1 + \frac{1}{4}YZ\right)$$
$$= (j0.0538)[1 + (-0.0053 + j0.9947)]$$
$$= -0.000054 + j0.0535$$
$$B = Z = 0.0746 + j0.394$$

The voltage at the receiving-end bus is taken as equal to the rated voltage of one per unit at zero degree phase shift, i.e., $1 < 0°$. The receiving-end power is, therefore,

$$V_2 I_2^*$$
$$(1 < 0°)(1_2 < 25.8°)$$
$$0.9 + j0.436$$

It is required to calculate the sending-end voltage and current:

$$V_1 = AV_2 + BI_2$$
$$= (0.989 + j0.002)(1 < 0°) + (0.0746 + j0.394)(1 < 25.8°)$$
$$= 1.227 + j0.234$$

The sending-end voltage is

$$|V_1| = |1.269| < 14.79°$$

The sending-end current is

$$I_1 = CV_2 + DI_2$$
$$(-0.000054 + j0.0535) + (0.989 + j0.002)(1 < -25.8°)$$
$$= 0.8903 - j0.3769$$
$$= 0.9668 < -22.944°$$

The sending-end power is

$$V_1 I_1^* = (1.269 < 14.79°)(0.9668 < 22.944°)$$
$$= 0.971 + j0.75$$

Thus, the active power loss is 0.071 per unit and the reactive power loss is 0.314 per unit. The reactive power loss increases as the PF of the load becomes more lagging.

The power supplied based on known sending-end and receiving-end voltages can also be calculated using the following equations.

The sending-end active power is

$$[V_1^2 - V_1 V_2 \cos(\theta_1 - \theta_2)]g_{12} - [V_1 V_2 \sin(\theta_1 - \theta_2)]b_{12}$$
$$[1.269^2 - 1.269\cos(14.79°)]0.464 - [1.269\sin(14.79°)](-2.450)$$
$$= 0.971 \text{ as before}$$

In this calculation, a prior calculated sending-end voltage of 1.269 < 14.79 for supply of per unit current at 0.9 lagging PF is used. *For a given load neither the sending-end voltage nor the receiving-end current is known.*

The sending-end reactive power is

$$
\begin{aligned}
Q_{12} &= [-V_1V_2 \sin(\theta_1 - \theta_2)]g_{12} - [V_1^2 - V_1V_2\cos(\theta_1 - \theta_2)]b_{12} - \frac{Y}{2}V_1^2 \\
&= [-1.269 \sin 14.79°]0.464 - [1.269^2 - 1.269\cos(14.79°)](-2.450) - (0.0269)(1.269^2) \\
&= 0.75 \text{ as before}
\end{aligned}
$$

The receiving-end active power is

$$
\begin{aligned}
P_{21} &= [V_2^2 - V_2V_1\cos(\theta_2 - \theta_1)]g_{12} - [V_2V_1 \sin(\theta_2 - \theta_1)]b_{12} \\
&= [1 - 1.269\cos(-14.79°)]0.464 - [1.269\sin(-14.79°)](-2.450) \\
&= 0.9
\end{aligned}
$$

and the receiving-end reactive power is

$$
\begin{aligned}
Q_{21} &= [-V_2V_1 \sin(\theta_2 - \theta_1)]g_{12} - [V_2^2 - V_2V_1\cos(\theta_2 - \theta_1)]b_{21} - \frac{Y}{2}V_2^2 \\
&= [-1.269\sin(-14.79°)]0.464 - [1 - 1.269\cos(-14.79°)](-2.450) - (0.0269)(1) = 0.436
\end{aligned}
$$

These calculations are merely a verification of the results. If the approximate relations of Equation 10.30 are applied,

$$
\begin{aligned}
P_{12} &\approx P_{21} = 0.627 \\
Q_{12} &\approx Q_{21} = 0.68
\end{aligned}
$$

This is not a good estimate for power flow, especially for the reactive power.

Example 10.7

Repeat the calculation in Example 10.6, with a long line distributed parameter model, with the same data. Calculate the sending-end current and voltage and compare the results.

$$
\begin{aligned}
Z &= 0.0746 + j0.394 \\
Y &= 0.0538
\end{aligned}
$$

The product ZY is

$$ZY = 0.021571 < 169.32°$$

α is the attenuation constant in nepers:

$$\alpha = |\gamma| \cos\left[\frac{1}{2}\tan^{-1}\left(-\frac{R}{X}\right)\right]$$

and β is the phase shift constant in radians:

$$\beta = |\gamma| \sin\left[\frac{1}{2}\tan^{-1}\left(-\frac{R}{X}\right)\right]$$

Thus,

$$\alpha = 0.0136687 \text{ Np}$$
$$\beta = 0.146235 \text{ rad}$$

The hyperbolic functions cosh γ and sinh γ are given by

$$\begin{aligned}\cosh\gamma &= \cosh\alpha \ \cos\beta + j\sinh\alpha \ \sin\beta \\ &= (1.00009)(0.987362) + j(0.013668)(0.14571) \\ &= 0.990519 < 0.1152^\circ \\ \sinh\gamma &= \sinh\alpha \ \cos\beta + j\cosh\alpha \ \sin\beta \\ &= (0.013668)(0.987362) + j(1.00009)(0.14571) \\ &= 0.1463449 < 84.698^\circ\end{aligned}$$

The sending-end voltage is thus

$$\begin{aligned}V_1 &= \cosh\gamma V_2 + Z_0 \sinh\gamma I_2 \\ &= (0.990519 < 0.1152^\circ)(1 < 0^\circ) + (2.73 < -5.361^\circ)(0.1463449 < 84.698^\circ) \\ &\quad (1 < -25.842^\circ) \\ &\quad 1.2697 < 14.73^\circ\end{aligned}$$

This result is fairly close to the one earlier calculated by using the Π model.
The sending-end current is

$$\begin{aligned}I_2 &= \left(\frac{1}{Z_0}\right)(\sinh\gamma)V_2 + (\cosh\gamma)I_2 \\ &= (0.3663 < 5.361^\circ)(0.146349 < 84.698^\circ) + (0.990519 < 0.1152^\circ)(1 < -25.84^\circ) \\ &= 0.968 < -22.865^\circ\end{aligned}$$

Again there is a good correspondence with the earlier calculated result using the equivalent Π model of the transmission line. The parameters of the transmission line shown in this example are for a 138 kV line of length approximately 120 mi. For longer lines, the difference between the calculations using the two models will diverge. This similarity of calculation results can be further examined.

Equivalence between Π and Long Line Model

For equivalence between long line and Π-models, *ABCD* constants can be equated. Thus, equating the *B* and *D* constants,

$$\begin{aligned}Z_s &= Z_0 \sinh\gamma l \\ 1 + \frac{1}{2}YZ &= \cosh\gamma l\end{aligned} \tag{10.87}$$

Thus,

$$Z_s = \sqrt{\frac{z}{y}}\sinh\gamma l = zl\frac{\sinh\gamma l}{l\sqrt{yz}} = Z\left[\frac{\sinh\gamma l}{\gamma l}\right] \tag{10.88}$$

i.e., the series impedance of the Π network should be increased by a factor ($\sinh\gamma l/\gamma l$).

$$1 + \frac{1}{2}YZ_c\sinh\gamma l = \cosh\lambda l$$

This gives the shunt element Y_p

$$\frac{1}{2}Y_p = \frac{Y}{2}\left[\frac{\tanh\gamma l/2}{\gamma l/2}\right] \tag{10.89}$$

i.e., the shunt admittance should be increased by $[(\tanh\gamma l/2)/(\gamma l/2)]$. For a line of medium length, both the series and shunt multiplying factors are ≈1.

10.9 Circle Diagrams

Consider a two-node two-bus system, similar to that of Figure 10.2. Let the sending-end voltage be $V_s < \theta°$ and the receiving-end voltage $V_r < \theta°$. Then

$$I_r = \frac{1}{B}V_s - \frac{A}{B}V_r \tag{10.90}$$

$$I_s = \frac{D}{B}V_s - \frac{1}{B}V_r \tag{10.91}$$

Constants A, B, C, and D can be written as

$$A = |A| < \alpha,\quad B = |B| < \beta,\quad D = |D| < \alpha \quad (A = D) \tag{10.92}$$

The receiving-end and sending-end currents can be written as

$$I_r = \left|\frac{1}{B}\right||V_s| < (\theta - \beta) - \left|\frac{A}{B}\right||V_r| < (\alpha - \beta) \tag{10.93}$$

$$I_s = \left|\frac{D}{B}\right||V_s| < (\alpha + \theta - \beta) - \left|\frac{1}{B}\right||V_r| < (\alpha - \beta) \tag{10.94}$$

The receiving-end power is

$$\begin{aligned} S_r &= V_r I_r^* \\ &= \frac{|V_r||V_s|}{|B|} < (\beta - \theta) - \left|\frac{A}{B}\right||V_r|^2 < (\beta - \alpha) \end{aligned} \tag{10.95}$$

The sending-end power is

$$S_s = \left|\frac{D}{B}\right| |V_s|^2 < (\beta - \alpha) - \frac{|V_s||V_r|}{|B|} < (\beta + \theta) \tag{10.96}$$

The real and imaginary parts are written as follows:

$$P_r = \frac{|V_s||V_r|}{|B|} \cos(\beta - \delta) - \left|\frac{A}{B}\right| |V_r|^2 \cos(\beta - \alpha) \tag{10.97}$$

$$Q_r = \frac{|V_s||V_r|}{|B|} \sin(\beta - \delta) - \left|\frac{A}{B}\right| \| V_r \|^2 \sin(\beta - \alpha) \tag{10.98}$$

Here, δ, the phase angle *difference*, is substituted for θ (the receiving-end voltage angle was assumed to be 0°). Similarly, the sending-end active and reactive powers are

$$P_s = \left|\frac{D}{B}\right| |V_s|^2 \cos(\beta - \alpha) - \frac{|V_s||V_r|}{|B|} \cos(\beta + \delta) \tag{10.99}$$

$$Q_s = \left|\frac{D}{B}\right| |V_s|^2 \sin(\beta - \alpha) - \frac{|V_s||V_r|}{|B|} \sin(\beta + \delta) \tag{10.100}$$

The received power is maximum at $\beta = \delta$:

$$P_r(\max) = \frac{|V_s||V_r|}{|B|} - \frac{|A||V_r|^2}{|B|} \cos(\beta - \alpha) \tag{10.101}$$

and the corresponding reactive power for maximum receiving-end power is

$$Q_r = -\frac{|A||V_r|^2}{|B|} \sin(\beta - \alpha) \tag{10.102}$$

The leading reactive power must be supplied for maximum active power transfer. For a short line, the equations reduce to the following:

$$P_r = \frac{|V_s||V_r|}{|Z|} \cos(\varepsilon - \delta) - \frac{|V_r|^2}{|Z|} \cos\varepsilon \tag{10.103}$$

$$Q_r = \frac{|V_s||V_r|}{|Z|} \sin(\varepsilon - \delta) - \frac{|V_r|^2}{|Z|} \sin\theta \tag{10.104}$$

$$P_s = \frac{|V_s|^2}{|Z|} \cos\theta - \frac{|V_s||V_r|}{|Z|} \cos(\varepsilon + \delta) \tag{10.105}$$

$$Q_s = \frac{|V_s|^2}{|Z|} \sin\theta - \frac{|V_s||V_r|}{|Z|} \sin(\varepsilon + \delta) \tag{10.106}$$

where ε is the impedance angle.

The center of the receiving-end circle is located at the tip of the phasor:

$$-\left|\frac{A}{B}\right||V_r|^2 < (\beta - \alpha) \tag{10.107}$$

In terms of the rectangular coordinates, the center of the circle is located at

$$-\left|\frac{A}{B}\right||V_r|^2 \cos(\beta - \alpha) \quad \text{MW} \tag{10.108}$$

$$-\left|\frac{A}{B}\right||V_r|^2 \sin(\beta - \alpha) \quad \text{Mvar} \tag{10.109}$$

The radius of the receiving-end circle is

$$\frac{|V_s||V_r|}{|B|} \quad \text{MVA} \tag{10.110}$$

The receiving-end circle diagram is shown in Figure 10.11. The operating point P is located on the circle by received real power P_r. The corresponding reactive power can be read immediately as Q_r. The torque angle can be read from the reference line as shown. For a constant receiving-end voltage V_r the center C is fixed and concentric circles result for varying V_s, Figure 10.12a. For constant V_s and varying V_r, the centers move along line OC and have radii in accordance with V_sV_r/B, Figure 10.12b.

The sending-end circle diagram is constructed on a similar basis and is shown in Figure 10.13. The center of the sending-end circle is located at the tip of the phasor:

$$\left|\frac{D}{B}\right||V_s|^2 < (\beta - \alpha) \tag{10.111}$$

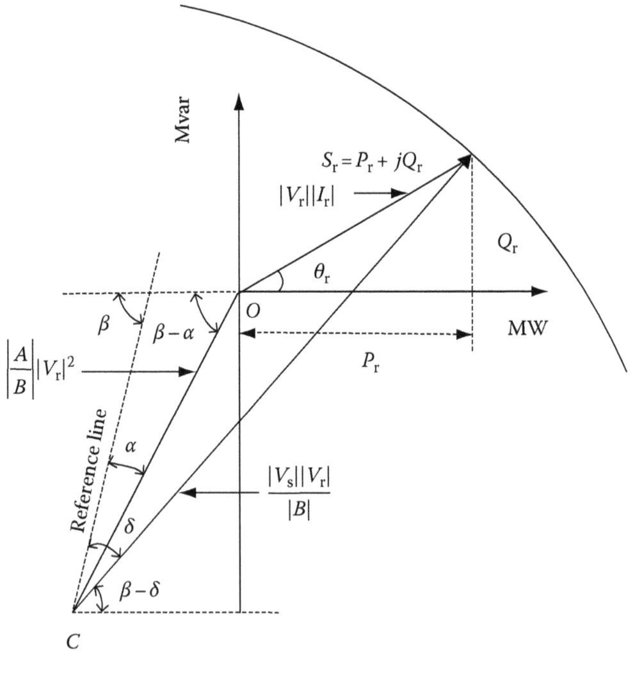

FIGURE 10.11
Receiving-end power circle diagram of a transmission line.

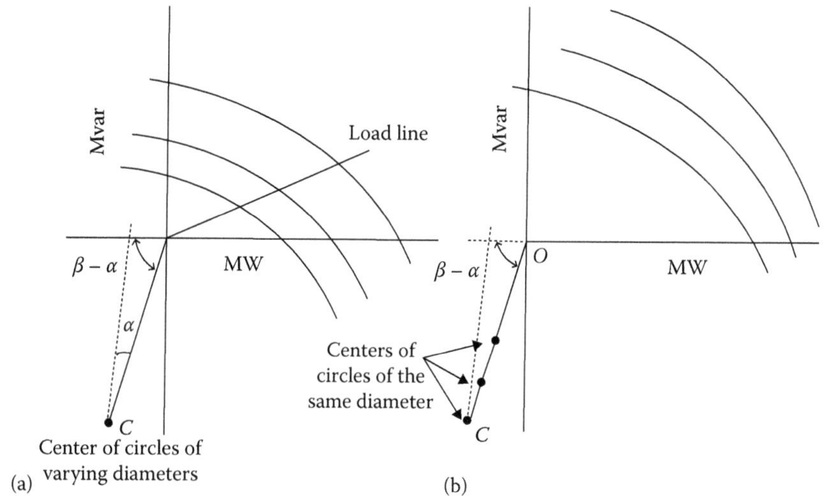

FIGURE 10.12
(a) Receiving-end power circles for different sending-end voltages and constant receiving-end voltage; (b) receiving-end power circles for constant sending-end voltage and varying receiving-end voltage.

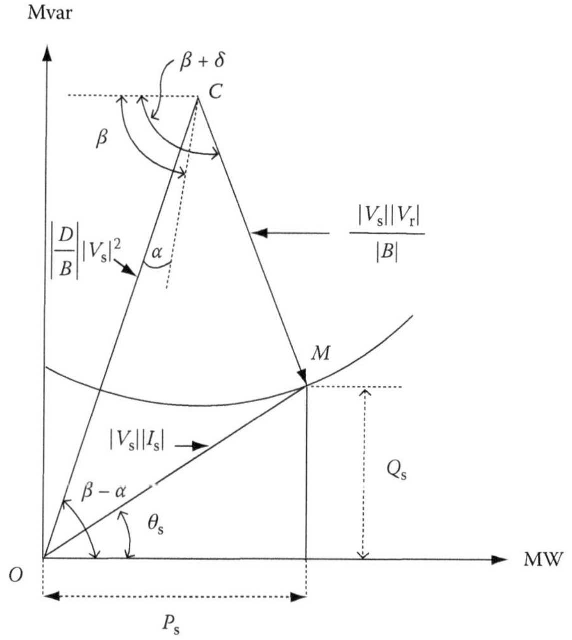

FIGURE 10.13
Sending-end power circle diagram.

or in terms of rectangular coordinates at

$$\left|\frac{D}{B}\right||V_s|^2\cos(\beta-\alpha)\text{MW}\quad \left|\frac{D}{B}\right||V_s|^2\sin(\beta-\alpha)\quad \text{Mvar} \tag{10.112}$$

The center of the sending-end circle is

$$\frac{|V_s||V_r|}{|B|} \tag{10.113}$$

Example 10.8

Consider Examples 10.6 and 10.7, solved for π and long line models. Draw a circle diagram and calculate (a) the sending-end voltage, (b) the maximum power transferred in MW if the sending-end voltage is maintained equal to the calculated voltage in (a), (c) the value of a shunt reactor required to maintain the receiving-end voltage equal to the sending-end voltage $=1$ per unit at no load, (d) the leading reactive power required at the receiving end when delivering 1.4 per unit MW load at 0.8 PF, the sending-end and receiving-end voltages maintained as in (a).

The calculated A and B constants from Example 10.6 are

$$A = 0.989 + j0.002 = |A| < \alpha = 0.989 < 0.116^\circ$$

$$B = 0.076 + j0.394 = |B| < \beta = 0.40 < 79^\circ$$

Thus, the coordinates of center C are

$$(-0.4732, \ -2.426)$$

The receiving-end power is $0.9 + j0.436$. The circle diagram is shown in Figure 10.14. CP by measurement $= 3.2 = |V_s||V_r|/|B|$. This gives $V_s = 1.28$, as calculated in Example 10.6.

(b) The maximum power transfer in MW is equal to 2.75 per unit, equal to O′O″ in Figure 10.14.

(c) The diameter of the circle for per unit sending- and receiving-end voltages is 2.5, with C as the center. The shunt reactor required ≈0.08 per unit, given by QQ′ in Figure 10.14.

(d) Following the construction shown in Figure 10.14, a new load line is drawn and the leading reactive power required ≈0.82 per unit, given by $Q_L Q'_L$.

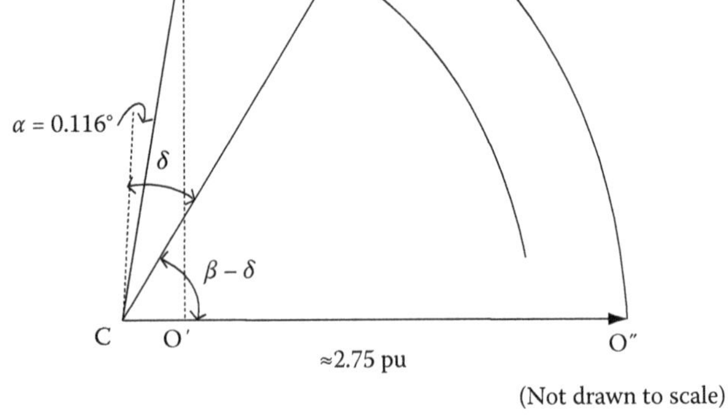

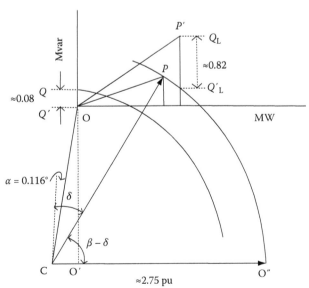

FIGURE 10.14
Circle diagram for solution of Example 10.8.

10.10 Modal Analysis

From Equation 10.64 for a perfect earth, we can write

$$\overline{L}\overline{C} = \frac{\overline{I}}{v^2} = \overline{M} \tag{10.114}$$

For kth conductor:

$$\frac{\partial V_k}{\partial x^2} = \frac{1}{v^2}\frac{\partial V_k}{\partial t^2} \tag{10.115}$$

This shows that the wave equation of each conductor is independent of the mutual influence of other conductors in the system—a result which is only partially true. However, for lightning surges and high frequencies, the depth of the current image almost coincides with the perfect earth. This may not be true for switching surges. The M matrix in (10.114) is not a diagonal matrix. We will use decoupling to diagonalize the matrix, with respect to transmission lines, this is called *Modal Analysis* as it gives different modes of propagation.

Consider a matrix of modal voltages $\overline{V}_{\mathrm{m}}$ and a transformation matrix $\overline{T}$, transforming conductor voltage matrix $\overline{V}$:

$$\overline{V} = \overline{T}\,\overline{V}_{\mathrm{m}} \tag{10.116}$$

Then the wave equation can be decoupled and can be written as

$$\begin{aligned}\frac{\partial^2 \overline{V}_{\mathrm{m}}}{\partial x^2} &= \overline{T}^{-1}\left[\overline{L}\overline{C}\right]\overline{T}\frac{\partial^2 \overline{V}_{\mathrm{m}}}{\partial t^2} \\ &= \overline{T}^{-1}\overline{M}^{-1}\overline{T}\frac{\partial^2 \overline{V}_{\mathrm{m}}}{\partial t^2} \\ &= \overline{\lambda}\frac{\partial^2 \overline{V}_m}{\partial t^2}\end{aligned} \tag{10.117}$$

For decoupling matrix $\overline{\lambda} = \overline{T}^{-1}\overline{M}^{-1}\overline{T}$ must be diagonal. This is done by finding the eigenvalues of $\overline{M}$ from the solution of its characteristic equation. The significance of this analysis is that for n conductor system, the matrices are of order n, and n number of modal voltages are generated which are independent of each other. Each wave travels with a velocity

$$v_n = \frac{1}{\lambda_k}, \quad k = 1, 2, \ldots, n \tag{10.118}$$

And the actual voltage on the conductors is given by (10.117).

For a three-phase line,

$$\overline{T} = \begin{vmatrix} 1 & 1 & 1 \\ -1 & 0 & 1 \\ 0 & -1 & 1 \end{vmatrix} \tag{10.119}$$

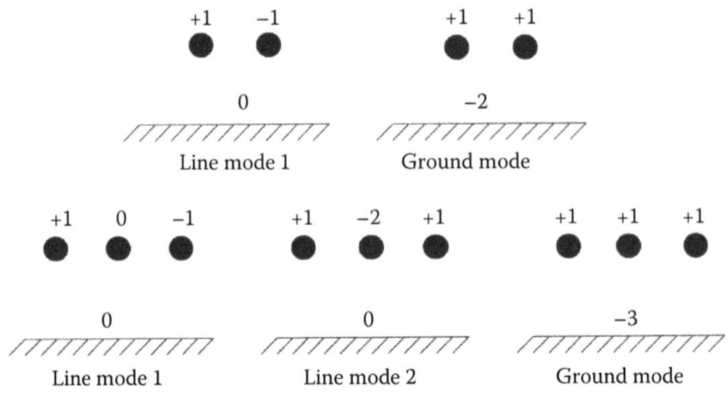

FIGURE 10.15
Mode propagation two-conductor and three-conductor transmission lines.

Figure 10.15 shows modes for two-conductor and three-conductor lines. For a two-conductor line, there are two modes. In the line mode, the voltage and current travel over one conductor returning through the other, none flowing through the ground. In the ground mode, modal quantities travel over both the conductors and return through the ground. For a three-conductor line there are two line modes and one ground mode.

The line-to-line modes of propagation are close to the speed of light and encounter less attenuation and distortion compared to ground modes. The ground mode has more resistance and hence more attenuation and distortion. The resistance of the conductors and earth resistivity plays an important role.

Mode propagation gives rise to further distortion of multi-conductor lines. For voltages below corona inception voltage the velocity of propagation of the line components is close to the velocity of light and the distortion is negligible. For ground currents the charges induced are near the surface, while the return current is well below the surface, at a depth depending upon the frequency and earth's resistivity. For a perfectly conducting earth, the current will flow at the earth's surface at velocity equal to the velocity of light. The concept of modal analysis is necessary for transient analysis and wave propagation.

10.11 Corona on Transmission Lines

The corona discharges form when the electrical field intensity exceeds the breakdown strength of air and local ionization occurs. There are a number of factors at which the corona forms, which include conductor diameter, conductor type (the smoother the conductor surface, the higher the critical disruptive voltage assuming the same diameter), line configuration, and weather. Corona will form when an electrode is charged to sufficiently high voltage from a power frequency source. The mechanism of corona formation is different for the positive and negative half cycles. The negative cycle tends to produce short-duration pulses of current up to 1 mA, lasting for 0.03 μs. The corona noise generated by negative charged electrode tends to be of low level compared to the noise generated by a positive charged electrode. The corona inception voltage is given by

$$V_c = \frac{2\pi r \varepsilon_0 E_c}{C \times 10^6} \tag{10.120}$$

where
r is the radius of conductor (m)
ε_0 is the permittivity of free space
E_c is the corona inception electric field (kV/m) on the conductor surface
C is the capacitance of the overhead line (μF/m)

The corona onset voltage is, generally, 30%–40% above the rated operating voltage.
Using Peek's equation, V_c can be expressed as

$$V_c = \frac{1.66 \times 10^{-3} n m k_d^{2/3} r \left(1 + \frac{0.3}{\sqrt{r}}\right)}{k_b} \tag{10.121}$$

where
n is the number of sub-conductors in a bundle
m is the conductor surface factor which varies from 0.7 to 0.8
k_d is relative air density
r is the conductor radius in cm
C is the capacitance as above in μF/m
k_b is the ratio of maximum to average surface gradient for bundle conductors

This is given by the expression

$$k_b = 1 + 2(n-1)\sin\left(\frac{\pi}{n}\right)\left(\frac{r}{A}\right) \tag{10.122}$$

where A is the distance between adjacent sub-conductors in cm.
The corona has three major effects:

1. Corona loss: A method of estimating the corona loss is established by Burgsdorf. If the line is designed to have appropriate RIV (radio interference voltage), the corona loss from the line will be relatively low in dry weather condition. Under rainfall or when the water droplets are still present on the conductors, this loss will increase substantially. As an example, a 500 kV line, four bundle conductors, spaced 45 cm centers, conductor radius 1.288 cm, gives a loss of 1.7 kW/km for fair-weather conditions and a loss of 33.6 kW/km for wet-weather conditions.
2. Radio interference voltages (RIV): During design of an HV transmission line, RIV levels are considered taking into account the signal strengths of local transmissions and density of population.
3. Effect on wave propagations and flashovers: Corona can reduce overvoltages on open-circuited lines by attenuating the lightning and switching surges. The capacitance of a conductor increases due to corona effect and by increasing electrostatic coupling between shield wires and phase conductors, corona during lightning strokes to shield wires or towers reduces the voltage across the insulator strings and the probability of flashovers.

10.12 System Variables in Load Flow

In the above analysis, currents have been calculated on the basis of system impedances or admittances and assumed voltages. The complex powers have been calculated on the basis of voltages. In load flow situations, neither the voltages nor the currents at load buses are known. The required power balance is known. A bus can connect system admittances, shunt admittances, generators, loads, or reactive power sources. The consumer demand at any instant is uncontrollable in terms of active and reactive power requirements. For a two-node two-port circuit, this is represented by a *four-dimensional disturbance* vector p:

$$p = \begin{vmatrix} p_1 \\ p_2 \\ p_3 \\ p_4 \end{vmatrix} = \begin{vmatrix} P_{D1} \\ Q_{D1} \\ P_{D2} \\ Q_{D2} \end{vmatrix} \tag{10.123}$$

where P_{D1}, P_{D2}, Q_{D1}, and Q_{D2} are the active and reactive power demands at the two buses.

The magnitudes of voltages and their angles at buses 1 and 2 can be called *state variables* and represented by

$$x = \begin{vmatrix} x_1 \\ x_2 \\ x_3 \\ x_4 \end{vmatrix} = \begin{vmatrix} \theta_1 \\ |V_1| \\ \theta_2 \\ |V_2| \end{vmatrix} \tag{10.124}$$

Lastly, the active and reactive power generation at buses 1 and 2 may be called *control variables*:

$$u = \begin{vmatrix} u_1 \\ u_2 \\ u_3 \\ u_4 \end{vmatrix} = \begin{vmatrix} P_{G1} \\ Q_{G1} \\ P_{G2} \\ Q_{G2} \end{vmatrix} \tag{10.125}$$

Thus, for a two-bus system we have 12 variables. For large power systems, there will be thousands of variables and the load flow must solve for interplay between these variables. This will be examined as we proceed along with load flow calculations in the following chapters. In a steady-state operation,

$$f(x, p, u) = 0 \tag{10.126}$$

Problems

10.1 Mathematically derive *ABCD* constants for nominal T models shown in Table 10.1.

10.2 A 500 mi (800 km) long three-phase transmission line operating at 230 kV supplies a load of 200 MW at 0.85 PF. The line series impedance $Z = 0.35 + j0.476\ \Omega/\text{mi}$ and shunt admittance $y = j5.2 \times 10^{-6}$ S/mi. Find the sending-end current, voltage, power, and power loss, using short-line model, nominal Π model, and long line model. Find the line regulation in each case. What is the SIL loading of the line?

10.3 A 400 kV transmission line has the following A and B constants: $A = 0.9 < 2°$; $B = 120 < 75°$. Construct a circle diagram and ascertain (a) the sending-end voltage for a receiving-end load of 200 MW at 0.85 PF, (b) the maximum power that can be delivered for sending an end voltage of 400 kV, and (c) the Mvar required at the receiving-end to support a load of 400 MW at 0.85 PF, the sending-end voltage being held at 400 kV.

10.4 Plot the incident and reflected currents for the line of Example 10.4 at midpoint of the line.

10.5 A 230 kV transmission line has the following line constants: $A = 0.85 < 4°$; $B = 180 < 65°$. Calculate the power at unity PF that can be supplied with the same sending-end and receiving-end voltage of 230 kV. If the load is increased to 200 MW at 0.9 PF, with the same sending-end and receiving-end voltage of 230 kV, find the compensation required in Mvar at the receiving end.

10.6 Draw the current and voltage profile of a symmetrical 230 kV, 300 mi long line at no load and at 100 MW, 0.8 PF load. Use the following line constants: $L = 1.98$ mH/mi; $C = 0.15$ μF/mi.

10.7 An underground cable has an inductance of 0.45 μH/m and a capacitance of 80 pF/m. What is the velocity of propagation along the cable?

10.8 Derive an expression for the maximum active power that can be transmitted over a short transmission line of impedance $Z = R + jX$. Find the maximum power for an impedance of $0.1 + j1.0$, line voltage 4160 V, and regulation not to exceed 5%.

10.9 A loaded long line is compensated by shunt power capacitors at the midpoint. Draw a general vector diagram of the sending-end and receiving-end voltages and currents.

10.10 Derive the *ABCD* constants of a transmission line having resistance 0.1 Ω/mi, reactance 0.86 Ω/mi, and capacitance 0.04 μF/mi. What is the electrical length of the line?

Bibliography

Anderson, J.G. *Transmission Reference Book.* New York: Edison Electric Company, 1968.

Beweley, L.V. *Traveling Waves on Transmission Systems*, 2nd edn. New York: John Wiley, 1951.

Blackwell, W.A. and L.L. Grigsby. *Introductory Network Theory*. Boston, MA: PWS Engineering, 1985.

Chowdhri, P. *Electromagnetic Transients in Power Systems*. Somerset, England: Research Study Press, 1996.

Das, J.C. *Transients in Electrical Systems*. New York: McGraw Hill, 2010.

EHV Transmission. *IEEE Trans.* (special issue), PAS-85-1966(6): 555–700, 1966.

EPRI. Transmission Line Reference Book—345 kV and Above. Palo Alto, CA: EPRI, 1975.

Gary, C., D. Crotescu, and G. Dragon. Distortion and attenuation of traveling waves caused by transient corona, CIGRE Study Committee Report 33, 1989.

Gonen, T. Electrical Power Transmission System Engineering, Analysis and Design, 2nd edn. Boca Raton, FL: CRC Press, 2009.

Kimberk, E.W. *Direct Current Transmission*, vol. 1. New York: John Wiley, 1971.

Naidu, M.S. and V. Kamaraju. *High Voltage Engineering*, 2nd edn. New York: McGraw Hill, 1999.
Rao, S. *EHV-AC, HVDC Transmission and Distribution Engineering*. New Delhi: Khanna Publishers, 2004.
Stott, B. Review of load-flow calculation methods. *Proc. IEEE* 62(7), July 1974.
Westinghouse Electric Corporation. *Electrical Transmission and Distribution Reference Book*, 4th edn. East Pittsburgh, PA: Westinghouse Electric Corporation, 1964.

11

Load Flow Methods: Part I

The Y-matrix iterative methods were the very first to be applied to load flow calculations on the early generation of digital computers. They required minimum storage but, however, may not converge on some load flow problems. This deficiency in Y-matrix methods led to Z-matrix methods, which had a better convergence, but required more storage and slowed down on large systems. These methods of load flow are discussed in this chapter.

We discussed the formation of bus impedance and admittance matrices in Chapter 3. For a general network with $n+1$ nodes, the admittance matrix is

$$\overline{Y} = \begin{vmatrix} Y_{11} & Y_{12} & \cdot & Y_{1n} \\ Y_{21} & Y_{22} & \cdot & Y_{2n} \\ \cdot & \cdot & \cdot & \cdot \\ Y_{n1} & Y_{n2} & \cdot & Y_{nn} \end{vmatrix} \tag{11.1}$$

where each admittance $Y_{ii}(i=1, 2, 3, 4, \ldots)$ is the self-admittance or driving point admittance of node i, given by the diagonal elements, and it is equal to the algebraic sum of all admittances terminating at that node. $Y_{ik}(i, k=1, 2, 3, 4, \ldots)$ is the mutual admittance between nodes i and k or transfer admittance between nodes i and k and is equal to the negative of the sum of all admittances directly connected between those nodes. In Chapter 3, we discussed how each element of this matrix can be calculated. The following modifications can be easily implemented in the bus admittance matrix:

1. Changing a branch admittance from y_{sr} to $y_{\text{sr}}+\Delta y_{\text{sr}}$ between buses S and R leads to

$$\begin{aligned} Y_{\text{ss}} &\rightarrow Y_{\text{ss}} + \Delta y_{\text{sr}} \\ Y_{\text{sr}} &\rightarrow Y_{\text{sr}} - \Delta y_{\text{sr}} \\ Y_{\text{rr}} &\rightarrow Y_{\text{rr}} + \Delta y_{\text{sr}} \end{aligned} \tag{11.2}$$

2. For addition of a new branch of admittance y_{a} between *existing* buses S and R, add to matrix $\overline{Y}$ change matrix $\Delta\overline{Y}$, given by:

$$\Delta\overline{Y} = \begin{array}{c|cc|} & S & R \\ S & y_{\text{a}} & -y_{\text{a}} \\ R & -y_{\text{a}} & y_{\text{a}} \end{array} \tag{11.3}$$

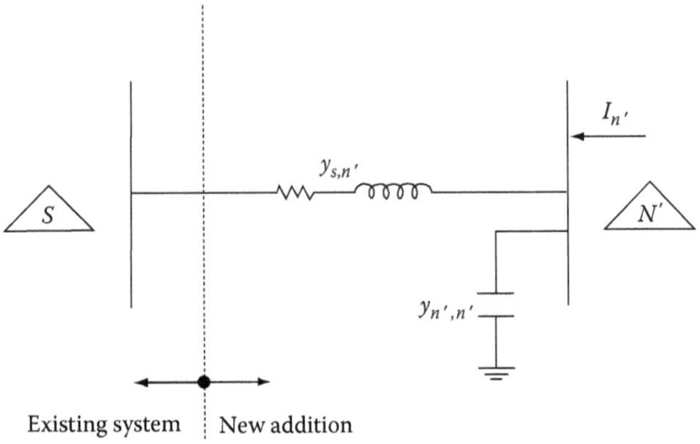

FIGURE 11.1
Adding a new branch between an existing bus S and a new bus N' in a Y matrix.

3. Addition of a new branch between an existing bus S and a *new bus* is shown in Figure 11.1. Let the new bus be designated N'. The order of the Y matrix is increased by one, from n to $n+1$. The injected current at bus N' is

$$\begin{aligned} I_{n'} &= Y_{n's}V_s + Y_{n'n'}V_{n'} \\ Y_{n's} &= -y_{s,n'} \\ Y_{n'n'} &= y_{n'n'} + y_{sn'} \end{aligned} \tag{11.4}$$

All equations remain unchanged except the bus S equation:

$$I_s = Y_{s1}V_1 + \cdots + \left(Y_{ss} + y_{s,n'}\right)V_s + \cdots + Y_{sn'}V_{n'} \tag{11.5}$$

Thus,

$$\begin{aligned} Y_{ss} &\rightarrow Y_{ss} + y_{sn'} \\ Y_{sn'} &\rightarrow -y_{sn'} \end{aligned} \tag{11.6}$$

11.1 Modeling a Two-Winding Transformer

A transformer can be modeled by its leakage impedance as in short-circuit calculations; however, in load flow calculations, a transformer can act as a control element:

- Voltage control is achieved by adjustments of taps on the windings, which change the turns' ratio. The taps can be adjusted under load, providing an automatic control of the load voltage (Appendix C). The under-load taps generally provide ±10%–20% voltage adjustments around the rated transformer voltage, in 16 or 32 steps. Off-load taps provide ±5% voltage adjustment, two taps of 2.5% below the rated voltage and two taps of 2.5% above the rated voltage. These off-load taps must be set at optimum level before commissioning, as these cannot be adjusted under load.

- Transformers can provide phase-shift control to improve the stability limits.
- The reactive power flow is related to voltage change and voltage adjustments indirectly provide reactive power control.

We will examine each of these three transformer models at appropriate places. The model for a ratio-adjusting type transformer is discussed in this chapter. The impedance of a transformer will change with tap position. For an autotransformer, there may be 50% change in the impedance over the tap adjustments. The reactive power loss in a transformer is significant, and the X/R ratio should be correctly modeled in load flow.

Consider a transformer of ratio 1:n. It can be modeled by an ideal transformer in series with its leakage impedance Z (the shunt magnetizing and eddy current loss circuit is neglected), as shown in Figure 11.2a. With the notations shown in this figure,

$$V_s = \frac{(V_r - ZI_r)}{n} \tag{11.7}$$

As the power through the transformer is invariant,

$$V_s I_s^* + (V_r - ZI_r)I_r^* = 0 \tag{11.8}$$

Substituting $I_s/(-I_r) = n$, the following equations can be written as

$$I_s = [n(n-1)Y]V_s + nY(V_s - V_r) \tag{11.9}$$

$$I_r = nY(V_r - V_s) + [(1-n)Y]V_r \tag{11.10}$$

Equations 11.9 and 11.10 give the equivalent circuits shown in Figure 11.2b and c, respectively. For a phase-shifting transformer, a model is derived in Chapter 12.

The equivalent Π circuits of the transformer in Figure 11.2c lead to

$$Y_{\text{ss}} = n^2 Y_{\text{sr}} \quad Y_{\text{rs}} = Y_{\text{sr}} = -nY_{\text{sr}} \quad Y_{\text{rr}} = Y_{\text{sr}} \tag{11.11}$$

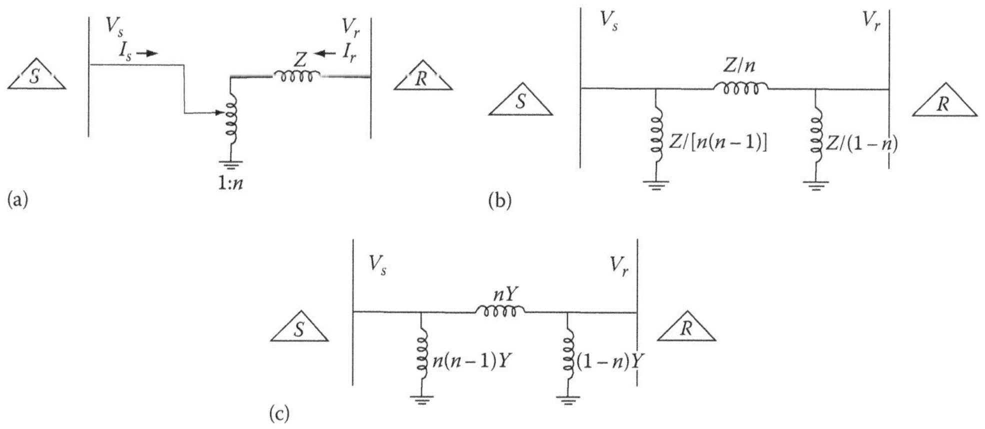

FIGURE 11.2
(a) Equivalent circuit of a ratio-adjusting transformer, ignoring shunt elements; (b) equivalent Π impedance network; (c) equivalent Π admittance network.

If the transformer ratio changes by Δn,

$$\begin{aligned} Y_{ss} &\rightarrow Y_{ss} + [(n+\Delta n)^2 - n^2]Y_{sr} \\ Y_{sr} &\rightarrow Y_{sr} - \Delta n Y_{sr} \\ Y_{rr} &\rightarrow Y_{rr} \end{aligned} \quad (11.12)$$

Example 11.1

A network with four buses and interconnecting impedances between the buses, on a per unit basis, is shown in Figure 11.3a. It is required to construct a Y matrix. The turns' ratio of the transformer should be accounted for according to the equivalent circuit developed above. All impedances are converted into admittances, as shown in Figure 11.3b, and the Y matrix is then written ignoring the effect of the transformer turns' ratio settings as

$$\overline{Y}_{bus} = \begin{vmatrix} 2.176 - j7.673 & -1.1764 + j4.706 & -1.0 + j3.0 & 0 \\ -1.1764 + j4.706 & 2.8434 - j9.681 & -0.667 + j2.00 & -1.0 + j3.0 \\ -1.0 + j3.0 & -0.667 + j2.00 & 1.667 - j8.30 & j3.333 \\ 0 & -1.0 + j3.0 & j3.333 & 1.0 - j6.308 \end{vmatrix}$$

The turns' ratio of the transformer can be accommodated in two ways: (1) by equations and (2) by the equivalent circuit of Figure 11.2c. Let us use Equation 11.2 to modify the elements of the admittance matrix:

$$\begin{aligned} Y_{33} &\rightarrow Y_{33} + \left[(n+\Delta n)^2 - n^2\right] Y_{34} \\ &= 1.667 - j8.30 + [1.1^2 - 1][-j3.333] \\ &= 1.667 - j7.60 \end{aligned}$$

$$\begin{aligned} &Y_{34} \rightarrow Y_{34} - \Delta n Y_{34} \\ &j3.333 - (1.1 - 1)(-j3.333) \\ &\quad = j3.666 \end{aligned}$$

The modified Y bus matrix for the transformer turns' ratio is

$$\overline{Y}_{bus,modified} = \begin{vmatrix} 2.1764 - j7.673 & -1.1764 + j4.706 & -1.0 + j3.0 & 0 \\ -1.1764 + j4.706 & 2.8434 - j9.681 & -0.667 + j2.00 & -1.0 + j3.0 \\ -1.0 + j3.0 & -0.667 + j2.00 & 1.667 - j7.60 & j3.667 \\ 0 & -1.0 + j3.0 & j3.667 & 1.0 - j6.308 \end{vmatrix}$$

As an alternative, the equivalent circuit of the transformer is first modified for the required turns' ratio, according to Figure 11.2c. This modification is shown in Figure 11.3c, where the series admittance element between buses 3 and 4 is modified and additional shunt admittances are added at buses 3 and 4. After the required modifications, the admittance matrix is formed, which gives the same results as those arrived at above.

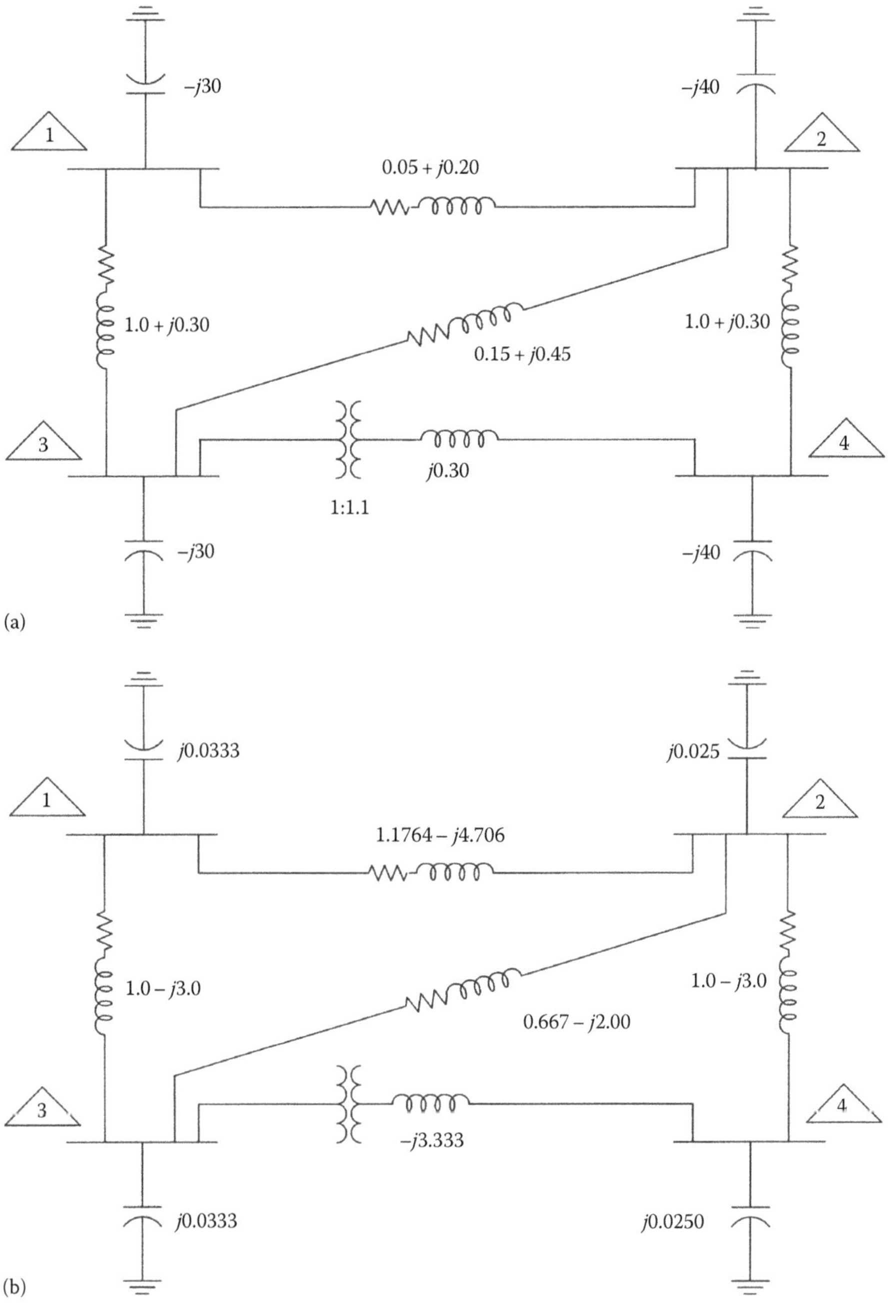

FIGURE 11.3
(a) Impedance data for a four-bus network with a ratio-adjusting transformer; (b) the four-bus system with impedance data converted to admittance;

(continued)

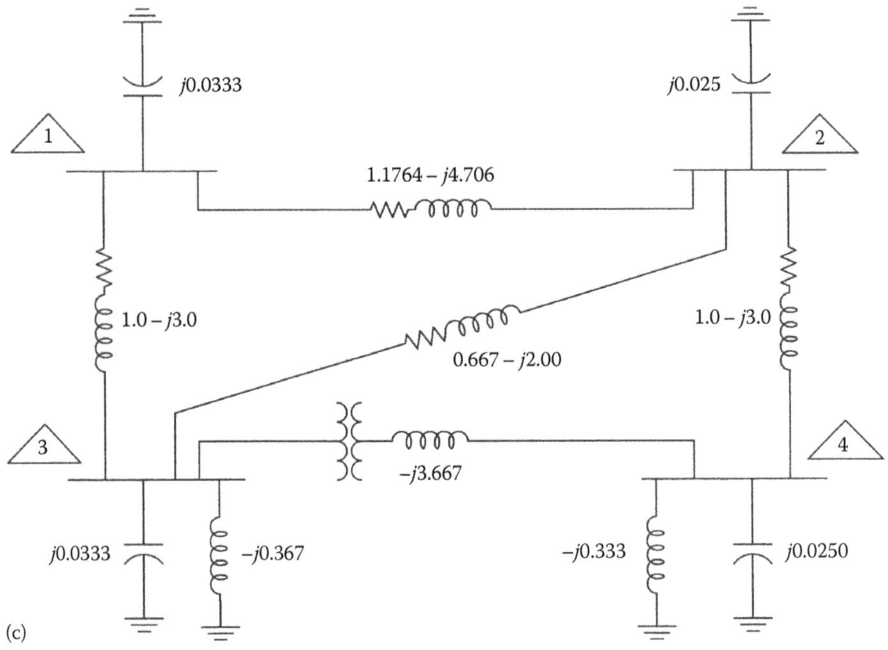

FIGURE 11.3 (continued)
(c) equivalent circuit adjusted for transformer turns' ratio.

11.2 Load Flow—Bus Types

In a load flow calculation, the voltages and currents are required under constraints of bus power. The currents are given by

$$\bar{I} = \overline{Y}\,\overline{V} \tag{11.13}$$

and the power flow equation can be written as

$$S_s = P_s + jQ_s = V_s I_s^* \quad s = 1, 2, \ldots, n \tag{11.14}$$

where n is the number of buses.

From Equations 11.13 and 11.14, there are $2n \times 2$ equations and there are $4n$ variables ($2n$ current variables and $2n$ voltage variables) presenting a difficulty in solution. The current can be eliminated in the above formation:

$$S_s = P_s + jQ_s = V_s \left(\sum_{r=1}^{n} Y_{sr} V_r \right)^* \quad r, s = 1, 2, \ldots, n \tag{11.15}$$

TABLE 11.1

Load Flow—Bus Types

Bus Type	Known Variable	Unknown Variable
PQ	Active and reactive power (P, Q)	Current and voltage (I, V)
PV	Active power and voltage (P, V)	Current and reactive power (I, Q)
Swing	Voltage	Current, active and reactive power (I, P, and Q)

Now there are $2n$ equations and $2n$ variables. This does not lead to an immediate solution. The constant power rather than constant current makes the load flow problem nonlinear. The loads and generation may be specified but the sum of loads is not equal to the sum of generation as network losses are indeterminable until the bus voltages are calculated. Therefore, the exact amount of total generation is unknown. One solution is to specify the power of all generators except one. The bus at which this generator is connected is called the *swing bus* or slack bus. A utility source is generally represented as a swing bus or slack bus, as the consumers' system is much smaller compared to a utility's system. After the load flow is solved and the voltages are calculated, the injected power at the swing bus can be calculated.

Some buses may be designated as PQ buses while the others are designated as PV buses. At a PV bus the generator active power output is known and the voltage regulator controls the voltage to a specified value by varying the reactive power output from the generator. There is an upper and lower bound on the generator reactive power output depending on its rating, and for the specified bus voltage, these bounds should not be violated. If the calculated reactive power exceeds generator Q_{max}, then Q_{max} is set equal to Q. If the calculated reactive power is lower than the generator Q_{min}, then Q is set equal to Q_{min}.

At a PQ bus, neither the current nor the voltage is known, except that the load demand is known. A "mixed" bus may have generation and also directly connected loads. The characteristics of these three types of buses are shown in Table 11.1.

The above description shows that the load flow problem is essentially nonlinear and iterative methods are used to find a solution.

11.3 Gauss and Gauss–Seidel *Y*-Matrix Methods

The principal of Jacobi iteration is shown in Figure 11.4. The program starts by setting initial values of voltages, generally equal to the voltage at the swing bus. In a well-designed power system, voltages are close to rated values and in the absence of a better estimate all the voltages can be set equal to 1 per unit. From node power constraint, the currents are known and substituting back into the *Y*-matrix equations, a better estimate of voltages is obtained. These new values of voltages are used to find new values of currents. The iteration is continued until the required tolerance on power flows is obtained. This is diagrammatically illustrated in Figure 11.4. Starting from an initial estimate of x_0, the final value of x^* is obtained through a number of iterations. The basic flowchart of the iteration process is shown in Figure 11.5.

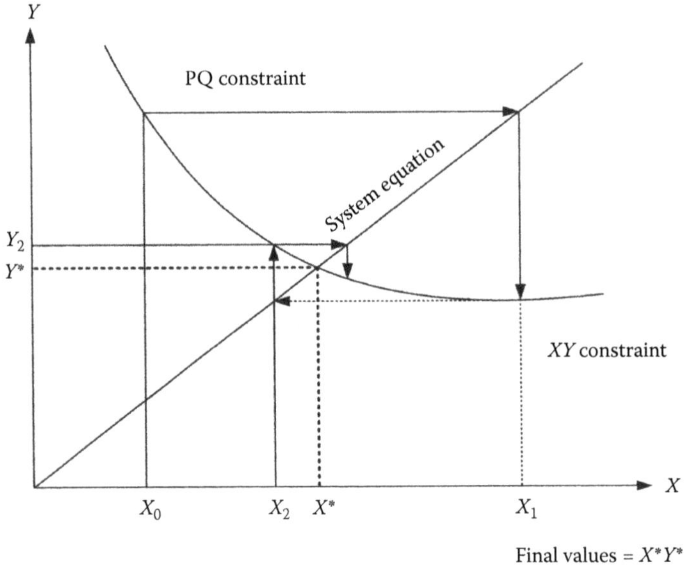

FIGURE 11.4
Illustration of numerical iterative process for final value of a function.

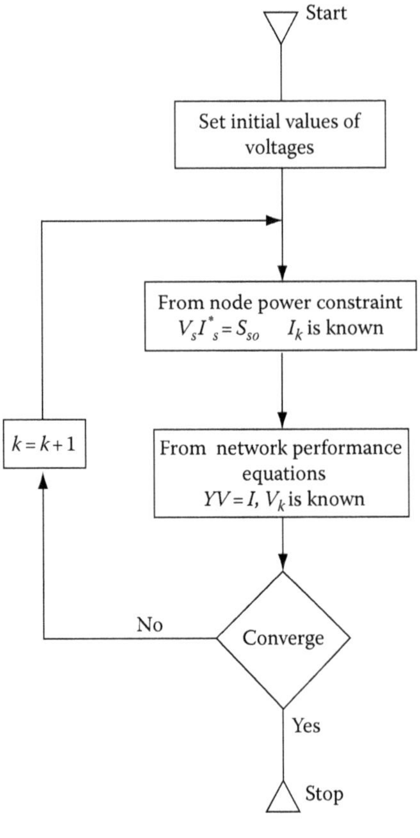

FIGURE 11.5
Flowchart of basic iterative process of Jacobi-type iterations.

11.3.1 Gauss Iterative Technique

Consider that n linear equations in n unknowns $(x_1, \ldots, x_n)$ are given. The a coefficients and b dependent variables are known:

$$\begin{aligned} a_{11}x_1 + a_{12}x_2 + \cdots + a_n x_n &= b_1 \\ a_{21}x_1 + a_{22}x_2 + \cdots + a_2 n x_n &= b_2 \\ \cdots \\ a_{n1}x_1 + a_{n2}x_2 + \cdots + a_{nn}x_n &= b_n \end{aligned} \tag{11.16}$$

These equations can be written as

$$\begin{aligned} x_1 &= \frac{1}{a_{11}}(b_1 - a_{12}x_2 - a_{13}x_3 - \cdots - a_{1n}x_n) \\ x_2 &= \frac{1}{a_{22}}(b_2 - a_{21}x_1 - a_{23}x_3 - \cdots - a_{2n}x_n) \\ \cdots \\ x_n &= \frac{1}{a_{nn}}(b_n - a_{n1}x_1 - a_{n2}x_2 - \cdots - a_{n,n-1}x_{n-1}) \end{aligned} \tag{11.17}$$

An initial value for each of the independent variables $x_1, x_2, \ldots, x_n$ is assumed. Let these values be denoted by

$$x_1^0,\ x_2^0,\ x_3^0,\ \ldots,\ x_n^0 \tag{11.18}$$

The initial values are estimated as

$$\begin{aligned} x_1^0 &= \frac{y_1}{a_{11}} \\ x_2^0 &= \frac{y_2}{a_{22}} \\ \cdots \\ x_n^0 &= \frac{y_n}{a_{nn}} \end{aligned} \tag{11.19}$$

These are substituted into Equation 11.17, giving

$$\begin{aligned} x_1^0 &= \frac{1}{a_{11}}\left[b_1 - a_{12}x_2^0 - a_{13}x_3^0 - \cdots a_{1n}x_n^0\right] \\ x_2^0 &= \frac{1}{a_{22}}\left[b_2 - a_{21}x_1^0 - a_{23}x_3^0 - \cdots a_{2n}x_n^0\right] \\ \cdots \\ x_n^0 &= \frac{1}{a_{nn}}\left[b_{nn} - a_{n1}x_1^0 - a_{n2}x_2^0 - \cdots a_{n,n-1}x_{n-1}^0\right] \end{aligned} \tag{11.20}$$

These new values of

$$x_1^1, x_2^1, \ldots, x_n^1 \tag{11.21}$$

are substituted into the next iteration. In general, at the kth iteration,

$$\begin{aligned}
x_1^k &= \frac{1}{a_{11}}\left[b_1 - a_{12}x_2^{k-1} - a_{13}x_3^{k-1} - \cdots - a_{1n}x_n^{k-1}\right] \\
x_2^k &= \frac{1}{a_{22}}\left[b_2 - a_{21}x_1^{k-1} - a_{23}x_3^{k-1} - \cdots - a_{2n}x_n^{k-1}\right] \\
&\cdots \\
x_n^k &= \frac{1}{a_{nn}}\left[b_n - a_{n1}x_2^{k-1} - a_{n2}x_2^{k-1} - \cdots - a_{n,n-1}x_{n-1}^{k-1}\right]
\end{aligned} \tag{11.22}$$

Example 11.2

Consider a three-bus network with admittances as shown in Figure 11.6. The currents at each of the buses are fixed, which will not be the case in practice, as the currents are dependent on voltages and loads. Calculate the bus voltages for the first five iterations by the Gauss iterative technique.

The matrix equation is

$$\begin{vmatrix} 3 & -1 & 0 \\ -1 & 4 & -3 \\ 0 & -3 & 6 \end{vmatrix} \begin{vmatrix} v_1 \\ v_2 \\ v_3 \end{vmatrix} = \begin{vmatrix} 3 \\ 4 \\ 2 \end{vmatrix}$$

Thus,

$$\begin{aligned}
3v_1 - v_2 &= 3 \\
-v_1 + 4v_2 - 3v_3 &= 4 \\
-3v_2 + 6v_3 &= 2
\end{aligned}$$

There are three equations and three unknowns. Let $v_1 = v_2 = v_3 = 1$ be the initial values. This relationship is generally valid as a starting voltage estimate in load flow, because the voltages

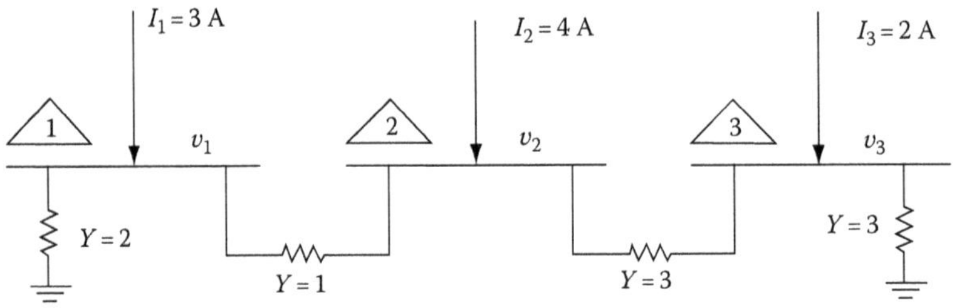

FIGURE 11.6
A three-bus system for Examples 11.2, 11.3, and 11.5.

TABLE 11.2

Gauss Iterative Solution of Example 11.2

Voltage	$k=0$	$k=1$	$k=2$	$k=3$	$k=4$	$k=5$
$v_{1(k)}$	1	1.333	1.667	1.653	1.806	1.799
$v_{2(k)}$	1	2.000	1.958	2.417	2.397	2.608
$v_{3(k)}$	1	0.8333	1.333	1.312	1.542	1.532

throughout the network must be held close to the rated voltages. The following equations can then be written according to the Gauss iterative method:

$$\begin{aligned} v_1^{k+1} &= 1 + \frac{1}{3}v_2^k \\ v_2^{k+1} &= 1 + \frac{1}{4}v_1^k + \frac{3}{4}v_3^k \\ v_3^{k+1} &= \frac{1}{3} + \frac{1}{2}v_2^k \end{aligned} \tag{11.23}$$

The values of v_1, v_2, and v_3 at $k=1$ are

$$\begin{aligned} k &= 1 \\ v_1 &= 1 + (0.3333)(1) = 1.333 \\ v_2 &= 1 + 0.25(1) + 0.75(1) = 2.000 \\ v_3 &= 0.3333 + (0.25)(1) = 0.833 \end{aligned}$$

$$\begin{aligned} k &= 2 \\ v_1 &= 1 + (0.3333)(2) = 1.667 \\ v_2 &= 1 + 0.25(1.333) + 0.75(0.8333) = 1.958 \\ v_3 &= 0.3333 + 0.5(2.000) = 1.333 \end{aligned}$$

The results for the first five iterations are presented in Table 11.2.

11.3.2 Gauss–Seidel Iteration

Instead of substituting the $k-1$ approximations into all the equations in the kth iteration, the kth iterations are immediately used as soon as these are found. This will hopefully reduce the number of iterations.

$$\begin{aligned} x_1^k &= \frac{1}{a_{11}}\left[b_1 - a_{12}x_2^{k-1} - a_{13}x_3^{k-1} - \cdots - a_{1n}x_n^{k-1}\right] \\ x_2^k &= \frac{1}{a_{22}}\left[b_2 - a_{21}x_1^{k} - a_{23}x_3^{k-1} - \cdots - a_{2n}x_n^{k-1}\right] \\ x_3^k &= \frac{1}{a_{33}}\left[b_3 - a_{31}x_1^{k} - a_{32}x_2^{k} - \cdots - a_{3n}x_n^{k-1}\right] \\ &\cdots \\ x_n^k &= \frac{1}{a_{nn}}\left[b_n - a_{n1}x_1^{k} - a_{n2}x_2^{k} - \cdots - a_{n,n-1}x_{n-1}^{k}\right] \end{aligned} \tag{11.24}$$

TABLE 11.3

Gauss–Seidel Iterative Solution of Example 11.3

Voltage	$k=0$	$k=1$	$k=2$	$k=3$	$k=4$	$k=5$
v_1	1	1.333	1.694	1.818	1.875	1.901
v_2	1	2.083	2.455	2.625	2.703	2.739
v_3	1	1.375	1.561	1.646	1.685	1.703

Example 11.3

Solve the three-bus system of Example 11.2 by Gauss–Seidel iteration.

We again start with the assumption of unity voltages at all the buses, but the new values of voltages are immediately substituted into the downstream equations. From Equation 11.23, the first iteration is as follows:

$k = 1$

$v_1 = 1 + 0.3333(1) = 1.333$. This value of v_1 is immediately used

$v_2 = 1 = 0.25(1.333) + 0.75(1) = 2.083$. This value of v_2 is immediately used

$v_3 = 0.3333 + 0.5(2.083) = 1.375$

The result of the first five iterations is shown in Table 11.3. A better estimate of the final voltages is obtained in fewer iterations.

11.3.3 Convergence

The convergence can be defined as

$$\varepsilon^k = \left|V_s^{k-1} - V_s^k\right| < \varepsilon \quad \text{(specified value)} \tag{11.25}$$

The calculation procedure is repeated until the specified tolerance is achieved. The value of ε is arbitrary. For Gauss–Seidel iteration, $\varepsilon = 0.00001$–0.0001 is common. This is the largest allowable voltage change on any bus between two successive iterations before the final solution is reached. Approximately 50–150 iterations are common, depending on the number of buses and the impedances in the system.

11.3.4 Gauss–Seidel *Y*-Matrix Method

In load flow calculations, the system equations can be written in terms of current, voltage, or power at the kth node. We know that the matrix equation in terms of unknown voltages, using the bus admittance matrix for $n+1$ nodes, is

$$\begin{vmatrix} I_1 \\ I_2 \\ \cdots \\ I_k \end{vmatrix} = \begin{vmatrix} Y_{11} & Y_{12} & \cdots & Y_{1n} \\ Y_{21} & Y_{22} & \cdots & Y_{2n} \\ \cdots & \cdots & \cdots & \cdots \\ Y_{n1} & Y_{n2} & \cdots & Y_{nn} \end{vmatrix} \begin{vmatrix} V_{01} \\ V_{02} \\ \cdots \\ V_{0n} \end{vmatrix} \tag{11.26}$$

where 0 is the common node. The significance of zero or common node or slack bus is discussed in Chapter 3.

Although the currents entering the nodes from generators and loads are not known, these can be written in terms of P, Q, and V:

$$I_k = \frac{P_k - jQ_k}{V_k^*} \tag{11.27}$$

The convention of the current and power flow is important. Currents entering the nodes are considered positive, and thus the power *into* the node is also positive. A load *draws* power out of the node and thus the active power and inductive vars are entered as $-P - j(-Q) = -P + jQ$. The current is then $(-P + jQ)/V^*$. The nodal current at the kth node becomes

$$\frac{P_k - jQ_k}{V_k^*} = Y_{k1}V_1 + Y_{k2}V_2 + V_{k3}Y_3 + \cdots + Y_{kk}V_k + \cdots + Y_{kn}V_n \tag{11.28}$$

This equation can be written as

$$V_k = \frac{1}{Y_{kk}}\left[\frac{P_k - jQ_k}{V_k^*} - Y_{k1}V_1 - Y_{k3}V_3 - \cdots - Y_{kn}V_n\right] \tag{11.29}$$

In general, for the kth node,

$$V_k = \frac{1}{Y_{kk}}\left[\frac{P_k - jQ_k}{V_k^*} - \sum_{i=1}^{i=n} Y_{ki}V_i\right] \quad \text{for } i \neq k \tag{11.30}$$

The kth bus voltage at $k+1$ iteration can be written as

$$V_k^{k+1} = \frac{1}{Y_{kk}}\left[\frac{P_k - jQ_k}{V_k^*} - \sum_{i=1}^{k-1} Y_{ki}V_i^{k+1} - \sum_{i=k+1}^{n} Y_{ki}V_i^k\right] \tag{11.31}$$

The voltage at the kth node has been written in terms of itself and the other voltages. The first equation involving the swing bus is omitted, as the voltage at the swing bus is already specified in magnitude and phase angle.

The Gauss–Seidel procedure can be summarized for PQ buses in the following steps:

1. Initial phasor values of load voltages are assumed, the swing bus voltage is known, and the controlled bus voltage at generator buses can be specified. Though an initial estimate of the phasor angles of the voltages will accelerate the final solution, it is not necessary and the iterations can be started with zero degree phase angles or the same phase angle as the swing bus. A *flat voltage start* assumes $1 + j0$ voltages at all buses, except the voltage at the swing bus, which is fixed.
2. Based on the initial voltages, the voltage at a bus in the first iteration is calculated using Equation 11.30, i.e., at bus 2:

$$V_{21} = \frac{1}{Y_{22}}\left[\frac{P_2 - jQ_2}{V_{20}^*} - Y_{21}V_{10} - \cdots - Y_{2n}V_{n0}\right] \tag{11.32}$$

3. The estimate of the voltage at bus 2 is refined by repeatedly finding new values of V_2 by substituting the value of V_2 into the right-hand side of the equation.
4. The voltage at bus 3 is calculated using the latest value of V_2 found in step 3 and similarly for other buses in the system.

This completes one iteration. The iteration process is repeated for the entire network till the specified convergence is obtained.

A generator bus is treated differently; the voltage to be controlled at the bus is specified and the generator voltage regulator varies the reactive power output of the generator within its reactive power capability limits to regulate the bus voltage:

$$Q_k = -I_m\left[V_k^*\left\{\sum_{i=1}^{i=n} Y_{ki}V_i\right\}\right] \tag{11.33}$$

where I_m stands for the imaginary part of the equation. The revised value of Q_k is found by substituting the most updated value of voltages:

$$Q_k^{k+1} = -I_m\left[V_k^{k*}\sum_{i=1}^{i=n} Y_{ki}V_i^{k+1} + V_k^{k*}\sum_{i=1}^{i=n} Y_{ki}V_i^{k}\right] \tag{11.34}$$

The angle δ_k is the angle of the voltage in Equation 11.30:

$$\begin{aligned}\delta_k^{k+1} &= \angle \text{ of } V_k^{k+1}\\ &= \angle \text{ of }\left[\frac{P_k - jQ_k^{k+1}}{Y_{kk}\left(V_k^k\right)^*} - \sum_{i=1}^{k-1}\frac{Y_{ki}}{Y_{kk}}V_i^{k+1} - \sum_{i=k+1}^{n}\frac{Y_{ki}}{Y_{kk}}V_i^{k}\right]\end{aligned} \tag{11.35}$$

For a PV bus the upper and lower limits of var generation to hold the bus voltage constant are also given. The calculated reactive power is checked for the specified limits:

$$Q_{k(\min)} < Q_k^{k+1} < Q_{k(\max)} \tag{11.36}$$

If the calculated reactive power falls within the specified limits, the new value of voltage V_k^{k+1} is calculated using the specified voltage magnitude and δ_k^k. This new value of voltage V_k^{k+1} is made equal to the specified voltage to calculate the new phase angle δ_k^{k+1}.

If the calculated reactive power is outside the specified limits, then,

$$\text{If } Q_k^{k+1} > Q_{k(\max)} \qquad Q_k^{k+1} = Q_{k(\max)} \tag{11.37}$$

$$\text{If } Q_k^{k+1} < Q_{k(\max)} \qquad Q_k^{k+1} = Q_{k(\max)} \tag{11.38}$$

This means that the specified limits are not exceeded and beyond the reactive power bounds, the PV bus is treated like a PQ bus. A flowchart is shown in Figure 11.7.

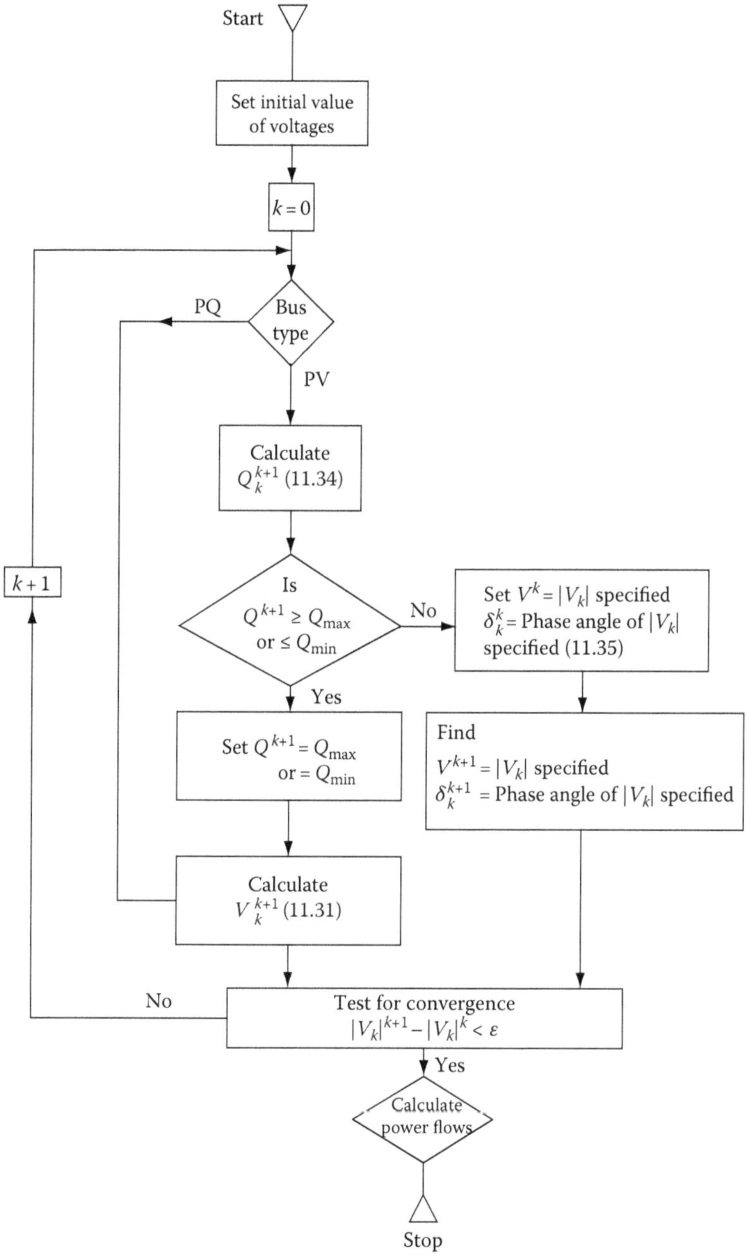

FIGURE 11.7
Flowchart for Gauss–Seidel method of load flow.

Example 11.4

A three-bus radial system of distribution is shown in Figure 11.8 consisting of three branch impedances shown in per unit on a 100 MVA base. Bus 1 is a swing bus with voltage $1 < 0°$ in per unit. Buses 2 and 3 are load buses, loads as shown, and bus 4 is a generation bus, i.e., a PV bus with voltage of 1.05 per unit, phase angle unknown. The generator is a 25 MVA unit, rated

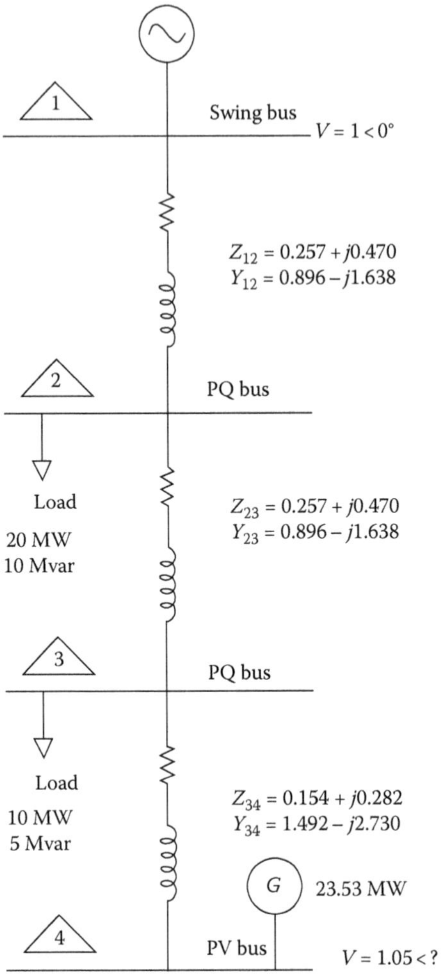

FIGURE 11.8
Four-bus network with generator for load flow (Examples 11.4, 11.6, and 11.9).

power factor of 0.8 (15 Mvar output at rated MW output of 20). Though the generator will be connected through a step-up transformer, its impedance is ignored. Also, the capacitance to ground of the transmission lines is ignored. Calculate the load flow by the Gauss–Seidel iterative method.

The impedances are converted into admittances and a Y matrix is formed:

$$\overline{Y}_{bus} = \begin{vmatrix} 0.896 - j1.638 & -0.896 + j1.638 & 0 & 0 \\ -0.896 + j1.638 & 1.792 - j3.276 & -0.896 + j1.638 & 0 \\ 0 & -0.896 + j1.638 & 2.388 - j4.368 & -1.492 + j2.730 \\ 0 & 0 & -1.492 + j2.730 & 1.492 - j2.730 \end{vmatrix}$$

Let the initial estimate of voltages at buses 2 and 3 be $1 < 0°$ and $1 < 0°$, respectively. The voltage at bus 2, V_2, is

$$V_2^1 = \frac{1}{Y_{22}}\left[\frac{P_2 - jQ_2}{V_2^{0*}} - Y_{21}V_1^0 - Y_{23}V_3^0Y_{24}V_4^0\right]$$

$$= \frac{1}{1.792 - j3.276}\left[\frac{0.2 + j0.1}{1.0 + j0} - (-0.896 + j1.638)(1.0 < 0°) - (-0.896 + j1.638)(1.0 < 0°)\right]$$

$$= 0.9515 < -2.059° \tag{11.39}$$

Substitute this value back into Equation 11.39 and recalculate:

$$V_2^{1(2)} = \frac{1}{1.792 - j3.276}\left[\frac{-0.2 + j0.1}{0.9515 < 2.059°} + 1.792 - j3.276\right]$$

$$= 0.948 < -2.05°$$

Another closer estimate of V_2 is possible by repeating the process. Similarly,

$$V_3^1 = \left[\frac{1}{2.388 - j4.368}\right]$$

$$\times \left[\frac{-0.1 + j0.05}{1.0 + j0} - (-0.896 + j1.638)(0.948 < -2.05°) - (-1.492 + j2.730)(1.05 < 0°)\right]$$

$$= 0.991 < -1.445°$$

Note that the newly estimated value of bus 2 voltage is used in this expression. The bus 3 voltage after another iteration is

$$V_{3_{12}} = 0.993 < -1.453°$$

For the generator bus 4 the voltage at the bus is specified as $1.05 < 0°$. The reactive power output from the generator is estimated as

$$Q_4^0 = -I_m\left[V_{4_0}^*\{Y_{41}V_1 + Y_{42}V_{2_{1(2)}} + Y_{43}V_{3_{1(2)}} + Y_{44}V_{4_0}\}\right] \tag{11.40}$$

Substituting the numerical values,

$$Q_4^0 = -I_m[1.05 < 0°[(-1.492 + j2.730)(0.991 < -1.453°) + (1.492 - j2.730)(1.05 < 0°)]]$$

This is equal to 0.125 per unit. The generator reactive power output at rated load is 0.150 per unit.

The angle δ of the voltage at bus 4 is given by

$$\angle V_4^1 = \angle\frac{1}{Y_{44}}\left[\frac{P_4 - jQ_{4_0}}{V_4^{0*}} - Y_{41}V_1 - Y_{42}V_2^{1(2)} - Y_{43}V_3^{1(2)}\right] \tag{11.41}$$

The latest values of the voltage are used in Equation 11.41. Substituting the numerical values and the estimated value of the reactive power output of the generator:

$$\angle V_4^1 = \angle\frac{1}{1.492 - j2.730}\left[\frac{0.2 - j0.125}{1.05 < 0°} - (-1.492 + j2.730)(0.993 < -1.453°)\right]$$

This is equal to $<0.54°$.

V_4^1 is given as $1.05 < 0.54°$; the value of voltage found above can be used for further iteration and estimate of reactive power.

$$Q_4^1 = -I_m[1.05 < -0.54°\{(-1.492 + j2.730)(0.993 < -1.453°) + (1.492 - j2.730)(1.05 < 0.54°)\}]$$

This gives 0.11 per unit reactive power. One more iteration can be carried out for accuracy. The results of the first and subsequent iterations are shown in Table 11.4.

Now suppose that for the specified voltage of 1.05 per unit, the generator reactive power requirement exceeds its maximum specified limit of 0.15, i.e., a generator rated at 0.9 power factor will have a reactive capability at full load equal to 0.109 per unit. The 1.05 per unit voltage cannot then be maintained at bus 3. The maximum reactive power is substituted and the bus voltage calculated as in a PV bus.

11.4 Convergence in Jacobi-Type Methods

Example 11.4 shows that in the very first iteration the calculated results are close to the final results. Most power system networks are well conditioned and converge. Generally, the convergence is better ensured when the diagonal elements of the Y matrix are larger than the off-diagonal elements. This may not always be the case. In systems that have a wide variation of impedances, oscillations may occur without convergence.

11.4.1 Ill-Conditioned Network

Consider two linear equations:

$$1{,}000x_1 + 20{,}001x_2 = 40{,}003$$
$$x_1 + 2x_2 = 4$$

The solution is

$$x_1 = -2, \quad x_2 = 3$$

Let the coefficient of x_1 in the first equation change by −0.1%, i.e., the coefficient is 999. The new equations are

$$999x_1 + 20{,}001x_2 = 40{,}003$$
$$x_1 + 2x_2 = 4$$

The solution is

$$x_1 = -\frac{2}{3}, \quad x_2 = \frac{7}{3}$$

TABLE 11.4

Gauss–Seidel Iterative Solution of Network of Example 11.4

	Bus 2		Bus 3		Bus 4				
Iteration	**Voltage**	**Load Current**	**Voltage**	**Load Current**	**Voltage**	**Generator Current**	**Generator Active Power**	**Generator Reactive Power**	**Converge *d* for ε =**
1	0.948 < −2.05	0.236	0.993 < −1.453	0.1126	1.05 < 0.6	0.2179	0.2	0.12	
2	0.944 < −2.803	0.2370	0.991 < −1.268	0.1128	1.05 < 0.863	0.2194	0.2	0.1144	0.12
3	0.943 < −2.717	0.2373	0.991 < −1.125	0.1129	1.05 < 0.999	0.22	0.2	0.1155	0.0025
4	0.942 < −2.647	0.2373	0.990 < −1.012	0.1129	1.050 < 1.111	0.22	0.2	0.1157	0.002
10	0.942 < −2.426	0.2373	0.990 < −0.645	0.1129	1.050 < 1.482	0.2199	0.2	0.1155	0.0006

Now let the coefficient of x_1 in the first equation change by +1%, i.e., it is 10,001. The new equations are

$$10{,}001x_1 + 20{,}001x_2 = 40{,}003$$
$$x_1 + 2x_2 = 4$$

The solution of these equations is

$$x_1 = 2, \quad x_2 = 1$$

This is an example of an *ill-conditioned* network. A small perturbation gives rise to large oscillations in the results.

11.4.2 Negative Impedances

Negative impedances may be encountered while modeling certain components, i.e., duplex reactors and sometimes the three-winding transformers. When negative elements are present, the Gauss–Seidel method of load flow will not converge.

11.4.3 Convergence Speed and Acceleration Factor

The convergence speed of the Y-matrix method corresponds to the indirect solution of linear simultaneous equations:

$$\overline{A}\overline{x} = \overline{b} \tag{11.42}$$

The matrix $\overline{A}$ can be written as

$$\overline{A} = D - \overline{L}_l - \overline{U}_u \tag{11.43}$$

The indirect process solution can be expressed as

$$\begin{aligned}
x_1^{k+1} &= \left(\frac{b_1 - a_{12}x_2^k - a_{13}x^k \cdots - a_{1n}x_n^k}{a_{11}}\right)\\
x_1^{k+1} &= x_1^k + \alpha\left(x_1^{k+1} - x_1^k\right)\\
x_2^{k+1} &= \left(\frac{b_2 - a_{21}x_1^{k+1} - a_{23}x_3^k \cdots - a_{2n}x_n^k}{a_{22}}\right)\\
x_2^{k+1} &= x_2^k + \alpha\left(x_2^{k+1} - x_2^k\right)\\
&\cdots
\end{aligned} \tag{11.44}$$

In matrix form:

$$\begin{aligned}
\overline{x}^{k+1} &= \overline{D}^{-1}\left[\overline{b} + \overline{L}_l\overline{x}^{k+1} + \overline{U}_u\overline{x}^k\right]\\
\overline{x}^{k+1} &= \overline{x}^k + \alpha\left(\overline{x}^{k+1} - \overline{x}^k\right)
\end{aligned} \tag{11.45}$$

From these equations,

$$\overline{x}^{k+1} = \overline{M}\overline{x}^{k} + \overline{g}$$

where

$$\begin{aligned}\overline{M} &= (\overline{D} - \alpha\overline{L}_l)^{-1}\left[(1-\alpha)\overline{L}_l + \overline{U}_u\right] \\ \overline{g} &= \alpha(\overline{D} - \alpha\overline{L}_l)^{-1}\overline{b}\end{aligned} \tag{11.46}$$

For Gauss method, $\alpha = 1$.

Equation 11.46 can be transformed as

$$\begin{aligned}\overline{x}^{k+1} &= \overline{M}\overline{x}^{k} + \overline{g} \\ &= \overline{M}(\overline{M}\overline{x}^{k-1} + \overline{g}) + \overline{g} \\ &= \overline{M}^{2}\overline{x}^{k+1} + (\overline{M}+1)\overline{g} \\ &\cdots \\ &= \overline{M}^{k+1}\overline{x}^{0} + (\overline{M}^{k} + \overline{M}^{k-1} + \cdots \overline{M} + 1)\overline{g}\end{aligned} \tag{11.47}$$

If $\lambda_1, \lambda_2, \lambda_3, \ldots$ are the eigenvalues of a square matrix $\overline{M}$, and $\lambda_i \neq \lambda_j$ $(i \neq j)$ and $x_1, x_2, x_3, \ldots$ are the eigenvectors, and then the eigenvectors are linearly independent. An arbitrary vector can be expressed as a linear combination of eigenvectors:

$$\overline{x} = c_1\overline{x}_1 + c_2\overline{x}_2 + \cdots + c_n\overline{x}_n \tag{11.48}$$

As $\overline{M}\overline{x} = \lambda\overline{x}$, the above equation can be written as

$$\begin{aligned}M\overline{x} &= C_1\lambda_1\overline{x}_1 + C_2\lambda_2\overline{x}_2 + \cdots + C_n\lambda_n\overline{x}_n \\ \overline{M}^{2}\overline{x} &= C_1\lambda_1^{2}\overline{x}_1 + C_2\lambda_2^{2}\overline{x}_2 + \cdots + C_n\lambda_n^{2}\overline{x}_n \\ \overline{M}^{k+1}\overline{x} &= C_1\lambda_1^{k+1}\overline{x}_1 + C_2\lambda_2^{k+1}\overline{x}_2 + \cdots + C_n\lambda_n^{k+1}\overline{x}_n\end{aligned} \tag{11.49}$$

If

$$|\lambda_i| < 1 \quad (i = 1, 2, 3, \ldots, n)$$

then the first term of Equation 11.49 converges to zero as $k \to \infty$. Otherwise, it diverges. The speed of convergence is dominated by the maximum value of λ_i. One method of reducing the value of λ_i is to increase α, i.e., if the Δx between iterations is not small enough, it is multiplied by a numerical factor α to increase its value; α is called the acceleration factor. The value of α generally lies between

$$0 < \alpha < 2$$

An α value of 1.4–1.6 is common. A value of $\alpha < 1$ is called a decelerating constant.

The larger the system, the more time it takes to solve the load flow problem, depending on the value of λ.

Example 11.5

Solve the system of Example 11.3 using the Gauss–Seidel method, with an acceleration factor of 1.6.

$$V_1^{(k+1)'} = 1 + \frac{1}{3}V_2^k$$

$$V_1^{(k+1)} = V_1^k + 1.6\left(V_1^{(k+1)'} - V_1^k\right)$$

$$V_2^{(k+1)'} = 1 + \frac{1}{4}V_1^{k+1} + \frac{3}{4}V_3^k$$

$$V_2^{(k+1)} = V_2^k + 1.6\left(V_2^{(k+1)'} - V_2^k\right)$$

$$V_3^{(k+1)'} = \frac{1}{3} + \frac{1}{2}V_2^{k+1}$$

$$V_3^{(k+1)} = V_3^k + 1.6\left(V_3^{(k+1)'} - V_3^k\right)$$

Using the above equations, the values of voltages for $k = 1$ are calculated as follows:

$$V_1^{1'} = 1 + \frac{1}{3}(1) = 1.333$$

$$V_1^1 = 1 + 1.6(1.333 - 1) = 1.6(0.333) = 1.538$$

$$V_2^{2'} = 1 + \frac{1}{4}(1.538) + \frac{3}{4}(1) = 2.1345$$

$$V_2^2 = 1 + 1.6(2.1345 - 1) = 2.8152$$

$$V_3^{3'} = \frac{1}{3} + \frac{1}{2}(2.8152) = 1.7406$$

$$V_3^3 = 1 + 1.6(1.7406 - 1) = 2.1849$$

The results of the first five iterations are shown in Table 11.5. These can be compared to the results in Table 11.3. The mismatch between bus voltages progressively reduces.

TABLE 11.5

Solution of Example 11.5, with Gauss–Seidel Iteration and Acceleration Factor of 1.6

Voltage	$k=0$	$k=1$	$k=2$	$k=3$	$k=4$	$k=5$
$V_1^{k'}$	1	1.333	1.938	2.135	1.912	1.839
V_1^k	1	1.538	2.178	2.109	1.794	1.866
$V_2^{k'}$		2.135	3.183	2.986	2.598	2.686
V_2^k	1	2.815	3.404	2.735	2.516	2.788
$V_3^{k'}$		1.741	2.035	1.700	1.591	1.727
V_3^k		2.185	1.945	1.533	1.626	1.788

Example 11.6

Repeat the calculations of Example 11.4, with an acceleration factor of 1.6.

The voltage at bus 2 is calculated in the first two iterations as $0.948 < -2.05°$ (Example 11.4). Using an acceleration factor of 1.6,

$$V_2^1 = V_2^0 + 1.6\left(V_2^{1'} - V_2^0\right)$$
$$= 1 < 0° + 1.6(0.948 < -2.05° - 1 < 0°)$$
$$= 0.916 - j0.054$$

Use this voltage to calculate the initial voltage on bus 3:

$$V_3^{1'} = \frac{1}{2.388 - j4.368}\left[\frac{-0.1 + j0.05}{1.0 + j0} - (-0.896 + j1.638)(0.916 - j0.054) - (-1.493 + j2.730)(1.05 < 0°)\right]$$
$$= 0.979 - j0.032$$

Calculate the bus 3 voltage with the acceleration factor:

$$V_3^1 = V_3^0 + 1.6\left(V_3^{1'} - V_3^0\right)$$
$$= 1 < 0° + 1.6(0.979 - j0.032 - 1 + j0)$$
$$= 0.966 - j0.651$$

For bus 4 calculate the reactive power, as in Example 11.4, using the voltages found above. The results are shown in Table 11.6, which shows that there is initial oscillation in the calculated results, which are much higher than the results obtained in Table 11.4. However, at the 10th iteration, the convergence is improved compared to the results in Table 11.4. Selection of an unsuitable acceleration factor can result in a larger number of iterations and oscillations and a lack of convergence.

11.5 Gauss–Seidel Z-Matrix Method

From Chapter 3, we know how a bus impedance matrix can be formed. We will revisit bus impedance matrix from load flow considerations. The voltage at the swing bus is defined and need not be included in the matrix equations. The remaining voltages are

$$\begin{vmatrix} V_2 \\ V_3 \\ \cdot \\ V_n \end{vmatrix} = \begin{vmatrix} Z_{22} & Z_{23} & \cdots & Z_{2n} \\ Z_{32} & Z_{33} & \cdots & Z_{3n} \\ \cdot & \cdot & \cdot & \cdot \\ Z_{n2} & Z_{n3} & \cdots & Z_{nn} \end{vmatrix} \begin{vmatrix} I_2 \\ I_3 \\ \cdot \\ I_n \end{vmatrix} \tag{11.50}$$

Therefore, we can write

$$V_k = Z_{k2}I_2 + Z_{k3}I_3 + \cdots + Z_{kn}I_n \tag{11.51}$$

TABLE 11.6

Solution of Example 11.6, with an Acceleration Factor of 1.6

	Bus 2		Bus 3		Bus 4				
Iteration	Voltage	Load Current	Voltage	Load Current	Voltage	Generator Current	Generator Reactive Power	Generator Active Power	Convergence
1	0.917 < −3.403	0.2439	0.970 < −3.105	0.1153	1.023	0.23	0.124	0.2	−0.034
2	0.939 < −3.814	0.2380	0.974 < −3.134	0.1148	1.045	0.2252	0.124	0.2	0.0094
3	0.929 < −3.587	0.2406	0.988 < −2.049	0.1132	1.053	0.2236	0.124	0.2	0.01295
4	0.947 < −2.836	0.2360	0.998 < −1.083	0.1103	1.050	0.2100	0.0927	0.2	0.0140
10	0.943 < −2.326	0.2372	0.991 < −0.485	0.1129	1.050 < 1.650	0.2193	0.1142	0.2	−0.00053

The current at the kth branch can be written as

$$I_k = \frac{P_k - jQ_k}{V_k^*} - y_k V_k \tag{11.52}$$

where y_k is the admittance of bus k to a common reference bus. If the kth branch is a generator, which supplies real and reactive power to the bus, P and Q are entered as positive values. If the kth branch draws power from the network, then P and Q are entered as negative values.

The general procedure is very similar to the Y admittance method:

$$\begin{aligned} V_k &= Z_{kl}I_1 + \cdots + Z_{kn}I_n \\ &= Z_{kl}\left[\frac{P_1 - jQ_1}{V_1^*} - y_1V_1\right] + \cdots + Z_{kn}\left[\frac{P_n - jQ_n}{V_n^*} - y_nV_n\right] \\ V_k &= \sum_{i=1}^{i-n} Z_{ki}\left[\frac{P_i - jQ_i}{V_i^*} - y_iV_i\right] \end{aligned} \tag{11.53}$$

The iteration process is as follows:

1. Assume initial voltages for the n buses:

$$V_1^0, V_2^0, \ldots, V_n^0 \tag{11.54}$$

2. Calculate V_1 from Equation 11.53 in terms of the initial assumed voltages and substitute back into the same equation for a new corrected value:

$$V_1^1 = \sum_{i=1}^{n} Z_{ki}\left(\frac{Pi - jQ_i}{V_i^{0*}} - yiV_i^0\right) \tag{11.55}$$

3. Repeat for bus 2, making use of corrected values of V_1 found in step 2. Iterate to find a corrected value of V_2 before proceeding to the next bus.
4. When all the bus voltages have been evaluated, start all over again for the required convergence.

At a generator bus, the reactive power is not known and an estimate of reactive power is necessary. This can be made from

$$Q_k = -I_m \frac{V_k^*}{Z_{kk}}\left[V_k - \sum_{i=1, i\neq k}^{i-n} Z_{ki}\frac{P_i - jQ_i}{V_i^*}\right] \tag{11.56}$$

A Z-impedance matrix has strong convergence characteristics, at the expense of larger memory requirements and preliminary calculations. Modifications due to system changes are also comparatively difficult.

11.6 Conversion of Y to Z Matrix

A number of techniques for formation of a bus-impedance matrix are outlined in Chapter 3. We will examine one more method by which a bus-impedance matrix can be constructed from an admittance matrix by a step-by-step pivotal operation. The following equations are supported:

$$\text{new } Y_{kk} = \frac{1}{Y_{kk}} (Y_{kk} = \text{pivot}) \quad (11.57)$$

$$\text{new } Y_{kj} = \frac{Y_{kj}}{Y_{kk}} \quad j = 1, \ldots, n(j \neq k) \quad (11.58)$$

$$\text{new } Y_{ik} = -\frac{Y_{ik}}{Y_{kk}} \quad i = 1, \ldots, n(j \neq k) \quad (11.59)$$

$$\text{new } Y_{ij} = Y_{ij} - \left[\frac{Y_{ik}Y_{kj}}{Y_{kk}}\right] \quad (i \neq k, j \neq k) \quad (11.60)$$

The choice of a pivot is arbitrary. The new nonzero elements can be avoided by a proper choice (Appendix D). The procedure can be illustrated by an example.

Example 11.7

Consider a hypothetical four-bus system, with the admittances as shown in Figure 11.9. Form a Y matrix and transform to a Z matrix by pivotal manipulation. Check the results with the step-by-step buildup method of the bus-impedance matrix described in Chapter 3.

The Y bus matrix is easily formed by examination of the network,

$$\bar{Y} = \begin{vmatrix} 7 & -2 & 0 & -3 \\ -2 & 5 & -3 & 0 \\ 0 & -3 & 4 & 0 \\ -3 & 0 & 0 & 6 \end{vmatrix}$$

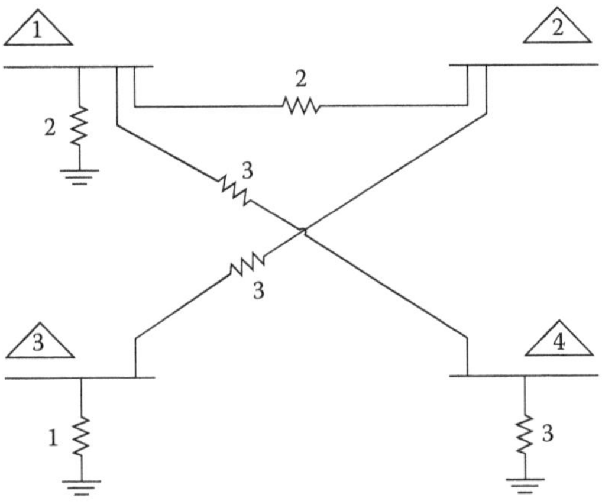

FIGURE 11.9
Four-bus network for Example 11.7.

pivot (3, 3),

$$Y_{31} = \frac{Y_{31}}{Y_{33}} = 0$$

$$Y_{32} = \frac{Y_{32}}{Y_{33}} = -\frac{3}{4} = -0.75$$

$$Y_{33} = \frac{1}{Y_{33}} = 0.25$$

$$Y_{34} = \frac{Y_{34}}{Y_{33}} = 0$$

$$Y_{13} = -\frac{Y_{13}}{Y_{33}} = 0$$

$$Y_{23} = -\frac{Y_{23}}{Y_{33}} = -\left(-\frac{3}{4}\right) = 0.75$$

$$Y_{43} = -\frac{Y_{43}}{Y_{33}} = 0$$

$$Y_{22} = Y_{22} - \frac{Y_{23}Y_{32}}{Y_{33}} = 5 - \frac{(-3)(-3)}{4} = 2.75$$

etc.

The transformed matrix is

$$\begin{vmatrix} 7 & -2 & 0 & -3 \\ -2 & 2.75 & 0.75 & 0 \\ 0 & -0.75 & 0.25 & 0 \\ -3 & 0 & 0 & 6 \end{vmatrix}$$

Next, use the pivot (4, 4). The transformed matrix is

$$\begin{vmatrix} 5.5 & -2 & 0 & 0.5 \\ -2 & 2.75 & 0.75 & 0 \\ 0 & -0.75 & 0.25 & 0 \\ -0.5 & 0 & 0 & 0.167 \end{vmatrix}$$

Next, use the pivot (2, 2),

$$\begin{vmatrix} 4.045 & 0.727 & 0.545 & 0.5 \\ -0.727 & 0.364 & 0.273 & 0 \\ 0.545 & 0.273 & 0.455 & 0 \\ -0.5 & 0 & 0 & 0.167 \end{vmatrix}$$

and finally the pivot (1, 1) gives the Z matrix

$$\overline{Z} = \begin{vmatrix} 0.2470 & 0.1797 & 0.1350 & 0.1236 \\ 0.1797 & 0.4950 & 0.3709 & 0.0898 \\ 0.1350 & 0.3709 & 0.528 & 0.0674 \\ 0.1236 & 0.0898 & 0.0674 & 0.2289 \end{vmatrix}$$

Step-by step buildup

Chapter 3 showed a step-by-step building method for a Z matrix.

Node 1 to ground and adding branch 1 to 2 and 2 to 3 give the Z matrix as

$$\begin{vmatrix} 0.5 & 0.5 & 0.5 \\ 0.5 & 1 & 1 \\ 0.5 & 1 & 1.33 \end{vmatrix}$$

Add the 1 ohm branch between node 3 and ground. The new Z matrix is

$$\begin{vmatrix} 0.5 & 0.5 & 0.5 \\ 0.5 & 1 & 1 \\ 0.5 & 1 & 1.33 \end{vmatrix} - \frac{1}{2.33}\begin{vmatrix} 0.5 \\ 1 \\ 1.33 \end{vmatrix}\begin{vmatrix} 0.5 & 1 & 1.33 \end{vmatrix} = \begin{vmatrix} 0.393 & 0.285 & 0.215 \\ 0.285 & 0.571 & 0.429 \\ 0.215 & 0.429 & 0.571 \end{vmatrix}$$

Add branch 1 to 4:

$$\bar{Z} = \begin{vmatrix} 0.393 & 0.285 & 0.215 & 0.393 \\ 0.285 & 0.571 & 0.429 & 0.285 \\ 0.215 & 0.429 & 0.571 & 0.215 \\ 0.393 & 0.285 & 0.215 & 0726 \end{vmatrix}$$

Finally, add the 0.333 ohm branch between node 4 and ground. The final Z matrix is

$$\begin{vmatrix} 0.393 & 0.285 & 0.215 & 0.393 \\ 0.285 & 0.571 & 0.429 & 0.285 \\ 0.215 & 0.429 & 0.571 & 0.215 \\ 0.393 & 0.285 & 0.215 & 0.726 \end{vmatrix} - \frac{1}{1.059}\begin{vmatrix} 0.393 \\ 0.285 \\ 0.215 \\ 0.726 \end{vmatrix}$$

$$\begin{vmatrix} 0.393 & 0.285 & 0.215 & 0.726 \end{vmatrix} = \begin{vmatrix} 0.247 & 0.1797 & 0.135 & 0.1236 \\ 0.1797 & 0.495 & 0.3709 & 0.0898 \\ 0.1350 & 0.3709 & 0.528 & 0.0674 \\ 0.1236 & 0.0898 & 0.0674 & 0.2289 \end{vmatrix}$$

This is the same result as the one arrived at by pivotal operation.

Example 11.8

Solve the network of Example 11.2 by the Z-matrix method.

The Y matrix of Example 11.2 is

$$\bar{Y} = \begin{vmatrix} 3 & -1 & 0 \\ -1 & 4 & -3 \\ 0 & -3 & 6 \end{vmatrix}$$

Form the Z matrix,

$$\overline{Z} = \begin{vmatrix} 0.385 & 0.154 & 0.077 \\ 0.154 & 0.462 & 0.231 \\ 0.071 & 0.231 & 0.282 \end{vmatrix}$$

Then

$$\begin{vmatrix} v_1 \\ v_2 \\ v_3 \end{vmatrix} = \overline{Z} \begin{vmatrix} 3 \\ 4 \\ 2 \end{vmatrix} = \begin{vmatrix} 1.923 \\ 2.769 \\ 1.718 \end{vmatrix}$$

No more iterations are required, as the currents are given. These values exactly satisfy the original equations. Compare these results obtained after five iterations in Tables 11.2, 11.3, and 11.5, which are still approximate and require more iterations for better accuracy.

Example 11.9

Solve Example 11.4 with a Z bus matrix of load flow.

Using the same initial voltage estimate on the buses as in Example 11.4,

$$\begin{aligned} I_2^0 &= \left[\frac{P_2 + jQ_2}{V_2^0}\right]^* - Y_{21}V_1 \\ &= \left[\frac{-0.2 - j0.1}{1.0}\right]^* - (-0.896 + j1.638) \times 1 < 0° \\ &= 0.696 - j1.538 \end{aligned}$$

The equation for the swing bus need not be written:

$$\begin{aligned} I_3^0 &= \left[\frac{P_3 + jQ_3}{V_3^0}\right]^* - Y_{31}V_1 \\ &= [(0.1 \quad j0.05)1 < 0°]^* \quad (0)1 < 0° - -0.1 + j0.05 \end{aligned}$$

The reactive power at bus 4 is

$$Q_4^0 = \text{Im}\left[V_4^0\left(Y_{41}V_1^0 + Y_{42}V_2^0 + Y_{43}V_3^0 + Y_{44}V_4^0\right)^*\right]$$

$$\text{Im}\left[1.05 < 0°[(-1.492 + j2.730)(1 < 0°) + (1.492 - j2.730)(1.05 < 0°)]^*\right] = j0.144$$

This is within the generator rated reactive power output.

$$\begin{aligned} I_4^0 &= \left[\frac{P_4 + jQ_4}{V_4^0}\right]^* - Y_{41}V_1 \\ &= 0.19 - j0.137 \end{aligned}$$

TABLE 11.7

Comparison of Y and Z Matrix Methods for Load Flow

No.	Compared Parameter	Y Matrix	Z Matrix	Remarks
1	Digital computer memory requirements	Small	Large	Sparse matrix techniques easily applied to Y matrix
2	Preliminary calculations	Small	Large	Software programs can basically operate from the same data input
3	Convergence characteristics	Slow, may not converge at all	Strong	Both methods may slow down on large systems
4	System modifications	Easy	Slightly difficult	See text

Thus, the voltages are given by

$$\begin{vmatrix} V_2 \\ V_3 \\ V_4 \end{vmatrix} = \overline{Z} \begin{vmatrix} I_2 \\ I_3 \\ I_4 \end{vmatrix}$$

where the Z matrix is formed from the Y matrix of Example 11.4. The equation for the swing bus is omitted.

$$\overline{Z} = \begin{vmatrix} 0.257 + j0.470 & 0.257 + j0.470 & 0.257 + j0.470 \\ 0.257 + j0.470 & 0.514 + j0.940 & 0.514 + j0.940 \\ 0.257 + j0.470 & 0.514 + j0.940 & 0.668 + j1.222 \end{vmatrix}$$

Therefore,

$$\begin{vmatrix} V_2 \\ V_3 \\ V_4 \end{vmatrix} = \overline{Z} \begin{vmatrix} 0.696 - j1.538 \\ -0.1 + j0.05 \\ 0.19 - j0.137 \end{vmatrix} = \begin{vmatrix} 0.966 - j0.048 \\ 1.03 - j0.028 \\ 1.098 + j0.004 \end{vmatrix} = \begin{vmatrix} 0.983 < -2.84^\circ \\ 1.030 < -1.56^\circ \\ 1.098 < 0.21^\circ \end{vmatrix}$$

The calculations are repeated with the new values of voltages. Note that V_4 is higher than the specified limits in the first iteration. Since V_4 is specified as 1.05, in the next iteration we will use $V_4 = 1.05 < 0.21^\circ$ and find the reactive power of the generator.

A comparison with the Y-matrix method is shown in Table 11.7.

11.7 Triangular Factorization Method of Load Flow

A commercial program of load flow used triangular factorization. A matrix can be factored into $\overline{L}$ and $\overline{U}$ (Appendix A). Here we factor $\overline{Y}$ bus matrix into $\overline{L}$ and $\overline{U}$ form

$$\overline{L}\,\overline{U} = \overline{Y}_{\text{bus}} \tag{11.61}$$

Then

$$\overline{L}\,\overline{U}\,\overline{V} = \overline{I} \tag{11.62}$$

As an intermediate step define a new voltage $\overline{V}'$, so that

$$\begin{aligned} \overline{L}\,\overline{V}' &= \overline{I} \\ \overline{U}\,\overline{V} &= \overline{V}' \end{aligned} \tag{11.63}$$

The lower triangular system is first solved for the vector $\overline{V}'$ by forward substitution. Then $\overline{V}$ is found by backward substitution. When $\overline{Y}$ is symmetric only $\overline{L}$ is needed as it can be used to produce the back substitution as well as the forward substitution. When the sparsity is exploited by optimal ordering (Appendix D), and the computer program processes and stores only the nonzero terms, the triangular factorization can give efficient direct solution.

Example 11.10

To illustrate the principle, consider a bus admittance matrix:

$$\overline{Y}_B = \begin{vmatrix} -j & 2j & j \\ 2j & -j3 & j3 \\ j & j3 & -4j \end{vmatrix}$$

The current injection vector is

$$\begin{vmatrix} j1.75 \\ j2.0 \\ 0 \end{vmatrix}$$

Then the voltages at each of the three nodes can be calculated:

$$\begin{vmatrix} V_1 \\ V_2 \\ V_3 \end{vmatrix} = \overline{Y}_B^{-1}\overline{I} = \begin{vmatrix} 0.973 \\ 0.902 \\ 0.92 \end{vmatrix} \tag{11.64}$$

Now calculate matrices $\overline{L}$ and $\overline{U}$ based upon the techniques discussed in Appendix A. These matrices are

$$\overline{L} = \begin{vmatrix} -j & 0 & 0 \\ j2 & j & 0 \\ j & j5 & -j28 \end{vmatrix}$$

$$\overline{U} = \begin{vmatrix} 1 & -2 & -1 \\ 0 & 1 & 5 \\ 0 & 0 & 1 \end{vmatrix}$$

The product can be verified to give the original bus admittance matrix. Then we can write

$$\begin{vmatrix} -j & 0 & 0 \\ j2 & j & 0 \\ j & j5 & -j28 \end{vmatrix} \begin{vmatrix} V_1' \\ V_2' \\ V_3' \end{vmatrix} = \begin{vmatrix} j1.75 \\ j2.0 \\ 0 \end{vmatrix}$$

This gives

$$\begin{vmatrix} V_1' \\ V_2' \\ V_3' \end{vmatrix} = \begin{vmatrix} -1.75 \\ 5.5 \\ 0.92 \end{vmatrix}$$

Then

$$\begin{vmatrix} 1 & -2 & -1 \\ 0 & 1 & 5 \\ 0 & 0 & 1 \end{vmatrix} \begin{vmatrix} V_1 \\ V_2 \\ V_3 \end{vmatrix} = \begin{vmatrix} -1.75 \\ 5.5 \\ 0.92 \end{vmatrix}$$

or

$$\begin{vmatrix} V_1 \\ V_2 \\ V_3 \end{vmatrix} = \begin{vmatrix} 0.973 \\ 0.902 \\ 0.92 \end{vmatrix}$$

This is same as the one arrived at in Equation 11.64. The calculated voltages V_1, V_2, and V_3 are much below-rated voltages. The example is for illustrative purpose only; the voltages at the buses are maintained close to the rated voltages. In a practical distribution system it is desirable to operate with a voltage slightly higher than the rated voltages.

Note that the bus admittance matrix in the above example is symmetrical. Referring to Appendix A, for a symmetric matrix, the calculation can be further simplified. The matrix $\overline{L}$ can be written as

$$\overline{L} = \overline{U}^{t}\overline{D} = \begin{vmatrix} 1 & 0 & 0 & 0 \\ \dfrac{Y_{21}}{Y_{11}} & 1 & 0 & 0 \\ \dfrac{Y_{31}}{Y_{11}} & \dfrac{Y_{32}}{Y_{22}} & 1 & 0 \\ \dfrac{Y_{41}}{Y_{11}} & \dfrac{Y_{42}}{Y_{22}} & \dfrac{Y_{43}}{Y_{33}} & 1 \end{vmatrix} \begin{vmatrix} Y_{11} & \cdot & \cdot & \cdot \\ \cdot & Y_{22} & \cdot & \cdot \\ \cdot & \cdot & Y_{33} & \cdot \\ \cdot & \cdot & \cdot & Y_{44} \end{vmatrix} \tag{11.65}$$

Then

$$\overline{Y}_{B} = \overline{U}^{t}\overline{D}\,\overline{U} \tag{11.66}$$

Equation 11.66 can be solved by denoting

$$\begin{aligned} \overline{U}^{t}\overline{V}'' &= \overline{I} \\ \overline{D}\,\overline{V} &= \overline{V}'' \\ \overline{U}\,\overline{V} &= \overline{V}' \end{aligned} \tag{11.67}$$

Example 11.11

Solve example using Equation 11.65.
From Example 11.10,

$$\overline{U}^{t}\overline{V}'' = \overline{I}$$

or

$$\begin{vmatrix} 1 & 0 & 0 \\ -2 & 1 & 0 \\ -1 & 5 & 1 \end{vmatrix} \begin{vmatrix} V_1'' \\ V_2'' \\ V_3'' \end{vmatrix} = \begin{vmatrix} j1.75 \\ j2.0 \\ 0 \end{vmatrix}$$

This gives

$$\begin{vmatrix} V_1'' \\ V_2'' \\ V_3'' \end{vmatrix} = \begin{vmatrix} j1.75 \\ j5.5 \\ -j25.75 \end{vmatrix}$$

Also

$$\overline{D}\overline{V}' = \overline{V}''$$

or

$$\begin{vmatrix} -j & 0 & 0 \\ 0 & j & 0 \\ 0 & 0 & -j28 \end{vmatrix} \begin{vmatrix} V_1' \\ V_2' \\ V_3' \end{vmatrix} = \begin{vmatrix} V_1'' \\ V_2'' \\ V_3'' \end{vmatrix} = \begin{vmatrix} j1.75 \\ j5.5 \\ -j25.75 \end{vmatrix}$$

From which

$$\begin{vmatrix} V_1' \\ V_2' \\ V_3' \end{vmatrix} = \begin{vmatrix} -1.75 \\ 5.5 \\ 0.92 \end{vmatrix}$$

This is the same result as that of Example 11.10.
Finally

$$\overline{U}\overline{V} = \overline{V}'$$

as in Example 11.10. Thus, it is a three-step calculation.

Problems

(The problems can be solved without modeling the systems on a digital computer.)

11.1 In Example 11.1 consider that the line between buses 2 and 3 is removed. Form the Y-impedance matrix. Modify this matrix by reconnecting the removed line.

11.2 Figure 11.P.1 shows a distribution system with data as shown. Convert all impedances to per unit base and form a Y matrix.

11.3 Calculate the currents and voltages in the system of Figure 11.P.1 by (1) Gauss iterative method, (2) Gauss–Seidel iterative method, and (3) Gauss–Seidel iterative method with an acceleration factor of 1.6. Calculate to first two iterations in each case.

11.4 Convert the Y matrix of Problem 11.2 into a Z matrix by (1) pivotal manipulation, and (2) step-by-step buildup.

11.5 Calculate currents and voltages by the Z-matrix method in Problem 11.3.

11.6 Figure 11.P.2 is a modification of the circuit in Example 11.5, with a tap-adjusting transformer. Calculate load flow with the transformer at rated voltage tap and at 1:1.1 tap. How does the reactive power flow change with the transformer tap adjustment? Calculate to first iteration.

11.7 Solve for voltages in Examples 11.8 and 11.9 using triangular factorization method and compare the results.

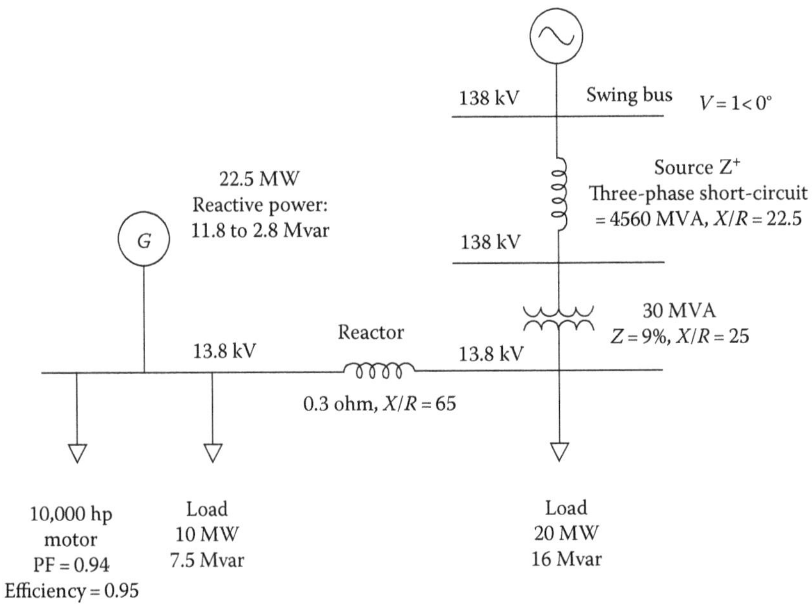

FIGURE 11.P.1
Network with impedance and generation data for load flow problems (Problems 11.2 through 11.5).

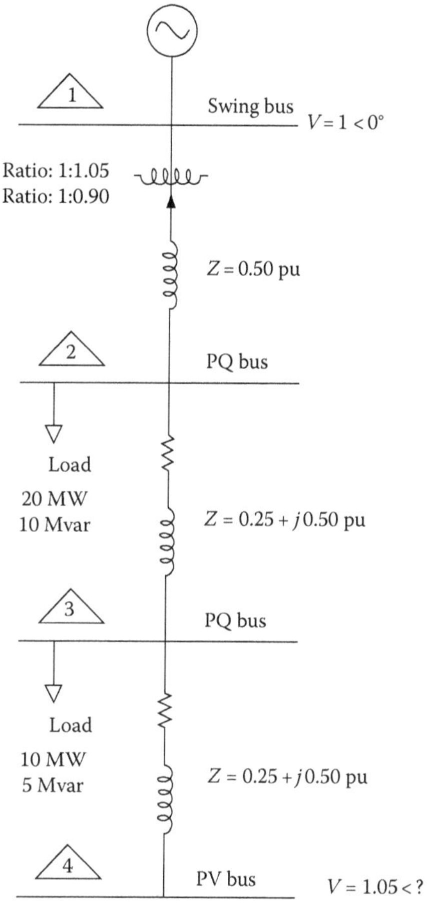

FIGURE 11.P.2
Four-bus system with ratio-adjusting transformer for Problem 11.6.

Bibliography

Arrillaga, J., J.C.P. Arnold, and B.J. Harker. *Computer Modeling of Electrical Power Systems.* New York: John Wiley, 1983.

Bergen, A.R. and V. Vittal. *Power System Analysis*, 2nd edn. Upper Saddle River, NJ: Prentice Hall, 1999.

Brown, H.E. *Solution of Large Networks by Matrix Methods.* New York: John Wiley, 1972.

Brown, H.E., G.K. Carter, H.H. Happ, and C.E. Person. Power flow solution by impedance iterative methods. *IEEE Trans. PAS* 2: 1–10, 1963.

Granger, J.J. and W.D. Stevenson. *Power System Analysis.* New York: McGraw-Hill, 1994.

Hubert, F.J. and D.R. Hayes. A rapid digital computer solution for power system network load flow. *IEEE Trans. PAS* 90: 934–940, 1971.

IEEE. *Power System Analysis*, Standard 399-1990.

Laughton, M.A. Decomposition techniques in power systems network load flow analysis using the nodal impedance matrix. *Proc. IEEE* 115(4): 539–542, 1968.

Powell, L. *Power System Load Flow Analysis*. New York: McGraw Hill, 2004.
Shipley, R.B. *Introduction to Matrices and Power Systems*. New York: John Wiley, 1976.
Stagg, G.W. and A.H. El-Abiad. *Computer Methods in Power System Analysis*. New York: McGraw-Hill, 1968.
Tinny, W.F. and J.W. Walker. Direct solution of sparse network equations by optimally ordered triangular factorization. *Proc. IEEE* 55: 1801–1809, November 1967.
Wang, X.-F., Y. Song, and M. Irving. *Modern Power System Analysis*. New York: Springer, 2008.

12

Load Flow Methods: Part II

The Newton–Raphson (NR) method has powerful convergence characteristics, though computational and storage requirements are heavy. The sparsity techniques and ordered elimination (discussed in Appendix D) led to its earlier acceptability and it continues to be a powerful load flow algorithm even in today's environment for large systems and optimization [1]. A lesser number of iterations are required for convergence, as compared to the Gauss–Seidel method, provided that the initial estimate is not far removed from the final results, and these do not increase with the size of the system [2]. The starting values can even be first estimated using a couple of the iterations with the Gauss–Seidel method for load flow and the results input into the NR method as a starting estimate. The modified forms of the NR method provide even faster algorithms. Decoupled load flow and fast decoupled solution methods are offshoots of the NR method.

12.1 Function with One Variable

Any function of x can be written as the summation of a power series, and Taylor's series of a function $f(x)$ is

$$y = f(x) = f(a) + f'(a)(x-a) + \frac{f''(a)}{2!}(x-a)^2 + \cdots + \frac{f^n(a)(x-a)^n}{n!} \tag{12.1}$$

where $f'(a)$ is the first derivative of $f(a)$. Neglecting the higher terms and considering only the first two terms, the series is

$$y = f(x) \approx f(a) + f'(a)(x-a) \tag{12.2}$$

The series converges rapidly for values of x near to a. If x_0 is the initial estimate, then the tangent line at $(x_0, f(x_0))$ is

$$y = f(x_0) + f'(x_0)(x_1 - x_0) \tag{12.3}$$

where x_1 is the new value of x, which is a closer estimate. This curve crosses the x *axis* (Figure 12.1) at the new value x_1. Thus

$$\begin{aligned} 0 &= f(x_0) + f'(x_0)(x_1 - x_0) \\ x_1 &= x_0 - \frac{f(x_0)}{f'(x_0)} \end{aligned} \tag{12.4}$$

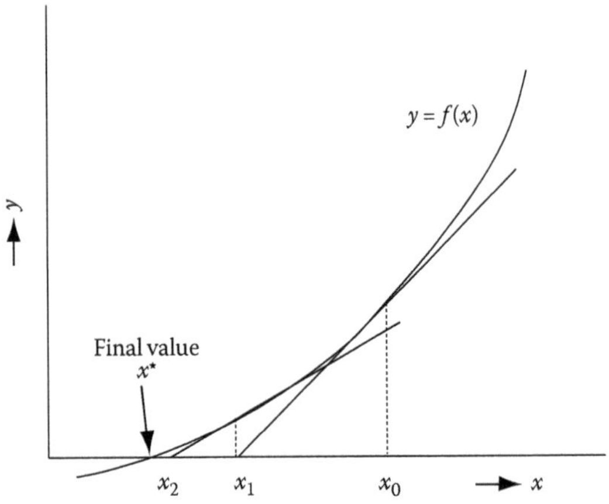

FIGURE 12.1
Progressive tangents to a function $y = f(x)$ at x_0, x_1, and x_2, showing the crossings with the x axis, each crossing approaching closer to the final value x^*.

In general

$$x_{k+1} = x_k - \frac{f(x_k)}{f'(x_k)} \tag{12.5}$$

Example 12.1

Consider a function

$$f(x) = x^3 - 2x^2 + 3x - 5 = 0$$

Find the value of x.

The derivative is

$$f'(x) = 3x^2 - 4x + 3$$

Let the initial value of $x = 3$, then $f(x) = 13$ and $f'(x) = 18$. For $k = 1$, from Equation 12.4, $x_1 = 3 - (13/18) = 2.278$. Table 12.1 is compiled to $k = 4$ and gives $x = 1.843$. As a verification, substituting this value into the original equation, the identity is approximately satisfied.

TABLE 12.1
Iterative Solution of a Function of One Variable (Example 12.1)

k	x_k	$f(x_k)$	$f'(x_k)$
0	3	13	18
1	2.278	3.277	9.456
2	1.931	0.536	6.462
3	1.848	0.025	5.853
4	1.844	≈0	

12.2 Simultaneous Equations

The Taylor series, Equation 12.1, is applied to n nonlinear equations in n unknowns, $x_1, x_2, \ldots, x_n$:

$$\begin{aligned} y_1 &\cong f_1(x_1, x_2, \ldots, x_n) + \Delta x_1 \frac{\partial f_1}{\partial x_1} + \Delta x_2 \frac{\partial f_1}{\partial x_2} + \cdots + \Delta x_n \frac{\partial f_1}{\partial x_n} \\ &\cdots\cdots\cdots\cdots \\ y_n &\cong f_n(x_1, x_2, \ldots, x_n) + \Delta x_1 \frac{\partial f_n}{\partial x_1} + \Delta x_2 \frac{\partial f_n}{\partial x_2} + \cdots + \Delta x_n \frac{\partial f_n}{\partial x_n} \end{aligned} \tag{12.6}$$

As a first approximation, the unknowns represented by the initial values $x_1^0, x_2^0, x_3^0, \ldots$ can be substituted into the above equations, that is,

$$y_n = f_n\left(x_1^0, x_2^0, \ldots, x_n^0\right) + \Delta x_1^0 \left.\frac{\partial f_n}{\partial x_1}\right|_0 + \cdots + \Delta x_n^0 \left.\frac{\partial f_n}{\partial x_n}\right|_0 \tag{12.7}$$

where $x_1^0, x_2^0, x_3^0, \ldots, x_n^0$ are the first estimates of n unknowns. On transposing,

$$\begin{aligned} y_1 - f_1^0 &= \left.\frac{\partial f_1}{\partial x_1}\right|_0 \Delta x_1^0 + \left.\frac{\partial f_1}{\partial x_2}\right|_0 \Delta x_2^0 + \cdots + \left.\frac{\partial f_1}{\partial x_n}\right|_0 \Delta x_n^0 \\ &\cdots\cdots\cdots \\ y_1 - f_1^0 &= \left.\frac{\partial f_1}{\partial x_1}\right|_0 \Delta x_1^0 + \left.\frac{\partial f_n}{\partial x_2}\right|_0 \Delta x_2^0 + \cdots + \left.\frac{\partial f_n}{\partial x_{n(0)}}\right|_0 \Delta x_n^0 \end{aligned} \tag{12.8}$$

where $f_n\left(x_1^0, x_2^0, \ldots, x_n^0\right)$ is abbreviated as f_n^0.

The original nonlinear equations have been reduced to linear equations in

$$\Delta x_1^0, \Delta x_2^0, \ldots, \Delta x_n^0 \tag{12.9}$$

The subsequent approximations are

$$\begin{aligned} x_1^1 &= x_1^0 + \Delta x_1^0 \\ x_2^1 &= x_2^0 + \Delta x_2^0 \\ &\cdots\cdots \\ x_n^1 &= x_n^0 + \Delta x_n^0 \end{aligned} \tag{12.10}$$

or in matrix form

$$\begin{vmatrix} y_1 - f_1^0 \\ y_2 - f_2^0 \\ \cdots \\ y_n - f_n^0 \end{vmatrix} = \begin{vmatrix} \dfrac{\partial f_1}{\partial x_1} & \cdots & \dfrac{\partial f_1}{\partial x_n} \\ \dfrac{\partial f_2}{\partial x_1} & \cdots & \dfrac{\partial f_2}{\partial x_n} \\ \cdots & \cdots & \cdots \\ \dfrac{\partial f_n}{\partial x_1} & \cdots & \dfrac{\partial f_n}{\partial x_n} \end{vmatrix} \begin{vmatrix} \Delta x_1^0 \\ \Delta x_2^0 \\ \cdots \\ \Delta x_n^0 \end{vmatrix} \tag{12.11}$$

The matrix of partial derivatives is called a Jacobian matrix. This result is written as

$$x^{k+1} = x^k - \left[J(x^k)\right]^{-1} f(x^k) \tag{12.12}$$

This means that determination of unknowns requires inversion of the Jacobian.

Example 12.2

Solve the two simultaneous equations:

$$f_1(x_1, x_2) = x_1^2 + 2x_2 - 3 = 0$$
$$f_2(x_1, x_2) = x_1 x_2 - 3x_2^2 + 2 = 0$$

Let the initial values of x_1 and x_2 be 3 and 2, respectively.

$$\bar{J}(x) = \begin{vmatrix} \dfrac{\partial f_1}{\partial x_1} & \dfrac{\partial f_1}{\partial x_2} \\ \dfrac{\partial f_2}{\partial x_1} & \dfrac{\partial f_2}{\partial x_2} \end{vmatrix} = \begin{vmatrix} 2x_1 & 2 \\ x_2 & x_1 - 6x_2 \end{vmatrix}$$

Step 0:

$$x^1 = x^0 - \begin{vmatrix} 6 & 2 \\ 2 & -9 \end{vmatrix}^{-1} f(x^0)$$

$$= \begin{vmatrix} 3 \\ 2 \end{vmatrix} - \begin{vmatrix} \dfrac{9}{58} & \dfrac{2}{58} \\ \dfrac{2}{58} & -\dfrac{6}{58} \end{vmatrix} \begin{vmatrix} 10 \\ 4 \end{vmatrix} = \begin{vmatrix} 1.586 \\ 1.241 \end{vmatrix}$$

Step 1:

$$x^2 = x^1 - \begin{vmatrix} 3.172 & 2 \\ 1.241 & -5.86 \end{vmatrix}^{-1} f(x^1)$$

$$= \begin{vmatrix} 1.586 \\ 1.241 \end{vmatrix} - \begin{vmatrix} 0.278 & 0.095 \\ 0.095 & -0.151 \end{vmatrix} \begin{vmatrix} 1.997 \\ -0.652 \end{vmatrix} = \begin{vmatrix} 1.092 \\ 1.024 \end{vmatrix}$$

Step 2:

$$x^3 = x^2 - \begin{vmatrix} 2.184 & 2 \\ 1.025 & -5.058 \end{vmatrix}^{-1} f(x^2)$$

$$= \begin{vmatrix} 1.092 \\ 1.025 \end{vmatrix} - \begin{vmatrix} 0.386 & 0.153 \\ 0.078 & -0.167 \end{vmatrix} \begin{vmatrix} 0.242 \\ -0.033 \end{vmatrix} = \begin{vmatrix} 1.004 \\ 1.001 \end{vmatrix}$$

Thus, the results are rapidly converging to the final values of x_1 and x_2 which are 1 and 1, respectively.

12.3 Rectangular Form of Newton–Raphson Method of Load Flow

The power flow equation at a PQ node is

$$S_s^0 = P_s^0 + jQ_s^0 = V_s \sum_{r=1}^{r=n} (Y_{sr}V_r)^* \tag{12.13}$$

Voltages can be written as

$$\begin{aligned} V_s &= e_s + jh_s \\ V_r &= e_r - jh_r \end{aligned} \tag{12.14}$$

Thus, the power is

$$\begin{aligned} &(e_s + jh_s)\left[\sum_{r=1}^{r=n} (G_{sr} - jB_{sr})(e_r - jh_r)\right] \\ &= (e_s + jh_s)\left[\sum_{r=1}^{r=n} (G_{sr}e_r - B_{sr}h_r) - j\sum_{r=1}^{r=n} (G_{sr}h_r - B_{sr}e_r)\right] \end{aligned} \tag{12.15}$$

Equating the real and imaginary parts, the active and reactive power at a PQ node is

$$P_s = e_s\left[\sum_{r=1}^{r=n} (G_{sr}e_r - B_{sr}h_r)\right] + h_s\left[\sum_{r=1}^{r=n} (G_{sr}h_r + B_{sr}e_r)\right] \tag{12.16}$$

$$Q_s = e_s\left[\sum_{r=1}^{r=n} (-G_{sr}h_r - B_{sr}e_r)\right] + h_s\left[\sum_{r=1}^{r=n} (G_{sr}e_r - B_{sr}h_r)\right] \tag{12.17}$$

where P_s and Q_s are functions of e_s, e_n h_s, and h_r. Starting from the initial values, new values are found which differ from the initial values by ΔP_s and Δq_s:

$$\Delta P_s^0 = P_s - P_s^0 \quad \text{(first iteration)} \tag{12.18}$$

$$\Delta Q_s^0 = Q_s - Q_s^0 \quad \text{(first iteration)} \tag{12.19}$$

For a PV node (generator bus) voltage and power are specified. The reactive power equation is replaced by a voltage equation:

$$|V_g|^2 = e_g^2 + h_g^2 \tag{12.20}$$

$$\Delta|V_g^0|^2 = |V_g|^2 - |V_g^0|^2 \tag{12.21}$$

Consider the four-bus distribution system of Figure 11.8. It is required to write the equations for voltage corrections. Buses 2 and 3 are the load (PQ) buses, bus 1 is the slack bus, and bus 4 is the generator bus. In Equation 12.11 the column matrix elements on the left consisting of $(y_1 - f_1^0)\ldots(y_n - f_n^0)$ are identified as ΔP, ΔQ, etc. The unknowns of the matrix equation are Δe and Δh. The matrix equation for voltage corrections is written as

$$\begin{vmatrix} \Delta P_2 \\ \Delta Q_2 \\ \Delta P_3 \\ \Delta Q_3 \\ \Delta P_4 \\ \Delta V_4^2 \end{vmatrix} = \begin{vmatrix} \partial P_2/\partial e_2 & \partial P_2/\partial h_2 & \partial P_2/\partial e_3 & \partial P_2/\partial h_3 & \partial P_2/\partial e_4 & \partial P_2/\partial h_4 \\ \partial Q_2/\partial e_2 & \partial Q_2/\partial h_2 & \partial Q_2/\partial e_3 & \partial Q_2/\partial h_3 & \partial Q_2/\partial e_4 & \partial Q_2/\partial h_4 \\ \partial P_3/\partial e_2 & \partial P_3/\partial h_2 & \partial P_3/\partial e_3 & \partial P_3/\partial h_3 & \partial P_3/\partial e_4 & \partial P_3/\partial h_4 \\ \partial Q_3/\partial e_2 & \partial Q_3/\partial h_2 & \partial Q_3/\partial e_3 & \partial Q_3/\partial h_3 & \partial Q_3/\partial e_4 & \partial Q_3/\partial h_4 \\ \partial P_4/\partial e_2 & \partial P_4/\partial h_2 & \partial P_4/\partial e_3 & \partial P_4/\partial h_3 & \partial P_4/\partial e_4 & \partial P_4/\partial h_4 \\ \partial V_4^2/\partial e_2 & \partial V_4^2/\partial h_2 & \partial V_4^2/\partial e_3 & \partial V_4^2/\partial h_3 & \partial V_4^2/\partial e_4 & \partial V_4^2/\partial h_4 \end{vmatrix} \begin{vmatrix} \Delta e_2 \\ \Delta h_2 \\ \Delta e_3 \\ \Delta h_3 \\ \Delta e_4 \\ \Delta h_4 \end{vmatrix} \quad (12.22)$$

In the rectangular form there are two equations per load (PQ) and generator (PV) buses. The voltage at the swing bus is known and thus there is no equation for the swing bus.

Equation 12.22 can be written in the abbreviated form as

$$\bar{g} = \bar{J}\bar{x} \quad (12.23)$$

$$\begin{aligned} \Delta e_s &= e_s(\text{new}) - e_s(\text{old}) \\ \Delta P_s &= P_{so} - P_s(e_2, h_2, e_3, h_3, e_4, h_4) \\ \Delta Q_s &= Q_{so} - Q_s(e_2, h_2, e_3, h_3, e_4, h_4) \\ \Delta V_s^2 &= V_{so}^2 - (e_s^2 + h_s^2) \\ \Delta h_s &= h_s(\text{new}) - h_s(\text{old}) \end{aligned} \quad (12.24)$$

where Δe_s and Δh_s are voltage corrections.

Equation 12.12 can be written as

$$x^{k+1} - x^k = -\left[\bar{J}(x^k)\right]^{-1} h(x^k) \quad (12.25)$$

or

$$\begin{vmatrix} \Delta e_2 \\ \Delta h_2 \\ \cdot \\ \cdot \end{vmatrix} = \bar{J}^{-1} \begin{vmatrix} -P_2(e_2, h_2, \ldots) + P_{20} \\ -Q_2(e_2, h_2, \ldots) + Q_{20} \\ \cdot \\ \cdot \end{vmatrix} \quad (12.26)$$

The partial coefficients are calculated numerically by substituting assumed initial values into partial derivative equations:

Off-diagonal elements, $s \neq r$:

$$\frac{\partial P_s}{\partial e_r} = -\frac{\partial Q_s}{\partial h_r} = G_{sr}e_s + B_{sr}h_s \quad (12.27)$$

$$\frac{\partial P_s}{\partial h_r} = \frac{\partial Q_s}{\partial e_r} = -B_{sr}e_s + G_{sr}h_s \quad (12.28)$$

$$\frac{\partial V_s^2}{\partial e_r} = \frac{\partial V_s^2}{\partial h_r} = 0 \quad (12.29)$$

Diagonal elements:

$$\frac{\partial P_s}{\partial e_s} = \sum_{r=1}^{r=n} (G_{sr}e_r - B_{sr}h_r) + G_{ss}e_s + B_{ss}h_s \tag{12.30}$$

$$\frac{\partial P_s}{\partial h_s} = \sum_{r=1}^{r=n} (G_{sr}h_r - B_{sr}e_r) - B_{ss}e_s + G_{ss}h_s \tag{12.31}$$

$$\frac{\partial Q_s}{\partial e_s} = \sum_{r=1}^{r=n} -(G_{sr}h_r - B_{sr}e_r) - B_{ss}e_s + G_{ss}h_s \tag{12.32}$$

$$\frac{\partial Q_s}{\partial h_s} = \sum_{r=1}^{r=n} (G_{sr}e_r - B_{sr}h_r) - G_{ss}e_s - B_{ss}h_s \tag{12.33}$$

$$\frac{\partial V_s^2}{\partial e_s} = 2e_s \tag{12.34}$$

$$\frac{\partial V_s^2}{\partial h_s} = 2h_s \tag{12.35}$$

12.4 Polar Form of Jacobian Matrix

The voltage equation can be written in polar form

$$V_s = V_s(\cos\theta_s + j\sin\theta_s) \tag{12.36}$$

Thus, the power is

$$V_s\left(\sum_{r=1}^{r=n} Y_{sr}V_{sr}\right)^* = V_s(\cos\theta_s - j\sin\theta_s)\sum_{r=1}^{r=n}(G_{sr} - jB_{sr})V_r(\cos\theta_r - j\sin\theta_r) \tag{12.37}$$

Equating real and imaginary terms:

$$P_s = V_s \sum_{r=1}^{r=n} V_r[(G_{sr}\cos(\theta_s - \theta_r) + B_{sr}\sin(\theta_s - \theta_r)] \tag{12.38}$$

$$Q_s = V_s \sum_{r=1}^{r=n} V_r[(G_{sr}\sin(\theta_s - \theta_r) - B_{sr}\cos(\theta_s - \theta_r)] \tag{12.39}$$

The Jacobian in polar form for the same four-bus system is

$$\begin{vmatrix} \Delta P_2 \\ \Delta Q_2 \\ \Delta P_3 \\ \Delta Q_3 \\ \Delta P_4 \end{vmatrix} = \begin{vmatrix} \partial P_2/\partial\theta_2 & \partial P_2/\partial V_2 & \partial P_2/\partial\theta_3 & \partial P_2/\partial V_3 & \partial P_2/\partial\theta_4 \\ \partial Q_2/\partial\theta_2 & \partial Q_2/\partial V_2 & \partial Q_2/\partial\theta_3 & \partial Q_2/\partial V_3 & \partial Q_2/\partial\theta_4 \\ \partial P_3/\partial\theta_2 & \partial P_3/\partial V_2 & \partial P_3/\partial\theta_3 & \partial P_3/\partial V_3 & \partial P_3/\partial\theta_4 \\ \partial Q_3/\partial\theta_2 & \partial Q_3/\partial V_2 & \partial Q_3/\partial\theta_3 & \partial Q_3/\partial V_3 & \partial Q_3/\partial\theta_4 \\ \partial P_4/\partial\theta_2 & \partial P_4/\partial V_2 & \partial P_4/\partial\theta_3 & \partial P_4/\partial V_3 & \partial P_4/\partial\theta_4 \end{vmatrix} \begin{vmatrix} \Delta\theta_2 \\ \Delta V_2 \\ \Delta\theta_3 \\ \Delta V_3 \\ \Delta\theta_4 \end{vmatrix} \tag{12.40}$$

The slack bus has no equation, because the active and reactive power at this bus are unspecified and the voltage is specified. At PV bus 4, the reactive power is unspecified and there is no corresponding equation for this bus in terms of the variable ΔV_4.

$$\Delta P_s = P_s^0 - P_s(\theta_2 V_2 \theta_3 V_3 \theta_4) \tag{12.41}$$

$$\Delta Q_s = Q_s^0 - Q_s(\theta_2 V_2 \theta_3 V_3 \theta_4) \tag{12.42}$$

The partial derivatives can be calculated as follows:

Off-diagonal elements, s ≠ r:

$$\frac{\partial P_s}{\partial \theta_r} == G_{sr} V_s V_r \sin(\theta_s - \theta_r) - B_{sr} V_s V_r \cos(\theta_s - \theta_r) \tag{12.43}$$

$$\begin{aligned}\frac{\partial P_s}{\partial V_r} &= G_{sr} V_s \cos(\theta_s - \theta_r) + B_{sr} \sin(\theta_s - \theta_r) \\ &= -\left(\frac{1}{V_r}\right)\frac{(\partial Q_s)}{(\partial \theta_r)}\end{aligned} \tag{12.44}$$

$$\frac{\partial Q_s}{\partial \theta_r} = -G_{sr} V_s V_r \cos(\theta_s - \theta_r) - B_{sr} V_s V_r \sin(\theta_s - \theta_r) \tag{12.45}$$

$$\begin{aligned}\frac{\partial Q_s}{\partial V_r} &= G_{sr} V_s \sin(\theta_s - \theta_r) - B_{sr} V_s \cos(\theta_s - \theta_r) \\ &= \left(\frac{1}{V_r}\right)\left(\frac{\partial P_s}{\partial \theta_r}\right)\end{aligned} \tag{12.46}$$

Diagonal elements:

$$\begin{aligned}\frac{\partial P_s}{\partial \theta_s} &= V_s \sum_{r=1}^{r=n} V_r[-G_{sr} \sin(\theta_s - \theta_r) + B_{sr} \cos(\theta_s - \theta_r)] - V_s^2 B_{ss} \\ &= -Q_s - V_s^2 B_{ss}\end{aligned} \tag{12.47}$$

$$\begin{aligned}\frac{\partial P_s}{\partial V_s} &= \sum_{r=1}^{r=n} V_r[G_{sr} \cos(\theta_s - \theta_r) + B_{sr} \sin(\theta_s - \theta_r)] - V_s G_{ss} \\ &= \left(\frac{P_s}{V_s}\right) + V_s G_{ss}\end{aligned} \tag{12.48}$$

$$\begin{aligned}\frac{\partial Q_s}{\partial \theta_s} &= V_s \sum_{r=1}^{r=n} V_r[G_{sr} \cos(\theta_s - \theta_r) + B_{sr} \sin(\theta_s - \theta_r)] - V_s^2 G_{ss} \\ &= P_s - V_s^2 G_{ss}\end{aligned} \tag{12.49}$$

$$\begin{aligned}\frac{\partial Q_s}{\partial V_s} &= \sum_{r=1}^{r=n} V_r[G_{sr} \sin(\theta_s - \theta_r) - B_{sr} \cos(\theta_s - \theta_r)] - V_s B_{ss} \\ &= \left(\frac{Q_s}{V_s}\right) - V_s B_{ss}\end{aligned} \tag{12.50}$$

12.4.1 Calculation Procedure of Newton–Raphson Method

The procedure is summarized in the following steps, and flowcharts are shown in Figures 12.2 and 12.3.

- Bus admittance matrix is formed.
- Initial values of voltages and phase angles are assumed for the load (PQ) buses. Phase angles are assumed for PV buses. Normally, the bus voltages are set equal to the slack bus voltage, and phase angles are assumed equal to 0°, that is, a *flat* start.

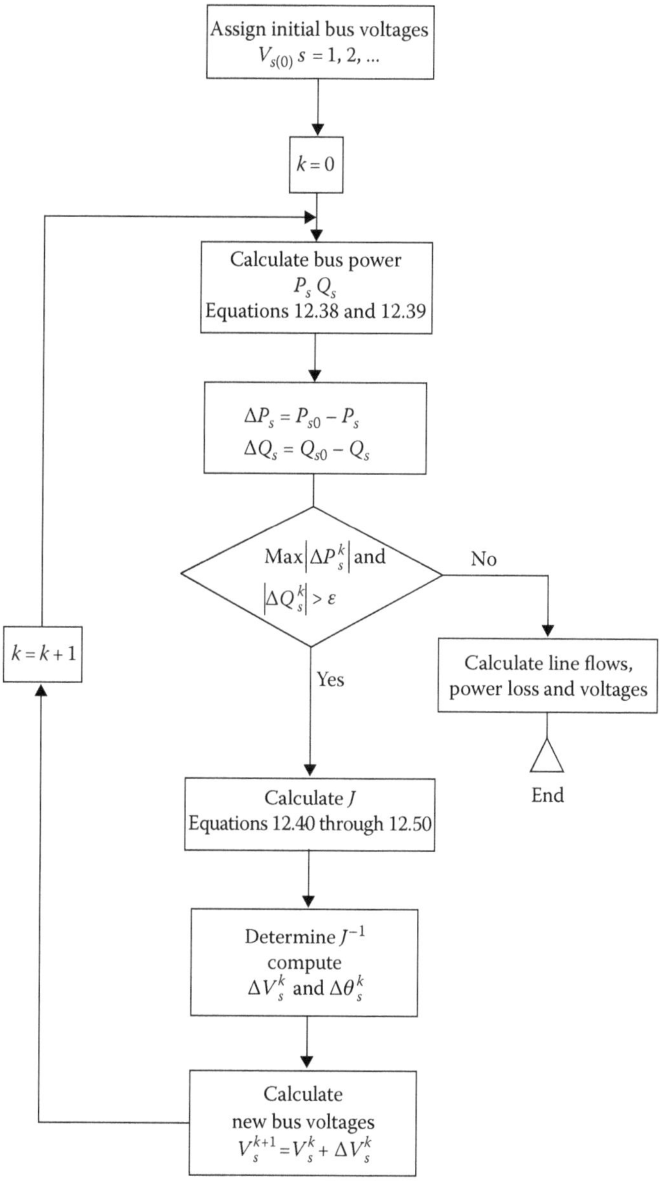

FIGURE 12.2
Flowchart for NR method of load flow for PQ buses.

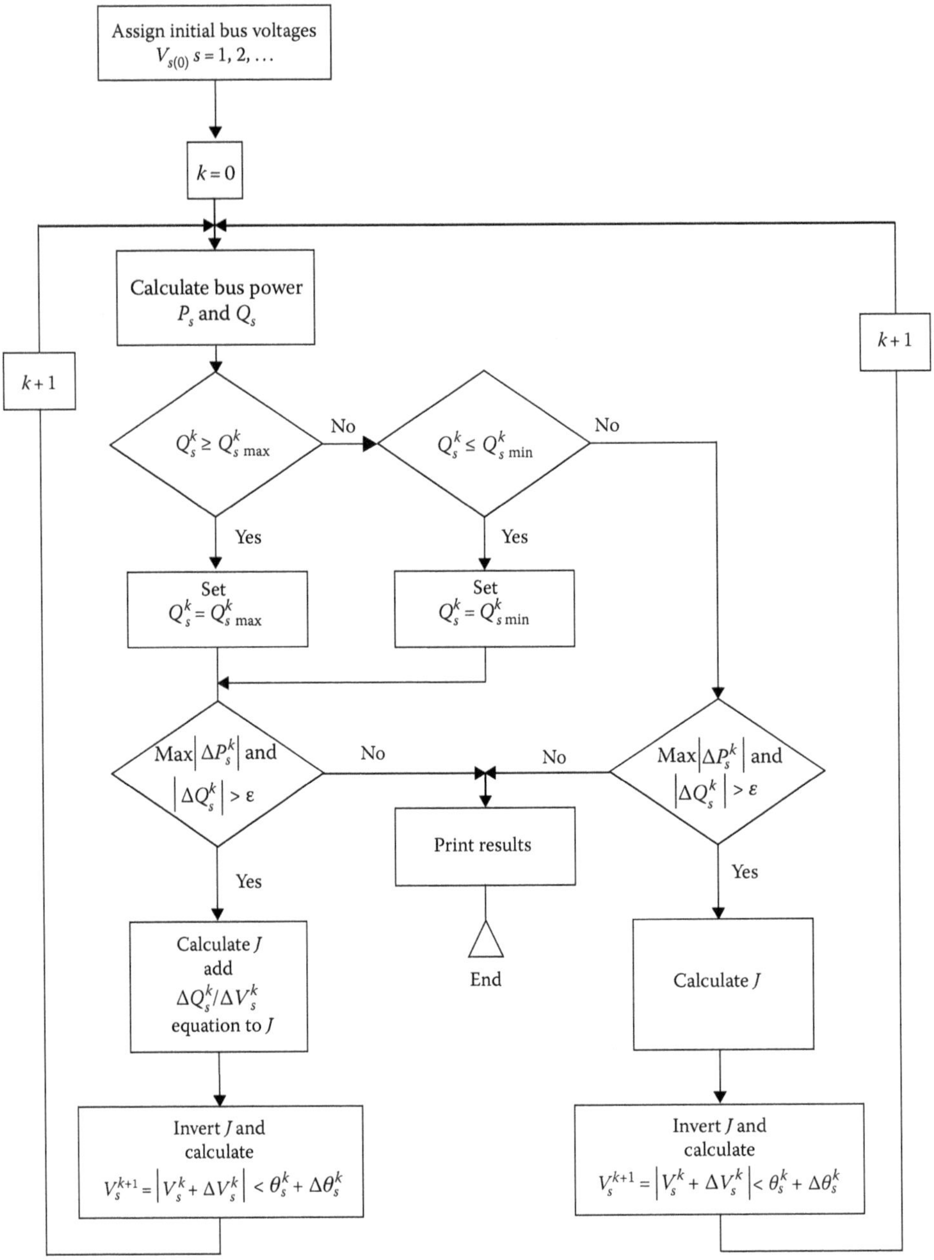

FIGURE 12.3
Flowchart for NR method of load flow for PV buses.

- Active and reactive powers, P and Q, are calculated for each load bus.
- ΔP and ΔQ can, therefore, be calculated on the basis of the given power at the buses.
- For PV buses, the exact reactive power is not specified, but its limits are known. If the calculated value of the reactive power is within limits, only ΔP is calculated. If the calculated value of reactive power is beyond the specified limits, then an appropriate limit is imposed and ΔQ is also calculated by subtracting the calculated value of the reactive power from the maximum specified limit. The bus under consideration is now treated as a PQ (load) bus.
- The elements of the Jacobian matrix are calculated.
- This gives $\Delta\theta$ and $\Delta|V|$.
- Using the new values of $\Delta\theta$ and $\Delta|V|$, the new values of voltages and phase angles are calculated.
- The next iteration is started with these new values of voltage magnitudes and phase angles.
- The procedure is continued until the required tolerance is achieved. This is generally 0.1 kW and 0.1 kvar.

Example 12.3

Consider a transmission system of two 138 kV lines, three buses, each line modeled by an equivalent Π network, as shown in Figure 12.4a, with series and shunt admittances as shown. Bus 1 is the swing bus (voltage 1.02 per unit), bus 2 is a PQ bus with load demand of 0.25 + *j*0.25 per unit, and bus 3 is a voltage-controlled bus with bus voltage of 1.02 and a load of 0.5 *j*0 per unit all on 100 MVA base. Solve the load flow using the NR method, polar axis basis.

First, form a *Y* matrix as follows:

$$\bar{Y} = \begin{vmatrix} 0.474 - j2.428 & -0.474 + j2.45 & 0 \\ -0.474 + j2.45 & 1.142 - j4.70 & -0.668 + j2.297 \\ 0 & -0.668 + j2.297 & 0.668 - j2.272 \end{vmatrix}$$

The active and reactive power at bus I (swing bus) can be written from Equations 12.38 and 12.39 as

$$\begin{aligned} P_1 &= 1.02 \times 1.02[0.474\cos(0.0 - 0.0) + (-2.428)\sin(0.0 - 0.0)] \\ &\quad + 1.02V_2[(-0.474)\cos(0.0 - \theta_2) + 2.45\sin(0.0 - \theta_2) \\ &\quad + 1.02 \times 1.02[0.0\cos(0.0 - \theta_3) + 0.0\sin(0.0 - \theta_3)] \end{aligned}$$

$$\begin{aligned} Q_1 &= 1.02 \times 1.02[0.474\sin(0.0 - 0.0) - (-2.428)\cos(0.0 - 0.0)] \\ &\quad + 1.02V_2[(-0.474)\sin(0.0 - \theta_2) - 2.45\cos(0.0 - \theta_2) \\ &\quad + 1.02 \times 1.02[0.0\sin(0.0 - \theta_3) - 0.0\cos(0.0 - \theta_3)] \end{aligned}$$

These equations for the swing bus are not immediately required for load flow, but can be used to calculate the power flow from this bus, once the system voltages are calculated to the required tolerance.

Similarly, the active and reactive powers at other buses are written

$$P_2 = V_2 \times 1.02[(-0.474)\cos(\theta_2 - 0) + 2.45\sin(\theta_2 - 0)] + V_2$$
$$\times V_2[1.142\cos(\theta_2 - \theta_2) + (-4.70)\sin(\theta_2 - \theta_2)] + V_2$$
$$\times 1.02[(-0.668)\cos(\theta_2 - \theta_3) + 2.297\sin(\theta_2 - \theta_3)]$$

Substituting the initial values ($V_2 = 1$, $\theta_2 = 0°$), $P_2 = -0.0228$.

$$Q_2 = V_2 \times 1.02[(-0.474)\sin(\theta_2 - 0.0) - 2.45\cos(\theta_2 - 0.0)] + V_2$$
$$\times V_2[1.142\sin(\theta_2 - \theta_2) - (-4.70)\cos(\theta_2 - \theta_2)] + V_2$$
$$\times 1.02[(-0.668)\sin(\theta_2 - \theta_3) + 2.297\cos(\theta_2 - \theta_3)]$$

Substituting the numerical values, $Q_2 = -0.142$.

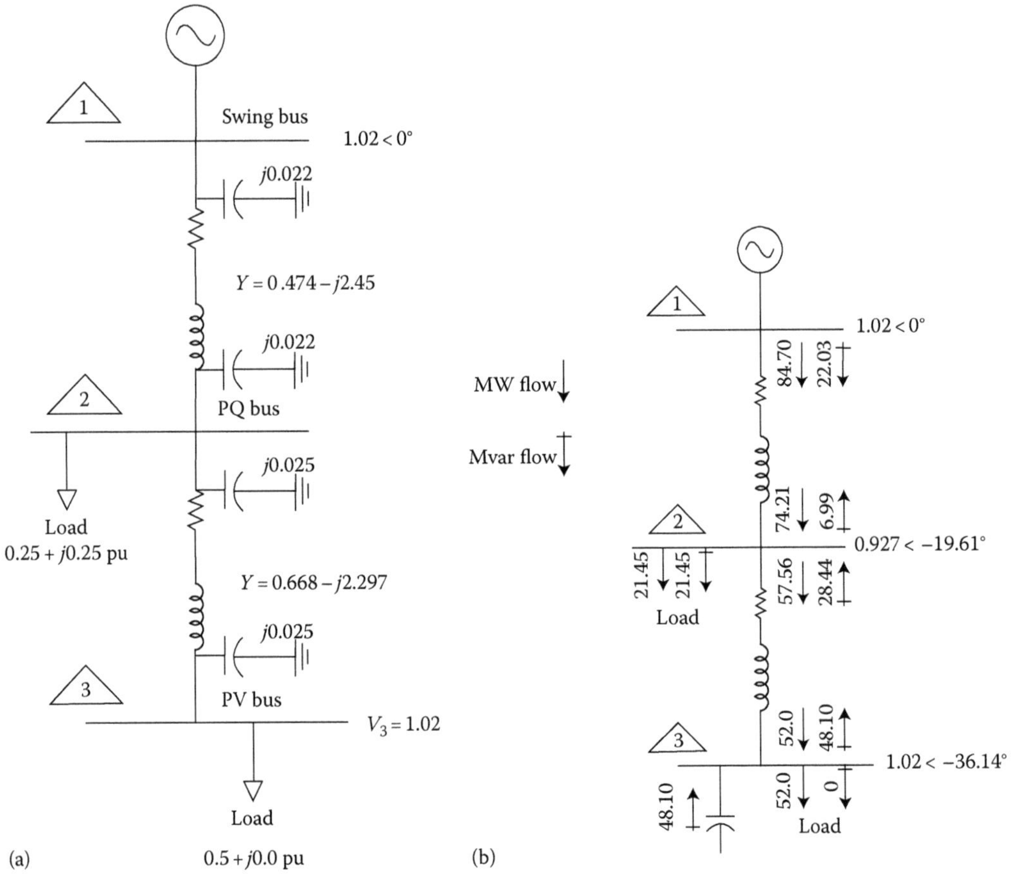

FIGURE 12.4
(a) System of Example 12.3 for load flow solution; (b) final converged load flow solution with reactive power injection at PV bus 3;

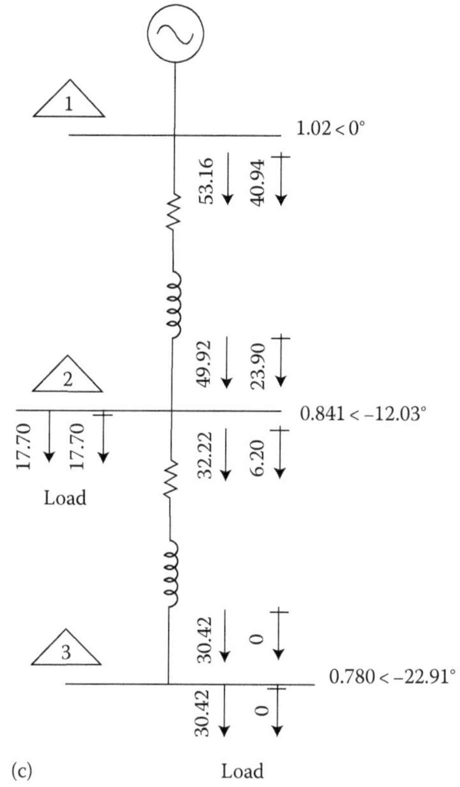

FIGURE 12.4 (continued)
(c) converged load flow with bus 3 treated as a PQ bus.

$$\begin{aligned}P_3 &= 1.02 \times 1.02[0.0\cos(\theta_3 - 0.0) + 0.0\sin(\theta_3 - 0.0)] + 1.02\\ &\quad \times V_2[(-0.668)\cos(\theta_3 - \theta_2) + 2.297\sin(\theta_3 - \theta_2)] + 1.02\\ &\quad \times 1.02[0.668\cos(\theta_3 - \theta_3) + (-2.047)\sin(\theta_3 - \theta_3)]\end{aligned}$$

Substituting the values, $P_3 = 0.0136$.

$$\begin{aligned}Q_3 &= 1.02 \times 1.02[0.0\sin(\theta_3 - 0.0) - 0.0\cos(\theta_3 - 0.0)] + 1.02\\ &\quad \times V_2[(-0.668)\sin(\theta_3 - \theta_2) - 2.297\cos(\theta_3 - \theta_2) + 1.02\\ &\quad \times 1.02[0.668\sin(\theta_3 - \theta_3) - (-2.272)\cos(\theta_3 - \theta_3)]\end{aligned}$$

Substituting initial values, $Q_3 = -0.213$.

The Jacobian matrix is

$$\begin{vmatrix}\Delta P_2\\ \Delta Q_2\\ \Delta P_3\end{vmatrix} = \begin{vmatrix}\partial P_2/\partial\theta_2 & \partial P_2/\partial V_2 & \partial P_2/\partial\theta_3\\ \partial Q_2/\partial\theta_2 & \partial Q_2/\partial V_2 & \partial Q_2/\partial\theta_3\\ \partial P_3/\partial\theta_2 & \partial P_3/\partial V_2 & \partial P_3/\partial\theta_3\end{vmatrix}\begin{vmatrix}\Delta\theta_2\\ \Delta V_2\\ \Delta\theta_3\end{vmatrix}$$

The partial differentials are found by differentiating the equations for P_2, Q_2, P_3, etc.

$$\frac{\partial P_2}{\partial \theta_2} = 1.02[V_2(0.474)\sin\theta_2 + 2.45\cos\theta_2] + 1.02[V_2(0.668)\sin(\theta_2-\theta_3) + V_2(2.297)\cos(\theta_2-\theta_3)]$$
$$= 4.842$$

$$\frac{\partial P_2}{\partial \theta_3} = V_2(1.02)[(-0.668)\sin(\theta_2-\theta_3) - 2.297\cos(\theta_2-\theta_3)]$$
$$= -2.343$$

$$\frac{\partial P_2}{\partial V_2} = 1.02[(-0.474)\cos\theta_2 + 2.45\sin\theta_2]$$
$$+ 2V_2(1.142) + 1.02[(-0.608)\cos(\theta_2-\theta_3) + 2.297\sin(\theta_2-\theta_3)]$$
$$= 1.119$$

$$\frac{\partial Q_2}{\partial \theta_2} = 1.02[-V_2(0.474)\cos\theta_2 + 2.45\sin\theta_2] + 1.02[V_2(-0.668)\cos(\theta_2-\theta_3) + 2.297\sin(\theta_2-\theta_3)]$$
$$= -1.1648$$

$$\frac{\partial Q_2}{\partial V_2} = 1.02[(-0.474)\sin\theta_2 - 2.45\cos\theta_2] + 2V_2(4.28)$$
$$+ 1.02[(-0.668)\sin(\theta_2-\theta_3) - 2.297\cos(\theta_2-\theta_3)]$$
$$= 4.56$$

$$\frac{\partial Q_2}{\partial \theta_3} = 1.02V_2[0.668\cos(\theta_2-\theta_3) - 2.297\sin(\theta_2-\theta_3)]$$
$$= 0.681$$

$$\frac{\partial P_3}{\partial \theta_2} = 1.02V_2[(0.668)\sin(\theta_3-\theta_2) - 2.297\cos(\theta_3-\theta_2)]$$
$$= 2.343$$

$$\frac{\partial P_3}{\partial V_2} = 1.02[(-0.668)\cos(\theta_3-\theta_2) + 2.297\sin(\theta_3-\theta_2)]$$
$$= 0.681$$

$$\frac{\partial P_3}{\partial \theta_3} = 1.02[0.668\sin(\theta_3-\theta_2) + 2.297\cos(\theta_3-\theta_2)]$$
$$= 2.343$$

Therefore, the Jacobian is

$$\bar{J} = \begin{vmatrix} 4.842 & 1.119 & -2.343 \\ -1.165 & 4.56 & 0.681 \\ -2.343 & 0.681 & 2.343 \end{vmatrix}$$

The system equations are

$$\begin{vmatrix} \Delta\theta_2^1 \\ \Delta V_2^1 \\ \Delta\theta_3^1 \end{vmatrix} = \begin{vmatrix} 4.842 & 1.119 & -2.343 \\ -1.165 & 4.56 & 0.681 \\ -2.343 & 0.681 & 2.343 \end{vmatrix}^{-1} \begin{vmatrix} -0.25-(-0.0228) \\ 0.25-(-0.142) \\ -0.5-0.0136 \end{vmatrix}$$

Inverting the Jacobian gives

$$\begin{vmatrix} \Delta\theta_2^1 \\ \Delta V_2^1 \\ \Delta\theta_3^1 \end{vmatrix} = \begin{vmatrix} 0.371 & -0.153 & 0.4152 \\ 0.041 & 0.212 & -0.021 \\ 0.359 & -0.215 & 0.848 \end{vmatrix} \begin{vmatrix} -0.2272 \\ 0.392 \\ -0.5136 \end{vmatrix} = \begin{vmatrix} -0.357 \\ -0.084 \\ -0.601 \end{vmatrix}$$

The new values of voltages and phase angles are

$$\begin{vmatrix} \theta_2^1 \\ V_2^1 \\ \theta_3^1 \end{vmatrix} = \begin{vmatrix} 0 \\ 1 \\ 0 \end{vmatrix} + \begin{vmatrix} -0.357 \\ 0.084 \\ -0.601 \end{vmatrix} = \begin{vmatrix} -0.357 \\ 1.084 \\ -0.729 \end{vmatrix}$$

This completes one iteration. Using the new bus voltages and phase angles the power flow is recalculated. Thus, at every iteration, the Jacobian matrix changes and has to be inverted.

In the first iteration, we see that the bus 2 voltage is 8.4% higher than the rated voltage; the angles are in radians. The first iteration is no indication of the final results. The hand calculations, even for a simple three-bus system, become unwieldy. The converged load flow is shown in Figure 12.4b. A reactive power injection of 48.10 Mvar is required at bus 3 to maintain a voltage of 1.02 per unit and supply the required active power of 0.5 per unit from the source. There is a reactive power loss of 48.68 Mvar in the transmission lines themselves, and the active power loss is 11.25 MW. The bus phase angles are high with respect to the swing bus, and the bus 2 operating voltage is 0.927 per unit, that is, a voltage drop of 7.3% under load. Thus, even with voltages at the swing bus and bus 3 maintained above rated voltage, the power demand at bus 2 cannot be met and the voltage at this bus dips. The load demand on the system is too high, it is lossy, and requires augmentation or reduction of load. A reactive power injection at bus 2 will give an entirely different result.

If bus 3 is treated as a load bus, the Jacobian is modified by adding a fourth equation of the reactive power at bus 3. In this case the bus 3 voltage dips down to 0.78 per unit, that is, a voltage drop of 22%; the converged load flow is shown in Figure 12.4c. At this lower voltage of 0.78 per unit, bus 3 can support an active load of only 0.3 per unit. This is not an example of a practical system, but it illustrates the importance of reactive power injection, load modeling, and its effect on the bus voltages.

The load demand reduces proportionally with reduction in bus voltages. This is because we have considered a constant impedance type of load, that is, the load current varies directly with the voltage as the load impedance is held constant. The load types are discussed further.

12.5 Simplifications of Newton–Raphson Method

The NR method has quadratic convergence characteristics; therefore, the convergence is fast and solution to high accuracy is obtained in the first few iterations. The number of iterations does not increase appreciably with the size of the system. This is in contrast to the Gauss–Seidel method of load flow that has slower convergence even with appropriately applied acceleration factors. The larger the system, the larger are the number of iterations; 50–150 iterations are common.

The NR method, however, requires more memory storage and necessitates solving a large number of equations in each iteration step. The Jacobian changes at each iteration and must be evaluated afresh. The time required for one iteration in the NR method may be 5–10 times that of the Gauss–Seidel method. Some simplifications that can be applied are as follows.

From Equation 12.22, the first equation is

$$\Delta P_2 = \left(\frac{\partial P_2}{\partial e_2}\right)\Delta e_2 + \left(\frac{\partial P_2}{\partial h_2}\right)\Delta h_2 + \left(\frac{\partial P_2}{\partial e_3}\right)\Delta e_3 + \left(\frac{\partial P_2}{\partial h_3}\right)\Delta h_3 + \left(\frac{\partial P_2}{\partial e_4}\right)\Delta e_4 + \left(\frac{\partial P_2}{\partial h_4}\right)\Delta h_4 \tag{12.51}$$

The second term of this equation $(\partial P_2/\partial h_2)\Delta h_2$ denotes the change in bus 2 active power for a change of h_2 to $h_2+\Delta h_2$. Similarly, the term $(\partial P_2/\partial e_2)\Delta e_2$ indicates change in bus 2 active power for a change of e_2 to $e_2+\Delta e_2$.

The change in power at bus 2 is a function of the voltage at bus 2, which is a function of the voltages at other buses. Considering the effect of e_2 only

$$\Delta P_2 \doteq \left(\frac{\partial P_2}{\partial e_2}\right)\Delta e_2 \tag{12.52}$$

Thus, the Jacobian reduces to only diagonal elements:

$$\begin{vmatrix} \Delta P_2 \\ \Delta Q_2 \\ \Delta P_3 \\ \Delta Q_3 \\ \Delta P_4 \\ \Delta V_4 \end{vmatrix} = \begin{vmatrix} \partial P_2/\partial e_2 & \cdot & \cdot & \cdot & \cdot & \cdot \\ \cdot & \partial Q_2/\partial h_2 & \cdot & \cdot & \cdot & \cdot \\ \cdot & \cdot & \partial P_3/\partial e_3 & \cdot & \cdot & \cdot \\ \cdot & \cdot & \cdot & \partial Q_3/\partial h_3 & \cdot & \cdot \\ \cdot & \cdot & \cdot & \cdot & \partial P_4/\partial e_4 & \cdot \\ \cdot & \cdot & \cdot & \cdot & \cdot & \partial v_4^2/\partial h_4 \end{vmatrix} \begin{vmatrix} \Delta e_2 \\ \Delta h_2 \\ \Delta e_3 \\ \Delta h_3 \\ \Delta e_4 \\ \Delta h_4 \end{vmatrix} \tag{12.53}$$

Method 2 reduces the Jacobian to a lower triangulation matrix:

$$\begin{aligned} \Delta P_2 &\doteq \left(\frac{\partial P_2}{\partial e_2}\right)\Delta e_2 \\ \Delta Q_2 &\doteq \left(\frac{\partial Q_2}{\partial e_2}\right)\Delta e_2 + \left(\frac{\partial Q_2}{\partial h_2}\right)\Delta h_2 \end{aligned} \tag{12.54}$$

Thus, the Jacobian matrix is

$$\begin{vmatrix} \Delta P_2 \\ \Delta Q_2 \\ \Delta P_3 \\ \Delta Q_3 \\ \Delta P_4 \\ \Delta V_4^2 \end{vmatrix} = \begin{vmatrix} \partial P_2/\partial e_2 & \cdot & \cdot & \cdot & \cdot & \cdot \\ \partial Q_2/\partial e_2 & \partial Q_2/\partial h_2 & \cdot & \cdot & \cdot & \cdot \\ \partial P_3/\partial e_2 & \partial P_3/\partial h_2 & \partial P_3/\partial 3e_3 & \cdot & \cdot & \cdot \\ \partial Q_3/\partial e_2 & \partial Q_3/\partial h_2 & \partial Q_3/\partial h_3 & \partial Q_3/\partial h_3 & \cdot & \cdot \\ \partial P_4/\partial e_2 & \partial P_4/\partial h_2 & \partial P_4/\partial e_3 & \partial P_4/\partial h_3 & \partial P_4/\partial e_4 & \cdot \\ \partial V_4^2/\partial e_2 & \partial V_4^2/\partial h_2 & \partial V_4^2/\partial e_3 & \partial V_4^2/\partial h_3 & \partial V_4^2/\partial e_4 & \partial V_4^2/\partial h_4 \end{vmatrix} \times \begin{vmatrix} \Delta e_2 \\ \Delta h_2 \\ \Delta e_3 \\ \Delta h_3 \\ \Delta e_4 \\ \Delta h_4 \end{vmatrix} \tag{12.55}$$

Method 3 relates P_2 and Q_2 to e_2 and h_2, P_3 and Q_3 to e_3 and h_3, etc. This is the Ward–Hale method. The Jacobian is

$$\begin{vmatrix} \Delta P_2 \\ \Delta Q_2 \\ \Delta P_3 \\ \Delta Q_3 \\ \Delta P_4 \\ \Delta V_4^2 \end{vmatrix} = \begin{vmatrix} \partial P_2/\partial e_2 & \partial P_2/\partial e_2 & \cdot & \cdot & \cdot & \cdot \\ \partial Q_2/\partial e_2 & \partial Q_2/\partial h_2 & \cdot & \cdot & \cdot & \cdot \\ \cdot & \cdot & \partial P_3/\partial e_3 & \partial P_3/\partial h_3 & \cdot & \cdot \\ \cdot & \cdot & \partial Q_3/\partial e_3 & \partial Q_3/\partial h_3 & \cdot & \cdot \\ \cdot & \cdot & \cdot & \cdot & \partial P_4/\partial e_4 & \partial P_4/\partial h_4 \\ \cdot & \cdot & \cdot & \cdot & \partial V_4^2/\partial e_4 & \partial v_4^2/\partial h_4 \end{vmatrix} \times \begin{vmatrix} \Delta e_2 \\ \Delta h_2 \\ \Delta e_3 \\ \Delta h_3 \\ \Delta e_4 \\ \Delta h_4 \end{vmatrix} \quad (12.56)$$

Method 4: A combination of methods 2 and 3:

$$\begin{vmatrix} \Delta P_2 \\ \Delta Q_2 \\ \Delta P_3 \\ \Delta Q_3 \\ \Delta P_4 \\ \Delta V_4^2 \end{vmatrix} = \begin{vmatrix} \partial P_2/\partial e_2 & \partial P_2/\partial h_2 & \cdot & \cdot & \cdot & \cdot \\ \partial Q_2/\partial e_2 & \partial Q_2/\partial h_2 & \cdot & \cdot & \cdot & \cdot \\ \partial P_3/\partial e_2 & \partial P_3/\partial h_2 & \partial P_3/\partial e_3 & \partial P_3/\partial h_3 & \cdot & \cdot \\ \partial Q_3/\partial e_2 & \partial Q_3/\partial h_2 & \partial Q_3/\partial eh_3 & \partial Q_3/\partial h_3 & \cdot & \cdot \\ \partial P_4/\partial e_2 & \partial P_4/\partial h_2 & \partial P_4/\partial e_3 & \partial P_4/\partial h_3 & \partial P_4/\partial e_4 & \partial P_4/\partial h_4 \\ \partial V_4^2/\partial e_2 & \partial V_4^2/\partial h_2 & \partial V_4^2/\partial e_3 & \partial V_4^2/\partial h_3 & \partial V_4^2/\partial e_4 & \partial V_4^2/\partial h_4 \end{vmatrix} \times \begin{vmatrix} \Delta e_2 \\ \Delta h_2 \\ \Delta e_3 \\ \Delta h_3 \\ \Delta e_4 \\ \Delta h_4 \end{vmatrix} \quad (12.57)$$

Method 5 may give the least iterations for a value of $\beta < 1$, a factor somewhat akin to the acceleration factor in the Gauss–Seidel method (>1). The Jacobian is of the form LDU (Appendix A):

$$\begin{vmatrix} \Delta P_2 \\ \Delta Q_2 \\ \Delta P_3 \\ \Delta Q_3 \\ \Delta P_4 \\ \Delta V_4 \end{vmatrix} = (L + D) \begin{vmatrix} \Delta e_2^k \\ \Delta h_2^k \\ \Delta e_3^k \\ \Delta h_3^k \\ \Delta e_4^k \\ \Delta h_4^k \end{vmatrix} + \beta U \begin{vmatrix} \Delta e_2^{k-1} \\ \Delta h_2^{k-1} \\ \Delta e_3^{k-1} \\ \Delta h_3^{k-1} \\ \Delta e_4^{k-1} \\ \Delta h_4^{k-1} \end{vmatrix} \quad (12.58)$$

12.6 Decoupled Newton–Raphson Method

It has already been demonstrated that there is strong interdependence between active power and bus voltage angle and between reactive power and voltage magnitude. The active power change ΔP is less sensitive to changes in voltage magnitude, and changes in reactive power ΔQ are less sensitive to changes in angles. In other words, the coupling between P and bus voltage magnitude is weak and between reactive power and phase angle is weak.

The Jacobian in Equation 12.22 can be rearranged as follows:

$$\begin{vmatrix} \Delta P_2 \\ \Delta P_3 \\ \Delta P_4 \\ \Delta Q_2 \\ \Delta Q_3 \end{vmatrix} = \begin{vmatrix} \partial P_2/\partial\theta_2 & \partial P_2/\partial\theta_3 & \partial P_2/\partial\theta_4 & \partial P_2/\partial V_2 & \partial P_2/\partial V_3 \\ \partial P_3/\partial\theta_2 & \partial P_3/\partial\theta_3 & \partial P_3/\partial\theta_4 & \partial P_3/\partial V_2 & \partial P_3/\partial V_3 \\ \partial P_4/\partial\theta_2 & \partial P_4/\partial\theta_3 & \partial P_4/\partial\theta_4 & \partial P_4/\partial V_2 & \partial P_4/\partial V_3 \\ \partial Q_2/\partial\theta_2 & \partial Q_2/\partial\theta_3 & \partial Q_2/\partial\theta_4 & \partial Q_2/\partial V_2 & \partial Q_2/\partial V_3 \\ \partial Q_3/\partial\theta_2 & \partial Q_3/\partial\theta_3 & \partial Q_3/\partial\theta_4 & \partial Q_3/\partial V_2 & \partial Q_3/\partial V_3 \end{vmatrix} \begin{vmatrix} \Delta\theta_2 \\ \Delta\theta_3 \\ \Delta\theta_4 \\ \Delta V_2 \\ \Delta V_3 \end{vmatrix} \tag{12.59}$$

Considering that

$$G_{sr} \lll B_{sr} \tag{12.60}$$

$$\sin(\theta_s - \theta_r) \lll 1 \tag{12.61}$$

$$\cos(\theta_s - \theta_r) \simeq 1 \tag{12.62}$$

The following inequalities are valid:

$$\left|\frac{\partial P_s}{\partial\theta_r}\right| \ggg \left|\frac{\partial P_s}{\partial V_r}\right| \tag{12.63}$$

$$\left|\frac{\partial Q_s}{\partial\theta_r}\right| \lll \left|\frac{\partial Q_s}{\partial V_r}\right| \tag{12.64}$$

Thus, the Jacobian is

$$\begin{vmatrix} \Delta P_2 \\ \Delta P_3 \\ \Delta P_4 \\ \Delta Q_2 \\ \Delta Q_3 \end{vmatrix} = \begin{vmatrix} \partial P_2/\partial\theta_2 & \partial P_2/\partial\theta_3 & \partial P_2/\partial\theta_4 & \cdot & \cdot \\ \partial P_3/\partial\theta_2 & \partial P_3/\partial\theta_3 & \partial P_3/\partial\theta_4 & \cdot & \cdot \\ \partial P_4/\partial\theta_2 & \partial P_4/\partial\theta_3 & \partial P_4/\partial\theta_4 & & \\ \cdot & \cdot & \cdot & \partial Q_2/\partial V_2 & \partial Q_2/\partial V_3 \\ \cdot & \cdot & \cdot & \partial Q_3/\partial V_2 & \partial Q_3/\partial V_3 \end{vmatrix} \begin{vmatrix} \Delta\theta_2 \\ \Delta\theta_3 \\ \Delta\theta_4 \\ \Delta V_2 \\ \Delta V_3 \end{vmatrix} \tag{12.65}$$

This is called P–Q decoupling.

12.7 Fast Decoupled Load Flow

Two synthetic networks, P–θ and P–V, are constructed. This implies that the load flow problem can be solved separately by these two networks, taking advantage of P–Q decoupling [3].

In a P–θ network, each branch of the given network is represented by conductance, the inverse of series reactance. All shunt admittances and transformer off-nominal voltage taps which affect the reactive power flow are omitted, and the swing bus is grounded. The bus conductance matrix of this network is termed $\overline{B}^{\theta}$.

The second model is called a Q–V network. It is again a resistive network. It has the same structure as the original power system model, but voltage-specified buses (swing bus and PV buses) are grounded. The branch conductance is given by

$$Y_{sr} = -B_{sr} = \frac{x_{sr}}{x_{sr}^2 + r_{sr}^2} \tag{12.66}$$

These are equal and opposite to the series or shunt susceptance of the original network. The effect of phase-shifter angles is neglected. The bus conductance matrix of this network is called $\overline{B}^{v}$.

The equations for power flow can be written as

$$\frac{P_s}{V_s} = \sum_{r=1}^{r=n} V_r[G_{sr}\cos(\theta_s - \theta_r) + B_{sr}\sin(\theta_s - \theta_r)] \tag{12.67}$$

$$\frac{Q_s}{V_s} = \sum_{r=1}^{r=n} V_r[G_{sr}\sin(\theta_s - \theta_r) - B_{sr}\cos(\theta_s - \theta_r)] \tag{12.68}$$

and partial derivatives can be taken as before. Thus, a single matrix for load flow can be split into two matrices as follows:

$$\begin{vmatrix} \Delta P_2/V_2 \\ \Delta P_3/V_3 \\ \cdot \\ \Delta P_n/V_n \end{vmatrix} = \begin{vmatrix} B_{22}^{\theta} & B_{23}^{\theta} & \cdot & B_{2n}^{\theta} \\ B_{32}^{\theta} & B_{33}^{\theta} & \cdot & B_{3n}^{\theta} \\ \cdot & \cdot & \cdot & \cdot \\ B_{n2}^{\theta} & B_{n3}^{\theta} & \cdot & B_{nn}^{\theta} \end{vmatrix} \begin{vmatrix} \Delta\theta_2 \\ \Delta\theta_3 \\ \cdot \\ \Delta\theta_n \end{vmatrix} \tag{12.69}$$

The correction of phase angle of voltage is calculated from this matrix:

$$\begin{vmatrix} \Delta Q_2/V_2 \\ \Delta Q_3/V_3 \\ \cdot \\ \Delta Q_n/V_n \end{vmatrix} = \begin{vmatrix} B_{22}^{v} & B_{23}^{v} & \cdot & B_{2n}^{v} \\ B_{32}^{v} & B_{33}^{v} & \cdot & B_{3n}^{v} \\ \cdot & \cdot & \cdot & \cdot \\ B_{n2}^{v} & B_{n3}^{v} & \cdot & B_{nn}^{v} \end{vmatrix} \begin{vmatrix} \Delta V_2 \\ \Delta V_3 \\ \cdot \\ \Delta V_n \end{vmatrix} \tag{12.70}$$

The voltage correction is also calculated from this matrix.

These matrices are real, sparse, and contain only admittances; these are constants and do not change during successive iterations. This model works well for $R/X \ll 1$. If this is not true, this approach can be ineffective. If phase shifters are not present, then both the matrices are symmetrical. Equations 12.69 and 12.70 are solved alternately with the most recent voltage values. This means that one iteration implies one solution to obtain $|\Delta\theta|$ to update θ and then another solution for $|\Delta V|$ to update V.

Example 12.4

Consider the network of Figure 12.5a. Let bus 1 be a swing bus, buses 2 and 3 PQ buses, and bus 4 a PV bus. The loads at buses 2 and 3 are specified as is the voltage magnitude at bus 4. Construct P–θ and Q–V matrices.

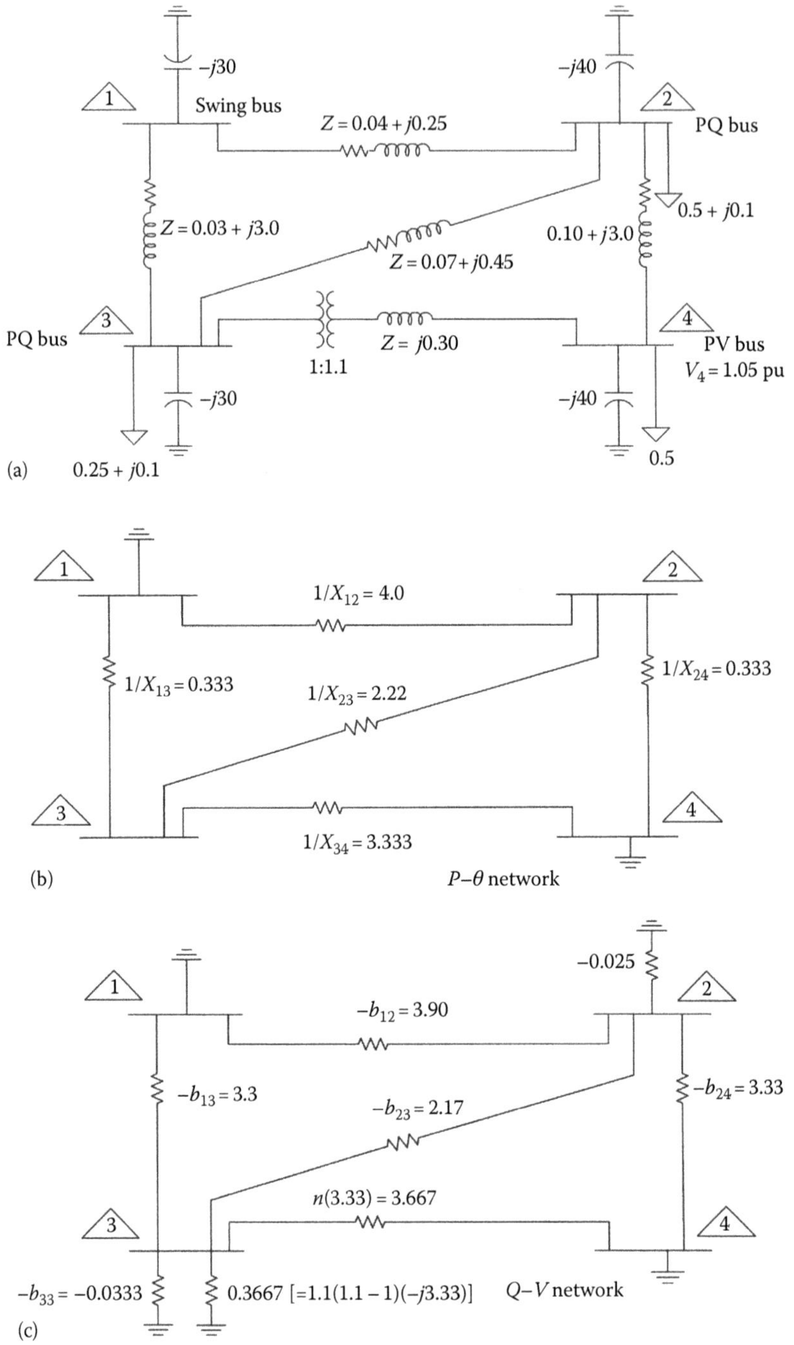

FIGURE 12.5
(a) Four-bus system with voltage tap adjustment transformer; (b) decoupled P–θ network; (c) Q–V network.

***P–θ* Network**

First construct the *P–θ* network shown in Figure 12.5b. Here the elements are calculated as follows:

$$B_{sr}^{\theta} = -\frac{1}{X_{sr}} = -B_{sr}\left(1 + \left(\frac{G_{sr}}{B_{sr}}\right)^2\right)$$

$$B_{ss}^{\theta} = \sum_{r}\left(\frac{1}{X_{sr}}\right) = \sum_{r} B_{sr}\left(1 + \left(\frac{G_{sr}}{B_{sr}}\right)^2\right)$$

All shunt susceptances are neglected and the swing bus connected to ground.

The associated matrix is

$$\overline{B}^{\theta} = \begin{vmatrix} 6.553 & -2.22 & -0.333 \\ -2.22 & 5.886 & -3.333 \\ -0.333 & -3.333 & 3.666 \end{vmatrix} \begin{vmatrix} \Delta\theta_2 \\ \Delta\theta_3 \\ \Delta\theta_4 \end{vmatrix} = \begin{vmatrix} \Delta P_2/V_2 \\ \Delta P_3/V_3 \\ \Delta P_4/V_4 \end{vmatrix}$$

***Q–V* Network**

This has the structure of the original model, but voltage-specified buses, that is, swing bus and PV bus are directly connected to ground.

$$B_{sr}^{V} = -B_{sr} = \frac{X_{sr}}{R_{sr}^2 + X_{sr}^2}$$

The *Q–V* network is shown in Figure 12.5c. The associated matrix is

$$\overline{B}^{V} = \begin{vmatrix} 9.345 & -2.17 \\ -2.17 & 9.470 \end{vmatrix} \begin{vmatrix} \Delta V_2 \\ \Delta V_3 \end{vmatrix} = \begin{vmatrix} \Delta Q_2/V_2 \\ \Delta Q_3/V_3 \end{vmatrix}$$

The power flow equations can be written as in Example 12.3.

12.8 Model of a Phase-Shifting Transformer

A model of a transformer with voltage magnitude control through tap changing under load or off-load tap operation was derived in Section 11.1. The voltage drop in a transmission line is simulated in a line drop compensator, which senses the remote secondary voltage and adjusts the voltage taps. The voltage taps, however, do not change the phase angle of the voltages appreciably. A minor change due to change of the transformer impedance on account of tap adjustment and the resultant power flow through it can be ignored. The real power control can be affected through phase-shifting of the voltage. A phase-shifting transformer changes the phase angle without appreciable change in the voltage magnitude; this is achieved by injecting a voltage at right angle to the corresponding line-to-neutral voltage (Figure 12.6a).

The *Y* bus matrix for load flow is modified. Consider the equivalent circuit representation of Figure 12.6b. Let the regulating transformer be represented by an ideal transformer

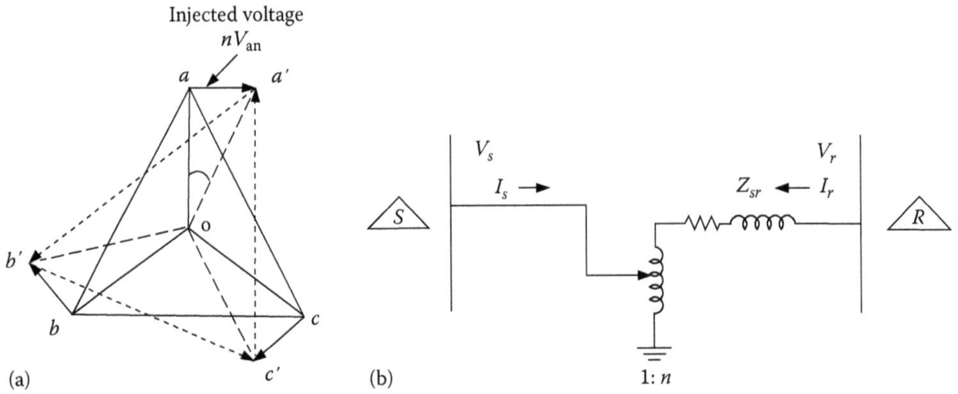

FIGURE 12.6
(a) Voltage injection vector diagram of a phase-shifting transformer; (b) schematic diagram of phase-shifting transformer.

with a series impedance or admittance. Since it is an ideal transformer, the complex power input equals the complex power output, and for a voltage adjustment tap changing transformer we have already shown that

$$I_s = n^2yV_s - nyV_r \tag{12.71}$$

where n is the ratio of the voltage adjustment taps (or currents). Also,

$$I_r = y(V_r - nV_s) \tag{12.72}$$

These equations cannot be represented by a bilateral network. The Y matrix representation is

$$\begin{vmatrix} I_s \\ I_r \end{vmatrix} = \begin{vmatrix} n^2y & -ny \\ -ny & y \end{vmatrix} \begin{vmatrix} V_s \\ V_r \end{vmatrix} \tag{12.73}$$

If the transformer has a phase-shifting device, n should be replaced with $n = Ne^{j\phi}$ and the relationship is

$$\begin{vmatrix} I_s \\ I_r \end{vmatrix} = \begin{vmatrix} N^2y & -N^*y \\ -Ny & y \end{vmatrix} \begin{vmatrix} V_s \\ V_r \end{vmatrix} \tag{12.74}$$

where N^* is a conjugate of N. In fact, we could write

$$N = n\varepsilon^{j0} \quad \text{for transformer without phase shifting} \tag{12.75}$$

$$N = n\varepsilon^{j\phi} \quad \text{for transformer with phase shifting} \tag{12.76}$$

The phase shift will cause a redistribution of power in load flow, and the bus angles solved will reflect such redistribution. A new equation must be added in the load flow to calculate the new phase-shifter angle based on the latest bus magnitudes and angles [4]. The specified active power flow is

$$P_{(\text{spec})} = V_s V_r \sin(\theta_s - \theta_r - \theta_\alpha) b_{sr} \tag{12.77}$$

where θ_α is the phase-shifter angle. As the angles are small:

$$\sin(\theta_s - \theta_r - \theta_\alpha) \approx \theta_s - \theta_r - \theta_\alpha \tag{12.78}$$

Thus,

$$\theta_\alpha = \theta_s - \theta_r - \frac{P_{(\text{spec})}}{V_s V_r b_{sr}} \tag{12.79}$$

θ_α is compared to its maximum and minimum limits. Beyond the set limits the phase shifter is tagged nonregulating. Incorporation of a phase-shifting transformer makes the Y matrix nonsymmetric.

Consider the four-bus circuit of Example 11.1. Let the transformer in line 3 to 4 be replaced with a phase-shifting transformer, with phase shift $\phi = -4°$. The Y bus can then be modified as follows.

Consider the elements related to line 3 to 4, *ignoring the presence of the transformer*. The submatrix is

	3	4
3	$1.667 - j8.30$	$j3.333$
4	$j3.333$	$1.0 - j6.308$

With phase shifting the modified submatrix becomes

	3	4
3	$1.667 - j8.30$	$e^{j4}\, j3.333$
4	$e^{-j4}(j3.333)$	$1.0 - j6.308$

12.9 DC Load Flow Models

The concept of decoupling and *P–θ and P–V models* is the key to the dc circuit models. All voltages are close to the rated voltages, the X/R ratio is >3.0, the angular difference between the tie-line flow is small, and the transformer taps are near 1.0 pu. With all these assumptions a high speed of calculation results, with some sacrifice of accuracy. The dc load flow was attractive to reduce the CPU (central processing unit in digital computers) time, and in modern times with high speed computing there is a demise of dc load flow. Its efficacy is to conduct examination of possible outage conditions and sensitivity analysis, for a very large number of system configurations.

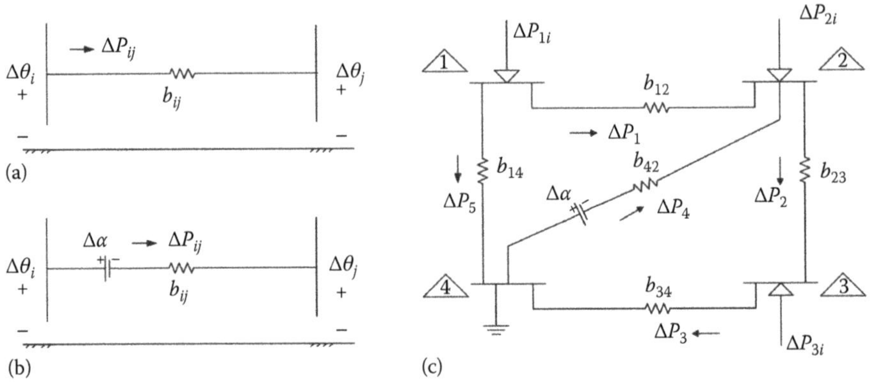

FIGURE 12.7
(a) DC circuit P–θ model of a tie-line; (b) dc circuit P–θ model of a voltage tap adjusting transformer; (c) circuit diagram for P–θ network equations.

12.9.1 P–θ Network

The P–θ circuit model of a series element is shown in Figure 12.7a. The active power flow between a bilateral node can be written as

$$\Delta p_{ji} \approx (\Delta\theta_j - \Delta\theta_i)b_{ij} \tag{12.80}$$

When a phase adjustment element (like a phase-shifting transformer) is provided in series as in the model shown in Figure 12.7b, the relation is

$$\Delta p_{ji} \approx (\Delta\theta_j - \Delta\theta_i + \Delta\alpha)b_{ij} \tag{12.81}$$

The partial derivatives $\partial P_{ij}/\partial V_i, \partial P_{ij}/\partial V_j$ are ignored.

Consider the four-bus circuit of Figure 12.7c. In terms of branch flows, the following equations can be written from branch relationships:

$$\begin{aligned}
\Delta\theta_2 &= \Delta\theta_1 - \frac{\Delta P_1}{b_{12}} \\
\Delta\theta_3 &= \Delta\theta_2 - \frac{\Delta P_2}{b_{23}} \\
\Delta\theta_4 &= \Delta\theta_3 - \frac{\Delta P_3}{b_{34}} \\
\Delta\alpha + \Delta\theta_2 &= \Delta\theta_4 - \frac{\Delta P_4}{b_{24}} \\
\Delta\theta_4 &= \Delta\theta_1 - \frac{\Delta P_5}{b_{14}}
\end{aligned} \tag{12.82}$$

From the node relationships

$$\begin{aligned}
\Delta P_1 + \Delta P_5 &= \Delta P_{1i} \\
-\Delta P_1 + \Delta P_2 - \Delta P_4 &= \Delta P_{2i} \\
-\Delta P_2 + \Delta P_3 &= \Delta P_{3i}
\end{aligned} \tag{12.83}$$

Or in the matrix form

$$\begin{vmatrix} -1/b_{12} & 0 & 0 & 0 & 0 & \vdots & 1 & -1 & 0 \\ 0 & -1/b_{23} & 0 & 0 & 0 & \vdots & 0 & 1 & -1 \\ 0 & 0 & -1/b_{34} & 0 & 0 & \vdots & 0 & 0 & 1 \\ 0 & 0 & 0 & -1/b_{24} & 0 & \vdots & 0 & -1 & 0 \\ 0 & 0 & 0 & 0 & -1/b_{14} & \vdots & 1 & 0 & 0 \\ \cdots & \cdots & \cdots & \cdots & \cdots & \cdots & \cdots & \cdots & \cdots \\ 1 & 0 & 0 & 0 & 1 & \vdots & 0 & 0 & 0 \\ -1 & 1 & 0 & -1 & 0 & \vdots & 0 & 0 & 0 \\ 0 & -1 & 1 & 0 & 0 & \vdots & 0 & 0 & 0 \end{vmatrix} \begin{vmatrix} \Delta P_1 \\ \Delta P_2 \\ \Delta P_3 \\ \Delta P_4 \\ \Delta P_5 \\ \cdots \\ \Delta\theta_1 \\ \Delta\theta_2 \\ \Delta\theta_3 \end{vmatrix} = \begin{vmatrix} 0 \\ 0 \\ 0 \\ \Delta\alpha \\ 0 \\ \cdots \\ \Delta P_{1i} \\ \Delta P_{2i} \\ \Delta P_{3i} \end{vmatrix} \quad (12.84)$$

This can be written as

$$\begin{vmatrix} -\bar{z} & \bar{C} \\ \bar{C}^t & \bar{Y} \end{vmatrix} \begin{vmatrix} \Delta P \\ \Delta\theta \end{vmatrix} = \begin{vmatrix} \Delta\alpha \\ \Delta P_i \end{vmatrix} \quad (12.85)$$

Or

$$\begin{vmatrix} \Delta P \\ \Delta\theta \end{vmatrix} = \begin{vmatrix} \bar{y} & \bar{T} \\ \bar{T}^t & \bar{Z} \end{vmatrix} \begin{vmatrix} \Delta\alpha \\ \Delta P_i \end{vmatrix}$$

$$= |\bar{S}_\theta| \begin{vmatrix} \Delta\alpha \\ \Delta P_i \end{vmatrix} \quad (12.86)$$

where

- $\bar{S}_\theta$ is sensitivity matrix for P–θ network
- $\bar{C}$ is the connection matrix relating incidence nodes to branches
- $\bar{Y}$ is the submatrix of shunt elements if present (normally $=0$)
- $\bar{z}$ is the submatrix of branch impedances

In Equation 12.86, $\bar{y}$ is the submatrix of terms relating a change in branch flow ΔP to a change in phase-shift angle $\Delta\alpha$, $\bar{T}$ is the submatrix of terms relating a change in branch flow ΔP to a change in node power injection ΔP_i, $\bar{T}^t$ is the submatrix relating a change in phase angle $\Delta\theta$ to a change in phase-shifter angle $\Delta\alpha$, and $\bar{Z}$ is the submatrix of terms relating a change in node phase angle $\Delta\theta$ to node injection ΔP_i.

The sensitivity factor relating an MW change to generation at bus i to an MW change in flow of line j is

$$\frac{\Delta P_j}{\Delta P_{Gi}} = (Z_{mi} - Z_{ni})b_j \quad (12.87)$$

where m and n are respective terminal buses of line j.

The sensitivity factor relating a phase angle change in a phase shifter in line l to an MW change in flow of line j is

$$\frac{\Delta P_j}{\Delta \alpha_l} = [(Z_{ms} - Z_{mr} - Z_{ns} + Z_{nr})b_l]b_j \tag{12.88}$$

where buses m and n are terminal buses of line j and s and r are terminal buses of phase-shifter circuit l.

12.9.2 Q–V Network

The dc formation of a transmission line (Figure 12.8a) is given by

$$\begin{aligned}\Delta q_{ij} &\approx (\Delta V_i - \Delta V_j)b_{ij} - 2b_{sh}\Delta V_i \\ &= \Delta q_{ij}b_{ij} - 2b_{sh}V_i\end{aligned} \tag{12.89}$$

In Equation 12.89, b_{sh} is one-half of the total shunt susceptance, that is, the total line shunt susceptance $= 2b_{sh}$. We obtain twice the normal susceptance in each leg of the Π model

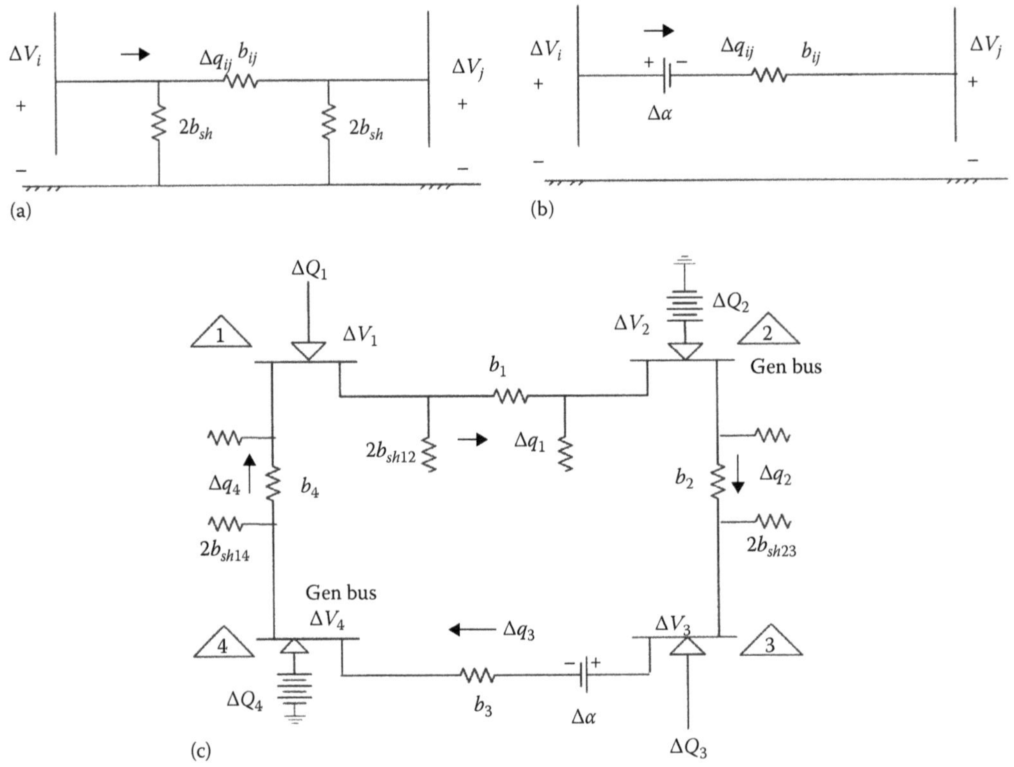

FIGURE 12.8
(a) Dc circuit Q–V model of a tie-line; (b) dc circuit Q–V model of phase-shifting transformer; (c) circuit diagram for Q–V network equations.

because of partial differentiation of the power flow equation with respect to voltage. Looking from the receiving end:

$$\begin{aligned} \Delta q_{ji} &\approx (\Delta V_j - \Delta V_i)b_{ji} - 2b_{sh}\Delta V_j \\ &= \Delta q_{ji}b_{ji} - 2b_{sh}V_j \end{aligned} \tag{12.90}$$

The dc model of a voltage tap adjusting transformer (Figure 12.8b) is

$$\begin{aligned} \Delta q_{ij} &= (\Delta V_i - \Delta V_j - \Delta\alpha)b_{ij} \\ \Delta q_{ji} &= (\Delta V_j - \Delta V_i - \Delta\alpha)b_{ji} \end{aligned} \tag{12.91}$$

In the circuit of Figure 12.8c, the generator terminal voltages ΔV_2 and ΔV_4 are considered as control parameters and ΔQ_2 and ΔQ_4 are the dependent parameters. Following equations can be written based on branch relations:

$$\begin{aligned} \Delta V_1 - \Delta V_2 &= \frac{\Delta q_1}{b_1} \\ \Delta V_2 - \Delta V_3 &= \frac{\Delta q_2}{b_2} \\ \Delta V_3 - \Delta V_4 - \Delta\alpha &= \frac{\Delta q_3}{b_3} \\ \Delta V_4 - \Delta V_1 &= \frac{\Delta q_4}{b_4} \end{aligned} \tag{12.92}$$

From the node relations

$$\begin{aligned} &\Delta q_1 + \Delta q_4 - 2(b_{sh1} + b_{sh4})\Delta V_1 = \Delta Q_1 \\ &-\Delta q_1 + \Delta q_2 - \Delta Q_2 = 2(b_{sh1} + b_{sh2})\Delta V_2 \\ &-\Delta q_2 + \Delta q_3 - 2b_{sh2}\Delta V_3 = \Delta Q_3 \\ &-\Delta q_4 - \Delta q_3 - \Delta Q_4 = 2(b_{sh4})\Delta V_4 \end{aligned} \tag{12.93}$$

Or in the matrix form

$$\left|\begin{array}{cccc:cccc} -1/b_1 & 0 & 0 & 0 & 1 & 0 & 0 & 0 \\ 0 & -1/b_2 & 0 & 0 & 0 & 0 & -1 & 0 \\ 0 & 0 & -1/b_3 & 0 & 0 & 0 & 1 & 0 \\ 0 & 0 & 0 & -1/b_4 & 0 & 0 & 0 & 0 \\ \cdots & \cdots & \cdots & \cdots & \cdots & \cdots & \cdots & \cdots \\ 1 & 0 & 0 & 1 & -2b_{sh1} - 2b_{sh4} & 0 & 0 & 0 \\ -1 & 1 & 0 & 0 & 0 & -1 & 0 & 0 \\ 0 & -1 & 1 & 0 & 0 & 0 & -2b_{sh2} & 0 \\ 0 & 0 & -1 & -1 & 0 & 0 & 0 & -1 \end{array}\right| \left|\begin{array}{c} \Delta q_1 \\ \Delta q_2 \\ \Delta q_3 \\ \Delta q_4 \\ \cdots \\ \Delta V_1 \\ \Delta Q_2 \\ \Delta V_3 \\ \Delta Q_4 \end{array}\right|$$

$$= \begin{vmatrix} 1 & 0 & 0 & 0 & \vdots & 0 & 1 & 0 & 0 \\ 0 & 1 & 0 & 0 & \vdots & 0 & -1 & 0 & 0 \\ 0 & 0 & 1 & 0 & \vdots & 0 & 0 & 0 & 1 \\ 0 & 0 & 0 & 1 & \vdots & 0 & 0 & 0 & 1 \\ \cdots & \cdots & \cdots & \cdots & \cdots & \cdots & \cdots & \cdots & \cdots \\ 0 & 0 & 0 & 0 & \vdots & 1 & 0 & 0 & 0 \\ 0 & 0 & 0 & 0 & \vdots & 0 & 2b_{sh1}+2b_{sh2} & 0 & 0 \\ 0 & 0 & 0 & 0 & \vdots & 0 & 0 & 1 & 0 \\ 0 & 0 & 0 & 0 & \vdots & 0 & 0 & 0 & 2b_{sh4} \end{vmatrix} \begin{vmatrix} 0 \\ 0 \\ \Delta\alpha \\ 0 \\ \cdots \\ \Delta Q_1 \\ \Delta V_2 \\ \Delta Q_3 \\ \Delta V_4 \end{vmatrix} \tag{12.94}$$

Or in the abbreviated form

$$\bar{J}\Delta x = H\Delta u \tag{12.95}$$

The sensitivity factor relating an Mvar change in injection at bus i to an Mvar change in flow of line j is given by

$$\frac{\Delta q_j}{\Delta q_i} = (Z_{mi} - Z_{ni})b_j \tag{12.96}$$

where buses m and n are the respective buses of line j.

The sensitivity factor relating an Mvar change in injection at bus i to a voltage change at bus k is given by

$$\frac{\Delta V_k}{\Delta Q_j} = Z_{ki} \tag{12.97}$$

The sensitivity factor relating a change in voltage at the generator bus i to an Mvar change in flow of line j is

$$\frac{\Delta q_j}{\Delta V_i} = [(Z_{ms} - Z_{ns})b_{is} + (Z_{mr} - Z_{nr})b_{ir}]b_{mn} \tag{12.98}$$

where the generator is connected to bus i with terminal buses s and r.

The sensitivity factor relating a change in voltage at a generator bus i to a voltage change at a bus k is

$$\frac{\Delta V_k}{\Delta V_i} = Z_{ks}b_{is} + Z_{kr}b_{ir}$$

12.10 Second Order Load Flow

In Equation 12.2 Taylor's series we neglected second order terms. Also from Equation 12.22 and including second order terms we can write

$$\begin{vmatrix} \Delta P \\ \Delta Q \end{vmatrix} = \bar{J} \begin{vmatrix} \Delta h \\ \Delta e \end{vmatrix} + \begin{vmatrix} SP \\ SQ \end{vmatrix} \tag{12.99}$$

Equation 12.99 has been written in rectangular coordinates. It can be shown that second order terms can be simplified as [5]

$$\begin{aligned} SP_i &= P_i^{\text{cal}}(\Delta e, \Delta h) \\ SQ_i &= Q_i^{\text{cal}}(\Delta e, \Delta h) \end{aligned} \tag{12.100}$$

No third or higher order terms are present. If the Equation 12.99 is solved an exact solution or near to exact solution can be expected in the very first iteration. The second order terms can be estimated from previous iteration, and $\Delta P, \Delta Q$ are corrected by the second order term as

$$\begin{vmatrix} \Delta P - SP \\ \Delta Q - SQ \end{vmatrix} = \bar{J} \begin{vmatrix} \Delta h \\ \Delta e \end{vmatrix} \tag{12.101}$$

Or

$$\begin{vmatrix} \Delta h \\ \Delta e \end{vmatrix} = \bar{J}^{-1} \begin{vmatrix} \Delta P - SP \\ \Delta Q - SQ \end{vmatrix} \tag{12.102}$$

Initially the vectors SP and SQ are set to zero. And P^{cal}, Q^{cal} are estimated. For a PV bus, the voltage magnitude is fixed, and therefore the increments must satisfy

$$e_i \Delta e_i + h_i \Delta h_i = V_i \Delta V_i \tag{12.103}$$

The voltages are updated:

$$\begin{aligned} e^{k+1} &= e^k + \Delta e \\ h^{k+1} &= h^k + \Delta h \end{aligned} \tag{12.104}$$

The second order terms SP and SQ are calculated using $\Delta P, \Delta Q$, and $P^{\text{cal}}, Q^{\text{cal}}$ are reestimated. The number of iterations and CPU time decrease with large systems over NR method.

Twenty five years back there was, practically, no power system software available for PC (personal computers) use. Some of the commercial power system software available today claim unlimited bus numbers and number of iterations is not so much of a consideration. The programs allow more than one load flow algorithm to be chosen. The computing requirements and storage for PC may be heavy and the project files may run into hundreds of MBs (Megabytes).

12.11 Load Models

Load modeling has a profound impact on load flow studies. Figure 12.9 shows the effect of change of operating voltage on constant current, constant MVA, and constant impedance load types. Heavy industrial motor loads are approximately constant MVA loads, while commercial and residential loads are mainly constant impedance loads. Classification into commercial, residential, and industrial is rarely adequate and one approach has been to divide the loads into individual load components. The other approach is based on measurements. Thus, the two approaches are

- Component-based models
- Models based on measurements

A component-based model is a *bottom-up* approach in the sense that different load components comprising the loads are identified. Each load component is tested to determine the relations between real and reactive power requirements versus voltage and frequency. A load model in exponential or polynomial form can then be developed from the test data.

The measurement approach is a *top-down* approach in the sense that the model is based on the actual measurement. The effect of the variation of voltage on active and reactive power consumption is recorded and, based on these, the load model is developed.

A composite load, that is, a consumer load consisting of heating, air-conditioning, lighting, computers, and television is approximated by combining load models in certain proportions based on load surveys. This is referred to as a *load window*.

Construction of the load window requires certain data, that is, load saturation, composition, and diversity data. Any number of load windows can be defined.

The load models are normalized to rated voltage, rated power, and rated frequency and are expressed in per unit. The exponential load models are

$$\frac{P}{P_n} = \left|\frac{V}{V_n}\right|^{\alpha_v} \left|\frac{f}{f_n}\right|^{\alpha_f} \tag{12.105}$$

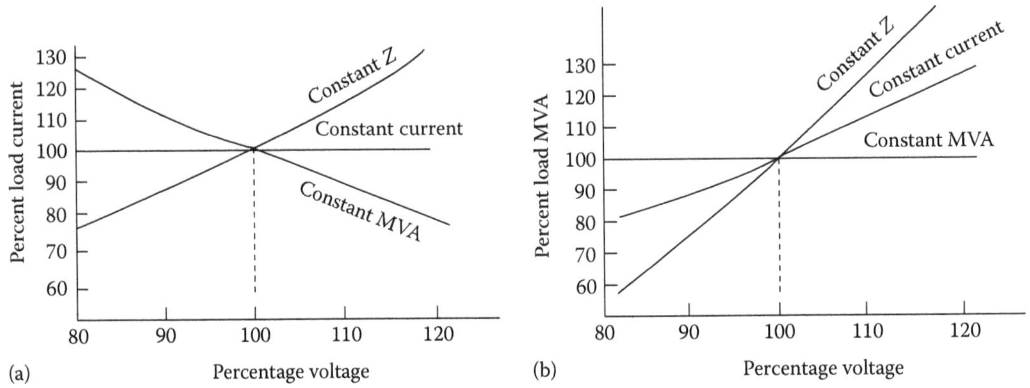

FIGURE 12.9
(a) Behavior of constant current, constant MVA, and constant impedance loads with respect to current loading as a function of voltage variations; (b) behavior of constant current, constant MVA, and constant impedance loads with respect to MVA loading as a function of voltage variations.

$$\frac{Q}{P_n} = \frac{Q_0}{P_0}\left|\frac{V}{V_n}\right|^{\beta_v}\left|\frac{f}{f_n}\right|^{\beta_f} \tag{12.106}$$

where

V is the initial value of the voltage
P_0 is the initial value of power
V_n is the adjusted voltage
P_n is the power corresponding to this adjusted voltage

The exponential factors depend on the load type. The frequency dependence of loads is ignored for load flow studies and is applicable to transient and dynamic stability studies. Another form of load model is the polynomial model.

Consider the equation

$$P = V^n \tag{12.107}$$

For all values of n, $P=0$ when $V=0$, and $P=1$ when $V=1$, as it should be. Differentiating

$$\frac{\mathrm{d}P}{\mathrm{d}V} = nV^{n-1} \tag{12.108}$$

For $V=1$, $\mathrm{d}P/\mathrm{d}V=n$. The value of n can be found by experimentation if $\mathrm{d}P/\mathrm{d}V$, that is, change in active power with change in voltage, is obtained. Also, by differentiating $P=VI$.

$$\frac{\mathrm{d}P}{\mathrm{d}V} = I + V\frac{\mathrm{d}I}{\mathrm{d}V} \tag{12.109}$$

For V and n equal to unity, the exponential n from Equations 12.108 and 12.110 is

$$n = 1 + \frac{\Delta I}{\Delta V} \tag{12.110}$$

The exponential n for a composite load can be found by experimentation if the change of current for a change in voltage can be established. For a constant power load $n=0$, for a constant current load $n=1$, and for a constant MVA load $n=2$. The composite loads are a mixture of these three load types and can be simulated with values of n between 0 and 2. The following are some quadratic expressions for various load types [6]. An EPRI Electric Power Research Institute report [7] provides more detailed models.

Air conditioning:

$$P = 2.18 + 0.298V - 1.45V^{-1} \tag{12.111}$$

$$Q = 6.31 - 15.6V + 10.3V^2 \tag{12.112}$$

Fluorescent lighting:

$$P = 2.97 - 4.00V + 2.0V^2 \tag{12.113}$$

$$Q = 12.9 - 26.8V + 14.9V^2 \tag{12.114}$$

Induction motor loads:

$$P = 0.720 + 0.109V + 0.172V^{-1} \tag{12.115}$$

$$Q = 2.08 + 1.63V - 7.60V^2 + 4.08V^3 \tag{12.116}$$

12.12 Induction Motor Models

We will discuss the induction motor model in some detail, because of its importance. The equivalent circuit of the induction motor, shown in Figure 12.10, is derived in many texts. Also see Section 6.11. The power transferred across the air gap is

$$P_g = I_2^2 \frac{r_2}{s} \tag{12.117}$$

Referring to Figure 12.10, r_2 is the rotor resistance, s the motor slip, and I_2 the rotor current. Mechanical power developed is the power across the air gap minus copper loss in the rotor, that is,

$$(1 - s)P_g$$

Thus, the motor torque T in Newton meters can be written as

$$T = \frac{1}{\omega_s} I_2^2 \frac{r_2}{s} \approx \frac{1}{\omega_s} \frac{V_1^2(r_2/s)}{(R_1 + r_2/s)^2 + (X_1 + x_2)^2} \tag{12.118}$$

where

R_1 is the stator resistance

V_1 is the terminal voltage

ω_s is the synchronous angular velocity $= 2\pi f/p$, p being the number of *pairs* of poles

From Equation 12.118 the motor torque varies approximately as the square of the voltage. Also, if the load torque remains constant and the voltage dips, there has to be an increase in the current.

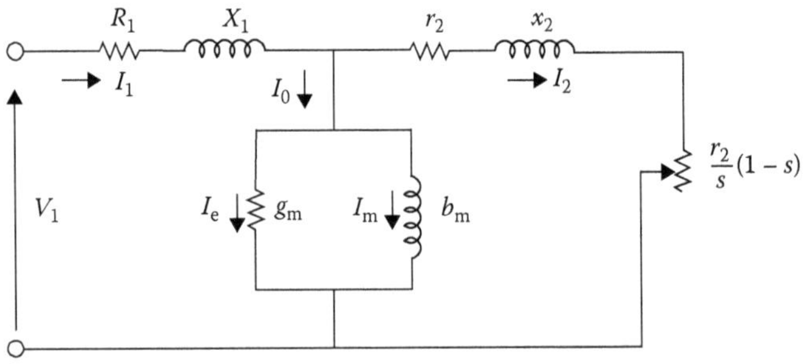

FIGURE 12.10
Equivalent circuit of an induction motor for balanced positive sequence voltages.

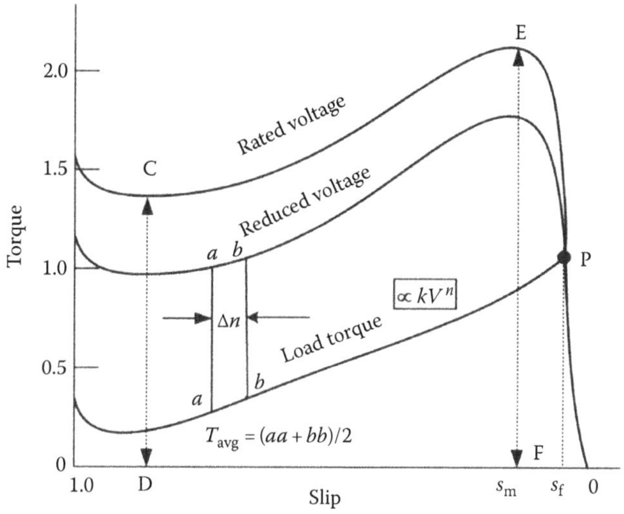

FIGURE 12.11
Torque-speed characteristics of an induction motor at rated voltage and reduced voltage with superimposed load characteristics.

Figure 12.11 shows typical torque-speed characteristics of an induction motor at rated voltage and reduced voltage. Note the definitions of locked rotor torque, minimum accelerating torque, and breakdown torque during the starting cycle. Referring to Figure 12.11, there is a cusp or reverse curvature in the accelerating torque curve, which gives the minimum accelerating torque or breakaway torque (=CD). The maximum torque, called the breakdown torque in the curve, occurs at slip s_m and is given by EF. The normal operating full-load point and slip s_f is defined by operating point P. The starting load characteristics shown in this figure are for a fan or blower, and varies widely depending on the type of load to be accelerated. Assuming that the load torque remains constant, that is, a conveyor motor, when a voltage dip occurs, the slip increases and the motor torque will be reduced. It should not fall below the load torque to prevent a stall. Considering a motor breakdown torque of 200%, and full load torque, the maximum voltage dip to prevent stalling is 29.3%. Figure 12.12 shows torque-speed characteristics for National Electrical Manufacturer's Association (NEMA) design motors A, B, C, D, and F [8].

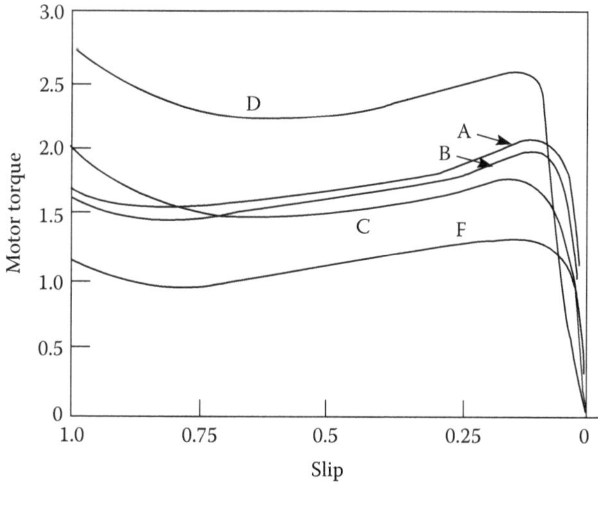

FIGURE 12.12
Torque speed characteristics of NEMA designs A, B, C, D, and F motors.

12.12.1 Double Cage Rotors

The maximum torque is proportional to the square of the applied voltage; it is reduced by stator resistance and leakage reactance, but is independent of rotor resistance. The slip for maximum torque is obtained from above equation by $dT/ds = 0$, which gives

$$s = \frac{r_2}{\sqrt{R_1^2 + (X_1 + x_2)^2}} \approx \frac{r_2}{(X_1 + x_2)} \tag{12.119}$$

The maximum torque is obtained by inserting the value of s in Equation 12.118

$$T_m = \frac{0.5V_1^2}{\omega_s\left[\sqrt{\left[R_1^2 + (X_1 + x_2)\right]^2} + R_1\right]} \tag{12.120}$$

The slip at the maximum torque is directly proportional to the rotor resistance (Equation 12.108). In many applications it is desirable to produce a higher starting torque at the time of starting. External resistances can be introduced in wound rotor motors which are short-circuited as the motor speeds up. In squirrel cage designs deep rotor bars or double cage rotors are used. The top bar is of lower cross section and has higher resistance than the bottom bar of larger cross section. The flux patterns make the leakage resistance of the top cage negligible compared to the lower cage. At starting, the slip frequency is equal to the supply system frequency and most of the starting current flows in the top cage, and the motor produces high starting torque. As the motor accelerates, the slip frequency decreases and the lower cage takes more current because of its low leakage reactance compared to the resistance. Figure 12.13a shows the equivalent circuit of a double cage induction motor, and Figure 12.13b its torque speed characteristics.

The effects of voltage and frequency variations on induction motor performance are shown in Table 12.2 [9]. We are not much concerned about the effect of frequency variation in load flow analysis, though this becomes of importance in harmonic analysis. The electromagnetic force (EMF) of a three-phase ac winding is given by

$$V = 4.44K_w f T_{ph}\Phi \tag{12.121}$$

where

V is the phase EMF
K_w is the winding factor
f is the system frequency
T_{ph} are the turns per phase
Φ is the mutual rotating flux

Maintaining the voltage constant, a variation in frequency results in an inverse variation in the flux. Thus, a lower frequency results in overfluxing the motor and its consequent derating. In variable-frequency drive systems V/f is kept constant to maintain a constant flux relation.

We discussed the negative sequence impedance of an induction motor for calculation of an open conductor fault in Example 2.5. A further explanation is provided with respect to Figure 12.14 which shows the negative sequence equivalent circuit of an induction motor.

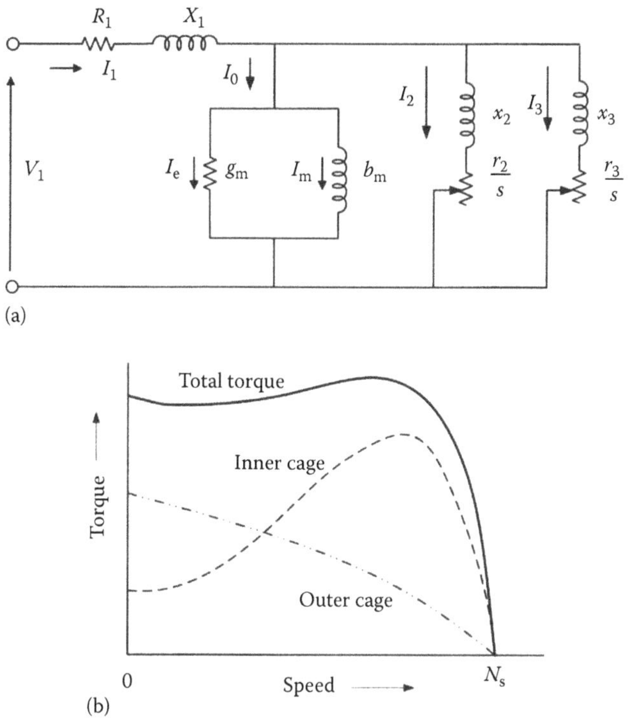

FIGURE 12.13
(a) Equivalent circuit of a double age induction motor; (b) torque speed characteristics.

TABLE 12.2
Effects of Voltage Variations on Induction Motor Performance

Characteristics of Induction Motor	Variation with Voltage	Performance at Rated Voltage (1.0 per Unit) and Other Than Rated Voltages				
		0.80	0.95	1.0	1.05	1.10
Torque	$=V^2$	0.64	0.90	1.0	1.10	1.21
Full-load slip	$=1/V^2$	1.56	1.11	1.0	0.91	0.83
Full-load current	$\approx 1/V$	1.28	1.04	1.0	0.956	0.935
Full-load efficiency		0.88	0.915	0.92	0.925	0.92
Full-load power factor		0.90	0.89	0.88	0.87	0.86
Starting current	$=V$	0.80	0.95	1.0	1.05	1.10
No load losses (watts)	$=V^2$	0.016	0.023	0.025	0.028	0.030
No load losses (vars)	$=V^2$	0.16	0.226	0.25	0.276	0.303

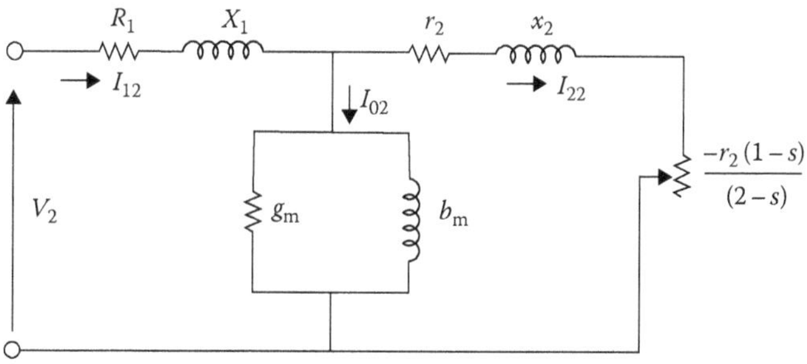

FIGURE 12.14
Equivalent circuit of an induction motor for negative sequence voltage.

When a negative sequence voltage is applied, the mmf wave in the air gap rotates backwards at a slip of 2.0 per unit. The slip of the rotor with respect to the backward rotating field is 2 – s. This results in a retarding torque component and the net motor torque reduces to

$$T = \frac{r_2}{\omega_s}\left(\frac{I_2^2}{s} - \frac{I_{22}^2}{2-s}\right) \tag{12.122}$$

where I_{22} is the current in the negative sequence circuit.

From equivalent circuits of Figures 12.10 and 12.14, we can write the approximate positive and negative sequence impedances of the motor:

$$\begin{aligned} Z_1 &= \left[\left(R_1 + \frac{r_2}{s}\right)^2 + (X_1 + x_2)^2\right]^{1/2} \\ Z_2 &= \left[\left(R_1 + \frac{r_2}{(2-s)^2} + (X_1 + x_2)^2\right]^{1/2} \end{aligned} \tag{12.123}$$

Therefore, approximately, the ratio $Z_1/Z_2 = I_s/I_f$, where I_s is the starting current or the locked rotor current of the motor and I_f is the full load current ($s = 1$ at start). For an induction motor with a locked rotor current of six times the full-load current the negative sequence impedance is one-sixth of the positive sequence impedance. A 5% negative sequence component in the supply system will produce 30% negative sequence current in the motor, which gives rise to additional heating and losses. Equations 12.123 are a simplification; the rotor resistance will change with respect to high rotor frequency and rotor losses are much higher than the stator losses. A 5% voltage unbalance may give rise to 38% negative sequence current with 50% increase in losses and 40°C higher temperature rise as compared to operation on a balanced voltage with zero negative sequence component. Also, the voltage unbalance is not equivalent to the negative sequence component. The NEMA definition of percentage voltage unbalance is maximum voltage deviation from the average voltage divided by the average voltage as a percentage. Operation above 5% unbalance is not recommended.

The zero sequence impedance of motors, whether the windings are connected in wye or delta formation, is infinite. The motor windings are left ungrounded.

From load flow considerations, it is conservative to assume that a balanced reduction in voltage at the motor terminals gives rise to a balanced increase in line current, inversely proportional to the reduced voltage. More accurate models may use part of the motor load as a constant kVA load and part as the constant impedance load.

Table 12.3 shows approximate load models for various load types. A synchronous motor is modeled as a synchronous generator with negative active power output and positive reactive power output, akin to generators, assuming that the motor has a leading rated power factor. The reactive power supplied into the system depends on the type of excitation and control system. As a load flow provides a static picture, the time constants applicable with control devices are ignored, that is, if a load demands a certain reactive power output, which is within the capability of a generator, it is available instantaneously. The drive systems can be assumed to have a constant active power demand with the reactive power demand varying with the voltage.

TABLE 12.3

Representation of Load Models in Load Flow

Load Type	*P* (Constant kVa)	*Q* (Constant kVa)	*P* (Constant *Z*)	*Q* (Constant *Z*)	Generator Power	Generator Reactive min/max
Induction motor running	$+P$	Q (lagging)				
Induction or synchronous starting			$+P$	Q (lagging)		
Generator					$+P$	$Q_{(\text{max})}/Q_{(\text{min})}{}^*$
Synchronous motor					$-P$	Q (leading)
Power capacitor				Q (leading)		
Rectifier	$+P$			Q (lagging)		
Lighting			$+P$	Q (lagging)		

$Q_{\text{min}}{}^*$ for generator can be leading.

12.13 Impact Loads and Motor Starting

Load flow presents a frozen picture of the distribution system at a given instant, depending on the load demand. While no idea of the transients in the system for a sudden change in load application or rejection or loss of a generator or tie-line can be obtained, a steady-state picture is presented for the specified loading conditions. Each of these transient events can be simulated as the initial starting condition, and the load flow study rerun as for the steady-state case. Suppose a generator is suddenly tripped. Assuming that the system is stable after this occurrence, we can calculate the redistribution of loads and bus voltages by running the load flow calculations afresh, with generator omitted. Similarly, the effect of an outage of a tie-line, transformer, or other system component can be studied.

12.13.1 Motor Starting Voltage Dips

One important application of the load flow is to calculate the *initial* voltage dip on starting a large motor in an industrial distribution system. From the equivalent circuit of an induction motor in Figure 12.10 and neglecting the magnetizing and eddy current loss circuit, the starting current or the locked rotor current of the motor is

$$I_{\text{lr}} = \frac{V_1}{(R_1 + r_2) + j(X_1 + x_2)} \tag{12.124}$$

The locked rotor current of squirrel-cage induction motors is, generally, six times the full-load current across the line starting, that is, the full rated voltage applied across the motor terminals with the motor at standstill. Higher or lower values are possible, depending on motor design.

Wound rotor motors may be started with an external resistance in the circuit to reduce the starting current and increase the motor starting torque. Synchronous motors are asynchronously started and their starting current is generally 3–4.5 times the rated full load current across the line starting. The starting currents are at a low power factor and may give rise to unacceptable voltage drops in the system and at motor terminals. On large voltage dips, the stability of running motors in the same system may be jeopardized, the motors may stall, or the magnetic contactors may drop out. The voltage tolerance limit of solid-state devices is much lower and a voltage dip >10% may precipitate a shutdown. As the system impedances or motor reactance cannot be changed, an impedance may be introduced in the motor circuit to reduce the starting voltage at motor terminals. This impedance is short-circuited as soon as the motor has accelerated to approximately 90% of its rated speed. The reduction in motor torque, acceleration of load, and consequent increase in starting time are of consideration. Some methods of reduced voltage starting are reactor starting, autotransformer starting, wye–delta starting (applicable to motors which are designed to run in delta), capacitor starting, and electronic soft starting. Other methods for large synchronous motors may be part-winding starting or even variable-frequency starting as discussed in Section 12.13.3. The starting impact load varies with the method of starting and needs to be calculated carefully along with the additional impedance of a starting reactor or autotransformer introduced into the starting circuit.

12.13.2 Snapshot Study

This will calculate only the initial voltage dip and no idea of the time-dependent profile of the voltage is available. To calculate the starting impact, the power factor of the starting current is required. If the motor design parameters are known, this can be calculated from Equation 12.124; however, rarely, the resistance and reactance components of the locked rotor circuit will be separately known. The starting power factor can be taken as 20% for motors under 1000 hp and 15% for motors >1000 hp. The manufacturer's data should be used when available.

Consider a 10,000 hp, four-pole synchronous motor, rated voltage 13.8 kV, rated power factor 0.8 leading, and full-load efficiency 95%. It has a full load current of 410.7 A. The starting current is four times the full-load current at a power factor of 15%. Thus, the starting impact is 5.89 MW and 38.82 Mvar.

During motor starting, generator transient behavior is important. On a simplistic basis the generators may be represented by a voltage behind a transient reactance, which for motor starting may be taken as the generator transient reactance. Prior to starting the voltage behind this reactance is simply the terminal voltage plus the voltage drop caused by the load through the transient reactance, that is, $V_t = V + jIX_d$, where I is the load current. For a more detailed solution the machine reactances change from subtransient to transient to synchronous, and open-circuit subtransient, transient, and steady-state time constants should be modeled with excitation system response.

Depending on the relative size of the motor and system requirements more elaborate motor starting studies may be required. The torque speed and accelerating time of the motor is calculated by step-by-step integration for a certain interval, depending on the accelerating torque, system impedances, and motor and loads' inertia. A transient stability program can be used to evaluate the dynamic response of the system during motor starting [10,11]; also EMTP type programs can be used.

12.13.3 Motor Starting Methods

Table 12.4 shows a summary of the various starting methods. System network stiffness, acceptable voltage drops, starting reactive power limitation, and motor and load

TABLE 12.4
Starting Methods of Motors

Starting Method	Reference Figure No.	*I* Starting	*T* Starting	Cost Ratio	Qualifications
Full voltage starting	a	1	1	1	Simplest starting method giving highest starting efficiency, provided voltage drops due to inrush currents are acceptable.
Reactor starting	b	α	α^2	2.5	Simple switching closed transition, torque per kVA is lower as compared to auto-transformer starting. A single reactor can be used to start a number of motors.
Krondrofer starting	c	α^2	α^2	3.5	Closed circuit transition requires complex switching, (a) close Y, then S; (b) open Y, (c) close R, (d) open S. Applicable to weak electrical systems, where the reduced starting torque of motor can still accelerate the loads.
Part-winding starting	d	α	α^2	Varies	Inrush current depends upon design of starting winding. Closed circuit transition by switching the parallel stator winding by closing R. The starting torque cannot be varied and fixed at the design stage. Start: close S. Run: close R.
Shunt capacitor starting	e	Varies	Varies	3	May create harmonic pollution (harmonic filters are required). Capacitors switched off by voltage/current control when the speed approaches approximately 95% of the full load speed, (bus voltage rises as the motor current falls). Start: close S1, S2. Run: open S2.
Shunt capacitor in conjunction with other reduced voltage starting	f	Varies	Varies	Varies	A reactor starting with shunt capacitors has two-fold reduction in starting current, due to reactor and capacitor. The motor terminal voltage and torque is increased as compared to reactor start.
Low frequency starting	g	Varies (150%–200% of load current)	Slightly > load torque	6–7	May create harmonic pollution, not generally used due to high cost. Gives a smooth acceleration. The current from supply system can be reduced to 150%–200% of the full load current. One low frequency starting equipment can be used to start a number of motors in succession by appropriate switching.
Pony Motor	h	—	—	Varies	Not generally used. The starting motor is high torque intermittent rated machine. Applicable when load torque during acceleration is small. Synchronizing required at S.

parameters need to be considered when deciding upon a starting method. Limitation of stating voltage dips is required for

- The utilities will impose restrictions on the acceptable voltage dips in their systems on starting of motors.
- The starting voltage dips will impact the other running motors in service. The stability of motors in service may be jeopardized on excessive voltage dips.
- The motor starter contactors may drop out on excessive voltage dips. Low voltage motor starter contactors may drop out in the first cycle of the voltage dip any where from 15% to 30%. Medium voltage dc motor contactors may ride through voltage dips of 30% or more for a couple of cycles depending upon the trapped magnetism, but the auxiliary relays in the motor starting circuit may drop out first.
- The drive and electronic loads will be more sensitive to the voltage dips and may set an upper limit to the acceptable voltage dip.
- On restoration of the voltage, the induction motors will reaccelerate, increasing the current demand from the supply system, which will result in more voltage drops. This phenomenon is cumulative and the resulting large inrush currents on reacceleration may cause a shutdown. This situation can be evaluated with a transient stability type program with dynamic load and generation models.
- A large starting voltage dip may reduce the net accelerating torque of the motor below the load torque during acceleration and result in lockout and unsuccessful starting or synchronizing.

The characteristics of starting methods in Table 12.4, and the associated diagram of starting connections (Figure 12.15) are summarized as follows:

1. **Full voltage starting, Figure 12.15a**
This is the simplest method for required starting equipment and controls and is the most economical. It requires a stiff supply system because of mainly reactive high starting currents. The motor terminal voltage during starting and, therefore, the starting torque will be reduced, depending upon the voltage drop in the impedance of the supply system. It is the preferred method, especially when high inertia loads requiring high breakaway torques are required to be accelerated. Conversely, a too fast run up of low inertia loads can subject the drive shaft and coupling to high mechanical torsional stresses.

2. **Reactor starting, Figure 12.15b**
A starting reactor tapped at 40%, 65%, and 80% and rated on an intermittent duty basis is generally used. It allows a smooth start with an almost unobservable disturbance in transferring from the reduced to full voltage (when the bypass breaker is closed). The reactor starting equipment and controls are simple, though the torque efficiency is poor. Reactance of the reactor for a certain tap can be approximately calculated by the expression: $X = (1 - R)/R$ where R is the tap ratio or the voltage ratio in per unit of the rated voltage. Thus, the starting current at a particular tap is given by

$$I_s = \frac{I}{Z_s + Z_m + Z_L} \tag{12.125}$$

It is in per unit based upon the motor rated voltage. Z_s, Z_m, and Z_L are the system, motor, and reactor impedance, reduced to a common base, say motor starting kVA base, and are

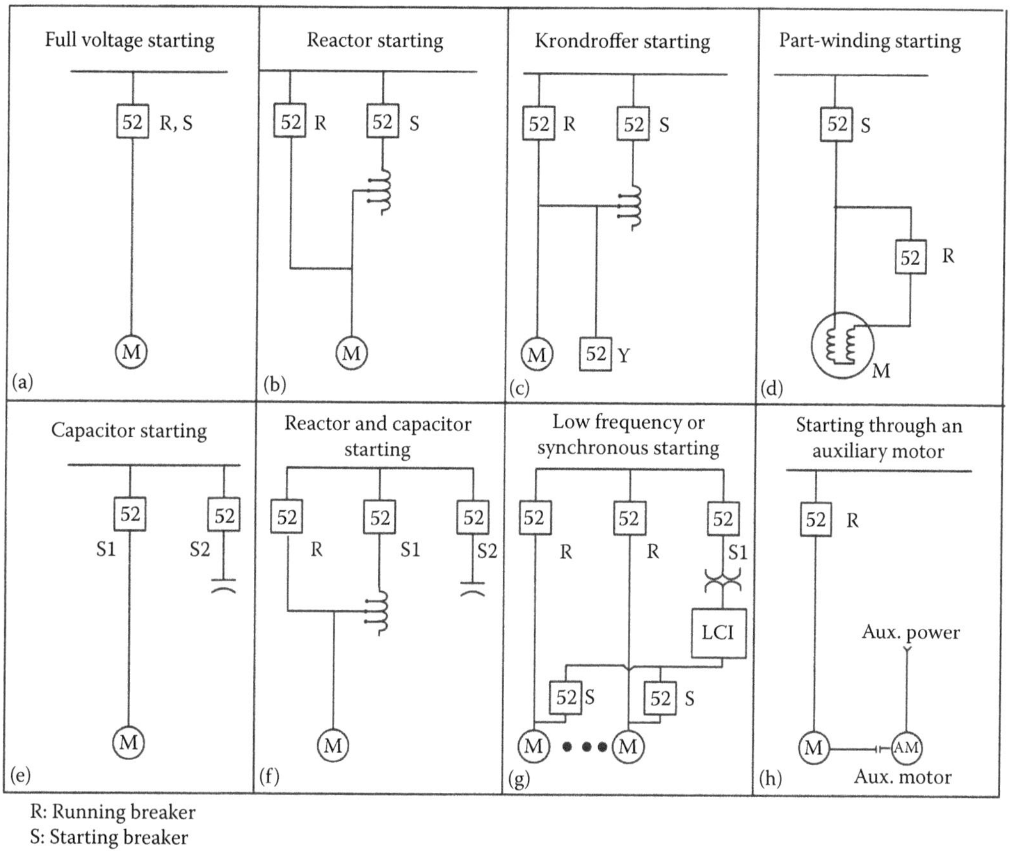

FIGURE 12.15
Starting methods of motors.

vector quantities; they also give the starting power factor. As these impedances act as a potential divider, the starting voltage at the motor terminals is simply I_sZ_m.

The motor starting impedance can be calculated by $(V/\sqrt{3} \times I_s)$ where V is the motor rated line to line voltage and I_s is the starting current at the motor at starting power factor.

3. Krondroffer starting, Figure 12.15c

An advantage of this method of starting is high torque efficiency, at the expense of additional control equipment and three switching devices per starter. Current from the supply system is further reduced by factor α, for the same starting torque obtained with a reactor start. Though close transition reduces the transient current and torque during transition, a quantitative evaluation is seldom attempted. The motor voltage during transition may drop significantly. For weak supply systems, this method of starting at 30%–45% of the voltage can be considered, provided the reduced asynchronous torque is adequate to accelerate the load.

4. Part winding starting, Figure 12.15d

The method is applicable to large synchronous motors of thousands of horsepower, designed for part-winding starting. These have at least two parallel circuits in the stator winding. This may add 5%–10% to the motor cost. The two windings cannot be exactly symmetrical in fractional slot designs and the motor design becomes specialized regarding winding pitch, number of slots, coil groupings, etc. The starting winding may be designed for a higher temperature rise during starting. Proper sharing of the current between paralleled windings, limiting temperature rises, and avoiding hot spots becomes a design consideration. Though no external reduced voltage starting devices are required and the controls are inherently simple; the starting characteristics are fixed and cannot be altered. Part-winding starting has been applied to large TMP (thermo-mechanical pulping) synchronous motors of 10,000 hp and above, yet some failures have been known to occur.

5. Capacitor starting, Figure 12.15e and f

The power factor of the starting current of even large motors is low, rarely exceeding 0.25. Starting voltage dip is dictated by the flow of starting reactive power over mainly inductive system impedances. Shunt connected power capacitors can be sized to meet a part of the starting reactive kvar, reducing the reactive power demand from the supply system. The voltage at the motor terminals improves and, thus, the available asynchronous torque. The size of the capacitors selected should ensure a certain starting voltage across the motor terminals, considering the starting characteristics and the system impedances. As the motor accelerates and the current starts falling, the voltage will increase and the capacitors are switched off at 100%–104% of the normal voltage, sensed through a voltage relay. A redundant current switching is also provided. For infrequent starting having shunt capacitors rated at 60% of the motor voltage are acceptable. When harmonic resonance is of concern, shunt capacitor filters can be used.

Capacitor and reactor starting can be used in combination (Figure 12.15f). A reactor reduces the starting inrush current and a capacitor compensates part of the lagging starting kvar requirements. These two effects in combination further reduce the starting voltage dip [12].

6. Low frequency starting or synchronous starting, Figure 12.15g

Cycloconverters and load commutated inverters (LCIs) have also been used for motor starting [13,14]. During starting and at low speeds the motor does not have enough back EMF to commute the inverter thyristors and auxiliary means must be provided. A synchronous motor runs synchronized at low speed, with excitation applied, and accelerates smoothly as LCI frequency increases. The current from the supply system can be reduced to 150%–200% of the full load current, and the starting torque need only be slightly higher than the load torque. The disadvantages are cost, complexity, and large dimensions of the starter. Tuned capacitor filters can be incorporated to control harmonic distortion and possible resonance problems with the load generated harmonics. Large motors require a coordinated starting equipment design.

Figure 12.15g shows that one starter can be switched to start a number of similar rated motors. In fact, any of the reduced voltage systems can be used to start a number of similar motors, though additional switching complexity is involved.

7. Starting through an auxiliary motor, Figure 12.15f

It is an uncommon choice, yet a sound one in some circumstances involving a large machine in relation to power supply system capabilities. No asynchronous torque is required for acceleration and the motor design can be simplified. The starting motor provides an economical solution only when the load torque during acceleration is small. Disadvantages are increased shaft length and the rotational losses of the starting motor, after the main motor is synchronized.

12.13.3.1 Number of Starts and Load Inertia

NEMA Ref. [8] specifies two starts in succession with motor at ambient temperature and for the WK^2 of load and the starting method for which motor was designed and one start with motor initially at a temperature not exceeding its rated load temperature. If the motor is required to be subjected to more frequent starts, it should be so designed. The starting time is approximately given by the expression

$$t = \frac{2.74 \sum WK^2 N_s^2 10^5}{P_r} \int \frac{dn}{T_a - T_l} \tag{12.126}$$

where
- t is accelerating time in s
- P_r is rated output
- other terms have been described before

As per Equation 12.126, accelerating time is solely dependent upon the load inertia and inversely proportional to the accelerating torque. The normal inertia loads of the synchronous and induction motors are tabulated in [8]. For a synchronous machine it is defined by the following equation:

$$WK^2 = \frac{0.375(\text{hp rating})^{1.15}}{(\text{Speed rpm}/1000)^2}\ \text{lb-ft}^2 \tag{12.127}$$

The heat produced during starting is given by the following expression:

$$h = 2.74 \sum WK^2 N_s^2 10^{-6} \int \frac{T_a}{T_a - T_l} s\,ds \tag{12.128}$$

where h is the heat produced in kW-s and, again, it depends on load inertia. The importance of load inertia on the starting time and accelerating characteristics of the motor is demonstrated.

The inertia constant H is often used in the dynamic motor starting studies. It is defined as

$$H = \frac{(0.231)(WK^2)(r/\min)^2 \times 10^{-6}}{\text{kVA}} s \tag{12.129}$$

where WK^2 is the combined motor and load inertia in lb-ft^2. The inertial constant does not vary over large values. Reference [15] provides some typical values for synchronous generators. The accelerating torque curve (motor torque-load torque) can be plotted from standstill to motor full load speed. The motor torque-speed curve can be adjusted for the expected voltage dip during starting (Figure 12.11). A severe voltage dip or reduced voltage starting methods may reduce motor torque below the load torque, and the motor cannot be started. The accelerating torque curve thus calculated can be divided into a number of incremental steps and average accelerating torque calculated for each incremental speed (Figure 12.11). Then the accelerating time Δt can be calculated on a step-by-step basis for each speed incremental:

$$\Delta t = \frac{WK^2 \Delta n}{307T} s \tag{12.130}$$

where
WK^2 is in lb-ft^2
T is in lb-ft

12.13.4 Starting of Synchronous Motors

Synchronous motors of the revolving field type can be divided into two main catagories: Salient solid pole synchronous motors, which are provided with concentrated field windings on the pole bodies; and cylindrical rotor synchronous motors, which have distributed phase winding over the periphery of the rotor.

The salient pole synchronous motor can be solid pole type and has poles in one piece of solid mass, made of cast steel, forged steel, or alloy steel. The starting torque is produced by eddy currents induced in the pole surface. Sometimes, slots are provided in the pole surface to absorb thermal stresses.

The salient laminated pole type has poles of punched sheet steel, laminated, compressed, and formed. The pole head may be equipped with starting windings of the deep-slot squirrel-cage type or double squirrel-cage type.

Synchronous induction motors, can again be of a salient pole or cylindrical rotor type. These motors can be designed to operate as induction motors for a short period when they fall out of step, that is, due to a sudden reduction in system voltage or excessive load torque and resynchronize automatically on reduction of load torque or restoration of voltage.

The starting of synchronous motors should consider similar factors as discussed for the induction motors, with the addition of synchronization, pulling out of step, and resynchronization.

The synchronous motor design permits lower starting currents for a given starting torque as compared with a squirrel-cage induction motor, typically, in ratio of 1:1.75. This may become an important system design consideration, when the starting voltage dips in the power system have to be limited to acceptable levels.

Figure 12.16 depicts the starting characteristics of two basic types of synchronous motors: salient pole and laminated pole. Salient pole motors have higher starting torque (curve a) and pulsating or oscillating torque (curve b). The cylindrical rotor machines have starting torque characteristics akin to induction motors (curve c) and smaller oscillating torque (curve d). Salient laminated pole construction with damper windings may produce a characteristic somewhere in-between the solid pole and cylindrical rotor designs.

Referring to Figure 12.16, T_a is the asynchronous torque of the motor analogous to the induction motor starting torque, as a synchronous motor is started asynchronously. The rotor saliency causes another torque T_p, which pulsates at a frequency equal to twice the slip frequency. Considering a supply system frequency of f Hz, and a rotor slip of s, the magnetic field in the rotor has a frequency of sf. Due to saliency it can be divided into two components: (a) a forward revolving field at frequency sf in the same direction as the rotor, and (b) a field revolving in reverse direction at sf. Since the rotor revolves at a speed $(1-f)s$, the forward field revolves at $(1-s)f+sf$, that is, at fundamental frequency with respect to stator, in synchronism with the rotating field produced by the stator three-phase windings.

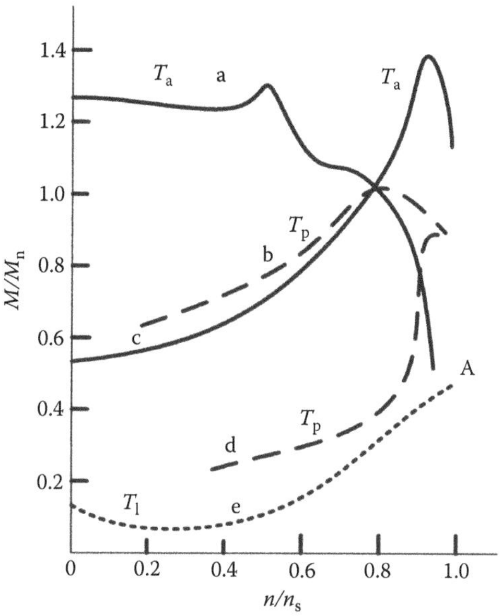

FIGURE 12.16
Starting characteristics of synchronous motors.

The interaction of this field with the stator field produces torque T_a in a fixed direction. Negative sequence field revolves at $(1-f)s - sf = (1-2f)s$. This is as viewed from the stator and, thus, it has a slip of $f - (1-2s)f = 2sf$ with respect to the field produced by the exciting current. The total torque produced by the synchronous motor is therefore

$$T_m = T_a + T_p \cos 2s\omega t \tag{12.131}$$

The currents corresponding to the positive and negative fields are I_a and I_p. The current I_a is at fundamental frequency and I_p is the pulsating current at $(1-2s)f$ Hz. The total current is given by the expression

$$I = I_a \cos(\omega t - \phi) + I_p \cos(1-2s)\omega t \tag{12.132}$$

where ϕ is the power factor angle at starting.

Due to electrical and magnetic asymmetry, T_a develops a sudden change near 50% speed, resulting in a singular point. This is called Gorges Phenomenon [16]. The shape and size of this dip are governed by machine data and the power system. If resistance is zero, the saddle disappears. Solid poles motor without damper windings and slip-rings will give a more pronounced saddle, while the saddle is a minimum in cylindrical rotor and laminated pole constructions with damper windings. The external resistances during asynchronous starting impact both, T_a and T_p and the saddle. At slip $s = 0.5$, the reverse rotating field comes to a halt as viewed from the stator and torque T_p becomes zero, so also the pulsating current I_p. For slip less than 0.5, the direction of the torque changes; this produces characteristic saddle. The average starting torque and the starting current can be calculated based upon the direct axis and quadrature-axis equivalent circuits of the synchronous machine [17].

Figure 12.16 also shows the load torque T_l during starting (curve e). The accelerating torque is $T_a - T_l$ and the motor will accelerate till point A, close to the synchronous speed. At this moment, pull-in torque refers to the maximum load torque under which the motor will pull into synchronism while overcoming inertia of the load and the motor when the excitation is applied. The torque of the motor during acceleration as induction motor at 5% slip at rated voltage and frequency is called nominal pull-in torque. Pull-in and pull-out torques of salient pole synchronous motors for normal values of load inertia are specified in NEMA standards [8]. For motors of 1250 hp and above, speeds 500–1800 rpm, the pull-in torque should not be less than 60% of the full load torque. The maximum slip at pull-in can be approximately determined from the following inequality:

$$s < \frac{242}{N_s}\sqrt{\frac{P_m}{\sum WK^2 f}} \tag{12.133}$$

where

N_s is the synchronous speed in rpm
f is the system frequency
P_m is the maximum output of the synchronous motor with excitation applied, when pulling into synchronism
WK^2 is the inertia of the motor and driven load in kg-m^2

It is necessary to ascertain the load torque and operating requirements before selecting a synchronous motor application. For example, wood chippers in the paper industry have large pull-out torque requirements of 250%–300% of the full load torque. Banbury mixers require starting, pull-in, and pull-out torques of 125%, 125%, and 250% respectively. Thus, the load types are carefully considered in applications. Load inertia impacts the heat produced during starting and the starting time. NEMA [8] gives the starting requirements of pumps and compressors and inertia ratios.

Example 12.5

A 3500 hp induction motor is started in a system configuration as shown in Figure 12.17a. The only load on the transformer secondary is the 3500 hp motor. This is a typical starting arrangement

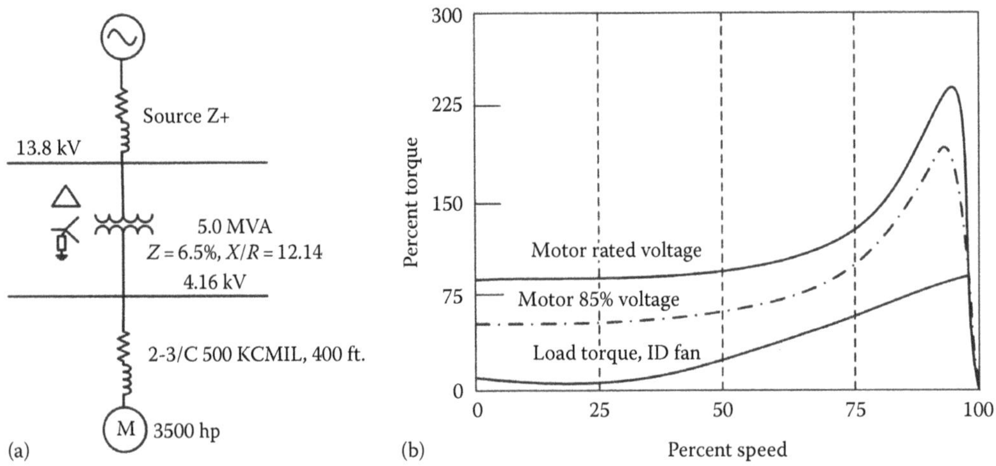

FIGURE 12.17
(a) System configuration and impedances for starting of 3500 hp motor; (b) motor and load torque speed characteristics.

for large motors. While a motor connected to a dedicated transformer may be able to tolerate and accelerate its load, the other loads if connected to the same bus may shut down or experience unacceptable voltage dips during starting. The motor is four-pole, rated voltage 4 kV, and has a full load power factor of 92.89%, efficiency = 94.23%, and full load current is 430.4 A. Again the motor rated voltage is slightly lower than the rated bus voltage of 4.16 kV. The locked rotor current is 576% at 19.47% power factor. This gives a starting impact of 3.344 MW, 17.35 Mvar, equivalent to 17.67 MVA. This starting impact is calculated at rated voltage. As the starting impact load is a constant impedance load, if the starting voltage is lower the starting impact load will reduce proportionally. The transformer is rated 5.00 MVA. Short-term loading of the transformer is acceptable, but if more frequent starts are required, the transformer rating must be carefully considered. The 5 MVA transformer of 6.5% impedance can take a three-phase short-circuit current of 12.3 kA for 2 s according to ANSI/IEEE [18] and also NEMA [8] allows two starts per hour, one with the motor at ambient temperature and the other when the motor has attained its operating temperature.

The motor torque speed characteristics, power factor, and slip are shown in Figure 12.17b. It has a locked rotor torque of 88.17% and breakdown torque of 244.5% at rated voltage. The standard load inertial according to NEMA for this size of motor is 8700 lb-ft^2. The motor drives a boiler ID (Induced Draft) fan of twice the NEMA inertia = 17400 lb-ft^2 ($H = 4.536$). To this must be added the inertia of the motor and coupling, say the total load inertia is 19790 lb-ft^2 ($H = 5.16$). The speed torque curve of the load is shown in Figure 12.17b. For a dynamic motor starting study this data must be obtained from the manufacturer.

The system configuration shown in Figure 12.17a has a source impedance (positive sequence) of $0.166 + j0.249$ ohm. The transformer has an impedance of 6.5%, $X/R = 12.14$ and the motor is connected through two cables in parallel per phase of 350 KCMIL, 400 ft long.

The starting current, slip, and motor terminal voltage are shown in Figure 12.18b. It is seen that there is a starting voltage dip of 16% and the starting current is reduced approximately

(a)

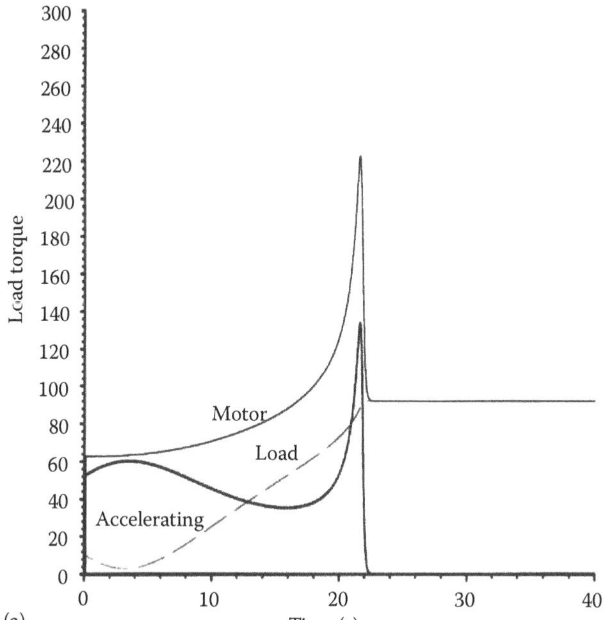

FIGURE 12.18
(a) Motor, load, and accelerating torque;

(continued)

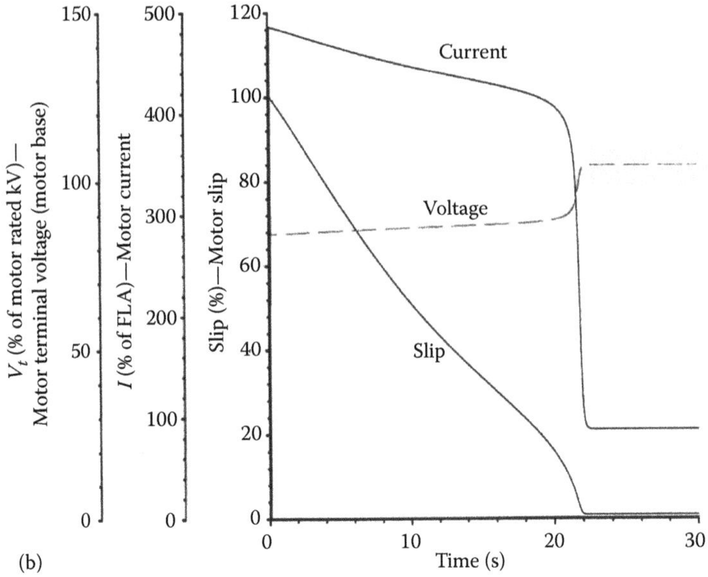

(b)

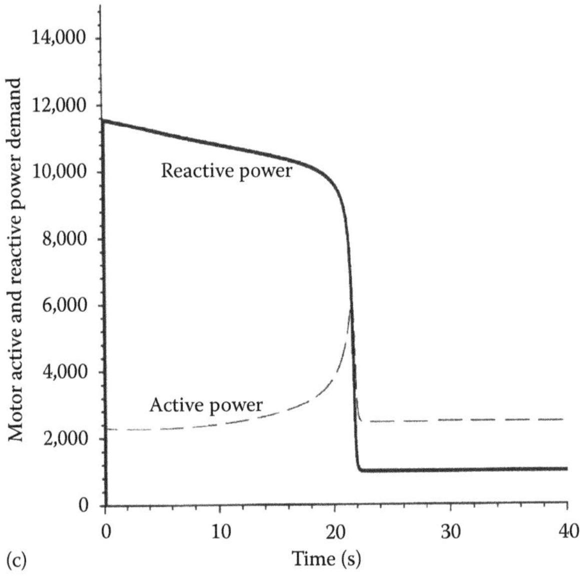

(c)

FIGURE 12.18 (continued)
(b) motor current, terminal voltage, and slip; (c) starting active and reactive power demand, Example 12.5.

proportional to the voltage dip. The voltage dip is shown based upon the rated motor voltage of 4 kV. The motor takes approximately 22 s to start. It is not unusual to see a starting time of 40–50 s for boiler ID fan motors. Figure 12.18a shows the starting torque, the load torque, and the accelerating torque, and Figure 12.18c shows the active and reactive power drawn by the motor during starting.

TABLE 12.5

Calculation of Starting Time of Motor (Example 12.6)

		Average Torque		
No.	**Speed Range Δn**	**%**	**lb-ft²**	**Time Δt (s)**
1	0%–25%	50.3	5274	5.5
2	25%–50%	44.70	4687	6.18
3	50%–75%	39.0	4089	7.09
4	75% to full load	65	6816	4.25

$\Sigma\Delta t = 23.02$.

Example 12.6

Calculate the starting time using the data in Example 12.5 by hand calculations.

As per procedure described above first calculate the voltage dip at the motor terminals. This is a simple one step load flow problem. Applying a starting impact load of 3.344 MW, 17.35 Mvar to all the impedances shown in series in Figure 12.17 gives an approximate voltage dip of 15%. Assuming that this voltage dip remains practically throughout the starting period, reduce the motor torque speed characteristics by $(0.85)^2$, that is by 72% (Figure 12.17b). For this example divide the starting period into four sections and calculate the accelerating torque in each strip. The result is shown in Table 12.5. Calculate the average torque in each period. Then using Equation 12.130, Δt for each incremental Δn is calculated. The summation gives 23.02 s, which is approximately close to the calculation result in Example 12.5. Smaller the steps Δn, more accurate the result.

12.14 Practical Load Flow Studies

The requirements for load flow calculations vary over a wide spectrum, from small industrial systems to large automated systems for planning, security, reactive power compensation, control, and on-line management. The essential requirements are

- High speed, especially important for large systems
- Convergence characteristics, which are of major consideration for large systems, and the capability to handle ill-conditioned systems
- Ease of modifications and simplicity, that is, adding, deleting, and changing system components, generator outputs, loads, and bus types
- Storage requirement, which becomes of consideration for large systems

The size of the program in terms of number of buses and lines is important. Practically, all programs will have data reading and editing libraries, capabilities of manipulating system variables, adding or deleting system components, generation, capacitors, or slack buses. Programs have integrated databases, that is, the impedance data for short-circuit or load flow calculations need not be entered twice, and graphic user interfaces. Which type of algorithm will give the speediest results and converge easily is difficult to predict precisely. Table 11.7 shows a comparison of earlier Z and Y matrix methods. Most programs will incorporate more than one solution method. While the Gauss–Seidel method with acceleration is still an option for smaller systems, for large systems some form of the NR decoupled method and fast load flow algorithm are commonly used, especially for optimal power flow studies. Speed can be accelerated by optimal ordering (Appendix D). In fast decoupled load flow the convergence is geometric, and less than five iterations are required for practical accuracies. If differentials are calculated efficiently the speed of the fast decoupled method can be even five times that of the NR method. Fast decoupled load flow is employed in optimization studies and in contingency evaluation for system security.

The preparation of data, load types, extent of system to be modeled, and specific problems to be studied are identified as a first step. The data entry can be divided into four main categories: bus data, branch data, transformers and phase shifters, and generation and load data. Shunt admittances, that is, switched capacitors and reactors in required steps, are represented as fixed admittances. Apart from voltages on the buses, the study will give branch power flows, identify transformer taps, phase-shifter angles, loading of generators and capacitors, power flow from swing buses, load demand, power factors, system losses, and overloaded system components.

Example 12.7

Consider a 15-bus network shown in Figure 12.19. It is solved by the fast decoupled load flow method, using a computer program. The impedance data are shown in Table 12.6. The bus types and loading input data are in Table 12.7. The bus voltages on PV buses and generator maximum and minimum var limits are specified in this table. All loads are modeled as *constant kVA loads*. There are 100 Mvar of shunt power capacitors in the system; this is always a constant impedance load.

The results of the load flow calculation are summarized in Tables 12.8 through 12.10. Table 12.8 shows bus voltages, Table 12.9 shows power flows, losses, bus voltages, and voltage drops, and Table 12.10 shows the overall load flow summary. The following observations are of interest:

- Voltages on a number of buses in the distribution are below acceptable level. The voltage on the 138 kV bus 10 is 12% below rated voltage and at bus 4 is 6.1% below rated voltage. There is a 5.85% voltage drop in transmission line L_8. The voltages at buses 2 and 3 are 4.2 and 6.6%, respectively, below rated voltages. Bus 9 voltage cannot be maintained at 1.04 per unit with the generator supplying its maximum reactive power. When corrective measures are to be designed, it is not necessary to address all the buses having low operating voltages. In this example, correcting voltages at 230 kV buses 2 and 3 will bring up the voltages in the rest of the system. Bus 10 will require additional compensation.
- As a constant kVA load model is used, the load demand is the sum of the total load plus the system losses (Table 12.10). It is 431.35 MW and 176.88 Mvar, excluding the Mvar supplied by shunt capacitors.
- There is a total of 100 Mvar of power capacitors in the system. These give an output of 84.14 Mvar. The output is reduced as a square of the ratio of operating voltage to rated voltage. The behavior of shunt power capacitors is discussed in Chapter 13.
- The generators operate at their rated active power output. However, there is little reactive power reserve as the overall operating power factor of the generators is 88.90 lagging.

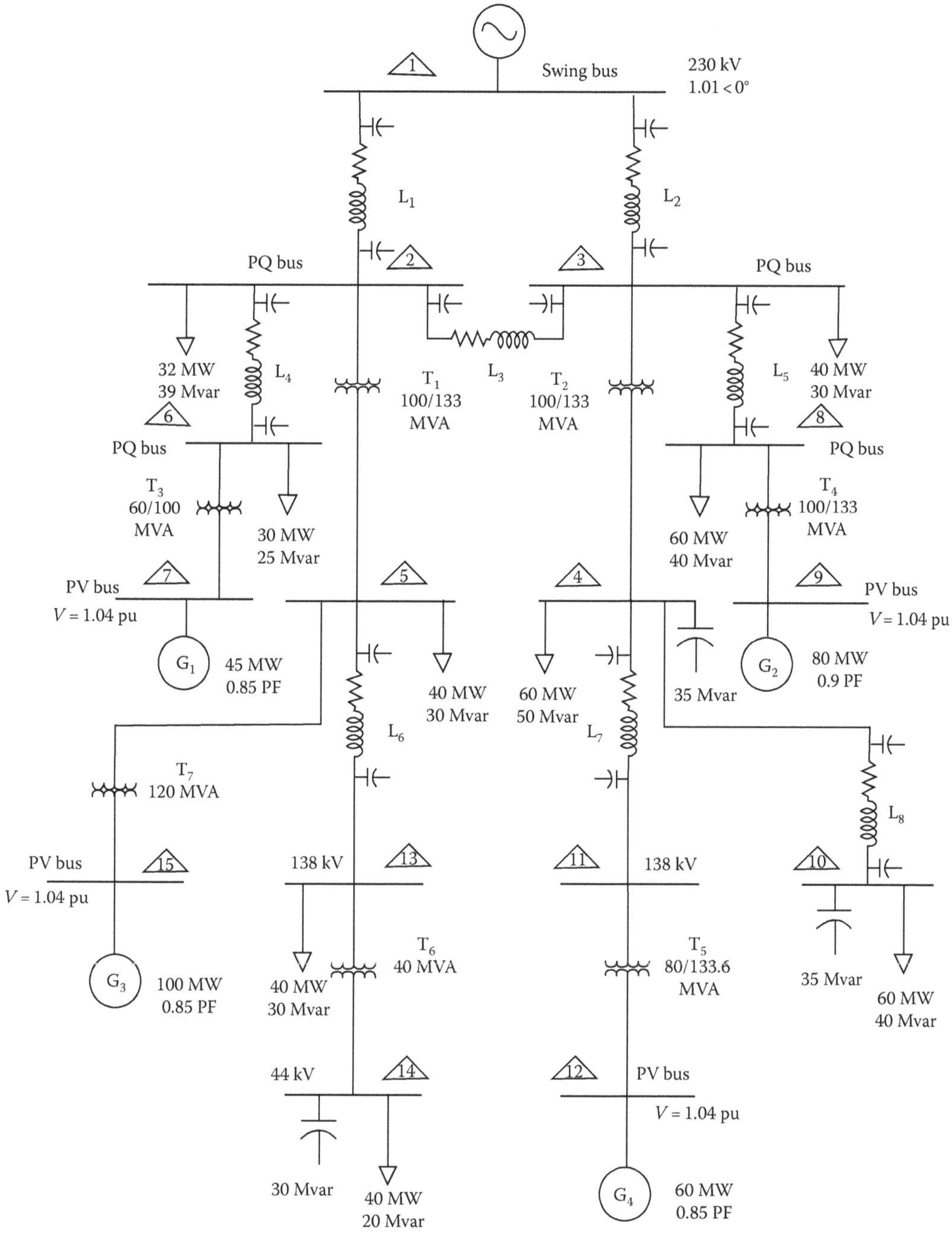

FIGURE 12.19
A 15-bus power system for Examples 12.5 and 12.6.

TABLE 12.6

Examples 12.7 and 12.8: Impedance Data

From Bus	To Bus	Connection Type	Impedance Data in per Unit on a 100 MVA Base, Unless Specified Otherwise
1	2	L_1: 230 kV transmission line	$Z = 0.0275 + j0.1512$, Π-model, $y = j0.2793$
1	3	L_2: 230 kV transmission line	$Z = 0.0275 + j0.1512$, Π-model, $y = j0.2793$
2	3	L_3: 230 kV transmission line	$Z = 0.0413 + j0.2268$, Π-model, $y = j0.4189$
2	6	L_4: 230 kV transmission line	$Z = 0.0157 + j0.008$, Π-model, $y = j0.1396$
3	8	L_5: 230 kV transmission line	$Z = 0.0157 + j0.008$, Π-model, $y = j0.1396$
4	11	L_6: 138 kV transmission line	$Z = 0.0224 + j0.1181$, Π-model, $y = j0.0301$
4	10	L_7: 138 kV transmission line	$Z = 0.0224 + j0.1181$, Π-model, $y = j0.0301$
5	13	L_8: 138 kV transmission line	$Z = 0.0373 + j0.1968$, Π-model, $y = j0.0503$
2	5	T_1: 100/133 MVA two-winding transformer, 230–138 kV	$Z = 10\%$ on 100 MVA base, $X/R = 34.1$, tap 1:1.02
3	4	T_2: 100/133 MVA two-winding transformer, 230–138 kV	$Z = 10\%$ on 100 MVA base, $X/R = 34.1$, tap 1:1.03
6	7	T_3: 60/100.2 MVA transformer, 230–13.8 kV (generator step-up transformer)	$Z = 10\%$ on 60 MVA base, $X/R = 34.1$, tap 1:1.03
8	9	T_4: 100/133 MVA transformer, 230–13.8 kV (generator step up transformer)	$Z = 10\%$ on 100 MVA base, $X/R = 34.1$, tap 1:1.04
11	12	T_5: 80/133.6 MVA two-winding transformer, 138–13.8 kV (generator step up transformer)	$Z = 9\%$ on 80 MVA base, $X/R = 34.1$, tap 1:1.05
13	14	T_6: 40 MVA two-winding transformer, 138–44 kV	$Z = 8\%$ on a 40 MVA base, $X/R = 27.3$, tap 1:1.01
5	15	T_7: 120 MVA two-winding transformer (generator step-up transformer), 138–13.8 kV	$Z = 11\%$ on a 120 MVA base, $X/R = 42$, tap 1:1.02

TABLE 12.7

Examples 12.7 and 12.8: Bus, Load, and Generation Data

Bus Identification	Type of Bus	Nominal Voltage (kV)	Specified Voltage (per Unit)	Bus Load (MW/Mvar)	Shunt Capacitor (Mvar)	Generator Rating (MW) Power Factor, and max/min Var Output (Mvar)
1	Swing	230	1.01 < 0	—		—
2	PQ	230	—	52/39		
3	PQ	230	—	40/30		
4	PQ	138	—	60/50	35	
5	PQ	138	—	40/30		
6	PQ	230	—	30/25		
7	PV	13.8	1.04 < ?	—	—	45 MW, 0.85 pF (28/3)
8	PQ	230	—	60/40		
9	PV	13.8	1.04 < ?	—		80 MW, 0.9 pF (38/3)
10	PQ	138	—	60/40	35	
11	PQ	138	—	—		
12	PV	13.8	1.04 < ?	—	—	60 MW, 0.85 pF (37/3)
13	PQ	138	—	40/30	—	—
14	PQ	44		40/30	30	
15	PV	13.8	1.04 < ?			100 MW, 0.85 pF (62/6)

TABLE 12.8

Example 12.7: Bus Voltages, Magnitudes, and Phase Angles (Constant kVA Loads)

Bus #	Voltage	Bus #	Voltage	Bus #	Voltage
1	1.01 < 0.0°	6	0.9650 < −5.0°	11	0.9696 < −6.4°
2	0.9585 < −5.7°	7	1.04 < −0.7°	12	1.04 < −2.4°
3	0.9447 < −6.6°	8	0.9445 < −5.5°	13	0.9006 < −13.0°
4	0.9393 < −10.7°	9	1.0213 < −0.6°	14	0.9335 < −18.7°
5	0.9608 < −7.1°	10	0.8808 < −18.6°	15	1.03.54 < −1.8°

Some reserve spinning power capability should be available to counteract sudden load demand or failure of a parallel line or tie circuit.

- There are two devices in tandem that control the reactive power output of the generators—the voltage-ratio taps on the transformer and the generator voltage regulators. The role of these devices in controlling reactive power flow is discussed in Chapter 13.
- System losses are 9.337 MW and 42 Mvar. An explanation of the active power loss is straightforward; the reactive power loss mainly depends on the series reactance and how far the reactive power has to travel to the load (Chapter 13). Table 12.9 identifies branch flows from-and-to between buses and the losses in each of the circuits. The high-loss circuits can be easily identified.

TABLE 12.9

Example 12.7: Active and Reactive Power Flows, Losses, Bus Voltages, and Percentage Voltage Drops (Constant kVA Loads)

	Connected Buses		From–To Flow		To–From Flow		Losses		Bus Voltage (%)		Voltage
Circuit	From Bus #	To Bus #	MW	Mvar	MW	Mvar	kW	kvar	From	To	Drop (%)
L_1	1	3	68.174	10.912	−66.748	−30.160	1426.0	−19,248.6	101.00	95.85	5.15
L_2	1	2	78.177	19.253	−76.219	−35.239	1957.7	−15,986.0	101.00	94.47	6.53
L_3	2	3	6.819	−14.602	−6.788	−23.169	30.6	−37,770.9	95.85	94.47	1.38
L_4	2	6	−14.794	−11.240	14.835	−1.466	41.3	−12,705.9	95.85	96.50	0.65
T_1	2	5	22.732	17.002	−22.697	−16.126	25.7	876.3	95.85	96.08	0.23
L_5	3	8	−19.786	−2.021	19.857	−10.072	71.9	−12,092.4	94.47	94.45	0.01
T_2	3	4	62.792	30.428	−62.632	−24.975	159.9	5,453.6	94.47	93.93	0.53
L_7	4	11	−58.956	−12.124	59.867	14.183	910.9	2,059.9	93.93	96.96	3.03
L_8	4	10	61.685	18.025	−59.904	−12.790	1781.3	5,234.9	93.93	88.08	5.85
L_6	5	13	82.402	35.807	−80.421	−27.964	1981.2	7,842.5	96.08	90.06	6.01
T_7	5	15	−99.704	−49.680	−99.997	61.999	293.3	12,319.0	96.08	103.54	7.46
T_3	6	7	−44.865	−23.418	44.999	28.000	134.4	4,581.9	96.50	103.99	7.49
T_4	8	9	−79.761	−29.872	79.999	38.000	238.3	8,127.8	94.45	102.13	7.68
T_5	11	12	−59.867	−14.183	60.000	18.711	132.8	4,527.8	96.96	104.00	7.04
T_6	13	14	40.422	−2.035	−40.274	6.072	147.8	4,036.1	90.06	93.35	3.29

TABLE 12.10

Example 12.7: Load Flow Summary (Constant kVA Loads)

Item	MW	Mvar	MVA	Percentage Power Factor
Swing bus	146.35	30.16	149.43	97.9 lagging
Generators	285.00	146.71	320.55	88.90 lagging
Total demand	431.35	176.875	466.21	92.5 lagging
Total load modeled	422.01 MW, constant kVA	303 Mvar, constant kVA	519.99	81.2 lagging
Power capacitors output		−84.18		
Total losses	9.337	42.742		
System mismatch	0.010	0.003		

Example 12.8

The load flow of Example 12.7 is repeated with constant impedance representation of loads. All other data and models remain unchanged.

Table 12.11 shows bus voltages, Table 12.12 branch load flow and losses, and Table 12.13 an overall load summary. The results can be compared with those of the corresponding tables for the constant kVA load model of Example 12.5. The following observations are of interest:

- In Example 12.7, with a constant kVA load, the overall demand (excluding the reactive power supplied by capacitors) is 431.35 MW and 176.88 Mvar. With the constant impedance load model, "off-loading" occurs with voltage reduction, and the load demand reduces to 397.29 MW and 122.66 Mvar. This helps the system to recover and the resulting voltage drops throughout are smaller as compared to those of a constant kVA load.
- The bus voltages improve in response to the reduced load demand. These are 2%–5% higher as compared to a constant kVA load model.
- The generators reactive output reduces in response to increased bus voltage. The redistribution of active power changes the phase angles of the voltages.
- The active power loss decreases and the reactive power loss increases. This can be explained in terms of improved bus voltages, which allow a greater transfer of power between buses and hence a larger reactive power loss.

TABLE 12.11

Example 12.8: Bus Voltages, Magnitudes, and Phase Angles (Constant Impedance Loads)

Bus #	Voltage	Bus #	Voltage	Bus #	Voltage
1	$1.01 < 0.0°$	6	$0.9750 < -3.5°$	11	$0.9841 < -4.0°$
2	$0.9717 < -4.3°$	7	$1.04 < 0.8°$	12	$1.04 < 0.0°$
3	$0.9659 < -5.0°$	8	$0.9646 < -3.8°$	13	$0.9312 < -10.2°$
4	$0.9692 < -8.2°$	9	$1.04 < 0.9°$	14	$0.9729 < -15.2°$
5	$0.9765 < -5.1°$	10	$0.9375 < -14.7°$	15	$1.04 < 0.2°$

TABLE 12.12

Example 12.8: Active and Reactive Power Flows, Losses, Bus Voltages, and Percentage Voltage Drops (Constant Impedance Type Loads)

	Connected Buses		From–To Flow		To–From Flow		Losses		Bus Voltage (%)		Voltage
Circuit	From Bus #	To Bus #	MW	Mvar	MW	Mvar	kW	kvar	From	To	Drop (%)
L_1	1	3	51.949	3.668	−51.134	−26.625	815.4	−22,957.8	101.00	97.17	3.83
L_2	1	2	60.344	6.662	−59.240	−27.892	1103.8	−21,230.4	101.00	96.59	4.41
L_3	2	3	5.512	−18.254	−5.498	−20.993	14.3	−39,247.6	97.17	96.59	0.58
L_4	2	6	−16.287	−7.255	16.331	−5.751	44.1	−13,005.9	97.17	97.50	0.33
T_1	2	5	12.807	15.307	−12.794	−14.885	12.4	421.6	97.17	97.65	0.47
L_5	3	8	−23.729	−2.245	23.826	−10.291	97.6	−12,536.2	96.59	96.64	0.05
T_2	3	4	51.148	23.141	−51.049	−19.764	99.0	3,376.6	96.59	96.92	0.33
L_7	4	11	−59.046	−0.277	59.877	1.785	830.8	1,508.3	96.92	98.41	1.49
L_8	4	10	53.830	5.994	−52.652	−4.345	1178.1	1,648.2	96.92	93.75	3.17
L_6	5	13	74.394	25.049	−72.930	−20.070	1463.3	4,978.4	97.65	93.12	4.53
T_7	5	15	−99.738	−38.769	−100.00	49.774	262.0	11,005.2	97.65	104.00	6.35
T_3	6	7	−44.880	−17.905	45.00	24.301	120.0	4,091.6	97.50	104.00	6.50
T_4	8	9	−79.773	−27.016	80.00	21.996	222.7	7,592.4	96.64	104.00	7.36
T_5	11	12	−59.878	−1.785	60.00	34.608	122.2	4,167.1	98.41	104.00	5.59
T_6	13	14	38.248	−5.943	−38.121	9.396	126.5	3,453.3	94.12	97.29	4.17

TABLE 12.13

Example 12.8: Load Flow Summary (Constant Impedance-Type Load)

Item	MW	Mvar	MVA	Percentage Power Factor
Swing bus	112.29	10.39	112.77	99.6 lagging
Generators	285.00	112.33	306.34	93.0 lagging
Total demand	397.293	122.660	415.80	95.5 lagging
Total load demand	390.785	189.40	434.26	89.9 lagging
Total constant impedance-type load modeled	422.01	303 (and 100 Mvar capacitors) = :202 Mvar		0
Total losses	6.51	66.74		
System mismatch	0.011	0.001		

12.14.1 Contingency Operation

Consider now that the largest generator of 100 MW on bus 15 is tripped. Assuming that there is no widespread disruption, the operating voltages on buses 13–15 dip by 12%–15%. Also, the voltage on the 230 kV bus 9 is down by approximately 9%. Even if the swing bus has all the spinning reserve and the capability to supply the increased demand due to loss of the 100 MW generator, and none of the system components, transformers, or lines is overloaded, the system has potential voltage problems.

While this load flow is illustrative of what to expect in load flow results, the number of buses in a large industrial distribution system may approach 300 and in the utility systems 5000 or more. Three-phase models may be required. The discussions are continued in the chapters to follow.

Problems

(These problems can be solved without modeling on a digital computer.)

12.1 Solve the following equations, using the NR method, to the second iteration. Initial values may be assumed as equal to zero.

$$9x_1 - x_1x_2 - 6 = 0$$
$$x_1 + 6x_2 - x_3^2 - 10 = 0$$
$$x_2x_3^2 - 10x_3 + 4 = 0$$

12.2 Write power flow equations of the network shown in Figure P12.1, using the NR method in polar form.

12.3 Solve the three-bus system of Figure P12.1, using the NR method to two iterations.

12.4 Solve Problem 12.3, using the decoupled NR method, to two iterations.

12.5 Construct P–θ and Q–V networks of the system shown in Figure P12.1.

12.6 Solve the network of Figure P12.1, using the fast decoupled method, to two iterations.

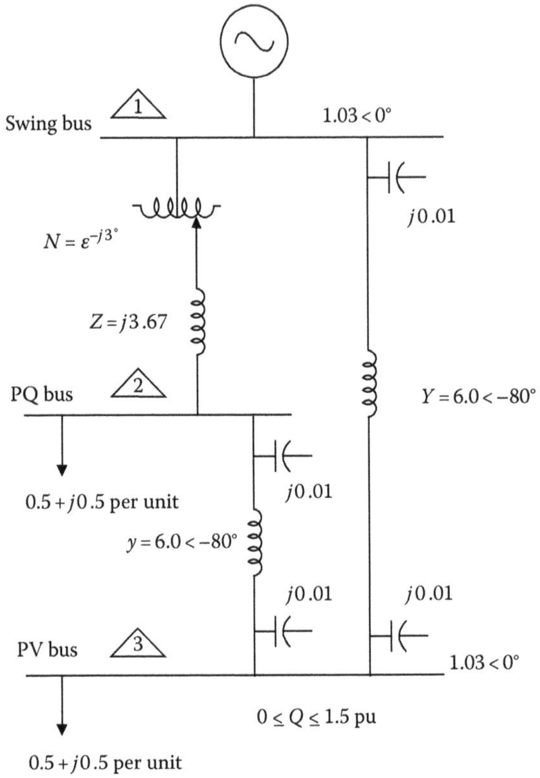

FIGURE P12.1
A three-bus system for Problems 12.2, 12.3, 12.5 through 12.7.

12.7 Solve the network of Figure P12.1 using the Ward-Hale method.

12.8 Solve Example 12.3 for one more iteration and show that the bus 2 voltage dips below rated system voltage.

12.9 What are the major differences in constant kVA, constant impedance, and constant current loads? How will each of these behave under load flow?

12.10 How can the initial starting drop of a motor be calculated using a load flow program? Is a specific algorithm necessary? What is the effect of generator models?

12.11 A 1500 hp, 4 kV, four-pole induction motor has a slip of 1.5%, a locked rotor current six times the full load current, and 94% efficiency. Considering that $X_m = 3$ per unit, and stator reactance = rotor reactance = 0.08 per unit, draw the equivalent positive and negative sequence circuit. If the voltage has a 5% negative sequence component calculate the positive sequence and negative sequence torque.

12.12 What are the dimensions of the Y matrix in Example 12.7? What is the percentage of populated elements with respect to the total elements?

12.13 Calculate the starting time, current, voltage, and torque profiles by hand calculations for 2000 hp, 2.4 kV, four-pole motor, full load slip = 1.75%, full load power factor = 0.945, full load efficiency = 93.5%, starting current = 6 times the full load current at a power factor of 0.15, and total load inertia including that of motor and coupling = 12,000 lb-ft^2. Apply the load and motor torque speed characteristics as shown in Figure 12.17b. The total impedance in the motor circuit at the time of starting = 2.5 pu on 100 MVA base.

12.14 In Problem 12.13, a drive system load can tolerate a maximum 10% voltage dip. Design a motor starting strategy applying at least two starting methods from Table 12.4.

References

1. R.J. Brown and W.F. Tinney. Digital solution of large power networks. *Trans. AIEE PAS* 76: 347–355, 1957.
2. W.F. Tinney and C.E. Hart. Power flow solution by Newton's method. *Trans. IEEE PAS* 86: 1449–1456, 1967.
3. B. Stott and O. Alsac. Fast decoupled load flow. *Trans. IEEE PAS* 93: 859–869, 1974.
4. N.M. Peterson and W.S. Meyer. Automatic adjustment of transformer and phase-shifter taps in the Newton power flow. *Trans. IEEE* 90: 103–108, 1971.
5. M.S. Sachdev and T.K.P. Medicheria. A second order load flow technique. *IEEE Trans. PAS* 96: 189–195, 1977.
6. M.H. Kent, W.R. Schmus, F.A. McCrackin, and L.M. Wheeler. Dynamic modeling of loads in stability studies in Stability of Large Electrical Power Systems, Editors: R.T. Byerly and E.W. Kimberk, IEEE Press, 1974, NJ.
7. EPRI. Load Modeling for Power Flow and Transient Stability Computer Studies. Report EL-5003, 1987.
8. NEMA. Large Machines-Induction Motors. Standard MG1 Part 20, 1993.
9. J.R. Linders. Effect of power supply variations on AC motor characteristics. *IEEE Trans. Ind. Applic.* 8: 383–400, July/August 1972.
10. J.C. Das. Effects of momentary voltage dips on the operation of induction and synchronous motors. *Trans. IEEE Ind. Appl. Soc.* 26: 711–718, 1990.
11. J.C. Das and J. Casey. Characteristics and analysis of starting of large synchronous motors. In: *Conf. Record, IEEE I&CPS Technical Conference*, Sparks, NV, May 1999.
12. G.S. Sangha. Capacitor-reactor start of large synchronous motor on a limited capacity network. *IEEE Trans. Ind. Appl.* 20(5): 1337–1343, September/October 1984.
13. J. Langer. Static frequency changer supply system for synchronous motors driving tube mills. *Brown Boveri Rev.* 57: 112–119, March 1970.
14. C.P. LeMone. Large MV motor starting using AC inverters. In: *ENTELEC Conf. Record*, Houston, TX, May 1984.
15. Westinghouse. *Transmission and Distribution Handbook*, 4th edn. East Pittsburg, PA: Westinghouse Electric and Manufacturing Co., 1964.
16. J. Bredthauer and H. Tretzack. HV synchronous motors with cylindrical rotors. *Siemens Power Eng.* V(5): 241–245, September/October 1983.
17. H.E. Albright. Applications of large high speed synchronous motors. *IEEE Trans. Ind. Appl.* 16(1): 134–143, January/February 1980.
18. IEEE Std. C37.91. IEEE Guide for Protecting Power Transformers, 2008.

13

Reactive Power Flow and Control

Chapters 11 and 12 showed that there is a strong relationship between voltage and reactive power flow, though real power flow on loaded circuits may further escalate the voltage problem. The voltages in a distribution system and to the consumers must be maintained within a certain plus–minus band around the rated equipment voltage, ideally from no load to full load, and under varying loading conditions. Sudden load impacts (starting of a large motor) or load demands under contingency operating conditions, when one or more tie-line circuits may be out of service, result in short-time or prolonged voltage dips. High voltages may occur under light running load or on sudden load throwing and are of equal considerations, though low voltages occur more frequently. ANSI C84.1 [1] specifies the preferred nominal voltages and operating voltage ranges A and B for utilization and distribution equipment operating from 120 to 34,500 V. For transmission voltages over 34,500 V, only nominal and maximum system voltage is specified. Range B allows limited excursions outside range A limits. As an example, for a 13.8 kV nominal voltage, range A = 14.49 – 12.46 kV and range B = 14.5 – 13.11 kV. Cyclic loads, for example, arc furnaces giving rise to flicker, must be controlled to an acceptable level. The electrical apparatuses have a certain maximum and minimum operating voltage range in which normal operation is maintained, that is, induction motors are designed to operate successfully under the following conditions [2]:

1. Plus or minus 10% of rated voltage, with rated frequency.
2. A combined variation in voltage and frequency of 10% (sum of absolute values) provided that the frequency variations do not exceed ±5% of rated frequency.
3. Plus or minus 5% of frequency with rated voltage.

Motor torque, speed, line current, and losses vary with respect to the operating voltage, as shown in Table 12.2. Continuous operation beyond the designed voltage variations is detrimental to the integrity and life of the electrical equipment.

A certain balance between the reactive power consuming and generating apparatuses is required. This must consider losses, which may be a considerable percentage of the reactive load demand.

When the reactive power is transported over mainly reactive elements of the power system, the reactive power losses may be considerable and these add to the load demand (Example 12.7). This reduces the active power delivery capability of most electrical equipment rated on a kVA base. As an example, consider the reactive power flow through a 0.76 ohm reactor. For a 70 Mvar input the output is 50 Mvar and 20 Mvar is lost in the reactor itself. If the load voltage is to be maintained at 1.0 per unit, the source side voltage should be raised to 1.28 per unit, representing a voltage drop of 28% in the reactor. Figure 13.1 shows reactive power loss and voltage drops in lumped reactance.

In a loaded transmission line, when power transfer is below the surge impedance loading, the charging current exceeds the reactive line losses and this excess charging

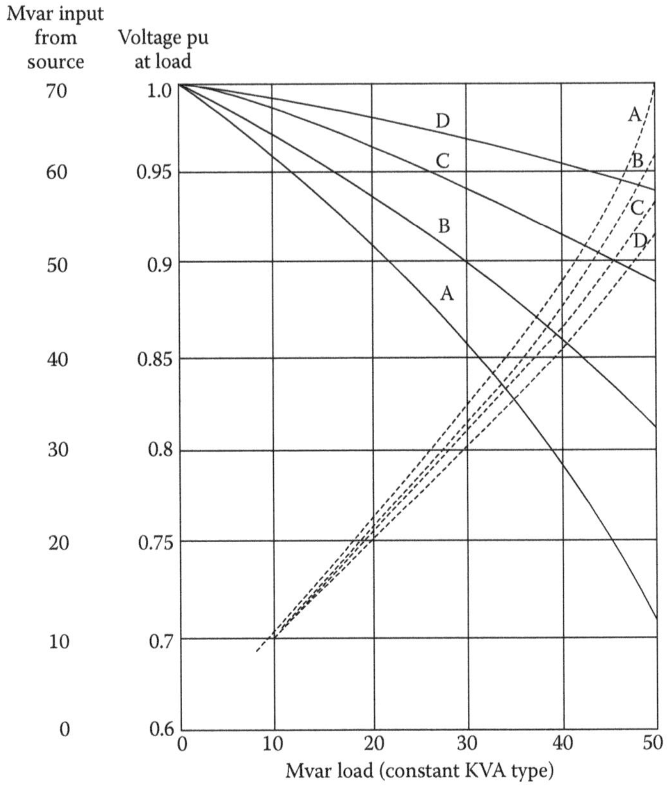

FIGURE 13.1
Voltage drop and reactive power loss in lumped reactance: A: 0.76 ohm; B: 0.57 ohm; C: 0.38 ohm; D: 0.19 ohm.

current must be absorbed by shunt reactors and generators. Above surge impedance loading the reactive power must be supplied to the line. At 1.5 times the surge loading an increase of 150 MW will increase the reactive power losses in a 500 kV transmission line by about 95 Mvar, or about 50% of the line charging. At twice surge loading, approximately 1800 MW (surge impedance 277 ohm), a 100 MW load increase will increase reactive losses by 100 Mvar.

A *V–Q* control may necessitate addition of leading or lagging reactive power sources, which may be passive or dynamic in nature. Shunt reactors and capacitors are examples of passive devices. The static var controller (SVC) is an example of a dynamic device. It should be ensured that all plant generators operate within their reactive power capability limits and remain stable. The on-load tap changing transformers must be able to maintain an acceptable voltage within their tap setting range.

Assessment of voltage problems in a distribution system under normal and contingency load-flow conditions, therefore, requires investigations of the following options:

- Location of reactive power sources, that is, series and shunt capacitors, synchronous condensers, voltage regulators, overexcited synchronous motors, and SVCs with respect to load.
- Control strategies of these reactive power sources.
- Provision of on-load tap changing equipment on tie transformers.

- Undervoltage load shedding.
- Stiffening of the system, that is, reduction of system reactance, which can be achieved by bundle conductors, duplicate feeders, and additional tie-lines.
- Redistribution of loads.

13.1 Voltage Instability

Consider the power flow on a mainly inductive tie-line, Figure 13.2a. The load demand is shown as $P+jQ$, the series admittance $Y_{sr}=g_{sr}+b_{sr}$ and $Z=R_{sr}+jX_{sr}$. A similar circuit is considered in Section 10.2. The power flow equation from the source bus (an infinite bus) is given by

$$
\begin{aligned}
P+jQ &= V_r e^{-j\theta}[(V_s - V_r e^{j\theta})(g_{sr}+jb_{sr})] \\
&= [(V_s V_r \cos\theta - V_r^2)g_{sr} + V_s V_r b_{sr}\sin\theta] + j[(V_s V_r \cos\theta - V_r^2)b_{sr} - V_s V_r g_{sr}\sin\theta]
\end{aligned}
\tag{13.1}
$$

If resistance is neglected,

$$P = V_s V_r b_{sr}\sin\theta \tag{13.2}$$

$$Q = \left(V_s V_r \cos\theta - V_r^2\right) b_{sr} \tag{13.3}$$

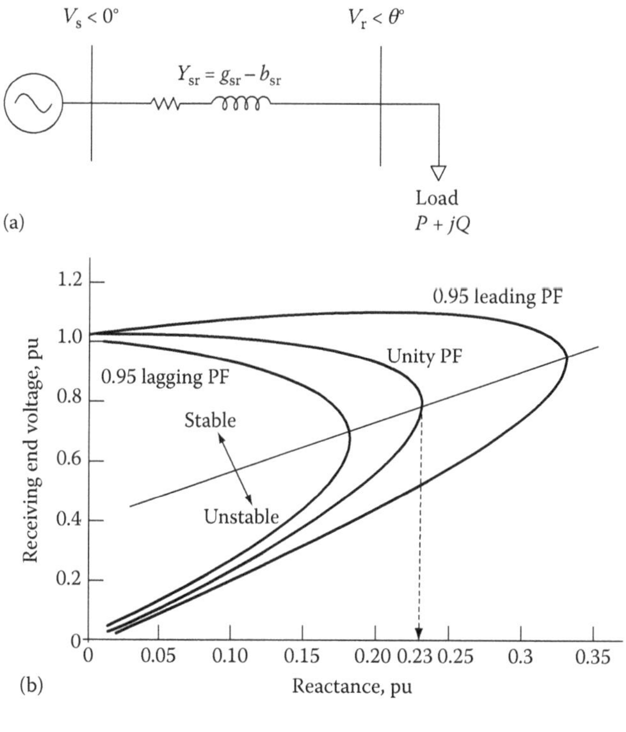

FIGURE 13.2
(a) Active and reactive power flow through a predominantly reactive two-port connector; (b) receiving end voltage per unit versus system reactance in per unit for different power factors and constant power output. Stable and unstable regions and critical reactance are shown.

These are the same equations as derived in Equations 10.22 and 10.23. If the receiving end load changes by a factor $\Delta P + \Delta Q$, then

$$\Delta P = (V_s b_{sr} \sin\theta)\Delta V + (V_s V_r b_{sr} \cos\theta)\Delta\theta \tag{13.4}$$

and

$$\Delta Q = (V_s \cos\theta - 2V_r) b_{sr} \Delta V - (V_s V_r b_{sr} \sin\theta)\Delta\theta \tag{13.5}$$

where
ΔV_r is the scalar change in voltage V_r
$\Delta\theta$ is the change in angular displacement

If θ is eliminated from Equation 13.1 and resistance is neglected, a dynamic voltage equation of the system is obtained as follows:

$$V_r^4 + V_r^2(2QX_{sr} - V_s^2) + X_{sr}^2(P^2 + Q^2) = 0 \tag{13.6}$$

The positive real roots of this equation are

$$V_r = \left[\frac{-2QX_{sr} + V_s^2}{2} \pm \frac{1}{2}\sqrt{(2QX_{sr} - V_s^2)^2 - 4X_{sr}^2(P^2 + Q^2)}\right]^{1/2} \tag{13.7}$$

This equation shows that, in a lossless line, the receiving end voltage V_r is a function of the sending end voltage V_s, series reactance X_{sr}, and receiving end real and reactive power. Voltage problems are compounded when reactive power flows over heavily loaded active power circuits. If reactive power is considered as zero and the sending end voltage is 1 per unit, then Equation 13.7 reduces to

$$V_r = \left[\frac{1}{2} \pm \frac{\sqrt{1 - 4X_{sr}^2 P^2}}{2}\right]^{1/2} \tag{13.8}$$

For two equal real values of V_r,

$$(1 - 4X_{sr}^2 P^2)^{1/2} = 0, \quad \text{i.e., } X_{sr} = \frac{1}{2}P \tag{13.9}$$

This value of X_{sr} may be called a critical reactance. Figure 13.2b shows the characteristics of receiving end voltage, against system reactance for constant power flow, at lagging and leading power factors, with the sending end voltage maintained constant. This figure shows, for example, that at a power factor of 0.9 lagging, voltage instability occurs for any reactance value exceeding 0.23 per unit. This reactance is the critical reactance. For a system reactance less than the critical reactance, there are two values of voltages, one higher and the other lower. The lower voltage represents unstable operation, requiring large amount of source current. For a system reactance close to the critical reactance, voltage instability can occur for a small positive excursion in the power demand. As the power factor improves, a higher system reactance is permissible for the power transfer.

The voltage instability can be defined as the limiting stage beyond which the reactive power injection does not elevate the system voltage to normal. The system voltage can only be adjusted by reactive power injection until the system voltage stability is maintained. From Equation 13.7, the critical system reactance at voltage stability limit is obtained as follows:

$$\begin{aligned} \left(2QX_{\mathrm{sr}} - V_{\mathrm{s}}^2\right)^2 &= 4X_{\mathrm{sr}}^2(P^2 + Q^2) \\ 4X_{\mathrm{sr}}^2 P^2 + 4X_{\mathrm{sr}} Q V_{\mathrm{s}}^2 - V_{\mathrm{s}}^4 &= 0 \end{aligned} \tag{13.10}$$

The solution of this quadratic equation gives

$$\begin{aligned} X_{\mathrm{sr(critical)}} &= \frac{V_{\mathrm{s}}^2}{2}\left[\frac{Q \pm \sqrt{Q^2 - P^2}}{P^2}\right] \\ &= \frac{V_{\mathrm{s}}^2}{2P}(-\tan\phi + \sec\phi) \end{aligned} \tag{13.11}$$

where ϕ is the power factor angle.

Enhancing the thermal capacity of radial lines by use of shunt capacitors is increasingly common. However, there is a limit to which capacitors can extend the load-carrying capability [3].

In practice the phenomenon of collapse of voltage is more complex. Constant loads are assumed in the above scenario. At lower voltages, the loads may be reduced, though this is not always true, that is, an induction motor may not stall until the voltage has dropped more than 25%, even then the magnetic contactors in the motor starter supplying power to the motor may not drop out. This lock out of the motors and loss of some loads may result in voltage recovery, which may start the process of load interruption afresh, as the motors try to reaccelerate on the return voltage.

From Equation 13.5 as θ is normally small:

$$\frac{\Delta Q}{\Delta V} = \frac{V_{\mathrm{s}} - 2V_{\mathrm{r}}}{X_{\mathrm{sr}}} \tag{13.12}$$

If the three phases of the line connector are short-circuited at the receiving end, the receiving end short-circuit current is

$$I_{\mathrm{r}} = \frac{V_{\mathrm{s}}}{X_{\mathrm{sr}}} \tag{13.13}$$

This assumes that the resistance is much smaller than the reactance. At no load $V_{\mathrm{r}} = V_{\mathrm{s}}$; therefore,

$$\frac{\partial Q}{\partial V} = -\frac{V_{\mathrm{r}}}{X_{\mathrm{sr}}} = -\frac{V_{\mathrm{s}}}{X_{\mathrm{sr}}} \tag{13.14}$$

Thus,

$$\left|\frac{\partial Q}{\partial V}\right| = \text{short-circuit current} \tag{13.15}$$

Alternatively, we could say that

$$\frac{\Delta V_s}{V} \approx \frac{\Delta V_r}{V} = \frac{\Delta Q}{S_{sc}} \tag{13.16}$$

where S_{sc} is the short-circuit level of the system. This means that the voltage regulation is equal to the ratio of the reactive power change to the short-circuit level. This gives the obvious result that the receiving end voltage falls with the decrease in system short-circuit capacity, or increase in system reactance. A stiffer system tends to uphold the receiving end voltage.

Example 13.1

Consider the system of Figure 13.3. Bus C has two sources of power, one transformed from the 400 kV bus A and connected through a transmission line and the other from the 230 kV bus B. These sources run in parallel at bus C. The voltages at buses A and B are maintained equal to the rated voltage. A certain load demand at bus C dips the voltage by 10 kV. What is the reactive power compensation required at bus C to bring the voltage to its rated value?

An approximate solution is given by Equation 13.16. Based on the impedance data shown in Figure 13.3, calculate the short-circuit current at bus C. This is equal to 7.28 kA. Therefore,

$$\frac{\partial Q}{\partial V} = 13.73 \text{ Mvar/kV}$$

A reactive power injection of 13.73 Mvar is required to compensate the voltage drop of 10 kV.

The voltage-reactive power stability problem is more involved than portrayed above. Power systems have a hierarchical structure: power generation, transmission lines, subtransmission level, distribution level, and finally consumer level. As reactive power does not travel well over long distances, it becomes a local problem. The utility companies operate with an agreed voltage level at inter-tie connections, and provide for their own reactive power compensation to meet this requirement. It can be said that there are no true hierarchical structures in terms of reactive power flow.

13.1.1 Relation with Real Power Instability

The voltage instability is not a single phenomenon and is closely linked to the electromagnetic stability and shares all aspects of active power stability, though there are differences.

Consider the equations of active and reactive power flow (Equations 10.21 and 10.22) for a short line. Let the voltages be fixed. The angle δ (phasor difference between the sending end and receiving end voltage vectors) varies with receiving bus power, as shown in Figure 13.4a. Beyond the maximum power drawn, there is no equilibrium point. Below the maximum power drawn there are two equilibria, one in the stable state and the other in the unstable state. A load flow below the maximum point is considered statically stable.

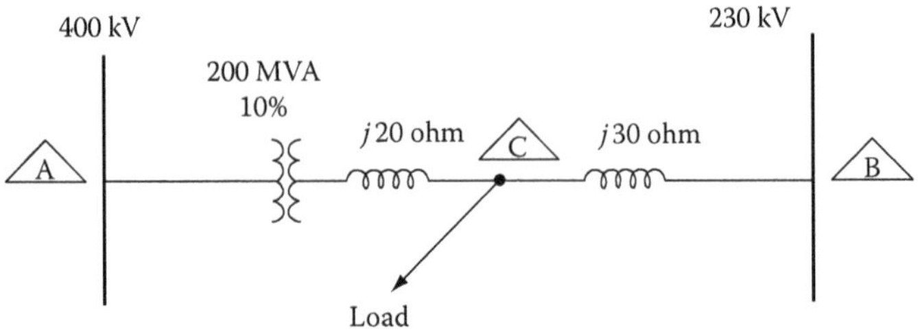

FIGURE 13.3
System for calculation of reactive power injection on voltage dip (Example 13.1).

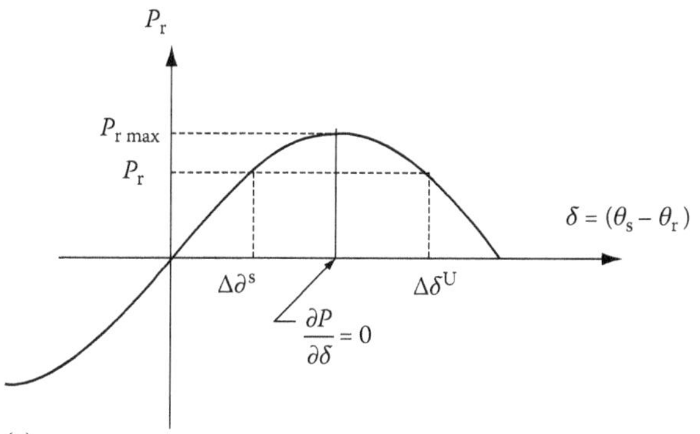

(a)

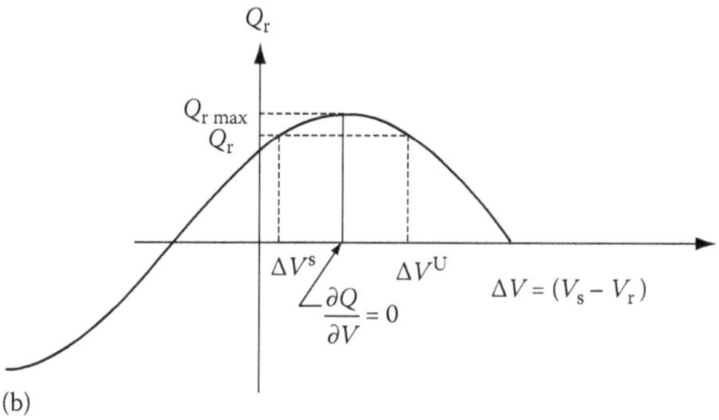

(b)

FIGURE 13.4
(a) To illustrate real power steady-state instability; (b) to illustrate reactive power steady-state instability.

A similar curve can be drawn for reactive power flow, as in Figure 13.4b. The angles are fixed, and the bus voltage magnitude changes. The character of this curve is the same as that of the active power flow curve. A reactive power demand above the maximum reactive power results in the nonexistence of a load-flow situation. The point ΔV^U is statically unstable, corresponding to $\Delta\delta^U$. If we define,

$$\left|P_{(\max)} - P\right| < \varepsilon \tag{13.17}$$

$$\left|Q_{(\max)} - Q\right| < \varepsilon \tag{13.18}$$

then, however small ε may be, there will always be two equilibrium points. Such an equilibrium point may be called a bifurcation point. The derivatives $\partial P/\partial\delta$ and $\partial Q/\partial V$ are zero at the static stability limit.

13.2 Reactive Power Compensation

The need for reactive power compensation is obvious from the aforementioned discussions [4,5]. An acceptable, if not ideally flat, voltage profile has to be maintained. Concurrently the stability is improved and the maximum power limit that can be transmitted increases.

13.2.1 Z_0 Compensation

It was shown in Chapter 10 that a flat voltage profile can be achieved with a SIL loading of V^2/Z_0. The surge impedance can be modified with addition of capacitors and reactors so that the desired power transmitted is given by the ratio of the square of the voltage and modified surge impedance:

$$P_{\text{new}} = \frac{V^2}{Z_{\text{modified}}} \tag{13.19}$$

The load can suddenly change, and ideally, the compensation should also adjust instantaneously according to Equation 13.19. This is not a practical operating situation and stability becomes a consideration. Passive and active compensators are used to enhance stability. The passive compensators are shunt reactors, shunt capacitors, and series capacitors. The active compensators generate or absorb reactive power at the terminals where these are connected. Their characteristics are described further.

13.2.2 Line Length Compensation

The line length of a wavelength of 3100 miles arrived at in Chapter 10 for a 60 Hz power transmission is too long. Even under ideal conditions, the natural load (SIL) cannot be transmitted $>\lambda/4$ ($=775$ miles). Practical limits are much lower. As $\beta = \omega\sqrt{LC}$, the inductance can be reduced by series capacitors, thereby reducing β. The phase-shift angle between the sending end and receiving end voltages is also reduced and the stability limit is, therefore, improved.

13.2.3 Compensation by Sectionalization of Line

The line can be sectionalized, so that each section is independent of others, that is, meets its own requirements of flat voltage profile and load demand. This is compensation by sectionalizing, achieved by connecting constant-voltage compensators along the line. These are active compensators, that is, thyristor-switched capacitors (TSCs), thyristor-controlled reactors (TCRs), and synchronous condensers. All three types of compensating strategies may be used in a single line.

Consider that a *distributed* shunt inductance L_{shcomp} is introduced. This changes the effective value of the shunt capacitance as

$$\begin{aligned} j\omega C_{\text{comp}} &= j\omega C + \frac{1}{j\omega L_{\text{shcomp}}} \\ &= j\omega C(1 - K_{\text{sh}}) \end{aligned} \tag{13.20}$$

where

$$K_{\text{sh}} = \frac{1}{\omega^2 L_{\text{shcomp}} C} = \frac{X_{\text{sh}}}{X_{\text{shcomp}}} = \frac{b_{\text{shcomp}}}{b_{\text{sh}}} \tag{13.21}$$

where K_{sh} is the degree of shunt compensation. It is negative for a shunt capacitance addition.

Similarly, let a distributed *series* capacitance C_{srcomp} be added. The degree of series compensation is given by K_{sc}:

$$K_{\text{sc}} = \frac{X_{\text{srcomp}}}{X_{\text{sr}}} = \frac{b_{\text{sr}}}{b_{\text{srcomp}}} \tag{13.22}$$

The series or shunt elements added are distinguished by subscript "comp" in Equation 13.22. Combining the effects of series and shunt compensations:

$$\begin{aligned} Z_{0\text{comp}} &= Z_{\text{o}}\sqrt{\frac{1-K_{\text{sc}}}{1-K_{\text{sh}}}} \\ P_{0\text{comp}} &= P_{\text{o}}\sqrt{\frac{1-K_{\text{sh}}}{1-K_{\text{sc}}}} \end{aligned} \tag{13.23}$$

Also,

$$\beta_{\text{comp}} = \beta\sqrt{(1-K_{\text{sh}})(1-K_{\text{sc}})} \tag{13.24}$$

The effects are summarized as follows:

- Capacitive shunt compensation increases β and power transmitted and reduces surge impedance. Inductive shunt compensation has the opposite effect, reduces β and power transmitted and increases surge impedance. A 100% inductive shunt capacitance will theoretically increase the surge impedance to infinity. Thus, at no load, shunt reactors can be used to cancel the Ferranti effect.
- Series capacitive compensation decreases surge impedance and β and increases power transfer capacity. Series compensation is applied more from the steady state and transient stability consideration rather than from power factor improvement. It provides better load division between parallel circuits, reduced voltage drops, and better adjustment of line loadings. It has practically no effect on the load power factor. Shunt compensation, on the other hand, directly improves the load power factor. Both types of compensations improve the voltages and, thus, affect the power transfer capability. The series compensation reduces the large shift in voltage that occurs between the sending and receiving ends of a system and improves the stability limit.

The performance of a symmetrical line was examined in Section 10.7. At no load, the midpoint voltage of a symmetrical compensated line is given by:

$$V_{\text{m}} = \frac{V_{\text{s}}}{\cos(\beta l/2)} \tag{13.25}$$

Therefore, series capacitive and shunt inductive compensation reduce the Ferranti effect, while shunt capacitive compensation increases it.

The reactive power at the sending and receiving ends of a symmetrical line was calculated in Equation 10.86 and is reproduced as follows:

$$Q_s = -Q_r = \frac{\sin\theta}{2}\left[Z_0 I_m^2 - \frac{V_m^2}{Z_0}\right]$$

This equation can be manipulated to give the following equation in terms of natural load of the line and $P_m = V_m I_m$ and $P_0 = V_0^2/Z_0$:

$$Q_s = -Q_r = P_0\frac{\sin\theta}{2}\left[\left(\frac{PV_0}{P_0 V_m}\right)^2 - \left(\frac{V_m}{V_0}\right)^2\right] \quad (13.26)$$

For $P = P_0$ (natural loading) and $V_m = 1.0$ per unit, $Q_s = Q_r = 0$.

If the terminal voltages are adjusted so that $V_m = V_0 = 1$ per unit:

$$Q_s = -Q_r = P_0\frac{\sin\theta}{2}\left[\left(\frac{P}{P_0}\right)^2 - 1\right] \quad (13.27)$$

At no load,

$$Q_s = -Q_r = -P_0 \tan\frac{\theta}{2} \approx -P_0\frac{\theta}{2} \quad (13.28)$$

If the terminal voltages are adjusted so that for a certain power transfer, $V_m = 1$ per unit, then the sending end voltage is

$$V_s = V_m\left(1 - \sin^2\frac{\theta}{2}\left[1 - \left(\frac{P}{P_0}\right)^2\right]\right)^{1/2} = -V_r \quad (13.29)$$

When series and shunt compensation are used, the reactive power requirement at *no load* is approximately given by

$$Q_s = -P\frac{\beta l}{2}(1 - K_{sh}) = -Q_r \quad (13.30)$$

If K_{sh} is zero, the reactive power requirement of a series compensated line is approximately the same as that of an uncompensated line, and the reactive power handling capability of terminal synchronous machines becomes a limitation. Series compensation schemes, thus, require SVCs or synchronous condensers/shunt reactors.

13.2.4 Effect on Maximum Power Transfer

The power transfer for an uncompensated lossless line under load, phase angle δ is

$$P = \frac{V_s V_r}{Z_0 \sin\beta l}\sin\delta = \frac{V_s V_r}{Z_0 \sin\theta}\sin\delta \quad (13.31)$$

This can be put in a more familiar form by assuming that $\sin\theta = \theta$:

$$\beta l = \theta = \omega l\sqrt{LC}$$

$$Z_0\theta = \omega l\sqrt{LC}\left(\sqrt{\frac{L}{C}}\right) = \omega lL = X_{sc} \tag{13.32}$$

$$P = \frac{V_s V_r}{Z_0 \sin\theta}\sin\delta \approx \frac{P_0}{\sin\theta}\sin\delta \approx \frac{P_0}{\theta}\delta \tag{13.33}$$

We know that $\delta = 90°$ for a theoretical steady-state limit. A small excursion or change of power transmitted or switching operations will bring about instability, Figure 13.4a. Practically, an uncompensated line is not operated at >30° load angle.

With compensation the ratio of powers becomes:

$$\frac{P_{comp}}{P} = \frac{1}{\sqrt{\frac{1-K_{sc}}{1-K_{sh}}}\sin\left[\beta x\sqrt{(1-K_{sc})(1-K_{sh})}\right]} \tag{13.34}$$

Series compensation has a more pronounced effect on P_{max}. Higher values of K_{sc} can give rise to resonance problems and in practice $K_{sc} \not> 0.8$. Series compensation can be used with a line of any length and power can be transmitted over a larger distance than is otherwise possible.

Example 13.2

A 650 mile line has $\beta = 0.116°$/mile.

1. Calculate P_{max} as a function of P_0 for an uncompensated line.
2. What are the limitations of operating under these conditions?
3. Considering a load angle of 30°; calculate the sending end voltage if the midpoint voltage is held at 1 per unit. Also, calculate the sending end reactive power of the uncompensated line.
4. Recalculate these parameters with 80% series compensation.

1. The electrical length is $\theta = \beta_1 = 0.116 \times 650 = 75.4°$. Therefore, the maximum power transfer as a function of natural load is $1/\sin 75.4° = 1.033P_0$.
2. An uncompensated transmission line is not operated close to the steady-state limit.
3. If the midpoint voltage is maintained at 1 per unit, by adjustment of the sending end and receiving end voltages, the sending end voltage from Equation 13.29 is

$$(1)\left[1 - (\sin 37.25)^2(1 - 0.5^2)\right]^{1/2} = 0.852 \text{ pu}$$

The ratio $P/P_0 = 0.5$ for load angle $\delta = 30°$. Thus, the voltages are too low. The reactive power input required from Equation 13.27 is 0.36 kvar/kW of load transferred at each end, that is, a total of 0.72 kvar/kW transmitted.

4. With 80% series compensation, the maximum power transfer from Equation 13.34 is

$$\frac{P_{comp}}{P} = \frac{1}{\sqrt{1-0.8}\sin\left[75.4\sqrt{1-0.8}\right]} = 4.0278P_0$$

From Equation 13.23 the surge impedance with compensation is $0.447Z_0$. Therefore, the natural loading $= 2.237P_0$. If the line is operated at this load, a flat voltage profile is obtained. The new electrical length of the line is

$$\theta' = \sin^{-1}\left(\frac{P'_0}{P'_{max}}\right) = 33.7°$$

13.2.5 Compensation with Lumped Elements

The distributed shunt or series compensation derived earlier is impractical and a line is compensated by lumped series and shunt elements at the midpoint or in sections, using the same methodology. The problem usually involves steady state as well as dynamic and transient stability considerations.

Figure 13.5a shows a midpoint compensation [4]. Each half-section of the line is shown as an equivalent Π model. The circuit of Figure 13.5a can be redrawn as shown in Figure 13.5b. The phasor diagram is shown in Figure 13.5c. The degree of compensation for the central half, κ_m, is

$$\kappa_m = \frac{b_{shcomp}}{0.5b_{sh}} \tag{13.35}$$

For equal sending and receiving end voltages:

$$P = \frac{V^2}{X_{sr}(1-s)}\sin\delta \tag{13.36}$$

where

$$s = \frac{X_{sr}}{2}\frac{b_{sh}}{4}(1-\kappa_m) \tag{13.37}$$

The midpoint voltage can be expressed as

$$V_m = \frac{V\cos(\delta/2)}{1-s} \tag{13.38}$$

The equivalent circuit of the line with compensation is represented in Figure 13.5d. For $s < 1$, the midpoint compensation increases the midpoint voltage, which tends to offset the series voltage drop in the line. If the midpoint voltage is controlled by the variation of the susceptance of the compensating device, then using the relationship in Equation 13.38, Equation 13.36 can be written as

$$P = \frac{V^2}{X_{sr}(1-s)}\sin\delta = \frac{V_m V}{X_{sr}\cos(\delta/2)}\sin\delta = 2\frac{V_m V}{X_{sr}}\sin\frac{\delta}{2} \tag{13.39}$$

If the midpoint voltage is equal to the sending end voltage and equal to the receiving end voltage,

$$P = \frac{2V^2}{X_{sr}} \sin\frac{\delta}{2} \tag{13.40}$$

This is shown in Figure 13.5e. Thus, the power transmission characteristics are a sinusoid whose amplitude varies as s varies. Each sinusoid promises ever higher power transfer. For $P > \sqrt{2}P_{max}$, angle $\delta > \pi/2$. When the transmission angle increases, the compensator responds by changing the susceptance to satisfy Equation 13.40. The economic limit of

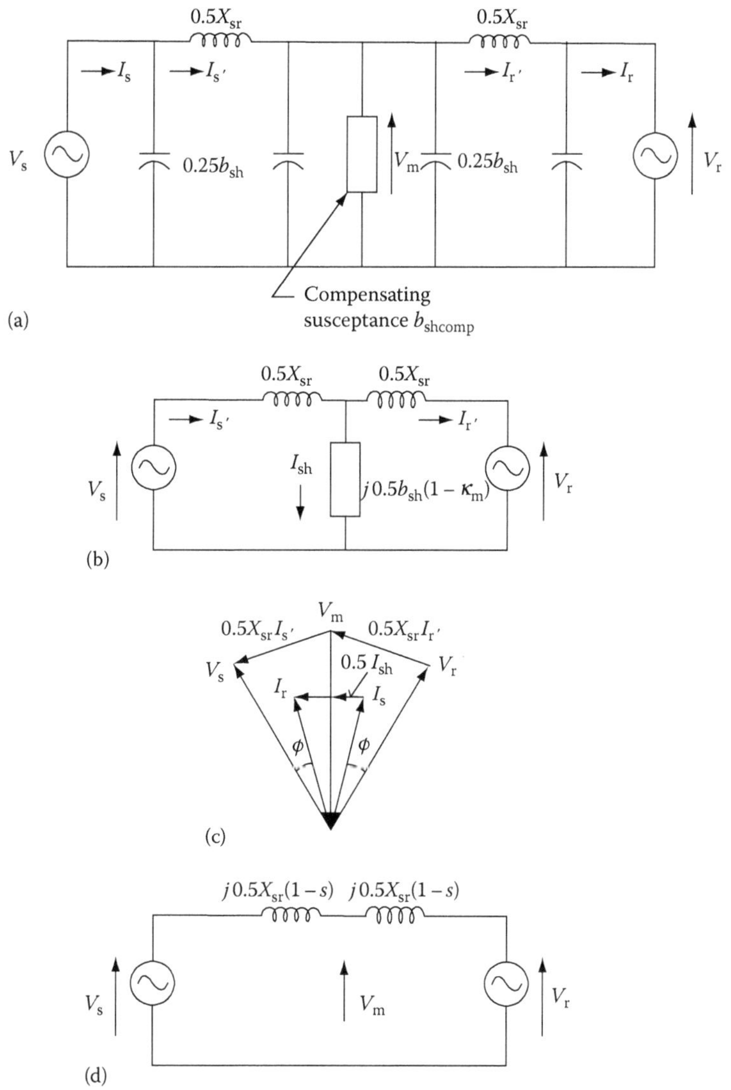

FIGURE 13.5
(a) Midpoint compensation of a transmission line; (b) circuit of (a) redrawn; (c) phasor diagram; (d) equivalent circuit;

(continued)

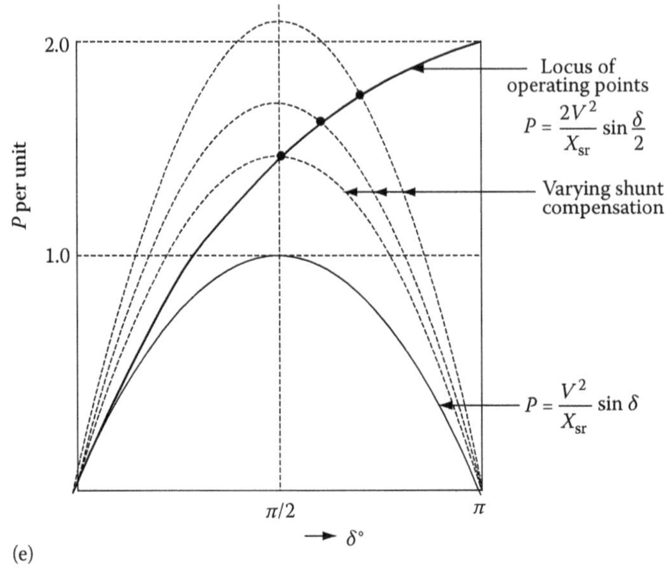

FIGURE 13.5 (continued)
(e) P–δ characteristics and dynamic response operation of a midpoint compensator.

a compensator to put effective capacitive susceptance may be much lower than the maximum power transfer characteristics given by Equation 13.40. When the compensator limit is reached it does not maintain a constant voltage and acts like a fixed capacitor.

The shunt compensation to satisfy Equation 13.38 can be calculated from the following equation:

$$b_{\text{shcomp}} = -\frac{4}{X_{\text{sr}}}\left[1 - \frac{V}{V_{\text{m}}}\cos\frac{\delta}{2}\right] + \frac{b_{\text{sh}}}{2} \tag{13.41}$$

In Chapter 12, we have seen that reactive power cannot be transferred over long distances. Due to the inherent problem of flow of reactive power connected with reduction of voltage and usable voltage band, voltage controlled (PV) buses must be scattered throughout the power system.

13.3 Reactive Power Control Devices

Some of the devices and strategies available for reactive power compensation and control are as follows:

- Synchronous generators
- Synchronous condensers
- Synchronous motors (overexcited)
- Shunt capacitors and reactors

- Static var controllers
- Series capacitors with power system stabilizers (PSSs)
- Line dropping
- Undervoltage load shedding
- Voltage reduction
- Under load tap changing transformers
- Setting lower transfer limits

13.3.1 Synchronous Generators

Synchronous generators are primary voltage-control devices and primary sources of spinning reactive power reserve. Figure 13.6 shows the reactive capability curve of a generator. To meet the reactive power demand, generators can be rated to operate at 0.8 power factor, at a premium cost. Due to thermal time constants associated with the generator exciter, rotor, and stator, some short-time overload capability is available and can be usefully utilized. On a continuous basis, a generator will operate successfully at its treated voltage and with a power factor at a voltage not more than 5% above or below the rated voltage, but not necessarily in accordance with the standards established for operation at the rated voltage [6]. The generator capability curves at voltages other than the rated voltage may differ.

Referring to Figure 13.6, the portion SPQ is limited by the generator megawatt output, the portion QN by the stator current limit, and the portion NM by the excitation current limit and ST by end-iron heating limit. *Q* is the rated load and power factor operating point. In the leading reactive power region the minimum excitation limit (MEL) and the

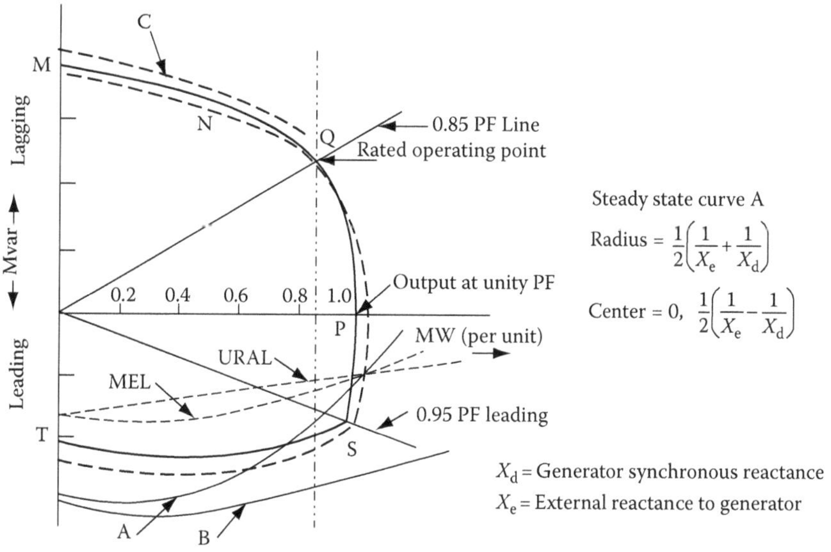

FIGURE 13.6
Reactive capability curve of a 0.85 power factor rated synchronous generator with stability limit curves. Curve A, classical stability limit; curve B, enhanced stability limit with high-response excitation system; curve C, 20 min overload reactive capability, dotted reactive capability curve shows operation at higher than rated voltage.

normal manufacturer's underexcited reactive ampère limit (URAL) are imposed. MEL plays an important role in voltage control. The MEL curve rises slightly above the pull-out curve also called classical static stability curve A, providing an extra margin of stability and preventing stator end-turn heating from low excitation. Where cables or a large shunt capacitors' bank can remain in service after a large disturbance, the capability of the generators to absorb capacitive current comes into play and appropriate settings of MEL are required. Curve B shows the improvement in the stability achieved by high-response excitation systems. The dotted curve shows the effect of operating at a higher than rated voltage, and curve C shows the short-time overload capability in terms of reactive power output.

The reactive output of a generator decreases if the system voltage increases and conversely it increases if the system voltage dips. This has a *stabilizing* effect on the voltage. The increased or decreased output acts in a way to counteract the voltage dip or voltage rise.

13.3.2 Synchronous Condensers

A synchronous condenser is a dynamic compensator and is characterized by its large synchronous reactance and heavy field windings. It develops a zero power factor leading current with least expenditure in active power losses. Figure 13.7 shows a *V* curve for a constant output voltage. The rated reactive power output for a 0.8 leading power factor synchronous motor is also shown for comparison. In a synchronous condenser at 100% excitation, full-load leading kvars are obtained. At about 10% of the excitation current, the leading kvar output falls to a minimum corresponding to the losses. The rated field current is defined as the current for rated machine output at rated voltage. The lagging kvar is usually limited to approximately one-third of the maximum leading kvar to prevent loss of synchronism on a disturbance. The full-load power factor may be as low as 0.02. The current drawn by a synchronous condenser can be varied from leading to lagging by change of its excitation, that is, machine internal voltage behind its synchronous reactance (steady state) and transient reactance (for transient stability) characteristics.

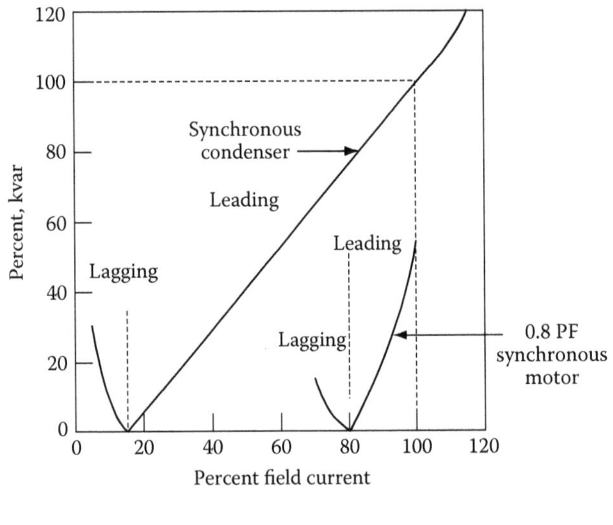

FIGURE 13.7
V curves for a synchronous condenser and 0.8 leading power factor synchronous motor.

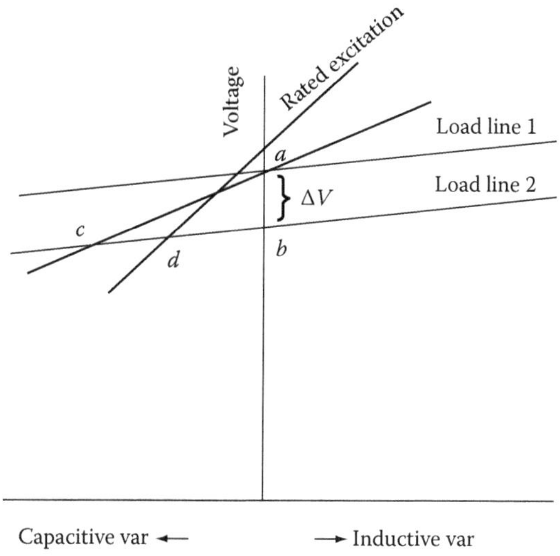

FIGURE 13.8
Operation of a synchronous condenser on sudden voltage dip.

A synchronous condenser provides stepless control of reactive power in both underexcited and overexcited regions. Synchronous condensers have been used for normal voltage control, high-voltage dc (HVDC) applications, and voltage control under upset conditions with voltage regulators. The transient open circuit time constant is high (of the order of few seconds) even under field forcing conditions, yet synchronous condensers have been applied to improve the transient stability limits on voltage swings. At present, these are being replaced with static var compensators, which have faster responses.

Figure 13.8 shows the response characteristics of a synchronous condenser. Consider that the condenser is initially operating at point *a*, neither providing inductive nor capacitive reactive power. A sudden drop in voltage (ΔV), forces the condenser to swing to operating point *c* along the load line 2. As it is beyond the rated operating point *d* at rated excitation, the operation at *c* is limited to a short time, and the steady-state operating point will be restored at *d*.

13.3.3 Synchronous Motors

In an industrial distribution system, synchronous motors can be economically selected, depending on their speed and rating. Synchronous motors are more suitable for driving certain types of loads, for example, refiners and chippers in the pulp and paper industry, Banbury mixers, screw-type plasticators and rubber mill line shafts, compressors, and vacuum pumps. The power factor of operation of induction motors deteriorates at low speeds due to considerable overhang leakage reactance of the windings. Synchronous motors will be more efficient in low-speed applications and can provide leading reactive power. Higher initial cost as compared to an induction motor is offset by the power savings. A careful selection of induction and synchronous motors in an industrial environment can obviate the problems of reactive power compensation and maintain an appropriate voltage profile at the load centers. Synchronous motors for such applications are normally rated to operate at 0.8 power factor leading at full load and need a higher rated excitation system as compared to unity power factor motors. An evaluation generally calls for synchronous motors versus induction motors with power factor improvement capacitors.

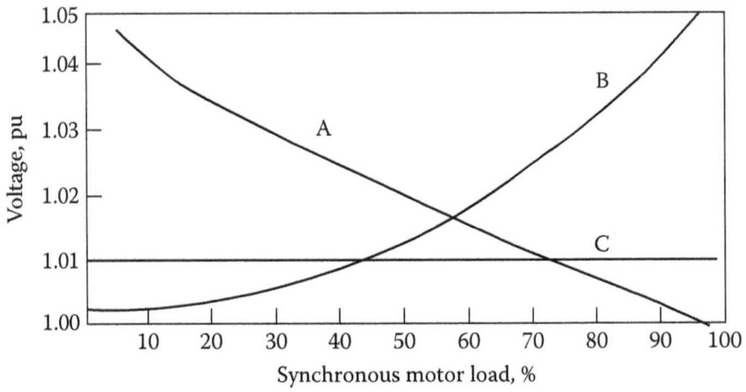

FIGURE 13.9
Effect of synchronous motor controller type on voltage regulation. A, constant current or unregulated controller; B, constant power-factor controller; C, constant kvar controller.

The type of synchronous motor excitation control impacts its voltage control characteristics. The controllers can be constant current, constant power factor, or constant kvar type, though many uncontrolled excitation systems are still in use. A constant current controller compensates for decrease in motor field current due to field heating; the reactive power output from the motor increases at no load and consequently the voltage rises at no load and falls as the motor is loaded. A constant power factor controller increases the reactive power output with load, and the voltage rises with the load. A cyclic load variation will result in cyclic variation of reactive power and thus the voltage. A constant reactive power controller will maintain a constant voltage profile, irrespective of load, but may give rise to instability when a loaded motor is subjected to a sudden impact load. Constant var and power factor controllers are common. Figure 13.9 shows the voltage control characteristics of these three types of controllers.

13.3.4 Shunt Power Capacitors

Shunt capacitors are extensively used in industrial and utility systems at all voltage levels. As we will see, the capacitors are the major elements of flexible ac transmission systems (FACTS). Much effort is being directed in developing higher power density, lower cost improved capacitors, and an increase in energy density by a factor of 100 is possible. These present a constant impedance type of load, and the capacitive power output varies with the square of the voltage:

$$\text{kvar}_{V2} = \text{kvar}_{V1}\left|\frac{V_2}{V_1}\right|^2 \tag{13.42}$$

where
kvar_{V1} is output at voltage V_1
kvar_{V2} is output at voltage V_2

As the voltage reduces, so does the reactive power output, when it is required the most. This is called the *destabilizing* effect of the power capacitors. (Compare with the stabilizing effect of a generator excitation system.)

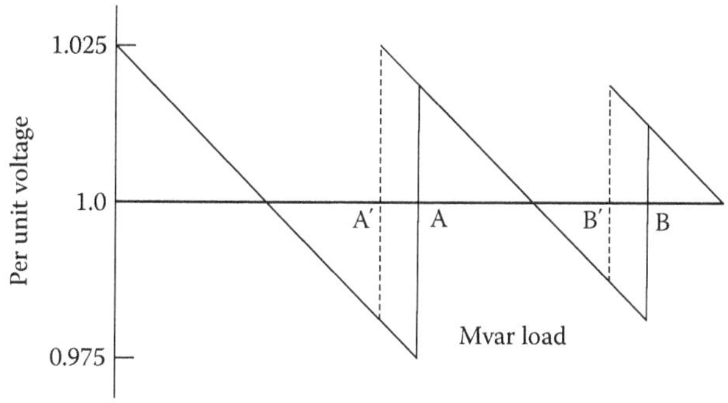

FIGURE 13.10
Sequential switching of capacitors in two steps with reactive power flow control to maintain voltage within a certain acceptable band. The no-load voltage is assumed to be 1.025 per unit.

Capacitors can be switched in certain discrete steps and do not provide a stepless control. Figure 13.10 shows a two-step sequential reactive power switching control to maintain voltage within a certain band. As the reactive power demand increases and the voltage falls (shown linearly in Figure 13.10 for simplicity), the first bank is switched at A, which compensates the reactive power and suddenly raises the voltage. The second step switching is, similarly, implemented at B. On reducing demand the banks are taken out from service at A′ and B′. A time delay is associated with switching in either direction to override transients. Voltage or power factor dependent switching controls are also implemented. Power capacitors have a wide range of applications:

- Power factor improvement capacitors switched with motors at low- and medium-voltage levels or in multistep power factor improvement controls in industrial systems
- Single or multiple banks in industrial distribution systems at low- and medium-voltage substations
- As harmonic filters in industrial distribution systems, arc furnaces, steel mills, and HVDC transmission
- Series or shunt devices for reactive power compensation in transmission systems
- Essential elements of SVC and FACTS controllers

Some of the problems associated with the shunt power capacitors are as follows:

- Switching inrush currents at higher frequencies and switching overvoltages
- Harmonic resonance problems
- Limited overvoltage withstand capability
- Limitations of harmonic current loadings
- Possibility of self-excitation of motors when improperly applied as power factor improvement capacitors switched with motors
- Prolonging the decay of residual motor voltage on disconnection, and trapped charge on disconnection, which can increase the inrush current on reswitching and lead to restrikes
- Requirements of *definite purpose* switching devices

These are offset by low capital and maintenance costs, modular designs varying from very small to large units, and fast switching response.

13.3.5 Static Var Controllers

The var requirements in transmission lines swing from lagging to leading, depending on the load. Shunt compensation by capacitors and reactors is one way. However, it is slow, and power circuit breakers have to be derated for frequent switching duties. The IEEE and CIGRE definition of a static var generator (SVG) embraces different semiconductor power circuits with their internal control enabling them to produce var output proportional to an input reference. An SVG becomes an SVC when equipped with external or system controls, which derive its reference from power system operating requirements and variables. SVCs can be classified into following categories [7]:

TCR (Figure 13.11a)

TSC (Figure 13.11b)

Fixed capacitor and TCR (FC-TCR) (Figure 13.11c)

TSC and TCR (TSC-TCR) (Figure 13.11d)

Figure 13.12 shows the circuit diagram of an FC-TCR. The arrangement provides discrete leading vars from the capacitors and continuously lagging vars from TCRs. The capacitors are used as tuned filters, as considerable harmonics are generated by thyristor control (Chapter 17).

The steady-state characteristics of an FC-TCR are shown in Figure 13.13. The control range is AB with a positive slope, determined by the firing angle control.

$$Q_{\alpha} = \left| b_{c} - b_{i(\alpha)} \right| V^{2} \tag{13.43}$$

where

b_c is the susceptance of the capacitor

$b_{i(\alpha)}$ is the susceptance of the inductor at firing angle a

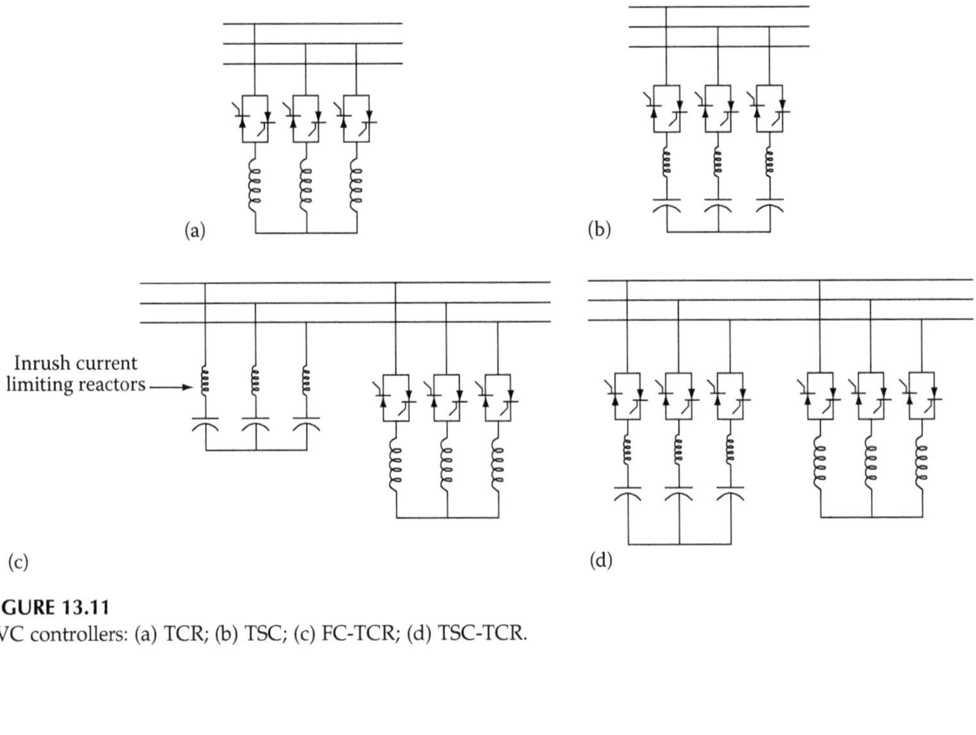

FIGURE 13.11
SVC controllers: (a) TCR; (b) TSC; (c) FC-TCR; (d) TSC-TCR.

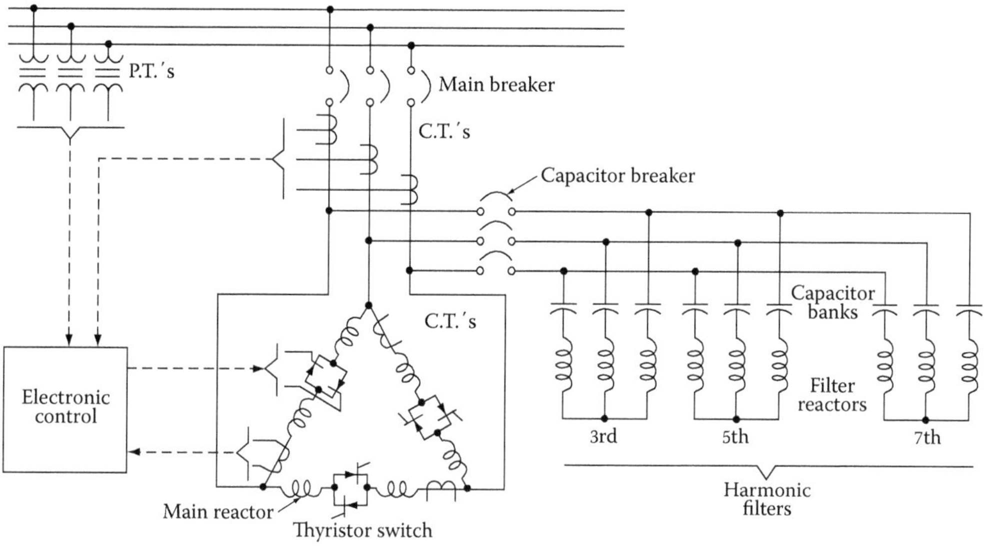

FIGURE 13.12
Circuit diagram of an FC-TCR, with switched capacitor filters.

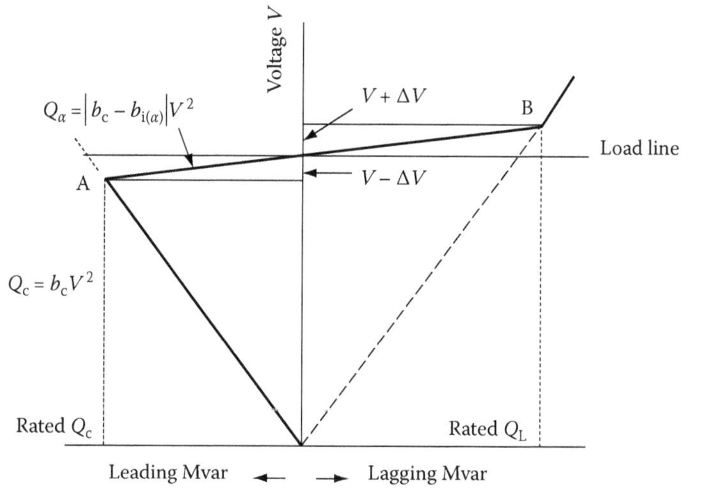

FIGURE 13.13
Steady-state *V*–*Q* characteristics of an FC-TCR.

As the inductance is varied, the susceptance varies over a large range. The voltage varies within limits $V \pm \Delta V$. Outside the control interval AB, the FC-TCR acts like an inductor in the high-voltage range and like a capacitor in the low-voltage range. The response time is of the order of 1 or 2 cycles. The compensator is designed to provide emergency reactive and capacitive loading beyond its continuous steady-state rating.

A TSC-TCR provides thyristor control for the reactive power control elements, capacitors, and reactors. Improved performance under large system disturbances and lower power loss are obtained. Figure 13.14 shows the *V*–*I* characteristics. A certain short-time

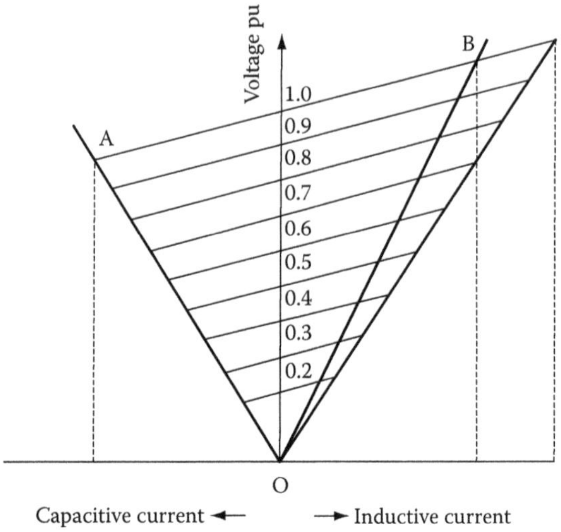

FIGURE 13.14
V–I characteristics of an SVC (TSC-TCR) showing reduced output at lower system voltages.

overload capability is provided both in the maximum inductive and capacitive regions (shown for the inductive region in Figure 13.14).

Voltage regulation with a given slope can be achieved in the normal operating range. The maximum capacitive current decreases linearly with the system voltage, and the SVC becomes a fixed capacitor when the maximum capacitive output is reached. The voltage support capability decreases with decrease in system voltage.

SVCs are ideally suited to control the varying reactive power demand of large fluctuating loads (i.e., rolling mills and arc furnaces) and dynamic overvoltages due to load rejection and are used in HVDC converter stations for fast control of reactive power flow.

Compensation by sectionalizing is based on a midpoint dynamic shunt compensator. With a dynamic compensator at the midpoint the line tends to behave like a symmetrical line. The power transfer equation for equal sending and receiving end voltages is Equation 13.40. The advantage of static compensators is apparent. The midpoint voltage will vary with the load, and an adjustable midpoint susceptance is required to maintain constant voltage magnitude. With rapidly varying loads, it should be possible for the reactive power demand to be rapidly corrected, with least overshoot and voltage rise. Figure 13.15 shows transient power angle curves for an uncompensated line, with phase angle shift, shunt compensation, and series compensation. With the midpoint voltage held constant, the angles between the two systems can each approach 90°, for a total static stability limit angle of 180°.

The power system oscillation damping can be obtained by rapidly changing the output of the SVC from capacitive to inductive so as to counteract the acceleration and deceleration of interconnected machines. The transmitted electrical power can be increased by capacitive vars when the machines accelerate and it can be decreased by reactive vars when the machines decelerate.

13.3.6 Series Capacitors

An implementation schematic of the series capacitor installation is shown in Figure 13.16. The performance under normal and fault conditions should be considered. Under fault conditions, the voltage across the capacitor rises, and unlike a shunt capacitor, a series

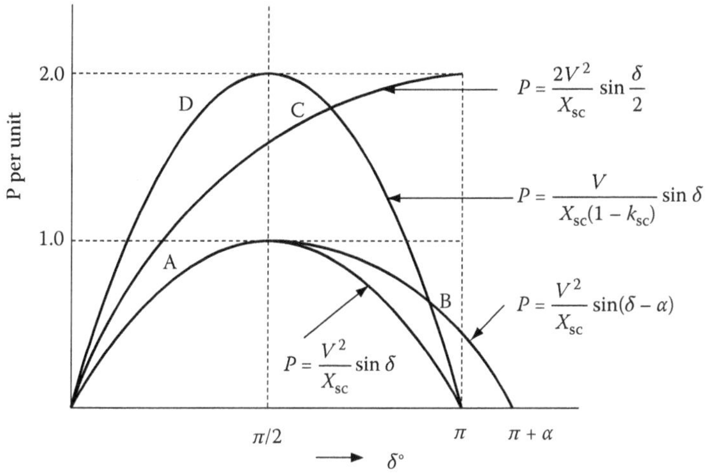

FIGURE 13.15
P–δ characteristics. A, uncompensated line; B, with phase-angle regulator; C, with midpoint shunt compensation; D, with series compensation.

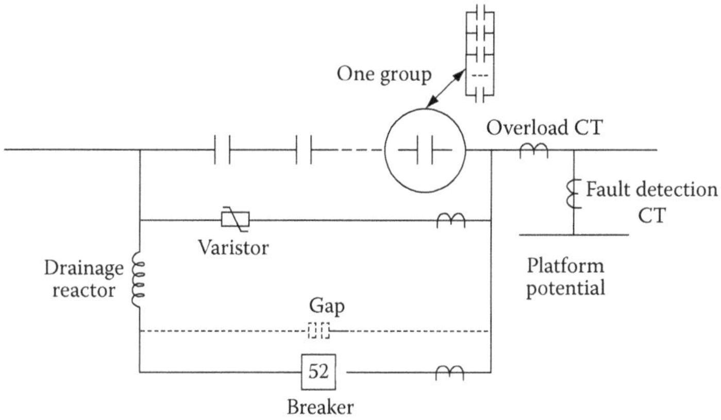

FIGURE 13.16
Schematic diagram of a series capacitor installation.

capacitor experiences many times its rated voltage due to fault currents. A zinc oxide varistor in parallel with the capacitor may be adequate to limit this voltage. Thus, in some applications the varistor will reinsert the bank immediately on termination of a fault. For locations with high fault currents a parallel fast acting triggered gap is introduced, which operates for more severe faults. When the spark gap triggers it is followed by closure of the bypass breaker. Immediately after the fault is cleared, to realize the beneficial effect of series capacitor on stability, it should be reinserted quickly, and the main gap is made self-extinguishing. A high-speed reinsertion scheme can reinsert the series capacitors in a few cycles. The bypass switch must close at voltages in excess of nominal, but not at levels too low to initiate main gap spark-over.

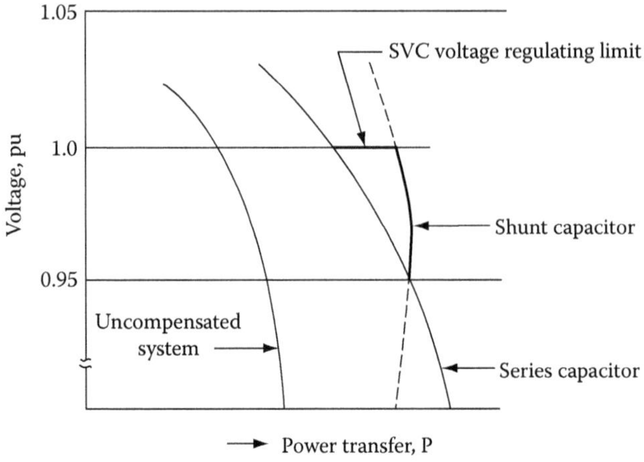

FIGURE 13.17
Series versus shunt compensation; power transfer characteristics for a mid-compensated line.

The discharge reactor limits the magnitude and frequency of the current through the capacitor when the gap sparks over. This prevents damage to the capacitors and fuses. A series capacitor must be capable of carrying the full line current. Its reactive power rating is

$$I^2X_c \text{ per phase} \tag{13.44}$$

and, thus, the reactive power output varies with the load current.

Figure 13.17 shows the impact of series versus shunt compensation at the midpoint of a transmission line. Both systems are, say, designed to maintain 95% midpoint voltage. The midpoint voltage does not vary much when an SVC is applied, but with series compensation it varies with load. However, for a transfer of power higher than the SVC control limit, the voltage falls rapidly as the SVC hits its ceiling limit, while series compensation holds the midpoint voltage better.

A series capacitor has a natural resonant frequency given by

$$f_n = \frac{1}{2\pi\sqrt{LC}} \tag{13.45}$$

where f_n is usually less than the power system frequency. At this frequency the electrical system may reinforce one of the frequencies of the mechanical resonance, causing *subsynchronous resonance* (SSR). If f_r is the SSR frequency of the compensated line, then at resonance,

$$\begin{aligned} 2\pi f_r L &= \frac{1}{2\pi f_r C} \\ f_r &= f\sqrt{K_{sc}} \end{aligned} \tag{13.46}$$

This shows that the SSR occurs at frequency f_r, which is equal to normal frequency multiplied by the square root of the degree of compensation. The transient currents at subharmonic frequency are superimposed upon the power frequency component and may be damped out within a few cycles by the resistance of the line. Under certain conditions subharmonic currents can have a destabilizing effect on rotating machines. A dramatic voltage rise can occur if the generator quadrature axis reactance and the system capacitive

TABLE 13.1

Characteristics of Reactive Power Sources

Equipment	Characteristics
Generators	Primary source of reactive power reserve, continuously adjustable reactive power in inductive and capacitive regions limited by generator reactive power-reactive capability curve. Response time depending on excitation system, fast response excitation systems of 0.1 s possible. Limited capacitive reactive power on underexcitation
Synchronous condensers	Continuously adjustable reactive power in inductive and capacitive regions, slow response (1 s). Limited inductive reactive power
SVC	Same as synchronous condensers, wider range of control, fast response, 1–2 cycles, better range of reactive power capability. Harmonic pollution. SSR of consideration for series capacitors
Power capacitors	Switchable in certain steps, only capacitive reactive power, response dependent on control system, of the order of a couple of cycles with circuit breaker control
Shunt reactors	Single unit per line, inductive reactive power only, switching response dependent on control system, of the order of a couple of cycles with power circuit breaker controls
Load and line dropping	Emergency measures. Load dropping also reduces active power
ULTC transformers	In certain steps only, slow response does not generate reactive power, only reroutes it

reactance are in resonance. There is no field winding or voltage regulator to control quadrature axis flux in a generator. Magnetic circuits of transformers can be driven to saturation and surge arresters can fail. The inherent dominant subsynchronous frequency characteristics of the series capacitor can be modified by a parallel-connected TCR.

Table 13.1 shows the comparative characteristics and applications of reactive power compensating devices.

13.4 Some Examples of Reactive Power Flow

Three examples of the reactive power flow are considered to illustrate reactive power/voltage problems in transmission [8], generating, and industrial systems.

Example 13.3

Figure 13.18a shows three sections of a 230 kV line, with a series impedance of $0.113 + j0.80\ \Omega$ per mile and a shunt capacitance of 0.2 MΩ per mile. Each section is 150 miles long. A voltage-regulating transformer is provided at each bus. The reactive power flow is considered under the following conditions:

- The rated voltage applied at slack bus 1 at no load.
- A load of 40 Mvar applied at furthermost right bus 4. No tap adjustment on transformers and no reactive power injection are provided.
- The transformer's tap adjustment raises the voltage at each bus to approximately 230 kV.
- A 30 Mvar capacitive injection is provided at bus 3.
- Transformer tap adjustment and reactive power injection of 5 Mvar each at buses 2 and 3.

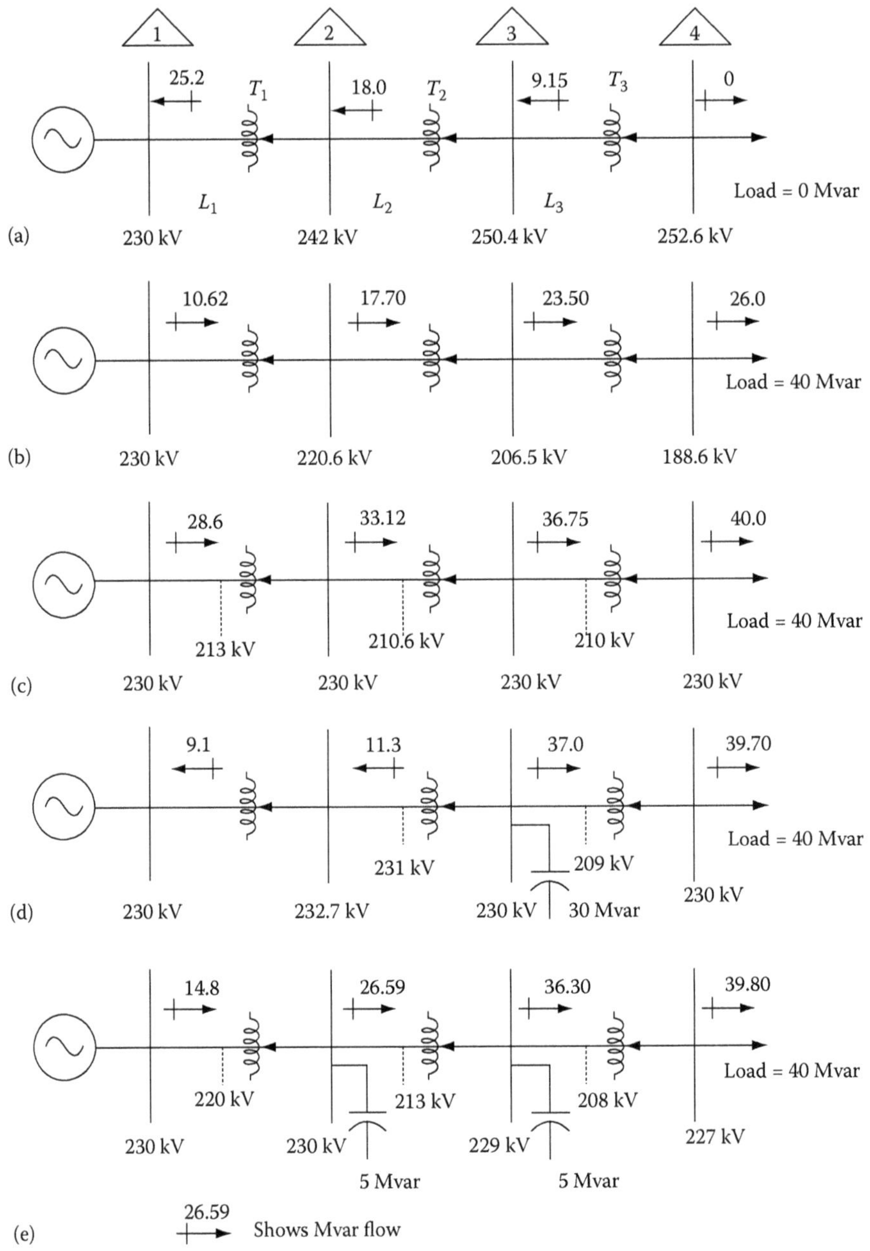

FIGURE 13.18
(a) A three-section 230 kV line at no load; (b) with 40 Mvar load at bus 4, no tap adjustments, and no reactive power injection; (c) with transformer tap adjustment to compensate for the line voltage drop; (d) with reactive power injection of 30 Mvar at bus 3; (e) with reactive power injection of 5 Mvar at buses 2 and 3 and transformer tap adjustments;

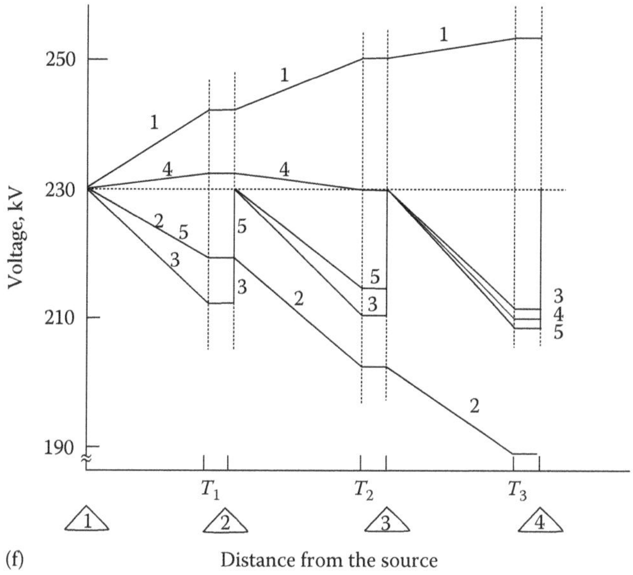

FIGURE 13.18 (continued)
(f) voltage profile along the line in all five cases of study.

1. At no load the voltage rises at the receiving end due to charging capacitance current. Each line section is simulated by a Π network. The voltage at bus 4 is 252.6 kV. The charging current of each section flows cumulatively to the source, and 25.2 Mvar capacitive power must be supplied from the slack bus.
2. The line sections in series behave like a long line and the voltage will rise or fall, depending on the loading, unless a section is terminated in its characteristic impedance. On a lossless line the voltage profile is flat and the reactive power is limited. On a practical lossy line and termination in characteristic impedance the voltage will fall at a moderate rate, and when not terminated in characteristic impedance, the voltage will drop heavily, depending on the reactive power flow. Figure 13.18b shows progressive fall of voltage in sections, and the voltage at bus 4 is 188.06 kV, that is, 81.7% of nominal voltage. At this reduced voltage only 26 Mvar of load can be supplied.
3. Each tap changer raises the voltage to the uniform level of the starting section voltage. Note that the charging kvar of the line, modeled as shunt admittance, acts like a reactive power injection at the buses, augmenting the reactive power flow from the upstream bus. The voltages on the primary of the transformer, that is, at the termination of each line section, are shown in Figure 13.8c.
4. A reactive power injection of 30 Mvar at bus 3 brings the voltages of buses 2 and 3 close to the rated voltage of 230 kV. However, the source still supplies capacitive charging power of the transmission line system and the reactive power to the load flows from the nearest injection point. Thus, reactive power injections in themselves are not an effective means to transmit reactive power over long distances.
5. This shows the effect of tap changers as well as reactive power injection. Reactive power injection at each bus with tap changing transformers can be an effective way to transmit reactive power over long distance, but the injections at each bus may soon become greater than the load requirement. This shows that the compensation is best provided as close to the load as practicable.

Figure 13.18e shows the voltage profile in each of the aforementioned cases.

Example 13.4

This example illustrates that relative location of power capacitors and the available tap adjustments on the transformers profoundly affect the reactive power flow.

Figure 13.19 illustrates effects of off-load and on-load tap changing with shunt capacitors located at the load or at the primary side of the transformer. Figure 13.19a through c shows the load flow when there are no power capacitors and only a tap changing transformer. Figure 13.19d through f shows capacitors located on the load bus. Figure 13.19g through i shows capacitors located at the transformer primary side. In each of these cases, the objective is to maintain load voltage equal to rated voltage for a primary voltage dip of 10%. In each case a constant power load of 30 MW and 20 Mvar is connected to the transformer secondary. Only reactive power flows are shown in these figures for clarity.

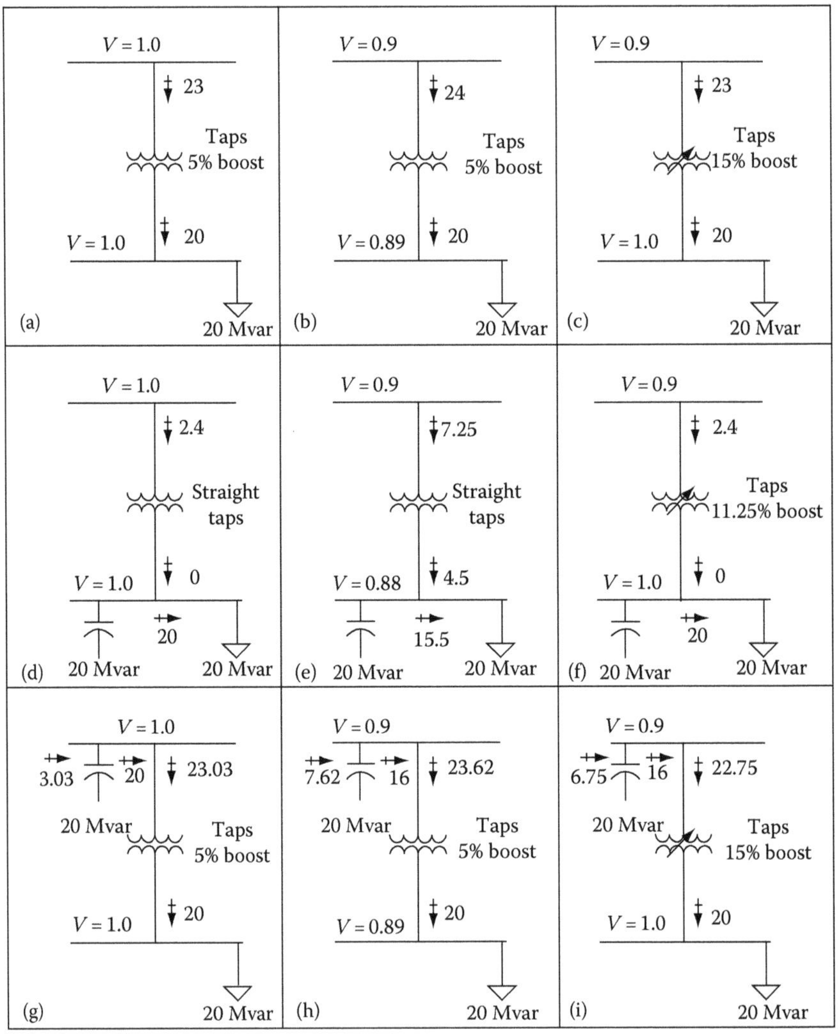

FIGURE 13.19
Reactive power flow with power capacitors and transformer tap changing.

Figure 13.19a shows that with the transformer primary voltage at rated level, taps on the transformer provide a 5% secondary voltage boost to maintain the load voltage at 1.00 per unit and counteract the voltage drop in the transformer. If the primary voltage dips by 10%, the load voltage will be 0.89 per unit and the reactive power input to transformer increases by 1.0 Mvar. An underload adjustment of taps must provide a 15% voltage boost to maintain the load voltage equal to the rated voltage, as in Figure 13.19c.

For load flow in Figure 13.19d the transformer secondary voltage is at rated level, as capacitors directly supply 20 Mvar of load demand and no secondary voltage boost is required from the transformer tap settings. Yet, 2.4 Mvar is drawn from the source, which represents reactive power losses in the transformer itself. A 10% dip in the primary voltage results in a load voltage of 0.88 per unit and the reactive power requirement from the source side increases to 7.25 Mvar. Thus, load capacitors have not helped the voltage as compared to the scenario in Figure 13.19b. The combination of power capacitor and on-load tap changing to maintain the rated secondary voltage is shown in Figure 13.19f.

Figure 13.19g through i shows the voltages, tap settings, and reactive power flows, when the power capacitors of the same kvar rating are located on the primary side of the transformer. The source has to supply higher reactive power as compared to the capacitors located on the load bus.

Example 13.5

Figure 13.20 shows a distribution system with a 71 MVA generator, a utility tie transformer of 40/63 MVA, and loads lumped on the 13.8 kV bus. A 0.38 ohm reactor in series with the generator limits the short-circuit currents on the 13.8 kV system to acceptable levels. However, it raises the generator operating voltage and directly reduces its reactive power capability due to losses. The total plant load including system losses is 39 MW and 31 Mvar. The excess generator power is required to be supplied into the utility's system and also it should be possible to run the plant on an outage of the generator. The utility transformer is sized adequately to meet this contingency.

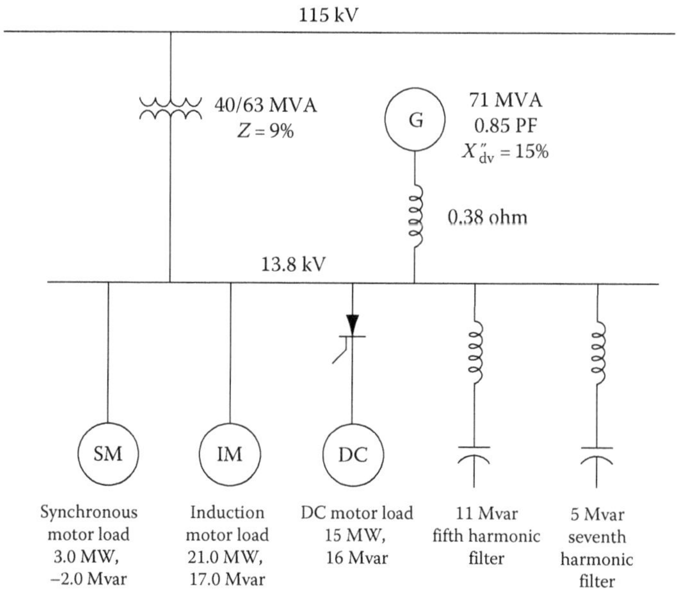

FIGURE 13.20
A 13.8 kV industrial cogeneration facility (Example 13.5).

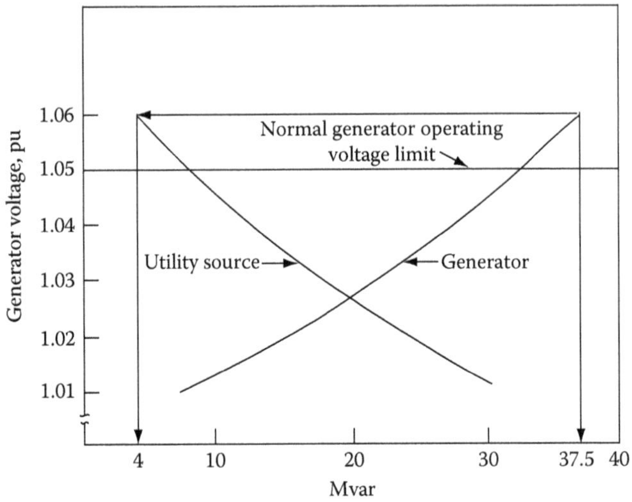

FIGURE 13.21
Reactive power sharing between generator and utility system versus voltage of generation for Example 13.5.

In order to maintain voltages at acceptable levels in the plant distribution, it is required to hold the 13.8 kV bus voltage within ±2% of the rated voltage. Also, the initial voltage dip on sudden outage of the generator should be limited to 12% to prevent a process loss.

Figure 13.21 shows reactive power sharing of load demand from the utility source and generator, versus generator operating voltage. To meet the load reactive power demand of 31 Mvar, the generator must operate at 1.06 per unit voltage. It delivers its rated Mvar output of 37.4. Still, approximately 4 Mvar of reactive power should be supplied by the utility source. A loss of 10.4 Mvar occurs in the generator reactor and transformer. To limit the generator operating voltage to 1.05, still higher reactive power must be supplied from the utility's source.

If reactive power compensation is provided in the form of shunt harmonic filters (Chapter 19), the performance is much improved. With an 11 Mvar filter alone, the generator can be operated at 1.05 per unit voltage. Its reactive power output is 28.25 Mvar, the source contribution is 1 Mvar, and the loss is 9 Mvar. Addition of another 5 Mvar of the seventh harmonic filter results in a flow of 6 Mvar into the utility's system; the generator operates at 30.25 Mvar, approximately at 81% of its rated reactive power capability at rated load, and the remaining reactive power capability of the generator cannot be normally utilized with its maximum operating voltage of 1.05 per unit. It serves as a reserve in case of a voltage dip.

On the sudden loss of a generator, initial voltage dips of 17.5%, 5.5%, and 4%, respectively, occur with no harmonic filter, with only the fifth harmonic filter, and with both fifth and seventh harmonic filters in service. This is another important reason to provide reactive power compensation. It will prevent a process loss, and to bring the bus voltage to its rated level, some load shedding is required.

13.5 Flexible AC Transmission Systems

The concept of FACTS was developed by EPRI and many FACTS operating systems are already implemented [9–12]. The world's first thyristor-controlled series capacitor was put in service on the Bonneville Power Authority's 500 kV line in 1993.

We have examined the problem of power flow over transmission lines and the role of SVCs' series and shunt compensation to change the impedance of the line or its transmission angle. Advances in recent years in power electronics, software, microcomputers, and fiber-optic transmitters that permit signals to be sent to and fro from high-voltage levels make possible the design and use of fast FACTS controllers. For example, the use of thyristors makes it possible to switch capacitors' order of magnitude faster than with circuit breakers and mechanical devices.

Another thrust leading to FACTS is the use of electronic devices in processes, industry, and home, which has created an entirely new demand for power quality that the energy providers must meet. Power quality problems include voltage sags and swells, high-frequency line-to-line surges, steep wave fronts, or spikes caused by switching of loads, harmonic distortions, and outright interruptions, which may extend over prolonged periods. An impulse is a unidirectional pulse of 10 ms in duration. A voltage sag is a reduction in nominal voltage for more than 0.01 s and less than 2.5 s. A swell is an increase in voltage for more than 0.01 s and less than 2.5 s. Low and high voltages are reduction and increase in voltage for more than 2.5 s. All these events are observable. A problem that is not easily detected is the common-mode noise, which occurs on all conductors of an electrical circuit at the same time. The tolerance of processes to these power quality problems is being investigated more thoroughly; however, the new processes and electronic controls are more sensitive to power quality problems. It is estimated that the power quality problems cost billions of dollars per year to the American industry. Though this book is not about mitigating the power quality problems and detailed discussions of this subject, two figures of interest are Figures 13.22 and 13.23. Figure 13.22 categorizes the various power quality problems and the technologies that can be applied to mitigate these problems [13]. Figure 13.23 is the new ITI (Information Technology Industry Council, earlier CBEMA) curve, which replaces the earlier CBEMA (Computer and Business Equipment Manufacturer's Association) [14] curve for acceptable power quality of computers and electronic equipment. The voltage dips and swells and their time duration become of importance. The other requirements of voltage balance, transients, phase angle displacements, and frequency variations are not discussed.

The FACTS use voltage source bridge, rather than current source configuration (Chapter 17). The requirements of FACTS controllers are as follows:

- The converter should be able to act as an inverter or rectifier with leading or lagging reactive power, that is, four-quadrant operation is required, as compared to current source line commutated converter, which has two-quadrant operations.
- The active and reactive power should be independently controllable with the control of phase angle.

The basic electronic devices giving rise to this thrust in the power quality are listed in Table 13.2. FACTS devices control the flow of ac power by changing the impedance of the transmission line or the phase angle between the ends of a specific line. Thyristor controllers can provide the required fast control, increasing or decreasing the power flow on a specific line and, responding almost instantaneously to stability problems. FACTS devices can be used to dampen the subsynchronous oscillations, which can be damaging to rotating equipment, that is, generators. These devices and their capabilities are only briefly discussed in this chapter.

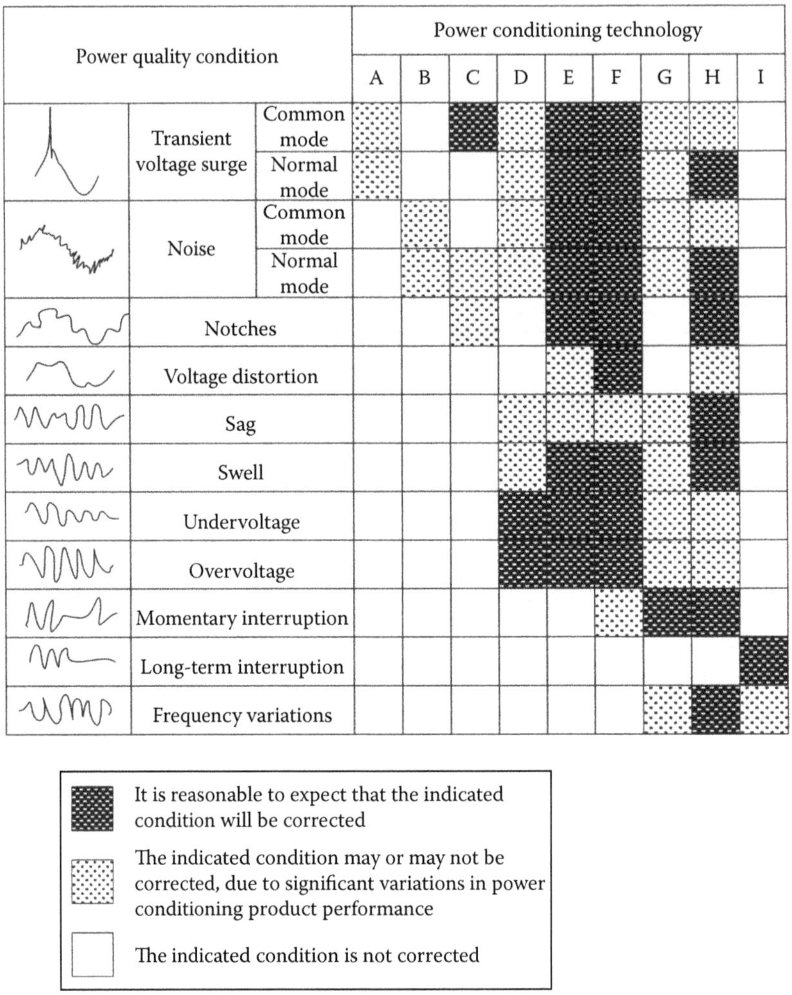
Power quality condition
Power conditioning technology
A
B
C
D
E
F
G
H
I
Transient voltage surge
Common mode
Normal mode
Noise
Common mode
Normal mode
Notches
Voltage distortion
Sag
Swell
Undervoltage
Overvoltage
Momentary interruption
Long-term interruption
Frequency variations
It is reasonable to expect that the indicated condition will be corrected
The indicated condition may or may not be corrected, due to significant variations in power conditioning product performance
The indicated condition is not corrected

A = TVSS
B = EMI/RFI filter
C = Isolation transformer
D = Electronic voltage regulator
E = Ferroresonane voltage regulator
F = Motor generator
G = Standby power system
H = Uninterruptible power supply (UPS)
I = Standby engine generator

FIGURE 13.22
Power quality problems and their mitigation technologies.

13.5.1 Synchronous Voltage Source

A solid-state synchronous voltage source (SS) can be described as analogous to a synchronous machine. A rotating condenser has a number of desirable characteristics, that is, high capacitive output current at low voltages and a source impedance that does not cause harmonic resonance with the transmission network. It has a number of shortcomings too, for example, slow response, rotational instability, and high maintenance.

An SS can be implemented with a voltage source inverter using gate turn-off thyristors (GTOs). An elementary six-pulse voltage source inverter with a dc voltage source can produce a balanced set of three quasi-square waveforms of a given frequency. The output

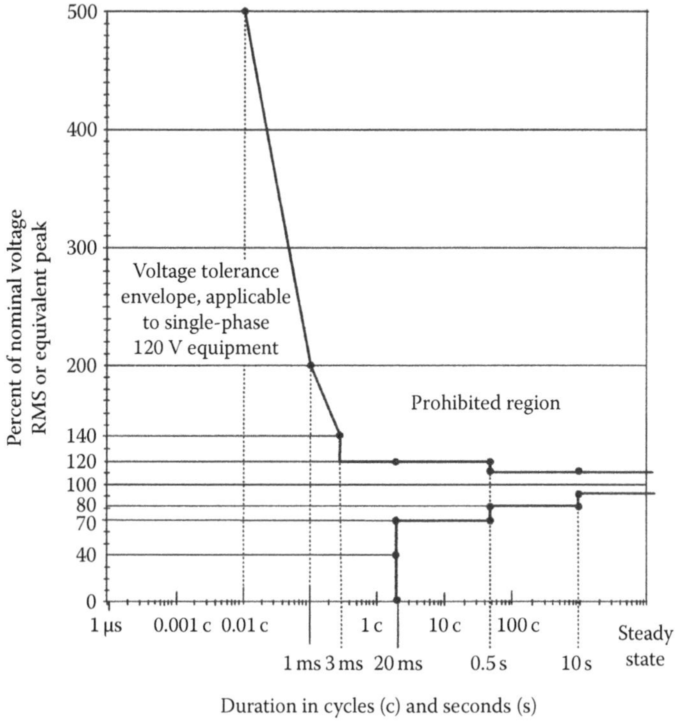

FIGURE 13.23
ITI curve, 2000.

TABLE 13.2
FACTS Devices[a]

Application	STATCOM or STATCON	SPFC (SSSC)	UPFC	NGH-SSR Damper
Voltage control	X	X	X	
Var compensation	X		X	
Series impedance control		X	X	X
MW flow control		X	X	
Transient stability	X	X	X	X
System isolation		X	X	
Damping of oscillations	X	X	X	X

[a] Each of these devices has varying degrees of control characteristics broadly listed in Table 13.2.

STATCOM, STATCON, static synchronous compensator or static condenser; SPFC, SSSC, series power flow controller; static series synchronous compensator; UPFC, unified power flow controller; NGH-SSR, Narain G. Hingorani subsynchronous resonance damper [9].

of the six-pulse inverter will contain harmonics of unacceptable level for transmission line application, and a multipulse inverter can be implemented by a variety of circuit arrangements (see Chapter 17).

The reactive power exchange between the inverter and the ac system can be controlled by varying the amplitude of the three-phase voltage produced by the SS. Similarly, the real

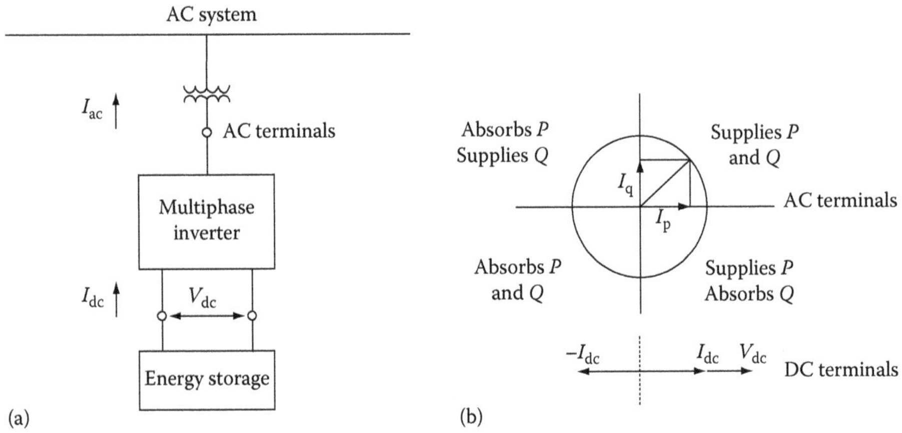

FIGURE 13.24
(a) A shunt connected synchronous voltage source; (b) possible modes of operation for real and reactive power generation.

power exchange between the inverter and the ac system can be controlled by phase shifting the output voltage of the inverter with respect to the ac system. Figure 13.24a shows the coupling transformer, the inverter, and an energy source, which can be a dc capacitor, battery, or superconducting magnet.

The reactive and real power generated by the SS can be controlled independently and any combination of real power generation/absorption with var generation and absorption is possible, as shown in Figure 13.24b. The real power supplied/absorbed must be supplied by the storage device, while the reactive power exchanged is internally generated in the SS.

The reactive power exchange is controlled by varying the amplitude of three-phase voltage. For a voltage greater than the system voltage, the current flows through the reactance from the inverter into the system, that is, the capacitive power is generated. For an inverter voltage less than the system voltage the inverter absorbs reactive power. The reactive power is, thus, exchanged between the system and the inverter, and the real power input from the dc source is zero. In other words, the inverter simply interconnects the output terminals in such a way that the reactive power currents can freely flow through them.

The real power exchange is controlled by phase shifting the output voltage of the inverter with respect to the system voltage. If the inverter voltage leads the system voltage, it provides real power to the system from its storage battery. This results in a real component of the current through tie reactance, which is in phase opposition to the ac voltage. Conversely, the inverter will absorb real power from the system to its storage device if the voltage is made to lag the system voltage.

This bidirectional power exchange capability of the SS makes complete temporary support of the ac system possible.

13.5.2 Static Synchronous Compensator

Figure 13.25 shows the schematic of an inverter-based shunt static synchronous compensator (STATCOM), sometimes called a static condenser (STATCON). It is a shunt reactive power compensating device, shown in Figure 13.24a. It can be considered as an SS with a storage device as a dc capacitor. A GTO-based power converter produces an ac voltage in

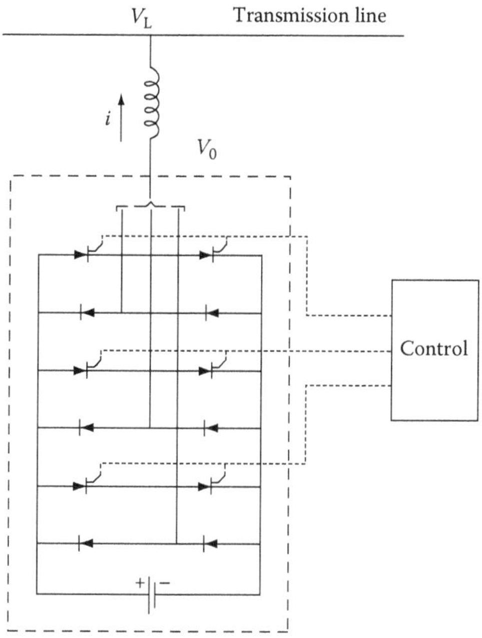

FIGURE 13.25
Schematic of a STATCOM or STATCON inverter-based static shunt compensator.

phase with the transmission line voltage. When the voltage source is greater than the line voltage ($V_L < V_0$), leading vars are drawn from the line and the equipment appears as a capacitor; when voltage source is less than the line voltage ($V_L > V_0$), a lagging reactive current is drawn. The basic building block is a voltage-source inverter, which converts dc voltage at its terminals into three-phase ac voltage. A STATCON may use many six-pulse inverters, output phase shifted and combined magnetically to give a pulse number of 24 or 48 for the transmission systems. Using the principal of harmonic neutralization, the output of n basic six-pulse inverters, with relative phase displacements, can be combined to produce an overall multiphase system. The output waveform is nearly sinusoidal and the harmonics present in the output voltage and input current are small (see Chapter 20). This ensures waveform quality without passive filters.

The *V–I* characteristics are shown in Figure 13.26. As compared to the characteristics of an SVC, the STATCON is able to provide rated reactive current under reduced voltage conditions. It also has transient overload capacity both in the inductive and capacitive regions, the limit being set by the junction temperature of the semiconductors. By contrast an SVC can only supply diminishing output current with decreasing system voltage.

The ability to produce full capacitive current at low voltages makes it ideally suitable for improving the first swing (transient) stability. The dynamic performance capability far exceeds that of a conventional SVC. It has been shown that the current system can transition from full rated capacitive to full rated inductive vars in approximately a *quarter-cycle*.

STATCON, just like SVC, behaves like an ideal midpoint compensator until the maximum capacitive output current is reached. The reactive power output of a STATCOM varies linearly with the system voltage while that of the SVC varies with the square of the voltage. In an SVC, TCRs produce high harmonic content, as the current waveform is chopped off in the phase-controlled rectifiers, and passive filters are required. A STATCON uses phase multiplication or pulse-width modulation and the harmonic generation is a minimum.

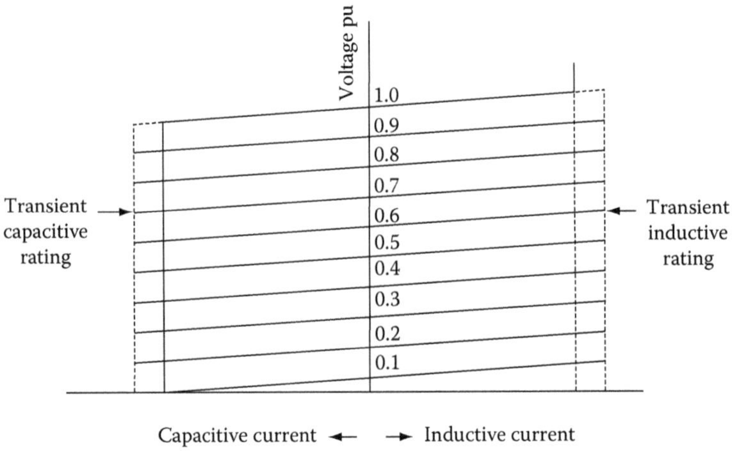

FIGURE 13.26
V–I characteristics of a STATCON.

Summarizing, the following advantages are obtained:

- Interface with real power sources
- Higher response to system changes
- Mitigation of harmonics as compared to an FC-TCR
- Superior low-voltage performance

13.5.3 Static Series Synchronous Compensator

A static series synchronous compensator (SSSC) may also be called a series power flow controller (SPFC). The basic circuit is that of an SS in *series* with the transmission line (Figure 13.27). We have observed that conventional series compensation can be considered as a reactive impedance in series with the line, and the voltage across it is proportional to the line current. A series capacitor increases the transmitted power by a certain percentage, depending on the series compensation for a given δ. In contrast, an SSSC injects a

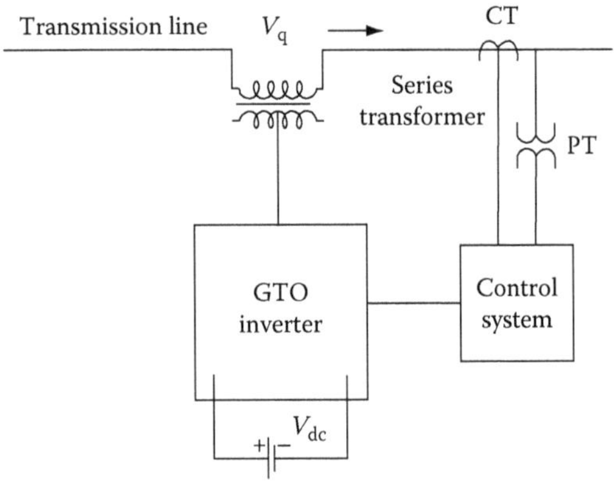

FIGURE 13.27
Series connected synchronous voltage source, SPFC, or SSSC.

compensating voltage, V_q, in series with the line *irrespective of the line current*. The SSSC increases the maximum power transfer by a fraction of the power transmitted, nearly independent of δ:

$$P_q = \frac{V^2}{X_{sc}} \sin\delta + \frac{V}{X_{sc}} V_q \cos\left(\frac{\delta}{2}\right) \tag{13.47}$$

While a capacitor can only increase the transmitted power, the SSSC can decrease it by simply reversing the polarity of the injected voltage. The reversed voltage adds directly to the reactive power drop in the line and the reactive line impedance is increased. If this reversed polarity voltage is larger than the voltage impressed across the line by sending and receiving end systems, the power flow will reverse.

$$|V_q| = |V_s - V_r + IX_L| \tag{13.48}$$

Thus, stable operation of the system is possible with positive and negative power flows, and due to the response time being less than 1 cycle, the transition from positive to negative power flow is smooth and continuous. Figure 13.28 shows P–δ curves of a series capacitor and an SSSC.

The SSSC can negotiate both the reactive and active power with an ac system, simply by controlling the angular position of the injected voltage with respect to the line current. One important application is simultaneous compensation of both reactive and resistive elements of the line impedance. By applying series compensation, the X/R ratio decreases. As R remains constant the ratio is $(X_L - X_c)/R$. As a result the reactive component of the current supplied by the receiving end progressively increases while the real component of the current transmitted to the receiving end progressively decreases. An SSSC can inject a component of voltage in antiphase to that developed by the line resistance drop to counteract the effect of resistive voltage drop on the power transmission.

The dynamic stability can be improved, as the reactive line compensation with simultaneous active power exchange can damp power system oscillations. During periods of angular

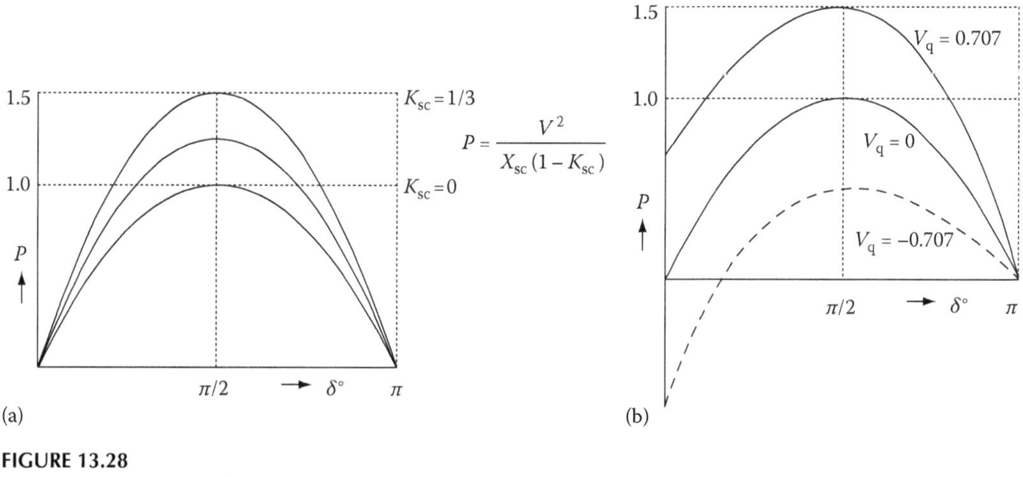

FIGURE 13.28
(a) P–δ characteristics of conventional series capacitive compensation; (b) P–δ characteristics of an SSSC as a function of the compensating voltage V_q.

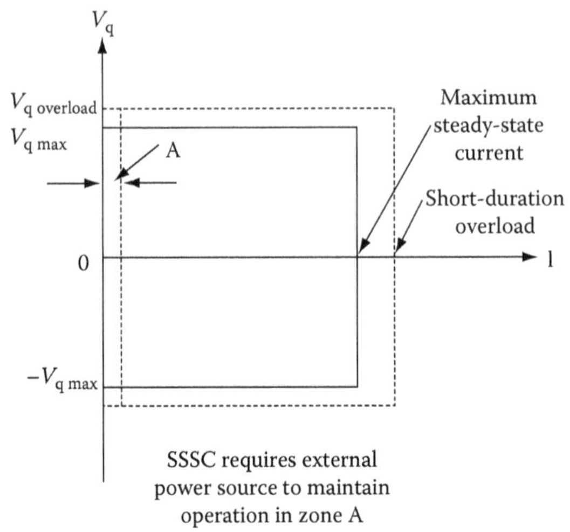

FIGURE 13.29
Range of SSSC voltage versus current, with overload capacity shown in dotted lines.

acceleration (increase of torque angle of the machines), an SSSC with suitable energy storage can provide maximum capacitive line compensation to increase the active power transfer and also absorb active power, acting like a damping resistor in series with the line.

The problems of SSR stated for a series capacitor in Section 13.3.6 are avoided. The SSSC is essentially an ac voltage source, which operates only at the fundamental frequency and its output impedance at other frequencies can be theoretically zero, though SSSC does have a small inductive impedance of the coupling transformer. An SSSC does not form a series resonant circuit with the line inductance, rather it can damp out the subsynchronous oscillations that may be present due to existing series capacitor installations.

Figure 13.29 shows the characteristics of an SSSC. The VA (volt-ampere) rating is simply the product of maximum line current and the maximum series compensating voltage. Beyond the maximum rated current, the voltage falls to zero. The maximum current rating is practically the maximum steady-state line current. In many practical applications only capacitive series line compensation is required and an SSSC can be combined with a fixed series capacitor.

If the device is connected to a short line with infinite buses, unity voltages, and constant phase angle difference, the characteristic can be represented by a circle in the P–Q plane with

$$\text{Center} = \frac{S_0 Z^*}{2R} \tag{13.49}$$

$$\text{Radius} = \left|\frac{S_0 Z^*}{2R}\right| \tag{13.50}$$

where

$S_0 = P_0 + jQ_0 =$ uncompensated power flow
Z is the series impedance of the line $= R_{sc} + jX_{sc}$
Z^* is complex conjugate of Z

The operating characteristics are defined by the *edge* of the circle only. Figure 13.30 shows the power transfer capability of a variable series capacitor with an SPFC controller. The portion of the characteristics to the left of the origin shows power reversal capability.

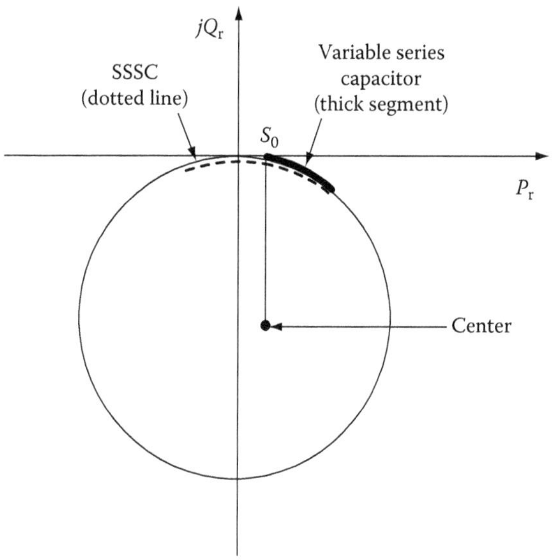

FIGURE 13.30
Operating characteristics of an SSSC in *P–Q* plane, showing comparison with variable series capacitor.

13.5.4 Unified Power Flow Controller

A unified power controller consists of two voltage source switching converters, a series and a shunt converter, and a dc link capacitor (Figure 13.31). The arrangement functions as an ideal ac-to-ac power converter in which real power can flow in either direction between ac terminals of the two converters, and each inverter can independently generate or absorb

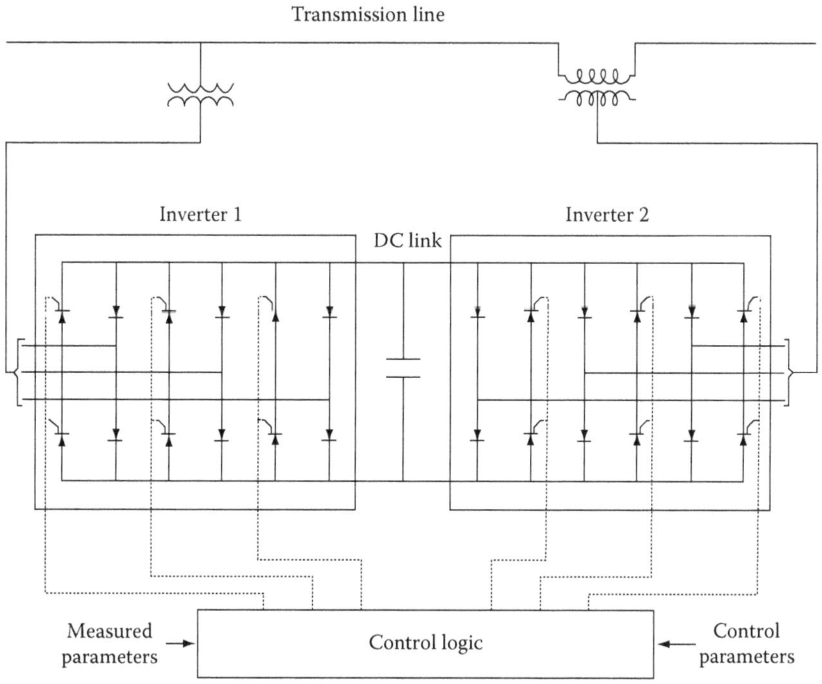

FIGURE 13.31
Schematic diagram of a UPFC.

reactive power at its own terminals. Inverter 2 injects an ac voltage V_q with controllable magnitude and phase angle (0°–360°) at the power frequency in series with the line. This injected voltage can be considered as an SS.

The real power exchanged at the ac terminal is converted by inverter into dc power, which appears as dc link voltage. The reactive power exchanged is generated internally by the inverter.

The basic function of inverter 1 is to absorb the real power demanded by inverter 2 at the dc link. This dc link power is converted back into ac and coupled with the transmission line through a shunt transformer. Inverter 1 can also generate or absorb controllable reactive power. The power transfer characteristics of a short transmission line with a unified power flow controller (UPFC) connecting two infinite buses, unity voltage, and a constant phase angle, can be represented by a circle on the *P–Q* plane:

$$(P_R - P_0)^2 + (Q_R - Q_0)^2 = \left(\frac{V_i}{Z}\right)^2 \tag{13.51}$$

where

P_0 and Q_0 are the line uncompensated real and reactive power
V_i is the magnitude of the injected voltage
Z is the line series impedance

The center is at the uncompensated power level S_0 and the radius is V_i/Z. Consider a UPFC with 0.25 per unit (pu) voltage limit. Let the series reactance of the line be 1.0 pu and the uncompensated receiving end power be $1 + j0$ pu. The UPFC can then control the receiving end power within a circle of 2.5 pu. With its center being at (1, 0), power transfer could be controlled between +3.5 and −1.5 pu.

The allowable operating range with the UPFC is anywhere inside the circle, while the SPFC operating range is the circle itself. The portion of the UPFC circle inside the SPFC circle represents operation of the UPFC with real power transfer from the transmission system to the series inverter to the shunt inverter, while the portion of the UPFC circle outside the SPFC circle represents the transfer of real power from the shunt inverter to the series inverter to the transmission system (Figure 13.32). A comparison with phase angle

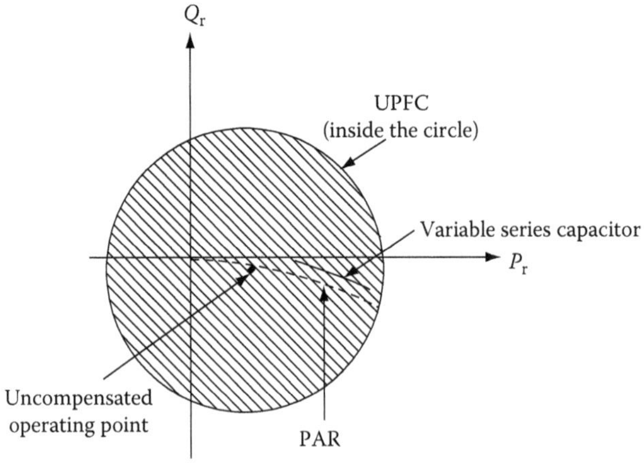

FIGURE 13.32
Comparison of characteristics of UPFC, PAR, and series capacitor in *P–Q* plane.

regulator (PAR) and series capacitors is superimposed in this figure. These devices provide one-dimensional control, while the UPFC provides simultaneous and independent P and Q control over a wide range.

Table 13.2 also shows characteristics and applications of FACTS controllers.

Problems

13.1 Mathematically derive Equations 13.6 and 13.37.

13.2 The voltage under load flow at a certain bus in an interconnected system dips by 10%, while the voltages on adjacent buses are held constant. The available short-circuit current at this bus is 21 kA, and the voltage is 230 kV. Find the reactive power injection to restore the voltage to its rated value of 230 kV.

13.3 In Example 13.1, the line is compensated by a shunt compensation of $K_{sh}=0.3$. Calculate all the parameters of Example 13.2. Repeat for series and shunt compensation of $K_{sh}=K_{sc}=0.3$.

13.4 A 100 MVA load at 0.8 power factor and 13.8 kV is supplied through a short transmission line of 0.1 per unit reactance (100 MVA base). Calculate the reactive power loss, and the load voltage. Size a capacitor bank at load terminals to limit the voltage drop to 2% at full load. What are the capacitor sizes for three-step switching to maintain the load voltage no more than 2% below the rated voltage as the load varies from zero to full load?

13.5 A 200 MW, 18 kV, 0.85 power factor generator is connected through a 200 MVA step-up transformer of 18–500 kV and of 10% reactance to a 500 kV system. The transformer primary windings are provided with a total of five taps, two below the rated voltage of 2.5% and 5% and two above the rated voltage of 2.5% and 5%. Assuming that initially the taps are set at a rated voltage of 18–500 kV, what is the generation voltage to take full rated reactive power output from the generator? If the operating voltage of the generator is to be limited to rated voltage, find the tap setting on the transformer. Find the reactive power loss through the transformer in each case. Neglect resistance.

13.6 A 100 mile (160 km) long line has $R=0.3$ ohm, $L=3$ mH, and $C=0.015$ μF. It delivers a load of 100 MVA at 0.85 power factor at 138 kV. If the sending end voltage is maintained at 145 kV, find the Mvar rating of the synchronous condenser at the receiving end at no load and at full load.

13.7 Derive an equation for the load line of the synchronous condenser shown in Figure 13.8.

13.8 Plot P–δ curves of (a) 230 kV uncompensated line, (b) with midpoint shunt compensation of 0.6, and (c) with series compensation of 0.6. The line is 200 miles long, has a series reactance of 0.8 ohm/mile, and a susceptance $y=5.4\times10^{-6}$ S/mile.

13.9 Compare the performance of conventional SVC devices with FACT controllers.

13.10 Compare UPFC with STATCON and SSSC.

References

1. ANSI. *Voltage Rating for Electrical Power Systems and Equipments (60 Hz)*, Standard C84.1, 1988.
2. NEMA. *Large Machines-Induction Motors*, Standard MG1-Part 20, 1993.
3. P. Kessel and R. Glavitsch. Estimating voltage stability of a power system. *Trans. IEEE PWRD* 1: 346–352, 1986.
4. T.J.E. Miller. *Reactive Power Control in Electrical Power Systems*. New York: John Wiley, 1982.
5. IEEE Committee Report. Var management-problem and control. *Trans. IEEE* PAS 103: 2108–2116, 1984.
6. IEEE. *A IEEE Standard Rated IONIVA and above, for Cylindrical Rotor Synchronous Generators*, Standard C50.13, 2005.
7. CIGRE WG 30-01, Task Force No. 2. In: I.A. Erinmez (ed.), *Static Var Compensators*. Paris, France: CIGRE, 1986.
8. T.M. Kay, P.W. Sauer, R.D. Shultz, and R.A. Smith. EHV and UHV line loadability dependent on var supply capability. *Trans. IEEE PAS* 101: 3586–3575, 1982.
9. N.G. Hingorani. Flexible AC transmission. *IEEE Spectrum* 30 (4): 40–45, 1993.
10. C. Schauder, M. Gernhard, E. Stacey, T. Lemak, L. Gyugi, T.W. Cease, and A. Edris. Development of a 100 Mvar static condenser for voltage control of transmission systems. *Trans. IEEE PWRD* 10: 1486–1496, 1995.
11. R.J. Nelson, J. Bain, and S.L. Williams. Transmission series power flow control. *IEEE Trans. PD* 10 (1), pp. 504–510, January 1995.
12. N.G. Hingorani and L. Gyugyi. *Understanding FACTS*. New York: IEEE Press, 2000.
13. J.C. Das. Industrial power systems. In: *Encyclopedia of Electrical and Electronics Engineering*, Vol. 10. New York: John Wiley, 2002.
14. IEEE. *IEEE Recommended Practice for Emergency and Standby Power Systems for Industrial and Commercial Applications*, Std. 446 1995.

14

Three-Phase and Distribution System Load Flow

Normally, three-phase systems can be considered as balanced. Though some unbalance may exist, due to asymmetry in transmission lines, machine impedances, and system voltages, these are often small and may be neglected. A single-phase positive sequence network of a three-phase system is adequate for balanced systems. During faults, though the systems are balanced, voltages are not. Unbalance voltages and unbalance in three-phase networks can also occur simultaneously. The unbalances cannot be ignored in every case, i.e., a distribution system may serve considerable single-phase loads. In such cases, three-phase models are required. A three-phase network can be represented both in the impedance and the admittance forms.

Equation 3.32 for a three-phase network can be written as

$$\begin{vmatrix} V_{pq}^{a} \\ V_{pq}^{b} \\ V_{pq}^{c} \end{vmatrix} + \begin{vmatrix} e_{pq}^{a} \\ e_{pq}^{b} \\ e_{pq}^{c} \end{vmatrix} = \begin{vmatrix} Z_{pq}^{aa} & Z_{pq}^{ab} & Z_{pq}^{ac} \\ Z_{pq}^{ba} & Z_{pq}^{bb} & Z_{pq}^{bc} \\ Z_{pq}^{ca} & Z_{pq}^{cb} & Z_{pq}^{cc} \end{vmatrix} \begin{vmatrix} i_{pq}^{a} \\ i_{pq}^{b} \\ i_{pq}^{c} \end{vmatrix} \tag{14.1}$$

The equivalent three-phase circuit is shown in Figure 14.1a, and its single-line representation in Figure 14.1b. In the condensed form, Equation 14.1 is

$$V_{pq}^{abc} + e_{pq}^{abc} = Z_{pq}^{abc} i_{pq}^{abc} \tag{14.2}$$

Similarly, for a three-phase system, Equation 3.33 in the admittance form is

$$\begin{vmatrix} i_{pq}^{a} \\ i_{pq}^{b} \\ i_{pq}^{c} \end{vmatrix} + \begin{vmatrix} j_{pq}^{a} \\ j_{pq}^{b} \\ j_{pq}^{c} \end{vmatrix} = \begin{vmatrix} y_{pq}^{aa} & y_{pq}^{ab} & y_{pq}^{ac} \\ y_{pq}^{ba} & y_{pq}^{bb} & y_{pq}^{bc} \\ y_{pq}^{ca} & y_{pq}^{cb} & y_{pq}^{cc} \end{vmatrix} \begin{vmatrix} V_{pq}^{a} \\ V_{pq}^{b} \\ V_{pq}^{c} \end{vmatrix} \tag{14.3}$$

In the condensed form, we can write

$$i_{pq}^{abc} + j_{pq}^{abc} = y_{pq}^{abc} V_{pq}^{abc} \tag{14.4}$$

A three-phase load flow study is handled much like a single-phase load flow. Each voltage, current, and power becomes a three-element vector and each single-phase admittance element is replaced by a 3 × 3 admittance matrix.

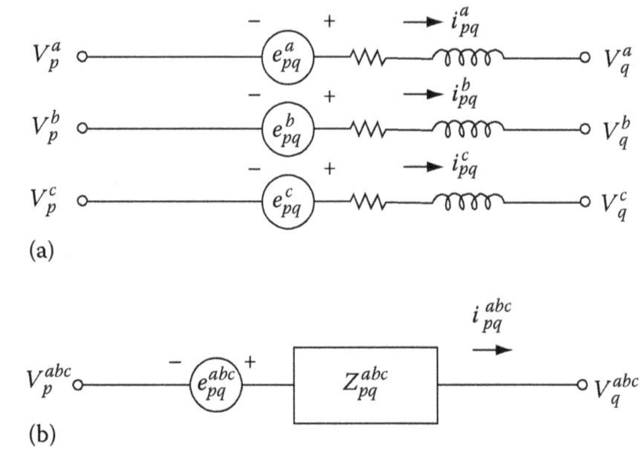

FIGURE 14.1
(a) Three-phase network representation, primitive impedance matrix; (b) single-line representation of three-phase network.

14.1 Phase Coordinate Method

In Chapter 1 we showed that a symmetrical system, when transformed by symmetrical component transformation, is decoupled and there is advantage in arriving at a solution using this method. The assumptions of a symmetrical system are not valid when the system is unbalanced. Untransposed transmission lines, large single-phase traction loads, and bundled conductors are some examples. Unbalanced currents and voltages can give rise to serious problems in the power system, i.e., negative sequence currents have a derating effect on generators and motors and ground currents can increase the coupling between transmission line conductors. Where the systems are initially coupled (Example B.1), then even after symmetrical component transformation the equations remain coupled. The method of symmetrical components does not provide an advantage in arriving at a solution. By representing the system in phase coordinates, i.e., phase voltages, currents, impedances, or admittances, the initial physical identity of the system is maintained. Using the system in the phase frame of reference, a generalized analysis of the power system network can be developed for unbalance, i.e., short-circuit or load flow conditions [1–3]. The method uses a nodal Y admittance matrix and, due to its sparsity, optimal ordering techniques are possible. Series and shunt faults and multiple unbalanced faults can be analyzed. The disadvantage is that it takes more iterations to arrive at a solution.

Transmission lines, synchronous machines, induction motors, and transformers are represented in greater detail. The solution technique can be described in the following steps:

- The system is represented in phase frame of reference.
- The nodal admittance matrix is assembled and modified for any changes in the system.
- The nodal equations are formed for solution.

The nodal admittance equation is the same as for a single-phase system:

$$\bar{Y}\bar{V} = \bar{I} \tag{14.5}$$

Each node is replaced by three equivalent separate nodes. Each voltage and current is replaced by phase-to-ground voltages and three-phase currents; I and V are column vectors of nodal phase currents and voltages. Each element of $\overline{Y}$ is replaced with a 3×3 nodal admittance sub-matrix. Active sources such as synchronous machines can be modeled with a voltage source in series with passive elements. Similarly, transformers, transmission lines, and loads are represented on a three-phase basis. The system base Y matrix is modified for the conditions under study, e.g., a series fault on opening a conductor can be simulated by Y-matrix modification. The shunt faults, i.e., single phase to ground, three phase to ground, two phase to ground, and their combinations can be analyzed by the principal of superimposition.

Consider a phase-to-ground fault at node k in a power system. It is equivalent to setting up a voltage V_f at k equal in magnitude but opposite in sign to the prefault voltage of the node k. The only change in the power system that occurs due to fault may be visualized as the application of a fault voltage V_f at k and the point of zero potential. If the effect of V_f is superimposed upon the prefault state, the fault state can be analyzed. To account for effect of V_f, all emf sources are replaced by their internal admittances and converted into equivalent admittance based on the prefault nodal voltage. Then, from Equation 14.5:

$$\begin{aligned} &I_i = 0, \quad \text{and} \quad V_k = \text{prefault voltage} \\ &i = 1, 2, \ldots, N, \quad i \neq k \end{aligned} \tag{14.6}$$

The fault current is

$$i_k = \sum_{i=1}^{i=N} Y_{ki} E_i \tag{14.7}$$

where E_i is the net *postfault* voltage.

For two single line-to-ground faults occurring at two different nodes p and q,

$$\begin{aligned} &I_i = 0, \quad i = 1, 2, \ldots, N \\ &i \neq p, q \end{aligned} \tag{14.8}$$

where V_p and V_q are equal to prefault values. Nodes p and q may represent any phase at any busbar. The currents I_p and I_q are calculated from

$$I_k = \sum_{i=1}^{N} Y_{ki} E_i \quad k = p, q \tag{14.9}$$

Thus, calculation of multiple unbalanced faults is as easy as a single line-to-ground fault, which is not the case with the symmetrical component method.

14.2 Three-Phase Models

Three-phase models of cables and conductors are examined in Appendix B. Y matrices of three-phase models are examined in this chapter, mainly for use in the factored Gauss–Y admittance method of load flow.

14.2.1 Conductors

A three-phase conductor with mutual coupling between phases and ground wires has an equivalent representation shown in Figure 14.2a and b (see Appendix B for details) and the following equations are then written for a line segment:

$$\begin{vmatrix} V_a - V'_a \\ V_b - V'_b \\ V_c - V'_c \end{vmatrix} = \begin{vmatrix} Z_{aa'-g} & Z_{ab'-g} & Z_{ac'-g} \\ Z_{ba'-g} & Z_{bb'-g} & Z_{bc'-g} \\ Z_{ca'-g} & Z_{cb'-g} & Z_{cc'-g} \end{vmatrix} \begin{vmatrix} I_a \\ I_b \\ I_c \end{vmatrix} \tag{14.10}$$

See also Equation B.22.

In the admittance form, Equation 14.10 can be written as

$$\begin{vmatrix} I_a \\ I_b \\ I_c \end{vmatrix} = \begin{vmatrix} Y_{aa-g} & Y_{ab-g} & Y_{ac-g} \\ Y_{bc-g} & Y_{bb-g} & Y_{bc-g} \\ Y_{ca-g} & Y_{cb-g} & Y_{cc-g} \end{vmatrix} \begin{vmatrix} V_a - V'_a \\ V_b - V'_b \\ V_c - V'_c \end{vmatrix} \tag{14.11}$$

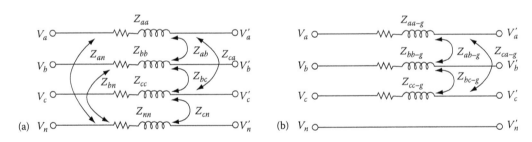

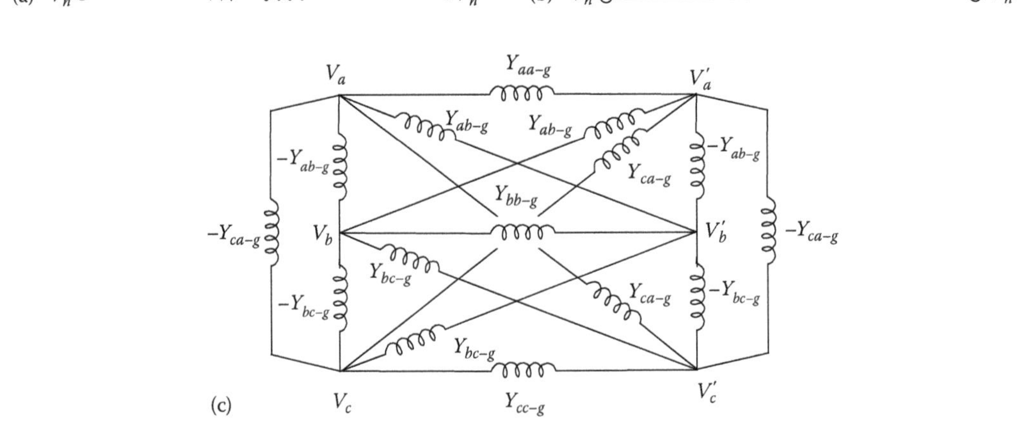

FIGURE 14.2
(a) Mutual couplings between a line section with ground wire in the impedance form; (b) transformed network in impedance form; (c) equivalent admittance network of a series line section.

Equation 14.11 can be rearranged as follows:

$$
\begin{aligned}
I_a &= Y_{aa-g}(V_a - V_a') + Y_{ab-g}(V_b - V_b') + Y_{ac-g}(V_c - V_c') \\
&= Y_{aa-g}(V_a - V_a') + Y_{ab-g}(V_a - V_b') + Y_{ac-g}(V_a - V_c') \\
&\quad - Y_{ab-g}(V_a - V_b) - Y_{ac-g}(V_a - V_c) \\
I_b &= Y_{bb-g}(V_b - V_b') + Y_{bb-g}(V_b - V_a') + Y_{bc-g}(V_b - V_c') \\
&\quad - Y_{ab-g}(V_c - V_a) - Y_{bc-g}(V_b - V_c) \\
I_b &= Y_{cc-g}(V_c - V_c') + Y_{cb-g}(V_c - V_b') + Y_{ca-g}(V_c - V_a') \\
&\quad - Y_{cb-g}(V_c - V_b) - Y_{ca-g}(V_c - V_a)
\end{aligned}
\tag{14.12}
$$

The same procedure can be applied to nodes V_a', V_b', and V_c'. This gives the equivalent series circuit of the line section as shown in Figure 14.2c. The effect of coupling is included in this diagram. Therefore, in the nodal frame we can write the three-phase Π model of a line as

$$
\begin{vmatrix} I_a \\ I_b \\ I_c \\ I_a' \\ I_b' \\ I_c' \end{vmatrix} = \begin{vmatrix} \mathcal{Y}^{abc} + \frac{1}{2}Y_{\text{sh}} & -\mathcal{Y}^{abc} \\ -\mathcal{Y}^{abc} & \mathcal{Y}^{abc} + \frac{1}{2}Y_{\text{sh}} \end{vmatrix} \begin{vmatrix} V_a \\ V_b \\ V_c \\ V_a' \\ V_b' \\ V_c' \end{vmatrix}
\tag{14.13}
$$

where

$$
\overline{Y}^{abc} = \overline{Z}^{-1,abc}
\tag{14.14}
$$

There is a similarity between the three-phase and single-phase admittance matrix, each element being replaced by a 3×3 matrix.

The shunt capacitance (line charging) can also be represented by current injection. Figure 14.3a shows the capacitances of a feeder circuit and Figure 14.3b shows current injection. The charging currents are

$$
\begin{aligned}
I_a &= -\frac{1}{2}[Y_{ab} + Y_{ac} + Y_{an}]V_a + \frac{Y_{ab}}{2}V_b + \frac{Y_{ac}}{2}V_c \\
I_b &= -\frac{1}{2}[Y_{ab} + Y_{ac} + Y_{an}]V_b + \frac{Y_{ab}}{2}V_a + \frac{Y_{bc}}{2}V_c \\
I_c &= -\frac{1}{2}[Y_{ab} + Y_{ac} + Y_{an}]V_c + \frac{Y_{ac}}{2}V_a + \frac{Y_{bc}}{2}V_b
\end{aligned}
\tag{14.15}
$$

14.2.2 Generators

The generators can be modeled by an internal voltage behind the generator transient reactance. This model is different from the power flow model of a generator, which is specified with a power output and bus voltage magnitude.

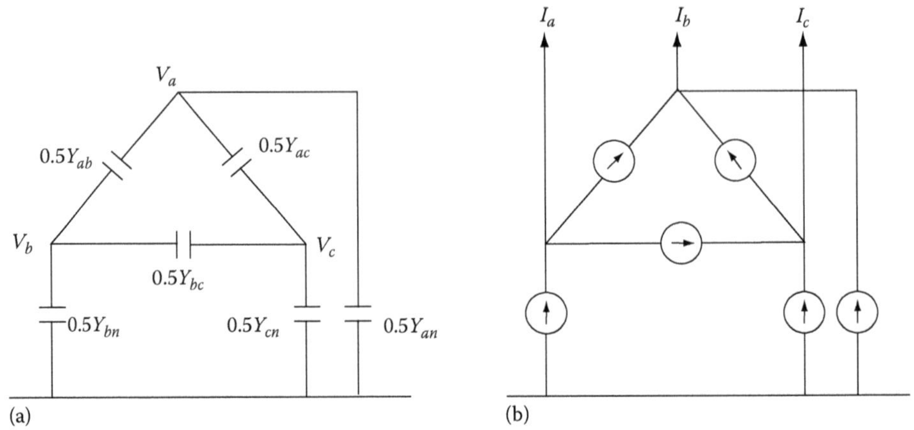

FIGURE 14.3
(a) Capacitances in a three-phase circuit; (b) equivalent current injections.

The positive, negative, and zero sequence admittances of a generator are well identified. The zero sequence admittance is

$$Y_0 = \frac{1}{R_0 + jX_0 + 3(R_g + jX_g)} \tag{14.16}$$

where
R_0 and X_0 are the generator zero sequence resistance and reactance
R_g and X_g are the resistance and reactance added in the neutral grounding circuit R_g and X_g are zero for a solidly grounded generator

Similarly,

$$Y_1 = \frac{1}{X'_d} \tag{14.17}$$

$$Y_2 = \frac{1}{X_2} \tag{14.18}$$

where
X'_d is the generator direct axis transient reactance
X_2 is generator negative sequence reactance (resistances ignored)

These sequence quantities can be related to the phase quantities as follows:

$$\bar{Y}^{abc} = \begin{vmatrix} Y_{11} & Y_{12} & Y_{13} \\ Y_{21} & Y_{22} & Y_{23} \\ Y_{31} & Y_{32} & Y_{33} \end{vmatrix}$$

$$= \frac{1}{3}\bar{T}_s\bar{Y}^{012}\bar{T}_s^t = \frac{1}{3}\begin{vmatrix} Y_0 + Y_1 + Y_2 & Y_0 + aY_1 + a^2Y_2 & Y_0 + a^2Y_1 + aY_2 \\ Y_0 + a^2Y_1 + aY_2 & Y_0 + Y_1 + Y_2 & Y_0 + aY_1 + a^2Y_2 \\ Y_0 + aY_1 + a^2Y_2 & Y_0 + a^2Y_1 + aY_2 & Y_0 + Y_1 + Y_2 \end{vmatrix} \tag{14.19}$$

where a is vector operator $1 < 120°$.

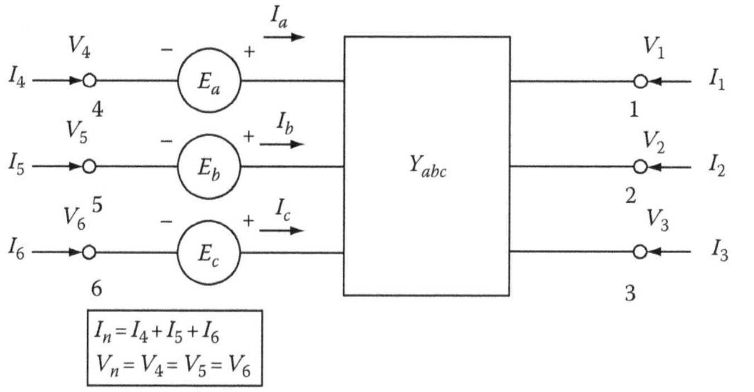

FIGURE 14.4
Norton equivalent circuit of a generator.

The machine model suitable for unbalance loading and neutral current flow is written as

$$\begin{vmatrix} I_1 \\ I_2 \\ I_3 \\ S^*/E_1^* \\ I_n \end{vmatrix} = \begin{vmatrix} Y_{11} & Y_{12} & Y_{13} & -Y_1 & -Y_0 \\ Y_{21} & Y_{22} & Y_{23} & -a^2Y_1 & Y_0 \\ Y_{31} & Y_{32} & Y_{33} & -aY_1 & -Y_0 \\ -Y_1 & -aY_1 & -a^2Y_1 & 3Y_1 & 0 \\ -Y_0 & -Y_0 & -Y_0 & 0 & 3Y_0 \end{vmatrix} \begin{vmatrix} V_1 \\ V_2 \\ V_3 \\ E_1 \\ V_n \end{vmatrix} \tag{14.20}$$

Referring to Figure 14.4, we can write

$$I_a = \frac{S_1^*}{E_1^*} \quad I_b = \frac{S_2^*}{E_2^*} \quad I_c = \frac{S_c^*}{E_3^*} \tag{14.21}$$

$$I_a = \frac{S_1^*}{E_1^*} \quad I_b = \frac{S_2^*}{aE_1^*} \quad I_c = \frac{S_c^*}{a^2E_1^*} \tag{14.22}$$

$$I_a + I_b + I_c = \frac{S^*}{E_1^*} = \frac{S_1^* + S_2^* + S_3^*}{E_1^*} \tag{14.23}$$

where
- S_1, S_2, and S_3 are the individual phase powers
- S is the total power
- E_1 is the positive sequence voltage behind the transient reactance

For a solidly grounded system the neutral voltage is zero. The internal machine voltages E_1, E_2, and E_3 are balanced; however, the terminal voltages V_1, V_2, and V_3 depend on internal machine impedances and unbalance in machine currents, I_a, I_b, and I_c. Because of unbalance, each phase power is not equal to one-third of the total power. I_1, I_2, and I_3 are injected currents and I_n is the neutral current. Equation 14.20 can model unbalances in the machine inductances and external circuit [2].

14.2.3 Generator Model for Cogeneration

The cogenerators in distribution system load flow are not modeled as PV type machines, i.e., to control the bus voltage, as discussed in Chapters 11 and 12. They are controlled to

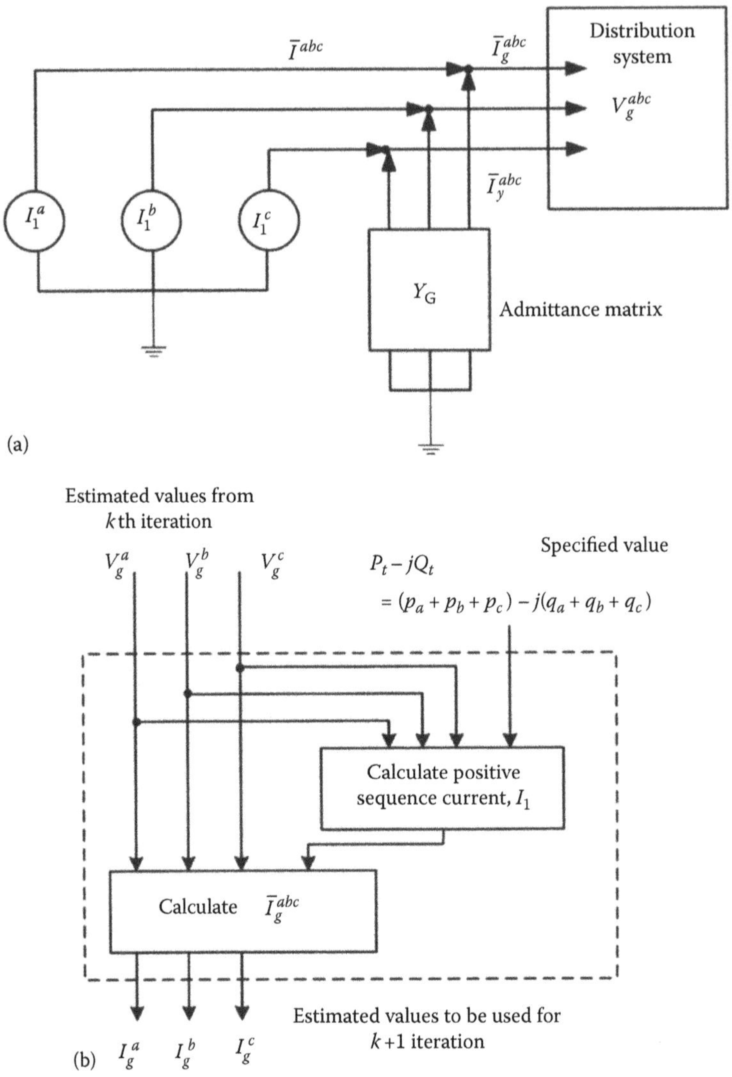

FIGURE 14.5
(a) Norton equivalent circuit of a generator for distribution systems; (b) circuit for calculations of load flow.

maintain a constant power and power factor, and power factor controllers may be required. Thus, for load flow the synchronous generators can be modeled as constant complex power devices. The induction generators require reactive power which will vary with the terminal voltage. Assuming a voltage close to rated voltage, these can also be modeled as *P–Q* devices. Figure 14.5a shows the Norton equivalent of the generator and Figure 14.5b shows the load flow calculation procedure [4]. The generator is represented by three injected currents. For the short-circuit calculations, the generator model is the same, except that I_1 is kept constant, as the internal voltage of the generator can be assumed not to change immediately after the fault.

14.2.4 Three-Phase Transformer Models

Three-phase transformer models considering winding connections, and turns ratio, are described in this section [5]. Consider a 12-terminal coupled network, consisting of three primary windings and three secondary windings mutually coupled through the transformer core (Figure 14.6). The short-circuit primitive matrix for this network is

$$\begin{vmatrix} i_1 \\ i_2 \\ i_3 \\ i_4 \\ i_5 \\ i_6 \end{vmatrix} = \begin{vmatrix} y_{11} & y_{12} & y_{13} & y_{14} & y_{15} & y_{16} \\ y_{21} & y_{22} & y_{23} & y_{24} & y_{25} & y_{26} \\ y_{31} & y_{32} & y_{33} & y_{34} & y_{35} & y_{36} \\ y_{41} & y_{42} & y_{43} & y_{44} & y_{45} & y_{46} \\ y_{51} & y_{52} & y_{53} & y_{54} & y_{55} & y_{56} \\ y_{61} & y_{62} & y_{63} & y_{64} & y_{65} & y_{66} \end{vmatrix} \begin{vmatrix} v_1 \\ v_2 \\ v_3 \\ v_4 \\ v_5 \\ v_6 \end{vmatrix} \tag{14.24}$$

This ignores tertiary windings. It becomes a formidable problem for calculation if all the Y elements are distinct. Making use of the symmetry, the Y matrix can be reduced to

$$\begin{vmatrix} y_p & -y_m & y'_m & y''_m & y'_m & y''_m \\ -y_m & y_s & y''_m & y'''_m & y''_m & y'''_m \\ y'_m & y''_m & y_p & -y_m & y'_m & y''_m \\ y''_m & y'''_m & -y_m & y_s & y''_m & y'''_m \\ y'_m & y''_m & y'_m & y''_m & y_p & -y_m \\ y''_m & y'''_m & y''_m & y'''_m & -y_m & y_s \end{vmatrix} \tag{14.25}$$

This considers that windings 1, 3, and 5 are primary windings and windings 2, 4, and 6 are secondary windings with appropriate signs for the admittances. The primed elements are all zero if there are no mutual couplings, e.g., in the case of three single-phase transformers.

$$\begin{vmatrix} i_1 \\ i_2 \\ i_3 \\ i_4 \\ i_5 \\ i_6 \end{vmatrix} = \begin{vmatrix} y_p & -y_m & 0 & 0 & 0 & 0 \\ -y_m & y_s & 0 & 0 & 0 & 0 \\ 0 & 0 & y_p & -y_m & 0 & 0 \\ 0 & 0 & -y_m & y_s & 0 & 0 \\ 0 & 0 & 0 & 0 & y_p & -y_m \\ 0 & 0 & 0 & 0 & -y_m & y_s \end{vmatrix} \begin{vmatrix} v_1 \\ v_2 \\ v_3 \\ v_4 \\ v_5 \\ v_6 \end{vmatrix} \tag{14.26}$$

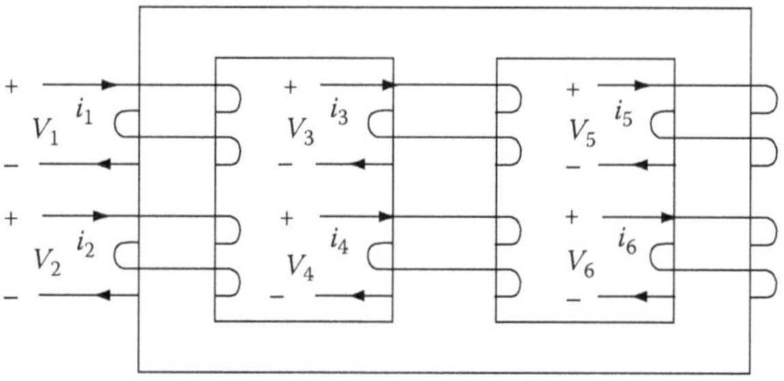

FIGURE 14.6
Elementary circuit of a three-phase transformer showing 12-terminal coupled primitive network.

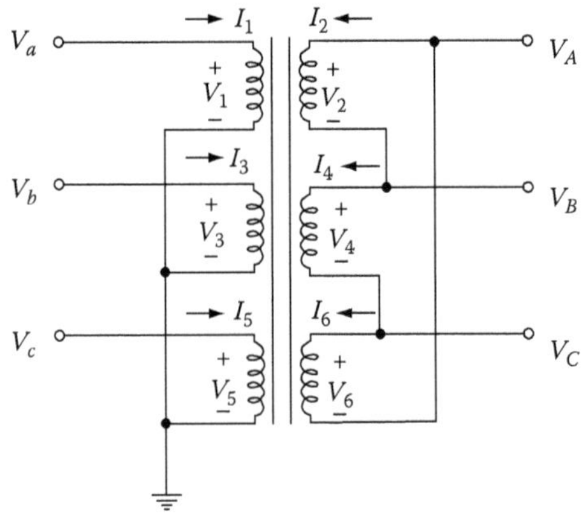

FIGURE 14.7
Circuit of a grounded wye–delta transformer with voltage and current relations for derivation of connection matrix.

Consider a three-phase wye–delta transformer (Figure 14.7). The branch and node voltages in this figure are related by the following connection matrix

$$\begin{vmatrix} v_1 \\ v_2 \\ v_3 \\ v_4 \\ v_5 \\ v_6 \end{vmatrix} = \begin{vmatrix} 1 & 0 & 0 & 0 & 0 & 0 \\ 0 & 0 & 0 & 1 & -1 & 0 \\ 0 & 1 & 0 & 0 & 0 & 0 \\ 0 & 0 & 0 & 0 & 1 & -1 \\ 0 & 0 & 1 & 0 & 0 & 0 \\ 0 & 0 & 0 & -1 & 0 & 1 \end{vmatrix} \begin{vmatrix} V_a \\ V_b \\ V_c \\ V_A \\ V_B \\ V_C \end{vmatrix} \tag{14.27}$$

or we can write

$$\bar{v}_{\text{branch}} = \overline{N}\,\overline{V}_{\text{node}} \tag{14.28}$$

$\overline{N}$ is the connection matrix.

Kron's transformation [6] is applied to the connection matrix $\overline{N}$ to obtain the node admittance matrix

$$\overline{Y}_{\text{node}} = \overline{N}^t \overline{Y}_{\text{prim}} \overline{N} \tag{14.29}$$

The node admittance matrix is obtained in phase quantities as

$$\overline{Y}_{\text{node}} = \begin{vmatrix} y_s & y'_m & y'_m & -(y_m + y''_m) & (y'_m + y''_m) & 0 \\ y'_m & y_s & y'_m & 0 & -(y_m + y''_m) & (y_m + y''_m) \\ y'_m & y'_m & y_s & (y_m + y''_m) & 0 & -(y_m + y''_m) \\ -(y_m + y''_m) & 0 & (y_m + y''_m) & 2(y_s - y'''_m) & -(y_s - y'''_m) & -(y_s - y'''_m) \\ (y_m + y''_m) & -(y_m + y''_m) & 0 & -(y_s - y'''_m) & 2(y_s - y'''_m) & -(y_s - y'''_m) \\ 0 & (y_m + y''_m) & -(y_m + y''_m) & -(y_s - y'''_m) & -(y_s - y'''_m) & 2(y_s - y'''_m) \end{vmatrix} \tag{14.30}$$

The primed y_m vanish when the primitive admittance matrix of three-phase bank is substituted in Equation 14.29. The primitive admittances are considered on per unit basis, and both primary and secondary voltages are 1 per unit. But a wye–delta transformer so obtained must consider a turns ratio of $\sqrt{3}$, so that wye and delta node voltages are still 1.0 per unit. The node admittance matrix can be divided into submatrices as follows:

$$\bar{Y}_{\text{node}} = \begin{vmatrix} \bar{Y}_{\text{I}} & \bar{Y}_{\text{II}} \\ Y_{\text{II}}^t & \bar{Y}_{\text{III}} \end{vmatrix} \tag{14.31}$$

where each 3 × 3 submatrix depends on the winding connections, as shown in Table 14.1. The submatrices in this table are defined as follows:

$$\bar{Y}_{\text{I}} = \begin{vmatrix} y_t & 0 & 0 \\ 0 & y_t & 0 \\ 0 & 0 & y_t \end{vmatrix} \quad \bar{Y}_{\text{II}} = \frac{1}{3}\begin{vmatrix} 2y_t & -y_t & -y_t \\ -y_t & 2y_t & -y_t \\ -y_t & -y_t & 2Y_t \end{vmatrix} \quad \bar{Y}_{\text{III}} = \frac{1}{\sqrt{3}}\begin{vmatrix} -y_t & y_t & 0 \\ 0 & -y_t & y_t \\ y_t & 0 & -y_t \end{vmatrix} \tag{14.32}$$

Here, y_t is the leakage admittance per phase in per unit. Note that all primed y_m are dropped. Normally, these primed values are much smaller than the unprimed values. These are considerably smaller in magnitude than the unprimed values and the numerical values of y_s; y_p and y_m are equal to the leakage impedance y_t obtained by short-circuit test, Appendix C. It is assumed that all three transformer banks are identical.

Table 14.1 can be used as a simplified approach to the modeling of common core-type three-phase transformer, in an unbalanced system. With more complete information the model in Equation 14.30 can be used for benefits of accuracy.

TABLE 14.1

Submatrices of Three-Phase Transformer Connections

Winding Connections		Self-Admittance		Mutual Admittance	
Primary	**Secondary**	**Primary**	**Secondary**	**Primary**	**Secondary**
Wye–G	Wye–G	$\bar{Y}_{\text{I}}$	$\bar{Y}_{\text{I}}$	$-\bar{Y}_{\text{I}}$	$-\bar{Y}_{\text{I}}$
Wye–G Wye Wye	Wye Wye–G Wye	$\bar{Y}_{\text{II}}$	$\bar{Y}_{\text{II}}$	$-\bar{Y}_{\text{II}}$	$-\bar{Y}_{\text{II}}$
Wye–G	Delta	$\bar{Y}_{\text{I}}$	$\bar{Y}_{\text{II}}$	$\bar{Y}_{\text{III}}$	$\bar{Y}_{\text{III}}^t$
Wye	Delta	$\bar{Y}_{\text{II}}$	$\bar{Y}_{\text{II}}$	$\bar{Y}_{\text{III}}$	$\bar{Y}_{\text{III}}^t$
Delta	Wye	$\bar{Y}_{\text{II}}$	$\bar{Y}_{\text{II}}$	$\bar{Y}_{\text{III}}^t$	$\bar{Y}_{\text{III}}$
Delta	Wye–G	$\bar{Y}_{\text{II}}$	$\bar{Y}_{\text{I}}$	$\bar{Y}_{\text{III}}^t$	$\bar{Y}_{\text{III}}$
Delta	Delta	$\bar{Y}_{\text{II}}$	$\bar{Y}_{\text{II}}$	$-\bar{Y}_{\text{II}}$	$-\bar{Y}_{\text{II}}$

Y_{III}^t is transpose of Y_{III}.

14.2.4.1 Symmetrical Components of Three-Phase Transformers

As stated in earlier chapters, in most cases it is sufficient to assume that the system is balanced. Then, symmetrical components models can be arrived at by using symmetrical component transformations. Continuing with wye–ground–delta transformer of Figure 14.7 first consider the self-admittance matrix. The transformation is

$$\overline{Y}_{012}^{pp} = \overline{T}_s^{-1} \begin{vmatrix} y_p & y'_m & y'_m \\ y'_m & y_p & y'_m \\ y'_m & y'_m & y_p \end{vmatrix} \overline{T}_s = \begin{vmatrix} y_p + 2y'_m & 0 & 0 \\ 0 & y_p - y'_m & 0 \\ 0 & 0 & y_p - y'_m \end{vmatrix} \tag{14.33}$$

Note that zero sequence admittance is different from the positive and negative sequence impedances. If y'_m is neglected all three are equal.

Similarly,

$$\overline{Y}_{012}^{ss} = \frac{1}{3}\overline{T}_s^{-1} \begin{vmatrix} 2(y_s - y'''_m) & -(y_s - y'''_m) & -(y_s - y'''_m) \\ -(y_s - y'''_m) & 2(y_s - y'''_m) & -(y_s - y'''_m) \\ -(y_s - y'''_m) & -(y_s - y'''_m) & 2(y_s - y'''_m) \end{vmatrix} \overline{T}_s = \begin{vmatrix} 0 & 0 & 0 \\ 0 & (y_s - y'''_m) & 0 \\ 0 & 0 & (y_s - y'''_m) \end{vmatrix} \tag{14.34}$$

As expected there is no zero sequence self-admittance in the delta winding.

Finally, the mutual admittance matrix gives

$$\overline{Y}_{012}^{ps} = \frac{1}{\sqrt{3}}\overline{T}_s^{-1} \begin{vmatrix} -(y_m + y''_m) & 0 & (y_m + y''_m) \\ (y_m + y''_m) & -(y_m + y''_m) & 0 \\ 0 & (y_m + y''_m) & -(y_m + y''_m) \end{vmatrix} \overline{T}_s$$

$$= \begin{vmatrix} 0 & 0 & 0 \\ 0 & -(y_m + y''_m) < 30^0 & 0 \\ 0 & 0 & -(y_m + y''_m) < -30^0 \end{vmatrix} \tag{14.35}$$

A phase shift of 30° occurs in the positive sequence network and a negative phase shift of −30° occurs in the negative sequence. Similar results were arrived at in Section 2.5.3. No zero sequence currents can occur between wye and delta side of a balanced wye–delta transformer. Based upon above equations the positive, negative, and zero sequence admittances of the transformer are shown in Figure 14.8.

If the phase shift is ignored

- $y'_m = y''_m = y'''_m$, which are zero in a three-phase bank
- $y_p - y_m$ is small admittance, as y_p is only slightly $>y_m$ and $y_p - y_m$ and $y_s - y_m$ as open-circuit admittances

Then the simplified model returns to that shown in Table 14.1.

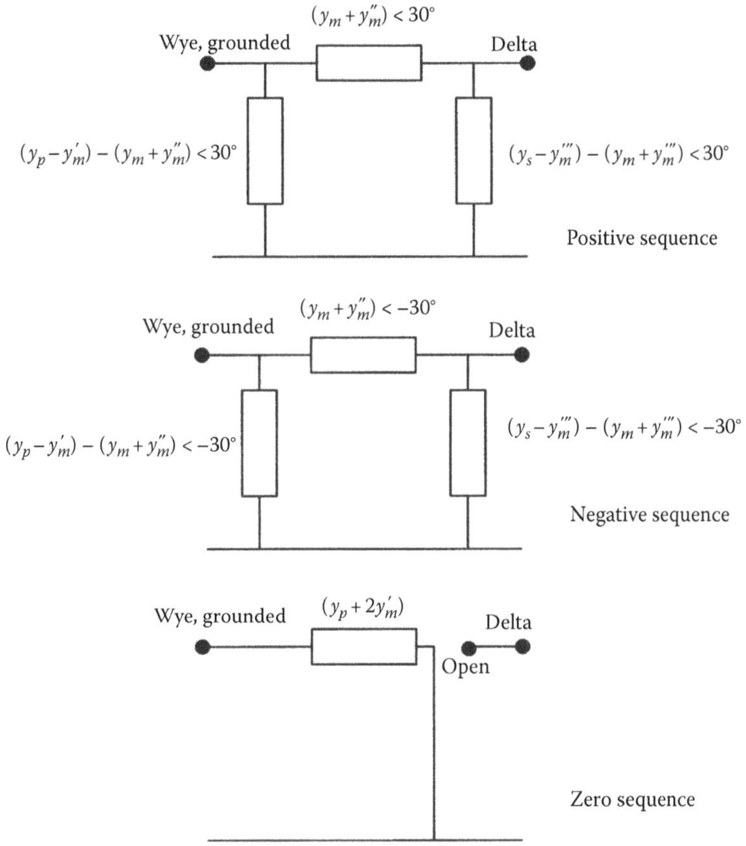

FIGURE 14.8
Sequence impedance circuits of a wye-grounded delta two-winding transformer.

If the off-nominal tap ratio between primary and secondary windings is α: β, where α and β are the taps on the primary and secondary side, respectively, in per unit, then the submatrices are modified as follows:

- Divide self-admittance of primary matrix by α^2
- Divide self-admittance of secondary matrix by β^2
- Divide mutual admittance matrixes by $\alpha\beta$

Consider a wye-grounded transformer. Then, from Table 14.1:

$$\bar{Y}^{abc} = \begin{vmatrix} \bar{Y}_{\mathrm{I}} & -\bar{Y}_{\mathrm{I}} \\ -\bar{Y}_{\mathrm{I}} & \bar{Y}_{\mathrm{I}} \end{vmatrix} = \begin{vmatrix} y_t & 0 & 0 & -y_t & 0 & 0 \\ 0 & y_t & 0 & 0 & -y_t & 0 \\ 0 & 0 & y_t & 0 & 0 & -y_t \\ -y_t & 0 & 0 & y_t & 0 & 0 \\ 0 & -y_t & 0 & 0 & y_t & 0 \\ 0 & 0 & -y_t & 0 & 0 & y_t \end{vmatrix} \tag{14.36}$$

For off-nominal taps the matrix is modified as

$$\overline{Y}^{abc} = \begin{vmatrix} \frac{y_t}{\alpha^2} & 0 & 0 & \frac{y_t}{\alpha\beta} & 0 & 0 \\ 0 & \frac{y_t}{\alpha^2} & 0 & 0 & \frac{y_t}{\alpha\beta} & 0 \\ 0 & 0 & \frac{y_t}{\alpha^2} & 0 & 0 & \frac{y_t}{\alpha\beta} \\ \frac{y_t}{\alpha\beta} & 0 & 0 & \frac{y_t}{\beta^2} & 0 & 0 \\ 0 & \frac{y_t}{\alpha\beta} & 0 & 0 & \frac{y_t}{\beta^2} & 0 \\ 0 & 0 & \frac{y_t}{\alpha\beta} & 0 & 0 & \frac{y_t}{\beta^2} \end{vmatrix} = |\overline{Y}_{yg-y}| \tag{14.37}$$

A three-phase transformer winding connection of delta primary and grounded-wye secondary is commonly used. From Table 14.1, its matrix equation is

$$\overline{Y}^{abc} = \begin{vmatrix} \frac{2}{3}y_t & -\frac{1}{3}y_t & -\frac{1}{3}y_t & -\frac{y_t}{\sqrt{3}} & \frac{y_t}{\sqrt{3}} & 0 \\ -\frac{1}{3}y_t & \frac{2}{3}y_t & -\frac{1}{3}y_t & 0 & -\frac{y_t}{\sqrt{3}} & \frac{y_t}{\sqrt{3}} \\ -\frac{1}{3}y_t & -\frac{1}{3}y_t & \frac{2}{3}y_t & \frac{y_t}{\sqrt{3}} & 0 & -\frac{y_t}{\sqrt{3}} \\ -\frac{y_t}{\sqrt{3}} & 0 & \frac{y_t}{\sqrt{3}} & y_t & 0 & 0 \\ \frac{y_t}{\sqrt{3}} & -\frac{y_t}{\sqrt{3}} & 0 & 0 & y_t & 0 \\ 0 & \frac{y_t}{\sqrt{3}} & -\frac{y_t}{\sqrt{3}} & 0 & 0 & y_t \end{vmatrix} \tag{14.38}$$

where y_t is the leakage reactance of the transformer.

For an off-nominal transformer, the Y matrix is modified as shown below:

$$\overline{Y}^{abc} = \begin{vmatrix} \frac{2}{3}\frac{y_t}{\alpha^2} & -\frac{1}{3}\frac{y_t}{\alpha^2} & -\frac{1}{3}\frac{y_t}{\alpha^2} & -\frac{y_t}{\sqrt{3}\alpha\beta} & \frac{y_t}{\sqrt{3}\alpha\beta} & 0 \\ -\frac{1}{3}\frac{y_t}{\alpha^2} & \frac{2}{3}\frac{y_t}{\alpha^2} & -\frac{1}{3}\frac{y_t}{\alpha^2} & 0 & -\frac{y_t}{\sqrt{3}\alpha\beta} & \frac{y_t}{\sqrt{3}\alpha\beta} \\ -\frac{1}{3}\frac{y_t}{\alpha^2} & -\frac{1}{3}\frac{y_t}{\alpha^2} & \frac{2}{3}\frac{y_t}{\alpha^2} & \frac{y_t}{\sqrt{3}\alpha\beta} & 0 & -\frac{y_t}{\sqrt{3}\alpha\beta} \\ -\frac{y_t}{\sqrt{3}\alpha\beta} & 0 & \frac{Y_t}{\sqrt{3}\alpha\beta} & \frac{y_t}{\beta^2} & 0 & 0 \\ \frac{y_t}{\sqrt{3}\alpha\beta} & -\frac{y_t}{\sqrt{3}\alpha\beta} & 0 & 0 & \frac{y_t}{\beta^2} & 0 \\ 0 & \frac{y_t}{\sqrt{3}\alpha\beta} & -\frac{y_t}{\sqrt{3}\alpha\beta} & 0 & 0 & \frac{y_t}{\beta^2} \end{vmatrix} \tag{14.39}$$

where α and β are the taps on the primary and secondary side in per unit.

In a load-flow analysis, the equation of a wye-grounded delta transformer and $\alpha = \beta = 1$ can be written as

$$\begin{vmatrix} I_A \\ I_B \\ I_C \\ I_a \\ I_b \\ I_c \end{vmatrix} = \begin{vmatrix} y_t & 0 & 0 & -\frac{1}{\sqrt{3}}y_t & \frac{1}{\sqrt{3}}y_t & 0 \\ 0 & y_t & 0 & 0 & -\frac{1}{\sqrt{3}}y_t & \frac{1}{\sqrt{3}}y_t \\ 0 & 0 & y_t & \frac{1}{\sqrt{3}}y_t & 0 & -\frac{1}{\sqrt{3}}y_t \\ -\frac{1}{\sqrt{3}}y_t & 0 & \frac{1}{\sqrt{3}}y_t & \frac{2}{3}y_t & -\frac{1}{3}y_t & -\frac{1}{3}y_t \\ \frac{1}{\sqrt{3}}y_t & -\frac{1}{\sqrt{3}}y_t & 0 & -\frac{1}{3}y_t & \frac{2}{3}y_t & -\frac{1}{3}y_t \\ 0 & \frac{1}{\sqrt{3}}y_t & -\frac{1}{\sqrt{3}}y_t & -\frac{1}{3}y_t & -\frac{1}{3}y_t & \frac{2}{3}y_t \end{vmatrix} \begin{vmatrix} V_A \\ V_B \\ V_C \\ V_a \\ V_b \\ V_c \end{vmatrix} \tag{14.40}$$

Here, the currents and voltages with capital subscripts relate to primary and those with lowercase subscripts relate to secondary. In the condensed form we will write it as

$$\bar{I}_{ps} = \bar{Y}_{Y-\Delta}\bar{V}_{ps} \tag{14.41}$$

Using symmetrical component transformation:

$$\begin{vmatrix} \bar{I}_p^{012} \\ \bar{I}_s^{012} \end{vmatrix} = \begin{vmatrix} \bar{T}_s & 0 \\ 0 & \bar{T}_s \end{vmatrix}^{-1} \bar{Y}_{y-\Delta} \begin{vmatrix} \bar{T}_s & 0 \\ 0 & \bar{T}_s \end{vmatrix} \begin{vmatrix} \bar{V}_p^{012} \\ \bar{V}_s^{012} \end{vmatrix} \tag{14.42}$$

Expanding:

$$\begin{vmatrix} \bar{I}_p^{012} \\ \bar{I}_s^{012} \end{vmatrix} = \begin{vmatrix} y_t & 0 & 0 & 0 & 0 & 0 \\ 0 & y_t & 0 & 0 & y_t < -30^\circ & 0 \\ 0 & 0 & y_t & 0 & 0 & y_t < 30^\circ \\ 0 & 0 & 0 & 0 & 0 & 0 \\ 0 & y_t < 30^\circ & 0 & 0 & y_t & 0 \\ 0 & 0 & y_t < -30^\circ & 0 & 0 & y_t \end{vmatrix} \begin{vmatrix} \bar{V}_p^{012} \\ \bar{V}_s^{012} \end{vmatrix} \tag{14.43}$$

The positive sequence equations are

$$\begin{aligned} I_{p1} &= y_t V_{p1} - y_t < -30^\circ V_{s1} \\ I_{s1} &= y_t V_{s1} - y_t < 30^\circ V_{p1} \end{aligned} \tag{14.44}$$

The negative sequence equations are

$$\begin{aligned} I_{p2} &= y_t V_{p2} - y_t < 30^\circ V_{s2} \\ I_{s2} &= y_t V_{s2} - y_t < -30^\circ V_2 \end{aligned} \tag{14.45}$$

The zero sequence equation is

$$\begin{aligned} I_{p0} &= y_t V_{p0} \\ I_{s0} &= 0 \end{aligned} \tag{14.46}$$

For a balanced system only the positive sequence component needs to be considered. The power flow on the primary side:

$$\begin{aligned} S_{ij} &= V_i I_{ij}^* = V_i\left(y_t^* V_i^* - y_t^* < 30^\circ V_j^*\right) \\ &= [y_t V_i^2 \cos\theta_{yt} - y_t|V_i V_j|\cos\left(\theta_i - \theta_{yt} - (\theta_j + 30^\circ)\right)] \\ &\quad + j\left[-y_t V_i^2 \sin\theta_{yt} - y_t|V_i V_j|\sin\left(\theta_i - \theta_{yt} - (\theta_j + 30^\circ)\right)\right] \end{aligned} \tag{14.47}$$

and on the secondary side:

$$\begin{aligned} S_{ji} &= V_i I_{ji}^* = V_j\left(y_t^* V_j^* - y_t^* < -30^\circ V_i^*\right) \\ &= \left[y_t V_j^2 \cos\theta_{yt} - y_t|V_j V_i|\cos\left(\theta_j - \theta_{yt} - (\theta_i - 30^\circ)\right)\right] \\ &\quad + j\left[-y_t V_j^2 \sin\theta_{yt} - y_t|V_j V_i|\sin\left(\theta_j - \theta_{yt} - (\theta_i - 30^\circ)\right)\right] \end{aligned} \tag{14.48}$$

14.2.5 Load Models

For a distribution system, the load window concept is discussed in Chapter 12. Based on test data a detailed load model can be derived, and the voltage/current characteristics of the models are considered. The models are not entirely three-phase balanced types and single-phase loads give rise to unbalances. A load window can be first constructed and percent of each load type allocated, Figure 14.9. By testing the power/voltage, characteristics of most load types is known. As an example fluorescent lighting the power requirement reduces when the voltage dips and increases as the voltage is restored to the operating voltage. Conversely for air conditioning loads, the power requirement increases as the voltage rises and also as the voltage dips below rated, giving a U-shaped curve [7]. A typical three-phase load is shown in Figure 14.10. The unbalance is allowed by load current injections.

Incandescent lighting	Fluorescent lighting	Space heating	Dryer	Refrig freezer	Elect. range	TV	Others	Total = 100%

FIGURE 14.9
Load window showing composition of various load types.

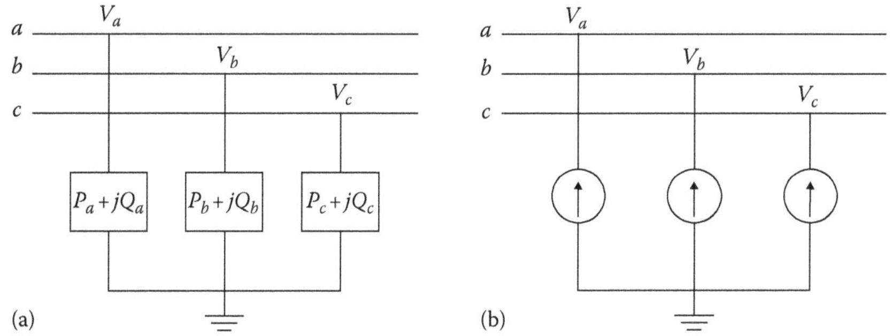

FIGURE 14.10
(a) Three-phase load representation; (b) equivalent current injection.

14.3 Distribution System Load Flow

The distribution system analysis requirements are as follows:

- It should be capable of modeling 4–44 kV primary distribution feeders and networks, 120/208 V secondary networks, and isolated 277/480 V systems simultaneously. These systems become very large and may consist of 4,000 secondary distribution buses (three-phase models = 12,000 buses) and 2,000 primary distribution buses. The system must be, therefore, capable of handling 20,000 buses. As an example, Southern California Edison serves 4.2 million customers over a territory of 50,000 square miles and there are 600 distribution substations, 3,800 distribution circuits over 38,000 circuit miles, 61,000 switches, 800 automatic reclosures, and 7,600 switched capacitor banks. Data generation and modeling become very time consuming.
- The system is inherently unbalanced. In the planning stage, it may be adequate to model the system on a single-phase basis, but the operation requires three-phase modeling. The capability of modeling three-phase systems, line segments, and transformers with phase shifts is a must. The core and copper losses need to be considered. A nonlinear model of the core losses is appropriate.
- The cogenerators should be capable of being modeled on primary and secondary systems.
- Contingency analysis is required to study the effect of outage of feeders.
- Short-circuit calculations are performed using the same database. The load currents may not be small compared to short-circuit currents and cannot be neglected. Prefault voltages and current injections obtained in load flow are input into the short-circuit calculations. Contributions of the cogenerators to faults must be included. Figure 14.11 shows a flowchart.
- A feeder has several interfeeder switches to link with other feeders. Under heavy loading or contingency conditions, normally open switches are closed to prevent system overloads. Optimal interfeeder switching decisions are required to be made in a distribution system. As any switching operation must be carried out to prevent overloads, the problem of optimal switching translates into a problem of

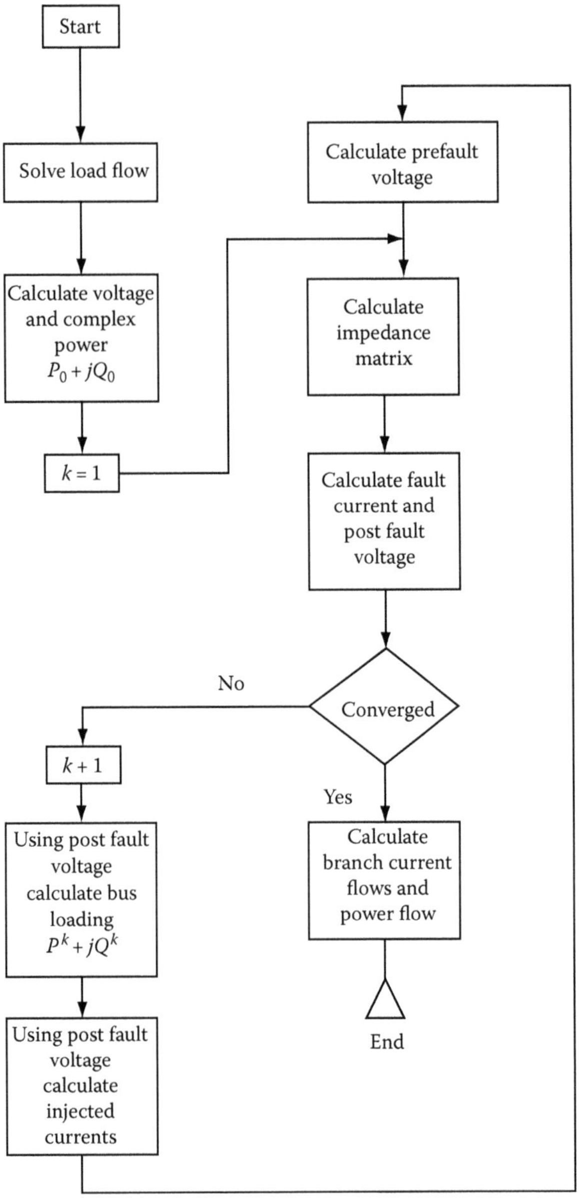

FIGURE 14.11
Flowchart for short-circuit calculations considering prefault voltages and currents.

dispatching currents through alternate routes to meet the load demand to prevent overloads and achieve phase balance. The radial nature of the network is preserved and the number of switching operations is minimized.

- Optimal capacitor location on distribution feeders for voltage control and energy loss reduction is required.

In today's environment, distribution networks are being influenced by energy savings and improvements in power quality. Regenerative methods of generation and storage, e.g., fuel cells, are receiving impetus. Optimization of alternative sources of energy and maximization of network utilization without overloading a section are required [8].

14.3.1 Methodology

The Gauss and Newton–Raphson (NR) methods are applicable. In the NR approach, because of the low X/R ratio of conductors associated with distribution system line segments compared to transmission systems, the Jacobian cannot be decoupled. A Gauss method using a sparse bifactored bus admittance matrix flowchart is shown in Figure 14.12.

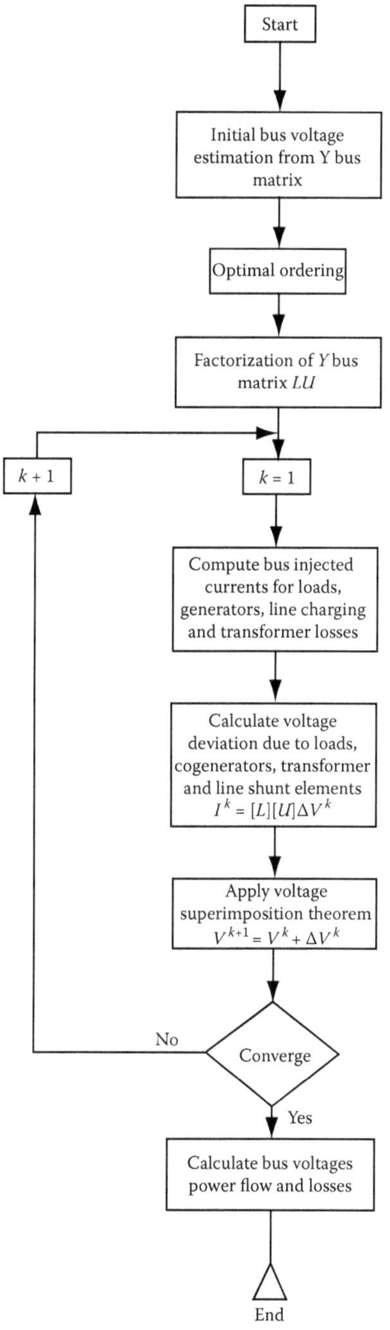

FIGURE 14.12
Flowchart for distribution system load flow using Gauss factored Y matrix.

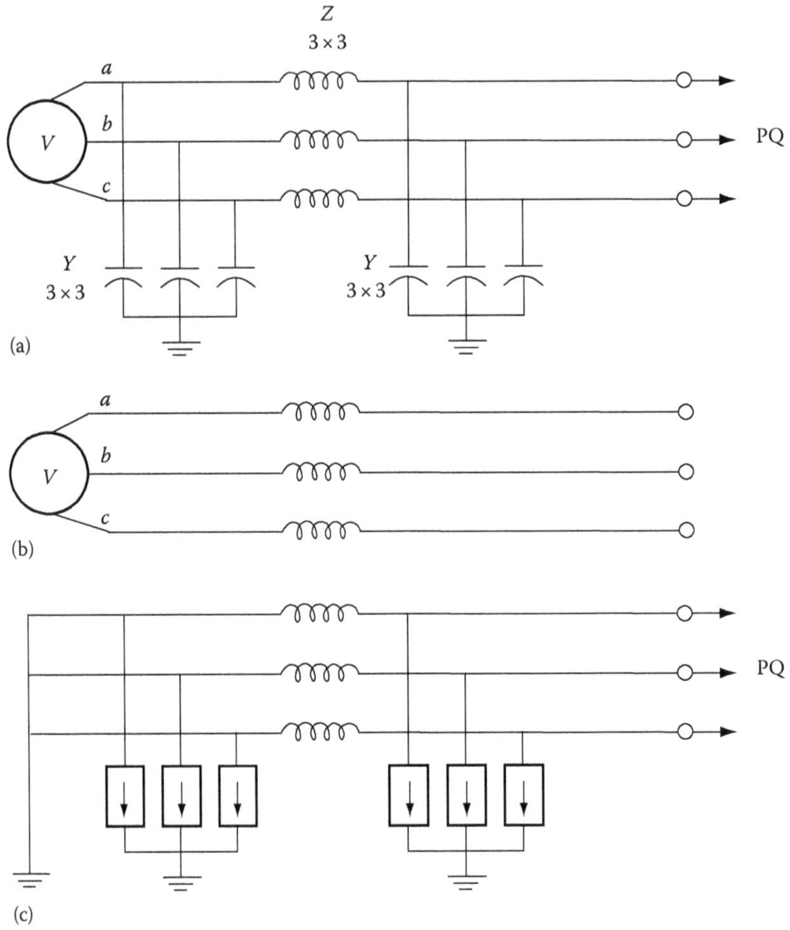

FIGURE 14.13
(a) Original system for load flow; (b) only swing bus activated; (c) swing bus grounded, only current injections activated.

As the voltage-specified bus in the system is only the swing bus, the rate of convergence is comparable to that of the NR method.

The voltage at each bus can be considered as having contributions from two sources, and the theorem of superimposition is applied. These two sources are the voltage-specified station bus and the load and generator buses (Figure 14.13a). The loads and cogenerators are modeled as current injection sources. The shunt capacitance currents are also included in the current injections. Using the superimposition principal, only one source is active at a time. When the swing bus voltage is activated, all current sources are disconnected (Figure 14.13b). When the current sources are activated, the swing bus is short-circuited to ground (Figure 14.13c).

14.3.2 Distribution System as a Ladder Network

A distribution system forms a ladder network, with loads teed-off in a radial fashion (Figure 14.14a). It is a nonlinear system, as most loads are of constant kW and kvar. However, linearization can be applied.

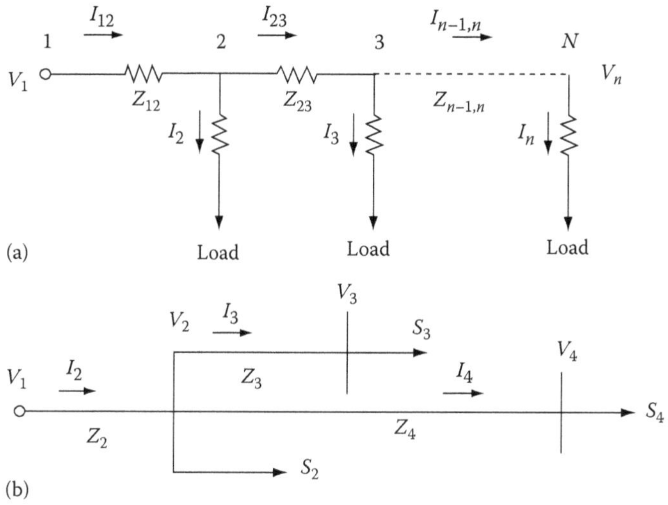

FIGURE 14.14
(a) Distribution system as a ladder network; (b) load flow with a teed-off lateral.

For a linear network assume that the line and load impedances are known and the source voltage is known. Starting from the last node, and assuming a node voltage of V_n, the load current is given by

$$I_n = \frac{V_n}{Z_{1n}} \tag{14.49}$$

Note that I_n is also the line current, as this is the last node. Therefore, the voltage at node $n-1$ can be obtained simply by adding the voltage drop:

$$V_{n-1} = V_n + I_{n-1,n}Z_{n-1,n} \tag{14.50}$$

This process can be carried out until the sending end node is reached. The calculated value of the sending end voltage will be different from the actual applied voltage. Since the network is linear, all the line and load currents and node voltages can be multiplied by the ratio

$$\frac{V_{\text{actual}}}{V_{\text{calculated}}} = \frac{V_{\text{s}}}{V_{\text{calculated}}} \tag{14.51}$$

Actually the node current must be calculated on the basis of complex power at the load:

$$I_{\text{node}} = \left(\frac{S_{\text{node}}}{V_{\text{node}}}\right)^* \tag{14.52}$$

Starting from the last node, the sending end voltage is calculated in the *forward sweep*, as in the linear case. This voltage will be different from the sending end voltage. Using this voltage, found in the first iteration, a *backward sweep* is performed, i.e., the voltages are recalculated starting from the first node to the nth node. This new voltage is used to recalculate the currents and voltages at the nodes in the second forward sweep. The process can be repeated until the required tolerance is achieved.

A lateral circuit (Figure 14.14b), can be handled as follows [9,10]:

- Calculate the voltage at node 2, starting from node 4, ignoring the lateral to node 3. Let this voltage be V_2.
- Consider that the lateral is isolated and is an independent ladder. Now the voltage at node 3 can be calculated and therefore current I_3 is known.
- The voltage at node 2 is calculated back, i.e., voltage drop I_3Z_3 is added to voltage V_3. Let this voltage be V_2'. The difference between V_2 and V_2' must be reduced to an acceptable tolerance. The new node 3 voltage is $V_{3(\text{new})} = V_3 - (V_2 - V_2')$. The current I_3 is recalculated and the calculations iterated until the desired tolerance is achieved.

14.4 Optimal Capacitor Locations

The optimum location of capacitors in a distribution system is a complex process [11,12]. The two main criteria are voltage profiles and system losses. Correcting the voltage profile will require capacitors to be placed toward the end of the feeders, while emphasis on loss reduction will result in capacitors being placed near load centers. An automation strategy based on intelligent customer meters, which monitor the voltage at consumer locations and communicate this information to the utility, can be implemented [13].

We will examine the capacitor placement algorithm based on loss reduction and energy savings using dynamic programming concepts [14]. A reader may first pursue Chapter 15 before going through this section. We can define the objective functions as follows:

- Peak power loss reduction
- Energy loss reduction
- Voltage and harmonic control
- Capacitor cost

The solved variables are fixed and switched capacitors, and their number, size, location, and switched time. Certain assumptions in this optimization process are the loading conditions of the feeders, type of feeder load, and capacitor size based on the available standard ratings and voltages. Consider the placement of a capacitor at node K (Figure 14.15). The peak power loss reduction of the segment is

$$PL_k = 3R_k\left(2I_{LK}I_{CK} - I_{CK}^2\right) \tag{14.53}$$

where

R_k is the resistance of segment K

I_{LK} is the peak reactive current in segment K before placement of the capacitor

I_{CK} is the total capacitor current flowing through segment K

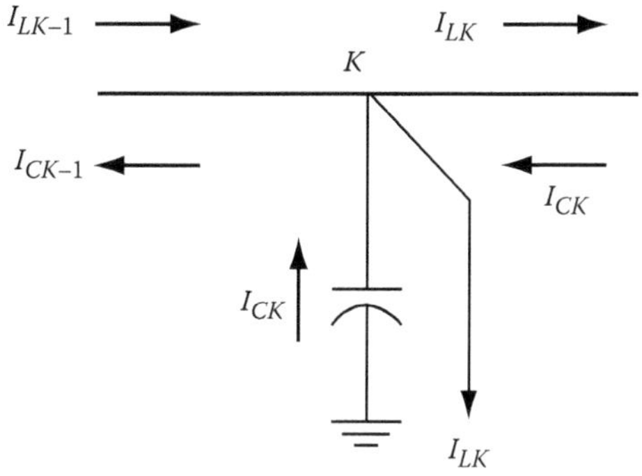

FIGURE 14.15
Load and capacitance current flows at node k with a capacitor bank.

The power loss reduction under average load can be written as

$$PLA_K = 3R_K\left(2I_{CK}I_{LK}LF - I_{CK}^2\right) \tag{14.54}$$

where all the terms are as defined before and LF is the load factor. The overall objective function for loss reduction can then be written as

$$F = \sum_{k=1}^{N} [F_p PL_K + F_{av} PLA_K] \tag{14.55}$$

where
F_p is the monetary conversion factor for power loss reduction under peak load in \$/kW/year
F_{av} is the monetary conversion factor for power loss under average load in \$/kW/year
N is the number of line segments

An optimal value function can be defined as follows:

$$S(x, y_x) = Max\left[\sum_{k=1}^{N} [F_p PL_K + F_{av} PLA_K]\right] \tag{14.56}$$

where $S(x, y_x)$ is the maximum value of the objective function calculated from substation to line segment x. If the capacitive current flowing through segment x is y_x. Also,

$$I_{CX} = I_{base} y_x \tag{14.57}$$

We can define some constraints in the optimization process:

The power factor at the substation outlet is to be maintained lagging, i.e., the load current is greater than the capacitive current:

$$I_{L1} > I_{C1} \tag{14.58}$$

This requires that the total capacitive current flowing through $(k-1)$ segment be at least equal to the total capacitive current flowing into node k:

$$I_{CK-1} \geq I_{CK} \tag{14.59}$$

The values of y_x are defined over an adequate range of discrete values.

A dynamic programming formulation seeks the minimum (or maximum) path subject to constraints and boundary conditions. The recursive relation is given by

$$S(x, y_x) = MAX_{y_{x-1} \geq y_x}[F_p PL_X(I_{\text{base}} y_x) + F_{\text{av}} PLA_x(I_{\text{base}} y_x) + S(x-1, y_{x-1})] \tag{14.60}$$

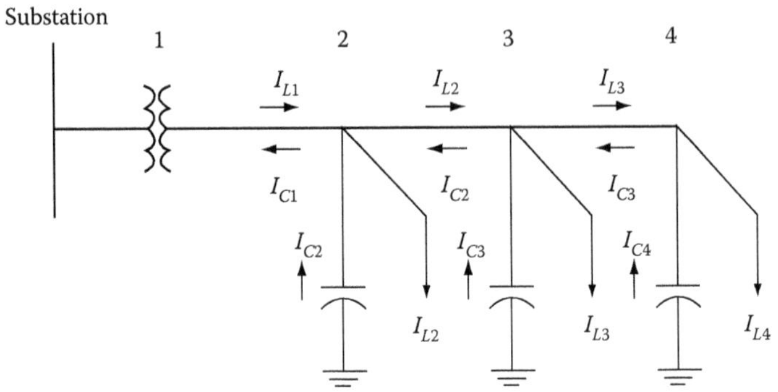

FIGURE 14.16
Three-section distribution system with capacitor compensation.

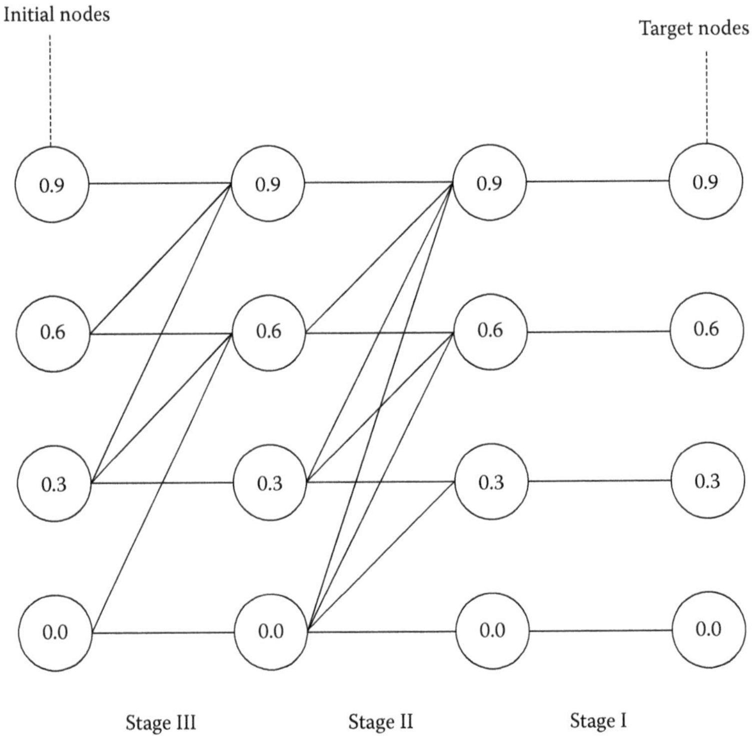

FIGURE 14.17
Three-stage flowchart of dynamic programming.

The boundary condition is given by

$$S(0, y_0) = 0 \quad \text{for all } y_0 \tag{14.61}$$

i.e., stage 0 means that the substation is encountered and no capacitors are installed beyond this point. The recursive procedure is illustrated by an example:

Consider the three-segment system in Figure 14.16. Assume that the value of $I_{L1} = 1.0$ per unit and the smallest capacitor size is 0.3 per unit with an upper limit of 0.9 per unit. In Figure 14.17 stage 1 shows that a target node can be constructed with every possible value of I_{c1}. A path is added for each target node and the cost is calculated using Equation 14.54 for $N = 1$:

$$\text{cost} = 3R_1(2 \times 1.0 \times 0.9 - 0.9^2)F_p \times I_{\text{base}}^2 + 3R_1(2 \times 1.0 \times 0.9 - 0.9^2)F_{\text{av}}I_{\text{base}}^2$$

The graph can be constructed as shown in Figure 14.17 for the other two stages also, and the cost of the path connecting the two nodes can be calculated in an identical manner. Once the cost associated with every path is calculated, the problem reduces to finding the maximum cost path from one of the initial nodes to the target nodes [14]. The voltage and switching constraints can be added to the objective function.

References

1. L. Roy. Generalized polyphase fault analysis program: Calculation of cross country faults. *Proc. IEE (Part B)* 126 (10): 995–1000, 1979.
2. M.A. Loughton. The analysis of unbalanced polyphase networks by the method of phase coordinates, Part 1. System representation in phase frame reference. *Proc. IEE* 115 (8): 1163–1172, 1968.
3. M.A. Loughton. The analysis of unbalanced polyphase networks by the method of phase coordinates, Part 2. System fault analysis. *Proc. IEE* 1165 (5): 857–865, 1969.
4. T.H. Chen. Generalized distribution analysis system. PhD dissertation, University of Texas at Arlington, Arlington, TX, May 1990.
5. M. Chen and W.E. Dillon. Power system modeling. *Proc. IEEE PAS* 62: 901–915, 1974.
6. G. Kron. *Tensor Analysis of Networks*. London: McDonald, 1965.
7. R.B. Adler and C.C. Mosher. Steady state voltage power characteristics for power system loads. In: R.T. Byerly and E.W. Kimbark (eds.), *Stability of Large Power Systems*. New York: IEEE Press, 1974.
8. H.L. Willis, H. Tram, M.V. Engel, and L. Finley. Selecting and applying distribution optimization methods. *IEEE Comp. Appl. Power* 9(1): 12–17, 1996.
9. W.H. Kersting and D.L. Medive. An application of ladder network to the solution of three-phase radial load flow problems. In: *IEEE PES Winter Meeting*, New York, 1976.
10. W.H. Kersting. *Distribution System Modeling and Analysis*, 2nd edn. Boca Raton, FL: CRC Press, 2006.
11. B.R. Williams and D.G. Walden. Distribution automation strategy for the future. *IEEE Comp. Appl. Power* 7(3): 16–21, 1994.
12. J.J. Grainger and S.H. Lee. Optimum size and location of shunt capacitors for reduction of losses on distribution feeders. *IEEE Trans. PAS* 100 (3): 1105–1118, 1981.
13. S.H. Lee and J.J. Gaines. Optimum placement of fixed and switched capacitors of primary distribution feeders. *IEEE Trans. PAS* 100: 345–352, 1981.
14. S.K. Chan. Distribution system automation. PhD dissertation, University of Texas at Arlington, Arlington, TX, 1982.

15

Optimization Techniques

The application of computer optimization techniques in power systems is reaching new dimensions with improvements in algorithm reliability, speed, and applicability. Let us start with a simple situation. Optimization can be aimed at reducing something undesirable in the power system, for example, the system losses or cost of operation, or maximizing a certain function, for example, efficiency or reliability. Such maxima and minima are always subject to certain constraints, that is, tap settings on transformers, tariff rates, unit availability, fuel costs, etc. The problem of optimization is thus translated into the problem of constructing a reliable mathematical model aimed at maximizing or minimizing a certain function, within the specified constraints.

It is possible to model a wide range of problems in planning, design, control, and measurement. Traditionally, optimization has been used for the economic operation of fossil-fueled power plants, using an economic dispatch approach. In this approach, inequality constraints on voltages and power flows are ignored and real power limits on generation and line losses are accounted for. A more complicated problem is system optimization over a period of time.

The optimization techniques are often applied off-line. For many power system problems, an off-line approach is not desirable, because optimal solution is required for immediate real-time implementation and there is a need for efficient and reliable methods. Table 15.1 shows the interaction of various levels of system optimization.

Linear programming [1–3] deals with situations where a maximum or minimum of a certain set of linear functions is desired. The equality and inequality constraints define a region in the multidimensional space. Any point in the region or boundary will satisfy all the constraints; thus, it is a region enclosed by the constraints and not a discrete single value solution. Given a meaningful mathematical function of one or more variables, the problem is to find a maximum or minimum, when the values of the variables vary within some certain allowable limits. The variables may react with each other or a solution may be possible within some acceptable violations or may not be possible at all.

Mathematically, we can minimize

$$f(x_1, x_2, \ldots, x_n) \tag{15.1}$$

subject to

$$\begin{aligned} &g_1(x_1, x_2, \ldots, x_n) \le b \\ &g_2(x_1, x_2, \ldots, x_n) \le b_2 \\ &\ldots \\ &g_m(x_1, x_2, \ldots, x_n) \le b_n \end{aligned} \tag{15.2}$$

The linear programming is a special case of an objective function (Equation 15.1) when all the constraints (Equation 15.2) are linear.

TABLE 15.1
Various Levels of System Optimization

Time Duration	Control Process	Optimized Function
Seconds	Automatic generation control	Minimize area control error, subject to system dynamic constraints
Minutes	Optimal power flow	Minimize instantaneous cost of operation or other indexes, for example, pollution
Hours and days	Unit commitment, hydrothermal	Minimize cost of operation
Weeks	Grid interchange coordination	Minimize cost with reliability constraints
Months	Maintenance scheduling	Minimize cost with reliability constraints
Years	Generation planning	Minimize expected investment and operational costs

15.1 Functions of One Variable

A function $f(x)$ has its global minima at a point x^* if $f(x) \leq f(x^*)$ for all values of x over which it is defined. Figure 15.1 shows that the function may have relative maxima or minima. A *stationary point*, sometimes called a *critical point*, is defined where

$$f'(x) = \frac{df(x)}{dx} = 0 \tag{15.3}$$

The critical point and stationary point are used interchangeably. Sometimes the critical point is defined as the any point that could be global optima. This is discussed further. The function $f(x)$ is said to have a weak relative maximum at x_0 if there exists a $\in$, $0 < \in < \delta$, such that $f(x) \leq f(x^0)$ and there is at least one point x in the interval $[x^0 - \in, x^0 + \in]$ such that $f(x) = f(x^0)$. The relative minimum of a function occurs at a point where its derivative is zero. For a vector,

$$\begin{aligned} g_1 &= \left.\frac{\partial f}{\partial x_1}\right|_{x=x_0} = 0 \\ &\ldots \\ g_n &= \left.\frac{\partial f}{\partial x_n}\right|_{x=x_0} = 0 \end{aligned} \tag{15.4}$$

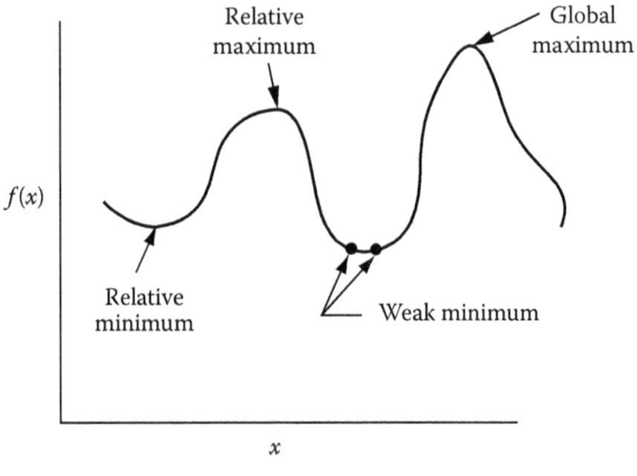

FIGURE 15.1
A single value function showing relative maxima and minima.

The derivative condition is necessary, but not sufficient, as the derivative can occur at maxima or saddle points. Additional conditions are required to ascertain that a minimum has been obtained.

15.2 Concave and Convex Functions

Important characteristics of the functions related to the existence of maxima and minima are the convexity and concavity. A function $f(x)$ is concave for some interval in x, if for any two points x_1 and x_2 in the interval and all values of λ, $0 \leq \lambda \leq 1$

$$f[\lambda x_2 + (1-\lambda)x_1] \leq \lambda f(x_2) + (1-\lambda)f(x_1) \tag{15.5}$$

The definition of the concave function is

$$f[\lambda x_2 + (1-\lambda)x_1] \geq f(x_2) + (1-\lambda)f(x_1) \tag{15.6}$$

Figure 15.2 shows convexity and concavity. Some functions may not be definitely convex or concave. The function $f(x) = x^3$ in Figure 15.2d is concave in the interval $(-\infty, 0)$ and convex in the interval $(0, \infty)$.

The convexity plays an important role. If it can be shown that the objective function is convex and the constraint set is convex, then the problem has a unique solution. It is not easy to demonstrate this. The optimal power flow problem (Chapter 16) is generally non-convex. Therefore, multiple minima may exist, which may differ substantially.

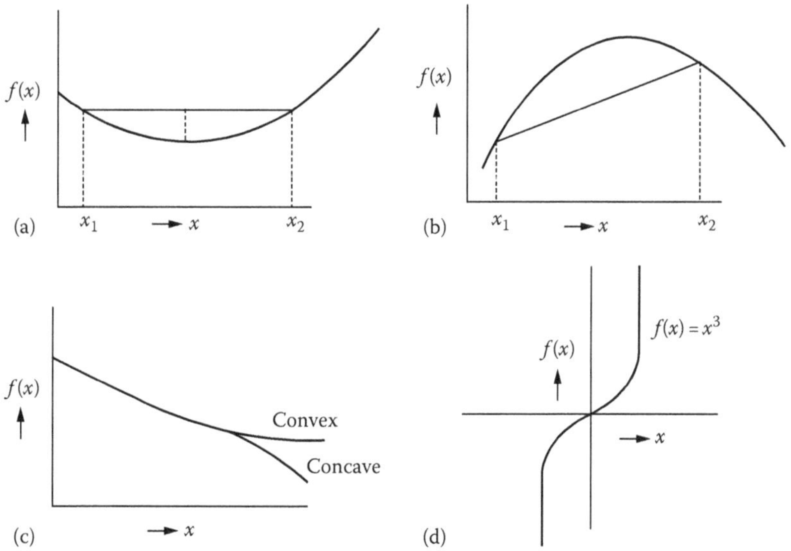

FIGURE 15.2
(a) A convex function; (b) concave function; (c) convex and concave function; (d) graph of function $f(x) = x^3$.

15.3 Taylor's Theorem

If $f(x)$ is continuous and has a first derivative, then for any two points x_1 and x_2, where $x_2 = x_1 +$ incremental h, there is a θ, $0 \le \theta \le 1$, so that

$$f(x_2) = f(x_1) + hf'[\theta x_1 + (1-\theta)x_2] \tag{15.7}$$

Extending

$$f(x_2) = f(x_1) + hf'(x_1) + \frac{h^2}{2!}f'(x_1) + \cdots + \frac{h^n}{n!}f^{(n)}[\theta x_1 + (1-\theta)x_2] \tag{15.8}$$

A function $f(x)$ has a relative maximum or minimum at x^* only if n is even, where n is the order of the first nonvanishing derivative at x^*. It is a maximum if $f^{(n)}(x^*) < 0$ and a minimum if $f^{(n)}(x^*) > 0$.

Example 15.1

Consider $f(x) = (x-1)^4$:

$$\frac{df(x)}{dx} = 4(x-1)^3, \quad \frac{d^2f(x)}{dx^2} = 12(x-1)^2, \quad \frac{d^3f(x)}{dx^3} = 24(x-1), \quad \frac{d^4f(x)}{dx^4} = 24$$

The fourth derivative is even and is the first nonvanishing derivative at $x = 1$; therefore, $x = 1$ is the minimum.

For a continuous differentiable function, in a small interval, the maximum and minimum can be determined as shown earlier. These are relative or local, and most of the time we will be interested in finding the global maxima or minima for $a \le x \le b$. A procedure for this is as follows.

Compute $f(a)$ and $f(b)$, compute $f'(x)$, find roots of $f'(x) = 0$, If there are no roots in $[a, b]$, z^* is larger of $f(a)$ and $f(b)$. If there are roots in $[a, b]$, then z^* is the largest of $f(a)$, $f(b)$, $f(x_1), \ldots, f(x_k)$.

Example 15.2

Find global maxima of

$$f(x) = x^3 + x^2 - x + 4, \quad 0 \le x \le 2$$

Here

$$f(0) = 4$$
$$f(2) = 14$$

Compute $f'(x)$

$$f'(x) = 3x^2 + 2x - 1$$

The roots are 1/3 and −1. The root −1 is not in interval [0, 2]; therefore, ignore it. Then

$$z^* = \text{Max}\left[f(0),\ f(2),\ f\left(\frac{1}{3}\right)\right] = 14$$

Optima of Concave and Convex Functions

- The convexity of a function assumes great importance for minimization. If $f(x)$ is convex over a closed interval $a \le x \le b$, then any relative minimum of $f(x)$ is also the global minima.
- The global maximum of a convex function $f(x)$ over a closed interval $a \le x \le b$ is taken on either $x = a$ or $x = b$ or both.
- If the functions $f_k(\overline{x})$, $k = 1, 2, \ldots, p$ are convex functions over some convex set X in Euclidean space E^n, then function $f(\overline{x}) = \sum_{k=1}^{n} f_k(\overline{x})$ is also convex over X.
- The sum of convex functions is a convex function. The sum of concave functions is a concave function.
- If $f(\overline{x})$ is a convex function over a closed convex set X in E^n, then any local minimum of $f(\overline{x})$ is also the global minimum of $f(\overline{x})$ over X.

Functions of Multivariables

Equation 15.8 can be written as

$$f(\overline{x}_2) = f(\overline{x}_1) + \overline{\nabla} f[\theta \overline{x}_1 + (1 - \theta)\overline{x}_2]h \tag{15.9}$$

where $\overline{\nabla} f$ is the gradient vector:

$$\overline{\nabla} f = \left(\frac{\partial f}{\partial x_1}, \frac{\partial f}{\partial x_2}, \ldots, \frac{\partial f}{\partial x_n}\right) \tag{15.10}$$

Matrix $\overline{H}$ is of $m \times n$ dimensions and is called a Hessian. It consists of second partial derivatives of $f(x)$.

$$H_{ij} = \frac{\partial^2 f(x_1, \ldots, x_n)}{\partial x_i \partial x_j} \tag{15.11}$$

A sufficient condition for $f(x)$ to have a relative minimum at point x^* is that $\overline{H}$ be positive definite. Also, if the solution to the gradient set of equations is unique, then the solution is a global minimum.

Example 15.3

Consider the solution of a function

$$3x_1^2 - 9x_1 + 4x_2^2 - 3x_1x_2 - 3x_2$$

The gradient vector is

$$g = \begin{vmatrix} 6x_1 - 9 - 3x_2 \\ 8x_2 - 3x_1 - 3 \end{vmatrix}$$

Setting it to zero and solving for x_1 and x_2 give

$$\begin{vmatrix} x_1 \\ x_2 \end{vmatrix} = \begin{vmatrix} 3/13 \\ -1/13 \end{vmatrix}$$

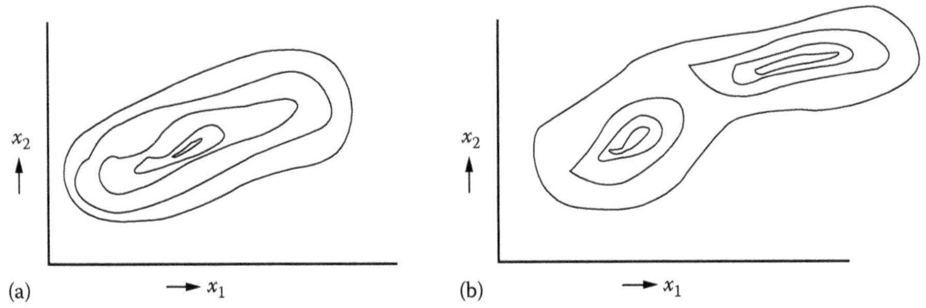

FIGURE 15.3
(a) Function showing strong unimodality; (b) nonunimodal function.

This solution is unique. The Hessian is given by

$$\overline{H} = \begin{vmatrix} \dfrac{\partial^2 f}{\partial x_1^2} & \dfrac{\partial^2 f}{\partial x_1 \partial x_2} \\ \dfrac{\partial^2 f}{\partial x_2 \partial x_1} & \dfrac{\partial^2 f}{\partial x_2^2} \end{vmatrix} = \begin{vmatrix} 6 & -3 \\ -3 & 8 \end{vmatrix}$$

Thus, $\overline{H}$ is a positive definite, and, therefore, the above solution is a global solution.

A property of a function can be defined as unimodality. A function is unimodal if there is a path from every point x to the optimal point along which the function continuously increases or decreases. Figure 15.3a shows a strongly unimodal function, while Figure 15.3b shows a nonunimodal function. A strictly unimodal function will have just one local optimum, which corresponds to the global optimum.

15.4 Lagrangian Method: Constrained Optimization

Suppose a function

$$f(x_1 x_2) = k \tag{15.12}$$

has to be minimized subject to constraint that

$$g(x_1 x_2) = b \tag{15.13}$$

The function $f(x_1, x_2)$ increases until it just touches the curve of $g(x_1, x_2)$. At this point, the slopes of f and g will be equal. Thus,

$$\begin{aligned} \frac{dx_1}{dx_2} &= -\frac{\partial f/\partial x_2}{\partial f/\partial x_1} \quad \text{slope of } f(x_1, x_2) \\ \frac{dx_1}{dx_2} &= -\frac{\partial g/\partial x_2}{\partial g/\partial x_1} \quad \text{slope of } g(x_1, x_2) \end{aligned} \tag{15.14}$$

Therefore,

$$\frac{\partial f/\partial x_2}{\partial f/\partial x_1} = \frac{\partial g/\partial x_2}{\partial g/\partial x_1} \tag{15.15}$$

or

$$\frac{\partial f/\partial x_2}{\partial f/\partial x_1} = \frac{\partial g/\partial x_2}{\partial g/\partial x_1} = \lambda \tag{15.16}$$

This common ratio λ is called the Lagrangian multiplier. Then

$$\frac{\partial f}{\partial x_1} - \lambda \frac{\partial g}{\partial x_1} = 0 \tag{15.17}$$

$$\frac{\partial f}{\partial x_2} - \lambda \frac{\partial g}{\partial x_2} = 0 \tag{15.18}$$

The Lagrangian function is defined as

$$F(x_1, x_2, \lambda) = f(x_1, x_2) + \lambda[b - g(x_1, x_2)] \tag{15.19}$$

Differentiation of Equation 15.19 with respect to x_1, x_2, and λ and equating to zero will give Equations 15.17 and 15.18. These are the same conditions as if a new unconstrained function h of three variables is minimized:

$$h(x_1 x_2 \lambda) = f(x_1 x_2) - \lambda g(x_1 x_2) \tag{15.20}$$

Thus,

$$\begin{aligned} \frac{\partial h}{\partial x_1} &= 0 \\ \frac{\partial h}{\partial x_2} &= 0 \\ \frac{\partial h}{\partial \lambda} &= 0 \end{aligned} \tag{15.21}$$

Example 15.4

Minimize the function

$$f(x_1 x_2) = x_1 + 2x_1 x_2 + 9.5x_2 = k$$

for

$$g(x_1 x_2) = x_1^2 + x_2 - 12$$

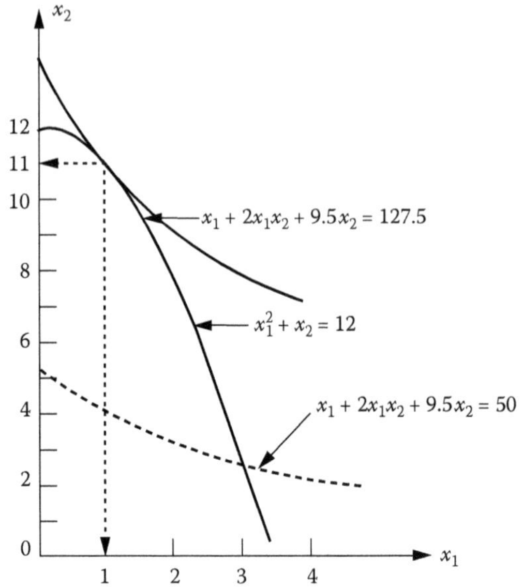

FIGURE 15.4
Functions in Example 15.3.

Form the function with the Lagrangian multiplier as in Equation 15.20,

$$h(x_1x_2) = x_1 + 2x_1x_2 + 9.5x_2 - \lambda(x_1^2 + x^2 - 12)$$

$$\frac{\partial h}{\partial x_1} = 1 + 2x^2 - 2\lambda x_1 = 0$$

$$\frac{\partial h}{\partial x_2} = 2x_1 + 9.5 - \lambda = 0$$

$$\frac{\partial h}{\partial \lambda} = -x_1^2 - x^2 + 12 = 0$$

This gives $x_1 = 1$, $x_2 = 11$, and $\lambda = 11.5$. The value of the function is 127.5. However, this is not the global minimum.

Care has to be exercised in applying this method. A convex and a concave function may be tangential to each other while the absolute minimum may be somewhere else, as in this example. Figure 15.4 shows that the calculated value is not really the minimum.

15.5 Multiple Equality Constraints

The function

$$f(x_1, x_2, \ldots, x_n) \tag{15.22}$$

subject to m equality constraints

$$\begin{aligned} &g_1(x_1, x_2, \ldots, x_n)\\ &g_2(x_1, x_2, \ldots, x_n)\\ &\ldots\\ &g_m(x_1, x_2, \ldots, x_n) \end{aligned} \tag{15.23}$$

can be minimized by forming a function

$$\begin{aligned} &h(x_1, x_2, \ldots, x_n, \lambda_1, \lambda_2, \ldots, \lambda_m) \\ &\quad = f(x_1, x_2, \ldots, x_n) + \lambda_1 g_1(x_1, x_2, \ldots, x_n) + \cdots + \lambda_m g_m(x_1, x_2, \ldots, x_n) \end{aligned} \tag{15.24}$$

where $n+m$ partial derivatives with respect to x_i, $i=1$ to n, and λ_j, $j=1$ to m, are obtained. The simultaneous equations are solved and the substitution gives the optimum value of the function.

15.6 Optimal Load Sharing between Generators

We will apply Lagrangian multipliers to the optimal operation of generators, ignoring transmission losses. The generators can be connected to the same bus, without appreciable impedance between these, which will be a valid system for ignoring losses. The cost of fuel impacts the cost of real power generation. This relationship can be expressed as a quadratic equation:

$$C_i = \frac{1}{2} a_n P_{Gn}^2 + b_n P_{Gn} + w_n \tag{15.25}$$

where
a_n, b_n, and w_n are constants
w_n is independent of generation

The slope of the cost curve is the incremental fuel cost (IC):

$$\frac{\partial C_1}{\partial P_{Gn}} = (\mathrm{IC})_n = a_n P_{Gn} + b_n \tag{15.26}$$

or inversely, the generation can be expressed as a polynomial of the form:

$$P_{Gn} = \alpha_n + \beta_n \left(\frac{\mathrm{d}C_n}{\mathrm{d}P_{Gn}}\right) + \gamma_n \left(\frac{\mathrm{d}C_n}{\mathrm{d}P_{Gn}}\right)^2 + \cdots \tag{15.27}$$

Considering spinning reserve, the total generation must exceed power demand. The following inequality must be strictly observed:

$$\sum P_\mathrm{G} > P_\mathrm{D} \tag{15.28}$$

where
P_G is the real rated power capacity
P_D is the load demand

The load on each generator is constrained between upper and lower limits:

$$P_{\min} \leq P \leq P_{\max} \tag{15.29}$$

The operating cost should be minimized, so that various generators optimally share the load:

$$\begin{aligned} C &= \sum_{i=1}^{i=n} C_i P_{Gi} \text{ is minimum when} \\ &\sum_{i=1}^{i=n} P_{Gi} - P_D = 0 \end{aligned} \tag{15.30}$$

Further, the loading of each generator is constrained in Equation 15.29.

This is a nonlinear programming (NLP) problem as the cost index C is nonlinear. If the inequality constraint of Equation 15.28 is ignored, a solution is possible by Lagrangian multipliers:

$$\Gamma = C - \lambda\left[\sum_{i=1}^{i=n} P_{Gi} - P_D\right] \tag{15.31}$$

where λ is the Lagrangian multiplier.

Minimization is achieved by the condition that

$$\frac{\partial \Gamma}{\partial P_{Gn}} = 0 \tag{15.32}$$

Since C is a function of P only, the partial derivative becomes a full derivative:

$$\begin{aligned} &\frac{dC_n}{dP_{Gn}} = \lambda, \quad \text{that is,} \\ &\frac{dC_1}{dP_{G1}} = \frac{dC_2}{dP_{G2}} = \cdots = \frac{dC_n}{dP_{Gn}} = \lambda \end{aligned} \tag{15.33}$$

That is, all units must operate at the same incremental cost. Figure 15.5 shows the graphic iteration of λ starting from an initial value of λ_0. The three different curves for C represent three different polynomials given by Equation 15.25. An initial value of $IC_0 = \lambda_0$ is assumed and the outputs of the generators are computed. If

$$\sum_{i=1}^{n} P_{Gi} = P_D \tag{15.34}$$

the optimum solution is reached, otherwise increment λ by $\Delta\lambda$ and recalculate the generator outputs.

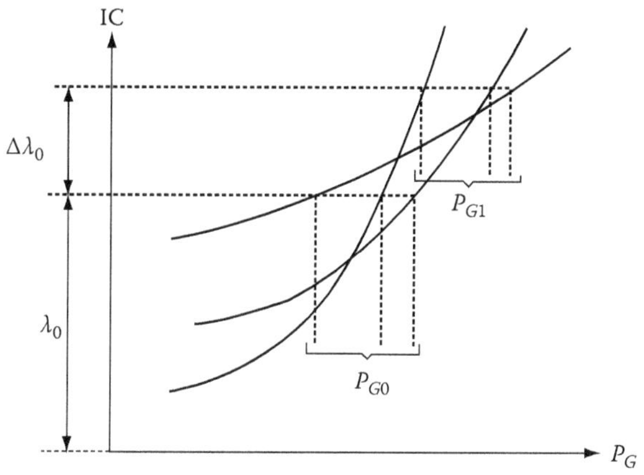

FIGURE 15.5
Graphical representation of Lagrangian multiplier iteration.

Example 15.5

Consider that two generators are required to share a load of 500 MW. The incremental costs of these two generators are given by

$$\frac{dC_a}{dP_{Ga}} = \lambda_a = 0.003P_a + 3.0\$/\text{MW h}$$

$$\frac{dC_b}{dP_{Gb}} = \lambda_b = 0.005P_b + 2.2\$/\text{MW h}$$

If P_a and P_b are the loads on each generator, then

$$P_a + P_b = 500$$

and from Equation 15.33

$$\lambda_a = \lambda_b$$

Solution of these equations gives

$$P_a = 212.5 \text{ MW}$$
$$P_b = 287.5 \text{ MW}$$
$$\lambda = \lambda_a = \lambda_b = 3.637(\$/\text{MW h})$$

15.7 Inequality Constraints

A function $f(x_1, x_2)$ subject to inequality constraint $g(x_1, x_2) \geq 0$ can be minimized by adding a nonnegative Z^2 to the inequality constraints; $g(x_1, x_2)$ takes a value other than zero, only if the constraint $g(x_1, x_2) \geq 0$ is violated. If $g(x_1, x_2) < 0$, Z^2 takes a value required to satisfy the equation $g(x_1, x_2) + Z^2 = 0$. If $g(x_1, x_2) \geq 0$, then $Z^2 = 0$:

$$h(x_1, x_2) = f(x_1, x_2) - \mu\left[g(x_1, x_2) + Z^2\right] = 0 \qquad (15.35)$$

The function h now has four variables, x_1 x_2, μ, and Z. The partial derivatives of h are obtained with respect to these variables and are equated to zero:

$$\begin{aligned}
\frac{\partial h}{\partial x_1} &= \frac{\partial f}{\partial x_1} - \mu\frac{\partial g}{\partial x_1} = 0 \\
\frac{\partial h}{\partial x_2} &= \frac{\partial f}{\partial x_2} - \mu\frac{\partial g}{\partial x_2} = 0 \\
\frac{\partial h}{\partial \mu} &= -g(x_1, x_2) - Z^2 = 0 \\
\frac{\partial h}{\partial Z} &= -2\mu Z = 0
\end{aligned} \tag{15.36}$$

The last condition means that either μ or Z or both μ and Z must be equal to zero.

Example 15.6

Minimize

$$f(x_1, x_2) = 2x_1 + 2x_1x_2 + 3x_2 = k$$

subject to the inequality constraint

$$g(x_1, x_2) = x_1^2 + x_2 \geq 0$$

First form the unconstrained function

$$\begin{aligned}
h &= f(x_1, x_2) - \mu\left[g(x_1, x_2) + Z^2\right] \\
&= 2x_1 + 2x_1x_2 + 3x_2 - \mu\left(x_1^2 + x_2 - 3 + Z^2\right) = k
\end{aligned}$$

This gives

$$\begin{aligned}
\frac{\partial h}{\partial x_1} &= 2 + 2x_2 - 2\mu x_1 = 0 \\
\frac{\partial h}{\partial x_1} &= 2x_1 + 3 - \mu = 0 \\
\frac{\partial h}{\partial \mu} &= x_1^2 + x_2 - 3 + Z^2 = 0 \\
\frac{\partial h}{\partial Z} &= 2\mu Z
\end{aligned}$$

From the last equation, either Z or μ or both are zero. Assume $Z = 0$, then solving the equations gives $x_1 = 0.76$ and $x_2 = 2.422$. Again, it can be shown that this is the relative minimum and not the global minimum.

15.8 Kuhn–Tucker Theorem

This theorem makes it possible to solve an NLP problem with several variables, where the variables are constrained to satisfy certain equality and inequality constraints. The minimization problem with constraints for control variables can be stated as

$$\min f(\overline{x}, \overline{u}) \tag{15.37}$$

subject to the equality constraints

$$g(\overline{x}, \overline{u}, \overline{p}) = 0 \tag{15.38}$$

and inequality constraints:

$$\overline{u} - \overline{u}_{\max} \leq 0 \tag{15.39}$$

$$\overline{u}_{\min} - \overline{u} \leq 0 \tag{15.40}$$

Assuming convexity of the functions defined above, the gradient

$$\overline{\nabla}\Gamma = 0 \tag{15.41}$$

where Γ is the Lagrangian formed as

$$\Gamma = f(\overline{x}, \overline{u}) + \lambda^t g(\overline{x}, \overline{u}, \overline{p}) + \alpha^t_{\max}(\overline{u} - \overline{u}_{\max}) + \alpha^t_{\min}(\overline{u}_{\min} - \overline{u}) \tag{15.42}$$

and

$$\begin{aligned} &\alpha^t_{\max}(\overline{u} - \overline{u}_{\max}) = 0 \\ &\alpha^t_{\min}(\overline{u}_{\min} - \overline{u}) = 0 \\ &\alpha_{\max} \geq 0 \\ &\alpha_{\min} \geq 0 \end{aligned} \tag{15.43}$$

Equation 15.43 is an *exclusion* equation. Multiples $\alpha_{\max}$ and $\alpha_{\min}$ are dual variables associated with the upper and lower limits on control variables, somewhat similar to λ. The superscript t stands for transpose (see Appendix A). If u_i violates a limit, it can be either the upper or lower limit and not both. Thus, either of the two inequality constraints (Equation 15.39 or 15.40) is active at any one time, that is, either $\alpha_{\max}$ or $\alpha_{\min}$ exists at one time and not both.

The gradient equation can be written as

$$\frac{\partial \Gamma}{\partial x} = \frac{\partial f}{\partial x} + \left(\frac{\partial g}{\partial x}\right)^t \lambda = 0 \tag{15.44}$$

$$\frac{\partial \Gamma}{\partial u} = \frac{\partial f}{\partial u} + \left(\frac{\partial g}{\partial u}\right)^t \lambda + \alpha_i = 0 \tag{15.45}$$

In Equation 15.45,

$$\begin{aligned}
&\alpha_i = \alpha_{i,\max} \quad \text{if } u_i - u_{i,\max} > 0 \\
&\alpha_i = -\alpha_{i,\min} \quad \text{if } u_{i,\min} - u_i > 0 \\
&\frac{\partial \Gamma}{\partial \lambda} = g(\overline{x}, \overline{u}, \overline{p}) = 0
\end{aligned} \tag{15.46}$$

Thus, α given by Equation 15.45 for any feasible solution, with λ computed from Equation 15.44, is a negative gradient with respect to $\overline{u}$:

$$\alpha = -\frac{\partial \Gamma}{\partial u} \tag{15.47}$$

At the optimum, α must also satisfy the exclusion equations:

$$\begin{aligned}
&\alpha_i = 0 \quad \text{if } u_{i(\min)} < u_i < u_{i(\max)} \\
&\alpha_i = \alpha_{i(\max)} \geq 0 \quad \text{if } u_i < u_{i(\max)} \\
&\alpha_i = -\alpha_{i(\min)} \leq 0 \quad \text{if } u_i < u_{i(\min)}
\end{aligned} \tag{15.48}$$

Using Equation 15.47, these equations can be written as

$$\begin{aligned}
&\frac{\partial \Gamma}{\partial u_i} = 0 \quad \text{if } u_{i(\min)} < u_i < u_{i(\max)} \\
&\frac{\partial \Gamma}{\partial u_i} \leq 0 \quad \text{if } u_i = u_{i(\max)} \\
&\frac{\partial \Gamma}{\partial u_i} \geq 0 \quad \text{if } u_i = u_{i(\min)}
\end{aligned} \tag{15.49}$$

15.9 Search Methods

In the above discussions, the functions must be continuous, differentiable, or both. In practice, very little is known about a function to be optimized. Many functions may defy simple characterization, convex or concave. The search methods can be classified as follows:

1. The unconstrained one-dimensional search methods, which can be simultaneous or sequential.

   The simultaneous search method can be again subdivided into the following:

   Exhaustive search

   Random search

   The random methods can be divided into the following:

   Dichotomous search

   Equal interval search

   Fibonacci search

2. Multidimensional search methods, which can also be simultaneous or sequential. Again the simultaneous search methods are exhaustive search and random search, while sequential methods are as follows:

   Multivariate grid search

   Univariate search

   Powell's method

   Method of steepest decent

   Fletcher Powell method

   Direct search

One-dimensional search methods place no constraints on the function, that is, continuity or differentiability. An exhaustive one-dimensional search method, for example, subdivides the interval [0, 1] into $\Delta x/2$ equally spaced intervals, and the accuracy of the calculations will depend on the selection of Δx. In practice, the functions have more than one variable, maybe thousands of variables, which may react with each other, and, hence, multidimensional methods are applied. We will discuss and analyze some of the search methods stated earlier.

15.9.1 Univariate Search Method

A univariate search changes one variable at a time so that the function is maximized or minimized. From a starting point, with a reasonable estimate of the solution x_0, find the next point x_1 by performing a maximization (or minimization) with respect to the variable

$$\bar{x}_1 = \bar{x}_0 + \lambda_1 \bar{e}_1 \tag{15.50}$$

where $e_1 = [1, 0, \ldots, 0]$ and λ_1 is a scalar. This can be generalized so that

$$f(\bar{x}_k + \lambda_{k+1}\bar{e}_{k+1}) \quad k = 0, 1, \ldots, n-1 \tag{15.51}$$

is maximized. The process is continued until $|\lambda_k|$ is less than some tolerance value.

Table 15.2 for two functions illustrates this procedure for minimization. The advantage is that only one function is minimized at a time, while the others are held constant. The new values of the other functions are found, one at a time, by substituting the value of the first function and minimizing. The advantage is that it does not require calculations of derivatives. The search method is ineffective when the variables have interaction, or geometrically there are deep valleys and/or ridges.

TABLE 15.2

Univariate Search Method—Function of Two Variables

Calculated Points	Minimized Function	Value Found	Best Current Estimate
(6, 6)	$f(x_1, 6)$	$x_1 = 5.2$	5.2, 6
(5.2, 6)	$f(5.2, x_2)$	$x_2 = 2.9$	5.2, 2.9
(5.2, 2.9)	$f(x_1, 2.9)$	$x_1 = 1.8$	1.8, 2.9
(1.8, 2.9)	$f(1.8, x_2)$	$x_2 = 0.8$	1.8, 0.8

15.9.2 Powell's Method of Conjugate Directions

If a quadratic function to be minimized is

$$h(\overline{x}) = \overline{x}'\overline{A}\overline{x} + \overline{b}'\overline{x} + c \tag{15.52}$$

The direction $\overline{p}$ and $\overline{q}$ are then defined as conjugate directions [3] if

$$\overline{p}'\overline{A}\overline{q} = 0 \tag{15.53}$$

For a unique minimum, it is necessary that matrix $\overline{A}$ is positive definite. Each iterative step begins with a one-dimensional search in n linearly independent directions. If these directions are called

$$\overline{r}_1, \overline{r}_2, \ldots, \overline{r}_n \tag{15.54}$$

and we assume that we start at point x_0, and then initially these directions are chosen to be the coordinates:

$$r_1 = (x_1, 0, \ldots, 0), \quad r_2 = (0, x_2, \ldots, 0), \ldots, r_n = (0, 0, \ldots, x_n) \tag{15.55}$$

The first iteration corresponds to the univariate method, in which one variable is changed at a time. Each iteration develops a new direction. If a positive definite quadratic function is being minimized, then after n iterations all the directions are mutually conjugate.

15.10 Gradient Methods

Starting with an initial value, a sequence of points can be generated so that each subsequent point makes

$$f(x^0) > f(x^1) > f(x^2) \cdots \tag{15.56}$$

$$f(x^0 + \Delta x) = f(x^0) + [\nabla f(x^0)]^t \Delta x + \cdots \tag{15.57}$$

Close to x^0, vector $\nabla f(x^0)$ is in a direction to increase $f(x)$. Thus, to minimize it, $-\nabla f(x^0)$ is used. If a gradient vector is defined as

$$\overline{g}_k \doteq \overline{\nabla} f(x^k) \tag{15.58}$$

then,

$$\overline{x}^{k+1} = \overline{x}^k + h_k(-\overline{g}_k) \tag{15.59}$$

Example 15.7

Consider a function

$$f(x_1, x_2) = 2x_1^2 + 2x_1x_2 + 3x_2^2$$

Gradient vector g is

$$\nabla f = \begin{vmatrix} \dfrac{\partial f}{\partial x_1} \\ \dfrac{\partial f}{\partial x_2} \end{vmatrix} = \begin{vmatrix} 4x_1 + 2x_2 \\ 2x_1 + 6x_2 \end{vmatrix}$$

The successive calculations to $k = 5$ are shown in Table 15.3. The initial assumed values are $x_1 = 4.000$ and $x_2 = 0.000$.

15.10.1 Method of Optimal Gradient

The method is also called the method of steepest decent [1,4], Determine h_k so that

$$h_k = \text{minimum } f(\overline{x}^k - h\overline{g}_k) \tag{15.60}$$

The procedure is as follows:

- Set initial value of $x(= x^0)$
- Iteration count $k = 0$
- Calculate gradient vector g_k
- Find h_k to minimize Equation 15.60

TABLE 15.3

Example 15.6: Minimization of a Function with Gradient Method

k	x_1, x_2	$f(x_1, x_2)$	g_k	$\lvert g_k \rvert$	$\dfrac{g_k'}{\lvert g_k \rvert}$
0	4.000	32.000	16.000	17.888	0.895
	0.000		8.000		0.447
1	3.553	22.670	13.318	13.942	0.955
	−0.447		4.124		0.296
2	2.598	11.294	8.906	8.936	0.997
	−0.743		0.738		0.083
3	1.601	4.528	4.752	5.065	0.938
	−0.826		−1.754		0.346
4	0.663	0.934	1.692	2.297	0.737
	−0.480		−1.554		−0.676
5	−0.074	0.0972	0.096	1.032	0.093
	0.196		1.028		0.996

then

$$\bar{x}^{k+1} = \bar{x}^k - h_k \bar{g}_k \tag{15.61}$$

The convergence is reached if

$$\left| f(\bar{x}^{k+1}) - f(\bar{x}^k) \right| < \in \tag{15.62}$$

TABLE 15.4

Example 15.7: Minimization with Optimal Gradient Method

k	x^k	f_k	g_k	$f(x^k - hg_k)$	h_k
0	4.000	32	16.000	$960.00h^2 - 320.00h + 32$	0.1758
	0.000		8.000		
1	1.1872	5.4176	1.936	$94.330h^2 - 40.5196h + 5.4176$	0.2148
	−1.4064		−6.064		
2	0.7713	1.0646	2.883	$16.623h^2 - 5.4027h + 1.0646$	0.1863
	−0.101		0.937		
3	0.2342	0.208	0.386	$3.593h^2 - 1.5524h + 0.208$	0.2160
	−0.2755		−1.185		
4	0.1508	0.0407	0.564	$0.9543h^2 - 0.3519h + 0.0407$	0.18615
	−0.0196		0.184		
5	0.0458	0.01780	—	—	—
	0.0538				

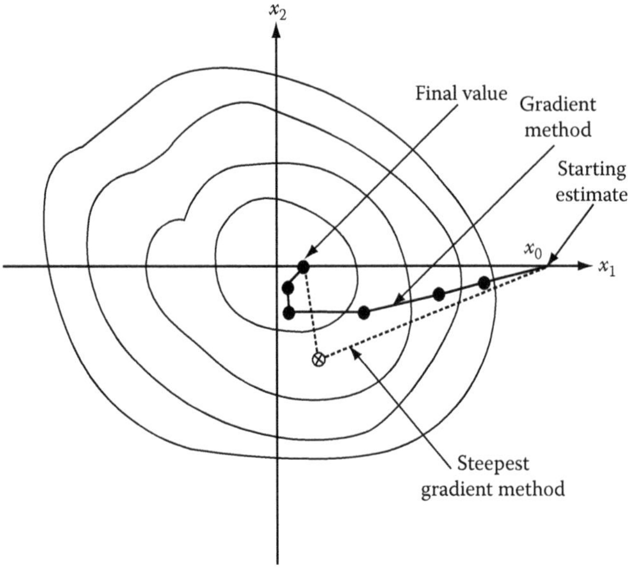

FIGURE 15.6
Convergence in gradient and optimal gradient methods.

Example 15.8

The previous example is solved by the optimal gradient method to five iterations. The results of the calculations are shown in Table 15.4. The minimum is being approached faster in lesser iterations. A comparison between the gradient method and the optimal gradient method is shown in Figure 15.6.

All iterative methods of minimization, whether quadratically convergent or not, locate h as the limit of sequence $x_0, x_1, x_2, \ldots,$ where x_0 is the initial approximation to the position of the minimum and where for each subsequent iteration, x_i is the position of the minimum with respect to variations along the line through x_i in some specified direction p_i. The method of steepest decent uses the directions of the negative gradient of $f(x)$ at x_i, and the method of alternate directions uses cyclically the directions of the n coordinate axes. Methods that calculate each new direction as a part of iteration cycles are more powerful than those in which the directions are assigned in advance [5].

15.11 Linear Programming—Simplex Method

A linear programming problem [1] can be defined as

$$\begin{aligned}
&\text{Max } z = \sum_{j=1}^{n} c_j x_j \\
&\text{st} \\
&\sum_{j=1}^{i=n} \alpha_{ij} x_j = b_i \quad i = 1, 2, \ldots, m \\
&x_j \geq 0 \quad j = 1, 2, \ldots, n
\end{aligned} \tag{15.63}$$

where "st" stands for "subject to." The coefficients C_j, and a_{ij} are real scalars and may vary. The objective function

$$f(\overline{x}) = \sum_{j=1}^{n} c_j x_j \tag{15.64}$$

is in linear form in variables x_j, whose values are to be determined. The inequalities can be converted by addition of either nonnegative slack or surplus variables to equalities. In matrix form, Equation 15.63 can be written as

$$\begin{aligned}
\max z &= \overline{c}'\overline{x} \\
\overline{A}\overline{x} &= \overline{b} \\
\overline{x} &\geq \overline{0}
\end{aligned} \tag{15.65}$$

Define the following:

A *feasible solution* to the linear programming is a vector that satisfies all the constraints of the problem. The X feasible solutions are defined as

$$X = \{\overline{x} | A\overline{x} = \overline{b}, \quad \overline{x} \geq \overline{0}\} \tag{15.66}$$

A *basic feasible solution* is a feasible solution with no more than m positive x values. These positive x values correspond to linearly independent columns of matrix $\overline{A}$. A nondegenerate basic feasible solution is a basic feasible solution with exactly m positive x_j values. It has fewer than m positive x_j values.

The following can be postulated:

- Every basic feasible solution corresponds to an extreme point of the convex set of feasible solutions of X.
- Every extreme point of X has m linearly independent vectors (columns of matrix $\overline{A}$) associated with it.
- There is some extreme point at which the objective function z takes its maximum or minimum value.

The matrix $\overline{A}$ can be partitioned into a basic matrix $\overline{B}$ and a nonbasic matrix $\overline{N}$. Similarly, the vector $\overline{x}$ can be partitioned and the following equation can be written:

$$\overline{A}\overline{x} = (\overline{B}, \overline{N})\begin{bmatrix}\overline{x}_B \\ \overline{x}_N\end{bmatrix} = \overline{b} \tag{15.67}$$

Therefore,

$$\overline{B}\overline{x}_B + \overline{N}\overline{x}_N = \overline{b} \tag{15.68}$$

For the basic solution, $\overline{x}_N$ should be zero; $\overline{x}_B = B^{-1}\overline{b}$ is the basic solution. Variables $\overline{x}_B$ are called the basic variables. If the vector $\overline{c}$ is partitioned as $\overline{c}_B$, $\overline{c}_N$, then the optimum function $z = \overline{c}_B\,\overline{x}_B$. From the basic properties of vectors in E^n, any combination of the columns of the $\overline{A}$ matrix can be written as a linear combination of columns in the matrix $\overline{B}$.

$$\bar{a}_j = \sum_{i=1}^{m} y_{ij}\bar{a}_i = B\overline{y}_i \tag{15.69}$$

or

$$\overline{y}_j = B^{-1}\bar{a}_j \tag{15.70}$$

If a new basic feasible solution is required, then one or more vectors $\bar{a}_i$ should be removed from $\overline{B}$, and substituted with some other vectors $\bar{a}_j$ from $\overline{N}$. In the simplex method of Dantzig, only one vector at a time is removed from the basis matrix and replaced with the new vector that enters the basis [6]. As each feasible solution corresponds to an *extreme point* of the convex set of solutions, by replacing one vector at a time, *the movement is from one extreme point to an adjacent extreme point of the convex set*. The interior point method basically differs in this concept.

Consider a vector $\bar{a}_r$ from the set $\bar{a}_j$, which is to enter the basis, so that $\overline{y}_{rj} \neq 0$; then, from Equation 15.69,

$$\bar{a}_r = \frac{1}{y_{ri}}\bar{a}_j - \sum_{i=1}^{m}\frac{y_{ij}}{y_{rj}}\bar{a}_i \quad (i \neq r) \tag{15.71}$$

The substitution gives the new basic solution. One more condition that needs to be satisfied is that x_{Br} is nonnegative. The column r to be removed from the basis can be chosen by the equation

$$\frac{x_{Br}}{y_{rj}} = \min\left\{\frac{x_{Bi}}{y_{ij}}, \quad y_{ij} > 0\right. \tag{15.72}$$

The minimum in Equation 15.72 may not be unique. In this case, one or more variables in the new basic solution will be zero and the result will be a degenerate basic solution. The vector to enter the basis can be selected so that

1. The new objective function = old objective function $+(x_{Br}/y_{rj})(c_j - z_j)$.
2. In the absence of degeneracy, $(x_{Br}/y_{rj}) > 0$. Therefore, $(c_j - z_j)$ should be selected to be greater than zero.

Example 15.9

Maximize

$$\begin{aligned} z &= 2x_1 - 5x_2 \\ &\text{st} \\ &2x_1 + 4x_2 \geq 16 \\ &3x_1 + 9x_2 \leq 30 \\ &\quad x_1, x_2 \geq 0 \end{aligned}$$

Convert inequalities to equalities by adding slack or surplus variables:

$$\begin{aligned} 2x_1 + 4x_2 - x_3 &= 16 \\ 3x_1 + 9x_2 + x_4 &= 30 \\ x_1, x_2, x_3, x_4 &\geq 0 \end{aligned}$$

Define matrix $\overline{A}$ and its vectors $\overline{a}_p$, $j = 1, 2, 3$, and 4:

$$\overline{A} = \begin{vmatrix} 2 & 4 & -1 & 0 \\ 3 & 9 & 0 & 1 \end{vmatrix} \qquad \overline{b} = \begin{vmatrix} 16 \\ 30 \end{vmatrix}$$

Therefore,

$$\overline{a}_1 = \begin{vmatrix} 2 \\ 3 \end{vmatrix} \qquad \overline{a}_2 = \begin{vmatrix} 4 \\ 9 \end{vmatrix} \qquad \overline{a}_3 = \begin{vmatrix} -1 \\ 0 \end{vmatrix} \qquad \overline{a}_4 = \begin{vmatrix} 0 \\ 1 \end{vmatrix}$$

Consider that $\overline{a}_1$ and $\overline{a}_4$ form the initial basis; then

$$x_1\overline{a}_1 + x_4\overline{a}_4 = \overline{b}$$

$$x_1\begin{vmatrix} 2 \\ 3 \end{vmatrix} + x_4\begin{vmatrix} 0 \\ 1 \end{vmatrix} = \begin{vmatrix} 16 \\ 30 \end{vmatrix}$$

This gives $x_1 = 8$, $x_4 = 6$, and $x_3 = x_2 = 0$. Therefore, the initial feasible solution is

$$\bar{x}_B = \begin{vmatrix} 8 \\ 6 \end{vmatrix}$$

Express vectors $\bar{a}_2$ and $\bar{a}_3$, not in the basis in terms of vectors $\bar{a}_1$ and $\bar{a}_4$:

$$\bar{a}_2 = y_{12}\bar{a}_1 + y_{42}\bar{a}_4$$
$$\bar{a}_3 = y_{13}\bar{a}_1 + y_{43}\bar{a}_4$$

Therefore,

$$\begin{vmatrix} 4 \\ 9 \end{vmatrix} = y_{12}\begin{vmatrix} 2 \\ 3 \end{vmatrix} + y_{42}\begin{vmatrix} 0 \\ 1 \end{vmatrix}$$

Solving $y_{12} = 2$, $y_{42} = 7$; therefore, $\bar{y}_2 = |2,7|$.
Similarly,

$$\begin{vmatrix} -1 \\ 0 \end{vmatrix} = y_{13}\begin{vmatrix} 2 \\ 3 \end{vmatrix} + y_{43}\begin{vmatrix} 0 \\ 1 \end{vmatrix}$$

Solving $y_{13} = -1/2$, $y_{43} = 3/2$; therefore, $\bar{y}_3 = |-1/2, 3/2|$.
Compute z for $\bar{y}_2 = |2, 7|$ and $\bar{y}_3 = |-1/2, 3/2|$:

$$z_j = \bar{c}'_B\bar{y}_j$$

$\bar{c}'_B = |2, 0|$, $c_2 = -5$, $c_3 = 0$, therefore

$$z_2 = |2 \quad 0|\begin{vmatrix} 2 \\ 7 \end{vmatrix} = 4$$

$$z_3 = |2 \quad 0|\begin{vmatrix} -1/2 \\ 3/2 \end{vmatrix} = -1$$

where $z_2 - c_2 = 4 + 5 = 9$, and $z_3 - c_3 = -1 - 0 = -1$.
Since $z_3 - c_3 < 0$, $\bar{a}_3$ will enter the basis. The vector to leave the basis can only be $\bar{a}_1$ or $\bar{a}_4$. The criterion is given by Equation 15.72:

$$\min\left\{\frac{x_{Bi}}{y_{13}}, y_{13} > 0\right\} \quad \text{i.e.,} \quad \frac{x_1}{y_{13}} \quad \text{or} \quad \frac{x_4}{y_{43}} > 0$$

but $y_{13} = -1/2$; therefore, $\bar{a}_4$ leaves the basis. The original value of the objective function is

$$z = \bar{c}'_B\bar{x}_B = |2, 0|\begin{vmatrix} 8 \\ 6 \end{vmatrix} = 16$$

The new value is given by

$$\hat{z} = z + \frac{x_4}{y_{43}}(c_3 - z_3) = 16 + \frac{6}{3/2}(1) = 20$$

where $\hat{z}$ denotes the new value

Compute the new value of $\bar{x}_B$:

$$x_1\bar{a}_1 + x_3\bar{a}_3 = \bar{b}$$

$$x_1\begin{vmatrix} 2 \\ 3 \end{vmatrix} + x_3\begin{vmatrix} -1 \\ 0 \end{vmatrix} = \begin{vmatrix} 16 \\ 30 \end{vmatrix}$$

Solving for $x_1 = 10$, $x_3 = 4$:

$$\hat{z} = |2, 0|\begin{vmatrix} 10 \\ 4 \end{vmatrix} = 20$$

$\bar{a}_1$ was in the basis, $\bar{a}_4$ left the basis, and $\bar{a}_3$ entered the basis. Calculate $\bar{a}_2$ and $\bar{a}_4$ in terms of $\bar{a}_1$ and $\bar{a}_3$:

$$\bar{a}_2 = y_{12}\bar{a}_1 + y_{32}\bar{a}_3$$

$$\bar{a}_4 = y_{14}\bar{a}_1 + y_{34}\bar{a}_3$$

$$\begin{vmatrix} 4 \\ 9 \end{vmatrix} = y_{12}\begin{vmatrix} 2 \\ 3 \end{vmatrix} + y_{32}\begin{vmatrix} -1 \\ 0 \end{vmatrix}$$

$$\begin{vmatrix} 0 \\ 1 \end{vmatrix} = y_{14}\begin{vmatrix} 2 \\ 3 \end{vmatrix} + y_{34}\begin{vmatrix} -1 \\ 0 \end{vmatrix}$$

$$\bar{y}_2 => |3,\ 2| \quad \text{and} \quad \bar{y}_4 = |1/3,\ 2/3|.$$

Compute Z_j:

$$z_2 = |2,\ 0|\begin{vmatrix} 3 \\ 2 \end{vmatrix} = 6 \qquad z_4 = |2,\ 0|\begin{vmatrix} 1/3 \\ 2/3 \end{vmatrix} = \frac{2}{3}$$

$$z_2 - c_2 = 6 + 5 > 0$$

$$z_4 - c_4 = \frac{2}{3} - 0 > 0$$

Therefore, the optimal solution is $x_1 = 10$, $x_2 = 0$, $x_3 = 4$, $x_4 = 0$, and $z = 20$.

15.12 Quadratic Programming

The NLP solves the optimization problems that involve a nonlinear objective and constraint functions. Problems with nonlinearity confined to objective function only can be solved as an extension of the simplex method. Sensitivity and barrier methods (Chapter 16) are considered fairly generalized to solve NLP problems. Quadratic programming (QP) is a special case where the objective function is quadratic, involving square or cross products of two variables. These are solution methods with the additional assumption that the function is convex. The type of problem is referred to as quadratic optimization. It can be applied to maintaining required voltage profile, maximizing power flow and minimizing generation costs. The problem is of the following form:

$$\begin{aligned}&\text{Maximize } f(\overline{x}) = \overline{c}'\overline{x} + \overline{x}'P\overline{x}\\&\text{st}\\&A\overline{x} \leq \overline{b} \quad \text{and} \quad \overline{x} \geq 0\end{aligned} \tag{15.73}$$

There are m constraints and n variables; $\overline{A}$ is an $m \times n$ matrix, P is an $n \times n$ matrix, $\overline{c}$ is an n component vector, and $\overline{b}$ is an m component vector, all with known linear elements. The vector $\overline{x}' = (x_1, x_2, \ldots, x_n)$ is the vector of unknowns. The nonlinear part of the problem is the objective function second term $\overline{x}'P\overline{x}$. Because it is of quadratic form, the matrix P is symmetric. We can postulate the following:

1. The function $f(\overline{x})$ is concave if P is a negative semidefinite or negative definite.
2. The function $f(\overline{x})$ is convex if P is a positive semidefinite or positive definite.
3. There are no computationally feasible methods for obtaining global maximum to QP unless $f(\overline{x})$ is concave. If the function is concave then Kuhn–Tucker conditions are satisfied by a global maxima:

$$F(\overline{x}\overline{\lambda}) = \overline{c}'\overline{x} + \overline{x}'P\overline{x} + \sum_{i=1}^{m} \lambda_i \left[b_i - \sum_{j=1}^{n} a_{ij}x_j \right] \tag{15.74}$$

Therefore, the global maximum to QP is any point $\overline{x}_0$ that satisfies the following:

$$\begin{aligned}&\overline{x}_{0j} > 0\\&\frac{\partial F}{\partial x_j} = c_j + 2\sum_{k=1}^{n} p_{jk}x_k - \sum_{i=1}^{m} \lambda_i a_{ij} = 0 \quad \text{at } \overline{x} = \overline{x}_0\end{aligned} \tag{15.75}$$

$$\begin{aligned}&\overline{x}_{0j} = 0\\&\frac{\partial F}{\partial x_j} = c_j + 2\sum_{k=1}^{n} p_{jk}x_k - \sum_{i=1}^{m} \lambda_i a_{ij} \leq 0 \quad \text{at } \overline{x} = \overline{x}_0\end{aligned} \tag{15.76}$$

$$\begin{aligned}&\lambda_{0j} > 0\\&\frac{\partial F}{\partial \lambda_i} = b_i - \sum_{j=1}^{n} a_{ij}x_j = 0 \quad \text{at } \overline{x} = \overline{x}_0\end{aligned} \tag{15.77}$$

$$\begin{aligned}&\lambda_{0j} = 0\\&\frac{\partial F}{\partial \lambda_i} = b_i - \sum_{j=1}^{n} a_{ij}x_j \leq 0 \quad \text{at } \overline{x} = \overline{x}_0\end{aligned} \tag{15.78}$$

These relations can be rewritten by adding slack variables $y_j, j = 1, 2, \ldots, n, x_{si}, i = 1, 2, \ldots, n$ to Equations 15.75 and 15.77.

$$2\sum_{k=1}^{n} p_{jk}x_k - \sum_{i=1}^{m} \lambda_i a_{ij} + y_j = -c_j$$
$$\sum_{j=1}^{n} a_{ij}x_j + x_{si} = b_i \tag{15.79}$$

Note that

$$x_j y_j = 0 \quad j = 1, 2, \ldots, n$$
$$\lambda_i x_{si} = 0 \quad i = 1, 2, \ldots, n \tag{15.80}$$

thus, a point $\overline{x}_0$ will have a global maximum solution only if there exist nonnegative numbers $\lambda_1, \lambda_2, \ldots, \lambda_m, x_{s1}, x_{s2}, \ldots, x_{sm}, y_1, y_2, \ldots, y_n$ so that Equations 15.78 through 15.80 are satisfied. Except for constraints in Equation 15.73, these are all linear and simplex method can be applied.

A general equation for QP with only inequality constraint is

$$\text{minimize } c^t x + \frac{1}{2} x^t \overline{G} x$$
$$\text{st}$$
$$\overline{A}x > b \tag{15.81}$$

$\overline{G}$ and $\overline{A}$ are n-square symmetric matrices and x and c are n-vectors.

15.13 Dynamic Programming

Dynamic programming (DP) can be described as a computational method of solving optimization problems without reference to particularities, that is, linear or nonlinear programming. It can be used to solve problems where variables are continuous and discrete or for optimization of a definite integral. It is a multistage decision process, where the techniques of linear and nonlinear programming optimization are not applicable.

A system at discrete points can be represented by state vectors:

$$X_i = [x_1(i),\ x_2(i),\ \ldots,\ x_n(i)] \tag{15.82}$$

A *single* decision (*di*) is made out of a number of choices to start; at every stage, a single decision is taken, and then at stage $i+1$, state vector is $X(i+1)$. DP is applicable when

$$X(i+1) = G[X(i),\ d(i), i] \tag{15.83}$$

At each stage, a return function is described:

$$R(i) = R[X(i),\ d(i),\ i] \tag{15.84}$$

Certain constraints are imposed at each stage:

$$\phi[X(i), d(i), i] \leq 0 \geq 0 = 0 \tag{15.85}$$

If there are N stages and denote total return by I, then

$$I = \sum_{i=1}^{N} R(i) \tag{15.86}$$

that is, choose sequences $d(1)$, $d(20), \ldots,$ $d(N)$, which result in optimization of I. This is called discrete time multistage process. If any of the functions, G, $R, \ldots$ defining such a system does not involve chance elements, the system is referred to as deterministic, otherwise it is called probabilistic system.

15.13.1 Optimality

It is based on the principal of optimality, which states that a subpolicy of an optimal policy must in itself be an optimal subpolicy. The essential requirement is that the objective function must be separable.

The problem with n variables is divided into n subproblems, each of which contains only one variable. These are solved sequentially, so that the combined optimal solution of n problems yields the optimal solution to the original problem of n variables.

Let initial stage be C, and first decision $d(i)$ is taken arbitrarily and then corresponding to $d(i)$:

$$X(2) = G[C, d(1)] \tag{15.87}$$

The maximum return over the remaining $N-1$ stages is

$$f_{N-1}[G\{C, d(1)\}] \tag{15.88}$$

and the total return is sum of Equations 15.87 and 15.88. Then using principal of optimality,

$$f_N(C) = R[C, d_N(c)] + f_{N-1}[G\{C, d_N(C)\}] \tag{15.89}$$

Consider that an objective function is separable and can be divided into n subfunctions—let us say three, for example. Then,

$$f(x) = f_1(x_1) + f_2(x_2) + f_3(x_3) \tag{15.90}$$

The optimal value of $f(x)$ is then the maximum of

$$\max\,[f_1(x_1) + f_2(x_2) + f_3(x_3)] \tag{15.91}$$

taken over all the nonnegative integers of the constraint variables. If the optimal solution is described by $\hat{x}_1$, $\hat{x}_2$, and $\hat{x}_3$ and assuming that $\hat{x}_2$ and $\hat{x}_3$ are known, then the problem reduces to that of a single variable:

$$\max\,\{f_1 x_1 + [f_2\hat{x}_2 + f_3\hat{x}_3]\} \tag{15.92}$$

We do not know $\hat{x}_2$ and $\hat{x}_3$, but these must satisfy the constraints, that is, the principal of optimality is applied in stages, till the entire system is optimal.

As an example, considering that the total number of units in a generating station, their individual cost characteristics, and load cycle on the station are known, the unit commitment can be arrived at by DP.

Let the cost function be defined as

$F_N(x) =$ minimum cost of generating x MW from N units

$f_N(y) =$ cost of generating y MW by Nth unit

$f_{N-1}(x-y) =$ Minimum cost of generating $(x-y)$ MW by $N-1$ units DP gives the following recursive relationship:

$$F_N(x) = \text{Min}[f_N(y) + F_{N-1}(x-y)] \tag{15.93}$$

Let t units be generated by N units. As a first step, arbitrarily choose *any one* unit out of the t units. Then $F_1(t)$ is known from Equation 15.93. Now $F_2(t)$ is the minimum of

$$\begin{aligned} &[f_2(0) - F_1(t)] \\ &[f_2(1) - F_1(t-1)] \\ &\ldots \\ &[f_2(t) - F_1(0)] \end{aligned} \tag{15.94}$$

This will give the most economical *two units* to share the total load. The cost curve of these two units can be combined into a single unit, and the third unit can be added. Similarly calculate the minima of $F_3(t), F_4(t), \ldots, F_n(t)$.

Example 15.10

Consider that three thermal units, characteristics, and cost indexes as shown in Table 15.5 are available. Find the unit commitment for sharing a load of 200 MW. The indexes a and b for each unit in cost equation (Equation 15.25) are specified in Table 15.5 and indexes w are zero. Use a second degree polynomial as

$$C_i = \frac{1}{2}a_i P_{Gi}^2 + b_i P_{Gi} + w_i \quad \$/\text{h} \tag{15.95}$$

TABLE 15.5

Example 15.9: Dynamic Programming—A System with Three Thermal Units

	Generation Capability		Cost Indexes		
Unit Number	Minimum	Maximum	*a*	*b*	*w*
1	50	200	0.02	0.50	0
2	50	200	0.04	0.60	0
3	50	200	0.03	0.70	0

We select unit 1 as the first unit:

$$F_1(200) = f_1(200) = \frac{1}{2}(0.02)(200)^2 + (0.50)(200) = \$500/\text{h}$$

We will consider a step of 50 MW to illustrate the problem. Practically it will be too large. From Equation 15.94,

$$F_2(200) = \min\{[f_2(0) + F_1(200)],\ [f_2(50) + F_1(150)],\ [f_2(100) + F_1(100)][f_2(150) + F_1(50)], [f_2(200) + F_1(0)]\}$$

The minimum of these expressions is given by

$$[f_2(50) + F_1(150)] = \$380/\text{h}$$

Similarly calculate

$$F_2(150) = \min\{[f_2(0) + F_1(150)],\ [f_2(50) + F_1(100)],\ [f_2(100) + F_1(50)],\ [f_2(150) + F_1(0)]\}$$
$$F_2(100) = \min\{[f_2(0) + F_1(100)],\ [f_2(50) + F_1(50)],\ [f_2(100) + F_1(0)]\}$$
$$F_2(50) = \min\{[f_2(0) + F_1(50)],\ [f_2(50) + F_1(0)]\}$$

The minimum values are

$$F_2(150) = [f_2(50) + F_1(100)] = \$230/\text{h}$$
$$F_2(100) = [f_2(50) + F_1(50)] = \$130/\text{h}$$
$$F_2(50) = [f_2(0) + F_1(50)] = \$50/\text{h}$$

Bring the third unit in

$$F_3(200) = \min\{[f_3(0) + F_2(200)],\ [f_3(50) + F_2(150)][f_3(100) + F_2(100)][f_3(150) + F_2(50)][f_3(200) + F_2(0)]\}$$

The steps as for unit 2 are repeated, which are not shown. The minimum is given by

$$F_3(200) = [f_3(50) + F_2(150)] = \$302.50/\text{h}$$

Therefore, the optimum load sharing on units 1, 2, and 3 is 100, 50, and 50 MW, respectively. If we had started with any of the three units as the first unit, the results would have been identical. The dimensions for large systems are of major consideration, and often DP is used as a subprocess within an optimization process.

A final value problem can be converted into an initial value problem.

Example 15.11

Maximize

$$x_1^2 + x_2^2 + x_3^2$$
$$\text{st}$$
$$x_1 + x_2 + x_3 \geq 10$$

Let

$$u_3 = x_1 + x_2 + x_3$$
$$u_2 = x_1 + x_2 = u_3 - x_3$$
$$u_1 = x_1 = u_2 - x_2$$

Then

$$\begin{aligned} F_3(u_3) &= \text{Max}_{x_3}\left[x_3^2 + F_2(u_2)\right] \\ F_2(u_2) &= \text{Max}_{x_2}\left[x_2^2 + F_1(u_2)\right] \\ F_1(u_1) &= \left[x_1^2\right] = (u_2 - x_2)^2 \end{aligned} \tag{15.96}$$

Substituting

$$F_2(u_2) = \text{Max}_{x_2}\left[x_2^2 + (u_2 - x_2)^2\right] \tag{15.97}$$

Take the partial derivative and equate to zero for maxima:

$$\begin{aligned} \frac{\partial F_2(u_2)}{\partial x_2} &= 2x_2 - 2(u_2 - x_2) = 0 \\ x_2 &= 2u_2 \end{aligned} \tag{15.98}$$

Therefore,

$$\begin{aligned} F_3(u_3) &= \text{Max}_{x_3}\left[x_3^2 + F_2(u_2)\right] \\ &= \text{Max}_{x_2}\left[x_3^2 + 5(u_3 - x_3)^2\right] \end{aligned} \tag{15.99}$$

Taking the partial derivative

$$\begin{aligned} \frac{\partial F_3(u_3)}{\partial x_3} &= 2x_3 - 10(u_3 - x_3) = 0 \\ x_3 &= \left(\frac{5}{6}\right)u_3 \end{aligned} \tag{15.100}$$

Hence,

$$F_3(u_3) = \left(\frac{5}{6}\right)u_3^2 \qquad u_3 > 10$$

that is, $F_3(u_3)$ is maximum for $u_3 = 10$. Back substituting gives $x_3 = -8.55$, $x_2 = 3.34$, $x_1 = -1.67$. The maximum value of the function is 87.03.

15.14 Integer Programming

Many situations in power systems are discrete, that is, a capacitor bank is online or off-line. A generator is synchronized: on at some time (status = 1) and off (status = 0) at other times. If all the variables are of integer type, the problem is called integer programming. If some variables are of continuous type, the problem is called mixed integer programming. These problems have a non-convex feasible region and a linear interpolation between feasible points (status 0 and 1) gives an infeasible solution. Non-convexity makes these problems more difficult to solve than those of smooth continuous formulation. The two mathematical approaches are as follows:

- Branch and bound methods
- Cutting plane methods

In the branch and bound method, the problem is divided into subproblems based on the values of the variables, that is, consider three variables and their *relaxed* noninteger solution as $x_1 = 0.85$, $x_2 = 0.45$, and $x_3 = 0.92$. Here, x_2 is far removed from the integer values; thus, a solution is found with $x_2 = 0$ and $x_2 = 1$. This leads to two new problems, with alternate possibilities of $x_2 = 1$ or $x_2 = 0$. The process can be continued, where the possible solutions with all integer values will be displayed. Bounding is used to cut off whole sections of the tree, without examining them, and this requires an incumbent solution, which can be the best solution found so far in the process, or an initial solution using heuristic criteria.

In the cutting plane method, additional constraints, called cutting planes, are introduced, which create a sequence of continuous problems. The solution of these continuous problems is driven toward the best integer solution.

Optimization applied to power systems is a large-scale problem. Given a certain optimization problem, is there a guarantee that an optimal solution can be found and the method will converge? Will the optimized system be implementable? How can it be tested? A study [6] shows that optimal power flow results are sensitive to ULTC (under load tap changing) operation and load models. The optimized system did not converge on load flow and had transient stability problems. High-fidelity mathematical models and accurate robust methods of solution are required for real solutions to real problems and to avoid the dilemma of real solutions to nonreal problems or nonsolutions to real world problems [7].

Problems

15.1 Find minimum of the function $2x_1^2 + 3x_2^2 + 3x_3^2 - 6x_1 - 8x_2 + 20x_3 + 45$

15.2 Find global maximum of

$$f(\bar{x}) = 5(x_1 - 2)^2 - 7(x_2 + 3)^2 + 7x_1x_2 \qquad \bar{a} \leq \bar{x} \leq \bar{b}$$

$$\bar{a} = [0, 0] \qquad \bar{b} = [10, 4]$$

15.3 Minimize $z = 3x_1^2 + 2x_2^2$, subject to $3x_1 - 2x_2 = 8$

15.4 If two functions are convex, is it true that their product will also be convex?

15.5 Find maximum of $x_1^2x_2^2x_3^2x_4^2$, given that $x_1^2 + x_2^2 + x_3^2 + x_4^2 = c^2$

15.6 Minimize $f(x_1, x_2) = 2.5x_1^2 + 3x_1^2x_2 + 2.5x_2^2$ by gradient method.

15.7 Repeat Problem 15.6, with optimal gradient method.

15.8 Maximize $z = 4x_1 + 3x_2$

st

$$2x_1 + 3x_2 \leq 16$$
$$4x_1 + 2x_2 \leq 10$$
$$x_1, x_2 \geq 0$$

15.9 Three generators each have a minimum and maximum generation capability of 10 and 50 MW, respectively. The cost functions associated with these four generators can be represented by Equation 15.95. Assume that the cost factors of generators 1, 2, and 3 are the same as in Table 15.5. Consider a step of 10 MW.

15.10 The incremental costs of two generators in \$/MWh are given by

$$\frac{dC_1}{dP_{G1}} = 0.02P_{G1} + 4$$
$$\frac{dC_1}{dP_{G1}} = 0.015P_{G1} + 3$$

Assuming that the total load varies over 100–400 MW, and maximum and minimum loads on each generator are 250 and 50 MW, respectively, how the load should be shared between the two units as it varies?

References

1. P.E. Gill, W. Murray, and M.H. Wright. *Practical Optimization*. New York: Academic Press, 1984.
2. D.G. Lulenberger. *Linear and Non-Linear Programming*. Reading, MA: Addison Wesley, 1984.
3. R.J. Vanderbei. *Linear Programming: Foundations and Extensions*, 2nd edn. Boston, MA: Kluwer, 2001.
4. R. Fletcher and M.J.D. Powell. A rapidly convergent descent method for minimization. *Comp. J.* 5(2): 163–168, 1962.
5. R. Fletcher and C.M. Reeves. Function minimization by conjugate gradients. *Comp. J.* 7(2): 149–153, 1964.
6. G.B. Dantzig. *Linear Programming and Extensions*. Princeton, NJ: Princeton University Press, 1963.
7. E. Yaahedi and H.M.Z. El-Din. Considerations in applying optimal power flow to power system operation. *IEEE Trans. PAS* 4(2): 694–703, 1989.

16

Optimal Power Flow

In the load-flow problem, the system is analyzed in a symmetrical steady state. The specified variables are: Real and reactive power at PQ buses, real powers and voltages at PV buses, and voltages and angles at slack buses. The reactive power injections can also be determined, based on the upper and lower limits of the reactive power control. However, this does not immediately lead to optimal operating conditions, as infinitely variable choices exist in specifying a balanced steady-state load-flow situation.

The constraints on the control variables are not arbitrary, that is, the power generation has to be within the constraints of load demand and transmission losses. These demands and limits are commonly referred to as equality and inequality constraints. There are a wide range of control values for which these constraints may be satisfied, and it is required to select a performance that will minimize or maximize a desired performance index (Chapter 15).

16.1 Optimal Power Flow

The optimal power flow (OPF) problem was defined in early 1960, in connection with the economic dispatch of power [1]. Traditionally, the emphasis in performance optimization has been on the cost of generation; however, this problem can become fairly complex when the hourly commitment of units, hourly production of hydroelectric plants, and cogeneration and scheduling of maintenance without violating the needs for adequate reserve capacity are added. The OPF problem can be described as the cost of minimization of real power generation in an interconnected system where real and reactive power, transformer taps, and phase-shift angles are controllable and a wide range of inequality constraints are imposed. It is a static optimization problem of minute-by-minute operation. It is a nonlinear optimization problem. In load flow, we linearize the network equations in terms of given constraints about an assumed starting point and then increment it with Δ, repeating the process until the required tolerance is achieved. Today, any problem involving the steady state of the system is referred to as an OPF problem. The time span can vary (Table 15.1). In the OPF problem, the basic definitions of state variable, control vector, and input demand vector are retained. OPF requires solving a set of nonlinear equations, describing optimal operation of the power system:

$$\begin{aligned}&\text{Minimize } F(\overline{x},\overline{u})\\&\text{with constraints } g(\overline{x},\overline{u}) = 0\\&h(\overline{x},\overline{u}) \le 0\end{aligned} \tag{16.1}$$

TABLE 16.1

Constraints, Controls, and Objective Functions in OPF

Inequality and Equality Constraints	Controls	Objectives
Power flow equations	Real and reactive power generation	Minimize generation cost
Limits on control variables		Minimize transmission losses
Circuit loading active and reactive	Voltage profiles and Mvar generation at buses	Minimize control shifts
		Minimize number of controls rescheduled
Net area active and reactive power generation	LTC transformer tap positions	
		Optimize voltage profile
Active and reactive power flow in a corridor	Transformer phase shifts	Minimize area active and reactive power loss
	Net interchange	
Unit Mvar capability	Synchronous condensers	Minimize shunt reactive power compensation
Active and reactive reserve limits	SVCs, capacitors, and reactor banks	
		Minimize load shedding
Net active power export	Load transfer	Minimize air pollution
Bus voltage magnitudes and angle limits	HVDC line MW flows	
	Load shedding	
Spinning reserve	Line switching	
Contingency constraints	Standby start-up units	
Environmental and security constraints		

Note: Dependent variables are all variables that are not control functions.

where

$g(\overline{x}, \overline{u})$ represent nonlinear equality constraints and power flow constraints

$h(\overline{x}, \overline{u})$ are nonlinear inequality constraints, limits on the control variables, and the operating limits of the power system

Vector $\overline{x}$ contains dependent variables. The vector $\overline{u}$ consists of control variables; $F(\overline{x}, \overline{u})$ is a scalar objective function.

Constraints, controls, and objectives are listed in Table 16.1.

16.1.1 Handling Constraints

The constraints are converted into equality constraints by including slack variables or including constraints at the binding limits. Let $B(\overline{x}, \overline{u})$ be the binding constraint set, comprising both equalities and enforced inequalities $h(\overline{x}, \overline{u})$ with $u_{\min} \leq u \leq u_{\max}$.

The Lagrangian function of Equation 16.1 is then

$$\Gamma(\overline{x}, \overline{u}) = f(\overline{x}, \overline{u}) + \lambda^t B(\overline{x}, \overline{u}) \tag{16.2}$$

The binding constraints cannot be definitely known beforehand and identification during optimization becomes necessary. A back-off mechanism is necessary to free inequality constraints if later on these become nonbinding.

Another approach is the penalty modeling, which penalizes the cost function if the functional inequality constraint is violated. Its Lagrangian function is

$$\Gamma(\overline{x},\overline{u}) = f(\overline{x},\overline{u}) + \lambda^t B(\overline{x},\overline{u}) + W(\overline{x},\overline{u}) \tag{16.3}$$

where

$$W(\overline{x},\overline{u}) = \sum_i r_i h_i^2(\overline{x},\overline{u}) \tag{16.4}$$

where
- i is the violated constraint
- r_i is the penalty weight
- $h_i(\overline{x}, \overline{u})$ is the ith constraint

Kuhn–Tucker conditions should be met for the final $\overline{x}$, $\overline{u}$, and λ. Squaring the constraints increases the nonlinearity and decreases the sparsity.

The augmented Lanrangian function [2] is a convex function with suitable values of ρ and is written as

$$\Gamma(\overline{x},\overline{u},\lambda,\rho) = f(\overline{x},\overline{u}) - \lambda^t B(\overline{x},\overline{u}) + \frac{1}{2}\rho B(\overline{x},\overline{u})^t B(\overline{x},\overline{u}) \tag{16.5}$$

where

$$\begin{aligned} &B(\overline{x}, \overline{u}) = 0 \\ &(x,u)_{\text{min}} \leq (x,u) \leq (x,u)_{\text{max}} \end{aligned} \tag{16.6}$$

A large value of ρ may impair the solution approach and a small value of ρ may make the function concave. Thus, the augmented Lagrangian method minimizes Equation 16.5, subject to constraints (Equation 16.6).

16.2 Decoupling Real and Reactive OPF

It was shown in Chapter 13 that the real and reactive subsets of variables and constraints are weakly coupled. For high X/R systems, the effect of real power flow on voltage magnitude and of reactive power flow on voltage phase-angle change is relatively negligible. Thus, during a real power OPF subproblem, the reactive power control variables are kept constant, and in reactive power OPF, the real power controls are held constant at their

TABLE 16.2

Controls and Constraints of Decoupled Active and Reactive Power OPF

OPF	Constraints	Controls
Active power OPF	Network power flow	MW generation
	Bus voltage angles	Transformer phase shifter positions
	Circuit loading MW	Area MW interchange
	MW branch flow	HVDC line MW flows
	MW reserve margins	Load shedding
	Area MW interchange	Line switching
	MW flow on a corridor	Load transfer
	Limits on controls	Fast start-up units
Reactive power OPF	Network reactive power flow	Generator voltages and Mvars
	Bus voltage magnitude	LTC tap positions
	Mvar loading	Capacitor, reactor, SVC synchronous
	Unit Mvar capability	Condenser statuses
	Area Mvar generation	
	Corridor Mvar flow	
	Mvar reserve margins	
	Limits on controls	

previously set values [3]. A summary of control and constraints of decoupled subproblem are shown in Table 16.2. The advantages are as follows:

- Decoupling greatly improves computational efficiency.
- Different optimization techniques can be used.
- A different optimization cycle for each subproblem is possible.

Real power controls are set to optimize operating costs, while reactive power controls are optimized to provide secure postcontingency voltage level or reactive power dispatch. The solution method for the real and reactive power OPF subproblem can be different. Figure 16.1 shows the flowchart for the decoupled solution.

16.3 Solution Methods of OPF

OPF methods can be broadly classified into two optimization techniques:

- Linear programming (LP)–based methods
- Nonlinear programming–based methods

LP is reliable for solving specialized OPF problems characterized by linear separable objectives. The nonlinear programming techniques are as follows:

- Sequential quadratic programming
- Augmented Lagrangian methods

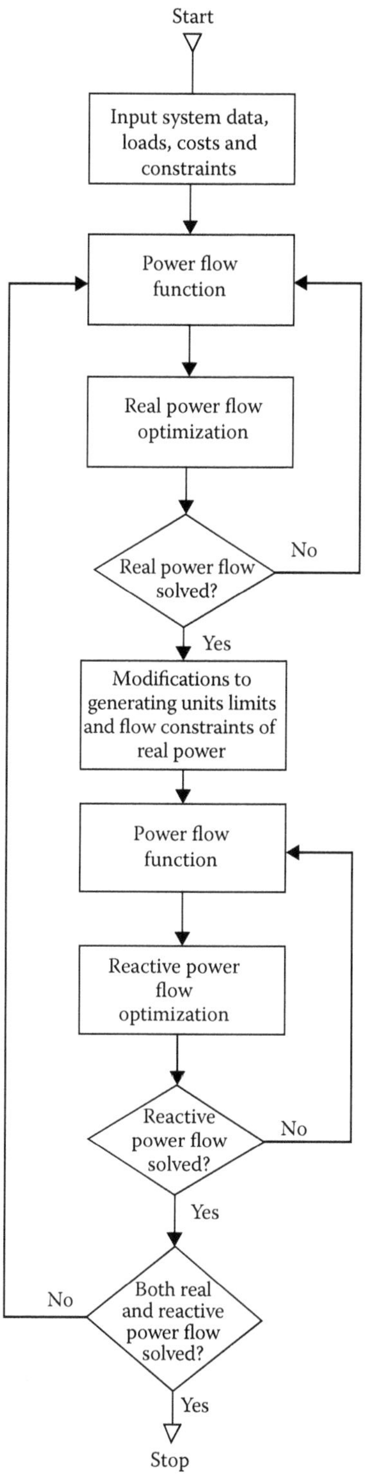

FIGURE 16.1
Decoupled optimal power flow.

- Generalized reduced gradient methods
- Projected augmented Lagrangian
- Successive linear programming
- Interior point (IP) methods

The new projective scaling Karmarkar's algorithm (see Section 16.9) for LP has the advantage of speed for large-scale problems by as much as 50:1 when compared to simplex methods. The variants of interior-point methods are dual affine, barrier, and primal affine algorithm.

16.4 Generation Scheduling Considering Transmission Losses

In Chapter 15, we considered optimizing generation, ignoring transmission losses. Generation scheduling, considering transmission line losses, can be investigated as follows.

The objective is to minimize total cost of generation at any time and meet the constraints of load demand with transmission losses, that is, minimize the function

$$C = \sum_{i=1}^{i=k} C_i(P_{Gi}) \tag{16.7}$$

Total generation = total demand plus total transmission losses. Therefore,

$$\sum_{i=1}^{k} P_{Gi} - P_D - P_L = 0 \tag{16.8}$$

where

k is the number of generating plants
P_L is the total transmission loss
P_D is the system load (total system demand)

To minimize the cost solve for the Lagrangian as

$$\Gamma = \sum_{i=1}^{k} C_i(P_{Gi}) - \lambda\left[\sum_{i=1}^{k} P_{Gi} - P_D - P_L\right] \tag{16.9}$$

If power factor at each bus is assumed to remain constant, the system losses are a function of generation and the power demand is unpredictable; thus, the power generation is the only control function. For optimum real power dispatch,

$$\frac{\partial \Gamma}{\partial P_{Gi}} = \frac{dC_i}{dP_{Gi}} - \lambda + \lambda \frac{\partial P_L}{\partial P_{Gi}} = 0 \quad i = 1, 2, \ldots, k \tag{16.10}$$

Rearranging

$$\lambda = \frac{dC_i}{dP_{Gi}} L_i \quad i = 1, 2, \ldots, k \tag{16.11}$$

where

$$L_i = \frac{1}{(1 - \partial P_L / \partial P_{Gi})} \tag{16.12}$$

is called the penalty factor of the ith plant. This implies that the minimum fuel cost is obtained when the incremental fuel cost of each plant multiplied by its penalty factor is the same for all the plants. The incremental transmission loss (ITL) associated with the ith plant is defined as $\partial P_L / \partial P_{Gi}$. Thus,

$$\frac{dC_i}{dP_{Gi}} = \lambda [1 - (\mathrm{ITL})_i] \tag{16.13}$$

This equation is referred to as the *exact coordination equation*.

16.4.1 General Loss Formula

The exact power flow equations should be used to account for transmission loss. Commonly, the transmission loss is expressed in terms of active power generation only. This is called the B-coefficient method. The transmission losses using B coefficients are given by

$$P_L = \sum_{m=1}^{k} \sum_{n=1}^{k} P_{Gm} B_{mn} P_{Gn} \tag{16.14}$$

where
P_{Gm} and P_{Gn} are the real power generation at m and n plants
B_{mn} are loss coefficients, which are constant under certain assumed conditions

In matrix form,

$$\overline{P}_L = \overline{P}_G^t \overline{B}\, P_G \tag{16.15}$$

$$\overline{P}_G = \begin{vmatrix} P_{G1} \\ P_{G2} \\ \cdots \\ P_{Gk} \end{vmatrix} \quad \overline{B} = \begin{vmatrix} B_{11} & B_{12} & \cdots & B_{1k} \\ B_{21} & B_{22} & \cdots & B_{2k} \\ \cdots & \cdots & \cdots & \cdots \\ B_{k1} & B_{k2} & \cdots & B_{kk} \end{vmatrix} \tag{16.16}$$

The B coefficients are given by

$$B_{mn} = \frac{\cos(\theta_m - \theta_n)}{|V_m||V_n| \cos\phi_m \cos\phi_n} \sum_p I_{pm} I_{pn} R_p \tag{16.17}$$

where

θ_m and θ_n are the phase angles of the generator currents at m and n with respect to a common reference

$\cos\phi_m$ and $\cos\phi_n$ are the power factors of the load currents at m and n plants

I_{pm} and I_{pn} are the current distribution factors, that is, the ratio of load current to total load current

R_p is the resistance of the pth branch

In order that B_{mn} do not vary with the load, the assumptions are as follows:

- Ratios I_{pm} and I_{pn} remain constant.
- Voltage magnitudes remain constant.
- The power factor of loads does not change, that is, the ratio of active to reactive power remains the same.
- Voltage phase angles are constant.

From Equations 16.15 and 16.16, for a three-generator system, the loss equation will be

$$P_L = B_{11}P_{G1}^2 + B_{22}P_{G2}^2 + B_{33}P_{G3}^2 + 2B_{12}P_{G1}P_{G2} + 2B_{23}P_{G2}P_{G3} + 2B_{31}P_{G3}P_{G1} \quad (16.18)$$

Due to simplifying assumptions, a more accurate expression for loss estimation may be required. The loss equation based on bus impedance matrix and current vector is

$$\bar{I}_{bus} = \bar{I}_p + j\bar{I}_q = \begin{vmatrix} I_{p1} \\ I_{p2} \\ \cdots \\ I_{pn} \end{vmatrix} + j\begin{vmatrix} I_{q1} \\ I_{q2} \\ \cdots \\ I_{qn} \end{vmatrix}$$

$$\begin{aligned} \overline{P}_L + j\overline{Q}_L &= \bar{I}_{bus}^t \overline{Z}_{bus} \bar{I}_{bus}^* \\ &= (\bar{I}_p + j\bar{I}_q)^t(\overline{R} + j\overline{X})(\bar{I}_p - j\bar{I}_q) \end{aligned} \quad (16.19)$$

The $\overline{Z}_{bus}$ is symmetrical; therefore, $\overline{Z}_{bus} = \overline{Z}_{bus}^t$. Expanding and considering the real part only,

$$\overline{P}_L = \bar{I}_p^t \overline{R}\,\bar{I}_p + \bar{I}_p^t \overline{X}\,\bar{I}_q + \bar{I}_q^t \overline{R}\,\bar{I}_q - \bar{I}_q^t \overline{X}\,\bar{I}_p$$

Again, as $\overline{X}$ is symmetrical, $\bar{I}_p^t \overline{X}\,\bar{I}_q + \bar{I}_q^t \overline{X}\,\bar{I}_p$, which gives

$$\overline{P}_L = \bar{I}_p^t \overline{R}\,\bar{I}_p + \bar{I}_q^t \overline{R}\,\bar{I}_q \quad (16.20)$$

The currents at a bus in terms of bus voltage and active and reactive power are

$$I_{pi} = \frac{1}{|V_i|}(P_i\cos\theta_i + Q_i\sin\theta_i) \quad (16.21)$$

$$I_{qi} = \frac{1}{|V_i|}(P_i\sin\theta_i - Q_i\cos\theta_i) \quad (16.22)$$

Substituting these current values into Equation 16.20 and simplifying, the loss equation is

$$P_{\mathrm{L}}=\sum_{j=1}^{n}\sum_{k=1}^{n}\left[\frac{R_{ki}\cos(\theta_i-\theta_k)}{|V_k||V_i|}(P_kP_i+Q_kQ_i)+\frac{R_{ki}\sin(\theta_i-\theta_k)}{|V_k||V_i|}(P_kQ_i-Q_kP_i)\right] \tag{16.23}$$

For small values of θ_k and θ_i, the second term of Equation 16.23 can be ignored:

$$P_{\mathrm{L}}=\sum_{j=1}^{n}\sum_{k=1}^{n}\left[\frac{R_{ki}\cos(\theta_i-\theta_k)}{|V_k||V_i|}(P_kP_i+Q_kQ_i)\right] \tag{16.24}$$

Equation 16.24 can be put in matrix form. Let

$$C_{ki}=\frac{R_{ki}\cos(\theta_i-\theta_k)}{|V_k||V_i|} \tag{16.25}$$

$$D_{ki}=\frac{R_{ki}\sin(\theta_i-\theta_k)}{|V_k||V_i|} \tag{16.26}$$

then Equation 16.24 is

$$P_{\mathrm{loss}}=\sum_{k=1}^{n}\sum_{i=1}^{n}(P_kC_{ki}P_i+Q_kC_{ki}Q_i+P_kD_{ki}Q_i-Q_kD_{ki}P_i) \tag{16.27}$$

or in matrix form:

$$P_{\mathrm{loss}}=|P_1,P_2,\ldots,P_n\quad Q_1,Q_2,\ldots,Q_n|$$

$$\times\begin{vmatrix} C_{11} & C_{12} & \cdots & C_{1n} & D_{11} & D_{12} & \cdots & D_{1n}\\ C_{21} & C_{22} & & C_{2n} & D_{21} & D_{22} & & D_{2n}\\ \cdots & & & & & & & \\ -D_{11} & -D_{12} & & -D_{1n} & C_{11} & C_{12} & & C_{1n}\\ -D_{21} & -D_{22} & & -D_{2n} & C_{21} & C_{22} & & C_{2n}\\ \cdots & & & & & & & \\ -D_{n1} & -D_{n2} & & -D_{nn} & C_{n1} & C_{n2} & & C_{nn}\end{vmatrix}\begin{vmatrix}P_1\\P_2\\\cdots\\P_n\\Q_1\\Q_2\\\cdots\\Q_n\end{vmatrix} \tag{16.28}$$

or in partitioned form:

$$\overline{P}_{\mathrm{loss}}=|\overline{P}_l\overline{Q}_l|\begin{vmatrix}\overline{C} & \overline{D}\\ -\overline{D} & \overline{C}\end{vmatrix}\begin{vmatrix}\overline{P}\\ \overline{Q}\end{vmatrix} \tag{16.29}$$

where $\overline{D}$ is zero for the approximate loss equation.

16.4.2 Solution of Coordination Equation

The solution of the coordination equation is an iterative process. For the nth plant,

$$\frac{\mathrm{d}C_n}{\mathrm{d}P_n}+\lambda\frac{\partial P_{\mathrm{L}}}{\partial P_n}=\lambda \tag{16.30}$$

As C_n is given by Equation 15.25,

$$(\mathrm{IC})_n = \frac{\mathrm{d}C_n}{\mathrm{d}P_n} = a_n P_n + b_n \ \$/\mathrm{MWh} \tag{16.31}$$

From Equation 16.14,

$$\frac{\partial P_\mathrm{L}}{\partial P_n} = 2\sum_{m=1}^{k} P_m B_{mn} \tag{16.32}$$

or

$$\frac{\partial P_\mathrm{L}}{\partial P_k} = 2\sum_{i=1}^{n} (C_{ki}P_i + D_{ki}Q_i) \tag{16.33}$$

In the approximate form,

$$\frac{\partial P_\mathrm{L}}{\partial P_k} = 2\sum_{i=1}^{n} C_{ki}P_i \tag{16.34}$$

Thus, substituting in (16.31),

$$\begin{aligned} a_n P_n + b_n + 2\lambda \sum_{m=1}^{k} 2B_{mn}P_m &= \lambda \\ P_n &= \frac{1 - (b_n/\lambda) - \sum_{m=1\, m\neq n}^{k} 2B_{mn}P_m}{(a_n/\lambda) + 2B_{mn}} \end{aligned} \tag{16.35}$$

The iterative process is enumerated as follows:

1. Assume generation at all buses except the swing bus and calculate bus power and voltages based on load flow.
2. Compute total power loss P_L. Therefore, ITL can be computed based on Equation 16.33.
3. Estimate initial value of λ, and calculate $P_1, P_2, \ldots, P_n$ based on equal incremental cost.
4. Calculate generation at all buses; Equation 16.34 or polynomial (15.26) can be used.
5. Check for ΔP at all generator buses:

$$\Delta P = \left| \sum_{i=1}^{n} P_{\mathrm{G}i}^{k} - P_\mathrm{D} - P_\mathrm{L}^{k} \right| \leq \varepsilon_1 \tag{16.36}$$

6. Is $\Delta P < \varepsilon$?

If no, update λ. If yes, is the following inequality satisfied?

$$P_{Gi}^{k} - P_{Gi}^{k-1} \leq \varepsilon_2 \tag{16.37}$$

7. If no, advance iteration count by 1 to $k+1$.

The flowchart is shown in Figure 16.2.

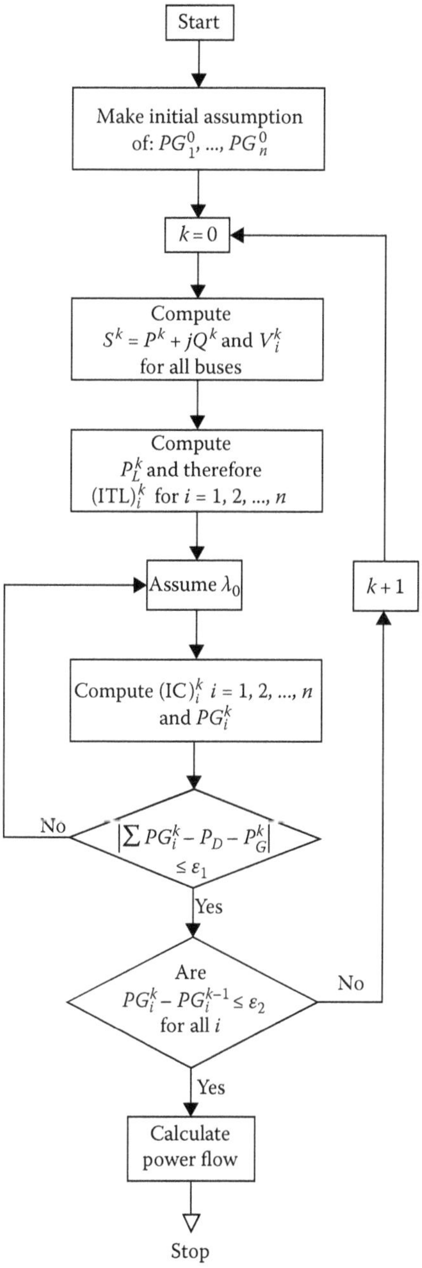

FIGURE 16.2
Flowchart: iterative generator scheduling considering transmission losses.

Example 16.1

Consider two generators of Example 15.4. These supply load through transmission lines as shown in Figure 16.3. The line impedances are converted to a load base of 500 MVA for convenience. The power factor of the load is 0.9 lagging. From Equation 16.34 and ignoring coefficients D_{ki},

$$\frac{\partial P_L}{\partial P_1} = 2(C_{11}P_1 + C_{12}P_2) = 2C_{11}P_1$$

where coefficient C_{12} represents the resistance component of the mutual coupling between the lines. As we are not considering coupled lines, $C_{12}=0$. If the sending-end and receiving-end voltages are all assumed to be equal to the rated voltage and $\theta_1=\theta_2=0$, then

$$\frac{\partial P_L}{\partial P_1} = 2C_{11}P_1 = 2R_{11}P_1 = 2 \times 0.005P_1 = 0.01P_1$$

Similarly,

$$\frac{\partial P_L}{\partial P_2} = 2C_{22}P_2 = 2 \times 0.0065P_2 = 0.013P_2$$

As an initial estimate, assume a value of λ, as calculated for load sharing without transmission line losses, as equal to 3.825. Set both $\lambda_1=\lambda_2=3.825$ and calculate the load sharing:

$$\lambda = \lambda_1 \frac{1}{1 - \partial P_L/\partial P_1}$$
$$= \lambda_2 \frac{1}{1 - \partial P_L/\partial P_2}$$

Here,

$$\lambda_1 = 0.003(500)P_1 + 3.3 = 1.5P_1 + 3.3$$
$$\lambda_2 = 0.005(500)P_2 + 2.2 = 2.5P_2 + 2.2$$

Thus,

$$3.825(1 - 0.01P_1) = 1.5P_1 + 3.3$$
$$3.825(1 - 0.013P_2) = 2.5P_2 + 2.2$$

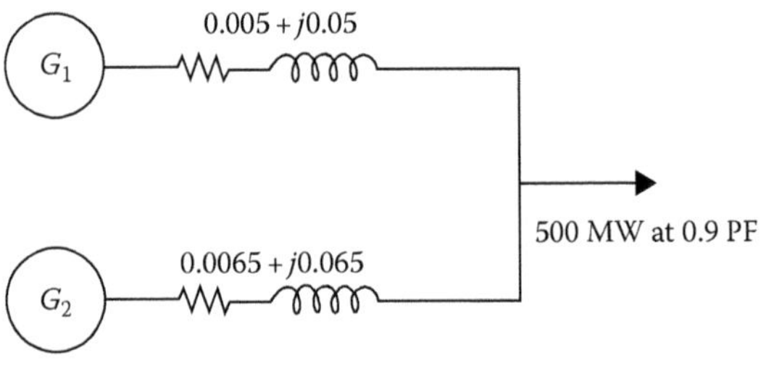

FIGURE 16.3
Configuration for Example 16.1.

This gives

$$P_1 = 0.341, \quad P_2 = 0.6373, \quad P_1 + P_2 = 0.9783 \text{ s}$$

The sum of powers should be >1.0, as some losses occur in transmission. The initial value of λ is low. Consider a demand of approximately 1.02 (2% losses) and adjust the value of λ. A value of 3.87 gives $P_1 = 0.37$ (=185 MW) and $P_2 = 0.655$ (=327.5 MW). The total demand is, therefore, 1.025, representing 2.5% losses.

A load flow is now required with outputs of generator 1 and generator 2 limited to the values calculated above. The load-flow results give a load voltage of $0.983 < -0.93°$ at the load bus and a reactive power input of 156 and 106 Mvar, respectively, from generators 1 and 2 (reactive power loss of 30 Mvar and an active power loss of 3 MW, which is approximately 0.6%). Thus, the total system demand is 1.006. New values of loss coefficients and λ values are found by the same procedure.

Practically, the generator scheduling considers the following constraints in the optimization problem [4–7].

- System real power balance
- Spinning reserve requirements
- Unit generation limits
- State of the unit, that is, it may be on or off
- Thermal unit minimum and maximum starting up/down times
- Ramp rate limits as the unit ramps up and down
- Fuel constraints
- System environmental (emission) limits, that is, SO_2 and NO_x limits
- Area emission limits

In addition, the following reactive power limits are imposed:

- Reactive power operating reserve requirements
- Reactive power generation limits and load bus balance
- System voltage and transformer tap limits

16.5 Steepest Gradient Method

The independent variables are represented by u. These are the variables controlled directly, say, voltage on a PV bus or generation. Dependent, state, or basic variables, which depend on the independent variables, are donated by x, that is, the voltage and angle on a PQ bus. Fixed, constant, or nonbasic variables may be in either of the above two classes and have reached an upper or lower bound and are being held at that bound, that is, when the voltage on a PV bus hits its limit. These are donated by p.

The optimization problem can be stated as minimizations:

$$\begin{aligned} &\min\ c(\overline{x},\overline{u}) \\ &\text{st} \\ &f(\overline{x},\overline{u},\overline{p})=0 \end{aligned} \tag{16.38}$$

Define the Lagrangian function:

$$\Gamma(\overline{x},\overline{u},\overline{p})=c(\overline{x},\overline{u})+\lambda f^t(\overline{x},\overline{u},\overline{p}) \tag{16.39}$$

The conditions for minimization of the unconstrained Lagrangian function are

$$\frac{\partial\Gamma}{\partial x}=\frac{\partial C}{\partial x}+\left[\frac{\partial f}{\partial x}\right]^t\lambda=0 \tag{16.40}$$

$$\frac{\partial\Gamma}{\partial u}=\frac{\partial C}{\partial u}+\left[\frac{\partial f}{\partial u}\right]^t\lambda=0 \tag{16.41}$$

$$\frac{\partial\Gamma}{\partial\lambda}=f(\overline{x},\overline{u},\overline{p})=0 \tag{16.42}$$

Equation 16.42 is the same as the equality constraint, and $\partial f/\partial x$ is the same as the Jacobian in the Newton–Raphson method of load flow.

Equations 16.40 through 16.42 are nonlinear and can be solved by the steepest gradient method. The control vector $\overline{u}$ is adjusted so as to move in the direction of steepest descent. The computational procedure is as follows:

An initial guess of $\overline{u}$ is made, and a feasible load-flow solution is found. The method iteratively improves the estimate of x:

$$\overline{x}^{k+1}=\overline{x}^k+\overline{\Delta}x \tag{16.43}$$

where $\overline{\Delta}x$ is obtained by solving the following equation:

$$\overline{\Delta}x=-\left[J(\overline{x}^k)\right]^{-1}f(\overline{x}^k) \tag{16.44}$$

Equation 16.40 is solved for λ:

$$\lambda=-\left[\left(\frac{\partial f}{\partial x}\right)^t\right]^{-1}\frac{\partial C}{\partial x} \tag{16.45}$$

λ is inserted into Equation 16.41 and the gradient is calculated:

$$\overline{\nabla}\Gamma=\frac{\partial C}{\partial u}+\left[\frac{\partial f}{\partial u}\right]^t\lambda \tag{16.46}$$

if $\nabla\Gamma \rightarrow 0$ is within the required tolerances, then the minimum is reached, otherwise find a new set of variables:

$$\overline{u}_{\text{new}} = \overline{u}_{\text{old}} + \overline{\Delta}u \tag{16.47}$$

where

$$\overline{\Delta}u = -\alpha\overline{\nabla}\Gamma \tag{16.48}$$

Here, Δu is the step in the negative direction of the gradient. The choice of α in Equation 16.48 is important. The step length is optimized. Too small a value slows the rate of convergence and too large a value may give rise to oscillations.

16.5.1 Adding Inequality Constraints on Control Variables

The control variables are assumed to be unconstrained in the above discussions. Practically these will be constrained, that is, generation has to be within certain upper and lower bounds:

$$\overline{u}_{\min} \leq \overline{u} \leq \overline{u}_{\max} \tag{16.49}$$

If Δu in Equation 16.48 causes u_i to exceed one of the limits, then it is set corresponding to that limit.

$$\begin{aligned} u_{i,\text{new}} &= u_{i,\max} \quad \text{if } u_{i,\text{old}} + \Delta u_i > u_{i,\max} \\ u_{i,\text{new}} &= u_{i,\min} \quad \text{if } u_{i,\text{old}} + \Delta u_i < u_{i,\min} \end{aligned} \tag{16.50}$$

Otherwise $u_{i,\text{new}} = u_{i,\text{old}} + \Delta u_i$

The conditions for minimizing Γ under constraint are (Chapter 15):

$$\begin{aligned} \frac{\partial\Gamma}{\partial u_i} &= 0 \quad \text{if } u_{i\min} < u_i < u_{i,\max} \\ \frac{\partial\Gamma}{\partial u_i} &\leq 0 \quad \text{if } u_i = u_{i,\max} \\ \frac{\partial\Gamma}{\partial u_i} &\geq 0 \quad \text{if } u_i < u_{i,\min} \end{aligned} \tag{16.51}$$

Thus, the gradient vector must satisfy the optimality constraints in Equation 16.51.

16.5.2 Inequality Constraints on Dependent Variables

The limits on dependent variables are an upper and lower bound, for example, on a PQ bus the voltage may be specified within an upper and a lower limit. Such constraints can be

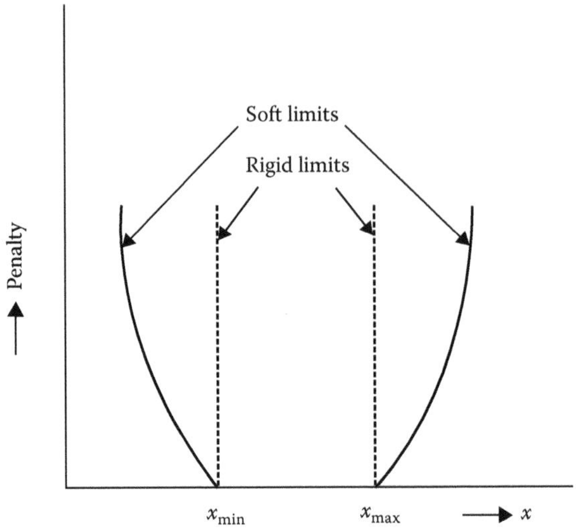

FIGURE 16.4
Penalty function: rigid and soft limits.

handled by the penalty function method (Chapter 15). The objective function is augmented by penalties for the constraint violations. The modified objective function can be written as

$$C' = C(\overline{x}, \overline{u}) + \sum W_j \tag{16.52}$$

where W_j is added for each violated constraint. A suitable penalty function is described by

$$\begin{aligned} W_j &= \frac{S_i}{2}(x_j - x_{j\,\max})^2 \quad \text{when } x_j > x_{j\,\max} \\ W_j &= \frac{S_i}{2}(x_j - x_{j\,\min})^2 \quad \text{when } x_j < x_{j\,\min} \end{aligned} \tag{16.53}$$

The plot is shown in Figure 16.4. This shows how the rigid bounds are replaced by soft limits. The higher the value of S_i, the more rigidly the function will be enforced [8].

16.6 OPF Using the Newton Method

We will examine OPF using the Newton method, through an example [9,10]. Consider a four-bus system with the following variables:

- Voltage magnitudes V_1, V_2, V_3, and V_4
- Voltage phase angles θ_1, θ_2, θ_3, and θ_4
- Real power injections p_1 and p_2 from controllable generators 1 and 2
- t_{23} and t_{34}, the tap ratios

The optimally ordered vector $\overline{y}$ for all variables is

$$\overline{y}^t = [p_1, p_2, t_{23}, t_{34}, \theta_1, v_1, \theta_2, v_2, \theta_3, v_3, \theta_4, v_4] \tag{16.54}$$

The objective function is the sum of cost functions of the active power outputs of the generators:

$$F_1(y) = \frac{1}{2}G_1p_1^2 + b_1p_1 + w_1 \tag{16.55}$$

$$F_2(y) = \frac{1}{2}G_2p_2^2 + b_2p_2 + w_2 \tag{16.56}$$

$$F(y) = F_1(y) + F_2(y) \tag{16.57}$$

The inequality constraints are the upper and lower bounds on the following:

- Power outputs of generators 1 and 2
- Phase-shift angle
- Voltages of the four buses
- Tap ratios of two transformers

16.6.1 Functional Constraints

The functional constraints consist of equalities that are always active and inequalities that are made active only when necessary to maintain feasibility. Real and active power loads are examples of functional equality constraints that are always active. The upper and lower bounds on the real and reactive power outputs of the generators are examples of functional inequality constraints that may be active or inactive.

The given equalities for buses 3 and 4 are

$$CP_i = P_i(y) - p_i \tag{16.58}$$

$$CQ_i = Q_i(y) - q_i \quad \text{for } i = 3, 4 \tag{16.59}$$

where
CP_i is the real power mismatch
CQ_i is the reactive power mismatch
P_i is the schedule real power load
q_i is the schedule reactive power load

The real power injection at buses 1 and 2 is controlled. The sum of real power flow at buses 1 and 2, including injections, is zero. The equations for this constraint are

$$CP_i = P_i(y) - p_i, \quad \text{for } i = 1, 2 \tag{16.60}$$

The reactive power output of two generators is constrained:

$$q_{i\,\min} \leq Q_i(y) \leq q_{i\,\max} \tag{16.61}$$

When $Q_i(y)$ is feasible, the mismatch in Equation 16.62 for CQ_i is made inactive. When $Q_i(y)$ needs to be enforced, CQ_i is made active:

$$CQ_i = [Q_i(y) - q_i] \tag{16.62}$$

16.6.2 Lagrangian Function

The Lagrangian for any problem similar to the example can be written in the following form:

$$\Gamma = F - \sum \lambda_{pi} CP_i - \sum \lambda_{qi} CQ_i \quad (16.63)$$

where
F is the objective function
λ_{pi} is the Lagrangian multiplier for CP_i
λ_{qi} is the Lagrangian for CQ_i

Equation 16.63 implies active constraints for real and reactive power injections.

The matrix equation of the linear system for minimizing the Lagrangian for any OPF by Newton method is

$$\begin{vmatrix} \overline{H}(y,\lambda) & -\overline{J}^t(y) \\ -\overline{J}(y) & \overline{0} \end{vmatrix} \begin{vmatrix} \overline{\Delta y} \\ \overline{\Delta \lambda} \end{vmatrix} = \begin{vmatrix} -\overline{g}(y) \\ -\overline{g}(\lambda) \end{vmatrix} \quad (16.64)$$

where
$\overline{H}(y,\lambda)$ is the Hessian matrix
$\overline{J}(y)$ is the Jacobian matrix
$\overline{g}(y)$ is the gradient w.r.t. y
$\overline{g}(\lambda)$ is the gradient w.r.t. λ

Equation 16.64 can be written as

$$\overline{W}\Delta z = -\overline{g} \quad (16.65)$$

The diagonal and upper triangle of the symmetric $\overline{W}$ matrix is shown in Figure 16.5, when all bus mismatch constraints are active. Some explanations are as follows: (1) the row and columns for $\Delta\theta_1$ is inactive as bus 1 is a swing bus, (2) reactive power injections at buses 1 and 2 can take any value, and, thus, (3) row/column $\Delta\lambda_{q1}$ and $\Delta\lambda_{q2}$ are inactive.

Gradient vector $\overline{g}$ is composed of first derivatives of the form $\partial\Gamma/\partial yi$ or $\partial\Gamma/\partial\lambda i$. As an example,

$$\frac{\partial \Gamma}{\partial p_2} = \frac{\partial}{\partial p_2}[F_2 - \lambda_{p2} CF_2] \quad (16.66)$$

$$= \frac{\partial}{\partial p_2}\left[\frac{1}{2}G_2 p_2^2 + b_2 p_2 + w_2 - \lambda_{p2}(y) - p_2\right]$$

$$= G_2 p_2 + b_2 + \lambda_{p2} \quad (16.67)$$

and

$$\frac{\partial \Gamma}{\partial \theta_2} = \frac{\partial}{\partial \theta_2}\{-\lambda_{pi} CP_i - \lambda_{qi} CQ_i - \lambda_{p2} CP_{2i} - \lambda_{q2} CQ_2\} \quad (16.68)$$

The Jacobian matrix is dispersed throughout the matrix W. Its elements are

$$\frac{\partial^2 \Gamma}{\partial y_i \partial \lambda_j} = \frac{\partial^2 \Gamma}{\partial \lambda_j \partial y_i} \quad (16.69)$$

	0	1	2	3	4	ΔZ		$-g$
	G1	J 1				Δp_1		$-\partial\Gamma/\partial p_1$
	G2		J 1			Δp_2		$-\partial\Gamma/\partial p_2$
0	H		H H J J	H H J J		Δt_{23}		$-\partial\Gamma/\partial t_{23}$
	H			H H J J	H H J J	Δt_{34}		$-\partial\Gamma/\partial t_{34}$
		H H J J	H H J J	H H J J		$\Delta\Theta_1$		$-\partial\Gamma/\partial \theta_1$
		H J J	H H J J	H H J J		ΔV_1		$-\partial\Gamma/\partial v_1$
1		0 0	J J 0 0	J J 0 0		$\Delta\lambda_{p1}$		$-\partial\Gamma/\partial \lambda_{p1}$
		0	J J 0 0	J J 0 0		$\Delta\lambda_{q1}$		$-\partial\Gamma/\partial \lambda_{q1}$
			H H J J	H H J J	H H J J	$\Delta\Theta_2$		$-\partial\Gamma/\partial \theta_2$
			H J J	H J J	H H J J	ΔV_2		$-\partial\Gamma/\partial v_2$
2			0 0	0 0	J J 0 0	$\Delta\lambda_{p2}$	=	$-\partial\Gamma/\partial \lambda_{p2}$
			0	0	J J 0 0	$\Delta\lambda_{q2}$		$-\partial\Gamma/\partial \lambda_{q2}$
				H H J J	H H J J	$\Delta\Theta_3$		$-\partial\Gamma/\partial \theta_3$
				H J J	H H J J	ΔV_3		$-\partial\Gamma/\partial v_3$
3				0 0	J J 0 0	$\Delta\lambda_{p3}$		$-\partial\Gamma/\partial \lambda_{p3}$
				0	J J 0 0	$\Delta\lambda_{q3}$		$-\partial\Gamma/\partial \lambda_{q3}$
					H H J J	$\Delta\Theta_4$		$-\partial\Gamma/\partial \theta_4$
					H J J	ΔV_4		$-\partial\Gamma/\partial v_4$
4					0 0	$\Delta\lambda_{p4}$		$-\partial\Gamma/\partial \lambda_{p4}$
					0	$\Delta\lambda_{q4}$		$-\partial\Gamma/\partial \lambda_{q4}$

FIGURE 16.5
Optimal load flow: matrix W.

where y_i could be θ_1, v_i t_{ij}, or p_i and λ_j can be λ_{pi} or λ_{qi}. These second-order partial derivatives are also first-order partial derivatives of the following form:

$$\frac{\partial P_i}{\partial \theta_i} = \frac{\partial Q_i}{\partial v_i} \tag{16.70}$$

These are elements of the Jacobian matrix J.

16.6.3 Hessian Matrix

Each element of H is a second-order partial derivative of the following form:

$$\frac{\partial^2 \Gamma}{\partial y_i \partial y_j} = \frac{\partial^2 \Gamma}{\partial y_j \partial y_i} \tag{16.71}$$

The elements of the Hessian for the example objective function are

$$\begin{aligned} \frac{\partial^2 F_1}{\partial p_1^2} &= G_1 \\ \frac{\partial^2 F_1}{\partial p_1 \partial \lambda_{p1}} &= 1 \\ \frac{\partial^2 F_1}{\partial p_2^2} &= G_2 \end{aligned} \tag{16.72}$$

Other elements of H are sums of several second-order partial derivatives, that is,

$$\frac{\partial^2\Gamma}{\partial\theta_2\partial v_4} = -\lambda_{p2}\frac{\partial^2 CP_2}{\partial\theta_2\partial v_4} - \lambda_{q2}\frac{\partial^2 CQ_2}{\partial\theta_2\partial v_4} - \lambda_{p4}\frac{\partial^2 CP_4}{\partial\theta_2\partial v_4} - \lambda_{q4}\frac{\partial^2 CQ_4}{\partial\theta_2\partial v_4} \tag{16.73}$$

The representative blocks 3–3 of the matrix W in Figure 16.5 are

$$\begin{vmatrix} \frac{\partial^2\Gamma}{\partial\theta_3^2} & \frac{\partial^2\Gamma}{\partial\theta_3\partial v_3} & \frac{-\partial^2\Gamma}{\partial\theta_3\partial\lambda_{p3}} & \frac{-\partial^2\Gamma}{\partial\theta_3\partial\lambda_{q3}} \\ 0 & \frac{\partial^2\Gamma}{\partial v_3^2} & \frac{-\partial^2\Gamma}{\partial v_3\partial\lambda_{p3}} & \frac{-\partial^2\Gamma}{\partial v_3\partial\lambda_{q3}}0 \\ 0 & 0 & 0 & 0 \\ 0 & 0 & 0 & 0 \end{vmatrix} \tag{16.74}$$

and block 3–4 is

$$\begin{vmatrix} \frac{\partial^2\Gamma}{\partial\theta_3\partial\theta_4} & \frac{\partial^2\Gamma}{\partial\theta_3\partial v_4} & \frac{-\partial^2\Gamma}{\partial\theta_3\partial\lambda_{p4}} & \frac{-\partial^2\Gamma}{\partial\theta_3\partial\lambda_{q4}} \\ \frac{\partial^2\Gamma}{\partial v_3\partial\theta_4} & \frac{\partial^2\Gamma}{\partial v_3\partial v_4} & \frac{-\partial^2\Gamma}{\partial v_3\partial\lambda_{p4}} & \frac{-\partial^2\Gamma}{\partial v_3\partial\lambda_{q4}} \\ \frac{-\partial^2\Gamma}{\partial\lambda_{p3}\partial\theta_4} & \frac{-\partial^2\Gamma}{\partial\lambda_{p3}\partial v_4} & 0 & 0 \\ \frac{-\partial^2\Gamma}{\partial\lambda_{q3}\partial\theta_4} & \frac{-\partial^2\Gamma}{\partial\lambda_{q3}\partial v_4} & 0 & 0 \end{vmatrix} \tag{16.75}$$

16.6.4 Active Set

The active set consists of variables that must be enforced for a solution. This set includes unconditional variables and the functions that would violate the constraints if the bounds were not enforced. The following variables are enforced unconditionally:

$$\theta_1 = 0$$

Power mismatch equations for buses 3 and 4, that is, CP_i and CQ_i
The mismatch equations for controllable real power (Equation 16.60)
The values of the following variables and functions have inequality constraints:

- Variables that will violate their bounds, p_1, p_2, t_{23}, t_{34}, and all bus voltages
- Mismatch equations for reactive power, where reactive power injections would violate one of their bounds. In this case, CQ_1 and CQ_2.

16.6.5 Penalty Techniques

A penalty can be modeled as a fictitious controllable quadratic function, aimed at increasing the optimized cost if the constraint is violated. A penalty is added when a lower or upper bound is violated. If the penalty function is added to Γ and taken into account for evaluation of $\overline{W}$ and $\overline{g}$, it effectively becomes a part of the objective function and creates a high cost of departure of the variable in the solution $\overline{W}\Delta z = -\overline{g}$. For large values of S_i, the function will be forced close to its bound. To modify W and g, the first and second derivatives of the function are computed, in a similar way to the other function in F:

$$\begin{aligned}\frac{\mathrm{d}Ey_i}{\mathrm{d}y_i} &= S_i\left(y_i^0 - \overline{y}_i\right)\\ \frac{\mathrm{d}^2Ey_i}{\mathrm{d}^2y_i} &= S_i\end{aligned} \tag{16.76}$$

where $\overline{y}_1$ is the upper bound and y_1^0 is the current value of y_1. The first derivative is added to $\partial\Gamma/\partial y_1$ of g and the second derivative to $\partial^2\Gamma/\partial y_{12}$. It is assumed that $\overline{W}$ and $\overline{g}$ are not factorized. However, practically, $\overline{W}$ will be factorized and its factors must be modified to reflect addition of penalties. The imposition of a penalty on a variable then requires a change in factors to reflect a change in one diagonal element of $\overline{W}$; $\overline{g}$ must also be modified. When a penalty is added or removed by modifying factors of $\overline{W}$, this effects irritative correction of the variable S_i depending on the computer word size. The larger the value of S_i, the more accurate the enforcement of the penalty.

To test whether a penalty is still needed, it is only necessary to test the sign for its Lagrange multiplier μ:

$$\mu_i = -S_i(y_i - \overline{y}_i) \tag{16.77}$$

If $\overline{y}_i$ is the upper bound and $\mu < 0$ or if $\overline{y}_i$ is the lower bound and μ is >0, the penalty is still needed. Otherwise, it can be removed.

16.6.6 Selecting Active Set

The problem is to find a good active set for solving

$$\overline{W}^k \Delta z^{k+1} = -\overline{g}^k \tag{16.78}$$

An active set that is correct at z^k may be wrong at z^{k+1}, because some inequality constraints will be violated and some enforced inequality constraints will not be necessary. The active set has to be adjusted in the iteration process until an optimum is found.

16.6.7 Algorithm for the Coupled Newton OPF

Initialize:

Vector $\overline{z}$ of variables y and Lagrange multipliers λ are given initial values; λ_{pi} can be set equal to unity and λ_{qi} equal to zero. All other variables can have initial values as in a load-flow program.

Select active set;
Evaluate g_k;

Test for optimum, if all the following conditions are satisfied:

- $\bar{g}_k = 0$
- λ_i and μ_i for all inequalities pass the sign test
- The system is feasible

If all these conditions are satisfied, the optimum has been reached. Exit.
Primary iteration. Evaluate and factorize $W(z_k)$.
Solve:

$$\overline{W}(z^k)\Delta z^k + 1 = -\bar{g}(z^k) \tag{16.79}$$

Compute new state. The flowchart is shown in Figure 16.6.
Some of the characteristics of this method are as follows:

- Each iteration is a solution for the minimum of a quadratic approximation of the Lagrangian function.
- Corrections in variables and Lagrangian multipliers that minimize the successive quadratic approximations are obtained in one simultaneous solution of the sparse linear matrix equation.
- All variables (control, state, or bounded) are processed identically.
- Penalty techniques are used to activate and deactivate the inequality constraints. The use of penalties is efficient and accuracy is not sacrificed.
- Solution speed is proportional to network size and is not much affected by the number of free variables or inequality constraints.

16.6.8 Decoupled Formation

From Equation 16.65, a decoupled formation of the Newton OPF can be written as

$$\overline{W}'\Delta z' = -\bar{g}'$$
$$\overline{W}''\Delta z'' = \bar{g}''$$

where $\overline{W}'$ and $\overline{W}''$ pertain to $P\theta$ and Pv subsystems, and similarly $\bar{g}'$ and $\bar{g}''$. The decoupling divides the problem into $P\theta$ and Pv subsystems. The elements of H omitted in decoupling are relatively small, while the elements of J may not be negligible. The decoupled W matrices have 2×2 formation in Equations 16.74 and 16.75. The factorization of $\overline{W}'$ and $\overline{W}''$ requires approximately the same computational effort as the nonfactorized version of Newton OPF [9].

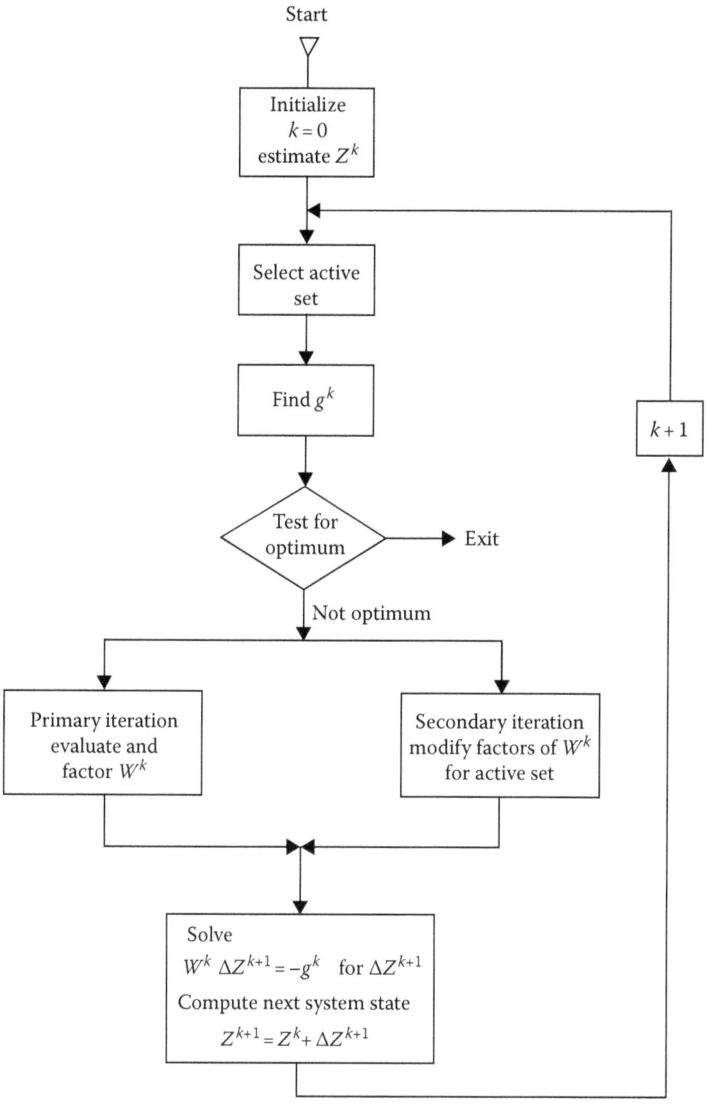

FIGURE 16.6
Flowchart of Newton method: optimal power flow.

16.7 Sequential Quadratic Programming

The method solves the problem of Equation 16.1 by repeatedly solving a quadratic programming approximation, which is a special case of nonlinear programming [11,12]. The objective function is quadratic and the constraints are linear (Chapter 15). The objective function $f(\overline{x}, \overline{u})$ is replaced by a quadratic approximation:

$$q^k(\overline{D}) = \nabla f(\overline{x},\overline{u})^k \overline{D} + \frac{1}{2}\overline{D}^t \nabla^2 \Gamma\left[(\overline{x},\overline{u})^k, \lambda^k\right]\overline{D} \quad (16.80)$$

The step D^k is calculated by solving the quadratic programming subproblem:

$$\begin{aligned}&\min q^k(\overline{D})\\&\text{st}\\&G(\overline{x},\overline{u})^k + J(\overline{x},\overline{u})^k\overline{D} = 0\\&H(\overline{x},\overline{u})^k + I(\overline{x},\overline{u})^k\overline{D} = 0\end{aligned} \tag{16.81}$$

where J and I are the Jacobian matrices corresponding to constraints G and H. The Hessian of the Lagrangian, $[\nabla^2\Gamma\ (\overline{x}, \overline{u})^k, \lambda^k]$, appearing in Equation 16.80 is calculated using quasi-Newton approximation. (Computation of the Hessian in the Newton method is time-consuming. Quasi-Newton methods provide an approximation of the Hessian at point k, using the gradient information from the previous iterations.) After D^k is computed by solving Equation 16.81, (x, u) is updated using

$$(\overline{x},\overline{u})^{k+1} = (\overline{x},\overline{u})^k + \alpha^k\overline{D}^k \tag{16.82}$$

where α^k is the step length. Ascertaining the step length in constrained systems must be chosen to minimize the objective function as well as to constrain violations. These two criteria are often conflicting, and a merit function is employed that reflects the relative importance of these two aims. There are several approaches to select merit functions. SQP has been implemented in many commercial packages.

16.8 Successive Linear Programming

Successive LP can accommodate all types of constraints quite easily and offers flexibility, speed, and accuracy for specific applications. The approach uses linearized programming solved by using different variants of the simplex method. The dual relaxation method and primal method upper bounds are more successfully implemented.

The primal approach tries to obtain an initial value of the objective function based on all the constraints and then modify the objective function by sequentially exchanging constraints on limits with those not on limits. The dual relaxation method finds an initial value of the objective function, which optimally satisfies n number of control variables. It then satisfies the remaining violated constraints by relaxing some binding constraints [13]. The method has simpler, less time-consuming initialization and can detect infeasibility at an early stage of the optimization process, when only a few of the constraints are considered binding in the solution.

Figure 16.7 shows the flowchart [14]. The method allows one critical violated constraint into tableau form at each step. One of the existing binding constraints must leave the basis. The currently violated constraint may enter the basis arbitrarily; however, the constraint to leave the basis must be optimally selected.

The constraints in the basis are tested for eligibility. This test is the *sensitivity* between the incoming constraint and existing binding constraint k:

$$S_k = \frac{\Delta r_{in}}{\Delta r_k} \tag{16.83}$$

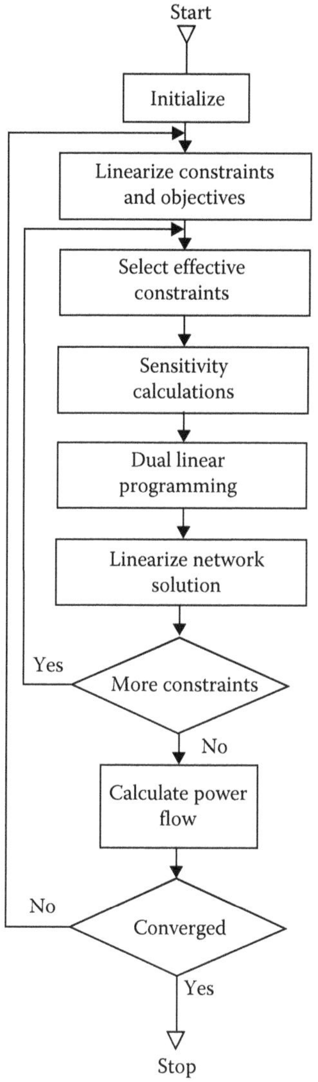

FIGURE 16.7
Flowchart: dual relaxation method.

where

Δr_{in} is the amount by which the violating branch flow or generation will be corrected by entering its constraint into the basis

Δr_k is the amount by which existing binding constraint k will change when freed

Constraint k is eligible if

- This constraint and the incoming constraint are both on the upper or lower limit and S_k is positive
- Both the constraints are on the opposite limits and S_k is negative

The *ratio test* involving the incremental cost y_k of each binding constraint is stated as follows:

The binding constraint to be freed from the basis *among all the eligible constraints* is the one for which

$$\left|\frac{y_k}{S_k}\right| = \text{minimum} \tag{16.84}$$

If no eligible constraints are found, the problem is infeasible.

16.9 Interior Point Methods and Variants

In 1984, Karmarkar [15] announced the polynomially bounded algorithm, which was claimed to be 50 times faster than the simplex algorithm. Consider the LP problem defined as

$$\begin{aligned}&\min c^{-t}\overline{x}\\&\text{st } \overline{A}\overline{x} = \overline{b}\\&\overline{x} \geq 0\end{aligned} \tag{16.85}$$

where

$\overline{c}$ and $\overline{x}$ are n-dimensional column vectors
$\overline{b}$ is an m-dimensional vector
$\overline{A}$ is an $m \times n$ matrix of rank m, $n \geq m$

The conventional simplex method requires 2^n iterations to find the solution. The polynomial-time algorithm is defined as an algorithm that solves the LP problem in $O(n)$ steps. The problem of Equation 16.85 is translated into

$$\begin{aligned}&\min\ c^{-t}\overline{x}\\&\text{st } \overline{A}\overline{x} = 0\\&e^{-t}\overline{x} = 1\\&\overline{x} \geq 0\end{aligned} \tag{16.86}$$

where $n \geq 2$, $\overline{e} = (1, 1, \ldots, 1)^t$ and the following holds:

- The point $x^0 = (1/n, 1/n, \ldots, 1/n)^t$ is feasible in Equation 16.86
- The objective value of Equation 16.86 = 0
- Matrix $\overline{A}$ has full rank of m

The solution is based on projective transformations followed by optimization over an inscribed sphere, which creates a sequence of points converging in polynomial time.

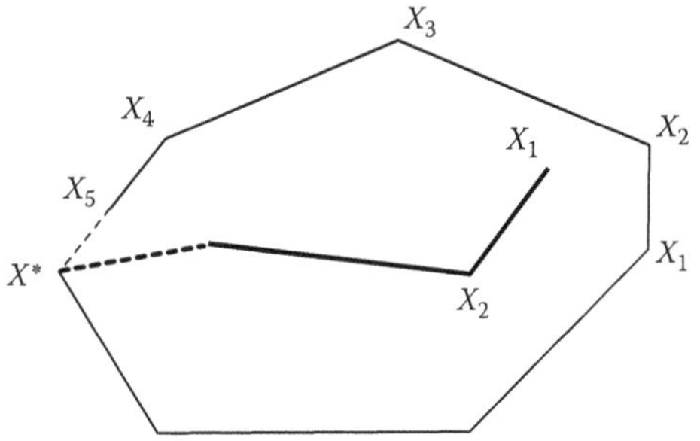

FIGURE 16.8
Iterations in simplex and IP method.

A projective transformation maps a polytope $P \subseteq R^n$ and a strictly IP $a \in P$ into another polytope P' and a point $a' \in P'$. The ratio of the radius of the largest sphere contained in P' with the same center a' is $O(n)$. The method is commonly called an IP method due to the path it follows during solution. The number of iterations required is not dependent on the system size.

We noted in Chapter 15 that, in the simplex method, we go from one extreme point to another extreme point (Example 15.8). Figure 16.8 shows a comparison of the two methods. The simplex method solves an LP problem, starting with one extreme point on the boundary of the feasible region, then going to a better neighboring extreme point along the boundary, and finally stopping at the optimum extreme point. The IP method stays in the interior of the polytope.

16.9.1 Karmarkar Interior Point Algorithm

The algorithm creates a sequence of points $x^0, x^1, \ldots, x^k$ in the following steps:

1. Let x^0 = center of simplex.
2. Compute next point $x^{k+l} = h(x^k)$. Function $\phi = h(a)$ is defined by the following steps:
3. Let $\overline{D} = \text{diag}(a_1, a_2, \ldots, a_n)$ be the diagonal matrix.
4. Augment $\overline{A}\,\overline{D}$ with rows of 1s

$$\overline{B} = \begin{vmatrix} \overline{A}\,\overline{D} \\ \overline{e}^t \end{vmatrix}$$

5. Compute orthogonal projection of Dc into the null space of B:

$$\overline{c}_p = \left[1 - \overline{B}^t(\overline{B}\,\overline{B}^t)^{-1}\overline{B}\right]\overline{D}c$$

6. The unit vector in the direction of $\overline{c}_p$ is

$$\overline{c}_u = \frac{\overline{c}_p}{|c_p|}$$

7. Take a step of length ωr in the direction of c_u:

$$\overline{z} = \overline{a} - \omega r \overline{c}_u \quad \text{where } r = \frac{1}{\sqrt{n(n-1)}}$$

8. Apply reverse protective transformation to z:

$$\overline{\phi} = \frac{\overline{D}\overline{z}}{\overline{e}^t\overline{D}\overline{z}}$$

Return ϕ to x^{k+l}.

The potential function is defined as follows:

$$f(\overline{x}) = \sum_{i=1}^{n} \text{In}\left(\frac{c^t x}{x_i}\right) \tag{16.87}$$

16.9.1.1 Check for Infeasibility

There should be certain improvement in the potential function at each step. If the improvement is not achieved, it can be concluded that the objective function must be positive. This forms a test of the feasibility.

16.9.1.2 Check for Optimality

The check involves going from the current IP to an extreme point without increasing the value of the objective function, and testing the extreme point for optimality. The check is carried out periodically.

It can be shown that in $O[n(q + \log n)]$ steps, a feasible point x is found such that

$$c^{-t}\overline{x} = 0 \quad \text{or} \quad \frac{c^{-t}\overline{x}}{c^{-t}a_0} \leq 2^{-q} \quad \text{where } a_0 = \left(\frac{1}{n}\right)\overline{e} \tag{16.88}$$

16.9.2 Barrier Methods

One year after Karmarkar announced his method, Gill et al. presented an algorithm based on projected Newton logarithmic barrier methods [16]. Gill's work [17] showed that Karmarkar work can be viewed as a special case of a barrier–function method for solving nonlinear programming problems. One drawback of the Karmarkar algorithm is that it does not generate dual solutions, which are of economic significance. Todd's work [18], an extension of Karmarkar's algorithm, generates primal and dual solutions with objective values converging to a common optimal primal and dual value. Barrier–function methods treat inequality constraints by creating a barrier function, which is a combination of original objective function and weighted sum of functions with positive singularity at the boundary. As the weight assigned to the singularities approaches zero, the minimum of barrier function approaches the minimum of original function.

The following are variants of Karmarkar's IP method:

- Projective scaling methods
- Primal and dual affine methods [19]
- Barrier methods
- Extended quadratic programming using IP

16.9.3 Primal–Dual IP Method

The algorithms based on Karmarkar's projective method have polynomial–time complexity requiring $O(nL)$ iterations. These algorithms do not appear to perform well in practice. The algorithms based on dual affine scaling methods exhibit good behavior to real world problems. Most primal–dual barrier path following methods have been shown to require $O(\sqrt{n}L)$ iterations at the most. We will discuss a primal–dual IP method for LP [20,21].

The algorithm is based on

- Newton method for solving nonlinear equations
- Lagrange method for optimization with inequality constraints
- Fiacco and McCormick barrier method for optimization with inequality constraints

Consider an LP problem (Equation 16.1), and let it be a primal problem. The dual problem is

$$\begin{aligned}&\max\ \overline{b}^t\overline{y}\\&\text{st}\ \overline{A}^t\overline{y}+\overline{z}=\overline{c}\\&\overline{z}\geq 0\end{aligned}\tag{16.89}$$

First Lagranges are formed with barriers:

$$\begin{aligned}\Gamma_p(\overline{x},\overline{y})&=\overline{c}^t\overline{x}-\mu\sum_{j=1}^{n}\text{In}(x_j)-y^t(\overline{A}\overline{x}-\overline{b})\\\Gamma_p(\overline{x},\overline{y},\overline{z})&=\overline{b}^t\overline{y}-\mu\sum_{j=1}^{n}\text{In}(z_j)-\overline{x}^T(\overline{A}^t\overline{y}+\overline{z}-\overline{c})\end{aligned}\tag{16.90}$$

The first-order necessary conditions for (16.90) are as follows:

$$\begin{aligned}&\overline{A}\overline{x}=\overline{b}\quad\text{(primal feasibility)}\\&\overline{A}^t\overline{y}+\overline{z}=\overline{c}\quad\text{(dual feasibility)}\\&\overline{x}_j\overline{z}_j=\mu\quad\text{for } j=1,2,\ldots,n\end{aligned}\tag{16.91}$$

If $\mu=0$, then the last expression in Equation 16.91 corresponds to ordinary complimentary slackness. In barrier methods, μ starts at some positive value and approaches zero, as $x(\mu)\rightarrow x^*$ (the constrained minima). Using Newton method to solve Equation 16.91, we have

$$\begin{aligned}&\overline{A}\,\overline{\Delta}x=\overline{b}-\overline{A}x^0=d\overline{P}\\&\overline{A}^t\overline{\Delta}y+\overline{\Delta}z=\overline{c}-\overline{z}^0-\overline{A}^ty^0=-d\overline{D}\\&\overline{Z}\,\overline{\Delta}x+\overline{X}\,\overline{\Delta}z=\mu\overline{e}-\overline{X}\,\overline{Z}\overline{e}\end{aligned}\tag{16.92}$$

where

$$\begin{aligned}\overline{X} &= \text{diag}(x_1^0,\ldots,x_n^0)\\ \overline{Z} &= \text{diag}(z_1^0,\ldots,z_n^0)\\ \overline{e}^t &= (1,1,\ldots,1)\end{aligned} \tag{16.93}$$

From Equation 16.92, the following equations are obtained:

$$\begin{aligned}\overline{\Delta y} &= (\overline{A}\,\overline{Z}^{-1}\overline{X}\,\overline{A}^t)^{-1}(\overline{b} - \mu\overline{A}\,\overline{Z}^{-1}\overline{e} - \overline{A}\,\overline{Z}^{-1}\overline{X}d\overline{D})\\ \overline{\Delta z} &= -d\overline{D} - \overline{A}^t\overline{\Delta y}\\ \overline{\Delta x} &= \overline{Z}^{-1}[\mu\overline{e} - \overline{X}\,\overline{Z}\overline{e} - \overline{X}\,\overline{\Delta z}]\end{aligned} \tag{16.94}$$

x, y, and z are then updated:

$$\begin{aligned}x^1 &= x^0 + \alpha_{\text{p}}\Delta x\\ y^1 &= y^0 + \alpha_{\text{d}}\Delta y\\ z^1 &= z^0 + \alpha_{\text{d}}\Delta z\end{aligned} \tag{16.95}$$

where α_{p} and α_{d} are step sizes for primal and dual variables, chosen to preserve $x > 0$ and $z > 0$.

This completes one iteration. Instead of taking several steps to converge for a fixed value of μ, μ is reduced from step to step. Monteiro and Adler [21] proposed a scheme for updating μ as follows:

$$\mu^{k+1} = \mu^k\left(1 - \frac{t}{\sqrt{n}}\right) \quad 0 < t < \sqrt{n} \tag{16.96}$$

and proved that convergence is obtained in a maximum of $O(\sqrt{n})$ iterations. The convergence criterion is

$$\frac{\overline{c}^t\overline{x} - \overline{b}^t\overline{y}}{1 + |\overline{b}^t\overline{y}|} < \in \tag{16.97}$$

16.10 Security and Environmental Constrained OPF

Security constrained OPF, written as SCOPF for abbreviation, considers outage of certain equipment or transmission lines [22,23]. The security constraints were introduced in early 1970 and online implementation became a new challenge. In conventional OPF, the insecurity of the system during contingency operations is not addressed. Traditionally, security has relied on preventive control, that is, the current-operating point is feasible in the event of occurrence of a given subset of the set at all possible contingencies. This means that the base case variables are adjusted to satisfy postcontingency constraints, which can be added to Equation 16.1:

$$\begin{aligned} &\min f(\bar{z}) \\ &\text{st } g(\bar{z}) = 0 \\ &h(\bar{z}) = 0 \\ &\phi_i(\bar{u}_i - \bar{u}_0) \leq \Theta_i \end{aligned} \tag{16.98}$$

Here, $f(\bar{z}) = f(\bar{x}, \bar{u})$ and the last constraint, called the coupling constraint, reflects the rate of change in the control variables of the base case. Θ_i is the vector of upper bounds reflecting ramp-rate limits. Without the coupling constraints, the problem is separable into $N+1$ subproblems.

The environmental constraints are implemented. The Clean Air Act requires utilities to reduce SO_2 and NO_x emissions. These are expressed as separable quadratic functions of the real power output of individual generating units.

The environmental constraints $e(Z) \leq 0$ can be added to Equation 16.98. This makes the problem of SCOPF rather difficult to solve by conventional methods, and decomposition strategies are used in the solution of OPF. Talukdar–Giras decomposition [17] is an extension of the sequential programming method. Benders decomposition [19] is a variable partitioning method, where certain variables are held fixed while the others are being solved. The values of the complicating variables are adjusted by the master problem, which contains the basic set and a subproblem, which has the extended OPF problem. Iterations between the master problem and subproblem continue until the original problem is solved. Table 16.3 is adapted from Ref. [23] and shows various levels of SCOPF. References [24–33] provide further reading.

TABLE 16.3

Power System Security Levels

Security Level	Description	Description	Control Actions
1	Secure	All loads supplied and no operating limits violated. The system is adequate, secure, and economical	
2	Correctly secure	All loads supplied, no operating limits violated, except under contingency conditions, which can be corrected without loss of load by contingency-constrained OPF	
3	Alert	All loads supplied, no operating limits violated, except that some violations caused by contingencies cannot be corrected without loss of load	Preventive rescheduling can raise it to levels 1 and 2
4	Correctable emergency	All loads supplied but operating limits are violated	Can be corrected by remedial actions and moves to levels 2 or 3
5	Noncorrectable emergency	All loads supplied, but operating limits violated. Can be corrected by remedial actions with some loss of load	Remedial actions result in some loss of load
6	Restorative	No operating limits violated, but there is loss of load	

References

1. J. Carpentier. Contribution á létude du Dispatching Economique. *Bull. Soc. Francaise Electr.* 3: 431–447, 1962.
2. A. Santos Jr. A dual augmented Lagrangian approach for optimal power flow. *IEEE Trans. Power Syst.* 3(3): 1020–1025, 1988.
3. R.R. Shoults and D.I. Sun. Optimal power flow based flow based upon P-Q decomposition. *IEEE Trans.* PAS 101: 397–405, 1982.
4. M. Piekutowski and I.R. Rose. A linear programming method for unit commitment incorporating generation configurations, reserve and flow constraints. *Power Apparatus Syst. IEEE Trans.* PAS-104(12): 3510–3516, December 1985.
5. D.I. Sun. Experiences with implementing optimal power flow for reactive scheduling in Taiwan power system. *IEEE Trans Power Syst.* 3(3): 1193–1200, 1988.
6. G.A. Maria and J.A. Findlay. A Newton optimal flow program for Ontario hydro EMS. *IEEE Trans Power Syst.* 2(3): 576–584, 1987.
7. C. Wang and S.M. Shahidehpour. A decomposition approach to nonlinear multi-area generation scheduling with tie-line constraints using expert systems. *IEEE Trans Power Syst.* 7(4): 1409–1418, 1992.
8. H.W. Dommel and W.F. Tinney. Optimal power flow solutions. *IEEE Trans. PAS* PAS-87: 1866–1876, 1968.
9. D.I. Sun, B.T. Ashley, B.J. Brewer, B.A. Hughes, and W.F. Tinney. Optimal power flow by Newton approach. *IEEE Trans* PAS 103: 2864–2880, 1984.
10. W.F. Tinney and D.I. Sun. Optimal power flow: Research and code development. EPRI Research Report EL-4894, February 1987.
11. S.N. Talukdar and T.C. Giras. A fast and robust variable matrix method for optimal power flows. *IEEE Trans.* PAS 101(2): 415–420, 1982.
12. R.C. Burchett, H.H. Happ, and K.A. Wirgau. Large scale optimal power flow. *IEEE Trans.* PAS 101(10): 3722–3732, 1982.
13. B. Sttot and J.L. Marinho. Linear programming for power system network security applications. *IEEE* PAS 98: 837–848, 1979.
14. D.S. Kirschen and H.P. Van Meeten. MW/voltage control in a linear programming based optimal load flow. *IEEE Trans. Power Syst.* 3: 782–790, May 1988.
15. N. Karmarkar. A new polynomial-time algorithm for linear programming. *Combinatorica* 4(4): 373–395, 1984.
16. P.E. Gill, W. Murray, M.A. Saunders, J.A. Tomlin, and M.H. Wright. On projected Newton barrier methods for linear programming and an equivalence to Karmarkar's projective method. *Math. Program.* 36: 183–209, 1986.
17. P.E. Gill. On projected Newton barrier methods for linear programming and an equivalence to Karmarkar's protective method. *Math. Program.* 36: 183–209, 1986.
18. M.J. Todd and B.P. Burrell. An extension of Karmarkar's algorithm for linear programming using dual variables. *Algorithmica* 1: 409–424, 1986.
19. R.E. Marsten. Implementation of dual affine interior point algorithm for linear programming. *ORSA Comput.* 1:287–297, 1989.
20. I.J. Lustig, R.E. Marsen, and N.D.F. Shanno. IP methods for linear programming computational state of art technical report. Computational Optimization Center, School of Industrial and System Engineering, Georgia Institute of Technology, Atlanta, GA, 1992.
21. R.D.C. Monteiro and I. Adler. Interior path following primal-dual algorithms. Part 1: Linear programming, Part II: Convex quadratic programming. *Math. Program.* 44: 27–66, 1989.
22. B. Stott, O. Alsac, and A.J. Monticelli. Security analysis and optimization. *Proc IEEE* 75(2): 1623–1644, 1987.
23. A.J. Monticelli, M.V.F. Pereira, and S. Granville. Security constrained optimal power flow with post-contingency corrective rescheduling. *IEEE Trans. Power Syst.* 2(1): 175–182, 1987.

24. R.C. Burchett, H.H. Happ, and D.R. Vierath. Quadratically convergent optimal power flow. *PAS* 103: 3267–3275, 1984.
25. A.M. Sasson. Optimal power flow solution using the Hessian matrix. *IEEE Trans* PAS 92: 31–41, 1973.
26. I.J. Lustig. Feasibility issues in an interior point method for linear programming. *Math. Program.* 49: 145–162, 1991.
27. IEEE Power Engineering Society. IEEE Tutorial Course–Optimal Power Flow, Solution Techniques, Requirements and Challenges–IEEE. Document Number 96TP 111–0, 1996.
28. J.F. Benders. Partitioning for solving mixed variable programming problems. *Numer. Math.* 4: 283–252, 1962.
29. S.N. Talukdar and V.C. Ramesh. A multi-agent technique for contingency constrained optimal power flows. *IEEE Trans. Power Syst.* 9(2): 885–861, May 1994.
30. I. Adler. An implementation of Karmarkar's algorithm for solving linear programming problems. *Math. Program.* 44: 297–335, 1989.
31. IEEE Committee Report. IEEE Reliability Test System. *IEEE Trans PAS* PAS-98: 2047–2054, 1979.
32. IEEE Working Group Report. Description and Bibliography of Major Economy-Security Functions Part I and Part II. *IEEE Trans.* 100: 211–235, 1981.
33. J.A. Momoh. Application of quadratic interior point algorithm to optimal power flow. EPRI Final Report, RP 2473–36 II, 1992.

17

Harmonics Generation

Harmonics in power systems can be studied under five distinct sections:

- Generation of harmonics
- Effects of harmonics
- Harmonic propagation, modeling, and analysis
- Mitigation of harmonics, passive and active filters
- Measurements of harmonics (not covered in this book)

Harmonics cause distortions of voltage and current waveforms, which have adverse effects on electrical equipment. Harmonics are one of the major power quality concerns. The estimation of harmonics from nonlinear loads is the first step in a harmonic analysis and this may not be straightforward. There is an interaction between the harmonic-producing equipment, which can have varied topologies, and the electrical system. Over the course of recent years, much attention has been focused on the analysis and control of harmonics, and standards have been established for permissible harmonic current and voltage distortions.

In this chapter, we will discuss the nature of harmonics and their generation by electrical equipment. Harmonics can have varied amplitudes and frequencies. The most common harmonics in power systems are sinusoidal components of a periodic waveform that have frequencies that can be resolved into some multiples of the fundamental frequency. Fourier analysis is the mathematical tool employed for such analysis, and Appendix E provides an overview. It is recommended that the reader becomes familiarized with Fourier analysis before proceeding with the subject of harmonics. Power systems also have harmonics that are noninteger multiples of the fundamental frequency and have aperiodic waveforms. The generation of harmonics in power system occurs from two distinct types of loads:

1. Linear time-invariant loads are characterized so that an application of a sinusoidal voltage results in a sinusoidal flow of current. These loads display a constant steady-state impedance during the applied sinusoidal voltage. If the voltage is increased, the current also increases in direct proportion. Incandescent lighting is an example of such a load. Transformers and rotating machines, under normal loading conditions, approximately meet this definition, though the flux wave in the air gap of a rotating machine is not sinusoidal. Tooth ripples and slotting may produce forward and reverse rotating harmonics. Magnetic circuits can saturate and generate harmonics. As an example, saturation in a transformer on abnormally high voltage produces harmonics, as the relationship between magnetic flux density B and the magnetic field intensity H in the transformer core is not linear.

The inrush current of a transformer contains odd and even harmonics, including a dc component. Yet, under normal operating conditions these effects are small. Synchronous generators in power systems produce sinusoidal voltages and the loads draw nearly sinusoidal currents. For the sinusoidal input voltages, the harmonic pollution produced, due to these types of loads, is small.

2. The second category of loads is described as nonlinear. In a nonlinear device, the application of a sinusoidal voltage does not result in a sinusoidal flow of current. These loads do not exhibit constant impedance during the entire cycle of applied sinusoidal voltage. *Nonlinearity is not the same as the frequency dependence of impedance*, i.e., the impedance of a reactor changes in proportion to the applied frequency, but it is linear at each applied frequency. On the other hand, nonlinear loads draw a current that may even be discontinuous or flow in pulses for a part of the sinusoidal voltage cycle. Some examples of nonlinear loads are as follows:
   - Adjustable drive systems
   - Cycloconverters
   - Arc furnaces
   - Switching mode power supplies
   - Computers, copy machines, and television sets
   - Static var compensators (SVCs)
   - HVDC transmission
   - Electric traction
   - Wind and solar power generation
   - Battery charging and fuel cells
   - Slip recovery schemes of induction motors
   - Fluorescent lighting and electronic ballasts

The distortion produced by nonlinear loads can be resolved into a number of categories:

- A distorted waveform having a Fourier series with fundamental frequency equal to power system frequency, and a periodic steady state exists. This is the most common case in harmonic studies.
- A distorted waveform having a submultiple of power system frequency and a periodic steady state exists. Certain types of pulsed loads and integral cycle controllers produce these types of waveforms.
- The waveform is aperiodic, but perhaps almost periodic. A trigonometric series expansion may still exist. Examples are arcing devices, for example, arc furnaces, and fluorescent, mercury, and sodium vapor lighting. The process is not periodic in nature, and a periodic waveform is obtained if the conditions of operation are kept constant for a length of time.

The components that are not an integral multiple of the power frequency are called noninteger harmonics.

The arc furnace loads are highly polluting and cause phase unbalance, flicker, impact loading, harmonics, and resonance, and may give rise to torsional vibrations in rotating machines.

17.1 Harmonics and Sequence Components

In a three-phase balanced system under non-sinusoidal conditions, the hth-order harmonic voltage (or current) can be expressed as

$$V_{ah} = V_h \sin(h\omega_0 t + \theta_h) \tag{17.1}$$

$$V_{bh} = V_h \sin\left(h\omega_0 t - \frac{2h\pi}{3} + \theta_h\right) \tag{17.2}$$

$$V_{ch} = V_h \sin\left(h\omega_0 t + \frac{2h\pi}{3} + \theta_h\right) \tag{17.3}$$

Under balanced conditions, the hth harmonic (frequency of harmonic $= h$ times the fundamental frequency) of phase b lags h times 120° behind that of the same harmonic in phase a. The hth harmonic of phase c lags h times 240° behind that of the same harmonic in phase a. In the case of triplen harmonics, shifting the phase angles by three times 120° or three times 240° results in cophasial vectors. Table 17.1 shows the sequence of harmonics, and the pattern is clearly positive–negative–zero. We can write

$$\text{Harmonics of the order } 3h+1 \text{ have positive sequence} \tag{17.4}$$

$$\text{Harmonics of the order } 3h+2 \text{ have negative sequence} \tag{17.5}$$

$$\text{and harmonics of the order } 3h \text{ are of zero sequence} \tag{17.6}$$

All triplen harmonics generated by nonlinear loads are zero sequence phasors. These add up in the neutral. In a three-phase four-wire system, with perfectly balanced single-phase loads between the phase and neutral, all positive and negative sequence harmonics will cancel out leaving only the zero sequence harmonics. In an unbalanced single-phase load, the neutral carries zero sequence and the residual unbalance of positive and negative sequence currents. Even harmonics are absent in the line because of phase symmetry (Appendix E) and unsymmetrical waveforms will add even harmonics to the phase conductors.

TABLE 17.1
Sequence of Harmonics

Harmonic Order	Sequence of the Harmonic
1	+
2	−
3	0
4	+
5	−
6	0
7	+
8	−
9	0
10, 11, 12	+, −, 0

17.2 Increase in Nonlinear Loads

Nonlinear loads are continually on the increase. It is estimated that during the next 10 years, 60% of the loads on utility systems will be nonlinear. Concerns for harmonics originate from meeting a certain power quality, which leads to the related issues of (1) effects on the operation of electrical equipment, (2) harmonic analysis, and (3) harmonic control. A growing number of consumer loads are sensitive to poor power quality and it is estimated that power quality problems cost U.S. industry tens of billion of dollars per year. While the expanded use of consumer automation equipment and power electronic controls is leading to higher productivity, these very loads are a source of electrical noise, and harmonics and are less tolerant to poor power quality. For example, ASDs (adjustable speed drives) are less tolerant to voltage sags and swells, and a voltage dip of 10% of certain duration may precipitate a shutdown.

17.3 Harmonic Factor

An index of merit has been defined as a harmonic distortion factor [1] (harmonic factor). It is the ratio of the root-mean square of the harmonic content to the root-mean square value of the fundamental quantity, expressed as a percentage of the fundamental:

$$\mathrm{DF} = \sqrt{\frac{\sum \text{of squares of amplitudes of all harmonics}}{\text{square of the amplitude of the fundamental}}} \times 100\% \tag{17.7}$$

Voltage and current harmonic distortion indexes, defined in Appendix F, are the most commonly used indexes. THD (total harmonic distortion) in common use is the same as DF.

17.4 Three-Phase Windings in Electrical Machines

The armature windings of a machine consist of phase coils, which span approximately a pole pitch. A phase winding consists of a number of coils connected in series, and the EMF generated in these coils is time displaced in phase by a certain angle. The air gap is bounded on either side by iron surfaces and provided with slots and duct openings and is skewed. Simple methods of estimating the reluctance of the gap to carry a certain flux across the gap are not applicable and the flux density in the air gap is not sinusoidal. Figure 17.1 shows that armature reaction varies between a pointed and flat-topped trapezium for a phase spread of $\pi/3$. Fourier analysis of the pointed waveform in Figure 17.1 gives

$$F = \frac{4}{\pi} F_m \cos \omega t \left[\sum_{n=1}^{n=\infty} \frac{1}{n} k_{mn} \sin nx \right] \tag{17.8}$$

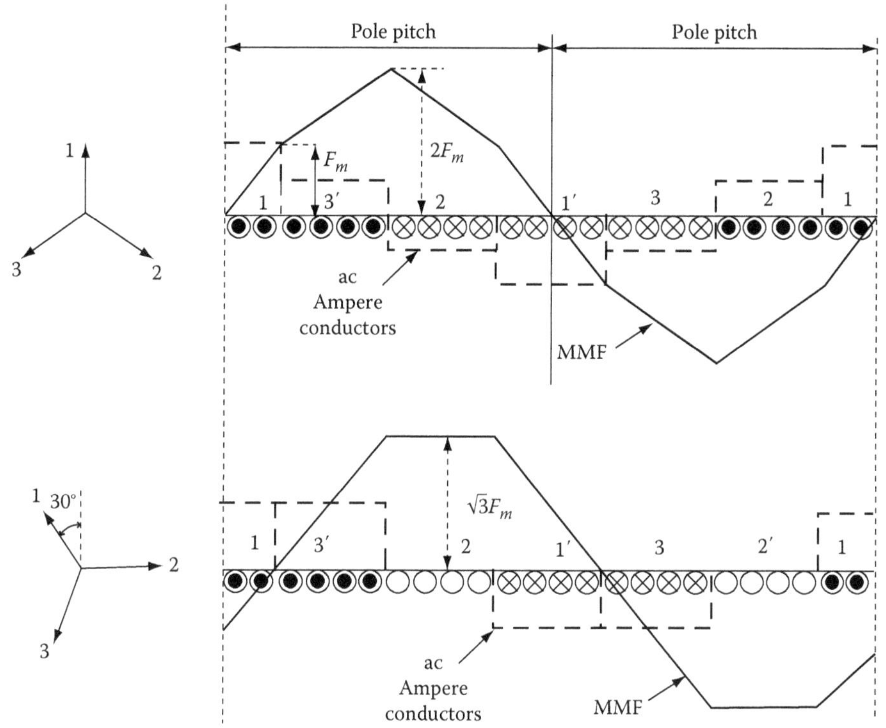

FIGURE 17.1
Armature reaction of a three-phase winding spanning a pole pitch.

$$k_{mn} = \frac{\sin(1/2)n\sigma}{g' \sin(1/2)(n\sigma/g')} \tag{17.9}$$

where
- k_{mn} is a winding distribution factor for the nth harmonic
- g' is the number of slots per pole per phase
- σ is the phase spread

The MMF (magneto motive force) of three phases will be given by considering the time displacement of currents and space displacement of axes as

$$\begin{aligned} F_t = {} & \frac{4}{\pi} F_m \cos\omega t \left[\sum_{n=1}^{n=\infty} \frac{1}{n} k_{mn} \sin nx \right] + \frac{4}{\pi} F_m \cos\left(\omega t - \frac{2}{3}\pi\right) \\ & \times \left[\sum_{n=1}^{n=\infty} \frac{1}{n} k_{mn} \sin n\left(x - \frac{2}{3}\pi\right) \right] + \frac{4}{\pi} F_m \cos\left(\omega t - \frac{4}{3}\pi\right) \\ & \times \left[\sum_{n=1}^{n=\infty} \frac{1}{n} k_{mn} \sin n\left(x - \frac{4}{3}\pi\right) \right] \end{aligned} \tag{17.10}$$

This gives

$$F_t = \frac{6}{\pi}F_{m1}\left[F_{m1}\sin(x-\omega t)+\frac{1}{5}k_{m5}\sin(5x-\omega t)-\frac{1}{7}k_{m7}\sin(7x-\omega t)+\cdots\right] \tag{17.11}$$

where k_{m5} and k_{m6} are harmonic winding factors.

The MMF has a constant fundamental, and harmonics of the order of 5, 7, 11, 13,..., or $6m \pm 1$, where m is any positive integer. The third harmonic and its multiples (triplen harmonics) are absent, though in practice some triplen harmonics are produced. The harmonic flux components are affected by phase spread, fractional slotting, and coil span. The pointed curve is obtained when $\sigma = 60°$ and $\omega t = 0$. The flat-topped curve is obtained when $\omega t = \pi/6$.

17.4.1 Cogging and Crawling of Induction Motors

Parasitic magnetic fields are produced in an induction motor due to harmonics in the MMF originating from

- Windings
- Certain combination of rotor and stator slotting
- Saturation
- Air gap irregularity
- Unbalance and harmonics in the supply system voltage

The harmonics move with a speed reciprocal to their order, either with or against the fundamental. Harmonics of the order of $6m+1$ move in the same direction as the fundamental field while those of $6m-1$ move in the opposite direction.

These harmonics can be considered to produce, by an additional set of rotating poles, rotor EMFs, currents, and harmonic torques akin to the fundamental frequency; but synchronous speeds depending upon the order of the harmonics. The resultant speed torque curve will be then a combination of the fundamental and harmonic torques. This produces a saddle in the torque speed characteristics and the motor can crawl at the lower speed of 1/7th of the fundamental.

This harmonic torque can be augmented by stator and rotor slotting. In n-phase winding, with g slots per pole per phase, EMF distribution factors of the harmonics are

$$n = 6Ag \pm 1 \tag{17.12}$$

where A is any integer, 0, 1, 2, 3, ... The harmonics of the order $6Ag+1$ rotate in the same direction as the fundamental, while those of order $6Ag-1$ rotate in the opposite direction.

Consider 24 slots in the stator of a four-pole machine. Then $g=2$ and 11th and 13th harmonics will be produced strongly. The harmonic induction torque thus produced can be augmented by the rotor slotting. Consider a rotor with 44 slots. The 11th harmonic has 44 half waves, each corresponding to a rotor bar. This will accentuate 11th harmonic torque and produce strong vibrations.

If the numbers of stator slots are equal to the number of rotor slots, the motor may not start at all, a phenomenon called cogging.

The rotor MMF has a harmonic content and with a certain combination of the stator and rotor slots, it is possible to get a stator and rotor harmonic torque producing a harmonic *synchronizing* torque. There will be a tendency to develop sharp synchronizing torque at some sub-synchronous speed.

The rotor slotting will produce harmonics of the order of

$$n = \frac{S_2}{p} \pm 1 \tag{17.13}$$

where S_2 is the number of rotor slots. Here, the plus sign means rotation with the machine. Consider a motor with $S_1 = 24$ and with $S_2 = 28$. The stator produces reversed 11th harmonic (reverse going) and 13th harmonic (forward going). The rotor develops a reversed 13th and forward 15th harmonic. The 13th harmonic is produced both by stator and rotor but of opposite rotation. The synchronous speed of the 13th harmonic is 1/13 of the fundamental synchronous speed. Relative to rotor it becomes

$$-\frac{(n_s - n_r)}{13} \tag{17.14}$$

where

n_s is the synchronous speed
n_r is the rotor speed

The rotor, therefore, rotates its own 13th harmonic at a speed of

$$-\frac{(n_s - n_r)}{13} + n_r \tag{17.15}$$

relative to the stator. The stator and rotor 13th harmonic fall into step when

$$+\frac{n_s}{13} = -\frac{(n_s - n_r)}{13} + n_r \tag{17.16}$$

This gives $n_r = n_s/7$, i.e., torque discontinuity is produced not by 7th harmonic but by 13th harmonic in the stator and rotor rotating in opposite directions.

The harmonic torques are avoided in the design of machines by proper selection of the rotor and stator slotting.

17.5 Tooth Ripples in Electrical Machines

Tooth ripples in electrical machinery are produced by slotting as these affect air-gap permeance. Figure 17.2 shows ripples in the air-gap flux distribution (exaggerated) because of variation in gap permeance. The frequency of flux pulsations corresponds to the rate at which slots cross the pole face, i.e., it is given by $2gf$, where g is the number of slots per pole and f is the system frequency. This stationary pulsation may be regarded as two waves of fundamental space distribution rotating at angular velocity $2g\omega$ in forward and backward directions. The component fields will have velocities of $(2g \pm 1)\omega$ relative to the armature

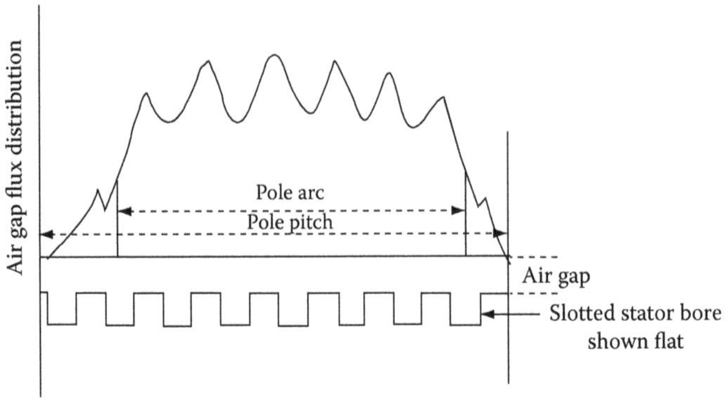

FIGURE 17.2
Gap flux distribution due to tooth ripples.

winding and will generate harmonic EMFs of frequencies $(2g \pm 1)f$ cycles per second. However, this is not the main source of tooth ripples. Since the ripples are due to slotting, these do not move with respect to conductors. Therefore, these cannot generate an EMF of pulsation. With respect to the rotor, the flux waves have a relative velocity of $2g\omega$ and generate EMFs of $2gf$ frequency. Such currents superimpose an MMF variation of $2gf$ on the resultant pole MMF. These can be again resolved into forward and backward moving components with respect to the *rotor*, and $(2g \pm 1)\omega$ with respect to the *stator*. Thus, stator EMFs at frequencies $(2g \pm 1)f$ are generated, which are the principal tooth ripples.

17.6 Synchronous Generators

The terminal voltage wave of synchronous generators must meet the requirements of NEMA, which states that the deviation factor of the open line-to-line terminal voltage of the generator shall not exceed 0.1.

Figure 17.3 shows a plot of a hypothetical generated wave, superimposed on a sinusoid, and the deviation factor is defined as

$$F_{\mathrm{DEV}} = \frac{\Delta E}{E_{\mathrm{OM}}} \tag{17.17}$$

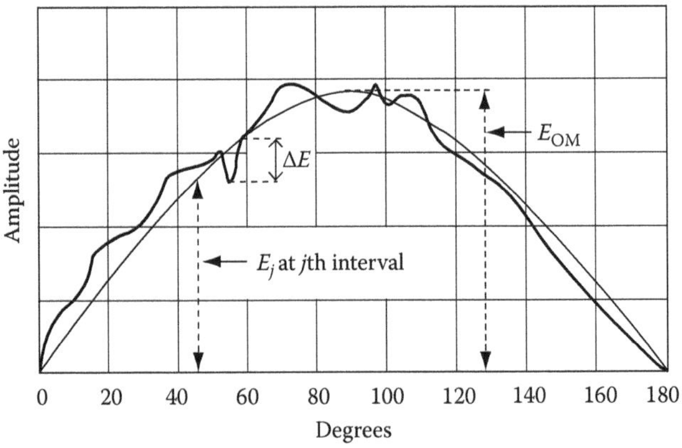

FIGURE 17.3
Measurements of deviation factor of a generator voltage.

where E_{OM} is calculated from a number of samples of instantaneous values:

$$E_{OM} = \sqrt{\frac{2}{J}\sum_{j=1}^{J} E_j^2} \tag{17.18}$$

The deviation from a sinusoid is very small.

Generator neutrals have predominant third harmonic voltages. In a wye-connected generator, with the neutral grounded through high impedance, the third harmonic voltage for a ground fault increases toward the neutral, while the fundamental frequency voltage decreases. The third harmonic voltages at line and neutral can vary considerably with load.

17.7 Transformers

Harmonics in transformers originate as a result of saturation, switching, high-flux densities, and winding connections. The following summarizes the main factors with respect to harmonic generation:

1. For economy in design and manufacture, transformers are operated close to the knee point of saturation characteristics of magnetic materials. Figure 17.4 shows a *B–H* curve and the magnetizing current waveform. A sinusoidal flux wave, required by sinusoidal applied voltage, demands a magnetizing current with a harmonic content. Conversely, with a sinusoidal magnetizing current, the induced EMF is peaky and the flux is flat topped.

   An explanation of the generation of the peaky magnetizing current considering the third harmonic is provided in Figure 17.5. A sinusoidal EMF E_a. generates a sinusoidal current flow, I_a, in lagging phase quadrature with E_a. These set up a flat-topped flux wave, ϕ_1, which can be resolved into two components: ϕ_α the fundamental flux wave, and ϕ_3 the third harmonic flux wave (higher harmonics are

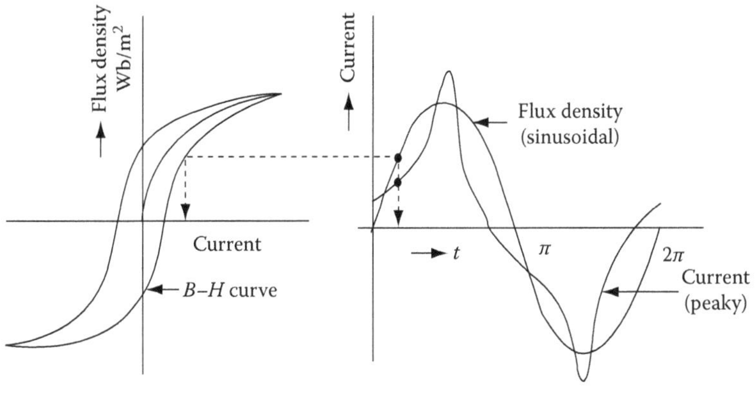

FIGURE 17.4
B–H curve of magnetic material and peaky transformer magnetizing current.

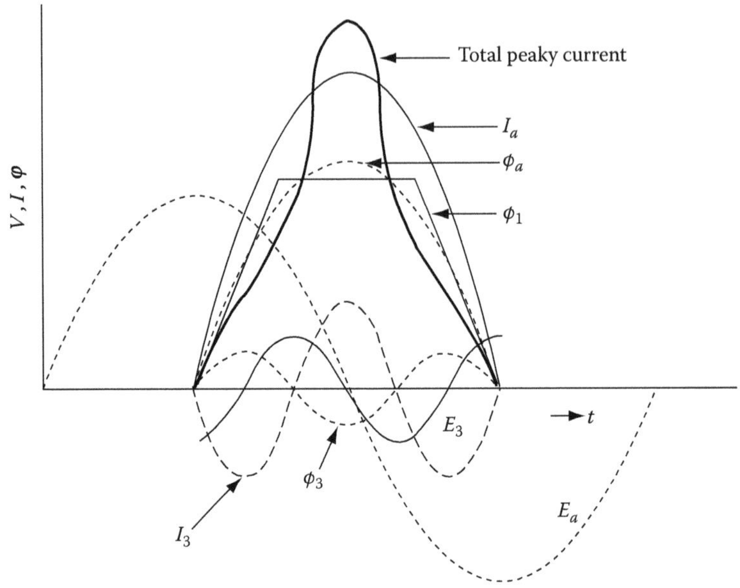

FIGURE 17.5
Origin of flat-topped flux wave in a transformer, third-harmonic current, and overall peaky magnetizing current.

neglected). The third harmonic flux can be supposed to produce a third harmonic EMF E_3 and a corresponding third harmonic current I_3, which when summed with I_a makes the total current peaky.

2. In a system of three-phase balanced voltages the 5th, 7th, 11th,... produce voltages displaced by 120° mutually, while the triplen harmonic voltages are cophasial. If the impedance to the third harmonic is negligible, only a very small third harmonic EMF is required to circulate a magnetizing current additive to the fundamental frequency, so as to maintain a sinusoidal flux. This is true if the transformer windings are delta connected. In wye–wye-connected transformers with isolated neutrals, as all the triplen harmonics are either directed inward or outward, these cancel between the lines, no third harmonic currents flow, and the flux wave in the transformer is flat topped. The effect on a wye-connected point is to make it oscillate at three times the fundamental frequency, giving rise to distortion of the phase voltages (Figure 17.6). Tertiary delta-connected windings are included in wye–wye-connected transformers for neutral stabilization.

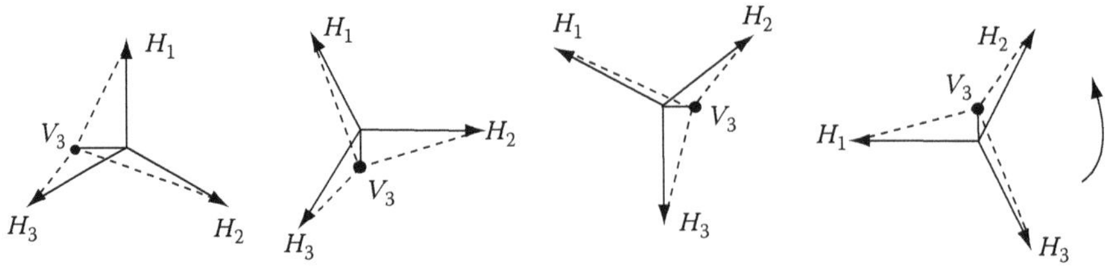

FIGURE 17.6
Phenomena of neutral oscillation in wye–wye-connected transformer, due to third-harmonic voltages.

3. Three-phase core-type transformers have magnetically interlinked phases, and the return paths of triplen harmonic fluxes lie outside the core, through the tank and transformer fluid, which have high reluctance. In five-limb transformers, the end limbs provide return paths for triplen harmonics.

   It can be said that power transformers generate very low levels of harmonic currents in steady-state operation, and the harmonics are controlled by design and transformer winding connections. The higher-order harmonics, i.e., the fifth and seventh, may be less than 0.1% of the transformer full-load current.

4. Energizing a power transformer does generate a high order of harmonics including a dc component. Figure 17.7 shows three conditions of energizing of a power transformer: (a) the switch closed at the peak value of the voltage, (b) the switch closed at zero value of the voltage, and (c) energizing with some residual trapped flux in the magnetic core due to retentivity of the magnetic materials. Figure 17.7d shows the spectrum of magnetizing inrush current, which resembles a rectified current and its peak value may reach 8–15 times the transformer full-load current, mainly depending on the transformer size. The asymmetrical loss due to conductor and core heating rapidly reduces the flux wave to symmetry about the time axis and typically the inrush currents last for a short duration (0.1 s).

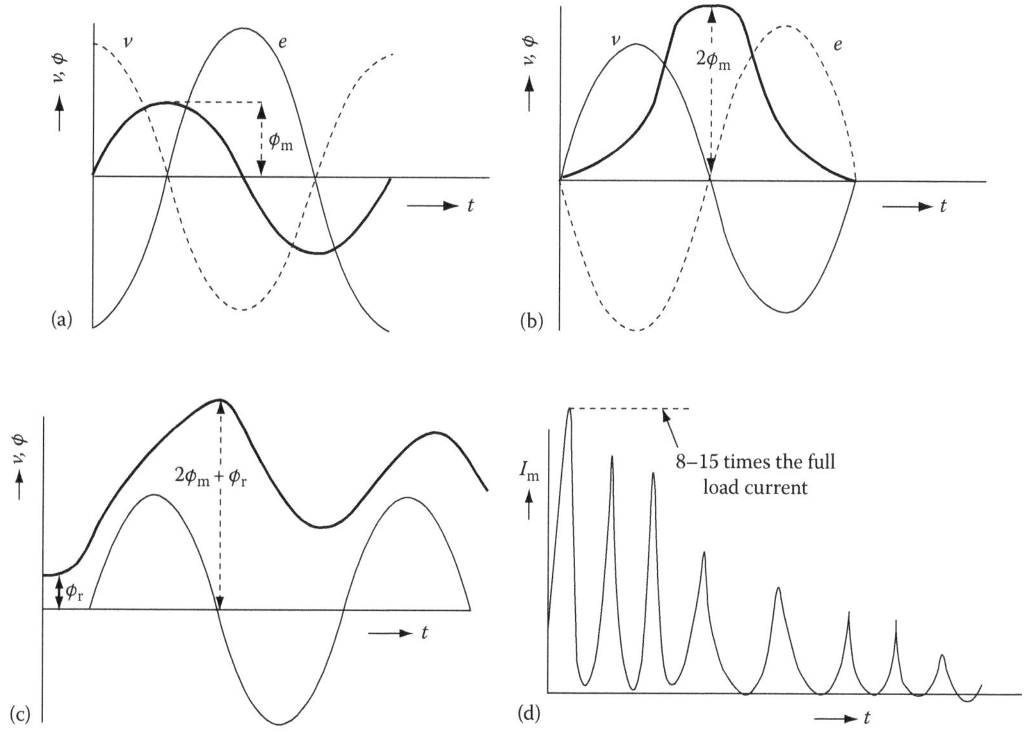

FIGURE 17.7
(a) through (d) Switching inrush current transients in a transformer (see text).

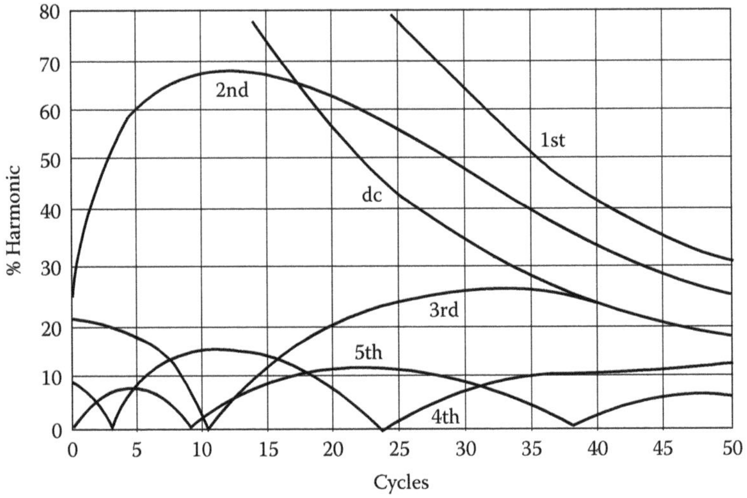

FIGURE 17.8
Harmonic components of the inrush current of a transformer.

Typical harmonics generated by the transformer inrush current are shown in Figure 17.8. Overexcitation of transformers in steady-state operation can produce harmonics. The generated fundamental frequency EMF is given by

$$V = 4.44 f T_{ph} B_m A_c \tag{17.19}$$

where
- T_{ph} is the number of turns in a phase
- B_m is the flux density (consisting of fundamental and higher-order harmonics)
- A_c is the area of core

Thus, the factor V/f is a measure of the overexcitation, though these currents do not normally cause a wave distortion of any significance. Exciting currents increase rapidly with voltage, and transformer standards specify application of 110% voltage without overheating the transformer. Under certain system upset conditions, the transformers may be subjected to even higher voltages and overexcitation. ANSI protective device number 24, volts per hertz relay is used for overexcitation protection.

17.8 Saturation of Current Transformers

Saturation of current transformers under fault conditions produces harmonics in the secondary circuits. Accuracy classification of current transformers is designated by one letter, C or T, depending on current transformer construction [2]. Classification C covers bushing-type transformers with uniformly distributed windings, and the leakage flux has a negligible effect on the ratio within the defined limits. A transformer with relaying accuracy class C200 means that the percentage ratio correction will not exceed 10% at

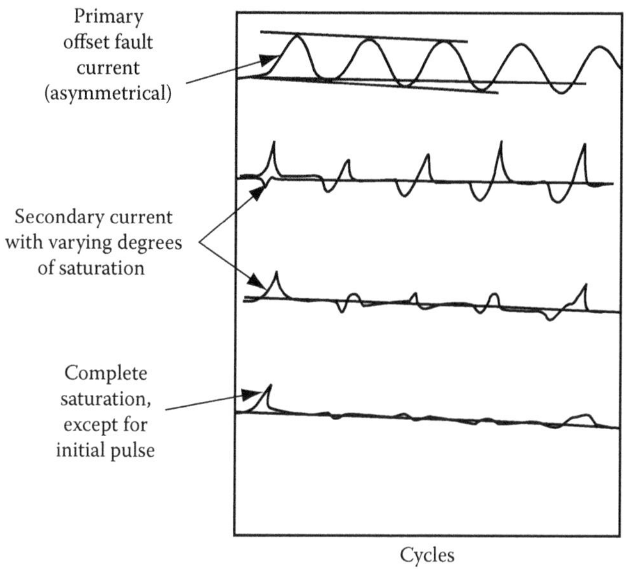

FIGURE 17.9
Saturation of a current transformer on asymmetrical fault current and origin of harmonics.

any current from 1 to 20 times the rated secondary current at a standard burden of 2.0 ohm, which will generate 200 V. The secondary voltage, as given by maximum fault current reflected on the secondary side multiplied by connected burden $(R+jX)$, should not exceed the assigned C accuracy class. When current transformers are improperly applied, saturation can occur, as shown in Figure 17.9 [3]. A completely saturated CT does not produce a current output, except during the first pulse, as there is a finite time to saturate and desaturate. The *transient* performance should consider the dc component of the fault current and its X/R ratio as it has far more effect in producing severe saturation of the current transformer.

As the CT saturation increases, so does the secondary harmonics, before the CT goes into a completely saturated mode. Harmonics of the order of 50% third, 30% fifth, 18% seventh, and 15% ninth and higher order may be produced. These can cause improper operation of the protective devices. This situation can be avoided by proper selection and application of current transformers.

17.9 Shunt Capacitors

High frequencies of inrush currents on switching of shunt capacitors were discussed in Chapter 5 and these are given by Equations 5.18 and 5.20 for single bank and back-to-back switching. The frequency of the system transient is typically less than 1 kHz for an isolated capacitor bank and less than 5 kHz for back-to-back switching. Series filter reactors and switching inrush current limiting reactors reduce these frequencies. The fast front of the switching surge voltage can cause a part-winding resonance and harmonic generation in a transformer, if the frequency coincides with the transformer's natural frequency, which is of the order of 10–100 kHz, the first resonance occurring in the range 7–15 kHz. The likelihood of exciting a part-winding resonance on switching is remote, but switching overvoltages are of major concern (see Chapter 18).

The power capacitors do not generate harmonics by themselves, but are the main cause of amplification of the harmonics due to resonance and increased harmonic distortion. These can also reduce harmonic distortion, when applied as filters. This important aspect of power capacitor is discussed in the chapters to follow.

17.10 Sub-Harmonic Frequencies

Chapter 13 shows that series compensation of transmission lines with capacitors can generate sub-harmonic frequencies and how these can be damped with some FACTS devices. Switching of long lines close to a generator can cause oscillations. Power oscillations can be described as

- *Inter-area mode oscillations*: These oscillations occur between one set of machines swinging against other set of machines in a different area of the transmission system. The oscillations are typically in the range 0.2–0.5 Hz.
- *Local mode oscillations*: These oscillations occur between one or more machines in a plant swinging against a large power source or network. The oscillations are typically in the range 0.7–2.0 Hz.
- *Interunit mode oscillations*: These oscillations occur when one machine swings against another machine in the same area in the same power plant. These oscillations are typically in the range 1.5–3.0 Hz.

Power system stabilizers in the excitation systems of the machines are used to stabilize the oscillations.

It can, generally, be said that the harmonics in the power systems from sources other than nonlinear loads are comparatively small, though these cannot always be ignored. Major sources of harmonies are nonlinear loads [4].

17.11 Static Power Converters

The primary sources of harmonics in the power system are power converters, rectifiers, inverters, and adjustable speed drives. The *characteristic* harmonics are those produced by the power electronic converters during normal operation and these harmonics are integer multiples of the fundamental frequency of the power system. *Noncharacteristic* harmonics are usually produced by sources other than power electronic equipment and may be at frequencies other than the integer multiple of the fundamental power frequency. The converters do produce some noncharacteristic harmonics, as ideal conditions of commutation and control are not achieved in practice. The ignition delay angles may not be uniform, and there may be unbalance in the supply voltages and the bridge circuits.

17.11.1 Single-Phase Bridge Circuit

The single-phase rectifier full-bridge circuit of Figure 17.10 is first considered. It is assumed that there is no voltage drop or leakage current, the switching is instantaneous, the voltage source is sinusoidal, and the load is resistive. For full wave conduction, the waveforms of input and output currents are then as shown in Figure 17.10b and c. The *average* dc current is

$$I_{dc} = \frac{2}{2\pi}\int_0^{\pi}\frac{E_m}{R}\sin\omega t\, d\omega t = \frac{2E_m}{\pi R} \tag{17.20}$$

and the rms value or the effective value of the output current, including all harmonics, is

$$I_{rms} = \sqrt{\frac{2}{2\pi}\int_0^{\pi}\left(\frac{E_m}{R}\right)^2\sin^2\omega t\, d\omega t} = \frac{E_m}{\sqrt{2}R} \tag{17.21}$$

The *input current has no harmonics*. The average dc voltage is given by

$$E_{dc} = \frac{2E_m}{\pi} \tag{17.22}$$

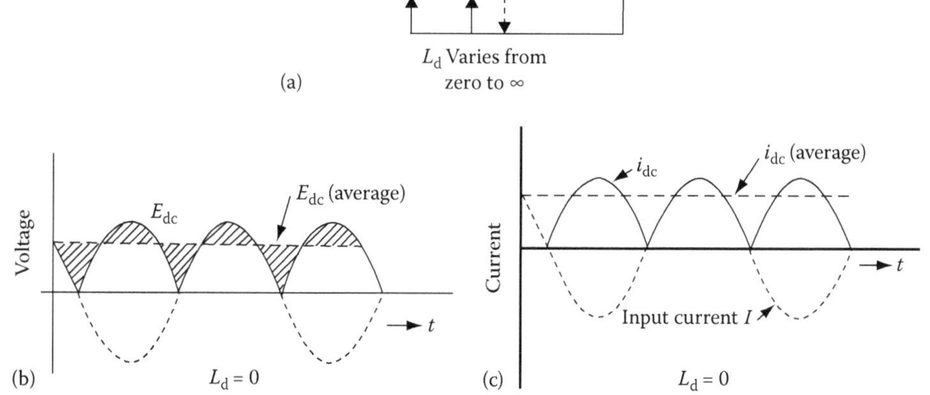

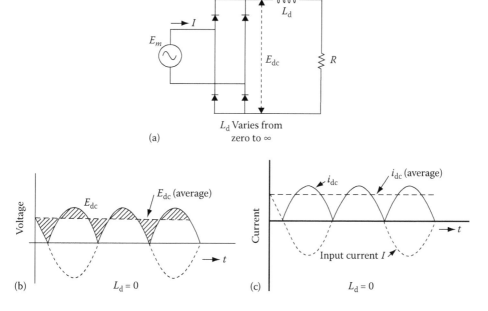

FIGURE 17.10
(a) A single-phase full rectifier bridge circuit, with resistive load; (b) and (c) waveforms with zero dc reactor.

The output ac power is defined as

$$P_{ac} = E_{rms}I_{rms} = \frac{(0.707E_{rms})^2}{R} \tag{17.23}$$

where E_{rms} considers the effect of harmonics on the output. The dc output power is

$$P_{dc} = E_{dc}I_{dc} = \left(\frac{2E_m}{\pi}\right)\left(\frac{2E_m}{\pi R}\right) = \frac{0.405E_m^2}{R} \tag{17.24}$$

The efficiency of rectification is given by P_{dc}/P_{ac} (81%). The *form factor* is a measure of the shape of the output voltage or current and it is defined as

$$FF = \frac{I_{rms}}{I_{dc}} = 1.11 \tag{17.25}$$

The *ripple factor*, which is a measure of the ripple content of the output current or voltage, is defined as the rms value of output voltage or current, including all harmonics, divided by the average value:

$$RF = \sqrt{\left(\frac{I_{rms}}{I_{dc}}\right)^2 - 1} = \sqrt{FF^2 - 1} \tag{17.26}$$

For the single-phase bridge circuit with resistive load, the ripple factor is

$$RF = \sqrt{\left(\frac{I_{rms}}{I_{dc}}\right)^2 - 1} = 0.48 \tag{17.27}$$

This shows that the ripple content of the dc output voltage is high (Figure 17.10b). This is not acceptable even for the simplest of applications. Let a series reactor be added in the dc circuit. The load current is no longer a sine wave but the average current is still equal to $2E_m/\pi R$. The ac line current is no longer sinusoidal, but approximates a poorly defined square wave with superimposed ripples (Figure 17.11a and b). The inductance has reduced the harmonic content of the load current by increasing the harmonic content of the ac line current. When the inductance is large, the ripple across the load is insignificant, and can be assumed constant, and the ac current wave is now a square wave (Figure 17.11c and d).

17.11.1.1 Phase Control

A silicon-controlled rectifier (SCR) can be turned on by applying a short pulse to its gate and turned off due to natural or line commutation. The term thyristor pertains to the family of semiconducting devices for power control. The angle by which the conduction is delayed after the input voltage starts to go positive until the thyristor is fired is called the delay angle. Figure 17.12b shows waveforms with a large dc reactor, and Figure 17.12c shows waveform with no dc reactor but identical firing angle. Thyristors 1 and 2 and 3 and 4 are fired in pairs as shown in Figure 17.12b. Even when the polarity of the voltage is reversed, the current keeps flowing in thyristors 1 and 2 until thyristors 3 and 4 are fired,

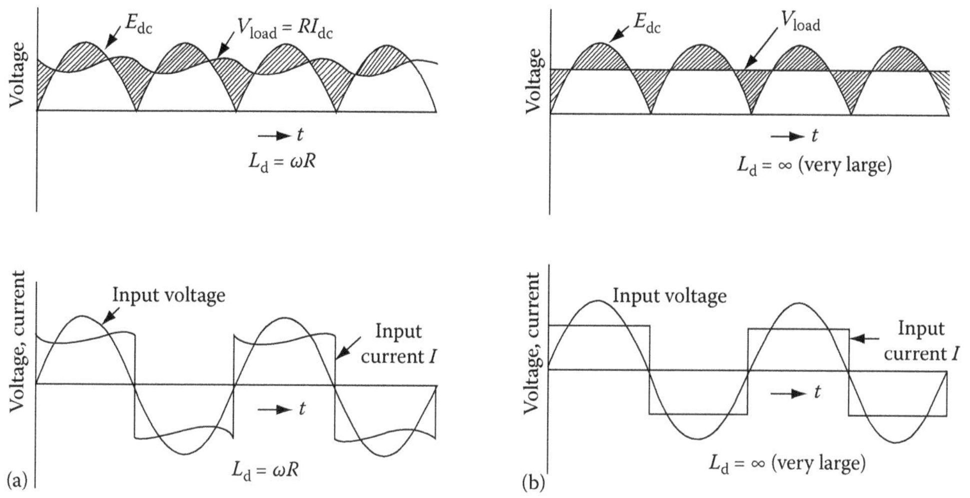

FIGURE 17.11
(a) Waveforms of a single-phase full rectifier bridge with small dc output reactor; (b) with large dc output reactor.

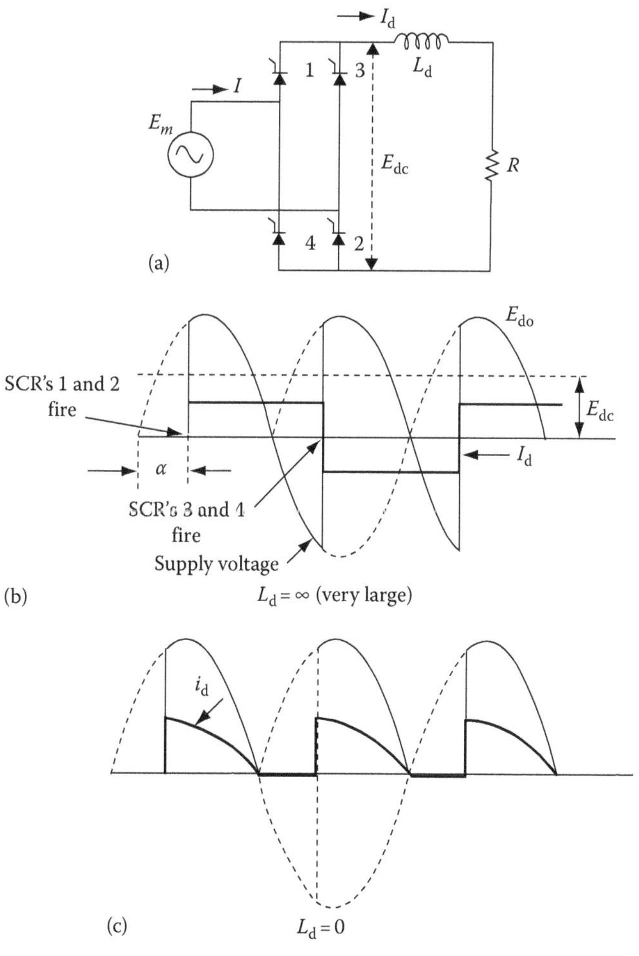

FIGURE 17.12
(a) Circuit of a single-phase fully controlled bridge; (b) and (c) waveforms with large dc reactor and with zero dc reactor.

Figure 17.12a. The firing of thyristors 3 and 4 reverse biases thyristors 1 and 2 and turns them off (this is referred to as class F type forced commutation or line commutation). The average dc voltage is

$$E_{dc} = \frac{2}{2\pi} \int_{\alpha}^{\pi+\alpha} E_m \sin \omega t \, d\omega(\omega t) = \frac{2E_m}{\pi} \cos \alpha \tag{17.28}$$

and the Fourier analysis of the rectangular current wave in Figure 17.12b gives

$$\begin{aligned} a_h &= -\frac{4I_a}{h\pi} \sin h\alpha, \quad h = 1,3,5,\ldots \\ &= 0 \quad h = 2,4,6,\ldots \end{aligned} \tag{17.29}$$

$$\begin{aligned} b_h &= \frac{4I_a}{h\pi} \cos h\alpha \quad h = 1,3,5,\ldots \\ &= 0 \quad h = 2,4,6,\ldots \end{aligned} \tag{17.30}$$

Since

$$I = \sum_{h=1,2,\ldots}^{\infty} [a_h \cos(h\omega t) + b_h \sin(h\omega t)] \tag{17.31}$$

The rms input current is given by

$$I = \frac{4}{\pi} I_d \left[\sin(\omega t - \alpha) + \frac{1}{3} \sin 3(\omega t - \alpha) + \frac{1}{5} \sin 5(\omega t - \alpha) + \cdots \right] \tag{17.32}$$

Triplen harmonics are present. Figure 17.13 shows harmonics as a function of the delay angle for a resistive load. The overlap angle (defined further) decreases the magnitude of the harmonics. When the output reactor is small, the current goes to zero, the input current wave is no longer rectangular, and the line harmonics increase.

17.11.1.2 Power Factor, Distortion Factor, and Total Power Factor

For sinusoidal voltages and currents, the power factor is defined as kW/kVA and the power factor angle ϕ is

$$\phi = \cos^{-1} \frac{\text{kW}}{\text{kVA}} = \tan^{-1} \frac{\text{kvar}}{\text{kW}} \tag{17.33}$$

The power factor of a converter is made up of two components: displacement and distortion. The effect of the two is combined in total power factor. The displacement component is the ratio of active power of the fundamental wave in watts to apparent power of fundamental wave in volt-amperes. This is the power factor as seen by the watt-hour and var-hour meters. The distortion component is that part associated with harmonic voltages and currents.

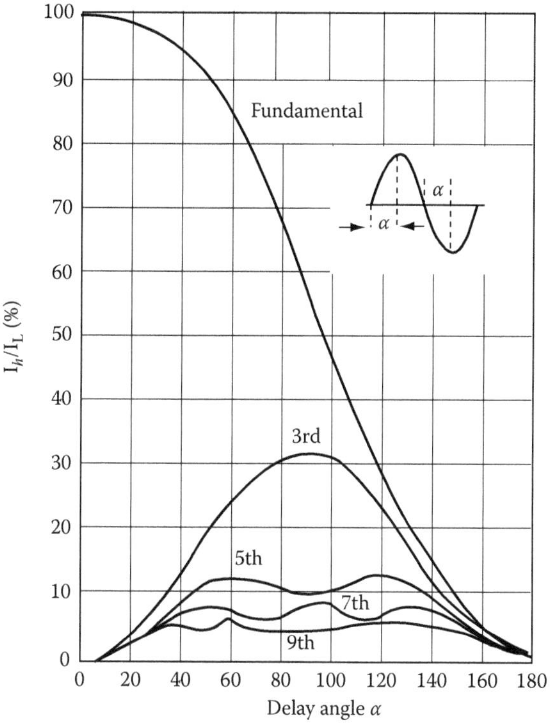

FIGURE 17.13
Harmonic generation as a function of phase-angle control, of delay angle, and of resistive load.

$$PF_t = PF_f \times PF_{\text{distortion}} \tag{17.34}$$

At fundamental frequency the displacement power factor will be equal to the total power factor, as the displacement power factor does not include kVA due to harmonics, while the total power factor does include it. For harmonic generating loads the total power factor will always be less than the displacement power factor. The discussion is continued in Section 19.7.

The fundamental input power factor angle is equal to the firing angle α. For the single-phase bridge circuit the input active and reactive power is

$$\text{Active power} = \frac{4}{2\pi} I_d E_m \cos\alpha \tag{17.35}$$

$$\text{Reactive power} = \frac{4}{2\pi} I_d E_m \sin\alpha \tag{17.36}$$

The power factor becomes depressed for large firing angles. This is the case whenever large phase control is used in the converter circuits. The maximum reactive power input for a half-controlled bridge will be one-half of that of a fully controlled bridge. The reactive power requirements of converters becomes important in many installations and this can be limited by limiting amount of phase control, reducing reactance of converter transformers, which limits μ and sequential control of converters, which has been popular in HVDC transmission systems. In the sequential control two or more converter sections can operate in series, with one section fully phased on and the other sections adding or subtracting from the voltage of the first section (Figure 17.14).

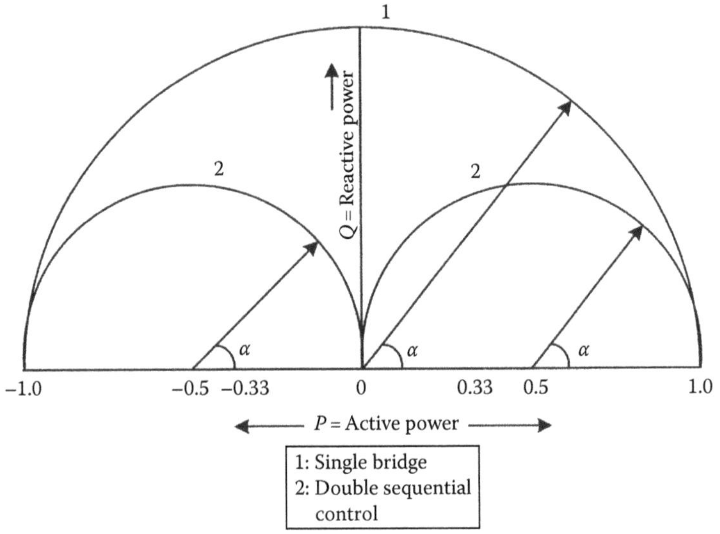

FIGURE 17.14
Reduction in reactive power with sequential control of converters.

Where controlled reactive power supply is not required, the reactive power consumed by the converters can be supplied by shunt capacitors and filters. In these cases, line commutated converters are economical and for HVDC transmission, line commutated converters have been used exclusively [5]. The thyristor-based converters become economical in large power handling capabilities as per device basis, the thyristors can handle two to three times more power than GTOs (gate turn-off thyristors), IGCTs (integrated gate bipolar transistors), and MTOs (MOS turn-off thyristors). There are variations in the current source converters like resonant converters, hybrid converters, and artificial commutation converters, which are not discussed. New converter topologies are being developed. With GTOs, the forced commutation can improve the power factor and reduce the input harmonic levels. The forced commutation techniques are

- Extinction angle control
- Pulse-width modulation, and sinusoidal pulse-width modulation (see section 17.17)
- Symmetrical angle control

Reference [6] describes a pulse-width modulation (PWM) converter with sinusoidal ac currents and minimum filter requirements. A full range four-quadrant operation is described, with control of input power factor. Reference [7] describes experimental test results on a three-phase bipolar transistor controlled current PWM with leading power factor. See Chapter 20 for some near-to-unity power factor converter topologies [8].

17.11.1.3 Harmonics on Output Side

For a single phase 2 pulse circuit in Figure 17.10a the output waveform (on the dc side) contains even harmonics of the input frequency and the Fourier expansion is

$$E_{d0} = E_{dc} + e_2 \sin 2\omega t + e_2' \cos 2\omega t + e_4 \sin 4\omega t + e_4' \cos 4\omega t + \cdots \quad (17.37)$$

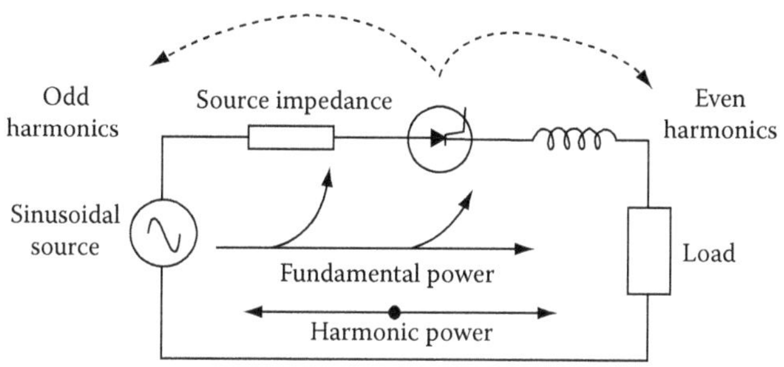

FIGURE 17.15
Single-phase 2 pulse controlled rectifier as a source of harmonic generation and harmonic power flow.

where e_m and $e'_m(m = 2, 4, 6, \ldots)$ are given by

$$e_m = \frac{2E_m}{\pi}\left[\frac{\sin(m+1)\alpha}{m+1} - \frac{\sin(m-1)\alpha}{m-1}\right] \tag{17.38}$$

$$e'_m = \frac{2E_m}{\pi}\left[\frac{\cos(m+1)\alpha}{m+1} - \frac{\cos(m-1)\alpha}{m-1}\right] \tag{17.39}$$

The converter can be considered as a harmonic current source; the even harmonics go into the load and the odd harmonics into the supply source. The harmonics fed into the supply system propagate into the power system. These can either be magnified or attenuated and are the subject of study in this book. The harmonics in the load circuit have an adverse effect on the loads, but are, generally, localized to the loads to which these connect. There is harmonic power associated with harmonic currents, which is a function of relative system impedances and load side impedance. Figure 17.15 shows this action. For a three-phase fully controlled 6-phase bridge circuit (converter) described in Section 17.11.2 (Figure 17.16a) only triplen harmonics of the input frequency are present in the output dc voltage. The harmonic order will be 6th, 12th, 18th...

If we replace the controlled rectifiers in the bottom half of Figure 17.16a with diodes, it is three-phase half controlled 3 pulse bridge circuit. Then the harmonics are 3rd, 6th, 9th...

These harmonics vary depending upon the firing angle α and overlap angle μ. The overlap angle is discussed in section 17.11.2.2.

It follows that as the pulse number of the converter increases, the ripple content in dc output voltage reduces.

17.11.2 Three-Phase Bridge Circuit

The three-phase fully controlled bridge is described, as it is most commonly used.

Figure 17.16a shows a three-phase fully controlled bridge circuit, and Figure 17.16b shows its current and voltage waveforms. The firing sequence of thyristors is shown in Table 17.2. At any time two thyristors are conducting. The firing frequency is six times the fundamental frequency and the firing angle can be measured from point O shown in Figure 17.16b. With a large output reactor, the output dc current is continuous and the input current is a rectangular pulse of $2\pi/3$ duration and amplitude i_d. The average dc voltage is

$$E_d = 2\left[\frac{3}{2\pi}\int_{-\pi/3+\alpha}^{\pi/3+\alpha} E_m \cos\omega t \, d(\omega t)\right] = \frac{3\sqrt{3}}{\pi}E_m \cos\alpha \tag{17.40}$$

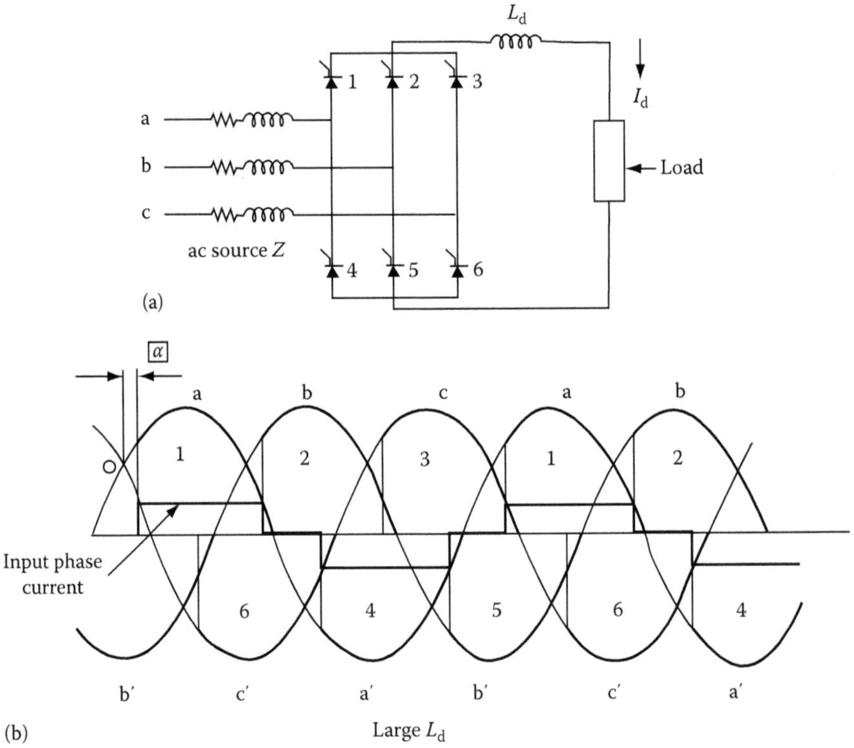

FIGURE 17.16
(a) Circuit of a three-phase fully controlled bridge, with large dc output reactor; (b) voltage and current waveforms for a certain delay angle α.

TABLE 17.2
Firing Sequence of Thyristors in Six-Pulse Full Converter

Conducting thyristors	5, 3	1, 5	6, 1	2, 6	4, 2	3, 4
Thyristor to be fired	1	6	2	4	3	5
Thyristor turning off	3	5	1	6	2	4

where E_m is the peak value of line to neutral voltage. For firing angles $>\pi/2$, the circuit can work as an inverter, i.e., dc power is fed back into the ac system. This requires that a dc source with opposite polarity is connected at the output. The power factor is lagging for rectifier operation and leading for inverter operation.

Figure 17.17 shows a connection diagram and waveforms for a three-phase fully controlled bridge, with delta–delta connection of the rectifier transformer and firing angle $\alpha = 0$; the input current is rectangular and its Fourier analysis gives

$$i_a = \frac{2\sqrt{3}}{\pi} I_d \left[\cos\omega t - \frac{1}{5}\cos 5\omega t + \frac{1}{7}\cos 7\omega t - \frac{1}{11}\cos 11\omega t + \frac{1}{3}\cos 13\omega t\right] \tag{17.41}$$

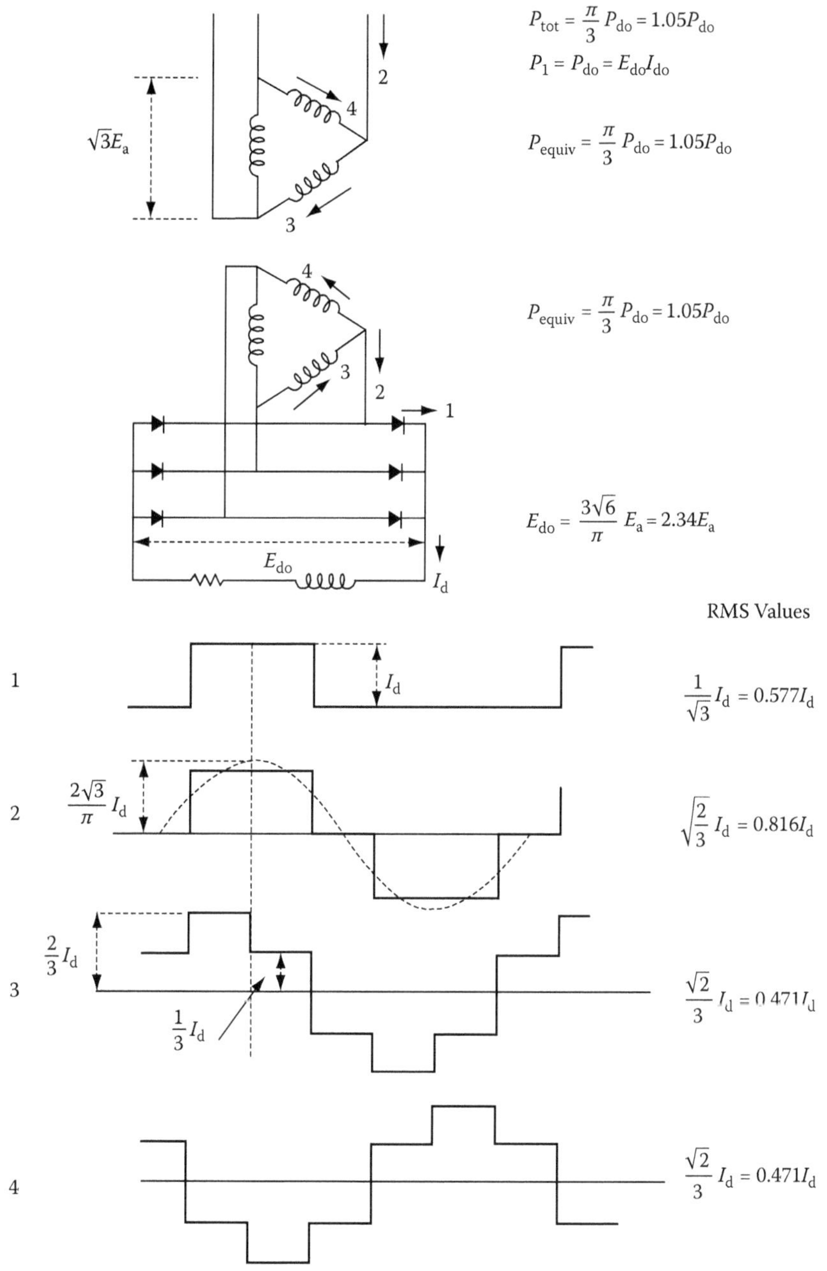

FIGURE 17.17
A six-pulse bridge circuit, zero delay angle, large output reactor, and delta–delta input transformer. Voltage and current relations.

Thus, maximum fundamental frequency current:

$$\frac{2\sqrt{3}}{\pi} I_{\mathrm{d}} \text{ peak} = \frac{\sqrt{6}}{\pi} I_{\mathrm{d}} \text{ rms}$$

Figure 17.18 shows a similar connection diagram and waveforms for delta–wye rectifier transformer connections. The input current is stepped and the resulting Fourier series for the current waveform is

$$i_a = \frac{2\sqrt{3}}{\pi} I_{\mathrm{d}} \left(\cos \omega t + \frac{1}{5} \cos 5\omega t - \frac{1}{7} \cos 7\omega t - \frac{1}{11} \cos 11\omega t + \frac{1}{13} \cos 13\omega t - , \ldots \right) \quad (17.42)$$

From these equations, the following observations can be made:

1. The line harmonics are on the order

$$h = pm \pm 1, \quad m = 1, 2, \ldots \quad (17.43)$$

   where p is the pulse number. The pulse number is defined as the total number of successive non-simultaneous commutations occurring within the converter circuit during each cycle when operating without phase control. This relationship also holds for a single-phase bridge converter, as the pulse number for a single-phase bridge circuit is 2. The harmonics given by Equation 17.43 are an integer of the fundamental frequency and are called characteristic harmonics, while all other harmonics are called noncharacteristic.
2. The triplen harmonics are absent. This is because an ideal rectangular wave shape and instantaneous transfer of current at the firing angle are assumed. In practice, some noncharacteristic harmonics are also produced.
3. The rms magnitude of the nth harmonic is I_f/h, i.e., the fifth harmonic is a maximum of 20% of the fundamental:

$$I_h = \frac{I_f}{h} \quad (17.44)$$

4. Fourier series of the input current is given by Equation 17.31

$$\begin{aligned} a_n &= -\frac{4I_{\mathrm{d}}}{h\pi} \sin\frac{n\pi}{3} \sin(h\alpha) \quad h = 1, 3, 5 \\ &= 0 \quad h = 2, 4, 6 \\ b_n &= -\frac{4I_{\mathrm{d}}}{h\pi} \sin\frac{h\pi}{3} \cos(h\alpha) \quad h = 1, 3, 5 \\ &= 0 \quad h = 2, 4, 6 \end{aligned} \quad (17.45)$$

   Therefore,

$$I = I_h \sin(h\omega t + \phi_h) \quad \text{where } \phi_h = \tan^{-1}\frac{a_h}{b_h} = -h\alpha \quad (17.46)$$

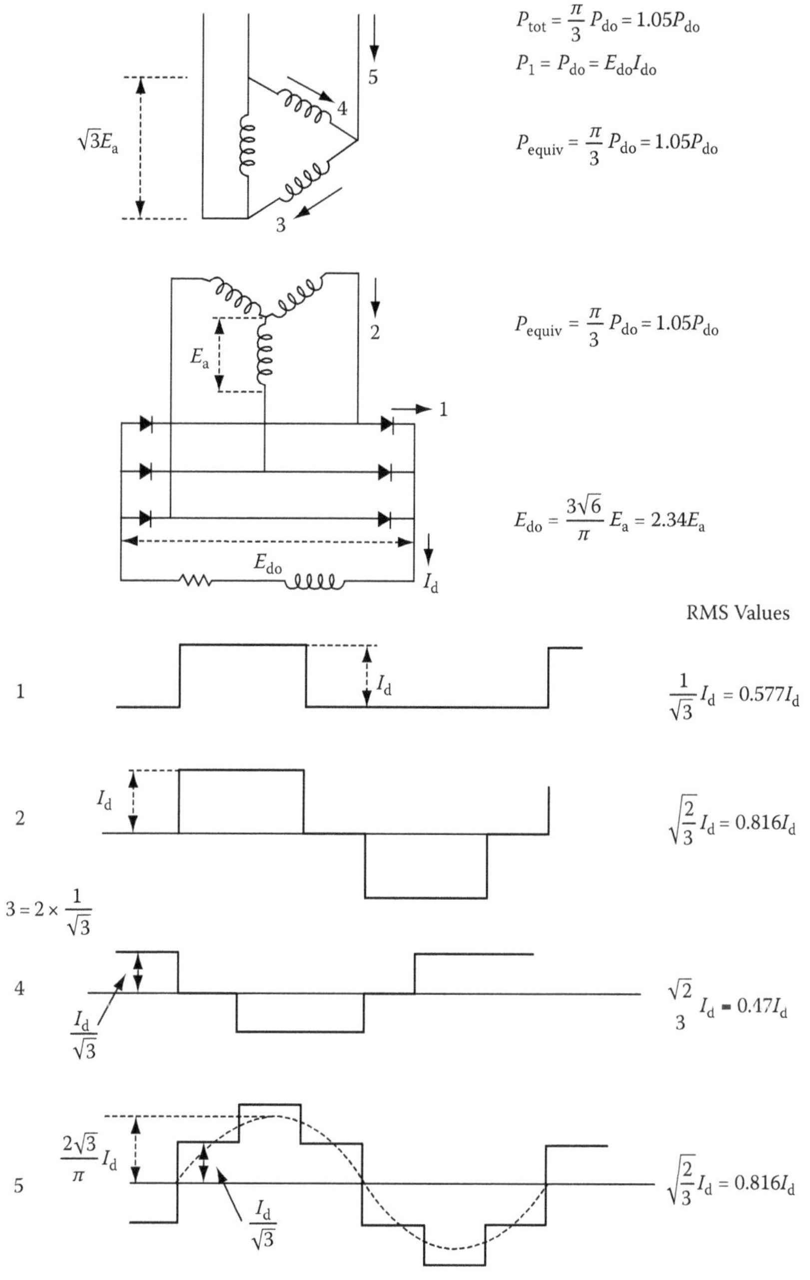

FIGURE 17.18
A six-pulse bridge circuit, zero delay angle, large output reactor, and delta–wye input transformer. Voltage and current relations.

The rms value of the nth harmonic input current is

$$I_h = \left(a_h^2 + b_h^2\right)^{1/2} = \frac{2\sqrt{2}}{h\pi}\sin\frac{h\pi}{3} \tag{17.47}$$

The rms value of the fundamental current is

$$I_1 = \frac{\sqrt{6}}{\pi}I_{\mathrm{d}} = 0.779I_{\mathrm{d}} \tag{17.48}$$

The rms input current (including harmonics) is

$$\left[\frac{2}{\pi}\int_{(-\pi/3)+\alpha}^{(\pi/3)+\alpha} I_{\mathrm{d}}^2 \mathrm{d}(\omega t)\right]^{1/2} = I_{\mathrm{d}}\sqrt{\frac{2}{3}} = 0.8165I_{\mathrm{d}} \tag{17.49}$$

5. The ripple factor of a six-pulse converter, with zero firing angle, is 0.076 and the lowest harmonic in the output of the converter is the sixth. As the pulse number of the converter increases, the ripple in the dc output voltage and the harmonic content in the input current are reduced. Also, for a given voltage and firing angle, the average dc voltage increases with the pulse number.
6. From Equation 17.49 rms current including harmonics is $0.8165I_{\mathrm{d}}$ and from Equation 17.41 the fundamental current is $0.7797I_{\mathrm{d}}$. Then the total harmonic current is

$$I_h = \left[(0.8165I_{\mathrm{d}})^2 - (0.7797I_{\mathrm{d}})^2\right]^{1/2} = 0.24I_{\mathrm{d}} \tag{17.50}$$

In the above analysis we assumed that the commutation is instantaneous. It may take some time before the current is commutated, through the inductive circuit of the ac system. This is discussed in the next section.

17.11.2.1 Cancellation of Harmonics Due to Phase Multiplication

Equations 17.41 and 17.42 show that the harmonics 5th, 7th, 17th are of opposite sign. We know that there is a 30° phase shift between the primary and secondary voltage vectors of a delta–wye transformer, while for a delta–delta- or wye–wye-connected transformer, this phase shift is 0°. If the load is equally divided on two transformers, one with delta–delta connections and the other with wye–delta or delta–wye connections, harmonics of the order of 5, 7, 17, ... are eliminated, and the system behaves like a 12-pulse circuit. This is called phase multiplication. The circuit is shown in Figure 17.19a and the waveform in the time domain in Figure 17.19b. Extending this concept, 24-pulse operation can be achieved with four transformers with 15° mutual phase shifts. As the magnitude of the harmonic is inversely proportional to the pulse number, the troublesome lower-order harmonics of larger magnitude are eliminated. This cancellation of harmonics, though, is not 100% as the ideal conditions of operation are rarely met in practice. The transformers should have exactly the same ratios and same impedances, the loads should be equally divided and converters should have exactly the same delay angle. Approximately 75% cancellation may be achieved in practice, and in harmonic analysis studies 25% residual harmonics are modeled.

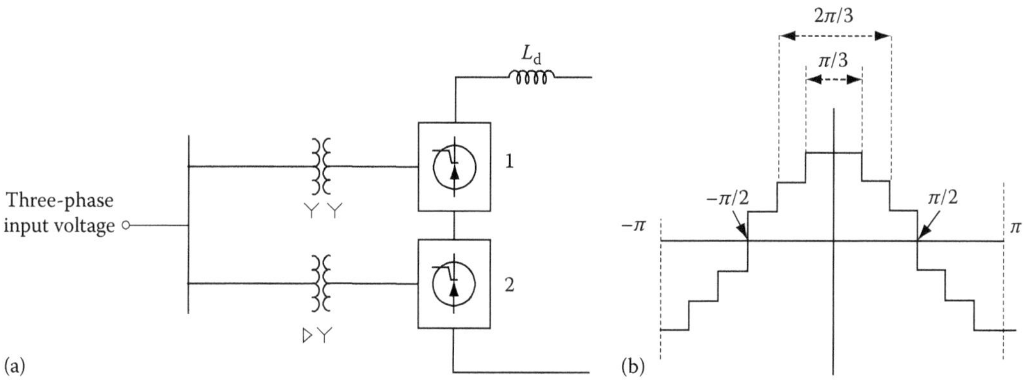

FIGURE 17.19
(a) Harmonic elimination with phase multiplication: circuit diagram; (b) input current waveform.

17.11.2.2 Effect of Source Impedance

The commutation of current from one SCR to another will take place instantaneously if the source impedance is zero. The commutation is delayed by an angle μ due to source inductance, and during this period a short circuit occurs through the conducting devices, the ac circulating current being limited by the source impedance; μ is called the overlap angle. When α is zero, the short-circuit conditions are those corresponding to maximum asymmetry and μ is large, i.e., slow initial rise. At $\alpha = 90°$, the conditions are of zero asymmetry with its fast rate of rise of current. Commutation produces two primary notches per cycle and four secondary notches of lesser amplitude, which are due to notch reflection from the other legs of the bridge (Figure 17.20). For a purely inductive source impedance, the output average dc voltage is reduced and is given by

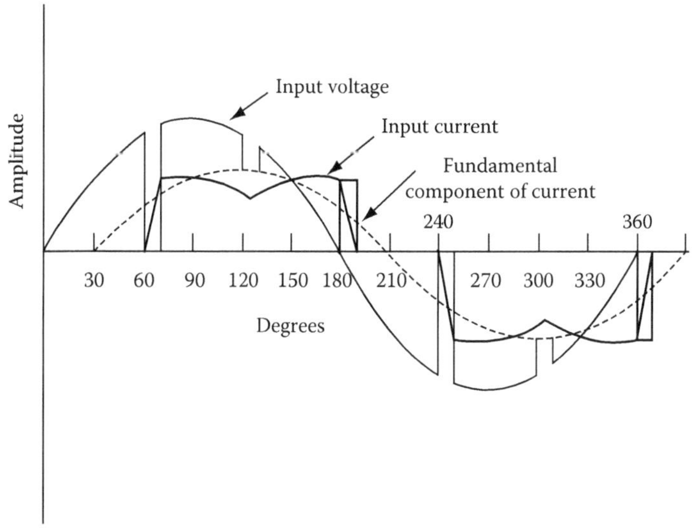

FIGURE 17.20
Voltage notching due to commutation in a six-pulse fully controlled bridge with dc output reactor.

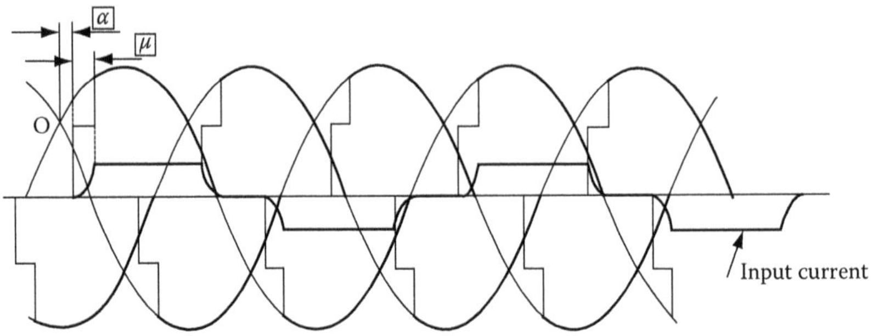

FIGURE 17.21
Effect of overlap angle on the current waveform.

$$E_d = E_{do} - \frac{3\omega L_s}{\pi} I_d \tag{17.51}$$

where
- L_s is the source inductance, and for a six-pulse fully controlled bridge
- E_{do} is given by Equation 17.40 (designated as E_d in this Equation) and it is called the internal voltage of the rectifier

Figure 17.21 shows that the overlap helps in reducing the harmonic content in the input current wave, which is rounded off and is more close to a sinusoid. Alternating-current harmonics at overlap are given by [1]:

$$I_h = I_{dc}\left[\sqrt{\frac{6}{\pi}}\frac{\sqrt{A^2 + B^2 - 2AB\cos(2\alpha + \mu)}}{h[\cos\alpha - \cos(\alpha + \mu)]}\right] \tag{17.52}$$

where

$$A = \frac{\sin[(h-1)(\mu/2)]}{h-1} \tag{17.53}$$

$$B = \frac{\sin[(h+1)(\mu/2)]}{h+1} \tag{17.54}$$

The depth of the voltage notch is calculated by the *IZ* drop and is a function of the impedance. The width of the notch is the commutation angle:

$$\mu = \cos^{-1}[\cos\alpha - (X_s + X_t)I_d] - \alpha \tag{17.55}$$

$$\cos\mu = 1 - \frac{2E_x}{E_{do}} \tag{17.56}$$

where
- X_s is the system reactance in per unit on converter base
- X_t is the transformer reactance in per unit on converter base
- I_d is the dc current in per unit on converter base
- E_x is the dc voltage drop caused by commutating reactance

TABLE 17.3

Theoretical and Typical Harmonic Spectrum of Six-Pulse Full Converters with Large dc Reactor

h	5	7	11	13	17	19	23	25
$1/h$	0.200	0.143	0.091	0.077	0.059	0.053	0.043	0.040
Typical	0.175	0.111	0.045	0.029	0.015	0.010	0.009	0.008

Notches cause EMI (electromagnetic interference) problems and misoperation of electronic devices, which sense the true zero crossing of the voltage wave.

As a six-pulse converter is most frequently used in industry, Appendix G provides graphical/analytical methods for the estimation of harmonics, with varying overlap angles. The assumption of a flat-topped wave is not correct. The actual wave and its effect on line harmonics are discussed in this appendix. Table 17.3 shows the theoretical magnitude of harmonics given by Equation 17.43 and a typical harmonic spectrum, assuming ripple-free dc output current and instantaneous commutation.

17.11.2.3 Effect of Output Reactor

The foregoing treatment of six-pulse converters assumes large dc inductance, so that the output dc current is continuous. At large phase-control angles and low output value of the reactor the current will be discontinuous, giving rise to increased line harmonics. See Appendix G for further discussions.

17.11.2.4 Effect of Load with Back EMF

Harmonic magnitude will be largely affected if the output dc load is active, for example, a battery charge. A dc motor has low inductance as well as a back EMF. The value of inductance at which the load current becomes discontinuous can be calculated by writing a differential equation of the form

$$L_{\mathrm{d}}\frac{\mathrm{d}i_{\mathrm{d}}}{\mathrm{d}t} + i_{\mathrm{d}}R_{\mathrm{d}} = E_m\cos(\omega t + \alpha) \tag{17.57}$$

and solving for i_{d}, equating it to zero, and evaluating L_{d}.

When the load has a back EMF, the load current waveform is not only decided by the firing angle, but also by the opposing voltage of the load. Figure 17.22 shows the waveforms of output voltage and the load current for a single-phase fully controlled circuit feeding a battery charger, neglecting source impedance; β is called the conduction angle. The harmonics are increased. At large control angles and with dc motor loads, having a back EMF, the discontinuous nature of the dc voltage and ac current, for a six-pulse converter, may give rise to higher harmonics and the fifth harmonic can reach peak levels up to three times that of the rectangular wave.

17.11.2.5 Inverter Operation

From Equation 17.40, the dc voltage is zero for a 90° delay angle. As the delay angle is further increased, the average dc voltage becomes negative (this can be examined by

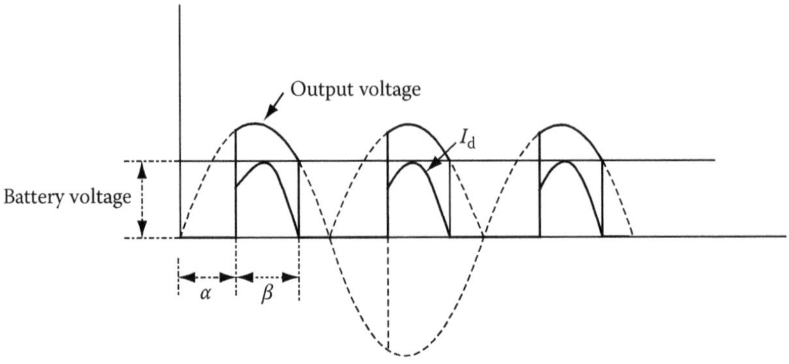

FIGURE 17.22
Current waveform with delay angle and a load with back EMF.

drawing output voltage waveform with advancing delay angle α in Figure 17.21). If instantaneous commutation is assumed then at 180° the negative voltage is as large as that for a rectifier at zero delay angle. However, operation at 180° is not possible, as some delay, denoted by γ, is needed for commutation of current and some additional time denoted by, angle δ is needed for the outgoing thyristors to turn off before voltage across it reverses. Otherwise the outgoing thyristor will commutate the current back, and lead to commutation failure. In HVDC transmission two modes of operation are recognized. In mode 1, the firing angle is varied and current remains constant. The duration for which the thyristor is reverse biased will reach a minimum value at some level of internal voltage and any further increase will result in commutation failure. Thus, in the second mode, the inverter operates at a constant margin angle or extinction angle. Inverter operation and angles γ and δ are not shown. See Ref. [9] for further details.

17.11.3 Diode Bridge Converter

Converters with an output dc reactor and front-end thyristors follow the dc link voltage. A full converter controls the amount of dc power from zero to full dc output. The voltage and current waveforms of this type of converter are discussed in Section 17.11.2. The harmonic injection into the supply system may be represented by a Norton equivalent. This type of converter is used at the front end of current source inverters.

The full-wave diode bridge with capacitor load, as shown in Figure 17.23, is the second type of converter. It converts from ac to dc and does not control the amount of dc power. This type of converter does not cause line notching, but the current drawn is more like a pulse current rather than the approximate square-wave current of the full converter. The voltage and current waveform are shown in Figure 17.23b. This circuit is better represented by a Thévenin equivalent and the source impedance has a greater impact.

Typical current harmonics for comparison are shown in Table 17.4. In the diode converter with dc link capacitor the fifth harmonic is higher by a factor of 3–4 and the seventh harmonic by a factor of 3. This type of converter with dc link capacitor is used in voltage source inverters (VSIs). Sometimes, a controlled bridge may replace the diode bridge preceding the dc link capacitor.

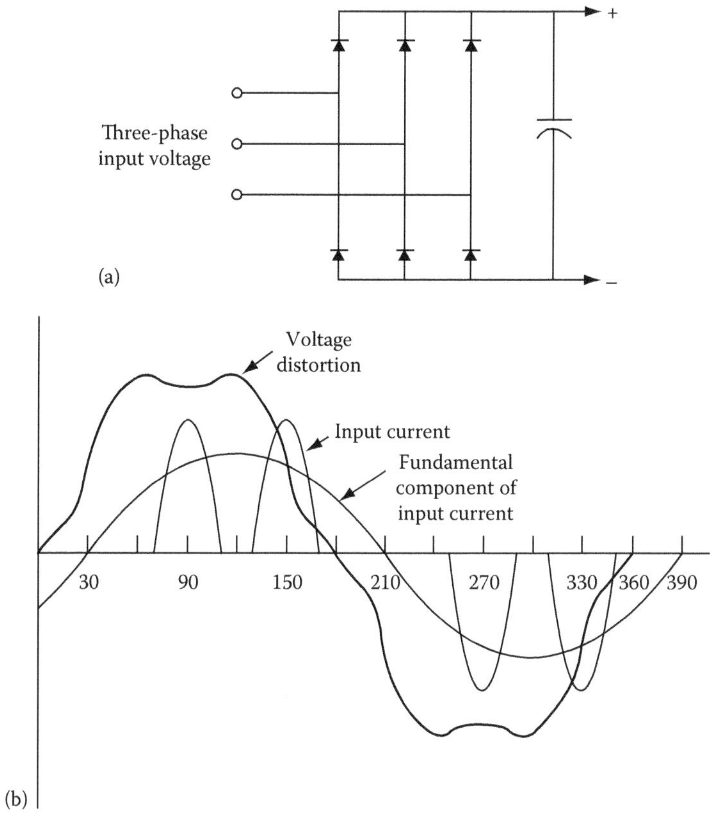

FIGURE 17.23
(a) Circuit of a diode bridge with dc link bus capacitor; (b) input current waveform.

TABLE 17.4

Typical Current Harmonics for a Six-Pulse Converter and Diode Bridge Converter, as a Percentage of Fundamental Frequency Current

Current Harmonic	Six-Pulse Converter	Diode Bridge Converter
5	17.94	64.5
7	11.5	34.6
11	4.48	5.25
13	2.95	5.89

17.12 Switch-Mode Power (SMP) Supplies

Single-phase rectifiers are used for power supplies in copiers, computers, TV sets, and household appliances. In these applications, the rectifiers use a dc filter capacitor and draw impulsive current from the ac supply. The harmonic current is worse than that given by Equation 17.32. Figure 17.24a and b shows conventional and switch mode power

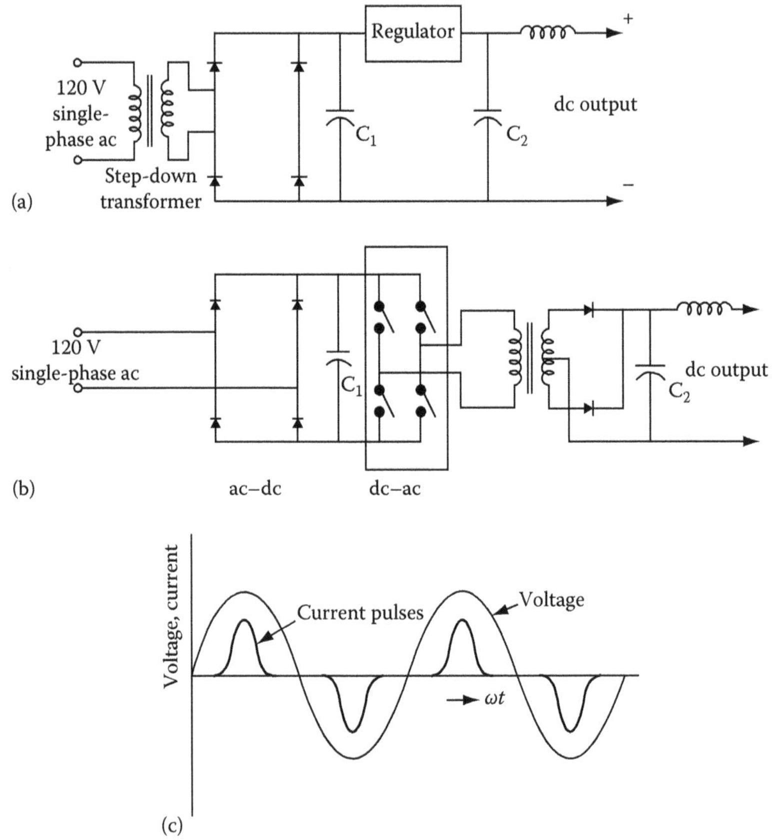

FIGURE 17.24
(a) Conventional power supply circuit; (b) switch mode power supply circuit; (c) input current pulse waveform.

supplies (SMPSs). In the conventional power supply system, the main ripple frequency is 120 Hz, and the current drawn is relatively linear. Capacitors C_1 and C_2 and inductor act as a passive filter. In the SMPSs, the incoming voltage is rectified at line voltage and the high dc voltage is stored in capacitor C_1. The transistorized switcher and controls switch the dc voltage from C_1 at a high rate (10–100 kHz). These high-frequency pulses are stepped down in a transformer and rectified. The switcher eliminates the series regulator and its losses in conventional power supplies. There are four common configurations used with the switched mode operation of the dc-to-ac conversion stage, and these are fly back, push–pull, half

TABLE 17.5
Spectrum of Typical Switch Mode Power Supply

Harmonic	Magnitude	Harmonic	Magnitude
1	1.000	9	0.157
3	0.810	11	0.024
5	0.606	13	0.063
7	0.370	15	0.07

Source: Reproduced from IEEE, *IEEE Recommended Practices and Requirements for Harmonic Control in Power Systems*, Standard 519, 1992.

bridge, and full bridge. The input current wave for such an SMPS is highly nonlinear, flowing in pulses for part of the sinusoidal ac voltage cycle (Figure 17.24c). The spectrum of an SMPS is given in Table 17.5 and shows high magnitude of the third and fifth harmonics.

17.13 Arc Furnaces

Arc furnaces may range from small units of a few ton capacity, power rating 2–3 MVA, to larger units having 400 t capacity and power requirement of 100 MVA. The harmonics produced by electric arc furnaces are not definitely predicted due to variation of the arc feed material. The arc current is highly nonlinear, and reveals a continuous spectrum of harmonic frequencies of both integer and noninteger order. The arc furnace load gives the worst distortion, and due to the physical phenomenon of the melting with a moving electrode and molten material, the arc current wave may not be same from cycle to cycle. The low-level integer harmonics predominate over the noninteger ones. There is a vast difference in the harmonics produced between the melting and refining stages. As the pool of molten metal grows, the arc becomes more stable and the current becomes steady with much less distortion. Figure 17.25 shows erratic rms arc current in a supply phase during the scrap melting cycle, and Table 17.6 shows typical harmonic content of two

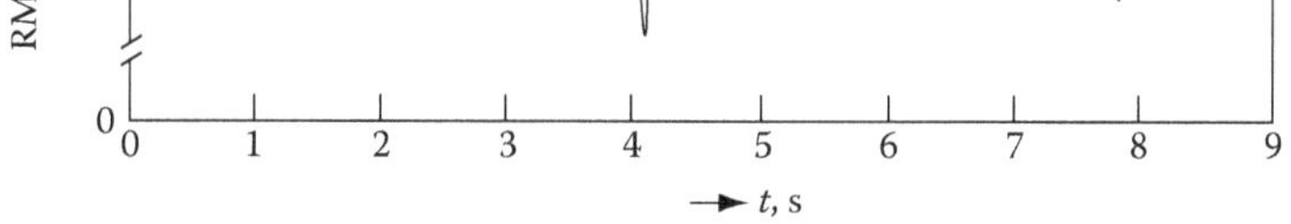

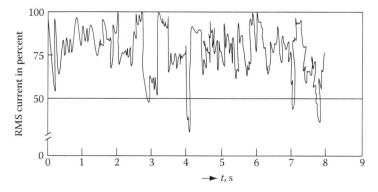

FIGURE 17.25
Erratic melting current in one-phase supply circuit of an arc furnace.

TABLE 17.6
Harmonic Content of Arc Furnace Current as Percentage of Fundamental

h	Initial Melting	Refining
2	7.7	0.0
3	5.8	2.0
4	2.5	0.0
5	4.2	2.1
7	3.1	

Source: Reproduced from IEEE, *IEEE Recommended Practices and Requirements for Harmonic Control in Power Systems*, Standard 519, 1992.

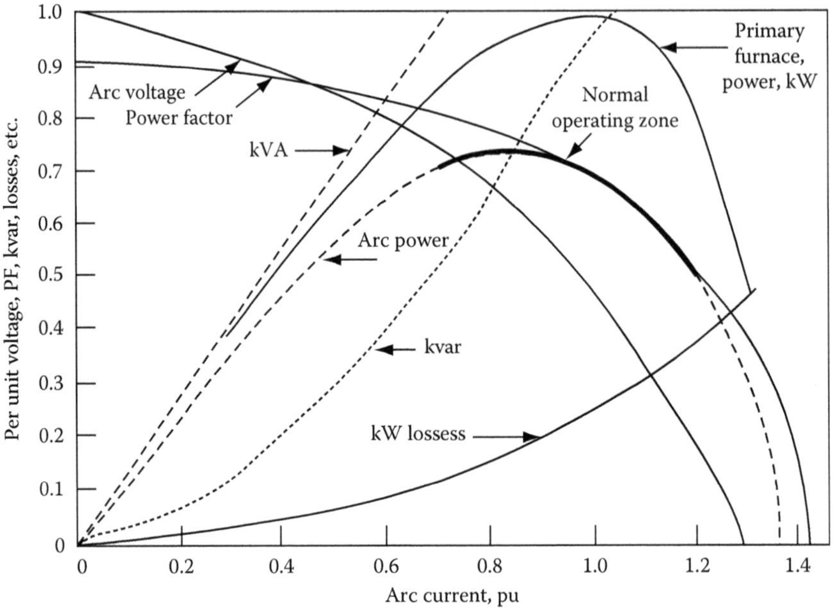

FIGURE 17.26
Typical performance curves for an arc furnace, showing the normal operating zone in thick line.

stages of the melting cycle in a typical arc furnace. The values shown in this table cannot be generalized. Both odd and even harmonics are produced. Arc furnace loads are harsh loads on the supply system, with attendant problems of phase unbalance, flicker, harmonics, impact loading, and possible resonance.

Figure 17.26 shows that the arc furnace presents a load of low lagging power factor. Large erratic reactive current swings cause voltage drops across the reactive impedance of the ac system, resulting in irregular variation of the terminal voltage. These voltage variations cause variation in the light output of the incandescent lamps and are referred to as flicker, based on the sensitivity of the human eye to the perception of variation in the light output of the incandescent lamps.

Appendix F describes the flicker limits. SVCs are used to control flicker and sometimes these may themselves produce interphase harmonics.

17.14 Cycloconverters

Cycloconverters are used in a wide spectrum of applications from ball-mill and linear motor drives to SVGs. The range of application for synchronous or induction motors varies from 1,000 to 50,000 hp and the speed control in the ratio of 50 to 1. Figure 17.27a shows the circuit of a three-phase single-phase cycloconverter, which synthesizes a 12 Hz output, and Figure 17.27c shows the output voltage waveform with resistive load. The positive converter operates for half the period of the output frequency and the negative converter operates for the other half. The output voltage is made of segments of input voltages (Figure 17.27b), and the average value of a segment depends on the delay angle for that

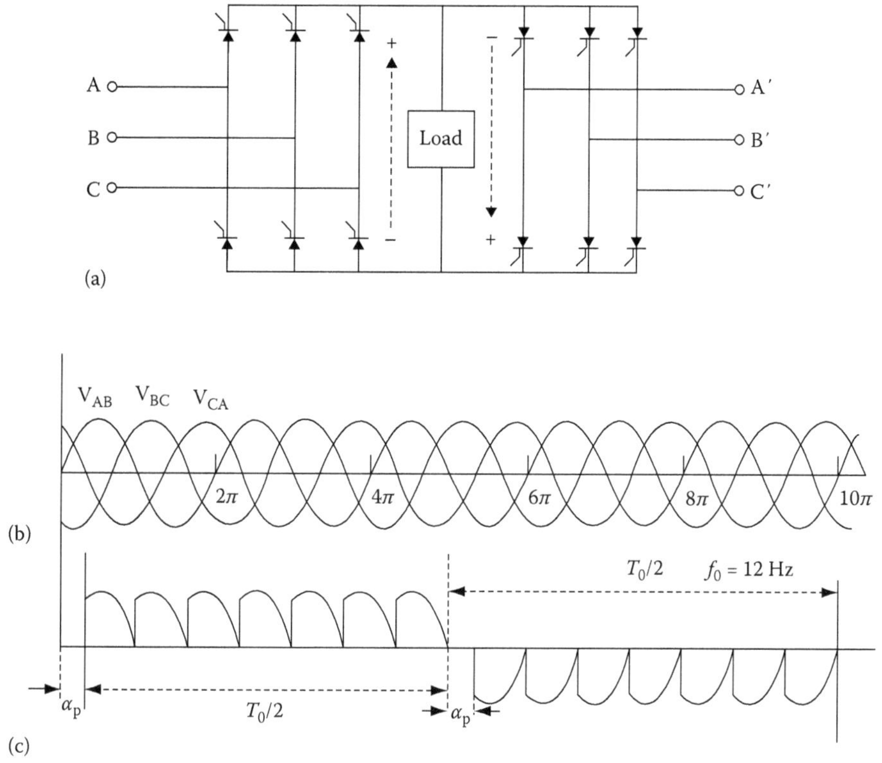

FIGURE 17.27
To illustrate principal of a cycloconverter.

segment; α_p is the delay angle of the positive converter and $\pi - \alpha_p$ is the delay angle of the negative converter. The output voltage contains harmonics and the input power factor is poor. For three-phase, three systems (Figure 17.27a), i.e., a total of 36 thyristors are required.

The voltage of a segment depends on the delay angle. If the delay angles of the segments are varied so that the average value of the segment corresponds as closely as possible to the variations in the desired sinusoidal output voltage, the harmonics in the output are minimized. Such delay angles can be generated by comparing a cosine signal at source frequency with an output sinusoidal voltage.

Cycloconverters have characteristic harmonic frequency of

$$f_h = (pm \pm 1)f \pm 6nf_0 \tag{17.58}$$

where

f_0 is the output frequency of the cycloconverter, $m, n = 1, 2, 3, \ldots$
p is the pulse number

Because of load unbalance and asymmetry between phase voltages and firing angles, noncharacteristic harmonics are also generated:

$$f_h = (pm \pm 1)f \pm 2nf_0 \tag{17.59}$$

As the output frequency varies, so does the spectrum of harmonics. Therefore, control of harmonics with single tuned filters becomes ineffective (Chapter 20). Reference [10] is entirely devoted to cycloconverters.

17.15 Thyristor-Controlled Reactor

Consider a TCR, controlled by two thyristors in an antiparallel circuit as shown in Figure 17.28a. If both thyristors are gated at maximum voltage, there are no harmonics and the reactor is connected directly across the voltage, producing a 90° lagging current, ignoring the losses. If the gating is delayed, the waveforms as shown in Figure 17.28b result. The instantaneous current through the reactor is

$$i = \sqrt{2}\frac{V}{X}(\cos\alpha - \cos\omega t) \quad \text{for } \alpha < \omega t < \alpha + \beta \tag{17.60}$$

$$= 0 \quad \text{for } \alpha + \beta < \omega t < \alpha + \pi \tag{17.61}$$

where
- V is the line-to-line fundamental rms voltage
- α is the gating angle
- β is the conduction angle

The fundamental component can be written as

$$I_{\mathrm{f}} = \frac{\beta - \sin\beta}{\pi X} V \tag{17.62}$$

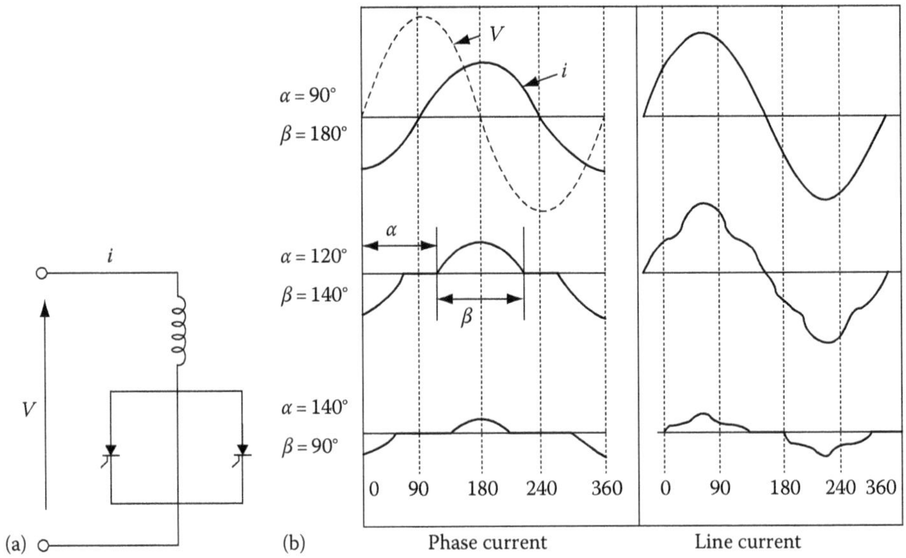

FIGURE 17.28
(a) Circuit of a thyristor-controller reactor; (b) current waveforms due to varying firing and conduction angles.

Appreciable amount of harmonics are generated. Assuming balanced gating angles, only odd harmonics are produced. The rms value is given by

$$I_h = \frac{4V}{\pi X}\left[\frac{\sin(h+1)\alpha}{2(h+1)} + \frac{\sin(h-1)\alpha}{2(h-1)} - \cos\alpha\frac{\sin h\alpha}{h}\right] \tag{17.63}$$

where $h = 3, 5, 7, \ldots$ Unequal conduction angles will produce even harmonics including a dc component.

17.16 Thyristor-Switched Capacitors

A capacitor traps charge at the maximum voltage when current is zero. This makes thyristor switching of capacitors difficult, as a possibility exists that maximum ac peak voltage can be applied to a capacitor charged to a maximum negative peak. The use of thyristors to switch capacitors is limited to allowing conduction for an integral number of half cycles and point of wave switching, i.e., gating angles of >90° are not used. The thyristors are gated at the peak of the supply voltage, when $dV/dt = 0$, and the capacitors are already charged to the peak of the supply voltage. In practice, all TSC circuits have some inductance, and oscillatory switching transients result. For a transient-free switching

$$\cos\alpha = 0 \tag{17.64}$$

$$V_c = \pm V\frac{X_c/X_L}{(X_c/X_L) - 1} \tag{17.65}$$

i.e., the capacitors are gated at supply voltage peak and the capacitors are charged to a higher than the supply voltage prior to switching. Since it is generally difficult to guarantee the second condition, it is difficult to prevent oscillatory transients. The transients on gating with $V = 0$ and with $dV/dt = 0$ are shown in Figure 17.29.

The synchronous SVC, described in Chapter 13, produces much lower harmonics. With pulse-width control (see Section 17.17), the ac side harmonics are controlled. Thyristor-controlled series compensation produces harmonics in series with the line (Chapter 13).

17.17 Pulse-Width Modulation

Over the years the pulse switching time of the power devices has been drastically reduced:

- SCR (fast thyristor) : 4 μs
- GTO (gate turn-off thyristor) : 1.0 μs
- GTR (giant transistor) : 0.8 μs
- IGBT (insulated gate bipolar) : 0.2 μs
- IGBT (power plate type) : 0.1 μs

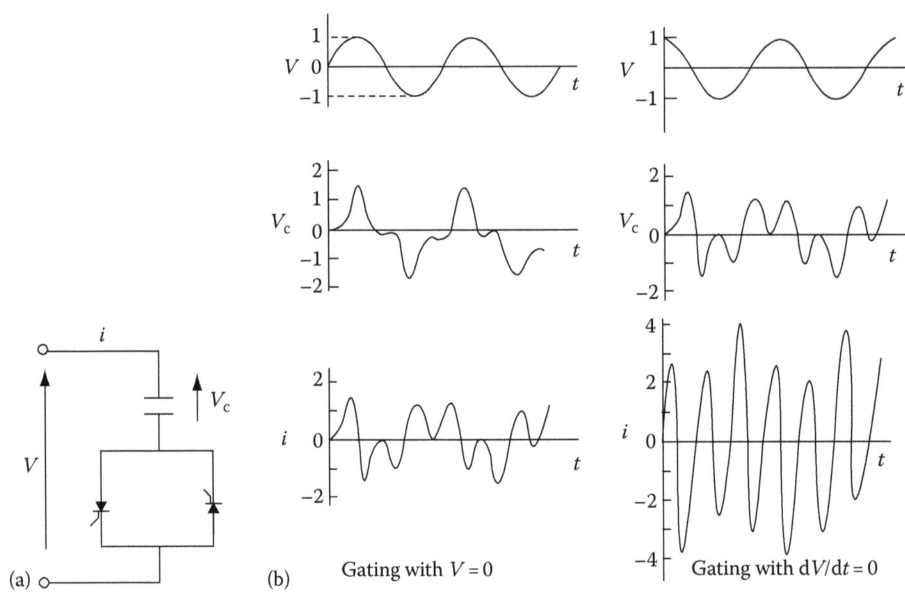

FIGURE 17.29
(a) Circuit of a thyristor-switched capacitor; (b) switching transients with $V = 0$ and with $dV/dt = 0$.

The voltage source invertors using IGBTs operate from a dc link bus. The inverter synthesizes a variable voltage, of variable-frequency waveform ($V/f =$ constant), by switching the dc bus voltage at high frequencies (10–20 kHz). The inverter output line-to-line voltage is a series of voltage pulses with constant amplitude and varying widths. IGBTs have become popular, as these can be turned on and off from simple low-cost driver circuits. Motor low-speed torque can be increased and improved low-speed stability is obtained. The high-frequency switching results in high dV/dt and the effects on motor insulation, connecting cables, and EMI are discussed in Chapter 18. Recent trends in soft switching technology reduce the rise time (Figure 17.30f and g).

Techniques of PWM are

- Single-pulse-width modulation
- Multiple-pulse-width modulation
- Sinusoidal-pulse-width modulation
- Modified sinusoidal-pulse-width modulation

In a single-pulse-width modulation technique, there is one pulse per half cycle and the width of the pulse is varied to control the inverter output voltage (Figure 17.30a and b). By varying A_r from 0 to A_c, the pulse width δ can be varied from 0° to 180°. The modulation index is defined as A_r/A_c. The harmonic content is high but can be reduced by using several pulses in each half cycle of the output voltage. The gating signal is generated by comparing a reference signal with a triangular carrier wave. This type of modulation is also known as uniform pulse-width modulation. The number of pulses per half cycle is $N = f_c/2f_0$, where f_c is the carrier frequency and f_0 is the output frequency.

In sinusoidal PWM, the pulse width is varied in proportion to the amplitude of the sine wave at the center of the pulse (Figure 17.30c and d). The distortion factor and the lower-order harmonic magnitudes are reduced considerably. The gating signals are

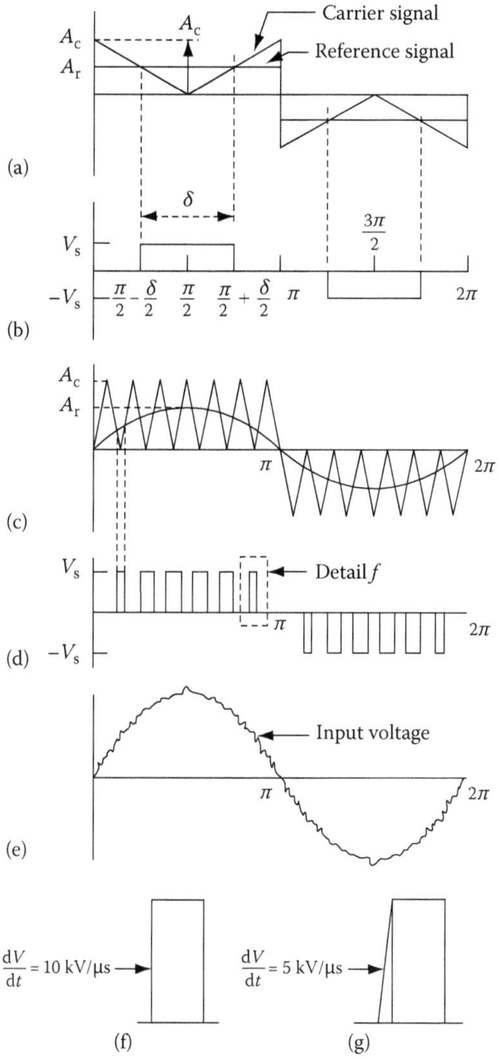

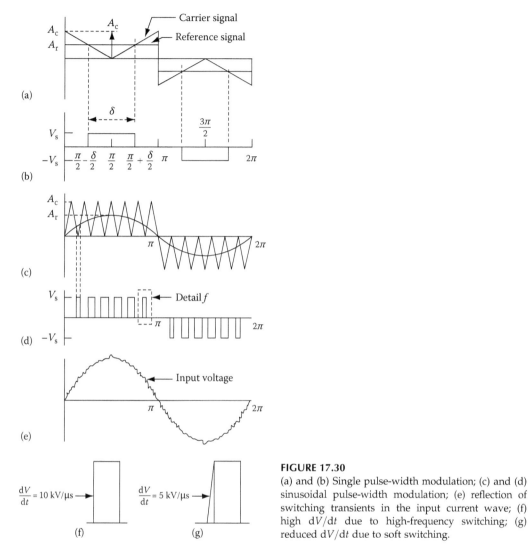

FIGURE 17.30
(a) and (b) Single pulse-width modulation; (c) and (d) sinusoidal pulse-width modulation; (e) reflection of switching transients in the input current wave; (f) high dV/dt due to high-frequency switching; (g) reduced dV/dt due to soft switching.

generated by comparing a reference sinusoidal signal with a triangular carrier wave of frequency f_c. The frequency of the reference signal f_r determines the inverter output frequency f_0, and its peak amplitude A_r controls the modulation index and output voltage V_0. This type of modulation eliminates harmonics and generates a nearly sinusoidal voltage wave. The input current waveform has a sinusoidal shape due to pulse-width shaping; however, harmonics at switching frequency are superimposed (Figure 17.30e). Soft switching can reduce high dV/dt (Figures 17.30f and g).

17.18 Adjustable Speed Drives

Adjustable speed drives account for the largest percentage of nonlinear loads in the industry. A comparison of electronic drive systems with type of motor, horse power rating, and drive system topology is shown in Table 17.7. Most drive systems require that the

TABLE 17.7

Adjustable-Speed Drive Systems

Drive Motor	Horse Power	Normal Speed Range	Converter Type
DC	1–10,000	50:1	Phase controlled, line commutated
Squirrel-cage induction	100–4,000	10:1	Current link, force commutated
Squirrel-cage induction	1–1,500	10:1	Voltage link, force commutated
Wound rotor	500–20,000	3:1	Current link, line commutated
Synchronous (brushless excitation)	1,000–60,000	50:1	Current link, load commutated
Synchronous or squirrel cage	1,000–60,000	50:1	Phase controlled, line commutated

incoming ac power supply be converted into dc. The dc power is then inverted back to ac at a frequency demanded by the speed reference of the ac variable-frequency drive or the dc feeds directly to dc drive systems through two- or four-quadrant converters. The fully controlled bridge circuit with output reactor and three-phase diode bridge circuit discussed above form the basic front-end input circuits to drive systems.

17.19 Pulse Burst Modulation

Typical applications of pulse burst modulation (PBM) are ovens, furnaces, heaters, and spot welders [4]. Three-phase PBM circuits can inject dc currents into the system, even when the load is purely resistive. A solid-state switch is kept turned on for an integer number γn of half cycles out of a total of n cycles (Figure 17.31). The control ratio $0 < \gamma < 1$ is adjusted by feedback control. The integral cycle control minimizes EMI, yet the circuit may inject significant dc currents into the power system. Neutral wire carries pulses of current at switch-off and switch-on, which have high harmonic content, depending on the control ratio γ. Harmonics in the 100–400 Hz band can reach 20% of the line current. Loading of neutrals with triplen and fifth harmonics is a concern. The spectrum is deficient in high-order harmonics.

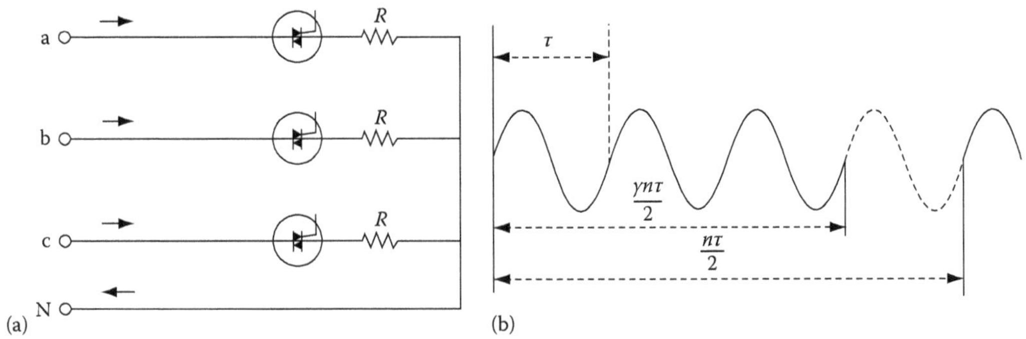

FIGURE 17.31
(a) Circuit for pulse burst modulation; (b) pulse burst modulation control, current waveform.

17.20 Chopper Circuits and Electric Traction

The dc traction power supply is obtained in the rectifier substations by unsmoothed rectification of utility ac power supply, and 12-pulse bridge rectifiers are common. Switching transients from commutation occur and harmonics are injected into the supply system. Auxiliary converters in the traction vehicles also generate harmonics, while EMI radiation is produced from fast current and voltage changes in the switching equipment.

A chopper with high inductive load is shown in Figure 17.32; the input current is pulsed and assumed as rectangular. The Fourier series is

$$i_c(t) = kI_a + \frac{I_a}{n\pi}\sum_{h=1}^{\infty} \sin 2h\pi k \cos 2h\pi f_c t + \frac{I_a}{h\pi}\sum_{h=1}^{\infty}(1 - \cos 2h\pi k)\sin 2h\pi f_c t \tag{17.66}$$

where

f_c is the chopping frequency

k is the mark-period ratio (duty cycle of the chopper $= t_1/T$)

The fundamental component is given for $h = 1$. In railway dc fed traction drives, thyristor choppers operate up to about 400 Hz. The chopper circuit is operated at fixed frequency and the chopping frequency is superimposed on the line harmonics. An input low-pass filter (Chapter 20) is normally connected to filter out the chopper-generated harmonics and to control the large ripple current. The filter has physically large dimensions, as it has a low resonant frequency. The worst-case harmonics occur at a mark-period ratio of 0.5. Imperfections and unbalances in the chopper phases modify the harmonic distribution and produce additional harmonics at all chopper frequencies. Transient inrush current to filter occurs when the train starts and may cause interference if it contains critical frequencies.

In a VSI-fed induction motor traction drive from a *dc traction system*, the inverter fundamental frequency increases from zero to about 120 Hz as the train accelerates. Up to the motor base speed, the inverter switches many times per cycle. To calculate source current harmonics this variable-frequency operation must be considered in addition to three-phase operation of the inverter. Harmonics in the dc link current depend on the spectrum of the switching function. Optimized PWM with quarter-wave symmetry is used in traction converters and though each dc link current component contains both odd and even harmonics, the positive and negative sequence components cancel in the dc link waveform, leaving only zero sequence and triplen harmonics in the spectrum.

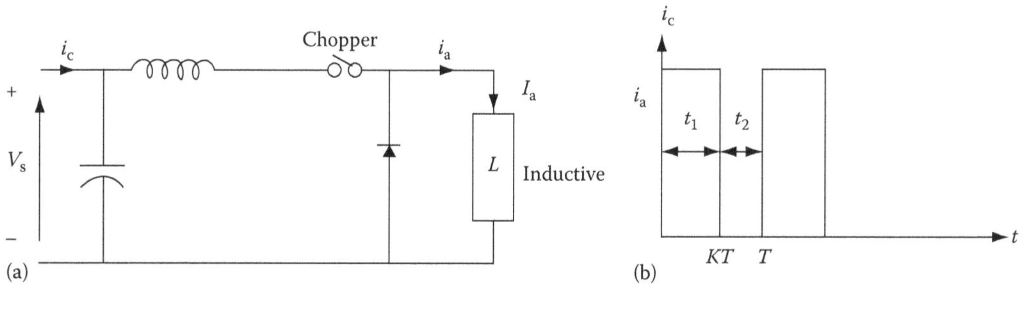

FIGURE 17.32
(a) A chopper circuit with input filter; (b) current waveform.

Multilevel VSI drives or step-down chopper drives are used with GTOs. VSIs may use a different control strategy, other than PWM, such as torque band control with asynchronous switching, and the disadvantage is that relatively high $6h$ harmonics are produced, together with third harmonic and some sub-harmonics. In the chopper inverter drive, the harmonics due to chopper and inverter combine. The input harmonic current spectrum consists of multiples of chopper frequency with side bands at six times the inverter frequency:

$$f_h = kf_c \pm 6hf_i \tag{17.67}$$

where
- k and h are positive integers
- f_c is the chopper frequency
- f_i is the inverter frequency

In ac-fed traction drives, the harmonics can be calculated depending on the drive system topology. Drives with dual semi-controlled converters and dc motors are rich sources of harmonics. In drives fed with a line pulse converter and voltage or current source inverter, the line converter is operated with PWM to regulate the demand to the VSI, while maintaining nearly unity power factor. The source current harmonics are mainly derived from the line operation of the pulse converter.

17.21 Slip Frequency Recovery Schemes

The slip frequency power of large induction motors can be recovered and fed back into the supply system. Figure 17.33 shows an example of a sub-synchronous cascade. The rotor slip frequency voltage is rectified, and the power taken by the rotor is fed into the supply system through a line commutated inverter. The speed of the induction motor can be adjusted as desired throughout the sub-synchronous range, without losses, though the reactive power consumption of the motor cannot be corrected in the arrangement shown in Figure 17.33.

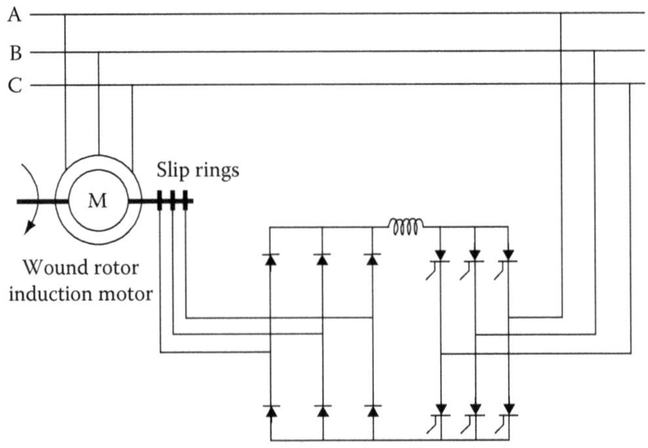

FIGURE 17.33
A slip recovery scheme for a wound rotor induction motor.

Such a system can cause sub-harmonics in the ac system. For six-pulse rectification, the power returned to the system pulsates at six times the rotor slip frequency. Torsional oscillations can be excited if the first or second natural torsional frequency of the mechanical system is excited, resulting in shaft stresses [11]. The ac harmonics for this type of load cannot be reduced by phase multiplication as dc current ripple is independent of the rectifier ripple.

17.22 Lighting Ballasts

Lighting ballasts may produce large harmonic distortions and third harmonic currents in the neutral. The newer rapid start ballast has a much lower harmonic distortion. The current harmonic limits for lighting ballasts are given in Tables 17.8 and 17.9. Table 17.8 shows that the limits for the newer ballasts are much lower as compared to earlier ballasts (Table 17.9). This also compares distortion produced by the lighting ballasts with other office equipment [12,13].

This chapter shows that it may not be always possible to estimate clearly harmonic emission. The topologies of harmonic-producing equipment are changing fast, and manufacturer's data can be used, wherever practical. The system impedance plays a major role. Consider that load currents are highly distorted but the system impedance is low. Thus, the

TABLE 17.8

Current Harmonic Limits for Lighting Ballasts

Harmonic	Maximum Value (%)
Fundamental	100
Second harmonic	5
Third harmonic	30
Individual harmonics >11th	7
Odd triplens	30
Harmonic factor	32

TABLE 17.9

THD Ranges for Different Types of Lighting Ballasts

Device Type	THD (%)
Older rapid start magnetic ballast	10–29
Electronic IC-based ballast	4–10
Electronic discrete-based ballast	18–30
Newer rapid start electronic ballast	<10
Newer instant start electronic ballast	15–27
High-intensity discharge ballast	15–27
Office equipment	50–150

Source: Reproduced from R. Arthur and R.A. Sanhan, *Neutral Currents in Three-Phase Wye Systems, Square D Company*. August 1996.

voltages will not be much distorted. Appendix F shows ANSI limits for harmonic currents that a user can inject into utility system, and these consider the system impedance. The limits are specified on the basis of the I_s/I_r ratio, where I_s is the short-circuit current and I_r is the load demand current. Further discussions continue in Appendix F.

17.23 Voltage Source Converters

The FACTS use voltage source bridges. The current source converters have been extensively used for HVDC transmission. In the 1990s, HVDC transmission using voltage source converters with PWM was introduced commercially called HVDC-*light* [5]. The requirements of FACT controllers are as follows:

- The converter should be able to act as an inverter or rectifier with leading or lagging reactive power, i.e., four-quadrant operation is required, as compared to current source line commutated converter, which has two-quadrant operations.
- The active and reactive power should be independently controllable with control of phase angle.

The principle is illustrated with respect to single valve operation (Figure 17.34). Consider that dc voltage remains constant and the turn-off device is turned on by gate control. Then the positive of dc voltage is applied to terminal A, and the current flows from $+V_d$ to A, i.e., inverter action. If the current flows from A to $+V_d$, *even when device 1 is turned on,* it will flow through the parallel diode, rectifier action. Thus, the power can flow in either direction. A valve with a combination of turn-off device and diode can handle power in either direction.

17.23.1 Three-Level Converter

Figure 17.35 shows the circuit of a three-level converter and associated waveforms. In Figure 17.35a, each half of the phase leg is split into two series connected circuits, midpoint connected through diodes, which ensure better voltage sharing between the two sections. Waveforms in Figures 17.35b through e are obtained corresponding to one three-phase leg. Waveform (Figure 17.35b) is obtained with 180° conduction of the devices. Waveform (Figure 17.35c) is obtained if 1 is turned off and 2A is turned on at an angle α *earlier* than for 180° conduction. The ac voltage V_a is clamped to zero with respect to midpoint N of the two capacitors. This occurs because devices 1A and 2A conduct and in combination with diodes clamp the voltage to zero. This continues for a period of 2α, till 1A is turned off and 2 is turned on and voltage is now $-V_d/2$, with both the 2 and 2A turned off and 1 and 1A

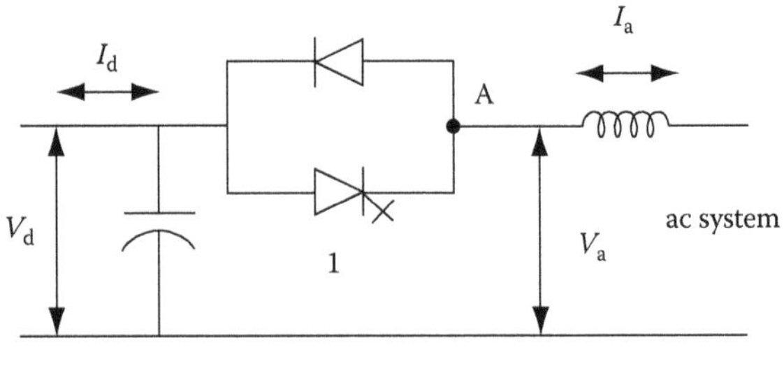

FIGURE 17.34
Principal of a voltage source converter.

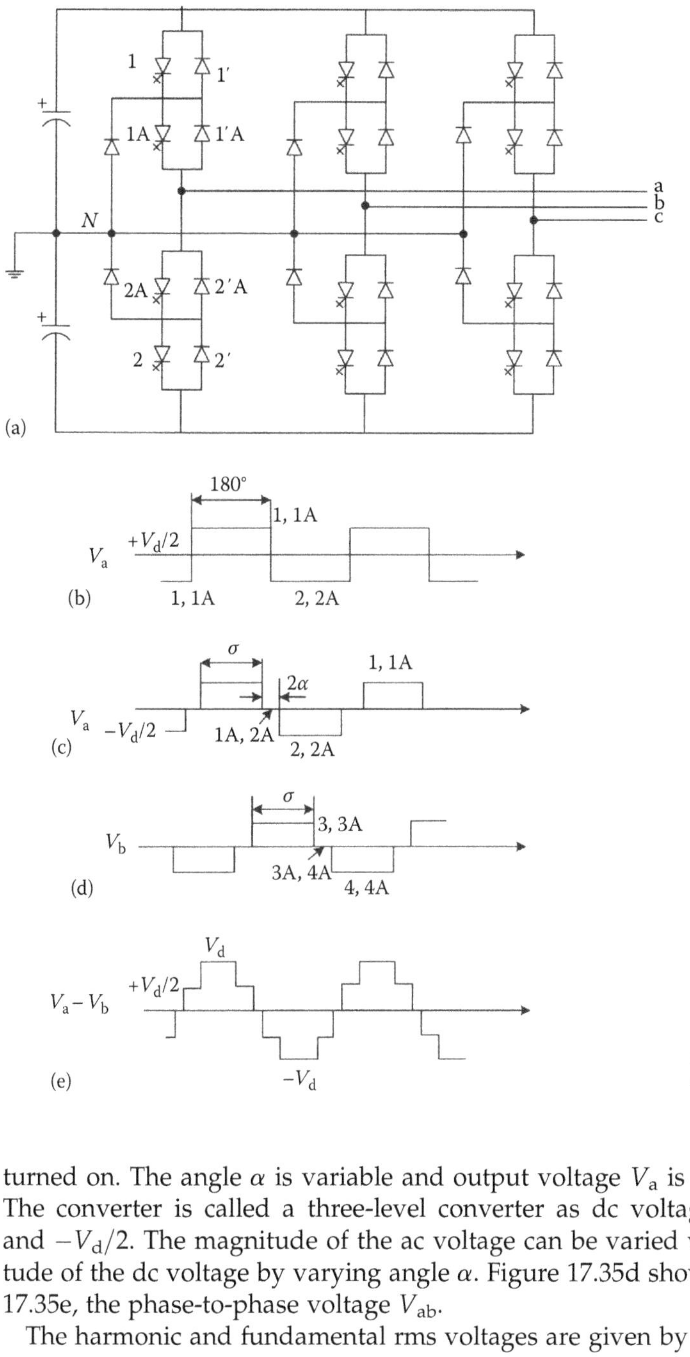

FIGURE 17.35
(a) A three-level three-phase voltage source converter. (b) through (e) Operational waveforms, see text.

turned on. The angle α is variable and output voltage V_a is square waves $\sigma = 180° - 2\alpha°$. The converter is called a three-level converter as dc voltage has three levels, $V_d/2$, 0, and $-V_d/2$. The magnitude of the ac voltage can be varied without changing the magnitude of the dc voltage by varying angle α. Figure 17.35d shows the voltage V_b and Figure 17.35e, the phase-to-phase voltage V_{ab}.

The harmonic and fundamental rms voltages are given by

$$V_h = \frac{2\sqrt{2}}{\pi}\left(\frac{V_d}{2}\right)\frac{1}{2}\sin\frac{h\alpha}{2}$$
$$V_f = \frac{2\sqrt{2}}{\pi}\left(\frac{V_d}{2}\right)\sin\frac{\alpha}{2} \tag{17.68}$$

$V_h = 0$ at $\alpha = 0°$ and maximum at $\alpha = 180°$.

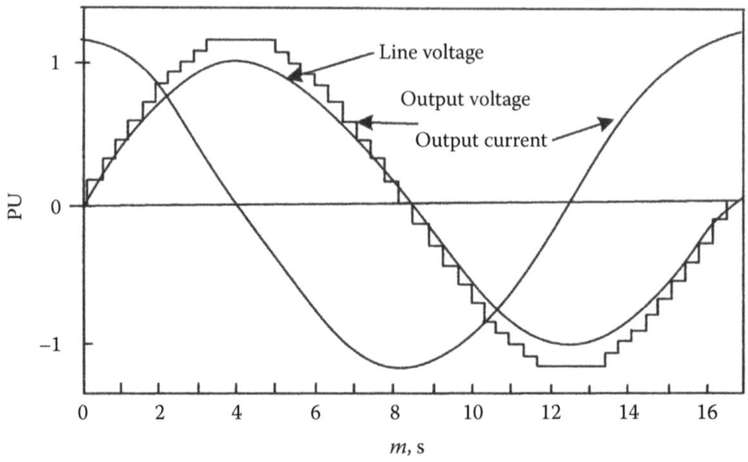

FIGURE 17.36
24-pulse operation for harmonic emission control in a STATCOM.

A STATCON may use many six-pulse converters outputs phase shifted and combined magnetically to give a pulse number of 24 or 48 for the transmission systems (Chapter 13). The output waveform is nearly sinusoidal and the harmonics present in the output current and voltage are small. This ensures waveform quality without passive filters. Figure 17.36 shows the output voltage and current waveform of a 48-pulse STATCON generating reactive power.

The PWM signals are produced using two carrier triangular signals; two different offsets of opposite sign are added to the triangular waveform to give the required carriers. These are compared with the reference voltage waveform and the output and its inverse is used for the gating signals—four gating signals one for each IGBT in a leg. The five-level converters are discussed in Refs. [14,15]. Harmonic emission from wind power generation is discussed in Chapter 22.

17.24 Inter-Harmonics

Inter-harmonics in power systems have attracted much attention in recent years [16]. Cycloconverters and arcing loads are the major sources. The integral cycle control (Section 17.19), induction motor cascade (Section 17.21), and low-frequency power line carrier (ripple control) are other sources. IEC [17] defines inter-harmonics as follows: "between the harmonics of the power frequency voltage and current, further frequencies can be observed, which are not an integer of the fundamental. They appear as discrete frequencies or as a wide-band spectrum." The term sub-harmonic is popular in the engineering community, but it has no official definition. A definition is sub-synchronous frequency component. Impact and limitation of inter-harmonics are described in Appendix F. Reference [4] has many references for further reading.

This chapter is indicative of the wide variations in the harmonic emissions, depending upon topologies. Thousands of such topologies exist, while every year some new ones

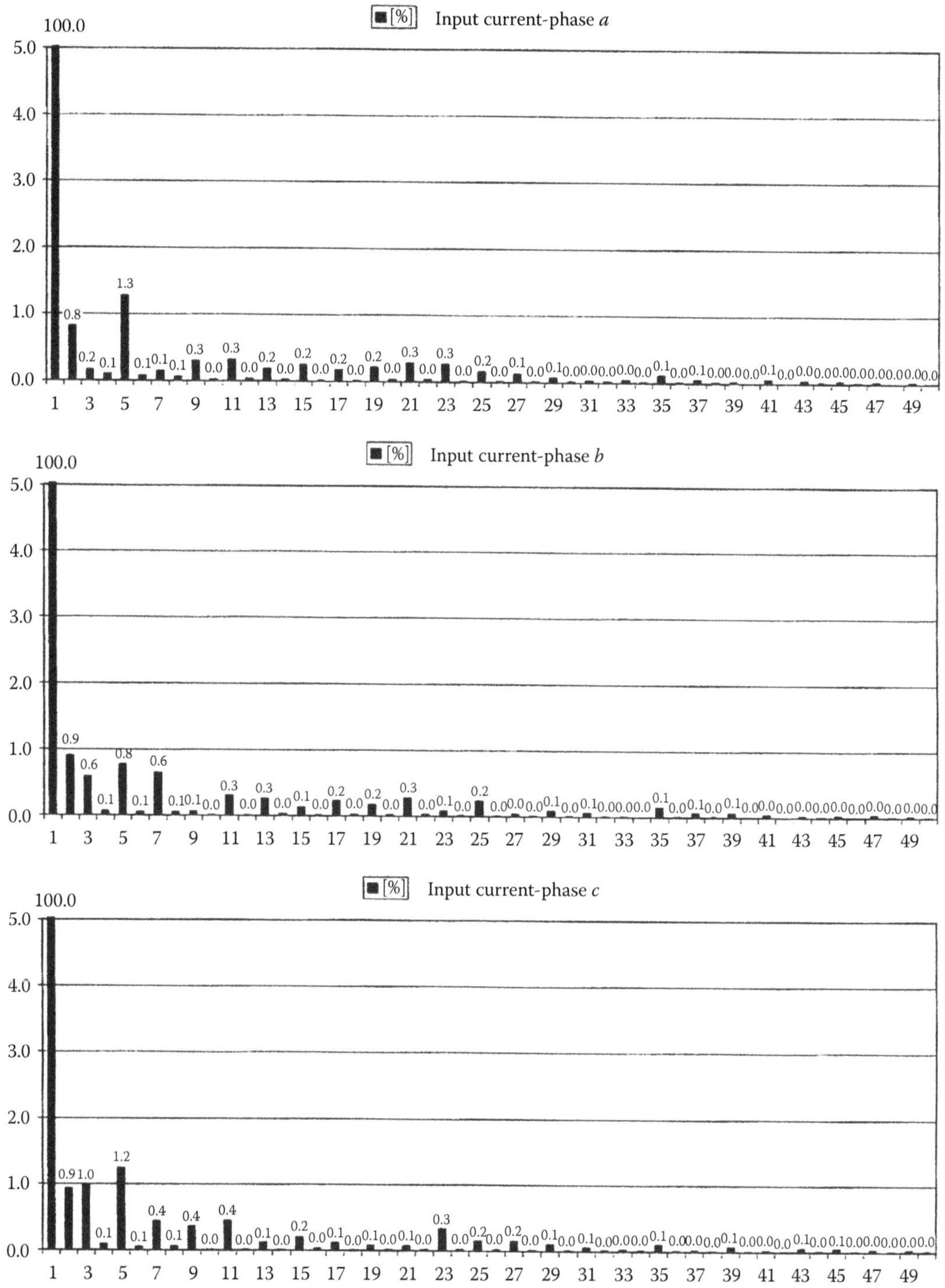

FIGURE 17.37
Harmonic emission, phases *a*, *b*, and *c*; 24-pulse PWM converter for a medium voltage drive system.

are added [18]. For the purpose of harmonic load flow and harmonic filter designs in Chapters 19 and 20, the harmonic emission from loads should be first estimated. This is not an easy task due to varied topologies. Mostly, appropriate spectrums and the harmonic angles should be ascertained from the manufacturers. Furthermore, the emission is also a function of load as well as system source impedance (Chapter 20). Figure 17.37 shows calculated harmonic current spectrum, based on specified system impedance at the point of interconnection. The drive system is PWM, 24-pulse. Note the difference in the harmonic emission in phases *a*, *b*, and *c*. Generally, for harmonic load flow three-phase models are not used; though it may be necessary to do so where harmonic phase unbalance exists. The emission is small and meets the requirements of IEEE 519 [1] without additional filters.

Note that *all* sources of harmonic generation are not described, for example, ferroresonance gives rise to overvoltages and harmonics (Appendix C).

Problems

17.1 Derive Equations 17.41 and 17.42, using Fourier series.

17.2 A six-pulse fully controlled converter operates at a three-phase, 480 V, 60 Hz system. The output current is 10 A and the firing angle is $\pi/4$. Calculate the input current, the harmonic amplitude of the output voltage, and the output voltage ripple factor.

17.3 Calculate the load resistance, the source inductance, and the overlap angle in Problem 17.2.

17.4 What are noncharacteristic harmonics? How are these generated?

17.5 Give an example of aperiodic harmonic generation.

17.6 Distinguish between displacement factor and power factor. A single-phase full bridge supplies a motor load. Assuming that the motor dc current is ripple-free, determine the input current (using Fourier analysis), harmonic factor, distortion factor, and power factor for an ignition delay angle of α.

17.7 What order of harmonics can be expected in the inrush current of a transformer? How long do these last? What gives rise to these harmonics? If the transformer core has trapped residual magnetism, will it reduce or increase the harmonics?

17.8 Draw a spectrum of line harmonics for a 12-pulse converter. Assume a rectangular current wave shape, zero overlap angle, and a large dc reactor to give ripple-free output current. Calculate the harmonic factor to the 29th harmonic.

17.9 Tabulate all possible two-winding transformer connections that will convert two equally loaded six-pulse converter circuits into a 12-pulse circuit.

17.10 Calculate the Fourier series for an input current to a six-pulse converter, with a firing angle of α. Calculate the harmonic spectrum for $\alpha = 15°$, 45°, and 60°. What will be the effect of doubling the source reactance?

17.11 Describe an appropriate model for a current source and voltage source inverter.

17.12 High ac system reactance increases overlap angle and a waveform closer to a sinusoid results. The ac system voltage distortion is reduced. Is this a correct statement?

17.13 Calculate the total power factor of a six-pulse converter, using Equations 17.52 and 17.55. The pertinent data are maximum rating 5 MVA, load voltage = 2.4 kV, load current = 500 A, angle $\alpha = 30°$, 13.8 kV rectifier transformer rating 5 MVA, percentage impedance 5.5%, X/R ratio = 8. What is the distortion power factor?

17.14 Plot a graph of fundamental and harmonic voltages of a three-level voltage converter.

17.15 Determine the harmonic factor, displacement factor, and the input power factor for the input current to a six-pulse current source converter, assuming a continuous load current (large reactance). Calculate these values if the rms input voltage is 480 V, and $\alpha = \pi/3$.

References

1. IEEE. *IEEE Recommended Practices and Requirements for Harmonic Control in Power Systems*, Standard 519, 1992.
2. ANSI/IEEE. *Requirements for Instrument Transformers*, Standard C57.13, 2005.
3. J.C. Das and J.R. Linders. Power system relaying. *Encyclopedia of Electronics and Electrical Engineers*, vol. 17. New York: John Wiley, 2002.
4. IEEE Working Group on Power System Harmonics. Power system harmonics: An overview. *IEEE Trans. Power Apparatus Syst. PAS* 102: 2455–2459, 1983.
5. J. Arrillaga. *High Voltage Direct Current Transmission*, 2nd edn. Piscataway, NJ: IEEE Press, 1998.
6. L. Malesani and P. Tenti. Three-phase AC/DC PWM converter with sinusoidal AC currents and minimum filter requirements. *IEEE Trans. Indust. Appl.* IA-23(1): 71–78, January/February 1987.
7. B.T. Ooi, J.C. Salmon, J.W. Dixon, and A. Kulkarni. A three-phase controlled-current PWM converter with leading power factor. *IEEE Trans. Indust. Appl.* IA-23(1): 78–84, February 1987.
8. J. Cardosa and T. Lipo. Current stiff converter topologies with resonant snubbers. *IEEE Industry Application Society Annual Meeting*, New Orleans, LA, 1997, pp. 1322–1329.
9. M.H. Rashid. *Power Electronics*. Upper Saddle River, NJ: Prentice Hall, 1988.
10. B.R. Pelly. *Thyristor Phase-Controlled Converters and Cycloconverters, Operation Control and Performation*. New York: John Wiley, 1971.
11. H. Flick. Excitation of sub-synchronous torsional oscillations in turbine generator sets by a current-source converter. *Siemens Power Eng.* IV(2): 83–86, 1982.
12. R. Arthur and R.A. Sanhan. *Neutral Currents in Three-Phase Wye Systems*. Oshkosh, WI: Square D Company, 0104ED9501R8/96, August 1996.
13. National Electrical Code. NFPA 70, 2008.
14. N. Hatti, Y. Kondo, and H. Akagi. Five level diode-clamped PWM converters connected back-to-back for motor drives. *IEEE Trans. Indust. Appl.* 44(4): 1268–1276, July/August 2008.
15. H. Akagi, S. Fujita, S. Yonetani, and Y. Kondo. A 6.6 kV transformer-less STATCOM based on a five level diode-clamped PWM converter: System design and experimentation of 200-V, 10 kVA laboratory model. *IEEE Trans. Indust. Appl.* 44(2): 672–680, March/April 2008.
16. R. Yacamini. Power system harmonics, Part 4—Interharmonics. *Power Eng. J.* 10(4): 185–193, 1996.
17. CEI/IEC 1000-2-1:1990. Electromagnetic Compatibility, Part 2: Environment, Sect. 1: Description of the Environment—Electromagnetic Environment for Low-Frequency Conducted Disturbances and Signaling in Public Power Supply Systems, 1st edn. 1990-05, IEC Standard.
18. F.L. Luo and H. Ye. *Power Electronics*. Boca Raton, FL: CRC Press, 2010.

18

Effects of Harmonics

Harmonics have deleterious effects on electrical equipment. These can be itemized as follows (see Ref. [1]):

1. Capacitor bank failure because of reactive power overload, resonance, and harmonic amplification. Nuisance fuse operation
2. Excessive losses, heating, harmonic torques, and oscillations in induction and synchronous machines, which may give rise to torsional stresses
3. Increase in negative sequence current loading of synchronous generators, endangering the rotor circuit and windings
4. Generation of harmonic fluxes and increase in flux density in transformers, eddy current heating, and consequent derating
5. Overvoltages and excessive currents in the power system, resulting from resonance
6. Derating of cables due to additional eddy current heating and skin effect losses
7. Inductive interference with telecommunication circuits
8. Signal interference and relay malfunctions, particularly in solid-state and microprocessor-controlled systems
9. Interference with ripple control and power line carrier systems, causing misoperation of the systems, which accomplish remote switching, load control, and metering
10. Unstable operation of firing circuits based upon zero voltage crossing detection and latching
11. Interference with large motor controllers and power plant excitation systems

In Chapter 5, we discussed the high-frequency inrush currents on switching of shunt capacitor banks. Nonlinear loads in presence of capacitors can bring about a resonant condition with one of the load-generated harmonics as stated above and discussed further in Section 18.5. This also has the following additional effects:

- Increase the transient inrush current of the transformers and prolong its decay rate [2]
- Increase the duty on the switching devices (see Chapter 5)
- Possibility of part-winding resonance exists if the predominant frequency of the transient coincides with natural frequency of the transformer (see Section 18.5.3)

18.1 Rotating Machines

18.1.1 Pulsating Fields and Torsional Vibrations

Harmonics will produce elastic deformation, that is, shaft deflection, parasitic torques, vibrations noise, additional heating, and lower the efficiency of rotating machines.

The movement of the harmonics is with or against the direction of the fundamental. Criterion of forward or reverse rotation is established from $h = 6m \pm 1$, where h is the order of harmonic and m is any integer. If $h = 6m + 1$, the rotation is in the forward direction, but at $1/h$ speed. Thus, 7, 13, 19, ... harmonics rotate in the same direction as the fundamental. Harmonics of the order 1, 4, 7, 10, 13, ... are termed positive sequence harmonics (Table 17.1).

If $h = 6m - 1$, the harmonic rotates in a reverse direction to the fundamental. Thus, 5, 11, 17, ... are the reverse rotating harmonics. Harmonics of the order 2, 5, 8, 11, 14 ... are the negative sequence harmonics.

In a synchronous machine, the frequency induced in the rotor is net rotational difference between fundamental frequency and the harmonic frequency. Fifth harmonic rotates in reverse with respect to stator and the induced frequency is that of sixth harmonic. Similarly, the forward rotating seventh harmonic with respect to stator produces sixth harmonic in the rotor. The interaction of these fields produces a pulsating torque at 360 Hz and results in the oscillations of the shaft. Similarly, the harmonic pair 11 and 13 produces a rotor harmonic of 12th. If the frequency of the mechanical resonance exists close to these harmonics during starting, large mechanical forces can occur.

The same phenomena occur in induction motors. Considering slip of the induction motors, the positive sequence harmonics, $h = 1, 4, 7, 10, 13, \ldots$, produce a torque of $(h - 1 + s)\omega$ in the direction of rotation, and the negative sequence harmonics, $h = 2, 5, 8, 11, 14, \ldots$, produce a torque of $-(h + 1 - s)\omega$ opposite to that of rotation. Here s is the slip of the induction motor.

It is possible that harmonic torques are magnified due to certain combinations of stator and rotor slots and cage rotors are more prone to circulation of harmonic currents as compared with the wound rotors.

The zero sequence harmonics ($h = 3, 6, \ldots$) do not produce a net flux density. These produce ohmic losses.

All parasitic fields produce noise and vibrations. The harmonic fluxes superimposed upon the main flux may cause tooth saturation, and zigzag leakage can generate unbalanced magnetic pull, which moves around the rotor. As a result, the rotor shaft can deflect and run through a critical resonant speed amplifying the torque pulsations.

Torque ripples may exist at various frequencies. If the inverter is a six-step type, then a sixth harmonic torque ripple is created, which would vary from 36 to 360 Hz when the motor is operated over the frequency range of 6–60 Hz. At low speeds, such torque ripple may be apparent as observable oscillations of the shaft speed or as torque and speed pulsations, usually termed *cogging*. It is also possible that some speeds within the operating range may correspond to natural mechanical frequencies of the load or support structure. At such frequencies, amplification can occur, giving rise to large dynamic stresses. Operation other than momentary, that is, during starting, should be avoided at these speeds.

The oscillating torques in synchronous generators can simulate the turbine generator into complex-coupled mode of vibration that results in torsional oscillations of rotor elements and flexing of turbine buckets. If the frequency of a harmonic coincides with the turbine-generator torsional frequency, it can be amplified by the rotor oscillation.

A documented case of failure of a large generator is described in Ref. [3]. A control loop within an SVC unit in a nearby steel mill resulted in modulation of 60 Hz waveform. This created upper and lower sidebands, producing 55 and 65 Hz current components. The reverse phase rotation manifested itself as a 115 Hz stimulating frequency on the rotor, which drifted between 114 and 118 Hz. This excited sixth-mode natural frequency of the rotor shaft, creating large torsional stresses.

The NGH-SSR (after Narain Hingorani sub-synchronous resonance suppressor) [4,5] scheme can minimize sub-synchronous electrical torque and hence mechanical torque and shaft twisting, limit buildup of oscillations due to sub-synchronous resonance, and protect series capacitors from overvoltages.

18.1.2 Sub-Harmonic Frequencies and Sub-Synchronous Resonance

We discussed sub-harmonic frequencies in Chapter 13 in conjunction with a series compensation of transmission lines. Generally, the transient currents excited by sub-harmonic resonant frequencies damp out quickly due to positive damping. This is a stable sub-harmonic mode. Under certain conditions this can become unstable. We know that in a synchronous machine, the positive sequence sub-harmonic frequencies will set up a flux, which rotates in the same direction as the rotor, and its slip frequency $=f_e-f$, where f_e is the frequency of the sub-harmonic. As f_e is $<f$, it is a negative slip and contributes to the negative damping. The synchronous machine can convert mechanical energy into electrical energy associated with sub-harmonic mode. If the negative damping is large, it can swap the positive resistance damping in the system and a small disturbance can result in large levels of currents and voltages. The sub-harmonic torque brought about by the difference frequency f_e-f rotates in a backward direction with respect to the main field and if this frequency coincides with one of natural torsional frequencies of the machine rotating system, damaging torsional oscillations can be excited. This phenomenon is called sub-synchronous resonance.

18.1.3 Increase of Losses

The effect of the harmonics on motor losses should consider the subdivision of losses into windage and friction, stator copper loss, core loss, rotor copper loss, and stray loss in the core and conductors and the effect of harmonics on each of these components. The effective rotor and stator leakage inductance decreases and the resistance increases with frequency. The effects are similar as discussed for negative sequence in Chapter 13, but these will be more pronounced at higher frequencies. The rotor resistance may increase 4–6 times the dc value, while leakage reactance may reduce to a fraction of the fundamental frequency value. The stator copper loss increases in proportion to the square of the total harmonic current plus an additional increase due to higher resistance at skin effect at higher frequencies. Harmonics contribute to magnetic saturation and the effect of distorted voltage on core losses may not be significant. Major loss components influenced by harmonics are stator and rotor copper loss and stray losses. A harmonic factor of 11% gives approximately 25% derating of general purpose motors [6].

18.1.4 Effect of Negative Sequence Currents

Synchronous generators have both a continuous and a short-time unbalanced current capabilities, which are shown in Tables 18.1 and 18.2 [7,8]. These capabilities are based

TABLE 18.1

Requirements of Unbalanced Faults on Synchronous Machines

Type of Synchronous Machine	Permissible $I_2^2 t$
Salient pole generator	40
Synchronous condenser	30
Cylindrical rotor generators	
Indirectly cooled	30
Directly cooled (0–800 MVA)	10
Directly cooled (801–1600 MVA)	10 – (0.00625)(MVA-800)

TABLE 18.2

Continuous Unbalance Current Capability of Generators

Type of Generator and Rating	Permissible I_2 (%)
Salient pole, with connected amortisseur windings	10
Salient pole, with nonconnected amortisseur windings	5
Cylindrical rotor, indirectly cooled	10
Cylindrical rotor, directly cooled to	
960 MVA	8
961–1200 MVA	6
1201–1500 MVA	5

upon 120 Hz negative sequence currents induced in the rotor due to continuous unbalance or unbalance under fault conditions. In the absence of harmonics unbalance loads and impedance asymmetries (i.e., non-transposition of transmission lines) require that the generators should be able to supply some unbalance currents. When these capabilities are exploited for harmonic loading, the variations in loss intensity at different harmonics versus 120 Hz should be considered. The following expression can be used for equivalent heating effects of harmonics translated into negative sequence currents:

$$I_{2\,\text{eqiv}} = \left[\left(\frac{6f}{120}\right)^{1/2}(K_{5,7})(I_5 + I_7)^2 + \left(\frac{12f}{120}\right)^{1/2}(K_{11,13})(I_{11} + I_{13})^2 + \cdots\right]^{1/2} \quad (18.1)$$

where

$K_{5,7}$, $K_{11,13}$, ... are correction factors to convert from maximum rotor surface loss intensity to average loss intensity [9] (from Figure 18.1)

f is the fundamental frequency

I_5, I_7 are the harmonic current in pu values

Example 18.1

Consider a synchronous generator, with continuous unbalance capability of 0.10 pu (Table 18.2). It is subjected to fifth and seventh harmonic loading of 0.07 and 0.04 pu, respectively. Is the unbalance capability exceeded?

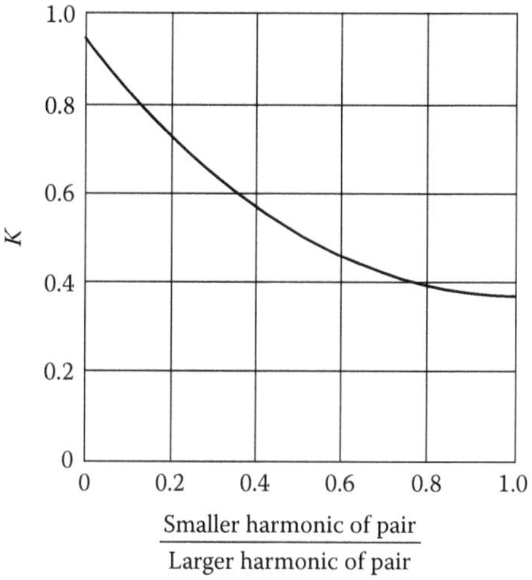

FIGURE 18.1
Ratio K for average loss to maximum loss based on harmonic pair. (From Ross, M.D. and Batchelor, J.W., *AIEE Trans.*, 62, 667, 1943.)

From Figure 18.1 and harmonic ratio 0.04/0.07 = 0.57, $K_{5,7} = 0.43$. From Equation 18.1,

$$I_{2\text{eqv}} = \left[\sqrt{3}(0.43)(0.07 + 0.04)\right]^{1/2} = 0.095$$

The continuous negative sequence capability is not exceeded. The example shows that the sum of harmonic currents in pu can exceed the generator continuous negative sequence capability, that is, in this case 11%, which exceeds 10% capability. Thus, calculation with simple summation is inaccurate.

18.1.5 Insulation Stresses

The high-frequency operation of modern PWM converters with IGBTs is discussed in Chapter 17. It subjects the motors to high dv/dt. It has an adverse effect on the motor insulation and contributes to the motor-bearing currents and shaft voltages. The rise time of the voltage pulse at the motor terminals influences the voltage stresses on the motor windings. As the rise time of the voltage becomes higher, the motor windings behave like a network of capacitive elements in series. The first coils of the phase windings are subject to overvoltages, as shown in Figure 18.2, which shows ringing. There has been a documented increase in the insulation failure rate caused by turn-to-turn shorts or phase-to-ground faults due to high dv/dt stress [10]. The common remedies are to provide inverter grade insulation or to add filters. The soft switching slows initial rate of rise as discussed in Chapter 17.

NEMA [6] has established limitations on voltage rise for general-purpose NEMA Design A and B induction motors and definite purpose-inverter-fed motors. Windings designed for definite purpose inverter grade motors use magnet wires with increase build and these polyester-based wires exhibit higher breakdown strength.

- The stator winding insulation system of general-purpose motors rated ≤600 V shall withstand $V_{peak} = 1$ kV and rise time ≥2 μs and for motors rated >600 V these limits are $V_{peak} \leq 2.5$ pu and rise time ≥1 μs.

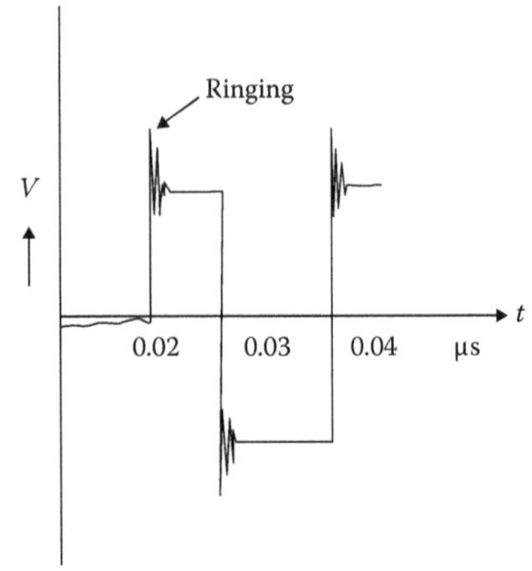

FIGURE 18.2
Pulse-width modulation, ringing in output voltage waveform due to traveling wave phenomena in interconnecting cables.

- For definite-purpose inverter-fed motors with base voltage rating ≤600 V, $V_{peak} \leq$ 1600 V and rise time is ≥0.1 μs. For motors with base voltage rating >600 V, $V_{peak} \leq 2.5$ pu and rise time ≥0.1 μs. V_{peak} is of single amplitude and 1 pu is the peak of the line-to-ground voltage at the maximum operating speed point.

The derating due to harmonic factor, effect on motor torque, starting current, and power factor are described in NEMA [6].

The motor windings can be exposed to higher than normal voltages due to neutral shift and common mode voltages [11], and in some current source inverters it can be as high as 3.3 times the crest of the nominal sinusoidal line-to-ground voltage.

Harmonics also impose higher dielectric stresses on the insulation of other electrical apparatuses. Harmonic overvoltages can lead to corona, void formation, and degradation.

18.1.6 Bearing Currents and Shaft Voltages

PWM inverters give rise to additional shaft voltages due to voltage and current spikes superimposed on the phase quantities during inverter operation. These will cause currents to flow through the bearings. If shaft voltages higher than 300 mV peak occur, the motor should be equipped with insulated bearings or shaft should be grounded.

18.1.7 Effect of Cable Type and Length

When the motor is connected through long cables, the high dv/dt pulses generated by PWM inverters cause traveling wave phenomena on the cables, resulting in reinforcement of incident and reflected waves due to impedance discontinuity at the motor terminals. And the voltages can reach twice the inverter output voltage. An anology can be drawn with long transmission lines and traveling wave phenomena. The incident traveling wave is reflected at the motor terminals, and reinforcement of incident and reflected waves occurs. Due to dielectric losses and cable resistance, damped ringing occurs as the wave is reflected from one end of the cable to the other. The ringing frequency is a function of cable length and wave propagation velocity, and is of the order of 50 kHz to 2 MHz [12]. Figure 18.2 is a representation of the ringing frequency. The type of cable between motor and the drive system is important.

By adding output filters, the cable charging current as well as dielectric stresses on the motor insulation can be reduced. The common filter types are

- Output line inductors
- Output limit filter
- Sine-wave filter
- Motor termination filter

An output inductor reduces dv/dt at the inverter and the motor. The ringing and overshoot may also be reduced. Output limit filters may consist of laminated core inductor or ferrite core inductors. A sine wave filter is a conventional low-pass filter formed from an output inductor, capacitor, and a damping resistor. Motor termination filters are first-order resistor/capacitor filters (Chapter 20).

18.2 Transformers

A transformer supplying nonlinear load may have to be derated. Harmonics effect transformer losses and eddy current loss density. The upper limit of the current distortion factor is 5% of the load current and the transformer should be able to withstand 5% overvoltage at rated load and 10% at no load. The harmonic currents in the applied voltage should not exceed these limits.

In addition to derating due to increased harmonic-current-induced eddy current loss, a drive system transformer may be subjected to severe current cycling and load demand, depending upon the drive system. The following calculations are based upon Ref. [13].

The losses in a transformer can be divided into (1) no-load losses and (2) load losses. The load losses consist of copper loss in windings and stray load losses. The stray load losses can be subdivided into the losses in the windings, and losses in the non-winding components of the transformer, that is, core clamps, structures, and tank. The total transformer loss P_{LL} is

$$P_{LL} = P + P_{EC} + P_{OSL} \tag{18.2}$$

where
P is I^2R loss
P_{EC} is the winding eddy current loss
P_{OSL} is other stray loss

The other stray load loss will increase proportional to the square of the current. However, these will not increase proportional to the square of the frequency, as in the winding eddy losses. The studies show that eddy current loss in bus bars, connections, and structural parts increase by a harmonic exponential factor of 0.8 or less. The effect of these losses varies depending upon the type of transformer.

For the liquid-immersed transformers, the top-oil rise θ_{TO} will increase as the total load losses increase due to harmonic loading. Unlike dry-type transformers where P_{OSL} is ignored, it must be considered for oil-immersed transformers as it impacts the top-oil temperature.

If the rms value of the current including harmonics is the same as the fundamental current, I^2R loss will be maintained the same. If the rms value due to harmonics increases, so does the I^2R loss.

$$I_{(\mathrm{pu})} = \left[\sum_{h=1}^{h=\max} (I_{h(\mathrm{pu})})^2 \right]^{1/2} \tag{18.3}$$

The eddy current loss P_{EC} is assumed to vary in proportion to the square of the electromagnetic field strength. Square of the harmonic current or the square of the harmonic number may be considered to be representative of it. Due to skin effect, the electromagnetic flux may not penetrate conductors at high frequencies. The leakage flux has its maximum concentration between the interfaces of the two windings:

$$P_{\mathrm{EC(pu)}} = P_{\mathrm{EC\text{-}R(pu)}} \sum_{h=1}^{h=\max} I_{h(\mathrm{pu})}^2 h^2 \tag{18.4}$$

where

$P_{\mathrm{EC\text{-}R}}$ is the winding eddy current loss under rated conditions

$I_{h(\mathrm{pu})}$ is the per unit rms current at harmonic h

To facilitate actual field measurements define winding eddy current loss at measured current and the power frequency by another term $P_{\mathrm{EC\text{-}O}}$. Then we can write the equation

$$P_{\mathrm{EC}} = P_{\mathrm{EC\text{-}O}} \times \frac{\sum_{h=1}^{h=h_{\max}} I_h^2 h^2}{I^2} = P_{\mathrm{EC\text{-}O}} \times \frac{\sum_{h=1}^{h=h_{\max}} I_h^2 h^2}{\sum_{h=1}^{h=h_{\max}} I_h^2} \tag{18.5}$$

where I is the rms load current. Define harmonic loss factor F_{HL} for the windings as $P_{\mathrm{EC}}/P_{\mathrm{EC\text{-}O}}$ in Equation 18.5.

$$F_{\mathrm{HL}} = \frac{\sum_{h=1}^{h=h_{\max}} [I_h/I]^2 h^2}{\sum_{h=1}^{h=h_{\max}} [I_h/I]^2} \tag{18.6}$$

In the above equation I can be substituted with I_1, where I_1 is the rms fundamental load current. It can be shown that *whether we normalize with respect to I or* I_1, the F_{HL} calculation gives the same results.

The heating due to other stray losses is not a consideration for dry-type transformers, since the heat generated is dissipated by the cooling air. With $P_{\mathrm{OSL}} = 0$, all the stray loss is assumed to occur in the windings. P_{LL} can be written as

$$P_{\mathrm{LL(pu)}} = \sum_{h=1}^{h=\max} I_{h(\mathrm{pu})}^2 \times (1 + F_{\mathrm{HL}} P_{\mathrm{EC\text{-}R(pu)}})\mathrm{pu} \tag{18.7}$$

To adjust the per unit loss density in the individual windings, the effect of F_{HL} must be known on each winding. The per unit value of the non-sinusoidal current for the dry-type transformers, which will make the result of Equation 18.7 equal to the design value of the loss density in the highest loss region for rated frequency and for rated current, is given by the following equation:

$$I_{\max(\text{pu})} = \left[\frac{P_{\text{LL-R(pu)}}}{1 + [F_{\text{HL}} \times P_{\text{EC-R(pu)}}]}\right]^{1/2} \tag{18.8}$$

18.2.1 Calculations from Transformer Test Data

The calculations of $P_{\text{EC-R}}$ can be made from the transformer test data. The maximum eddy current loss density is assumed 400% of the average value for that winding. The division of eddy current loss between the windings is

- Sixty percent in inner winding and 40% in outer winding in all transformers having a self-cooled rating of <1000 A, regardless of turns ration.
- Sixty percent in inner winding and 40% in outer winding in all transformers for all transformers having turns ratio of 4:1 or less.
- Seventy percent in the inner winding and 30% in outer winding for all transformers having turns ratio >4:1 and also having one or more windings with a maximum self-cooled rating of >1000 A.
- The eddy current loss distribution within each winding is assumed nonuniform.

The stray loss component of the load loss is calculated by the following expression:

$$\begin{aligned} P_{\text{TSL-R}} &= \text{Total load loss} - \text{copper loss} \\ &= P_{\text{LL}} - K\left[I_{(1-R)}^2 R_1 + I_{(2-R)}^2 R_2\right] \end{aligned} \tag{18.9}$$

In this expression for copper loss, R_1 and R_2 are the resistances measured at the winding terminals (i.e., H1 and H2 or X1 and X2) and should not be confused with winding resistances of each phase. $K=1$ for single-phase transformers, and 1.5 for the three-phase transformers.

Sixty-seven percent of the stray loss is assumed to be winding eddy losses for dry-type transformers:

$$P_{\text{EC-R}} = 0.67 P_{\text{TSL-R}} \tag{18.10}$$

Thirty-three percent of the total stray loss is assumed to be winding eddy losses for liquid immersed

$$P_{\text{EC-R}} = 0.33 P_{\text{TSL-R}} \tag{18.11}$$

The other stray losses are given by

$$P_{\text{OSL-R}} = P_{\text{TSL-R}} - P_{\text{EC-R}} \tag{18.12}$$

As the low-voltage winding is the inner winding, maximum $P_{\text{EC-R}}$ is given by

$$\max P_{\text{EC-R(pu)}} = \frac{K_1 P_{\text{EC-R}}}{K(I_{(2-R)})^2 R_2}\,\text{pu} \tag{18.13}$$

where K_1 is the division of eddy current loss in the inner winding, equal to 0.6 or 0.7, multiplied by the maximum eddy current loss density of 4.0 per unit, that is, 2.4 or 2.8, depending upon the transformer turns ratio and the current rating and K has already been defined depending on the number of phases.

Example 18.2

A delta–wye-connected dry-type isolation transformer of 13.8–2.4 kV is required for a 2.3 kV 3000 hp drive motor connected to a load commutated inverter (LCI), with the following current spectrum:

$$I = 709\ \text{A},\quad I_5 = 123\ \text{A},\quad I_7 = 79\ \text{A},\quad I_{11} = 31\ \text{A},\quad I_{13} = 20\ \text{A},\quad I_{17} = 11\ \text{A},$$
$$I_{19} = 7\ \text{A},\quad I_{23} = 6\ \text{A},\quad I_{25} = 5\ \text{A}$$

Calculate the harmonic loading. The following data are supplied by the manufacturer:
$R_1 = 1.052\ \Omega$, $R_2 = 0.0159\ \Omega$ and total load loss = 23,200 W at 75°C. Calculate the derating when supplying the same harmonic current spectrum.
The primary 13.8 kV winding current = 71.9 A. The copper loss in the windings is

$$1.5[(1.052)(71.19)^2 + (0.0159)(709)^2] = 19{,}986.3\ \text{W}$$

From the given loss data

$$P_{\text{TSL-R}} = 23{,}200 - 19{,}986.3 = 3{,}213.7\ \text{W}$$

The winding eddy loss is then

$$P_{\text{EC-R}} = 3213.7 \times 0.67 = 2153.2\ \text{W}$$

The transformer has a ratio of >4:1, but secondary current is <1000 A. Then, from Equation 18.13 the maximum $P_{\text{EC-Rmax}}$ is

$$P_{\text{EC-Rmax}} = \frac{2.4 \times 2153.2}{1.5(0.0159)(709)^2} = 0.431\ \text{pu}$$

Table 18.3 is constructed, based upon the transformer fundamental current of 709 A as the base current, the steps of calculation are obvious. The first five columns of this table apply to the calculations in this example. Then $F_{\text{HL}} = 2.7964/1.04 = 2.69$

$$P_{\text{LL(pu)}} = 1.04 \times (1 + 0.43 \times 2.69) = 2.24$$

Then from Equation 18.8, the maximum permissible value of non-sinusoidal current is

$$I_{\max\,(\text{pu})} = \sqrt{\frac{1.43}{1 + 2.69 \times 0.43}} = 0.81\ \text{pu}$$

Thus, the transformer capability with given non-sinusoidal load current harmonic composition is approximately 81% of its sinusoidal load current capability = 574.3 A. To properly apply a transformer for this application raises the transformer kVA by approximately 24%.

TABLE 18.3

Calculations of Derating of a Transformer Due to Harmonic Loads

h	I_h/I	$(I_h/I)^2$	h^2	$(I_h/I)^2\ h^2$	$h^{0.8}$	$(I_h/I)^2 h^{0.8}$
1	2	3	4	5	6	7
1	1.0	1.0	1.0	1.0	1.0	1.0
5	0.171	0.029	25	0.725	3.62	0.105
7	0.111	0.012	49	0.588	4.74	0.057
11	0.043	0.0018	121	0.2178	6.81	0.012
13	0.028	0.0006	169	0.1014	7.78	0.005
17	0.015	0.00022	289	0.0636	9.65	0.002
19	0.010	0.00010	361	0.0361	10.54	0.001
23	0.008	0.000064	529	0.0339	12.28	0.0008
25	0.007	0.000049	625	0.0306	13.13	0.0006
Σ		1.04		2.7964		1.183

18.2.2 Liquid-Filled Transformers

Liquid-filled transformers are similar to dry-type transformers except that effects of all stray losses are considered. For self-cooled OA (ONAN) mode, the top-oil temperature rise above ambient is given by

$$\theta_{\text{TO}} = \theta_{\text{TO-R}}\left[\frac{P_{\text{LL}} + P_{\text{NL}}}{P_{\text{LL-R}} + P_{\text{NL}}}\right]^{0.8} {}^\circ\text{C} \tag{18.14}$$

where

$\theta_{\text{TO-R}}$ is the top-oil temperature rise over ambient under rated conditions
P_{NL} is the no-load loss
P_{LL} is the load loss under rated conditions

Also

$$P_{\text{LL}} = P + F_{\text{HL}}P_{\text{EC}} + F_{\text{HL-STR}}P_{\text{OSL}} \tag{18.15}$$

where

$F_{\text{HL-STR}}$ is the harmonic loss factor for the other stray loss
P is I^2R portion of the load loss

The winding hottest spot conductor rise is given by

$$\theta_g = \theta_{g-R}\left(\frac{1 + F_{\text{HL}}P_{\text{EC-R(pu)}}}{1 + P_{\text{EC-R(pu)}}}\right)^{0.8} \tag{18.16}$$

where θ_g and θ_{g-R} are the hottest spot conductor rise over top oil under harmonic loading and under rated conditions, respectively.

For liquid-immersed transformers Equation 18.16 becomes

$$\theta_{g1} = \theta_{g1-R}\left(\frac{1 + 2.4F_{\text{HL}}P_{\text{EC-R(pu)}}}{1 + 2.4P_{\text{EC-R(pu)}}}\right)^{0.8} {}^\circ\text{C} \tag{18.17}$$

or

$$\theta_{g1} = \theta_{g1-R}\left(\frac{1 + 2.8F_{HL}P_{EC\text{-}R(pu)}}{1 + 2.8P_{EC\text{-}R(pu)}}\right)^{0.8} \,^{\circ}\mathrm{C} \tag{18.18}$$

where

θ_{g1} is the hottest-spot HV conductor rise over top-oil temperature

θ_{g1-R} is the hottest-spot HV conductor rise over top-oil under rated conditions

Example 18.3

Consider that the transformer in Example 18.2 is liquid immersed, ONAN, self-cooled. The top-oil temperature rise and winding hottest spot conductor rise are required to be calculated. In addition to the data in Example 18.2, the no-load loss of the transformer is 5000 W.

From Example 18.2:

$$P_{OSL\text{-}R} = P_{TSL\text{-}R} - P_{EC\text{-}R} = 3213.7 - 2153.2 = 1060.5\ \mathrm{W}$$

Assume the following temperature rises, which are normal for 55°C rated transformers.

HV and LV windings average = 55°C

Top-oil rise = 55°C

Hottest-spot conductor rise = 65°C

The harmonic loss factor is 2.69 as calculated in Example 18.2. Columns 6 and 7 are added to Table 18.3 and harmonic loss factor for the other stray loss is summation of column 7 divided by summation of column 3 = 1.183/1.04 = 1.1375. Assuming that harmonic load distribution is equal to 100% of the fundamental current, the calculated losses are

No load = 5,000 W

I^2R loss = 19,986.3 W

Winding eddy = 2,153.2 W

Other stray = 1,060.5 W

The total gives a rated loss of 28,200 W. This checks with the given loss data by the manufacturer = 5,000 + 23,200 = 28,200 W.

From Table 18.3, the rms current is given by square root of third column summation, that is, $\sqrt{1.04} = 1.02$. The total losses must be corrected to reflect rms current and also load factor. Here assume that load factor is 100%. Therefore, correction for the rms current is

$$P_{LL} = I^2_{rms\text{-}pu}L_f^2 = 1.02^2 \times 1^2 = 1.04$$

Correct the losses using harmonic multipliers calculated above:

Winding eddy = 1.04 × 2.69 × 2153.2 = 6023.8 W

Other stray loss = 1.04 × 1.1375 × 1060.5 = 1254.6 W

I^2R loss = 1.04 × 19,986.4 = 20,785.9 W

Thus, the total losses are now = 33,064 W

Then the top-oil temperature form Equation 18.14 is

$$\theta_{\mathrm{TO}} = 55 \times \left(\frac{33{,}064}{28{,}200}\right)^{0.8} = 62.47°\mathrm{C}$$

The rated inner winding losses corrected for rms current are

$$1.04 \times 1.5(0.0159)(709)^2 = 12{,}468\,\mathrm{W}$$

The low-voltage winding currents are less than 1000 A. Sixty percent of the winding eddy loss is assumed to occur in the LV winding and the maximum eddy loss at the hottest region is assumed to be four times average eddy loss. The hottest-spot conductor rise over top-oil temperature is calculated using Equation 18.17.

$$\theta_g = (65 - 55) \times \left(\frac{12{,}468 + 6{,}023.8 \times 2.4}{11{,}988 + 2{,}153 \times 2.4}\right)^{0.8} = 14.34°\mathrm{C}$$

Then the hottest-spot temperature is $62.47 + 14.34 = 76.81°\mathrm{C}$.

This exceeds the maximum of 65°C by 11.81°C. The transformer on continuous operation will be seriously damaged. This demonstrates that liquid-filled or air-cooled transformers cannot be loaded to their rated fundamental frequency rating when supplying nonlinear loads, and the derating must be calculated in each case.

18.2.3 UL *K*-Factor of Transformers

The UL (Underwriter's Laboratories) standards [14,15] also specify transformer derating (*K*-factors) when carrying non-sinusoidal loads. The UL *K*-factor is given by the expression

$$K\text{-factor} = \sum_{h=1}^{h=h_{\max}} I_{h-\mathrm{pu}}^2 h^2 \tag{18.19}$$

$I_{h-\mathrm{pu}}$ is rms current at harmonic h.

This can be shown to be equal to

$$K\text{-factor} = \left(\frac{\sum_{h=1}^{h=\max} I_h^2}{I_R^2}\right) F_{\mathrm{HL}} \tag{18.20}$$

I_R is rated fundamental rms current.

The harmonic loss factor is a function of the harmonic current distribution and independent of its relative magnitude. UL *K*-factor is dependent upon both the magnitude and distribution of the harmonic current. For a transformer with harmonic currents specified as per unit of the rated transformer secondary currents, the *K*-factor and harmonic loss factor will have the same numerical values. That is, the *K*-factor is equal to harmonic factor only when square root of the sum of harmonic currents squared equals the rated secondary current of the transformer.

There is marked difference between the derating in ANSI/IEEE recommendations and UL *K*-factor. If a transformer has 3% eddy current loss and a *K*-factor of 5, then the eddy current loss increases to $3\% \times 5\% = 15\%$. The UL derating ignores eddy current loss gradient. Figure 18.3 shows higher derating with the ANSI method of calculation for a six-pulse harmonic current spectrum, assuming a theoretical magnitude of harmonics of $1/h$, for a *K*-factor load of approximately 9.

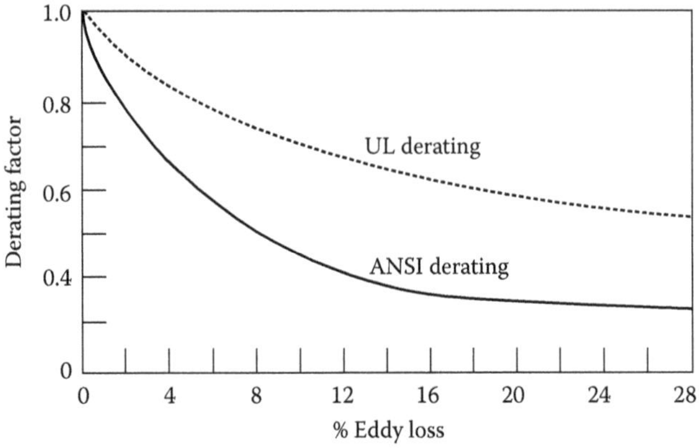

FIGURE 18.3
Relative derating of transformers: ANSI method and UL K-factor.

Example 18.4

Calculate the UL K-factor for the transformer loading harmonic spectrum of Example 18.1:
From Table 18.3, the UL K-factor = 2.7964. K-factor varies with the transformer harmonic load.

18.3 Cables

A non-sinusoidal current in a conductor causes additional losses. The ac conductor resistance is changed due to skin and proximity effects. Both these effects are dependent upon frequency, conductor size, cable construction, and spacing. We observed that even at 60 Hz the ac resistance of conductors is higher than the dc resistance (Table 7.12). With harmonic currents these effects are more pronounced. The ac resistance is given by

$$\frac{R_{ac}}{R_{dc}} = 1 + Y_{cs} + Y_{cp} \tag{18.21}$$

where
Y_{cs} is due to conductor resistance resulting from skin effect
Y_{cp} is due to proximity effect

The skin effect is an ac phenomenon, where the current density throughout the conductor cross section is not uniform and the current tends to flow more densely near the outer surface of the conductor than toward the center. This is because an ac flux results in induced EMFs, which are greater at the center than at the circumference, so that potential difference tends to establish currents that oppose the main current at the center and assist it at the circumference. The result is that the current is forced to the outside, reducing the effective area of the conductor. The effect is utilized in high ampacity hollow conductors and tubular bus bars, to save material costs. The skin effect is given by [16]

$$Y_{cs} = F(x_s) \tag{18.22}$$

where
Y_{cs} is due to skin effect losses in the conductor
$F(x_s)$ is the skin effect function

$$x_s = 0.875\sqrt{f\frac{k_s}{R_{dc}}} \tag{18.23}$$

where the factor k_s depends upon the conductor construction.

The proximity effect occurs because of distortion of current distribution between two conductors in close proximity. This causes concentration of current in parts of the conductors or bus bars closest to each other (currents flowing in forward and return paths). The expressions and graphs for calculating the proximity effect are given in Ref. [16]. The increased resistance of conductors due to harmonic currents can be calculated, and the derated capacity is

$$\frac{1}{1+\sum_{h=1}^{h=\max} I_h^2 R_h} \tag{18.24}$$

where
I_h is the harmonic current
R_h is the ratio of the resistance at that harmonic with respect to the conductor dc resistance

For a typical six-pulse harmonic spectrum, the derating is approximately 3%–6%, depending upon the cable size (Figure 18.4). See also Ref. [17].

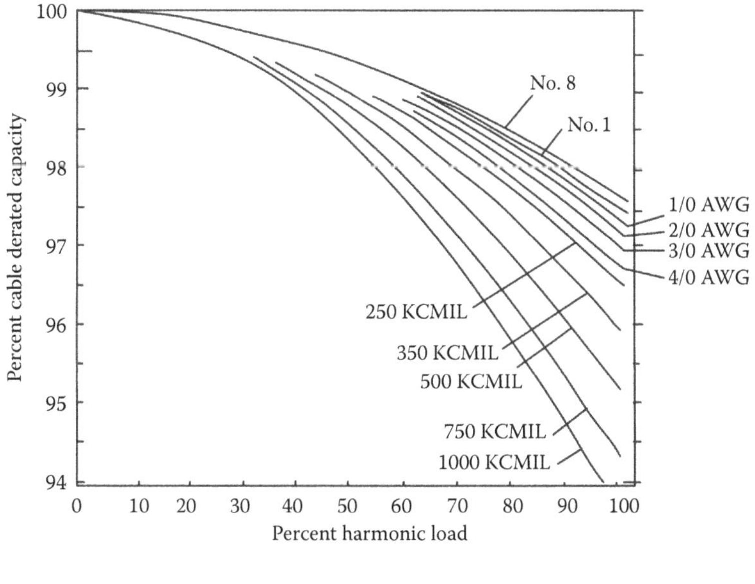

FIGURE 18.4
Derating of cables with typical six-pulse converter circuit.

18.4 Capacitors

The main effect of harmonics on capacitors is that a resonance condition can occur with one of the load-generated harmonics (see Section 18.5). Capacitors may be located in an industrial plant or close to such a plant that has significant harmonic producing loads. This location is very possibly subject to harmonic resonance and should be avoided or the capacitors can be used as harmonic filters. On a sub-transmission system where the capacitors are located far from the harmonic producing loads, the propagation of harmonics through the interconnecting systems need to be studied. The linear loads served from a common feeder, which also serves nonlinear loads of some other consumers, may become susceptible to harmonic distortion. There have been two approaches, one to consider capacitor placement from a reactive power consideration and then study the harmonic effects, and the second to study the fundamental frequency voltages, reactive power, and harmonic effects simultaneously. A consumer system that does not have harmonic-producing loads can be subject to harmonic pollution due to harmonic loads of other consumers in the system.

The capacitors can be severely overloaded due to harmonics, especially under resonant conditions and can even be damaged. The capacitors are intended to be operated at or below their rated voltage and frequency. The capacitors are capable of continuous operation under contingency system and capacitor bank conditions, provided that none of the following is exceeded [18,19].

110% of the rated rms voltage. If harmonics are present, this means

$$V_{\text{rms}} \leq 1.1 = \left[\sum_{h=1}^{h=h_{\max}} V_h^2\right]^{1/2} \tag{18.25}$$

The crest not to exceed $1.2 \times \sqrt{2}$ times the rated rms voltage, including harmonics but excluding transients.

$$V_{\text{crest}} \leq 1.2 \times \sqrt{2} \sum_{h=1}^{h=h_{\max}} V_h \tag{18.26}$$

The rms current should not exceed 135% of nominal rms current based upon rated kvar and rated voltage, including fundamental and harmonic currents.

$$I_{\text{rms}} \leq 1.35 = \left[\sum_{h=1}^{h=h_{\max}} I_h^2\right]^{1/2} \tag{18.27}$$

135% of rated kvar, that is, if harmonics are present:

$$\text{k var}_{\text{pu}} \leq 1.35 = \left[\sum_{h=1}^{h=h_{\max}} (V_h I_h)\right] \tag{18.28}$$

Unbalances within a capacitor bank, due to capacitor element failures and or individual fuse operations result in overvoltages on some capacitor units. Typically, individual

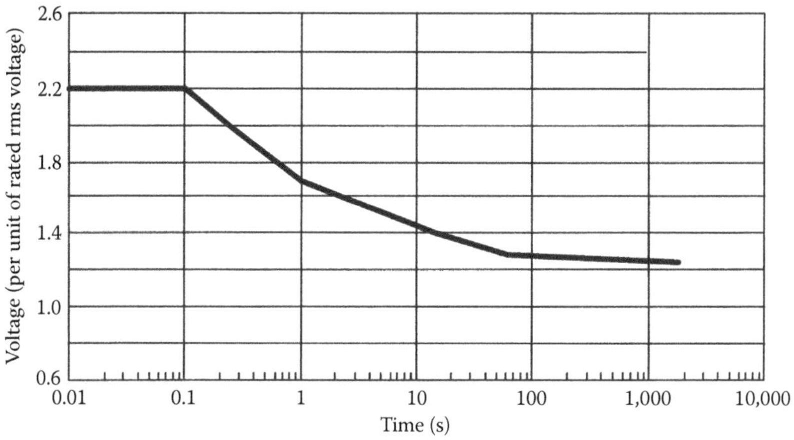

FIGURE 18.5
Maximum contingency power frequency overvoltage capability of capacitor units. (From IEEE, IEEE Standard for Shunt Power Capacitors, 2002, Standard 18.)

capacitor voltage is allowed to increase by 10% before unbalance detection removes capacitor banks from service. In a filter bank, failure of a capacitor unit will cause detuning. The limitations of the capacitor bank loadings become of importance in the design of capacitor filters (Chapter 20).

A capacitor tested according to IEEE Std. 18 will withstand a combined total of 300 applications of power frequency terminal-to-terminal overvoltages without superimposed transients or harmonic content; the magnitude and duration are shown in Figure 18.5 [19].

The capacitor unit is also expected to withstand transient currents inherent in the operation of power systems, which include infrequent high lightning currents and discharge currents due to nearby faults (see chapter 20). For frequent back-to-back switching the peak capacitor unit current should be held to a value lower than that shown in Figure 18.6.

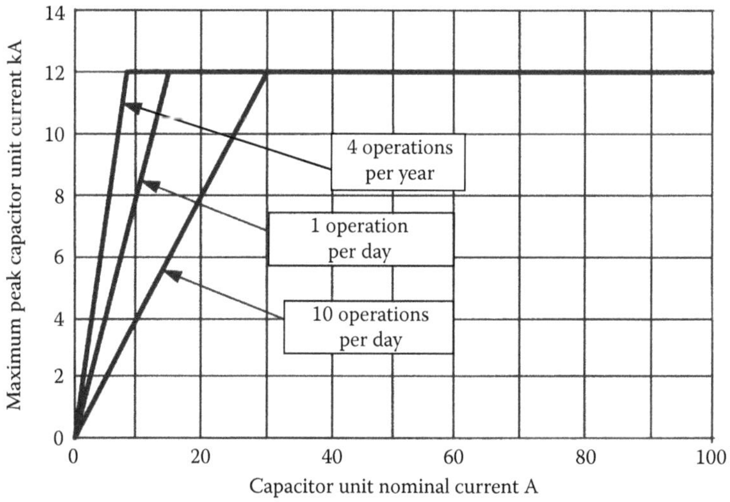

FIGURE 18.6
Transient current capability of capacitor units for regularly occurring transients.

The capacitor bank current is the number of capacitor unit current multiplied by the number of capacitor units or strings in parallel. In this figure, the curves are based upon straight lines from origin to 12 kA, and the slopes are 1500 times the nominal current for four operations per year, 800 times the nominal current for one operation per day, and 400 times the nominal current for 10 operations per day.

18.5 Harmonic Resonance

18.5.1 Parallel Resonance

When shunt capacitors are connected for power factor improvement, voltage support, or reactive power compensation, these act in parallel with the system impedance. Ignoring resistance, the impedance of the parallel combination is

$$\frac{j\omega L(1/j\omega C)}{(j\omega L + 1/j\omega C)} \tag{18.29}$$

This means that the system impedance is considered solely inductive in parallel with a capacitor. Assuming that L and C remain invariant with frequency, resonance occurs at a frequency where the inductive and capacitive reactance is equal and the denominator of Equation 18.29 is zero, that is, the impedance of the combination is infinite for a lossless system. The impedance angle changes abruptly as the resonant frequency is crossed (Figure 18.7). The inductive impedance of the power source and distribution (utility, transformers, generators, and motors) as seen from the point of application of the capacitors equals the capacitive reactance of the power capacitors at the resonant frequency

$$j2\pi f_n L = \frac{1}{j2\pi f_n C} \tag{18.30}$$

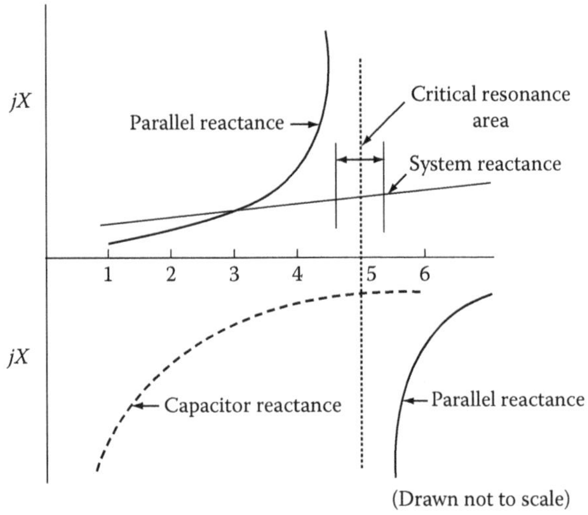

FIGURE 18.7
Parallel resonance in a lossless system with lumped inductance and capacitance.

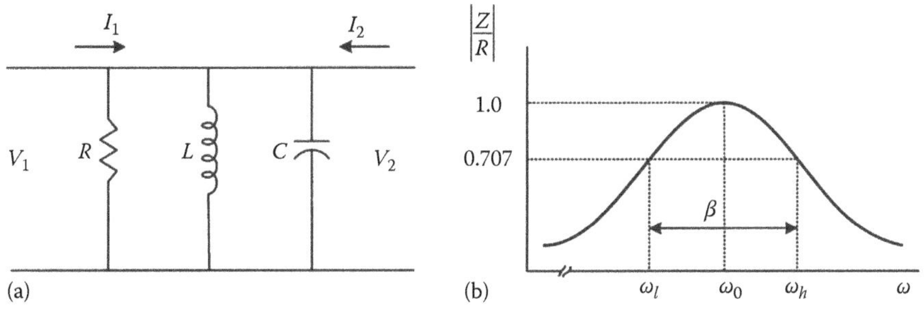

FIGURE 18.8
(a) A circuit for parallel resonance. (b) The impedance versus frequency plot.

where f_n is the resonant frequency. Depending upon the relative values of the inductance and capacitance, it may happen that this frequency coincides with one of the load-generated harmonics, say the fifth. If the resonant frequency is 300 Hz, $5j\omega L = 1/(5j\omega C)$, where ω pertains to the fundamental frequency. The capacitors form one branch of the tuned resonant circuit, while the rest of the system acts like the other parallel branch as seen from the point of harmonic current source. When excited at the resonant frequency, a harmonic current magnification occurs in the parallel tuned circuit, though the exciting input current is small. This magnification can be many times the exciting current and can even exceed the fundamental frequency current. It overloads the capacitors, may result in nuisance fuse operation and sever amplification of the harmonic currents and resulting distortion in the power system, which has consequent deleterious effects on the power system components.

Consider now a parallel RLC circuit (Figure 18.8a). The impedance is high at resonant frequency:

$$y = \frac{1}{R} + \frac{1}{j\omega L} + j\omega C \tag{18.31}$$

At resonance,

$$-\frac{1}{\omega L} + \omega C = 0 \quad \text{or} \quad \omega = \omega_{\alpha} = \frac{1}{\sqrt{LC}} \tag{18.32}$$

The magnitude Z/R with respect to frequency is plotted in Figure 18.8b.

The bandwidth β shown in Figure 18.8b is given by

$$\beta = \frac{\omega_a}{Q_a} \tag{18.33}$$

where the quality factor for the parallel circuit is given by

$$Q_a = \frac{R}{\omega_a L} = \omega_a RC = R\sqrt{\frac{C}{L}} \tag{18.34}$$

Resonance with one of the load-generated harmonics is the major effect of harmonics in presence of power capacitors. This condition has to be avoided in any application of power capacitors. More conveniently, the resonant condition in the power system can be expressed as

$$h = \frac{f_n}{f} = \sqrt{\frac{\text{KVA}_{\text{sc}}}{\text{kvar}_{\text{c}}}} \tag{18.35}$$

where

h is the order of harmonics
f_n is the resonant frequency
f is the fundamental frequency
KVA_{sc} is the short-circuit duty at the point of application of the shunt capacitors
kvar_{c} is the shunt power capacitor rating in kvar

Consider that the short-circuit level at the point of application of power capacitors is 500 MVA. Resonance at 5th, 7th, 11th, 13th, ..., harmonics will occur for a shunt capacitor size of 20, 10.2, 4.13, 2.95, ..., Mvar, respectively. The smaller the size of the capacitor, the higher is the resonant frequency.

The short-circuit level in a system is not a fixed entity. It varies with the operating conditions. In an industrial plant these variations may be more pronounced than in the utility systems, as a plant generator or part of the plant rotating loads may be out of service, depending upon the variations in processes and operation. The resonant frequency in the system will *float around*.

The lower-order harmonics are more troublesome from a resonance point of view. As the order of harmonics increases, their magnitude reduces (Chapter 17). Sometimes, a harmonic analysis study is limited to 25th or 29th harmonic only. This may not be adequate and possible resonances at higher frequencies may be missed. Table 18.4 shows the harmonic current spectrum at the secondary of a lightly loaded 480-V transformer, with a capacitor bank of 40 kvar. Resonance occurs at 29th and 31st harmonics and current distortion at these harmonics is 17.7% and 17.5%, respectively. The total THD is 25.4. Table 18.5 shows the effect of switching more capacitor units as the load develops, with a 160-kvar bank. Now the resonance occurs at 11th and 13th harmonics and the THD is 7.72%. These distortion limits are beyond the acceptable limits (see Appendix F). This shows that the higher-order harmonics can be equally troublesome and give rise to higher distortions, if a resonant condition occurs.

It can be concluded that

- The resonant frequency will swing around depending upon the changes in the system impedance, e.g., switching a tie circuit on or off, operation at reduced load. Some sections of the capacitors may be switched out altering the resonant frequency.
- An expansion or reorganization of the distribution system may bring out a resonant condition where none existed before.
- Even if the capacitors in a system are sized to escape current resonant condition, immunity from future resonant conditions cannot be guaranteed owing to system changes, e.g., increase in the short-circuit level of the utility system.

TABLE 18.4

Measurements of Harmonic Spectrum of a 480 V Substation with 40 kvar Capacitor Bank, Fundamental Current 35.4 A

Harmonic	Percentage	Phase Angle (°)	Harmonic	Percentage	Phase Angle (°)
5	0.6	146	22	0.2	−98
7	0.2	−59	24	0.2	−98
9	0.2	−138	26	0.2	−142
11	1.1	4	28	0.3	171
13	1.5	−9	30	0.6	28
17	0.4	51	32	0.3	84
19	1.8	51	36	0.2	37
21	0.5	26			
23	1.4	−88			
25	1.5	−96			
27	1.4	83			
29	17.7	31			
31	17.5	−106			
35	1.8	30			
37	2.3	59			
41	1.6	−28			
43	0.4	148			
Σ Odd	25.3%		Σ Even	0.9	

Note: THD: 25.42%.

TABLE 18.5

Measurements of Harmonic Spectrum of a 480 V Substation with 160 kvar Capacitor Bank, Fundamental Current 148.5 A

Harmonic	Percentage	Phase Angle (°)	Harmonic	Percentage	Phase Angle (°)
5	0.8	141	14	0.2	21
7	0.3	−63	16	0.2	−62
9	0.3	−125			
11	2.7	24			
13	6.7	63			
15	0.9	−10			
17	0.6	−151			
19	0.6	−179			
21	0.2	168			
25	0.2	34			
29	1.5	128			
31	1.5	−17			
37	0.2	82	Σ Even	0.4	
41	0.2	−7			
Σ Odd	7.7%				

Note: THD: 7.72%.

18.5.2 Series Resonance

In a series resonance circuit (Figure 18.9a), the resistance capacitance and inductance are in series and the inductive reactance at the resonance is again equal to the capacitive reactance; however, the circuit has low impedance at the resonant frequency equal to the resistance.

$$z = R + j\left(\omega L - \frac{1}{\omega C}\right) \tag{18.36}$$

At a certain frequency, say f_n, z is minimum when

$$\omega L - \frac{1}{\omega C} = 0 \quad \text{or} \quad \omega = \omega_n = \frac{1}{\sqrt{LC}} \tag{18.37}$$

Figure 18.9b shows the frequency response; the capacitive reactance inversely proportional to frequency is higher at low frequencies, while the inductive reactance directly proportional to frequency is higher at higher frequencies. Thus, the reactance is capacitive and angle of z negative below f_n and above f_n the circuit is inductive and angle z is positive (Figure 18.9c). The voltage transfer function $H_v = V_2/V_1 = R/Z$ is shown in Figure 18.9d. This curve is reciprocal of Figure 18.9b

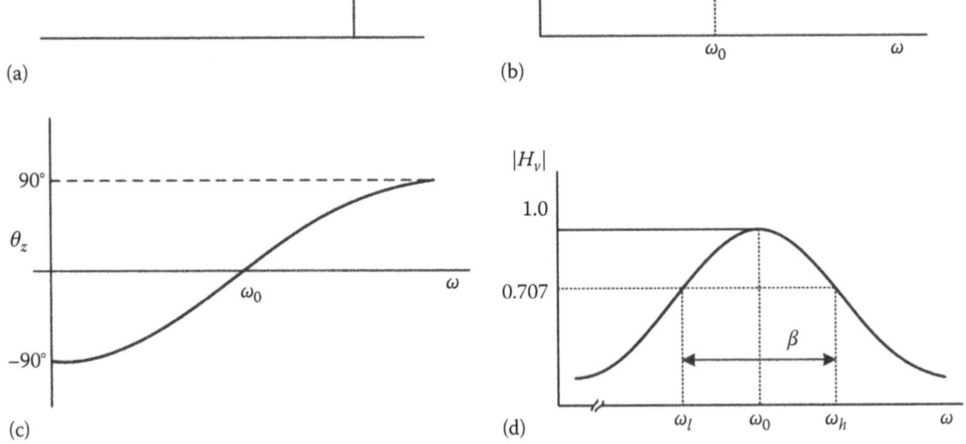

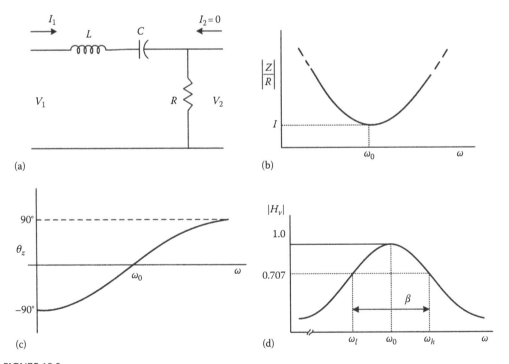

FIGURE 18.9
(a) A circuit for series resonance. (b) Impedance plot versus frequency. (c) Frequency versus impedance angle. (d) Transfer function versus frequency.

The bandwidth β shown in Figure 18.9d is given by

$$\beta = \frac{R}{L} = \frac{\omega_n}{Q_0}$$

or (18.38)

$$Q_0 = \frac{\omega_n L}{R}$$

The half power frequencies can also be expressed as

$$\omega_h = \frac{R}{2L} + \sqrt{\left(\frac{R}{2L}\right)^2 + \frac{1}{LC}} = \omega_0\left(\sqrt{1 + \frac{1}{4Q_0^2}} + \frac{1}{2Q_0}\right)$$

$$\omega_l = -\frac{R}{2L} + \sqrt{\left(\frac{R}{2L}\right)^2 + \frac{1}{LC}} = \omega_0\left(\sqrt{1 + \frac{1}{4Q_0^2}} - \frac{1}{2Q_0}\right) \quad (18.39)$$

The quality factor Q is defined as

$$Q_0 = 2\pi\left(\frac{\text{maximum energy stored}}{\text{energy dissipated per cycle}}\right) \quad (18.40)$$

Compare this with parallel circuit.

A series combination of the capacitor bank and line or transformer reactance may form a low impedance path for harmonic currents and can generate high voltages. Capacitive impedances of cables and long lines cause resonance in the range 5–15 kHz. Series compensation of power lines is another example. A transformer with a capacitor connected on its secondary can form a resonant circuit.

18.5.3 Part-Winding Resonance

Transformers have frequency-dependent impedance characteristics. The steady-state behavior can be determined by location of zeros (natural frequencies, eigenvalues) and poles (mode shapes). The zeros of the impedance function coincide with the natural frequencies. These can be determined by calculations or by measurements. In Figure 18.10, curve A shows impedance versus frequency plot as seen from the high-voltage terminal H_1 of a transformer. This shows a number of resonant frequencies. Now consider that a unit voltage is applied at terminal H_1, then the voltage at 75% winding tap is shown by the curve B, which shows much voltage escalation. The switching inrush current frequency for capacitor banks varies from 1 to 6 kHz, and depending upon the system configuration, traveling wave phenomena, a part-winding resonance can be excited. This will give rise to high overvoltages and insulation stresses and possibility of transformer winding failures [20–22].

18.6 Voltage Notching

The commutation notches in Figure 17.20 show that the voltage waveform has more than one zero crossing. Much electronic equipment uses crossover of the waveform to detect frequency or generate a reference for the firing angle. Such items of equipment are timers,

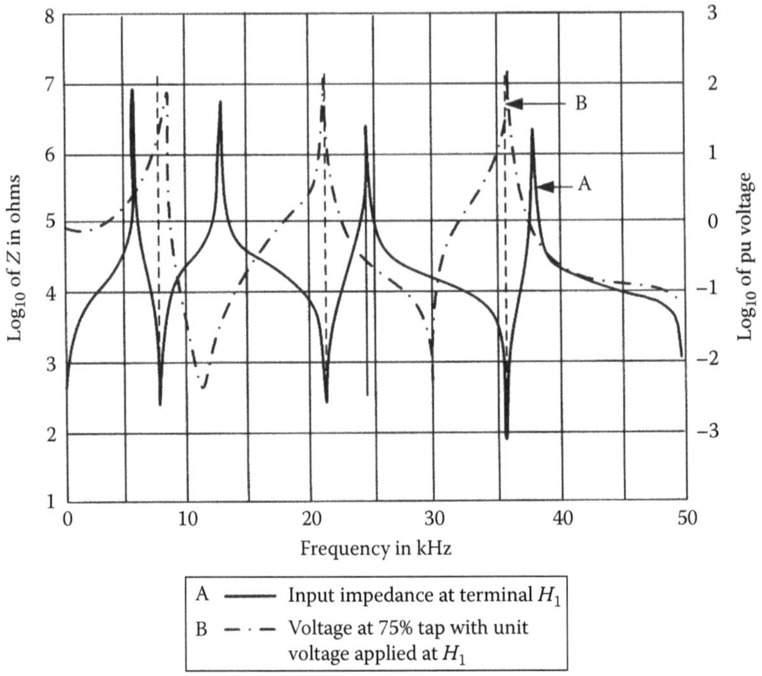

FIGURE 18.10
Input impedance plot of a transformer as seen from the high-voltage terminal H_1, showing a number of natural resonant frequencies.

domestic clocks, and UPS (uninterruptible power supply) systems. Many controllers using an inverse cosine type of control circuit will be susceptible to zero crossing. Automatic voltage regulators of generators are another example. The control of firing angle and hence the dc voltage and excitation will be affected. All drives using rectifier front end can be affected, and converter disruption can occur. Line harmonics seem to affect the accuracy and operation of magnetic devices and peripheral equipment also.

The second significant feature is the oscillations caused by the excitation of system inductance and capacitance. The capacitances may be those of power factor correction capacitors or cables. The frequency range is from 5 to 60 Hz. These can cause interference with telecommunications through coupling of the power line with the telephone line.

If the firing angle is near the crest of the wave, overvoltages can occur, as commutation notches near the peak of the voltage can drive it more than the normal crest voltage. The effect of high dv/dt may manifest in the failure of solid-state devices and electromagnetic interference (EMI).

18.7 Electromagnetic Interference

Disturbances generated by switching devices such as BJTs (bipolar junction transistors), IGBTs (insulated gate bipolar transistors), and high-frequency PWM modulation systems, and voltage notching due to converters, generate high-frequency switching harmonics.

Also, a short rise and fall time of 0.5 μs or less occurs due to the commutation action of the switches. This generates sufficient energy levels in the radio-frequency range in the form of damped oscillations between 10 kHz and 1 GHz. The following approximate classification of electromagnetic disturbance by frequency can be defined:

Below 60 Hz	: sub-harmonic
60–2 kHz	: harmonics
16–20 kHz	: acoustic noise
20–150 kHz	: range between acoustic and radio-frequency disturbance
150 kHZ to 30 MHz	: conducted radio-frequency disturbance
30 MHz to 1 GHz	: radiated disturbance

The high-frequency disturbances are referred to as EMI. The radiated form is propagated in free space as electromagnetic waves, and the conducted form is transmitted through the power lines, especially at distribution level. Conducted EMI is much higher than radiated noise. Two modes of conducted EMI are recognized, symmetrical or differential mode and asymmetrical or common mode. The symmetrical mode occurs between two conductors that form a conventional return circuit and common mode propagation takes place between a group of conductors and ground or other group of conductors.

The noise at frequencies around a few kilohertz can interfere with audiovisual equipment and electronic clocks. Shielding and proper filtering are the preventive measures. Lightning surges, arcing type of faults and operation of circuit breakers and fuses also produce EMI.

18.8 Overloading of Neutral

Figure 17.24 shows that the line current of switched mode power supplies flows in pulses. Also, the PBM technique (see Section 17.19) gives rise to neutral currents. At low level of current the pulses are nonoverlapping in a three-phase system, that is, only one phase of a three-phase system carries current at any one time. The only return path is through the neutral and as such the neutral can carry the summed currents of the three phases. Its rms value is, therefore, 173% of the line current (Figure 18.11). As the load increases, the pulses in the neutral overlap and the neutral current as a percentage of the line current reduces. The third harmonic is the major contributor to the neutral current; other triplen harmonics have insignificant contributions. A minimum of 33% third harmonic is required to produce 100% neutral current in a balanced wye system.

The nonlinear loads may not be perfectly balanced and single-phase unbalanced loads can create higher unbalance currents in the neutral. The National Electric Code (NEC), published by National Fire Protection Association (NFPA) [23], recommends that where the major portion of loads consists of nonlinear loads, the neutral shall be considered as a current-carrying conductor. Normally, a reduced neutral cross section as compared to phase conductors is specified in NEC. In some installations, the neutral current may exceed the maximum phase current. The single-phase branch circuits can be run with a separate neutral for each phase rather than using multiwire branch circuit with a shared neutral.

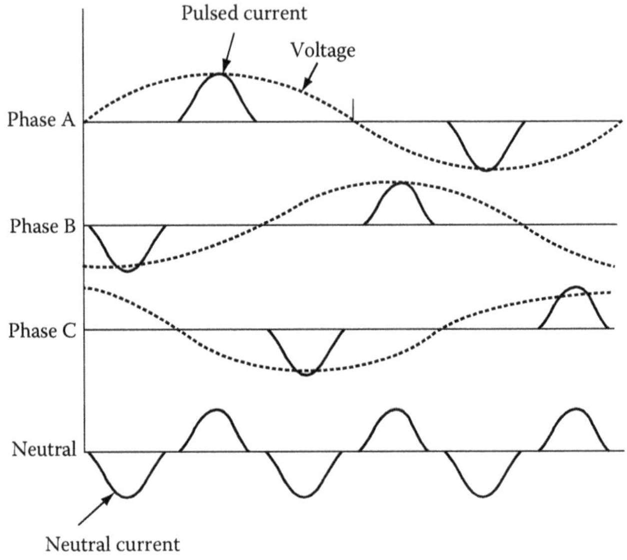

FIGURE 18.11
Summation of neutral current in a three-phase four-wire system.

18.9 Protective Relays and Meters

Harmonics may result in possible relay mis-operation. Relays that depend upon crest voltage and/or current or voltage zeros are affected by harmonic distortion on the wave. The excessive third harmonic zero sequence current can cause ground relays to false trip. A Canadian study [1] documents the following effects:

1. Relays exhibited a tendency to operate slower and/or with higher pickup values rather than to operate faster and/or with lower pickup values.
2. Static underfrequency relays were susceptible to substantial changes in the operating characteristics.
3. In most cases the changes in the operating characteristics were small over a moderate range of distortion during normal operation.
4. Depending upon the manufacturer, the overcurrent and overvoltage relays exhibited various changes in the operating characteristics.
5. Depending upon harmonic content, the operating torque of the relays could be reversed.
6. Operating time could vary widely as a function of the frequency mix in the metered quantity.
7. Balanced beam distance relays could exhibit both underreach and overreach.
8. Harmonics could impair the operation of high-speed differential relays.

Harmonic levels of 10%–20% are generally required to cause problems with relay operation. These levels are much higher than will be tolerated in the power systems.

Reference [1] states the impossibility of completely defining relay responses because of the variety of relays in use and the variations in the nature of distortions that can occur, even if the discussions is limited to 6-pulse and 12-pulse converters. Not only the harmonic magnitudes and predominant harmonic orders vary, but relative phase angles can also vary. The relays will respond differently to two wave shapes that have the same characteristics magnitudes but different phase angles relative to the fundamental.

Metering and instrumentation is affected by harmonics. Close to resonance, higher harmonic voltages may cause appreciable errors. A 20% fifth harmonic content can produce a 10%–15% error in a two-element three-phase watt transducer. The error due to harmonics can be positive, negative, or smaller with third harmonics, depending upon the type of the meter. The presence of harmonics reduces the reading on power factor meters.

18.10 Circuit Breakers and Fuses

Harmonic components can affect the current interruption capability of circuit breakers. The high di/dt at current zero can make interruption process more difficult. Figure 17.16 shows that for a six-pulse converter, the current zero is extended and di/dt at current zero is very high. There are no definite standards in the industry for derating. One method is to arrive at a maximum di/dt of the breaker, based upon interrupting rating at fundamental frequency, and then translate it into maximum harmonic levels assuming that the harmonic in question is in phase with the fundamental [1].

A typical reduction in current-carrying capacity of molded case circuit breakers, when in use at high-frequency sine wave currents is shown in Table 18.6. The effect on larger breakers can be more severe, since the phenomena are also related to skin effect and proximity effect.

Harmonics will reduce the current-carrying capacity of fuses. Also, the time current characteristics can be altered and the melting time will change. Harmonics also affect the interrupting rating of the fuses. Excessive transient overvoltages from current-limiting fuses, forcing current to zero before a natural zero crossing, may be generated. This can cause surge arrester operation and capacitor failures.

18.11 Telephone Influence Factor

Harmonic currents and voltages can produce electrical and magnetic fields that will impair the performance of communication systems. Due to proximity, there will be inductive coupling with the communication systems. Relative *weights* have been established by tests

TABLE 18.6

Reduction in Current-Carrying Capability (%) of Molded Case Circuit Breakers, 40°C Temperature Rise

Breaker Size	Sinusoidal Current	
	300 Hz	420 Hz
70 A	9	11
225 A	14	20

Note: Based on published data of one manufacturer.

for the various harmonic frequencies that indicate disturbance to voice frequency communication. This is based on disturbance produced by injection of a signal of the harmonic frequency relative to that produced by a 1 kHz signal similarly injected. The TIF (Telephone Influence Factor) weighting factor is a combination of C-message weighting characteristics, which account for relative interference effects of various frequencies in the voice band, and a capacitor that produces weighting that is directly proportional to the frequency to account for an assumed coupling function [24]. It is the dimensionless quantity that is indicative of the waveform and not the amplitude. It is given by

$$\mathrm{TIF} = \frac{\sqrt{\sum (X_f W_f)^2}}{X_\mathrm{t}} \tag{18.41}$$

$$\mathrm{TIF} = \sqrt{\sum \left[\frac{(X_f W_f)}{X_\mathrm{t}}\right]^2} \tag{18.42}$$

where
X_t is the total rms voltage or current
X_f is the single-frequency rms current or voltage at frequency f
W_f is the single-frequency weighting at frequency f

The TIF weighting function, which reflects C-message weighting and coupling normalized to 1 kHz, is given by

$$W_f = 5P_f f \tag{18.43}$$

where
P_f is the C-message weighting at frequency f
f is the frequency under consideration

The single-frequency TIF values are listed in Table 18.7. The weighting is high in the frequency range 2–3.5 kHz in which human hearing is most sensitive.

TABLE 18.7
1960 Single-Frequency TIF Values

Freq. (Hz)	TIF	Freq. (Hz)	TIF	Freq. (Hz)	TIF	Freq. (Hz)	TIF
60	0.5	1,020	5,100	1,860	7,820	3,000	9,670
180	30	1,080	5,400	1,980	8,330	3,180	8,740
300	225	1,140	5,630	2,100	8,830	3,300	8,090
360	400	1,260	6,050	2,160	9,080	3,540	6,730
420	650	1,380	6,370	2,220	9,330	3,660	6,130
540	1,320	1,440	6,560	2,340	9,840	3,900	4,400
660	2,260	1,500	6,680	2,460	10,340	4,020	3,700
720	2,760	1,620	6,970	2,580	10,600	4,260	2,750
780	3,360	1,740	7,320	2,820	10,210	4,380	2,190
900	4,350	1,800	7,570	2,940	9,820	5,000	840
1,000	5,000						

Source: Reproduced from IEEE Standard 519-1992 (*IEEE Recommended Practice and Requirements for Harmonic Control in Electrical Systems*) 1992.

TABLE 18.8

Balanced IT Guidelines for Converter Installation Tie Lines

Category	Description	IT
I	Levels unlikely to cause interference	Up to 10,000
II	Levels that might cause interference	10,000–50,000
III	Levels that probably will cause interference	>50,000

Source: Reproduced from IEEE Standard 519-1992 (*IEEE Recommended Practice and Requirements for Harmonic Control in Electrical Systems*) 1992.

Telephone interference is often expressed as a product of current and TIF, that is, IT product where I is the rms current in amperes. Alternatively, it is expressed as a product of voltage and TIF weighting, where the voltage is in kV, that is, kV-T product.

Table 18.8 [24] gives balanced IT guidelines for converter installations. The values are for circuits with an exposure between overhead systems, both power and telephone. Within an industrial plant or commercial building, the interference between power cables and twisted pair telephone cables is low, and interference is not normally encountered. Telephone circuits are particularly susceptible to the influence of ground return currents.

The effect of harmonics can be felt at a distance from their point of generation. Sometimes it eludes intuition until a rigorous study is conducted. Reference [24] details some abnormal conditions of harmonic problems. These are natural resonance of transmission lines, overexcitation of transformers, and harmonic resonance in the zero sequence circuits (see Chapter 19).

Problems

18.1 A synchronous generator has an I_2^2t of 30. Calculate the time duration for which the generator can tolerate the typical harmonic current spectrum of six-pulse converter given in Table 17.3, as a percentage of its rated full-load current, which is equal to the fundamental frequency rms current.

18.2 Explain how the fifth and seventh harmonic pair produces a sixth harmonic in the rotor circuit of a synchronous generator. What is the rotation of this harmonic?

18.3 Draw a table showing the forward reverse and zero sequence harmonics in a machine up to 37th harmonic.

18.4 Calculate the derating of a 5 MVA dry-type, 13.8–4.16 kV transformer, when subjected to typical harmonic spectrum of Table 17.3, as per unit of its rated current. The transformer data are $R_1 = 0.2747\ \Omega$, $R_2 = 0.02698\ \Omega$, total load loss = 30,000 W.

18.5 In Problem 18.4, the transformer is an oil-immersed transformer. Assume the temperatures as in Example 18.3. The no-load loss is 6000 W. The transformer supplies its rated fundamental frequency current and the harmonic spectrum is as in Problem 18.4. Calculate the top-oil temperature and hot-spot temperature.

18.6 In Problem 18.5, the transformer is loaded to 75% of its rating, and the spectrum currents based upon 75% of the fundamental current. Calculate the top-oil temperature and hot-spot temperature.

18.7 Calculate UL *K*-factor in Problem 18.4.

18.8 The system short-circuit level at a certain point in a distribution system is 1000 MVA. Calculate the kvar rating of the shunt capacitor banks that will give resonance at 300 and 420 Hz.

18.9 In a series resonant circuit $X_c = 300\ \Omega$ and resonance occurs at 5th harmonic. Find the value of the reactor and plot its impedance of the series circuit as the frequency varies from 60 to 600 Hz. Consider a Q factor of 50.

18.10 Repeat Problem 18.9 for a parallel resonant circuit.

18.11 Calculate half-power frequencies in Problem 18.9. What is the significance of these frequencies?

18.12 Repeat Problem 18.10 with a Q of 30.

References

1. IEEE. A report prepared by load characteristics task force. The effects of power system harmonics on power system equipment and loads. *IEEE PAS* 104: 2555–2561, 1985.
2. J.F. Witte, F.P. DeCesaro, and S.R. Mendis. Damaging long term overvoltages on industrial capacitor banks due to transformer energization inrush currents. *IEEE Trans. Ind. Appl.* 30(4): 1107–1115, July/August 1994.
3. IEEE. Working Group J5 of Rotating Machinery Protection subcommittee, Power System Relaying Committee. The impact of large steel mill loads on power generating units. *IEEE Trans. Power Deliv.* 15: 24–30, 2000.
4. N.G. Hingorani. A new scheme for subsynchronous resonance damping of torsional oscillations and transient torques—Part I. In: *IEEE PES Summer Meeting*, Paper no. 80 SM687-4, Minneapolis, MN, 1980.
5. N.G. Hingorani and K.P. Stump. A new scheme for subsynchronous resonance damping of torsional oscillations and transient torques—Part II. *IEEE PES Summer Meeting*, Paper no. 80 SM688-2, Minneapolis, MN, 1980.
6. NEMA. Motors and Generators, Parts 30 and 31, 1993. Standard MG-1.
7. ANSI. Rotating Electrical Machinery—Synchronous Machines, 1990. Standard C50.10.
8. ANSI. Requirements for Cylindrical Rotor 50 Hz and 60 Hz Synchronous Generators rated 10 MVA and above, 2005. Standard C50.13.
9. M.D. Ross and J.W. Batchelor. Operation of non-salient-pole–type generators supplying a rectifier load. *AIEE Trans.* 62: 667–670, 1943.
10. A.H. Bonnett. Available insulation systems for PWM inverter fed motors. *IEEE Ind. Appl. Mag.* 4: 15–26, 1998.
11. J.C. Das and R.H. Osman. Grounding of AC and DC low-voltage and medium-voltage drive systems. *IEEE Trans. Ind. Appl.* 34: 205–216, 1998.
12. J.M. Bentley and P.J. Link. Evaluation of motor power cables for PWM AC drives. *IEEE Trans. Ind. Appl.* 33: 342–358, 1997.
13. IEEE. IEEE Recommended Practice for Establishing Liquid-Filled and Dry-Type Power and Distribution Transformer Capability When Supplying Nonsinusoidal Load Currents, Standard C57-110, 2008.
14. U.L. Dry–Type General Purpose and Power Transformers, 1994. Standard UL 1561.
15. UL. Transformers Distribution, Dry-Type over 600V, 1994. Standard UL 1562.

16. J.H. Neher and M.H. McGrath. The calculation of the temperature rise and load capability of cable systems. *AIEE Trans.* 76(3): 752–764, October 1957.
17. A. Harnandani. Calculations of cable ampacities including the effects of harmonics. *IEEE Ind. Appl. Mag.* 4: 42–51, 1998.
18. IEEE. IEEE Standard for Shunt Power Capacitors, 2002. Standard 18.
19. IEEE. Draft Guide for the Application of Shunt Power Capacitors, 2006. P1036/D13a.
20. R.S. Bayless, J.D. Selmen, D.E. Traux, and W.E. Reid. Capacitor switching and transformer transients. *IEEE Trans. PWRD* 3(1): 349–357, Jan. 1988.
21. J.C. Das. Analysis and control of large shunt capacitor bank switching transients. *IEEE Trans. Ind. Appl.* 41(6): 1444–1451, Nov./Dec. 2005.
22. J.C. Das. Surge transference through transformers. *IEEE Ind. Appl. Mag.* 9(5): 24–32, Oct. 2003.
23. NFPA. National Electric Code 2009. NFPA 70.
24. IEEE. *IEEE Recommended Practice and Requirements for Harmonic Control in Electrical Systems,* Standard 519, 1992.

19

Harmonic Analysis

The purpose of harmonic analysis is to ascertain the distribution of harmonic currents, voltages, and harmonic distortion indices in a power system. This analysis is then applied to the study of resonant conditions and harmonic filter designs and also to study other effects of harmonics on the power system, i.e., notching and ringing, neutral currents, saturation of transformers, and overloading of system components. On a simplistic basis, harmonic simulation is much like a load flow simulation. The impedance data from a short-circuit study can be used and modified for the higher frequency effects. In addition to models of loads, transformers, generators, motors, etc., the models for harmonic injection sources, arc furnaces, converters, SVCs, etc., are included. These are not limited to characteristics harmonics, and a full spectrum of load harmonics can be modeled. These harmonic current injections will be at different locations in a power system. As a first step, a frequency scan is obtained, which plots the variation of impedance modulus and phase angle at a selected bus with variation of frequency or generates *R*–*X* plots of the impedance. This enables the resonant frequencies to be ascertained. The harmonic current flows in the lines are calculated, and the network, assumed to be linear at each step of the calculations with added constraints, is solved to obtain the harmonic voltages. The calculations may include the following:

- Calculation of harmonic distortion indices
- Calculation of TIF, KVT, and IT (see Section 18.11)
- Induced voltages on communication lines
- Sensitivity analysis, i.e., the effect of variation of a system component

This is rather a simplistic approach. The rigorous harmonic analysis gets involved because of interaction between harmonic-producing equipment and the power system, the practical limitations of modeling each component in a large power system, the extent to which the system should be modeled for accuracy, and the types of component and nonlinear source models. Furnace arc impedance varies erratically and is asymmetrical. Large power high-voltage dc (HVDC) converters and FACTS devices have large nonlinear loads, and superimposition is not valid. Depending on the nature of the study, *simplistic methods may give erroneous results.*

19.1 Harmonic Analysis Methods

There are a number of methodologies for the calculation of harmonics and effects of nonlinear loads. Direct measurements can be carried out, using suitable instrumentation. In a noninvasive test, the existing waveforms are measured. An EPRI project describes two methods, and Refs. [1,2] provide a summary of the research project sponsored by EPRI and BPA (Bonneville Power Administration) on HVDC system interaction from ac harmonics. One method uses harmonic sources on the ac network, and the other injects harmonic currents into the ac system using HVDC converter. In the later case, the system harmonic impedance is a ratio of the harmonic voltage and current. In the first case, a switchable shunt impedance such as a capacitor or filter bank is required at the point where the measurement is taken. By comparing the harmonic voltages and currents before and after switching the shunt device, the network impedance can be calculated. This research project implements (1) development of data acquisition system to measure current and voltage signals, (2) development of data processing package, (3) calculation of harmonic impedances, and (4) development of a computer model for impedance calculations. The analytical analysis can be carried out in the frequency and time domains. Another method is to model the system to use a state-space approach. The differential equations relating current and voltages to system parameters are found through basic circuit analysis. We will discuss frequency- and time-domain methods [3,4].

19.1.1 Frequency-Domain Analysis

For calculations in the frequency domain, the harmonic spectrum of the load is ascertained, and the current injection is represented by a Norton's equivalent circuit. Harmonic current flow is calculated throughout the system for each of the harmonics. The system impedance data are modified to account for higher frequency and reduced to their Thévenin equivalent. The principal of superposition is applied. If all nonlinear loads can be represented by current injections, the following matrix equations are applicable:

$$\overline{V}_h = \overline{Z}_h \overline{I}_h \tag{19.1}$$

$$\overline{I}_h = \overline{Y}_h \overline{V}_h \tag{19.2}$$

The formation of bus impedance and admittance matrices has already been discussed. The distribution of harmonic voltages and currents is no different for networks containing one or more sources of harmonic currents. During the steady state, the harmonic currents entering the network are considered as being produced by ideal sources that operate without repercussion. The entire system can then be modeled as an assemblage of passive elements. Corrections will be applied to the impedance elements for dynamic loads, e.g., generators and motors' frequency-dependent characteristics at each incremental frequency chosen during the study can be modeled. The system harmonic voltages are calculated by direct solution of the linear matrix equations (Equations 19.1 and 19.2).

In a power system, the harmonic injection will occur only on a few buses. These buses can be ordered last in the Y matrix, and a reduced matrix can be formed. For n-nodes and $n-j+1$ injections, the reduced Y matrix is

$$\begin{vmatrix} I_j \\ \cdot \\ I_n \end{vmatrix} = \begin{vmatrix} Y_{jj} & \cdot & Y_{jn} \\ \cdot & \cdot & \cdot \\ Y_{nj} & \cdot & Y_{nn} \end{vmatrix} \begin{vmatrix} V_i \\ \cdot \\ V_n \end{vmatrix} \tag{19.3}$$

where diagonal elements are the self-admittances, and the off-diagonal elements are transfer impedances as in the case of load flow calculations.

Linear transformation techniques are discussed in Chapter 3. The admittance matrix is formed from a primitive admittance matrix by transformation:

$$\overline{Y}_{abc} = \overline{A}'\overline{Y}_{\text{prim}}\overline{A} \tag{19.4}$$

and the symmetrical component transformation is given by

$$\overline{Y}_{012} = \overline{T}_s^{-1}\overline{Y}_{abc}\overline{T}_s \tag{19.5}$$

The vector of nodal voltages is given by

$$\overline{V}_h = \overline{Y}_h^{-1}\overline{I}_h = \overline{Z}_h\overline{I}_h \tag{19.6}$$

For the injection of a unit current at bus k,

$$\overline{Z}_{kk} = \overline{V}_k \tag{19.7}$$

where Z_{kk} is the impedance of the network seen from bus k. The current flowing in branch j_k is given by

$$\overline{I}_{jk} = \overline{V}_{jk}(\overline{V}_j - \overline{V}_k) \tag{19.8}$$

where Y_{jk} is the nodal admittance matrix of the branch connected between j and k.

Variation of the bus admittance matrix, which is produced by a set of modifications in the change of impedance of a component, can be accommodated by modifications to the Y-bus matrix as discussed before. For harmonic analysis, the admittance matrix must be built at each frequency of interest, for component level RLC parameters for circuit models of lines, transformers, cables, and other equipment. Thus, the harmonic voltages can be calculated. A new estimate of the harmonic injection currents is then obtained from the computed harmonic voltages, and the process is iterative until the convergence on each bus is obtained. Under resonant conditions, large distortions may occur, and the validity of assumption of linear system components is questionable. From Equation 19.1, we see that the harmonic impedance is important in the response of the system to harmonics. There can be interaction between harmonic sources throughout the system, and if these are ignored, the single-source model and the superposition can be used to calculate the harmonic distortion factors and filter designs. The assumption of constant system impedance is not valid, as the system impedance always changes; say due to switching conditions, operation, or future additions. These impedance changes in the system may have a more profound effect on the ideal current source modeling than the interaction between harmonic sources. A weak ac/dc interconnection defined with a short-circuit ratio (short-circuit capacity of the ac system divided by the dc power injected by the converter into converter bus) of <3 may have voltage and power instabilities, transient and dynamic overvoltages, and harmonic overvoltages [5].

19.1.2 Frequency Scan

A frequency scan is merely a repeated application in certain incremental steps of some initial value of frequency to the final value, these two values spanning the range of harmonics to be considered. The procedure is equally valid whether there are single or multiple harmonic sources in the system, so long as the principal of superimposition is

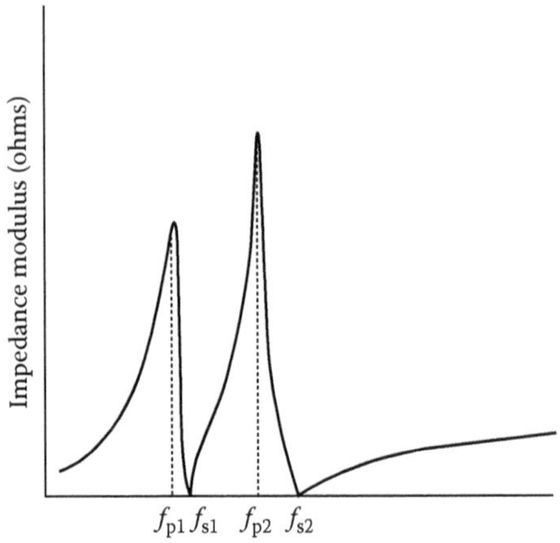

FIGURE 19.1
Frequency scan showing parallel and series resonance frequencies.

held valid. Then, for unit current injection, the calculated voltages give the driving point and transfer impedance, both modulus and phase angle. The Y_{bus} contains only linear elements for each frequency. Varying the frequency gives a series of impedances, which can be plotted to provide an indication of the resonant conditions. Figure 19.1 shows a frequency scan of impedance modulus versus frequency. The parallel resonances occur at peaks, which give the maximum impedances and the series resonances at the lowest points of the impedance plots. Figure 19.1 shows parallel resonance at two frequencies f_{p1} and f_{p2} and series resonance at f_{s1} and f_{s2}. We will see in Chapter 20 that such a frequency scan is obtained with two single-tuned shunt filters. Multiphase frequency scans can identify the harmonic resonance caused by single-phase capacitor banks.

19.1.3 Voltage Scan

A voltage scan may similarly be carried out by applying unit voltage to a node and calculating the voltages versus frequency in the rest of the system. The resulting voltages represent the voltage-transfer function to all other nodes in the system. This analysis is commonly called a voltage-transfer function study. The peaks in the scan identify the frequencies at which the voltages will be magnified, and the lowest points indicate frequencies where these will be attenuated.

19.1.4 Phase Angle of Harmonics

For simplicity, all the harmonics may be considered cophasial. This does not always give the most conservative results, unless the system has one predominant harmonic, in which case only harmonic magnitude can be represented. The phase angles of the current sources are functions of the supply voltage phase angle and are expressed as

$$\theta_h = \theta_{h,\text{spectrum}} + h(\theta_1 - \theta_{1,\text{spectrum}}) \tag{19.9}$$

where

θ_1 is the phase angle obtained from fundamental frequency load flow solution

$\theta_{h,\text{spectrum}}$ is the typical phase angle of harmonic current source spectrum (see also Appendix G)

The phase angles of a three-phase harmonic source are rarely 120° apart, as even a slight unbalance in the fundamental frequency can be reflected in a considerable unbalance in the harmonic phase angle.

19.1.5 Newton–Raphson Method

The Newton–Raphson method can be applied to harmonic current flow [6,7]. This is based on the balance of active power and reactive volt-amperes, whether at fundamental frequency or at harmonics. The active and reactive power balance is forced to zero by the bus voltage iterations.

Consider a system with $n+1$ buses, bus 1 is a slack bus, buses 2 through $m-1$ are conventional load buses, and buses m to n have nonsinusoidal loads. It is assumed that the active power and the reactive volt-ampere balance are known at each bus and that the nonlinearity is known. The power balance equations are constructed so that ΔP and ΔQ at all nonslack buses are zero for all harmonics. The form of ΔP and ΔQ as a function of bus voltage and phase angle is the same as in conventional load flow, except that Y_{bus} is modified for harmonics. The specified active and reactive powers are known at buses 2 through $m-1$, but only active power is known at buses m through n. Two additional parameters are required: the current balance and volt-ampere balance. The current balance for fundamental frequency is written, and the equation is modified for buses with harmonic injections. The harmonic response of the buses 1 through $m-1$ is modeled in admittance bus matrix.

The third equation is the apparent volt-ampere balance at each bus:

$$S_{\text{L}}^2 = \sum_{s} P_{\text{L}}^2 + \sum_{s} Q_{\text{L}}^2 + \sum D_{\text{L}}^2 \tag{19.10}$$

where the third term of the equation denotes distortion power at bus L, which is not considered as an independent variable, as it can be calculated from real and imaginary components of currents (see Section 19.7).

The final equations for the harmonic power flow become

$$\begin{vmatrix} \Delta W \\ \Delta I^1 \\ \Delta I^5 \\ \Delta I^7 \\ \cdots \end{vmatrix} = \begin{vmatrix} J^1 & J^5 & J^7 & \cdots & 0 \\ YG^{1,1} & YG^{1,5} & YG^{1,7} & \cdots & H^1 \\ YG^{5,1} & YG^{5,5} & YG^{5,7} & \cdots & H^5 \\ YG^{7,1} & YG^{7,5} & YG^{7,7} & \cdots & H^7 \\ \cdots & \cdots & \cdots & \cdots & \cdots \end{vmatrix} \tag{19.11}$$

where all elements in Equation 19.11 are subvectors and submatrices partitioned from ΔM (apparent mismatches), J, and ΔU, i.e., $\Delta M = J\Delta U$.

ΔW = mismatch active and reactive volt-amperes

ΔI^1 = mismatch fundamental current

ΔI^k = mismatch harmonic current at kth harmonic

J^1 = conventional power flow Jacobian

J^k = Jacobian at harmonic k

$$(YG)^{k,j} = Y^{k,k} + G^{k,k} \quad (k = j)$$
$$= G^{k,j} \quad (k \neq j)$$

where
- $Y^{k,k}$ is an array of partial derivatives of injection currents at the kth harmonic with respect to the kth harmonic voltage
- $G^{k,j}$ are the partial derivatives of the kth harmonic load current with respect to the jth harmonic supply voltage
- H^k is the partial derivative of nonsinusoidal loads for real and imaginary currents with respect to α and β

A flowchart is shown in Figure 19.2. Reference [8] discusses the impedance matrix method of harmonic analysis. Also, see Refs. [9,10].

19.1.6 Time-Domain Analysis

The simplest harmonic model is a rigid harmonic source and linear system impedance. A rigid harmonic source produces harmonics of a certain order and constant magnitude and phase, and the linear impedances do not change with frequencies. Multiple harmonic sources are assumed to act in isolation and the principal of superimposition applies. These models can be solved by iterative techniques, and the accuracy obtained will be identical to that of time-domain methods. For arc furnaces and even electronic converters under resonant conditions, an ideal current injection may cause significant errors. The nonlinear and time-varying elements in the power system can significantly change the interaction of the harmonics with the power system. Consider the following:

1. Most harmonic devices that produce uncharacteristic harmonics as terminal conditions are in practice not ideal, e.g., converters operating with unbalanced voltages.
2. There is interaction between ac and dc quantities, and there are interactions between harmonics of different order, given by switching functions (defined later).
3. Gate control of converters can interact with harmonics through synchronizing loops.

Time-domain analyses have been used for transient stability studies, transmission lines, and switching transients. It is possible to solve a wide range of differential equations for the power system using computer simulation and to build up a model for harmonic calculations, which could avoid many approximations inherent in the frequency-domain approach. Harmonic distortions can be directly calculated and making use of FFT (fast Fourier transform, see Appendix E). These can be converted into frequency domain. The graphical results are waveforms of zero crossing, ringing, high dv/dt, and commutation notches. The transient effects can be calculated, e.g., the part-winding resonance of a transformer can be simulated. The synchronous machines can be simulated with accurate models to represent saliency, and the effects of frequency can be dynamically simulated. EMTP is the most widely used program for simulation in the time domain.

For analysis in the time domain, a part of the system of interest may be modeled in detail. This detailed model consists of three-phase models of system components, transformers, harmonic sources, and transmission lines; and it may be coupled with a network model of

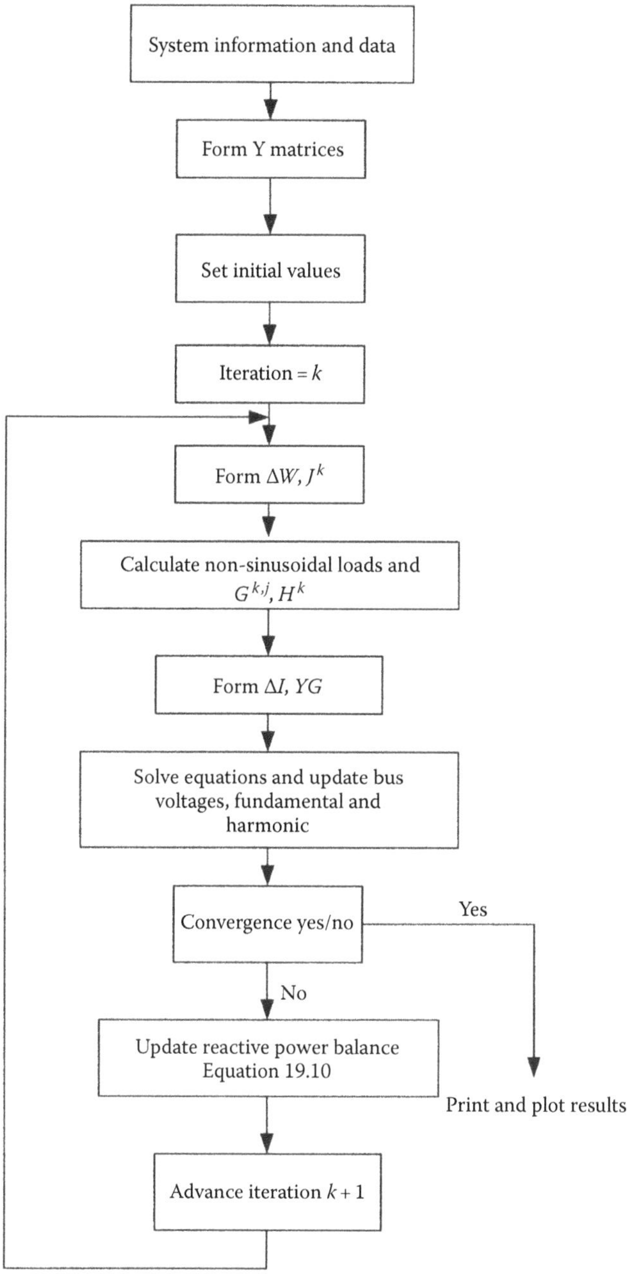

FIGURE 19.2
Harmonic power flowchart, Newton–Raphson method.

lumped RLC branches at interconnection buses to represent the driving point and transfer impedances of the selected buses. The overall system to be studied is considerably reduced in size, and time-domain simulation is simplified.

19.1.7 Switching Function

Switching function is a steady-state concept for the study of interactions between the ac and dc sides of a converter. The terminal properties of many converters can be

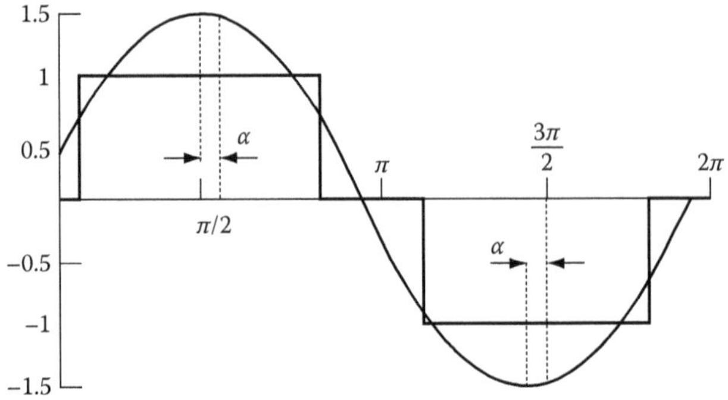

FIGURE 19.3
Switching function of a six-pulse converter.

approximated in the time domain by the converter switching function. The modulation/demodulation properties of the converter account for interaction between harmonics, generation of noncharacteristic harmonics, and propagation of dc harmonics on the ac side, and operation under unbalanced voltage or current. Consider the switching function of a converter, as shown in Figure 19.3. The switching function is 1 when dc current flows in the positive direction, −1 when it flows in the negative direction, and zero otherwise. The switching functions of three phases are symmetrical and balanced, and, in steady state, these lag the system voltage by a converter delay angle. The ac current output of phase *a* is the product of the switching function and the dc current in phase *a*:

$$I_a = I_{dc} S_a \tag{19.12}$$

$$I_b = I_{dc} S_b \tag{19.13}$$

$$I_c = I_{dc} S_c \tag{19.14}$$

$$V_{dc} = V_a S_a + V_b S_b + V_c S_c \tag{19.15}$$

The converter appears as a current source from the ac side and a voltage source from the dc side. The switching functions allow one to conduct studies for harmonic interaction between two or more converters in near proximity. The switching function assumes that the control system operates perfectly and delivers ignition pulses at regular intervals. Practically, there is interaction between the network harmonics and converter controls, which can be detected by modeling synchronizing loops.

19.2 Harmonic Modeling of System Components

19.2.1 Transmission Lines

The transmission line models are in Appendix B, which gives impedance calculations for a mutually coupled three-phase line, with ground wires, bundled conductors, and symmetrical component transformations. The transmission line model to be used is determined by the wavelength of the highest frequency of interest. Long-line effects should be represented for lines of length $150/h$ miles, where h is the harmonic number. The effect of higher frequencies is to increase the skin effect and proximity effects. A frequency-dependent model of the resistive component becomes important, though the effect on the reactance is ignored. The resistance can be multiplied by a factor $g(h)$ [11]:

$$R(h) = R_{dc}g(h) \tag{19.16}$$

$$g(h) = 0.035X^2 + 0.938X > 2.4 \tag{19.17}$$

$$= 0.35X + 0.3X \leq 2.4 \tag{19.18}$$

where

$$X = 0.3884\sqrt{\frac{f_h}{f}}\sqrt{\frac{h}{R_{dc}}} \tag{19.19}$$

where
f_h is the harmonic frequency
f is the system frequency

Another equation for taking into account the skin effect is:

$$R = R_e\left(\frac{j\mu\omega}{2\pi a}\frac{J_z(r)}{\partial J_z(r)/\partial r|_{r=a}}\right) \tag{19.20}$$

where
$J_z(r)$ is the current density
a is the outside radius of the conductor

Transmission Line Equations with Harmonics
In presence of harmonics, from Equation 10.56:

$$V_{s(h)} = V_{r(h)}\cosh(\gamma l_{(h)}) + I_{r(h)}Z_{0(h)}\sinh(\gamma l_{(h)})$$

$$= V_{r(h)}\cosh\left(\sqrt{Z_{(h)}Y_{(h)}}\right) + I_{r(h)}\sqrt{\frac{Z_{(h)}}{y_{(h)}}}\sinh\left(\sqrt{Z_{(h)}Y_{(h)}}\right) \tag{19.21}$$

$$I_{s(h)} = \frac{V_{r(h)}}{Z_{0(h)}}\sinh\left(\sqrt{Z_{(h)}Y_{(h)}}\right) + I_{r(h)}\cosh\left(\sqrt{Z_{(h)}Y_{(h)}}\right) \tag{19.22}$$

Similarly from

$$\begin{vmatrix} V_r \\ I_r \end{vmatrix} = \begin{vmatrix} \cosh(\gamma l) & -Z_0\sinh(\gamma l) \\ -\dfrac{\sinh(\gamma l)}{Z_0} & \cosh\gamma l \end{vmatrix} \begin{vmatrix} V_s \\ I_s \end{vmatrix} \tag{19.23}$$

Equations of receiving end current and voltages can be written.

The variation of the impedance of the transmission line with respect to frequency is of much interest. We derived the equivalent π model in Chapter 10, Equations 10.87 and 10.89. Repeating this equation:

$$Z_s = Z_0\sinh\gamma l$$
$$Y_p = \frac{Y}{2}\left[\frac{\tanh\gamma l/2}{\gamma l/2}\right] \tag{19.24}$$

Here, we have denoted the series element of the π-model with Z_s and the shunt element with Y_p. In the presence of harmonics, we can write

$$
\begin{aligned}
Z_{s(h)} &= Z_{0(h)} \sinh \gamma l_{(h)} = Z_{(h)} \frac{\sinh \gamma l_{(h)}}{\gamma l_{(h)}} \\
Y_{p(h)} &= \frac{\tanh(\gamma l_{(h)}/2)}{Z_{0(h)}} = \frac{Y_{(h)}}{2} \frac{\tanh(\gamma l_{(h)}/2)}{\gamma l_{(h)}/2}
\end{aligned}
\tag{19.25}
$$

Also

$$
\begin{aligned}
\gamma_{(h)} &= \sqrt{z_{(h)} y_{(h)}} \approx \frac{h\omega}{l}\sqrt{LC} \\
Z_{0(h)} &= \sqrt{z_{(h)}/y_{(h)}} \approx \sqrt{\frac{L}{C}}
\end{aligned}
\tag{19.26}
$$

Recall from Chapter 10 that

$$
\begin{aligned}
\lambda_{(h)} &= \frac{2\pi}{\beta_{(h)}} \approx \frac{l}{hf\sqrt{LC}} \\
v_{(h)} &= f\lambda_h \approx \frac{l}{\sqrt{LC}} \\
f_{osc(h)} &= \frac{v_{(h)}}{l} \approx \frac{1}{\sqrt{LC}}
\end{aligned}
\tag{19.27}
$$

Thus, the characteristic impedance, velocity of propagation, and frequency of oscillations are all independent of h while wavelength varies inversely with h.

A specimen impedance plot of a 230 kV line is shown in Figure 19.4. The series and shunt reactances are predominant. The shunt resistance normally considered zero in a pi model

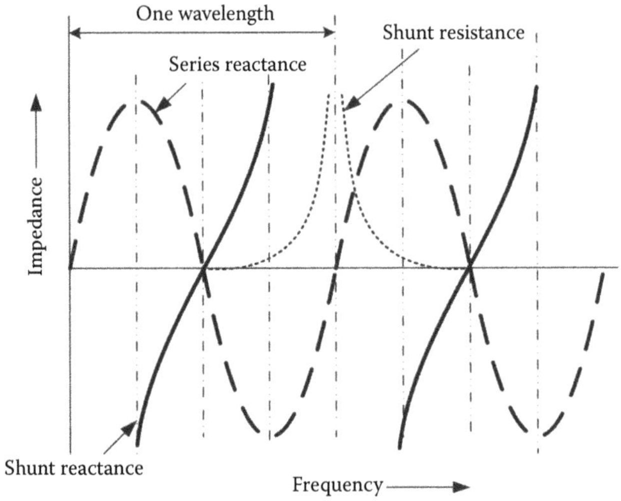

FIGURE 19.4
Impedance plot of a 230 kV line.

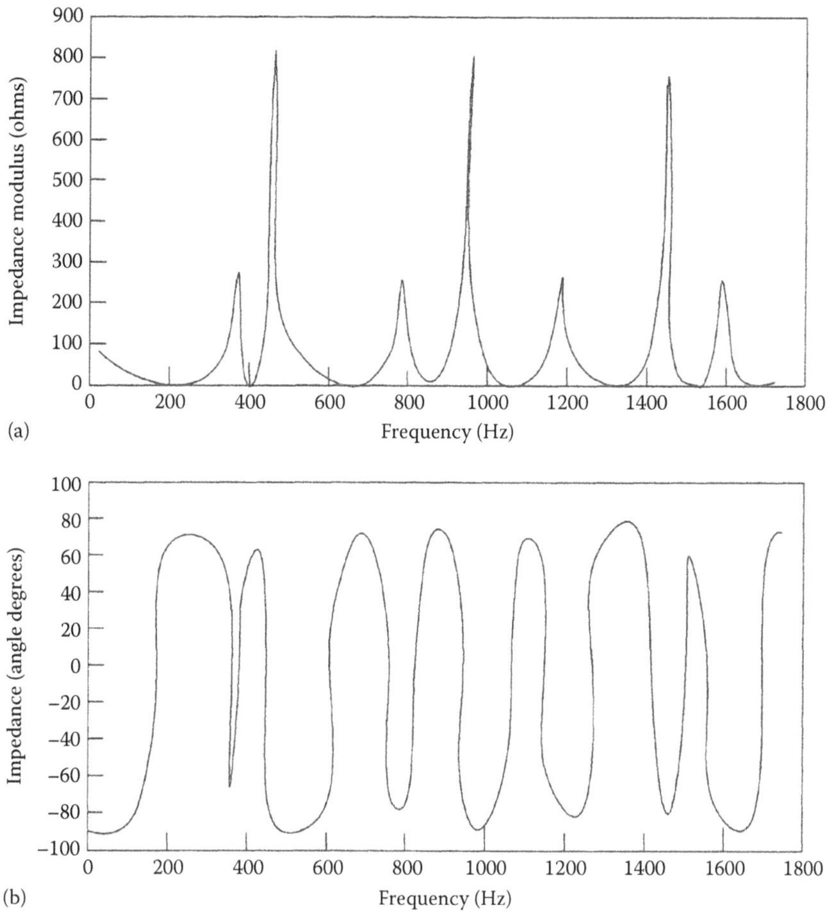

FIGURE 19.5
(a) Frequency scan of a 400 kV line with bundle conductors; (b) corresponding phase angle of the impedance modulus.

has considerable effect on resonant frequency. The impedance variations are similar to a series and parallel resonating circuits.

Figure 19.5a shows the calculated impedance modulus and Figure 19.5b, the impedance angle versus frequency plots of a 400 kV line, consisting of four bundled conductors of 397.5 KCMIL ACSR per phase in horizontal configuration. The average spacing between bundles is 35 ft, and the height above the ground is 50 ft. Two ground wires at ground potential are considered in the calculations, and the earth resistivity is 100 Ω m. Each conductor has a diameter of 0.806 in. (0.317 cm), and bundle conductors are spaced 6 in. (0.15 m) center to center.

The plots show a number of resonant frequencies. The impedance angle changes abruptly at each resonant frequency.

19.2.2 Underground Cables

Appendix B gives cable models. Cables have more significant capacitance than overhead lines. An estimate of length where long-line effects are modeled is $90/h$ miles.

19.2.3 Filter Reactors

The frequency-dependent Q of the filter reactors is especially important, as it affects the sharpness of tuning (Chapter 20). Resistance at high frequencies can be calculated by the following expressions:

$$R_h = \left[\frac{0.115h^2 + 1}{1.15}\right] R_f \quad \text{for aluminum reactors} \tag{19.28}$$

$$R_h = \left[\frac{0.055h^2 + 1}{1.055}\right] R_f \quad \text{for copper reactors} \tag{19.29}$$

19.2.4 Transformers

The single-phase and three-phase transformer models are discussed in load flow and Appendix C. A linear conventional T-circuit model of the transformer is shown in Appendix C, and fundamental frequency values of resistance and reactance can be found by a no-load and short-circuit tests on the transformer. The resistance of the transformer can be modified with increase in frequency according to Figure 19.6. While the resistance increases with frequency, the leakage inductance reduces. The magnetizing branch in the transformer model is often omitted if the transformer is not considered a source of the harmonics. This simplified model may not be accurate, as it does not model the nonlinearity in the transformer on account of:

- Core and copper losses due to eddy current and hysteresis effects. The core loss is a summation of eddy current and hysteresis loss; both are frequency dependent:

$$P_c = P_e P_h = K_e B^2 f^2 + K_h B^s f \tag{19.30}$$

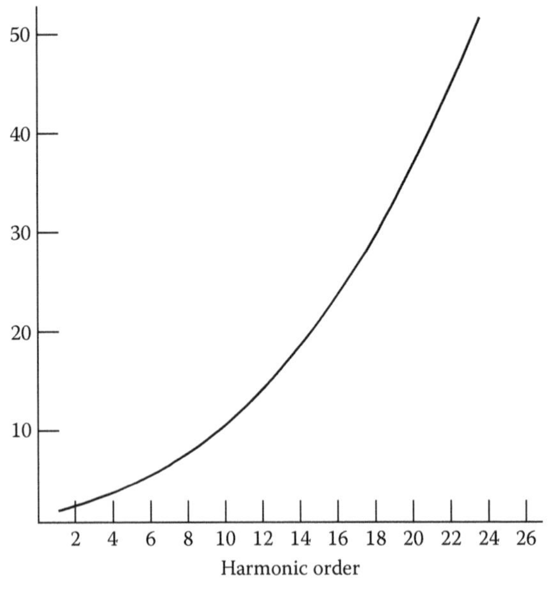

FIGURE 19.6
Increase in transformer resistance with frequency.

where

B is the peak flux density
s is the Steinmetz constant (typically 1.5–2.5, depending on the core material)
f is the frequency
K_e and K_h are constants

- Leakage fluxes about the windings, cores, and surrounding medium.
- Core magnetization characteristics, which are the primary source of transformer nonlinearity.

A number of approaches can be taken to model the nonlinearities. EMTP transformer models, *Satura* and *Hysdat*, are discussed in Appendix C. These models consider only core magnetization characteristics and neglect nonlinearities or frequency dependence of core losses and winding effects.

Capacitances of the transformer windings are generally omitted for harmonic analysis of distribution systems; however, for transmission systems, capacitances are included. Surge models of transformers are discussed in Appendix C. A simplified model with capacitances is shown in Figure 19.7; C_{bh} is the high-voltage bushing capacitance, C_b is the secondary bushing capacitance, C_{hg} and C_{lg} are distributed winding capacitances, and C_{hl} is the distributed winding capacitance between the windings. Typical capacitance values for core-type transformers are shown in Table 19.1.

Converter loads may draw dc and low-frequency currents through the transformers, i.e., a cycloconverter load. Geomagnetically induced currents flow on the earth's surface due to

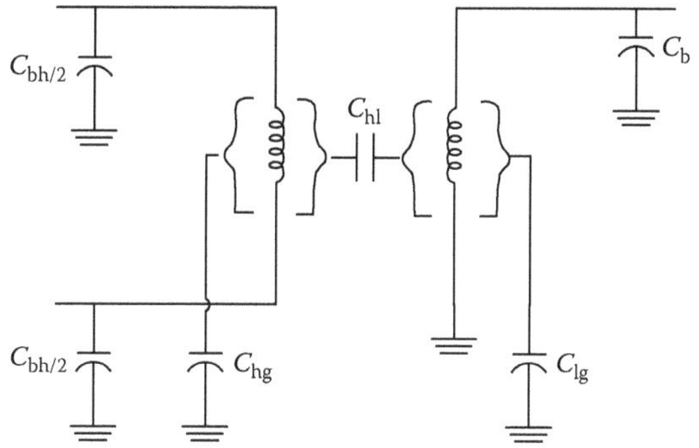

FIGURE 19.7
Simplified capacitance model of a two-winding transformer.

TABLE 19.1
Capacitance of Core-Type Transformers in nF

MVA Rating of Transformer	C_{hg}	C_{hl}	C_{lg}
1	1.2–14	1.2–17	3.1–16
2	1.4–16	1–18	3–16
5	1.2–14	1.1–20	5.5–17
10	4–7	4–11	8–18
25	2.8–4.2	2.5–18	5.2–20
50	4–6.8	3.4–11	3–24
75	3.5–7	5.5–13	2.8–30
100	3.3–7	5–13	4–40

geomagnetic disturbance are typically at low frequencies (0.001–0.1 Hz), reaching a peak value of 200 A. These can enter the transformer windings through grounded wye neutrals and bias the cores to cause half-cycle saturation (see Appendix C). Also, three-phase models of the transformers are discussed in detail in Chapter 14.

19.2.5 Induction Motors

Figure 12.10 shows the equivalent circuits of an induction motor. The shunt elements g_c and b_m are relatively large compared to R_1, r_2, X_1, and x_2. Generally, the locked rotor current of the motor is known and is given by Equation 12.125. At fundamental frequency, neglecting magnetizing and loss components, the motor reactance is

$$X_f = X_1 + x_2 \tag{19.31}$$

and the resistance is

$$R_f = R_1 + \frac{r_2}{s} \tag{19.32}$$

This resistance is not the same as used in short-circuit calculations. At harmonic frequencies, the reactance can be directly related to the frequency:

$$X_h = hX_f \tag{19.33}$$

though this relation is only approximately correct. The reactance at higher frequencies is reduced due to saturation. The stator resistance can be assumed to vary as the square root of the frequency:

$$R_{1h} = \sqrt{h} \cdot (R_1) \tag{19.34}$$

The harmonic slip is given by

$$s_h = \frac{h-1}{h} \quad \text{for positive sequence harmonics} \tag{19.35}$$

$$s_h = \frac{h+1}{h} \quad \text{for negative sequence harmonics} \tag{19.36}$$

The rotor resistance at harmonic frequencies is

$$r_{2h} = \frac{\sqrt{(1 \pm h)}}{s_h} \tag{19.37}$$

The motor impedance neglecting magnetizing resistance is infinite for triplen harmonics, as the motor windings are not grounded.

From Figure 12.10 motor impedance at harmonic h is:

$$R_1 + jhX_1 + \frac{jhx_m((r_2/s_h) + jhx_2)}{(r_2/s_h) + jh(x_m + x_2)} \tag{19.38}$$

where

$$s_h = 1 - \frac{n}{hn_s} \quad h = 3n + 1$$
$$s_h = 1 + \frac{n}{hn_s} \quad h = 3n - 1 \tag{19.39}$$

19.2.6 Generators

The synchronous generators do not produce harmonic voltages and can be modeled with shunt impedance at the generator terminals. An empirical linear model is suggested, which consists of a full subtransient reactance at a power factor of 0.2. The average inductance experienced by harmonic currents, which involve both the direct axis and quadrature axis reactances, is approximated by

$$\text{Average inductance} = \frac{L_d'' + L_q''}{2} \tag{19.40}$$

At harmonic frequencies, the fundamental frequency reactance can be directly proportioned. The resistance at harmonic frequencies is given by

$$R_h = R_{dc}\left[1 + 0.1\left(\frac{h_f}{f}\right)^{1.5}\right] \tag{19.41}$$

This expression can also be used for the calculation of harmonic resistance of transformers and cables having copper conductors. Model generators for harmonics with:

$$Z_{0(h)} = R_h + jhX_0 \quad h = 3, 6, 9\ldots$$
$$Z_{1(h)} = R_h + jhX_d'' \quad h = 1, 4, 7\ldots \tag{19.42}$$
$$Z_{2(h)} = R_h + jhX_2 \quad h = 2, 5, 8\ldots$$

X_0, X_d'', and X_2 are the generator zero sequence, subtransient, and negative sequence reactances at fundamental frequency.

19.3 Load Models

Figure 19. 8a shows a parallel *RL* load model. It represents bulk power load as an *RL* circuit connected to ground. The resistance and reactance components are calculated from fundamental frequency voltage, reactive volt-ampere, and power factor:

$$R = \frac{V^2}{S\cos\phi} L = \frac{V^2}{2\pi fS\sin\phi} \tag{19.43}$$

where S = three-phase apparent power.

The reactance is frequency dependent, and resistance may be constant or it can also be frequency dependent. Alternatively, the resistance and reactance may remain constant at all frequencies.

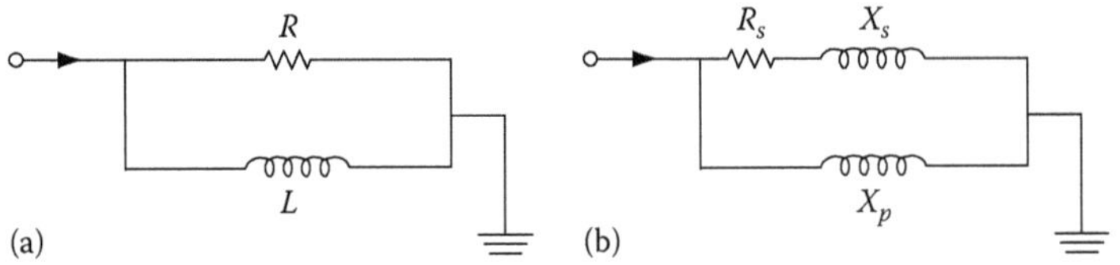

FIGURE 19.8
(a) Parallel *RL* load model; (b) CIGRE type-C model.

Figure 19.8b shows a CIGRE (Conference Internationale des Grands Reseaux Electriques à Haute Tension) type-C load model [12], which represents bulk power, valid between 5th and 30th harmonics. Here, the following relations are applicable:

$$R_s = \frac{V^2}{P} \qquad X_s = 0.073hR_s \qquad X_p = \frac{hR_s}{6.7(Q/P) - 0.74} \tag{19.44}$$

This load model was derived by experimentation.

For the distribution systems modeling,

- Cables are represented as equivalent π model
- For short lines model capacitance at the bus
- Transformers represented by their equivalent circuits
- Power factor correction capacitors modeled at their locations
- Harmonic filters, generators represented as discussed above
- All elements should be uncoupled using techniques discussed in this book
- Loads modeled as per their composition and characteristics

19.4 System Impedance

The system impedance to harmonics is not a constant number. Figure 19.9 shows the *R–X* plot of system impedance. The fundamental frequency impedance is inductive, its value representing the *stiffness* of the system. The resonances in the system make the *R–X* plots

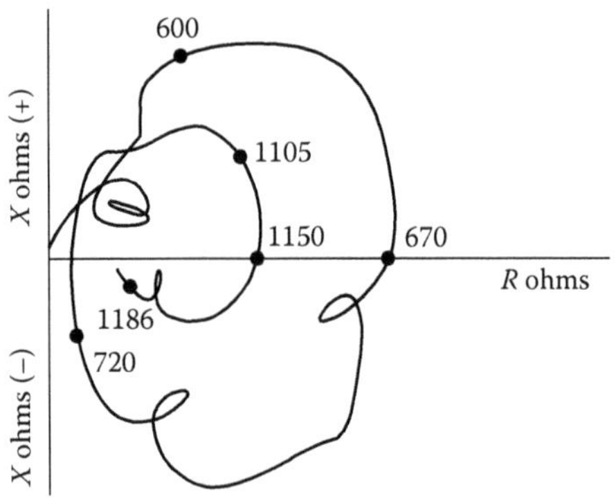

FIGURE 19.9
R–X plot of a supply system source impedance.

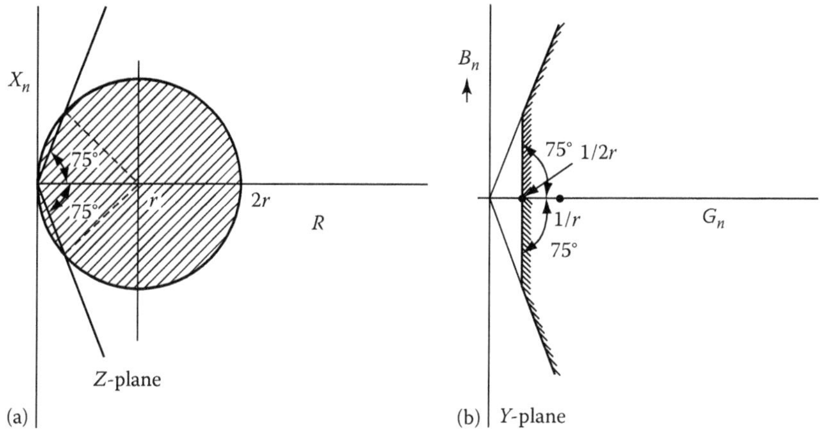

FIGURE 19.10
Generalized impedance plot (a) in R–X plane and (b) in Y plane.

a spiral shape, and the impedance may become capacitive. Such spiral-shaped impedances have been measured for high-voltage systems, and resonances at many frequencies are common. These frequencies at resonance points and also at a number of other points on the spiral-shaped curves are indicated as shown in Figure 19.9. At the resonance, the impedance reduces to a resistance. The system impedance can be ascertained by the following means:

- A computer solution can be used to calculate the harmonic impedances.
- In noninvasive measurements, the harmonic impedance can be calculated directly from the ratio of harmonic voltage and current reading.
- In another measurement method, shunt impedance is switched in the circuit, and the harmonic impedance is calculated by comparing the harmonic voltages and currents before and after switching.

The spiral-shaped impedance plots can be bounded in the Z plane by a circle on the right side of the Y axis and tangents to it at the origin at an angle of 75°, Figure 19.10a. This configuration can also be translated in the Y plane (Figure 19.10b) [13].

19.5 Three-Phase Models

The power system elements are not perfectly symmetrical. Asymmetry is involved in the circuit loading and mutual couplings and unbalanced self- and mutual impedances result. The problem is similar to the three-phase load flow and is compounded by the nonlinearities of the harmonic loads [14]. Single-phase models are not adequate when:

- Telephone interference is of concern. The influence of zero sequence harmonics is important, which gives rise to most of the interference with the communication circuits.
- There are single-phase capacitor banks in the system.

- There is single phase or unbalanced harmonic sources.
- Triplen harmonics are to be considered, ground currents are important, and significant unbalanced loading is present.

The network asymmetries and mutual couplings should be included, which require a 3×3 impedance matrix at each harmonic. Three-phase models of transformers with neutral grounding impedances, mutually coupled lines, distributed parameter transposed, and untransposed transmission lines should be possible in the harmonic analysis program, also see Chapter 14.

The networks can be divided into subsystems. A subsystem may be defined as any part of the network which is so divided that no subsystem has any mutual coupling between its constituent branches and those of the rest of the network. Following the example of a two-port network (Chapter 10):

$$\begin{vmatrix} \overline{V}_s \\ \overline{I}_s \end{vmatrix} = \begin{vmatrix} \overline{A} & \overline{B} \\ \overline{C} & \overline{D} \end{vmatrix} \begin{vmatrix} \overline{V}_r \\ \overline{I}_r \end{vmatrix} \tag{19.45}$$

that is, the current and voltages are now matrix quantities, the dimensions depending upon the section being considered, 3, 6, 9.... All sections contain the same number of mutually coupled elements, and all matrices are of the same order. Then, the equivalent model i matrix can be written as

$$\begin{vmatrix} \overline{I}_s \\ \overline{I}_r \end{vmatrix} = \begin{vmatrix} [\overline{D}][\overline{B}]^{-1} & \vdots & [\overline{C}] - [\overline{D}][\overline{B}]^{-1}[\overline{A}] \\ \cdots & \vdots & \cdots \\ [\overline{B}]^{-1} & \vdots & -[\overline{B}]^{-1}[\overline{A}] \end{vmatrix} \begin{vmatrix} \overline{V}_s \\ \overline{V}_r \end{vmatrix} \tag{19.46}$$

19.5.1 Uncharacteristic Harmonics

Uncharacteristic harmonics will be present. These may originate from variations in ignition delay angle. As a result, harmonics of the order $3q$ are produced in the direct voltage, and $3q \pm 1$ are produced in the ac line current, where q is an odd integer. Third harmonic and its odd multiples are produced in the dc voltage and even harmonics in the ac line currents. The uncharacteristic harmonics may be amplified, as these may shift the times of voltage zeros and result in more unbalance of firing angles. A firing angle delay of 1° causes approximately 1% third harmonic. The modeling of any user-defined harmonic spectrum and their angles should be possible in a harmonic analysis program.

On a simplistic basis for a 12-pulse converter model, noncharacteristic harmonics like 5, 7, 17, 19... at 15% of the level computed for a 6-pulse converter and model triplen harmonics 3, 9, 15,... as 1% of the fundamental. But this does not account for unbalance. If we write the unbalance voltages as

$$\begin{aligned} V_a &= V \sin \omega t \\ V_b &= V(1+d)\sin(\omega t - 120^\circ) \\ V_c &= V \sin(\omega t + 120^\circ) \end{aligned} \tag{19.47}$$

where d is the unbalance, then conduction intervals for the current are

$$\begin{aligned} t_a &= t_c = 120^\circ - e \\ t_b &= 120^\circ + 2e \end{aligned} \tag{19.48}$$

where

$$e = \tan^{-1}\frac{\sqrt{3}(1+d)}{3+d} - 30^\circ \tag{19.49}$$

The instantaneous currents ignoring overlap are

$$\begin{aligned} i_a &= \frac{4}{\pi}I_d\frac{(-1)^h}{h}\sin\left(\frac{ht_a}{2}\right)\sin(h\omega t) \\ i_b &= \frac{4}{\pi}I_d\frac{(-1)^h}{h}\sin\left(\frac{ht_b}{2}\right)\sin(h\omega t - 120^\circ - e/2) \\ i_c &= \frac{4}{\pi}I_d\frac{(-1)^h}{h}\sin\left(\frac{ht_c}{2}\right)\sin(h\omega t + 120^\circ - e) \end{aligned} \tag{19.50}$$

If there is 1% unbalance $d = 0.1$, then third harmonic currents are

$$\begin{aligned} i_{a3} &= 0.01586k < 180^\circ \\ i_{b3} &= 0.03169k < -2.36^\circ \\ i_{c3} &= 0.01586k < 175.27^\circ \end{aligned} \tag{19.51}$$

where

$$k = \frac{2\sqrt{3}I_d}{\pi} \tag{19.52}$$

The positive and negative sequence components of these currents will be injected into the network even if a delta connected transformer is used [15].

19.5.2 Converters

Chapter 17 discusses the harmonic generation from the converters. There is an interaction between the ac voltages and the dc current in weak ac/dc interconnections. Reference [5] describes voltage instabilities, control system instabilities, transient dynamic overvoltages, temporary and harmonic overvoltages, and low-order harmonic resonance. Reference [16] describes the inadequacy of commercial harmonic analysis programs to model transient system harmonics and development of converter models and nonlinear resistor models to represent mercury arc and sodium vapor lamps. The computational algorithm to determine the harmonic components of the converter current and reactive power drawn considers overlap angle, calculated V_d and I_d, and reiterates with overlap angle till $V_d I_d = P$, calculated the Fourier components of the input current and the reactive power drawn by the converter. A power series model is adopted for the nonlinear resistor.

The harmonic currents produced by nonlinear loads are derived on the assumption of a perfectly sinusoidal source that is a strong sinusoidal voltage system. These harmonic currents are then injected into the ac system to determine levels of voltage distortion. However, when the injected harmonic frequency is close to the parallel resonant frequency, the calculation algorithm often diverges and a transient converter simulation (TCS) is required. This requires ac system equivalents responding accurately to power and harmonic frequencies. An equivalent circuit consisting of number of tuned RLC branches has been proposed [17] as a solution to HVDC studies. Reference [10] describes frequency-dependent equivalent circuits suitable for integration in the time-domain solutions.

The converter from the dc side is a voltage source, and from the ac side it can be viewed as an admittances with a with current source, Figure 19.11a.

Many frequencies can be generated due to a single distortion. The three of these are important:

1. On dc side at frequency $h\omega_0$
2. On ac side, positive sequence $(h+1)\,\omega_0$ (19.53)
3. On ac side, negative sequence $(h-1)\,\omega_0$

This is shown in Figure 19.11(b).

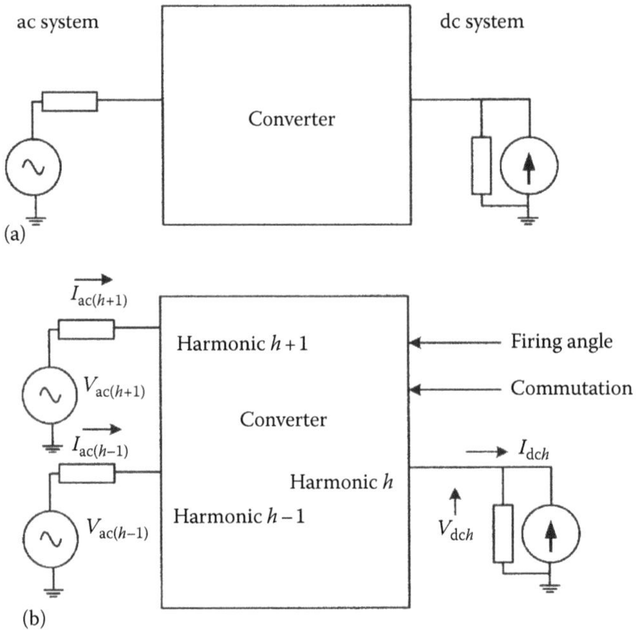

FIGURE 19.11
(a) Converter interconnecting ac and dc systems and (b) non-characteristics frequency interactions.

The elements of these matrices are derived by numerical techniques [18]. The frequency-dependent behavior of the converter may be defined as returned distortion (current or voltage) as a result of applied distortion at the same frequency. Manipulations of the Equation 19.53 lead to the interaction as shown in Figure 19.11b [10].

19.6 Modeling of Networks

The modeling is dependent on the network being studied. Networks vary in complexity and size, and, generally, it is not possible to include the detailed model of every component in the study. A high-voltage grid system may incorporate hundreds of generators, transmission lines, and transformers. Thus, the extent to which a system should be modeled has to be decided. The system to be retained and deriving an equivalent of the rest of the system can be addressed properly only through a sensitivity study. An example of the effect of the extent of system modeling is shown in Figure 19.12 for a 200 MW dc tie in a 230 kV system [4]. A 20-bus model shows resonance at the 5th and 12th harmonics, rather than at the 6th and 13th when larger numbers of buses are modeled. This illustrates the risk of inadequate modeling.

19.6.1 Industrial Systems

Industrial systems vary in size and complexity, and some industrial plants may generate their own power and have operating loads of 100 MW or more. The utility ties maybe at high voltages of 115, 138, or even 230 kV. It is usual to represent the utility source by its short-circuit impedance. There may be nearby harmonic sources or capacitors that will impact the extent of external system modeling. Generators and large rotating loads may be modeled individually, while an equivalent motor model can be derived connected through fictitious transformer impedance representing a number of transformers to which the

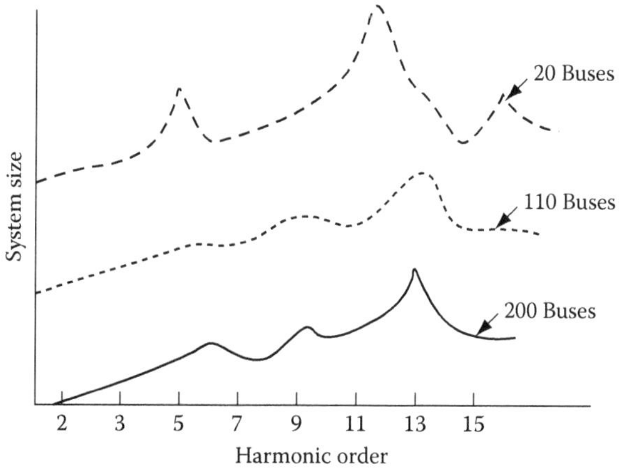

FIGURE 19.12
Errors introduced into a 200-bus system by inadequate modeling. (Adapted from IEEE Task Force on Harmonic Modeling and Simulation IEEE Trans. PD, 11, 452, 1996.)

motors are connected. This aggregating of loads is fairly accurate, provided that harmonic source buses and the buses having capacitor compensations are modeled in detail.

19.6.2 Distribution Systems

The primary distribution system voltage levels are 4–44 kV, while the secondary distribution systems are of low voltage (<600 V). A distribution system harmonic study is undertaken to investigate harmonic problems, resonances, and distortion in the existing system or to investigate the effect of adding another harmonic source. Single- or three-phase models can be used as demanded by the study. At low-voltage distribution level, the unbalance loads and harmonic sources are of particular importance, demanding a three-phase representation. The distribution systems are tied into the interconnected power network, and the system may be too large to be modeled. In most cases, it will be sufficiently accurate to represent a transmission network by its short-circuit impedance. This impedance may change, depending upon modifications to the system or system operation and is not a fixed entity. When capacitors or harmonic sources are present, a more detailed model will be necessary. The transmission system itself can be a significant source of harmonics in the distribution system, and measurement at points of interconnection may be necessary to ascertain it. The load models are not straightforward, if these are lumped with harmonic-producing loads. The motor loads can be segregated from the nonrotating loads, which can be modeled as an equivalent parallel *R–L* impedance. Again, it is not feasible to model each load individually, and feeder loads can be aggregated into large groups without loss of much accuracy.

The studies generally involve finite harmonic sources, and the background harmonic levels are often ignored. These can be measured, and analysis, generally, combines modeling and measurements for accuracy. A study shows that the harmonic currents at higher frequencies have widely varying phase angles, which result in their cancellation [19]. At lower frequencies up to the 13th, the cancellation is not complete. Unbalance loads on the feeder result in high-harmonic currents in the neutral and ground paths. At fifth and seventh harmonics, the loads can be modeled as (1) harmonic sources, (2) a harmonic source with a parallel *RL* circuit, and (3) a harmonic source with a series *RL* circuit. Radial distribution systems will generally exhibit a resonance or cluster of resonances between fifth and seventh harmonics. See also Ref. [20].

19.6.3 Transmission Systems

Transmission systems have higher X/R ratios and lower impedances, and the harmonics can be propagated over much longer distances. The capacitances of transformers and lines are higher, and these need to be included. The operating configuration range of a transmission system is much wider than that of a distribution system. A study may begin by identifying a local area, which must be modeled in detail. The distant portions of the system are represented as lumped equivalents. Equivalent impedance based on short-circuit impedance is one approach, the second approach uses a frequency versus impedance curve of the system, and there is a third intermediate area, whose boundaries must be carefully selected for accuracy. These can be based on geographical distance from the source bus. Series line impedance and the number of buses distant from the source are some other criteria. Figure 19.13 shows division of a network into areas of main study and external systems. Sensitivity methods provide a better analytical tool.

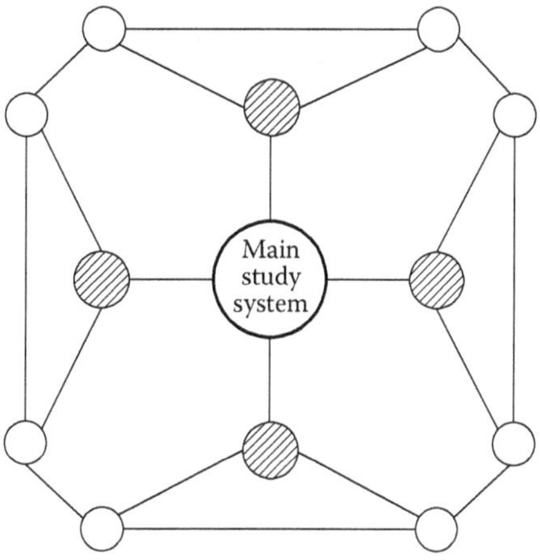

FIGURE 19.13
Division of a network into main study and external interconnected systems.

19.6.4 Sensitivity Methods

The purpose is to ascertain the sensitivity of the system response when a component parameter varies. Adjoint network analysis can be used. The network N consisting of linear passive elements is excited at the bus of primary interest by a unit current at the harmonic source bus and branch currents $I_1, I_2, \ldots, I_n$ are obtained. The transfer impedance T is defined as the voltage output across the bus of primary interest divided by the harmonic current of the input bus. The adjoint network N^*, which has the same topology as the original network, is excited by a unit current source from the output to obtain adjoint network branch currents $I_1^*, I_2^*, \ldots, I_n^*$ are obtained. The sensitivity of a transfer impedance T is defined as

$$S_x^T = \frac{\partial T}{\partial x}\left(\frac{x}{T}\right) \tag{19.54}$$

where x is any parameter R, L, or C at frequency denoted by S_x^T. The sensitivities can be calculated using

$$\frac{\partial T}{\partial x} = I_x \cdot I_x^* \tag{19.55}$$

where I_x and I_x^* are x element branch currents from two-network analysis of N and N^*, respectively, Figure 19.14a. The efficacy of the method is limited to small variations in the parameter. When large variations occur in external system equivalents, these can cause serious changes in the transfer function, and the bilinear theorem can be applied. These large changes in the transfer function of a two-port network to changes in an internal parameter are analyzed by pulling out Z of the network, effectively forming a three-port network (Figure 19.14b). For the transfer impedance, the following equation is obtained:

$$T = \frac{V_2}{I_1} = \frac{Z_{x\text{in}}T(0) + ZT(\infty)}{Z + Z_{x\text{in}}} \tag{19.56}$$

where
$T(0)$ is the transfer impedance when $Z = 0$
$T(\infty)$ is the transfer impedance when $Z = \infty$ (open circuited)
$Z_{x\text{in}}$ is the input impedance looking into the network from the nodes of Z [21]

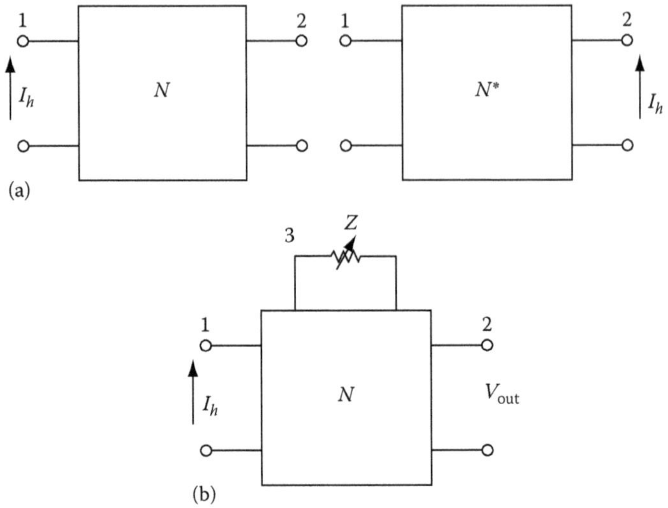

FIGURE 19.14
(a) Two-port N network, with a harmonic source and its adjoint network; (b) three-port network.

To summarize, for harmonic analysis program should be capable of the following:

- Represent load, transformer, generator, shunt capacitor, and induction motor models.
- Represent various types of load models.
- Form nodal impedance matrices for required range of frequencies.
- It should be possible to calculate harmonic impedances at any bus and harmonic frequency scans.
- Transmission lines, transformer, and reactor resistances must be corrected for skin/proximity effects.
- It should be possible to calculate harmonic current spectrum for line flows.
- Harmonic current injection at each harmonic source and its reaction with system should be considered.
- Sensitivity analysis should be possible.
- A powerful plotting interface is required.
- The parameters like TIF, KVT, and IT should be possible to be calculated.
- When harmonic filters are applied, their loading and harmonic load flow throughout should be calculated, with distortion indices, according to IEEE 519. The system and filter resonant frequencies should be capable of being plotted.
- The three-phase models should be accommodated.
- The impact of transformer-winding connections should be possible to be modeled.
- The harmonic-derating factors of the transformers should be possible to be calculated.

19.7 Power Factor and Reactive Power

The power factor of a converter is given by the following expression:

$$\text{Total PF} = \frac{q}{\pi}\sin\left(\frac{\pi}{q}\right) \quad (19.57)$$

where
q is the number of converter pulses
πq is the angle in radians

This ignores commutation overlap, no-phase overlap, and neglects transformer magnetizing current. For a six-pulse converter, the power factor is $3/\pi = 0.955$. A 12-pulse converter has a theoretical power factor of 0.988. With commutation overlap and phase retard, the power factor is given by [22]:

$$\text{PF} = \frac{E_d' I_d}{\sqrt{3}E_L I_L} = \frac{3}{\pi}\frac{1}{\sqrt{3f(\mu,\alpha)}}\left(\cos\alpha - \frac{E_x}{E_{do}}\right) \quad (19.58)$$

where
$E_d' = E_d + E_r + E_f$
E_d is the average direct voltage under load
E_r is the resistance drop
E_f is the total forward drop per circuit element
I_d is the dc load current in average amperes
E_L is the primary line-to-line ac voltage
I_L is the ac primary line current in rms amperes
α is the phase retard angle
μ is the angle of overlap or commutation angle
E_{do} is the theoretical dc voltage
E_x is the direct voltage drop due to commutation reactance

and

$$f(\mu,\alpha) = \frac{\sin\mu[2+\cos(\mu+2\alpha] - \mu[1+2\cos\alpha\cos(\mu+\alpha)]}{[2\pi\cos\alpha - \cos(\mu+\alpha)]^2} \quad (19.59)$$

The displacement power factor is

$$\cos\phi_1' = \frac{\sin^2\mu}{\sqrt{\mu^2+\sin^2\mu - 2\mu\sin\mu\cos\mu}} \quad (19.60)$$

This relationship neglects transformer magnetizing current. The correction for magnetizing current is approximately given by

$$\cos\phi_1 = \cos\left[\text{arc}\cos\phi_1' + \text{arc}\tan\left(\frac{I_{mag}}{I_1}\right)\right] \quad (19.61)$$

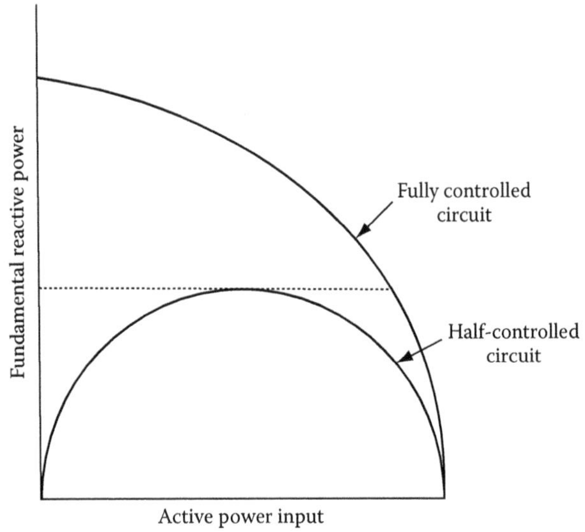

FIGURE 19.15
Reactive power requirements of fully controlled and half-controlled bridge circuits.

where cos ϕ_1' is the displacement power factor without transformer magnetizing current. The power factor of converters will vary with the type of converter and dc filter. In a PWM inverter, driven from a dc link voltage with a reactor and capacitor, the drive motor power factor is not truly reflected on the ac side and is compensated by the filter capacitor and reactor.

Line-commutated inverters require reactive power from the supply system. The closer is the operation to zero-voltage dc the more the reactive power required. Figure 19.15 shows the reactive power requirement of a fully controlled bridge circuit versus half-controlled bridge circuit. The maximum reactive power input for a half-controlled circuit is seen to be half of the fully controlled circuit. By sequential control, two converter sections are operated in series, one section fully phased on and the other section adding or subtracting from the voltage of the other section, see Figure 17.14.

In the case of sinusoidal voltage and current, the following relationship holds

$$S^2 = P^2 + Q^2 \tag{19.62}$$

where

P is the active power
Q is the reactive volt-ampere
S is the volt-ampere

This relationship has been amply explored in load flow section: $S = V_f I_f$, $Q = V_f I_f \sin(\theta_f - \delta_f)$, and $PF = P/S$.

In the case of nonlinear load or when the source has nonsinusoidal waveform, the reactive power Q can be defined as

$$Q = \sum_{h=1}^{h=\infty} V_h I_h \sin(\theta_h - \delta_h) \tag{19.63}$$

and the apparent power can be defined as

$$S = \sqrt{P^2 + Q^2 + D^2} \tag{19.64}$$

where D is the distortion power. Consider D^2 up to the third harmonic:

$$\begin{aligned} D^2 = {} & \left(V_0^2 + V_1^2 + V_2^2 + V_3^2\right)\left(I_0^2 + I_1^2 + I_2^2 + I_3^2\right) \\ & - \left(V_0 I_0 + V_1 I_1 \cos\theta_1 + V_2 I_2 \cos\theta_2 + V_3 I_3 \cos\theta_3\right)^2 \\ & - \left(V_1 I_1 \sin\theta_1 + V_2 I_2 \sin\theta_2 + V_3 I_3 \sin\theta_3\right)^2 \end{aligned} \tag{19.65}$$

An expression for distortion power factor can be arrived from current and voltage harmonic distortion factors. From the definition of these factors, rms harmonic voltages and currents can be written as

$$V_{\mathrm{rms}(h)} = V_f \sqrt{1 + \left(\frac{\mathrm{THD_V}}{100}\right)^2} \tag{19.66}$$

$$I_{\mathrm{rms}(h)} = I_f \sqrt{1 + \left(\frac{\mathrm{THD_I}}{100}\right)^2} \tag{19.67}$$

Therefore, the total power factor is

$$\mathrm{PF_{tot}} = \frac{P}{V_f I_f \sqrt{1 + (\mathrm{THD_V}/100)^2}\sqrt{1 + (\mathrm{THD_I}/100)^2}} \tag{19.68}$$

Neglecting the power contributed by harmonics and also voltage distortion, as it is generally small

$$\begin{aligned} \mathrm{PF_{tot}} &= \cos(\theta_f - \delta_f) \cdot \frac{1}{\sqrt{1 + (\mathrm{THD_I}/100)^2}} \\ &= \mathrm{PF_{displacement}} \mathrm{PF_{distortion}} \end{aligned} \tag{19.69}$$

The total power factor is the product of displacement power factor (which is the same as the fundamental power factor) and is multiplied by the distortion power factor as defined above.

19.8 Shunt Capacitor Bank Arrangements

Formation of shunt capacitor banks from small to large sizes and at various voltages is required for filter design and reactive power compensation. These can be connected in a variety of three-phase connections, which depend on the best utilization of the standard voltage ratings, fusing, and protective relaying. To meet certain kvar and voltage requirements, the banks are formed from standard unit power capacitors available in certain ratings and voltages. For high-voltage applications, these are outdoor rack mounted.

TABLE 19.2

Number of Series Groups in Y-Connected Capacitor Banks

		Available Capacitor Voltage; Kilovolt Per Unit												
V (kV)	*V* (kV)	21.6	19.92	14.4	13.8	13.28	12.47	9.96	9.54	8.32	7.96	7.62	7.2	6.64
500.0	288.7	14	15	20	21	22		29	30	35	36	38		
345.0	199.2		10			15	16	20	21	24	25	27		
230.0	132.8					10			14	16	17	18		20
161.0	92.9					7							13	14
138.0	79.7		4	6	6	6		8			10		11	12
115.0	66.4					5			7	8	9	9		10
69.0	39.8		2		3	3		4			5			6
46.0	26.56					2								4
34.5	19.92		1					2						3
24.9	14.4			1									2	
23.9	13.8				1									2
23.0	13.28					1								
14.4	8.32									1				
13.8	7.96										1			
13.2	7.62											1		
12.47	7.2												1	

For medium-voltage applications, a bank may be provided in an indoor or outdoor metal enclosure or it can be rack mounted outdoors. Table 19.2 [23] shows the number of series groups for wye-connected capacitor banks required for line operating voltages from 12.47 to 500 kV, i.e., for 500 kV application, 14 series strings of 21.6 kV rated capacitors, or 38 strings of 7.62 kV rated capacitors are required in a formation as shown in Figure 19.16. The unit sizes are, generally, limited to 100, 200, 300, and 400 kvar, and, in each string, a number of units are connected in parallel to obtain the required kvar. There are limitations in forming the series and parallel strings. A minimum number of units should be placed in parallel per series group to limit voltage on the remaining units to 110%, if any one unit goes out of service, say due to operation of its fuse. This is because capacitors are rated to withstand 10% overvoltage, though for filter designs, other considerations enter into this picture. One is that larger the size of the capacitor unit lost (say due to operation of its fuse), the greater will be the detuning effect. This may overload the units remaining in service, shift the design frequencies, and increase the distortion at the PCC (point of common coupling), beyond acceptable limits [see Chapter 20].

Individual capacitor-can fusing is selected to protect the rupture/current withstand rating of the can. The maximum clearing time curve of the fuse and the case rupture curve are plotted together. The case rupture characteristics vary with the design and size, and these data should be obtained from a manufacturer. The curves are for 10% and 50% probability boundary. The probability of case rupture can be defined as the opening of the case as a result of failure, from a mere cracked seam or bushing seal to a violent bursting of the case. Within safe zone, no greater damage than a slight swelling of the case will occur, though a case rupture is possible for low levels of short-circuit current flowing for extended period of time. To avoid such ruptures, the fuse link is coordinated so that it clears the fault in 300 s. The overvoltage duration that will occur on the capacitors remaining in service is of consideration. Also, the fuse must be selected for continuous current, switching inrush

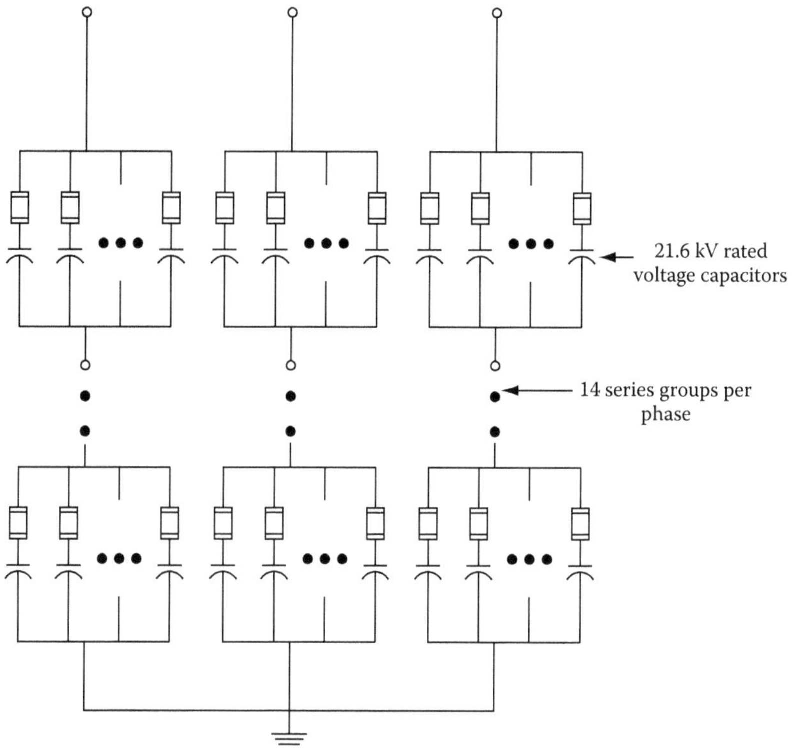

FIGURE 19.16
Formation of a 500 kV shunt capacitor bank with series groups.

current, and lightning surge current requirements. The available short-circuit and its clearing time at the point of application should be compared with the manufacturer's published data on short-circuit withstand. As the capacitor enclosures are grounded, a phase-to-ground fault can still occur, even if the capacitors are connected in ungrounded configuration. The maximum size of capacitor bank for group fusing is limited by the maximum size of the available fuse, current limiting, or T and K expulsion-type links. The expulsion-type fuses are generally used for outdoor rack-mounted capacitor banks. In an ungrounded wye-connected bank, the maximum short-circuit current is only three times the full load current, and a group fuse selection to meet all the criteria and fast clearance time becomes difficult. For externally fused banks, fuses should be fast to coordinate with the fast unbalance relay settings, but should not operate during switching or external faults.

Table 19.3 shows the minimum units in a parallel group [23]. Also, there is a limit to the number of parallel units connected in a series group when expulsion-type fuses are used for individual capacitor-can protection. The energy liberated and fed into the fault when a fuse operates is required to be limited, depending on the fuse characteristics (Figure 19.17). The energy release may be as follows:

$$E = 2.64 \text{ J per KVAC rated voltage} \tag{19.70}$$

$$E = 2.64(1.10)^2 \text{ J per KVAC 110\% voltage} \tag{19.71}$$

$$E = 2.64(1.20)^2 \text{ J per KVAC 120\% voltage} \tag{19.72}$$

TABLE 19.3

Minimum Number of Units in Parallel Per Series Group to Limit Voltage on Remaining Units to 110% with One Unit Out

Number of Series Groups	Grounded Y or Δ	Ungrounded Y	Double Y, Equal Sections
1	—	4	2
2	6	8	7
3	8	9	8
4	9	10	9
5	9	10	10
6	10	10	10
7	10	10	10
8	10	11	10
9	10	11	10
10	10	11	11
11	10	11	11
12 and over	11	11	11

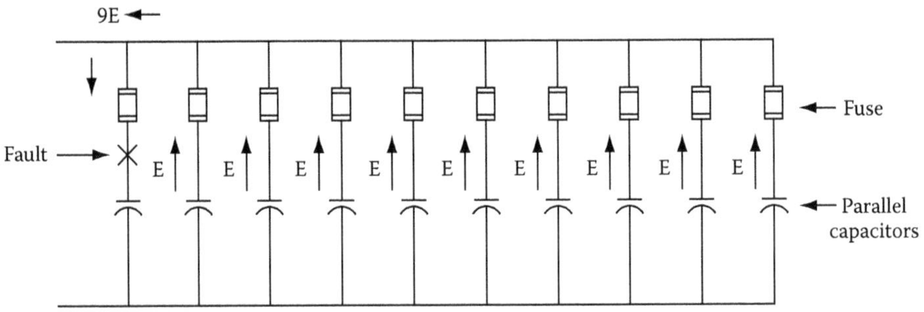

FIGURE 19.17
Energy fed into a fault from parallel capacitor units.

Normally, capacitors up to 3100 kvar can be connected in parallel when expulsion fuses are used. This limit can be exceeded if capacitor units are fused with current-limiting fuses (generally limited to indoor metal-enclosed installations).

Apart from the multiple series group grounded wye banks, the capacitors may be connected in:

- Ungrounded wye connection
- Ungrounded double wye neutrals
- Grounded double wye neutrals
- Delta connection

These connections are shown in Figure 19.18. Delta connection is common for low-voltage application with one-series group rated for line-to-line voltage. A wye-ungrounded group can be formed with one group per phase, when the required operating voltage corresponds to standard capacitor unit rating. The wye neutral is left ungrounded. Grounded Y neutrals

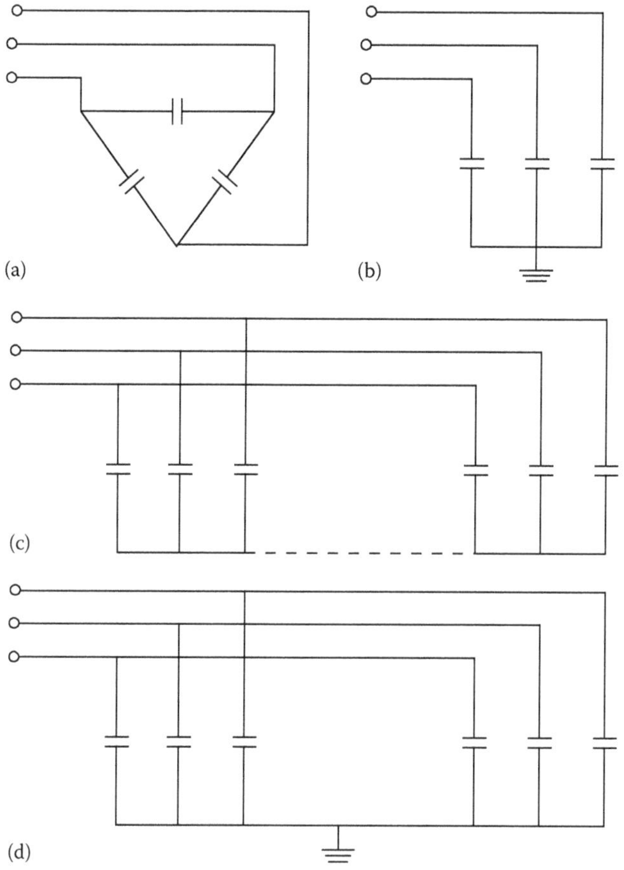

FIGURE 19.18
Three-phase connections of shunt capacitor banks. (a) Delta connection, (b) wye connection, wye ungrounded or grounded, (c) ungrounded double-wye (wye points connected or unconnected), and (d) grounded double-wye connection.

and multiple series group are used for voltages above 34.5 kV. Multiple series groups limit the fault current. Grounded capacitors provide a low impedance path for lightning surge currents and give some protection from surge voltages; however, third-harmonic currents can circulate, and these may overheat the system grounding impedances. In the case of an ungrounded multiple series group, wye-connected bank, third-harmonic currents do not flow, but the entire bank, including the neutral, should be insulated for the line voltage. Double-wye banks and multiple series groups are used when a capacitor bank becomes too large for the 3100 kvar per group for the expulsion type of fuses. The design of grounding grids and connection of neutral points of capacitor banks is of importance. Two methods of grounding are (1) single-point grounding and (2) peninsula grounding. With single-point grounding, the neutrals of capacitor banks of a given voltage are all connected through an insulted cable, which is connected to the grid at one point. There will be substantial voltage of the order of tens of kilovolts between the ends of the neutral bus and single-point ground during switching. The use of shielded cable will help reduce the voltage stress. In the event of a fault, high frequency currents can flow back into power system via the substation ground grid.

With peninsula grounding, the grounding grid is built under capacitor banks and bus work in a form resembling a series of peninsulas. In this arrangement, one or more ground conductors may be carried under the capacitor rack of each phase of each group and tied to main station ground at one point at the edge of capacitor area. All capacitor neutral connections are made to this isolated peninsula grounding grid conductors [23]. The capacitor bank and associated current transformer and voltage transformer potential will rise during capacitor switching, but the transients in the rest of the system will be reduced. Also, with peninsula grounding, all equipment at the neutral ends tends to rise to the same potential, and differential voltages can be avoided.

Figures 19.16 and 19.17 show externally fused capacitors. The selection of the fuse is determined by the coordination with the capacitor case rupture withstand curve and the transient inrush currents.

The capacitor banks may be internally fused, where one fuse is connected in series with each capacitor element. On a puncture or short circuit in a capacitor element, the current through the fuse increases in proportion to the number of units in parallel. This will melt the fuse in a short time of few milliseconds. Generally, the number of units that can be put in parallel is reduced and series groups increased compared to externally fused banks. Generally, at least there should be two capacitor units in parallel in each group.

The fuseless capacitor banks have an arrangement that does not use fuses, and it is not simply the elimination of fuses. When an 'all-film' capacitor has a dielectric short, the film burns away resulting in a weld between the foils—it creates a low resistance short, which does not generate large amount of heat or gas, as long as the current is limited through the short, allowing the capacitor to continue in operation indefinitely. These designs were not used with paper/film or all paper capacitors because dielectric shorts are more likely to have localized heating and gassing.

19.9 Unbalance Detection

An external fuse should be sized to meet a number of requirements, which include the protection of the case rupture withstand capability and withstanding the discharge current of a failed unit. Also, the fuse must withstand switching inrush currents, restriking current, and lightning surge currents. A failure of a fuse in a capacitor bank involving more than one series group per phase and in all ungrounded capacitor banks irrespective of the series groups will increase the voltage on the capacitors remaining in service. This magnitude can be calculated from the following expressions:

Grounded wye, double wye, or delta connected banks:

$$\%V = \frac{100S}{1 + (1 - \%R/100)(S - 1)} \tag{19.73}$$

where

S is the number of series groups >1

R is the percentage of capacitor units removed from one series group

$\%V$ is the percent of nominal voltage on the remaining units in affected series group

Say a series group has 5 capacitors in parallel, and there are three series groups per phase ($S = 3$), then failure of a fuse in one unit means 20% outage ($R = 20$), and $\%V = 115\%$. A capacitor cannot withstand higher than 10% voltage continuously, and this over voltage is not acceptable if operation for any length of time is required with one capacitor unit out of service.

Ungrounded wye or ungrounded double-wye neutrals are isolated.

$$\%V = \frac{100S}{S(1 - \%R/100) + 2/3(\%R/100)} \tag{19.74}$$

For the same configuration as for grounded bank, the $\%V$ will be 118.4%. The failure of a unit results in detuning in case of a filter bank; it is further discussed in Chapter 20. Generally, the voltage rise is limited to a maximum of 10% and sometimes, even lower. A bank can be designed so that on loss of a single unit the capacitor bank continues to operate in service and tripped immediately on loss of two units in the same group. The protection schemes to detect unbalance are not discussed; however, with modern microprocessor-based relays, it is possible to detect an unbalance as low as 1%–2% on failure of a capacitor unit in a banks of parallel and series groups.

19.10 Study Cases

At each step of the calculations in these examples, we will analyze the results and draw inferences as we proceed. Modifications to the original problem and its solution reveal the nature of the study and harmonic resonance problems. We will then continue with the same examples in Chapter 20 to address the resolution of harmonic problems that we discover in this section.

Example 19.1: A Substation with Single Harmonic Source

Consider a six-pulse drive system load and a 2000 hp induction motor connected as shown in Figure 19.19a; 10 and 5 MVA transformers have a percentage impedance of 6%. A 4.16 kV bus 2 is defined as the PCC. The explanation of PCC is included in Appendix F. It is required to calculate the harmonic voltages and currents into the 2000 hp motor and 10 MVA transformer T1 and also harmonic distortion factors starting with hand calculations. The source impedance of the supply system is first neglected in the hand calculations; it is relatively small compared to the transformer impedance. We will itemize the steps of the calculations and make observations as we proceed.

1. *Derive a harmonic current injection model*: The converter operates at a firing angle α of 15° and 5 MVA transformer T2 is fully loaded. The hand calculations are limited to the 17th harmonic. This is to demonstrate the procedure of calculations, rather than a rigorous solution of the harmonic flow problem. Harmonic flow calculations are generally carried to the 40th harmonic and preferably to the 50th. By limiting the calculation to lower frequencies, a possible higher frequency resonance can be missed, and serious errors in calculations of distortion factors can occur.

   The commutating reactance is $0.06\% + 0.03\% = 0.09\%$, approximately, on a 5 MVA base. The overlap angle is calculated using Equation 17.55, and it is ≈14.5°. From Appendix G, the percentage harmonic currents are 18%, 13%, 6.5%, 4.8%, and 2.8%, respectively, for the 5th, 7th, 11th, 13th, and 17th harmonics. In terms of the fundamental current of transformer T2 = 695 A, the injected harmonic currents are 125, 90, 76, 33, and 19 A, respectively, for the 5th, 7th, 11th, 13th, and 17th harmonics. The transformer must be capable of supplying these nonlinear loads.

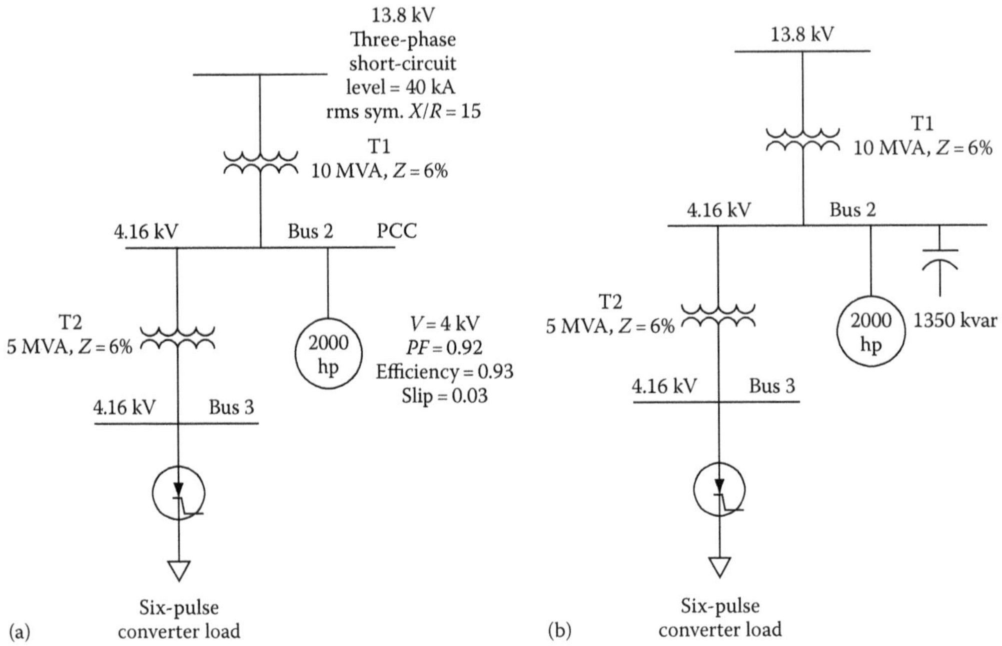

FIGURE 19.19
(a) System with harmonic load for Example 19.1; (b) with a shunt capacitor of 1.35 Mvar added at bus 2.

2. *Calculate bus harmonic impedances*: Table 19.4 shows the calculation of the transformer T1 resistance and reactance at harmonic frequencies. The resistance at fundamental frequency is multiplied by appropriate values read from the curve in Figure 19.6. The reactance at harmonic frequencies is directly proportional to the harmonic frequency, though this is an approximation.

   To calculate the harmonic impedances of the induction motor, its locked rotor reactance at rated voltage is taken as 16.7% (the same as in the short-circuit calculations). For a 2000 hp motor, this gives an X_m of 1.658 ohm. The fundamental frequency resistance is calculated from the X/R ratio of 30. This gives a resistance of 0.0522 ohm.

   Assume that 45% of the resistance is the stator resistance and 55% is the rotor resistance. The stator resistance is then 0.02487 ohm, and the rotor resistance is 0.03033 ohm. The slip

TABLE 19.4
Example 19.1: Calculation of Harmonic Impedances[a]

	Transformer T1			Motor			Bus 2		
Harmonic Order	*R*	*X*	*Z*	*R*	*X*	*Z*	*R*	*X*	*Z*
5	0.0277	0.5166	0.5173	0.1186	8.290	8.291	0.0225	0.4863	0.4868
7	0.0416	0.7231	0.7243	0.1525	11.606	11.607	0.0368	0.6807	0.6817
11	0.0832	1.1363	1.1393	0.1789	18.238	18.239	0.0568	1.0696	1.0711
13	0.0901	1.3420	1.3450	0.1987	21.554	21.555	0.0620	1.2614	1.2629
17	0.1730	1.7560	1.7645	0.2241	28.186	28.187	0.0976	1.653	1.6558

[a] All values are in ohms.

at fifth harmonic (negative sequence harmonic) is 1.2. The resistance at the fifth harmonic from Equations 19.34 through 19.37 is

$$0.02487\sqrt{5} + \frac{\sqrt{6}(0.03033)}{1.2} = 0.1186$$

Harmonic resistance and reactance for the motor, calculated in a likewise manner, are shown in Table 19.4.

The transformer and motor impedances are paralleled to obtain the harmonic bus impedances at each frequency. Here, the resistance and reactance are separately paralleled, akin to X/R calculation for the short-circuit currents. This method is not to be used for harmonic studies, and complex impedance calculation should be performed. The harmonic impedances calculated at bus 2, the PCC, are shown in Table 19.4.

3. *Calculate harmonic currents and voltages*: As the harmonic currents are known, the bus harmonic voltages can be calculated, by IZ multiplication. The harmonic currents in the motor and transformer T1 are then calculated, based on the harmonic voltages. Note that these do not exactly sum to the harmonic injected currents (Table 19.5).
4. *Calculate harmonic current and voltage distortion factors*: The limits of current and voltage distortion factors are discussed in Appendix F. The voltage harmonic distortion factor is given by

$$\frac{\sqrt{(60.85)^2 + (61.40)^2 + (81.40)^2 + (41.67)^2 + (31.46)^2}}{(4160/\sqrt{3})}$$

This gives 5.36% distortion. The calculation is not accurate as the higher order harmonics are ignored. The total demand distortion (TDD) is defined in Appendix F. Based on the current injected into the main transformer T1, it is

$$\frac{\sqrt{(117.36)^2 + (84.71)^2 + (71.45)^2 + (30.98)^2 + (17.80)^2}}{830}$$

Note that the TDD is based upon the total load demand, which includes the nonlinear load and the load current of the 2000 hp motor (see Appendix F). This gives a TDD of 19.9%. This exceeds the permissible levels at the PCC (Appendix F). This can be expected as the ratio of nonlinear load to total load is approximately 83%. When the nonlinear load exceeds about 35% of the total load, a careful analysis is required.

TABLE 19.5

Example 19.1: Harmonic Current Flow and Harmonic Voltages

Harmonic Order	Harmonic Current Injected (A)	Current into Transformer T1 (A)	Current into 2000 hp Motor (A)	Harmonic Voltage (V)
5	125	117.63	7.33	60.85
7	90	84.71	5.28	61.35
11	76	71.45	4.46	81.40
13	33	30.98	1.93	41.67
17	19	17.80	1.12	31.46

Harmonic voltage distortion factor = 5.36%.

5. *Reactive power compensation*: Consider now that a power factor improvement shunt capacitor bank at bus 2 in Figure 19.19b is added to improve the overall power factor of the supply system to 90%.

   The operating power factor of the converter is 0.828. A load flow calculation shows that 5.62 MW and 3.92 Mvar must be supplied from the 13.8 kV system to transformer T1. This gives a power factor of 0.82 at 13.8 kV, and there is 0.58 Mvar of loss in the transformers.

   A shunt capacitor bank of 1.4 Mvar at bus 2 will improve the power factor from the supply system to approximately 90%.

6. *Form a capacitor bank and decide its connections*: A capacitor rating of 2.77 kV and 150 kvar is standard. The reasons for using a voltage rating higher than the system voltage are discussed in Chapter 20. At the voltage of use, four units per phase, connected in ungrounded wye configuration, will give

$$3(600)\left[\frac{2.4}{2.77}\right]^2 = 1351.25\ \text{kvar}$$

   and

$$X_c = \frac{(\text{kV})^2 \times 10^3}{\text{kvar}} = \frac{2.77^2 \times 10^3}{600} = 12.788\ \text{ohm}$$

   or $c = 2.074\text{E-}4\text{F}$.

7. *Estimate resonance with capacitors*: The short-circuit level at bus 2 in Figure 19.19a is approximately 150 MVA and from Equation 18.20, parallel resonance can be estimated to coincide with the 11th harmonic. To ascertain the resonant frequency correctly, a frequency scan is made at small increments of frequency. To capture the resonance correctly, this increment of frequency should be as small as practical. A scan at five-cycle intervals will miss the resonance peak by four cycles.

8. *Limitations of hand calculations*: This brings a break point in the hand calculations. Assuming that the impedances are calculated at five cycle intervals, to cover the spectrum up to even the 17th harmonic, 192 complex calculations must be made in the first iteration. The hand calculations are impractical even for a small system.

9. *Frequency scan with capacitors*: Resorting to a computer calculation, the current injection is extended to the 35th harmonic, and the system short-circuit level of 40 kA is inputted.

   The results of frequency scan at two-cycle intervals show that the resonant frequency is between 630 and 632. (A two-cycle frequency step is used in the calculation.) This result was expected. The maximum angle is 88.59° and the minimum −89.93°; the impedance modulus is 122.70 ohm. The impedance modulus is shown in Figure 19.20a and phase angle in Figure 19.20b. If a frequency scan is made without the capacitors, it shows that the maximum impedance is 3 ohm, and it is a straight line of uniform slope (Figure 19.20a, dotted lines). The impedance at parallel resonance increases many fold. Figure 19.20b shows an abrupt change in impedance phase angle with the addition of capacitor.

10. *Harmonic study with capacitors*: The distribution of harmonic currents at each frequency is shown in Table 19.6. It shows that

    - The harmonic currents throughout the spectrum are amplified. This amplification is high at the 11th harmonic (close to the resonant frequency of 630–632 Hz). While the injected current is 76 A, the current in the capacitor bank is 884 A, in the supply transformer 753 A, and in the motor 55.7 A.
    - The 11th harmonic voltage is 1030 V. The harmonic voltage distortion is 43.3%, and the voltage–time wave shape in Figure 19.21 shows large distortion due to 11th harmonic.

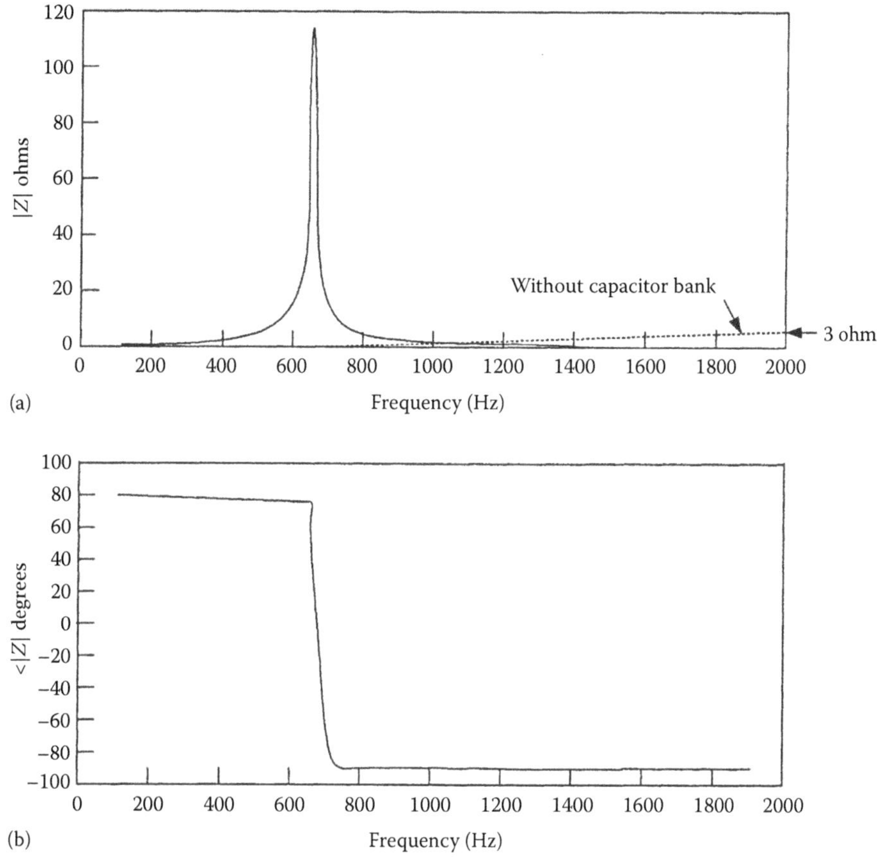

FIGURE 19.20
(a) Impedance modulus showing resonance at 11th harmonic with addition of 1.35 Mvar capacitor bank (Example 19.1); (b) impedance phase angle plot.

TABLE 19.6
Example 19.1: Harmonic Currents and Voltages with a Shunt Capacitor Bank

Order of the Harmonic	Injected Current (A)	Current in the Capacitor Bank (A)	Current in the Transformer T1, at PCC (A)	Current in the 2000 hp Motor (A)	Harmonic Voltage, Bus 2 (V)
5	125	36.4	150	11.1	93.5
7	90	71.5	150	11.1	13.9
11	76	884	753	55.7	1030
13	33	95	58	4.3	94.2
17	19	30	11	0.81	23.2
19	10.5	15.9	4.5	0.33	10.70
23	3.5	4.4	0.86	0.064	2.47
25	2.8	3.4	0.56	0.042	1.74
29	2	2.3	0.28	0.021	1.02
31	1.4	1.6	0.17	0.013	0.65
35	1.2	1.4	0.14	0.009	0.65

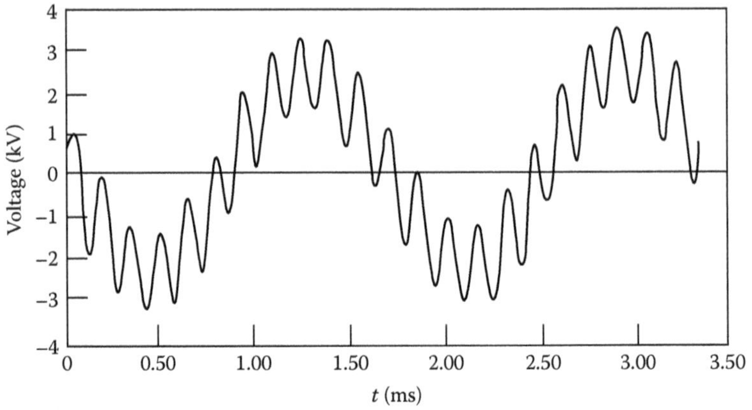

FIGURE 19.21
Voltage–time waveform, showing pollution from 11th harmonic.

The total harmonic current demand distortion factor is 94.52%. Appendix F shows that total TDD for this system should not exceed 8%.

- The harmonic loading on the capacitor bank is calculated as follows:

$$I_{\text{rms}} = \left[\sum_{h=1}^{h=35} I_h^2\right]^{1/2} = 806.6\text{ A}$$

The permissible rms current in the capacitors including harmonics is 1.35 times the rated current = 252.9 A. The current loading is exceeded. The rms voltage is calculated in a similar way to current, and it is 2614 V, V_{rms} rated = 2770 V, V_{rms} permissible = 3047 V. The voltage rating is not exceeded. The kvar loading is calculated from:

$$\text{kvar} = \sum_{h=1}^{h=35} I_h V_h = 2283$$

The rated kvar = 1351 and permissible kvar = 1.35 × 1351 = 1823. The rated kvar is exceeded.

11. *Sizing capacitor bank to escape resonance*: The capacitor bank is sized to escape the resonance with any of the load-generated harmonics. If six capacitors of 150 kvar, rated voltage 2.77 kV, are used in parallel per phase in wye configuration, the total three-phase kvar is 2026 at the operating voltage, and the resonant frequency will be approximately 515 Hz (Equation 18.20). As this frequency is not generated by the load harmonics and is also not a multiple of third harmonic frequencies, a repeat of harmonic flow calculation with this size of capacitor bank should considerably reduce distortion.

The results of calculation do show that the voltage distortion and TDD are 4% and 18%, respectively. Amplification of harmonic currents occurs noticeably at the fifth and seventh harmonics. Thus:

- The mitigation of the resonance problem by selecting the capacitor size to escape resonance may not minimize the distortion to acceptable levels.
- Current amplifications occur at frequencies adjacent to the resonant frequencies, though these may not exactly coincide with the resonant frequency of the system.

- The size of the capacitor bank has to be configured based on series parallel combinations of standard unit sizes and this may not always give the desired size. Of necessity, the capacitor banks have to be sized a step larger or a step smaller to adhere to configurations with available capacitor ratings and over- or under-compensation results.
- The resonant frequency swings with the change in system operating conditions, and this may bring about a resonant condition; however, carefully, the capacitors were sized in the initial phase. The resonant frequency is especially sensitive to change in the utility's short-circuit impedance.

The example is carried further in Chapter 20 for the design of passive filters.

Example 19.2: Large Industrial System

This example portrays a relatively large industrial distribution system with plant generation and approximately 18% of the total load consisting of six-pulse converters. The system configuration is shown in Figure 19.22. The loads are lumped on equivalent transformers—this is not desirable when harmonic sources are dispersed throughout the system. A 225-bus plant distribution system is reduced to an 9-bus system in Figure 19.22. Some reactive power compensation is provided by the power factor improvement capacitors switched with the medium-voltage motors at buses 6 and 7. The load flow shows that reactive power compensation of 7.2 and 6.3 Mvar at 12.47 kV buses 2 and 3, respectively, is required to maintain an acceptable voltage profile on loss of a plant generator. This compensation is provided in two-step switching. The power factor improvement capacitors at bus 3 are split into two sections of 4.5 and 2.7 Mvar, respectively. The capacitors at bus 2 are divided into two banks of 3.6 and 2.7 Mvar, respectively. The series reactors can be designed to turn the capacitor banks into shunt filters, but here the purpose is to limit inrush currents, especially on back-to-back switching, and also to reduce the switching duty on the circuit breakers (not shown in Figure 19.22).

Some medium-voltage motor load can be out of service, and their power factor improvement capacitors will also be switched out of service. The loads may be operated with one or two generators out of service, with some load shedding. The medium-voltage capacitors on buses 2 and 3 have load-dependent switching, and one or both banks may be out of service. This switching strategy is very common in industrial plants to avoid generation of excessive capacitive reactive power and also to prevent overvoltages at no load.

The following three operating conditions are studied for harmonic simulation:

1. All plant loads are operational with both generators running at their rated output, and all capacitors shown in Figure 19.22 are operational. Full converter loads are applied.
2. No. 2 generator is out of service. The motor loads are reduced to approximately 50%, and 2.7 Mvar capacitor banks at buses 2 and 3 are out of service, but the converter load is not reduced.
3. The effect of 30 MVA of bulk load and 18 MW of converter load connected through a 115 kV 75 mile transmission line (modeled with distributed line constants), which is ignored in cases 1 and 2, is added. These loads and transmission line model are superimposed on operating conditions 1.

The results of harmonic simulation are summarized in Tables 19.7 and 19.8. Table 19.7 shows harmonic current injection into the supply system at 115 kV and generators nos. 1 and 2. Table 19.8 shows the parallel and series resonant frequencies. The impedance modulus versus frequency plot of the 115 kV bus and 12.47 kV buses 2 and 3 for all three cases of study are shown in Figures 19.23 through 19.25. The R/X plots of the utility's supply system and impedance modulus versus frequency plots are shown in Figure 19.26.

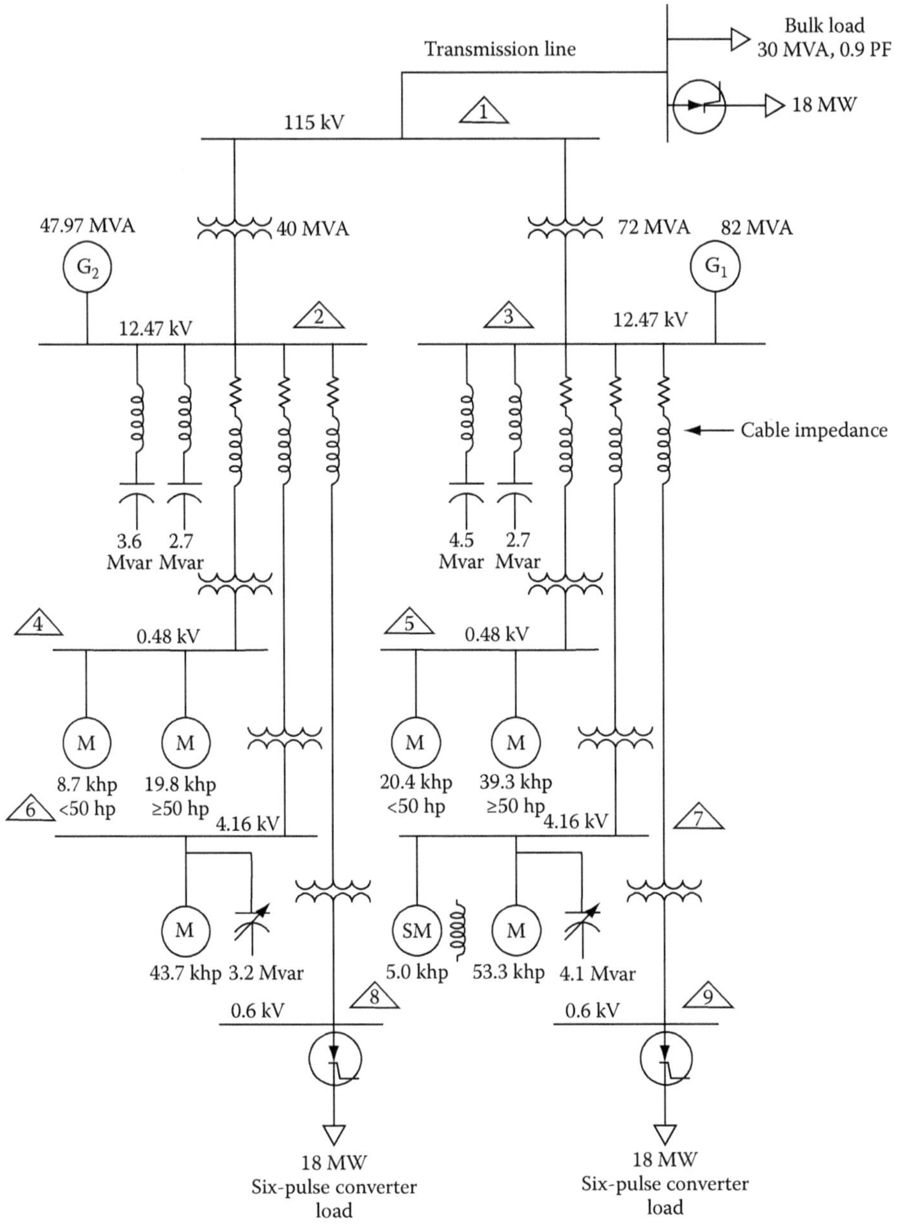

FIGURE 19.22
Single-line diagram of a large industrial plant loads aggregated for harmonic study (Example 19.2).

TABLE 19.7

Example 19.2: Harmonic Current Flow, Operating Conditions 1, 2, and 3

Harmonic Order	Current in Generator 1			Current in Generator 2			Current in Utility's System		
	1	2	3	1	2	3	1	2	3
5	187	257.00	167.00	209	Generator out of service	196.00	3.22	69.02	23.3
7	353	7.06	340	84.2		82.90	44.9	18.31	44.60
11	1.52	17.10	1.83	102		109.00	3.80	2.13	4.32
13	6.76	3.25	5.10	21.40		18.50	0.22	0.500	3.67
17	2.88	2.71	4.31	2.74		3.34	0.48	0.33	1.55
19	0.68	19.00	0.27	1.03		0.74	1.28	2.23	1.88
23	0.06	0.93	0.09	0.12		0.15	0.003	0.18	1.64
25	0.19	0.37	0.16	0.11		0.09	0.03	0.08	1.63

It is seen that the resonant frequencies vary over wide limits and so does the harmonic current flow. In operating condition 2 with partial loads and some capacitors out of service, additional resonant frequencies occur, which did not exist under operating condition 1. Under condition 3, the harmonic current injection at higher frequencies in the utility system increases appreciably. Condition 2 gives higher distortion at the PCC, though the distortion in generator 1 is reduced. The continuous negative sequence capability of generators ($I_2 = 10\%$) is exceeded in operating conditions 1 and 3.

The analysis is typical of large industrial plants where power factor improvement capacitors are provided in conjunction with nonlinear plant loads. The variations in loads and operating conditions result in large swings in the resonant frequencies. The example also illustrates the effect of power capacitors and nonlinear loads, which are located 75 miles away from the plant distribution. The need for mitigation of harmonics is obvious, and the example is continued further in Chapter 20 for the design of passive harmonic filters.

Example 19.3

Calculate TIF, IT, and kVT at the sending end of line L1 (Figure 19.22) operating condition 3 of Example 19.2.

IT and kVT are calculated from Equation 18.41, and TIF factors are given in Table 18.7. The harmonic currents or voltages in the line are known from harmonic analysis study. Table 19.9 shows the calculations. From this table,

$$\left[\sum_{h=1}^{h=49} I_h^2\right]^{1/2} = 0.158\,\text{kA}$$

The TIF-weighting factors are high in the frequency range 1620–3000, and, for accuracy, harmonic currents and voltages should be calculated up to about the 49th harmonic. From Table 19.9:

$$\left[\sum_{h=1}^{h=49} W_f^2 I_f^2\right]^{1/2} = \sqrt{4700.73} = 68.56$$

Therefore, IT = 68.56/0.159 = 360.85; kVT can be similarly calculated using harmonic voltages in kilovolt. TIF factors for industrial distribution systems are, normally, not a concern.

TABLE 19.8

Example 19.2: Resonant Frequencies and Impedance Modulus, Operating Conditions 1, 2, and 3

	1			2			3		
Bus ID	**Parallel Resonance**	**Series Resonance**	**Impedance Modulus**	**Parallel Resonance**	**Series Resonance**	**Impedance Modulus**	**Parallel Resonance**	**Series Resonance**	**Impedance Modulus**
Utility	403	424	239.6	373	385	318.6	400	425	213.6
				496	530				
Bus 2	586	1436	17.68	367	647	41.5	589	1436	28.26
				905					
Bus 3	400	635	33.76	493	860	26.24	401	635	12.89
	893	1343		1153	1699		892	1342	

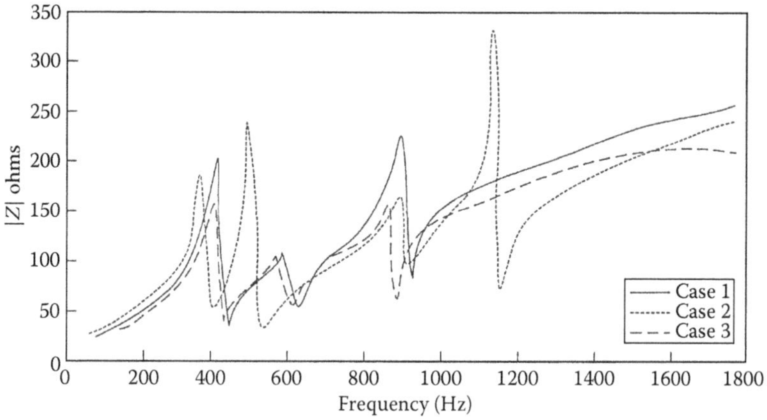

FIGURE 19.23
Impedance modulus versus frequency plot for three conditions of operation, 115 kV utility's bus (Example 19.2).

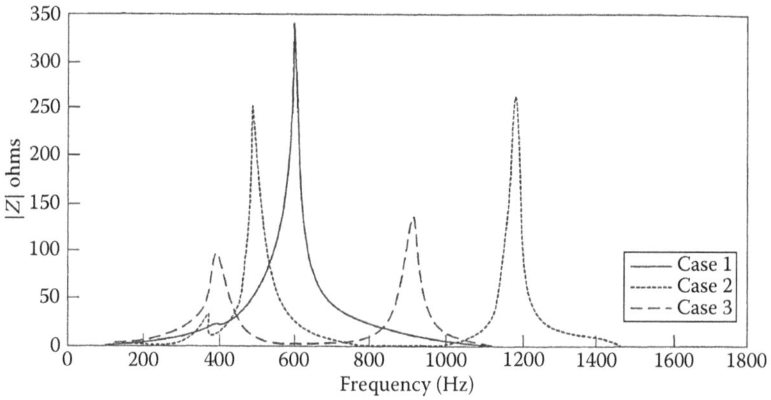

FIGURE 19.24
Impedance modulus versus frequency plot for three conditions of operation, 12.47 kV bus 2 (Example 19.2).

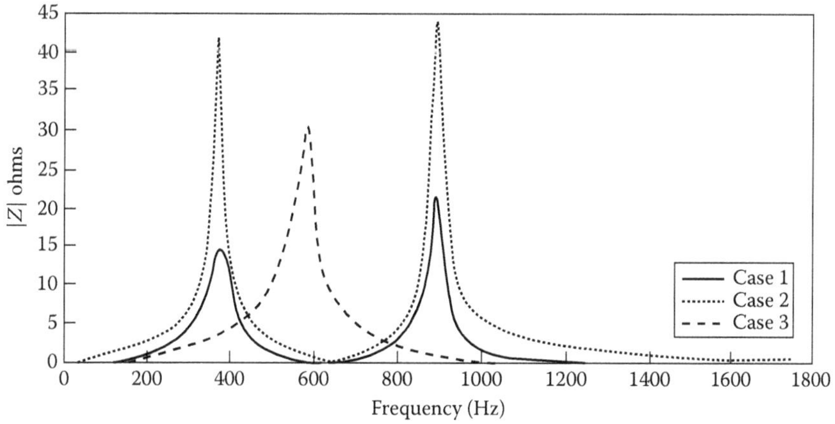

FIGURE 19.25
Impedance modulus versus frequency plot for three conditions of operation, 12.47 kV bus 3 (Example 19.2).

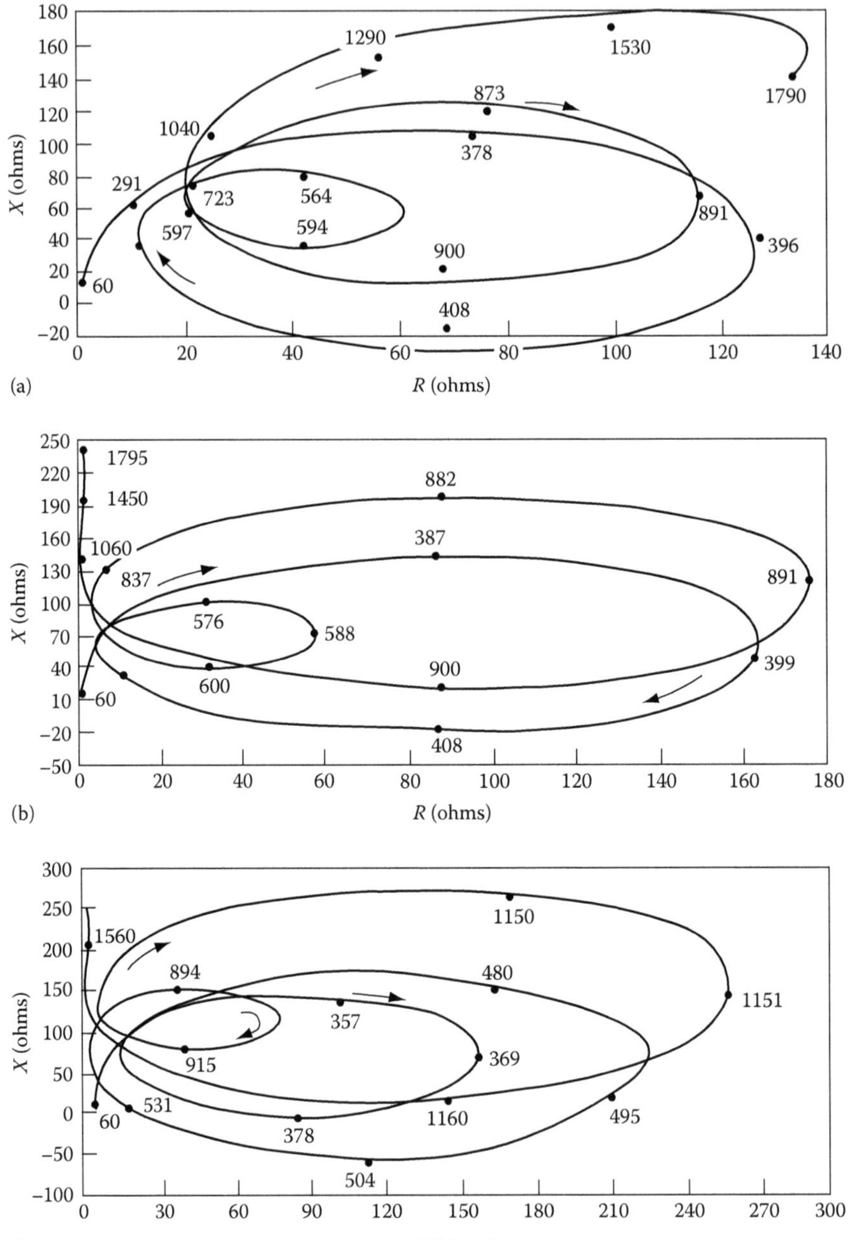

FIGURE 19.26
R–X plots of utility's source impedance under three conditions of operation (Example 19.2).

TABLE 19.9

Example 19.3: Calculation of IT Product, Sending End, Distributed Parameter Line L1 (Figure 19.22)

Harmonic	Frequency	TIF Weighting (W_f)	Harmonic Current in kA (I_f)	W_fI_f	$(W_fI_f)^2$
1	60	0.5	0.158	0.079	0.006
5	300	225	0.0127	2.857	8.165
7	420	650	0.00269	1.749	3.057
11	660	2,260	0.0110	24.860	618.020
13	780	3,360	0.00438	14.717	216.584
17	1,020	5,100	0.00431	21.980	483.164
19	1,140	5,630	0.00381	21.450	460.115
23	1,380	6,370	0.00257	16.371	268.006
25	1,500	6,680	0.00324	21.643	468.428
27	1,740	7,320	0.00292	21.374	456.865
31	1,860	7,820	0.00247	19.315	373.085
35	2,100	8,830	0.00175	15.453	238.780
37	2,220	9,330	0.00157	14.648	214.567
41	2,460	10,360	0.00146	15.126	228.784
43	2,580	10,600	0.00150	15.900	252.810
47	2,820	10,210	0.00150	15.315	234.549
49	2,940	9,820	0.00135	13.256	175.748

Example 19.4: Transmission System

Harmonic load flow in a transmission system network is illustrated in this example. The system configuration is shown in Figure 19.27. There are nine transmission lines, and the line constants are shown in Table 19.10. These are modeled with distributed parameters or π networks. Also, the 6-pulse and 12-pulse converters are modeled as ideal converters.

The frequency scan with *only harmonic injections* and without bulk linear loads shows numerous resonant frequencies with varying impedance modulus, e.g., Figure 19.28 shows these at the 230 kV bus 1. The resonance at higher frequencies is more predominant. The harmonic current flows in lines are shown in Table 19.11.

Figure 19.29 is a frequency scan with bulk loads applied. The resonant frequencies and a number of smaller resonance points are eliminated. The impedance modulus is reduced. Table 19.12 shows the cluster of resonant frequencies and impedance modulus in these two cases.

The effect on harmonic current flows in line sections 1-5 and 1-2 under harmonic injections only and under harmonic injections and loads is shown in Figures 19.30 and 19.31.

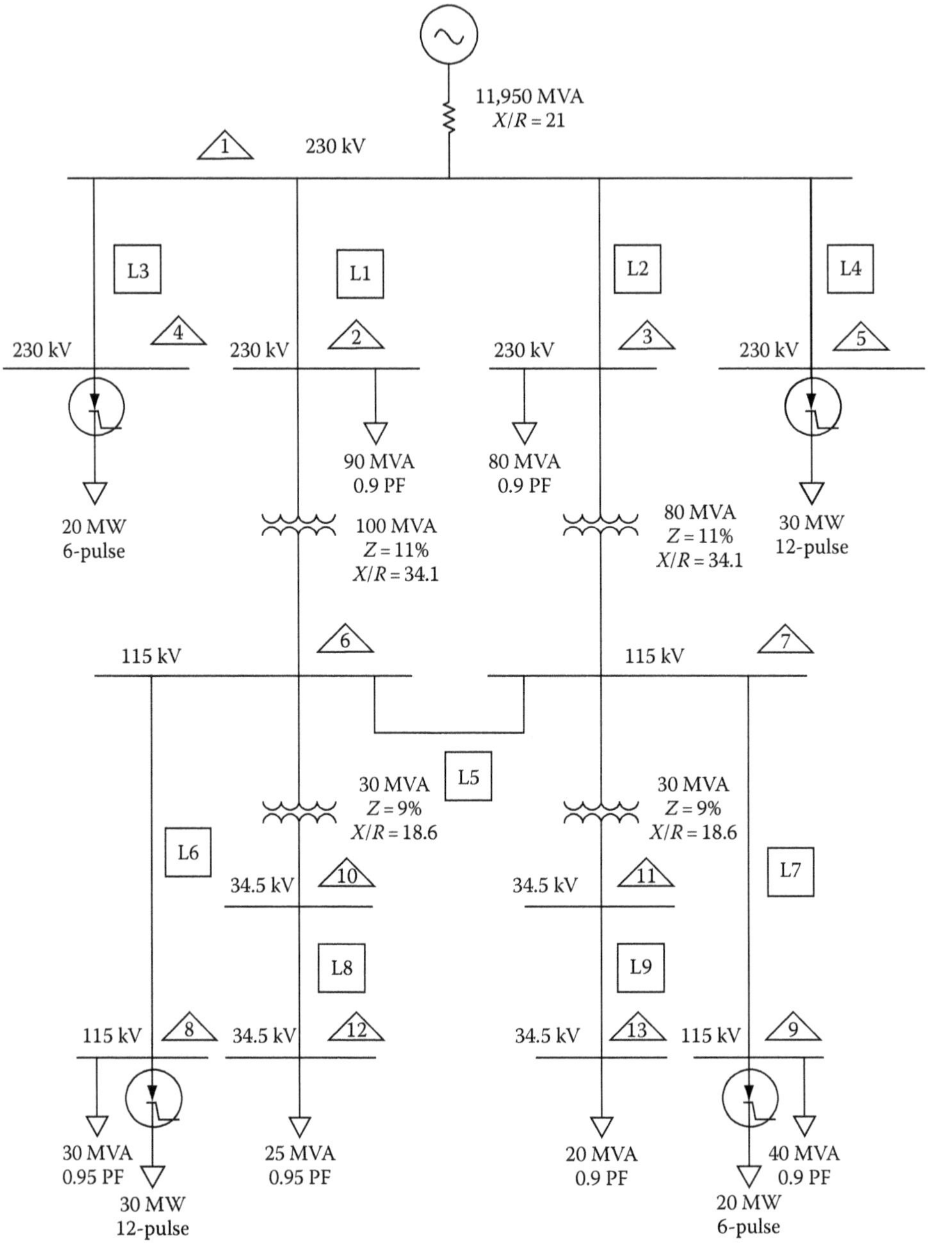

FIGURE 19.27
A transmission system for harmonic analysis study (Example 19.4).

TABLE 19.10

Example 19.4: Transmission Line Data and Models

Line No.	Voltage (kV)	Conductor ACSR KCMIL	Length (Miles)	GMD	*R* (Per Unit, 100 MVA Base)	X_L (Per Unit, 100 MVA Base)	X_c (Per Unit 100 MVA Base)	Model Type
L1	230	636	100	34	0.00003	0.08585	1.8290	Distributed
L2	230	636	120	34	0.00004	0.10302	1.5240	Distributed
L3	230	636	50	34	0.00002	0.04293	3.6580	Distributed
L4	230	636	40	34	0.00001	0.03434	4.5720	Distributed
L5	115	715	30	21	0.00004	0.08828	20.347	π model
L6	115	715	40	21	0.00005	0.11771	15.260	Distributed
L7	115	715	50	21	0.00006	0.14713	12.208	Distributed
L8	34.5	397.5	30	5	0.00072	0.62473	131.228	π model
L9	34.5	397.5	30	5	0.00072	0.62473	131.228	π model

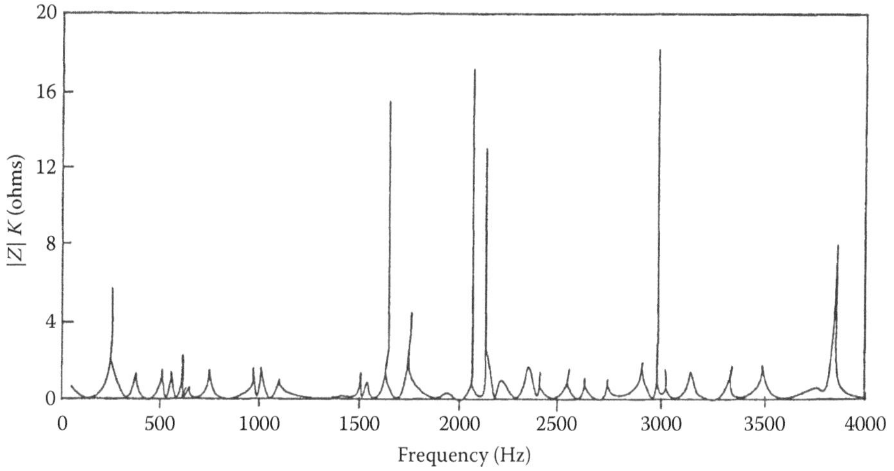

FIGURE 19.28
Cluster of resonant frequencies at 230 kV bus 1 without bulk load applied (Example 19.4).

Example 19.5: Three-Phase Modeling

Consider a 2.0 MVA, 2.4–0.48 kV delta–wye-connected transformer serving a mixed three-phase load and single-phase load, as shown in Figure 19.32. A single-phase 480–240 V, 500 kVA transformer serves lighting and switch-mode supply loads. The harmonic spectrum is modeled according to Chapter 17. The results of harmonic flow study are shown in Table 19.13, and the harmonic currents injected into the wye secondary of a 2 MVA, 2.4–0.48 kV transformer are shown. The capacitors at bus B amplify third harmonic currents. As the single-phase transformer is

TABLE 19.11

Harmonic Current Flow, with Harmonic Current Injection, Bulk Loads Not Applied

	Harmonic Current Flow (A)													
Line	5th	7th	11th	13th	17th	19th	23rd	25th	29th	31st	35th	37th	41st	43rd
L1														
1-2	5.71	7.16	7.73	2.55	6.17	0.90	0.52	18.5	7.92	2.71	0.41	42.81	0.04	1.31
2-1	4.39	1.50	0.06	1.39	2.79	0.95	0.47	0.32	7.91	2.24	0.40	46.20	0.08	0.89
#L2														
1-3	6.70	21.60	16.03	4.06	10.70	21.50	1.18	0.87	21.60	3.68	1.40	17.20	1.19	0.11
3-1	13.62	2.56	13.01	5.19	0.53	3.89	0.20	2.99	6.58	0.42	3.28	29.30	0.44	0.04
L3														
1-4	4.30	12.63	8.31	8.39	24.70	7.54	0.38	2.39	1.71	1.92	9.14	72.80	1.39	3.38
4-1	10.00	7.17	4.56	3.86	2.95	2.64	2.18	2.01	1.73	1.62	1.43	1.36	1.22	1.07
L4														
1-5	5.63	1.95	0.03	1.73	20.20	30.00	0.29	15.10	15.30	0.95	7.30	12.60	0.16	2.17
5-1	0.00	0.00	6.85	5.79	0.00	0.00	3.27	3.01	0.00	0.00	2.15	2.04	0.00	L60
L5														
6-7	7.59	4.45	11.31	16.10	6.75	0.07	2.68	9.01	1.24	0.30	23.6	148.00	1.61	0.25
7-6	6.37	5.67	7.88	15.10	6.45	0.87	3.20	18.50	19.40	3.43	22.0	173.00	0.88	4.34
L6														
6-8	0.84	1.00	12.50	13.10	0.60	0.86	10.30	22.40	24.10	5.10	33.2	250.00	1.79	1.24
8-6	0.00	0.00	13.70	11.60	0.00	0.00	6.55	6.02	0.00	0.00	4.30	4.70	0.00	3.24
L7														
7-9	33.20	11.10	32.40	24.40	4.47	6.09	4.64	8.46	2.52	2.05	30.5	239.00	1.78	4.40
9-7	20.10	14.30	9.13	7.72	5.91	5.28	4.37	4.02	3.46	3.24	2.87	2.71	2.45	2.14
L8														
10-12	1.20	1.50	3.62	0.07	1.73	3.26	22.2	46.70	23.5	3.09	5.25	3.31	0.00	2.61
21-10	0.00	0.00	0.00	0.00	0.00	0.00	0.00	0.00	0.00	0.00	0.00	0.00	0.00	0.00
L9														
11-13	1.40	1.19	5.29	3.63	3.03	2.72	6.13	13.50	12.50	1.81	6.17	23.8	0.00	0.08
13-11	0.00	0.00	0.00	0.00	0.00	0.00	0.00	0.00	0.00	0.00	0.00	0.00	0.00	0.00

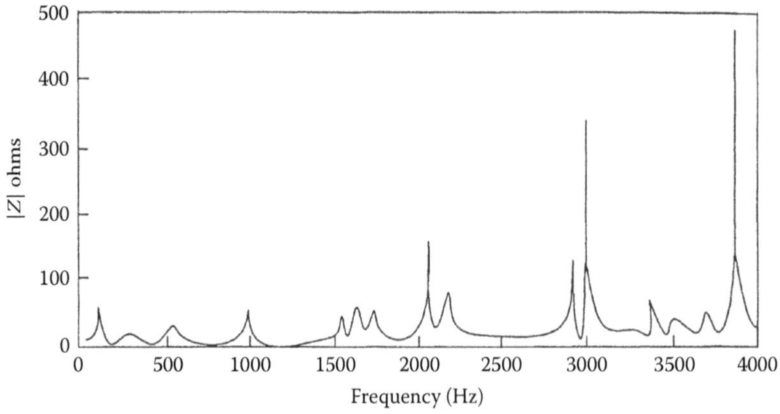

FIGURE 19.29
Shift in resonant frequencies at 230 kV bus 1 with bulk load applied (Example 19.4).

TABLE 19.12

Example 19.4: Resonant Frequencies and Maximum Impedance Modulus

	Without Load			With Load		
Bus ID	**Parallel Resonance (Hz)**	**Series Resonance (Hz)**	**Impedance Modulus (ohms)**	**Parallel Resonance (Hz)**	**Series Resonance (Hz)**	**Impedance Modulus (ohms)**
1	242, 362, 524, 581, 739, 958, 997, 1123, 1186, 1513, 1646, 1735, 1832, 2053, 2126, 2212, 2381, 2569, 2881, 2979	187, 340, 464, 572, 733, 872, 979, 1114, 1165, 1189, 1520, 1687, 1823, 1856, 2084, 2199, 2370, 2555, 2612, 2909	27,410	545, 974, 1742, 2074, 2153, 2582, 2902, 2990	870, 1166, 1874, 2114, 2574, 2610, 2934	3749
6	362, 580, 739, 958, 997, 1123, 1187, 1294, 1315, 1514, 1646, 1735, 1832, 2054, 2126, 2213, 2381, 2638	469, 643, 757, 967, 1022, 1129, 1279, 1312, 1346, 1538, 1685, 1819, 2008, 2056, 2149, 2372, 2636, 2872	46,300	1542, 1662, 1762, 2150, 2298, 2442, 2665	1558, 1702, 2014, 2166, 2394, 2657	1135
10	362, 580, 739, 958, 997, 1123, 1186, 1295, 1315, 1513, 1645, 2212	392, 589, 742, 962, 1004, 1126, 1243,1306, 1502, 1639, 1738, 2224	32,820	2306, 2402, 2622	2357, 2466, 2860	448.6

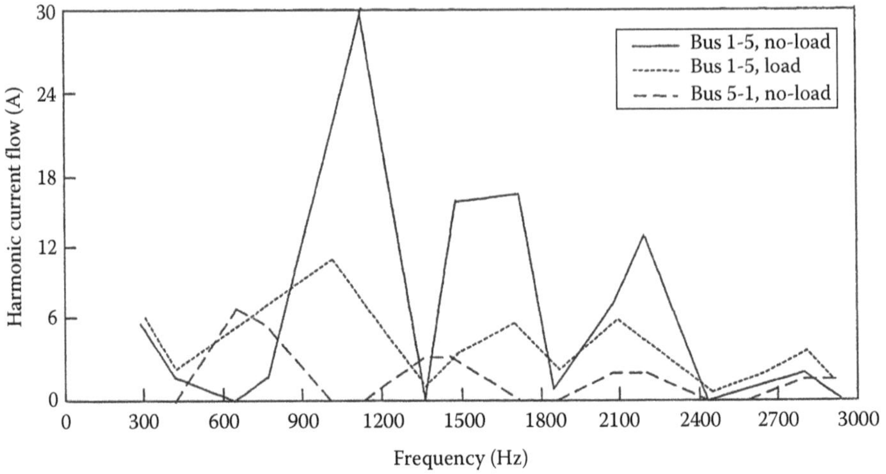

FIGURE 19.30
Harmonic current flow in bus tie 1-5 (Example 19.4).

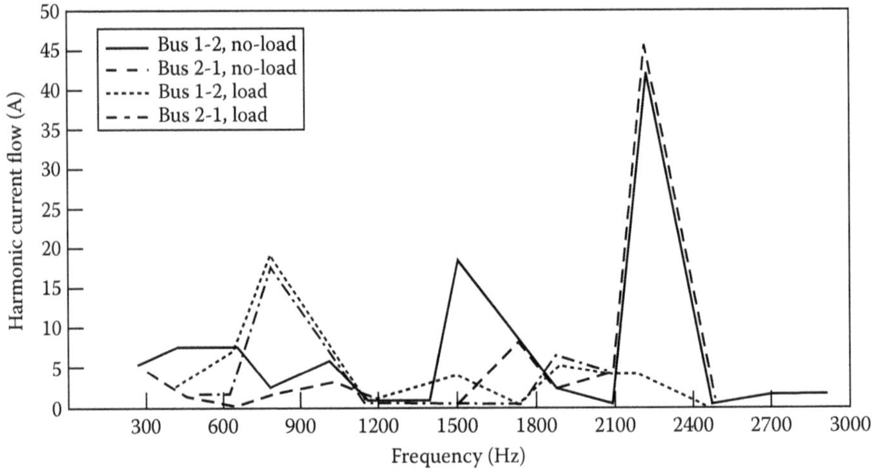

FIGURE 19.31
Harmonic current flow in bus tie 1-2 (Example 19.4).

connected between phases *a* and *b*, only these phases carry harmonic currents. The harmonic currents in phase *c* are zero. If the 350 kvar capacitors are removed, the harmonic-injection pattern changes. The currents at the second and third harmonics are reduced while those at the fourth, fifth, and seventh harmonics are increased, i.e., capacitors amplify lower order harmonics and attenuate higher order harmonics.

One solution of harmonic mitigation is to convert capacitors at bus B into filters. It may be necessary to increase the reactive power rating. A second possibility is to provide a harmonic filter on the load side of a single-phase transformer. A third possibility is to install zero-sequence traps on the wye side of the 2.4 kV transformer. We will discuss these in Chapter 20.

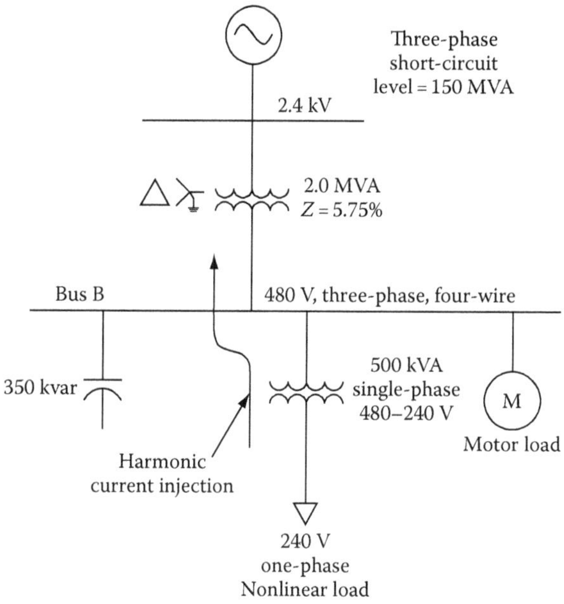

FIGURE 19.32
System with single-phase nonlinear load for three-phase simulation (Example 19.5).

TABLE 19.13
Example 19.5: Harmonic Current Injected Into Wye Secondary of Transformer T1 (in A), Figure 19.29

	With 350 kvar Capacitors on Bus B			350 kvar Capacitors on Bus B Removed		
h	Phase *a*	Phase *b*	Phase *c*	Phase *a*	Phase *b*	Phase *c*
2	141	141	0	46.3	46.3	0
3	456	456	(single-phase load)	241	241	(single-phase load)
4	8.8	8.8		150	150	
5	7.82	7.82		24	24	
6	6.30	6.30		18	18	
7	8.26	8.26		60	60	

Problems

19.1 Calculate overlap angle and harmonic spectrum (using equations in Appendix G) of a six-pulse converter operating in the following configuration:

Delay angle = 30°, supply source short-circuit level at 13.8 kV = 875 MVA, converter transformer 3000 kVA, percentage impedance = 5.5%, and converter load = 75% of the transformer kVA rating.

19.2 A 10 MVA transformer has a percentage impedance of 10% and X/R ratio of 10. Tabulate its resistance and reactance for harmonic spectrum of a six-pulse converter up to the 25th harmonic.

19.3 Repeat Problem 19.2 for an induction motor of a 2.3 kV, four-pole, 2500 hp, full-load power factor = 0.92, full-load efficiency = 0.93%, and locked rotor current = six times the full-load current at 20% power factor.

19.4 Calculate PF, DF, distortion power, and active and reactive power for the converter in Problem 19.1.

19.5 Harmonic propagation in a transmission system is to be studied. Describe how the extent of the system to be modeled will be decided. What models of lines and transformers will be used?

19.6 Form a capacitor bank of 15 Mvar, operating voltage 44 kV in wye configuration from unit capacitor sizes in Table 19.2. Expulsion fuses are to be used for the fusing of individual capacitor units.

19.7 Explain why IT and kVT will be different at the sending and receiving end of a transmission line.

19.8 Plot the impedance versus frequency across a transmission line using equivalent π model for a long line. Consider line parameters of Example 10.6.

References

1. G.D. Breuer et al. HVDC-AC interaction. Part 1: Development of harmonic measurement system hardware and software. *IEEE Trans. PAS* 101 (3): 701–706, March 1982.
2. G.D. Breuer et al. HVDC-AC interaction. Part II: AC system harmonic model with comparison of calculated and measured data. *IEEE Trans. PAS* 101 (3): 709–717, March 1982.
3. IEEE. Task Force on Harmonic Modeling and Simulation. Modeling and simulation of propagation of harmonics in electrical power systems. Part I: Concepts, models and simulation techniques. *IEEE Trans. Power Deliv.* 11: 452–465, 1996.
4. IEEE. Task Force on Harmonic Modeling and Simulation. Modeling and simulation of propagation of harmonics in electrical power systems. Part II: Sample systems and examples. *IEEE Trans. Power Deliv.* 11: 466–474, 1996.
5. C. Hatziadoniu and G.D. Galanos. Interaction between the AC voltages and Dc current in weak AC/DC interconnections. *IEEE Trans. Power Deliv.* 3: 1297–1304, 1988.
6. D. Xia and G.T. Heydt. Harmonic power flow studies. Part 1: Formulation and solution. *IEEE Trans. PAS* 101: 1257–1265, 1982.
7. D. Xia and G.T. Heydt. Harmonic power flow studies. Part II: Implementation and practical applications. *IEEE Trans PAS* 101: 1266–1270, 1982.
8. A.A. Mohmoud and R.D. Shultz. A method of analyzing harmonic distribution in AC power systems. *IEEE Trans PAS* 101: 1815–1824, 1982.
9. J. Arrillaga, D.A. Bradley, and P.S. Bodger. *Power System Harmonics*. New York: John Wiley, 1985.
10. J. Arrillaga, B.C. Smith, N.R. Watson, and A.R. Wood. *Power System Harmonic Analysis*. West Sussex, England: John Wiley, 1997.
11. EPRI. HVDC-AC system interaction for AC harmonics. EPRI Report 1: 7.2–7.3, 1983.
12. CIGRE. Working Group 36-05. Harmonic characteristic parameters, methods of study, estimating of existing values in the network. *Electra* 35–54, 1977.
13. E.W. Kimbark. *Direct Current Transmission*. New York: John Wiley, 1971.
14. M. Valcarcel and J.G. Mayordomo. Harmonic power flow for unbalance systems. *IEEE Trans. Power Deliv.* 8: 2052–2059, 1993.

15. D.J. Pileggi, N.H. Chandra, and A.E. Emanuel. Prediction of harmonic voltages in distribution systems. *IEEE Trans. PAS* 100 (3): 1307–1315, March 1981.
16. J.P. Tamby and V.I. John. Q'Harm: A harmonic power flow program for small power systems. *IEEE Trans. PS* 3 (3): 945–955, August 1988.
17. N.R. Watson and J. Arrillaga. Frequency dependent AC system equivalents for harmonic studies and transient converter simulation. *IEEE Trans. PD* 3 (3): 1196–1203, July 1988.
18. E.V. Larsen, D.H. Baker, and J.C. Mclver. Low order harmonic interaction on ac/dc systems. *IEEE Trans. PD* 4 (1): 493–501, 1989.
19. T. Hiyama, M.S.A.A. Hammam, and T.H. Ortmeyer. Distribution system modeling with distributed harmonic sources. *IEEE Trans. Power Deliv.* 4: 1297–1304, 1989.
20. M.F. McGranaghan, R.C. Dugan, and W.L. Sponsler. Digital simulation of distribution system frequency-response characteristics. *IEEE Trans. PAS* 100: 1362–1369, 1981.
21. M.F. Akram, T.H. Ortmeyer, and J.A. Svoboda. An improved harmonic modeling technique for transmission network. *IEEE Trans. Power Deliv.* 9: 1510–1516, 1994.
22. IEEE. *IEEE Recommended Practice and Requirements for Harmonic Control in Electrical Systems,* Standard 519, 1992.
23. ANSI/IEEE. *Guide for Protection of Shunt Capacitor Banks,* Standard C37.99, 2005.

20

Harmonic Mitigation and Filters

With the increase in consumer nonlinear load, the harmonics injected into the power supply system and their consequent effects are becoming of greater concern. Harmonic currents seeking a low impedance path or a resonant condition can travel through the power system and create problems for the consumers who do not have their own source of harmonic generation. IEEE standard 519 [1] stipulates who is responsible for what. A consumer can inject only a certain amount of harmonic current into the supply system, depending on his/her load, short-circuit current, and supply system voltage. The utility supply companies must ensure a certain voltage quality at the PCC (point of common coupling) with the consumer apparatus. These harmonic current and voltage limits [1,2] are discussed in Appendix F.

20.1 Mitigation of Harmonics

There are four major methodologies for mitigation of harmonics:

1. The equipment can be designed to withstand the effect of harmonics, for example, transformers, cables, and motors can be derated. Motors for PWM inverters can be provided with special insulation to withstand high du/dt, and the relays can be rms sensing.
2. Passive filters at suitable locations, preferably close to the source of harmonic generation, can be provided so that the harmonic currents are trapped at the source and the harmonics propagated in the system are reduced.
3. Active filtering techniques, generally, incorporated with the harmonic-producing equipment itself can reduce the harmonic generation at source. Hybrid combinations of active and passive filters are also a possibility.
4. Alternative technologies can be adopted to limit the harmonics at source, for example, phase multiplication, operation with higher pulse numbers, converters with interphase reactors, active wave-shaping techniques, multilevel converters, and harmonic compensation built into the harmonic-producing equipment itself to reduce harmonic generation.

What will be the most useful strategy in a given situation largely depends on the currents and voltages involved, the nature of loads, and the specific system parameters, for example, short-circuit level at the PCC.

20.2 Band-Pass Filters

The operation of a single-tuned (ST) shunt filter is explained with reference to Figure 20.1. (Any other type of filter connected in the shunt can be termed a shunt filter.) Figure 20.1a shows a system configuration with nonlinear load and Figure 21.1b shows the equivalent circuit. Harmonic current injected from the source, through impedance Z_c, divides into the filter and the system. The system impedance for this case, shown as Z_s, consists of the source impedance Z_u in series with the transformer impedance Z_t and paralleled with the motor impedance:

$$I_h = I_f + I_s \tag{20.1}$$

where
- I_h is the harmonic current injected into the system
- I_f is the current through the filter
- I_s is the current through the system impedance

Also,

$$I_f Z_f = I_s Z_s \tag{20.2}$$

that is, the harmonic voltage across the filter impedance (Z_f) equals the harmonic voltage across the equivalent power system impedance (Z_s).

$$I_f = \left[\frac{Z_s}{Z_f + Z_s}\right] I_h = \rho_f I_h \tag{20.3}$$

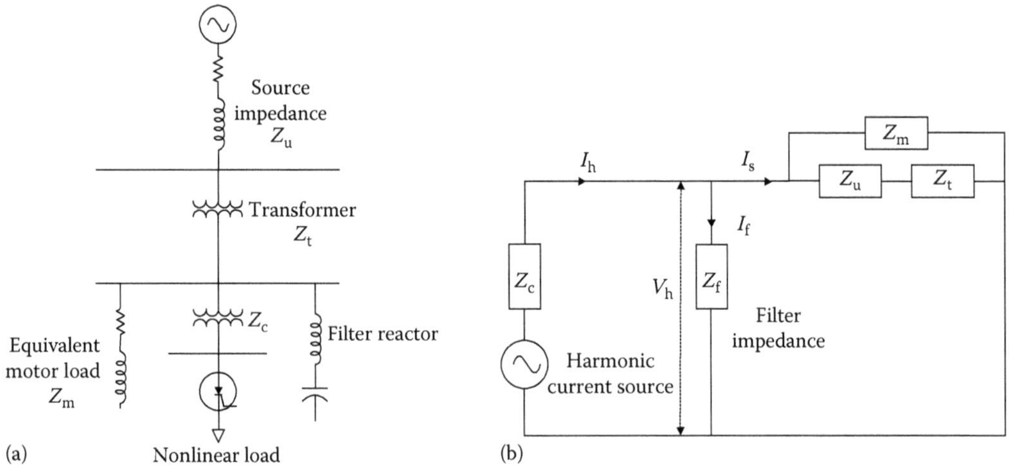

FIGURE 20.1
(a) Diagram of connections with an ST shunt filter and harmonic source; (b) equivalent circuit looking from harmonic injection as the source.

$$I_s = \left[\frac{Z_f}{Z_f + Z_s}\right] I_h = \rho_s I_h \tag{20.4}$$

where ρ_f and ρ_s are complex quantities, which determine the distribution of harmonic current in the filter and system impedance. These equations can also be written in terms of admittances.

A properly designed filter will have ρ_f close to unity, typically 0.995, and the corresponding ρ_s for the system will be 0.05. The impedance angles of ρ_f and ρ_s may be on the order of $-81°$ and $-2.6°$, respectively.

The harmonic voltages should be as low as possible. The equivalent circuit of Figure 20.1b shows that system impedance plays an important role in the harmonic current distribution. For an infinite system impedance, the filtration is perfect, as no harmonic current flows through the system impedance. Conversely, for a system of zero harmonic impedance, all the harmonic current will flow into the system and none in the filter. In the case where there is no filtration all the harmonic current passes on to the system. The lower the system impedance, that is, the higher the short-circuit current, the smaller is the voltage distortion, provided the filter impedance is lowered so that it takes most of the harmonic current.

In an ST filter, as the inductive and capacitive impedances are equal at the resonant frequency, the impedance is given by the resistance R.

$$Z = R + j\omega L + \frac{1}{j\omega C} = R \tag{20.5}$$

The following parameters can be defined:

ω_n is the tuned angular frequency in radians and is given by

$$\omega_n = \frac{1}{\sqrt{LC}} \tag{20.6}$$

X_0 is reactance of the inductor or capacitor at the tuned angular frequency. Here, $n = f_n/f$, where f_n is the filter-tuned frequency and f is the power system frequency.

$$X_0 = \omega_n L = \frac{1}{\omega_n C} = \sqrt{\frac{L}{C}} \quad \text{and} \quad \omega_n = \sqrt{\frac{1}{LC}} \tag{20.7}$$

The quality factor of the tuning reactor is defined as

$$Q = \frac{X_0}{R} = \frac{\sqrt{L/C}}{R} \tag{20.8}$$

It determines the sharpness of tuning. The pass band is bounded by frequencies at which

$$|Z_f| = \sqrt{2}R \tag{20.9}$$

$$\delta = \frac{\omega - \omega_n}{\omega_n} \tag{20.10}$$

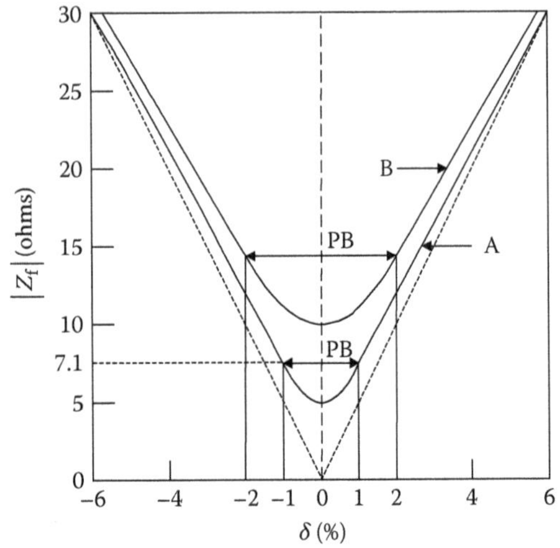

FIGURE 20.2
Response of an ST shunt filter, pass band, and asymptotes with varying Q factors.

At these frequencies, the net reactance equals resistance, capacitive on one side and inductive on the other. If it is defined as the deviation per unit from the tuned frequency, then for small frequency deviations, the impedance is approximately given by

$$|Z_f| = R\sqrt{1+4\delta^2Q^2} = X_0\sqrt{Q^{-2}+4\delta^2} \tag{20.11}$$

To minimize the harmonic voltage, Z_f should be reduced or the filter admittance should be high as compared to the system admittance.

The plot of the impedance is shown in Figure 20.2. The sharpness of tuning is dependent on R as well as on X_0 and the impedance of the filter at its resonant frequency can be reduced by reducing these. The asymptotes are at

$$|X_f| = \pm 2X_0|\delta| \tag{20.12}$$

The edges of the pass band are at $\delta = \pm 1/2Q$ and width $= 1/Q$. In Figure 20.2, curve A is for $R = 5$ ohm, $X_0 = 500$ ohm, and $Q = 100$, with asymptotes and pass band, as shown. Curve B is for $R = 10$ ohm, $X_0 = 500$ ohm, and $Q = 50$. These two curves have the same asymptotes. The resistance, therefore, affects sharpness of tuning.

20.2.1 Tuning Frequency

ST filter is not tuned exactly to the frequency of the harmonic it is intended to suppress. The system frequency may change, causing harmonic frequency to change. The tolerance on filter reactors and capacitors may change due to aging or temperature effects. The tolerance on commercial capacitor units is ±20% and on reactors ±5%. For filter applications it is necessary to adhere to closer tolerances on capacitors and reactors. Where a number of capacitor units are connected in series or parallel, these are carefully formed with tested values of the capacitance so that large phase unbalances do not occur. Any such unbalances between the phases will result in overvoltage stress; in addition, the neutral will not be at ground potential in ungrounded wye-connected banks. A tolerance of ±2.0% on reactors and a tolerance of +5% on capacitors (no negative tolerance)

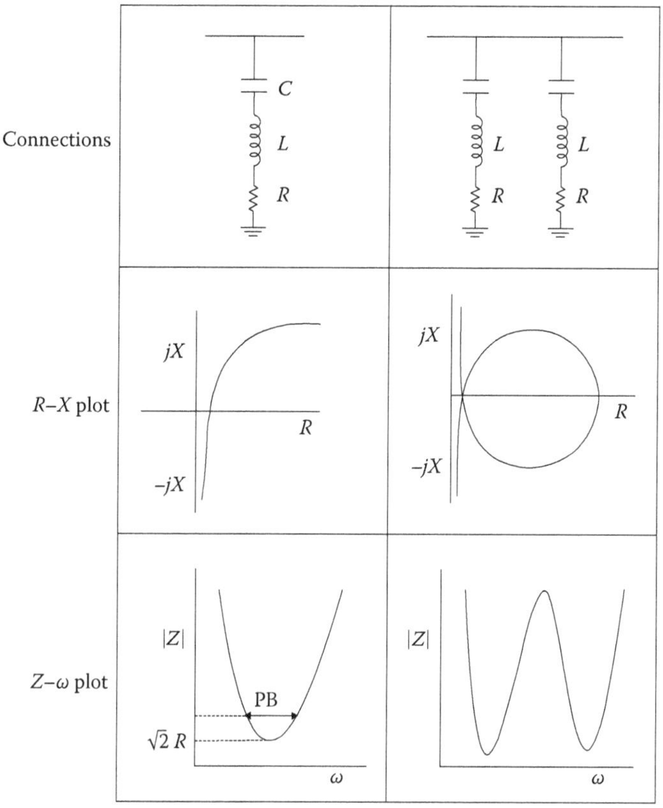

FIGURE 20.3
Circuit connections and R–X and Z–ω plots of an ST shunt filter in isolation and two ST filters in parallel.

in industrial environment are practical. Closer tolerances may be required for high-voltage dc (HVDC) applications.

A change in L or C of 2% causes the same detuning as a change of system frequency by 1% [3]:

$$\delta = \frac{\Delta_f}{f_n} + \frac{1}{2}\left(\frac{\Delta L}{L_n} + \frac{\Delta C}{C_n}\right) \tag{20.13}$$

Figure 20.3 shows the circuits, and R–X and Z–ω plots of ST filters in isolation and in parallel.

20.3 Practical Filter Design

Example 20.1

We will continue with Example 19.1 and go through the iterations of designing ST filters, so that the harmonic distortions are at acceptable limits.

Form an ST Shunt Filter

In Example 19.1, with the addition of a 1350 kvar capacitor bank, the resonant frequency is close to the 11th harmonic, which gives large magnification of harmonic currents and distortion factors. An ST filter design is generally started with the lowest harmonic, though the resonant harmonic may be higher. Consider that an ST filter tuned to the 4.7th harmonic is formed. Filter reactance is given by Equation 20.7, which gives filter $L = 0.682$ mH. Let the reactor X/R ratio be arbitrarily chosen as 40 at fundamental frequency. We will discuss Q of the filter reactors according to Equation 20.8, at the tuned frequency of the filter in Section 20.8.1. The results of the frequency scan and phase angle plot are shown in Figure 20.4a and b, respectively. These can be compared with those of Figure 19.20a and b. It is noted that

- The resonance is not eliminated. It will always shift to a frequency that is lower than the selected tuned frequency. This shifted parallel resonant frequency, in Figure 20.4a, is 257 Hz. It is given by

$$f_{11} = \frac{1}{2\pi}\sqrt{\frac{1}{(L_s + L)C}} \tag{20.14}$$

  where L_s is the system reactance.

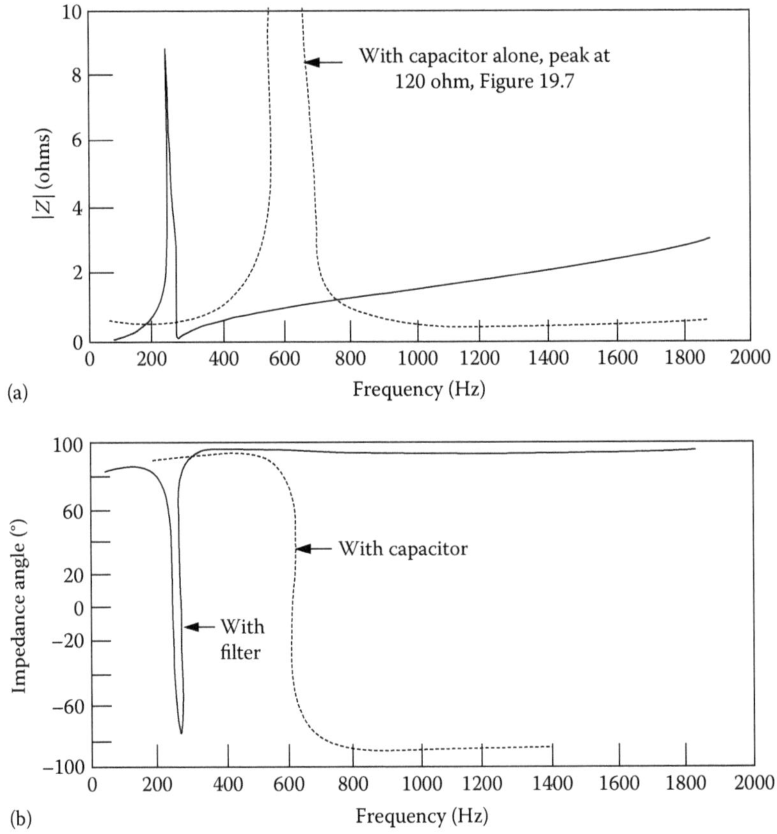

FIGURE 20.4
(a) Impedance modulus and (b) phase-angle plots with application of an ST fifth harmonic filter (Example 20.1); response compared with capacitors alone.

TABLE 20.1

Example 20.1: Harmonic Currents and Voltages with Fifth Harmonic ST Filter

Order of Harmonic	Current in ST Filter (A)	Current in Transformer T1 (A)	Current in 2000 hp Motor (A)	Harmonic Voltage, Bus 2(V)
5	79.56	42.30	3.13	26.3
7	24.07	61.40	4.54	53.4
11	14.95	56.8	4.20	77.76
13	6.18	25.00	1.85	40.38
17	3.38	14.50	1.08	30.7
19	1.93	8.44	0.63	19.9
23	0.60	2.70	0.20	7.71
25	0.48	2.16	0.16	6.71
29	0.34	1.54	0.11	5.57
31	0.24	1.08	0.08	4.17
35	0.20	0.95	0.06	3.26

- The resonance peak has its own value of Q given by

$$Q_1 = \frac{1}{(R_s + R)}\sqrt{\frac{(L_s + L)}{C}} \tag{20.15}$$

where R_s is the system resistance. Figure 20.4a shows that the impedance modulus is considerably reduced. It is 9.17 ohm as compared to 122.7 ohm with capacitor alone.

- The harmonic current flows are shown in Table 20.1. To calculate TDD (total demand distortion) according to the IEEE [1], short-circuit current and load-demand current at the PCC are required. From the data given in Figure 19.19a, short-circuit current is 20.82 kA. The average load current is 830 A. Thus, the ratio $I_{SC}/I_L = 25.08$. The permissible and calculated harmonic current distortion factors with the fifth harmonic ST filter are shown in Table 20.2. The TDD for 7th and 11th harmonics is above permissible limits and the total TDD is 11.87%. Though the total TDD is reduced from 94.52% (without filter reactor) to 11.87%, it is still above permissive limits.

Add Another ST Filter for Seventh Harmonic

As the distortion limits are not met, we can try splitting the 1350 kvar capacitor bank used to form the fifth harmonic ST filter into two equal parallel ST filters, one tuned to $n = 4.7$ as before and the other tuned to $n = 6.7$. We had used four units of 150 kvar, 2.77 kV per phase, giving an installed kvar of 1800 at 2.77 kV (=1350 kvar at 2.4 kV). This is split into two equal banks, 900 kvar, each at 2.77 kV, and ST filters are formed. The results of TDD calculation are again shown in Table 20.2. TDD for fifth harmonic worsens. The fifth harmonic current in the supply system is increased, while the seventh harmonic current is reduced. The harmonic current flow is shown in Table 20.3, case 1. The formation should meet requirements of Table 19.3.

Effect of Tuning Frequencies

As fifth harmonic current is increased, a sharper tuning of the fifth ST filter is attempted by shifting the resonant frequency to $n = 4.85$. TDD calculations are shown in Table 20.2. The harmonic current distributions are shown in Table 20.3, case 2. A closer tuning of the fifth ST filter increases the fifth harmonic loading of the filter, resulting in reduced TDD at these frequencies.

With the fifth ST filter tuned to $n = 4.85$ and the seventh to $n = 6.7$, the TDD for harmonics <11 meets the requirements; however, TDD for harmonics $11 < h < 17$ is still high. These point

TABLE 20.2

Example 20.1: Summary of TDD under Various Conditions Studied

Harmonic order → Operating condition ↓	<11		$11 \le h < 17$		$17 \le h < 23$		$23 \le h < 35$				$35 \le h$	Total TDD
IEEE limits	7.0		3.5		2.5		1.0				0.5	8.0
Harmonic order →	5	7	11	13	17	19	23	25	29	31	35	
No capacitor	14.172[a]	10.208[a]	8.608[a]	3.733[a]	2.144[a]	Not calculated (total TDD not accurate)						19.91%[a]
1350 kvar capacitors	18.07[a]	18.07[a]	90.723[a]	6.987[a]	1.325	0.542	0.103	0.067	0.033	0.020	0.017	94.52[a]
5th harmonic ST filter, $n = 4.7$	5.096	7.398[a]	6.843[a]	3.012	1.747	1.017	0.325	0.260	0.186	0.130	0.114	11.87[a]
5th and 7th harmonic ST filters, $n = 4.7$ and 6.7	8.72[a]	2.783	5.892[a]	2.662	1.578	0.922	0.296	0.237	0.169	0.119	0.100	11.03[a]
5th and 7th harmonic ST filters $n = 4.85$ and 6.7	5.422	2.759	5.855[a]	2.638	1.566	0.916	0.293	0.237	0.169	0.119	0.100	9.04[a]
5th, 7th, and 11th harmonic ST filters, $n = 4.85$, 6.7, and 10.6	3.572	2.700	1.658	2.110	2.082	1.42	0.571	0.504	0.425	0.320	0.238	6.85
5th, 7th, and 11th harmonic ST filters, with tolerances	8.735[a]	4.614	1.205	1.313	0.988	0.600	0.200	0.163	0.118	0.083	—	10.10[a]
One large 5th harmonic ST filter, $n = 4.85$	0.984	3.398	3.289	1.675	0.998	0.583	0.188	0.151	0.107	0.075	0.050	5.24

Note: $I_{SC}/I_L = 25$, $I_L = 830$ A = base current, six-pulse converter, 2.4 kV.

[a] Shows that TDD limits are exceeded. See Appendix F.

TABLE 20.3

Example 20.1: Harmonic Simulation, Fifth and Seventh ST Filters, and Effect of Tuning Frequency

Harmonic Order h ↓	Fifth ST Filter	Seventh ST Filter	2000 hp Motor	Supply System PCC
5	67.1[b]	30.2	5.36	72.4
	89.00[c]	12.23	3.33	45.00
7	45.2	60.7	1.71	23.1
	50.59	60.33	1.70	22.90
11	6.43	17.10	3.62	48.90
	6.91	16.89	3.60	48.6
13	2.73	6.56	1.63	22.1
	2.92	6.53	1.62	21.9
17	1.53	3.39	1.31	13.10
	1.62	3.37	0.96	13.00
19	0.87	1.91	0.57	7.65
	0.93	1.89	0.56	7.61
23	0.27	0.58	0.18	2.46
	0.29	0.58	0.18	2.44
25	0.22	0.46	0.15	1.97
	0.23	0.46	0.14	1.96
29	0.16	0.33	0.11	1.41
	0.16	0.33	0.10	1.41
31	0.11	0.23	0.07	0.99
	0.11	0.23	0.07	0.99
35	0.09	0.19	0.06	0.83
	0.09	0.19	0.06	0.83

[a] Harmonic current flow in ampères.
[b] Case 1: top row, fifth ST filter $n = 4.7$, seventh ST, filter $n = 6.7$.
[c] Case 2: bottom row, fifth ST filter $n = 4.85$, seventh ST filter $n = 6.7$.

to the necessity of adding another ST filter for the eleventh harmonic. Add another ST Filter for 11th Harmonic.

Another equal section of the capacitor bank is added and $n = 10.6$. The results of the simulation are shown in Table 20.4 and the TDD calculations are in Table 20.2. The TDD for each band of frequencies as well as total TDD now meets the requirements. A total of three ST filters of equal sections, each consisting of 900 kvar (300 kvar per phase) rated voltage of 2.77 kV, connected in ungrounded wye and provided with series tuning reactors $n = 4.85$, 6.7, and 10.6, respectively, meet the requirements. The frequency scan of the impedance and its phase angles, with all three filters in place, are plotted in Figure 20.5.

Minimum Filter

In this example, though additional reactive power is not required from fundamental frequency load flow conditions, yet to meet the TDD requirements, an additional 11th ST filter has become necessary. The converse may be also true, depending on the load power factor and percentage of nonlinear load to total load. Compromises are often required when filters must meet the requirements of certain reactive power compensation as well as TDD. A filter designed to control only the harmonic distortion, without the limitation of meeting a certain reactive power demand, is termed the *minimum filter*.

TABLE 20.4

Example 20.1: Harmonic Simulation 5th, 7th, and 11th ST Filters

$h\downarrow$	5th ST Filter	7th ST Filter	11th ST Filter	2000 hp Motor	Supply System PCC	Harmonic Voltage at PCC (Bus 2)
5	94.3	13.06	7.49	3.53	47.7	29.65
7	5.67	67.7	11.04	1.90	25.7	22.41
11	1.42	3.47	60.38	0.74	9.99	13.66
13	1.44	3.22	16.71	0.80	10.8	17.51
17	1.02	2.12	7.09	0.61	8.17	17.28
19	0.61	1.24	3.82	0.37	4.96	11.72
23	0.20	0.39	1.12	0.12	1.66	4.74
25	0.16	0.32	0.88	0.10	1.35	4.19
29	0.12	0.23	0.61	0.07	0.98	3.53
31	0.08	0.16	0.41	0.05	0.69	2.66
35	0.06	0.12	0.36	0.03	0.57	1.98

[a] Harmonic current flow in ampères; harmonic voltage in volts.

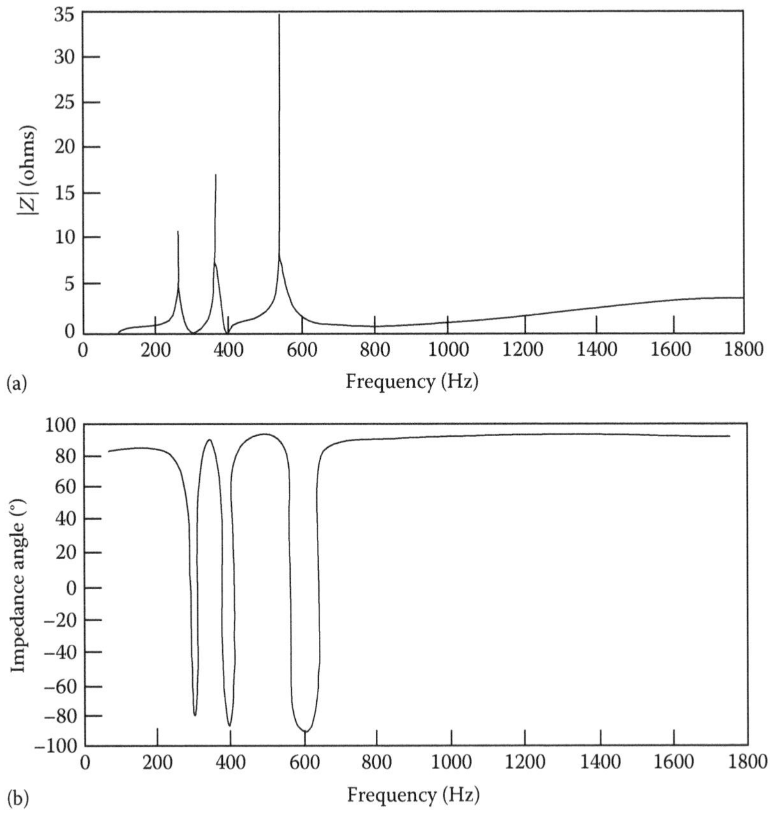

FIGURE 20.5
Example 20.1: (a) impedance modulus and (b) phase-angle plots with 5th, 7th, and 11th harmonic filters.

TABLE 20.5

Example 20.1: Current into PCC with One Large Fifth Harmonic ST Filter[a]

$h \rightarrow$	5	7	11	13	17	19	23	25	29	31	35
Harmonic current →	8.17	28.2	27.3	13.9	8.29	4.84	1.56	1.25	0.89	0.63	0.42

[a] Harmonic current flow in ampères.

One Large Fifth Harmonic ST Filter to Meet the TDD Requirements

As the relative harmonic loading of the filter and the system depends on the filter reactance, if the requirements of TDD are not met, the simplest artifice is to increase the size of the filter, that is, increase fundamental frequency Mvar rating and redesign the filter. Often this becomes a limitation of passive filters that to achieve the required TDD, the size of the filter is too large, resulting in overcompensation of the reactive power and giving rise to overvoltage when in service and reduced voltage when out of service.

In this example, we can iterate by trying increasing sizes of the fifth ST filter, without parallel seventh and eleventh filters, until the TDD requirements are met. Calculations show that a single fifth ST filter requires 9000 kvar of capacitors (3000 kvar per phase, that is, 10 units of 300 kvar, rated voltage 2.77 kV, connected in parallel) to satisfy the TDD requirements. The harmonic currents in the PCC with this filter are shown in Table 20.5 and TDD in Table 20.2.

The single large filter with 9.00 Mvar of capacitors is impractical. The example illustrates that *simply increasing the size of the filter may not give an optimum solution*.

Shifted Resonant Frequencies

Each parallel ST filter gives rise to a shifted resonant frequency, below its own tuned frequency. If the shifted resonance frequency coincides with one of the characteristics, noncharacteristic, or triplen harmonics present in the system, current magnification at these frequencies will occur. The switching inrush current of a transformer is rich in even and third harmonics (Figure 17.8). As the transformers are switched in and out, harmonic current injections into the system and filters will increase, though this will last for the switching duration of the transformers. It is possible that these currents are sufficiently magnified to give rise to large harmonic voltages. High overvoltages can occur if the system is sharply tuned to the harmonic that is being excited by the transformer inrush current, (second, third, fourth, and even harmonics). Capacitor banks could also fail prematurely. This places a constraint on the design of ST filters. The shifted resonance frequencies should have at least 30 cycle difference between the adjacent and odd or even harmonics. Even then, some amplification of the transformer switching inrush current will occur. In this example, with three ST filters, the shifted frequencies are 271, 367, and 549 Hz. Reference [4] illustrates switching transients when capacitors and transformers are switched together. These increase the decay time of the switching transient and harmonic resonance can occur.

The last two frequencies are close to the sixth and ninth harmonics. A wider band can be attempted by slightly lowering the tuning frequency of the 7th and 11th ST filters. The shifted frequencies can also be calculated from Equation 20.14.

Effect of Tolerances on Filter Components

The tolerances on capacitors and reactors will result in detuning. Consider that components of the following tolerances are selected:

- Capacitors: +5%
- Reactors: ±2%

Let the capacitance of the fifth and seventh filters increase by 5% and the inductance by 2%. This is quite a conservative assumption for checking the detuning effect and resulting current distribution. The series-tuned frequencies of the fifth and seventh filters will shift to a lower value.

TABLE 20.6

Harmonic Simulation with Tolerances on Filters[a] (See Text)

$h \downarrow$	5th ST Filter	7th ST Filter	11th ST Filter	2000 hp Motor	Supply System PCC
5	81.4	22.9	11.4	5.36	72.5
7	8.11	57.1	16.4	2.84	38.3
11	1.41	3.28	60.55	0.74	10.00
13	1.44	3.10	16.78	0.80	10.9
17	1.02	2.06	7.11	0.61	8.20
19	0.61	1.21	3.84	0.37	4.98
23	0.20	0.39	1.12	0.12	1.66
25	0.16	0.32	0.88	0.10	1.35
29	0.12	0.23	0.61	0.07	0.98
31	0.08	0.16	0.41	0.05	0.69

[a] Harmonic current flow in ampères.

The results of harmonic current flow are shown in Table 20.6. The harmonic distortion increases above the acceptable limits.

This points to the necessity of iterating the design with required tolerances and fine-tuning the selected tuning frequencies. Closer tolerances on the components are an option, but that may not be practical and economically justifiable. Capacitors with metalized film construction lose capacitance as they age, resulting in a gradual increase in the tuning frequency. Non-metalized electrode capacitors have a fairly stable capacitance. Tuning a harmonic filter more sharply than required to attain the desired performance unnecessarily stresses the components and generally makes the filter more prone to overload from other harmonic sources. Considerations should also be applied to the increase of the loads and consequent harmonics. Transformer energizing and clearing of nearby faults will lead to temporary surge of the filter harmonic current. The faults and energizing of transformer may result in saturation of transformers, which gives additional harmonic loading. If a harmonic filter is not going to be removed automatically during a system outage, it is desirable to do a system study to determine the filter performance during the outage. Switching transients are of concern and these are discussed in Chapter 5, though a more rigorous analysis will be required depending upon the application.

Outage of One of the Parallel Filters

Outage of one of the parallel ST filters should be considered. It will have the following effects:

- The current loading of remaining filters in service may increase substantially and the capacitors and reactors may be overloaded.
- The resonant frequencies will shift and may result in harmonic current amplification.
- The harmonic distortion will increase.

Table 20.7 shows the effect of outage of one of the three filters at a time. The harmonic distortion at the PCC increases in every case, though the filter components are not overloaded. According to IEEE Standard 519 [1], it is permissible to operate the system on a short-term basis with higher distortion limits at the PCC, provided that the faulty unit is placed back in service quickly after rectification.

Sometimes, the outage of a filter may result in overloading of the remaining filters in service. It then becomes necessary that parallel filters are also removed from service. The filter protection

TABLE 20.7

Example 20.1: Effect of Outage of One of the Parallel Filters[a]

	5th ST Out			7th ST Out			11th ST Out		
Harmonic Order h ↓	7th ST Filter	11th ST Filter	Utility (PCC)	7th ST Filter	11th ST Filter	Utility (PCC)	5th ST Filter	7th ST Filter	Utility (PCC)
5	53.28	30.57	194	85.48	6.79	43.2	89.00	12.23	45.00
7	72.2	11.78	27.5	22.9	44.57	104.00	50.59	60.33	22.90
11	3.5	61.53	10.2	1.48	63.27	10.5	6.91	16.89	48.60
13	3.37	17.47	11.3	1.59	1.85	1.20	2.92	6.53	21.90
17	2.24	7.49	8.64	1.14	7.97	9.20	1.62	3.37	13.00
19	1.31	4.05	5.25	0.68	4.31	5.59	0.93	1.89	7.61
23	0.42	1.19	1.76	0.22	1.27	1.87	0.29	0.58	2.44
25	0.34	0.93	1.43	0.18	0.98	1.52	0.23	0.46	1.96
29	0.24	0.64	1.04	0.13	0.68	1.11	0.16	0.33	1.41
31	0.19	0.44	0.73	0.09	0.47	0.78	0.11	0.23	0.99
35	0.14	0.40	0.46	0.07	0.41	0.48	0.07	0.14	0.46

[a] Harmonic current in ampères.

and switching scheme are designed so that, with the outage of a unit, the complete system is shut down. This brings another consideration, that is, redundancy is the filter application so that the harmonic emission at PCC is controlled within IEEE limits. Alternatively, enough spare parts and services should be available to bring the faulty unit in service in a short time.

Operation with Varying Loads

When load-dependent switching is required for reactive power compensation, multiple capacitor banks are switched in an ascending order, that is, 5th, 7th, and 11th. Generally, this will occur during start-up conditions; however, if sustained operation at reduced loads is required, it is necessary to control the harmonic distortion at each of the operating loads and switching steps. The harmonic loads may or may not decrease in proportion to the overall plant load. This adds another step in designing an appropriate passive filtering scheme to meet the TDD requirements.

Division of Reactive kvar between Parallel Filter Banks

When multiple parallel filters are required and the total kvar requirements are also known, it remains to find out the most useful distribution of kvar among the parallel filters. In the above example, the 5th, 7th, and 11th filters are based on equal kvar. This is too simplistic an approach, rarely implemented. As filters should be sized to handle the harmonic loading, one approach would be to divide the required kvar based on the percentage of harmonic current that each filter will carry. This will not be known in advance. The other method is to proportionate the filters with respect to harmonic current generation, that is, the lower order harmonics are higher in magnitude, so more kvar is allocated to a lower order filter. Again some iteration will be required to optimize the sizes initially chosen, based on the actual fundamental and harmonic current loadings and the desired reactive power compensation.

Losses in the Capacitors

The power capacitors have some active power loss component, though small. Figure 20.6 shows the average losses versus ambient temperature for capacitors. At an operating temperature of 40°C, the loss is approximately 0.10 W/kvar and increases to 0.28 W/kvar at −40°C. This loss should be considered in the filter design, by an equivalent series resistance inserted in the circuit.

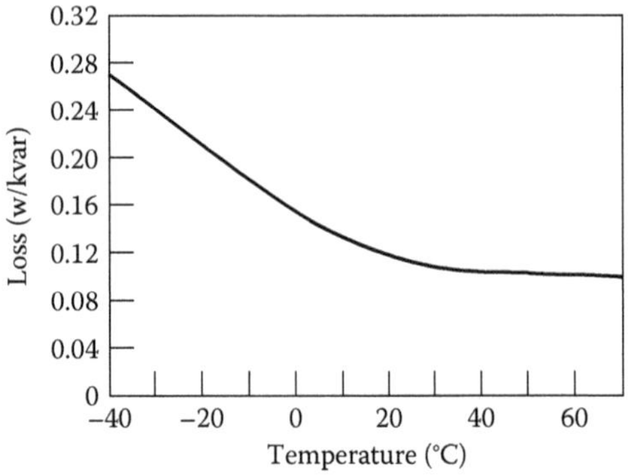

FIGURE 20.6
Average losses in film–foil capacitor units, with variation of temperature.

Harmonic Filter Detuning and Unbalance

The operation of an internal or external fuse or shorting of elements of a fuseless capacitor banks changes the capacitance of the filter and subjects it to higher overvoltage. Unbalance detection systems are applied and some considerations are as follows:

- The resonant frequency will change. Fuse operation in a parallel connected capacitor bank will decrease the capacitance and increase the resonant frequency. There is a possibility that shorting of elements of an externally fused capacitor bank increases the capacitance and decreases the resonant frequency. It is desirable to ascertain the maximum plus minus capacitance change that can be tolerated. The detuning may be a more stringent condition than the overvoltage on remaining units.
- Ambiguous indications are a possibility. For example, a negligible current will flow through a CT connecting the neutrals of a balanced ungrounded-double-wye bank, and this will not change if equal number of fuse failure or element short-outs occurs in the same phases of the wye banks. Where such a possibility exists, an alarm on first failure of a fuse or first failure of an element in fuseless bank is desirable to be provided.
- An arcing fault external to a capacitor unit can result in large change in filter capacitance and detuning. The unbalance protection may not always operate, depending upon filter configurations. A phase overcurrent relaying scheme can be designed according to IEEE Std. C37.99-2000 [5]. Furthermore, harmonic currents should not adversely affect trip or alarm level relays, and filters are required so that the operation is based on fundamental currents and voltages.
- To simulate the fuse failure in one capacitor unit in a phase and its effect on detuning, three-phase harmonic load flow modeling and analyses are required.

20.4 Relations in an ST Filter

The reactive power output of a capacitor at fundamental frequency is V^2/X_c. In the presence of a filter reactor, it is given by

$$\begin{aligned} S_f &= \frac{V^2}{X_L - X_c} \\ &= \frac{V^2}{X_c/n^2 - X_c} \\ &= \frac{n^2}{n^2 - 1} \times \text{(reactive power without reactor)} \end{aligned} \tag{20.16}$$

The reactive power output with a filter reactor tuned to, say, $4.85f$ is approximately 4% higher than that without the reactor. This is so because the voltage drop in the reactor is added to the capacitor voltage and its operating voltage is

$$V_c = V + V_L = V + \frac{V(j\omega L)}{\left(j\omega L - \frac{1}{j\omega C}\right)}$$
$$= V\left[\frac{n^2}{n^2 - 1}\right] \quad (20.17)$$

The capacitors in a fifth harmonic filter tuned to $4.85f$ operate at approximately 4% higher than the system voltage. While selecting the voltage rating on the filter capacitors, the following considerations are made:

- Higher operating voltage due to the presence of filter reactors.
- Sustained upward operating voltage of the utility power supply system. This may be due to location, for example, close to generating stations, or may be due to voltage adjustment tap changing on transformers.
- The higher voltages that will be imposed when one or two capacitor units in a parallel group go out of service. The neutral unbalance detection schemes may be set to alarm if a fuse operates on one of the parallel capacitor units and trip if the fuse operates on two units in the same phase.

The steady state fundamental frequency voltage can be

$$V_r = V\left(\frac{n^2}{n^2 - 1}\right) + \sum_{h=2}^{\infty} I_h X_{ch} \quad (20.18)$$

where
V_r is the rms rated voltage of the capacitor
I_h is the rms harmonic current at harmonic h
V is maximum system voltage across capacitor excluding voltage rise across the reactor
X_{ch} is the capacitive reactance at the harmonic h

For transient events, such as capacitor bank switching, circuit breakers restrikes, etc., the voltage rating is calculated from

$$V_r = \frac{V_{tr}}{\sqrt{2}k} \quad (20.19)$$

where
V_{tr} is the peak transient voltage
k factor is read from IEEE standard 1531-2003, figure 7 in per unit of peak of rated voltage, not reproduced here

If there are 100 transient events in a year, $k = 2.6$. For a 13.8 kV rated voltage $V_{tr} = 50.73$ kV peak, which is approximately 2.6 times the rated peak voltage. This implies that the capacitors can withstand 100 surges in a year of magnitude approximately 2.6 times the rated voltage. Higher voltage surges will be applicable if the frequency of surges is decreased.

For dynamic events generally lasting for a few fundamental cycles to several seconds, such as transformer energization and bus fault clearance, the voltage rating of the capacitor is calculated form the following equation:

$$V_{\mathrm{r}} = \frac{V_{\mathrm{d}}}{\sqrt{2}} \tag{20.20}$$

where V_{d} is the voltage across the capacitor reached during dynamic event. The power frequency short-time overvoltage capability is specified in Figure 8 of IEEE standard 1531-2003 [6], not reproduced here. A capacitor unit can withstand 2.2 pu of rated voltage for 0.01–0.1 s and at about 15 s the overvoltage withstand capability is reduced to 1.4 pu.

Power system studies, for example, EMTP switching transient studies, are required to establish the transient and dynamic overvoltages in a system. The rated voltage selected should be highest of the voltages given by Equations 20.18 through 20.20. Generally, an operating voltage 8%–10% higher than the nominal system voltage is selected. This reduces the reactive capability at the voltage of application as the square of the voltages (Equation 13.42).

Example 20.2

A 3 Mvar capacitor bank is required to be formed for an ST filter application at 13.8 kV. The ST filter is wye connected, tuned to 4.7f and ungrounded and has the harmonic loading $I_1 = 124$ A, $I_5 = 40$ A, $I_7 = 20$ A, and $I_{11} = 5$ A, higher order harmonics neglected for this example. The switching overvoltage studies show that these overvoltages are not more than 2.5 times the rated system voltage, and the estimated switching operations per year are 50. Calculate the rated voltage of the capacitor units required to form the bank.

$$X_{\mathrm{ch}} = \left(\frac{4.7^2}{4.7^2 - 1}\right)\left(\frac{13.8^2}{3}\right) = 66.49\ \Omega$$

From Equation 20.18, under steady-state operation, the capacitors will experience a voltage of

$$\begin{aligned} 124 \times 66.49 + 40 \times \frac{66.49}{5} + 20 \times \frac{66.49}{7} + 5 \times \frac{66.49}{11} &= 8244 + 531.9 + 190 + 30.22 \\ &= 8.90\ \mathrm{kV} \end{aligned}$$

For transient voltage capability and 50 switching operations, k as read from Figure 7 of IEEE standard 1531-2000, Ref. [6] is 2.75. Then the rms voltage rating of the capacitor form Equation 20.19 is

$$\frac{2.5 \times 13.8}{\sqrt{2} \times 2.75} = 8.872\ \mathrm{kV}$$

Check for the power frequency dynamic voltage capability from Figure 8 of Ref. [6]. This shows a voltage capability of 2.2 per unit of rated voltage, which is adequate based on the switching studies.

In the above example, if the switching frequency is raised to 500 per year, then k factor is 2.4 and the rated voltage of the capacitor unit should be 11.09 kV.

An 8.32 kV standard voltage rating capacitor unit has a continuous operating voltage capability of 9.152 kV. This voltage rating can be selected for the application; however, it is preferable to reserve 10% margin on the continuous rating of the capacitors for contingency operations. Alternatively, an immediate shutdown should be taken for replacement of failed unit or a blown fuse, as detected by unbalance detection equipment.

20.4.1 Number of Series Parallel Groups

Table 19.3 gives the minimum number of units in parallel per series group to limit voltage on remaining unit to 110% with one unit out of service. Equations 19.73 and 19.74 can be used for accurate calculations. The shift in tuning frequency and consequent overloading will be a more stringent condition, as compared to overvoltage limitation. Consider the 3000 kvar bank of Example 20.2; per phase 1000 kvar is required. The reactive power output of 8.32 kV capacitor will reduce by a factor of 0.917 when applied at 13.8 kV ungrounded wye configuration, and a minimum of four units per phase are required. A reference to the manufacturer's standard available unit sizes shows that four units of 300 kvar will give 1100 kvar per phase at the operating voltage of 13.8 kV, when connected in wye configuration, and a three-phase bank size of 3.3 Mvar. Thus, an adjustment in the required size has to be made based on the standard voltage and capacitor kvar size ratings. IEEE standard 519 allows that harmonic limits may be exceeded by 50% for shorter periods during start-ups or unusual conditions (Appendix F).

To meet the requirement of a certain reactive power output, a larger number of units are then required.

The fundamental loading of the capacitors is given by

$$\frac{V_{\mathrm{c}}^2}{X_{\mathrm{c}}}=\frac{V^2}{X_{\mathrm{c}}}\left[\frac{n^2}{n^2-1}\right]^2=S_{\mathrm{f}}\left[\frac{n^2}{n^2-1}\right] \tag{20.21}$$

where

$$S_f=\frac{V^2}{X_{\mathrm{c}}}\left[\frac{n^2}{n^2-1}\right]$$

and the harmonic loading is

$$\frac{I_{\mathrm{h}}^2X_{\mathrm{c}}}{h}=\frac{I_{\mathrm{h}}^2V^2}{Sh}\frac{h^2}{h^2-1} \tag{20.22}$$

When harmonic voltages and current flows are known from harmonic simulation, the harmonic loading can be found from

$$\sum_{h=2}^{h=\infty}V_hI_h \tag{20.23}$$

The fundamental frequency loading of the filter reactor is

$$\frac{V_{\mathrm{L}}^2}{X_{\mathrm{L}}}=\left[\frac{V_{\mathrm{c}}}{n^2}\right]^2\left[\frac{n^2}{X_{\mathrm{c}}}\right]=\frac{V_{\mathrm{c}}^2}{n^2X_{\mathrm{c}}}=\frac{S_{\mathrm{f}}}{n^2}\left[\frac{n^2}{n^2-1}\right] \tag{20.24}$$

The harmonic loading for the reactor is the same as for the capacitor.

The increase in bus voltage on switching a capacitor at a transformer secondary bus is approximately given by

$$\%\Delta V=\frac{\mathrm{kvar}_{\mathrm{capacitor}}Z_{\mathrm{t}}}{\mathrm{kVA}_{\mathrm{t}}} \tag{20.25}$$

A flowchart of the design of ST filters is thus shown in Figure 20.7.

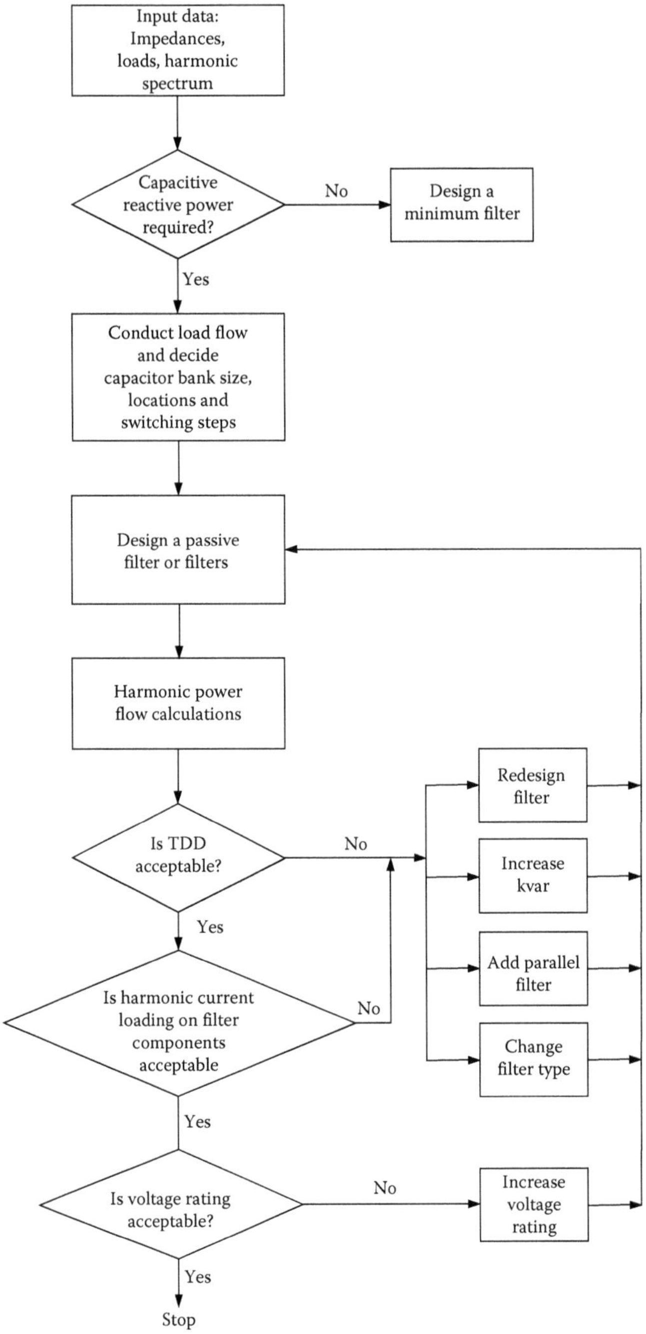

FIGURE 20.7
Flowchart for design of ST filters.

20.5 Filters for a Furnace Installation

Example 20.3

Figure 20.8 shows a furnace installation. The total operating load is 150 MVA. The PCC is the 230 kV side of 125/208 MVA transformer. A reactive power compensation of 120 Mvar is required, which is provided by four ST filters formed with 48, 24, 24 and 24 Mvar capacitors, respectively, for second, third, fourth, and fifth harmonics. These are connected at the main 34.5 kV bus. The capacitor banks are formed as follows:

- Second harmonic ST filter: double wye, grounded, three-series groups, each group containing eight units of 400 kvar capacitors of rated voltage 7.2 kV (Figure 20.9).
- Third, fourth, and fifth ST filters: single wye ungrounded, three-series groups, each group containing eight units of 400 kvar capacitors of rated voltage 7.2 kV.

The tuning frequencies for second, third, fourth, and fifth ST filters are 1.95, 2.95, 3.95, and 4.95 times the fundamental frequency, respectively.

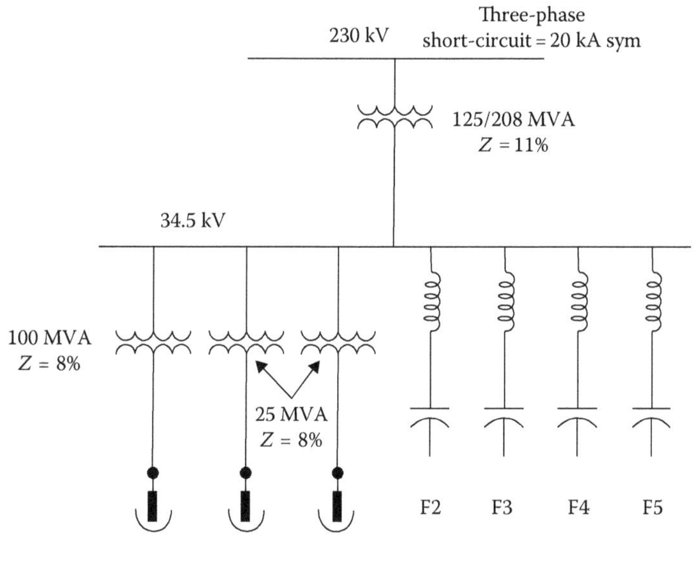

Filter	Harmonic	Filter Reactor	Capacitor Bank
F2	Second	1.694E-2 H, $Q = 100$	49 Mvar
F3	Third	1.480E-2 H $Q = 100$	24.5 Mvar
F4	Fourth	8.259E-3 H $Q = 100$	24.5 Mvar
F5	Fifth	5.260E-3 H $Q = 100$	24.5 Mvar

FIGURE 20.8
Single-line diagram of a furnace installation, showing ST filters.

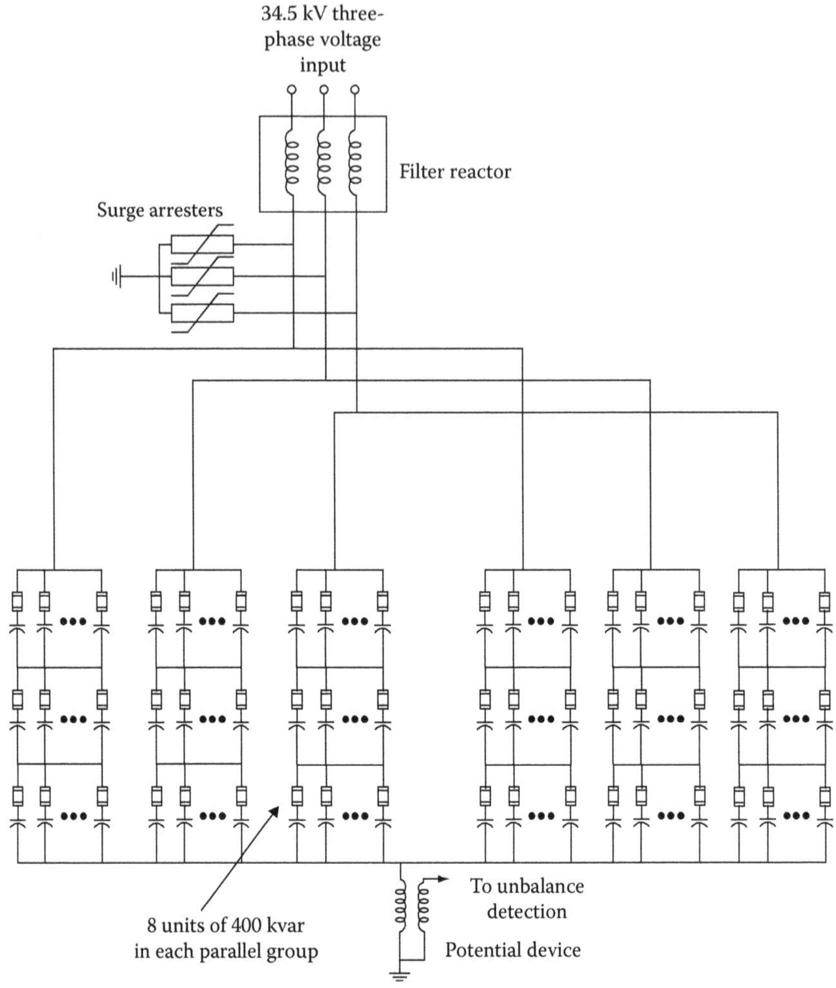

FIGURE 20.9
Formation of a 49 Mvar double-wye, 34.5 kV capacitor filter bank for second harmonic (Example 20.2).

The system is impacted with a harmonic spectrum during the melting cycle of the furnaces. The results of harmonic flow calculations and TDD at the PCC are shown in Table 20.8. This table shows harmonic current loading of the filters and harmonic currents fed into the 230 kV system. We observe that each ST filter operates effectively providing a low impedance path for the harmonic it is intended to shunt away. TDD at the 230 kV PCC is 1.32. Figure 20.9 shows formation of second harmonic filter.

Dynamic Stresses

Filters for furnace installations should receive special considerations with respect to transient surges. Consider that a second harmonic ST filter is not installed in a furnace installation, and the lowest harmonic order filter is the third. The shifted resonant frequency may coincide with one of the transformer inrush current harmonics or be close to it, and may increase the transformer inrush currents. Though these last for a short duration, these will stress the filter reactor and capacitors. Switching inrush currents were discussed

TABLE 20.8

Example 20.3: Harmonic Filters for Arc Furnace (Figure 20.8)[a]

Harmonic	Second Harmonic ST Filter	Third Harmonic ST Filter	Fourth Harmonic ST Filter	Fifth Harmonic ST Filter	Harmonic Currents at PCC (230 kV Bus)	Harmonic Voltages at PCC (230 kV Bus)
2	192.20	1.07	0.77	0.69	0.868	0.437
3	6.83	134.2	10.00	6.69	2.82	0.267
4	0.91	1.63	59.5	3.87	0.50	0.022
5	0.54	0.75	2.34	97.78	0.34	0.0337
7	5.08	6.18	13.39	28.89	3.50	8.37

Notes: Load demand = 350.8 A, three-phase short-circuit current at PCC = 20 kA, ratio $I_{SC}/I_L = 57$, permissible total TDD = 6.0%, permissible TDD for $h < 11 = 5\%$; calculated TDD = 1.32%, which is also total TDD.

[a] Harmonic current flow in ampères; harmonic voltages in kV.

in Section 5.8 and it was noted that the presence of filter reactors lowers the frequency as well as the magnitude of the inrush currents. However, their duration may increase, as the reactors are of high *Q*, giving less damping. Synchronous switching and resistance switching are the options. For normal switching, it will be necessary to calculate the effect of switching and of transformer inrush currents, and resulting harmonic voltages, and apply these to the specifications of filter reactors and capacitors.

Attenuation of Harmonics

In the above example, we declared a 230 kV bus as the PCC. If a 34.5 kV bus is declared as the PCC and the TDD is calculated, it will be higher than the TDD at the 230 kV bus. The calculated TDD at the 34.5 kV bus is 1.35. This shows attenuation of harmonics in propagation through system elements, in this case, the transformer impedance. The impact is the maximum at the point of injection and attenuation occurs as the harmonics are propagated into the system, unless there is amplification due to resonance. Partial resonances are common when capacitors are used.

Noninteger Harmonics

Furnace loads generate noninteger harmonics. Generally, for a furnace installation, the seventh harmonic filter is a high-pass filter (see Section 20.8).

20.6 Filters for an Industrial Distribution System

Example 20.4

Example 19.2 for application of power capacitors in an industrial plant showed that depending on the operating condition, resonant frequencies swing over a wide spectrum and the harmonic distortion at the PCC is high. The equivalent negative sequence current loading of generators is exceeded. Even without harmonics, a part of the generator negative sequence capability may be utilized due to unbalance loads and voltages and system asymmetries. The harmonic loading on the generators and harmonic distortion are reduced by turning capacitor banks at buses 2 and 3 into parallel fifth and seventh ST filters.

The following details are applicable for the final filter designs:

Bus 2, fifth harmonic ST filter: five units of 300 kvar, 7.2 kV, per phase, total capacitor Mvar = 4.5, connected in ungrounded wye configuration; $n=4.85$, $C=76.75$ μF, and $L=3.897$ mH.

Buses 2 and 3, seventh harmonic ST filter: three units of 300 kvar, 7.2 kV, per phase, giving 2.7 Mvar total, connected in ungrounded wye configuration; $n=6.75$, $C=46.05$ μF, and $L=3.353$ mH.

Bus 3, fifth harmonic ST filter: four units of 300 kvar, 7.2 kV, per phase, total capacitor Mvar = 3.6, $n=4.85$, $C=61.40$ μF, $L=4.872$ mH, and $Q=100$.

X/R ratio of reactors = 100 at fundamental frequency. The harmonic current flow is studied under the same three conditions as in Example 19.2 and TDD at the PCC is calculated. The normal load demand current $I_r=250$ A in cases 1 and 3, and 410 A in case 2, when No. 2 generator is out of service. The three-phase short-circuit current is 30.2 kA sym. The results of calculation are shown in Table 20.9. The following observations are of interest.

When the 75 mi 115 kV line and its harmonic load are modeled, TDD increases over normal operating condition 1. (See Example 19.2 for a description of the operating conditions for this system.) This shows the impact of harmonic loads that may be located at considerable distance from the consumer. These should invariably be considered in a harmonic analysis study.

In operating condition 2, seventh harmonic ST filters on buses 2 and 3 are out of service. TDD is slightly above the limits. This is acceptable for short-term operation.

The smaller generator G2 rated at 47.97 MVA has a higher harmonic loading as compared to the larger generator Gl of 82 MVA. The impedance modulus shows some interaction between ST filters and capacitors at motors. Generally, it is desirable to observe one strategy of reactive power compensation in a distribution system, due to the problem of secondary resonance, discussed in Section 20.7.

The resonant frequency varies by a maximum of 1.2% in the three cases. This can be compared to the much wider swings shown in Table 19.8, without filters. The resulting harmonic current flows through the system and filters change with switching operation, yet the TDD at the PCC remains within acceptable limits (Table 20.9). Also, the negative sequence loadings of the generators are at safe levels.

TABLE 20.9

Example 20.4: Harmonic Simulation with Filters[a]

Harmonic Order ↓	Current in Generator 1			Current in Generator 2			Current into the Utility's System (PCC)		
	1[b]	2	3	1	2	3	1	2	3
5	37	35.6	31.7	51.7	Generator out of service	48.7	6.78	7.05	12.8
7	24.4	104	27.3	24.9		26.3	4.13	25.81	5.44
11	2.03	42.3	2.3	25.7		27.2	1.24	5.45	3.70
13	3.76	3.99	2.87	17.6		15.2	1.17	0.21	4.15
17	2.83	1.76	4.0	10.0		12.1	0.75	0.56	0.90
19	2.03	1.60	8.20	7.13		4.96	0.54	0.45	2.13
23	1.94	1.81	2.89	7.71		9.53	0.54	0.47	0.95
25	1.75	1.73	1.30	7.44		6.19	0.48	0.44	1.80
Calculated TDD at PCC→							3.28	6.66	6.12
Permissible TDD at PCC→							7.5	6	7.5

[a] Harmonic current flow in ampères.
[b] 1, 2, 3 refer to study cases, see Example 19.2.

20.7 Secondary Resonance

In the case where there are secondary circuits that have resonant frequencies close to the switched capacitor bank, the initial surge can trigger oscillations in the secondary circuits that are much larger than the switched circuit. The ratio of these frequencies is given by

$$\frac{f_c}{f_m} = \sqrt{\frac{L_m C_m}{L_s C_s}} \tag{20.26}$$

where
- f_c is the coupled frequency
- f_m is the main circuit switching frequency
- L_s and C_s are the inductance and capacitance in the secondary circuit, respectively
- L_m and C_m are the inductance and capacitance in the main circuit, respectively

Figure 20.10a shows the circuit diagram and Figure 20.10b shows amplification of transient voltage in multiple capacitor circuits. The amplification effect is greater when the natural frequencies of the two circuits are almost identical. Damping ratios of the primary and coupled circuits will affect the degree of interaction between the two circuits [7].

Reference [4] shows an EMTP simulation of secondary resonance. A 220 kvar low-voltage capacitor remains connected to a 2 MVA, 13.8 kV–0.48 kV transformer, while a 6 Mvar capacitor bank is switched on 13.8 kV system to which the 2 MVA transformer is also connected. An overvoltage of 3.1 times the rated voltage occurs on 480 V secondary of the 2 MVA transformer and a peak current of 2200A flows through the 220 kvar capacitor. It is best not to apply capacitors at multi-voltage levels in a distribution system, as study and simulation of secondary resonances can be fairly involved.

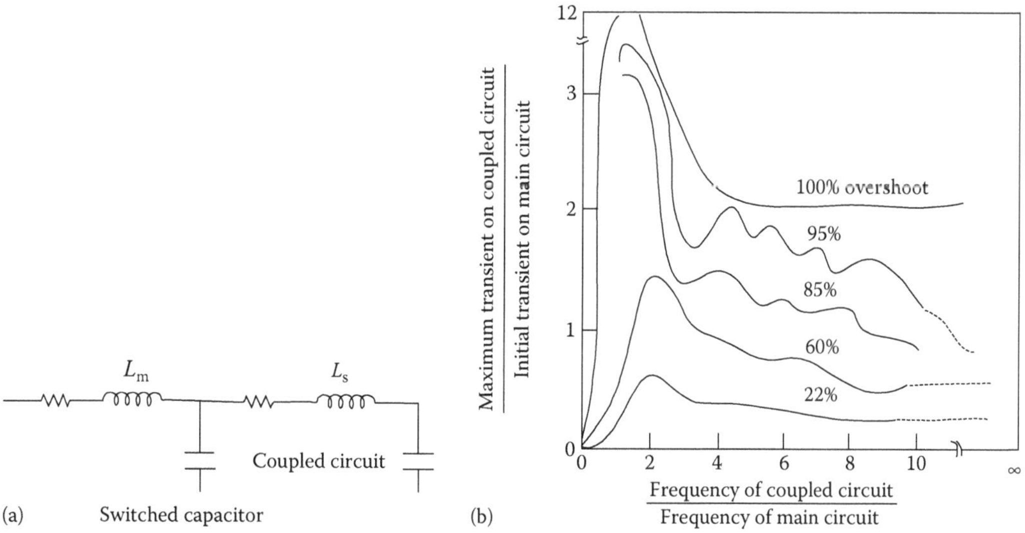

FIGURE 20.10
(a) Circuit of a secondary resonance; (b) overvoltages due to secondary resonance. (From Das, J.C., *Transients in Electrical Systems—Recognition Analysis and Mitigation*, McGraw Hill, New York, 2010.)

20.8 Filter Reactors

The filter reactors can be dry-type air-core or dry-type iron-core reactors, generally used in low- and medium-voltage applications, say, up to voltage levels of 30 kV. Fluid-filled iron-core reactors are used for medium-voltage applications. For voltages above 30 kV, air-core reactors are most popular. Air-core reactors can be designed for higher voltages and better tolerances as compared to iron-core reactors. For medium-voltage applications, single-phase iron-core reactors can be applied, and these have smaller dimensions and require much smaller magnetic clearances as compared to air-core reactors. Three-phase iron-core reactors are not generally used as it is difficult to adjust the reactance of one phase without impacting the reactance of the other phases. The change in magnetic material properties can give rise to wider fluctuations in the reactance value in iron-core reactors, though these are designed with lower flux densities and a series air gap and are smaller in dimensions. Reactors for filter applications are subjected to high harmonic frequencies. A harmonic current flow spectrum, based on the worst case operation, is normally required by a manufacturer for an appropriate design. The rated steady-state voltage is calculated as arithmetic sum of fundamental and harmonic voltages, similar to the capacitor:

$$V_{\mathrm{r}} = \sum_{h=1}^{h=\infty} I_h X_{\mathrm{R(h)}} \tag{20.27}$$

where $X_{\mathrm{R(h)}}$ is the reactance at the given harmonic order.

The earlier construction of air-insulated reactors, consisting of large conductors restrained in polyester or poured-in concrete, has given way to small parallel conductors, epoxy insulated, and encapsulated (Figure 20.11a and b for comparison). The parallel conductors share the current path and it is possible to use fiberglass epoxy composite to encapsulate the windings. The windings are, thus, completely supported along their lengths, and can better withstand compressive and tensile short-circuit forces. The filament fiberglass utilized in conjunction with epoxy resin to encapsulate has better tensile strength than steel. The reactors must withstand system-through fault symmetrical amperes and time duration of 3 s is generally specified. These should also withstand the mechanical stresses brought about by asymmetrical short-circuit currents. Dynamic stresses due to switching and transformer inrush currents have to be considered.

At medium-voltage levels, the harmonic filter is generally connected in ungrounded configuration, and when reactor is located on the source side of the capacitor, the available fault current is limited, while an iron-core reactor in the same situation may saturate, and will not decrease the short-circuit current. For high-voltage grounded banks, the filter reactor may be located on the neutral side. This may allow basic insulation level (BIL) of the reactor to be less than that of the system. The air-core coils of even large reactors can be vertically stacked, resulting in space saving. The middle phase is reverse wound to balance the magnetic fluxes.

The magnetic clearance required for air-core reactors is generally large, at least equal to half the diameter of the coil on the sides of the reactor and 1/4 the diameter of the coil for any magnetic material in the foundations, provided no close loops of the magnetic

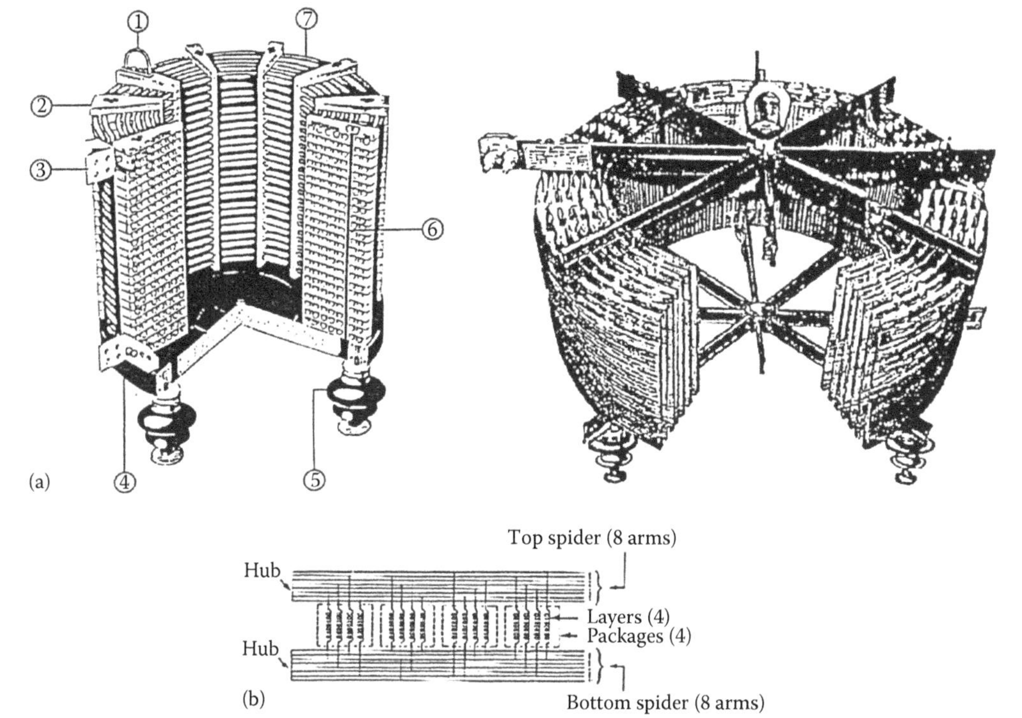

FIGURE 20.11
(a) Old and (b) new constructional features of reactors. 1, Lifting lug; 2, interlocked cleats; 3, terminal; 4, concrete or resin basin; 5, mounting insulator; 6, insulated tie rod; 7, large cable conductor.

materials exit. Foundation pedestals for the reactors can be designed with fiber glass reinforcements. Sometimes, where the required clearances cannot be maintained, 1/2 in. aluminum plates can be provided at the base and sides.

20.8.1 *Q* Factor

Apart from its impact on the filter performance, the Q factor determines the fundamental frequency losses and this could be an overriding consideration, especially when the reactors at medium-voltage level are required to be located indoors in metal or fiberglass enclosures and space is at a premium. Consider a second harmonic filter for the furnace installation in Figure 20.9. The inductor is 0.01694 mH, that is, inductive reactance = 6.386 ohm. An X/R of 50 gives a reactor resistance of 0.128 ohm. The rms current is 843 A. This gives a loss of approximately 272 kW/h (at fundamental), which is very substantial.

Equation 20.8 defines the filter Q based on the inductive or capacitive reactance at the tuned frequency (these are equal). The fundamental frequency losses and heat dissipation are of major consideration. This does not mean that the effect on filter performance can be ignored. The higher the value of Q, the more pronounced is the valley at the tuned frequency. For industrial systems, the value of R can be limited to the resistance built in the reactor itself.

Example 20.5

The effect of change in Q of the filter is examined in this example. We have an X/R of 100 at fundamental frequency for the filter reactors in example 20.4. Thus, the resistance values of the reactors are as follows:

Fifth harmonic filter at bus 2 = 0.01469 ohm

Fifth harmonic filter at bus 3 = 0.01836 ohm

Seventh harmonic filter at bus 2 and bus 3 = 0.012642 ohm

Figure 20.2 shows that the sharpness of tuning is dependent upon the resistance. Harmonic load flow of Example 20.3 is repeated with tuning reactors of $X/R = 10$ and the results are shown in Table 20.10. There is hardly an appreciable difference in the harmonic current flow. In industrial systems, the performance of ST filters will be, generally, indistinguishable for Q (Equation 20.8) = 20 to $Q = 100$.

The X/R of tuning reactors at 60 Hz is given by $3.07K^{0.377}$ where K is the three-phase kVA $= 3I^2X$ (I is the rated current in amperes and X is the reactance in ohms). X/R of a 1500 kVA reactor will be 50 while that of a 10 MVA reactor will be 100. High X/R reactors can be purchased at a cost premium. Thus, selection of X/R of the reactor depends upon

Initial capital investment

Active energy losses

Effectiveness of the filtering

The optimization of filter admittance and Q for the impedance angle of the network and δ *are required* for the transmission systems. The optimum value of Q is given by [3]:

$$Q = \frac{1 + \cos\phi_m}{2\delta_m + \sin\phi_m} \tag{20.28}$$

where ϕ_m is the network impedance angle. Consider a frequency variation of ±1%, a temperature coefficient of 0.02% per degree Celsius, and a temperature variation of ±30°C on the inductors and capacitors, then from Equation 20.13 $\delta = 0.006$. For an impedance angle $\phi_m = 80°$, the optimum Q from Equation 20.28 is 99.31. The higher the tolerances on components and frequency deviation, the lower the value of Q.

TABLE 20.10

Effect of Change of Q: Harmonic Simulation, Condition 1 of Example 20.4[a]

Harmonic	Current in Generator 1		Current in Generator 2		Current into Utility's System (PCC)	
Order	$X/R = 100$	$X/R = 7–10$	$X/R = 100$	$X/R = 7–10$	$X/R = 100$	$X/R = 7–10$
5	37	37.5	51.7	54.0	6.78	6.90
7	24.4	25.0	24.9	25.7	4.13	4.24
11	2.03	2.03	25.7	25.7	1.24	1.24
13	3.76	3.76	17.6	17.6	1.17	1.17
17	2.83	2.83	10.0	10.0	0.75	0.75
19	2.03	2.03	7.13	7.13	0.54	0.54
23	1.94	1.94	7.71	7.71	0.54	0.54
25	1.75	1.75	7.44	7.44	0.48	0.48

[a] Harmonic current flow in ampères.

20.9 Double-Tuned Filter

A double-tuned filter is derived from two ST filters and is shown in Figure 20.12. Its *R–X* plot and *Z–ω* plots are identical to that of two ST filters in parallel, as shown in Figure 20.3. The advantage with respect to two ST filters is that the power loss at fundamental frequency is less and one inductor instead of two is subjected to full impulse voltage. In Figure 20.12, the BIL on reactor L_2 is reduced while reactor L_1 sees the full impulse voltage. This is an advantage in high-voltage applications. The following equations [8] transform two ST filters of different frequencies into a single double-tuned filter:

$$C_1 = C_a + C_b \tag{20.29}$$

$$L_2 = \frac{(L_aC_a - L_bC_b)^2}{(C_a + C_b)^2(L_a + L_b)} \tag{20.30}$$

$$R_2 = R_a\left[\frac{a^2(1-x^2)}{(1+a)^2(1+x^2)}\right] - R_b\left[\frac{1-x^2}{(1+a)^2(1+x^2)}\right] + R_1\left[\frac{a(1-a)(1-x^2)}{(1+a)^2(1+x^2)}\right] \tag{20.31}$$

$$C_2 = \frac{C_aC_b(C_a + C_b)(L_a + L_b)^2}{(L_aC_a - L_bC_b)^2} \tag{20.32}$$

$$R_3 = -R_a\left[\frac{a^2x^4(1-x^2)}{(1+ax^2)^2(1+x^2)}\right] + R_b\left[\frac{(1-x^2)}{(1+ax^2)^2(1+x^2)}\right] + R_1\left[\frac{(1-x^2)(1-ax^2)}{(1-x^2)(1-ax^2)}\right] \tag{20.33}$$

$$L_1 = \frac{L_aL_b}{L_a + L_b} \tag{20.34}$$

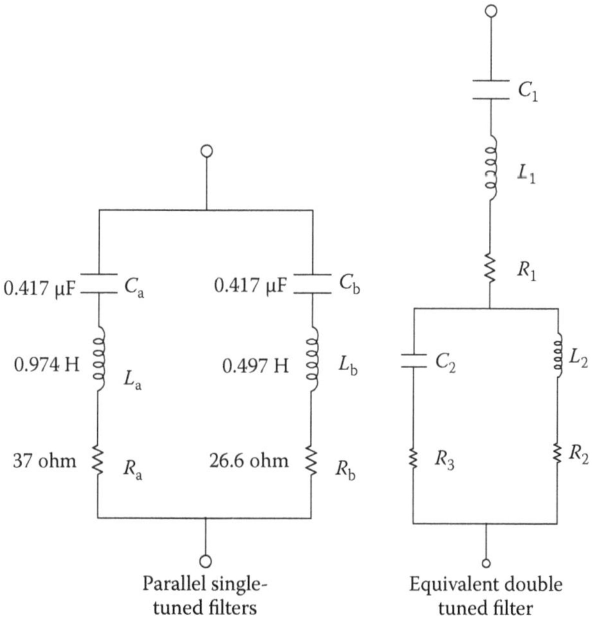

FIGURE 20.12
Equivalent circuits of two ST parallel filters and a single double-tuned filter.

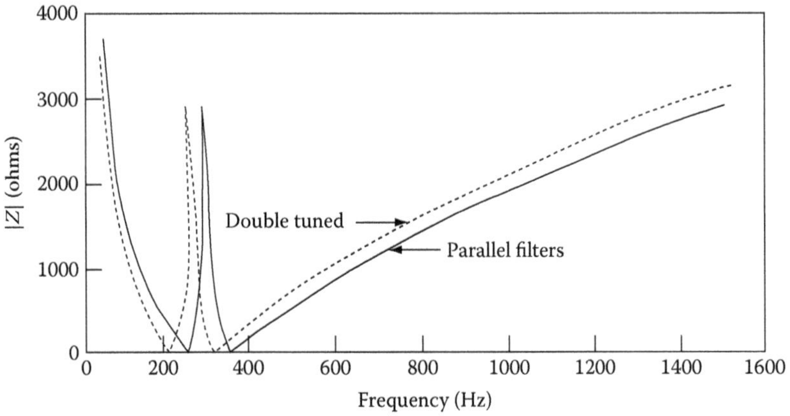

FIGURE 20.13
Z–ω plots of two parallel ST filters and equivalent double-tuned filter (Example 20.6).

where

$$a = \frac{C_a}{C_b} \quad x = \sqrt{\frac{L_b C_b}{L_a C_a}} \tag{20.35}$$

Generally, R_1 is omitted and R_2 and R_3 are modified so that the impedance near resonance is practically the same. Note that inductor L_1 will have some resistance, which is considered in the above equations.

Example 20.6

Consider the fifth and seventh filter section for a high-voltage application, at 50 Hz, numerical values as shown in Figure 20.12. It is converted into a double-tuned filter and the response is compared with the original ST parallel filters. The use of the above equations gives the following numerical values for the filter:

$$\begin{aligned} C_1 &= 0.34\ \mu\text{F} & R_1 &= 2.07\ \text{ohm} \\ C_2 &= 7.931\ \mu\text{F} & R_2 &= 1.527\ \text{ohm} \\ L_1 &= 0.329\ \text{H} & R_3 &= 1.232\ \text{ohm} \\ L_2 &= 0.039\ \text{H} \end{aligned}$$

The response of the two filters (without external connections) is superimposed in Figure 20.13.

20.10 Damped Filters

Figure 20.14 shows four types of damped filters. The first-order filter is not used as it has excessive loss at fundamental frequency and requires a large capacitor. The second-order high pass is generally used in composite filters for higher frequencies.

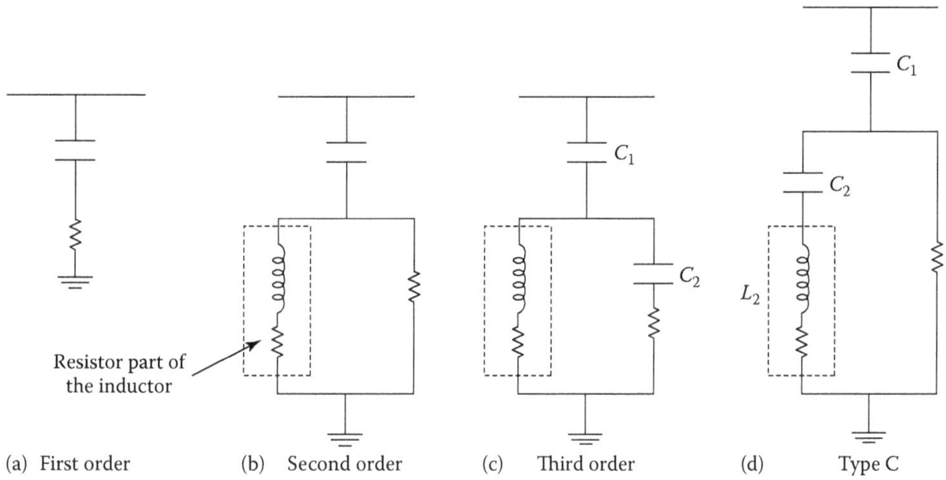

FIGURE 20.14
Circuits of damped filters: (a) first-order filter; (b) second-order filter; (c) third-order filter; (d) type-C filter.

If it were to be used for the full spectrum of harmonics, the capacitor size would become large and fundamental frequency losses in the resistor would be of consideration. This will be illustrated with an example. The filter is more commonly described as a second-order high-pass filter. The third-order filter has a substantial reduction in fundamental frequency losses, due to the presence of C_2, which increases the filter impedance; C_2 is very small compared to C_1. The filtering performance of type-C filters lies between that of second- and third-order filters. C_2 and L_2 are series tuned at fundamental frequency and the fundamental frequency loss is reduced. Also see Ref. [8].

The band-pass filters give rise to a shifted resonance frequency, while damped filters do not. This advantage of damped filters can be exploited and possible resonances at shifted frequencies can be avoided. Unlike ST parallel multiple filters, there are no parallel branches, yet the component sizing becomes comparatively large and it may not be possible to exploit this advantage in every system design. The performance and loading are less sensitive to tolerances. The behavior of damped filters can be described by the following two parameters [9]:

$$m = \frac{L}{R^2C} \tag{20.36}$$

$$f_0 = \frac{1}{2\pi CR} \tag{20.37}$$

The impedance can be expressed in the parallel equivalent form:

$$Y_f = G_f + jB_f \tag{20.38}$$

where

$$G_f = \frac{m^2x^4}{R_1[(1-mx^2)^2 + m^2x^2]} \tag{20.39}$$

$$B_f = \frac{x}{R_1}\left[\frac{1 - mx^2 + m^2x^2}{(1 - mx^2)^2 + m^2x^2}\right] \tag{20.40}$$

where

$$x = \frac{f}{f_0} \tag{20.41}$$

Considering that the filter is in parallel with an ac system of admittance $Y_a < \pm\phi_a$ (max), then the minimum total admittance as ϕ_a and Y vary is

$$Y = B_f \cos\phi_a + G_f \sin\phi_a \tag{20.42}$$

provided that the sign of each term is taken as positive and χ is less than the value that gives

$$|\cot\phi_f| = \left|\frac{G_f}{B_f}\right| = |\tan\phi_a| \tag{20.43}$$

For a given C, select parameters f_0 and m to obtain a sufficiently high admittance (low impedance) over the required frequency range. Values of m are generally between 0.5 and 2.

20.10.1 Second-Order High-Pass Filter

The characteristics of a second-order high-pass filter are shown in Figure 20.15, with its R–X and Z–ω plots. It has a low impedance above a corner frequency; thus, it will shunt

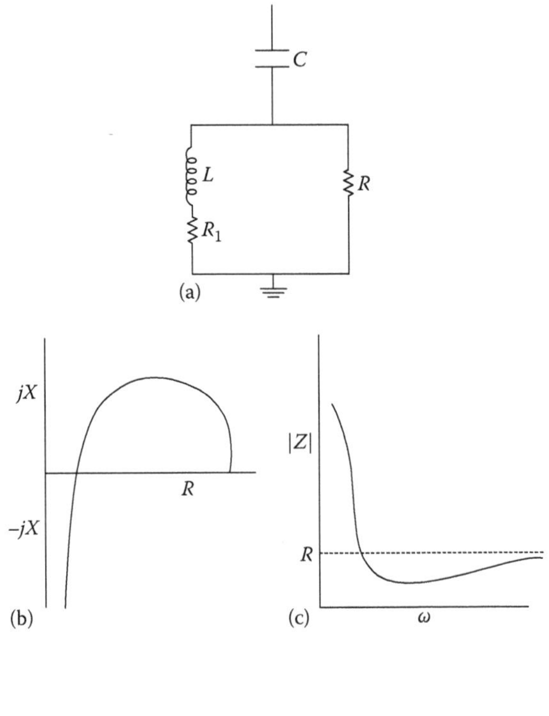

FIGURE 20.15
(a) Circuit and (b) R–X and (c) Z–ω plots of a second-order high-pass filter.

a large percentage of harmonics at or above the corner frequency. The sharpness of tuning in high-pass filters is the reciprocal of tuned filters:

$$Q = \frac{R}{(L/C)^{1/2}} = \frac{R}{X_L} = \frac{R}{X_c} \tag{20.44}$$

With high Q, the filtering action is more pronounced. Filter impedance is given by

$$Z = \frac{1}{j\omega C} + \left(\frac{1}{R} + \frac{1}{j\omega L}\right)^{-1} \tag{20.45}$$

The higher the resistance the greater is the sharpness of tuning. The Q value may vary from 0.5 to 2 and there is no optimum Q, unlike with band-pass filters.

The reactive power of the capacitor at fundamental frequency is the same as for an ST filter. The loading at harmonic h is

$$I_h^2 \frac{X_c}{h} = \frac{1}{S_f} \frac{I_h^2}{h} V^2 \left[\frac{n^2}{n^2 - 1}\right] \tag{20.46}$$

Thus, the total harmonic loading is

$$V^2 \frac{n^2}{S_f(n^2 - 1)} \sum_{h=\min}^{h=\max} \frac{I_h^2}{h} \tag{20.47}$$

The reactor loading at fundamental frequency can be calculated by assuming that current through the parallel resistor is zero, that is, current through the inductor is the same as through the capacitor; then, fundamental frequency loading is

$$I_L^2 X_L = I_c^2 \frac{X_c}{n^2} = \frac{S_f}{n^2}\left[\frac{n^2}{2} - 1\right] \tag{20.48}$$

At harmonic h, the harmonic current I_h divides into the resistance and inductance. The inductive component of the current is

$$I_{hL} = I_h \frac{R}{R + j\omega L} = I_h \frac{Q}{[Q^2 + (h/n)^2]^{1/2}} \tag{20.49}$$

The total harmonic loading is therefore

$$= Q^2 \frac{V^2}{S_f}\left[\frac{n^2}{n^2 - 1}\right] \sum_{n=\min}^{h=\max} \left[n \frac{I_h^2}{Q^2 n^2 + h^2}\right] \tag{20.50}$$

The loss in the resistor can be calculated as follows:

$$R = QhX_L \tag{20.51}$$

$$|I_R| = \frac{|I_L| X_L}{R} = \frac{I_L}{Qn} \tag{20.52}$$

Thus, the power loss is

$$I_R^2R = \frac{1}{Qn} I_L^2 X_L \tag{20.53}$$

$$= \frac{1}{Qn}(\text{Mvar loading}) \tag{20.54}$$

$$= \frac{S}{Qn^3}\left[\frac{n^2}{n^2-1}\right] \tag{20.55}$$

20.11 Design of a Second-Order High-Pass Filter

Example 20.7

In Example 20.1, we concluded that three ST filters for the 5th, 7th, and 11th harmonics designed around 900 kvar capacitors, rated voltage 2.77 kV (equivalent to 675 kvar at a system voltage of 2.4 kV) for each filter leg, provided adequate filtration and controlled TDD to acceptable values. A total of 2700 kvar was required. We also observed that if an ST filter for the fifth harmonic is used, it has to be of unpractically large size, requiring 9000 kvar of similarly rated capacitors. *Capacitor units rated at* 2.4 kV *could be used, depending on the system overvoltage profile.*

If a single second-order high-pass filter is designed to control TDD to acceptable levels in Example 20.1, its size will be still larger than the single fifth ST filter. This is so because a high-pass filter has a higher impedance at notch frequency, as compared to an ST filter. The application of a high-pass filter is, generally, for higher frequencies and notch reduction. Four cases of study are presented in Table 20.11.

TABLE 20.11

Example 20.6: High-Pass Filter[a]

Harmonic Order $h \downarrow$	Case 1	Case 2	Case 3	Case 4
5	322	145	79	46.2
7	173	60.8	33.5	21.5
11	31.6	36.7	20.3	25.6
13	9.21	14.3	7.81	13.0
17	3.05	6.80	3.65	7.96
19	1.44	3.61	1.93	4.66
23	0.33	0.98	0.52	1.49
25	0.23	0.73	0.39	1.19
29	0.13	0.46	0.24	0.84
31	0.08	0.30	0.16	0.58
35	0.06	0.21	0.09	0.43

[a] Harmonic currents in PCC, in ampères.

Case 1: Capacitor size 1.8 Mvar per phase; $C = 622$ μF, reactor $L = 0.4805$ mH (reactor $X/R = 100$), and $R = 0.18$ ohm.

Case 2: As case 1 except that filter $Q = 4.85$. For $n = 4.85$, $C = 622$ μF, $L = 0.4805$, and $R = 0.87$ ohm. A marked difference in the harmonic currents injected at the PCC occurs with higher Q. The TDD is still high.

Case 3: The size of capacitors is doubled, that is, 3.6 Mvar per phase and 10.8 Mvar total. The TDD at the PCC is still not in control, especially at lower harmonics.

The frequency scan of these three cases is shown in Figure 20.16. For large resistance, the filter reverts to an ST filter.

Case 4: An ST fifth harmonic filter of 300 kvar per phase ($C = 10.37$ μF, $L = 2.88$ mH, $Q = 100$, and $n = 4.85$) is paralleled with a high-pass filter; $C = 1.2$ Mvar per phase (3.6 Mvar total), $C = 414.9$ μF, $L = 0.346$ mH, filter $Q = 4.5$, and $n = 7$. The result of harmonic current flow into the PCC almost meets the TDD requirements.

This shows that three ST filters are the best design choice for Example 20.1. The circuit diagram of a high-pass filter in conjunction with parallel ST filters, and R–X and Z–ω plots, are shown in Figure 20.17.

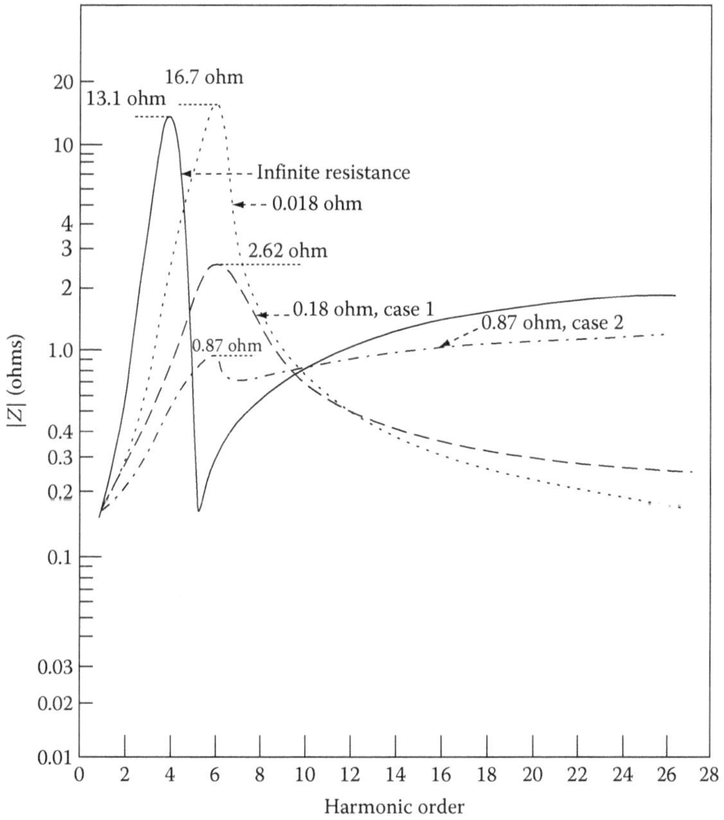

FIGURE 20.16
Z–ω plots of a second-order high-pass filter with varying Q and capacitor bank size (Example 20.6).

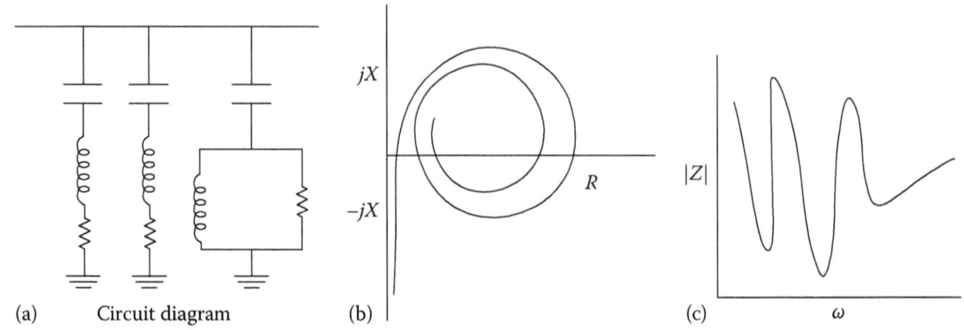

FIGURE 20.17
(a) High-pass filter for higher frequencies in parallel with two ST filters. (b) R–X and (c) Z–ω characteristics.

20.12 Zero-Sequence Traps

Zig-zag transformers and delta-wye transformers will act as zero-sequence traps when connected in the neutral circuit of a three-phase four-wire system. Figure 20.18 shows a delta-wye transformer serving single-phase nonlinear loads of switched-mode power supplies, PCs, printers, and fluorescent lighting. As discussed in Chapter 17, the neutral can carry excessive harmonic currents. A zig-zag or delta-wye transformers connected as shown in Figure 20.18 will reduce harmonic currents and voltage.

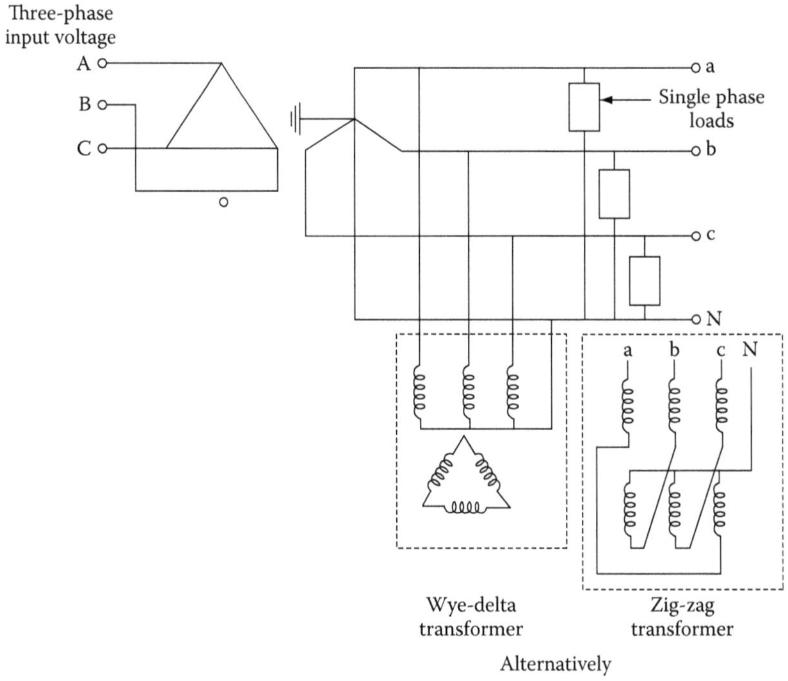

FIGURE 20.18
Delta-wye or zig-zag transformer used as neutral trap in a three-phase four-wire system serving nonlinear loads.

As we discussed in Chapter 2, the zero-sequence impedance of the core-type delta-wye transformer is low as the zero-sequence flux seeks a high reluctance path through air or the transformer tank. The delta winding carries the zero-sequence currents to balance the primary ampère turns. In an unbalanced system, the positive and negative sequence components will also be present, and these will not be suppressed. In a zig-zag transformer, all windings have the same number of turns, but each pair of windings on a leg is wound in the opposite direction. A zig-zag transformer has low zero-sequence impedance and works in the same manner as a delta-wye transformer.

Figure 20.18 shows that the three-phase four-wire system, with neutral solidly grounded, serves single-phase loads. The neutral currents have two parallel paths, both of low impedance, through the delta-wye or zig-zag transformer and also through the grounded neutral. The neutral voltage rise will be much less, though it will not be completely stable.

20.13 Limitations of Passive Filters

Passive filters have been widely applied to limit harmonic propagation, improve power quality, reduce harmonic distortion, and provide reactive power compensation simultaneously. These can be designed for large-current applications and high voltages. Many such filters are in operation for HVDC links, and passive filters are still the only choice when high voltages and currents are involved.

Some of the limitations of the passive filters are apparent from the examples [10,11]. These can be summarized as follows:

- Passive filters are not adaptable to the changing system conditions and once installed are *rigidly* in place. Neither the tuned frequency nor the size of the filter can be changed so easily. The passive elements in the filters are close tolerance components.
- A change in the system or operating condition can result in detuning and increased distortion. This can go undetected, unless there is online monitoring equipment in place.
- The design is largely affected by the system impedance. To be effective, the filter impedance must be less than the system impedance, and the design can become a problem for stiff systems. In such cases, a very large filter will be required. This may give rise to overcompensation of reactive power, and overvoltages on switching and undervoltages when out of service.
- Often, passive filters will require a number of parallel shunt branches. Outage of a parallel unit totally alters the resonant frequencies and harmonic current flows. This may increase distortion levels beyond permissible limits.
- Power losses in the resistance elements of passive filters can be very substantial for large filters.
- The parallel resonance between filter and the system (for single- or double-tuned filters) may cause amplification of currents of a characteristic or noncharacteristic harmonic. A designer has a limited choice in selecting the tuned frequency to avoid all possible resonances with the background harmonics. System changes

will alter this frequency to some extent, however carefully the initial design might have been selected.

- Damped filters do not give rise to a system parallel resonant frequency; however, these are not so effective as a group of ST filters. The impedance of a high-pass filter at its notch frequency is higher than the corresponding ST filter. The size of the filter becomes large to handle the fundamental and harmonic frequencies.
- The aging, deterioration, and temperature effects detune the filter in a random manner (though the effect of maximum variations can be considered in the design stage).
- If the converters feed back dc current into the system (with even harmonics), it can cause saturation of the filter reactor with resulting increase in distortion.
- Definite-purpose breakers are required. To control switching surges, special synchronous closing devices or resistor closing is required (see Chapter 4).
- The grounded neutrals of wye-connected banks provide a low-impedance path for the third harmonics. Third-harmonic amplification can occur in some cases.
- Special protective and monitoring devices (not discussed) are required.

20.14 Active Filters

By injecting harmonic distortion into the system, which is equal to the distortion caused by the nonlinear load, but of opposite polarity, the waveform can be corrected to a sinusoid. The voltage distortion is caused by the harmonic currents flowing in the system impedance. If a nonlinear current with opposite polarity is fed into the system, the voltage will revert to a sinusoid.

Active filters can be classified according to the way these are connected in the circuit [12,13]:

- In series connection
- In parallel shunt connection
- Hybrid connections of active and passive filters

20.14.1 Shunt Connection

As we have seen, the voltage in a weak system is very much dependent on current, while a stiff system of zero impedance will have no voltage distortion. Thus, provided that the system is not too stiff, a nonsinusoidal voltage can be corrected by injecting proper current. A harmonic current source is represented as a Norton equivalent circuit, and it may be implemented with a voltage-fed PWM inverter to inject a harmonic current of the same magnitude as that of the load into the system, but harmonics of opposite polarity. A shunt connection is shown in Figure 20.18a. The load current will be sinusoidal, so long as the load impedance is higher than the source impedance:

$$I_{\text{fl}} = I_{\text{h}} \quad \text{for} \quad |Z_{\text{L}}| \gg |Z_{\text{s}}| \quad I_{\text{h}} = 0 \tag{20.56}$$

In Chapter 17, we studied two basic types of converters—current and voltage. A converter with dc output reactor and constant dc current is a current harmonic source. A converter with a diode front end and dc capacitor has a highly distorted current depending on the ac source impedance, but the voltage at rectifier input is less dependent on ac impedance. This is a voltage harmonic source. It presents a low impedance and shunt connection will not be effective. A shunt connection is more suitable for current source controllers where the output reactor resists the change of current.

20.14.2 Series Connection

Figure 20.19b shows a series connection. A voltage V_f is injected in series with the line and it compensates the voltage distortion produced by a nonlinear load. A series active filter is more suitable for harmonic compensation of diode rectifiers where the dc voltage for the inverter is derived from a capacitor, which opposes the change of the voltage.

Thus, the compensation characteristics of the active filters are influenced by the system impedance and load. This is very much akin to passive filters; however, active filters have better harmonic compensation characteristics against the impedance variation and frequency variation of harmonic currents. The control systems of the active filters have a profound effect on the performance and a converter can have even a negative reactance. The active filters by themselves have the limitations that initial costs are high and do not constitute a cost-effective solution for nonlinear loads above approximately 1000 kW, though further developments will lower the costs and extend applicability.

20.14.3 Hybrid Connection

Hybrid connections of active and passive filters are shown in Figure 20.20. Figure 20.20a is a combination of shunt active and shunt passive filters. Figure 20.20b shows a combination of a series active filter and a shunt passive filter while Figure 20.20c shows an active filter in series with a shunt passive filter. The combination of shunt active and passive filters has already been applied to harmonic compensation of large steel mill drives [14]. The addition of a large shunt capacitor will reduce the load resistance and Equation 20.56 is no longer valid. The shunt passive filter will draw a large source current from a stiff system and may

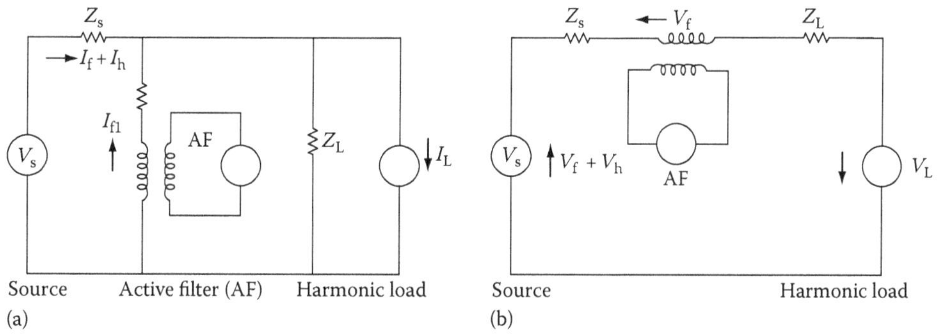

FIGURE 20.19
(a) Shunt connection of an active filter, (b) series connection of an active filter.

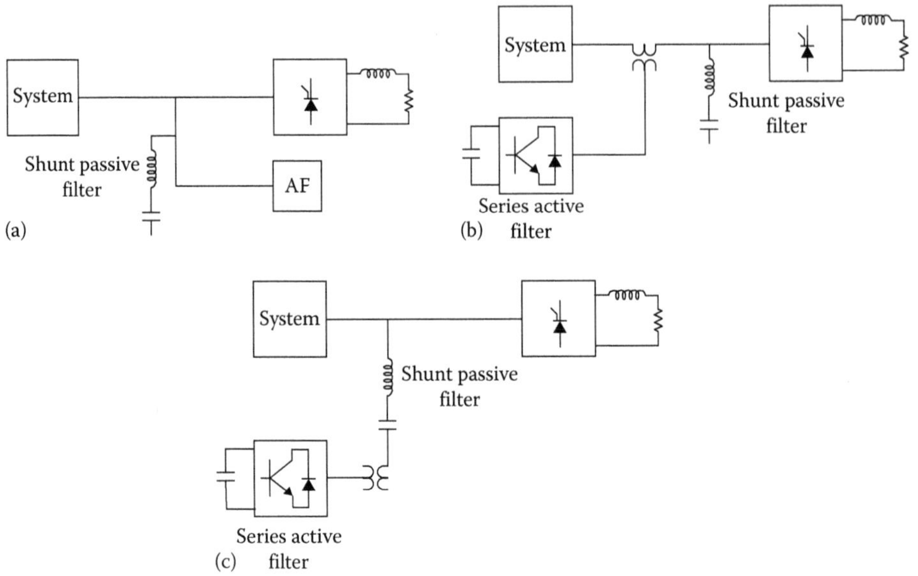

FIGURE 20.20
(a)–(c) Hybrid connections of active and passive filters.

act as a sink to the upstream harmonics. It is required that in a hybrid combination, the filters share compensation properly in the frequency domain.

In a series connection, the active filter is connected in series with the passive filter, both being in parallel with the load, as shown in Figure 20.20c. With suitable control of the active filter, it is possible to avoid resonance and improve filter performance. The active filter can be either voltage or current controlled. In current-mode control, the inverter is a voltage source to compensate for current harmonics. In voltage-mode control, the converter is a voltage-source inverter controlled to compensate for the voltage harmonics. The advantage is that the converter itself is far smaller, only about 5% of the load power. The active filter in such schemes regulates the effective source impedance as experienced by the passive filter, and the currents are forced to flow in the passive filter rather than in the system. This makes the passive filter characteristics independent of the actual source impedance and a consistent performance can be obtained.

20.14.4 Combination of Active Filters

A combination of series and shunt active filters is shown in Figure 20.21. This looks similar to the unified power controller discussed in Chapter 13, but its operation is different [14]. A series filter blocks harmonic currents flowing in and out of the distribution feeders. It detects the supply current and is controlled to present a zero impedance to the fundamental frequency and high resistance to the harmonics. The shunt filter absorbs the harmonics from the supply feeders and detects the bus voltage at the point of connection. It is controlled to present infinite impedance to the fundamental frequency and low impedance

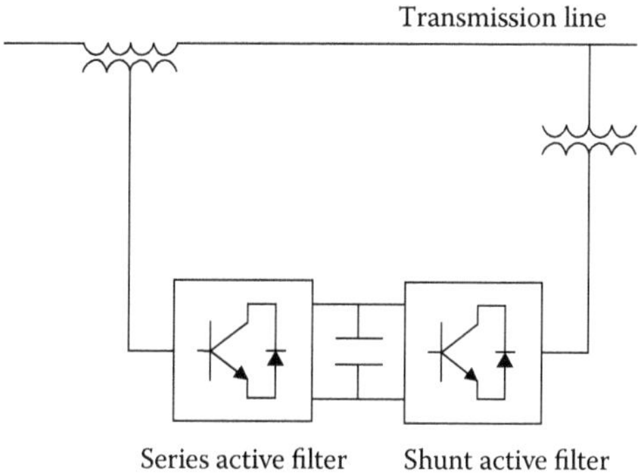

FIGURE 20.21
Connections of a unified power quality conditioner.

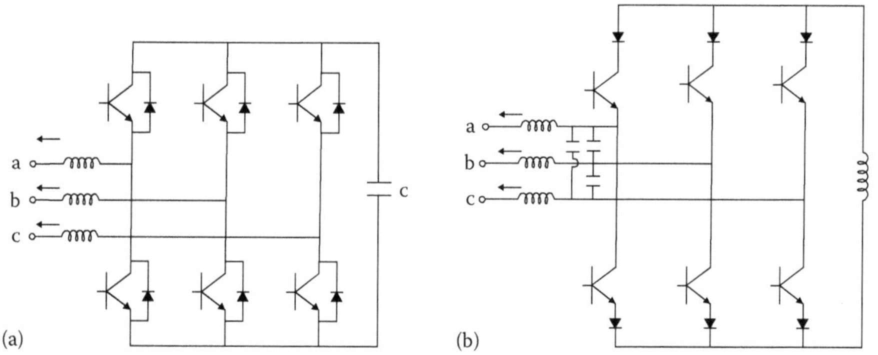

FIGURE 20.22
(a) Voltage-source inverter active filter; (b) current-source active inverter filter.

to the harmonics. The harmonic currents and voltages are extracted from the supply system in the time domain.

The electronics and power devices used in both types of converters for filters are quite similar. (Figure 20.22 shows three-phase voltage-source and current-source PWM converters.) The current-source active filter has a dc reactor with a constant dc current while the voltage-source active filter has a capacitor on the dc side with constant dc voltage. An output filter is provided to attenuate the inverter switching effects. In a current-source type, LC filters are necessary (Figure 20.22b). Transient oscillations can appear because of resonance between filter capacitors and inductors. The controls are implemented so that the inverter outputs a harmonic current equivalent but opposite to that of the load. The source side current is therefore sinusoidal, but the voltage will be sinusoidal only if the source does not generate any harmonics. Bipolar junction transistors are used with switching frequencies up to 50 kHz for modest ratings. SCRs and GTOs are used for higher power outputs.

A further classification is based on the control system, that is, time-domain and frequency-domain corrections.

20.15 Corrections in Time Domain

Corrections in the time domain are based on holding instantaneous voltage or current within reasonable tolerance of a sine wave. The error signal can be the difference between actual and reference waveforms. Time-domain techniques can be classified into three main categories [13]:

- Triangular wave
- Hysteresis
- Deadbeat

The error function can be instantaneous reactive power (IRP, described in Section 20.17) or EXT (extraction of fundamental frequency component). For EXT, the fundamental component of the distorted waveform is extracted through a 60 Hz filter and then the error function is $e(t) = f60(t) - f(t)$. For IRP, the error function is given by the difference between the instantaneous orthogonal transformation of *actual* and 60 Hz components of voltages and currents.

The triangular-wave method is easiest to implement and can be used to generate two-state or three-state switching functions. A two-state function can be connected positively or negatively, while a three-state function can be positive, negative, or zero (Figure 20.23).

In the two-state system, the inverter is always *on* (Figure 20.23a). The extracted error signal is compared to a high-frequency triangular carrier wave, and the inverter switches each time the waves cross. The result is an injected signal that produces equal and opposite distortion.

In a three-state system (hysteresis method), preset upper and lower limits are compared to an error signal (Figure 20.23b). So long as the error is within a tolerable band, there is no switching and the inverter is off.

The advantages of time-domain methods are fast response, though these are limited to one-node application, to which these are connected and take measurement from.

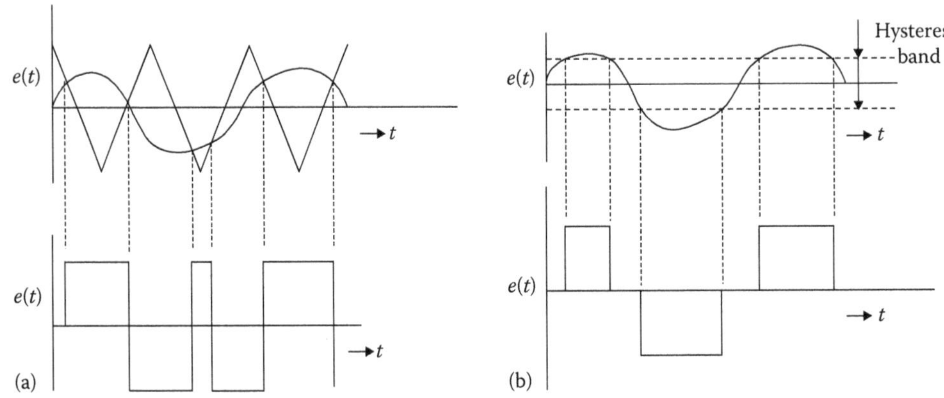

FIGURE 20.23
(a) Two-step switching function; (b) three-step switching function.

20.16 Corrections in the Frequency Domain

Fourier transformation is used to determine the harmonics to be injected. The error signal is extracted using a 60 Hz filter and the Fourier transform of the error signal is taken. The cancellation of M harmonics method allows for compensation up to the Mth harmonic, where M represents the highest harmonic to be compensated. A switching function is constructed by solving a set of nonlinear equations to determine the precise switching times and magnitudes. Quarter-wave symmetry is assumed to reduce the computations. Because an error function is used, the system can easily accommodate system changes but requires intense calculations and the time delays associated with it. The computations increase with M and the increased computational requirements are the main disadvantage, though these can be applied in dispersed networks.

The predetermined frequency method injects specific frequencies into the system, which are decided in the design stage of the system, much like passive harmonic filtering. This eliminates the need for real-time commutation of switching signals, but the harmonic levels present must be carefully evaluated beforehand and each filter is designed for the specific requirements. See also Refs. [15,16].

20.17 Instantaneous Reactive Power

The signal method of control generates an error signal based on input voltage or current and a reference sinusoidal waveform. A more elaborate function is the instantaneous power method, which calculates the desired current so that the instantaneous active and reactive power in a three-phase system is kept constant, that is, the active filter compensates for variation in instantaneous power [17]. By linear transformation the phase voltages e_a, e_b, e_c and load currents i_a, i_b, i_c are transformed into an α–β (two-phase) coordinate system:

$$\begin{vmatrix} e_\alpha \\ e_\beta \end{vmatrix} = \sqrt{\frac{2}{3}} \begin{vmatrix} 1 & -\frac{1}{2} & -\frac{1}{2} \\ 0 & \frac{\sqrt{3}}{2} & -\frac{\sqrt{3}}{2} \end{vmatrix} \begin{vmatrix} e_a \\ e_b \\ e_c \end{vmatrix} \tag{20.57}$$

and

$$\begin{vmatrix} i_\alpha \\ i_\beta \end{vmatrix} = \sqrt{\frac{2}{3}} \begin{vmatrix} 1 & -\frac{1}{2} & -\frac{1}{2} \\ 0 & \frac{\sqrt{3}}{2} & -\frac{\sqrt{3}}{2} \end{vmatrix} \begin{vmatrix} i_a \\ i_b \\ i_c \end{vmatrix} \tag{20.58}$$

The instantaneous real power p and the instantaneous imaginary power q are defined as

$$\begin{vmatrix} p \\ q \end{vmatrix} = \begin{vmatrix} e_\alpha & e_\beta \\ -e_\beta & e_\alpha \end{vmatrix} \begin{vmatrix} i_\alpha \\ i_\beta \end{vmatrix} \tag{20.59}$$

Here, p and q are not conventional watts and vars. p and q are defined by the instantaneous voltage in one phase and the instantaneous current in the other phase.

$$p = e_\alpha i_\alpha + e_\beta i_\beta = e_a i_a + e_b i_b + e_c i_c \tag{20.60}$$

To define IRP, the space vector of imaginary power is defined as

$$\begin{aligned} q &= e_\alpha i_\beta + e_\beta i_\alpha \\ &= \frac{1}{\sqrt{3}}[i_a(e_c - e_b) + i_b(e_a - e_c) + i_c(e_b - e_a)] \end{aligned} \tag{20.61}$$

Equation 20.59 can be written as

$$\begin{vmatrix} i_\alpha \\ i_\beta \end{vmatrix} = \begin{vmatrix} e_\alpha & e_\beta \\ -e_\beta & e_\alpha \end{vmatrix}^{-1} \begin{vmatrix} p \\ q \end{vmatrix} \tag{20.62}$$

These are divided into two kinds of currents:

$$\begin{vmatrix} i_\alpha \\ i_\beta \end{vmatrix} = \begin{vmatrix} e_\alpha & e_\beta \\ -e_\beta & e_\alpha \end{vmatrix}^{-1} \begin{vmatrix} p \\ 0 \end{vmatrix} + \begin{vmatrix} e_\alpha & e_\beta \\ -e_\beta & e_\alpha \end{vmatrix}^{-1} \begin{vmatrix} 0 \\ q \end{vmatrix} \tag{20.63}$$

This can be written as

$$\begin{vmatrix} i_\alpha \\ i_\beta \end{vmatrix} = \begin{vmatrix} i_{\alpha p} \\ i_{\beta p} \end{vmatrix} + \begin{vmatrix} i_{\alpha q} \\ i_{\beta q} \end{vmatrix} \tag{20.64}$$

where i_{ap} is the α-axis instantaneous active current:

$$i_{\alpha p} = \frac{e_\alpha}{e_\alpha^2 + e_\beta^2} p \tag{20.65}$$

$i_{\alpha q}$ is the α-axis instantaneous reactive current:

$$i_{\alpha q} = \frac{-e_\beta}{e_\alpha^2 + e_\beta^2} q \tag{20.66}$$

$i_{\beta p}$ is the β-axis instantaneous active current:

$$i_{\beta p} = \frac{e_\alpha}{e_\alpha^2 + e_\beta^2} p \tag{20.67}$$

and $i_{\beta q}$ is the β-axis instantaneous reactive current:

$$i_{\beta q} = \frac{e_\alpha}{e_\alpha^2 + e_\beta^2} q \tag{20.68}$$

The following equations exist:

$$\begin{aligned} p &= e_\alpha i_{\alpha P} + e_\beta i_{\beta P} \equiv P_{\alpha P} + P_{\beta P} \\ 0 &= e_\alpha i_{\alpha q} + e_\beta i_{\beta q} \equiv P_{\alpha q} + P_{\beta q} \end{aligned} \tag{20.69}$$

where the α-axis instantaneous active and reactive powers are

$$P_{\alpha p} = \frac{e_\alpha^2}{e_\alpha^2 + e_\beta^2} p \quad P_{\alpha q} = \frac{-e_\alpha e_\beta}{e_\alpha^2 + e_\beta^2} q \tag{20.70}$$

The β-axis instantaneous active and reactive power is

$$P_{\beta q} = \frac{e_\beta^2}{e_\alpha^2 + e_\beta^2} p \quad P_{\beta q} = \frac{e_\alpha e_\beta}{e_\alpha^2 + e_\beta^2} q \tag{20.71}$$

The sum of the instantaneous active powers in two axes coincides with the instantaneous real power in the three-phase circuit. The IRPs P_{aq} and $P_{\beta q}$ cancel each other and make no contribution to the instantaneous power flow from the source to the load.

Consider instantaneous power flow in a three-phase cycloconverter. The IRP on the source side is the IRP circulating between source and cycloconverter while the IRP on the output side is the IRP between the cycloconverter and the load. Therefore, there is no relationship between the IRPs on the input and output sides, and the instantaneous imaginary power on the input side is not equal to the instantaneous imaginary power on the output side. However, assuming zero active power loss in the converter, the instantaneous real power on the input side is equal to the real output power.

20.18 Harmonic Mitigation at Source

The harmonic mitigation at source, without filters, has attracted the attention of industry and researchers. This covers a wide field, spanning industrial applications to transmission systems and HVDC. We will briefly look at four systems from the harmonic mitigation viewpoint.

20.18.1 Phase Multiplication

The principle of harmonic elimination by phase multiplication is discussed in Chapter 17. Figure 20.24 shows a 2300 V medium-voltage drive system, where each motor phase is

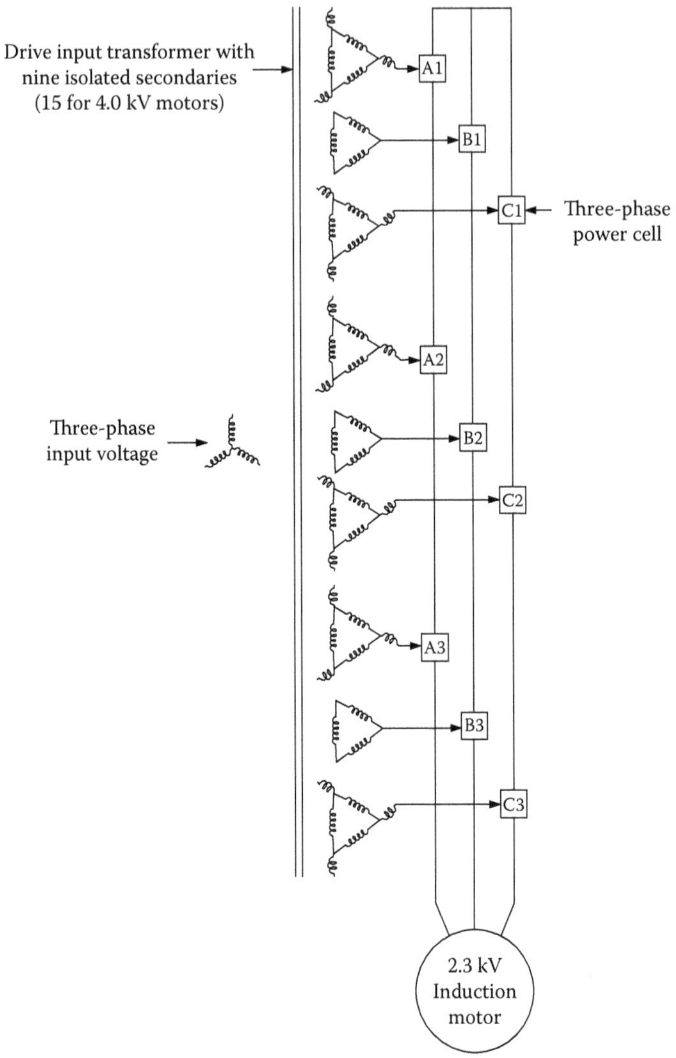

FIGURE 20.24
An ac drive system with secondary-phase multiplication and low harmonic distortion. (Courtesy of Robicon Corporation, New Kensington, PA.)

driven by three PWM cells. Each group of power cells is wye connected with a floating neutral and is powered by an isolated secondary winding of the drive input transformer. A greatly improved voltage waveform is obtained due to phase displacements in the transformer secondary windings, and the harmonic distortion meets IEEE limits [1] without filters. Another advantage is that the common mode voltages are eliminated.

20.18.2 Parallel Connected 12-Pulse Converters, with Interphase Reactor

Figure 20.25a shows the circuit of a conventional 12-pulse thyristor converter with interphase reactor and phase shift obtained through delta–delta and delta–wye input transformers. Figure 20.25b is a conventional stepped waveform of the 12-pulse converter. This can be rendered close to a sinusoid by superposition of a triangular current as shown in Figure 20.25c and the system has a better waveform than that of a 36-pulse thyristor converter [18].

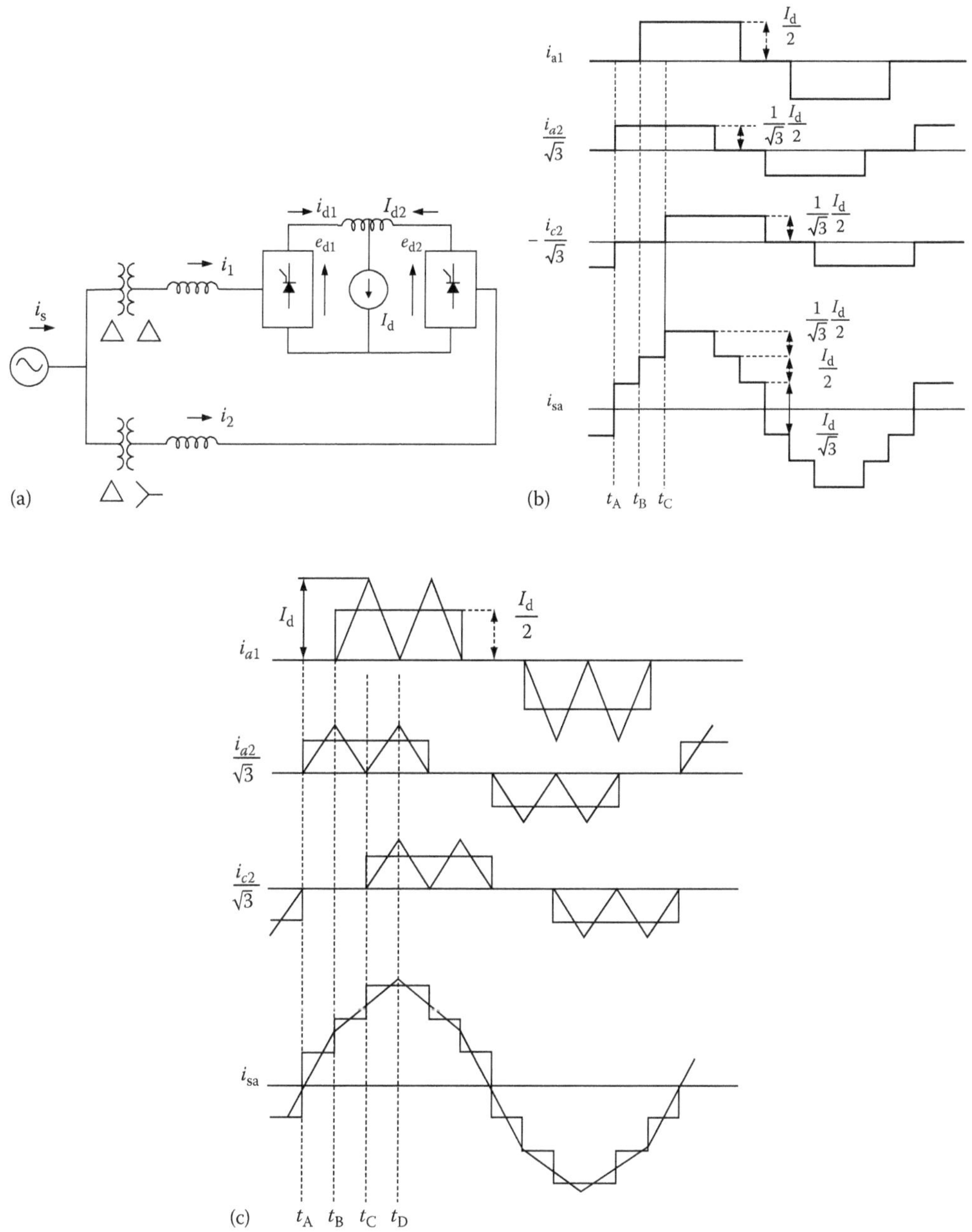

FIGURE 20.25
(a) A 12-pulse converter circuit with interphase reactor; (b) stepped voltage generation of a 12-pulse circuit; (c) improvement in voltage waveform with triangular-wave superimposition.

20.18.3 Active Current Shaping

By using proper control systems, the input current of converters can be forced to follow a sinusoid in phase with voltage, addressing the need for reactive power compensation as well as harmonic elimination [14]. The load current can be written as

$$I_L(t) = K\nu(t) + i_q(t) \tag{20.72}$$

where

$K\nu(t)$ is the active component of the load current
K is a coefficient that can be calculated in the control circuit
$i_q(t)$ is the nonactive component of the current

The nonactive current must be compensated to have maximum power factor and harmonic rejection. The desired reference current in the active filter is

$$I_q(t) = I_L(t) - k\nu(t) \tag{20.73}$$

20.19 Multilevel Converters

Multilevel converters are a new breed [19], which overcome some of the disadvantages of PWM inverters, namely,

The carrier frequency is high, usually between 2 and 20 kHz. In a normal PWM waveform, the pulse height is the dc-link voltage, dv/dt is high, and causes a strong EMI (see Figure 18.2). This introduces a number of harmonics and possible ringing. PWM needs rigorous switching conditions and resultant switching losses. The inverter control circuitry is complex.

The multilevel inverters emerged as a solution to high-power applications. The switching frequencies are low, equal to or only a few times the output frequency. The pulse heights are low. For an m-level inverter, the pulse height is V_m/m where V_m is the output voltage amplitude. This results in much smaller dv/dt and EMI as compared to PWM inverters. The harmonics and THD are further reduced. Smooth switching conditions are obtained with much lower switching power losses.

Mutilevel converters have been applied to HVDC, large motor drives, railway traction applications, UPFC, STATCON, and SVC (Chapter 13). The quality of output voltage increases as the number of voltage levels increases. The applications have been extended to active power filters, voltage sag compensators, and photovoltaic systems.

The various types of multilevel converters are as follows:

- Diode-clamped multilevel inverter (DCMI) was proposed by Nabae in 1980 [20]. It is also called the neutral point clamped (NPC) inverter, because the NPC inverter effectively doubles the device voltage without precise voltage matching.
- Capacitor-clamped (flying capacitors) multilevel inverters appeared in 1990s.
- Cascaded multilevel inverters with separate dc sources (CMIs) applications prevailed in mid-1990s for motor drives and utility applications. It has drawn great

interest for medium-voltage high-power inverters. It is also used for regenerative-type motor applications [21].

- Recently some new topologies of multilevel inverters have emerged, such as generalized multilevel inverters, hybrid multilevel inverters, and soft-switched multilevel inverters. These are applied at medium-voltage levels for mills, conveyors, pumps, fans, compressors, and so on. The applications are also extended to low-power applications [22,23].

Figure 20.26a shows a single-phase, five-level diode clamp circuit with four dc bus capacitors, C_1, C_2, C_3, and C_4. The staircase voltage wave is synthesized from several levels

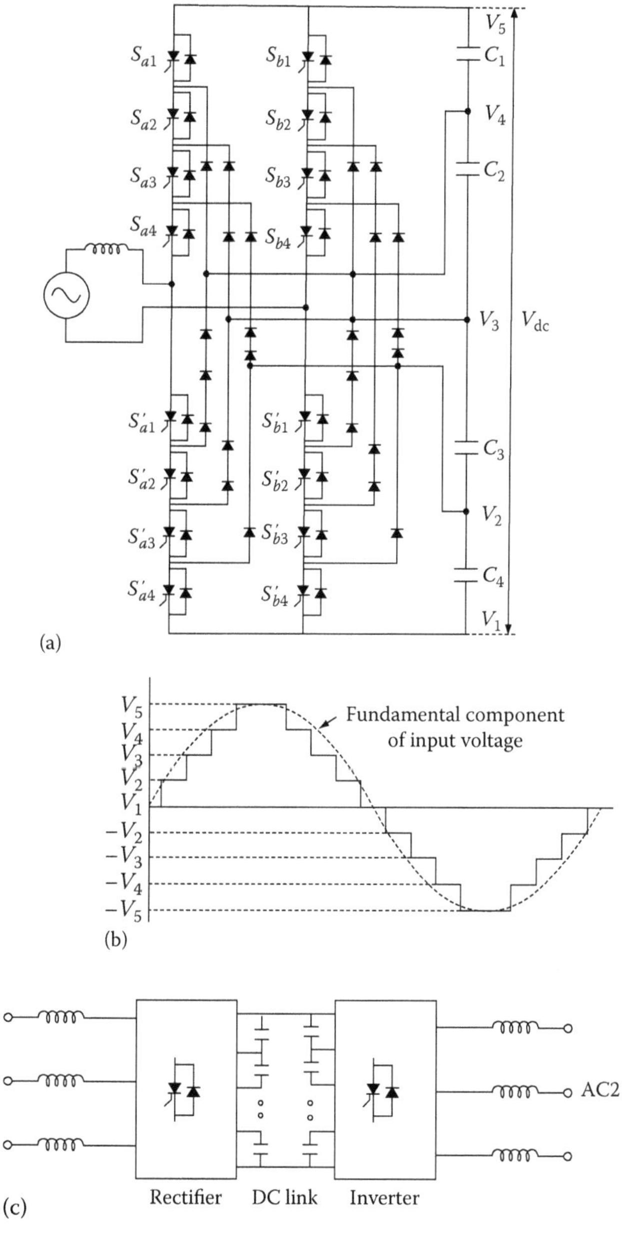

FIGURE 20.26
(a) Single-phase, full-bridge, five-level diode-clamp converter circuit; (b) stepped voltage generation resulting in low harmonic distortion; (c) two-diode clamp multilevel converters for back-to-back intertie system.

of dc capacitor voltages. An m-level diode clamp converter consists of $m-1$ capacitors on the dc bus and produces m levels of the phase voltage by appropriate switching. The voltage across each capacitor is $V_{dc}/4$. The staircase voltage shown in Figure 20.26b is generated by the five switch combinations shown in Figure 20.26a and the switching matrix shown below:

$$\begin{array}{ccccccccc} & S_{a1} & S_{a2} & S_{a3} & S_{a4} & S'_{a1} & S'_{a2} & S'_{a3} & S'_{a4} \\ V_4 & 1 & 1 & 1 & 1 & 0 & 0 & 0 & 0 \\ V_3 & 0 & 1 & 1 & 1 & 1 & 0 & 0 & 0 \\ V_2 & 0 & 0 & 1 & 1 & 1 & 1 & 0 & 0 \\ V_1 & 0 & 0 & 0 & 1 & 1 & 1 & 1 & 0 \\ V_0 & 0 & 0 & 0 & 0 & 1 & 1 & 1 & 1 \end{array} \quad (20.74)$$

With high switching levels, the harmonic content is low enough and filters are not needed. The disadvantages are large clamping diodes, unequal switching ratings, and real power control. The clamping diodes can be replaced with capacitors called flying capacitor-based control or multilevel converters using cascade converters with dc sources. An application in a back-to-back intertie connection is shown in Figure 20.26c. The resulting harmonic distortion is within IEEE limits without filters [19].

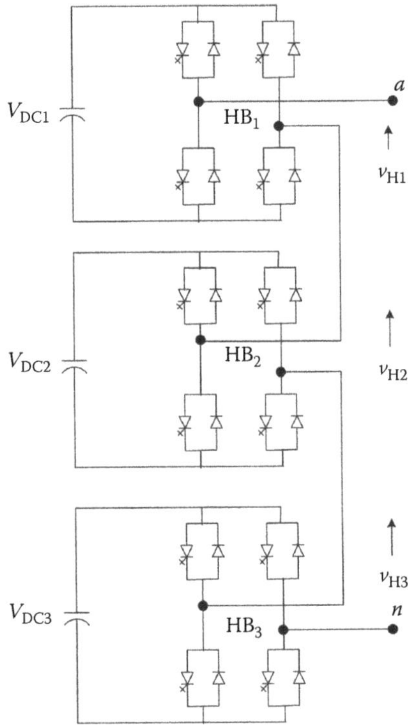

FIGURE 20.27
Multilevel inverter based on series connections of HBs.

TABLE 20.12

Operation of Multilevel Inverter (Figure 20.27)

E	v_{H1}	v_{H2}	v_{H3}
0	0	0	0
$+1E$	E	0	0
$+2E$	0	$2E$	0
$+3E$	E	$2E$	0
$+4E$	0	0	$4E$
$+5E$	E	0	$4E$
$+6E$	0	$2E$	$4E$
$+7E$	E	$2E$	$4E$
$-E$	$-E$	0	0
$-2E$	0	$-2E$	0
$-3E$	$-E$	$-2E$	0
$-4E$	0	0	$-4E$
$-5E$	$-E$	0	$-4E$
$-6E$	0	$-2E$	$-4E$
$-7E$	$-E$	$-2E$	$-4E$

Figure 20.27 shows the basic structure of multilevel inverters using H-Bridge converters. This shows one-phase leg with three HBs. In binary hybrid multilevel inverter, the dc-link voltage of HBi is

$$V_{DCi} = 2^{i-1}E \tag{20.75}$$

that is, for a three-level inverter of Figure 20.27, $V_{DC1} = E$, $V_{DC2} = 2E$, and $V_{DC3} = 4E$. The operation is listed in Table 20.12 and Figure 20.28 shows that the positive half output waveform has 15 levels. The negative half is identical. Note that the HB with higher dc link voltage has a lower number of commutations.

In a trinary hybrid inverter (THMI), the dc line voltages are

$$V_{DCi} = 3^{i-1}E \tag{20.76}$$

that is, for three-HB one phase leg, $V_{DC1} = E$, $V_{DC2} = 3E$, and $V_{DC3} = 9E$. The output voltage, as before is

$$v_{an} = \sum_{i=1}^{h} v_{Hi} = \sum_{i=1}^{h} F_i V_{DCi} \tag{20.77}$$

where F_i is a switching function. It can be shown that THMI has a greatest number of output voltage levels, and by proper switching angles, the odd and even harmonics can practically be eliminated. See Ref. [23].

The above techniques form an introduction, references are provided for the interested reader to probe further.

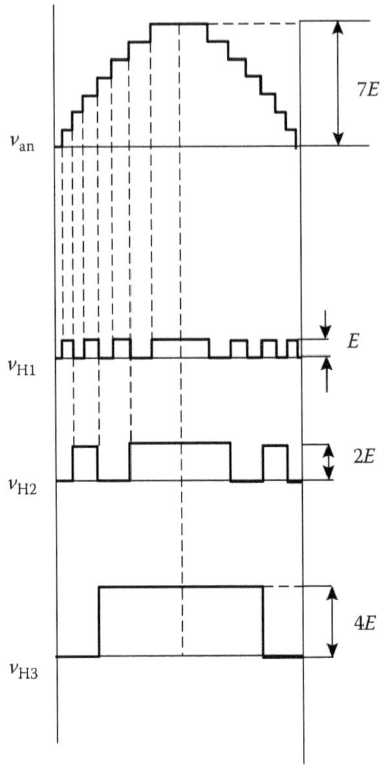

FIGURE 20.28
Positive half of waveform of BHMI, 15 levels.

References

1. IEEE. IEEE recommended practice and requirements for harmonic control in electrical systems, 1992. Standard 519.
2. IEC. Electromagnetic compatibility—Part 3: Limits—Section 2: Limits for harmonic current emission (equipment input current #16A per phase), 1995. Standard 61000-3-2.
3. E.W. Kimbark. *Direct Current Transmission*, Chap. 8. New York: John Wiley, 1971.
4. J.C. Das. *Transients in Electrical Systems—Recognition Analysis and Mitigation*. New York: McGraw Hill, 2010.
5. IEEE. IEEE guide for protection of the capacitor banks, 2000. Standard C37.99.
6. IEEE. IEEE guide for application and specifications of harmonic filters, 2003. Standard 1531.
7. J. Zaborszky and J.W. Rittenhouse. Fundamental aspects of some switching overvoltages in power systems. *IEEE Trans.* PAS 1727–1734, 1982.
8. J. Arrillaga, D.A. Bradley, and P.S. Bodger. *Power System Harmonics*. New York: John Wiley, 1985.
9. J.D. Anisworth. Filters, damping circuits and reactive voltamperes in HVDC converters. In: *High Voltage Direct Current Converters and Systems*. London, U.K.: Macdonald, 1965.
10. J.C. Das. Analysis and control of harmonic currents using passive filters. In: *TAPPI Proceedings, Atlanta Conference*, Atlanta, GA, 1999, pp. 1075–1089.
11. J.C. Das. Passive filters-potentialities and limitations. *IEEE Trans. Ind. Appl.* 40(1): 232–241, January/February 2004.
12. H. Akagi. Trends in active power line conditioners. *IEEE Trans. Power Electron.* 9: 263–268, 1994.
13. W.M. Grady, M.J. Samotyi, and A.H. Noyola. Survey of active line conditioning methodologies. *IEEE Trans. Power Deliv.* 5: 1536–1541, 1990.

14. H. Akagi. New trends in active filters for power conditioning. *IEEE Trans. Ind. Appl.* 32: 1312–1322, 1996.
15. A. Cavallini and G.C. Montanarion. Compensation strategies for shunt active filter control. *IEEE Trans. Power Electron.* 9: 587–593, 1994.
16. C.V. Nunez-Noriega and G.G. Karady. Five step-low frequency switching active filter for network harmonic compensation in substations. *IEEE Trans. Power Deliv.* 14: 1298–1303, 1999.
17. H. Akagi and A. Nabe. The p–q theory in three-phase systems under non-sinusoidal conditions. *ETEP* 3: 27–30, 1993.
18. T. Tanaka, N. Koshio, H. Akagi, and A. Nabae. Reducing supply current harmonics. *IEEE Ind. Appl. Mag.* 4: 31–35, 1998.
19. J.S. Lai and F.Z. Peng. Multilevel converters—A new breed of power converters. *IEEE Trans. Ind. Appl.* 32: 509–517, 1996.
20. A. Nabae, I. Takahashi, and H. Akagi. A neutral point clamped PWM inverter. *IEEE Trans. Ind. Appl.* 17: 518–523, 1981.
21. P.W. Hammond. New approach to enhance power quality for medium voltage AC drives. *IEEE Trans. Ind. Appl.* 33: 202–208, 1997.
22. A.M. Trzymadlowski. *Introduction to Modern Power Electronics*. New York: John Wiley, 1998.
23. F.L. Luo and H. Ye. *Power Electronics-Advanced Conversion Technologies*. Boca Raton, FL: CRC Press, 2010.

21

Arc Flash Hazard Analysis

In the past, industrial electrical systems in the United States were designed considering prevalent standards, i.e., ANSI/IEEE, NEC, OSHA, UL, NESC, and the like, and arc flash hazard was not a direct consideration for the electrical system designs. This environment is changing fast, and the industry is heading toward innovations in the electrical systems designs, equipment, and protection to limit the arc flash hazard, as it is detrimental to the workers' safety. This opens another chapter of the power system design, analysis, and calculations, hitherto not required. There is a spate of technical literature and papers on arc flash hazard, its calculation, and mitigation. References [1–8] describe arcing phenomena and arc flash calculations, sometimes commenting on the equations in IEEE standard 1584 [9]. This chapter addresses these issues, which have become of great importance in the power system planning, designs, and protective relay applications.

Arc flash is a dangerous condition associated with the unexpected release of tremendous amount of energy caused by an electric arc within electrical equipment [10]. This release is in the form of intense light, heat, and blast of arc products that may consist of vaporized components of enclosure material—copper, steel, or aluminum. Intense sound and pressure waves also emanate from the arc flash that resembles a confined explosion. Arcing occurs when the insulation between the live conductors breaks down, due to aging, surface tracking, treeing phenomena, and human error while maintaining electrical equipment in the energized state. The insulation systems are not perfectly homogeneous, and voids form due to thermal cycling. In non-self-restoring insulations, the treeing phenomenon starts with a discharge in a cavity, which enlarges over a period of time, and the discharge patterns resemble tree branches, hence the name "treeing" (Figure 21.1). As the treeing progresses, discharge activity increases, and ultimately the insulation resistance may be sufficiently weakened and breakdown occurs under electrical stress. The treeing phenomenon is of particular importance in XLPE and non-self-restoring insulations. Surface tracking occurs due to abrasion, irregularities, contamination, and moisture, which may lead to an arc formation between the line and ground. An example will be a contaminated insulator under humid conditions. Although online monitoring and partial discharge measurements are being applied as diagnostic tools, the randomness associated with a fault and insulation breakdown are well recognized, and a breakdown can occur at any time, jeopardizing a worker's safety, who may be in close proximity of the energized equipment. Arc temperatures are on the order of 35,000°F, about four times the temperature on the surface of the sun. An arc flash can therefore cause serious fatal burns.

21.1 Relating Short-Circuit Currents with Arc Flash and Personal Safety

The phenomenal progress made by the electrical and electronic industry since Thomas Edison propounded the principle of incandescent lighting in 1897 has sometimes been achieved at the cost of loss of human lives and disabilities. Although reference to electrical

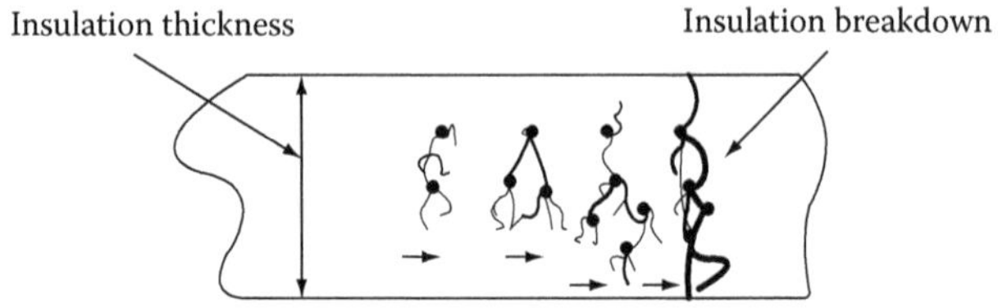

FIGURE 21.1
Progressive treeing in non-self-restoring insulation leading to breakdown.

safety can be found as early as about 1888, it was only in 1982 that Ralph Lee [11] correlated arc flash and body burns with short-circuit currents.

OSHA definition of a recordable injury, TRIR, for 1 year of exposure, is as follows:

$$\text{TRIR} = \frac{\text{Total number of recordable injuries and accidents}}{200{,}000\ \text{h}} \tag{21.1}$$

Most insurance companies accept this parameter of definition because there is a cost associated with these incidents.

An arcing phenomenon is associated with other hazards too, namely, arc blast, fire hazard, and shock hazard.

21.1.1 Arc Blast

As opposed to arc flash, which is associated with thermal hazard and burns, arc blast is associated with extreme pressure and rapid pressure buildup. Consider a person positioned directly in front of an event and high pressure impinging upon the chest and close to heart and the hazard associated with it.

A substance requires a different amount of physical space when it changes state, say from solid to vaporized particles. When the liquid copper evaporates, it expands 67,000 times. In documented instances, a motor terminal box exploded as a result of force created by the pressure buildup, parts flying across the room [12]. Pressure measurement of 2160 lb/ft^2 around the chest area and sound level of 141.5 dB at 2 ft have been made.

21.1.2 Fire Hazard and Electrical Shock

Fire hazard [13,14] and electrical shock are the other hazards. There are a number of ways the exposure to shock hazard occurs. The resistance of the contact point, the insulation of the ground under the feet, flow of current path through the body, the body weight, and the system voltage level are all important. The threshold perception level is 0.0010–0.0007 A, and, at about 40 mA, shock duration of 1 s can cause ventricular fibrillation and can be fatal. On average, one person is electrocuted in work places *every day in the United States*.

21.1.3 Time Motion Studies

Of necessity and for the continuity of processes, maintenance of electrical equipment in energized state has to be allowed for. If all maintenance work could be carried out in de-energized state, a short circuit cannot occur, and therefore there is no risk of arc flash hazard. However, the continuous process plants, where the shutdown of a process can result in colossal amount of loss, downtime, and restarting, it becomes necessary to maintain the equipment in the energized state. Prior to the institution of arc flash standards, this has been carried out for many years jeopardizing worker safety, and there are documented cases of injuries including fatal burns.

The time/motion studies show that human reaction time to sense, judge, and runaway from a hazardous situation varies form person to person. A typical time is on the order of 0.4 s (approximately 25 cycles). This means that 24 cycles is the shortest time in which a person can view a condition and BEGIN to move or act. In all other conditions, it is not possible to see a hazardous situation and move away from it. As will be further demonstrated, this reaction time is too large for a worker to move away and shelter himself from an arc flash hazard situation.

Thermal burns due to arc flash are only a part the picture for overall worker safety, training and establishing sound maintenance procedures and guidelines. Figure F.1 in [15] provides hazard/risk analysis procedure flowchart. It implies that each establishment must perform a number of tasks, training, and safety procedures that should be implemented for workers' safety. This chapter is confined to the analysis of arc flash thermal damage and calculation of arc flash boundary, subsequently defined.

21.2 Arc Flash Hazard Analysis

Currently, there are two major standards for arc flash calculations:

1. NFPA 70E, revised in 2009 [15]
2. IEEE 1584, 2000, which will undergo revisions [9]
3. IEEE 1584a, 2004, amendment 1 [16]
4. IEEE P1584b/D2 Draft 2, unapproved [17].

NFPA 70E, year 2009, in Annex D, Table D.1 provides limitations of various calculation methods. This is reproduced in Table 21.1. *The standard does not express any preference as to*

TABLE 21.1

Limitations of Calculation Methods

Source	Limitations/Parameters
Ralph Lee paper [11]	Calculates arc flash protection boundary for arc in open air; conservative over 600 V and becomes more conservative as voltage rises
Doughty/Neal paper [18]	Calculated incident energy for three-phase arc on systems rated 600 V and below applies to short-circuit currents between 16 and 50 kA
Raplph Lee paper [11]	Calculated incident energy for three-phase arc in open air on systems rated above 600 V becomes more conservative as voltage rises
IEEE Std. 1584 [9]	Calculates incident energy and arc flash boundary for 208 V–15 kV, three-phase 50–60 Hz; 700–106,000 A short-circuit currents and 13–152 mm conductor gaps[a]
ANSI/IEEE C2 [19]	Calculates incident energy for open air phase-to-ground arcs 1–500 kV for live line work

Source: NFPA-70-National electric code-2009.

[a] Note that IEEE 1584 contains a theoretically derived model applicable for any voltage.

which method should be used. Reference [17] recognizes the use of knowledge and experience of those who have performed studies as a guide in applying the standard.

It is recognized that to construct an *accurate* mathematical model of the arcing phenomenon is rather impractical. This is because of spasmodic nature of the fault caused by arc elongation blowout effects, physical flexing of cables and bus bars under short circuits, possible arc reignition, turbulent flow of plasma, high temperature gradients (the temperature at the core being on the order of 25,000 K while at the arc boundary on the order of 300–2000 K), and conductivity factors changing at a fast rate. For the arc interruption phenomenon in circuit breakers, Casey-Mayr model is frequently used and is considered fairly accurate (Chapter 4).

IEEE 1584 equations are empirical equations based upon laboratory test results, though the standard includes some Lee's equations too.

21.2.1 Ralph Lee's Equations

Ralph Lee equations from Ref. [11] are as follows:

Maximum power in a three-phase arc is

$$P = \mathrm{MVA_{bf}} \times 0.707^2 \ \mathrm{MW} \tag{21.2}$$

where $\mathrm{MVA_{bf}}$ is bolted fault MVA.

The distance in feet of a person from arc source for a just curable burn, i.e., skin temperature remains less than 80°C is

$$D_c = (2.65\mathrm{MVA_{bf}}t)^{1/2} \tag{21.3}$$

where t is the time of exposure (s).

The equation for the incident energy produced by a three-phase arc in open air on systems rated above 600 V is given by

$$E = \frac{793FVt_A}{D^2} \ \mathrm{cal/cm^2} \tag{21.4}$$

where

D is the distance from the arc source (in.)
F is the bolted fault short-circuit current (kA)
V is the system phase-to-phase voltage (kV)
t_A is the arc duration (s)

For the low-voltage systems of 600 V or below and for an arc in the open air, the estimated incident energy is

$$E_{MA} = 5271D_A^{-1.9593}t_A[0.0016F^2 - 0.0076F + 0.8938] \tag{21.5}$$

where

E_{MA} is the maximum open air incident energy (cal/cm^2)
F is short-circuit current (kA), range 16–50 kA
D_A is distance from arc electrodes (in.) (for distances 18 in. and greater)

The estimated energy for an arc in a cubic box of 20 in. open on one side is given by

$$E_{MB} = 1038.7D_B^{-1.4738}t_A[0.0016F^2 - 0.0076F + 0.8938] \quad (21.6)$$

where
E_{MB} is the incident energy
D_B is the distance from arc electrodes (in.) (for distances 18 in. and greater)

21.2.2 IEEE 1584 Equations

The IEEE equations are applicable for the electrical systems operating at 0.208 to 15 kV, three-phase, 50 or 60 Hz, available short-circuit current range 700–106,000 A, and conductor gap = 13–152 mm. For three-phase systems in open-air substations, open-air transmission systems, a theoretically derived model is available. For system voltage *below* 1 kV, the following equation is solved:

$$\log I_a = K + 0.662\log_{10} I_{bf} + 0.0966V + 0.000526G + 0.5588V(\log_{10} I_{bf}) - 0.00304G(\log_{10} I_{bf}) \quad (21.7)$$

where
I_a is the arcing current (kA)
G is the conductor gap, typical conductor gaps are specified in [9]
K is the −0.153 for open air arcs, −0.097 for arc in a box
V is the system voltage (kV)
I_{bf} is the bolted three-phase fault current (kA), rms symmetrical

For systems of 1 kV and higher solve, the following equation is solved:

$$\log_{10} I_a = 0.00402 + 0.983\log_{10} I_{bf} \quad (21.8)$$

This expression is valid for arcs both in open air and in a box. Use $0.85I_a$ to find a second arc duration. This second arc duration accounts for variations in the arcing current and the time for the overcurrent device to open. Calculate incident energy using both $0.85I_a$ and I_a and use the higher value.

Incident energy at working distance, an empirically derived equation, is given by

$$\log_{10} E_n = K_1 + K_2 + 1.081\log_{10} I_a + 0.0011G \quad (21.9)$$

The equation is based upon data normalized for an arc time of 0.2 s, where

E_n = incident energy (J/cm^2) normalized for time and distance
K_1 = 0.792 for open air and −0.555 for arcs in a box
K_2 = 0 for ungrounded and high-resistance-grounded systems and −0.113 for grounded systems

Note that resistance grounded, HR grounded, and ungrounded systems are all considered ungrounded for the purpose of the calculation of incident energy.

G is the conductor gap (mm), Table 21.4.
Conversion from normalized values gives the equation

$$E = 4.184 C_f E_n \left(\frac{t}{0.2}\right)\left(\frac{610^x}{D^x}\right) \tag{21.10}$$

where
E is the incident energy (J/cm^2)
C_f is the calculation factor = 1.0 for voltages above 1 kV and 1.5 for voltages at or below 1 kV
t is the arcing time (s)
D is the distance from the arc to the person, working distance, Table 21.2
x is the distance exponent as given in [9] and reproduced in Table 21.3

A theoretically derived equation can be applied for voltages above 15 kV or when gap is outside the range in the Table 21.4 from [9].

TABLE 21.2

Classes of Equipment and Typical Working Distances

Classes of Equipment	Typical Working Distance (mm)
15 kV switchgear	910
5 kV switchgear	910
Low-voltage switchgear	610
Low-voltage MCCs and panelboards	455
Cable	455
Other	To be determined in field

Source: IEEE, IEEE guide for performing arc-flash hazard calculations, Standard 1584, 2002.

TABLE 21.3

Factors for Equipment and Voltage Classes

System Voltage (kV)	Equipment Type	Typical Gap between Conductors	Distance × Factor
0.208–1	Open air	10–40	2.000
	Switchgear	32	1.473
	MCC and panels	25	1.641
	Cable	13	2.000
>1–5	Open air	102	2.000
	Switchgear	13–102	0.973
	Cable	13	2.000
>5–15	Open air	13–153	2.000
	Switchgear	153	0.973
	Cable	13	2.000

Source: IEEE, IEEE guide for performing arc-flash hazard calculations, Standard 1584, 2002.

TABLE 21.4

Classes of Equipment and Typical Bus Gaps

Classes of Equipment	Typical Bus Gaps (mm)
15 kV switchgear	153
5 kV switchgear	104
Low-voltage switchgear	32
Low-voltage MCCs and panelboards	25
Cable	13
Other	Not required

$$E = 2.142 \times 10^6 V I_{\text{bf}} \left(\frac{t}{D^2}\right) \tag{21.11}$$

The arc flash protection boundary, defined further, empirically derived equation is

$$D_{\text{B}} = \left[4.184 C_{\text{f}} E_{\text{n}} \left(\frac{t}{0.2}\right)\left(\frac{610^x}{E_{\text{B}}}\right)\right]^{1/x} \tag{21.12}$$

where E_{B} is the incident energy (J/cm^2) at the distance of arc flash protection boundary.

For the empirically derived equation,

$$D_{\text{B}} = \left[2.142 \times 10^6 V I_{\text{bf}} \left(\frac{t}{E_{\text{B}}}\right)\right]^{1/2} \tag{21.13}$$

Due to complexity of IEEE equations, the arc flash analysis is run on digital computers. Most commercially available programs analyze arc flash hazard as a subroutine to short-circuit calculations. It is obvious that the incident energy release and the consequent hazard depend upon the following:

- The available three-phase rms symmetrical short-circuit current. In low-voltage systems, the arc flash current will be 50%–60% of the bolted three-phase current, due to arc voltage drop. In medium and high-voltage systems, it will be only slightly lower than the bolted three-phase current. The short-circuit currents are accompanied by a dc component, whether it is the short circuit of a generator, motor, or utility source. However, for arc flash hazard calculations, the *dc component is ignored*. Also, any unsymmetrical fault currents, such as line-to-ground fault currents need not be calculated. As evident from the cited equations, only three-phase bolted fault current needs to be calculated.
- The time duration for which the event lasts. This is obviously the sum of protective relay (or any other protection device) operating time plus the opening time of the switching device. For example, if the relay operating time is 20 cycles and the interrupting time of the circuit breaker is 5 cycles, then the arc flash time is 25 cycles.
- The type of equipment, that is, switchgear or MCC and the operating voltage.
- The system grounding.

21.3 Hazard/Risk Categories

NFPA [Table 130.7(C)(11)] describes the personal protective equipment (PPE) characteristics for hazard/risk category of 0 and 1 through 4. These are shown in Table 21.5. There is a slight revision with respect to 2004 edition.

The standard ASTM F1506 [20] calls for every flame resistance garment to be labeled with an arc energy rating, ATPV (arc thermal performance exposure value). The rating of the garment is matched with the calculated incident energy release level. The test method of determining the ATPV states the incident energy on a multilayer system of materials that results in 50% probability that sufficient heat transfer through the test specimen is predicted to cause onset of second degree skin burn injury [15].

The maximum incident energy for which PPE is specified is 40 cal/cm^2 (167.36 J/cm^2). *It is not unusual to encounter energy levels much higher than 40 cal/cm^2 in actual electrical systems. Standards do not provide guidelines for higher incident energy levels*. Incident energy reduction techniques can be applied; otherwise, it is prudent not to maintain such equipment in energized state.

A category 4 PPE outfit looks like a space suit with face hood, eye shields, cover, and gloves. It restricts the mobility of a worker to perform delicate tasks, for example, maintenance work on terminals and wiring.

Thus, not only an accurate calculation of incident energy level but also its reduction in the planning and design stage and selection of appropriate protection and relaying of electrical systems are gaining importance [21].

21.3.1 Hazard Boundaries

The boundaries are defined in NFPA, and the following synopsis is relevant here, Figure 21.2.

The flash protection boundary is the distance at which threshold of second degree burns can occur, and the incident energy release is 1.2 cal/cm^2 (5.0 J/cm^2). This is the boundary

TABLE 21.5

Protective Clothing Characteristics

Hazard Category	Clothing Description	Arc Rating of PPE (cal/cm^2)
0	Nonmelting, flammable materials, i.e., untreated cotton, wool, rayon, or silk, or blends of these materials, with a fabric weight of 4.5 oz/yd^2	N/A
1	Arc-treated FR shirt and FR pants or FR overall	4
2	Arc-treated FR shirt and FR pants or FR overall	8
3	Arc-treated shirt and pants or FR overall, and arc suit selected, so that the system rating meets the required minimum	25
4	Arc-treated shirt and pants or FR overall, and arc flash suit selected, so that the system rating meets the required minimum	40

Source: NFPA 70E, Electrical safety in workplace, 2009.

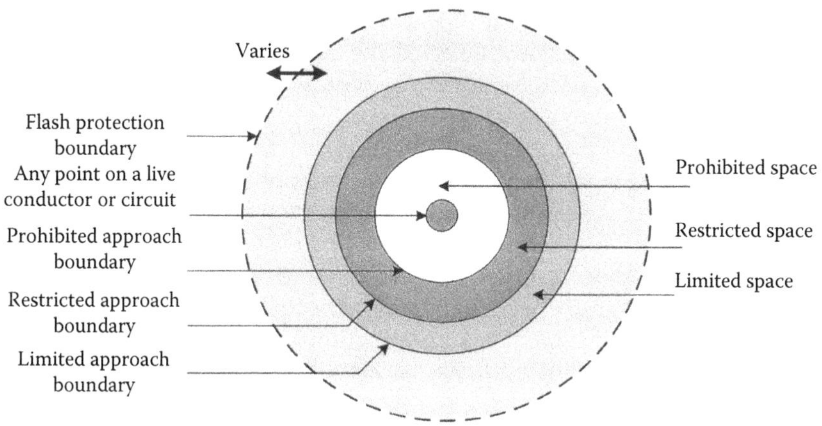

FIGURE 21.2
Hazard boundaries.

that is calculated by computer-based programs. Inside the boundary, the energy level will be higher. This boundary should not be crossed by any one, including a qualified person, without wearing the required PPE, Table 21.5. The PPE outfits are designed to minimize the risk of sustaining energy greater than 1.2/cal cm^2. *That is, threshold of second degree burns can still occur even with appropriate PPE, and these burns are considered curable.*

Unqualified persons, that is, those not specifically trained to carry out the required tasks are safe when they stay away from the energized part a certain distance, which is the limited-approach boundary. They should not cross the limited approach boundary and arc flash boundary unless escorted by a qualified person.

Crossing restricted approach boundary means that special shock prevention techniques and equipment are required, and an unqualified person is not allowed to cross this boundary.

Finally, the prohibited approach boundary establishes the space that can be crossed only, as if a live contact with exposed energized conductors or circuits was planned.

Working distance is defined as the closest distance to a worker's body excluding hands and arms. IEEE standard 1584 [9] specifies required working distances, Table 21.2. For example, for 15 kV switchgear, it is 36 in., while, for a 480 V MCC, it is 18 in. A larger working distance reduces the incident energy and therefore the HRC. Recommendations of IEEE for the working distances are followed.

The limited approach, restricted approach, and prohibited approach boundaries are all defined based upon the system voltage. *No calculations are required for establishing these boundaries.*

21.4 System Grounding: Impact on Incident Energy

Equation 21.9 for the calculation of incident energy has a factor K_2 for system grounding. It is zero for ungrounded and high-resistance-grounded systems and −0.113 for the solidly grounded systems. Thus, the incident energy release is higher for the ungrounded or high-resistance-grounded systems as compared to solidly grounded systems. The physical phenomenon leading to the release of incident energy for solidly grounded and ungrounded systems is not explicitly stated in the standard.

Lee's equations do not consider system grounding, and the incident energy will be same whether the system is grounded or ungrounded.

However, it is not the intent that solidly grounded systems should be adopted to reduce arc flash hazard. Much lower equipment damage and continuity of processes can be achieved with high-resistance-grounded systems. For low-voltage distributions, generally, the solidly grounded systems are not applied, and ungrounded systems are extinct, no more used. The solidly grounded systems versus high-resistance-grounding systems are amply discussed in the current literature (see Chapter 2). Approximately 70% of the faults in the electrical systems are line-to-ground faults. Sometimes, these may be self-clearing and of transient nature (e.g., in OH line systems) or may evolve into three-phase faults over a period of time.

Thus, the probability of a worker being subject to arc flash due to ground faults is much higher. The IEEE equations take a safe stance and state that if the hazard level is calculated for three-phase faults, it will be lower for the ground faults. In a single-phase to ground fault, the arc tends to extinguish at natural current zero. In three-phase systems, the phases are displaced 120 electrical degrees, and natural current zeros are similarly displaced. In industrial systems, the line-to-ground fault current in solidly grounded systems can even exceed three-phase bolted fault current, Chapter 2. A footnote in NFPA-90 reads as:

> …, high resistance grounding of low-voltage and 5 kV (nominal) systems, current limitations, and specifications of covered bus within equipment are techniques available to reduce the hazard of the system.

Example 21.1

Calculate the incident energy using IEEE and Ralph Lee's equations for a 30 kA three-phase bolted fault current in a 13.8 kV metal-clad switchgear, resistant grounded. The arcing time is 30 cycles for the arc fault current through the protective device.

Using IEEE equation (21.8), calculate the arcing current:

$$\log I_a = 0.00402 + 0.983 \log I_{bf}$$

This gives $I_a = 28.578$ kA. Note that the arc flash time is calculated based upon the arcing fault current through the protective device and not the bolted fault current. A further calculation for arc flash time is required at $0.85I_a$. Let us assume for this example that 30 cycles are applicable. Then, calculate normalized incident energy from (21.9):

$$\log\ E_n = (-0.555) + (-0.113) + 1.081 \log(28.578) + 0.0011(153)$$

Here, $K_1 = -0555$ (arc in a box), $K_2 = -0.113$, resistance-grounded system, and $G = 153$ mm, Table 21.2. This gives $\log E_n = 1.074$. Calculate incident energy in J/cm^2 using (21.10):

$$E = 4.184(1)(11.858)\left(\frac{0.5}{0.2}\right)\left(\frac{610^{0.973}}{910^{0.973}}\right)$$

Here, t is the arcing time = 0.5 s, the distance x is given in Table 21.3, and, for 15 kV, it is 0.973, and D the working distance from Table 21.2 is 910 mm.

This gives $E = 84.4$ J/cm^2 = 20.08 cal/cm^2.

Calculate the arc flash boundary (21.12)

$$D_B = \left[4.184(1)(11.858)\left(\frac{0.5}{0.2}\right)\left(\frac{610^{0.973}}{5}\right)\right]^{1/0.973}$$

Here, $C_f = 1$, E_n as calculated before is 11.858 J/cm^2 and $E_B = 5$ J/cm^2 by definition. This $D_B = 16541$ mm $= 54.267$ ft. This is rather a large distance at which a worker can get threshold of second degree burns.

Lee's Equations

The incident energy release given by (21.4) is 126.6 cal/cm^2 and from (21.3) distance in feet for a curable burn is 30.82 ft. There is a vast difference in these calculations. Note that Lee's equations do not consider system grounding. IEEE equations are normally used for arc flash evaluations.

21.5 Duration of an Arc Flash Event and Arc Flash Boundary

A maximum duration of 2 s for the total fault clearance time of an arc flash event is considered, though, in some cases, the fault clearance time can be higher.

Tables 21.6 and 21.7 show the arc flash boundary calculations according to IEEE 1584 equations for a bolted three-phase fault current of 30 kA; in 13.8 kV switchgear,

TABLE 21.6

Arc Flash Boundary and Incident Energy Release for 30 kA of Bolted Fault Current (28.58 kA rms Arc Flash Current) in 13.8 kV Switchgear, 13.8 kV System Resistance Grounded, Working Distance = 36 in., Gap = 153 mm

Arc Duration (s)	Arc Flash Boundary (in.)	Incident Energy
0.058	74	3.0
0.5	851	26
1.0	1736	52
1.5	2633	78
2	3539	104

TABLE 21.7

Arc Flash Boundary and Incident Energy Release for 30 kA of Bolted Fault Current (16.76 kA rms Arc Flash Current) in 480 V MCC, 480 V System High-Resistance Grounded, Working Distance = 18 in., Gap = 25 mm

Arc Duration (s)	Arc Flash Boundary (in.)	Incident Energy
0.050	36	3.8
0.5	147	38
1.0	225	75
1.5	288	113
2	343	151

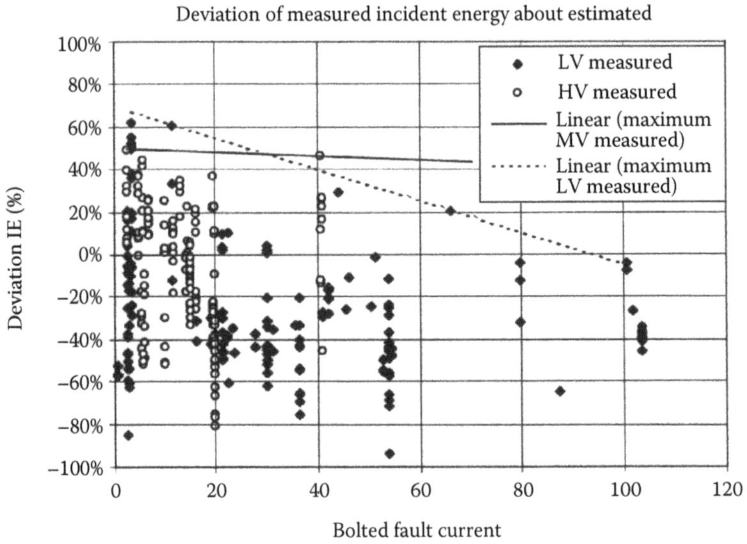

FIGURE 21.3
Variations in arc flash energy calculations from IEEE standard 1584.

13.8 kV system is resistance grounded and also, in a 480 V MCC, 480 V system is high-resistance grounded.

For a 30 kA, 2 s fault in 13.8 kV switchgear, the incident energy boundary is 3539 in. equal to 295 ft. For 1 s fault duration, it is 144.6 ft. This is rather a large distance. The variation of the incident energy boundary at which a worker can be exposed to 1.2 cal/cm^2 of incident energy and sustain threshold of second degree burns seems to be very large.

The working space in an electrical room switchgear installation in front of electrical equipment is limited and may be as low as 5–6 ft and must meet NEC requirements. Many companies, as a policy, keep the electrical rooms locked, and a worker must have the required PPE outfit before entering the electrical rooms.

Figure 21.3 shows deviation of measured incident energy from the calculated values for various bolted fault currents using IEEE 1584 equations.

A distribution system may be operated in various operating modes, that is, a tie line or a generator switched in or switched out of service. Bolted short-circuit currents in each of these operating modes are required to be calculated, and the arc flash hazard calculated for a variation between 100% and 85% of arcing current in each scenario.

21.5.1 Equipment Labeling

The equipment can then be labeled with respect to the maximum HRC calculated in each of the operating modes. Generally, the label contains equipment identification, system voltage, PPE required (categories 1, 2, 3, or 4), incident energy in cal/cm^2 or J/cm^2, arc flash boundary, restricted and prohibitive approach boundary, and the identification of the upstream protective device that cleared the fault. Such a label must be provided on each electrical equipment based upon the arc flash analysis study [15].

21.6 Protective Relaying and Coordination

Protective relaying is not covered in this book. A reader may consult the references [22–27]. For arc flash analysis, ground fault protection and its coordination need not be considered, as the calculations are based upon three-phase bolted faults, which will be cleared by the phase protection relays.

The elements in protective relaying systems include relays, direct-acting solid-state trip devices on low-voltage circuit breakers, and fuses. For medium and high-voltage systems, relayed breakers are exclusively used. A time–current coordination of these devices should insure selectivity and backup protection, so that minimum of unfaulted section of the distribution system is interrupted. A downstream protective device must operate faster than the next upstream device, though the magnitude of the short-circuit current flowing through these may be identical. While maximizing protection and minimizing the area of shutdown, it should be ensured that all system components, like transformers, reactors, cables, and rotating equipment, are protected with respect to continuous overloads and within their fault withstand capabilities and thermal damage capabilities. The coordination should also meet the specific requirements of the processes, for example, protection of a frequently started motor will be different as compared to a motor with 2 starts per hour, normal NEMA specifications. An equipment damage due to lack of protection can lead to more serious loss of production and downtime. Coordination of protective devices in a given situation is not a substitute for proper system planning and adequacy of system protection to perform required functions. To start with, the following data are required:

- A single-line diagram of the system, which shows all the protective devices to be coordinated on a time current basis. The switching conditions of breakers, alternate routing of power flow, and the switching devices that will be tripped or opened to isolate a fault must be earmarked.
- The load current flows under normal and contingency conditions and overloads if any and their time duration.
- Current transformer ratios, burdens, secondary resistances, accuracy classifications, and locations.
- Time current characteristics of all devices to be coordinated, total and minimum clearing time of fuses, let-through curves of fuses, and current limiters. Relay burdens, current setting ranges, and available functions are to be set.
- Short-circuit currents at the point of applications. Computer routines will calculate the arcing currents. The data required for short-circuit calculations is discussed in Chapter 7.
- Medium-voltage motor data, the thermal damage curves, heating and cooling time constants, embedded RTDs, number of starts, and blocking conditions.

This list is merely indicative. The coordination is carried out by providing a certain time margin between two protective devices to be coordinated in series. This is called the coordinating time interval (CTI) and varies with the type of relays and interrupting time of the breakers. This is further discussed in the appropriate applications in the following sections.

It is desirable that in a continuous process plant, maximum selectivity is obtained, and only the faulty section is disconnected from service. In the time–current coordination, this is achieved by progressively increasing the operating time of the protective devices toward the source. Thus, as the upstream sources of higher power handling capability are approached, the time to clear a fault increases—the selectivity and sensitivity normally pull in the opposite directions. For arc flash hazard reduction, it is desirable that the faults are cleared fast, and yet the selectivity is maintained. This is achieved by unit protection systems.

21.6.1 Unit Protection Systems, Differential Relaying

Differential relaying provides unit protection system, which means that the protection operates for a fault within the protected zone and is stable for any through fault currents. The differential protection can even be set lower than the load currents, and, thus, low level of fault currents can be cleared fast, reducing arc flash hazard and the equipment damage. Figure 21.4 shows overlapping zones of differential protections for transformer, 138 kV line, generator, and two sectionalized buses, which are achieved with suitable location of the current transformers. The transformer differential protection also protects the primary

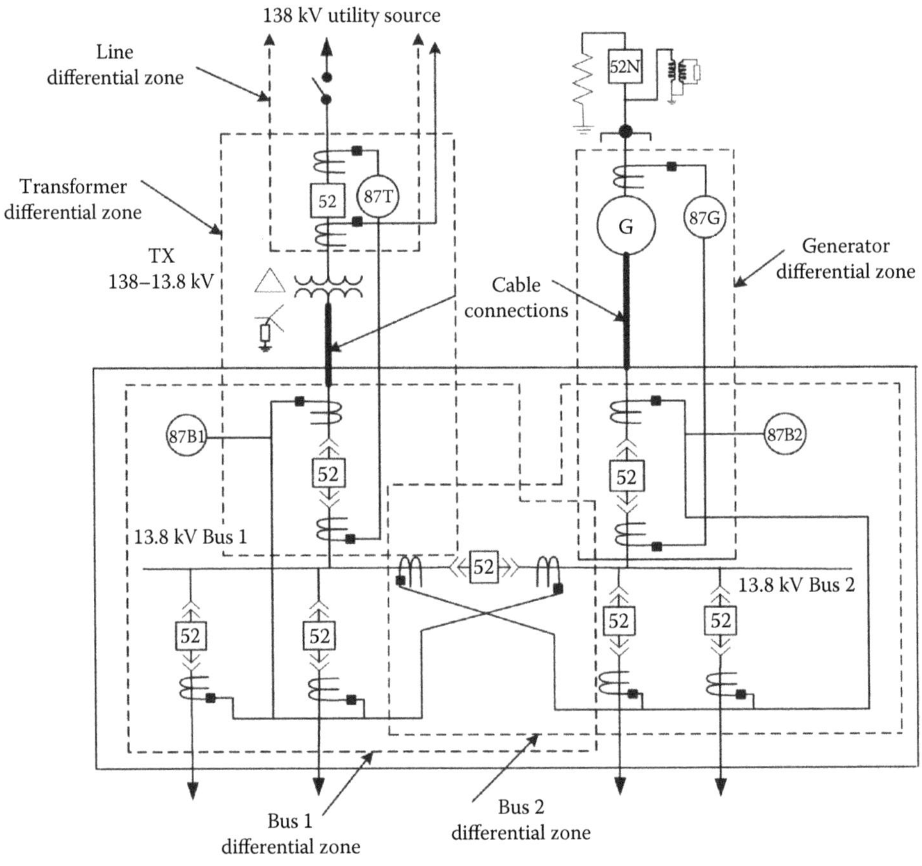

FIGURE 21.4
Overlapping zones of differential protection. *Note:* All other protections, for example, ground fault, directional, frequency, time overcurrent, etc., are not shown.

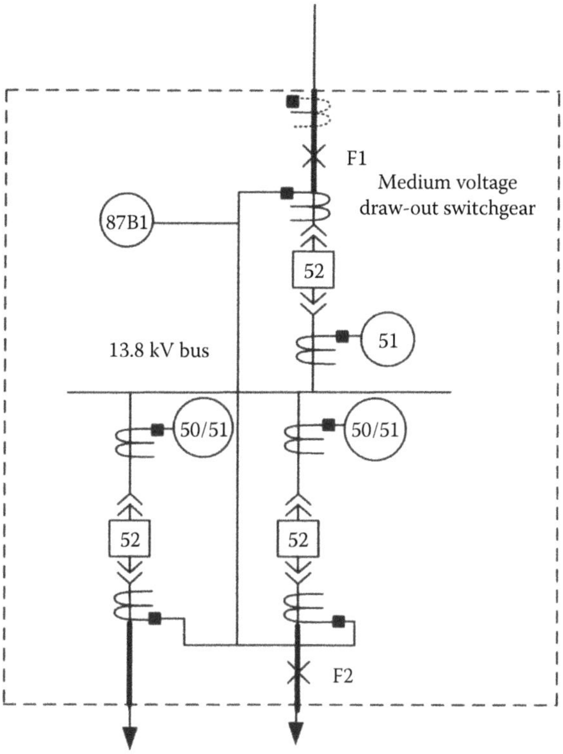

FIGURE 21.5
Unprotected areas outside differential relay bus protection zone, shown in bold.

and secondary cable connections of the transformers. Separate cable differential zone may be required for transformer secondary cables depending upon their lengths.

Note that differential protection covers the zones between the locations of the current transformers. There are unprotected areas when overlapping zones of differential relaying are not provided. Figure 21.5 shows bus differential protection, stand-alone, applied to medium-voltage switchgear. The window-type current transformers are normally located on breaker draw-out spouts. Thus, the area shown in bold on the incoming and feeder circuit breakers, which includes cable connections and some bus work inside the breaker cubicles, is outside the differential relay protection zone. The fault at F2 will be cleared by instantaneous overcurrent relay element, device 50, on the feeder breakers. Generally, with proper instantaneous overcurrent settings, low hazard risk category (HRC) levels, similar to that obtained with differential relays can be obtained. However, a fault on the incoming connections at F1 must be cleared by an upstream relay. The zone of the differential relay can be extended by locating the differential CT right at the cable entrances in the circuit breaker cubicles, CT shown dotted on the source side. Reference [28] describes an arc fault detection approach using higher harmonics and differential protection; this can serve as an early warning system for developing arc fault.

For bus differential protection, high-impedance-type voltage restraint relays have been popular. The problem of current transformer saturation is avoided by loading CTs with a high impedance unit. For external faults, there is high degree of error in CTs in the faulted circuit; however, for faults within protected zone, the CT errors are small. A nonlinear resistor limits the voltage developed across relay coil.

These have led to low-impedance current differential relays, which can accommodate CT mismatch and operate faster, in subcycle time. Though differential protection can be applied on all the medium-voltage buses, the cost factor can be a deterrent. The earlier pilot wire systems [22] to protect lines and cables, maximum length approximately 50 mi with metallic, or fiber pilot wires have been superseded by microprocessor-based multifunction programmable protective relays (MMPRs) of low-impedance current differential types. In earlier schemes, currents applied to sequence networks produce a composite current that is proportional to line current and has a polarity related to line current flow direction. Modern line current differential MMPRs can serve as pilot wire differential relays with fiber-optic interface, which can be continuously monitored and used for intertripping and waveform comparison of the currents at each end.

21.6.2 Arc Flash Detection Relays

These are designed for the protection of medium or low-voltage air-insulated switchgear. Loop-type fiber sensors, radial sensor, or lens-type sensors are used for arc detection, which can be supervised by adjustable overcurrent settings, or alternatively tripping can be arranged from light sensing only. From arc flash considerations, these provide even a faster fault detection time. The light-sensing fiber is run in all the compartments of switchgear and senses the radiant light of the arc flash. Compensation and settings in the relays are provided to override the ambient light. The manifestation of radiant light at a certain wave spectrum occurs first in an arc flash; the sensing and actuation of a digital breaker trip contact is fast, on the order of ¼ of a cycle. Figure 21.6 provides comparative operating times in the instantaneous zone of differential relays and light-sensing protection.

Protective relaying has been called an art as well as a science and is vast subject not covered in this book. Modern MMPRs have the capability of fault data capture, oscillography, monitoring, metering, self-diagnostic, and close tolerance programmable characteristics—even the front panel display and push buttons can be programmed. The selection of appropriate relay types and their settings has a profound effect on the ultimate arc hazard reduction. For practical relaying and coordination for arc flash analysis, this description is basic, and a reader may refer to the cited references.

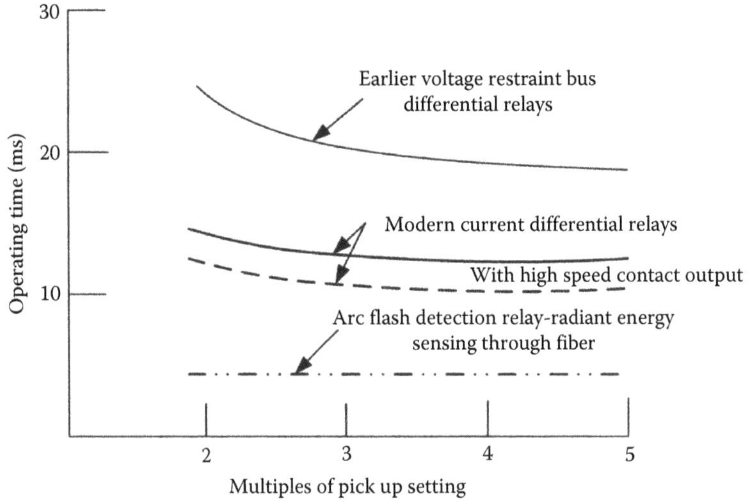

FIGURE 21.6
Comparative operating times of differential and light sensing relays.

21.7 Short-Circuit Currents

The examples and demonstrations in this chapter are based upon the following maximum short-circuit currents:

13.8 kV Systems
A maximum short-circuit current of 40 kA rms symmetrical is considered. Following is a real-world example of a large distribution system:

"Four utility tie transformers each of 30/50 MVA, 110–13.8 kV, six plant generators having a total installed capacity of 270 MVA, total running load 190 MVA—the excess power generated is supplied into the utility system, full stream production can be maintained on forced or maintenance outage of one or more generators and one utility tie transformer, yet, 13.8 kV circuit breakers with interrupting rating of 40.2 kA at 13.8 kV have been applied." At 13.8 kV primary distribution levels, the systems have been designed with earlier *K*-rated breakers with maximum 40.2 kA interrupting rating at 13.8 kV. Short-circuit current-limiting reactors have been used for the interconnections, and all the system is tied together at 13.8 kV primary distribution. This is, perhaps, a practical example of the maximum power distribution capability at 13.8 kV primary distribution voltages, which is the most popular voltage in the industry.

2.4 kV and 4.16 kV Systems
For these distribution systems, a maximum transformer rating of 7.5/10.5 MVA, 5.5% impedance on its own OA rating base is considered, which is conservative. The 7.5 MVA transformers serve mostly medium-voltage motor loads. On this basis, the maximum-calculated fault levels are 33 and 19 kA rms symmetrical, respectively, for 2.4 and 4.16 kV, respectively.

480 V Systems
Low-voltage switchgear rated up to 200 kA short-circuit interrupting rating and low-voltage power circuit breakers LVPCBs rated up to 5000 A are commercially available. This means a single transformer of 4000 kVA rating, 13.8–0.48 kV, percentage impedance 5.75% on 4.0 MVA base can be used. Assuming that the transformer serves mainly ac motor loads, the available short-circuit current will be approximately 100 kA, assuming that the transformer is connected to a 13.8 kV system with a short-circuit level of 40 kA rms symmetrical. For the purpose of calculations presented in this chapter, a maximum transformer size of 2500 kVA is considered.

21.7.1 Reducing Short-Circuit Currents

To reduce the arc flash hazard, the short-circuit levels in the electrical systems can be reduced by

1. The relative magnitudes of 33 kA at 2.4 kV and 19 kA at 4.16 kV, for the same equipment ratings, bring about a voltage selection criterion for arc flash reduction—that at the design stage select 4.16 kV rather than 2.4 kV for medium-voltage secondary distributions. This will have cost impacts depending upon the ratings, especially for induction motors.
2. For low-voltage systems of the same ratings and configurations, the short-circuit currents in a 575/600 V system will be approximately 83% of those in 480 V

systems. (Low-voltage cables of 1 or 2 kV voltage grade may be required for drive systems.)

3. Rating of the transformers plays another major role in reducing the short-circuit levels. For example, two 3.75 MVA medium-voltage transformers can be used compared to a single 7.5 MVA unit. Similarly, for low-voltage distributions, two 1.0 MVA transformers can be used versus a single 2.0 MVA transformer.
4. Double-ended substations with autoclosing of normally open bus section circuit breaker can be used. This configuration is not so popular, yet it is possible to transfer rotating loads (induction motors) without loss of processes. Though, synchronous motors should not be reconnected for autotransfer of power [29].
5. Current-limiting reactors are a possibility, but careful analysis and study are required. These may have adverse effect on voltage profiles and reactive power loss and are not recommended simply to lower short-circuit currents for the purpose of reducing HRC.

21.8 Arc Flash Calculations in Medium-Voltage Systems

Table 21.8 shows the maximum arc flash time, which is the sum of relay operating time plus breaker interrupting time for limiting the HRC to level 2 (8 cal/cm^2) and also level 4 (40 cal/cm^2) in the medium-voltage systems. This table shows that in order to limit HRC to 2, a 30 kA bolted short-circuit fault in a 13.8 kV system must be cleared in 0.15 s, and a 40 kA fault must be cleared in 0.11 s. Deducting 5 cycles for the breaker operating time, the time available for the downstream coordination is 0.06 and 0.027 s, respectively. This is too small for any downstream time–current coordination. *Thus, time–current coordination, howsoever implemented, cannot reduce the hazard level to 2 or lower for 13.8 kV distributions.*

Table 21.9 shows the calculations with differential relays and light-sensing relays, operating times as shown in Figure 21.6. For older version of differential relays, an operating time of 1.5 cycles is considered, while, for newer MMPRs with current differential, an

TABLE 21.8

Maximum Arcing Time for Limiting HRC to 2 and 4 for 13.8 kV, 4.16 kV, and 2.4 kV Systems, Resistance Grounded, Working Distance 36 in.

System Voltage (kV)	SC Current (kA rms sym.)	FCT (s)	Incident Energy (cal/cm^2)
13.8	40	0.11	8
		0.57	40
	30	0.15	8
		0.75	40
2.4	33	0.16	8
		0.78	40
4.16	19	0.28	8
	19	1.43	40

TABLE 21.9

Arc Flash Hazard Reduction in Medium-Voltage Systems through Differential and Arc Flash Light-Sensing Relays

System Voltage (kV)	Bus Bolted Fault (kA rms sym.)	Arc Fault (kA rms sym.)	Breaker Interrupting Time Cycles	Device Operating Time Cycles	Gap (mm)	Arc Flash Boundary (in.)	Incident Energy (cal/cm^2)	HRC	Remarks
13.8	40	37.92	5	1.5	153	241	7.6	2	1.5 cycles differential
				0.75		169	6.8	2	¾ cycle differential
				0.25		155	6.2	2	¼ cycle arc flash
			3	1.5		131	5.3	2	1.5 cycles differential
				0.75		110	4.4	2	¾ cycle differential
				0.25		94	3.8	1	¼ cycle arc flash
13.8	30	28.58	5	1.5	153	176	5.6	2	1.5 cycles differential
				0.75		124	5	2	¾ cycle differential
				0.25		113	4.6	2	¼ cycle arc flash
			3	1.5		96	3.9	1	1.5 cycles differential
				0.75		80	3.3	1	¾ cycle differential
				0.25		68	2.8	1	¼ cycle arc flash
4.16	19 (see text)	18.24	5	1.5	104	94	3	1	1.5 cycles differential
				0.75		66	2.7	1	¾ cycle differential
				0.25		60	2.5	1	¼ cycle arc flash
			3	1.5		51	2.1	1	1.5 cycles differential
				0.75		43	1.8	1	¾ cycle differential
				0.25		37	1.5	1	¼ cycle arc flash
2.4	33 (see text)	31.38	5	1.5	104	172	5.5	2	1.5 cycles differential
				0.75		121	4.9	2	¾ cycle differential
				0.25		110	4.5	2	¼ cycle arc flash
			3	1.5		94	3.8	1	1.5 cycles differential
				0.75		78	3.2	1	¾ cycle differential
				0.25		67	2.7	1	¼ cycle arc flash

operating time of ¾ of a cycle is considered. An operating time of ¼ cycle is considered for light-sensing relays supervised by current. These relays operate with either 3-cycle or 5-cycle circuit breakers. This table shows that even for 40 kA, short-circuit current in a 13.8 kV resistance grounded system HRC level of 1 can be obtained with light-sensing relays. The calculated HRC levels do not exceed 2 anywhere in this table.

21.8.1 Reduction of HRC through a Maintenance Mode Switch

As stated earlier, the application of differential protection on *all* medium-voltage-distribution systems will be expensive, and, practically, it is not implemented. It is common to apply differential or light-sensing relays at the main distribution switchgears of primary distribution voltage. With the proper application of a MMPR, with protective settings modified/brought in service through maintenance mode switch can reduce the hazard level, sometimes equivalent to that obtained by differential relays. The premise is that a fault may not coincide with the maintenance tasks, and, therefore, during maintenance, the selective coordination is sacrificed to some extent. If a fault does occur, the worker as well as the equipment is protected, but a wider area may be shutdown. In a MMPR changing or bringing into service, a protection function is done by forming an AND gate with the switch position and the protective settings, which actuate a discrete output. This can be wired in the trip circuit.

A typical 2.4 kV, MV motor control center is shown in Figure 21.7 and is served by a rather large transformer of 7.5/9.375 MVA, percentage impedance 5.5%. First, assume that relay R2 and the secondary breaker BK2 shown dotted in this figure are not provided. Also, there is no transformer primary fuse, and the transformer is protected by relay R1. This is permissible according to NEC [30] provided appropriate settings are provided on this relay and is typical of industrial installation systems. The time–current coordination (TCC) plot in Figure 21.8 illustrates that transformer ANSI frequent fault withstand curve is well protected by R1. The availability of maximum rating of current-limiting fuses for 13.8 kV applications for transformer primary protection is 300 E. Figure 21.8 shows that other unit-substation transformers are daisy chained, again typical of many industrial distribution systems.

Table 21.10 depicts the result of arc flash calculations. The incident energy for a fault at F4 on 2.4 kV MCC bus or fault at F3 is high, 75.6 cal/cm^2, while at all other fault locations, F1, F2, and F5, the hazard level is limited to 0–2.

For a fault at F5, downstream of NEMA E2 medium-voltage starter, with type R fuses, the HRC is zero. This has been calculated with a large motor and fuse size of 36 R. Table 21.11 shows that HRC with type R fuses can be limited to 2.

For all fault locations F1, with proper instantaneous overcurrent settings on relay R1, the HRC can be maintained below 2, as demonstrated in this case, with 40 kA symmetrical short-circuit current at 13.8 kV. Multiple HRC categories exist on the equipment depending upon the fault location.

The following strategies can be used for reducing the hazard level for fault at F4:

- Provide a main secondary breaker BK2 and bring in the protection relay R2 settings as shown in Figure 21.8, through a maintenance switch. This reduces the HRC to 2 at the bus, but the incoming cable compartment of the circuit breaker, cable terminations at the back of the circuit breaker and also the cables from the transformer to the switchgear (zone shown Z in Figure 21.7) are not protected by relay R2.

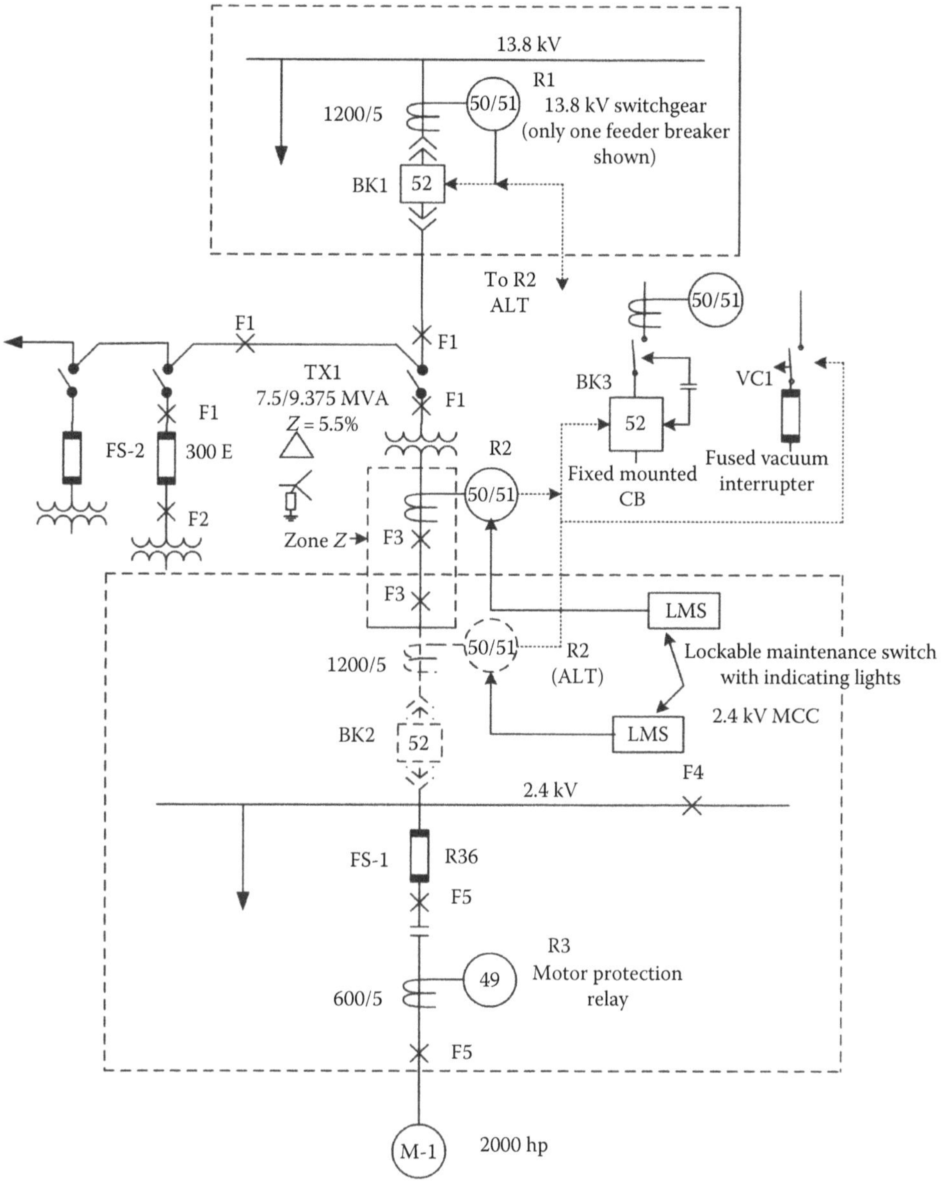

FIGURE 21.7
Primary unit substation configuration for arc flash hazard analysis.

- Eliminate BK2, provide CTs located in the secondary compartment or tank of the transformer, and connect relay R2 to these CTs. Now all the zone from transformer secondary to 2.4 kV MCC bus and also the connections to the individual motor starter on source side of motor fuse are protected. R2 should trip 13.8 kV circuit breaker BK1. This will result in a total shutdown of all the loads served from this feeder breaker.
- The industrial practice has been to provide transformer primary protection through current-limiting fuses. UL requires that, for indoor less-inflammable

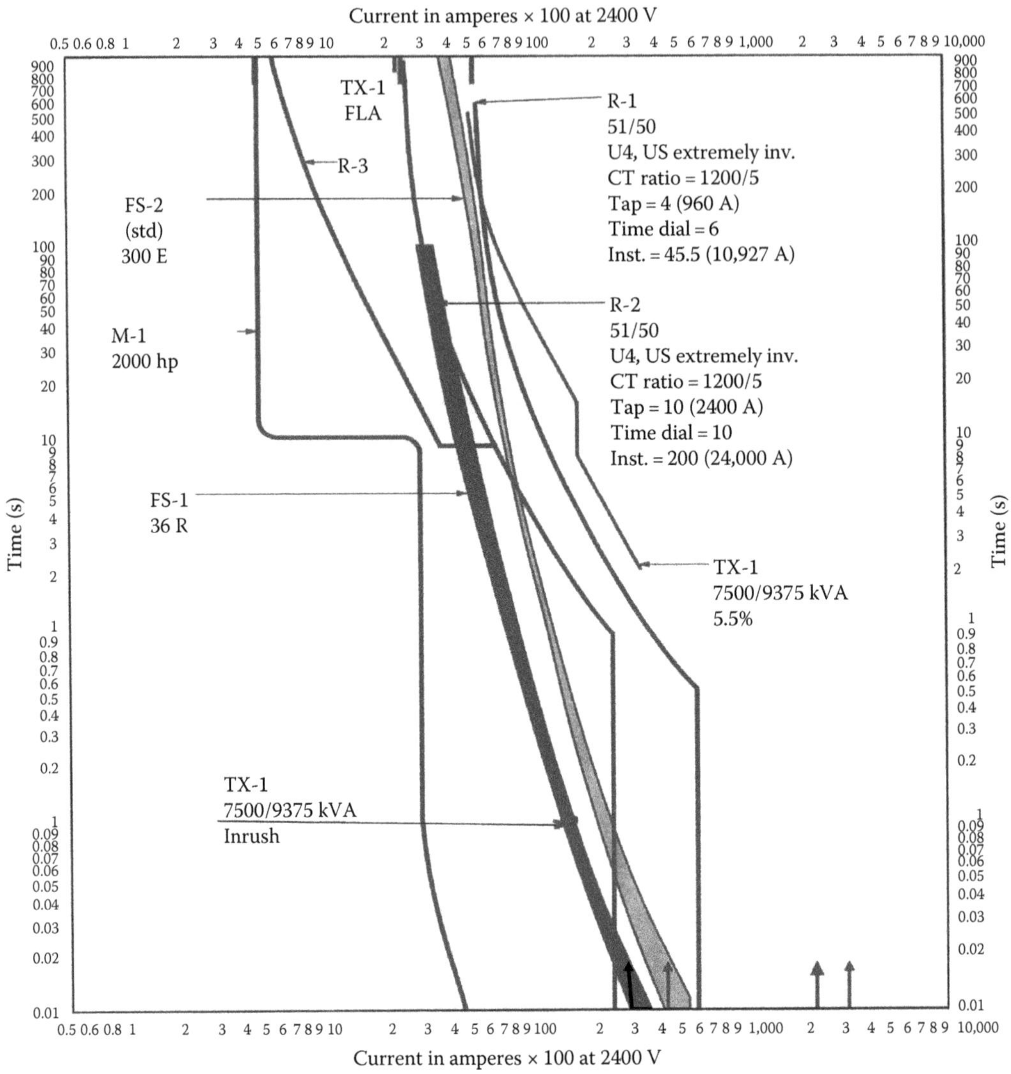

FIGURE 21.8
Time–current coordination in the system shown in Figure 21.7. R2 actuated through a maintenance mode switch reduces incident energy level to 4.3 cal/cm^2.

liquid-immersed transformers, current-limiting fuses must be provided to limit the energy let-through for an internal transformer fault and prevent explosion. Factory mutual, however, takes a different stance and specifies devices and relays for transformer protection. Fixed-mounted circuit breakers and cubical type of switchgear can be used for the primary protection of the transformers for cost savings.

- If the primary fused switch is replaced with fixed-mounted circuit breakers, which is tripped by relay R2, the area of shutdown is confined to the faulty substation only.
- Table 21.10 shows that with R2 set as shown in Figure 21.7, the incident energy is reduced from 75 to 4.3 cal/cm^2, HRC 2.

TABLE 21.10

Calculations of Arc Flash Hazard, System Configuration (Figure 21.7, TCC Plot; Figure 21.8, Working Distance 36 in.)

Configuration	Fault Location	Bolted Fault Current (kA)	Arc Flash Current (kA)	Arcing Time (s)	Arc Flash Boundary (in.)	Incident Energy (cal/cm^2)	HRC
No relay R2 and breaker BK2	F1	38.8 (R1)	36.6	0.093	159.6	6.4	2
	F2	38.8 (300 E fuse)	36.6	0.01	16.1	0.7	0
	F3	5.17 (R1)	4.93	1.637	2543	75.6	Danger
	F4	5.17 (R1)	4.93	1.637	2543	75.6	Danger
	F5	27.89 (36 R fuse)	26.61	0.02	21.9	0.9	0
Relay R2 and breaker BK2	F3	5.17 (R1)	4.93	1.637	2543	75.6	
	F4	28.41 (R2)	27.10	0.093	106.5	4.3	2
Relay R2, operated from CT in transformer	F3 or F4	28.41 (R2)	27.10	0.093	106.5	4.3	2

TABLE 21.11

HRC for Fault on Load Side of Medium-Voltage Motor Starters NEMA Type E2, Protected with Type R Current-Limiting Fuses, 2.4 and 4.16 kV Systems, Resistance Grounded

Bolted Fault Current (kA)	Type R Fuse	HRC
25–30	2 R–36 R	0
20	2 R–32 R	0
	36 R	1
15	2 R–24 R	0
	24 R–36 R	1
10	2 R–18 R	0
	24 R	1
	32 R–36 R	2

- Also, the settings on R2 coordinate well with the motor protection relay curve; allow starting of largest 2000 hp motor and supplying full load of the transformer.
- It may be inferred from Figure 21.7 that the settings on relay R1 can be changed to that of relay R2 through a maintenance switch. Though this avoids installation of another set of CTs, relay R2, and intertripping wiring, from the coordination perspective, adding R2 is a better choice.

21.8.2 Arc Resistant Switchgear

Arc resistance switchgear for medium-voltage and low-voltage applications will give zero incident energy outside the enclosure at the working distance. It must withstand an internal arcing fault and meet the requirements of IEEE C37.20.7-2007 [31].

Assembly type 1 has arc-resistant design for the front of the equipment only. Type 2 has freely accessible exterior for front back and sides. Accessibility suffixes A, B, C, and D further categorize the equipment, with respect to accessibility. For example, suffix "B" means that under normal operation of the equipment opening of the doors or cover of compartments specifically designed as low-voltage control or instrumentation compartments is permissible. This requires additional testing [31]. If any cover or panel is opened, which is not intended to be opened under normal operation, the equipment is no longer arc resistant.

The arc-resistant switchgear is specified for the maximum short-circuit duration, and IEEE standard [31] recommends a time duration of 0.5 s. Practically, the time duration can be reduced when fast differential or light-sensing protective features are provided. The arc flash duration withstand time has a major impact on the cost of the equipment.

The switchgear may be provided with pressure relief devices to exhaust the overpressure. Plenums may be run on top or rear of the assembly to conduct arc flash products outside the switchgear room. It is not the intention to lay down the specifications for applications, suffice to state that this is one means to reduce arc flash hazard. Nevertheless, all maintenance tasks cannot be carried out simply because the equipment is arc resistant. NFPA 70E-2009 [15] specifies a HRC of 4 for the arc-resistant switchgear, type 1 or 2 with clearing time of <0.5 s for the insertion or removal of circuit breaker from cubicles with door open. (This should be analyzed case by case depending upon the protection and system fault levels.) Of course, rarely, the circuit breakers will be locally racked out in modern work environment. Industrial establishments are resorting to remote racking, and presence of a worker near the equipment is not required except to engage the remote racking mechanisms.

21.9 Arc Flash Calculations in Low-Voltage Systems

Reference [32] states that annual exposures for low-voltage MCCs are 365, for other low-voltage equipments, annual exposures are 52, and for rest of the equipment at HV (>34 kV), MV (1–34 kV), and MV MCCs, they are total of 30. Therefore, the reduction of HRC for low-voltage MCCs is important. It also implies that LV MCCs are most frequently maintained equipment while energized.

According to NEC Article 240.85, for the low-voltage transformer protection, it is not necessary to provide a secondary main breaker, and protection with primary overcurrent protection devices is acceptable provided certain conditions stated in this article are met. Table 450-3(A) of NEC permits this omission of secondary protection device in supervised locations. Table 21.12 shows that, for the arc flash protection, this leads to a hazardous condition for a bus fault on the low-voltage switchgear that is directly connected to the transformer secondary. The arc fault current as seen by the current-limiting fuse, on the primary side, is reduced in the transformation ratio, that is, by a factor of 28.75, and the fuse takes a long time to operate resulting in release of large amount of incident energy. Even if the maximum arcing time is reduced to 2 s, the incident energy release is much above 40 cal/cm^2.

Before we discuss this situation and the measures to reduce arc flash hazards in the low-voltage systems, there have been many product innovations in the industry for controlling

TABLE 21.12

Low-Voltage Switchgear Directly Connected to Transformer Secondary, No Secondary Breaker, Incident Energy on a Bus Fault Cleared by Transformer Primary Fuse

Transformer (kVA)	Primary Current-Limiting Fuse	Bolted Three-Phase Fault Current (kA, sym.)	Arc Fault Current in the Fuse Reflected on 13.8 kV (kA)	Arcing Duration	Incident Energy (cal/cm^2)
1000	100E	21.3	0.289	461.96	7,082
1500	125E	43.8	0.549	121.26	3,723
2500	175E	51.3	0.598	374.72	12,672

the arc flash hazard in low-voltage distribution systems. Some of these and relevant aspects can be enumerated as follows.

1. *Zone interlocking*: Zone-selective interlocking is an old concept revisited for arc flash reduction. It can also be applied to medium-voltage systems and preserves the selective coordination between main, tie, and feeder breakers allowing fast tripping between device-desired zones. This is done through wired connections between trip units and relays. If a feeder detects a fault, it sends a restraint signal to the main breaker, but for a fault on the bus, the main breaker does not get a downstream restraint signal and trips without delay. The restraint logic is not instantaneous, and there is some time delay associated with it, so that there is no unrestrained tripping of the main. For conservatism, a delay of 20 ms can be added, though it varies from manufacturer to manufacturer. Also, care has to be exercised with motor loads. A motor load will contribute to the bus short-circuit current, and the feeder breaker should not send a restraint signal upstream when the motor contribution fault current flows through it.
2. *CPU-based protection and control*: Low-voltage power circuit breakers with redundant CPU-based protection, monitoring, control, and diagnostic functions, which replace discrete devices and hard wiring, are available. These provide enhanced reliability and arc flash reduction. Synchronization between two CPUs is maintained through a hardwired sync clock, and the CPUs and critical controls are powered through in-built redundant UPS systems. For arc fault reduction, bus differential algorithm with zone-based overcurrent protection is used. For work near the equipment, reduced energy let-through mode can be selected through a remotely located switch. This enables more sensitive protection settings to lower the arc flash incident energy. The maintenance-mode switch for the arc flash reduction system is being called with a variety of names (LMS in Figure 21.7); essentially, it acts upon the trip settings to modify these during maintenance operation.
3. *Current-limiting fuses*: A current-limiting fuse is designed to reduce equipment damage by interrupting the rising fault current before it reaches its peak value. Within its current-limiting range, the fuse operates within 1/4–1/2 cycles. By limiting the rising fault current, the I^2T let-through to the fault is reduced because of two counts: (1) high speed of fault clearance in 1/4 cycle typically in the current-limiting range and (2) fault current limitation. This reduces the fault damage [33]. See Chapter 5 for the characteristics of the current-limiting fuses. Manufacturers publish let-through curves to show the "threshold level" at which the current-limiting action starts. To utilize the current-limiting action of the fuse and

TABLE 21.13

Comparison of Incident Energy Release RK1 Fuses versus MCCBs, Low-Voltage Motor Starters

Bolted Fault Current (kA)	RK1 Fuse Size	MCCB	Incident Energy (cal/cm^2)	HRC
65	100–300	100–300	1.3	1
20	100–300	100–300	0.5	0
10	100–300	100–300	0.3	0

reduction of energy let through, arc flash current through the fuse must exceed the threshold level. Practically, (1) a large size fuse may be provided to achieve maximum coordination, and (2) some lack of coordination must be accepted. This is further illustrated in an example to follow.

The use of current-limiting fuses for low-voltage motor starters is being extolled in many publications. Table 21.13 is constructed for fault currents varying from 10 to 65 kA and RK1 fuses of 300, 200, 150, and 100 A versus the MCCBs of a certain manufacturers of similar current rating, and the magnetic pickup of MCCBs is set at eight times the current rating. This table shows that there is no difference in the incident energy release, fuses versus MCCBs. Fuses at motor starter level can give rise to single phasing and also need inventory control and replacements.

4. *Series connected devices*: Figure 21.9 is based upon NEC. The series ratings are applicable when the combinations are tested according to UL [34]. NEC article 240.89 [30] states that series ratings should not be used where
   a. Motors are connected to the load side of the higher-rated overcurrent device and on the line side of the lower-rated overcurrent device
   b. The sum of the motor full-load current exceeds 1% of the interrupting rating of lower-rated circuit breaker

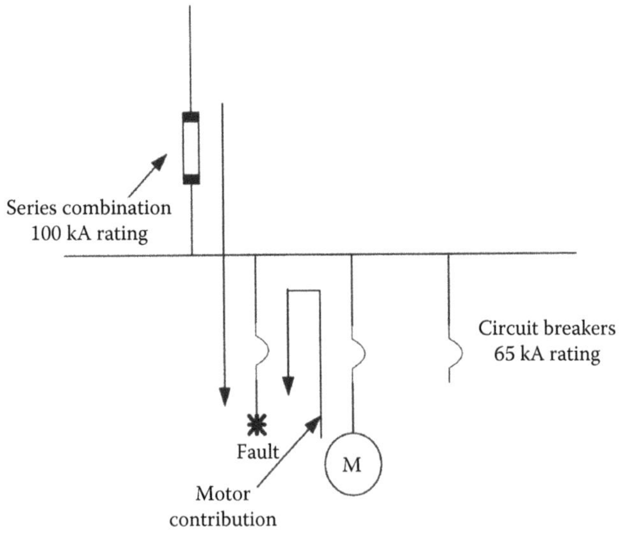

FIGURE 21.9
Example of an installation where the motor contribution exceeds NEC requirements (<1%) for the lowest rated circuit breaker.

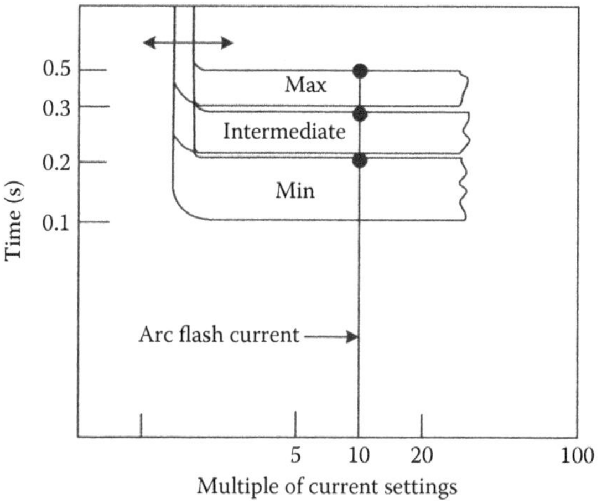

FIGURE 21.10
The ST bands of a low-voltage trip programmer. The arc flash times are considered at the top of the band for conservatism.

Integrally, fused circuit breakers have been used extensively in the industry to raise the short-circuit levels of the combinations, and most manufacturers are of the opinion that above NEC stipulations do not apply to integrally fused breakers—there is no motor contribution between the fuse and the breaker. However, for separately mounted devices, NEC requirements must be followed.

5. *Short time bands of LVPCBs*: LVPCBs have a short-circuit withstand capability of 30 cycles, MCCBs do not have any short-circuit withstand capability and must be provided with instantaneous protection, ICCBs may have short-circuit withstand capability of 15 cycles, yet these are provided with high-set instantaneous override. Thus, low-voltage breaker types can be an important criterion in selective coordination. There has been an attempt to split the 500 ms (30 cycles on 60 Hz basis) time withstand of LVPCBs in to much smaller short-time delay bands. Figure 21.10 illustrates that arc flash time duration is calculated based upon the maximum time of the ST delay band. This figure shows only three ST delay bands, minimum, intermediate, and maximum. The time associated with these bands is 0.2, 0.3, and 0.5 s, respectively. There is a small time gap between the bands, yet the coordination is achieved with this small gap or even when the bands seem to overlap.

   Table 21.14 shows a modern LVPCB, provided with an electronic trip device having seven short-time delay bands.

TABLE 21.14
Short Time Bands of Old versus New LVPCB Trip Devices

	Breaker A	Breaker B
Band	Max Clear (ms)	Max Clear (ms)
1	200	92
2	300	158
3	500	200
4		267
5		317
6		383
7		500

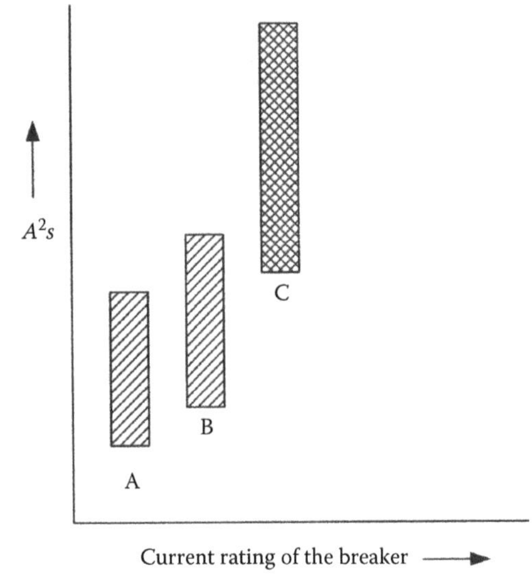

FIGURE 21.11
Coordination in the instantaneous zone based upon the I^2T let through. This may not be the sufficient criterion.

If a three-step coordination is considered, breaker type A (Table 21.14) will clear the fault in 200, 300, and 500 ms, respectively, while breaker type B will clear it in 92, 158, and 200 ms, respectively. The end result can be 2 levels of higher HRC with breaker type A.

6. *Coordination between instantaneous settings*: Coordination between two devices in series, which have low impedance between them and see the same magnitude of short-circuit current, is not so explicit. If the maximum I^2T let through of the downstream device is lower than the minimum I^2T of the upstream device, coordination is possible provided the two devices in series are tested at the same short-circuit power factor. Figure 21.11 shows that an upstream device C will coordinate with device A, but not with device B. For coordination with let-through current of current limiting devices see [35].

   Due to the importance of this aspect in arc flash reduction and coordination, recently, manufacturers have started publishing application tables of coordination between instantaneous settings of their particular breakers. According to one manufacturer, his LVPCB of 2000 rated amperes set at an instantaneous of 15 times the plug rating (instantaneous setting = 30 kA) will coordinate with his downstream MCCB frame 1200 A, magnetic setting not more 27 kA.

7. *Location of current sensors*: Location of current sensors becomes important with respect to the zone left unprotected, much akin to Figure 21.5, for the medium-voltage systems. The location of current sensors for main breaker is particularly important. Generally, the sensors in LVPCBs are located on the draw-out breaker element itself, location a or b in Figure 21.12. Even if the sensor is located on the source side of the breaker, at location a, the zone illustrated in this figure remains unprotected, and a fault in this zone must be cleared by transformer primary fuse. For a secondary transformer fault to be cleared by the transformer primary fuse, it will operate in the time delay zone releasing immense amount of incident energy due to delayed fault clearance.

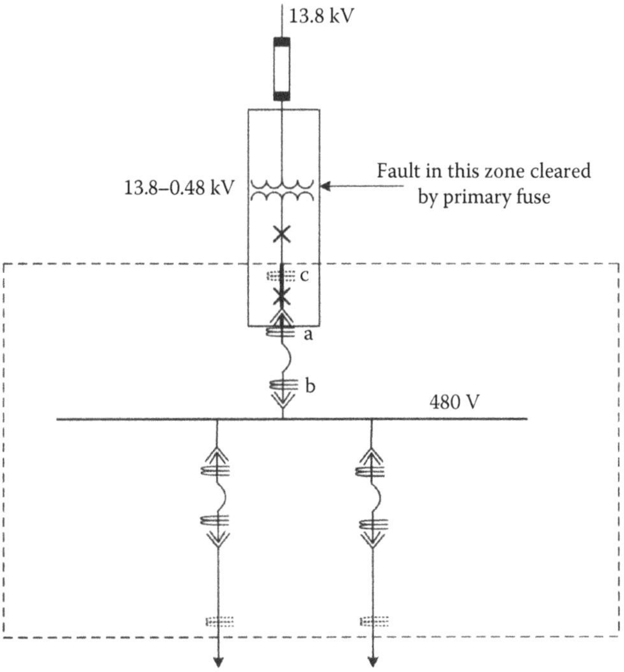

FIGURE 21.12
Alternative locations a, b, and c for the current sensors in a low-voltage switchgear.

If the sensors can be located in position c, right at the incoming and outgoing cables entrances to the low-voltage switchgear, and the assembly is provided with differential protection, the complete switchgear assembly is protected.

8. Innovations like close door remote racking, insulated bus bars, system redundancy, through the door voltage indicators, incorporation of safety shutters, arc flash protected infrared scanning windows, online insulation testing systems, and the like are strategies for worker safety.

Example 21.2

Table 21.15 shows the arc flash calculations for a three-level coordination, as shown in Figure 21.13 for a double-ended substation with parallel-running transformers. These coordination levels are between (1) MCC-molded case-circuit breakers and LVPCB feeder breakers, (2) feeder breaker and bus section breaker, and (3) bus section breaker and main breaker. HRC for a fault at F3 is included, and this fault is cleared by the transformer primary fuse. Two types of LVPCBs A and B as in Table 21.14 are considered.

The calculations in Table 21.15 are based upon a transformer impedance of 5.75% on its own OA base, 40 kA three-phase short-circuit level at 13.8 kV, motor load served from the transformers approximately equal to their kVA rating, and motor contributions removed after 5 cycles. The table is constructed with full three-step coordination between the faults at F1, F2 (bus section), and F2 (main breaker). It shows the following:

The incident energy release and HRC levels are much lower when a breaker trip device with closer ST delay bands is selected.

HRC 2 is obtained only in one case, with a 1000 kVA transformer and fault at MCC cleared in 92 ms at the feeder LVPCB.

If a fault is to be cleared by the primary transformer fuse, fault location F3, tremendous amount of incident energy is released and hazardous conditions exist. This area in the

TABLE 21.15

Arc Flash Hazard Calculations in Low-Voltage Systems and Various Transformer Ratings

				Breaker Type A Three ST Delay Bands, 200, 300, and 500 ms		Breaker Type B ST Delay Bands, Table 21.14, 92, 158, and 200 ms	
Tran. kVA (Total)	Primary Fuse	Fault at	Bolted Fault Current (kA)	Incident Energy (cal/cm^2)	HRC	Incident Energy (cal/cm^2)	HRC
2500	175 E	F1	49.44	22	3	11	3
		F2 (BS)	62.77	37	4	21	3
		F2 (M)	62.77	63	D[a]	26	4
		F3 (F)	62.77	b		244	D
2000	150 E	F1	41.40	19	3	9.4	3
		F2 (BS)	50.01	32	4	17	3
		F2 (M)	50.01	57	D	22	3
		F3 (F)	50.01	b		200	D
1500	125 E	F1	33.0	15	3	7.4	2
		F2 (BS)	38.48	24	3	14	3
		F2 (M)	38.48	40	4	17	3
		F3 (F)	38.48	b		155	D
1000	80 E	F1	23.55	12	3	7.1	2
		F2 (BS)	26.21	17	3	9.5	3
		F2 (M)	26.21	28	4	12	3
		F3 (F)	26.21	b		107	D

Note: F2(BS): Bus section breaker tripped, F2(M): main secondary breaker tripped, F3(F): transformer primary fuse operator. D[a]: abbreviated for Danger, b: arcing time limited to 2 s.

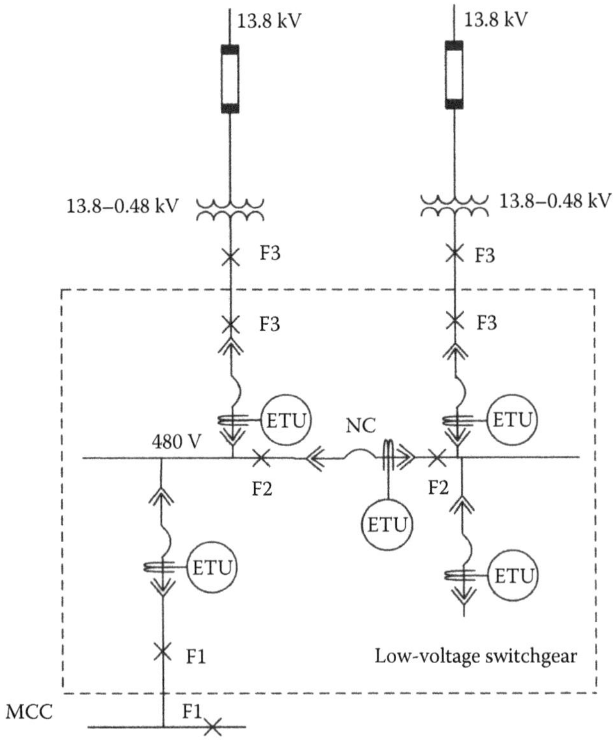

FIGURE 21.13
Low-voltage distribution system with three-step coordination.

low-voltage switchgear, which can remain unprotected due to location of sensors and trip devices, should not be ignored.

Some measures to reduce HRC are

1. Zone interlocking will effectively eliminate the time delay in clearing the fault in zones 2 and 3. Thus, there will be only one coordination interval, yet it gives HRC 3 for 65 kA fault.
2. A certain area as shown in Figure 21.12 can remain unprotected, and the fault in this zone must be cleared by the transformer primary fuse. This is not acceptable. An MMPR through a maintenance switch, similar to the configuration shown in Figure 21.7 for MV systems, can be used. The primary fuse protection can be replaced with a fixed-mounted circuit breaker and low-voltage main breaker eliminated. Again, this configuration is acceptable according to NEC.
3. The feeder LVPCBs can be set at instantaneous setting. The maximum time delay will be 3 cycles, and, in this mode, the HRC2 will be obtained for 65 kA bolted short-circuit current. This does not mean that coordination with downstream MCCBs or MCPs will be entirely lost, though some compromise will be made. Also, it is possible to select downstream devices to fully coordinate with instantaneous setting on feeder LVPCBs.
4. Current-limiting fuses can be used.

The calculation below illustrates some of these concepts to reduce HRC to 2. Figure 21.14 shows that there is no secondary breaker, and the protection to LV-switchgear bus and the cable

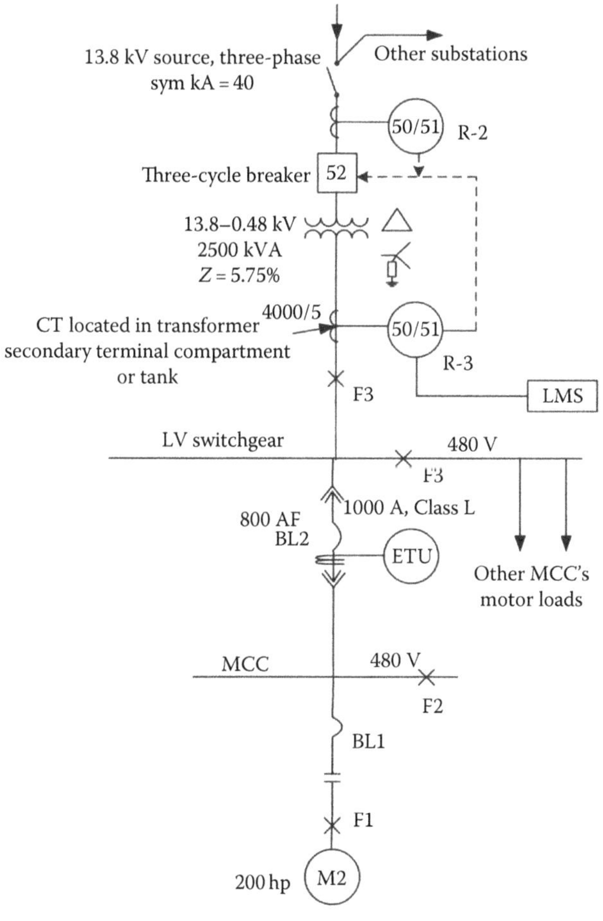

FIGURE 21.14
A low-voltage transformer with no secondary breaker, but primary circuit breaker tripped by secondary overcurrent relay.

connections from the transformer is provided by a CT-operated overcurrent relay that trips the transformer primary fixed-mounted breaker. The coordination of the protective devices is shown in Figure 21.15, and the results of arc flash calculation are shown in Table 21.16.

To further limit the HRC for a fault in the switchgear bus or incoming cables from the secondary of transformer fault location F3, a three-cycle primary breaker is required. Then, the arcing time, which is breaker operating time of 0.05 s plus relay operating time of 0.0167 cycles, gives incident energy of 6.4 cal/cm^2 and HRC2. For a 5-cycle breaker, the incident energy release increases to 8.6 cal/cm^2 and HRC of 3.

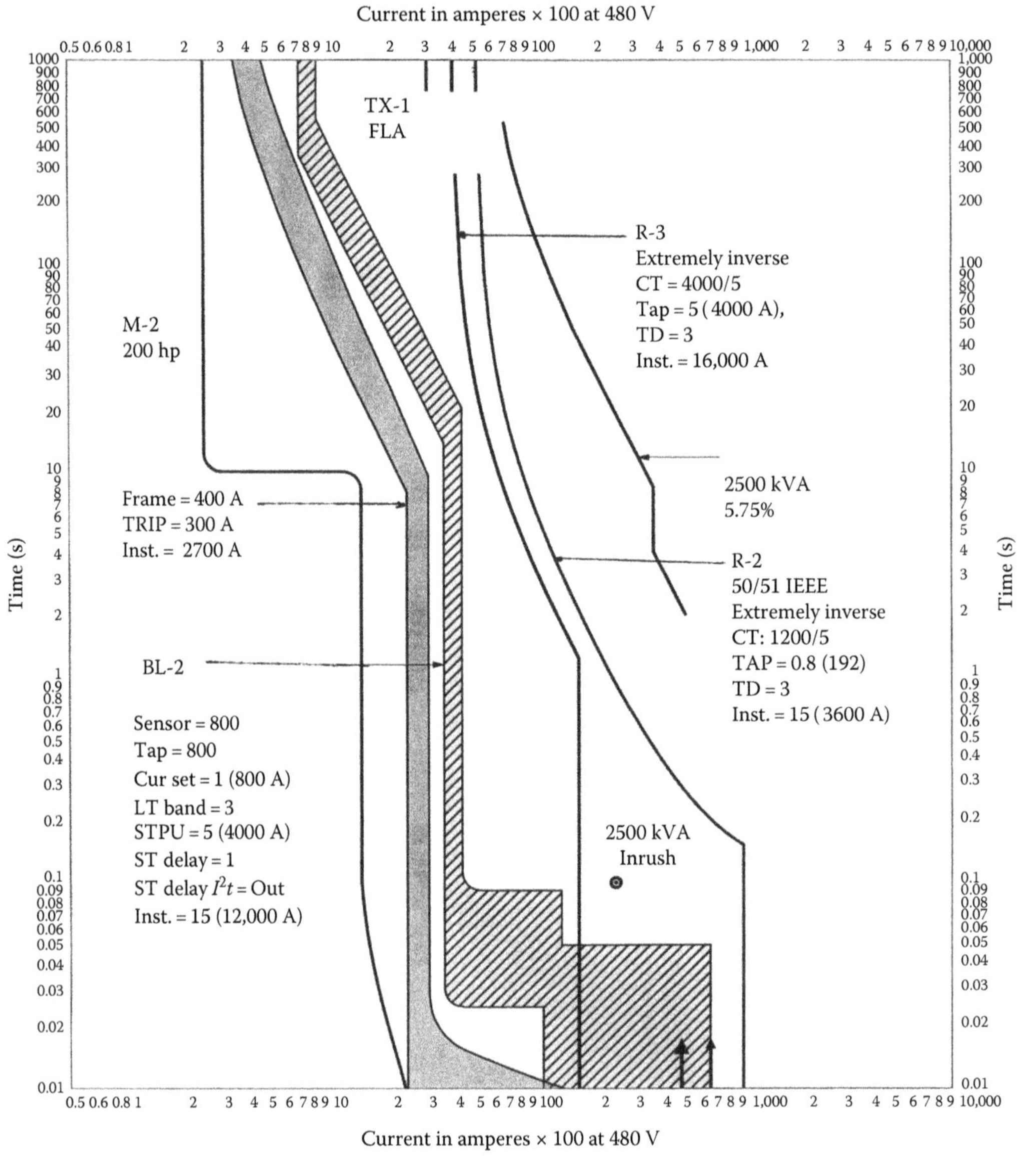

FIGURE 21.15
Coordination in the distribution system of Figure 21.14.

TABLE 21.16

Calculations of Arc Flash Hazard, Low-Voltage 480 V Distribution, 2500 kVA Transformer, System HR Grounded

System Configuration	TCC Plot	Fault at	Bolted kA, (Bus)	Arc Flash (kA)	FCT (s)	Working Distance (in.)	Incident Energy (cal/cm²)	HRC
No main secondary breaker, MMPR, with LMS, Figure 21.14	Figure 21.15	F1	53.76	22.99	0.01	18	0.90	0
		F2 No inst.	53.76	22.99	0.1	18	12	3
		F2 with inst.	53.76	22.99	0.05	18	6.4	2
		F3	65.61	26.87	0.0713[a]	24	6.4	2
Integrally fused main and secondary breakers, Figure 21.16	Figures 21.17 and 21.18	F2, MCC	53.26	22.91	0.008	18	1.1	0
		F3, no inst.	65.61	26.87	0.158	24	13	3
		F3, with inst.	65.61	26.87	0.05	24	5.3	2

[a] Primary breaker must be a three-cycle breaker.

The low-voltage MCC, fault at F2, is exposed to 12 cal/cm^2 of incident energy and HRC of 3, if no instantaneous settings on feeder breaker BL2 in Figure 21.14 are provided. With instantaneous settings, HRC is reduced to 2. Figure 21.15 illustrates that even with the instantaneous settings, reasonable coordination with the MCCB in low-voltage starters in MCC is achieved. For a fault downstream of the motor starter, the HRC is 0.

The instantaneous settings on R3 do not coordinate with feeder breaker settings. These can be brought into action through a maintenance switch. Then, under normal operation, the system is fully coordinated.

Maintenance switch should be possible to be remotely operated, should be provided with indicating lights, and should be key operated. These features can be provided when the maintenance switch operates on a MMPR, but not always with low-voltage circuit breaker trip programmers.

Next, the integrally fused breakers are considered, as illustrated in Figure 21.16. The main 4000 AF breaker is provided, with 4000 A class L limiter, and the 800 AF feeder breaker is provided with 1000 A class L limiter. The coordination with this arrangement is depicted in Figures 21.17 and 21.18. Note that the limiters do not coordinate well with the settings. Say, for a fault F2, exceeding 10 kA, 1000 A limiters will operate faster that the ST or instantaneous settings on the feeder breaker. This limits the incident energy, and HRC on MCC is reduced to zero.

The coordination in Figure 21.17 shows that 4000 A limiter does not help in limiting the HRC at LV switchgear, which is at category 3. This is so because the arc flash current of 26.87 kA is much lower than the threshold current level of 4000 A limiter, and, for this value of arc flash current, the short-time setting operated in 0.158 s gives HRC 3.

If some compromise is made or an instantaneous setting on main 4000 AF breaker is brought into action through a maintenance switch, the coordination can be altered with the addition of instantaneous as shown in Figure 21.18. The reduces the incident energy to 5.3 cal/cm^2 and HRC of 2.

No area within any equipment enclosure should remain unprotected. In Figure 21.16, it is necessary to add another CT-operated relay upstream of main secondary breaker, fault at F4 which should trip the transformer primary breaker. This comparative analysis shows that providing a MMPR with maintenance switch, with CTs located in transformer tank or secondary terminal box, covers the entire connections from the transformer to the bus, and also a fault on the bus and is more economical compared to other alternatives.

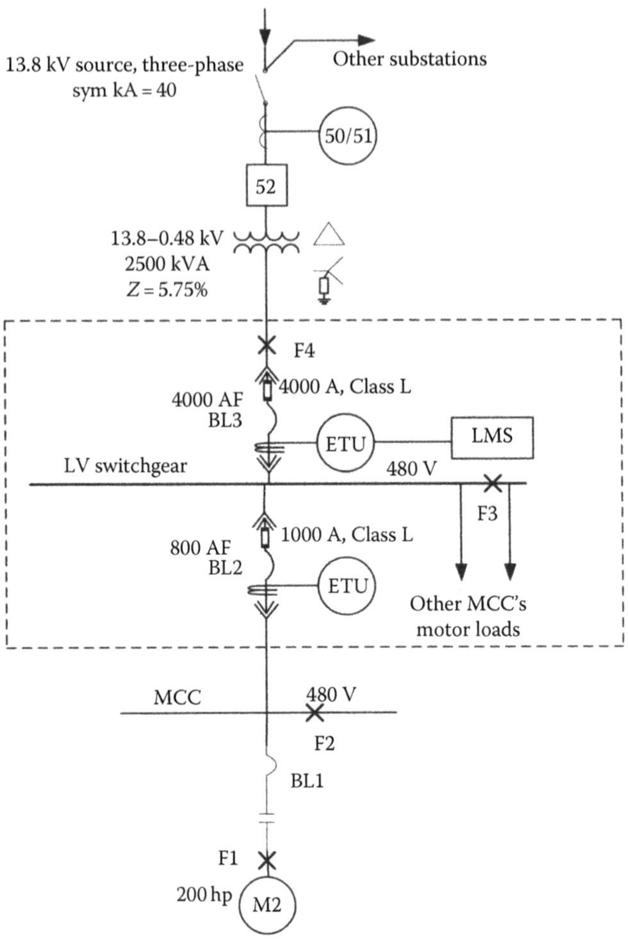

FIGURE 21.16
Low-voltage distribution with integrally fused primary and secondary breakers.

21.10 Accounting for Decaying Short-Circuit Currents

The application of short-circuit currents for calculations of switching devices short-circuit duties is quite different from their applications to arc flash calculations. To illustrate this point, rigorous analytical calculations of the short-circuit currents from generators and motors can be considered.

First, consider the fault decrement curve of the generator, Figure 6.6, which is reproduced as Figure 21.19 with ANSI/IEEE first cycle and interrupting duty short-circuit current superimposed. Note that the first cycle and interrupting duty ac components of the current are identical. Next short-circuit current of a 2.3 kV, 4-pole, full-load efficiency = 94.7%, full-load power factor = 88%, full-load current = 202 amperes, and locked rotor current = 1212 A is considered. The calculated parameters of this motor are stator resistance = 0.031 pu, rotor resistance = 0.015 pu, magnetizing reactance = 3.0 pu, magnetizing resistance = 100.0 pu, stator reactance = 0.0656 pu, rotor reactance = 0.0984 pu, motor transient reactance = 0.165 pu, ac short-circuit time constant = 0.295 s, and dc short-circuit time constant = 0.0142 s. Based upon this calculated data, the short-circuit profile with ANSI/IEEE values superimposed is shown in Figure 21.20. This shows

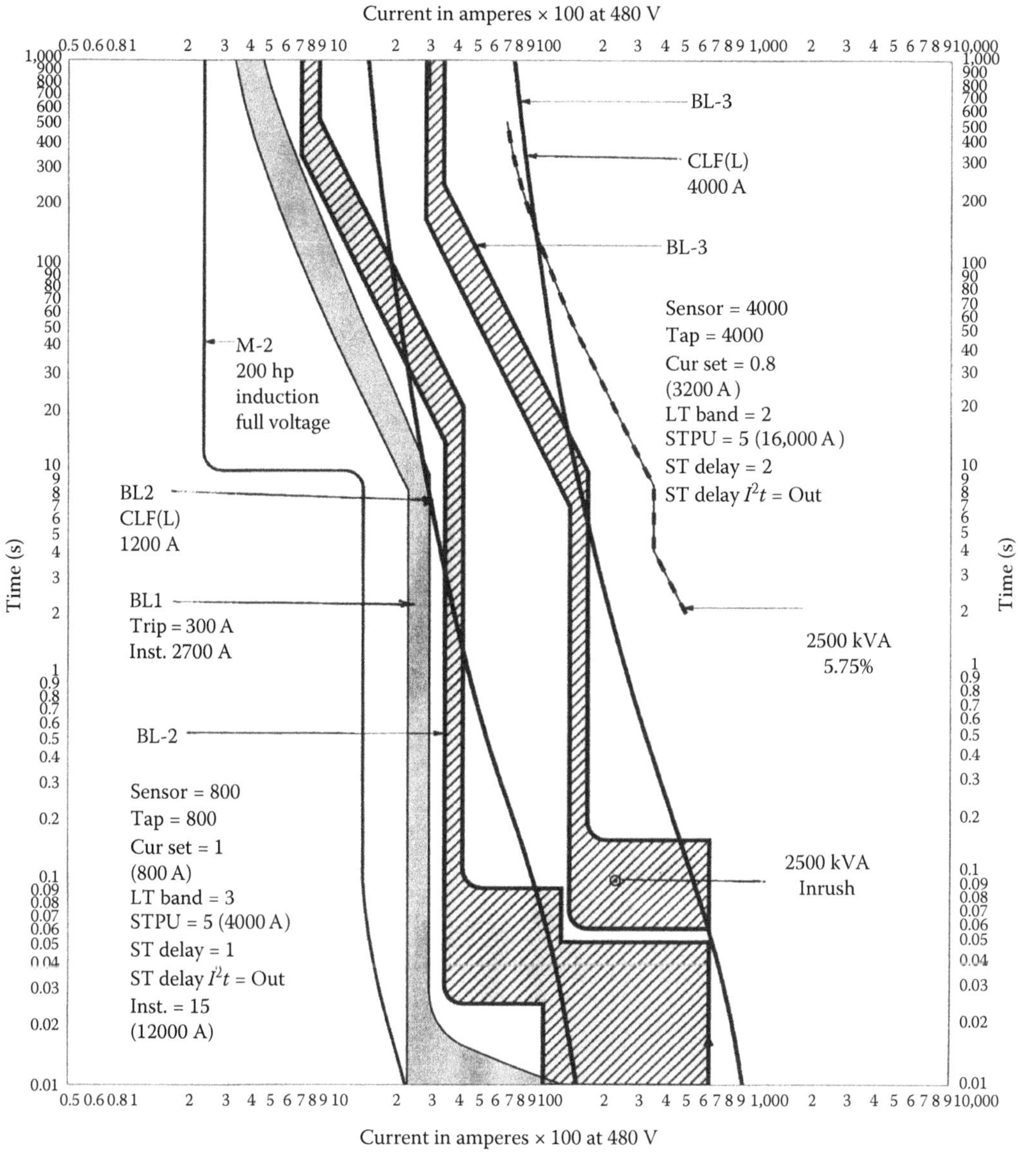

FIGURE 21.17
Coordination in the distribution system of Figure 21.16. Current limiters on main secondary breaker do not limit HRC.

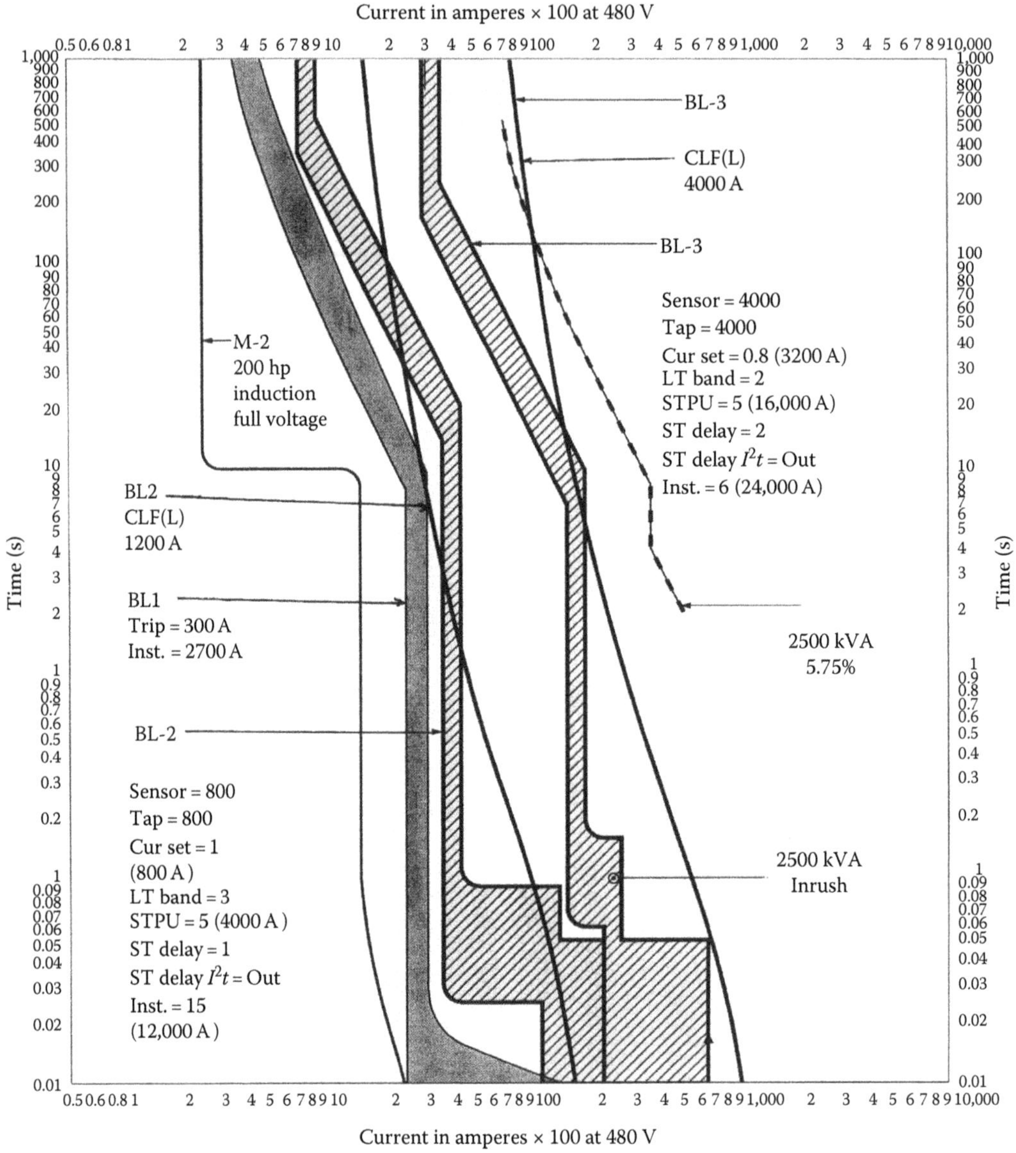

FIGURE 21.18
Coordination in Figure 21.17, modified by adding instantaneous setting on main secondary breaker, which gives HRC 2.

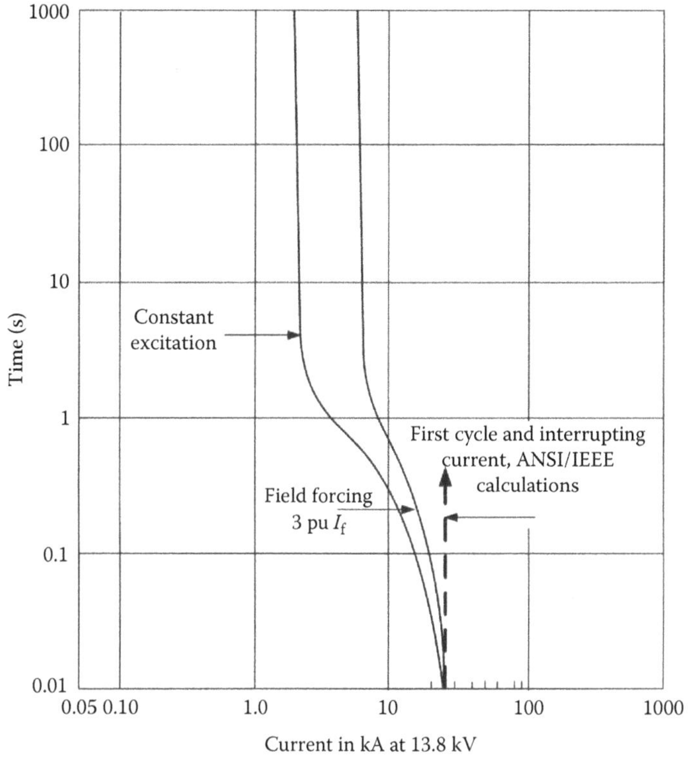

FIGURE 21.19
Short-circuit decrement curves of 100 MVA generator, Figure 6.6, with ANSI/IEEE first cycle and interrupting duty currents superimposed.

that ANSI/IEEE empirical calculations cannot represent accurately the short-circuit decay for arc flash calculations.

Consider a system bus that has multiple sources connected to it, which will contribute to the short-circuit currents. For example, Figure 21.21 shows a 13.8 kV bus, resistance-grounded system, which has a 40 MVA generator and a utility interconnection through a 45 MVA transformer. The generator operates in synchronism with a 138 kV utility source. A number of transformers carrying rotating ac motor load (induction and synchronous motors) and some static loads, not shown in Figure 21.21, are served from this bus. This is a typical industrial distribution cogeneration bus. The load distributions from this bus can be equivalence Thévenin impedances, as shown in this figure.

The calculated short-circuit current for a fault on the bus is 38 kA rms symmetrical. With respect to the calculations of short-circuit currents for arc flash analysis, using available arc flash calculation software programs, there are two situations:

1. No algorithm to account for the decay from the generators or motors is available. A user can select either the first cycle (momentary) or interrupting (1.5–5 cycle) symmetrical currents.
2. The program facilitates knocking out the motor contribution after a user-selectable time delay and similarly reduce the generator short-circuit current after a time delay.

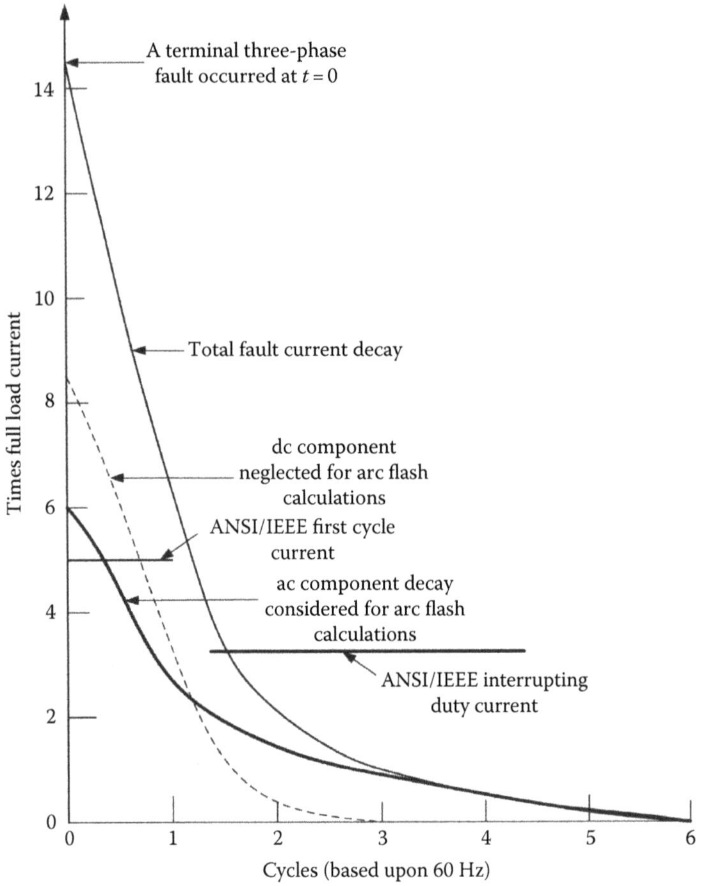

FIGURE 21.20
Calculated short-circuit currents for terminal fault of a 900 hp motor. The dc component is ignored for arc flash calculations.

The calculation to follow demonstrates that though situation 2 is better, it is still not accurate. Consider a fault on 13.8 kV bus of Figure 21.21. With no decay from the generator or motors, the incident energy accumulation profile is shown in Figure 21.22a. In this figure, for a bus fault, the generator breaker trips prior to opening of the utility tie breaker. The motor contributions continue for the entire period. This results in energy accumulation as shown.

Now, assume that the motor contributions are dropped in 6 cycles, and generator contribution is reduced to 300% of its full load current in 0.5 s. Then, this situation is depicted in Figure 21.22b. The total energy accumulated is reduced, as shown in the shaded area, compared to the situation shown in Figure 21.22a.

Yet, the total energy accumulations shown in Figure 21.22b are not accurate. In the real-world situation, it is not the step reduction but decay at a certain rate given by the transient parameters of the generators and motors, Figure 21.22c.

Therefore, incident energy calculated using the three figures will be

$$E_{\text{Figure 21.22a}} > E_{\text{Figure 21.22b}} > E_{\text{Figure 21.22c}}$$

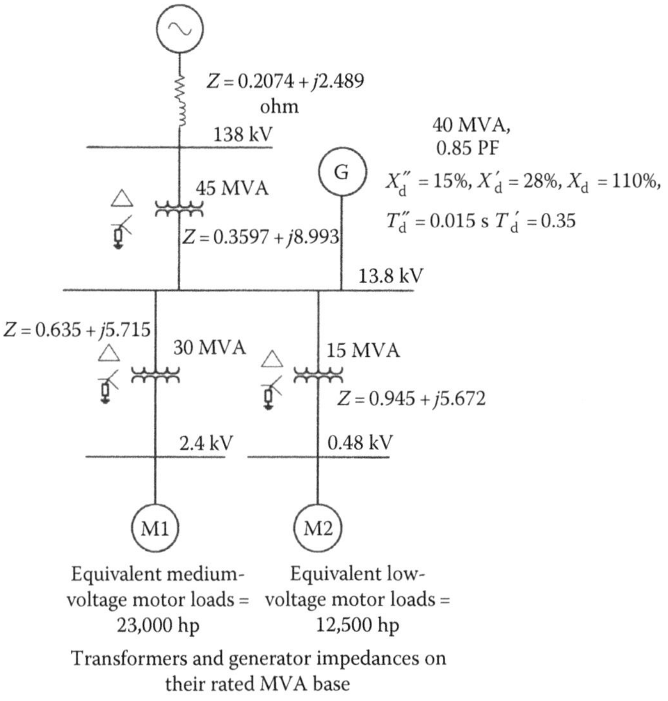

FIGURE 21.21
A 13.8 kV bus with multiple sources of short-circuit currents.

In other words, the calculations are overly conservative. In Figure 21.22b, it is only guesswork when the step change should be made.

There is no computer software that can simulate the results of Figure 21.22c. Theoretically, trapezoidal rule of integration or other step-by-step numerical techniques can be used. IEC standard [36] recommends integration over the period of short circuit to calculate the accumulated energy.

Example 21.3

Consider that the fault on bus A, fed from the generator and utility tie is removed simultaneously in 0.5 s. This is arbitrary to show the difference in calculations using the methodology shown in Figure 21.22a through c.

Calculation 1 (Figure 21.22a): No decay of short-circuit current from the generator or motors. The calculated results are shown in Table 21.17, row 1. Incident energy release = 36 cal/cm^2 and HRC = 4.

Calculation 2 (Figure 21.22b): Motor contribution knocked out in 8 cycles, and generator contribution reduced to 350% at 15 cycles, incident energy release = 30 cal/cm^2 and HRC = 4, Table 21.17, row 2.

Calculation 3 (Figure 21.22c): First, plot/calculate the overall current-time decrement curve of the short-circuit current. Utility source is considered non-decaying and motor and generator short-circuit contributions decay.

Plot decrement curves of the generator and motors. The symmetrical component of the generator short-circuit current, generator at no load at the rated voltage of the generator, using the generator parameters from Figure 21.21 can be plotted, Chapter 6. This fault decrement curve is plotted in Figure 21.23. The dc component is ignored. Similarly, the decrement curves of motors can be plotted. This gives the overall decrement curve by summation of the three components. The plot is extended to time 0.001 s (0.06 cycles).

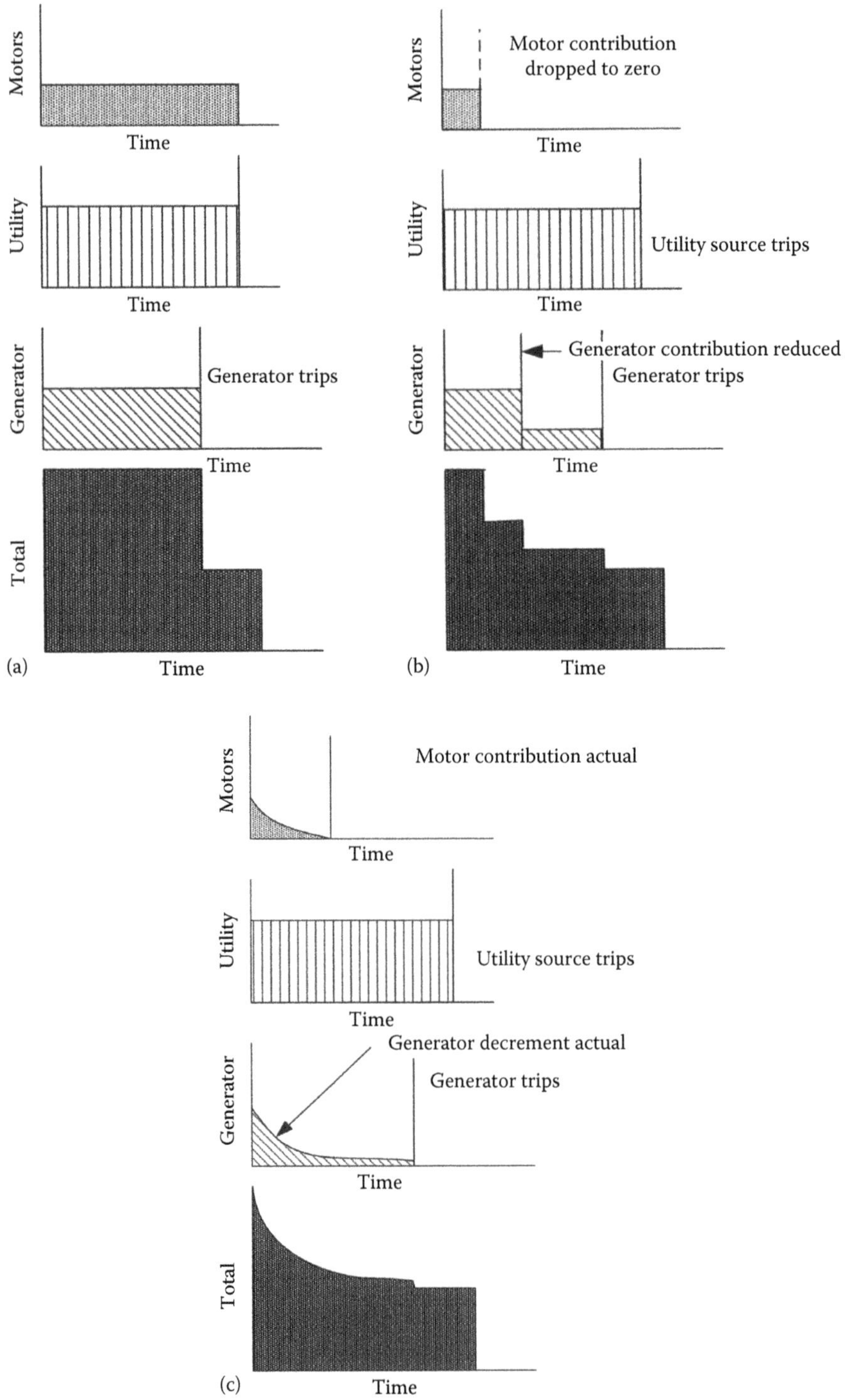

FIGURE 21.22
Accumulation of incident energy, see text.

TABLE 21.17

Variations in the Calculations of Incident Energy and HRC, 13.8 kV Switchgear, Resistance-Grounded system, Trip Delay 0.5 s, Breaker Opening Time 0.080 s (Five-Cycle Breaker)

Cal. No.	Bolted Current (kA rms)	Arcing Current (kA rms)	Incident Energy (cal/cm^2)	HRC
1	37.90	35.96	36	4
2	37.90	35.96	30	4
3			21.11	3

Note: Row 1, no decay; row 2, AC motor and generator decay, step change; row 3, accurate calculations.

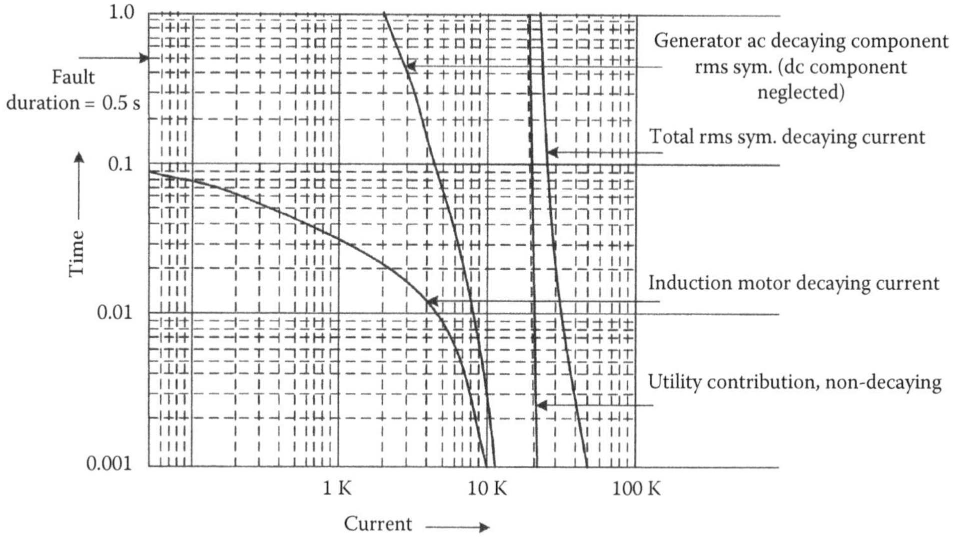

FIGURE 21.23
Calculation of short-circuit current decay profiles at 13.8 kV bus, Figure 21.21.

To calculate the incident energy, divide the 0.5 s interval into three intervals, arbitrary chosen for reasonable accuracy:

0.001–0.01 s

0.01–0.1 s

0.1–0.5 s

The average current in each interval and its time duration are shown in Table 21.18, with corresponding energy release. The summation gives 21.11 cal/cm^2 and HRC 3. There is a vast difference in the results obtained with calculating the total current decay, dividing it to number of sections, taking the average in each section, calculating the energy release in each section and then summing it up.

This shows that accounting for decay of short-circuit currents makes much difference in the ultimate results. Practically, with the hand calculations, it will be very time consuming to plot the decay at each bus, divide it into number of segments, and then calculate the incident energy and summate, though the calculation algorithm can be computerized.

TABLE 21.18

Calculations of Incident Energy by Plotting Overall Fault Decrement Curve

Bolted Current (kA)	Arcing Current (kA)	Duration (s)	Incident Energy (cal/cm^2)
35	33.25	0.01	0.61
29.2	27.64	0.09	4.5
23.5	22.48	0.40	16

The electrical distribution systems vary in complexity, power-handling capability, and interconnections. However, it is demonstrated that using the current technology, arc flash hazard can be reduced. The industry is moving in this direction. Each distribution system is unique in some way, yet the general guidelines and examples of calculations in this chapter, selection of appropriate protective devices, possible reduction of short-circuit currents in the design stage should be helpful in reducing HRC. A study of possible alternatives will be required, leading to an effective solution. See references [37 and 38] for further study.

Problems

21.1 Calculate the maximum arcing time to limit HRC to level 2 (incident energy less than or equal to 8 cal/cm^2), for a 13.8 kV resistance-grounded switchgear, three-phase bolted short-circuit current = 25 kA.

21.2 Repeat Problem 21.1 with solidly grounded system.

21.3 Repeat Problems 21.1 and 21.2 with 480 V system.

21.4 Calculate arc flash boundary in each case of Problems 21.1 through 21.3.

21.5 The generator of Example 6.1 runs in synchronism with a utility source, which gives 25 kA symmetrical short-circuit current at the 13.8 kV bus. Calculate the incident energy release for a 30-cycle fault at the 13.8 kV bus, contributed by the generator and utility source. Recalculate the incident energy with ANSI/IEEE generator short-circuit current contribution.

21.6 In Figure 21.7, the transformer TX1 of 7.5 MVA is replaced with a transformer of 3.75 MVA provided with a 300 E fuse, characteristics as shown in Figure 21.8. Recalculate the short-circuit currents and time–current coordination to ascertain the arcing time for a fault on 2.4 kV bus, fault location F4.

21.7 In Figure 21.14, replace 2500 kVA transformer with a 1000 kVA transformer of 5.5% impedance. Select appropriate transformer primary fuse protection. Replot the time–current coordination curve of Figure 21.15.

21.8 Design a 2.4 kV transformer substation to serve a load of 4000 hp motor load through one medium-voltage MCC connected to the secondary of the transformer. The largest motor is 1500 hp. The transformer is served from a 13.8 kV system with a symmetrical short-circuit level of 40 kA. Select all equipment ratings and protection systems applying commercially available standard equipment and protective devices. Run a time–current coordination plot. Calculate arc flash hazard levels. How these can be reduced to HRC 2?

21.9 Design a 480 V transformer substation to serve a load of 2000 hp motor of low-voltage motor loads, divided on four low-voltage motor control centers. The largest motor is 200 hp. The transformer is served from a 13.8 kV system with a symmetrical short-circuit level of 40 kA. Select all equipment ratings, and protection systems applying commercially available standard equipment and protective devices. Run a time–current coordination plot. Calculate arc flash hazard levels. How these can be reduced to HRC 2?

References

1. W.A. Brown and R. Shapiro. Incident energy reduction techniques. *IEEE Ind. Mag.* 15(3): 53–61, May/June 2009.
2. T. Gammon and J. Mathews. Conventional and recommended arc power and energy calculations and arc damage assessment. *IEEE Trans. Ind. Appl.* 39(3): 197–203, May/June 2003.
3. T. Gammon and J. Mathews. Instantaneous arcing fault models developed for building system analysis. *IEEE Trans. Ind. Appl.* 37(1): 197–203, January/February 2001.
4. V.V. Terzija and H.J. Koglin. On the modeling of long arc in still air and arc resistance calculations. *IEEE Trans. Power Deliv.* 19(3): 1012–1017, July 2004.
5. A.D. Stokes and D.K. Sweeting. Electrical arcing burn hazards. *IEEE Trans. Ind. Appl.* 42(1): 134–142, January/February 2006.
6. H.B. Land, III. Determination of the case of arcing faults in low-voltage switchboards. *IEEE Trans. Ind. Appl.* 44(2): 430–436, March/April 2008.
7. H.B. Land, III. The behavior of arcing faults in low-voltage switchboards. *IEEE Trans. Ind. Appl.* 44(2): 437–444, March/April 2008.
8. R. Wilkins, M. Allison, and M. Lang. Effects of electrode orientation in arc flash testing. In *Proc. IEEE IAS Annual Meeting*, Hong Kong, 2005, pp. 459–465.
9. IEEE. IEEE guide for performing arc-flash hazard calculations. Standard 1584, 2002.
10. J.C. Das. *Arc Flash Hazard*. New York: McGraw-Hill Year Book of Science and Technology, 18–20, 2008.
11. R. Lee. The other electrical hazard: Electrical arc blast burns. *IEEE Trans. Ind. Appl.* 1A-18(3): 246–251, May/June 1982.
12. M.G. Droouet and F. Nadeau. Pressure waves due to arcing faults in a substation. *IEEE Trans.* PAS-98(5):1632–1635, September 1979.
13. Society of Fire Protection Engineers(SPFE) Task Force. *Engineering Guide to Predicting First and Second Degree Skin Burns*, Boston, MA, 2000.
14. NFPA 921, Guide for fire and explosion investigations.
15. NFPA 70E—Electrical safety in workplace, 2009.
16. IEEE 1584a, IEEE guide for performing arc-flash hazard calculations—Amendment 1, 2004.
17. IEEE P 1584b/D2, Draft 2, Guide for performing arc-flash hazard calculations—Amendment, July 2009 (unapproved).
18. R.L. Doughty, T.E. Neal, and H.L. Floyd II. Predicting incident energy to better manage the electrical arc hazard on 600-V power distribution systems. *IEEE Trans. Ind. Appl.* 36(1): 257–269, January/February 2000.
19. ANSI/IEEE C2—National electric safety code, 2007.

20. ASTM F1506, Specifications for textile materials for wearing apparel for use by electrical workers exposed to momentary electrical arc and related thermal hazards, 1998.
21. J.C. Das, Design aspects of industrial distribution systems to limit arc flash hazard. *IEEE Trans. Ind. Appl.* 41(6): 1467–1482, November/December 2005.
22. Applied protective relaying, Westinghouse Electric Corporation, Newark, NJ, 1979.
23. ANSI/IEEE. IEEE recommended practice for protection and coordination of industrial and commercial power systems. Standard 242-1986.
24. J.L. Blackburn. *Protective Relaying*. Boca Raton, FL: CRC Press, 2007.
25. A.F. Silvia. *Protective Relaying*. Boca Raton, FL: CRC Press, 2009.
26. S.H. Horowitz (Ed.). *Power System Relaying*. New York: IEEE Press, 1980.
27. J.C. Das. Protective relay coordination—Ideal and practical. In *Conf. Record, IAS Annual Meeting*, San Diego, CA, 1989, pp. 1861–1874.
28. W.J. Lee, M. Sahni, K. Methaprasyoon, C. Kwan, Z. Ren, and M. Sheeley. A novel approach for arcing fault detection for medium-voltage/low-voltage switchgear. *IEEE Trans. Ind. Appl.* 45(4): 1475–1483, July/August 2009.
29. J.C. Das. Effects of momentary voltage dips on operation of induction and synchronous motors. *IEEE Trans. Ind. Appl.* 26(4):711–718, July/August 1990.
30. NFPA-70-National electric code-2009.
31. ANSI/IEEE C37.20.7. IEEE guide for testing metal enclosed switchgear rated upto 38 kV for internal arcing faults, 2007.
32. D.R. Doan, J.K. Slivka, and C.J. Bohrer. A summary of arc flash hazard assessment and safety improvements. *IEEE Trans. Ind. Appl.* 45(4): 1210–1216, July/August 2009.
33. R.L. Doughty, T.E. Neal, T.L. Macalady, V. Saportia, and K. Borgwald. The use of low-voltage current limiting fuses to reduce arc flash energy. *IEEE Trans. Ind. Appl.* 36(6):1741–1749, November/December 2000.
34. UL489-Molded case circuit breakers and circuit breaker enclosures, 1991.
35. M. Valeds, T. Papallo, and A. Crabtree. Method of determining selective capability of current limiting overcurrent devices using peak let-through current. In: *Conf. Record, IEEE Paper and Pulp Ind. Conf.*, Birmingham, AL, 2009, 145–153.
36. IEC 60909-0, Short-circuit currents in thee-phase AC systems, Part-0, Calculation of currents, 2001, Also IEC 60909-1, Factors for calculation.
37. J.C. Das. Protection planning and system design to reduce arc flash incident energy in a multi-voltage-level distribution system to 8 cal/cm^2 (HRC2) or less—Part I: Methodology: *IEEE Trans. Industry Appl.* 47(1), pp. 398–407, Jan./Feb. 2011.
38. J.C. Das. Protection planning and system design to reduce arc flash incident energy in a mutli-voltage-level disbtribution system to 8 cal/cm^2 (HRC2) or less—Part II: Analysis. *IEEE Trans. Industry Appl.* 47(1), pp. 408–420, Jan./Feb. 2011.

22

Wind Power

Renewable energy sources (RES) are becoming of increasing interest in view of rising energy costs and growing concerns of global climate changes and environmental effects. There is foreseeable scarcity of fossil fuels, while the energy demand is increasing in the emerging countries (Figure 1.16). There is energy dependence in the United States due to heavy imports. There seems to be an urgent need for the reduction of energy produced from fossil fuels and the increase in the share of energy production from renewable sources. This leads to a sensible energy mix and improvements in the generation, transmission, and consumption. More than 94 GW of wind power generation has been added worldwide by the end of 2007, out of which over 12 GW is in the United States and 22 GW in Germany alone. Looking at energy penetration levels (ratio of wind power delivered by total energy delivered), Denmark leads, reaching a level of 20% or more, followed by Germany. In some hours of the year, the wind energy penetration exceeds 100%, with excess sold to Germany and NordPool. Nineteen offshore projects operate in Europe producing 900 MW. U.S. offshore wind energy resources are abundant.

In the United States, wind power generation accounts for approximately 0.6% of the total, and Renewable Energy Laboratory (DOE/NREL) conducted an investigation as to how 20% of energy from wind will look like in 2030 (Figure 22.1) [1]. AEP (American Electric Power) produced a white paper that included 765 kV network overlay for U.S. power system that will increase reliability and allow for 400 GW of wind or other generation to be added. The political climate change in Washington can impact these decisions and accelerate exploitation of wind power. The modern wind power plants can be as large as 300 MW and are often located within a short distance of each other.

With respect to power system analysis and transient stability analyses, interconnections of wind power stations with grid systems pose even larger problems, which require a more thorough analysis of the interconnection and system isolation under disturbances.

A challenge of increasing the contributions from RES is that their outputs vary directly with the weather conditions. Figure 22.2a shows hourly wind production over 365 days and statistical nature of wind power production, and Figure 22.2b shows the statistical nature of wind power generation. The patterns will vary, depending upon the location, and these figures show the wide variations in wind power that can occur. Though the tools for predicting generation capabilities from RES have considerably improved, the scheduling of these units remains difficult. The power produced in offshore units may not be available even for days or weeks. With the exception of hydro and biomass, the RES have no storage capabilities and power generation follows the weather conditions. Increasing system flexibility will be required to manage the task of system balancing. The conventional power plants must accommodate the swings in power generation of RES. The conventional thermal plants have to exercise a greater amount of control, and will frequently operate inefficiently at partial load or will be shutdown and restarted. Thus, emission and fuel costs increase, besides increasing the cost of energy produced from these units. The capacity of existing storage power plants, mainly pumped hydro, is not adequate and system flexibility is needed for balancing functions.

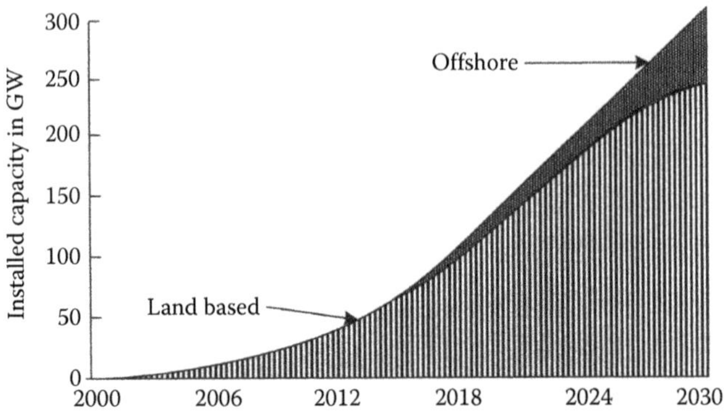

FIGURE 22.1
Twenty percent penetration of land-based and offshore wind power installed capacity.

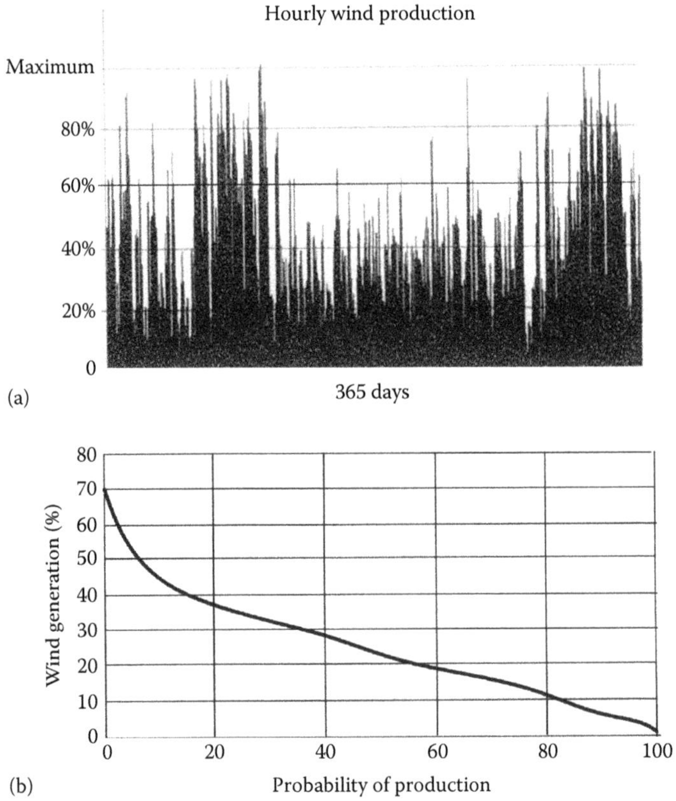

FIGURE 22.2
(a) Hourly variations of wind power, pattern depends upon location. (b) Probability of wind power production, which varies with location.

Transmission systems and distribution grids have been planned such that large power stations are as close as possible to the load centers. A large potential of RES is available far from the load centers. This leads to a new load-flow situation and bottlenecks already occur in transmission systems with large wind power. Contingency management serves short-term goals and in the long run appropriate reinforcement of the grids is a must; in the United States, an overlay of 765 kV ac transmission network is proposed (see Section 22.1).

These high-voltage lines become an issue due to ROW requirements and public opposition but become essential if large amounts of RES that are remotely located are to be integrated. The following aspects become important:

1. *Power flow control*: Large unexpected fluctuations in wind power can cause additional loop flows through transmission grids. When a large deviation has to be balanced by other sources, these may be located far away from the wind generation. The flow through parallel transmission paths may reverse, making it difficult to operate them. The preferred solution is to build high-voltage transmission systems.
2. *Congestion management*: This needs to be addressed from the market-side view point as well as from the technical viewpoint using FACTs devices. It is known that installed wind power in some grid areas has already reached a level that entails problems in grid management and grid control in strong wind periods caused by large wind fluctuations.
3. *Long-term and short-term voltage stability*: By replacing large synchronous generators with small generators integrated into sub-transmission systems or even distribution systems far from load centers, the amount of reactive power reserve in the system is considerably reduced. The reactive power contribution of the wind power plants is highly dependent upon the technology used, connection point, as well additional reactive power support; SVC, STATCON, switched capacitors or increasing the "must run" units. There is already reduction of long-term stability limit resulting in a reduction of maximum power transfer along interconnecting lines. The short-term voltage stability limit remains above long-term voltage stability limit and is not as critical, yet it should be addressed and planned for in the grid operation.
4. *Transient stability and low-voltage ride-through capability*: This is further discussed in Section 22.4.2. The stability is also a function of the generator type.
5. *Green house gas emissions*: The green house gas reduction range is in the range of 350–450 kg CO_2 per MWh of power generated by wind power plants.
6. *Reserve requirement*: The impact of wind power on tertiary control requirements can be estimated. Significant hourly fluctuations and forecast errors have to be managed. It is necessary to consider the effect of wind and solar energy together.
7. *Energy storage*: Large-scale storage is possible in pumped-hydro power stations. In the future compressed air energy storage is seen. The production and storage of hydrogen could be a solution, though the overall efficiency of hydrogen chain, production, storage, and combustion is low, on the order of 30%–35%. Developments in battery technology could lead to revolution in car technology.

22.1 AEP 765 kV Transmission Grid Initiative in the United States

The transmission grids in the United States are mature and heavily loaded. The constraints of thermal and voltage stability have been well documented on systems operating at voltages through 500 kV. Expanded EHV systems are required to provide long-term transmission capacity and integration of RES. HVDC can transmit more power over longer distances. In HVDC transmission, the transfer capability is influenced by the losses and not by reactive power or stability concerns as in ac transmission systems. There are two converter technologies used in the HVDC transmission: (1) current source converters (CSC) and (2) voltage source converters (VSC) (see Section 17.23). The CSC-based technology has been applied at ± 600 kV and 6300 MW on double bipoles (Brazil) and will be soon applied at ±800 kV up to 6400 MW (China). VSC-based HVDC transmission is available through ±320 kV and 1100 MW for symmetrical monopole configuration and up to 2200 MW at ±640 kV in bipolar configuration. References [2–4] describe HVDC transmission system configurations.

There is an interdependence of HVDC and HVAC transmission systems. For example, ac network where the CSC-based HVDC is connected must be strong relative to its transmitted power. CSC converters draw reactive power at the point of interconnection and the reactive power demand increases with loading. Harmonic filters and other reactive power compensation methods are required (Chapters 13 and 20). VSC-based systems do not require a strong ac network to operate properly and can serve to interconnect for smaller scale wind generation using induction generators. To support local system voltage, a VSC station can act as an SVC or STATCON.

AC transmission provides more accessibility and flexibility. AC transmission systems can be tapped more easily to serve loads or pickup resources over moderate distance. A system that is flexible enough to integrate new resources and additional future generation is required. In some cases, HVDC may be an advantage for point-to-point transmission of bulk power; it is an asynchronous link and short-circuit power is not transferred. For underground or submarine connections, HVDC cable systems provide high capacity over significant distances without reactive power compensation required by ac systems. This is not a comparison of HVDC vis-à-vis HVDC; the two systems must coexist. A technology comparison of EHV ac and HVDC systems is shown in Table 22.1 [5].

Figure 22.3 shows the line loadability versus distance adapted from [5]. For a 200 mi transmission distance, the number of ac lines required for 6000 MW capacity will be 2 765 kV lines, 6 500 kV lines, and 12 345 kV lines. These point to the reduction of ROW with 765 kV transmission lines. These numbers of lines are calculated without series compensation (Chapter 13). With series compensation, the lines required for 6000 MW are four and eight, respectively, for 500 kV and 345 kV, and one series-compensated 765 kV line can carry 4800 MW.

The energy efficiency is another important criterion and Figure 22.4 compares losses at full load, including line resistive and terminal equipment losses for the 6000 MW point-to-point transmissions. AEP is investigating larger 765 kV conductors in order to achieve greater

TABLE 22.1

EHV ac and HVDC Systems for 6000 MW Delivery

			Circuit					Conductor		
Voltage (kV)	No. of Lines	Circuits per Tower	SIL (MW)	Capacity (MW)	Series Comp.	Shunt React.	Shunt Caps.	Type	No. in Bundle	Area (kcmil)
345 ac	4	2	400	800	Yes	No	Yes	ACSR	2	1590
500 ac	4	1	900	1800	Yes	Yes	Yes	ACSR	3	1590
765 ac	2	1	2400	3100	No	Yes	Yes	ACSR/TW	6	957
±500 dc	2	1		3000				ACSR	3	2515
±800 dc	1	1		6000				ACSR	4	2515

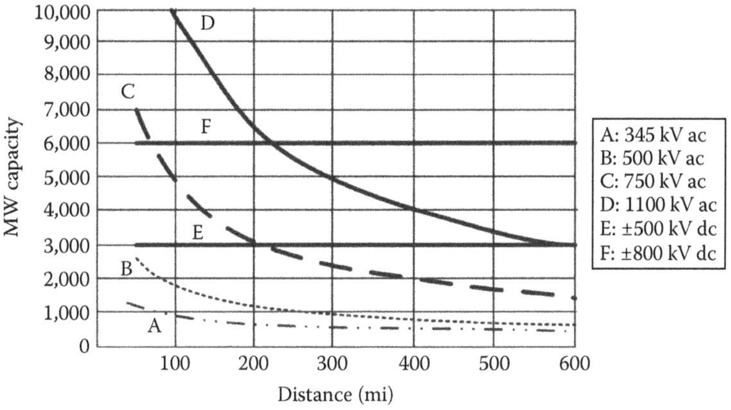

FIGURE 22.3
Line loadability versus distance, comparative for ac and dc transmission voltages as shown.

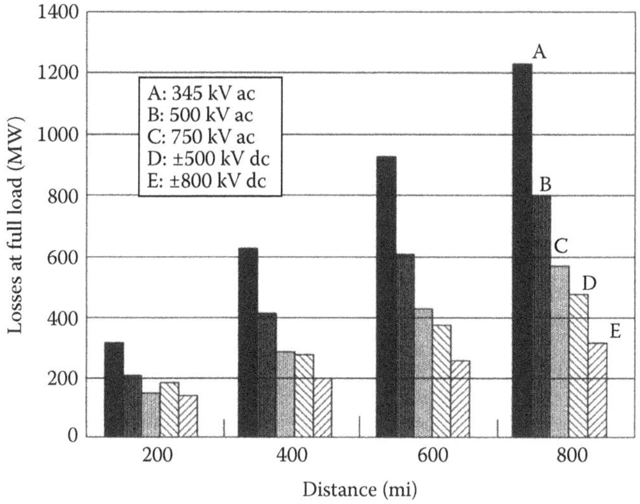

FIGURE 22.4
Comparative losses versus distance.

efficiency. It may be said that long-distance point-to-point transmission without intermediate accessibility to line (HVDC systems) is rather uncommon in much of the United States.

Figure 22.5 is an NREL map of wind power potentiality and illustrates an overlay of 765 kV ac transmission system and dc links. Note the potentiality of high wind power generation in the middle corridor of the country and coastal regions. As this figure is reproduced in gray tones, the differences in the wind power production areas may not be so clear. Joint studies will be necessary to determine final planned configurations. HVDC applications will be integrated depending upon the specific studies. The new grid concepts will be a strong integrator of systems and sources (see Refs. [6–11]).

22.1.1 Maximum Transfer Capability

In May 1965, NERC introduced a term transmission transfer capability (TTC) and later in 1996 approved a document entitled available transfer capability (ATC) definition and determination. According to this document, calculation of ATC requires calculation of several related terms including TTC, which is defined as the maximum power that can

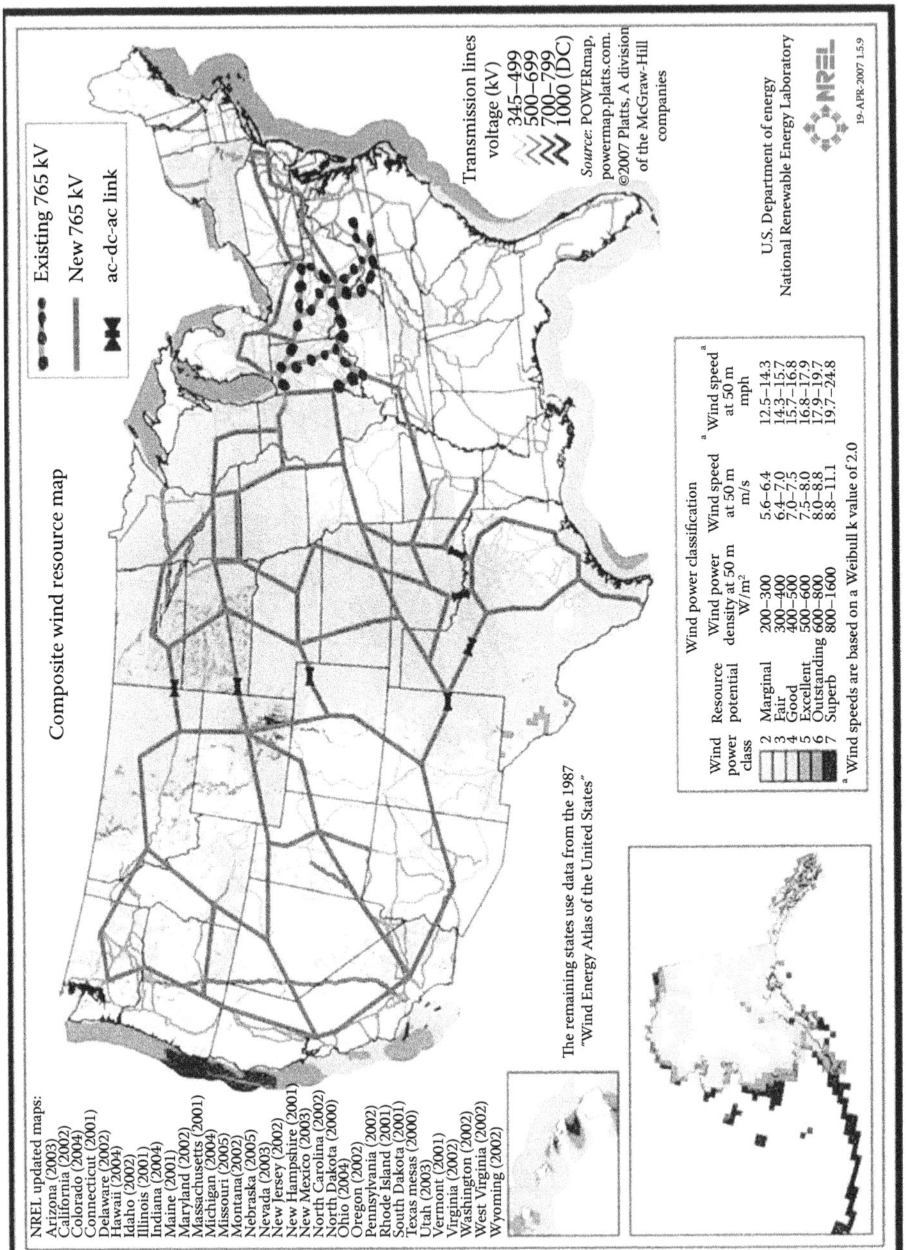

FIGURE 22.5
NERC map of interstate grid for wind power integration.

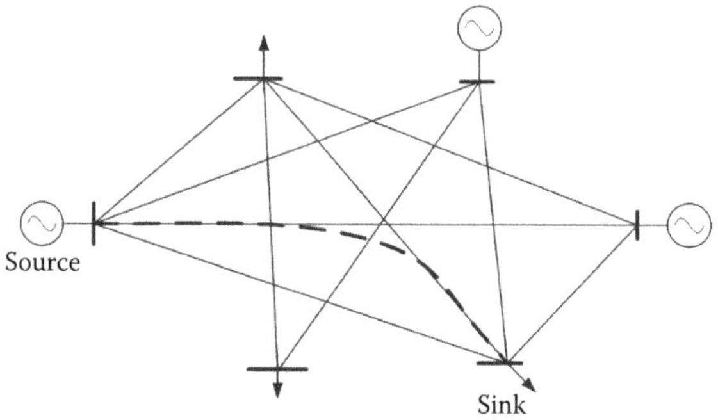

FIGURE 22.6
Representation of TTC (see text).

be transferred in a reliable manner between a pair of defined sources and sink locations in the interconnected system while meeting all predefined set of pre- and post contingency system conditions [12]. A number of contingency system conditions are considered and TTC is determined at each of these conditions. This maximum transferable power is commonly known as maximum transfer capability (MTC). A number of methods have been proposed for calculating it [13–17]—optimal power flow (OPF) (Chapter 16) is one such method. Consider Figure 22.6 and let the additional power be represented by scalar γ. This additional power is accounted for by modifying the active and reactive power demands at the sink bus. Assuming a constant power factor,

$$\begin{aligned} P &= P_0 + \gamma \cos \phi_0 \\ Q &= Q_0 + \gamma \sin \phi_0 \end{aligned} \tag{22.1}$$

where
P_0 and Q_0 and the base active and reactive powers
ϕ_0 is the power factor angle

The optimization is formulated as follows:

$$\begin{aligned} &\max\ \gamma \\ &\text{s.t.} \\ &G(x, \gamma) = 0 \\ &H(x, \gamma) \leq 0 \end{aligned} \tag{22.2}$$

where
x is a vector of system control and state variables
$G(x, \gamma) = 0$ are a set of equality constraints accounting the power balance equations
$H(x, \gamma) \leq 0$ are a set of inequality constraints according to system security limits

These can be

- Generation capacity
- Transient stability limit
- Voltage stability limit
- Voltage excursion limits
- Transmission line thermal capability

Then the solution of OPF gives MTC:

$$\text{MTC} = P_0 + \gamma_{\max}\phi_0 \tag{22.3}$$

Investigations of TTC with wind power generation become important. The varying nature of wind power generation and system voltage instability and transient stability are essential parameters. Monte Carlo methods are well suited to investigate these variables and effects of system performance. A simplified description of Monte Carlo methods is given in [18].

22.1.2 Power Reserves and Regulation

Different generation types behave differently with respect to regulation features.

Hydrogenation: Hydro power offers good power regulation. Approximately 40% of the capacity can be regulated in 1 min if the reservoir holds enough water. Hydro plants with pumped storage facilities offer good storage capability.

Nuclear power plants: These are operated at their maximum capacity throughout the year as the base load plants and do not offer much regulation capability. One minute regulation capability of boiling water nuclear reactors is generally on the order of 1%–3% or little higher.

Condensing power plants: Condensing plants using coal as fuel can be regulated by approximately 5% within 1 min. The regulation of peat-fueled plants is 3%. Starting time of a condensing plant is dependent upon time that elapsed from the last use. Warm plants can be started in some hours while cold plant starting time may be approximately 10 h.

Gas turbines: Gas turbines are capable of fast and reliable power regulation and can be started and synchronized to the grid in about 3 min. These can be regulated to approximately 40% of their capacity in 1 min.

Combined heat and power plants (*CHP*): These have different capabilities depending upon the operation. For example, bypassing a high-pressure preheater can increase electrical power in the range of 5%–10% and by bypassing the condenser in an extraction back-pressure plant it is possible to increase power quickly by 10%–15%.

Wind power: Wind power also offers from upward and downward regulation capability by rotor blade controls.

Load shedding: Load shedding is an important tool of regulation in case of disturbances, and the loads for fast active disturbances are disconnected to maintain transient stability. HVDC systems serve as reserves between two ac systems.

22.1.3 Congestion Management

Reliable operation means that the transmission system has to operate within constraints of thermal limits and contingency analyses. For large-scale wind generation, transmission system operators (TSO) and wind farm operators must coordinate the data in advance for operation planning, bottlenecks on the tie lines in advance, and congestion forecast. We talk about short-term, middle-term and long-term congestion forecasts. The main objective is to determine power system state in advance that fulfills the network security requirements. Network and market-based measures are adopted to which a third set of measures can also be added. Again it can be formulated as an optimization problem with an objective function to minimize

$$F_t = \sum_{a=1}^{N^A} \left(W_a^{\mathrm{R}} f_{t,a}^{\mathrm{R}} + W_a^{\mathrm{M}} f_{t,a}^{\mathrm{M}} \right) \tag{22.4}$$

where F_t is the objective function in time t, which is divided into number of intervals to be analyzed. Index a describes the control area of the transmission system, f^{R} describes power system requirements of current state, f^{M} describes total cost measures to avoid congestion, W^{R} is a weighting factor of sub-objective of network security requirements, W^{M} describes weight factor for sub-objective of total costs of the measures. The weights fulfill the following condition:

$$W_a^{\mathrm{R}} \gg W_a^{\mathrm{M}} \tag{22.5}$$

The criteria to determine a secure state can be maximum current on each line, maximum and minimum node voltages, and $n - 1$ security criteria. Change of topology and rerouting of power can be considered as possible actions to avoid congestions and the short-circuit currents for the new configuration are calculated. Criteria are modeled with penalty functions that become zero for secure power system state [19–21].

Market-based measures consider re-dispatch with minimal total cost. The forecast of wind generation is considered and wind farms cannot deliver more active power than predicted with the forecasted wind speed. To avoid congestion, only a reduction in the wind-power-projected output is possible, which depends upon regulation. The generation management is resorted to when network and market-based measures are not sufficient to avoid congestion.

22.2 Wind Energy Conversion

In the beginning of wind power development, the energy produced by small wind power turbines was expensive and subsidies were high. Today single units of 5 MW are on the market. Today's wind turbines all over the world have three-blade rotors, diameters ranging from 70–80 m, and mounted atop 60–80 m or higher towers. The tower heights up to 160 m are a technical feasibility. The typical turbine installed in the United States in 2006 can produce about 1.5 MW of power. Higher-rated units and offshore wind base plants may see a unit size of 5 MW or more by 2010. Figure 22.7 shows the developments of single unit wind power turbines in the United States.

Figure 22.8 is a schematic representation of electrical and mechanical features of a wind converter unit, with *upwind* rotor. The rotor speed that is on the order of 8–22 rpm is the input to gear box, and on the output side the speed is 1500–1800 rpm. The drive train dimensions are large, increasing the horizontal dimension of nacelle.

22.2.1 Drive Train

Several designs are under development to reduce the drive train weight and cost. One approach is to build direct drive permanent magnet generators that eliminates complexity of gear box. The slowly rotating generator will be larger in diameter, 4–10 m, and quite heavy. The decrease in cost and availability of rare earth permanent magnets is expected to significantly affect size and weight. The generator designs tend to be compact and lightweight and reduce electrical losses in windings as compared to wound rotor machines.

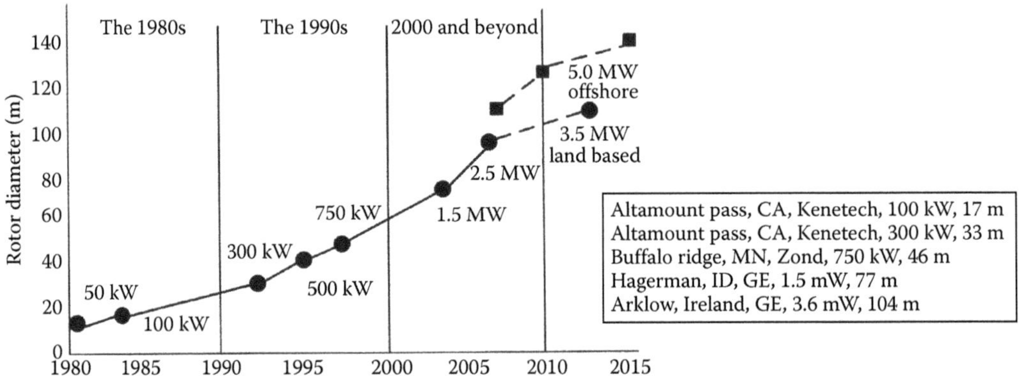

FIGURE 22.7
Development of single unit wind turbines in the United States.

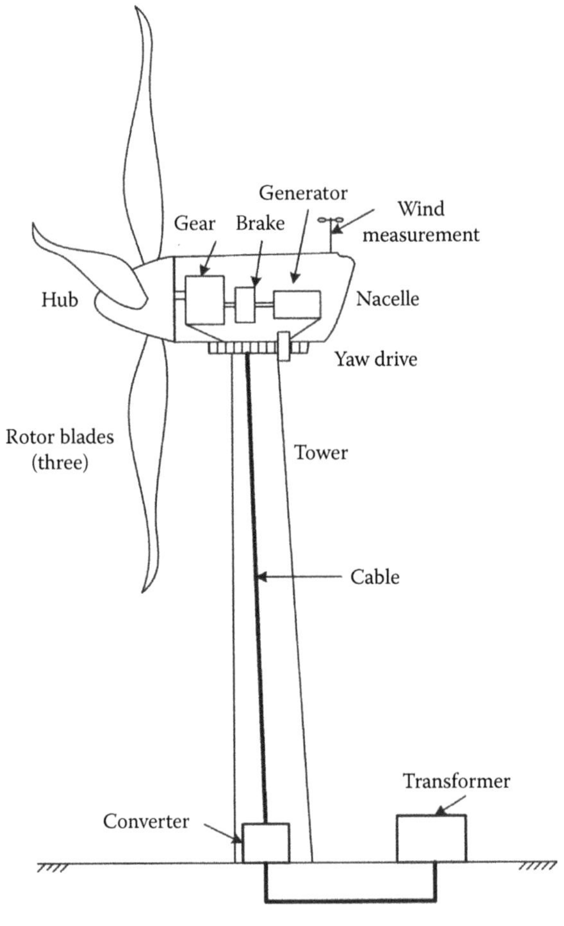

FIGURE 22.8
A schematic of wind power conversion unit, showing major components.

Prototypes have been built—a 1.5 MW design with 56 poles is only 4 m in diameter versus 10 m for a wound rotor design. It is undergoing testing at National Wind Technology Center.

A hybrid of direct drive approach uses low-speed generator. WindPACT drive train project has developed a single stage planetary drive operating at a gear box ratio of 9.16:1. The gear box drives a 72 pole PM generator and reduces the diameter of a 1.5 MW generator to 2 m [22]. Another development is the distributed drive train.

22.2.2 Towers

Depending upon the nacelle weight, the towers are constructed from steel or steel/concrete. The tower heights are up to 100 m. Wind speeds are height dependent, and lattice towers to 160 m height can be constructed. A relationship between tower mass and tower height shows sharp increase in the tower mass per meter height as turbine output increases, approximately 2000 kg/m for turbines in the output range of 1.5 MW. Large plants tend to have significantly greater mass per meter as the mast height increases. There is an ongoing effort to develop advanced tower designs that are easily transported and installed and are cost-effective.

The wind speeds are not uniform over the area of the rotor. The rotor profiles result in different wind speeds at the blades nearest to the ground level compared to the top of the blade travel. Gusts and changes in wind speed impact the rotor unequally. Influences due to tower shadow or windbreak effects cause fluctuations in power or torque.

22.2.3 Rotor Blades

As the wind turbines increase in output so does their blades, from 8 m in 1980 to 70 m for many land-based units. Improved blade designs have kept weight growth much lower than the geometric escalation of blade lengths. Work continues in the application of lighter and stronger carbon fiber in highly stressed areas to stiffen the blades, improve fatigue resistance, and simultaneously reduce the weight. Research continues in the development of lighter blades—carbon fiber and fiberglass.

22.3 Cube Law

Wind originates from difference in temperature and pressure in air mass. A change in these parameters alters the air density, ρ. The force exerted on volume of air V is given by

$$F = Vg\Delta\rho \tag{22.6}$$

where g is the gravitational constant. This produces kinetic energy:

$$E = \frac{1}{2}mv^2 \tag{22.7}$$

where

m is the mass of the air
v is the velocity being assumed constant
E is the wind power

This is rather an oversimplification. If we consider an air volume of certain cross section, a swirl free speed, upstream of the turbine, and downstream of the turbine, it will result in a reduction in speed, with a corresponding broadening of the cross-sectional area (wake decay). However, with this simplification of constant speed, the basic tenets of turbine output are still valid.

The following equation can be written for the air mass:

$$m = \rho A v \tag{22.8}$$

where A is the rotor area. Substituting (22.5) in (22.4), the theoretical power output is

$$P = \frac{1}{2}\rho A v^3 c_p \tag{22.9}$$

where the wind speed–dependent coefficient c_p describes the amount of energy converted by the wind turbine. It is on the order of 0.4–0.5.

All the energy in a moving stream of air cannot be captured. A block wall cannot be constructed, because some air must remain in motion after extraction. On the other hand, a device that does not slow the air will not extract any energy. The optimal blockage is called Betz limit, around 59%. The aerodynamic performance of blades has improved dramatically, and it is possible to capture about 80% of the theoretical limit. The new aerodynamic designs also minimize fouling due to dirt and bugs that accumulate at the leading edge and can reduce efficiency.

According to Betz, the maximum wind power turbine output is

$$P = \frac{16}{27} A_\mathrm{R} \frac{\rho}{2} v_1^3 \tag{22.10}$$

where

A_R is the air flow in the rotor area

v_1 is the wind velocity far upstream of the turbine

The maximum is obtained when

$$v_2 = \frac{2}{3} v_1 \quad \text{and} \quad v_3 = \frac{1}{3} v_1 \tag{22.11}$$

where

v_3 is the reduced velocity after the broadening of the air stream past the rotor

v_2 is the velocity in the rotor area

The ratio of power absorbed by turbine to that of moving air mass is

$$P_0 = A_\mathrm{R} \frac{\rho}{2} v_1^3 \tag{22.12}$$

and

$$c_p = \frac{P}{P_0} \tag{22.13}$$

Another way of defining c_p is with respect to λ, which is defined as follows:

$$\lambda = \frac{v_\mathrm{TS}}{v_\mathrm{RP}} = \frac{2\pi n r}{v_\mathrm{RP}} \tag{22.14}$$

where
v_{TS} is the blade tip speed
v_{RP} is the speed of rotor plane

v_{TS} can be written as follows:

$$v_{TS} = 2\pi nr \tag{22.15}$$

where
r is the rotor radius in m
n is the speed in s^{-1}

Typical values of λ are 8–10 and the tip speed ratio influences the power coefficient c_p. Also c_p is dependent upon wind speed. Combining these relations, Figure 22.9 shows power or torque versus the wind speed. (Torque is simply power divided by angular velocity ω.) Figure 22.10 shows performance coefficient as a function of tip speed ratio with blade pitch angle (see section below) as a parameter.

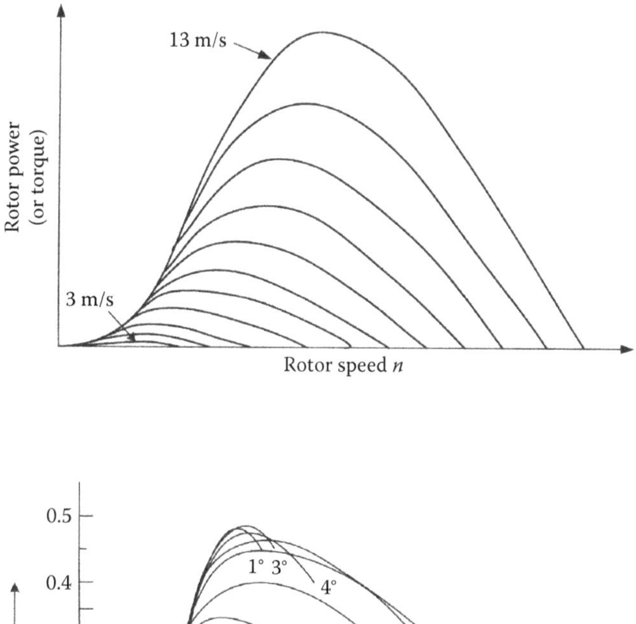

FIGURE 22.9
Rotor power of a wind turbine unit based upon rotor speed and wind velocity.

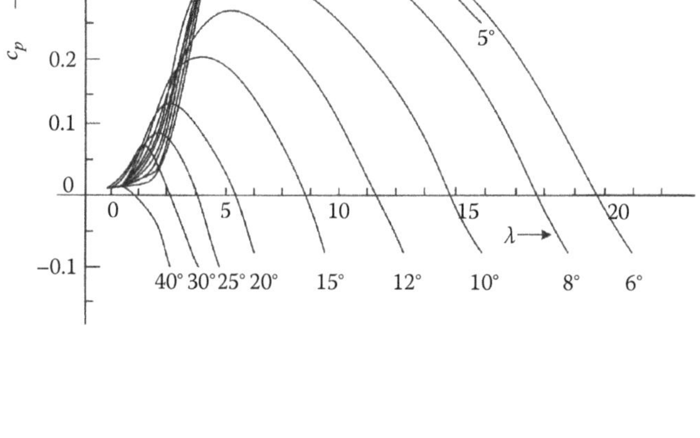

FIGURE 22.10
c_p–λ characteristics of a wind turbine unit with blade pitch angle.

22.4 Operation

Figure 22.11 shows power curves for typical modern turbine. The turbine output is controlled by rotating the blades about their long axis to change angle of attack called, "controlling the blade pitch." The turbine is pointed into the wind by rotating the nacelle about tower, which is called the "yaw control." Modern turbines operate with rotor positioned on the windward side of tower, which is referred to as an "upward rotor." A turbine generally starts producing in winds of about 12 mph, reaches a maximum power at about 28–30 mph, and shuts down "feather the blades" at about 50 mph. The amount of energy in the wind available for extraction by turbine increases with the cube of speed; thus, a 10% increase in speed means a 33% increase in energy. While the output increases proportional to rotor swept area, the volume of material and, thus, the cost increase as cube of the diameter. Controllers integrate signals from dozens of sensors to control rotor speed, blade pitch angle, and generator torque and power conversion voltage and phase angle.

Wind speed changes may occur over long periods of time or suddenly within a matter of seconds. The system has to be protected from the sudden gusts of wind. The mechanical loads are determined by dynamic forces and knowledge of dynamic wind behavior at a location is necessary for proper component ratings and the mechanical power acting on the turbine should be limited.

There are three methods to achieve it:

1. Stall
2. Active stall
3. Pitch control

The stall control exercises adjustable clutches on rotating blade tips to shut down. Under normal conditions, laminar air flow occurs on the blades. The lift values corresponding to angle of attack are achieved at low drag components. With wind speeds exceeding nominal

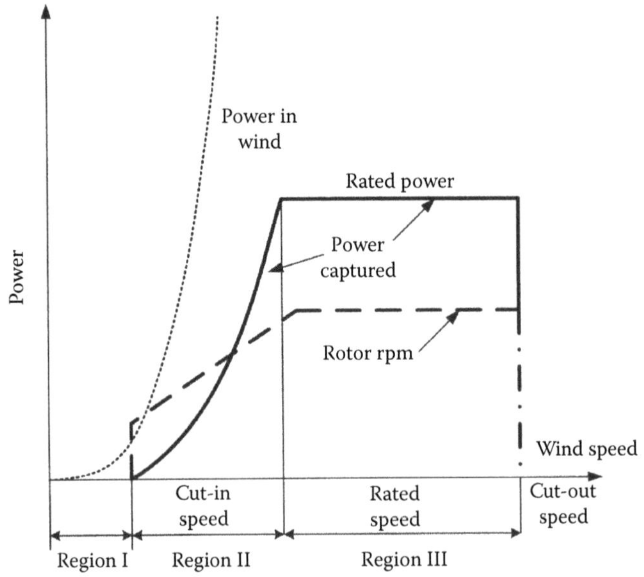

FIGURE 22.11
Typical wind generating unit operating curve, power versus wind speed.

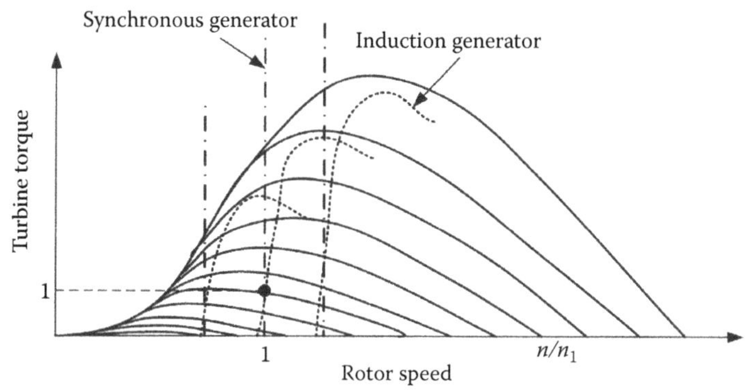

FIGURE 22.12
Wind turbine torque–speed characteristics by variation of generator frequency superimposed on wind power curves.

values at which the generator-rated output occurs, higher angles of attack and stalling occur. This is achieved by properly profiling the blades. The lift forces and lift coefficient are reduced and the drag forces and coefficients are increased. Stall-regulated machines are often designed with asynchronous generators of higher nominal output and rigid coupling with the grid is obtained. An active stall control is achieved by turnable rotor blades.

Variable blade pitch allows direct control of turbine. By varying the blade pitch, it is possible to control the torque of the turbine, and further adjustments bring about a stall condition. The control and regulation system is complex. In high-power applications, pitch control is used. For the design and control of pitch adjustments, a host of mechanical moments and forces must be considered. These include the forces and moments due to blade deflection, lift forces on blades, propeller moment, due to teetering, and frictional moments [23].

22.4.1 Speed Control

The curves of fixed and variable speed generators are marked in Figure 22.12. This shows that when the turbine is driven by a synchronous generator, varying the generator frequency at a certain wind speed will give operation at $n/n_1 = 1$, where n corresponds to the grid frequency. The turbine is constrained to follow the grid frequency. Sufficient turbine torque to drive the generator is available for wind speed above approximately 3.6 m/s for synchronous generators to about 3.8 m/s for asynchronous generators. Under variable frequency generator operation, the speed of rotation can be feely set within given limits. The turbine utilization of available wind power is optimized. A ramp control of the active power output is possible [23,24].

22.4.2 Behavior under Faults and Low-Voltage Ride Through

Figure 22.13 shows the recommendation of WECC (Western Electricity Coordinating Council) wind generation task force (WGTF) with respect to proposed voltage ride-through requirements for all wind generators [25]. A three-phase fault is cleared in nine cycles, and the post-fault voltage recovery dictates whether the wind power generating plant can

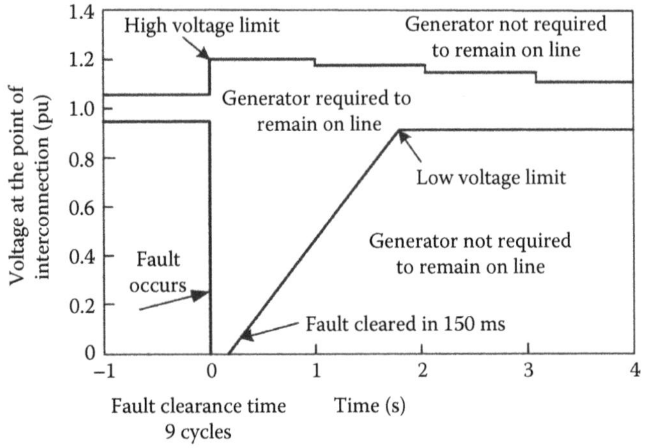

FIGURE 22.13
Proposed WECC voltage ride-through requirements of wind generators.

remain on line. The requirement does not apply to faults that will occur between the wind generator terminals and the high side of GSU (generator step up unit transformer) and the wind plants connected to transmission network via a radial line will not be required to ride through the fault on that line.

Figure 22.13 also addresses the high and low voltages. Generators based upon simple induction machines degrade power system voltage performance as these require excitation reactive power. The FERC (Federal Energy Regulatory Commission) orders 661 and 661A require new wind generators to have capability to control their reactive power within 0.95 leading to 0.95 lagging range. As this requirement can be expensive to comply, FERC requires it only if the interconnection study shows that it is needed. Modern wind generators provide this capability from power electronics that control the real power operation of the machine. Also see Refs. [26,27].

22.5 Wind Generators

22.5.1 Induction Generators

The operation and characteristics of induction motors are described in Chapter 12. The characteristics of an induction machine for negative slip are depicted in Figure 22.14. An induction motor will act as induction generator with negative slip. At $s = 0$, the induction motor torque is zero and if it is driven above its synchronous speed, the slip becomes negative and generator operation results. Using the same symbols as in Chapter 12, the maximum torque can be written as follows:

$$T_m = \frac{0.5V_1^2}{\omega_s\left[\sqrt{R_1^2 + (X_1 + x_2)^2} \pm R_1\right]} \tag{22.16}$$

The negative R_1 in (22.16) represents the maximum torque required to drive the machine as generator. The maximum torque is independent of the rotor resistance. For subsynchro-

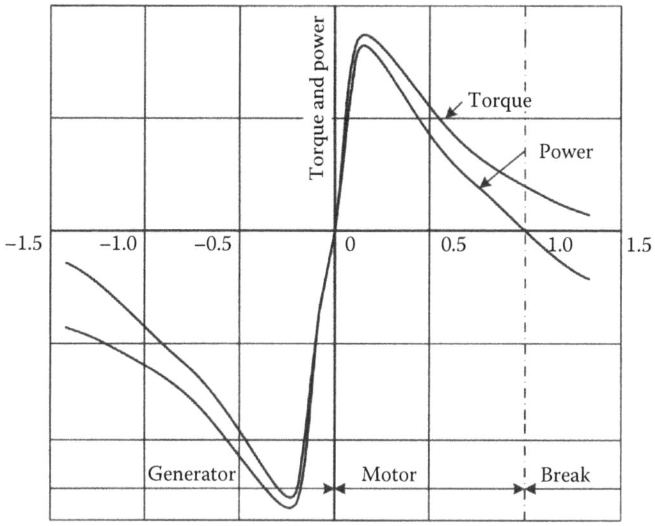

FIGURE 22.14
Torque and power characteristics of an induction machine versus slip induction motor, induction generator, and a brake.

nous operation, the rotor resistance does affect the slip at which the maximum torque occurs. For maximum torque at starting, $s = 1$, and

$$r_2 = \sqrt{R_1^2 + (X_1 + x_2)^2} \tag{22.17}$$

For super synchronous operation (generator operation), the maximum torque is independent of r_2, same as for motor operation, but increases with the reduction of both stator and rotor reactances. Therefore, we can write as follows:

$$\frac{T_{\text{m,gen}}(\text{supersyn.})}{T_{\text{m,motor}}(\text{subsyn.})} = \frac{\sqrt{R_1^2 + (X_1 + x_2)^2} + R_1}{\sqrt{R_1^2 + (X_1 + x_2)^2} - R_1} \approx \frac{X_1 + x_2 + R_1}{X_1 + x_2 - R_1} \tag{22.18}$$

The approximation holds as long as $R_1 \ll X_1$. The torque–speed characteristics of the machine above synchronism are similar to that for running as an induction motor. If the prime mover develops a greater driving torque than the maximum counter torque, the speed rises into an unstable region and the slip increases. At some high value of slip, the generating effect ceases and the machine becomes a brake.

Induction generators do not need synchronizing and can run in parallel without hunting and at any frequency, the speed variations of the prime mover are relatively unimportant. Thus, these machines are applied for wind power generation.

An induction generator must draw its excitation from the supply system, which is mostly reactive power requirement. On a sudden short circuit, the excitation fails, and with it the generator output; so in a way the generator is self-protecting.

As the rotor speed rises above synchronous speed, the rotor emf becomes in phase opposition to its subsynchronous position, because the rotor conductors are moving faster than the stator rotating field. This reverses the rotor current also and the stator component reverses. The rotor current locus is a complete circle. The stator current is clearly a leading current of definite phase angle. The output cannot be made to supply a lagging load.

An induction generator can be self-excited through a capacitor bank, without external dc source, but the frequency and generated voltage will be affected by speed, load, and capacitor rating. For an inductive load, the magnetic energy circulation must be dealt with by the capacitor bank as induction generator cannot do so.

For wind power applications, generators are either squirrel cage or wound rotor induction types with rotating field windings. The coupling with the grid, directly or through inverters, is of significance. Mostly induction generators are used.

The induction generator must draw its reactive power requirement from the grid source. When capacitors, SVCs, rotary phase shifters are connected, the operational capabilities can be parallel with synchronous machines, though resonance with grid inductance is a possibility. Induction generators produce harmonic and synchronous pulsating torques, akin to induction motors. A synchronous machine provides control of operating conditions, leading or lagging by excitation control. With respect to interconnection with the grid, following schemes exist.

22.5.2 Direct Coupled Induction Generator

The direct coupled induction machine is generally of 4-pole type; a gear box transforms the rotor speed to a higher speed for generator operation above synchronous speed. It requires reactive power from grid or ancillary sources, and starting after a blackout may be a problem. Wind-dependent power surges produce voltage drops and flicker. The connection to the grid is made through thyristor switches, which are bypassed after start. A wound rotor machine has the capability of adjusting the slip and torque characteristics by inserting resistors in the rotor circuit and the slip can be increased at an expense of more losses and heavier weight (Figure 22.15a). The system will not meet the current regulations of connection to grid and may be acceptable for isolated systems.

22.5.3 Induction Generator Connected to Grid through Full Size Converter

The induction generator is connected to the grid through two back-to-back VSC. Because of full power rating of the inverter, the cost of electronics is high. The wind-dependent power spikes are damped by the dc link. The grid side inverter need not be switched in and out so frequently and harmonic pollution occurs.

22.5.4 Doubly Fed Induction Generator

The stator of the induction machine is directly connected to the grid, while the rotor is connected through voltage source converter (Figure 22.15b). The energy flow over the converter in the rotor circuit is bidirectional. In subsynchronous mode, the energy flows to the rotor and in super synchronous mode it flows from rotor to the grid. The ratings of the converter are much reduced, generally 1/3rd of the full power, and depend upon the speed range of turbine. The power rating is

$$P = P_s \pm P_r \tag{22.19}$$

where P_s and P_r are the stator and rotor powers. But the rotor has only the slip frequency induced in its windings; therefore, we can write as follows:

$$P_r = P_a \times s \tag{22.20}$$

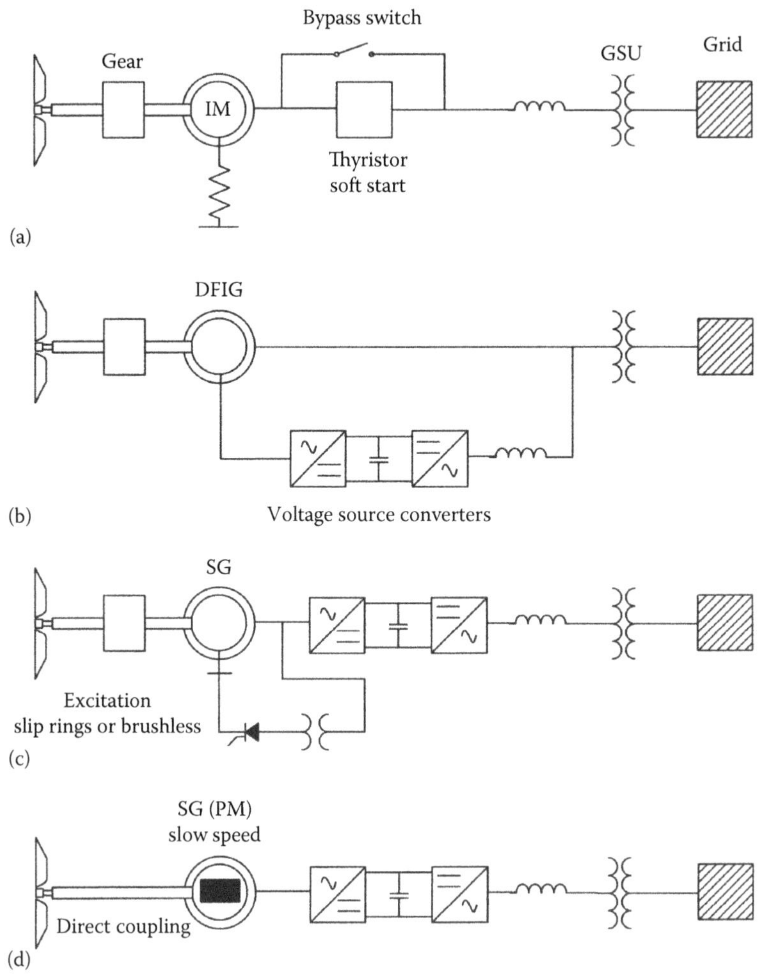

FIGURE 22.15
Grid connections of wind generators. (a) Direct connection of an induction generator, stall regulated. (b) Connections of a DFIG (Doubly-fed induction generator) variable speed generator, pitch regulated. (c) Synchronous generator, brush type or brushless type with voltage–source converters, pitch regulated. (d) Gearless connections of a low-speed permanent magnet generator.

where s is the slip. For a speed range of $\pm 30\%$, the slip is ± 0.3, and a third of converter power is required. Also we can write as follows:

$$n_s = \frac{f_r \pm f}{p} 120 \tag{22.21}$$

where p is the number of pair of poles.

Synchronous generators can be brush type or brushless type of permanent magnet excitation systems. These are also connected to the grid much alike asynchronous machines. The excitation power has to be drawn from the source, unless the generator is of permanent magnet type. Figure 22.15c and d shows typical connections.

22.6 Power Electronics

The entire output of variable voltage, variable frequency output of a wind generator, sometimes called the wild ac, is converted to direct current, which is then converted to utility quality ac power. The frequency converters condition the electrical energy from the wind generators and damp the influence on the grid connections. With respect to the topology of the electronic devices, we have

- Current-controlled rectifier
- Voltage source rectifier
- Voltage source inverters
- Current source inverters
- Newer technology of ac/dc/ac ZSI (Z-source inverters)

The rectifiers may be (1) uncontrolled, (2) bridges with dc/dc regulators, and (3) controlled. Again, pulse-controlled IGBT topology with phase multiplication to reduce harmonic generation is the preferred choice, as active and reactive control of power can be exercised (Chapter 13). Practically, all the installed variable speed synchronous generator wind turbines that employ fully rated converters use VSI topology. As the power of the converter rises, devices for PWM are needed to switch at higher frequencies. For such applications as an alternative CSI is proposed [28], however, harmonic performance suffers and var compensation is required.

Consider a synchronous generator grid connection. The constant voltage dc link is supplied by a controlled rectifier bridge and the rectifier is current controlled so that magnitude and phase angle of the generator current are controlled by triggering of the rectifier. By phase shifting the generator current in underexcited and overexcited regions, the generator voltage can be controlled and matched to the dc link. Also see Ref. [29].

Short-circuit current calculations according to machines integrated in industrial systems may not be always valid. For example, in a DFIM (Doubly-fed induction motor), the stator current for a nearby fault may be limited to nearly rated current if rotor power converter remains active. It may, however, be disabled by crowbar circuit for protection during fault, in which case the fault current will be several times the rated for a few cycles.

Switching of devices in converters gives rise to interference emission over a wide spectrum. The electromagnetic compatibility should be insured according to relevant standards.

22.6.1 ZS Inverters

CSI and VSI inverters suffer from some limitations. For example,

- In a VSI, ac output voltage is below the dc link voltage; it may be called a buck dc/ac inverter. In CSI, ac output voltage has to be greater than dc voltage feeding the inductor. It is a boost converter for dc/ac power conversion.
- In a VSI, upper and lower devices in each phase leg (Figure 17.35a) should not be gated simultaneously. An EMI noise could trigger these resulting in a shoot through that can destroy the devices. This is a major consideration for converter's reliability.

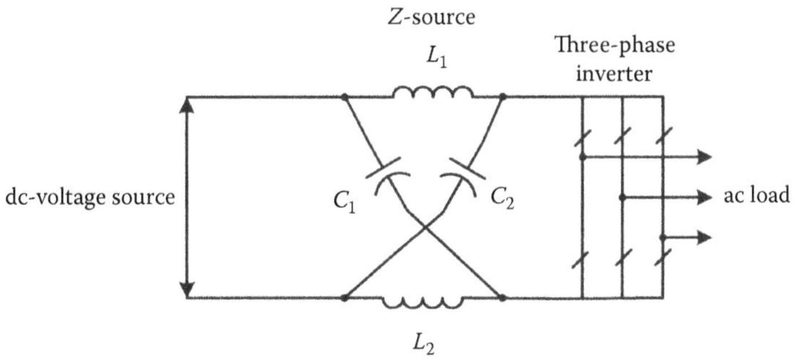

FIGURE 22.16
A ZSI converter, basic configuration.

- In CSI, at least one of the upper and one of the lower devices (Figure 17.16a) have to be gated on and maintained, otherwise an open circuit of dc inductor will occur, which could destroy the devices. Here also misgating due to EMI noise is a major concern.
- The circuits of VSI and CSI are not interchangeable.

ZS inverters overcome some of the above problems of traditional CSI and VSI technology. It employs an impedance network to couple the converter main circuit to the power source. Figure 22.16 shows a two port network, which consists of two inductors L_1 and L_2 and capacitors C_1 and C_2 connected in X shape. This is the impedance source coupling the converter to dc source. If the inductors are zero, the converter becomes a VSI and if the capacitors are zero the converter becomes CSI. The advantages are as follows:

- The converter is a buck-boost converter. There is no need for an additional dc/dc boost converter to obtain required dc.
- The *Z*-circuit restricts the overvoltages and overcurrents. The legs in the main bridge can be operated in open circuit or short circuit for a short time.
- ZSI has a function to suppress EMI noise. The shoot-through problem and misgating will not occur.

References [30,31] provide further reading. Reference [32] describes its application to a wind turbine, linking with the utility. Basically, a *Z*-network is introduced between the front-end diode rectifier and neutral point clamped inverter. This three-level diode clamped ac/dc/ACZSI has three-level output waveform, with the use of minimal passive components. Before the front-end diode rectifier, passive harmonic filters are placed and the neutral potential needed by the three-level inverter circuitry can be tapped from the wye-connected filter capacitors common point.

22.7 Reactive Power Control

Apart from the reactive power need of the induction machine itself, the passive connecting element consumes reactive power. These elements are transformers and cable connections to the point of grid. Then the reactive power is required by loads.

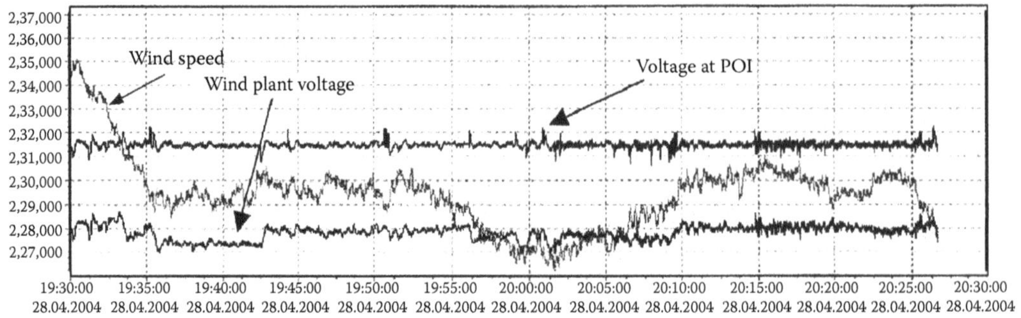

FIGURE 22.17
Wind plant voltage response and regulation at the point of connection.

Problems of transient low voltages can occur when the wind power generation is connected to relatively weak grid systems.

Some wind-generating plants have added SVCs and DSTATCOM (Chapter 13), which control the power factor to unity at the point of interconnection. The Argonne Mesa wind plant in New Mexico has a DSTATCOM, which controls the power factor to utility at the point of interconnection at Guadalupe 345 kV station bus. Four mechanically switched capacitor banks are located in the collector substation some two mi away from the interconnect substation. The DSTATCOM controls determine the required reactive power output based upon voltage and current measurements at 345 kV collector bus.

Figure 22.17 shows the impact of high winds on the 1.5 MW generators, connected to a 230 kV transmission line. Line drop compensation algorithms are utilized to synthesize voltage at the point of interconnection, located approximately 75 km from the wind plant. The flicker index of the voltage is less than 2% at the point of interconnection. In spite of considerable variation in the wind speed, the plant output is relatively stable.

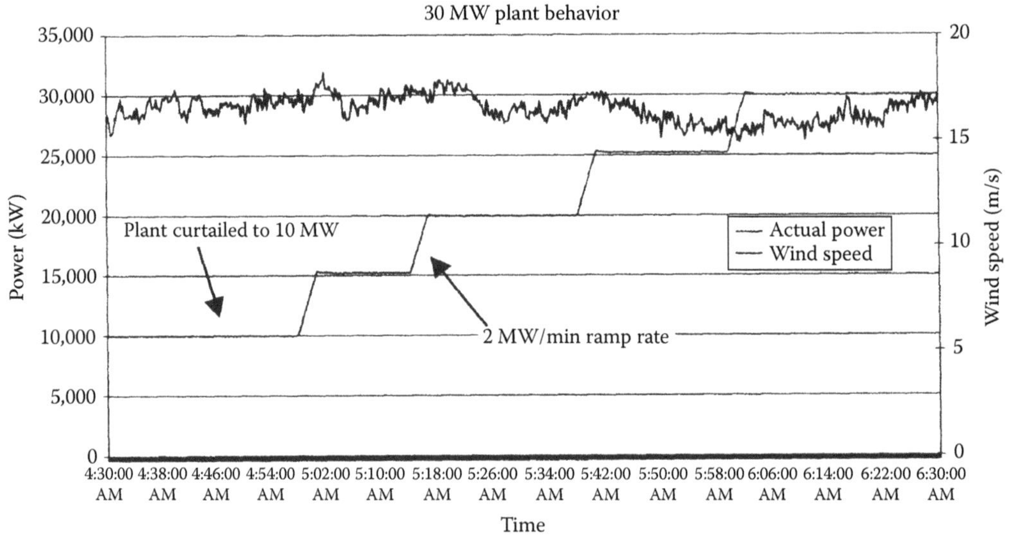

FIGURE 22.18
Active power response of a wind plant with ramp control.

Figure 22.18 shows field test results on an active power regulator and power rate limiter on an operating 30 MW wind plant. Initially the output is curtailed to 10 MW, and during the tests the active power command is raised in four 5 MW increments. The transition between each step ramp-rate is controlled to 2 MW/min.

22.8 Harmonics

Generation of harmonics and mitigation is discussed in Chapters 17 and 20. Even harmonics in wind generation can arise due to unsymmetrical half waves and may appear at fast load changes. Subharmonic can be produced due to periodical switching with variable frequency. Interharmonics can be generated when the frequency is not synchronized to the fundamental frequency, which may happen at low and high frequency switching.

The interharmonics due to back-to-back configuration of two converters can be calculated according to IEC [33].

$$f_{n,m} = [(p_1k_1) \pm 1]f \pm (p_2k_2)F \tag{22.22}$$

where

$f_{n,m}$ is the interharmonics frequency
f is the input frequency
F is the output frequency
p_1 and p_2 are pulse numbers of the two converters

Interharmonics are also generated due to speed-dependent frequency conversion between rotor and stator of DFIM and as side bands of characteristic harmonics of PWM converters. Noncharacteristic harmonics can be generated due to grid unbalance.

A topology shown in Figure 22.19 is advocated in Ref. [28]. Inverters 1 and 2 are series-connected bridge circuits that employ fully controllable switches with bidirectional blocking

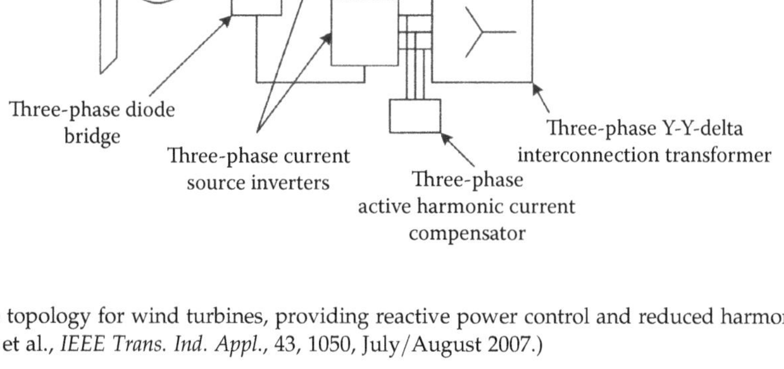

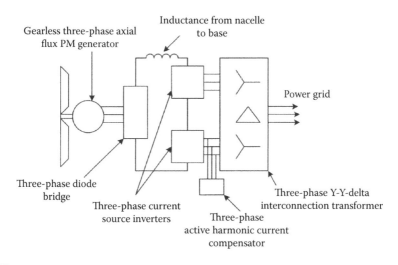

FIGURE 22.19
Current source topology for wind turbines, providing reactive power control and reduced harmonics. (Adapted from Tenca, P. et al., *IEEE Trans. Ind. Appl.*, 43, 1050, July/August 2007.)

capabilities. The bridges are switched at line frequency using phase-control techniques, which avoids PWM and the associated switching losses and allows the use of high power but relatively slow devices such as gate-controlled thyristors or even silicon-controlled thyristors. Using phase control, the dc link voltage is modulated and thus dc-link current, but the phase angle of phase-controlled inverter is tied to the firing angle (Chapter 17). By controlling the phase angle of the two converters so that $\phi_2 = -\phi_1$, inverter dc link voltage is modulated without affecting the fundamental power factor at the mains.

22.9 Computer Modeling

Modeling is required in the planning stage to conduct interconnection studies, grid reliability, and simulate energy capture for hybrid plants. The energy capability of the wind power generation for typical time history wind samples or wind probability density curve can be forecasted. Also simulation is required for aerodynamics, mechanical dynamics, and structures—this gives ideas of static and dynamic loads, predicted power curves, vibration modes, and control system response. Electrical transients in generators and power electronics need to be simulated. Excess staring currents, behavior under short circuit and under voltage transients, and isolation from grid can be studied in a time frame varying from a few microseconds to seconds. GE PSLF/PSDS and Siemens PTI/PSSE programs are designed for the study of large-scale interconnected systems; yet there is not much sharing of the data between consultants, utilities, and power planners. Engineering design models are implemented in three-phase simulation programs like EMTP and PSCAD [34–39].

22.9.1 Wind Turbine Controller

Classically, the wind turbine control principles can be depicted with reference to Figure 22.20. In zone 1, the system is operated at optimal rotor speed according to rotor aerodynamics in

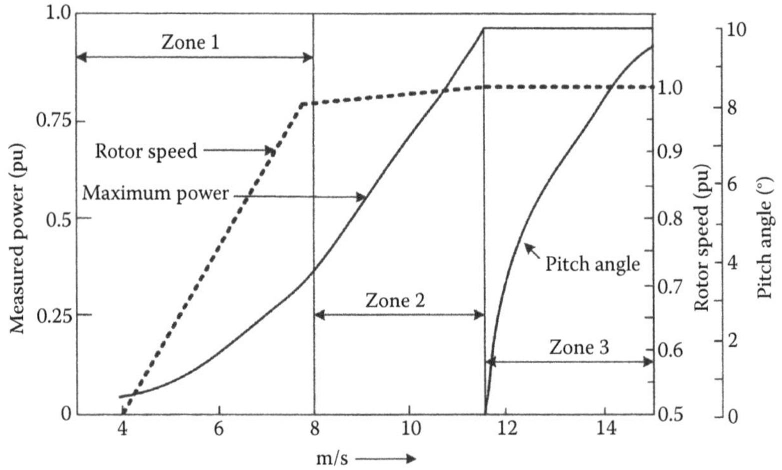

FIGURE 22.20
Wind turbine control zones.

order to extract maximum energy from the wind. A relation of power reference to rotor speed can be written as follows:

$$P_{\text{ref}} = f(\Omega_{\text{wt}})$$

and (22.23)

$$\Omega_{\text{wt-ref}} = f(P_{\text{meas}})$$

In zone 2, the rotor speed is limited to the nominal rotor speed, $\Omega_{\text{wt-nom}}$. For controller stability, a slope may be added to the power reference to rotor speed characteristics. In zone 3, the wind speed is high enough to allow nominal power production. In this zone, the objective is to maintain the rotor speed Ω_{wt} and the measured produced power at their nominal values.

Figure 22.21 shows one wind controller implementation strategy. The two quantities of interest are rotor speed Ω_{wt} and measured power P_{meas}. In order to regulate these two quantities, two control variables are available: the blade pitch angle β_{ref} and electromagnetic torque $T_{\text{em-ref}}$. The rotor speed is regulated by the pitch control angle and measured power is

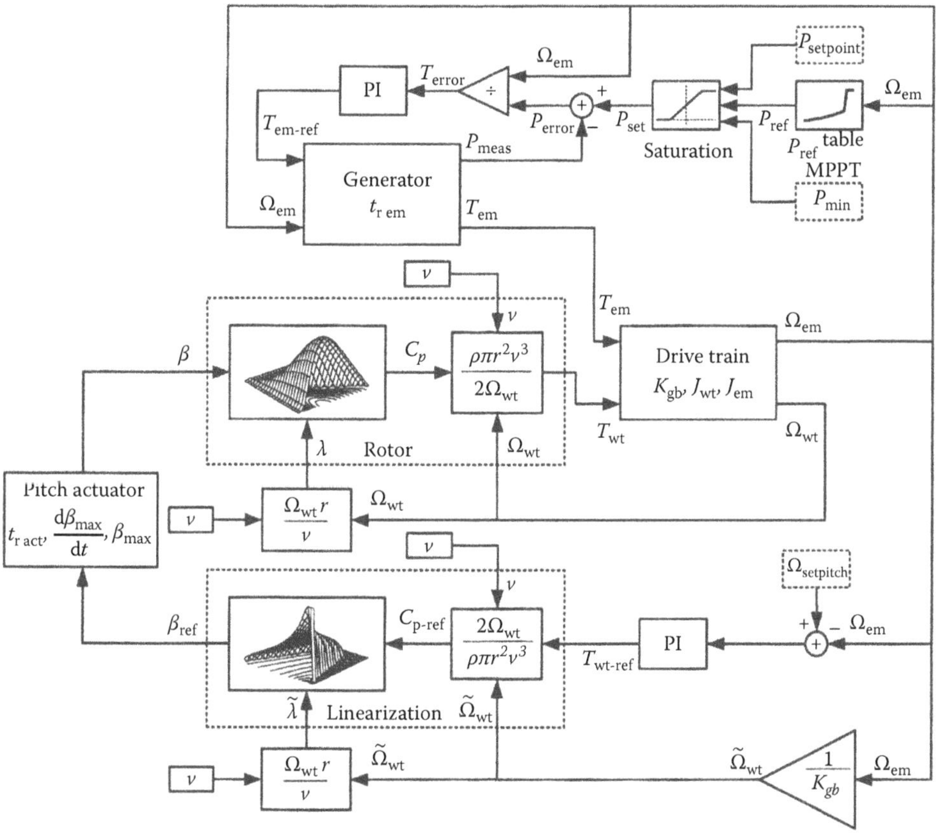

FIGURE 22.21
Block circuit diagram of a wind turbine controller. (From Venne, P. and Guillaud, X., Impact of turbine control strategy on deloaded operation, in: *CIGRE and IEEE PES Joint Symposium*, Calgary, Alberta, Canada, 2009.)

regulated by acting upon electromagnetic torque $T_{em\text{-}ref}$. A block circuit diagram is shown in Figure 22.21 [39]. This represents two control loops, one for the measured power P_{meas} and the other for the rotor speed Ω_{wt}. For the power control loop, P_{ref} is extracted from optimal power set point to rotor speed and an external set point $P_{setpoint}$ can be introduced in a saturation element to limit maximum value of P_{ref}. Then the internal power set point and the measured produced power are used to compute power error. From this, the torque error T_{error} is calculated using generator rotor speed Ω_{em}. Then the torque reference is used by a PI (proportional integral) controller to provide electromagnetic torque reference $T_{em\text{-}ref}$.

Analysis of any pitch-controlled turbine reveals a nonlinearity caused by the relation of wind turbine torque T_{wt} and pitch angle β. The capability to change power set point requires adjustment of the process gain depending upon wind *and* power set point as well. The principle of model inversion is used to tackle this problem and is called general gain scheduling. Each element of the process is inverted in the control algorithm, third loop in Figure 22.21. The pitch angle reference β_{ref} is obtained from power coefficient $c_{p\text{-}ref}$ using the approximated tip speed ratio $\tilde{\lambda}$. The power coefficient $c_{p\text{-}ref}$ is obtained from wind turbine torque reference $T_{wt\text{-}ref}$ provided by the PI controller using wind speed v, the approximated rotor speed $\tilde{\Omega}_{wt}$, and the mathematical expression of the coefficient reference $c_{p\text{-}ref}$ derived from the equation of rotor torque [40,41].

Though not exhaustive, this chapter has covered the major aspects of wind power generation and references provide further reading.

References

1. U.S. Department of Energy. Fact sheet for 20% wind energy report.
2. K.R. Padiyar. *Power Transmission Systems*. New York: John Wiley, 1990.
3. J. Arrillaga. *HVDC Transmission*, 2nd edn. Piscataway, NJ: IEEE Press, 1998.
4. J.C. Das. *Transients in Electrical Systems*. New York: McGraw-Hill, 2010.
5. J.A. Fleeman, R. Gutman, M. Heyeck, M. Baharman, and B. Normark. EHV and HVDC transmission working together to integrate renewable power. In *CIGRE and IEEE PES Joint Symposium*, Calgary, Alberta, Canada, 2009.
6. R.D. Dunlop, R. Gutman, and P.P. Marchenko. Analytical development of loadability characteristics for EHV and UHV transmission lines. *IEEE Trans.* PAS-98(2): 606–617, March/April 1979.
7. Interstate vision for wind integration. AEP 2007, www.aep.com
8. IEEE Transmission and Distribution Committee. An IEEE survey of US and Canadian overhead transmission outages at 230 kV and above. Paper no. WM054-7PWRD, *IEEE PES Winter Meeting*, Columbus, OH, January/February 1993.
9. H.C. Barnes and T.J. Nagel. AEP 765-kV system: General background relating to its development. *IEEE Trans.* PAS-88(9): 1313–1319, September 1969.
10. J.D. McDonald. The next generation grid: Energy infrastructure of the future. *IEEE Power Energy Mag.* 7(2), March/April 2009.
11. M.P. Bahrman and B.K. Johnson. The ABCs of HVDC transmission technologies. *IEEE Power Energy Mag.*, March/April 2007.
12. NERC. Transmission transfer capability task force. Available transfer capability definitions and determinations. Princeton, NJ, June 1996.
13. G.C. Ejebe, J.G. Waight, M. Saots-Nieto et al., Fast calculation of linear available transfer capability. *IEEE Trans.* PS-15: 1112–1116, August 2000.

14. M. Liang and A. Abur. Total transfer capability computation for multi-area power systems. *IEEE Trans.* PS-21, August 2006.
15. Y. Ou and C. Singh. Assessment of available transfer capability and margins. *IEEE Trans.* PS-21, August 2006.
16. PowerCon, R. Sun, Y. Fan, Y. Song, and Y. Sun. Development and application of software for ATC calculation, October 2006.
17. K. Audomvongseree and A. Yokoyama. Consideration of an appropriate TTC by probabilistic approach. *IEEE Trans.* PS-22, May 2002.
18. R. Billinton and L. Wenyuan. *Reliability Assessment of Electric Power Systems Using Monte Carlo Methods.* New York: Plenum, 1994.
19. ETSO. Counter measures for congestion management definition and basic concepts, June 2003, http://www.etso-net.org
20. A.F.K. Kamga and J.F. Verstege. A cross border congestion management system integrating AC and DC load flow models. In *WESC (Sixth World Energy System Conference)*, Turin, June 2006.
21. A.F.K. Kamga, S. Völler, and J.F. Verstege. Congestion management in transmission systems with large scale integration of wind energy. In: *CIGRE and IEEE PES Joint Symposium*, Calgary, Alberta, Canada, July 2009.
22. Following reports can be obtained: (i) D.A. Griffen. WindPACT turbine design scaling studies technical area 1—composite blades for 80–120 m rotor: March 21, 2000–March 15, 2001. NREL Rep. SR-500-29492. (ii) G. Bywaters, V. John, J. Lynch, P. Mattila, G. Norton, J. Stowell, M. Salata, O. Labath, A. Chertok, and D. Hablanian. Northe power systems WindPact drive train alternative design study report: Period of performance: April 12, 2001–January 31, 2005. NREL Rep. SR-500-35524. (iii) M.W. LaNier, LWST phase 1 conceptual design study: Evaluation of design and construction approaches for economical hybrid steel/concrete wind turbine towers: June 28, 2002–July 31, 2004. NREL Rep. SR-500-36777, www.nrel.gov/publications/
23. S. Heier, *Grid Integration of Wind Energy Conversion Systems*, 2nd edn. New York: John Wiley, 2009.
24. T. Burton, D. Sharpe, N. Jenkins, and E. Bossanyi. *Wind Power Handbook.* New York: John Wiley, 2002.
25. Western Electricity Coordinating Council disturbance monitoring reports.
26. IEEE. Guide for interfacing dispersed storage and generation facilities with electric facility systems, Standard 1001.1988.
27. IEEE. Standard for interconnecting distributed resources with electrical power systems. Standard 1547. 2003. Available with subject areas: Energy generation/Power Generation Smart Grid.
28. P. Tenca, A.A. Rockhill, and T.A. Lipo. Wind turbine current source converter providing reactive power control and reduced harmonics. *IEEE Trans. Ind. Appl.* 43: 1050–1060, July/August 2007.
29. R. Strzelecki and G. Benysek (Eds.) *Power Electronics in Smart Electrical Energy Networks.* Berlin, Germany: Springer, 2008.
30. F.L. Luo and H. Ye. *Power Electronics-Advanced Conversion Technologies.* Boca Raton, FL: CRC Press, 2010.
31. F.Z. Peng. Z-source inverter. *IEEE Trans. Ind. Appl.* 39: 504–510, 2003.
32. P.C. Loh, F. Gao, P.C. Tan, and F. Blaabjerg. Three-level AC-DC-AC Z-source converter using reduced passive component count. In: *Proceedings of IEEE PESC*, pp. 2691–2697, 2007.
33. IEC 61000-2-4. Electromagnetic compatibility, Part 2. Environmental section 4: Compatibility levels in industrial plants for low-frequency conducted disturbances.
34. Dynamic models for wind farms for power system studies, http://www.energy.sintef.no/wind/iea.asp
35. Dynamic model validation for the GE wind turbine, www.uwig.org UWIG Modeling User Group.
36. EPA, Renewable portfolio standards fact sheet, http://www.epa.gov/chp/state-policy/renewable_fs.html

37. IEC-61400-21. Wind turbine generator systems. Part 21: Measurement and assessment of power quality characteristics of grid connected wind turbines, 2001.
38. E.A. DEMeo, W. Grant, M.R. Milligan, and M.J. Schuerger. Wind power integration. *IEEE Power Energy Mag*. 3(6): 38–46, 2005.
39. GE energy and AWS true wind, Ontario Wind Integration Study, http://www.ieso.ca/imoweb/pubs/marketreports/opa-report-200610-1.pdf
40. P. Venne and X. Guillaud. Impact of turbine control strategy on deloaded operation. In: *CIGRE and IEEE PES Joint Symposium*, Calgary, Alberta, Canada, 2009.
41. T. Ackermann. (Ed.). *Wind Power in Power Systems*. New York: Wiley-Interscience, 2005.

Appendix A: Matrix Methods

A.1 Review Summary

A.1.1 Sets

A set of points is denoted by

$$S = (x_1, x_2, x_3) \tag{A.1}$$

This shows a set of three points, x_1, x_2, and x_3. Some properties may be assigned to the set

$$S = \{(x_1, x_2, x_3) | x_3 = 0\} \tag{A.2}$$

Equation A.2 indicates that the last component of the set $x_3 = 0$. Members of a set are called elements of the set. If a point x, usually denoted by $\bar{x}$, is a member of the set, it is written as

$$\bar{x} \in S \tag{A.3}$$

If we write

$$\bar{x} \notin S \tag{A.4}$$

then point x is not an element of set S. If all the elements of a set S are also the elements of another set T, then S is said to be a subset of T, or S is contained in T:

$$S \subset T \tag{A.5}$$

Alternatively, this is written as

$$T \supset S \tag{A.6}$$

The intersection of two sets S_1 and S_2 is the set of all points $\bar{x}$ such that $\bar{x}$ is an element of both S_1 and S_2. If the intersection is denoted by T, we write

$$T = S_1 \cap S_2 \tag{A.7}$$

The intersection of n sets is

$$T \equiv S_1 \cap S_2 \cap \cdots \cap S_n \equiv \bigcap_{i=1}^{n} S_i \tag{A.8}$$

The union of two sets S_1 and S_2 is the set of all points $\overline{x}$ such that $\overline{x}$ is an element of either S_1 or S_2. If the union is denoted by P, we write

$$P = S_1 \cup S_2 \tag{A.9}$$

The union of n sets is written as

$$P \equiv S_1 \cup S_2 \cup \cdots \cup S_n \equiv \bigcup_{i=1}^{n} S_i \tag{A.10}$$

A.1.2 Vectors

A vector is an ordered set of numbers, real or complex. A matrix containing only one row or column may be called a vector:

$$\overline{x} = \begin{vmatrix} x_1 \\ x_2 \\ \cdot \\ x_n \end{vmatrix} \tag{A.11}$$

where $x_1, x_2, \ldots, x_n$ are called the constituents of the vector. The transposed form is

$$\overline{x}' = |x_1, x_2, \ldots, x_n| \tag{A.12}$$

Sometimes, the transpose is indicated by a superscript letter t. A null vector $\overline{0}$ has all its components equal to zero and a sum vector $\overline{1}$ has all its components equal to 1.

The following properties are applicable to vectors

$$\begin{aligned}
\overline{x} + \overline{y} &= \overline{y} + \overline{x} \\
\overline{x} + (\overline{y} + \overline{z}) &= (\overline{x} + \overline{y}) + \overline{z} \\
\alpha_1(\alpha_2\overline{x}) &= (\alpha_1\alpha_2)\overline{x} \\
(\alpha_1 + \alpha_2)\overline{x} &= \alpha_1\overline{x} + \alpha_2\overline{x} \\
\overline{0}\overline{x} &= \overline{0}
\end{aligned} \tag{A.13}$$

Multiplication of two vectors of the same dimensions results in an inner or scalar product:

$$\begin{aligned}
\overline{x}'\overline{y} &= \sum_{i=1}^{n} x_i y_i = \overline{y}'\overline{x} \\
\overline{x}'\overline{x} &= |\overline{x}|^2 \\
\cos\phi &= \frac{\overline{x}'\overline{y}}{|\overline{x}||\overline{y}|}
\end{aligned} \tag{A.14}$$

where

ϕ is the angle between vectors

$|x|$ and $|y|$ are the geometric lengths

Two vectors $\overline{x}_1$ and $\overline{x}_2$ are orthogonal if

$$\overline{x}_1\overline{x}_2' = 0 \tag{A.15}$$

A.1.3 Matrices

1. A matrix is a rectangular array of numbers subject to certain rules of operation, and usually denoted by a capital letter within brackets [A], a capital letter in bold, or a capital letter with an overbar. The last convention is followed in this book. The dimensions of a matrix indicate the total number of rows and columns. An element a_{ij} lies at the intersection of row i and column j.
2. A matrix containing only one row or column is called a vector.
3. A matrix in which the number of rows is equal to the number of columns is a square matrix.
4. A square matrix is a diagonal matrix if all off-diagonal elements are zero.
5. A unit or identity matrix $\bar{I}$ is a square matrix with all diagonal elements $= 1$ and off-diagonal elements $= 0$.
6. A matrix is symmetric if, for all values of i and j, $a_{ij} = a_{ji}$.
7. A square matrix is a skew symmetric matrix if $a_{ji} = -a_{ji}$ for all values of i and j.
8. A square matrix whose elements below the leading diagonal are zero is called an upper triangular matrix. A square matrix whose elements above the leading diagonal are zero is called a lower triangular matrix.
9. If in a given matrix rows and columns are interchanged, the new matrix obtained is the transpose of the original matrix, denoted by $\bar{A}'$.
10. A square matrix $\bar{A}$ is an orthogonal matrix if its product with its transpose is an identity matrix:

$$\bar{A}\bar{A}' = \bar{I} \tag{A.16}$$

11. The conjugate of a matrix is obtained by changing all its complex elements to their conjugates, that is, if

$$\bar{A} = \begin{vmatrix} 1-i & 3-4i & 5 \\ 7+2i & -i & 4-3i \end{vmatrix} \tag{A.17}$$

then its conjugate is

$$\bar{A}^* = \begin{vmatrix} 1+i & 3+4i & 5 \\ 7-2i & i & 4+3i \end{vmatrix} \tag{A.18}$$

A square matrix is a unit matrix if the product of the transpose of the conjugate matrix and the original matrix is an identity matrix:

$$\bar{A}^{*\prime}\bar{A} = \bar{I} \tag{A.19}$$

12. A square matrix is called a Hermitian matrix if every i–j element is equal to the conjugate complex j–i element, that is,

$$\bar{A} = \bar{A}^{*\prime} \tag{A.20}$$

13. A matrix, such that

$$\overline{A}^2 = \overline{A} \tag{A.21}$$

is called an idempotent matrix.

14. A matrix is periodic if

$$\overline{A}^{k+1} = \overline{A} \tag{A.22}$$

15. A matrix is called nilpotent if

$$\overline{A}^{k} = 0 \tag{A.23}$$

where k is a positive integer. If k is the least positive integer, then k is called the index of nilpotent matrix.

16. Addition of matrices follows a commutative law:

$$\overline{A} + \overline{B} = \overline{B} + \overline{A} \tag{A.24}$$

17. A scalar multiple is obtained by multiplying each element of the matrix with a scalar. The product of two matrices $\overline{A}$ and $\overline{B}$ is only possible if the number of columns in $\overline{A}$ equals the number of rows in $\overline{B}$.

If $\overline{A}$ is an $m \times n$ matrix and $\overline{B}$ is $n \times p$ matrix, the product $\overline{A}\,\overline{B}$ is an $m \times p$ matrix where

$$c_{ij} = a_{i1}b_{1j} + a_{i2}b_{2j} + \cdots + a_{in}b_{nj} \tag{A.25}$$

Multiplication is not commutative:

$$\overline{A}\,\overline{B} \neq \overline{B}\,\overline{A} \tag{A.26}$$

Multiplication is associative if confirmability is assured:

$$\overline{A}(\overline{B}\,\overline{C}) = (\overline{A}\,\overline{B})\overline{C} \tag{A.27}$$

It is distributive with respect to addition:

$$\overline{A}(\overline{B} + \overline{C}) = \overline{A}\,\overline{B} + \overline{A}\,\overline{C} \tag{A.28}$$

The multiplicative inverse exists if $|A| \neq 0$. Also,

$$(\overline{A}\,\overline{B})' = \overline{B}'\overline{A}' \tag{A.29}$$

18. The transpose of the matrix of cofactors of a matrix is called an adjoint matrix. The product of a matrix $\overline{A}$ and its adjoint is equal to the unit matrix multiplied by the determinant of A. If A is a square matrix,

$$\overline{A}_{\text{adj}}\overline{A} = \overline{A}\,\overline{I} \tag{A.30}$$

This property can be used to find the inverse of a matrix (see Example A.4).

19. By performing elementary transformations, any nonzero matrix can be reduced to one of the following forms called the normal forms:

$$[I_r]\ [I_r \quad 0] \begin{vmatrix} I_r \\ 0 \end{vmatrix} \begin{vmatrix} I_r & 0 \\ 0 & 0 \end{vmatrix} \tag{A.31}$$

The number r is called the rank of matrix $\overline{A}$. The form

$$\begin{vmatrix} I_r & 0 \\ 0 & 0 \end{vmatrix} \tag{A.32}$$

is called the first canonical form of $\overline{A}$. Both row and column transformations can be used here. The rank of a matrix is said to be r if (1) it has at least one nonzero minor of order r, and (2) every minor of $\overline{A}$ of order higher than $r = 0$. Rank is a nonzero row (the row that does not have all the elements $= 0$) in the upper triangular matrix.

Example A.1

Find the rank of the matrix

$$\overline{A} = \begin{vmatrix} 1 & 4 & 5 \\ 2 & 6 & 8 \\ 3 & 7 & 22 \end{vmatrix}$$

This matrix can be reduced to an upper triangular matrix by elementary row operations (see the following):

$$\overline{A} = \begin{vmatrix} 1 & 4 & 5 \\ 0 & 1 & 1 \\ 0 & 0 & 12 \end{vmatrix}$$

The rank of the matrix is 3.

A.2 Characteristics Roots, Eigenvalues, and Eigenvectors

For a square matrix $\overline{A}$, the $\overline{A} - \lambda\overline{I}$ matrix is called the characteristic matrix; λ is a scalar, and $\overline{I}$ is a unit matrix. The determinant $|A - \lambda I|$ when expanded gives a polynomial, which is called the characteristic polynomial of $\overline{A}$, and the equation $|A - \lambda I| = 0$ is called the characteristic equation of matrix $\overline{A}$. The roots of the characteristic equation are called the characteristic roots or eigenvalues.

Some properties of eigenvalues are as follows:

- Any square matrix $\overline{A}$ and its transpose $\overline{A}'$ have the same eigenvalues.
- The sum of the eigenvalues of a matrix is equal to the trace of the matrix (the sum of the elements on the principal diagonal is called the trace of the matrix).
- The product of the eigenvalues of the matrix is equal to the determinant of the matrix. If

$$\lambda_1, \lambda_2, \ldots, \lambda_n$$

are the eigenvalues of $\overline{A}$, then the eigenvalues of

$$\begin{aligned} k\overline{A} &\quad \text{are } k\lambda_1, k\lambda_2, \ldots, k\lambda_n \\ \overline{A}^m &\quad \text{are } \lambda_1^m, \lambda_2^m, \ldots, \lambda_n^m \\ \overline{A}^{-1} &\quad \text{are } 1/\lambda_1, 1/\lambda_2, \ldots, 1/\lambda_n \end{aligned} \tag{A.33}$$

- Zero is a characteristic root of a matrix, only if the matrix is singular.
- The characteristic roots of a triangular matrix are diagonal elements of the matrix.
- The characteristics roots of a Hermitian matrix are all real.
- The characteristic roots of a real symmetric matrix are all real, as the real symmetric matrix will be Hermitian.

A.2.1 Cayley–Hamilton Theorem

Every square matrix satisfies its own characteristic equation:

$$\text{If } |\overline{A} - \lambda\overline{I}| = (-1)^n(\lambda^n + a_1\lambda^{n-1} + a_2\lambda^{n-2} + \cdots + a_n) \tag{A.34}$$

is the characteristic polynomial of an $n \times n$ matrix, then the matrix equation

$$\begin{gathered} \overline{X}^n + a_1\overline{X}^{n-1} + a_2\overline{X}^{n-2} + \cdots + a_n\overline{I} = 0 \\ \text{is satisfied by } \overline{X} = \overline{A} \\ \overline{A}^n + a_1\overline{A}^{n-1} + a_2\overline{A}^{n-2} + \cdots + a_n\overline{I} = 0 \end{gathered} \tag{A.35}$$

This property can be used to find the inverse of a matrix.

Example A.2

Find the characteristic equation of the matrix

$$\overline{A} = \begin{vmatrix} 1 & 4 & 2 \\ 3 & 2 & -2 \\ 1 & -1 & 2 \end{vmatrix}$$

and then the inverse of the matrix.

The characteristic equation is given by

$$\begin{vmatrix} 1-\lambda & 4 & 2 \\ 3 & 2-\lambda & -2 \\ 1 & -1 & 2-\lambda \end{vmatrix} = 0$$

Expanding, the characteristic equation is

$$\lambda^3 - 5\lambda^2 - 8\lambda + 40 = 0$$

and then by the Cayley–Hamilton theorem

$$\overline{A}^2 - 5\overline{A} - 8\overline{I} + 40\overline{A}^{-1} = 0$$
$$40\overline{A}^{-1} = -\overline{A}^2 + 5\overline{A} + 8\overline{I}$$

We can write

$$40A^{-1} = -\begin{vmatrix} 1 & 4 & 2 \\ 3 & 2 & -2 \\ 1 & -1 & 2 \end{vmatrix}^2 + 5\begin{vmatrix} 1 & 4 & 2 \\ 3 & 2 & -2 \\ 1 & -1 & 2 \end{vmatrix} + 8\begin{vmatrix} 1 & 0 & 0 \\ 0 & 1 & 0 \\ 0 & 0 & 1 \end{vmatrix}$$

The inverse is

$$A^{-1} = \begin{vmatrix} -0.05 & 0.25 & 0.3 \\ 0.2 & 0 & -0.2 \\ 0.125 & -0.125 & 0.25 \end{vmatrix}$$

This is not an effective method of finding the inverse for matrices of large dimensions.

A.2.2 Characteristic Vectors

Each characteristic root λ has a corresponding nonzero vector $\overline{x}$ that satisfies the equation $|\overline{A} - \lambda\overline{I}|\ \overline{x} = 0$. The nonzero vector $\overline{x}$ is called the characteristic vector or eigenvector. The eigenvector is, therefore, not *unique*.

A.3 Diagonalization of a Matrix

If a square matrix $\overline{A}$ of $n \times n$ has n linearly independent eigenvectors, then a matrix $\overline{P}$ can be found so that

$$\overline{P}^{-1}\overline{A}\,\overline{P} \tag{A.36}$$

is a diagonal matrix.

The matrix $\overline{P}$ is found by grouping the eigenvectors of $\overline{A}$ into a square matrix, that is, $\overline{P}$ has eigenvalues of $\overline{A}$ as its diagonal elements.

A.3.1 Similarity Transformation

The transformation of matrix $\overline{A}$ into $\overline{P}^{-1}\,A\overline{P}$ is called a *similarity transformation*. Diagonalization is a special case of similarity transformation.

Example A.3

$$\text{Let } \overline{A} = \begin{vmatrix} -2 & 2 & -3 \\ 2 & 1 & -6 \\ -1 & -2 & 0 \end{vmatrix}$$

Its characteristics equation is

$$\begin{vmatrix} -2-\lambda & 2 & -3 \\ 2 & 1-\lambda & -6 \\ -1 & -2 & 0-\lambda \end{vmatrix} = 0$$

$$(-2-\lambda)(-\lambda+\lambda^2-12)-2(-2\lambda-6)-3(-4+1-\lambda)=0$$

$$\lambda^3+\lambda^2-21\lambda-45=0$$
$$(\lambda-5)(\lambda-3)(\lambda-3)=0$$

$$\lambda=5,\,-3,\,-3$$

The eigenvector is found by substituting the eigenvalues. For $\lambda=5$

$$-\begin{vmatrix} -7 & 2 & -3 \\ 2 & -4 & -6 \\ -1 & -2 & -5 \end{vmatrix}\begin{vmatrix} x \\ y \\ z \end{vmatrix} = \begin{vmatrix} 0 \\ 0 \\ 0 \end{vmatrix}$$

By manipulation of the rows, this can be reduced to

$$\begin{vmatrix} -1 & -2 & -5 \\ 0 & 16 & 32 \\ 0 & 0 & 0 \end{vmatrix}\begin{vmatrix} x \\ y \\ z \end{vmatrix} = \begin{vmatrix} 0 \\ 0 \\ 0 \end{vmatrix}$$

Therefore

$$-x-2y-5z=0$$
$$16y+32z=0$$

As eigenvectors are not unique, by assuming that $z=1$, and solving, one eigenvector is

$$(-1,-2,1)^t$$

Similarly, eigenvectors for $\lambda = -3$ can be found. This gives the following values:

$$\begin{vmatrix} x_1 \\ x_2 \\ (1/3)(x_1 + 2x_2) \end{vmatrix}$$

Choose arbitrarily $x_1 = 2$, $x_2 = -1$ and then

$$(2, -1, 0)^t$$

is another eigenvector. Similarly,

$$(3, 0, 1)^t$$

can be the third eigenvector. A matrix formed of these vectors is

$$\overline{P} = \begin{vmatrix} -1 & 2 & 3 \\ -2 & -1 & 0 \\ 1 & 0 & 1 \end{vmatrix}$$

and the diagonalization is obtained:

$$\overline{P}^{-1}\overline{A}\,\overline{P} = \begin{vmatrix} 5 & 0 & 0 \\ 0 & -3 & 0 \\ 0 & 0 & -3 \end{vmatrix}$$

This contains the eigenvalues as the diagonal elements.

Now choose some other eigenvectors and form a new matrix, say,

$$\overline{P} = \begin{vmatrix} 1 & 1 & 3 \\ 2 & 1 & 0 \\ -1 & 1 & 1 \end{vmatrix}$$

Again with these values, $\overline{P}^{-1}A\overline{P}$ is

$$\begin{vmatrix} 1 & 1 & 3 \\ 2 & 1 & 0 \\ -1 & 1 & 1 \end{vmatrix}^{-1} \begin{vmatrix} -2 & 2 & -3 \\ 2 & 1 & -6 \\ -1 & -2 & 0 \end{vmatrix} \begin{vmatrix} 1 & 1 & 3 \\ 2 & 1 & 0 \\ -1 & 1 & 1 \end{vmatrix} - \begin{vmatrix} 5 & 0 & 0 \\ 0 & -3 & 0 \\ 0 & 0 & -3 \end{vmatrix}$$

This is the same result as before.

A.4 Linear Independence or Dependence of Vectors

Vectors $\overline{x}_1, \overline{x}_2, \ldots, \overline{x}_n$ are dependent if all vectors (row or column matrices) are of the same order, and n scalars $\lambda_1, \lambda_2, \ldots, \lambda_n$ (not all zeros) exist such that

$$\lambda_1\overline{x}_1 + \lambda_2\overline{x}_2 + \lambda_3\overline{x}_3 + \cdots + \lambda_n\overline{x}_n = 0 \tag{A.37}$$

Otherwise, they are linearly independent. In other words, if vector $\overline{x}_K + 1$ can be written as a linear combination of vectors $(\overline{x}_1, \overline{x}_2, \ldots, \overline{x}_n)$, then it is linearly dependent, otherwise it is linearly independent. Consider the vectors

$$\overline{x}_3 = \begin{vmatrix} 4 \\ 2 \\ 5 \end{vmatrix} \quad \overline{x}_1 = \begin{vmatrix} 1 \\ 0.5 \\ 0 \end{vmatrix} \quad \overline{x}_2 = \begin{vmatrix} 0 \\ 0 \\ 1 \end{vmatrix}$$

then

$$\overline{x}_3 = 4\overline{x}_1 + 5\overline{x}_2$$

Therefore, $\overline{x}_3$ is linearly dependent on $\overline{x}_1$ and $\overline{x}_2$.

A.4.1 Vector Spaces

If $\overline{x}$ is any vector from all possible collections of vectors of dimension n, then for any scalar α, the vector $\alpha\overline{x}$ is also of dimension n. For any other *n-vector* $\overline{y}$, the vector $\overline{x} + \overline{y}$ is also of dimension n. The set of all n-dimensional vectors are said to form a linear vector space E^n. Transformation of a vector by a matrix is a linear transformation:

$$\overline{A}(\alpha\overline{x} + \beta\overline{y}) = \alpha(\overline{A}\overline{x}) + \beta(\overline{A}\overline{y}) \tag{A.38}$$

One property of interest is

$$\overline{A}\overline{x} = 0 \tag{A.39}$$

that is, whether any nonzero vector $\overline{x}$ exists that is transformed by matrix $\overline{A}$ into a zero vector. Equation A.39 can only be satisfied if the columns of $\overline{A}$ are linearly dependent. A square matrix whose columns are linearly dependent is called a singular matrix, and a square matrix whose columns are linearly independent is called a nonsingular matrix. In Equation A.39 if $\overline{x} = 0$, then columns of $\overline{A}$ must be linearly independent. The determinant of a singular matrix is zero and its inverse does not exist.

A.5 Quadratic Form Expressed as a Product of Matrices

The quadratic form can be expressed as a product of matrices:

$$\text{Quadratic form} = \overline{x}'A\overline{x} \tag{A.40}$$

where

$$\overline{x} = \begin{vmatrix} x_1 \\ x_2 \\ x_3 \end{vmatrix} \quad \overline{A} = \begin{vmatrix} a_{11} & a_{12} & a_{13} \\ a_{21} & a_{22} & a_{23} \\ a_{31} & a_{32} & a_{33} \end{vmatrix} \tag{A.41}$$

Therefore,

$$\bar{x}'A\bar{x} = |x_1\ x_2\ x_3| \begin{vmatrix} a_{11} & a_{12} & a_{13} \\ a_{21} & a_{22} & a_{23} \\ a_{31} & a_{32} & a_{33} \end{vmatrix} \begin{vmatrix} x_1 \\ x_2 \\ x_3 \end{vmatrix}$$
$$= a_{11}x_1^2 + a_{22}x_2^2 + a_{33}x_3^2 + 2a_{12}x,\ x_2 + 2a_{23}x_2x_3 + 2a_{13}x_1x_3 \tag{A.42}$$

A.6 Derivatives of Scalar and Vector Functions

A scalar function is defined as

$$y \cong f(x_1, x_2, \ldots, x_n) \tag{A.43}$$

where $x_1, x_2, \ldots, x_n$ are n variables. It can be written as a scalar function of an n-dimensional vector, that is, $y = f(\bar{x})$, where $\bar{x}$ is an n-dimensional vector:

$$\bar{x} = \begin{vmatrix} x_1 \\ x_2 \\ \cdot \\ x_n \end{vmatrix} \tag{A.44}$$

In general, a scalar function could be a function of several vector variables, that is, $y = f(\bar{x}, \bar{u}, \bar{p})$, where $\bar{x}$, $\bar{u}$, and $\bar{p}$ are vectors of various dimensions. A vector function is a function of several vector variables, that is, $y = f(\bar{x}, \bar{u}, \bar{p})$.

A derivative of a scalar function with respect to a vector variable is defined as

$$\frac{\partial \bar{f}}{\partial x} = \begin{vmatrix} \partial f/\partial x_1 \\ \partial f/\partial x_2 \\ \cdot \\ \partial f/\partial x_n \end{vmatrix} \tag{A.45}$$

The derivative of a scalar function with respect to a vector of n dimensions is a vector of the same dimension. The derivative of a vector function with respect to a vector variable x is defined as

$$\frac{\partial \bar{f}}{\partial x} = \begin{vmatrix} \partial f_1/\partial x_1 & \partial f_1/\partial x_2 & \cdot & \partial f_1/\partial x_n \\ \partial f_2/\partial x_1 & \partial f_2/\partial x_2 & \cdot & \partial f_2/\partial x_n \\ \cdot & \cdot & \cdot & \cdot \\ \partial f_m/\partial x_1 & \partial f_m/\partial x_2 & \cdot & \partial f_m/\partial x_n \end{vmatrix} = \begin{vmatrix} [\partial f_1/\partial x_1]^t \\ [\partial f_2/\partial x_2]^t \\ \cdot \\ [\partial f_m/\partial x_n]^t \end{vmatrix} \tag{A.46}$$

If a scalar function is defined as

$$\begin{aligned}\bar{s} &= \lambda^t f(\bar{x},\bar{u},\bar{p}) \\ &= \lambda_1 f_1(\bar{x},\bar{u},\bar{p}) + \lambda_2 f_2(\bar{x},\bar{u},\bar{p}) + \cdots + \lambda_m f_m(\bar{x},\bar{u},\bar{p})\end{aligned} \tag{A.47}$$

then $\partial s/\partial\lambda$ is

$$\frac{\partial\bar{s}}{\partial\lambda} = \begin{vmatrix} f_1(\bar{x},\bar{u},\bar{p}) \\ f_2(\bar{x},\bar{u},\bar{p}) \\ \cdot \\ f_m(\bar{x},\bar{u},\bar{p}) \end{vmatrix} = f(\bar{x},\bar{u},\bar{p}) \tag{A.48}$$

and $\partial s/\partial x$ is

$$\frac{\partial\bar{s}}{\partial x} = \begin{vmatrix} \lambda_1\dfrac{\partial f_1}{\partial x_1} + \lambda_2\dfrac{\partial f_2}{\partial x_1} + \cdots + \lambda_m\dfrac{\partial f_m}{\partial x_1} \\ \lambda_1\dfrac{\partial f_1}{\partial x_2} + \lambda_1\dfrac{\partial f_2}{\partial x_2} + \cdots + \lambda_m\dfrac{\partial f_m}{\partial x_2} \\ \cdots \\ \lambda_1\dfrac{\partial f_1}{\partial x_n} + \lambda_1\dfrac{\partial f_2}{\partial x_n} + \cdots + \lambda_m\dfrac{\partial f_m}{\partial x_n} \end{vmatrix} = \begin{vmatrix} \dfrac{\partial f_1}{\partial x_1} + \dfrac{\partial f_2}{\partial x_1} + \cdots + \dfrac{\partial f_m}{\partial x_1} \\ \dfrac{\partial f_1}{\partial x_2} + \dfrac{\partial f_2}{\partial x_2} + \cdots + \dfrac{\partial f_m}{\partial x_2} \\ \cdots \\ \dfrac{\partial f_1}{\partial x_n} + \dfrac{\partial f_2}{\partial x_n} + \cdots + \dfrac{\partial f_m}{\partial x_n} \end{vmatrix} \begin{vmatrix} \lambda_1 \\ \lambda_2 \\ \cdot \\ \lambda_m \end{vmatrix} \tag{A.49}$$

Therefore,

$$\frac{\partial\bar{s}}{\partial x} = \left|\frac{\partial f}{\partial x}\right|^t \bar{\lambda} \tag{A.50}$$

A.7 Inverse of a Matrix

The inverse of a matrix is often required in the power system calculations, though it is rarely calculated directly. The inverse of a square matrix $\bar{A}$ is defined so that

$$\bar{A}^{-1}\bar{A} = \bar{A}\bar{A}^{-1} = \bar{I} \tag{A.51}$$

The inverse can be evaluated in many ways.

A.7.1 By Calculating the Adjoint and Determinant of the Matrix

$$\bar{A}^{-1} = \frac{\bar{A}_{\text{adj}}}{|A|} \tag{A.52}$$

Example A.4

Consider the matrix

$$\bar{A} = \begin{vmatrix} 1 & 2 & 3 \\ 4 & 5 & 6 \\ 3 & 1 & 2 \end{vmatrix}$$

Its adjoint is

$$\bar{A}_{\text{adj}} = \begin{vmatrix} 4 & -1 & -3 \\ 10 & -7 & 6 \\ -11 & 5 & -3 \end{vmatrix}$$

and the determinant of $\bar{A}$ is equal to -9.
Thus, the inverse of $\bar{A}$ is

$$\bar{A}^{-1} = \begin{vmatrix} -\frac{4}{9} & \frac{1}{9} & \frac{1}{3} \\ -\frac{10}{9} & \frac{7}{9} & -\frac{2}{3} \\ \frac{11}{9} & -\frac{5}{9} & \frac{1}{3} \end{vmatrix}$$

A.7.2 By Elementary Row Operations

The inverse can also be calculated by elementary row operations. This operation is as follows:

1. A unit matrix of $n \times n$ is first attached to the right side of matrix $n \times n$ whose inverse is required to be found.
2. Elementary row operations are used to force the augmented matrix so that the matrix whose inverse is required becomes a unit matrix.

Example A.5

Consider a matrix

$$\bar{A} = \begin{vmatrix} 2 & 6 \\ 3 & 4 \end{vmatrix}$$

It is required to find its inverse.

Attach a unit matrix of 2×2 and perform the operations as shown as follows:

$$\begin{vmatrix} 2 & 6 \\ 3 & 4 \end{vmatrix}\begin{vmatrix} 1 & 0 \\ 0 & 1 \end{vmatrix} \to \frac{R_1}{2}\begin{vmatrix} 1 & 3 \\ 3 & 4 \end{vmatrix}\begin{vmatrix} \frac{1}{2} & 0 \\ 0 & 1 \end{vmatrix} \to R_2 - 3R_1 \begin{vmatrix} 1 & 3 \\ 0 & -5 \end{vmatrix}\begin{vmatrix} \frac{1}{2} & 0 \\ \frac{-3}{2} & 1 \end{vmatrix} \to R_1 + \frac{5}{3}R_2$$

$$\begin{vmatrix} 1 & 0 \\ 0 & -5 \end{vmatrix}\begin{vmatrix} \frac{-2}{5} & \frac{3}{5} \\ \frac{-3}{2} & 1 \end{vmatrix} \to R_2 - \frac{1}{5}\begin{vmatrix} 1 & 0 \\ 0 & 1 \end{vmatrix}\begin{vmatrix} \frac{-2}{5} & \frac{3}{5} \\ \frac{3}{10} & \frac{-1}{5} \end{vmatrix}$$

Thus, the inverse is

$$\overline{A}^{-1} = \begin{vmatrix} \frac{-2}{5} & \frac{3}{5} \\ \frac{3}{10} & \frac{-1}{5} \end{vmatrix}$$

A.7.2.1 Properties of Inverse Matrices

Some useful properties of inverse matrices are the following:

The inverse of a matrix product is the product of the matrix inverses taken in reverse order, that is,

$$[\overline{A} \quad \overline{B} \quad \overline{C}]^{-1} = [\overline{C}]^{-1}[\overline{B}]^{-1}[\overline{A}]^{-1} \tag{A.53}$$

The inverse of a diagonal matrix is a diagonal matrix whose elements are the respective inverses of the elements of the original matrix:

$$\begin{vmatrix} A_{11} & & \\ & B_{22} & \\ & & C_{33} \end{vmatrix}^{-1} = \begin{vmatrix} \frac{1}{A_{11}} & & \\ & \frac{1}{B_{22}} & \\ & & \frac{1}{C_{33}} \end{vmatrix} \tag{A.54}$$

A square matrix composed of diagonal blocks can be inverted by taking the inverse of the respective submatrices of the diagonal block:

$$\begin{vmatrix} [\text{block } A] & & \\ & [\text{block } B] & \\ & & [\text{block } C] \end{vmatrix}^{-1} = \begin{vmatrix} [\text{block } A]^{-1} & & \\ & [\text{block } B]^{-1} & \\ & & [\text{block } C]^{-1} \end{vmatrix} \tag{A.55}$$

A.7.3 Inverse by Partitioning

Matrices can be partitioned horizontally and vertically, and the resulting submatrices may contain only one element. Thus, a matrix $\overline{A}$ can be partitioned as shown in the following:

$$\overline{A} = \left|\begin{array}{ccc:c} a_{11} & a_{12} & a_{13} & a_{14} \\ a_{21} & a_{22} & a_{23} & a_{24} \\ \hdashline a_{31} & a_{32} & a_{33} & a_{34} \\ a_{41} & a_{42} & a_{43} & a_{44} \end{array}\right| = \left|\begin{array}{cc} \overline{A}_1 & \overline{A}_2 \\ \overline{A}_3 & \overline{A}_4 \end{array}\right| \quad \text{(A.56)}$$

where

$$\overline{A}_1 = \left|\begin{array}{ccc} a_{11} & a_{12} & a_{13} \\ a_{21} & a_{22} & a_{23} \\ a_{31} & a_{32} & a_{33} \end{array}\right| \quad \text{(A.57)}$$

$$\overline{A}_2 = \left|\begin{array}{c} a_{14} \\ a_{24} \\ a_{34} \end{array}\right| \quad \overline{A}_3 = |a_{41} \quad a_{42} \quad a_{43}| \quad \overline{A}_4 = [a_{44}] \quad \text{(A.58)}$$

Partitioned matrices follow the rules of matrix addition and subtraction. Partitioned matrices $\overline{A}$ and $\overline{B}$ can be multiplied if these are confirmable and columns of $\overline{A}$ and rows of $\overline{B}$ are partitioned exactly in the same manner:

$$\left|\begin{array}{cc} \overline{A}_{11_{2\times 2}} & \overline{A}_{12_{2\times 1}} \\ \overline{A}_{21_{1\times 2}} & \overline{A}_{22_{1\times 1}} \end{array}\right| \left|\begin{array}{cc} \overline{B}_{11_{2\times 3}} & \overline{B}_{12_{2\times 1}} \\ \overline{B}_{21_{1\times 3}} & \overline{B}_{22_{1\times 1}} \end{array}\right| = \left|\begin{array}{cc} \overline{A}_{11}\overline{B}_{11} + \overline{A}_{12}\overline{B}_{21} & \overline{A}_{11}\overline{B}_{12} + \overline{A}_{12}\overline{B}_{22} \\ \overline{A}_{21}\overline{B}_{11} + \overline{A}_{22}\overline{B}_{21} & \overline{A}_{21}\overline{B}_{12} + \overline{A}_{22}\overline{B}_{22} \end{array}\right| \quad \text{(A.59)}$$

Example A.6

Find the product of two matrices $\overline{A}$ and $\overline{B}$ by partitioning

$$\overline{A} = \left|\begin{array}{ccc} 1 & 2 & 3 \\ 2 & 0 & 1 \\ 1 & 3 & 6 \end{array}\right| \quad \overline{B} = \left|\begin{array}{cccc} 1 & 2 & 1 & 0 \\ 2 & 3 & 5 & 1 \\ 4 & 6 & 1 & 2 \end{array}\right|$$

is given by

$$\overline{A}\overline{B} = \left|\begin{array}{cc} \left|\begin{array}{cc} 1 & 2 \\ 2 & 0 \end{array}\right| \left|\begin{array}{ccc} 1 & 2 & 1 \\ 2 & 3 & 5 \end{array}\right| + \left|\begin{array}{c} 3 \\ 1 \end{array}\right| |4\ 6\ 1| & \left|\begin{array}{cc} 1 & 2 \\ 2 & 0 \end{array}\right| \left|\begin{array}{c} 0 \\ 1 \end{array}\right| + \left|\begin{array}{c} 3 \\ 1 \end{array}\right| |2| \\ |1 \quad 3| \left|\begin{array}{ccc} 1 & 2 & 1 \\ 2 & 3 & 5 \end{array}\right| + |6|\ |4\ 6\ 1| & |1 \quad 3| \left|\begin{array}{c} 0 \\ 1 \end{array}\right| + |6|\ |2| \end{array}\right|$$

A matrix can be inverted by partition. In this case, each of the diagonal submatrices must be square. Consider a square matrix partitioned into four submatrices:

$$\overline{A} = \begin{vmatrix} \overline{A}_1 & \overline{A}_2 \\ \overline{A}_3 & \overline{A}_4 \end{vmatrix} \tag{A.60}$$

The diagonal submatrices $\overline{A}_1$ and $\overline{A}_4$ are square, though these can be of different dimensions. Let the inverse of $\overline{A}$ be

$$\overline{A}^{-1} = \begin{vmatrix} \overline{A}_1'' & \overline{A}_2'' \\ \overline{A}_3'' & \overline{A}_4'' \end{vmatrix} \tag{A.61}$$

then

$$\overline{A}^{-1}\overline{A} = \begin{vmatrix} \overline{A}_1'' & \overline{A}_2'' \\ \overline{A}_3'' & \overline{A}_4'' \end{vmatrix} \begin{vmatrix} \overline{A}_1 & \overline{A}_2 \\ \overline{A}_3 & \overline{A}_4 \end{vmatrix} = \begin{vmatrix} 1 & 0 \\ 0 & 1 \end{vmatrix} \tag{A.62}$$

The following relations can be derived from this identity:

$$\begin{aligned} \overline{A}_1'' &= \left[\overline{A}_1 - \overline{A}_2\overline{A}_4^{-1}\overline{A}_3\right]^{-1} \\ \overline{A}_2'' &= -\overline{A}_1''\overline{A}_2\overline{A}_4^{-1} \\ \overline{A}_4'' &= \left[-\overline{A}_3\overline{A}_1^{-1}\overline{A}_2 + \overline{A}_4\right]^{-1} \\ \overline{A}_3'' &= -\overline{A}_4''\overline{A}_3\overline{A}_1^{-1} \end{aligned} \tag{A.63}$$

Example A.7

Invert the following matrix by partitioning

$$\overline{A} = \begin{vmatrix} 2 & 3 & 1 \\ 1 & 1 & 3 \\ 1 & 2 & 4 \end{vmatrix}$$

$$\overline{A}_1 = \begin{vmatrix} 2 & 3 \\ 1 & 1 \end{vmatrix} \quad \overline{A}_2 = \begin{vmatrix} 0 \\ 3 \end{vmatrix} \quad \overline{A}_3 = |1 \ 2| \quad \overline{A}_4 = |4|$$

$$\overline{A}_1'' = \left[\begin{vmatrix} 2 & 3 \\ 1 & 1 \end{vmatrix} - \begin{vmatrix} 0 \\ 3 \end{vmatrix} \left|\frac{1}{4}\right| |1 \ 2|\right]^{-1} = \begin{vmatrix} \frac{2}{7} & \frac{12}{7} \\ \frac{1}{7} & -\frac{8}{7} \end{vmatrix}$$

$$\overline{A}_2'' = -\begin{vmatrix} \frac{2}{7} & \frac{12}{7} \\ \frac{1}{7} & -\frac{8}{7} \end{vmatrix} \begin{vmatrix} 0 \\ 3 \end{vmatrix} \left|\frac{1}{4}\right| = \begin{vmatrix} -\frac{9}{7} \\ \frac{6}{7} \end{vmatrix}$$

$$\overline{A}_3'' = -\begin{bmatrix} 1 \\ 7 \end{bmatrix} |1\ 2| \begin{vmatrix} -1 & 3 \\ 1 & -2 \end{vmatrix} = \begin{vmatrix} -\frac{1}{7} & \frac{1}{7} \end{vmatrix}$$

$$\overline{A}_4'' = \left[-|1\ 2| \begin{vmatrix} -1 & 3 \\ 1 & -2 \end{vmatrix} \begin{vmatrix} 0 \\ 3 \end{vmatrix} + [4] \right]^{-1} = \frac{1}{7}$$

$$\overline{A}^{-1} = \begin{vmatrix} \frac{2}{7} & \frac{12}{7} & -\frac{9}{7} \\ \frac{1}{7} & -\frac{8}{7} & \frac{6}{7} \\ -\frac{1}{7} & \frac{1}{7} & \frac{1}{7} \end{vmatrix}$$

A.8 Solution of Large Simultaneous Equations

The application of matrices to the solution of large simultaneous equations constitutes one important application in the power systems. Mostly, these are sparse equations with many coefficients equal to zero. A large power system may have more than 3000 simultaneous equations to be solved.

A.8.1 Consistent Equations

A system of equations is consistent if they have one or more solutions.

A.8.2 Inconsistent Equations

A system of equations that has no solution is called inconsistent, that is, the following two equations are inconsistent:

$$x + 2y = 4$$
$$3x + 6y = 5$$

A.8.3 Test for Consistency and Inconsistency of Equations

Consider a system of n linear equations:

$$\begin{aligned} a_{11}x_1 + a_{12}x_2 + \cdots + a_{1n}x_1 &= b_1 \\ a_{21}x_1 + a_{22}x_2 + \cdots + A_{2n}x_2 &= b_2 \\ \cdots \\ a_{n1}x_1 + a_{n2}x_2 + \cdots + a_{mn}x_{n=b_n} \end{aligned} \tag{A.64}$$

Form an augmented matrix $\overline{C}$,

$$\overline{C} = [\overline{A}, \overline{B}] = \begin{vmatrix} a_{11} & a_{12} & \cdot & a_{1n} & b_1 \\ a_{21} & a_{22} & \cdot & a_{2n} & b_2 \\ \cdot & \cdot & \cdot & \cdot & \cdot \\ a_{n1} & a_{n2} & \cdot & a_{nn} & b_n \end{vmatrix} \tag{A.65}$$

The following holds for the test of consistency and inconsistency:

- A unique solution of the equations exists if rank of $\overline{A}$ = rank of $\overline{C} = n$, where n is the number of unknowns.
- There are infinite solutions to the set of equations if rank of $\overline{A}$ = rank of $\overline{C} = r$, $r < n$.
- The equations are inconsistent if rank of $\overline{A}$ is not equal to rank of $\overline{C}$.

Example A.8

Show that the equations

$$\begin{aligned} 2x + 6y &= -11 \\ 6x + 20y - 6z &= -3 \\ 6y - 18z &= -1 \end{aligned}$$

are inconsistent.

The augmented matrix is

$$\overline{C} = \overline{A}\overline{B} = \begin{vmatrix} 2 & 6 & 0 & -11 \\ 6 & 20 & -6 & -3 \\ 0 & 6 & -18 & -1 \end{vmatrix}$$

It can be reduced by elementary row operations to the following matrix:

$$\begin{vmatrix} 2 & 6 & 0 & -11 \\ 0 & 2 & -6 & 30 \\ 0 & 0 & 0 & -91 \end{vmatrix}$$

The rank of A is 2 and that of C is 3. The equations are not consistent.

The equations (A.64) can be written as

$$\overline{A}\overline{x} = \overline{b} \tag{A.66}$$

where

$\overline{A}$ is a square coefficient matrix
$\overline{b}$ is a vector of constants
$\overline{x}$ is a vector of unknown terms

If $\overline{A}$ is nonsingular, the unknown vector $\overline{x}$ can be found by

$$\overline{x} = \overline{A}^{-1}\overline{b} \tag{A.67}$$

This requires calculation of the inverse of matrix $\overline{A}$. Large system equations are not solved by direct inversion, but by a sparse matrix techniques.

Example A.9

This example illustrates the solution by transforming the coefficient matrix to an upper triangular form (backward substitution). The equations

$$\begin{vmatrix} 1 & 4 & 6 \\ 2 & 6 & 3 \\ 5 & 3 & 1 \end{vmatrix} \begin{vmatrix} x_1 \\ x_2 \\ x_3 \end{vmatrix} = \begin{vmatrix} 2 \\ 1 \\ 5 \end{vmatrix}$$

can be solved by row manipulations on the augmented matrix as follows:

$$\begin{vmatrix} 1 & 4 & 6 \\ 2 & 6 & 3 \\ 5 & 3 & 1 \end{vmatrix} \begin{vmatrix} 2 \\ 1 \\ 5 \end{vmatrix} \rightarrow R_2 - 2R_1 = \begin{vmatrix} 1 & 4 & 6 \\ 0 & -2 & -9 \\ 5 & 3 & 1 \end{vmatrix} \begin{vmatrix} 2 \\ -3 \\ 5 \end{vmatrix} \rightarrow R_3 - 5R_1$$

$$= \begin{vmatrix} 1 & 4 & 6 \\ 0 & -2 & -9 \\ 0 & -17 & -29 \end{vmatrix} \begin{vmatrix} 2 \\ -3 \\ -5 \end{vmatrix} \rightarrow R_3 - \frac{17}{2}R_2 = \begin{vmatrix} 1 & 4 & 6 \\ 0 & -2 & -9 \\ 0 & 0 & 47.5 \end{vmatrix} \begin{vmatrix} 2 \\ -3 \\ 20.5 \end{vmatrix}$$

Thus,

$$\begin{aligned} 47.5x_3 &= 20.5 \\ -2x_2 - 9x_3 &= -3 \\ x_1 + 4x_2 + 6x_3 &= 2 \end{aligned}$$

which gives

$$\overline{x} = \begin{vmatrix} 1.179 \\ -0.442 \\ 0.432 \end{vmatrix}$$

A set of simultaneous equations can also be solved by partitioning

$$\left|\begin{array}{c:c} a_{11}, \ldots, a_{1k} & a_{1m}, \ldots, a_{1n} \\ \ldots & \ldots \\ a_{k1}, \ldots, a_{kk} & a_{km}, \ldots, a_{kn} \\ \hdashline a_{m1}, \ldots, a_{mk} & a_{mm}, \ldots, a_{mn} \\ \ldots & \ldots \\ a_{n1}, \ldots, a_{nk} & a_{nm}, \ldots, a_{nn} \end{array}\right| \begin{vmatrix} x_1 \\ \cdot \\ x_k \\ -- \\ x_m \\ \cdot \\ x_n \end{vmatrix} = \begin{vmatrix} b_1 \\ \cdot \\ b_k \\ -- \\ b_m \\ \cdot \\ b_n \end{vmatrix} \tag{A.68}$$

Equation A.68 is horizontally partitioned and rewritten as

$$\begin{vmatrix} \overline{A}_1 & \overline{A}_2 \\ \overline{A}_3 & \overline{A}_4 \end{vmatrix} = \begin{vmatrix} \overline{X}_1 \\ \overline{X}_2 \end{vmatrix} \begin{vmatrix} \overline{B}_1 \\ \overline{B}_2 \end{vmatrix} \tag{A.69}$$

Vectors $\overline{X}_1$ and $\overline{X}_2$ are given by

$$\overline{X}_1 = \left\lfloor \overline{A}_1 - \overline{A}_2\overline{A}_4^{-1}\overline{A}_3 \right\rfloor^{-1} \left\lfloor \overline{B}_1 - \overline{A}_2\overline{A}_4^{-1}\overline{B}_2 \right\rfloor \tag{A.70}$$

$$\overline{X}_2 = \left\lfloor \overline{A}_4^{-1}(\overline{B}_2 - \overline{A}_3\overline{X}_1) \right\rfloor \tag{A.71}$$

A.9 Crout's Transformation

A matrix can be resolved into the product of a lower triangular matrix $\overline{L}$ and an upper unit triangular matrix $\overline{U}$, that is,

$$\begin{vmatrix} a_{11} & a_{12} & a_{13} & a_{14} \\ a_{21} & a_{22} & a_{23} & a_{24} \\ a_{31} & a_{32} & a_{33} & a_{34} \\ a_{41} & a_{42} & a_{43} & a_{44} \end{vmatrix} = \begin{vmatrix} l_{11} & 0 & 0 & 0 \\ l_{21} & l_{22} & 0 & 0 \\ l_{31} & l_{32} & l_{33} & 0 \\ l_{41} & l_{42} & l_{43} & l_{44} \end{vmatrix} \begin{vmatrix} 1 & u_{12} & u_{13} & u_{14} \\ 0 & 1 & u_{23} & u_{24} \\ 0 & 0 & 1 & u_{34} \\ 0 & 0 & 0 & 1 \end{vmatrix} \tag{A.72}$$

The elements of $\overline{U}$ and $\overline{L}$ can be found by multiplication:

$$\begin{aligned} l_{11} &= a_{11} \\ l_{21} &= a_{21} \\ l_{22} &= a_{22} - l_{21}u_{12} \\ l_{31} &= a_{31} \\ l_{32} &= a_{32} - l_{31}u_{12} \\ l_{33} &= a_{33} - l_{31}u_{13} - l_{32}u_{23} \\ l_{41} &= a_{41} \\ l_{42} &= a_{42} - l_{41}u_{12} \\ l_{43} &= a_{43} - l_{41}u_{13} - l_{42}u_{23} \\ l_{44} &= a_{44} - a_{41}u_{14} - l_{42}u_{24} - l_{43}u_{3} \end{aligned} \tag{A.73}$$

and

$$\begin{aligned} u_{12} &= \frac{a_{12}}{l_{11}} \\ u_{13} &= \frac{a_{13}}{l_{11}} \\ u_{14} &= \frac{a_{14}}{l_{11}} \\ u_{23} &= \frac{(a_{23} - l_{21}u_{13})}{l_{22}} \\ u_{24} &= \frac{(a_{24} - l_{21}u_{14})}{l_{22}} \\ u_{34} &= (a_{34} - l_{31}u_{14} - l_{32}u_{24})l_{33} \end{aligned} \tag{A.74}$$

In general,

$$l_{ij} = a_{ij} - \sum_{k=1}^{k=j-1} l_{ik}u_{kj} \quad i \geq j \tag{A.75}$$

for $j = 1, \ldots, n$

$$u_{ij} = \frac{1}{l_{ii}}\left(a_{ij} - \sum_{k=1}^{k=j-1} l_{ik}u_{kj}\right) \quad i < j \tag{A.76}$$

Example A.10

Transform the following matrix into LU form

$$\begin{vmatrix} 1 & 2 & 1 & 0 \\ 0 & 3 & 3 & 1 \\ 2 & 0 & 2 & 0 \\ 1 & 0 & 0 & 2 \end{vmatrix}$$

From Equations A.75 and A.76,

$$\begin{vmatrix} 1 & 2 & 1 & 0 \\ 0 & 3 & 3 & 1 \\ 2 & 0 & 2 & 0 \\ 1 & 0 & 0 & 2 \end{vmatrix} = \begin{vmatrix} 1 & 0 & 0 & 0 \\ 0 & 3 & 0 & 0 \\ 2 & -4 & 4 & 0 \\ 1 & -2 & 1 & 2.33 \end{vmatrix} \begin{vmatrix} 1 & 2 & 1 & 0 \\ 0 & 1 & 1 & 0.33 \\ 0 & 0 & 1 & 0.33 \\ 0 & 0 & 0 & 1 \end{vmatrix}$$

The original matrix has been converted into a product of lower and upper triangular matrices.

A.10 Gaussian Elimination

Gaussian elimination provides a natural means to determine the LU pair:

$$\begin{vmatrix} a_{11} & a_{12} & a_{13} \\ a_{21} & a_{22} & a_{23} \\ a_{31} & a_{32} & a_{33} \end{vmatrix} \begin{vmatrix} x_1 \\ x_2 \\ x_3 \end{vmatrix} = \begin{vmatrix} b_1 \\ b_2 \\ b_3 \end{vmatrix} \tag{A.77}$$

First, form an augmented matrix,

$$\begin{vmatrix} a_{11} & a_{12} & a_{13} & b_1 \\ a_{21} & a_{22} & a_{23} & b_2 \\ a_{31} & a_{32} & a_{33} & b_3 \end{vmatrix} \tag{A.78}$$

1. Divide the first row by a_{11}. This is the only operation to be carried out on this row. Thus, the new row is

$$\begin{gathered} 1 \quad a'_{12} \quad a'_{13} \quad b'_1 \\ a'_{12} = \frac{a_{12}}{a_{11}}, \quad a'_{13} = \frac{a_{13}}{a_{11}}, \quad b'_1 = \frac{b_1}{a_{11}} \end{gathered} \tag{A.79}$$

This gives

$$l_{11} = a_{11}, \quad u_{11} = 1, \quad u_{12} = a'_{12}, \quad u_{13} = a'_{13} \tag{A.80}$$

2. Multiply new row 1 by $-a_{21}$ and add to row 2. Thus, a_{21} becomes zero.

$$\begin{gathered} 0 \quad a'_{22} \quad a'_{23} \quad a'_{33} \quad b'_2 \\ a'_{22} = a_{22} - a_{21}a'_{12} \\ a'_{23} = a_{23} - a_{21}a'_{13} \\ b'_2 = b_2 - a_{21}b'_1 \end{gathered} \tag{A.81}$$

Divide new row 2 by a'_{22}. Row 2 becomes

$$\begin{gathered} 0 \quad 1 \quad a''_{23} \quad b''_2 \\ a''_{23} = \frac{a'_{23}}{a'_{22}} \\ b''_2 = \frac{b'_2}{a'_{22}} \end{gathered} \tag{A.82}$$

This gives

$$l_{21} = a_{21}, \quad l_{22} = a'_{22}, \quad u_{22} = 1, \quad u_{23} = a'_{23} \tag{A.83}$$

3. Multiply new row 1 by $-a_{31}$ and add to row 3. Thus, row 3 becomes

$$\begin{gathered} 0 \quad a'_{32} \quad a'_{33} \quad b'_3 \\ a'_{32} = a_{32} - a_{32}a'_{12} \\ a'_{33} = a_{33} - a_{31}a'_{13} \end{gathered} \tag{A.84}$$

Multiply row 2 by $-a_{32}$ and add to row 3. This row now becomes

$$0 \quad 0 \quad a''_{33} \quad b''_3 \tag{A.85}$$

Divide new row 3 by a''_{33}. This gives

$$\begin{gathered} 0 \quad 0 \quad 1 \quad b'''_3 \\ b'''_3 = \frac{b''_3}{a''_{33}} \end{gathered} \tag{A.86}$$

From these relations,

$$l_{33} = a''_{33}, \quad l_{31} = a_{31}, \quad l_{32} = a'_{32}, \quad u_{33} = 1 \tag{A.87}$$

Thus, all the elements of $\overline{L}$, $\overline{U}$ have been calculated and the process of forward substitution has been implemented on vector $\overline{b}$.

A.11 Forward–Backward Substitution Method

The set of sparse linear equations

$$\overline{A}\overline{x} = \overline{b} \tag{A.88}$$

can be written as

$$\overline{L}\,\overline{U}\overline{x} = \overline{b} \tag{A.89}$$

or

$$\overline{L}\overline{y} = \overline{b} \tag{A.90}$$

where

$$\overline{y} = \overline{U}\,\overline{x} \tag{A.91}$$

$\overline{L}\overline{y} = \overline{b}$ is solved for $\overline{y}$ by forward substitution. Thus, $\overline{y}$ is known. Then, $\overline{U}\,\overline{x} = \overline{y}$ is solved by backward substitution.

Solve $\overline{L}\,\overline{y} = \overline{b}$ by forward substitution:

$$\begin{vmatrix} l_{11} & 0 & 0 & 0 \\ l_{21} & l_{22} & 0 & 0 \\ l_{31} & l_{32} & l_{33} & 0 \\ l_{41} & l_{42} & l_{43} & l_{44} \end{vmatrix} \begin{vmatrix} y_1 \\ y_2 \\ y_3 \\ y_4 \end{vmatrix} = \begin{vmatrix} b_1 \\ b_2 \\ b_3 \\ b_4 \end{vmatrix} \tag{A.92}$$

Thus,

$$\begin{aligned} y_1 &= \frac{b_1}{l_{11}} \\ y_2 &= \frac{(b_2 - l_{21}y_1)}{l_{22}} \\ y_3 &= \frac{(b_3 - l_{31}y_1 - l_{32}y_2)}{l_{33}} \\ y_4 &= \frac{(b_4 - l_{41}y_1 - l_{42}y_2 - l_{43}y_3)}{l_{44}} \end{aligned} \tag{A.93}$$

Now solve $\overline{U}\overline{x} = \overline{y}$ by backward substitution:

$$\begin{vmatrix} 1 & u_{12} & u_{13} & u_{14} \\ 0 & 1 & u_{23} & u_{24} \\ 0 & 0 & 1 & u_{34} \\ 0 & 0 & 0 & 1 \end{vmatrix} \begin{vmatrix} x_1 \\ x_2 \\ x_3 \\ x_4 \end{vmatrix} = \begin{vmatrix} y_1 \\ y_2 \\ y_3 \\ y_4 \end{vmatrix} \tag{A.94}$$

Thus,

$$\begin{aligned} x_4 &= y_4 \\ x_3 &= y_3 - u_{34}x_4 \\ x_2 &= y_2 - u_{23}x_3 - u_{24}x_4 \\ x_1 &= y_1 - u_{12}x_2 - u_{13}x_3 - u_{14}x_4 \end{aligned} \tag{A.95}$$

The forward–backward solution is generalized by the following equation:

$$\overline{A} = \overline{L}\overline{U} = (\overline{L}_{\mathrm{d}} + \overline{L}_1)(\overline{I} + \overline{U}_{\mathrm{u}}) \tag{A.96}$$

where
- $\overline{L}_{\mathrm{d}}$ is the diagonal matrix
- $\overline{L}_1$ is the lower triangular matrix
- $\overline{I}$ is the identity matrix
- $\overline{U}_{\mathrm{u}}$ is the upper triangular matrix

Forward substitution becomes

$$\begin{aligned} \overline{L}\overline{y} &= \overline{b} \\ (\overline{L}_{\mathrm{d}} + \overline{L}_1)\overline{y} &= \overline{b} \\ \overline{L}_{\mathrm{d}}\overline{y} &= \overline{b} - \overline{L}_1\overline{y} \\ \overline{y} &= \overline{L}_{\mathrm{d}}^{-1}(\overline{b} - \overline{L}_1\overline{y}) \end{aligned} \tag{A.97}$$

that is,

$$\begin{vmatrix} y_1 \\ y_2 \\ y_3 \\ y_4 \end{vmatrix} = \begin{vmatrix} 1/l_{11} & 0 & 0 & 0 \\ 0 & 1/l_{22} & 0 & 0 \\ 0 & 0 & 1/l_{33} & 0 \\ 0 & 0 & 0 & 1/l_{44} \end{vmatrix} \times \left[\begin{vmatrix} b_1 \\ b_2 \\ b_3 \\ b_4 \end{vmatrix} - \begin{vmatrix} 0 & 0 & 0 & 0 \\ l_{21} & 0 & 0 & 0 \\ l_{31} & l_{32} & 0 & 0 \\ l_{41} & l_{42} & l_{43} & l_{44} \end{vmatrix} \begin{vmatrix} y_1 \\ y_2 \\ y_3 \\ y_4 \end{vmatrix} \right] \tag{A.98}$$

Backward substitution becomes

$$\begin{aligned} (\overline{I} + \overline{U}_{\mathrm{u}})\overline{x} &= \overline{y} \\ \overline{x} &= \overline{y} - \overline{U}_{\mathrm{u}}\overline{x} \end{aligned} \tag{A.99}$$

that is,

$$\begin{vmatrix} x_1 \\ x_2 \\ x_3 \\ x_4 \end{vmatrix} = \begin{vmatrix} y_1 \\ y_2 \\ y_3 \\ y_4 \end{vmatrix} - \begin{vmatrix} 0 & u_{12} & u_{13} & u_{14} \\ 0 & 0 & u_{23} & u_{24} \\ 0 & 0 & 0 & u_{34} \\ 0 & 0 & 0 & 0 \end{vmatrix} \begin{vmatrix} x_1 \\ x_2 \\ x_3 \\ x_4 \end{vmatrix} \tag{A.100}$$

A.11.1 Bifactorization

A matrix can also be split into LU form by sequential operation on the columns and rows. The general equations of the bifactorization method are

$$\begin{aligned} l_{ip} &= a_1 p \quad \text{for } \geq p \\ u_{pj} &= \frac{a_{pj}}{a_{pp}} \quad \text{for } j > p \\ a_{ij} &= a_1 j - l_{ip}u_{pj} \quad \text{for } i > p, \quad j > p \end{aligned} \tag{A.101}$$

Here, the letter p means the path or the pass. This will be illustrated with an example.

Example A.11

Consider the matrix

$$\overline{A} = \begin{vmatrix} 1 & 2 & 1 & 0 \\ 0 & 3 & 3 & 1 \\ 2 & 0 & 2 & 0 \\ 1 & 0 & 0 & 2 \end{vmatrix}$$

It is required to convert it into LU form. This is the same matrix of Example A.10.

Add an identity matrix, which will ultimately be converted into a U matrix and the $\overline{A}$ matrix will be converted into an L matrix:

$$\begin{vmatrix} 1 & 2 & 1 & 0 \\ 0 & 3 & 3 & 1 \\ 2 & 0 & 2 & 0 \\ 1 & 0 & 0 & 2 \end{vmatrix} \begin{vmatrix} 1 & 0 & 0 & 0 \\ 0 & 1 & 0 & 0 \\ 0 & 0 & 1 & 0 \\ 0 & 0 & 0 & 1 \end{vmatrix}$$

First step, $p = 1$:

1			
0	3	3	0
2	−4	0	0
1	−2	−1	2

1	2	1	0
0	1	0	0
0	0	1	0
0	0	1	0

The shaded columns and rows are converted into L and U matrix column and row and the elements of $\overline{A}$ matrix are modified using Equation A.101, that is,

$$\begin{aligned} a_{32} &= a_{32} - l_{31}u_{12} \\ &= 0 - (2)(2) = -4 \\ a_{33} &= a_{33} - l_{31}u_{31} \\ &= 2 - (2)(1) = 0 \end{aligned}$$

Second step, pivot column 2, $p=2$:

1			
0	3	3	0
2	−4	4	1.32
1	−2	1	2.66

1	2	1	0
0	1	1	0.33
0	0	1	0
0	0	0	1

Third step, pivot column 3, $p=3$:

1	0	0	0
0	3	0	0
2	−4	4	0
1	−2	1	2.33

1	2	1	0
0	1	1	0.33
0	0	1	0.33
0	0	0	1

This is the same result as derived before in Example A.10.

A.12 LDU (Product Form, Cascade, or Choleski Form)

The individual terms of L, D, and U can be found by direct multiplication. Again, consider a 4×4 matrix

$$\begin{vmatrix} a_{11} & a_{12} & a_{13} & a_{14} \\ a_{21} & a_{22} & a_{23} & a_{24} \\ a_{31} & a_{32} & a_{33} & a_{34} \\ a_{41} & a_{42} & a_{43} & a_{44} \end{vmatrix} = \begin{vmatrix} 1 & 0 & 0 & 0 \\ l_{21} & 1 & 0 & 0 \\ l_{31} & l_{32} & 1 & 0 \\ l_{41} & l_{42} & l_{43} & 1 \end{vmatrix} \begin{vmatrix} d_{11} & 0 & 0 & 0 \\ 0 & d_{22} & 0 & 0 \\ 0 & 0 & d_{33} & 0 \\ 0 & 0 & 0 & d_{44} \end{vmatrix} \begin{vmatrix} 1 & u_{12} & u_{13} & u_{14} \\ 0 & 1 & u_{23} & u_{24} \\ 0 & 0 & 1 & u_{34} \\ 0 & 0 & 0 & 1 \end{vmatrix} \quad \text{(A.102)}$$

The following relations exist:

$$
\begin{aligned}
d_{11} &= a_{11} \\
d_{22} &= a_{22} - l_{21}d_{11}u_{12} \\
d_{33} &= a_{33} - l_{31}d_{11}u_{13} - l_{32}d_{22}u_{23} \\
d_{44} &= a_{44} - l_{41}d_{11}u_{14} - l_{42}d_{22}u_{24} - l_{43}d_{33}u_{34} \\
u_{12} &= a_{12}/d_{11} \\
u_{13} &= a_{13}/d_{11} \\
u_{14} &= a_{14}/d_{11} \\
u_{23} &= (a_{23} - l_{21}d_{11}u_{13})/d_{22} \\
u_{24} &= (a_{24} - l_{21}d_{11}u_{14})/d_{22} \\
u_{34} &= (a_{34} - l_{31}d_{11}u_{14} - l_{32}d_{22}u_{24})/d_{33} \\
l_{21} &= a_{21}/d_{11} \\
l_{31} &= a_{31}/d_{11} \\
l_{32} &= (a_{32} - l_{31}d_{11}u_{12})/d_{22} \\
l_{41} &= a_{41}/d_{11} \\
l_{42} &= (a_{42} - l_{41}d_{11}u_{12})/d_{22} \\
l_{43} &= (a_{43} - l_{41}d_{11}u_{13} - l_{42}d_{22}u_{23})/d_{33}
\end{aligned}
\tag{A.103}
$$

In general,

$$
\begin{aligned}
d_{ii} &= a_{11} - \sum_{j=1}^{i=1} l_{ij}d_{jj}u_{ji} \quad \text{for } i = 1, 2, \ldots, n \\
u_{ik} &= \left[a_{ik} - \sum_{j=1}^{i=1} l_{if}d_{jj}u_{jk} \right] / d_{ii} \quad \text{for } k = i + 1 \ldots, n \quad i = 1, 2, \ldots, n \\
l_{ki} &= \left[a_{ki} - \sum_{j=1}^{i=1} l_{kj}d_{jj}u_{ji} \right] / d_{ii} \quad \text{for } k = i + 1, \ldots, n \quad i = 1, 2, \ldots, n
\end{aligned}
\tag{A.104}
$$

Another scheme is to consider A as a product of sequential lower and upper matrices as follows:

$$
A = (L_1L_2, \ldots, L_n)(U_n, \ldots, U_2U_1) \tag{A.105}
$$

$$
\begin{vmatrix} a_{11} & a_{12} & a_{13} & a_{14} \\ a_{21} & a_{22} & a_{23} & a_{24} \\ a_{31} & a_{32} & a_{33} & a_{34} \\ a_{41} & a_{42} & a_{43} & a_{44} \end{vmatrix}
=
\begin{vmatrix} l_{11} & 0 & 0 & 0 \\ l_{21} & 1 & 0 & 0 \\ l_{31} & 0 & 1 & 0 \\ l_{41} & 0 & 0 & 1 \end{vmatrix}
\begin{vmatrix} 1 & 0 & 0 & 0 \\ 0 & a_{22_2} & a_{23_2} & a_{24_2} \\ 0 & a_{32_2} & a_{33_2} & a_{34_2} \\ 0 & a_{42_2} & a_{43_2} & a_{44_2} \end{vmatrix}
\begin{vmatrix} 1 & u_{12} & u_{13} & u_{14} \\ 0 & 1 & 0 & 0 \\ 0 & 0 & 1 & 0 \\ 0 & 0 & 0 & 1 \end{vmatrix}
\tag{A.106}
$$

Here the second step elements are denoted by subscript 2 to the subscript.

$$
\begin{aligned}
&l_{21} = a_{21} \quad l_{31} = a_{31} \quad l_{41} = a_{41} \\
&u_{12} = a_{12}/l_{11} \quad u_{13} = a_{13}/l_{11} \quad u_{14} = a_{14}/l_{11} \\
&a_{ij_2} = a_{1j} - l_{1i}u_{1j} \quad i, j = 2, 3, 4
\end{aligned}
\tag{A.107}
$$

All elements correspond to first step, unless indicated by subscript 2.

In general, for the *kth* step,

$$
\begin{aligned}
d_{kk}^k &= a_{kk}^k \quad k = 1, 2, \ldots, n-1 \\
l_{ik}^k &= a_{ik}^k / a_{kk}^k \\
u_{kj} &= a_{kj}^k / a_{kk}^k \\
a_{ij}^{k+1} &= \left(a_{ij}^k - a_{ik}^k a_{kj}^k\right) / a_{kk}^k \\
&\quad k = 1, 2, \ldots, n-1 \quad i, j = k+1, \ldots, n
\end{aligned}
\tag{A.108}
$$

Example A.12

Convert the matrix of Example A.10 into LDU form

$$
\begin{vmatrix} 1 & 2 & 1 & 0 \\ 0 & 3 & 3 & 1 \\ 2 & 0 & 2 & 0 \\ 1 & 0 & 0 & 2 \end{vmatrix} = l^1 \times l^2 \times l^3 \times D \times u^3 \times u^2 \times u^1
$$

The lower matrices are

$$
l^1 \times l^2 \times l^3 = \begin{vmatrix} 1 & 0 & 0 & 0 \\ 0 & 1 & 0 & 0 \\ 2 & 0 & 1 & 0 \\ 1 & 0 & 0 & 1 \end{vmatrix} \begin{vmatrix} 1 & 0 & 0 & 0 \\ 0 & 1 & 0 & 0 \\ 0 & -4/3 & 1 & 0 \\ 1 & -2/3 & 0 & 1 \end{vmatrix} \begin{vmatrix} 1 & 0 & 0 & 0 \\ 0 & 1 & 0 & 0 \\ 0 & 0 & 1 & 0 \\ 0 & 0 & 1/4 & 0 \end{vmatrix}
$$

The upper matrices are

$$
u^3 \times u^2 \times u^1 = \begin{vmatrix} 1 & 0 & 0 & 0 \\ 0 & 1 & 0 & 1/3 \\ 0 & 0 & 1 & 1/3 \\ 0 & 0 & 0 & 1 \end{vmatrix} \begin{vmatrix} 1 & 0 & 0 & 0 \\ 0 & 1 & 1 & 0 \\ 0 & 0 & 1 & 0 \\ 0 & 0 & 0 & 1 \end{vmatrix} \begin{vmatrix} 1 & 2 & 1 & 0 \\ 0 & 1 & 0 & 0 \\ 0 & 0 & 1 & 0 \\ 0 & 0 & 0 & 1 \end{vmatrix}
$$

The matrix D is

$$D = \begin{vmatrix} 1 & 0 & 0 & 0 \\ 0 & 3 & 0 & 0 \\ 0 & 0 & 4 & 0 \\ 0 & 0 & 0 & 7/3 \end{vmatrix}$$

Thus, the LDU form of the original matrix is

$$\begin{vmatrix} 1 & 0 & 0 & 0 \\ 0 & 1 & 0 & 0 \\ 2 & -4/3 & 1 & 0 \\ 1 & -2/3 & 1/4 & 1 \end{vmatrix} \begin{vmatrix} 1 & 0 & 0 & 0 \\ 0 & 3 & 0 & 0 \\ 0 & 0 & 4 & 0 \\ 0 & 0 & 0 & 7/3 \end{vmatrix} \begin{vmatrix} 1 & 2 & 1 & 0 \\ 0 & 1 & 1 & 1/3 \\ 0 & 0 & 1 & 1/3 \\ 0 & 0 & 0 & 1 \end{vmatrix}$$

If the coefficient matrix is symmetrical (for a linear bilateral network), then

$$[L] = [U]^t \tag{A.109}$$

Because

$$\begin{aligned} l_{ip}(\text{new}) &= a_{ip}/a_{pp} \\ u_{pi} &= a_{pi}/a_{pp} \ (a_{ip} = a_{pi}) \end{aligned} \tag{A.110}$$

The LU and LDU forms are extensively used in power systems.

Bibliography

Brown, H.E. *Solution of Large Networks by Matrix Methods*. New York: Wiley Interscience, 1975.

Corbeiller, P.L. *Matrix Analysis of Electrical Networks*. Cambridge, MA: Harvard University Press, 1950.

Lewis, W.E. and D.G. Pryce. *The Application of Matrix Theory to Electrical Engineering*. London, U.K.: E&F N Spon, 1965.

Shipley, R.B. *Introduction to Matrices and Power Systems*. New York: Wiley, 1976.

Stignant, S.A. *Matrix and Tensor Analysis in Electrical Network Theory*. London, U.K.: Macdonald, 1964.

Appendix B: Calculation of Line and Cable Constants

This appendix presents an overview of calculations of line and cable constants with an emphasis on three-phase models and transformation matrices. Practically, the transmission or cable system parameters will be calculated using computer-based subroutine programs. For simple systems, the data are available in tabulated form for various conductor types, sizes, and construction [1–4]. Nevertheless, the basis of these calculations and required transformations are of interest to a power system engineer. The models described are generally applicable to steady-state studies. Frequency-dependent models are required for transient analysis studies, not discussed in this book [5].

B.1 AC Resistance

As we have seen, the conductor ac resistance is dependent upon frequency and proximity effects, temperature, spiraling, and bundle conductor effects, which increase the length of wound conductor in spiral shape with a certain pitch. The ratio R_{ac}/R_{dc} considering proximity and skin effects is given in Chapter 18. The resistance increases linearly with temperature and is given by the following equation:

$$R_2 = R_1\left(\frac{T + t_2}{T + t_1}\right) \tag{B.1}$$

where

R_2 is the resistance at temperature t_2

R_1 is the resistance at temperature t_1

T is the temperature coefficient, which depends on the conductor material

It is 234.5 for annealed copper, 241.5 for hard drawn copper, and 228.1 for aluminum. The resistance is read from manufacturers' data, databases in computer programs, or generalized tables. The internal resistance and the ac-to-dc resistance ratios using Bessel's functions is provided in Ref. [6]. The frequency-dependent behavior can be simulated by dividing the conductor into n hollow cylinders with an internal reactance and conductivity, and the model replicated by an equivalent circuit consisting of series of connections of hollow cylinders, Ref. [7]. Table 7.12 shows ac-to-dc resistance of conductors at 60 Hz.

B.2 Inductance

The *internal* inductance of a solid, smooth, round metallic cylinder of infinite length is due to its internal magnetic field when carrying an alternating current and is given by

$$L_{\text{int}} = \frac{\mu_0}{8\pi}\ \text{H/m (Henry per meter)} \tag{B.2}$$

where μ_0 is the permeability $= 4\pi \times 10^{-7}$ (H/m). Its *external* inductance is due to the flux outside the conductor and is given by

$$L_{\text{ext}} = \frac{\mu_0}{2\pi} \ln\left(\frac{D}{r}\right)\ \text{H/m} \tag{B.3}$$

where
- D is any point at a distance D from the surface of the conductor
- r is the conductor radius

In most inductance tables, D is equal to 1 ft and adjustment factors are tabulated for higher conductor spacings. The total reactance is

$$L = \frac{\mu_0}{2\pi}\left[\frac{1}{4} + \ln\frac{D}{r}\right] = \frac{\mu_0}{2\pi}\left[\ln\frac{D}{e^{-1/4}r}\right] = \frac{\mu_0}{2\pi}\left[\ln\frac{D}{\text{GMR}}\right]\ \text{H/m} \tag{B.4}$$

where GMR is called the geometric mean radius and is $0.7788r$. It can be defined as the radius of a tubular conductor with an infinitesimally thin wall that has the same external flux out to a radius of 1 ft as the external and internal flux of a solid conductor to the same distance.

B.2.1 Inductance of a Three-Phase Line

We can write the inductance matrix of a three-phase line in terms of flux linkages λ_a, λ_b, and λ_c:

$$\begin{vmatrix} \lambda_a \\ \lambda_b \\ \lambda_c \end{vmatrix} = \begin{vmatrix} L_{aa} & L_{ab} & L_{ac} \\ L_{ba} & L_{bb} & L_{bc} \\ L_{ca} & L_{cb} & L_{cc} \end{vmatrix} \begin{bmatrix} I_a \\ I_b \\ I_c \end{bmatrix} \tag{B.5}$$

The flux linkage λ_a, λ_b, and λ_c are given by

$$\begin{aligned}
\lambda_a &= \frac{\mu_0}{2\pi}\left[I_a \ln\left(\frac{1}{\text{GMR}_a}\right) + I_b \ln\left(\frac{1}{D_{ab}}\right) + I_c \ln\left(\frac{1}{D_{ac}}\right)\right] \\
\lambda_b &= \frac{\mu_0}{2\pi}\left[I_a \ln\left(\frac{1}{D_{ba}}\right) + I_b \ln\left(\frac{1}{\text{GMR}_b}\right) + I_c \ln\left(\frac{1}{D_{bc}}\right)\right] \\
\lambda_c &= \frac{\mu_0}{2\pi}\left[I_a \ln\left(\frac{1}{D_{ca}}\right) + I_b \ln\left(\frac{1}{D_{cb}}\right) + I_c \ln\left(\frac{1}{\text{GMR}_c}\right)\right]
\end{aligned} \tag{B.6}$$

where
$D_{ab}, D_{ac}, \ldots$ are the distances between the conductor of a phase and conductors of b and c phases
L_{aa}, L_{bb}, and L_{cc} are the self-inductances of the conductors
L_{ab}, L_{bb}, and L_{cc} are the mutual inductances

If we assume a symmetrical line, that is, the GMR of all three conductors is equal and also the spacing between the conductors is equal. The equivalent inductance per phase is

$$L = \frac{\mu_0}{2\pi} \ln\left(\frac{D}{\text{GMR}}\right) \text{ H/m} \tag{B.7}$$

The phase-to-neutral inductance of a three-phase symmetrical line is the same as the inductance per conductor of a two-phase line.

B.2.2 Transposed Line

A transposed line is shown in Figure B.1. Each phase conductor occupies the position of two other phase conductors for one-third of the length. The purpose is to equalize the phase inductances and reduce unbalance. The inductance derived for a symmetrical line is still valid and the distance D in Equation B.7 is substituted by GMD (geometric mean distance). It is given by

$$\text{GMD} = (D_{ab}D_{bc}D_{ca})^{1/3} \tag{B.8}$$

A detailed treatment of transposed lines with rotation matrices is given in Ref. [8].

B.2.3 Composite Conductors

A transmission line with composite conductors is shown in Figure B.2. Consider that group X is composed of n conductors in parallel and each conductor carries $1/n$ of the line current. The group Y is composed of m parallel conductors, each of which carries $-1/m$ of the return current. Then L_x, the inductance of conductor group X, is

$$L_x = 2 \times 10^{-7} \ln \frac{\sqrt[nm]{(D_{aa'}D_{ab'}D_{ac'} \ldots D_{am}) \ldots (D_{na'}D_{nb'}D_{nc'} \ldots D_{nm})}}{\sqrt[n^2]{(D_{aa}D_{ab}D_{ac} \ldots D_{an}) \ldots (D_{na}D_{nb}D_{nc} \ldots D_{nn})}} \text{ H/m} \tag{B.9}$$

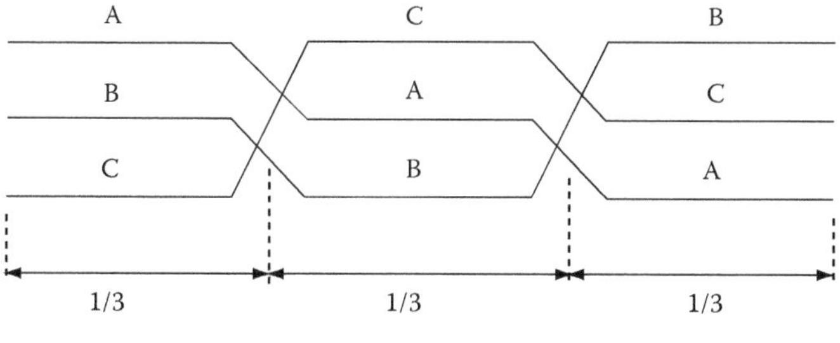

FIGURE B.1
Transposed transmission line.

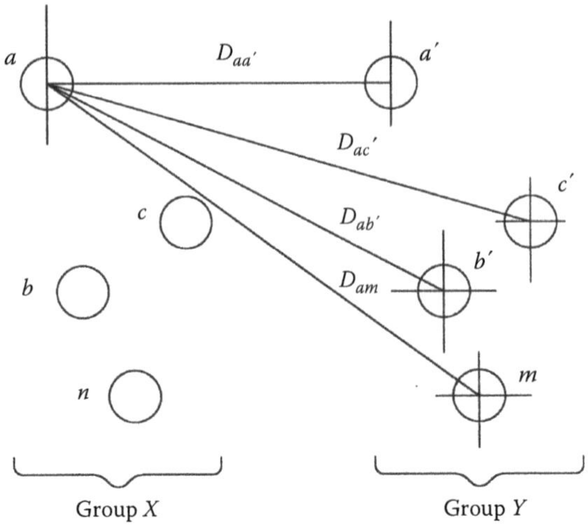

FIGURE B.2
Inductance of composite conductors.

We write Equation B.9 as

$$L_x = 2 \times 10^{-7} \ln\left(\frac{D_m}{D_{sx}}\right) \text{ H/m} \tag{B.10}$$

Similarly,

$$L_y = 2 \times 10^{-7} \ln\left(\frac{D_m}{D_{sy}}\right) \text{ H/m} \tag{B.11}$$

The total inductance is

$$L = (L_x + L_y) \text{ H/m} \tag{B.12}$$

B.3 Impedance Matrix

In Chapter 1, we decoupled a symmetrical three-phase line 3 × 3 matrix having equal self-impedances and mutual impedances (see Equation 1.37). We showed that the off-diagonal elements of the sequence impedance matrix are zero. In high-voltage transmission lines, which are transposed, this is generally true and the mutual couplings between phases are almost equal. However, the same cannot be said of distribution lines and these may have unequal off-diagonal terms. In many cases, the off-diagonal terms are smaller than the diagonal terms and the errors introduced in ignoring these will be small. Sometimes, an equivalence can be drawn by the equations

$$\begin{aligned} Z_s &= \frac{Z_{aa} + Z_{bb} + Z_{cc}}{3} \\ Z_m &= \frac{Z_{ab} + Z_{bc} + Z_{ca}}{3} \end{aligned} \tag{B.13}$$

that is, an average of the self- and mutual impedances can be taken. The sequence impedance matrix then has only diagonal terms. (See Example B.1.)

B.4 Three-Phase Line with Ground Conductors

A three-phase transmission line has couplings between phase-to-phase conductors and also between phase-to-ground conductors. Consider a three-phase line with two ground conductors, as shown in Figure B.3. The voltage V_a can be written as

$$\begin{aligned} V_a = {} & R_a I_a + j\omega L_a I_a + j\omega L_{ab} I_b + j\omega L_{ac} I_c + j\omega L_{aw} I_w + j\omega L_{av} I_v \\ & - j\omega L_{an} + V'_a + R_n I_n + j\omega L_n I_n - j\omega L_{an} I_a - j\omega L_{bn} I_b \\ & - j\omega L_{cn} I_c - j\omega L_{wn} I_w - j\omega L_{vn} I_v \end{aligned} \tag{B.14}$$

where

$R_a, R_b, \ldots, R_n$ are resistances of phases $a, b, \ldots, n$
$L_a, L_b, \ldots, L_n$ are the self-inductances
$L_{ab}, L_{ac}, \ldots, L_{an}$ are the mutual inductances

This can be written as

$$\begin{aligned} V_a = {} & (R_a + R_n)I_a + R_n I_b + R_n I_c + j\omega(L_a + L_n - 2L_{an})I_a \\ & + j\omega(L_{ab} + L_n - L_{an} - L_{bn})I_b + j\omega(L_{ac} + L_n - L_{an} - L_{cn})I_c + R_n I_w \\ & + j\omega(L_{aw} + L_n - L_{an} - L_{wn})I_w + R_n I_v + j\omega(L_{av} + L_n - L_{an} - L_{vn})I_v + V'_a \\ = {} & Z_{aa\text{-}g} I_a + Z_{ab\text{-}g} I_b + Z_{ac\text{-}g} I_c + Z_{aw\text{-}g} I_w + Z_{av\text{-}g} I_v + V'_a \end{aligned} \tag{B.15}$$

where

$Z_{aa\text{-}g}$ and $Z_{hb\text{-}g}$ are the self-impedances of a conductor with ground return
$Z_{ab\text{-}g}$ and $Z_{ac\text{-}g}$ are the mutual impedances between two conductors with common earth return

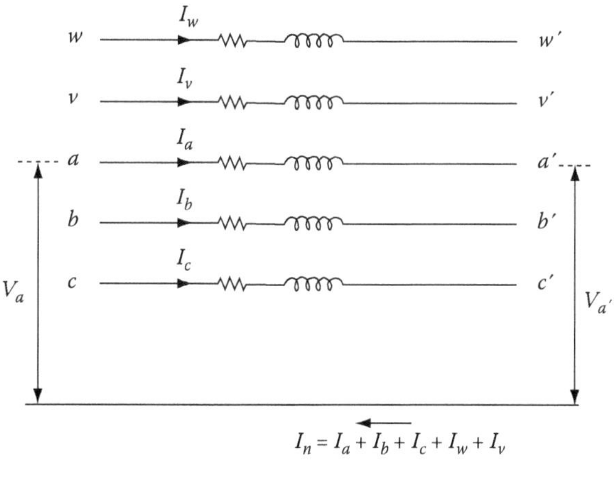

FIGURE B.3
Transmission line section with two ground conductors.

Similar equations apply to the voltages of other phases and ground wires. The following matrix then holds for the voltage differentials between terminals marked w, v, a, b, and c, and w', v', a', b', and c':

$$\begin{vmatrix} \Delta V_a \\ \Delta V_b \\ \Delta V_c \\ \Delta V_w \\ \Delta V_w \end{vmatrix} = \begin{vmatrix} Z_{aa\text{-}g} & Z_{ab\text{-}g} & Z_{ac\text{-}g} & Z_{aw\text{-}g} & Z_{av\text{-}g} \\ Z_{ba\text{-}g} & Z_{bb\text{-}g} & Z_{bc\text{-}g} & Z_{bw\text{-}g} & Z_{bv\text{-}g} \\ Z_{ca\text{-}g} & Z_{cb\text{-}g} & Z_{cc\text{-}g} & Z_{cw\text{-}g} & Z_{cv\text{-}g} \\ Z_{wa\text{-}g} & Z_{wb\text{-}g} & Z_{wc\text{-}g} & Z_{ww\text{-}g} & Z_{wv\text{-}g} \\ Z_{va\text{-}g} & Z_{vb\text{-}g} & Z_{vc\text{-}g} & Z_{vw\text{-}g} & Z_{vv\text{-}g} \end{vmatrix} \begin{vmatrix} I_a \\ I_b \\ I_c \\ I_w \\ I_v \end{vmatrix} \tag{B.16}$$

In the partitioned form, this matrix can be written as

$$\begin{vmatrix} \Delta\overline{V}_{abc} \\ \Delta\overline{V}_{wv} \end{vmatrix} = \begin{vmatrix} \overline{Z}_A & \overline{Z}_B \\ \overline{Z}_C & \overline{Z}_D \end{vmatrix} \begin{vmatrix} \overline{I}_{abc} \\ \overline{I}_{wv} \end{vmatrix} \tag{B.17}$$

Considering that the ground wire voltages are zero,

$$\begin{aligned} \Delta\overline{V}_{abc} &= \overline{Z}_A\overline{I}_{abc} + \overline{Z}_B\overline{I}_{wv} \\ 0 &= \overline{Z}_c\overline{I}_{abc} + \overline{Z}_D\overline{I}_{wv} \end{aligned} \tag{B.18}$$

Thus,

$$\overline{I}_{wv} = -\overline{Z}_D^{-1}\overline{Z}_C\overline{I}_{abc} \tag{B.19}$$

$$\Delta\overline{V}_{abc} = (\overline{Z}_A - \overline{Z}_B\overline{Z}_D^{-1}\overline{Z}_C)\overline{I}_{abc} \tag{B.20}$$

This can be written as

$$\Delta\overline{V}_{abc} = \overline{Z}_{abc}\overline{I}_{abc} \tag{B.21}$$

$$\overline{Z}_{abc} = \overline{Z}_A - \overline{Z}_B\overline{Z}_D^{-1}\overline{Z}_C = \begin{vmatrix} Z_{aa'\text{-}g} & Z_{ab'\text{-}g} & Z_{ac'\text{-}g} \\ Z_{ba'\text{-}g} & Z_{bb'\text{-}g} & Z_{bc'\text{-}g} \\ Z_{ca'\text{-}g} & Z_{cb'\text{-}g} & Z_{cc'\text{-}g} \end{vmatrix} \tag{B.22}$$

The five-conductor circuit is reduced to an equivalent three-conductor circuit. The technique is applicable to circuits with any number of ground wires provided that the voltages are zero in the lower portion of the voltage vector.

B.5 Bundle Conductors

Consider bundle conductors, consisting of two conductors per phase (Figure B.4). The original circuit of conductors a, b, and c and a', b', and c' can be transformed into an equivalent conductor system of a'', b'', and c''.

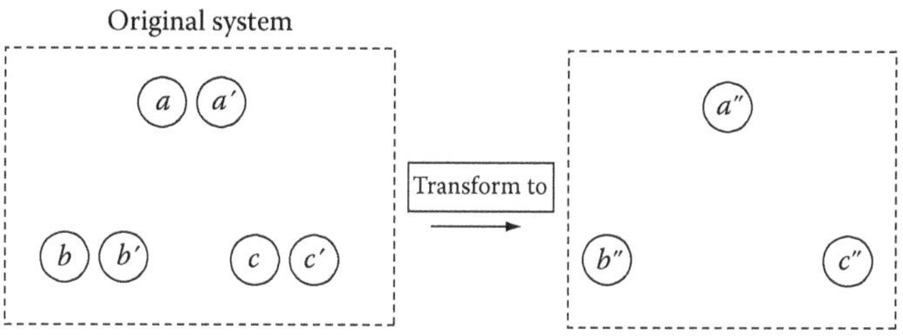

FIGURE B.4
Transformation of bundle conductors to equivalent single conductors.

Each conductor in the bundle carries a different current and has a different self- and mutual impedance because of its specific location. Let the currents in the conductors be I_a, I_b, and I_c, and I'_a, I'_b, and I'_c, respectively. The following primitive matrix equation can be written as follows:

$$\begin{vmatrix} V_a \\ V_b \\ V_c \\ -- \\ V'_a \\ V'_b \\ V'_c \end{vmatrix} \begin{vmatrix} Z_{aa} & Z_{ab} & Z_{ac} & \vdots & Z_{aa'} & Z_{ab'} & Z_{ac'} \\ Z_{ba} & Z_{bb} & Z_{bc} & \vdots & Z_{ba'} & Z_{bb'} & Z_{bc'} \\ Z_{ca} & Z_{cb} & Z_{cc} & \vdots & Z_{ca'} & Z_{cb'} & Z_{cc'} \\ -- & -- & -- & -- & -- & -- & -- \\ Z_{a'a} & Z_{a'b} & Z_{a'c} & \vdots & Z_{a'a'} & Z_{a'b'} & Z_{a'c'} \\ Z_{b'a} & Z_{b'b} & Z_{b'c} & \vdots & Z_{b'a'} & Z_{b'b'} & Z_{b'c'} \\ Z_{c'a} & Z_{c'b} & Z_{c'c} & \vdots & Z_{c'a'} & Z_{c'b'} & Z_{c'c'} \end{vmatrix} \begin{vmatrix} I_a \\ I_b \\ I_c \\ -- \\ I_{a'} \\ I_{b'} \\ I_{c'} \end{vmatrix} \tag{B.23}$$

This can be partitioned so that

$$\begin{vmatrix} \overline{V}_{abc} \\ \overline{V}_{a'b'c'} \end{vmatrix} = \begin{vmatrix} \overline{Z}_1 & \overline{Z}_2 \\ \overline{Z}_3 & \overline{Z}_4 \end{vmatrix} \begin{vmatrix} \overline{I}_{abc} \\ \overline{I}_{a'b'c'} \end{vmatrix} \tag{B.24}$$

for symmetrical arrangement of bundle conductors $\overline{Z}_1 = \overline{Z}_4$.

Modify so that the lower portion of the vector goes to zero. Assume that

$$\begin{aligned} V_a &= V'_a = V''_a \\ V_b &= V'_b = V''_b \\ V_c &= V'_c = V''_c \end{aligned} \tag{B.25}$$

The upper part of the matrix can then be subtracted from the lower part:

$$\begin{vmatrix} V_a \\ V_b \\ V_c \\ -- \\ 0 \\ 0 \\ 0 \end{vmatrix} = \begin{vmatrix} Z_{aa} & Z_{ab} & Z_{ac} & \vdots & Z_{aa'} & Z_{ab'} & Z_{ac'} \\ Z_{ba} & Z_{bb} & Z_{bc} & \vdots & Z_{ba'} & Z_{bb'} & Z_{bc'} \\ Z_{ca} & Z_{cb} & Z_{cc} & \vdots & Z_{ca'} & Z_{cb'} & Z_{cc'} \\ -- & -- & -- & -- & -- & -- & -- \\ Z_{a'a}-Z_{aa} & Z_{a'b}-Z_{ab} & Z_{a'c}-Z_{ac} & \vdots & Z_{a'a'}-Z_{aa'} & Z_{a'b'}-Z_{ab'} & Z_{a'c'}-Z_{ac'} \\ Z_{b'a}-Z_{ba} & Z_{b'b}-Z_{bb} & Z_{b'c}-Z_{bc} & \vdots & Z_{b'a'}-Z_{ba'} & Z_{b'b'}-Z_{bb'} & Z_{b'c'}-Z_{bc'} \\ Z_{c'a}-Z_{ca} & Z_{c'b}-Z_{cb} & Z_{c'c}-Z_{cc} & \vdots & Z_{c'a'}-Z_{ca'} & Z_{c'b'}-Z_{cb'} & Z_{c'c}-Z_{cc'} \end{vmatrix} \begin{vmatrix} I_a \\ I_b \\ I_c \\ -- \\ I_{a'} \\ I_{b'} \\ I_{c'} \end{vmatrix} \tag{B.26}$$

We can write it in the partitioned form as

$$\begin{vmatrix} \overline{V}_{abc} \\ 0 \end{vmatrix} = \begin{vmatrix} \overline{Z}_1 & \overline{Z}_2 \\ \overline{Z}_2^t - \overline{Z}_1 & \overline{Z}_4 - \overline{Z}_2 \end{vmatrix} \begin{vmatrix} \overline{I}_{abc} \\ \overline{I}_{a'b'c'} \end{vmatrix} \tag{B.27}$$

$$\begin{aligned} I''_a &= I_a + I'_a \\ I''_b &= I_b + I'_b \\ I''_c &= I_c + I'_c \end{aligned} \tag{B.28}$$

The matrix is modified as shown in the following:

$$\left|\begin{array}{cccc:cc} Z_{aa} & Z_{ab} & Z_{ac} & Z_{aa'} - Z_{aa} & Z_{ab'} - Z_{ab} & Z_{ac'} - Z_{ac} \\ Z_{ba} & Z_{bb} & Z_{bc} & Z_{ba'} + Z_{ba} & Z_{bb'} + Z_{bb} & Z_{bc'} - Z_{bc} \\ Z_{ca} & Z_{cb} & Z_{cc} & Z_{ca'} - Z_{ca} & Z_{cb'} - Z_{cb} & Z_{cc'} - Z_{cc} \\ \hdashline Z_{a'a} - Z_{aa} & Z_{a'b} - Z_{ab} & Z_{a'c} - Z_{ac} & Z_{a'a'} - Z_{aa'} - Z_{a'a} + Z_{aa} & Z_{a'b'} - Z_{ab'} - Z_{a'b} + Z_{ab} & Z_{a'c'} - Z_{ac'} - Z_{a'c} + Z_{ac} \\ Z_{b'a} - Z_{ba} & Z_{b'b} - Z_{bb} & Z_{b'c} - Z_{bc} & Z_{b'a'} - Z_{ba'} - Z_{b'a} + Z_{ba} & Z_{b'b'} - Z_{bb'} - Z_{b'b} + Z_{bb} & Z_{b'c'} - Z_{bc'} - Z_{b'c} + Z_{bc} \\ Z_{c'a} - Z_{ca} & Z_{c'b} - Z_{cb} & Z_{c'c} - Z_{cc} & Z_{c'a'} - Z_{ca'} - Z_{c'a} + Z_{ca} & Z_{c'b'} - Z_{cb'} - Z_{c'b} + Z_{cb} & Z_{c'c'} - Z_{cc'} - Z_{c'c} + Z_{cc} \end{array}\right| \left|\begin{array}{c} I_a + I'_a \\ I_b + I'_b \\ I_c + I'_c \\ \hdashline I'_a \\ I'_b \\ I'_c \end{array}\right| \tag{B.29}$$

or in partitioned form

$$\begin{vmatrix} \overline{V}_{abc} \\ 0 \end{vmatrix} = \begin{vmatrix} \overline{Z}_1 & \overline{Z}_2 - \overline{Z}_1 \\ \overline{Z}_2^t - \overline{Z}_1 & (\overline{Z}_4 - \overline{Z}_2) - \left(\overline{Z}_2^t - \overline{Z}_1\right) \end{vmatrix} \begin{vmatrix} \overline{I}''_{abc} \\ \overline{I}'_{a'b'c'} \end{vmatrix} \tag{B.30}$$

This can now be reduced to following 3×3 matrix as before:

$$\begin{vmatrix} V''_a \\ V''_b \\ V''_c \end{vmatrix} = \begin{vmatrix} Z''_{aa} & Z''_{ab} & Z''_{ac} \\ Z''_{ba} & Z''_{bb} & Z''_{bc} \\ Z''_{ca} & Z''_{cb} & Z''_{cc} \end{vmatrix} \begin{vmatrix} I''_a \\ I''_b \\ I''_c \end{vmatrix} \tag{B.31}$$

B.6 Carson's Formula

The theoretical value of $Z_{abc\text{-}g}$ can be calculated by Carson's formula (ca. 1926). This is of importance even today in calculations of line constants. For an n-conductor configuration, the earth is assumed as an infinite uniform solid with a constant resistivity. Figure B.5 shows image conductors in the ground at a distance equal to the height of the conductors above ground and exactly in the same formation, with the same spacings between the conductors. A flat conductor formation is shown in Figure B.5.

$$Z_{ii} = R_i + 4\omega P_{ii} G + j\left[X_i + 2\omega G \ln\frac{S_{ii}}{r_i} + 4\omega Q_{ii} G\right] \ \Omega/\text{mi} \tag{B.32}$$

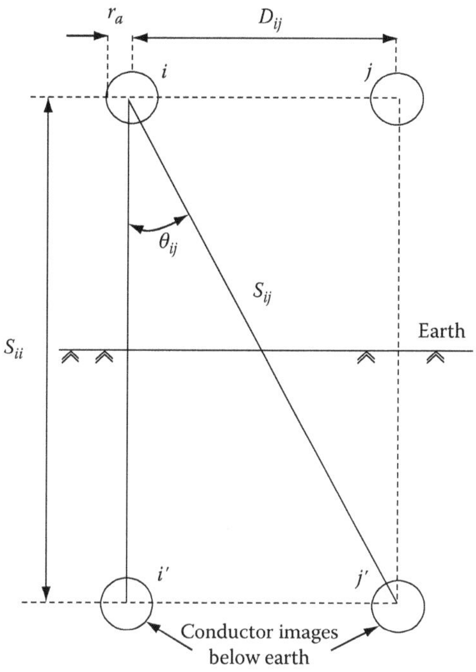

FIGURE B.5
Conductors and their images; (Carson's formula).

$$Z_{ij} = 4\omega P_{ii}G + j\left[2\omega G \ \ln\frac{S_{ij}}{D_{ij}} + 4\omega Q_{ij}G\right] \ \Omega/\text{mi} \tag{B.33}$$

where
- Z_{ii} is the self-impedance of conductor i with earth return (ohm/mi)
- Z_{ij} is the mutual impedance between conductors i and j (ohm/mi)
- R_i is the resistance of conductor (ohm/mi)
- S_{ii} is the conductor to image distance of the ith conductor to its own image
- S_{ij} is the conductor to image distance of the ith conductor to the image of the jth conductor
- D_{ij} is the distance between conductors i and j
- r_i is the radius of conductor (ft)
- ω is the angular frequency
- $G = 0.1609347 \times 10^{-7}$ ohm-cm
- GMR_i is the geometric mean radius of conductor i
- ρ is the soil resistivity
- θ_{ij} is the angle as shown in Figure B.5

Expressions for P and Q are

$$P = \frac{\pi}{8} - \frac{1}{3\sqrt{2}}k\ \cos\theta + \frac{k^2}{16}\cos 2\theta\left(0.6728 + \ln\frac{2}{k}\right) + \frac{k^2}{16}\theta\ \sin\theta + \frac{k^3\cos 3\theta}{45\sqrt{2}} - \frac{\pi k^4\cos 4\theta}{1536} \tag{B.34}$$

$$Q = -0.0386 + \frac{1}{2}\ln\frac{2}{k} + \frac{1}{3\sqrt{2}}\cos\theta - \frac{k^2\cos 2\theta}{64} + \frac{k^3\cos 3\theta}{45\sqrt{2}}$$

$$-\frac{k^4\sin 4\theta}{384} - \frac{k^4\cos 4\theta}{384}\left(\ln\frac{2}{k} + 1.0895\right) \quad \text{(B.35)}$$

where

$$k = 8.565 \times 10^4 S_{ij}\sqrt{\frac{f}{\rho}} \quad \text{(B.36)}$$

S_{ij} is in feet
ρ is soil resistivity (ohm-m)
f is the system frequency

This shows dependence on frequency as well as on soil resistivity.

B.6.1 Approximations to Carson's Equations

These approximations involve P and Q and the expressions are given by

$$P_{ij} = \frac{\pi}{8} \quad \text{(B.37)}$$

$$Q_{ij} = -0.03860 + \frac{1}{2}\ln\frac{2}{k_{ij}} \quad \text{(B.38)}$$

Using these assumptions, $f = 60$ Hz and soil resistivity = 100 ohm-m, the equations reduce to

$$Z_{ii} = R_i + 0.0953 + j0.12134\left(\ln\frac{1}{\text{GMR}_i} + 7.93402\right) \text{ ohm/mi} \quad \text{(B.39)}$$

$$Z_{ij} = 0.0953 + j0.12134\left(\ln\frac{1}{D_{ij}} + 7.93402\right) \text{ ohm/mi} \quad \text{(B.40)}$$

Equations B.39 and B.40 are of practical significance for calculations of line impedances.

Example B.1

Consider an unsymmetrical overhead line configuration, as shown in Figure B.6. The phase conductors consist of 556.5 kcmil (556,500 circular mils) of ACSR conductor consisting of 26 strands of aluminum, two layers, and seven strands of steel. From the properties of ACSR conductor tables, the conductor has a resistance of 0.1807 ohm at 60 Hz and its GMR is 0.0313 ft at 60 Hz; conductor diameter = 0.927 in. The neutral consists of 336.4 kcmil, ACSR conductor, resistance 0.259 ohm per mi at 60 Hz and 50°C and GMR 0.0278 ft, and conductor diameter 0.806 in. It is required to form a primitive Z matrix and convert it into a 3 × 3 Z_{abc} matrix and then to sequence impedance matrix Z_{012}.

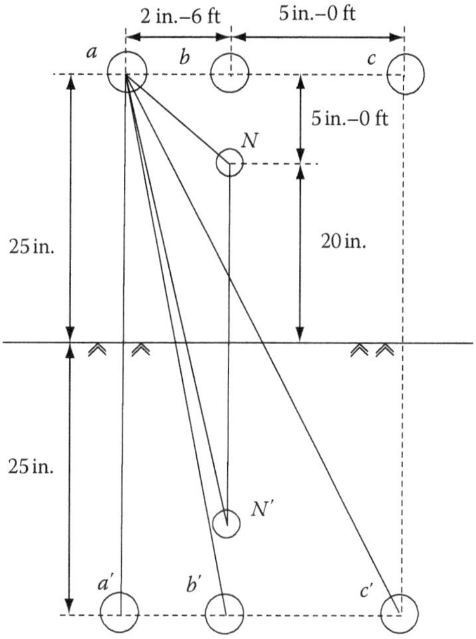

FIGURE B.6
Distribution line configuration for calculation of line parameters (Examples B.1 and B.3).

Using Equations B.39 and B.40,

$$Z_{aa} = Z_{bb} = Z_{cc} = 0.1859 + j1.3831$$
$$Z_{nn} = 0.3543 + j1.3974$$
$$Z_{ab} = Z_{ba} = 0.0953 + j0.8515$$
$$Z_{bc} = Z_{cb} = 0.0953 + j0.7674$$
$$Z_{ca} = Z_{ac} = 0.0953 + j0.7182$$
$$Z_{an} = Z_{na} = 0.0953 + j0.7539$$
$$Z_{bn} = Z_{nb} = 0.0953 + j0.7674$$
$$Z_{cn} = Z_{nc} = 0.0953 + j0.7237$$

Therefore, the primitive impedance matrix is

$$\bar{Z}_{prim} = \begin{vmatrix} 0.1859 + j1.3831 & 0.0953 + j0.08515 & 0.0953 + j0.7182 & 0.0953 + j0.7539 \\ 0.0953 + j0.8515 & 0.1859 + j1.3831 & 0.0953 + j0.7624 & 0.0953 + j0.7674 \\ 0.0953 + j0.7182 & 0.0953 + j0.7624 & 0.1859 + j1.3831 & 0.0953 + j0.7237 \\ 0.0953 + j0.7539 & 0.0953 + j0.7674 & 0.0953 + j0.7237 & 0.3543 + j1.3974 \end{vmatrix}$$

Eliminate the last row and column using Equation B.22

$$\bar{Z}_{abc} = \begin{vmatrix} 0.1846 + j0.9825 & 0.0949 + j0.4439 & 0.0921 + j0.3334 \\ 0.0949 + j0.4439 & 0.1864 + j0.9683 & 0.0929 + j0.3709 \\ 0.0921 + j0.3334 & 0.0929 + j0.3709 & 0.1809 + j1.0135 \end{vmatrix} \text{ ohm/mi}$$

Convert to Z_{012} by using the transformation equation (1.35)

$$\bar{Z}_{012} = \begin{vmatrix} 0.3705 + j1.7536 & 0.0194 + j0.0007 & -0.0183 + j0.0055 \\ -0.0183 + j0.0055 & 0.0907 + j0.6054 & -0.0769 - j0.0146 \\ 0.0194 + j0.0007 & 0.0767 + j0.0147 & 0.0907 + j0.6054 \end{vmatrix}$$

This shows the mutual coupling between sequence impedances. We could average out the self- and mutual impedances according to Equation B.13

$$Z_s = \frac{Z_{aa} + Z_{bb} + Z_{cc}}{3} = 0.184 + j0.9973$$

$$Z_m = \frac{Z_{ab} + Z_{bc} + Z_{ca}}{3} = 0.0933 + j0.38271$$

The matrix Z_{abc} then becomes

$$\bar{Z}_{abc} = \begin{vmatrix} 0.184 + j0.9973 & 0.933 + j0.3827 & 0.933 + j0.3827 \\ 0.933 + j0.3827 & 0.184 + j0.9973 & 0.933 + j0.3827 \\ 0.933 + j0.3827 & 0.933 + j0.3827 & 0.184 + j0.9973 \end{vmatrix} \text{ohm/mi}$$

and this gives

$$\bar{Z}_{012} = \begin{vmatrix} 0.3706 + j1.7627 & 0 & 0 \\ 0 & 0.0907 + j0.6146 & 0 \\ 0 & 0 & 0.0907 + j0.6146 \end{vmatrix} \text{ohm/mi}$$

Example B.2

Figure B.7 shows a high-voltage line with two 636,000 mils ACSR bundle conductors per phase. Conductor GMR = 0.0329 ft, resistance = 0.1688 ohm per mi, diameter = 0.977 in., and spacings are as shown in Figure B.7. Calculate the primitive impedance matrix, reduce it to a 3 × 3 matrix, and then convert it into a sequence component matrix.

From Equations B.39 and B.40 and the specified spacings in Figure B.7, matrix Z_1 is

$$\bar{Z}_1 = \begin{vmatrix} 0.164 + j1.3770 & 0.0953 + j0.5500 & 0.0953 + j0.4659 \\ 0.953 + j0.5500 & 0.164 + j1.3770 & 0.0953 + j0.5500 \\ 0.0953 + j0.4659 & 0.0953 + j0.5500 & 0.164 + j1.3770 \end{vmatrix}$$

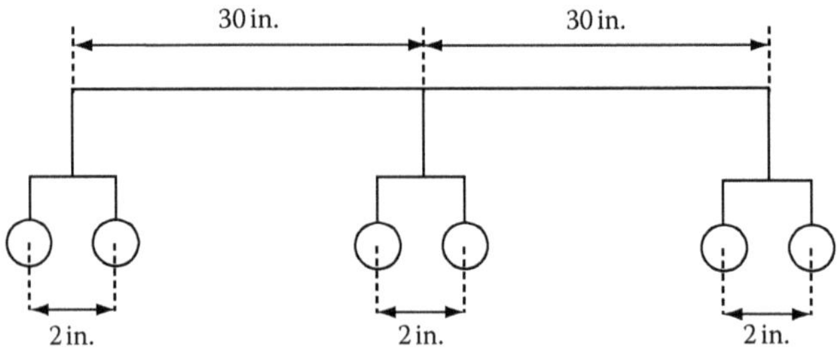

FIGURE B.7
Configuration of bundle conductors (Example B.2).

This is also equal to Z_4, as the bundle conductors are identical and symmetrically spaced. Matrix Z_2 of Equation B.27 is

$$\bar{Z}_2 = \begin{vmatrix} 0.0953 + j0.8786 & 0.0953 + j0.5348 & 0.0953 + j0.4581 \\ 0.953 + j0.5674 & 0.0953 + j0.8786 & 0.0953 + j0.5348 \\ 0.0953 + j0.4743 & 0.0953 + j0.08786 & 0.0953 + j0.8786 \end{vmatrix}$$

The primitive matrix is 6 × 6 given by Equation B.23 formed by partitioned matrices according to Equation B.24.Thus, from $\bar{Z}_1$ and $\bar{Z}_2$, the primitive matrix can be written. From these two matrices, we will calculate matrix equation (B.30)

$$\bar{Z}_1 - \bar{Z}_2 = \begin{vmatrix} 0.069 + j0.498 & j0.0150 & j0.0079 \\ -j0.0171 & 0.069 + j0.498 & j0.0150 \\ -j0.00847 & -j0.0170 & 0.069 + j0.498 \end{vmatrix}$$

and

$$\bar{Z}_k = (\bar{Z}_1 - \bar{Z}_2) - \left(\bar{Z}_2^t - \bar{Z}_1\right) = \begin{vmatrix} 0.138 + j0.997 & -j0.0022 & -j0.0005 \\ -j0.0022 & 0.138 + j0.997 & -j0.0022 \\ -j0.0005 & -j0.0022 & 0.138 + j0.997 \end{vmatrix}$$

The inverse is

$$\bar{Z}_k^{-1} = \begin{vmatrix} 0.136 - j0.984 & 0.000589 - j0.002092 & 0.0001357 - j0.0004797 \\ 0.0005891 - j0.002092 & 0.136 - j0.981 & 0.0005891 - j0.002092 \\ 0.0001357 - j0.0004797 & 0.0005891 - j0.002092 & 0.136 - j0.981 \end{vmatrix}$$

then, the matrix $(\bar{Z}_2 - \bar{Z}_1)\bar{Z}_k^{-1}\left(\bar{Z}_2^t - \bar{Z}_1\right)$ is

$$\begin{vmatrix} 0.034 + j0.2500 & -0.000018 - j0.000419 & 0.0000363 - j0.0003871 \\ -0.000018 - j0.000419 & 0.034 + j0.2500 & -0.000018 - j0.000419 \\ 0.0000363 - j0.000387 & -0.000018 - j0.000419 & 0.034 + j0.2500 \end{vmatrix}$$

Note that the off-diagonal elements are relatively small as compared to the diagonal elements. The required 3 × 3 transformed matrix is then Z_1 minus the above matrix:

$$\bar{Z}_{\text{transformed}} = \begin{vmatrix} 0.13 + j1.127 & 0.095 + j0.55 & 0.095 + j0.466 \\ 0.095 + j0.55 & 0.13 + j1.127 & 0.095 + j0.55 \\ 0.095 + j0.466 & 0.095 + j0.55 & 0.13 + j1.127 \end{vmatrix} \text{ ohm/mi}$$

Using Equation 1.35, the sequence impedance matrix is

$$\bar{Z}_{0.12} = \begin{vmatrix} 0.32 + j2.171 & 0.024 - j0.014 & -0.024 - j0.014 \\ -0.024 - j0.014 & 0.035 + j0.605 & -0.048 + j0.028 \\ 0.024 - j0.014 & 0.048 + j0.028 & 0.035 + j0.605 \end{vmatrix} \text{ ohm/mi}$$

B.7 Capacitance of Lines

The shunt capacitance per unit length of a two-wire, single-phase transmission line is

$$C = \frac{\pi\varepsilon_0}{\ln(D/r)} \text{ F/m (Farads per meter)} \tag{B.41}$$

where ε_0 is the permittivity of free space $= 8.854 \times 10^{-12}$ F/m, and other symbols are as defined before. For a three-phase line with equilaterally spaced conductors, the line-to-neutral capacitance is

$$C = \frac{2\pi\varepsilon_0}{\ln(D/r)} \text{ F/m} \tag{B.42}$$

For unequal spacings, D is replaced with GMD from Equation B.7. The capacitance is affected by the ground, and the effect is simulated by a mirror image of the conductors exactly at the same depth as the height above the ground. These mirror-image conductors carry charges that are of opposite polarity to conductors above the ground (Figure B.8). From this figure, the capacitance to ground is

$$C_n = \frac{2\pi\varepsilon_0}{\ln(\text{GMD}/r) - \ln\left(\sqrt[3]{S_{ab'}S_{bc'}S_{ca'}} / \sqrt[3]{S_{aa'}S_{bb'}S_{cc'}}\right)} \tag{B.43}$$

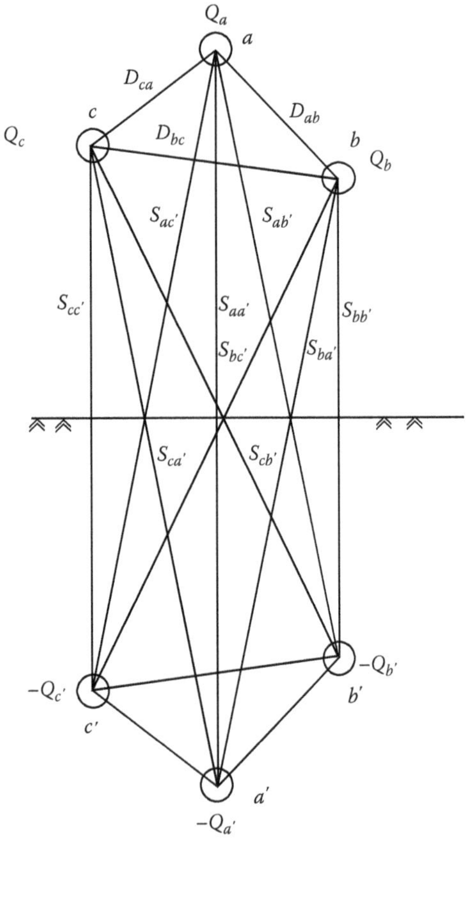

FIGURE B.8
Calculation of capacitances, conductors, mirror images, spacings, and charges.

Using the notations in Equation B.10, this can be written as

$$C_n = \frac{2\pi\varepsilon_0}{\ln(D_m/D_s)} = \frac{10^{-9}}{18\ln(D_m/D_s)}\ \text{F/m} \tag{B.44}$$

B.7.1 Capacitance Matrix

The capacitance matrix of a three-phase line is

$$\overline{C}_{abc} = \begin{vmatrix} C_{aa} & -C_{ab} & -C_{ac} \\ -C_{ba} & C_{bb} & -C_{bc} \\ -C_{ca} & -C_{cb} & C_{cc} \end{vmatrix} \tag{B.45}$$

This is diagrammatically shown in Figure B.9a. The capacitance between the phase conductors a and b is C_{ab}, and the capacitance between conductor a and ground is $C_{aa} - C_{ab} - C_{ac}$.

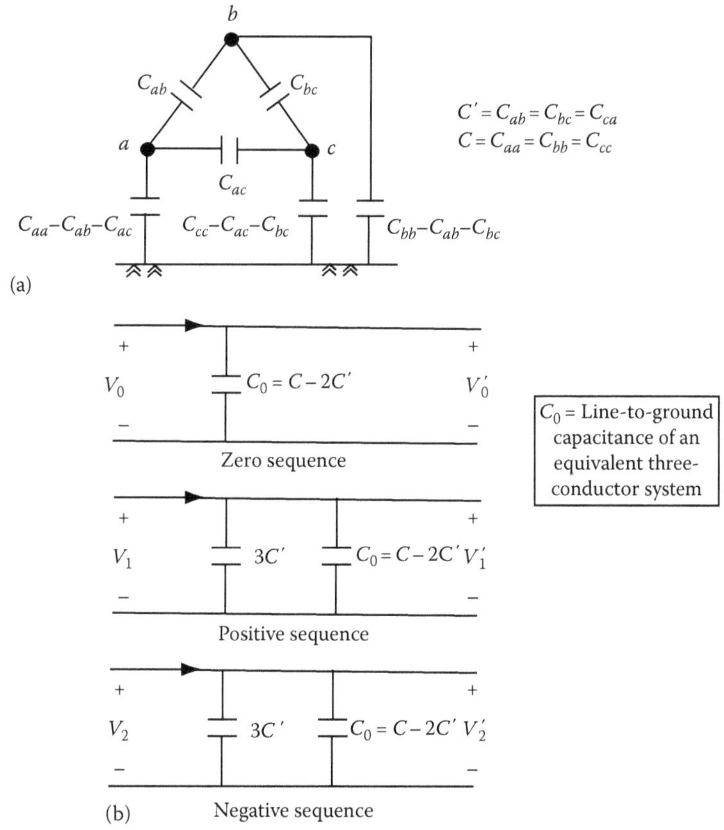

FIGURE B.9
(a) Capacitances of a three-phase line; (b) equivalent positive, negative, and zero sequence networks of capacitances.

If the line is perfectly symmetrical, all the diagonal elements are the same and all off-diagonal elements of the capacitance matrix are identical:

$$\overline{C}_{abc} = \begin{vmatrix} C & -C' & -C' \\ -C' & C & -C' \\ -C' & -C' & C \end{vmatrix} \quad \text{(B.46)}$$

Symmetrical component transformation is used to diagonalize the matrix

$$\overline{C}_{012} = \overline{T}_s^{-1}\overline{C}_{abc}\overline{T}_s = \begin{vmatrix} C-2C' & 0 & 0 \\ 0 & C+C' & 0 \\ 0 & 0 & C+C' \end{vmatrix} \quad \text{(B.47)}$$

The zero, positive, and negative sequence networks of capacitance of a symmetrical transmission line are shown in Figure B.9b. The eigenvalues are $C-2C'$, $C+C'$, and $C+C'$. The capacitance $C+C'$ can be written as $3C'+(C-2C')$, that is, it is equivalent to the line capacitance of a three-conductor system plus the line-to-ground capacitance of a three-conductor system.

In a capacitor, $V=Q/C$. The capacitance matrix can be written as

$$\overline{V}_{abc} = \overline{P}_{abc}\overline{Q}_{abc} = \overline{C}_{abc}^{-1}\overline{Q}_{abc} \quad \text{(B.48)}$$

where $\overline{P}$ is called the potential coefficient matrix, that is,

$$\begin{vmatrix} V_a \\ V_b \\ V_c \end{vmatrix} = \begin{vmatrix} P_{aa} & P_{ab} & P_{ac} \\ P_{ba} & P_{bb} & P_{bc} \\ P_{ca} & P_{cb} & P_{cc} \end{vmatrix} \begin{vmatrix} Q_a \\ Q_b \\ Q_c \end{vmatrix} \quad \text{(B.49)}$$

where

$$P_{ii} = \frac{1}{2\pi\varepsilon_0}\ln\frac{S_{ii}}{r_i} = 11.17689\ \ln\frac{S_{ii}}{r_i} \quad \text{(B.50)}$$

$$P_{ij} = \frac{1}{2\pi\varepsilon_0}\ln\frac{S_{ij}}{D_{ij}} = 11.17689\ \ln\frac{S_{ij}}{D_{ij}} \quad \text{(B.51)}$$

where

S_{ij} is the conductor-to-image distance below ground (ft)
D_{ij} is the conductor-to-conductor distance (ft)
r_i is the radius of the conductor (ft)
ε_0 is the permittivity of the medium surrounding the conductor $= 1.424 \times 10^{-8}$

For sine-wave voltage and charge, the equation can be expressed as

$$\begin{vmatrix} I_a \\ I_b \\ I_c \end{vmatrix} = j\omega \begin{vmatrix} C_{aa} & -C_{ab} & -C_{ac} \\ -C_{ba} & -C_{bb} & -C_{bc} \\ -C_{ca} & -C_{cb} & C_{cc} \end{vmatrix} \begin{vmatrix} V_a \\ V_b \\ V_c \end{vmatrix} \quad \text{(B.52)}$$

The capacitance of three-phase lines with ground wires and bundle conductors can be addressed as in the calculations of inductances. The primitive P matrix can be partitioned and reduces to a 3×3 matrix.

Example B.3

Calculate the matrices P and C for Example B.l. The neutral is 30 ft above ground and the configuration of Figure B.6 is applicable.

The mirror images of the conductors are drawn in Figure B.6. This facilitates calculation of the spacings required in Equations B.50 and B.51 for the P matrix. Based on the geometric distances and conductor diameter, the primitive P matrix is

$$\overline{P} = \begin{vmatrix} P_{aa} & P_{ab} & P_{ac} & P_{an} \\ P_{ba} & P_{bb} & P_{bc} & P_{bn} \\ P_{ca} & P_{cb} & P_{cc} & P_{cn} \\ P_{na} & P_{nb} & P_{nc} & P_{nn} \end{vmatrix}$$

$$= \begin{vmatrix} 80.0922 & 33.5387 & 21.4230 & 23.3288 \\ 33.5387 & 80.0922 & 25.7913 & 24.5581 \\ 21.4230 & 25.7913 & 80.0922 & 20.7547 \\ 23.3288 & 24.5581 & 20.7547 & 79.1615 \end{vmatrix}$$

This is reduced to a 3×3 matrix

$$P = \begin{vmatrix} 73.2172 & 26.3015 & 15.3066 \\ 26.3015 & 72.4736 & 19.3526 \\ 15.3066 & 19.3526 & 74.6507 \end{vmatrix}$$

Therefore, the required $\overline{C}$ matrix is inverse of $\overline{P}$, and $\overline{Y}_{abc}$ is

$$\overline{Y}_{abc} = j\omega\overline{P}^{-1} = \begin{vmatrix} j6.0141 & -j1.9911 & -j0.7170 \\ -j1.9911 & j6.2479 & -j1.2114 \\ -j0.7170 & -j1.2114 & j5.5111 \end{vmatrix} \mu\text{S/mi}$$

B.8 Cable Constants

The construction of cables varies widely; it is mainly a function of insulation type, method of laying, and voltage of application. For high-voltage applications above 230 kV, oil-filled paper-insulated cables are used, though recent trends see the development of solid dielectric cables up to 345 kV. A three-phase solid dielectric cable has three conductors enclosed within a sheath, and because the conductors are much closer to each other than those in an overhead line and the permittivity of insulating medium is much higher than that of air, the shunt capacitive reactance is much lower as compared to an overhead line. Thus, the use of a T or Π model is required even for shorter cable lengths.

The inductance per unit length of a single-conductor cable is given by

$$L = \frac{\mu_0}{2\pi} \ln\frac{r_1}{r_2} \text{ H/m} \tag{B.53}$$

where

r_1 is the radius of the conductor

r_2 is the radius of the sheath, that is, the cable outside diameter divided by 2

When single-conductor cables are installed in magnetic conduits, the reactance may increase by a factor of 1.5. Reactance is also dependent on conductor shape, that is, circular or sector, and on the magnetic binders in three-conductor cables.

B.8.1 Zero Sequence Impedance of the OH Lines and Cables

The zero sequence impedance of the lines and cables is dependent upon the current flow through a conductor and return through the ground, or sheaths and encounters the impedance of these paths. The zero sequence current flowing in one phase also encounters the currents arising out of that conductor self-inductance, from mutual inductance to other two-phase conductors, from the mutual inductance to the ground and sheath return paths, and the self-inductance of the return paths. Tables and analytical expressions are provided in [9]. As an example, the zero sequence impedance of a three-conductor cable with a solidly bonded and grounded sheath is given by

$$z_0 = r_c + r_e + j0.8382\frac{f}{60}\log_{10}\frac{D_e}{\text{GMR}_{3c}} \tag{B.54}$$

Where

r_c is the ac resistance of one conductor (ohm/mi)

r_e is the ac resistance of earth return (depending upon equivalent depth of earth return and soil resistivity, taken as 0.286 ohm/mi)

D_e is the distance to equivalent earth path (see [9])

GMR_{3c} = geometric mean radius of conducting path made up of three actual conductors taken as a group, in.:

$$\text{GMR}_{3c} = \sqrt[3]{\text{GMR}_{1c}S^2} \tag{B.55}$$

Where GMR_{1c} is the geometric mean radius of individual conductor and $S = (d + 2t)$, where d is the diameter of the conductor and t is the thickness of the insulation.

B.8.2 Concentric Neutral Underground Cable

We will consider a concentric neutral construction as shown in Figure B.10a. The neutral is concentric to the conductor and consists of a number of copper strands that are wound helically over the insulation. Such cables are used for underground distribution, directly buried or installed in ducts. Referring to Figure B.10a, d is the diameter of the conductor, d_0 is the outside diameter of the cable over the concentric neutral strands, and d_s is the

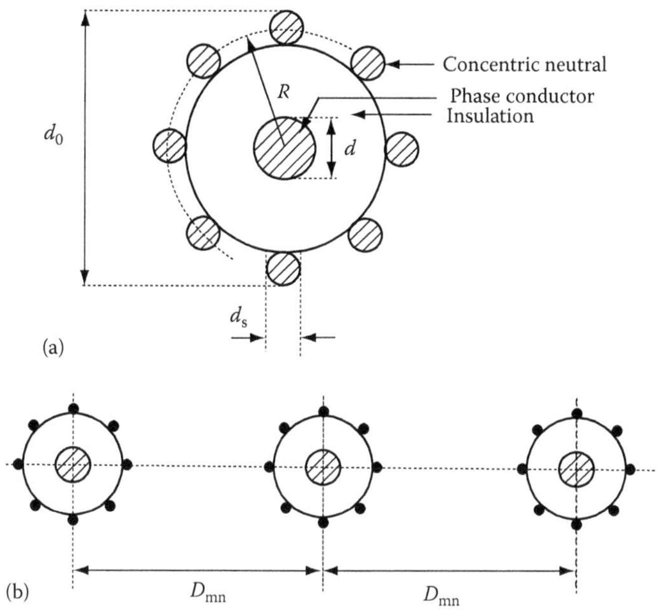

FIGURE B.10
(a) Construction of a concentric neutral cable; (b) configuration for calculation of cable series reactance (Example B.4).

diameter of an individual neutral strand. Three cables in flat formation are shown in Figure B.10b. The GMR of a phase conductor and a neutral strand is given by the expression

$$\mathrm{GMR_{cn}} = \sqrt[n]{\mathrm{GMR_s} n R^{n-1}} \tag{B.56}$$

where
- $\mathrm{GMR_{cn}}$ is the equivalent GMR of the concentric neutral
- $\mathrm{GMR_s}$ is the GMR of a single neutral strand
- n is the number of concentric neutral strands
- R is the radius of a circle passing through the concentric neutral strands (see Figure B.10a) $= (d_0 - d_s)/2$ (ft)

The resistance of the concentric neutral is equal to the resistance of a single strand divided by the number of strands.

The geometric mean distance between concentric neutral and adjacent phase conductors is

$$D_{ij} = \sqrt[n]{D_{\mathrm{mn}}^n - R^n} \tag{B.57}$$

where D_{ij} is the *equivalent* center-to-center distance of the cable spacing. Note that it is less than D_{mn}, the center-to-center spacing of the adjacent conductors, Figure B.10b. Carson's formula can be applied and the calculations are similar to those in Example B.1.

Example B.4

A concentric neutral cable system for 13.8 kV has a center-to-center spacing of 8 in. The cables are 500 kcmil, with 16 strands of #12 copper wires. The following data are supplied by the manufacturer:

GMR phase conductor = 0.00195 ft
GMR of neutral strand = 0.0030 ft
Resistance of phase conductor = 0.20 ohm/mi

Resistance of neutral strand = 10.76 ohm/mi. Therefore, the resistance of the concentric neutral = 10.76/16 = 0.6725 ohm/mi.

Diameter of neutral strand = 0.092 in.
Overall diameter of cable = 1.490 in.
Therefore, $R = (1.490 - 0.092)/24 = 0.0708$ ft.

The effective conductor phase-to-phase spacing is approximately 8 in., from Equation B.57.
The primitive matrix is a 6 × 6 matrix, similar to Equation B.16. In the partitioned form, Equation B.17, the matrices are

$$\bar{Z}_a = \begin{vmatrix} 0.2953 + j1.7199 & 0.0953 + j1.0119 & 0.0953 + j0.9278 \\ 0.0953 + j1.0119 & 0.2953 + j1.7199 & 0.0953 + j1.0119 \\ 0.0953 + j0.9278 & 0.0953 + j1.0119 & 0.2953 + j1.7199 \end{vmatrix}$$

The spacing between the concentric neutral and the phase conductors is approximately equal to the phase-to-phase spacing of the conductors. Therefore,

$$\bar{Z}_B = \begin{vmatrix} 0.0953 + j1.284 & 0.0953 + j1.0119 & 0.0953 + j0.9278 \\ 0.0953 + j1.0119 & 0.0953 + j1.284 & 0.0953 + j1.0119 \\ 0.0953 + j0.9278 & 0.0953 + j1.0119 & 0.0953 + j1.284 \end{vmatrix}$$

Matrix $\bar{Z}_c = \bar{Z}_B$ and matrix $\bar{Z}_D$ are given by

$$\bar{Z}_D = \begin{vmatrix} 0.7678 + j1.2870 & 0.0953 + j1.0119 & 0.0953 + j0.9278 \\ 0.0953 + j1.0119 & 0.7678 + j1.2870 & 0.0953 + j1.0119 \\ 0.0953 + j0.9278 & 0.0953 + j1.0119 & 0.7678 + j1.2870 \end{vmatrix} z$$

This primitive matrix can be reduced to a 3 × 3 matrix, as in other examples.

B.8.3 Capacitance of Cables

In a single-conductor cable, the capacitance per unit length is given by

$$C = \frac{2\pi\varepsilon\varepsilon_0}{\ln(r_1/r_2)} \text{ F/m} \tag{B.58}$$

Note that ε is the permittivity of the dielectric medium relative to air. The capacitances in a three-conductor cable are shown in Figure B.11. This assumes a symmetrical construction, and the capacitances between conductors and from conductors to the sheath are equal. The circuit of Figure B.11a is successively transformed and Figure B.11d shows that the net capacitance per phase = $C_1 + 3C_2$.

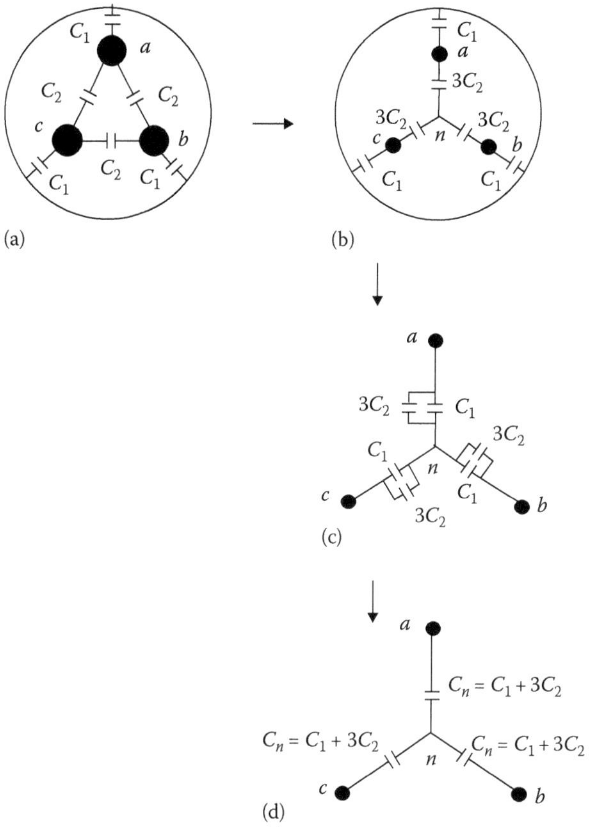

FIGURE B.11
(a) Capacitances in a three-conductor cable; (b) and (c) equivalent circuits; (d) final capacitance circuit.

TABLE B.1

Typical Values for Dielectric Constants of Cable Insulation

Type of Insulation	Permittivity (ε)
Polyvinyl chloride (PVC)	3.5–8.0
Ethylene-propylene insulation (EP)	2.8–3.5
Polyethylene insulation	2.3
Cross-linked polyethylene	2.3–6.0
Impregnated paper	3.3–3.7

By change of units, Equation B.58 can be expressed as

$$C = \frac{7.35\varepsilon}{\log(r_1/r_2)}\ \text{pF/ft} \tag{B.59}$$

This gives the capacitance of a single-conductor shielded cable. Table B.1 gives values of ε for various cable insulation types.

References

1. D.G. Fink (Ed.). *Standard Handbook for Electrical Engineers,* 10th edn. New York: McGraw-Hill, 1969.
2. T. Croft, C. Carr, and J.H. Watt. *American Electrician's Handbook,* 9th edn. New York: McGraw-Hill, 1970.
3. Central Station Engineers. *Electrical Transmission and Distribution Reference Book,* 4th edn. East Pittsburgh, PA: Westinghouse Corp., 1964.
4. The Aluminum Association. *Aluminum Conductor Handbook,* 2nd edn. Washington, DC, 1982.
5. J.C. Das. *Transients in Electrical Systems.* New York: McGraw-Hill, 2010.
6. W.D. Stevenson. *Elements of Power System Analysis,* 2nd edn. New York, McGraw Hill, 1962.
7. CIGRE, Working Group 33.02(Internal Voltages). *Guide Lines for Representation of Network Elements When Calculating Transients,* Paris.
8. P.M. Anderson. *Analysis of Faulted Systems.* Ames, IA: Iowa State University Press, 1973.
9. Westinghouse. *Transmission and Distribution Reference Book.* East Pittsburg, PA, 1964.

Appendix C: Transformers and Reactors

A power transformer is an important component of the power system. The transformation of voltages is carried out from generating voltage level to transmission, subtransmission, distribution, and consumer level. The installed capacity of the transformers in a power system may be seven or eight times the generating capacity. The special classes of transformers include furnace, converter, regulating, rectifier, phase shifting, traction, welding, and instrument (current and voltage) transformers. Large converter transformers are installed for HVDC transmission.

The transformer models and their characteristics are described in the relevant sections of this book in various chapters (Chapters 1, 2, 9, 14, etc.). This appendix provides basic concepts and discusses autotransformers, step-voltage regulators, and transformer models, not covered elsewhere in this book.

C.1 Model of a Two-Winding Transformer

We represented a transformer model by its series impedance in the load flow and short-circuit studies. We also developed models for tap changing, phase shifting, and reactive power flow control transformers. Concepts of leakage flux, total flux, and mutual and self-reactances in a circuit of two magnetically coupled coils are described in Chapter 6, and Equation 6.32 of a unit transformer is derived. These can be extended and a matrix model can be written as

$$\begin{vmatrix} v_1 \\ v_2 \\ \cdot \\ v_n \end{vmatrix} = \begin{vmatrix} r_{11} & r_{12} & \cdot & r_{1n} \\ r_{21} & r_{22} & \cdot & r_{2n} \\ \cdot & \cdot & \cdot & \cdot \\ r_{n1} & r_{n2} & \cdot & r_{nn} \end{vmatrix} \begin{vmatrix} i_1 \\ i_2 \\ \cdot \\ i_n \end{vmatrix} + \begin{vmatrix} L_{11} & L_{12} & \cdot & L_{1n} \\ L_{21} & L_{22} & \cdot & L_{2n} \\ \cdot & \cdot & \cdot & \cdot \\ L_{n1} & L_{n2} & \cdot & L_{nn} \end{vmatrix} \frac{\mathrm{d}}{\mathrm{d}t} \begin{vmatrix} i_1 \\ i_2 \\ \cdot \\ i_n \end{vmatrix} \tag{C.1}$$

A two-winding transformer model can be derived from the circuit diagram shown in Figure C.1a and the corresponding phasor diagram (vector diagram) shown in Figure C.2. The transformer supplies a load current I_2 at a terminal voltage V_2 and lagging power factor angle ϕ_2. Exciting the primary winding with voltage V_1 produces changing flux linkages. Though the coils in a transformer are tightly coupled by interleaving the windings and are wound on a magnetic material of high permeability, all the flux produced by primary windings does not link the secondary. The winding leakage flux gives rise to leakage reactances. In Figure C.2, Φ_m is the main or mutual flux, assumed to be constant. The EMF induced in the primary windings is E_1, which lags Φ_m by 90°. In the secondary winding, the ideal transformer produces an EMF E_2 due to mutual flux linkages. There has

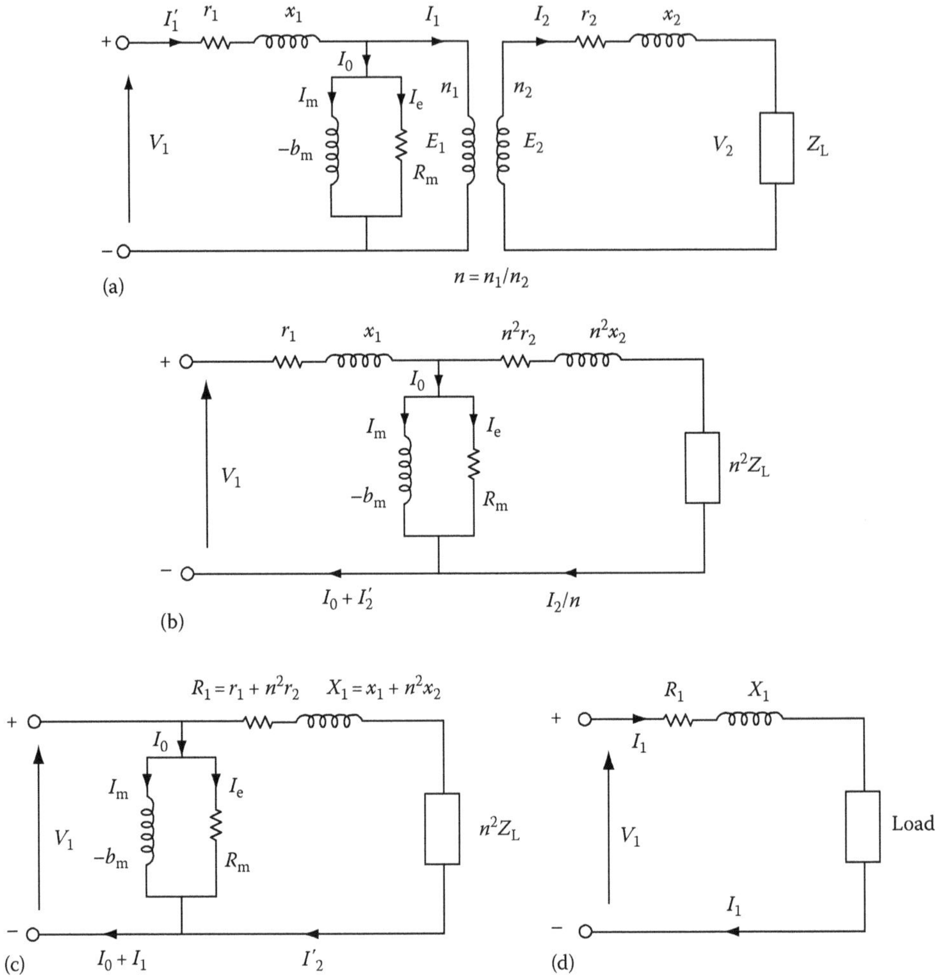

FIGURE C.1
(a) Equivalent circuit of a two-winding transformer; (b)–(d) simplifications to the equivalent circuit.

to be a primary magnetizing current even at no load, in a time phase with its associated flux, to excite the core. The pulsation of flux in the core produces losses. Considering that the no-load current is sinusoidal (which is not true under magnetic saturation; see Chapter 17), it must have a core loss component due to hysteresis and eddy currents:

$$I_0 = \sqrt{I_m^2 + I_e^2} \tag{C.2}$$

where
I_m is the magnetizing current
I_e is the core loss component of the current
I_0 is the no-load current

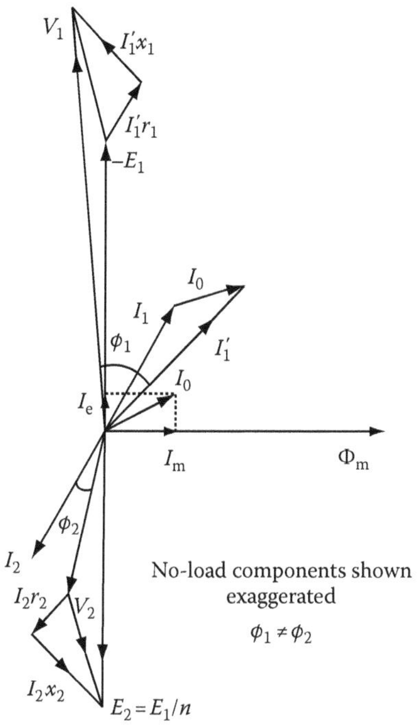

FIGURE C.2
Vector diagram of a two-winding transformer on load.

I_m and I_e are in phase quadrature. The generated EMF because of flux Φ_m is given by

$$E_2 = 4.44fn_2\Phi_m \tag{C.3}$$

where
E_2 is in volts when Φ_m is in Wb/m^2
n_2 is the number of secondary turns
f is the frequency

As primary ampère turns must be equal to the secondary ampère turns, that is, $E_1I_1 = E_2I_2$, we can write

$$\frac{E_1}{E_2} = \frac{n_1}{n_2} = n$$

and (C.4)

$$\frac{I_1}{I_2} \approx \frac{n_2}{n_1} = \frac{1}{n}$$

The current relation holds because the no-load current is small. The terminal relations can now be derived. On the primary side, the current is compounded to consider the no-load component of the current, and the primary voltage is equal to $-E_1$ (to neutralize the EMF of induction) and I_1r_1 and I_1x_1 drop in the primary windings. On the secondary side, the terminal voltage is given by the induced EMF E_2 less I_2r_2 and I_2x_2 drops in the secondary windings. The equivalent circuit is therefore as shown in Figure C.1a. The transformer is an ideal lossless transformer of turns ratio n.

In Figure C.1a, we can refer the secondary resistance and reactance to the primary side or vice versa. The secondary windings of n turns can be replaced with an equivalent winding referred to the primary, where the copper loss in the windings and the voltage drop in reactance are the same as in the actual winding. We can denote the resistance and reactance of the equivalent windings as r_2' and x_2', respectively:

$$\begin{aligned} I_1^2 r_2' &= I_2^2 r_2 \quad r_2' = r_2\left(\frac{I_2^2}{I_1^2}\right) \approx r_2\left(\frac{n_1}{n_2}\right)^2 = n^2 r_2 \\ x_2' &= x_2\left(\frac{I_2 E_1}{I_1 E_2}\right) \approx x_2\left(\frac{n_1}{n_2}\right)^2 = n^2 x_2 \end{aligned} \tag{C.5}$$

The transformer is a single phase ideal transformer with no losses and having a turns ratio of unity and no secondary resistance or reactance. By also transferring the load impedance to the primary side, the unity ratio ideal transformer can be eliminated and the magnetizing circuit is pulled out to the primary terminals without appreciable error (Figure C.1b and c). In Figure C.1d, the magnetizing and core loss circuit is altogether omitted. The equivalent resistance and reactances are

$$\begin{aligned} R_1 &= r_1 + n^2 r_2 \\ X_1 &= x_1 + n^2 x_2 \end{aligned} \tag{C.6}$$

Thus, on a simplified basis, the transformer positive or negative sequence model is given by its percentage reactance specified by the manufacturer, on the transformer natural cooled MVA rating base. This reactance remains fairly constant and is obtained by a short-circuit test on the transformer. The magnetizing circuit components are obtained by an open circuit test.

The expression for hysteresis loss is

$$P_h = K_h f B_m^s \tag{C.7}$$

where

K_h is a constant

s is the Steinmetz exponent, which varies from 1.5 to 2.5, depending on the core material; generally, it is $= 1.6$

The eddy current loss is

$$P_e = K_e f^2 B_m^2 \tag{C.8}$$

where K_e is a constant. Eddy current loss occurs in core laminations, conductors, tanks, and clamping plates. The core loss is the sum of the eddy current and hysteresis loss. In Figure C.2, the primary power factor angle ϕ_1 is $> \phi_2$.

C.1.1 Open Circuit Test

Figure C.3 shows no-load curves when an open circuit test at rated frequency and varying voltage is made on the transformer. The test is conducted with the secondary winding open circuited and rated voltage applied to the primary winding. For high-voltage transformers, the secondary winding may be excited and the primary winding opened. At constant frequency, B_m is directly proportional to applied voltage and the core loss is approximately proportional to B_m^2. The magnetizing current rises steeply at low flux densities, then more slowly as iron reaches its maximum permeability, and thereafter again steeply, as saturation sets in.

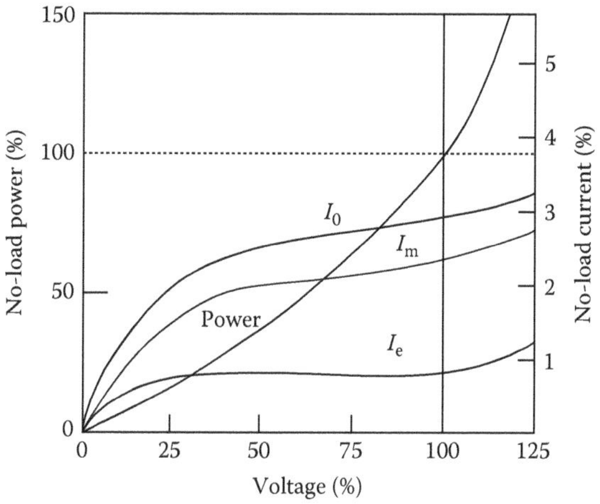

FIGURE C.3
No-load test on a transformer.

From Figure C.1, the open circuit admittance is

$$Y_{OC} = g_m - jb_m \tag{C.9}$$

This neglects the small voltage drop across r_1 and x_1. Then,

$$g_m = \frac{P_0}{V_1^2} \tag{C.10}$$

where
P_0 is the measured power
V_1 is the applied voltage

Also,

$$b_m = \frac{Q_0}{V_1^2} = \sqrt{\frac{S_0^2 - P_0^2}{V_1^2}} \tag{C.11}$$

where P_0, Q_0, and S_0 are measured active power, reactive power, and volt-ampères on open circuit, respectively. Note that the exciting voltage E_1 is not equal to V_1, due to the drop that no-load current produces through r_1 and x_1. Corrections can be made for this drop.

C.1.2 Short-Circuit Test

The short-circuit test is conducted at the rated current of the winding, which is shorted, and a reduced voltage is applied to the other winding to circulate a full-rated current:

$$P_{sc} = I_{sc}^2 R_1 = I_{sc}^2 (r_1 + n^2 r_2) \tag{C.12}$$

where
P_{sc} is the measured active power on short circuit
I_{sc} is the short-circuit current

$$Q_{sc} = I_{sc}^2 X_1 = I_{sc}^2 (x_1 + n^2 x_2) \tag{C.13}$$

Example C.1

The copper loss of a 2500 kA, 13.8–4.16 kV delta–wye connected three-phase transformer is 4 kW on the delta side and 3 kW on the wye side. Find R_1, r_1, r_2, and r_2' for phase values throughout. If the total reactance is 5.5%, find X_1, x_1 x_2, and x_2', assuming that the reactance is divided in the same proportion as resistance.

The copper loss per phase on the 13.8 kV side $=4/3=1.33$ kW and the current per phase $=$ 60.4 A. Therefore, $r_1=0.364$ ohms. Similarly for the 4.16 kV side, the copper loss $=3/3=1$ kW, the current $=346.97$ A, and $r_2=0.0083$ ohm per phase; r_2 referred to the primary side $r_2'=(0.0083)$ $(13.8\times 4.16)^2=0.273$ ohm, and $R_1=0.364+0.273=0.637$ ohm. A 5.5% reactance on a transformer MVA base of $2.5=4.19$ ohms referred 13.8 kV side, and then $x_1=(4.19)(0.364)/0.637=2.39$ ohm and $x_2'=1.79$ ohm. Referred to the 4.16 kV side $x_2=0.162$ ohm. The transformer X/R ratio $=6.56$.

Example C.2

The transformer of Example C.1 gave the following results on open circuit test: open circuit on the 4.16 kV side, rated primary voltage and frequency, input $=10$ kW, and no-load current $=2.5$ A. Find the magnetizing circuit parameters.

The active component of the current $I_e=3.33/13.8=0.241$ A per phase. Therefore,

$$g_m=\frac{10\times 10^3}{3\times(13.8\times 10^3)^2}=0.017\times 10^{-3}\text{ mhos}$$

The magnetizing current is

$$I_m=\sqrt{I_0^2-I_e^2}=\sqrt{1.44^2-0.241^2}=1.42\text{ A}$$

The power factor angle of the no-load current is 9.63°, and b_m from Equation C.11 is -0.103×10^{-3} mhos per phase.

C.2 Transformer Polarity and Terminal Connections

C.2.1 Additive and Subtractive Polarity

The relative direction of induced voltages, as appearing on the terminals of the windings, is dependent on the order in which these terminals are taken out of the transformer tank. As the primary and secondary voltages are produced by the same mutual flux, these must be in the same direction in each turn. The load current in the secondary voltage flows in a direction so as to neutralize the mmf of the primary voltage. How the induced voltages will appear as viewed from the terminals depends on the relative direction of the windings. The polarity refers to the definite order in which the terminals are taken out of the tank. Polarity may be defined as the voltage vector relations of transformer leads as brought out of the tank. Referring to Figure C.4a, the polarity is the relative direction of the induced voltage from H_1 to H_2 as compared with that from X_1 to X_2, both being in the same order. The order is important in the definition of polarity.

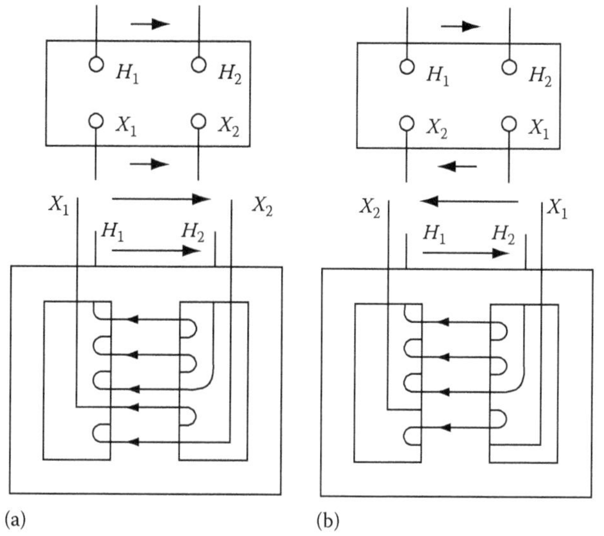

FIGURE C.4
(a) Polarity and polarity markings, subtractive polarity; (b) additive polarity.

When the induced voltages are in the opposite direction, as in Figure C.4b, the polarity is said to be additive, and when in the same direction, as in Figure C.4a, it is said to be subtractive. According to the ANSI standard, all liquid immersed power and distribution transformers have subtractive polarity. Dry-type transformers also have subtractive polarity.

When the terminals of any winding are brought outside the tank and marked so that H_1 and X_1 are adjacent, the polarity is subtractive. The polarity is additive when H_1 is diagonally located with respect to X_1 (Figure C.4b). The lead H_1 is brought out as the right-hand terminal of the high-voltage group as seen when facing the highest voltage side of the case. The polarity is often marked by dots on the windings. If H_1 is dotted, then X_1 is dotted for subtractive polarity. The currents are in phase. Angular displacement and terminal markings for three-phase transformers and autotransformers are discussed in Ref. [1].

C.3 Parallel Operation of Transformers

Transformers may be operated in parallel to supply increased loads, and for reliability, redundancy, and continuity of the secondary loads. Ideally, the following conditions must be satisfied:

- The phase sequence must be the same.
- The polarity must be the same.
- Voltage ratios must be the same.
- The vector group, that is, the angle of phase displacement between primary and secondary voltage vectors, should be the same.
- Impedance voltage drops at full load should be the same, that is, the percentage impedances based on the rated MVA rating must be the same.

It is further desirable that the two transformers have the same ratio of percentage resistance to reactance voltage drops, that is, the same X/R ratios.

With the conditions mentioned earlier met, the load sharing will be proportional to the transformer MVA ratings. It is basically a parallel circuit with two transformer impedances in parallel and a common terminal voltage:

$$I_1 = \frac{IZ_2}{Z_1 + Z_2}$$
$$I_2 = \frac{IZ_1}{Z_1 + Z_2} \quad \text{(C.14)}$$

where
I_1 and I_2 are the current loadings of each transformer
I is the total current

In terms of the total MVA load, S, the equations are

$$S_1 = \frac{SZ_2}{Z_1 + Z_2}$$
$$S_2 = \frac{SZ_1}{Z_1 + Z_2} \quad \text{(C.15)}$$

While the polarity and vector group are essential conditions, two transformers may be paralleled when they have

- Unequal ratios and equal percentage impedances
- Equal ratios and unequal percentage impedances
- Unequal ratios and unequal percentage impedances

It is not a good practice to operate transformers in parallel when

- Either of the two parallel transformers is overloaded by a significant amount above its rating
- The no-load circulating current exceeds 10% of the full-rated load
- The arithmetical sum of the circulating current and load current is >110%

The circulating current means the current circulating in the high- and low-voltage windings, excluding the exciting current.

Example C.3

A 10 MVA, 13.8–4.16 kV transformer has a per-unit resistance and reactance of 0.005 and 0.05, respectively. This is paralleled with a 5 MVA transformer of the same voltage ratio, and having per unit resistance and reactance of 0.006 and 0.04, respectively. Calculate how these will share a load of 15 MVA at 0.8 power factor lagging.

Convert Z_1 and Z_2 on any common MVA base and apply Equations C.14 and C.15. The results are as follows:

10 MVA transformer: $S_1 = 9.255 < -37.93°$

5 MVA transformer: $S_2 = 5.749 < -35.20°$

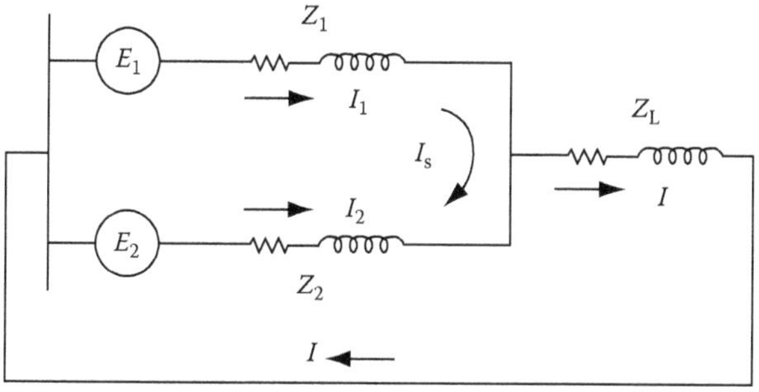

FIGURE C.5
Equivalent circuit of two parallel running transformers with different voltage ratios and percentage impedance.

The loads do not sum to 15 MVA because of different power factors. The MW and Mvar components of the transformer loads should sum to the total load components:

10 MVA transformer: 7.30 MW and 5.689 Mvar

5 MVA transformer: 4.698 MW and 3.314 Mvar

The total load is equal to an approximate load MW of 12 MW and 9.0 Mvar. The transformers do not share the load proportional to their ratings.

If the terminal voltages differ, there will be circulating current at no load. With reference to Figure C.5, the load sharing is given by the following equations:

$$I_1 = \frac{E_1 - V}{Z_1} \qquad I_2 = \frac{E_2 - V}{Z_2} \tag{C.16}$$

where V is the load voltage. This is given by

$$V = (I_1 + I_2)Z_L = \left(\frac{E_1 - V}{Z_1} + \frac{E_2 - V}{Z_2}\right)Z_L \tag{C.17}$$

This can be written as

$$V\left(\frac{1}{Z_1} + \frac{1}{Z_2} + \frac{1}{Z_L}\right) = \left(\frac{E_1}{Z_1} + \frac{E_2}{Z_2}\right) \tag{C.18}$$

For a given load, the calculation is iterative in nature, as shown in Example C.4.

Example C.4

In Example C.3, the transformers have the same percentage impedances and the same X/R ratios; the secondary voltage of the 10 MVA transformer is 4 kV and that of the 5 MVA transformer is 4.16 kV. Calculate the circulating current at no load.

We will work on per-phase basis. The 10 MVA transformer impedance referred to 4 kV secondary is $0.008 + j0.08$ ohm, and the 5 MVA transformer impedance at 4.16 kV secondary is $0.0208 + j0.1384$ ohm; $Z_1 + Z_2 = 0.0288 + j0.2184$. *Assume that the load voltage is 4 kV*; then, on a per-phase basis, the load is 5 MVA at 0.8 power factor and the load impedance is $0.853 + j0.64$ ohm.

$$\frac{E_1}{Z_1} + \frac{E_2}{Z_2} = \frac{4000}{\sqrt{3}(0.008 + j0.08)} + \frac{4160}{\sqrt{3}(0.0208 + j0.1384)} = 5409 - j45.550\,\text{kA}$$

Also,

$$\left(\frac{1}{Z_1}+\frac{1}{Z_2}+\frac{1}{Z_L}\right)=3.048-j20.003$$

From Equation C.18, the load voltage is 2260 – j74.84 volts phase to neutral. From Equation C.16, the 10 MVA transformer load current is 980.04 – j445.347 and that of the 5 MVA transformer is 672.17 – j874.42. The total load current is 1652.2 – j1319.75 and the single-phase load MVA is 3.645 MW and 3.112 Mvar. This is much different from the desired loading of 4 MW and 3 Mvar. This is due to assumption of the load voltage. The calculation can be repeated with a lower estimate of load voltage and recalculation of load impedance.

C.4 Autotransformers

The circuit of an autotransformer is shown in Figure C.6a. It has windings common to primary and secondary, that is, the input and output circuits are electrically connected. The primary voltage and current are V_1 and I_1, respectively, and the secondary voltage

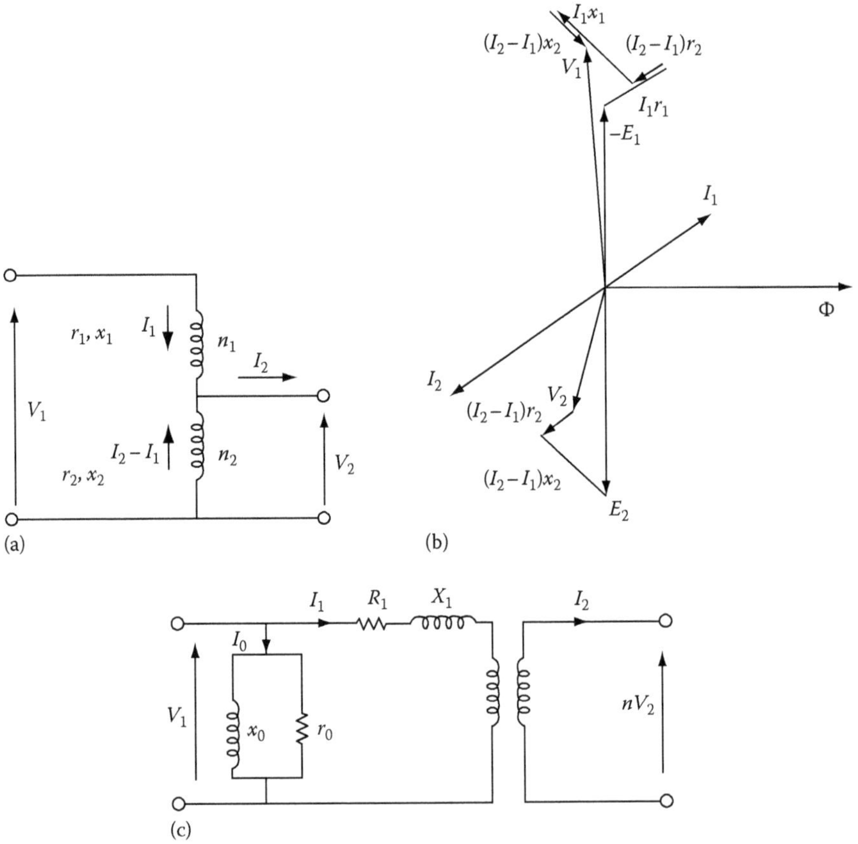

FIGURE C.6
(a) Circuit of an autotransformer, step-down configuration shown; (b) vector diagram of an autotransformer on load; (c) equivalent circuit of an autotransformer.

and current are V_2 and I_2, respectively. If the numbers of turns are n_1 and n_2, as shown, then neglecting losses

$$\frac{V_1}{V_2} = \frac{I_2}{I_1} = \frac{n_1 + n_2}{n_2} = n \tag{C.19}$$

The ampère turns $I_1 n_1$ oppose ampère turns $I_2 n_2$ and the common part of the winding carries a current of $I_2 - I_1$. Consequently, a smaller cross-section of the conductor is required. Conductor material in the autotransformer as a percentage of conductor material in a two-winding transformer for the same kVA output is

$$\frac{M_{\text{auto}}}{M_{\text{two-winding}}} = \frac{I_1(n_1) + (I_2 - I_1)n_2}{I_1(n_1 + n_2) + I_2 n_2}$$

$$= 1 - \frac{2}{(n) + (I_2/I_2)} = 1 - \frac{V_2}{V_1} \tag{C.20}$$

The savings in material cost are most effective for transformation voltages close to each other. For a voltage ratio of 2, approximately 50% savings in copper could be made.

The vector diagram is shown in Figure C.6b and the equivalent circuit in Figure C.6c. Neglecting the magnetizing current,

$$V_1 = -E_1 - I_1(r_1 \cos\phi + x_1 \sin\phi) + (I_2 - I_1)(r_2 \cos\phi + x_2 \sin\phi) \tag{C.21}$$

where ϕ is the load power factor. Note that the impedance drop in the common winding is added, because the net current is opposed to the direction of I_1. The equation for the secondary voltage is

$$V_2 = E_2 - (I_2 - I_1)(r_2 \cos\phi + x_2 \sin\phi) \tag{C.22}$$

Combining these two equations, we write

$$V_1 = nV_2 + I_1\left[\left(r_1 + (n-1)^2 r_2\right)\cos\phi + \left(x_1 + (n-1)^2 x_2\right)\sin\phi\right] \tag{C.23}$$

This means that the equivalent resistance and reactance corresponding to a two-winding transformer are

$$\begin{aligned} R_1 &= r_1 + (n-1)^2 r_2 \\ X_1 &= x_1 + (n-1)^2 x_2 \end{aligned} \tag{C.24}$$

An autotransformer can be tested for impedances exactly as a two-winding transformer. The resistance and reactance referred to the secondary side is

$$R_2 = \frac{R_1}{n^2} = r_2 + \frac{r_1}{n^2}$$
$$X_2 = \frac{X_1}{n^2} = x_2 + \frac{x_1}{n^2} \tag{C.25}$$

The kVA rating of the circuit with respect to the kVA rating of the windings is $(n-1)/n$. A 1-MVA, 33–22 kV autotransformer has an equivalent two-winding kVA of $(1.5-1)/1.5 = 0.333 \times 1000 = 333$ kVA. This results in considerable saving in winding capacity. The series impedance is less than that of a two-winding transformer. This is beneficial from the load-flow point of view, as the losses and voltage drop will be reduced; however, a larger contribution to short-circuit current results.

A three-phase autotransformer connection is shown in Figure C.7a. Such banks are usually Y-connected with a grounded neutral, and a tertiary winding is added for third-harmonic circulation and neutral stabilization (see Figure 17.6). This circuit is akin to that of a three-winding transformer, and the positive and zero sequence circuits are as shown. The T-circuit positive sequence parameters are calculated by shorting one set of terminals and applying positive sequence voltage to the other terminals and keeping the third set of terminals open circuited.

$$\begin{vmatrix} Z_H \\ Z_X \\ Z_Y \end{vmatrix} = \frac{1}{2}\begin{vmatrix} 1 & 1 & -1 \\ 1 & -1 & 1 \\ 1 & 1 & -1 \end{vmatrix}\begin{vmatrix} Z_{HX} \\ Z_{HY} \\ Z_{XY} \end{vmatrix} \text{pu} \tag{C.26}$$

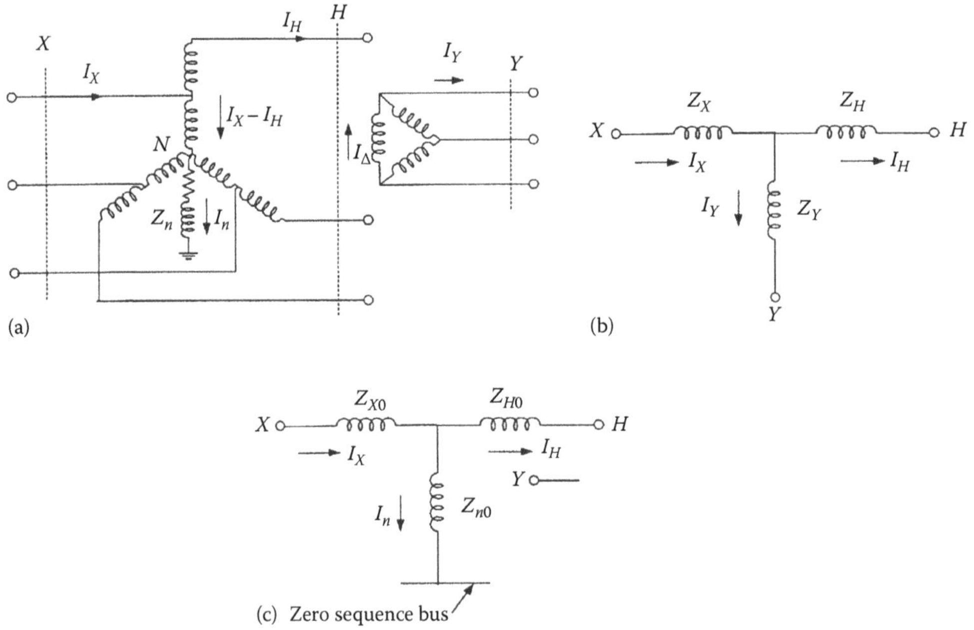

FIGURE C.7
(a) Circuit of a three-phase auto-transformer with tertiary delta; (b), (c) Positive and zero sequence circuits.

This is identical to Equation 1.54. All impedances are in pu.

The zero sequence impedance is given by

$$\begin{vmatrix} Z_{X0} \\ Z_{H0} \\ Z_{n0} \end{vmatrix} = \frac{1}{2} \begin{vmatrix} 1 & -1 & 1 & (n-1)/n \\ 1 & 1 & -1 & -(n-1)/n^2 \\ -1 & 1 & 1 & 1/n \end{vmatrix} \begin{vmatrix} Z_{HX} \\ Z_{HY} \\ Z_{XY} \\ 6Z_n \end{vmatrix} \quad \text{(C.27)}$$

where n is defined in Equation C.19. If the neutral of the autotransformer is ungrounded, $Z_n = \infty$ and all impedances in Equation C.27 become infinity. One way to solve this problem is to convert the Y equivalent to a delta equivalent and then take the limit as $Z_n = \infty$. The delta equivalent impedances thus calculated are called "resonant delta."

C.4.1 Scott Connection

Two autotransformers with suitable taps can be used in a Scott connection, for three-phase to two-phase conversion and flow of power in either direction. The arrangement is shown in Figure C.8. The line voltage V appears between terminals C and B and also between terminals A and B and A and C. The voltage between A and S is $(V\sqrt{3})/2$; the second autotransformer, called the teaser transformer, has $(\sqrt{3}/2)$ turns. The two secondaries having equal turns produce voltages equal in magnitude and phase quadrature. The neutral of the three-phase system can be located on the second or teaser transformer. The neutral must have a voltage of $V/\sqrt{3}$ to terminal A, that is, the neutral point can be trapped at $V[(\sqrt{3}/2) - 1/\sqrt{3}] = 0.288n_1$ turns from S. It can be shown that the three-phase side is balanced for a two-phase balanced load, that is, if the load is balanced on one side, it will be balanced on the other.

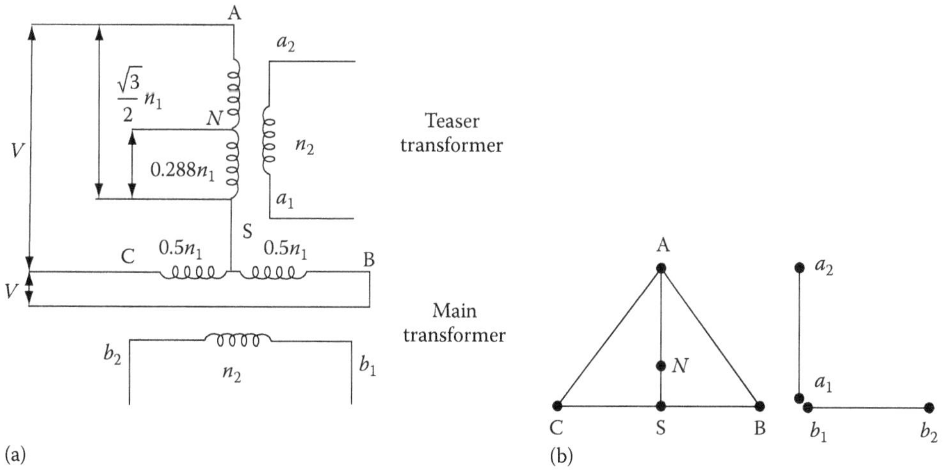

FIGURE C.8
(a) Circuit of a Scott-connected transformer; (b) three-phase primary and two-phase secondary voltages.

C.5 Step-Voltage Regulators

Step-voltage regulators [2] are essentially autotransformers. The most common voltage regulators manufactured today are of single-phase type with reactive switching, resulting in ±10% voltage regulation in 32 steps, 16 "boosting" and 16 "bucking." The rated voltages are generally up to 19,920 kV (line to neutral), 150 kV BIL (basic insulation level), and the current rating ranges from 5 to 2,000 A (not at all voltage levels). The general application is in distribution systems and three single-phase voltage regulators can be applied in wye or delta connection to a three-phase three-wire or three-phase four-wire system. The winding common to the primary and secondary is designated as a shunt winding, and the winding not common to the primary and secondary is designated as a series winding. The series winding voltage is 10% of the regulator-applied voltage. The polarity of this winding is changed with a reversing switch to accomplish buck or boost of the voltage. When the voltage regulation is provided on the load side, it is called a type-A connection (Figure C.9a). The core excitation varies as the shunt winding is connected across the source voltage. In a type-B connection (Figure C.9b), the regulation is provided on the load side and the source voltage is applied by way of series taps. Figure C.9d shows the schematic of a tap-changing circuit with current-limiting reactors and equalizer windings.

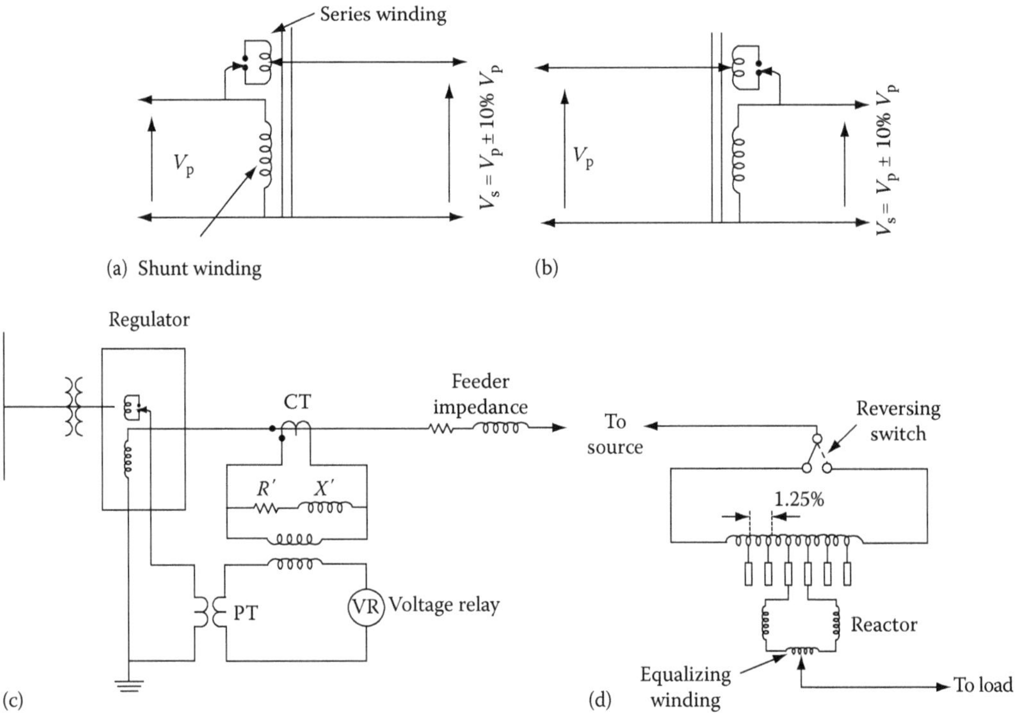

FIGURE C.9
Circuits of (a) type-A and (b) type-B connection step-voltage regulators; (c) schematic of line-drop compensator; (d) reactance-type tap changer with reversing switch.

C.5.1 Line Drop Compensator

The step regulators are controlled through a line drop compensator; its schematic circuit is shown in Figure C.9c. The voltage drop in the line from the regulator to the load is simulated in an R'–X' network in the compensator. The settings on these elements are decided on the basis of load flow prior to insertion of the regulator, that is, the voltage and current at the point of application give the system impedance to be simulated by R' and X' in the line drop compensator.

C.6 Extended Models of Transformers

A transient transformer model should address saturation, hysteresis, eddy current, and stray losses. Saturation plays an important role in determining the transient behavior of the transformer. Extended transformer models can be very involved and these are not required in every type of study. At the same time, a simple model may be prone to errors. As an example, in distribution system load flow, representing a transformer by series impedance alone and neglecting the shunt elements altogether may not be proper, as losses in the transformers may be considerable. For studies on switching transients, it is necessary to include capacitance of the transformers as high-frequency surges will be transferred more through electrostatic couplings rather than through electromagnetic couplings. For short-circuit calculations, capacitance and core loss effects can be neglected. Thus, the type of selected model depends on the nature of the study. There are many approaches to the models, some of which are briefly discussed.

The equivalent circuit of the shunt branch of a transformer for nonlinearity can be drawn as shown in Figure C.10. The excitation current has half-wave symmetry and contains only odd harmonics (see Appendix E). We may consider the excitation current as composed of two components: a fundamental frequency component and a distortion component. The fundamental frequency component is broken into two components, i_e and i_m, as discussed before, which give rise to shunt components g_m and $-b_m$ (Figure C.1a). The distortion component may be considered as a number of equivalent harmonic current sources in parallel with the fundamental frequency components, each of which can be represented in the phasor form, $I_{ei} < \theta_i$. To consider the effect of variation in the supply system voltages,

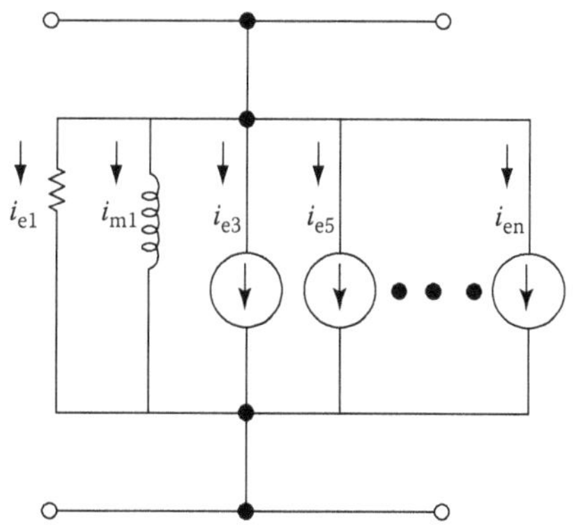

FIGURE C.10
Transformer shunt branch model considering nonlinearity.

the model parameters at three voltage levels of maximum, minimum, and rated voltage can be approximated by quadratic functions of the supply system voltage:

$$W = a + bV + cV^2$$
$$W = R_m, X_m, I_{e3}, I_{e5}, \ldots, \theta_3, \theta_5, \ldots \tag{C.28}$$

where $I_{e3}, I_{e5}, \ldots, \theta_3, \theta_5, \ldots$ are the harmonic currents and their angles. The coefficients a, b, and c can be found from

$$\begin{vmatrix} a \\ b \\ c \end{vmatrix} = \begin{vmatrix} 1 & V_{min} & V_{min}^2 \\ 1 & V_{rated} & V_{rated}^2 \\ 1 & V_{max} & V_{max}^2 \end{vmatrix} \begin{vmatrix} W_0 \\ W_1 \\ W_2 \end{vmatrix} \tag{C.29}$$

where W_0, W_1, W_2 are measured values of W for V_{min}, V_{rated}, and V_{max}, respectively.

C.6.1 Modeling the Hysteresis Loop

A model of the hysteresis loop can be constructed, based on measurements. The locus of the midpoints of the loop is obtained by measurements at four points and its displacement by a *consuming function,* whose maximum value is *ob*, and *ef* changes periodically by half-wave symmetry (Figure C.11). The consuming function can be written as $f(x) = -ob \sin \omega t$. The periphery can be then represented by 16 line segments [3]:

$$i = (i_k - m_k\phi_k) + m_k\phi - ob \sin(\omega t)$$
$$\phi_{k-1} < |\phi| \leq \phi_k \tag{C.30}$$
$$k = 1, 2, \ldots, 16$$

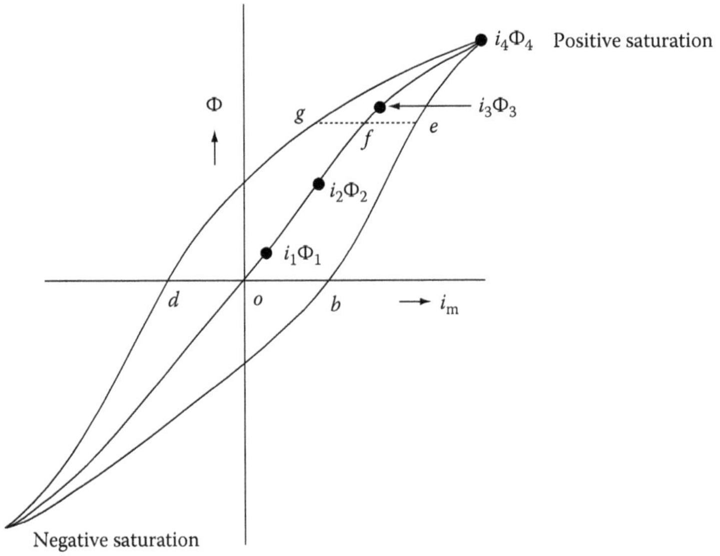

FIGURE C.11
Piecewise hysteresis loop curve fitting.

C.6.2 EMTP Models

Figure C.12a shows a single-phase model; R remains constant, and it is calculated from excitation losses. The nonlinear inductor is modeled from transformer excitation data and from its nonlinear V–I characteristics. In modern transformers, the cores saturate sharply and there is a well-defined knee. Often a two-slope piecewise linear inductor is adequate to model such curves. The saturation curve is not supplied as a flux–current curve, but as an rms voltage–rms current curve. The *satura* routine in EMTP [4] converts voltage–current input into flux–current data. Figure C.12b and c shows an example of this conversion for a 750 MVA, 420–27 kV five-leg, core type, wye–delta-connected transformer. The nonlinear inductance should be connected between the windings closest to the iron core. The input data are presented in per-unit values with regard to the winding connections and the base current and voltage. For the delta winding, the rms excitation current is $1/\sqrt{3}$ times the excitation current. Also rms excitation current in delta winding is approximated given by

$$I_{\text{m-w}} = \left(I_{\text{ex-w}}^2 - \left(\frac{P_{\text{ex}}}{3V_{\text{ex}}}\right)^2\right)^{1/2} \tag{C.31}$$

where

$I_{\text{m-w}}$ is winding magnetizing current
P_{ex} is the measured excitation power
V_{ex} is the excitation voltage

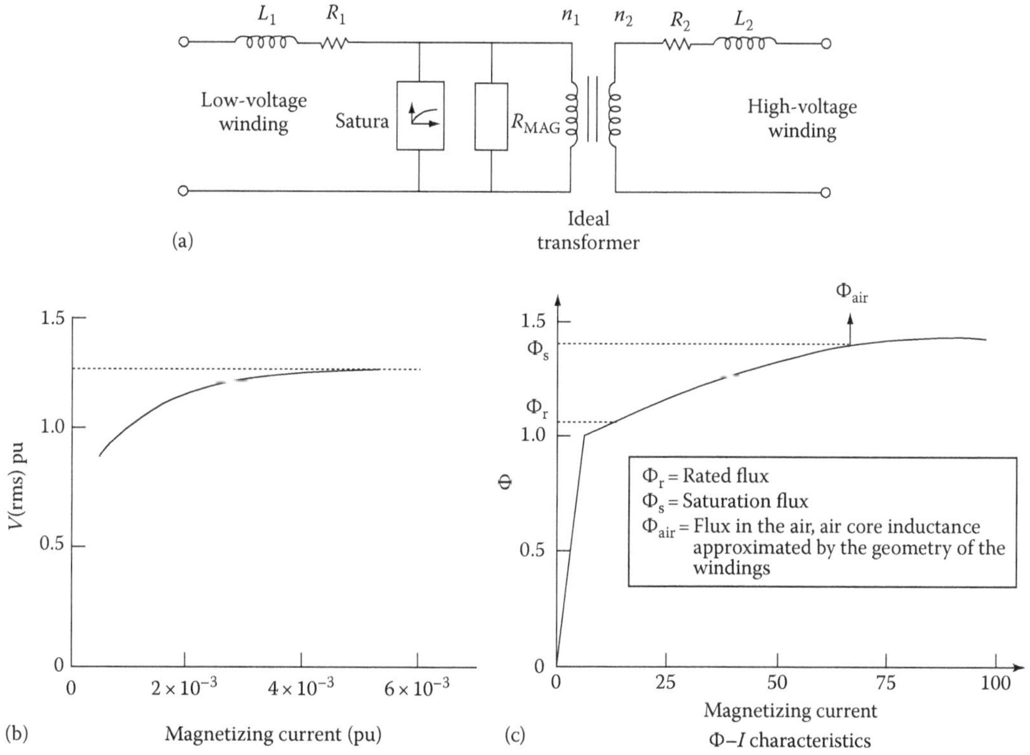

FIGURE C.12
(a) EMTP model *satura*; (b), (c) conversion of V–I characteristics into Φ–I characteristics.

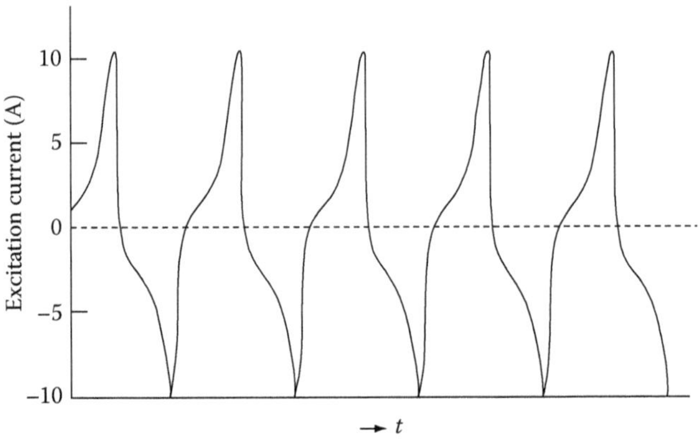

FIGURE C.13
Simulation of inrush current of a 50 kVA transformer.

Then the data are converted into unit values for the equivalent phase: S base = 250 MVA, V base = 27 kV, and I base = 9259 A. There is linear interpolation between the assumed values and finite-difference approximation to sinusoidal excitation. The hysteresis is ignored.

The EMTP model *hysdat* represents a hysteresis loop in 4–5 points to 20–25 points for a specific core material. The positive and negative saturation points, as shown in Figure C.11, need only to be specified. The geometric construction leading to positive saturation point A and negative saturation point is specified in Ref. [4]. The positive and negative saturation points are shown on the loop. Figure C.13 shows EMTP simulation of the excitation current in a single-phase, 50 kVA transformer. Other EMTP models are BECATRAN and FDBIT; the latter model can represent the behavior of transformer over wide range of frequencies, though the data for generating this model will not be readily available.

C.6.3 Nonlinearity in Core Losses

Figure C.14 shows a frequency domain approach and considers that winding resistance and leakage reactance remain constant and the nonlinearity is confined to the core characteristics [5]. The core loss is modeled as a superimposition of losses occurring in fictitious harmonic and eddy current resistors. The magnetizing characteristics of the transformer are defined by a polynomial expressing the magnetizing current in terms of flux linkages:

$$i_M = A_0 + A_1\lambda + A_2\lambda^2 + A_3\lambda^3 + \cdots \tag{C.32}$$

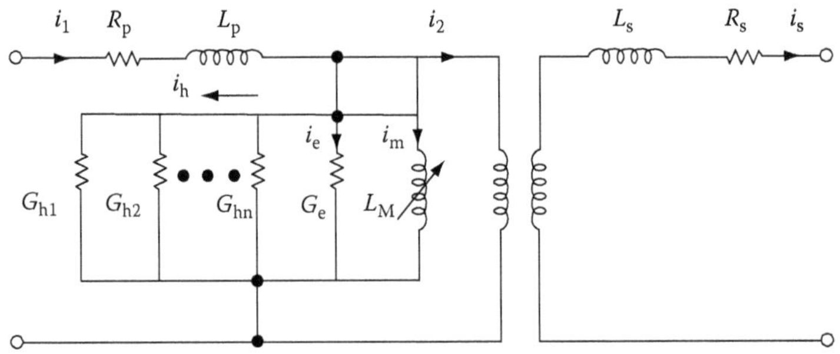

FIGURE C.14
Nonlinear shunt model with superimposition of harmonic currents in resistors.

Only a specific order of harmonic currents flow to appropriate G_h resistors in Figure C.14. From Equations C.7 and C.8, the core loss equation is

$$P_{fe} = P_h + P_e = K_h B^s f + K_e B^2 f^2 \tag{C.33}$$

For a sinusoidal voltage, this can be written as

$$P_{fe} = k_h f^{1-s} E^s + k_e E^2 \quad (k_h \neq K_h \text{ and } k_e \neq K_e) \tag{C.34}$$

This defines two-conductance G_h for hysteresis loss and G_e for eddy current loss, given by

$$G_h = k_h f^{1-s}, \quad G_e = k_e \tag{C.35}$$

C.7 High-Frequency Models

For the response of transformers to transients, a very detailed model may include each winding turn and turn-to-turn inductances and capacitances [6]. Consider a disk-layer winding or pancake sections as shown in Figure C.15a. Each numbered rectangular block

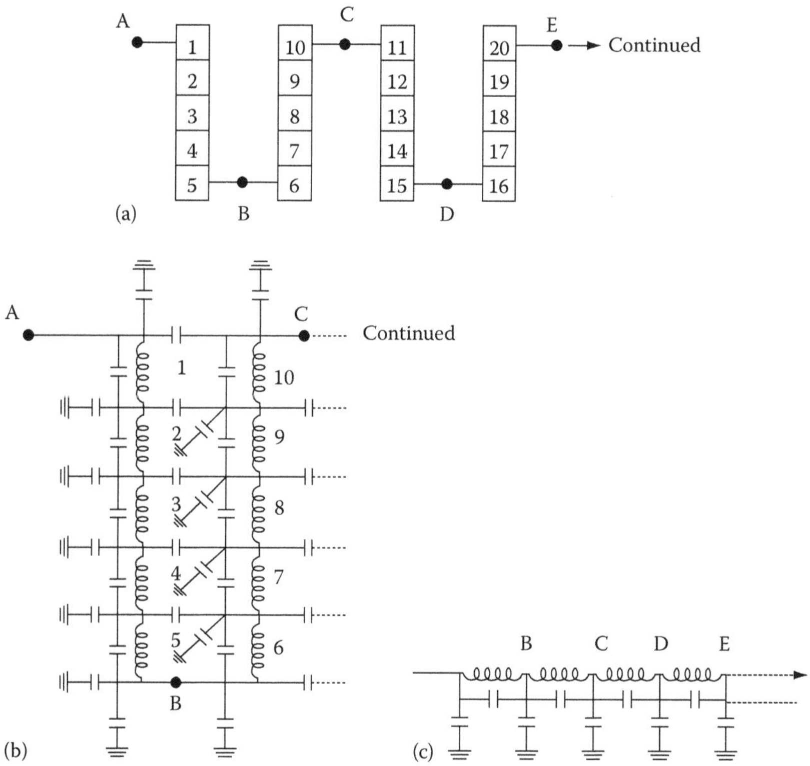

FIGURE C.15
(a) Winding turns in a pancake coil; (b) circuit of winding inductances and capacitances; (c) simplified circuit model.

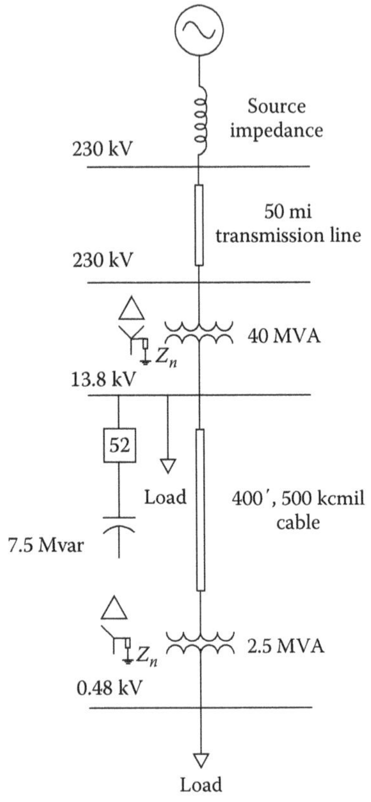

FIGURE C.16
Circuit for simulation of capacitor switching transient.

represents the cross section of a turn. The winding line terminal is at A and winding continues beyond E. Each section can be represented by a series of inductance elements with series and shunt capacitances as shown in Figure C.15b. Though the model looks complex, the mutual inductances are not shown, resistances are not represented, and no interturn capacitances are shown. This circuit will be formidable in terms of implementation. For most applications, representation of each turn is not justified, and, by successive lumping, a much simpler model is obtained (Figure C.15c).

C.7.1 Surges Transferred through Transformers

The lightning and switching surges can be transferred through transformer couplings [7]. Winding capacitive couplings predominate at high frequencies. For protection of insulation of transformers, primary and secondary surge arresters are necessary. Consider the circuit in Figure C.16. A 7.5 Mvar capacitor bank is switched at the 13.8 kV bus and the resulting switching overvoltages on the secondary of a 2.5 MVA, 13.8–0.48 kV transformer connected through 400 ft 500 KCMIL, 15 kV cable are simulated using *EMTP*. The transformer model shown in Figure 19.7 is used. The results are shown in Figure C.17. This figure shows high-frequency components, due to multiple reflections in the connecting cable, and the peak secondary. The secondary voltage is 3000 V. This is very high for a 480 V system. Secondary surge arresters and capacitors applied at transformer terminals will appreciably reduce this voltage and the high frequency of oscillations shown in this figure.

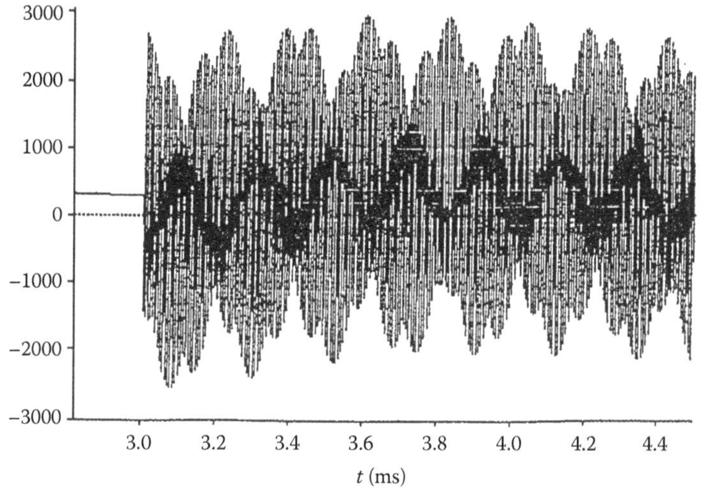

FIGURE C.17
EMTP simulation of transient voltage on 480 V secondary of 2.5 MVA transformer in Figure C.16.

The switching of transformers can give rise to voltage escalation inside the transformer windings. The windings have internal "ringing" frequencies and certain switching operations can excite these frequencies creating excessive intra-winding stresses. Sometimes, the repeated restrikes in switching devices can also give rise to high-rise transients. Surge arresters are ineffective to limit these voltages. A snubber circuit (usually a capacitor in series with a resistor) connected phase to ground can limit these voltage peaks (see also Refs. [8–11]).

C.8 Duality Models

Duality-based models can be used to represent transformers. These models are based on core topology and utilize the correspondence between electric and magnetic circuits, as expressed by the principle of duality. Voltage, current, and inductance in electrical circuits correspond to flux, mmf, and reluctance, respectively:

$$
\begin{aligned}
I &= \frac{V}{R} \\
\Phi &= \frac{\text{MMF}}{l/\mu\mu_0 a} = \frac{\text{MMF}}{S}
\end{aligned}
\tag{C.36}
$$

where
- l is the length of the magnetic path
- a is the cross-sectional area
- S is the reluctance, which is analogous to resistance in an electrical circuit and determines the magnetomotive force necessary to produce a given magnetic flux

Permeance is the reverse of reluctance.

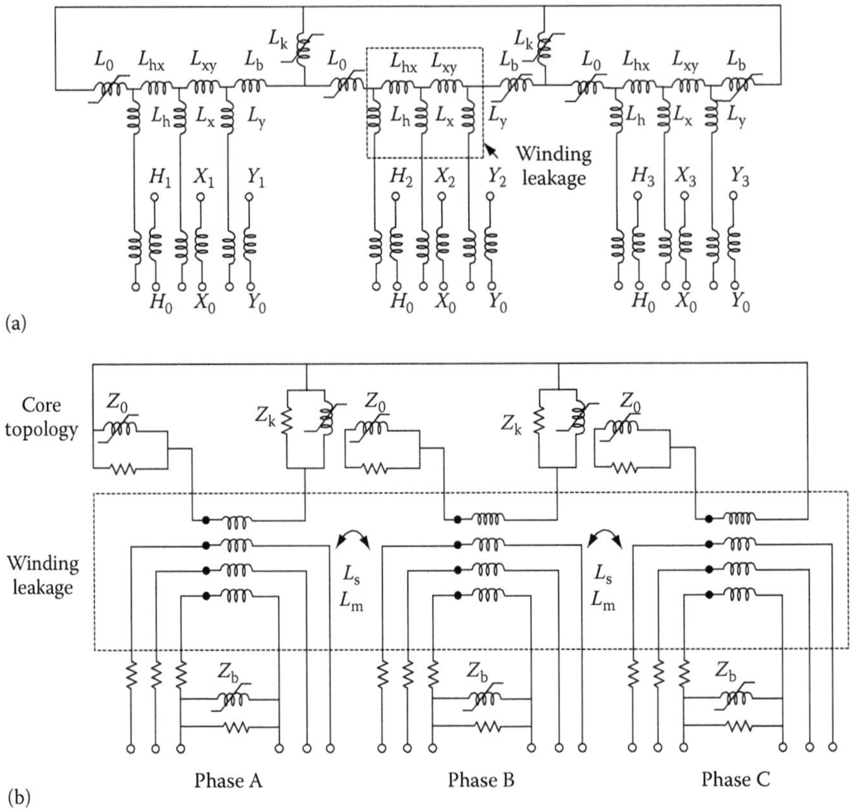

FIGURE C.18
(a) Duality-based circuit model of a core type, three-winding transformer; (b) simplified circuit derived from (a).

Figure C.18a shows electrical equivalent circuit of a three-winding core-type transformer portraying magnetic coupling in three- and five-limbed transformers [12]. Nonlinear inductances correspond to iron flux paths in the magnetic circuit, permitting each core limb to be modeled separately. Each L_k represents top and lower yokes and each L_b represents a wound limb; L_0 represents the flux path through the air, outside the core and around the windings. Finally, the ladder network between linear inductances L_0 and L_b represents winding leakages through air. Inductances L_h and L_y represent unequal flux linkages between turns due to finite winding radial build, and these are small compared to L_0 and L_b. This model is simplified as shown in Figure C.18b. The various inductances are calculated from short-circuit tests.

Duality models can be used for low-frequency transient studies, such as short circuits, inrush currents, ferroresonance, and harmonics [13].

C.9 GIC Models

Geomagnetically induced currents (GICs) flow through the earth's surface due to solar magnetic disturbances and these are typically 0.001–0.1 Hz and can reach peak values of 200 A. These can enter transformer windings through grounded neutrals (Figure C.19a),

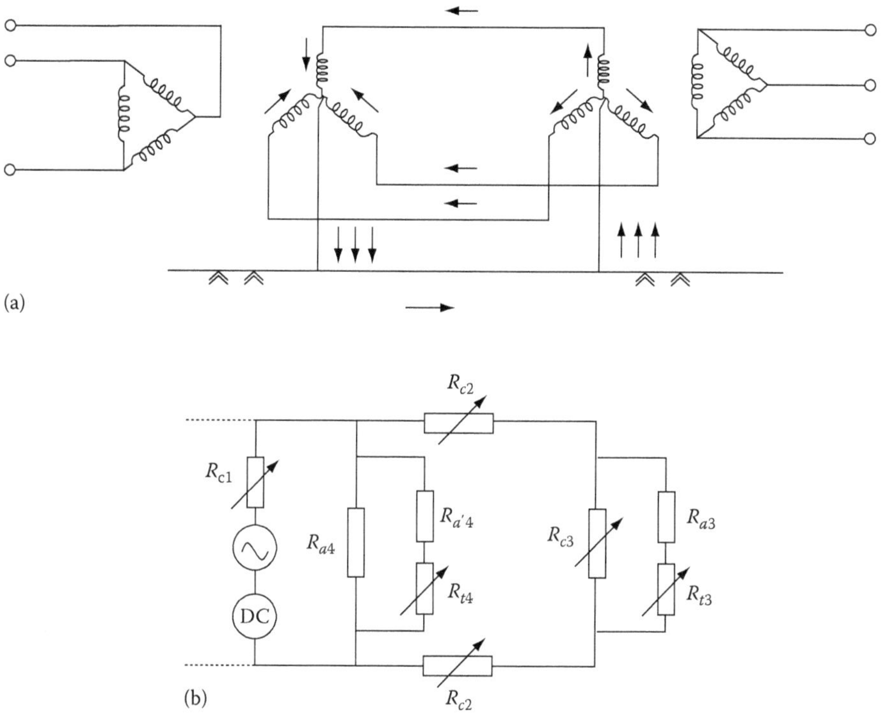

FIGURE C.19
(a) GIC entering the grounded neutrals of wye-connected transformers; (b) transformer model for GIC simulation.

and bias the transformer core to half-cycle saturation. As a result, the transformer magnetizing current is greatly increased. Harmonics increase and these could cause reactive power consumption, capacitor overload, false operation of protective relays, etc. [14].

A model for GIC is shown in Figure C.19b. Four major flux paths are included. All R elements represent reluctances in different branches. Subscripts c, a, and t stand for core, air, and tank, respectively; and 1, 2, 3, and 4 represent major branches of flux paths. Branch 1 represents the sum of core and air fluxes within the excitation windings, branch 2 represents the flux path in yoke, branch 3 represents the sum of fluxes entering the side leg, part of which leaves the side leg and enters the tank, and branch 4 represents flux leaving the tank from the center leg. An iterative program is used to solve the circuit of Figure C.19 so that nonlinearity is considered.

C.10 Ferroresonance

Ferroresonance can occur when a nonlinear inductance, like the iron-core reactance of a transformer, is excited in parallel with capacitance through high impedance. Overvoltages up to 3–4 times the nominal voltage can occur. Ferroresonance occurs because of reactance of a distribution transformer X_m and the capacitance from the bushings and underground cables form a resonant circuit. Highly distorted voltage waveforms are produced, which include a 60 Hz component. Considerable harmonics are generated. We may postulate, based upon the current literature, that

- High overvoltages of the order of 4.5 per unit can be created.
- Resonance can occur over a wide range of X_c/X_m, possibly in the range [15,16]:

$$0.1 < \frac{X_c}{X_m} < 40$$

- Resonance occurs only when the transformer is unloaded, or very lightly loaded. Transformers loaded to more than 10% of their rating are not susceptible to ferroresonance.

The capacitance of cables varies between 40–100 nF per 1000 ft, depending upon conductor size. However, the magnetizing reactance of a 35 kV transformer is several times higher than that of a 15 kV transformer; the ferroresonance can be more damaging at higher voltages. For delta-connected transformers, the ferroresonance can occur for less than 100 ft of cable. Therefore, the grounded wye–wye transformer connection has become the most popular in underground distribution system in North America. It is more resistant, though not totally immune to ferroresonance. During three-phase switching, the poles of the switch may not close simultaneously, or a current limiting fuse in one or two lines may be open, giving rise to ferroresonance (Figure C.20).

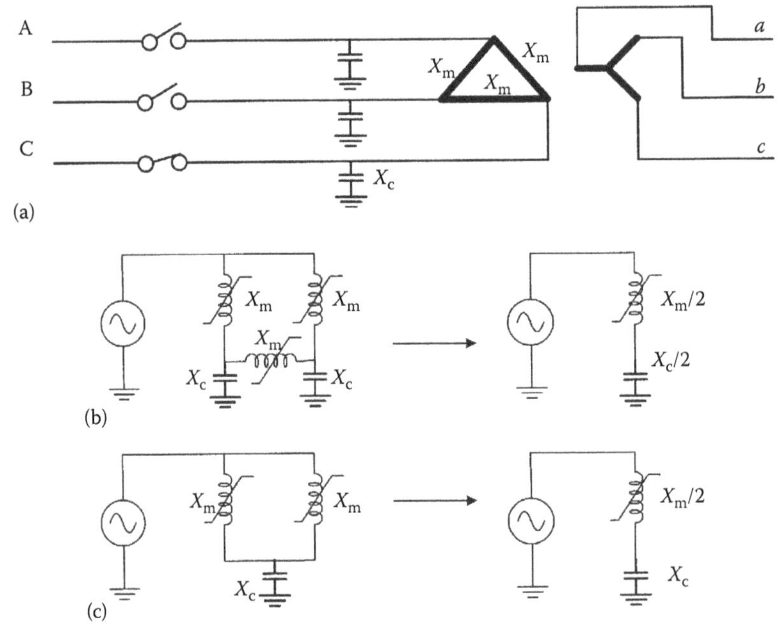

FIGURE C.20
(a) A circuit for possible ferroresonance. Equivalent circuits derived with (b) one phase closed (c) two phases closed.

C.11 Reactors

We have discussed the following applications of reactors in various chapters of this book:

- Current-limiting reactors, mainly from the standpoint of limiting the short-circuit currents in a power system. These can be applied in a feeder, in a tie-line, in synchronizing bus arrangements (Figure 7.13), and as generator reactors (Figure 13.20).
- Shunt reactors for reactive power compensation.
- Reactors used in static var controllers, that is, SVCs, TCRs, TSCs, and discharge reactors in a series capacitor (Chapter 13).
- Harmonic filter reactors and inrush current-limiting reactors.
- Line reactors to limit notching effects and dc reactors for ripple current limitation in drive systems.
- Neutral grounding reactor.

Further applications are as follows:

- Smoothing reactors are used in series with an HVDC line or inserted into a dc circuit to reduce harmonics on the dc side. Filter reactors are installed on the ac side of converters. Radio interference and power-line carrier filter reactors are used to reduce high-frequency noise propagation.
- Reactors are installed in series in a medium-voltage feeder (high-voltage side of the furnace transformer) to improve efficiency, reduce electrode consumption, and limit short-circuit currents.
- An arc suppression reactor, called a Peterson coil, is a single-phase variable reactor that is connected between the neutral of a transformer and ground for the purpose of achieving a resonant grounding system, though such grounding systems are not in common use in the United States but prevalent in Europe. The inductance is varied to cancel the capacitance current of the system for a single line-to-ground fault.
- Reactors are used in reduced-voltage motor starters to limit the starting inrush currents.
- Series reactors may be used in transmission systems to modify the power flow by changing the transfer impedance. Complexity of power grids results in power flow situations, where one area can be affected by switching, loading, and outage conditions occurring in other area. Strategic placement of reactors can increase power transfer capability and improve reliability.
- Shunt reactors are used in long high-voltage transmission lines, which generate a high amount of leading reactive power when lightly loaded. Conversely, these absorb a large amount of reactive power when heavily loaded (Chapter 13). This impacts the voltage profile. Shunt reactors absorb reactive power and lower the system voltage.

We will discuss a duplex reactor, which can sometimes be usefully applied to limit short-circuit currents and at the same time improve steady-state performance as compared to a

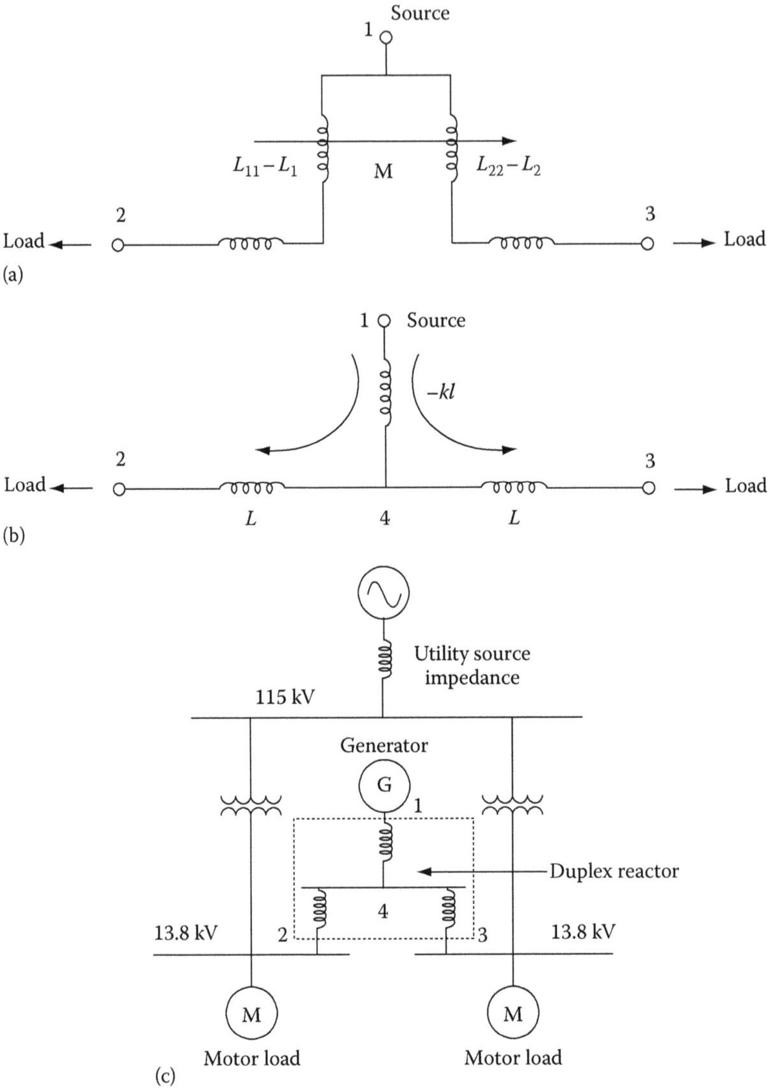

FIGURE C.21
(a), (b) Equivalent circuit of a duplex reactor, (c) a circuit showing application of a duplex reactor.

conventional reactor [17]. It consists of two magnetically coupled coils per phase. The magnetic coupling, which is dependent on the geometric proximity of the coils, is responsible for desirable properties of a duplex reactor under short-circuit and load-flow conditions. The equivalent circuit is shown in Figure C.21a and b and its application in Figure C.21c. The mutual coupling between coils is

$$L_m = \sqrt{(L_{11} - L_1)(L_{22} - L_2)} = k\sqrt{L_{11}L_{22}} = kL \quad \text{(C.37)}$$

where k is the coefficient of coupling. Note that the inductance of sections 1–4 in the T equivalent circuit, for direction of current flow from source to loads, becomes negative $(= -kL)$. Between terminals 2 and 4, and 3 and 4, it is L. The terminal 4 is fictitious and the source terminal is 1, while the load terminals are 2 and 3. Thus, the effective inductance between the source and a load terminal is $(1-k)L$, where L is the inductance of each winding and k is the coefficient of coupling. Effective reactance to load flow is reduced by a factor of k and voltage drops will also be reduced by the same factor. For a short circuit on any of the load buses, the currents in one of the windings reverse and the effective inductance is $2L(1+k)$. The short-circuit currents will be more effectively limited. The limitations are that k is dependent on the geometry of the coils and is almost independent of the current loading of the coils. Cancellation of the magnetic field under load flow occurs only when the second winding is loaded. The advantages of a duplex reactor are well exploited if the loads are split equally on two buses (Figure C.21c).

References

1. ANSI. Terminal markings and connections for distribution and power transformers, 1978 (Revised 1992). Standard C57.12.70.
2. IEEE. Standard requirements, Terminology and test code for step-voltage regulators, 1999. Standard C57.15.
3. C.E. Lin, J.B. Wei, C.L. Huang, and C.J. Huang. A new method for representation of hysteresis loops. *IEEE Trans. Power Deliv.* 4: 413–419, 1989.
4. Canadian/American EMTP User Group. *ATP RuleBook*. Portland, OR, 1987–1992.
5. J.D. Green and C.A. Gross. Non-linear modeling of transformers. *IEEE Trans. Ind. Appl.* 24: 434–438, 1988.
6. W.J. McNutt, T.J. Blalock, and R.A. Hinton. Response of transformer windings to system transient voltages. *IEEE Trans.* PAS-9: 457–467, 1974.
7. J.C. Das, Surge transference through transformers. *IEEE Ind. Mag.* 9(5), 24–32, October 2003.
8. P.T.M. Vaessen. Transformer model for high frequencies. *IEEE Trans. Power Deliv.* 3: 1761–1768, 1988.
9. T. Adielson, et al. Resonant overvoltages in EHV transformers-modeling and application. *IEEE Trans.* PAS-100: 3563–3572, 1981.
10. IEEE. IEEE application guide for high-voltage circuit breakers rated on symmetrical current basis, Standard C37.010-1999.
11. J.F. Perkins. Evaluation of switching surge overvoltages on medium voltage power systems. *IEEE Trans.* PAS-101: 1727–1734, 1982.
12. X. Chen and S.S. Venkta. A three-phase three-winding core-type transformer model for low-frequency transient studies. *IEEE Trans.* PD-12: 775–782, 1997.
13. A. Narang and R.H. Brierley. Topology based magnetic model for steady state and transient studies for three-phase core type transformers. *IEEE Trans.* PS-9: 1337–1349, 1994.
14. S. Lu, Y. Liu, and J.D.R. Ree. Harmonics generated from a DC biased transformer. *IEEE Trans.* PD-8: 725–731, 1993.
15. D.R. Smith, S.R. Swanson, and J.D. Borst. Overvoltages with remotely switched cable fed grounded Wye–Wye transformers. *IEEE Trans.* PAS-94, 1843–1853, 1975.
16. R.H. Hopkinson. Ferroresonance during single-phase switching of three-phase distribution transformer banks. *IEEE Trans.* PAS-84: 289–293, 1965.
17. J.C. Das, W.F. Robertson, and J. Twiss. Duplex reactor for large cogeneration distribution system-an old concept reinvestigated. In: *TAPPI, Engineering Conference*, Nashville, TN, 1991, pp. 637–648.

Appendix D: Sparsity and Optimal Ordering

It is seen that linear simultaneous equations representing a power system are sparse. These give sparse matrices. Consider a 500-node system, and assuming that, on average, there are two bus connections to a node, the admittance matrix A will have 500 diagonal elements and $500 \times 2 \times 2 = 2000$ off-diagonal elements. The total number of elements in A is $500 \times 500 = 250{,}000$. The population of nonzero elements is, therefore, 1.0%. The assumption of two buses per node in a transmission network is high. Typically, it will be 1–1.5, further reducing the percentage of nonzero elements.

When the matrix is factorized in LU or LDU form, nonzero elements are created where none existed before:

$$a_{ij}\,(\text{new}) = a_{ij}\,(\text{primitive}) - \frac{a_{ip}a_{pj}}{a_{pp}} \tag{D.1}$$

If a_{ij} is zero in the original matrix at the beginning of the pth step and both a_{ip} and a_{pj} are nonzero in value, then a_{ij} (new) becomes a new nonzero term.

$$\begin{array}{|c|c|c|c|}\hline & & & \\ \hline & a_{pp} & a_{pj} & \\ \hline & & & \\ \hline & a_{ip} & 0 & \\ \hline \end{array} \Rightarrow \begin{array}{|c|c|c|}\hline & & \\ \hline & a_{pp} & a_{pj} \\ \hline & & \\ \hline & a_{ip} & a_{ij} \\ \hline \end{array} \tag{D.2}$$

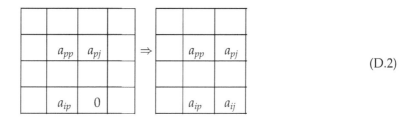

D.1 Optimal Ordering

Optimal ordering is the rearrangement of the order of sequence of columns and rows in the matrix to minimize nonzero terms. When a programming scheme that stores and processes only nonzero terms is used, saving in operations and computer memory can be achieved by keeping the stored tables of factors as short as possible. A truly optimal ordering scheme should produce the fewest nonzero elements. For large systems, the determination of such an optimal ordering itself may require a long computer time, which may not be justifiable practically, for the benefits in reducing the nonzero elements. An effective algorithm for absolute optimal ordering has not been developed; however, several schemes give near optimal solutions.

D.2 Flow Graphs

Power system matrices are symmetrical, and, therefore, their structures can be described by flow graphs. Optimal elimination then translates into a topological problem. The following theorems can be postulated:

1. Any graph that is obtained from another graph by a node-eliminating process is independent of the order in which the nodes are eliminated.
2. When any two successive nodes are adjacent in the original graph or in the graph generated by the elimination process, the node with the smaller *valence* occurs first.

The valency of a node connected in a graph is the number of new paths created or added as a result of elimination of the node. Hence, the valency of the node is defined as the number of new links added to the graph, that is, the new nonzero elements generated in the coefficient matrix because of elimination of the node.

In a connected graph, the nodes communicate between each other. A node can be eliminated if a path exists and the flow in the graph is not interrupted. In the graph of Figure D.1, node 14 can be eliminated. This will not add any new branches nor any new nonzero elements. Similarly, node 13 can be eliminated, as a path exists between node 12 and 5, and, in the coefficient matrix, element $a_{12,5}$ will be nonzero. However, elimination of node 4 will add a new link between 3 and 5 and hence the element $a_{3,5}$, which was zero earlier, will appear as a nonzero element in the matrix. Elimination of node 7 adds four new links, as shown. The new links are shown dotted.

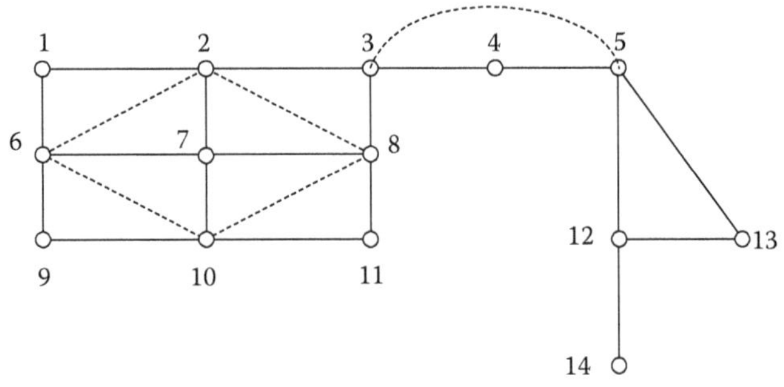

FIGURE D.1
A network for node elimination.

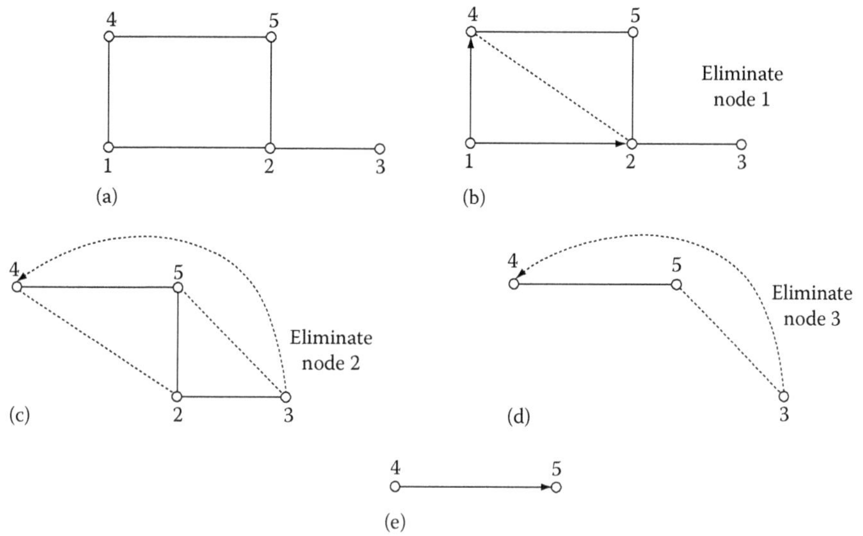

FIGURE D.2
(a) Graph of a five-node network; (b)–(e) graphs with successive elimination of nodes 1, 2, 3, and 4.

Consider the network of Figure D.2 and the ordered matrix as shown below, with nonzero terms. The ordering here is 1, 2, 3, 4, and 5:

	1	2	3	4	5
1	X	X		X	
2	X	X	X		X
3		X	X		
4	X			X	X
5		X		X	X

(D.3)

Eliminate node 1 (Figure D.2). Two new nonzero elements are introduced:

$$\left|\begin{array}{cc:ccc} X & X & & X & \\ \hdashline X & X & X & X_{\text{new}} & X \\ & X & X & & \\ X & X_{\text{new}} & & X & X \\ & X & & X & X \end{array}\right| \tag{D.4}$$

Eliminate node 2, and four new nonzero elements are produced. The matrix, after elimination of node 1 and introduction of new elements due to elimination of node 2, is

$$\left|\begin{array}{c|ccc} X & X & X_{\text{new}} & X \\ \hline X & X & X_{\text{new}} & X_{\text{new}} \\ X_{\text{new}} & X_{\text{new}} & X & X \\ X & X_{\text{new}} & X & X \end{array}\right| \tag{D.5}$$

Eliminate node 3. No new nonzero element is produced. There is already a path between 3 and 5 (dotted). The modified matrix after elimination of node 2 is

$$\left|\begin{array}{c|cc} X & X_{\text{new}} & X_{\text{new}} \\ \hline X_{\text{new}} & X & X \\ X_{\text{new}} & X & X \end{array}\right| \tag{D.6}$$

Node 3 followed by node 4 can now be eliminated.

Now consider the reordered scheme of elimination, shown in Figure D.3.

1. Node 3 is first eliminated; no new link or nonzero element is created.
2. Node 4 is next eliminated; creates two new nonzero elements.
3. Node 5 is next eliminated; creates no new nonzero element.
4. Node 2 can now be eliminated.

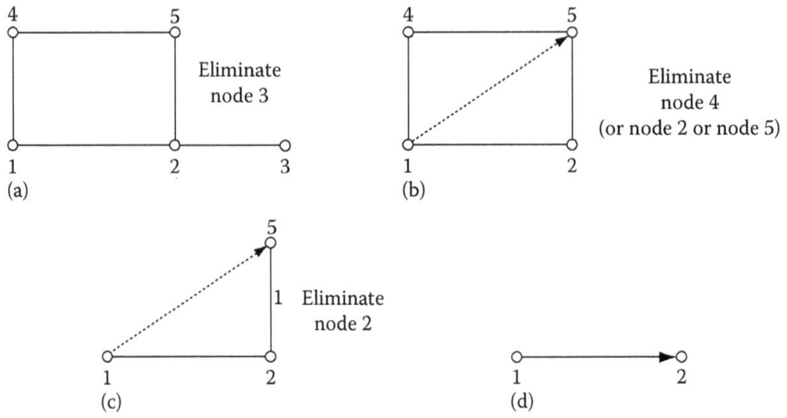

FIGURE D.3
(a)–(d) Graphs with alternate order of node elimination.

This gives the ordered original matrix as 3, 4, 5, 2, and 1:

	3	4	5	2	1
3	X			X	
4		X	X		X
5		X	X	X	
2	X		X	X	X
1		X		X	X

(D.7)

The number of nonzero elements created at the *p*th step is

$$N_{\text{nonzero}} = \frac{n(n-1)}{2} \tag{D.8}$$

where n is the number of nodes directly connected to the pivot node, but not directly connected to each other.

D.3 Optimal Ordering Schemes

A number of near optimal schemes have been developed:

Scheme 1

The rows of the coefficient matrix are numbered according to the number of nonzero off-diagonal elements, before elimination. The rows with the least number of off-diagonal terms come first, followed by the next rows in ascending order of nonzero off-diagonal terms. This method is simple and fast to execute but does not give the minimum number of nonzero terms. In other words, the scheme is to reorder the nodes so that the number of connecting branches to each node is in ascending order in the original network.

Scheme 2

The rows in the coefficient matrix are numbered so that, at each step, the next row to be operated upon is the one with fewest nonzero terms. If more than one row meets this criterion, select any one first. Thus, the nodes are numbered so that at each step of elimination, the next node to be eliminated is the one having the fewest connected branches, that is, the minimum degree. This method requires simulation of the elimination process to take into account the changes in the branch connections as each node is eliminated.

Scheme 3

At each step of the elimination process, select the node that produces the smallest number of new branches. From the network point of view, at each step, the next node to be eliminated is the one that produces the fewest row equivalents of every feasible alternative.

Scheme 4

If at any stage of elimination, more than one node has the same degree, then remove the one that creates the minimum off-diagonal zero terms, that is, new links in the system graph. This exploits the merits of Schemes 2 and 3.

Consider the network of Figure D.4a. Its matrix without prior ordering, in terms of numbered nodes in serial ascending order, is

	1	2	3	4	5	6	7	8	9	10	11
1	X	X			X						
2	X	X	X								
3		X	X	X							
4			X	X				X	X		
5	X				X	X				X	
6		X			X	X					
7							X	X			
8				X			X	X	X		X
9				X				X	X		
10					X					X	
11								X			X

(D.9)

An examination of the coefficient matrix shows that

Nodes 7, 10, and 11 have only one connection with the other nodes and thus one off-diagonal element. According to Scheme 1, the optimal ordering sequence can start with any of these nodes, in any order.

Nodes 1, 3, 6, and 9 come next. These have two connections with the other nodes in the original network and can be eliminated next, in any order.

Nodes 2, 4, and 5 have three connections and should be eliminated next in any order, while node 8 has four connections and should be eliminated last.

Thus, the optimal ordering according to Scheme 1 is 7, 10, 11, 1, 3, 4, 6, 9, 2, 4, 5, and 8. This is shown graphically in Figure D.4b and each step of elimination is marked by the side of the node in a triangle. This scheme generates four new elements.

Next, Scheme 2 is applied and it is shown in Figure D.4c. While optimal ordering could have been written straightaway, without examination of the modified graphs in Scheme 1, or straight from the examination of the coefficient matrix, the optimal ordering according

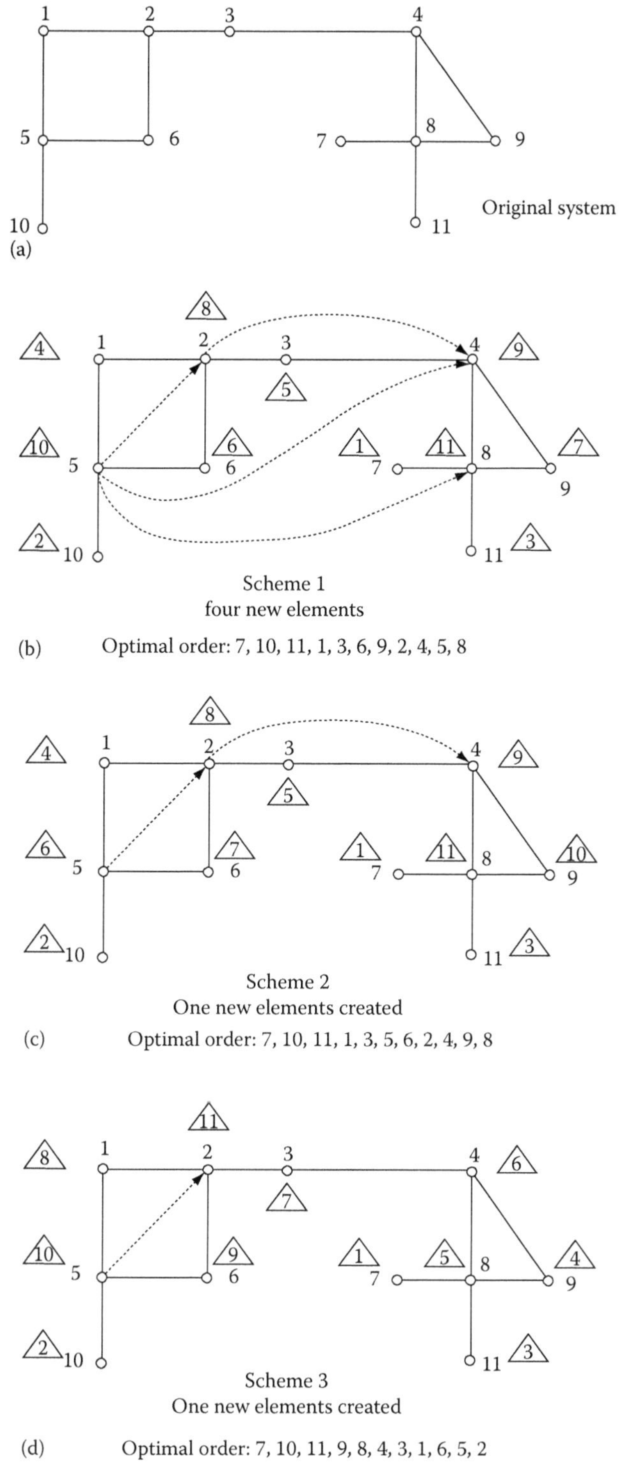

FIGURE D.4
(a) Graph of original 11-node network; (b)–(d) alternate optimal ordering Schemes 1, 2, and 3.

to Scheme 2 is not easily visualized, unless the modified graphs on successive node elimination are examined.

After nodes 7, 10, and 11 are eliminated, nodes 8, 9, 3, 1, 5, and 6 have two branches connected to them. Thus, there is a choice in elimination order. Figure D.4c shows the elimination in the order 1, 3, 5, and 6. Elimination of 1 and 3 give one new element each.

As the network should be examined after each elimination, the node 2 that had three branches has only one branch connected to it after these eliminations. Thus, it comes next in the order of elimination. This leaves nodes 4, 9, and 8, which have two branches connected to each, in a triangle. Thus, any of these can be eliminated first. This gives an optimal order 7, 10, 11, 1, 3, 5, 6, 2, 4, 9, and 8.

This order is not unique. Other orders will satisfy the Scheme 2 criteria.

Figure D.4d shows Scheme 3. Here, only one new element is produced. The optimal order of Scheme 2 could also give the same result. Consider that after elimination of nodes 7, 10, and 11 in Scheme 2, there is a choice to eliminate 8, 9, 5, 6, or 1. Let us eliminate 9. Node 8 then has only one path to node 4 and should be eliminated next. The next node to eliminate will be node 4 (single path) followed by node 3. Thus, Scheme 2 could also give only one new element. As an exercise, the reader can construct a graph and optimal order for Scheme 4.

Bibliography

El-Abiad, A.H. (Ed.). Solution of network problems by sparsity techniques, Chap. 1. In *Proceedings of the Arab School of Science and Technology*, Kuwait, 1983. Distributed by McGraw Hill, New York.

Happ, H.H. The solution of system problems by tearing. *IEEE Proc.* 62(7): 930–940, July 1974.

Sato, N. and W.F. Tinney. Technique for exploiting the sparsity of the network admittance matrix. *IEEE Trans.* PAS 82: 944–950, 1963.

Tinney, W.F. and J.W. Walker. Direct solutions of sparse network equations by optimally ordered triangular factorization. *IEEE Proc.* 55: 1801–1809, 1967.

Appendix E: Fourier Analysis

E.1 Periodic Functions

A function is said to be periodic if it is defined for all real values of t, and if there is a positive number T such that

$$f(t) = f(t+T) = f(t+2T) = f(t+nT) \tag{E.1}$$

then T is called the period of the function.

If k is any integer and $f(t+kT) = f(t)$ for all values of t and if two functions $f_1(t)$ and $f_2(t)$ have the same period T, then the function $f_3(t) = af_1(t) + bf_2(t)$, where a and b are constants, also has the same period T. Figure E.1 shows a periodic function.

E.2 Orthogonal Functions

Two functions $f_1(t)$ and $f_2(t)$ are orthogonal over the interval (T_1, T_2) if

$$\int_{T_1}^{T_2} f_1(t) f_2(t) = 0 \tag{E.2}$$

Figure E.2 shows two orthogonal functions over the period T.

E.3 Fourier Series and Coefficients

A periodic function can be expanded in a Fourier series. The series has the expression:

$$f(t) = a_0 + \sum_{n=1}^{\infty} \left(a_n \cos\left(\frac{2\pi nt}{T}\right) + b_n \sin\left(\frac{2\pi nt}{T}\right)\right) \tag{E.3}$$

where a_0 is the average value of function $f(t)$. It is also called the dc component, and a_n and b_n are called the coefficients of the series. A series such as Equation E.3 is called a trigonometric Fourier series. The Fourier series of a periodic function is the sum of sinusoidal components of different frequencies. The term $2\pi/T$ can be written as ω.

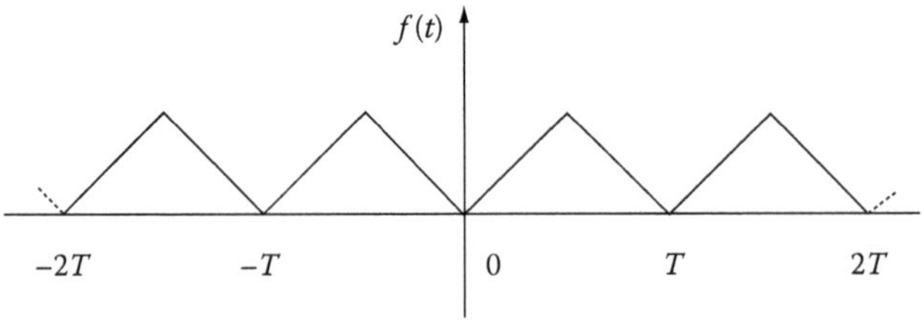

FIGURE E.1
A periodic function.

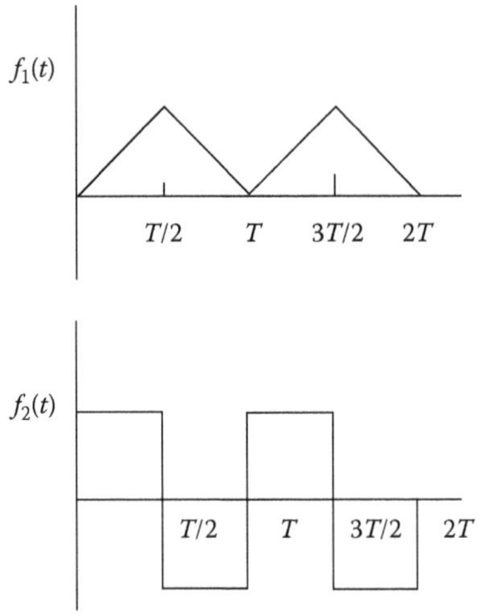

FIGURE E.2
Orthogonal functions.

The nth term $n\omega$ is then called the nth harmonic and $n = 1$ gives the fundamental; a_0, a_n, and b_n are calculated as follows:

$$a_0 = \frac{1}{T}\int_{-T/2}^{T/2} f(t)\mathrm{d}t \tag{E.4}$$

$$a_n = \frac{2}{T}\int_{-T/2}^{T/2} \cos\left(\frac{2\pi nt}{T}\right)\mathrm{d}t \quad \text{for } n = 1, 2, \ldots, \infty \tag{E.5}$$

$$b_n = \frac{2}{T}\int_{-T/2}^{T/2} \sin\left(\frac{2\pi nt}{T}\right)\mathrm{d}t \quad \text{for } n = 1, 2, \ldots, \infty \tag{E.6}$$

These equations can be written in terms of angular frequency:

$$a_0 = \frac{1}{2\pi}\int_{-\pi}^{\pi} f(x)\,\omega t\,\mathrm{d}\omega t \tag{E.7}$$

$$a_n = \frac{1}{\pi}\int_{-\pi}^{\pi} f(x)\omega t\cos(n\omega t)\,\mathrm{d}\omega t \tag{E.8}$$

$$b_n = \frac{1}{\pi} \int_{-\pi}^{\pi} f(x)\, \omega t \sin(n\omega t)\, d\omega t \tag{E.9}$$

This gives

$$x(t) = a_0 + \sum_{n=1}^{\infty} [a_n \cos(n\omega t) + b_n \sin(n\omega t)] \tag{E.10}$$

We can write

$$\begin{aligned} a_n \cos n\omega t + b_n \sin \omega t &= \left[a_n^2 + b_n^2\right]^{1/2} [\sin \phi_n \cos n\omega t + \cos \phi_n \sin n\omega t] \\ &= \left[a_n^2 + b_n^2\right]^{1/2} \sin(n\omega t + \phi_n) \end{aligned} \tag{E.11}$$

where

$$\phi_n = \tan^{-1} \frac{a_n}{b_n}$$

The coefficients can be written in terms of two separate integrals:

$$\begin{aligned} a_n &= \frac{2}{T} \int_0^{T/2} x(t) \cos\left(\frac{2\pi n t}{T}\right) dt + \frac{2}{T} \int_{-T/2}^{0} x(t) \cos\left(\frac{2\pi n t}{T}\right) dt \\ b_n &= \frac{2}{T} \int_0^{T/2} x(t) \sin\left(\frac{2\pi n t}{T}\right) dt + \frac{2}{T} \int_{-T/2}^{0} x(t) \sin\left(\frac{2\pi n t}{T}\right) dt \end{aligned} \tag{E.12}$$

Example E.1

Find the Fourier series of a function defined by

$$\begin{matrix} x + \pi & 0 \le x \le \pi \\ -x - \pi & -\pi \le x < 0 \end{matrix}$$

Find a_0, which is given by

$$a_0 = \frac{1}{\pi} \int_{-\pi}^{0} (-x - \pi)\, dx + \frac{1}{\pi} \int_0^{\pi} (x + \pi)\, dx = \pi$$

$$\begin{aligned} a_n &= \frac{1}{\pi} \int_{-\pi}^{0} (-x - \pi) \cos nx\, dx + \frac{1}{\pi} \int_0^{\pi} (x + \pi) \cos nx\, dx \\ &- \frac{4}{n^2 \pi} \quad \text{if } n \text{ is odd} \\ &= 0 \quad \text{if } n \text{ is even} \end{aligned}$$

b_n is given by

$$b_n = \frac{1}{\pi}\int_{-\pi}^{0}(-x-\pi)\sin nx\,dx + \frac{1}{\pi}\int_{0}^{\pi}(x+\pi)\sin nx\,dx$$

$$= \frac{4}{n} \quad \text{if } n \text{ is odd}$$

$$= 0 \quad \text{if } n \text{ is even}$$

Thus, the Fourier series is

$$f(x) = \frac{\pi}{2} - \frac{4}{\pi}\left(\frac{\cos x}{1^2} + \frac{\cos 3x}{3^2} + \cdots\right) + 4\left(\frac{\sin x}{1} + \frac{\sin 3x}{3} + \cdots\right)$$

E.4 Odd Symmetry

A function $f(x)$ is said to be an odd or skew symmetric function, if

$$f(-x) = -f(x) \tag{E.13}$$

The area under the curve from $-T/2$ to $T/2$ is zero. This implies that

$$a_0 = 0, \quad a_n = 0 \tag{E.14}$$

$$b_n = \frac{4}{T}\int_{0}^{T/2} f(t)\sin\left(\frac{2\pi nt}{T}\right)dt \tag{E.15}$$

Figure E.3a shows a triangular function, having odd symmetry. The Fourier series contains only sine terms.

E.5 Even Symmetry

A function $f(x)$ is even symmetric, if

$$f(-x) = f(x) \tag{E.16}$$

The graph of such a function is symmetric with respect to the y axis. The y-axis is a mirror of the reflection of the curve.

$$a_0 = 0, \quad b_n = 0 \tag{E.17}$$

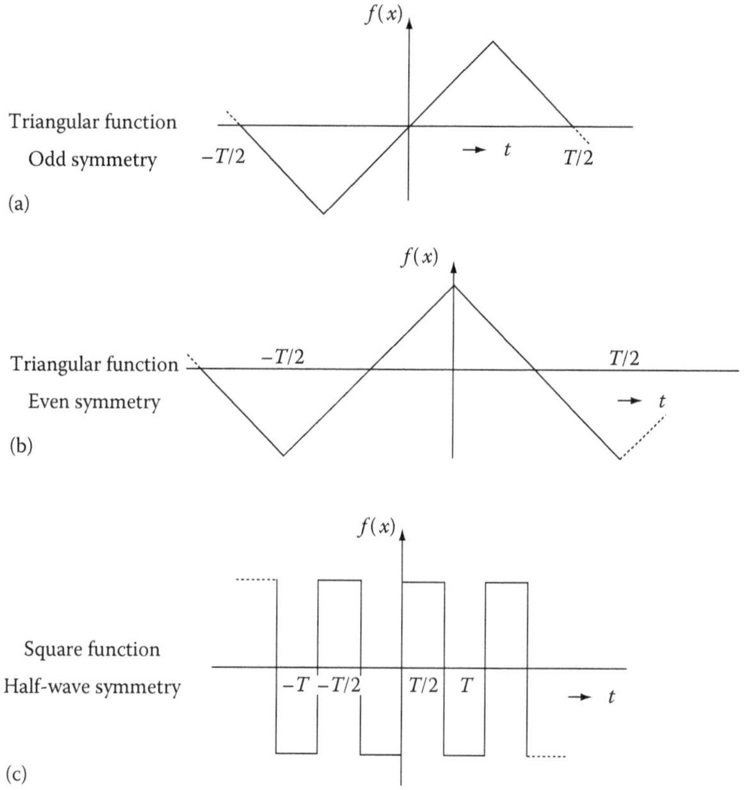

FIGURE E.3
(a) Triangular function with odd symmetry; (b) triangular function with even symmetry; (c) square function with half-wave symmetry.

$$a_n = \frac{4}{T}\int_0^{T/2} f(t)\cos\left(\frac{2\pi nt}{T}\right)\mathrm{d}t \tag{E.18}$$

Figure E.3b shows a triangular function with odd symmetry. The Fourier series contains only cosine terms. Note that the odd and even symmetry has been obtained with the triangular function, by shifting the origin.

E.6 Half-Wave Symmetry

A function is said to have half-wave symmetry if

$$f(x) = -f(x + T/2) \tag{E.19}$$

Figure E.3c shows a square-wave function that has half-wave symmetry, with respect to the period $-T/2$. The negative half-wave is the mirror image of the positive half, but phase shifted by $T/2$ (or π rad). Due to half-wave symmetry, the average value is zero. The function contains only odd harmonics.

If n is odd, then

$$a_n = \frac{4}{T}\int_0^{T/2} x(t)\cos\left(\frac{2\pi nt}{T}\right)dt \tag{E.20}$$

and $a_n = 0$ for $n =$ even. Similarly,

$$b_n = \frac{4}{T}\int_0^{T/2} x(t)\sin\left(\frac{2\pi nt}{T}\right)dt \tag{E.21}$$

for $n =$ odd, and it is zero for $n =$ even.

E.7 Harmonic Spectrum

The Fourier series of a square-wave function is

$$f(t) = \frac{4k}{\pi}\left(\frac{\sin\omega t}{1} + \frac{\sin 3\omega t}{3} + \frac{\sin 5\omega t}{5} + \cdots\right) \tag{E.22}$$

where k is the amplitude of the function. The magnitude of the nth harmonic is $1/n$, when the fundamental is expressed as one per unit. The construction of a square wave from the component harmonics is shown in Figure E.4a, and the plotting of harmonics as a percentage of the magnitude of the fundamental gives the harmonic spectrum of Figure E.4b. A harmonic spectrum indicates the relative magnitude of the harmonics with respect to the fundamental and is not indicative of the sign (positive or negative) of the harmonic, nor its phase angle.

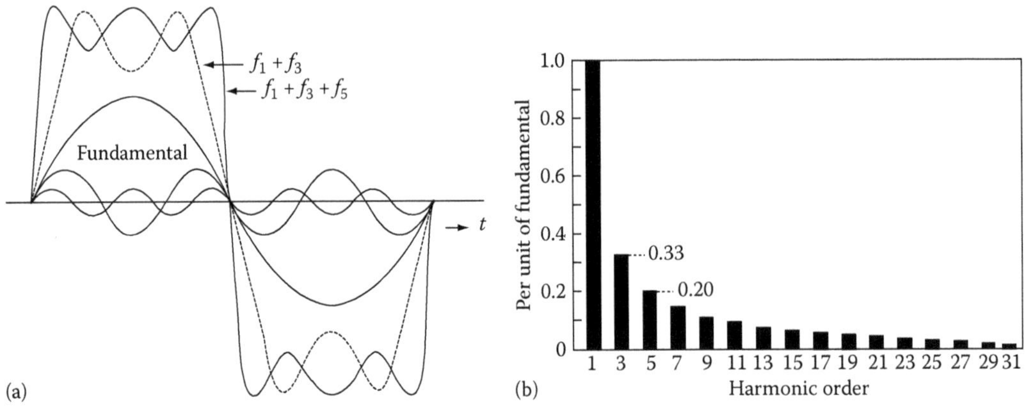

FIGURE E.4
(a) Construction of a square wave from its harmonic components; (b) harmonic spectrum of a square wave.

E.7.1 Constructing Fourier Series from Graphs and Tables

When the function is given as a graph or table of values, a Fourier series can be constructed as follows:

$$a_0 = \frac{1}{\pi}\int_0^{2\pi} f(x)\mathrm{d}x = \frac{2}{2\pi - 0}\int_0^{2\pi} f(x)\mathrm{d}x$$
$$= \text{twice mean value of } f(x), \quad 0 \text{ to } 2\pi \tag{E.23}$$

Similarly,

$$a_n = \text{twice mean value of } f(x)\cos nx, \quad 0 \text{ to } 2\pi \tag{E.24}$$

$$b_n = \text{twice mean value of } f(x)\sin nx, \quad 0 \text{ to } 2\pi \tag{E.25}$$

Example E.2

Construct the Fourier series for the function given in Table E.1. The step-by-step calculations are shown in Table E.2.

TABLE E.1
Example E.2

X	0°	30°	60°	90°	120°	150°	180°	210°	240°	270°	300°	330°
$f(x)$	3.45	4.87	6.98	2.56	1.56	1.23	0.5	2.6	3.7	4.8	3.2	1.1

TABLE E.2
Example E.2: Step-by-Step Calculations

x(°)	$f(x)$	$f(x)$ sin x	$f(x)$ sin 2x	$f(x)$ cos x	$f(x)$ cos 2x
0	3.45	0	0	3.45	3.45
30	4.87	2.435	4.217	4.217	2.435
60	6.98	6.07	6.07	3.49	−3.49
90	2.56	2.56	0	0	−2.56
120	1.56	1.351	−1.351	−0.78	−0.78
150	1.23	0.615	−1.015	−1.065	0.615
180	0.5	0	0	−0.5	0.5
210	2.6	−1.30	2.252	−2.252	1.3
240	3.7	−3.204	3.204	−1.85	−1.85
270	4.8	−4.8	0	0	−4.8
300	3.2	−2.771	−2.771	1.6	−1.6
330	1.1	−0.55	−0.957	0.957	0.55
Totals	36.55	0.406	9.649	7.276	−6.23

Note: $a_0 = 2 \times$ mean of $f(x) = (2)(36.55)/12 = 6.0917$.

$$a_0 = 2 \times \text{mean of } f(x)$$
$$a_1 = 2 \times \text{mean of } f(x) \cos x$$
$$b_1 = 2 \times \text{mean of } f(x) \sin x$$

$a_0 = 2 \times$ mean of $f(x) = (2)(36.55)/12 = 6.0917$.
From the summation of functions in Table E.2,

$$f(x) = a_0 + a_1 \cos x + a_2 \cos 2x + \cdots b_1 \sin x + b_2 \sin 2x + \cdots$$
$$= 6.092 + 1.212 \cos x - 1.038 \cos 2x + \cdots + 0.068 \sin x + 1.608 \sin 2x + \cdots$$

E.8 Complex Form of Fourier Series

A vector with amplitude A and phase angle θ with respect to a reference can be resolved into two oppositely rotating vectors of half the magnitude so that

$$|A| \cos \theta = |A/2| e^{j\theta} + |A/2| e^{-j\theta} \tag{E.26}$$

Thus,

$$a_n \cos n\omega t + b_n \sin n\omega t \tag{E.27}$$

can be substituted by

$$\cos(n\omega t) = \frac{e^{jn\omega t} + e^{-jn\omega t}}{2} \tag{E.28}$$

$$\sin(n\omega t) = \frac{e^{jn\omega t} - e^{-jn\omega t}}{2j} \tag{E.29}$$

Thus,

$$x(t) = \frac{a_0}{2} + \frac{1}{2} \sum_{n=1}^{n=\infty} (a_n - jb_n) e^{jn\omega t} + \frac{1}{2} \sum_{n=1}^{n=\infty} (a_n - jb_n) e^{-jn\omega t} \tag{E.30}$$

We introduce negative values of n in the coefficients, that is,

$$a_{-n} = \frac{2}{T} \int_{-T/2}^{T/2} x(t) \cos(-n\omega t)\, dt = \frac{2}{T} \int_{-T/2}^{T/2} x(t) \cos(n\omega t)\, dt = a_n \quad n = 1, 2, 3, \ldots \tag{E.31}$$

$$b_{-n} = \frac{2}{T} \int_{-T/2}^{T/2} x(t) \sin(-n\omega t)\, dt = -\frac{2}{T} \int_{-T/2}^{T/2} x(t) \sin(n\omega t)\, dt = -b_n \quad n = 1, 2, 3, \ldots \tag{E.32}$$

Hence,

$$\sum_{n=1}^{\infty} a_n e^{-jn\omega t} = \sum_{n=-1}^{\infty} a_n e^{jn\omega t} \quad \text{(E.33)}$$

and

$$\sum_{n=1}^{\infty} jb_n e^{-jn\omega t} = \sum_{n=-1}^{\infty} jb_n e^{jn\omega t} \quad \text{(E.34)}$$

Therefore, substituting in Equation E.30, we obtain

$$x(t) = \frac{a_0}{2} + \frac{1}{2}\sum_{n=-\infty}^{\infty} (a_n - jb_n)e^{jn\omega t} = \sum_{n=-\infty}^{\infty} c_n e^{jn\omega t} \quad \text{(E.35)}$$

This is the expression for a Fourier series expressed in exponential form, which is the preferred approach for analysis. The coefficient c_n is complex and given by

$$c_n = \frac{1}{2}(a_n - jb_n) = \frac{1}{T}\int_{-T/2}^{T/2} x(t)e^{-jn\omega t}\mathrm{d}t \quad n = 0, \pm 1, \pm 2, \ldots \quad \text{(E.36)}$$

E.8.1 Convolution

When two harmonic phasors of different frequencies are convoluted, the result will be harmonic phasors at sum and difference frequencies:

$$\begin{aligned} &|a_k| \sin(k\omega t + \theta_k)|b_m| \sin(m\omega t + \theta_m) \\ &= \frac{1}{2}|a_k||b_m|\left[\sin\left((k-m)\omega t + \theta_k - \theta_m + \frac{\pi}{2}\right) - \sin\left((k+m)\omega t + \theta_k + \theta_m + \frac{\pi}{2}\right)\right] \end{aligned} \quad \text{(E.37)}$$

Convolution is a process of correlating one time series with another time series that has been reversed in time.

E.9 Fourier Transform

Fourier analysis of a continuous periodic signal in the time domain gives a series of discrete frequency components in the frequency domain. The Fourier integral is defined by the following expression:

$$X(f) = \int_{\infty}^{-\infty} x(t)e^{-j2\pi ft}\mathrm{d}t \quad \text{(E.38)}$$

If the integral exists for every value of parameter f (frequency), then this equation describes the Fourier transform. The Fourier transform is a complex quantity:

$$X(f) = \text{Re}\,X(f) + j\text{Im}\,X(f) \tag{E.39}$$

where
Re $X(f)$ is the real part of the Fourier transform
Im $X(f)$ is the imaginary part of the Fourier transform

The amplitude or *Fourier spectrum of* $x(t)$ is given by

$$|X(f)| = \sqrt{R^2(f) + I^2(f)} \tag{E.40}$$

$\phi(f)$ is the phase angle of the Fourier transform and is given by

$$\phi(f) = \tan^{-1}\left[\frac{\text{Im}\,X(f)}{\text{Re}\,X(f)}\right] \tag{E.41}$$

The inverse Fourier transform or the backward Fourier transform is defined as

$$x(t) = \int_{-\infty}^{\infty} X(f)e^{j2\pi ft}\,df \tag{E.42}$$

Inverse transformation allows determination of a function of time from its Fourier transform. Equations E.38 and E.42 are a Fourier transform pair and the relationship can be indicated by

$$x(t) \leftrightarrow X(f) \tag{E.43}$$

Example E.3

Consider a function defined as

$$\begin{aligned} x(t) &= \beta e^{-\alpha t} \quad t > 0 \\ &= 0 \quad t < 0 \end{aligned} \tag{E.44}$$

It is required to write its forward Fourier transform.
From Equation E.38,

$$\begin{aligned} X(f) &= \int_{0}^{\infty} \beta e^{-\alpha t} e^{-j2\pi ft}\,dt \\ &= \frac{-\beta}{\alpha + j2\pi f} e^{-(\alpha + j2\pi f)t}\Big|_0^{\infty} \\ &= \frac{\beta}{\alpha + j2\pi f} = \frac{\beta\alpha}{\alpha^2 + (2\pi f)^2} - j\frac{2\pi f\beta}{\alpha^2 + (2\pi f)^2} \end{aligned}$$

This is equal to

$$\frac{\beta}{\sqrt{\alpha^2 + (2\pi f)^2}} e^{j \tan^{-1} [-2\pi f/\alpha]} \tag{E.45}$$

Example E.4

Convert the function arrived at in Example E.3 to $x(t)$.

The inverse Fourier transform is

$$x(t) = \int_{-\infty}^{\infty} X(f) e^{j2\pi ft} df$$

$$= \int_{-\infty}^{\infty} \left[\frac{\beta\alpha}{\alpha^2 + (2\pi f)^2} - j \frac{2\pi f\beta}{\alpha^2 + (2\pi f)^2} \right] e^{j2\pi ft} df$$

$$= \int_{-\infty}^{\infty} \left[\frac{\beta\alpha \cos(2\pi ft)}{\alpha^2 + (2\pi f)^2} + \frac{2\pi f\beta \sin(2\pi ft)}{\alpha^2 + (2\pi f)^2} \right] df + j \int_{-\infty}^{\infty} \left[\frac{\beta\alpha \sin(2\pi ft)}{\alpha^2 + (2\pi f)^2} + \frac{2\pi f\beta \cos(2\pi ft)}{\alpha^2 + (2\pi f)^2} \right] df$$

The imaginary term is zero, as it is an odd function.

This can be written as

$$x(t) = \frac{\beta\alpha}{(2\pi)^2} \int_{-\infty}^{\infty} \frac{\cos(2\pi tf)}{(\alpha/2\pi)^2 + f^2} df + \frac{2\pi\beta}{(2\pi)^2} \int_{-\infty}^{\infty} \frac{f \sin(2\pi tf)}{(\alpha/2\pi)^2 + f^2} df$$

As

$$\int_{-\infty}^{\infty} \frac{\cos \alpha x}{b^2 + x^2} dx = \frac{\pi}{b} e^{-ab}$$

and

$$\int_{-\infty}^{\infty} \frac{x \sin ax}{b^2 + x^2} dx = \pi e^{-ab}$$

$x(t)$ becomes

$$x(t) = \frac{\beta\alpha}{(2\pi)^2} \left[\frac{\pi}{\alpha/2\pi} e^{-(2\pi t)(\alpha-\pi)} \right] + \frac{2\pi\beta}{(2\pi)^2} [\pi e^{-(2\pi t)(\alpha-\pi)}]$$

$$= \frac{\beta}{2} e^{-\alpha t} + \frac{\beta}{2} e^{-\alpha t} = \beta e^{-\alpha t} \quad t > 0$$

that is,

$$\beta e^{-\alpha t} t > 0 \leftrightarrow \frac{\beta}{\alpha + j2\pi f} \tag{E.46}$$

Example E.5

Consider a rectangular function defined by

$$\begin{aligned} x(t) &= K \quad |t| \leq T/2 \\ &= 0 \quad |t| > T/2 \end{aligned} \tag{E.47}$$

The Fourier transform is

$$X(f) = \int_{-T/2}^{T/2} Ke^{-j2\pi ft}\, dt = KT\left[\frac{\sin(\pi fT)}{\pi fT}\right] \tag{E.48}$$

This is shown in Figure E.5. The term in parentheses in Equation E.48 is called the *sinc function*. The function has zero value at points $f = n/T$.

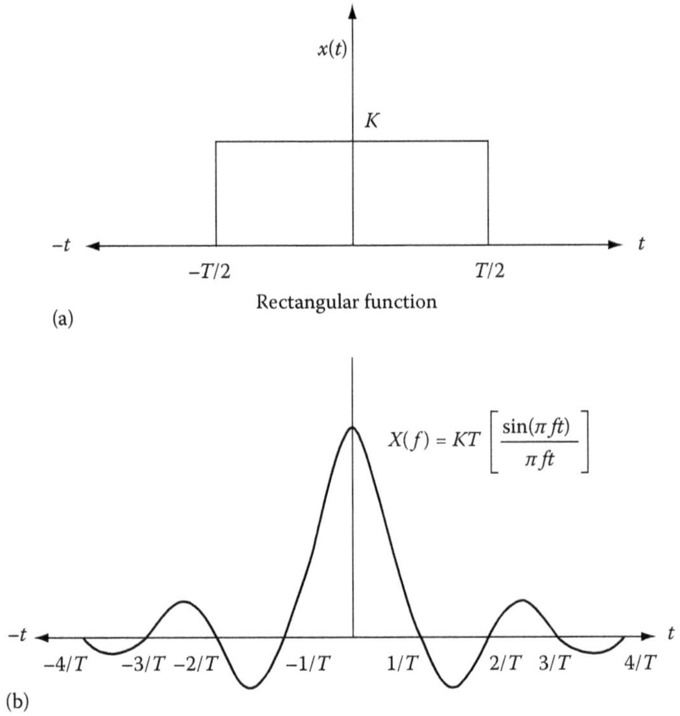

FIGURE E.5
(a) Bandwidth-limited rectangular function with even symmetry, amplitude *K*; (b) the sinc function, showing side lobes.

E.10 Sampled Waveform: Discrete Fourier Transform

The sampling theorem states that if the Fourier transform of a function $x(t)$ is zero for all frequencies greater than a certain frequency f_c, then the continuous function $x(t)$ can be uniquely determined by a knowledge of the sampled values. The constraint is that $x(t)$ is zero for frequencies greater than f_c, that is, the function is band limited at frequency f_c. The second constraint is that the sampling spacing must be chosen so that

$$T = \frac{1}{(2f_c)} \tag{E.49}$$

The frequency $1/T = 2f_c$ is known as the *Nyquist sampling rate.*

Aliasing means that the high-frequency components of a time function can impersonate a low frequency if the sampling rate is low. Figure E.6 shows a high frequency and a low frequency that share identical sampling points. Here, a high frequency is impersonating a low frequency for the same sampling points. The sampling rate must be high enough for the highest frequency to be sampled at least twice per cycle, $T = 1/(2f_c)$.

Often the functions are recorded as sampled data in the time domain, the sampling being done at a certain frequency. The Fourier transform is represented by the summation of discrete signals where each sample is multiplied by

$$e^{-j2\pi fnt_1} \tag{E.50}$$

that is,

$$X(f) = \sum_{n=-\infty}^{\infty} x(nt_1)e^{-j2\pi fnt_1} \tag{E.51}$$

Figure E.7 shows the sampled time domain function and the frequency spectrum for a discrete time domain function.

Where the frequency domain spectrum and the time domain function are sampled functions, the Fourier transform pair is made of discrete components:

$$X(f_k) = \frac{1}{N}\sum_{n=0}^{N-1} x(t_n)e^{-j2\pi kn/N} \tag{E.52}$$

$$X(t_n) = \sum_{k=0}^{N-1} X(f_k)e^{j2\pi kn/N} \tag{E.53}$$

Figure E.8 shows discrete time and frequency functions. *The discrete Fourier transform* approximates the continuous Fourier transform. However, errors can occur in the approximations involved. Consider a cosine function $x(t)$ and its continuous Fourier transform $X(f)$, which consists of two impulse functions that are symmetric about zero frequency (Figure E.9a).

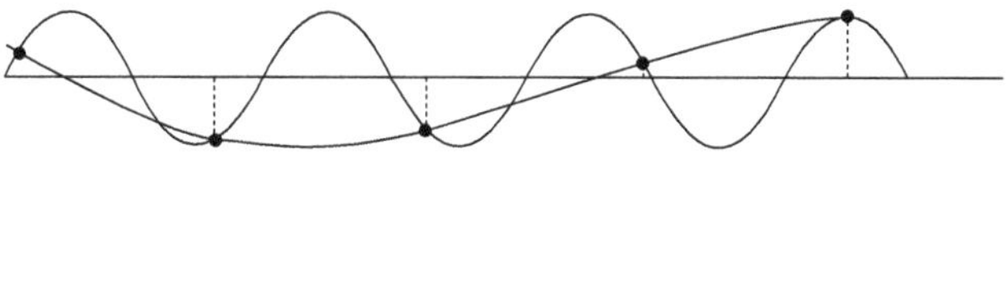

FIGURE E.6
High-frequency impersonating a low frequency—to illustrate aliasing.

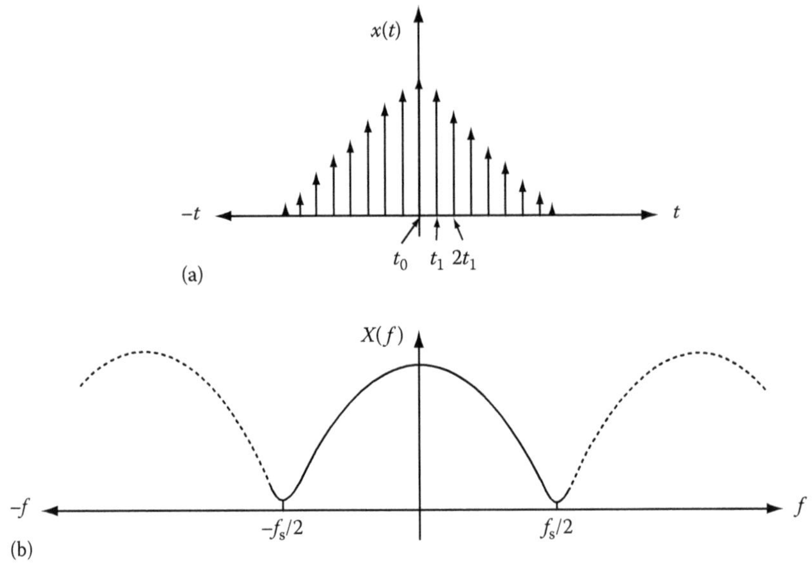

FIGURE E.7
(a) Sampled time domain function; (b) frequency spectrum for the discrete time domain function.

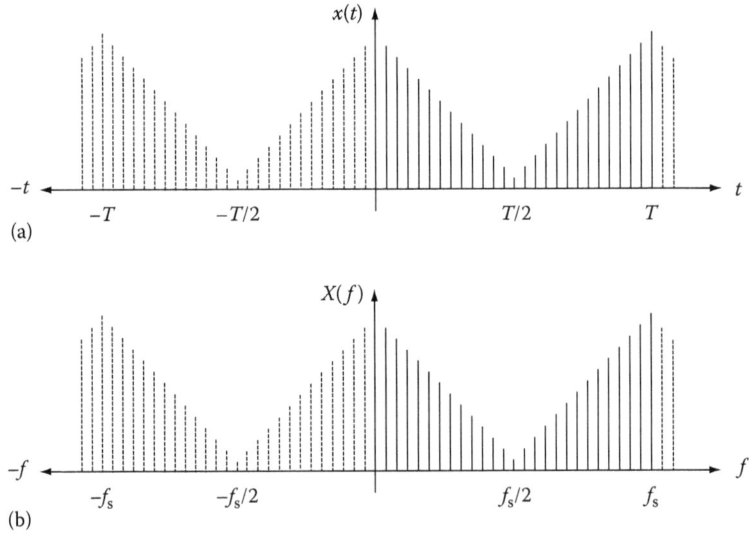

FIGURE E.8
Discrete time (a) and frequency (b) domain functions.

The finite portion of $x(t)$, which can be viewed through a unity amplitude window $w(t)$, and its Fourier transform $W(f)$, which has side lobes, are shown in Figure E.9b.

Figure E.9c shows that the corresponding convolution of two frequency signals results in blurring of $X(f)$ into two sine x/x-shaped pulses. Thus, the estimate of $X(f)$ is fairly corrupted.

The sampling of $x(t)$ is performed by multiplying with $c(t)$ (Figure E.9d); the resulting frequency domain function is shown in Figure E.9e.

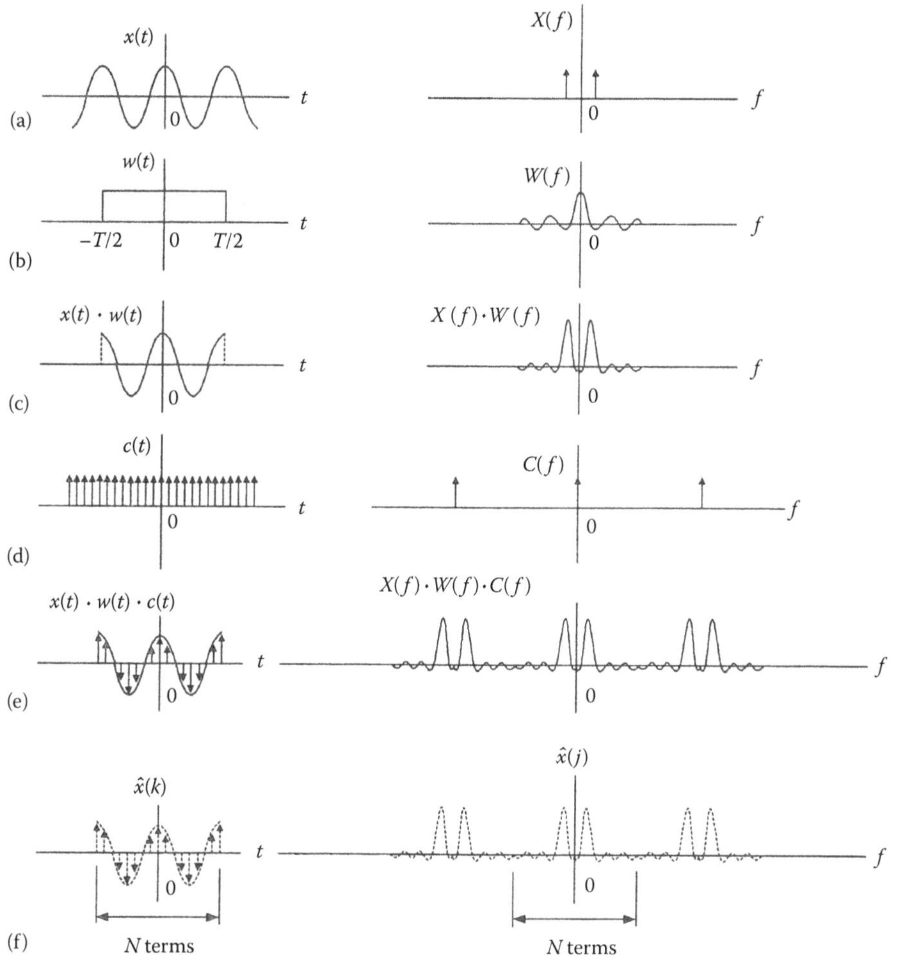

FIGURE E.9
Fourier coefficients of the discrete Fourier transform viewed as corrupted estimate of the continuous Fourier transform. (a) $x(t)$ and Fourier transform $X(f)$, (b) unit amplitude window $w(t)$ and $W(f)$, (c) convolution of $x(t)$ and $w(t)$, (d) discrete sampling function, (e) convolution $x(t)$, $w(t)$ and $c(t)$, (f) discrete band width limited function based upon (e). (From Bergland, G.D., *IEEE Spect.*, 7, 41, 1969.)

The continuous frequency domain function shown in Figure E.9e can be made discrete if the time function is treated as one period of a periodic function. This forces both the time domain and frequency domain functions to be infinite in extent, periodic, and discrete (Figure E.9f). *The discrete Fourier transform* is reversible mapping of N terms of the time function into N terms of the frequency function. Some problems are outlined in the following.

E.10.1 Leakage

Leakage is inherent in the Fourier analysis of any finite record of data. The function may not be localized on the frequency axis and has side lobes (Figure E.5). The objective is to localize the contribution of a given frequency by reducing the leakage through these side lobes. The usual approach is to apply a data window in the time domain, which has lower

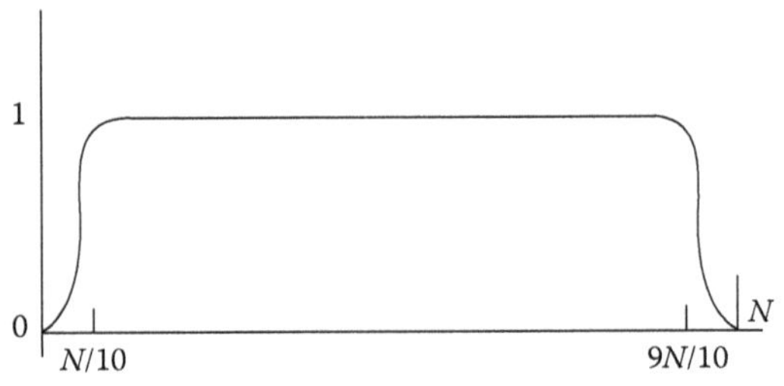

FIGURE E.10
An extended data window.

side lobes in the frequency domain, as compared to a rectangular data window. An extended cosine bell data window is shown in Figure E.10. A raised cosine wave is applied to the first and last 10% of the data and a weight of unity is applied for the middle 90% of the data. A number of other types of windows that give more rapidly decreasing side lobes have been described in the literature [1].

E.10.2 Picket-Fence Effect

The picket-fence effect can reduce the amplitude of the signal in the spectral windows, when the signal being analyzed falls in between the orthogonal frequencies, say, between the third and fourth harmonics. The signal will be experienced by both the third and fourth harmonic spectral windows, and in the worst case halfway between the computed harmonics. By analyzing the data with a set of samples that are identically zero, the fast Fourier transform (FFT) algorithm (Section E.11) can compute a set of coefficients with terms lying in between the original harmonics.

E.11 Fast Fourier Transform

The FFT is simply an algorithm that can compute the discrete Fourier transform more rapidly than any other available algorithm.

Define

$$W = e^{-j2\pi/N} \tag{E.54}$$

The frequency domain representation of the waveform is

$$X(f_k) = \frac{1}{N}\sum_{n=0}^{N=1} x(t_n)W^{kn} \tag{E.55}$$

The equation can be written in matrix form:

$$\begin{vmatrix} X(f_0) \\ X(f_1) \\ \cdot \\ X(f_k) \\ \cdot \\ X(f_{N-1}) \end{vmatrix} = \frac{1}{N}\begin{vmatrix} 1 & 1 & \cdot & 1 & \cdot & 1 \\ 1 & W & \cdot & W^k & \cdot & W^{N-1} \\ \cdot & \cdot & \cdot & \cdot & \cdot & \cdot \\ 1 & W^k & \cdot & W^{k2} & \cdot & W^{k(N-1)} \\ \cdot & \cdot & \cdot & \cdot & \cdot & \cdot \\ 1 & W^{N-1} & \cdot & W^{(N-1)k} & \cdot & W^{(N-1)^2} \end{vmatrix}\begin{vmatrix} x(t_0) \\ x(t_1) \\ \cdot \\ x(t_n) \\ \cdot \\ x(t_{N-1}) \end{vmatrix} \tag{E.56}$$

or in a condensed form:

$$[\overline{X}(f_k)] = \frac{1}{N}[\overline{W}^{kn}][\overline{x}(t_n)] \tag{E.57}$$

where $[\overline{X}(f_k)]$ is a vector representing N components of the function in the frequency domain, while $[\overline{x}(t_n)]$ is a vector representing N samples in the time domain. Calculation of N frequency components from N time samples therefore requires a total of $N \times TN$ multiplications.

For $N=4$:

$$\begin{vmatrix} X(0) \\ X(1) \\ X(2) \\ X(3) \end{vmatrix} = \begin{vmatrix} 1 & 1 & 1 & 1 \\ 1 & W^1 & W^2 & W^3 \\ 1 & W^2 & W^4 & W^6 \\ 1 & W^3 & W^6 & W^9 \end{vmatrix} \begin{vmatrix} x(0) \\ x(1) \\ x(2) \\ x(3) \end{vmatrix} \tag{E.58}$$

However, each element in matrix $[\overline{W}^{kn}]$ represents a unit vector with clockwise rotation of $2n/N$ ($n=0, 1, 2, \ldots, N-1$). Thus, for $N=4$ (i.e., four sample points), $2\pi/N=90°$. Thus,

$$W^0 = 1 \tag{E.59}$$

$$W^1 = \cos \pi/2 - j \sin \pi/2 = -j \tag{E.60}$$

$$W^2 = \cos \pi - j \sin \pi = -1 \tag{E.61}$$

$$W^3 = \cos 3\pi/2 - j \sin 3\pi/2 = j \tag{E.62}$$

$$W^4 = W^0 \tag{E.63}$$

$$W^6 = W^2 \tag{E.64}$$

Hence, the matrix can be written in the following form:

$$\begin{vmatrix} X(0) \\ X(1) \\ X(2) \\ X(3) \end{vmatrix} = \begin{vmatrix} 1 & 1 & 1 & 1 \\ 1 & W^1 & W^2 & W^3 \\ 1 & W^2 & W^0 & W^2 \\ 1 & W^3 & W^2 & W^1 \end{vmatrix} \begin{vmatrix} x(0) \\ x(1) \\ x(2) \\ x(3) \end{vmatrix} \tag{E.65}$$

This can be factorized into

$$\begin{vmatrix} X(0) \\ X(1) \\ X(2) \\ X(3) \end{vmatrix} = \begin{vmatrix} 1 & W^0 & 0 & 0 \\ 1 & W^2 & 0 & 0 \\ 0 & 0 & 1 & W^1 \\ 0 & 0 & 1 & W^3 \end{vmatrix} \begin{vmatrix} 1 & 0 & W^0 & 0 \\ 0 & 1 & 0 & W^0 \\ 1 & 0 & W^2 & 0 \\ 0 & 1 & 0 & W^2 \end{vmatrix} \begin{vmatrix} x(0) \\ x(1) \\ x(2) \\ x(3) \end{vmatrix} \tag{E.66}$$

Computation requires four complex multiplications and eight complex additions. Computation of Equation E.58 requires 16 complex multiplications and 12 complex additions. The computations are reduced.

While this forms an overview, an interested reader may like to probe further. Reference [1] lists 62 further references.

Reference

1. G.D. Bergland. A guided tour of the fast Fourier transform. *IEEE Spect.* 7: 41–52, 1969.

Appendix F: Limitation of Harmonics

F.1 Harmonic Current Limits

The limits of harmonic indices, current, and voltage are defined at the point of common coupling (PCC) [1]. This can be a point of metering or the point of connection of the consumer apparatus with the utility supply company, the point of interference. Within an industrial plant, *PCC is the point between the nonlinear load and other load*. Thus, if the nonlinear loads are dispersed throughout the distribution system, then PCCs are all the buses to which these loads are connected.

HVDC systems and SVCs owned and operated by the utility are excluded from the definition of the PCC. Harmonic measurements are recommended at PCC. It is assumed that the system is characterized by the short-circuit impedance and that the effect of capacitors is neglected. The recommended current distortion limits are concerned with total demand distortion (TDD), which is defined as the total root-sum square harmonic current distortion in a percentage of maximum demand load current (15 or 30 min demand).

The limits for the current distortion that a consumer must adhere to are shown in Tables F.1 and F.2 [1]. The ratio I_{SC}/I_L is the ratio of the short-circuit current available at the PCC to the maximum fundamental frequency current and is calculated on the basis of the average maximum demand for the preceding 12 months. As the size of user load decreases with respect to the size of the system, the percentage of the harmonic current that the user is allowed to inject into the utility system increases.

Tables F.1 and F.2 are applicable to six-pulse rectifiers. For higher pulse numbers, the limits of the characteristic harmonics are increased by a factor of $\sqrt{P/6}$ *provided* that the amplitudes of noncharacteristic harmonics are less than 25% of the limits specified in the tables.

An overriding article is provided that the transformer connecting the user to the utility system should not be subjected to harmonic currents in excess of 5% of the transformer-rated current. Where this requirement is not met, a higher rated transformer should be considered.

The injected harmonic currents may create resonance with the utility system, and a consumer must insure that harmful series and parallel resonances are not occurring. The utility source impedance may have a number of resonant frequencies (Figure 19.6) and it is necessary to model the utility system in greater detail for harmonic flow calculations.

The limit of harmonic current injection does not limit a user's choice of converters or selection of harmonic-producing equipment technology, nor does it lay down limits on the harmonic emission from equipment. It is left to the user how he would adhere to the limits of harmonic current injection, whether by choice of an alternative technology, use of passive or active filters, or any other harmonic mitigating device. This is in contrast to the IEC, which has laid out the maximum emission limits from the equipment [2].

TABLE F.1

Current Distortion Limits for General Distribution Systems (120 V–69 kV)

	Maximum Harmonic Current Distortion in Percentage of Fundamental					
	Harmonic Order (Odd Harmonics)[b]					
I_{sc}/I_L[a]	<11	$11 \leq h < 17$	$17 \leq h < 23$	$23 \leq h < 35$	$35 \leq h$	TDD
<20[c]	4.0	2.0	1.5	0.6	0.3	5.0
20–50	7.0	3.5	2.5	1.0	0.5	8.0
50–100	10.0	4.5	4.0	1.5	0.7	12.0
100–1000	12.0	5.5	5.0	2.0	1.0	15.0
>1000	15.0	7.0	6.0	2.5	1.4	20.0

Source: IEEE, Standard 519, IEEE recommended practice and requirements for harmonic control in electrical systems, 1992. Copyright 1992 IEEE. All rights reserved.

Notes: For PCCs from >69 to 161 kV, the limits are 50% of the limits above. Current distortions that occur in a dc offset, for example, half-wave converters, are not allowed.

[a] I_{sc}, maximum short-circuit current at PCC; I_L, maximum load current (fundamental frequency) at PCC.

[b] Even harmonics are limited to 25% of the odd harmonic limits above.

[c] All power generation equipment is limited to these values of current distortion regardless of I_{sc}/I_L.

TABLE F.2

Current Distortion Limits for General Transmission Systems (>161 kV), Dispersed Generation and Cogeneration

	Harmonic Order (Odd Harmonics)[b]					
I_{sc}/I_L[a]	<11	$11 \leq h < 17$	$17 \leq h < 23$	$23 \leq h < 35$	$35 \leq h$	THD
<50[c]	2.0	1.0	0.75	0.3	0.15	2.5
>50	3.0	1.5	1.15	0.45	0.22	3.75

Source: IEEE, IEEE recommended practice and requirements for harmonic control in electrical systems, 1992, Standard 519. Copyright 1992 IEEE. All rights reserved.

Notes: For PCCs from >69 to 161 kV, the limits are 50% of the limits above. Current distortions that occur in a dc offset, for example, half-wave converters, are not allowed.

[a] I_{sc}, maximum short-circuit current at PCC; I_L, maximum load current (fundamental frequency) at PCC.

[b] Even harmonics are limited to 25% of the odd harmonic limits above.

[c] All power generation equipment is limited to these values of current distortion regardless of I_{sc}/I_L.

The IEC standard [2] involves limits on harmonic current injected into a public distribution network by nonlinear appliances with an input current less than or equal to 16 A. This classifies such appliances into four classes (A–D). For each class, harmonic current emission limits are established up to the 39th harmonic. The classes are

Class A: Three-phase appliances

Class B: Portable appliances

Class C: Lighting appliances, including dimmer systems

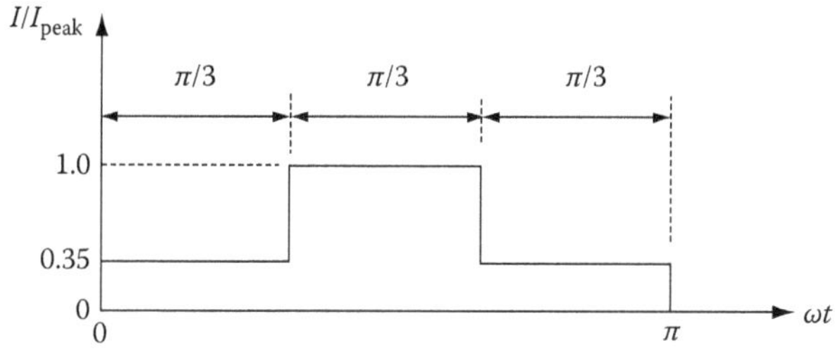

FIGURE F.1
Line current block for harmonic emission control. (From IEC, Electromagnetic compatibility—Part 3: Limits—Section 2: Limits for harmonic current emission (equipment input current #16A per phase), 1995, Standard 61000-3-2.)

TABLE F.3

Harmonic Emission Limits for Class D Appliances

Harmonic Order	Maximum Admissible Harmonic Current (mA/W)	Maximum Admissible Harmonic Current (A)
3	3.4	2.3
5	1.9	1.14
7	1.0	0.77
9	0.5	0.4
11	0.35	0.33
$13 \leq n \leq 39$	$3.85/n$	$0.15\ (15/n)$

Source: IEC, Electromagnetic compatibility—Part 3: Limits—Section 2: Limits for harmonic current emission (equipment input current #16A per phase), 1995, Standard 61000-3-2.

Class D: Appliances with input current with an assigned special wave shape and an active power input $P < 600$ W measured according to the method illustrated in the standard. This special wave shape is shown in Figure F.1 and the harmonic emission limits are presented in Table F.3.

The standard lays down methods for testing individual appliance harmonic emissions in the regulated frequency range. We have seen that the harmonic injection into a system is also a function of the system impedance. Unlike the IEEE [1], IEC limits apply to all systems irrespective of their stiffness.

F.2 Voltage Quality

While a user can inject only a certain amount of harmonic current into the utility system as discussed in Section F.1, the utilities and power producer must meet a certain voltage quality to the consumers. The recommended voltage distortion limits are given in Table F.4. The index used is THD (total harmonic voltage distortion) as a percentage of nominal fundamental frequency voltage. The limits are system design values for *worst case* for normal operation, conditions lasting for more than 1 h. For shorter periods, during startups or unusual conditions, the limits can be exceeded by 50%. If the limits are exceeded, harmonic mitigation through the use of filters or stiffening of the system through parallel feeders is recommended.

TABLE F.4

Harmonic Voltage Limits for Power Producers (Public Utilities or Cogenerators)

	Harmonic Distortion in Percentage at PCC		
	<69 kV	>69–161 kV	>161 kV
Maximum for individual harmonic	3.0	1.5	1.0
THD%	5.0	2.5	1.5

Source: IEEE, IEEE recommended practice and requirements for harmonic control in electrical systems, 1992, Standard 519. Copyright 1992 IEEE. All rights reserved.

Note: High-voltage systems can have up to 2.0% THD where the cause is an HVDC terminal that will attenuate by the time it is tapped for a user.

Example F.1

The harmonic current and voltage spectrum of a 12-pulse LCI inverter, operating at $\alpha = 15°$, are presented in first two columns of Table F.5. The demand current is 1200 A, which is also the inverter maximum operating current. The available short-circuit current at the PCC is 14 kA. Calculate the distortion limits. Are these exceeded?

Table F.5 is extended to show permissible limits of the TDD at each of the harmonics, and individual and total permissible current distortions are calculated from Table F.1 for $I_{SC}/I_L < 20$ (actual $I_{SC}/I_L = 11.67$). The noncharacteristic harmonics are reduced to 25% and the characteristic harmonics (Equation 17.43) are multiplied by a factor of the square root of p over 6, that is, $\sqrt{2}$.

Table F.5 shows that current distortion limits on a number of harmonics are exceeded. Also, the total THD is greater than permissible limits. The total THD voltage is below 5% limit, but the

TABLE F.5

Example F.1[a]

h	I_h (A)	Harmonic Distortion	IEEE Limits	V_h (V)
5	24.32	2.027	1.0	20.86
7	13.43	1.119	1.0	16.13
11	65.42	5.452	2.82	123.46
13	39.69	3.331	2.82	88.52
17	1.87	0.156	0.375	5.45
19	0.97	0.081	0.375	3.16
23	8.45	0.704	0.846	33.34
25	8.54	0.711	0.846	36.62
29	0.98	0.081	0.15	4.87
31	0.76	0.063	0.15	4.04
35	5.02	0.418	0.423	30.14
37	3.27	0.273	0.423	20.75
41	0.23	0.019	0.075	1.62
43	0.23	0.019	0.075	1.70
47	3.45	0.287	0.423	27.81
49	2.58	0.215	0.423	21.69
Total distortion		6.887%	5%	4.08%

[a] 12-pulse harmonic content, 15° firing angle, 4.16 kV bus; load current = base current = 1200 A; short-circuit current = 14 kA.

distortion at the 11th harmonic is 3.32%, which exceeds the maximum permissible limit of 3% on an individual harmonic.

In this example, the distortion limits exceed the permissible limits, though a 12-pulse converter is used, and the base load equals the converter load, that is, all the load is nonlinear. Generally, the harmonic-producing loads will be some percentage of the total load demand and this will reduce TDD as it is calculated on the basis of total fundamental frequency demand current. Also, note that if the short-circuit level of the system changes so does the harmonic distribution (Chapter 20).

F.3 Commutation Notches

Commutation notches are shown in Figure F.2a. For low-voltage systems, the notch depth, the total notch area of the line-to-line voltage at the PCC, and THD should be limited as shown in Table F.6. The notch area is given by

$$A_N = V_N t_N \tag{F.1}$$

where

A_N is the notch area in volt-microseconds
V_N is the depth of the notch in volts, line-to-line, (L-L) of the deeper notch in the group
t_N is the width of the notch in microseconds

Consider the equivalent circuit of Figure F.2b and let us define the following inductances:

L_t is the inductance of the drive transformer
L_L is the reactance of the feeder line
L_s is the inductance of the source

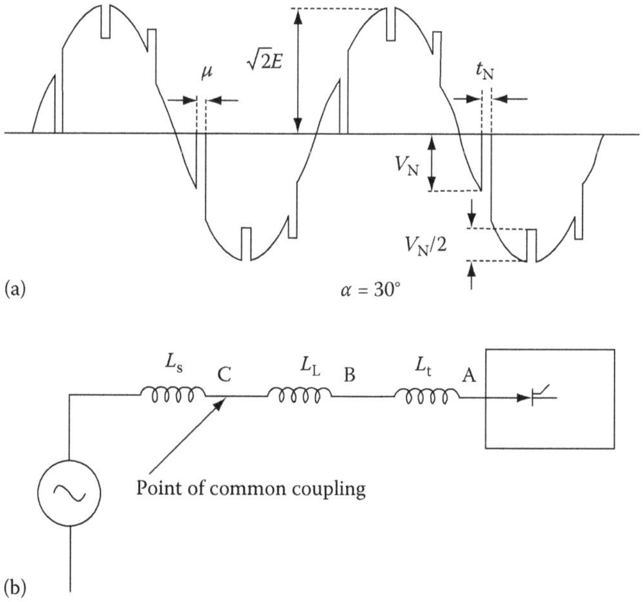

FIGURE F.2
(a) Voltage notches; (b) circuit diagram for calculation of notch area.

TABLE F.6

Low-Voltage System Classification and Distortion Limits

	Special Applications	General Systems	Dedicated Systems
Notch depth	10%	20%	50%
THD (voltage)	3%	5%	10%
Notch area (A_n)	16,400	22,800	36,500

Source: IEEE, IEEE recommended practice and requirements for harmonic control in electrical systems, 1992, Standard 519.

Notes: The notch area for voltages other than 480 V is given by multiplication by a factor $V/480$. Notch area is given in volts-microseconds at rated voltage and current. Special applications include hospitals and airports. A dedicated system is exclusively dedicated to the converter load.

The notch depth depends where we look into the system. The primary voltage goes to zero at the converter terminals, point A in Figure F.2b, and the depth of the notch is maximum. At B, the depth of the notch in per unit of the notch depth at A is given by

$$\frac{L_s + L_L}{L_s + L_t + L_L} \tag{F.2}$$

If point C in Figure F.2b is defined as the PCC, then the depth of the notch at the PCC in per unit with respect to notch depth at the converter is

$$V_N = \frac{L_s}{L_L + L_t + L_s} \tag{F.3}$$

The impedances in the converter circuit are acting a sort of potential divider. The width of the notch is given by the commutation angle μ and by the following expression:

$$t_N = \frac{2(L_L + L_t + L_s)I_d}{e} \tag{F.4}$$

where e is the instantaneous voltage (L-L) prior to the notch. The area of the notch at the converter terminals is given by

$$A_N = 2I_d(L_L + L_t + L_s) \tag{F.5}$$

A relationship between line notching and distortion factor is given by

$$V_h = \left[\sum_{h=5}^{h=\infty} V_h^2\right]^{1/2} = \left[2V_N^2 t_N + 4(V_N/2)^2 t_N\right]^{1/2} = \sqrt{3V_N^2 t_N f} \tag{F.6}$$

In Equation F.6, the two deeper and four less deep notches per cycle (Figure F.2a) are considered.

Example F.2

Consider a system configuration as shown in Figure F.3. It is required to calculate the notch depth and notch area at buses A, B, and C.

The inductances throughout the system are calculated and are shown in Figure F.3. The source inductance (reflected at 480 V), based on the given short-circuit data, is 0.64 μH. It is a stiff system at 13.8 kV, and the source reactance is small.

A 1-MVA transformer reactance referred to the 480 V side, based on the X/R ratio, $X_t = 5.638\% = 0.1299$ ohm. This gives an inductance of 34.4 μH at 480 V. Similarly, for a 1.5 MVA transformer, the inductance is 23.2 μH. The feeder inductance is given or it can be calculated, based on given cable/bus duct data. The notch depth at bus C as a percentage of depth at the converter = (34.4 + 0.64)/(34.4 + 0.64 + 30) = 54%. Referring to Table F.6, this exceeds the limits even for a dedicated system.

Consider that the converter is supplying a motor load of 500 hp at 460 V and the dc current is continuous and equal to 735 A. The notch area at the converter ac terminals is then given by

$$
\begin{aligned}
2I_d(L_s + L_{T1} + L_{F1}) &= 2 \times 735(0.64 + 34.3 + 30) \\
&= 95609\,\text{V-}\mu\text{s}
\end{aligned}
$$

Therefore, the notch area at bus C = 0.54 × 95,609 = 51,628 V-μs. From Table F.6, the limits for dedicated systems and general systems are 36,500 and 22,800, respectively. The calculated value is much above these limits.

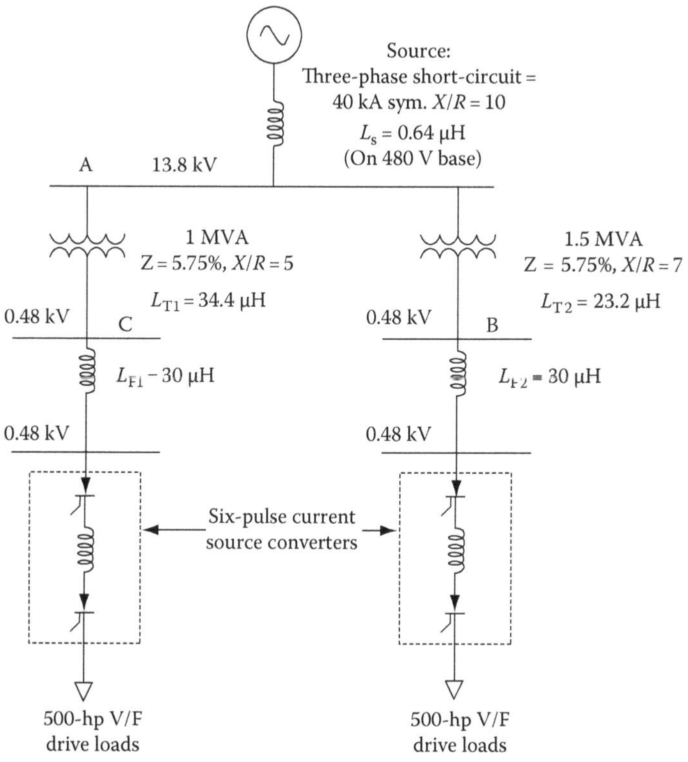

FIGURE F.3
System configuration with drive loads for calculation of notch area (Example F.2).

The notch area at bus A due to the converter at bus C is

$$\frac{0.64}{0.64 + 34.4 + 30}(95{,}609) = 940.8\,\text{V-}\mu\text{s}$$

In a likewise calculation, the notch depth at bus B = 44.28%, area = 35061 V-μs. The notch area at bus A due to the converter load at bus B = 940.8 V-μs.

The total notch area at bus A can be considered as a sum of the notch areas due to converter loads at buses B and C. This assumes that the notch widths due to commutation are additive, which is a very conservative assumption.

This example suggests that problems of notching can be mitigated by the following:

- Decreasing the source impedance behind the PCC bus.
- Increasing the impedance ahead of (on the load side) the PCC bus. Isolation transformers and line reactors are commonly used and serve the same purpose.

F.4 Interharmonics

We defined interharmonics in Section 17.24 and briefly discussed the type of nonlinear loads, which may give rise to interharmonics in power systems. The interharmonic frequency components greater than the power frequency cause heating effects that are similar to those caused by harmonics. The impact on light flicker (Section F.5) is important. Modulation of power system voltage with interharmonic voltage introduces variations in system voltage rms value. The IEC flicker meter is used to measure the light flicker indirectly by simulating the response of an incandescent lamp and human eye–brain response to visual stimuli [3]. The other impacts of concern are as follows:

- Excitation of subsynchronous conditions in turbo-generator shafts (see also Section 18.1.2).
- Interharmonic voltage distortions similar to other harmonics.
- Interference with low frequency power line carrier control signals.
- Overloading of conventional tuned filters. See Section 20.13 for the limitations of the passive filters, the tuning frequencies, the displaced frequencies, and possibility of a resonance with the series-tuned frequency. As the interharmonics vary with the operating frequency of a cycloconverter, a resonance can be brought where none existed before making the design of single tuned filters impractical.

IEC 61000-4-15 [4] has established a method of measurement of harmonics and interharmonics utilizing a 10- or 12-cycle window for 50 and 60 Hz systems. This results in a spectrum with 5 Hz resolution. The 5 Hz bins are combined to produce various groupings and components for which limits and guidelines can be referenced. The IEC limits the interharmonic voltage distortion to 0.2% for the frequency range from DC to 2 kHz.

There are yet no limits of interharmonics assigned in IEEE 519-1992 [1]. The following recommendations are from Ref. [5] and may be incorporated in this standard:

- Limit of 0.2% for frequencies less than 140 Hz to address flicker of incandescent lamps and fluorescent lamps.
- Limit individual interharmonic component distortion to less than 1% above 140 Hz up to some frequency, yet to be determined, to protect low frequency PLC (power line carrier), address sensitivity to light flicker within 8 Hz of harmonic frequencies, and account for resonances created by harmonic filters.
- For higher frequencies, limit interharmonic voltage component and total distortion to some percentage related to proposed frequency-dependent harmonic voltage limits to protect high frequency PLC and filter resonances. Alternatively, define a linear limit curve with increasing slope, to recognize the reduced impact on light flicker with increasing frequency.

F.5 Flicker

Voltage flicker occurs due to operation of rapidly varying loads, such as arc furnaces, which affect the system voltage. This can cause annoyance by causing visible light flicker on tungsten filament lamps. The human eye is most sensitive to light variations in the frequency range of 5–10 Hz, and voltage variations of less than 0.5% in this frequency can cause annoying flicker on tungsten lamps. The percentage pulsation of voltage related to frequency at which it is most perceptible, from various references, is included in Figure 10.3 of Ref. [1]. Though this figure has been in use for a long time, it was superseded in IEEE Std. 1453 [6]. The solid-state compensators and loads may produce modulation of the voltage magnitude that is more complex than what was envisaged in the original flicker curves. This standard adopts IEC standard 61000-4-15 [4] in total. Define

$$\mathrm{Plt} = \sqrt[3]{\frac{1}{12} \times \sum_{j=1}^{12} \mathrm{Pst}_j^3} \tag{F.7}$$

where Plt is a measure of long-term perception of flicker obtained for a 2 h period. This is made up of 12 consecutive Pst values, where Pst is a measure of short-term perception of flicker for 10 min interval. This value is the standard output of IEC flicker meter. Further qualification is that IEC flicker meter is suited to events that occur once per hour or more often. The curves in Figure 10.3 of Ref. [1] are still useful for infrequent events like a motor start, once per day or even as frequent as some residential air conditioning equipment. Figure F.4 depicts comparison of IEEE and IEC for flicker irritation.

For acceptance of flicker-causing loads to utility systems, IEC standards [3,4,7] are recommended. The application of shape factors allows the effect of loads with voltage fluctuations other than the rectangular to be evaluated in terms of Pst values. Further research is needed in issues related to the effect of interharmonics on flicker and flicker transfer coefficients from HV to LV electrical power systems [8].

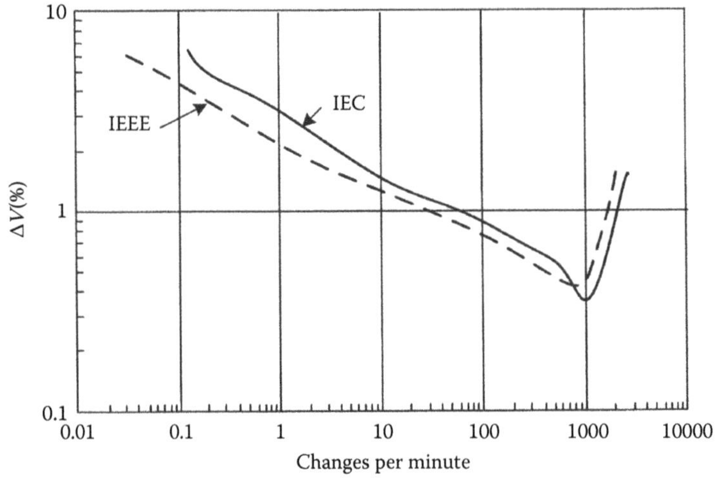

FIGURE F.4
Comparison of IEC and IEEE standards with respect to flicker tolerance. (From IEEE, Recommended practice for measurement and limits of voltage fluctuations and associated light flicker on AC power systems, Standard 1453, 2004.)

Two levels, planning level and compatibility level, are defined. Compatibility level is the specified disturbance level in a specified environment for coordination in setting the emission and immunity limits. Planning level, in a particular environment, is adopted as a reference value for limits to be set for the emissions from large loads and installations, in order to coordinate those limits with all the limits adopted for equipment intended to be connected to the power supply system.

As an example, planning levels for Pst and Plt in MV (voltages >1 kV and <35 kV), HV (voltages >35 kV and <230 kV), and EHV (voltages >230 kV) are shown in Table F.7; and compatibility levels for LV and MV power systems are shown in Table F.8.

TABLE F.7
Planning Levels for Pst and Plt in MV, HV, and EHV Power Systems

	Planning Levels	
Pst or Plt	**MV**	**HV–EHV**
Pst	0.9	0.8
Plt	0.7	0.6

TABLE F.8
Compatibility Levels for Pst and Plt in LV and MV Systems

Pst or Plt	**Compatibility Level**
Pst	1.0
Plt	0.8

Arc furnaces cause flicker because the current drawn during melting and refining periods is erratic and fluctuates widely and the power factor is low. There are other loads that can generate flicker, for example, large spot welding machines often operate close to the flicker perception limits. Industrial processes may comprise a number of motors having rapidly varying loads or starting at regular intervals, and even domestic appliances such as cookers and washing machines can cause flicker on weak systems. However, the harshest load for flicker is an arc furnace. During the melting cycle of a furnace, the reactive power demand is high. Figure 17.26 gives typical performance curves of an arc furnace, and Figure 17.25 shows that an arc furnace current is random and no periodicity can be assigned (thus Fourier analysis cannot be used), yet some harmonic spectrums have been established and Table 17.6 shows typical harmonics during melting and refining stage. Note that even harmonics are produced during melting stage. The high reactive power demand and poor power factor cause cyclic voltage drops in the supply system. Reactive power flow in an inductive element requires voltage differential between sending end and receiving ends and there is reactive power loss in the element itself. When the reactive power demand is erratic, it causes corresponding swings in the voltage dips, much depending upon the stiffness of the system behind the application of the erratic load. This voltage drop is proportional to the short-circuit MVA of the supply system and the arc furnace load.

For a furnace installation, the short-circuit voltage depression (SCVD) is defined as

$$\mathrm{SCVD} = \frac{2\mathrm{MW}_{\mathrm{furnace}}}{\mathrm{MVA}_{\mathrm{SC}}} \tag{F.8}$$

where

MW is the installed load of the furnace in $\mathrm{MW}_{\mathrm{furnace}}$

$\mathrm{MVA}_{\mathrm{SC}}$ is the short-circuit level of the utility supply system

This gives an idea whether potential problems with flicker can be expected. An SCVD of 0.02–0.025 may be in the acceptable zone between 0.03 and 0.035 in the borderline zone, and above 0.035 objectionable [9]. When there are multiple furnaces, these can be grouped into one equivalent MW. Example 20.2 describes the use of tuned filters to compensate for the reactive power requirements of an arc furnace installation. The worst flicker occurs during the first 5–10 min of each heating cycle and decreases as the ratio of the solid-to-liquid metal decreases.

The significance of $\Delta V/V$ and a number of voltage changes are illustrated with reference to Figure F.5 [4]. This shows a 50 Hz waveform, having a 1.0 average voltage with a relative voltage change $\Delta v/\bar{v} = 40\%$ and with 8.8 Hz rectangular modulation. It can be written as

$$v(t) = 1 \times \sin(2\pi \times 50t) \times \left\{1 + \frac{40}{100} \times \frac{1}{2} \times \mathrm{signum}[2\pi \times 8.8 \times t]\right\} \tag{F.9}$$

Each full period produces two distinct changes: one with increasing magnitude and one with decreasing magnitude. Two changes per period with a frequency of 8.8 Hz give rise to 17.6 changes per second.

F.5.1 Control of Flicker

The response of the passive compensating devices is slow. When it is essential to compensate load fluctuations within a few milliseconds, SVCs are required. Referring to Figure 13.12,

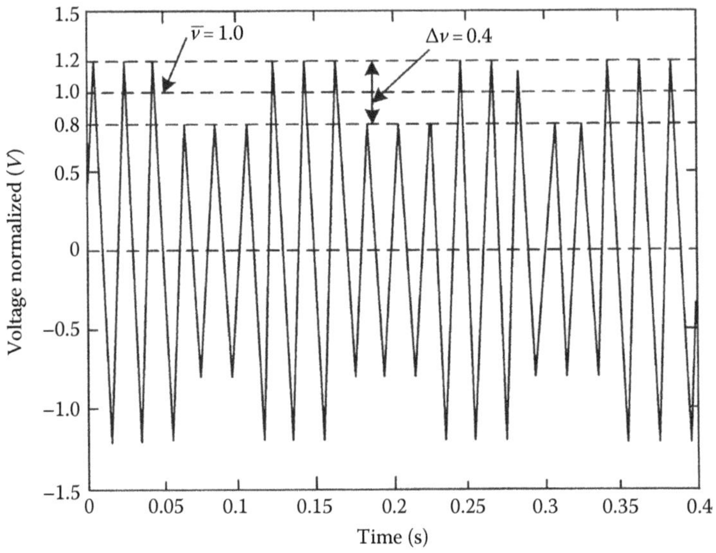

FIGURE F.5
Modulation with rectangular voltage change $\Delta V/V = 40\%$, 8.8 Hz. 17.6 changes per second. (From IEC, Electromagnetic compatibility (EMC)—Part 4: Testing and measurement techniques (Section 15: Flicker meter—Functional and design specifications), 2003.)

large TCR flicker compensators of 200 MW have been installed for arc furnace installations. Closed-loop control is necessary due to the randomness of load variations and complex circuitry is required to achieve response times of less than one cycle. Significant harmonic distortion may be generated, and harmonic filters will be required. TSCs have also been installed and these have inherently one cycle delay as the capacitors can only be switched when their terminal voltage matches the system voltage. Thus, the response time is slower. SVCs employing TSCs do not generate harmonics, but the resonance with the system and transformer needs to be checked.

We discussed STATCOM in Chapter 13. It has been long recognized that reactive power can be generated without the use of bulk capacitors and reactors and STATCOM makes it possible. As discussed in Chapter 13, it is capable of operating with leading or lagging power factors. With the design of high bandwidth control capability, STATCOM can be used to force three-phase currents of arbitrary waveshape through the line reactance. This means that it can be made to supply non-sinusoidal, unbalanced, randomly fluctuating currents demanded by the arc furnace. With a suitable choice of dc capacitor, it can also supply the fluctuating real power requirements, which cannot be achieved with SVCs. Harmonics are not of concern (Figure 17.36).

The instantaneous reactive power on the source side is the reactive power circulating between the electrical system and the device, while reactive power on the output side is the instantaneous reactive power between the device and its load. There is no relation between the instantaneous reactive powers on the load and source side, and the instantaneous imaginary power on the input is not equal to the instantaneous reactive power on the output (Chapter 15). The STATCOM for furnace compensation may use vector control based on the concepts of instantaneous active and reactive power, i_α and i_β (Chapter 20).

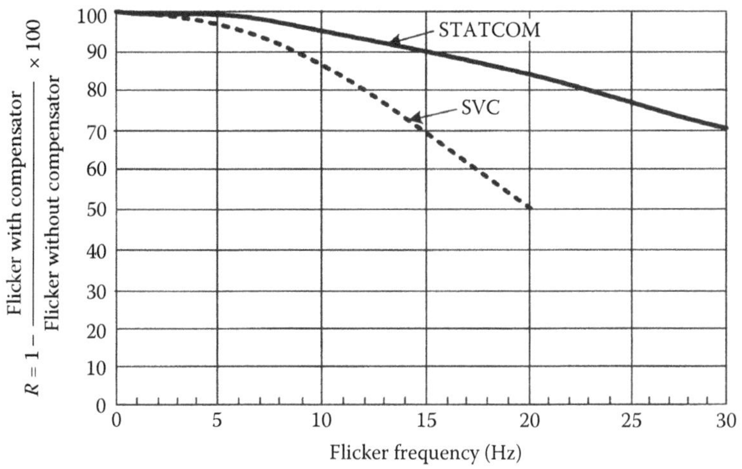

FIGURE F.6
Flicker factor R with STATCON and SVC.

Figure F.6 shows flicker reduction factor as a function of flicker frequency STATCOM versus SVC [10]. Flicker mitigation with a fixed reactive power compensator and an active compensator—a hybrid solution for welding processes—is described in [11].

References

1. IEEE. IEEE recommended practice and requirements for harmonic control in electrical systems, 1992. Standard 519.
2. IEC. Electromagnetic compatibility. Part 3: Limits (Section 2: Limits for harmonic current emission (equipment input current #16 A per phase)), 1995. Standard 61000-3-2.
3. IEC. Electromagnetic compatibility. Part 3: Limits (Section 3: Limitations of voltage fluctuations and flicker in low-voltage supply systems for equipment with rated current ≤16 A per phase and not subjected to conditional connection, Geneva), 2008. Standard 61000-3-3.
4. IEC. Electromagnetic compatibility (EMC). Part 4: Testing and measurement techniques (Section 15: Flicker meter—Functional and design specifications), 2003.
5. IEEE. IEEE interharmonics task force working document IH0101, 2000.
6. IEEE. Recommended practice for measurement and limits of voltage fluctuations and associated light flicker on AC power systems. Standard 1453, 2004.
7. IEC. Electromagnetic compatibility (EMC). Part 3-11: Limits: Limitation of voltage fluctuations and flicker in low-voltage supply systems for equipment with rated current <=75 A, Geneva, 2000. Standard 61000-3-11.
8. S.M. Halpin and V. Singhvi. Limits for interharmonics in the 1–100 Hz range based upon lamp flicker considerations. *IEEE Trans. Power Deliv*. 22(1): 270–276, January 2007.
9. S.R. Mendis, M.T. Bishop, and J.F. Witte. Investigation of voltage flicker in electric arc furnace power systems. *IEEE Ind. Mag*. 2: 28–34, 1996.
10. C.D. Schauder and L. Gyugyi, STATCOM for electric arc furnace compensation. EPRI Workshop, Palo Alto, CA, 1995.
11. M. Routimo, A. Makinen, M. Salo, R. Seesvuori, J. Kiviranta, and H. Tuusa. Flicker mitigation with hybrid compensator. *IEEE Trans. Ind. Appl*. 44(4): 1227–1238, July/August 2008.

Appendix G: Estimating Line Harmonics

The harmonic generation is a function of the topology of the harmonic-producing equipment (Chapter 17). Six-pulse current source converters are most commonly applied, and the harmonic estimation from these has been discussed in Chapter 17 and is visited again. Figure G.1 shows the line current waveforms. The theoretical or textbook waveform is rectangular and considers instantaneous commutation (Figure G.1a). The effect of commutation delay and firing angle still retains the flat-top assumption (Figure G.1b). The dc current is not flat topped and the actual waveform has a ripple (Figure G.1c). For lower values of the dc reactor and large phase-control angles, the current is discontinuous (Figure G.1d).

G.1 Waveform without Ripple Content

An estimation of the harmonics ignoring waveform ripple is provided by Equations 17.45 through 17.49, which are reproduced for ease of reference as follows:

$$I_h = I_d\sqrt{\frac{6}{\pi}}\sqrt{\frac{A^2 + B^2 - 2AB\cos(2\alpha + \mu)}{h[\cos\alpha - \cos(\alpha + \mu)}} \tag{G.1}$$

where

$$A = \frac{\sin\left[(h-1)\frac{\mu}{2}\right]}{h-1} \tag{G.2}$$

and

$$B = \frac{\sin\left[(h+1)\frac{\mu}{2}\right]}{h+1} \tag{G.3}$$

The angle μ is given by

$$\mu = \cos^{-1}[\cos\alpha - (X_S + X_t)I_d] - \alpha \tag{G.4}$$

We also defined X_s and X_t as the system and transformer reactance in per unit on a converter base and I_d as the dc current in per unit on a converter base. The calculation is illustrated with an example:

Consider a 2 MVA delta–wye, 13.8–0.48 kV transformer, impedance 5.75%, and a supply system short-circuit level of 500 MVA. The six-pulse converter operates with $\alpha = 45°$.

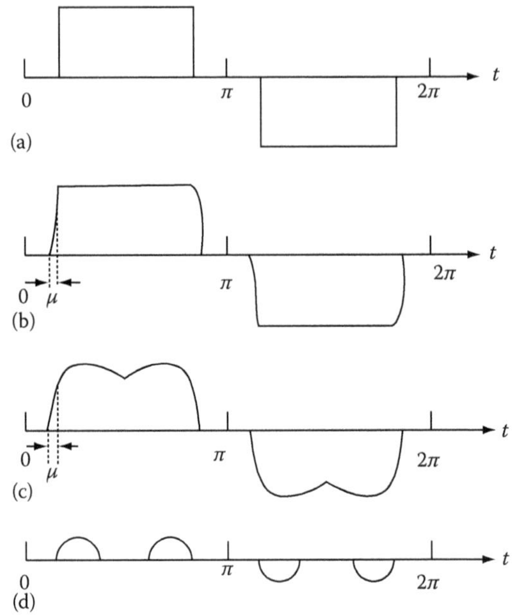

FIGURE G.1
(a) Rectangular current waveform; (b) waveform with commutation angle; (c) waveform with ripple content; (d) discontinuous waveform due to large delay angle control.

Considering a converter base equal to the rated kVA of the transformer and converting the transformer and source impedances to a converter base, we have $X_t = 0.0575$ and $X_s = 0.004$ in per unit. The resistance is ignored. From Equation G.4,

$$\mu = \cos^{-1}[\cos 45° - (0.0575 + 0.004)] - 45° = 4.8°$$

We will calculate the magnitude of the fifth harmonic. From Equation G.2, $A = 0.0417$, and from Equation G.3, $B = 0.0414$. Substituting these values into Equation G.1, we have a fifth harmonic current equal to 27.47% of the dc current.

For large gating angles and relatively small commutation angles, the current can be assumed to be trapezoidal. For this waveform,

$$\frac{I_h}{I_d} = 2\frac{\sqrt{2}}{\pi} \cdot \frac{\sin h\pi/3}{\pi} \cdot \frac{\sin h\mu/2}{h\mu/2} \quad \text{(G.5)}$$

or

$$\frac{I_h}{I_1} = \frac{1}{h} \cdot \frac{\sin h\mu/2}{h\mu/2} \cdot \frac{\sin h\pi/3}{h\pi/3} \quad \text{(G.6)}$$

where I is fundamental frequency current.

We will calculate the fifth harmonic current, based on Equations G.5 and G.6.

From Equation G.5,

$$I_h = \frac{2\sqrt{2}}{\pi} \frac{\sin 300°}{\pi} \frac{\sin 12°}{0.2094} = 0.2465 I_d$$

We are more interested in the harmonics as a percentage of the fundamental frequency current and, from Equation G.6, $I_h = 19.86\%$ of the fundamental current.

G.2 Waveform with Ripple Content

The above equations ignored ripple content. Figure G.2 considers a ripple content, which is sinusoidal, and a sine half-wave is superimposed on the trapezoidal waveform. The following equations are applicable [1].

$$I_{\mathrm{h}} = I_{\mathrm{c}}\frac{2\sqrt{2}}{\pi}\left[\frac{\sin\left(\frac{h\pi}{3}\right)\sin\frac{h\mu}{2}}{h^2\frac{\mu}{2}} + \frac{r_{\mathrm{c}}g_{\mathrm{h}}\cos\left(h\frac{\pi}{6}\right)}{1-\sin\left(\frac{\pi}{3}+\frac{\mu}{2}\right)}\right] \tag{G.7}$$

where

$$g_{\mathrm{h}} = \frac{\sin\left[(h+1)\left(\frac{\pi}{6}-\frac{\mu}{2}\right)\right]}{h+1} + \frac{\sin\left[(h-1)\left(\frac{\pi}{6}-\frac{\mu}{2}\right)\right]}{h-1} - \frac{2\sin\left[h\left(\frac{\pi}{6}-\frac{\mu}{2}\right)\sin\left(\frac{\pi}{3}+\frac{\mu}{2}\right)\right]}{h} \tag{G.8}$$

where

I_{c} is the value of the dc current at the end of the commutation

r_{c} is the ripple coefficient ($=\Delta i/I_{\mathrm{c}}$)

In Figure G.2, the time zero reference is at $\omega t' = 0$, at the center of the current block. This is even symmetry, and, therefore, only cosine terms are present. The instantaneous current is then

$$i_{\mathrm{h}} = I_{\mathrm{h}}\sqrt{2}\cos n\omega t' \tag{G.9}$$

The current harmonics referred to $\omega t' = 0$ (Figure G.2) are

$$i_{\mathrm{h}} = I_{\mathrm{h}}\sqrt{2}\sin\frac{n\pi}{2}\sin\left[n(\omega t - \phi_1)\right] \tag{G.10}$$

where ϕ_1 is the phase angle between fundamental current and source voltage ($=\alpha + \mu/2$).

We will continue with the example to illustrate the method of calculation. Consider that $V_{\mathrm{do}} = 2.34\ (480/\sqrt{3}) = 648$ V. The short-circuit current on the 480 V bus is designated as $I_{\mathrm{s}} = 39$ kA. Let $I_{\mathrm{c}} = 2$ kA, and $V_{\mathrm{d}}/V_{\mathrm{do}} = 0.8$, where V_{d} is the dc operating voltage and V_{do} is

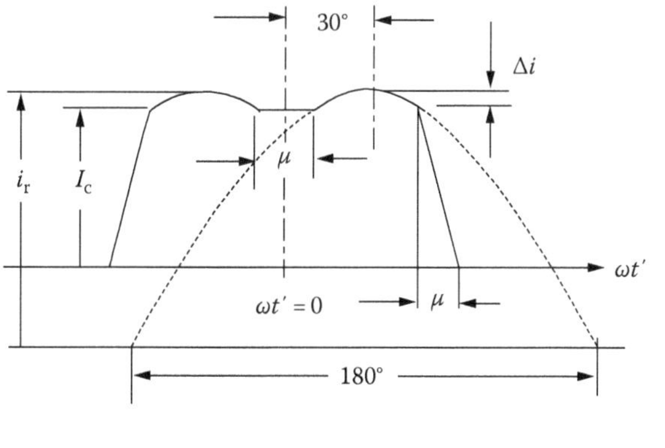

FIGURE G.2
Trapezoidal current waveform with superimposed sinusoidal ac ripple.

the no load voltage; then, $I_c/I_s\sqrt{2} = 0.036$. Using these values and from Figure G.3 for $a = 45°$, the overlap angle μ can be read as approximately 4.5°. Let us say it is 4.8°, as before.

Entering the values of μ and $V_d/V_{do} = 0.8$ in Figure G.4, we read $A_r = 0.1$, where A_r is the voltage-ripple integral or the ripple area.

Calculate the ripple current $\Delta i = A_r V_{do}/X_r$:

$$X_r = \omega L_d + 2X_c$$

where

X_r is ripple reactance
L_d is the inductance in the dc circuit in H
X_c is the commutating reactance

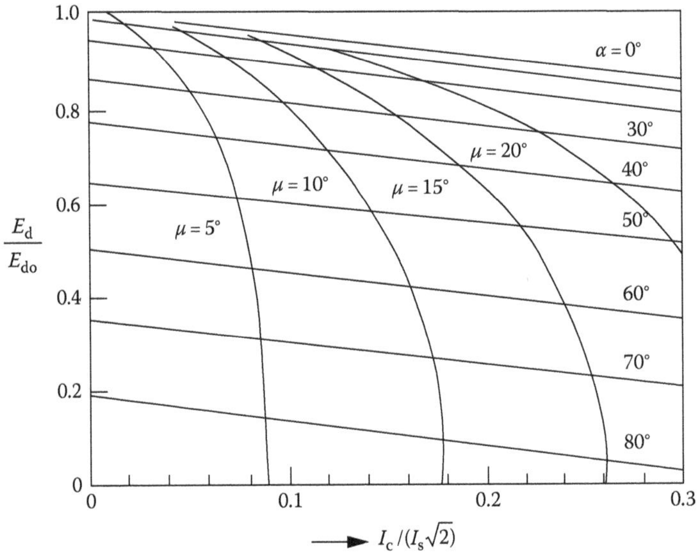

FIGURE G.3
Converter load curves for six-pulse bridge.

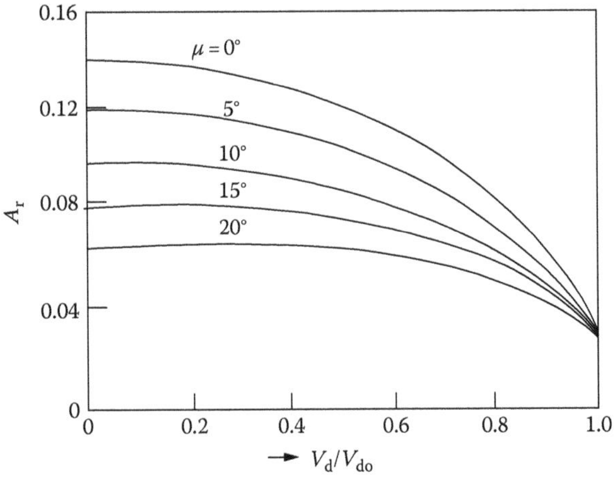

FIGURE G.4
Voltage-ripple integral for calculation of ripple current Δi.

First consider that there is no dc inductance, $X_r = 2X_c = 2 \times 0.0071$ ohm (the reactance of the transformer and the source referred to the 480 V side), and that $\Delta i = 4563$, then the ripple coefficient $r_c = \Delta i / I_c = 2.28$.

Now, the harmonics as a percentage of the fundamental current can be calculated graphically from Figure G.5a through h from Ref. [1]. Negative values indicate a phase shift of π, and I_1 is the fundamental frequency current.

Continuing with the example, and entering Figure G.5a for $\mu = 4.8$ and $r_c = 2.28$, we find that it is outside the range on the X-axis. An approximate value of 40% fifth harmonic can be read.

More accurately, Equations G.7 and G.8 can be used. From Equation G.8, and substituting the numerical values,

$$g_h = \frac{\sin 165.6°}{6} + \frac{\sin 110.4°}{4} - \frac{2(\sin 138° \ \sin 62.4°)}{5} = 0.03875$$

and from Equation G.7,

$$I_h = \frac{2\sqrt{2}}{\pi}\left[\frac{\sin 300° \ \sin 12°}{1.0472} + \frac{(2.28)(0.03857)\cos 150°}{0.113}\right] = -0.726 I_c$$

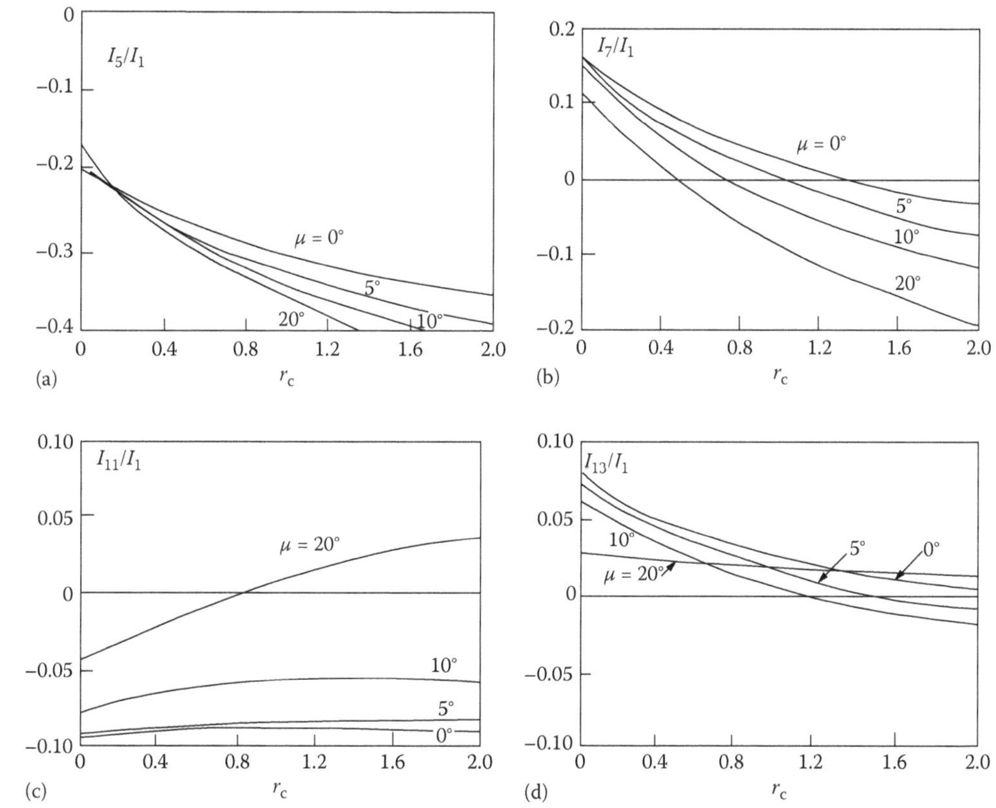

FIGURE G.5
(a)–(h) Harmonic currents in per unit of fundamental frequency current as a function of r_c.

(continued)

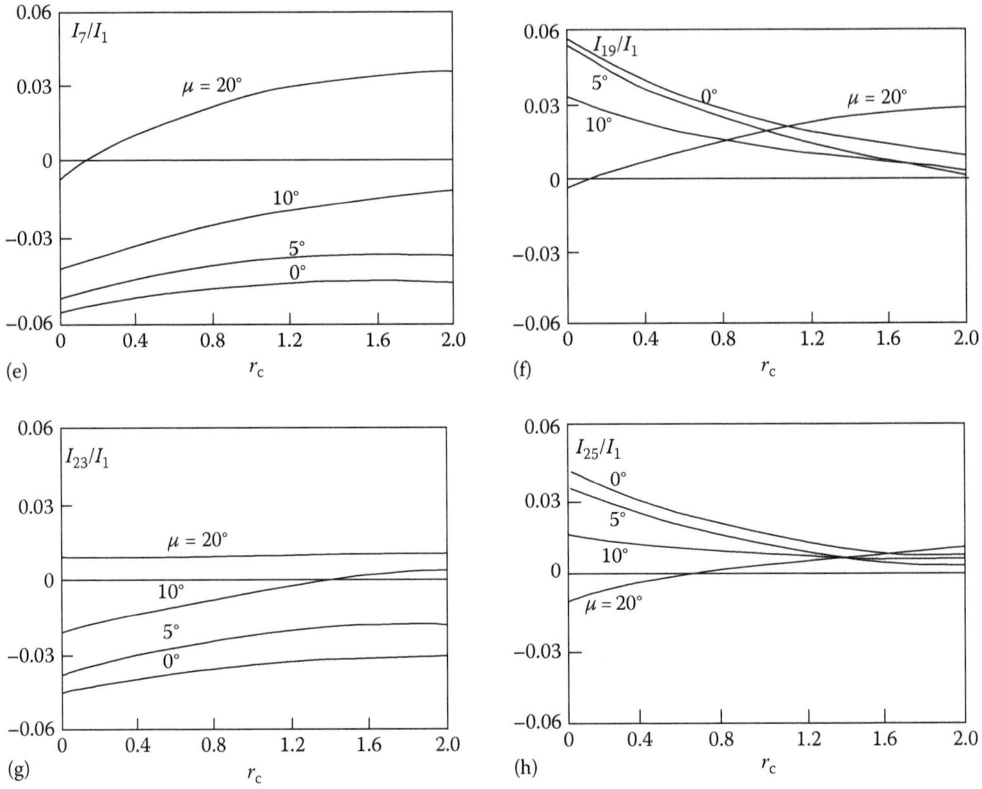

FIGURE G.5 (continued)

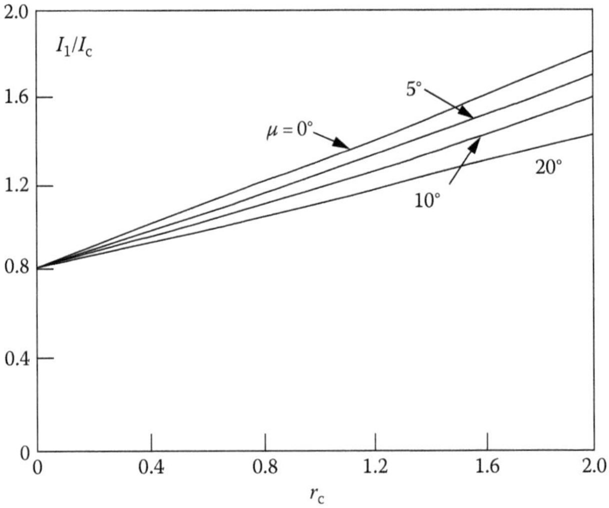

FIGURE G.6
Fundamental line current and commutation current as a function of r_c.

The ratio of currents I_1/I_c is given in Figure G.6. For $r_c = 2.28$, it is 1.8 and therefore the fifth harmonic in terms of the fundamental current is 42.3%. This is close to 40% estimated from the graphs.

If a dc inductor of 1 mH is added, then $X_r = 0.3912$ ohm, $\Delta i = 165.6$, the ripple coefficient $r_c = \Delta i / I_c = 0.08$, and the fifth harmonic reduces to 20% of the fundamental current. The reader may also refer to classical work in Ref. [2] and a recent work in Ref. [3].

G.3 Phase Angle of Harmonics

When a predominant harmonic source acts in isolation, it may not be necessary to model the phase angles of the harmonics. For multiple harmonic sources, phase angles should be modeled. Figure G.7a shows the time–current waveform of a six-pulse current source converter, when the phase angles are represented and it is recognizable as the line current of a six-pulse converter, with overlap and no ripple content; however, the waveform of

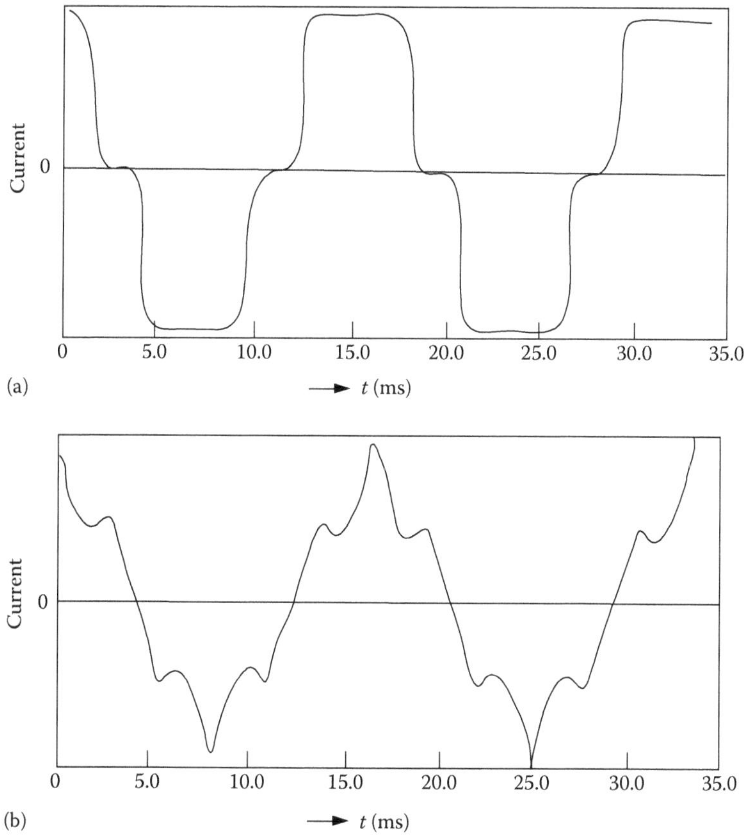

FIGURE G.7
(a) Six-pulse converter current, harmonics at proper angles; (b) all harmonics cophasial.

Figure G.7b has exactly the same spectra, but all the harmonics are cophasial. It can be shown that the harmonic current flow and the calculated distortions for a single-source harmonic current will be almost identical for these two waveforms. Assuming that the harmonics are cophasial in multisource current models does not always give the most conservative results.

Figure G.8a shows the time–current waveform of a pulse-width-modulated or voltage source converter, with harmonic phase angles. The harmonic spectrum and phase angles are shown in Table G.1. The phase angle of the fundamental is shown to be zero. Equation 19.9 applies and for a certain phase angle of the fundamental, the phase angle of the harmonics is calculated by shifting the angle column by $h\theta_1$ (harmonic order multiplied by fundamental frequency phase angle). Figure G.8b shows the waveform with all harmonics cophasial. Some observations as for Figure G.7a and b are applicable.

For multiple source assessment, the worst-case combination of the phase angles can be obtained by performing harmonic studies with one harmonic-producing element modeled at a time. The worst-case harmonic level, voltage, or current is the arithmetical summation of the harmonic magnitudes in each study. This will be a rather lengthy study. Alternatively, all the harmonic sources can be simultaneously modeled, with proper phase angles. The fundamental frequency angles are known by prior load flow and the angle of harmonics can be calculated using Equation 19.9.

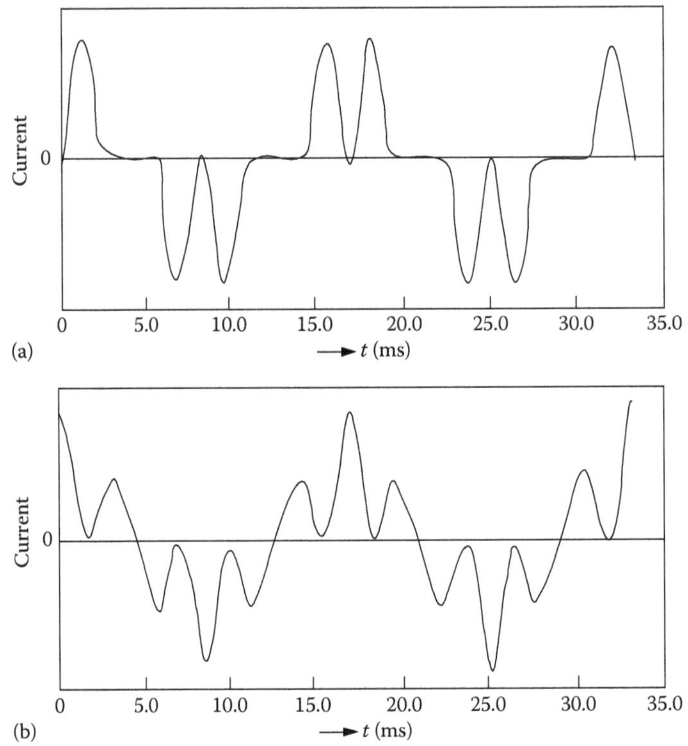

FIGURE G.8
(a) PWM voltage source converter current, harmonics with proper phase angles; (b) all harmonics cophasial.

TABLE G.1

Harmonic Spectrum of a Voltage Source or Pulse-Width-Modulated ASD

Harmonic Order	Magnitude	Phase Angle
1	100.0	0
3	3.1	−160
5	60.7	−178
7	41.2	−172
9	0.7	158
11	3.85	165
13	7.74	−177
14	0.3	−53
15	0.41	135
17	3.2	32
18	0.3	228
19	1.54	179
21	0.32	110
23	1.8	38
25	1.3	49
29	1.2	95
31	0.7	222

References

1. A.D. Graham and E.T. Schonholzer. Line harmonics of converters with DC-motor loads. *IEEE Trans. Ind. Appl.* 19: 84–93, 1983.
2. J.C. Read. The calculation of rectifier and inverter performance characteristics. *JIEE, UK,* 495–509, 1945.
3. M. Grotzbach and R. Redmann. Line current harmonics of VSI-fed adjustable-speed drives. *IEEE Trans. Ind. Appl.* 36: 683–690, 2000.

Index

D

I

J

M

N

U

V

W

www.ingramcontent.com/pod-product-compliance
Ingram Content Group UK Ltd.
Pitfield, Milton Keynes, MK11 3LW, UK
UKHW051835180425
457613UK00022B/1260

* 9 7 8 1 1 3 8 0 7 5 0 4 7 *